新纪元汉英林业词汇

New Era Chinese-English Forestry Glossary

刘琪璟　王辉民　编

中国林业出版社

内容简介

本书收录了林业及相关领域的汉英词汇共约 5.3 万条（英语词汇 4.9 万条），包括森林生态学、测树学、森林培育学、森林经理学、植物学、土壤学、气象学、野生动物、有害生物防治、林产化学工业、森林采运、木材加工、林业机械等。本书为双向检索，其中英汉部分以索引的形式编排。本书可供从事林业科学及生态学研究与教学、林业技术与资源管理人员使用。

图书在版编目（CIP）数据

新纪元汉英林业词汇 / 刘琪璟，王辉民编. —北京：中国林业出版社，2017.4

ISBN 978-7-5038-8988-2

I. ①新… II. ①刘… ②王… III. ①林业—词汇—英、汉 IV. ①S7-61

中国版本图书馆 CIP 数据核字（2017）第 094370 号

中国林业出版社·教育出版分社

策划编辑：肖基浒　　责任编辑：张　佳　肖基浒

电话：010-83143555　　传真：010-83143516

出版发行　中国林业出版社（100009　北京市西城区德内大街刘海胡同 7 号）
　　　　　Email: jiaocaipublic@163.com　电话（010）83143555
　　　　　http://lycb.forestry.gov.cn

经　　销　新华书店

印　　刷　北京中科印刷有限公司

版　　次　2017 年 4 月第 1 版

印　　次　2017 年 4 月第 1 次印刷

开　　本　880 mm × 1230 mm　1/32

印　　张　32

字　　数　1149 千字

定　　价　168.00 元

前　言

本书在编辑过程中参考了大量国内外相关工具书、期刊、国家行业标准、高等院校教材以及互联网资源。对不同来源的词汇经过汇总合并、甄别筛选，历时10年，最后编辑成书。

本书收录的是林业及相关领域的专业词汇，包括一部分在林业科学文献中比较常用的公共词汇。关于动植物名称，鉴于已经有相关的英拉汉对照工具书，而且包括的种类比较齐全，故本书基本上没有收录植物名称，动物也只收录少量常见的种类。林产化学方面有关化学物质的名称仅收录常见的种类，更多有关化学分子式的词汇请读者查阅化学方面的工具书。

本书为双向检索，但为了节省版面，英汉部分以索引的形式编排。汉英部分出现的英文词汇全部编入英文索引。

本书编辑过程中，下列人员（以拼音排序）参加校对：陈丽萍、胡杨、贾全全、林波、刘丽婷、刘文慧、罗春旺、罗梅、吕延杰、梅光义、孟盛旺、孟晓清、石宽、史景宁、孙翀、陶立超、王建军、王明初、徐倩倩、杨君珑、衣旭彤、于健、张博、张国春、周光、周华、庄会霞。特别感谢出版社编审人员对本书初稿的仔细认真审校。

尽管本书编辑过程中投入了大量的时间和精力，但仍然难以避免纰漏和错误，诚恳希望读者批评指正，以便再版时订正。

编　者

2016 年 10 月

目　录

使用说明

1. 本书词汇按汉语拼音顺序排列，同音汉字采用计算机系统默认排序。汉字索引给出的是汉语词汇的首汉字。

2. 关于多音字，只对每个检索词条的词首汉字进行处理。鉴于排版工作量的关系，词中间的多音字采用计算机默认排序。如“重”字，读音有“chóng”和“zhòng”。“重复”和“重量”中的“重”为不同汉字，而在“双重抽样”和“湿重”中尽管是不同读音，但未做分别处理，而是由系统按“zhòng”排序。再如“莎草”的词头按“suō”处理，而“纸莎草”中的“莎”则由系统按“shā”排序。

3. 中文词条中的圆括号“( )”表示解释或同义词。如“喜好(干湿或阴阳等)、下风(背风)”。

4. 中文词条中的方括号“[ ]”表示可以省略，如“偏途[演替]顶极”。

5. 英文词条之间用分号“;”隔开。

6. 英文词条中的圆括号“( )”表示可以省略或同义词，如“cutting (felling)”，其中的“felling”和“cutting”是同义词；再如“reinforcement (planting)”中的“planting”是对前面的词的解释。这方面很容易判断，故未单独做区分。

7. 英文词条的近义词或可替换词，在括号里加“*or*”表示，如“disk (*or* disc)”。

8. 对部分英文词条的复数不规则词尾在括号里用“*plur.*”标注，如“taxon (*plur.* taxa)”。

汉字索引

阿(ā)........1
埃(āi)........1
癌(ái)........1
矮(ǎi)........1
艾(ài)........2
爱(ài)........2
隘(ài)........2
安(ān)........2
桉(ān)........2
氨(ān)........3
鞍(ān)........3
岸(àn)........3
按(àn)........3
胺(àn)........3
案(àn)........3
暗(àn)........3
肮(āng)........4
凹(āo)........4
螯(áo)........4
坳(ào)........4
奥(ào)........4
澳(ào)........4
八(bā)........4
巴(bā)........5
扒(bā)........5
芭(bā)........5
疤(bā)........5
拔(bá)........5
把(bǎ)........5
靶(bǎ)........5
坝(bà)........5
霸(bà)........5
耙(bà)........5
白(bái)........5
百(bǎi)........6
柏(bǎi)........7
摆(bǎi)........7
败(bài)........7
拜(bài)........7
扳(bān)........7
班(bān)........7
斑(bān)........7
搬(bān)........7
板(bǎn)........7
版(bǎn)........9
半(bàn)........9
伴(bàn)........11
拌(bàn)........11
瓣(bàn)........11
绑(bǎng)........11
棒(bàng)........11
棓(bàng)........12
傍(bàng)........12
磅(bàng)........12
包(bāo)........12
苞(bāo)........12
孢(bāo)........12
胞(bāo)........13
雹(báo)........13
薄(báo)........13
饱(bǎo)........14
宝(bǎo)........14
保(bǎo)........14
报(bào)........15
抱(bào)........15
鲍(bào)........15
暴(bào)........15
爆(bào)........16
刨(bào)........16
曝(bào)........17
杯(bēi)........17
卑(bēi)........17
北(běi)........17
贝(bèi)........17
备(bèi)........17
背(bèi)........17
倍(bèi)........18
被(bèi)........18
焙(bèi)........19
锛(bēn)........19
本(běn)........19
苯(běn)........19
崩(bēng)........19
绷(bēng)........19
逼(bī)........19
鼻(bí)........19
比(bǐ)........19
吡(bǐ)........20
彼(bǐ)........20
俾(bǐ)........20
必(bì)........20
毕(bì)........20
闭(bì)........20
庇(bì)........20
荜(bì)........20
壁(bì)........20
避(bì)........21
臂(bì)........21
边(biān)........21
编(biān)........23
蝙(biān)........23
鞭(biān)........23
贬(biǎn)........23
扁(biǎn)........23
匾(biǎn)........23
苄(biàn)........23
变(biàn)........23
便(biàn)........24
遍(biàn)........24
辨(biàn)........24
辫(biàn)........24
标(biāo)........24
飑(biāo)........26
表(biǎo)........26
裱(biǎo)........27
别(bié)........27
瘪(biě)........27
玢(bīn)........27
宾(bīn)........27
滨(bīn)........27
槟(bīn)........27
濒(bīn)........27
冰(bīng)........27
兵(bīng)........29
丙(bǐng)........29
柄(bǐng)........29
蛃(bǐng)........29
并(bìng)........29
病(bìng)........29
拨(bō)........29
波(bō)........30
玻(bō)........30
钵(bō)........30
剥(bō)........30
播(bō)........31
伯(bó)........31
驳(bó)........31
泊(bó)........31
勃(bó)........31
铂(bó)........31
博(bó)........31
不(bú)........31
卟(bǔ)........32
补(bǔ)........32
捕(bǔ)........33
哺(bǔ)........33
不(bù)........33
布(bù)........35
步(bù)........35
部(bù)........36
簿(bù)........36
擦(cā)........36
材(cái)........36
财(cái)........37
裁(cái)........37
采(cǎi)........37
彩(cǎi)........39
参(cān)........40
餐(cān)........40
残(cán)........40
蚕(cán)........41
仓(cāng)........41
苍(cāng)........41
舱(cāng)........41

藏(cáng) 41
操(cāo) 41
槽(cáo) 41
草(cǎo) 41
侧(cè) 42
测(cè) 44
策(cè) 45
参(cēn) 45
层(céng) 45
叉(chā) 45
插(chā) 46
差(chā) 46
茬(chá) 46
茶(chá) 46
查(chá) 47
岔(chà) 47
拆(chāi) 47
柴(chái) 47
豺(chái) 47
觇(chān) 47
掺(chān) 47
缠(chán) 47
蟾(chán) 47
产(chǎn) 47
铲(chǎn) 48
颤(chàn) 48
菖(chāng) 48
常(cháng) 48
偿(cháng) 49
长(cháng) 49
厂(chǎng) 50
场(chǎng) 50
敞(chǎng) 50
抄(chāo) 50
超(chāo) 50
晁(cháo) 52
巢(cháo) 52
潮(cháo) 52
车(chē) 52
掣(chè) 52
尘(chén) 52
沉(chén) 52
陈(chén) 53
晨(chén) 53
衬(chèn) 53
称(chēng) 53
撑(chēng) 53
成(chéng) 53
呈(chéng) 55
承(chéng) 55
城(chéng) 56
乘(chéng) 56
程(chéng) 56
橙(chéng) 56
秤(chèng) 57
吃(chī) 57
池(chí) 57
迟(chí) 57
持(chí) 57
匙(chí) 57
尺(chǐ) 57
呎(chǐ) 57
齿(chǐ) 57
斥(chì) 58
赤(chì) 58
炽(chì) 58
翅(chì) 58
冲(chōng) 59
充(chōng) 59
虫(chóng) 59
重(chóng) 60
抽(chōu) 61
酬(chóu) 61
稠(chóu) 61
筹(chóu) 62
臭(chòu) 62
出(chū) 62
初(chū) 62
樗(chū) 64
除(chú) 64
厨(chú) 64
锄(chú) 64
雏(chú) 64
橱(chú) 64
储(chǔ) 64
处(chǔ) 65
畜(chù) 65
触(chù) 65
川(chuān) 65
氚(chuān) 65
穿(chuān) 65
传(chuán) 65
船(chuán) 66
椽(chuán) 66
串(chuàn) 66
疮(chuāng) .. 67
窗(chuāng) .. 67
床(chuáng) .. 67
创(chuàng) .. 67
吹(chuī) 67
垂(chuí) 67
槌(chuí) 68
锤(chuí) 68
春(chūn) 68
纯(chún) 68
莼(chún) 69
唇(chún) 69
醇(chún) 69
戳(chuō) 69
疵(cī) 69
瓷(cí) 69
磁(cí) 69
雌(cí) 69
次(cì) 70
刺(cì) 71
枞(cōng) 71
樬(cōng) 71
从(cóng) 71
丛(cóng) 72
粗(cū) 72
促(cù) 73
猝(cù) 73
醋(cù) 73
簇(cù) 73
催(cuī) 74
脆(cuì) 74
萃(cuì) 74
淬(cuì) 74
翠(cuì) 74
村(cūn) 74
存(cún) 74
搓(cuō) 74
锉(cuò) 74
错(cuò) 74
搭(dā) 74
达(dá) 75
打(dǎ) 75
大(dà) 76
呆(dāi) 78
代(dài) 78
带(dài) 79
贷(dài) 81
待(dài) 81
怠(dài) 81
袋(dài) 81
戴(dài) 81
丹(dān) 81
担(dān) 81
单(dān) 81
胆(dǎn) 87
淡(dàn) 87
弹(dàn) 87
蛋(dàn) 88
氮(dàn) 88
当(dāng) 88
挡(dǎng) 88
刀(dāo) 88
叨(dāo) 89
导(dǎo) 89
岛(dǎo) 90
捣(dǎo) 90
倒(dǎo) 90
倒(dào) 90
到(dào) 90
盗(dào) 90
道(dào) 90
稻(dào) 90
得(dé) 90
德(dé) 90
灯(dēng) 90
登(dēng) 91
等(děng) 91
凳(dèng) 92
低(dī) 92
堤(dī) 94
滴(dī) 94
狄(dí) 94
迪(dí) 94
敌(dí) 94
笛(dí) 94

镝(dí) 94
抵(dǐ) 94
底(dǐ) 94
地(dì) 95
递(dì) 99
第(dì) 99
蒂(dì) 100
缔(dì) 100
颠(diān) 100
典(diǎn) 100
点(diǎn) 100
碘(diǎn) 100
电(diàn) 100
玷(diàn) 103
垫(diàn) 103
淀(diàn) 103
靛(diàn) 103
凋(diāo) 103
貂(diāo) 103
雕(diāo) 103
吊(diào) 103
掉(diào) 104
调(diào) 104
跌(diē) 104
迭(dié) 104
叠(dié) 104
碟(dié) 104
蝶(dié) 104
丁(dīng) ... 104
酊(dīng) ... 105
钉(dīng) ... 105
顶(dǐng) ... 105
订(dìng) 106
钉(dìng) 106
定(dìng) 106
丢(diū) 108
东(dōng) ... 108
冬(dōng) ... 108
动(dòng) ... 108
冻(dòng) ... 110
栋(dòng) ... 110
洞(dòng) ... 110
兜(dōu) 110
抖(dǒu) 110
陡(dǒu) 110
斗(dǒu) 110
豆(dòu) 110
都(dū) 110
毒(dú) 110
独(dú) 111
读(dú) 111
犊(dú) 111
堵(dǔ) 111
杜(dù) 111
度(dù) 111
渡(dù) 111
镀(dù) 111
端(duān) ... 111
短(duǎn) ... 112
断(duàn) ... 113
椴(duàn) ... 113
煅(duàn) ... 113
锻(duàn) ... 113
堆(duī) 113
队(duì) 114
对(duì) 114
兑(duì) 115
吨(dūn) 115
钝(dùn) 115
盾(dùn) 115
顿(dùn) 116
多(duō) 116
夺(duó) 120
躲(duǒ) 120
垛(duò) 120
剁(duò) 120
舵(duò) 120
惰(duò) 120
俄(é) 120
鹅(é) 120
蛾(é) 120
额(é) 120
厄(è) 120
恶(è) 120
萼(è) 121
颚(è) 121
鳄(è) 121
恩(ēn) 121
蒽(ēn) 121
儿(ér) 121
耳(ěr) 121
饵(ěr) 121
二(èr) 121
发(fā) 123
伐(fá) 125
垡(fá) 127
阀(fá) 127
筏(fá) 127
法(fǎ) 127
珐(fà) 128
帆(fān) 128
番(fān) 128
蕃(fān) 128
翻(fān) 128
凡(fán) 128
矾(fán) 128
繁(fán) 128
反(fǎn) 129
返(fǎn) 130
泛(fàn) 130
范(fàn) 130
方(fāng) 130
芳(fāng) 131
防(fáng) 131
房(fáng) 134
仿(fǎng) 134
纺(fǎng) 134
放(fàng) 134
飞(fēi) 136
非(fēi) 136
菲(fēi) 139
绯(fēi) 139
鲱(fēi) 139
肥(féi) 139
榧(fěi) 139
废(fèi) 139
沸(fèi) 140
费(fèi) 140
分(fēn) 140
芬(fēn) 143
酚(fēn) 143
焚(fén) 143
粉(fěn) 143
粪(fèn) 144
丰(fēng) 144
风(fēng) 144
枫(fēng) 146
封(fēng) 146
疯(fēng) 146
峰(fēng) 146
葑(fēng) 146
锋(fēng) 146
蜂(fēng) 146
冯(féng) 147
缝(féng) 147
凤(fèng) 147
佛(fó) 147
夫(fū) 147
呋(fū) 147
跗(fū) 147
稃(fū) 147
孵(fū) 147
敷(fū) 147
弗(fú) 147
伏(fú) 147
扶(fú) 147
服(fú) 147
氟(fú) 147
浮(fú) 147
符(fú) 148
匐(fú) 148
幅(fú) 148
辐(fú) 148
福(fú) 149
抚(fǔ) 149
斧(fǔ) 149
俯(fǔ) 149
辅(fǔ) 149
腐(fǔ) 150
父(fù) 151
付(fù) 151
负(fù) 151
附(fù) 152
赴(fù) 152
复(fù) 152
副(fù) 154
赋(fù) 154
傅(fù) 154
富(fù) 155
腹(fù) 155

缚(fù) 155
覆(fù) 155
伽(gā) 155
改(gǎi) 155
钙(gài) 155
盖(gài) 156
概(gài) 156
甘(gān) 156
苷(gān) 156
柑(gān) 156
杆(gān) 156
干(gān) 156
杆(gǎn) 159
秆(gǎn) 159
赶(gǎn) 159
感(gǎn) 159
橄(gǎn) 159
干(gàn) 159
冈(gāng) 160
刚(gāng) 160
肛(gāng) 160
纲(gāng) 160
钢(gāng) 160
杠(gàng) 161
高(gāo) 161
睾(gāo) 164
镐(gǎo) 164
戈(gē) 164
疙(gē) 165
哥(gē) 165
鸽(gē) 165
搁(gē) 165
割(gē) 165
革(gé) 165
阁(gé) 165
格(gé) 165
隔(gé) 165
葛(gě) 166
个(gè) 166
各(gè) 166
铬(gè) 166
给(gěi) 167
根(gēn) 167
跟(gēn) 168
庚(gēng) 168
耕(gēng) 168
埂(gěng) 168
绠(gěng) 168
梗(gěng) 169
更(gèng) 169
工(gōng) 169
弓(gōng) 170
公(gōng) 170
功(gōng) 171
攻(gōng) 171
宫(gōng) 171
供(gōng) 171
拱(gǒng) 172
珙(gǒng) 172
共(gòng) 172
勾(gōu) 172
沟(gōu) 172
钩(gōu) 173
篝(gōu) 173
枸(gǒu) 173
构(gòu) 173
购(gòu) 173
估(gū) 173
咕(gū) 173
孤(gū) 174
菁(gū) 174
酤(gū) 174
箍(gū) 174
古(gǔ) 174
谷(gǔ) 175
股(gǔ) 175
骨(gǔ) 175
鼓(gǔ) 175
固(gù) 175
故(gù) 177
雇(gù) 177
刮(guā) 177
寡(guǎ) 177
挂(guà) 177
拐(guǎi) 178
怪(guài) 178
关(guān) . . . 178
观(guān) . . . 178
棺(guān) . . . 178
冠(guān) . . . 178
管(guǎn) . . . 178
贯(guàn) . . . 179
惯(guàn) . . . 179
灌(guàn) . . . 179
罐(guàn) . . . 180
光(guāng) . . 180
广(guǎng) . . 182
归(guī) 183
龟(guī) 183
规(guī) 183
硅(guī) 183
轨(guǐ) 184
柜(guì) 184
贵(guì) 184
桂(guì) 184
跪(guì) 184
辊(gǔn) 184
滚(gǔn) 184
棍(gùn) 184
锅(guō) 184
国(guó) 185
果(guǒ) 186
裹(guǒ) 186
过(guò) 186
哈(hā) 187
蛤(há) 187
胲(hǎi) 187
海(hǎi) 188
害(hài) 189
含(hán) 189
焓(hán) 189
寒(hán) 189
罕(hǎn) 190
汉(hàn) 190
汗(hàn) 190
旱(hàn) 190
焊(hàn) 190
夯(hāng) . . . 190
行(háng) . . . 190
绗(háng) . . . 190
航(háng) . . . 190
蒿(hāo) 191
豪(háo) 191
好(hǎo) 191
好(hào) 191
号(hào) 191
耗(hào) 191
禾(hé) 191
合(hé) 191
河(hé) 192
荷(hé) 193
核(hé) 193
盒(hé) 194
翮(hé) 194
赫(hè) 194
褐(hè) 194
鹤(hè) 194
壑(hè) 194
黑(hēi) 194
痕(hén) 195
亨(hēng) 195
恒(héng) 195
桁(héng) 195
鸻(héng) 195
横(héng) 195
烘(hōng) . . . 197
哄(hōng) . . . 197
红(hóng) . . . 197
宏(hóng) . . . 198
虹(hóng) . . . 198
洪(hóng) . . . 198
喉(hóu) 199
猴(hóu) 199
后(hòu) 199
厚(hòu) 200
候(hòu) 200
呼(hū) 200
狐(hú) 201
弧(hú) 201
胡(hú) 201
壶(hú) 201
葫(hú) 201
湖(hú) 201
槲(hú) 201
蝴(hú) 201
糊(hú) 201
虎(hǔ) 201
琥(hǔ) 201
互(hù) 201
户(hù) 202

护(hù)......202
戽(hù)......202
瓠(hù)......202
花(huā).....202
划(huá).....204
华(huá).....204
铧(huá).....204
滑(huá).....205
划(huà).....205
华(huà).....205
化(huà).....205
画(huà).....206
桦(huà).....206
怀(huái)....207
槐(huái)....207
踝(huái)....207
坏(huài)....207
还(huán)...207
环(huán)...207
缓(huǎn)...209
幻(huàn)...209
换(huàn)...209
荒(huāng)..209
皇(huáng)..210
黄(huáng)..210
蝗(huáng)..210
磺(huáng)..210
簧(huáng)..210
灰(huī).....210
挥(huī).....211
恢(huī).....211
辉(huī).....211
回(huí).....211
茴(huí).....212
毁(huǐ).....213
汇(huì).....213
会(huì).....213
荟(huì).....213
绘(huì).....213
喙(huì).....213
昏(hūn)....213
婚(hūn)....213
浑(hún)....213
混(hùn)....213
活(huó).....214
火(huǒ).....215
或(huò).....217
货(huò).....217
获(huò).....217
霍(huò).....217
击(jī)......217
机(jī)......217
肌(jī)......218
矶(jī)......218
鸡(jī)......219
唧(jī)......219
积(jī)......219
姬(jī)......219
基(jī)......219
畸(jī)......221
稽(jī)......221
激(jī)......221
几(jī)......221
吉(jí).......221
级(jí).......221
极(jí).......221
即(jí).......222
急(jí).......222
疾(jí).......222
棘(jí).......222
集(jí).......222
籍(jí).......224
几(jǐ)......224
己(jǐ)......224
挤(jǐ)......224
脊(jǐ)......225
计(jì).......225
记(jì).......226
纪(jì).......226
技(jì).......226
忌(jì).......226
季(jì).......226
剂(jì).......226
迹(jì).......226
既(jì).......226
继(jì).......226
寄(jì).......226
蓟(jì).......227
系(jì).......227
加(jiā).......227
家(jiā)......229
夹(jiā)......229
茄(jiā)......229
荚(jiá)......229
颊(jiá)......229
蛱(jiá)......229
甲(jiǎ)......229
钾(jiǎ)......230
假(jiǎ)......230
槚(jiǎ)......230
价(jià)......230
驾(jià)......231
架(jià)......231
嫁(jià)......231
尖(jiān)....231
坚(jiān)....232
间(jiān)....232
肩(jiān)....232
监(jiān)....232
兼(jiān)....232
拣(jiǎn)....232
茧(jiǎn)....232
捡(jiǎn)....232
检(jiǎn)....232
减(jiǎn)....233
剪(jiǎn)....233
简(jiǎn)....234
碱(jiǎn)....234
间(jiàn)....235
件(jiàn)....236
饯(jiàn)....236
建(jiàn)....236
剑(jiàn)....236
健(jiàn)....236
渐(jiàn)....236
腱(jiàn)....237
溅(jiàn)....237
鉴(jiàn)....237
键(jiàn)....237
箭(jiàn)....237
江(jiāng)...237
将(jiāng)...237
姜(jiāng)...237
浆(jiāng)...237
僵(jiāng)...238
讲(jiǎng)...238
奖(jiǎng)...238
桨(jiǎng)...238
降(jiàng)...238
交(jiāo).....238
郊(jiāo).....240
浇(jiāo).....240
胶(jiāo).....240
焦(jiāo).....242
礁(jiāo).....242
嚼(jiáo).....242
角(jiǎo).....242
绞(jiǎo).....242
铰(jiǎo).....243
矫(jiǎo).....243
脚(jiǎo).....243
搅(jiǎo).....243
缴(jiǎo).....243
较(jiào).....243
教(jiào).....243
窖(jiào).....243
酵(jiào).....243
校(jiào).....243
阶(jiē)......243
皆(jiē)......244
接(jiē)......244
揭(jiē)......245
街(jiē)......245
结(jiē)......245
孑(jié)......245
节(jié)......245
杰(jié)......246
洁(jié)......246
结(jié)......246
捷(jié)......246
截(jié)......247
解(jiě)......247
介(jiè)......248
芥(jiè)......248
界(jiè)......248
蚧(jiè)......248
借(jiè)......248
今(jīn).....248
金(jīn).....248
津(jīn).....249

筋(jīn) 249
紧(jǐn) 249
锦(jǐn) 249
尽(jìn) 249
进(jìn) 249
近(jìn) 250
劲(jìn) 251
浸(jìn) 251
禁(jìn) 251
茎(jīng) 252
京(jīng) 252
经(jīng) 252
荆(jīng) 254
晶(jīng) 254
精(jīng) 254
井(jǐng) 254
颈(jǐng) 254
景(jǐng) 255
警(jǐng) 255
径(jìng) 255
净(jìng) 256
胫(jìng) 257
痉(jìng) 257
竞(jìng) 257
静(jìng) 257
境(jìng) 258
镜(jìng) 258
纠(jiū) 258
鸠(jiū) 258
九(jiǔ) 259
久(jiǔ) 259
酒(jiǔ) 259
旧(jiù) 259
救(jiù) 259
厩(jiù) 259
就(jiù) 259
居(jū) 259
掬(jū) 259
局(jú) 259
菊(jú) 259
橘(jú) 259
咀(jǔ) 259
矩(jǔ) 259
举(jǔ) 260
巨(jù) 260
拒(jù) 260
具(jù) 260
剧(jù) 261
距(jù) 261
飓(jù) 261
锯(jù) 261
聚(jù) 263
捐(juān) 264
卷(juǎn) 264
绢(juàn) 264
眷(juàn) 264
孓(jué) 264
决(jué) 264
绝(jué) 264
掘(jué) 265
蕨(jué) 266
爵(jué) 266
攫(jué) 266
军(jūn) 266
均(jūn) 266
菌(jūn) 266
喀(kā) 267
卡(kǎ) 267
开(kāi) 268
揩(kāi) 269
凯(kǎi) 269
勘(kān) 269
看(kān) 269
坎(kǎn) 269
砍(kǎn) 269
康(kāng) . . . 269
糠(kāng) . . . 270
抗(kàng) . . . 270
考(kǎo) 271
栲(kǎo) 271
烤(kǎo) 271
靠(kào) 271
苛(kē) 271
柯(kē) 271
科(kē) 272
颗(kē) 272
髁(kē) 272
壳(ké) 272
可(kě) 272
克(kè) 274
刻(kè) 274
客(kè) 274
课(kè) 274
垦(kěn) 274
啃(kěn) 274
坑(kēng) 275
空(kōng) . . . 275
孔(kǒng) . . . 276
空(kòng) . . . 277
控(kòng) . . . 277
口(kǒu) 277
叩(kòu) 277
扣(kòu) 277
寇(kòu) 278
刳(kū) 278
枯(kū) 278
苦(kǔ) 278
库(kù) 278
酷(kù) 278
夸(kuā) 278
跨(kuà) 278
会(kuài) 278
块(kuài) 279
快(kuài) 279
宽(kuān) . . . 279
狂(kuáng) . . 280
矿(kuàng) . . 280
框(kuàng) . . 280
亏(kuī) 280
盔(kuī) 280
魁(kuí) 280
馈(kuì) 280
溃(kuì) 280
昆(kūn) 280
髡(kūn) 280
醌(kūn) 280
捆(kǔn) 280
困(kùn) 281
扩(kuò) 281
蛞(kuò) 281
阔(kuò) 281
廓(kuò) 282
垃(lā) 282
拉(lā) 282
喇(lǎ) 282
腊(là) 282
蜡(là) 282
辣(là) 282
来(lái) 282
莱(lái) 282
赖(lài) 282
兰(lán) 282
拦(lán) 283
栏(lán) 283
蓝(lán) 283
篮(lán) 283
缆(lǎn) 283
榄(lǎn) 283
滥(làn) 283
廊(láng) 283
朗(lǎng) 283
浪(làng) 283
莨(làng) 283
捞(lāo) 283
劳(láo) 284
老(lǎo) 284
烙(lào) 284
涝(lào) 284
耢(lào) 284
酪(lào) 284
乐(lè) 284
勒(lè) 284
雷(léi) 284
蕾(lěi) 285
累(lěi) 285
肋(lèi) 285
泪(lèi) 285
类(lèi) 285
棱(léng) 285
楞(léng) 285
冷(lěng) 286
狸(lí) 287
离(lí) 287
梨(lí) 288
犁(lí) 288
篱(lí) 288
李(lǐ) 288
里(lǐ) 288
浬(lǐ) 288
理(lǐ) 288

力(lì)........288
历(lì)........288
立(lì)........288
利(lì)........290
沥(lì)........291
例(lì)........291
栎(lì)........291
栗(lì)........291
砾(lì)........291
粒(lì)........291
连(lián).....291
帘(lián).....294
莲(lián).....294
涟(lián).....294
联(lián).....294
廉(lián).....294
镰(lián).....294
脸(liǎn).....294
链(liàn).....294
楝(liàn).....295
良(liáng)...295
凉(liáng)...295
梁(liáng)...295
椋(liáng)...295
粮(liáng)...295
两(liǎng)...295
亮(liàng)...296
量(liàng)...296
晾(liàng)...296
疗(liáo).....296
瞭(liǎo).....296
料(liào).....296
撂(liào).....297
列(liè)......297
劣(liè)......297
烈(liè)......297
猎(liè)......297
裂(liè)......297
鬣(liè)......298
邻(lín)......298
林(lín)......298
临(lín)......304
淋(lín)......304
磷(lín)......304
鳞(lín)......305
檩(lǐn).....305
灵(líng).....305
陵(líng).....305
菱(líng).....305
零(líng).....305
龄(líng).....305
岭(lǐng)....305
领(lǐng)....305
另(lìng).....305
令(lìng).....305
溜(liū)......305
留(liú)......305
流(liú)......306
硫(liú)......308
馏(liú)......308
榴(liú)......308
瘤(liú)......308
柳(liǔ)......308
六(liù)......308
鹨(liù)......309
龙(lóng)....309
隆(lóng)....309
拢(lǒng)....309
垄(lǒng)....309
楼(lóu).....309
蝼(lóu).....309
髅(lóu).....309
搂(lǒu).....309
镂(lòu).....309
漏(lòu).....309
露(lòu).....310
卢(lú)......310
芦(lú)......310
炉(lú)......310
卤(lǔ)......310
鲁(lǔ)......310
陆(lù)......310
录(lù)......310
鹿(lù)......310
路(lù)......310
露(lù)......311
孪(luán)....311
銮(luán)....311
卵(luǎn)....311
乱(luàn)....311
掠(luè).....311
略(luè).....311
伦(lún).....311
轮(lún).....311
罗(luó).....312
逻(luó).....312
螺(luó).....312
裸(luǒ).....313
洛(luò).....314
络(luò).....314
落(luò).....314
吕(lǚ)......314
旅(lǚ)......314
铝(lǚ)......314
履(lǚ)......314
率(lǜ)......315
绿(lǜ)......315
葎(lǜ)......315
氯(lǜ)......315
滤(lǜ)......316
麻(má).....316
马(mǎ).....316
玛(mǎ).....316
码(mǎ).....316
蚂(mǎ).....316
埋(mái).....317
霾(mái).....317
买(mǎi).....317
迈(mài).....317
麦(mài).....317
卖(mài).....317
脉(mài).....317
蛮(mán)....317
满(mǎn)....317
螨(mǎn)....317
蔓(màn)....317
漫(màn)....317
慢(màn)....318
芒(máng)...318
杧(máng)...318
盲(máng)...318
牻(máng)...318
猫(māo)....318
毛(máo)....318
矛(máo)....319
锚(máo)....319
铆(mǎo)....319
冒(mào)....319
贸(mào)....319
帽(mào)....319
玫(méi).....319
眉(méi).....319
梅(méi).....319
媒(méi).....319
楣(méi).....319
煤(méi).....319
酶(méi).....320
霉(méi).....320
每(měi).....320
美(měi).....320
镁(měi).....320
闷(mēn)....320
门(mén)....320
虻(méng)...321
萌(méng)...321
蒙(méng)...321
猛(měng)...321
锰(měng)...321
孟(mèng)...321
弥(mí)......321
迷(mí)......321
醚(mí)......321
米(mǐ)......321
觅(mì)......321
泌(mì)......321
密(mì)......322
幂(mì)......323
蜜(mì)......323
绵(mián)...323
棉(mián)...323
免(miǎn)...323
面(miàn)...323
苗(miáo)...323
描(miáo)...324
瞄(miáo)...324
灭(miè).....324
篾(miè).....324
民(mín)....324
敏(mǐn)....324
名(míng)...324

明(míng) . . . 325
鸣(míng) . . . 325
螟(míng) . . . 325
命(mìng) . . . 325
摹(mó) 325
模(mó) 325
膜(mó) 326
摩(mó) 326
磨(mó) 326
蘑(mó) 327
抹(mǒ) 327
没(mò) 327
磨(mò) 327
末(mò) 327
茉(mò) 327
沫(mò) 327
莫(mò) 327
墨(mò) 327
模(mú) 327
母(mǔ) 327
牡(mǔ) 328
木(mù) 328
目(mù) 336
苜(mù) 336
牧(mù) 336
幕(mù) 336
目(mù) 336
钠(nà) 336
奈(nài) 336
耐(nài) 336
萘(nài) 338
南(nán) 338
难(nán) 338
楠(nán) 338
囊(náng) . . . 338
挠(náo) 338
脑(nǎo) 339
内(nèi) 339
嫩(nèn) 341
能(néng) 341
尼(ní) 342
泥(ní) 342
拟(nǐ) 342
逆(nì) 343
腻(nì) 343
年(nián) 343
黏(nián) 344
捻(niǎn) 345
碾(niǎn) 345
念(niàn) 345
酿(niàng) . . . 345
鸟(niǎo) 345
尿(niào) 346
脲(niào) 346
捏(niē) 346
啮(niè) 346
镊(niè) 346
柠(níng) 346
凝(níng) 346
拧(nǐng) . . . 346
牛(niú) 346
扭(niǔ) 347
钮(niǔ) 347
农(nóng) . . . 347
浓(nóng) . . . 348
耨(nòu) 348
奴(nú) 348
努(nǔ) 348
暖(nuǎn) . . . 348
疟(nüè) 348
挪(nuó) 348
诺(nuò) 348
噢(ō) 348
欧(ōu) 348
偶(ǒu) 348
耦(ǒu) 349
藕(ǒu) 349
爬(pá) 349
耙(pá) 349
帕(pà) 349
拍(pāi) 349
排(pái) 349
牌(pái) 350
派(pài) 350
潘(pān) 350
攀(pān) 350
盘(pán) 351
磐(pán) 351
蟠(pán) 351
判(pàn) 351
庞(páng) . . . 351
旁(páng) . . . 351
抛(pāo) 351
跑(pǎo) 352
泡(pào) 352
炮(pào) 352
疱(pào) 352
胚(pēi) 352
陪(péi) 353
培(péi) 353
赔(péi) 353
配(pèi) 353
喷(pēn) 353
盆(pén) 355
棚(péng) 355
硼(péng) 355
篷(péng) 355
膨(péng) 355
碰(pèng) 355
批(pī) 355
坯(pī) 355
披(pī) 355
劈(pī) 355
皮(pí) 356
枇(pí) 356
疲(pí) 356
啤(pí) 357
蜱(pí) 357
偏(piān) 357
胼(pián) 357
片(piàn) 357
漂(piāo) 358
瓢(piáo) 358
漂(piǎo) 358
票(piào) 358
嘌(piào) 358
撇(piē) 358
拼(pīn) 358
贫(pín) 359
频(pín) 359
品(pǐn) 359
平(píng) 359
评(píng) 363
苹(píng) 363
凭(píng) 363
屏(píng) 363
瓶(píng) 363
坡(pō) 363
破(pò) 363
剖(pōu) 363
扑(pū) 364
铺(pū) 364
匍(pú) 364
菩(pú) 364
葡(pú) 364
蒲(pú) 364
脯(pǔ) 365
圃(pǔ) 365
普(pǔ) 365
谱(pǔ) 365
蹼(pǔ) 365
瀑(pù) 365
七(qī) 365
栖(qī) 365
桤(qī) 365
期(qī) 365
欺(qī) 366
漆(qī) 366
齐(qí) 366
奇(qí) 366
歧(qí) 366
脐(qí) 366
畦(qí) 366
崎(qí) 366
骑(qí) 366
棋(qí) 366
蛴(qí) 366
旗(qí) 366
鳍(qí) 366
企(qǐ) 366
杞(qǐ) 367
启(qǐ) 367
起(qǐ) 367
气(qì) 368
弃(qì) 371
汽(qì) 371
契(qì) 371
槭(qì) 371
器(qì) 371
卡(qiǎ) 371

千(qiān)....371
扦(qiān)....371
迁(qiān)....372
牵(qiān)....372
铅(qiān)....372
签(qiān)....372
前(qián)....372
钳(qián)....373
潜(qián)....373
浅(qiǎn)....374
欠(qiàn)....375
茜(qiàn)....375
嵌(qiàn)....375
歉(qiàn)....375
枪(qiāng)...375
腔(qiāng)...375
强(qiáng)...375
墙(qiáng)...376
蔷(qiáng)...376
强(qiǎng)...376
抢(qiǎng)...376
羟(qiǎng)...376
锹(qiāo)....376
敲(qiāo)....376
橇(qiāo)....376
乔(qiáo)....377
桥(qiáo)....377
樵(qiáo)....377
壳(qiào)....377
峭(qiào)....377
翘(qiào)....377
撬(qiào)....377
鞘(qiào)....377
切(qiē).....377
窃(qiè).....378
侵(qīn).....378
亲(qīn).....379
芹(qín).....379
琴(qín).....379
禽(qín).....379
青(qīng)....379
轻(qīng)....379
氢(qīng)....380
倾(qīng)....380
清(qīng)....381
蜻(qīng)....381
情(qíng)....381
晴(qíng)....381
氰(qíng)....381
穹(qióng)...381
琼(qióng)...381
丘(qiū).....381
秋(qiū).....382
蚯(qiū).....382
求(qiú).....382
球(qiú).....382
区(qū)......382
驱(qū)......383
屈(qū)......383
蛆(qū)......383
躯(qū)......383
趋(qū)......383
鸲(qú)......383
渠(qú)......383
蠼(qú)......384
曲(qǔ)......384
取(qǔ)......384
去(qù)......384
趣(qù)......385
圈(quān)...385
权(quán)...385
全(quán)...385
泉(quán)...387
拳(quán)...387
犬(quǎn)...387
缺(quē).....387
雀(què).....388
确(què).....388
裙(qún)....388
群(qún)....388
髯(rán).....390
燃(rán).....390
染(rǎn).....390
壤(rǎng)....391
让(ràng)....391
扰(rǎo).....391
绕(rào).....391
热(rè)......391
人(rén).....394
仁(rén).....396
稔(rěn).....396
刃(rèn).....396
认(rèn).....396
任(rèn).....396
韧(rèn).....396
妊(rèn).....396
日(rì)......396
茸(róng)....397
绒(róng)....397
容(róng)....397
溶(róng)....398
榕(róng)....398
熔(róng)....398
融(róng)....398
冗(rǒng)....398
柔(róu).....398
揉(róu).....399
鞣(róu).....399
肉(ròu).....399
如(rú)......399
蠕(rú)......399
乳(rǔ)......399
入(rù)......400
软(ruǎn)....400
蕊(ruǐ).....401
芮(ruì).....401
锐(ruì).....401
瑞(ruì).....401
闰(rùn).....401
润(rùn).....401
若(ruò).....401
弱(ruò).....401
洒(sǎ)......401
撒(sǎ)......401
萨(sà)......402
塞(sāi).....402
鳃(sāi).....402
赛(sài).....402
三(sān).....402
伞(sǎn).....404
散(sǎn).....404
散(sàn).....404
桑(sāng)....405
扫(sǎo).....405
色(sè)......405
森(sēn).....405
杀(shā).....410
沙(shā).....411
砂(shā).....411
沙(shā).....411
砂(shā).....411
沙(shā).....411
纱(shā).....412
砂(shā).....412
莎(shā).....412
筛(shāi)....412
晒(shài)....413
山(shān)....413
杉(shān)....414
珊(shān)....414
煽(shān)....414
闪(shǎn)....414
扇(shàn)....414
善(shàn)....414
伤(shāng)...414
商(shāng)...414
熵(shāng)...415
赏(shǎng)...415
上(shàng)...415
烧(shāo)....416
梢(shāo)....417
勺(sháo)....417
芍(sháo)....417
杓(sháo)....417
少(shǎo)....417
奢(shē).....417
赊(shē).....417
舌(shé).....417
蛇(shé).....417
设(shè).....417
社(shè).....418
射(shè).....418
涉(shè).....419
摄(shè).....419
麝(shè).....419
申(shēn)....419
伸(shēn)....419
砷(shēn)....419
深(shēn)....419
神(shén)....420

审(shěn)....420
肾(shèn)....420
渗(shèn)....420
椹(shèn)....421
升(shēng)...421
生(shēng)...421
声(shēng)...430
牲(shēng)...430
绳(shéng)...430
省(shěng)...431
圣(shèng)...431
盛(shèng)...431
剩(shèng)...431
尸(shī).....431
失(shī).....431
施(shī).....431
湿(shī).....432
蓍(shī).....433
十(shí).....433
石(shí).....433
时(shí).....435
识(shí).....435
实(shí).....435
拾(shí).....436
食(shí).....436
蚀(shí).....437
十(shí).....437
史(shǐ).....437
矢(shǐ).....437
使(shǐ).....437
始(shǐ).....437
示(shì).....438
世(shì).....438
市(shì).....438
势(shì).....438
事(shì).....438
饰(shì).....438
试(shì).....438
视(shì).....439
柿(shì).....439
适(shì).....439
室(shì).....440
释(shì).....441
嗜(shì).....441
噬(shì).....441
螫(shì).....441
收(shōu)....441
手(shǒu)....442
守(shǒu)....443
首(shǒu)....443
寿(shòu)....443
受(shòu)....443
狩(shòu)....444
授(shòu)....444
售(shòu)....444
兽(shòu)....444
瘦(shòu)....444
书(shū).....444
枢(shū).....444
梳(shū).....444
舒(shū).....444
疏(shū).....444
输(shū).....445
赎(shú).....446
熟(shú).....446
属(shǔ).....446
鼠(shǔ).....446
薯(shǔ).....446
曙(shǔ).....446
束(shù).....446
树(shù).....446
竖(shù).....450
数(shù).....450
刷(shuā)....451
衰(shuāi)...451
摔(shuāi)...451
甩(shuǎi)...451
闩(shuān)..451
栓(shuān)..452
双(shuāng).452
霜(shuāng).455
水(shuǐ)....455
税(shuì)....460
睡(shuì)....460
顺(shùn)....460
瞬(shùn)....461
说(shuō)....461
蒴(shuò)....461
司(sī)......461
丝(sī)......461
私(sī)......462
斯(sī)......462
撕(sī)......462
死(sǐ)......462
似(sì)......462
四(sì)......462
寺(sì)......464
饲(sì)......464
松(sōng)....464
凇(sōng)....465
送(sòng)....465
搜(sōu).....465
苏(sū)......465
酥(sū)......466
俗(sú)......466
素(sù)......466
速(sù)......466
嗉(sù)......466
塑(sù)......466
溯(sù)......466
宿(sù)......467
酸(suān)....467
算(suàn)....467
随(suí).....467
髓(suǐ).....468
岁(suì).....468
碎(suì).....468
隧(suì).....469
穗(suì).....469
损(sǔn).....469
隼(sǔn).....469
榫(sǔn).....469
莎(suō).....469
梭(suō).....469
蓑(suō).....469
羧(suō).....469
缩(suō).....469
所(suǒ).....469
索(suǒ).....469
锁(suǒ).....470
他(tā)......470
塌(tā)......470
塔(tǎ)......470
踏(tà)......470
胎(tāi)......471
台(tái)......471
抬(tái)......471
苔(tái)......471
薹(tái)......471
太(tài)......471
态(tài)......471
钛(tài)......471
泰(tài)......471
坍(tān).....471
摊(tān).....471
滩(tān).....471
弹(tán).....472
坛(tán).....472
檀(tán).....472
炭(tàn).....472
探(tàn).....472
碳(tàn).....473
羰(tāng)....473
搪(táng)....473
镗(táng)....473
糖(táng)....473
螳(táng)....474
躺(tǎng)....474
烫(tàng)....474
趟(tàng)....474
逃(táo).....474
桃(táo).....474
陶(táo).....474
淘(táo).....474
讨(tǎo).....474
套(tào).....474
特(tè)......474
誊(téng)....475
藤(téng)....475
剔(tī)......475
梯(tī)......475
锑(tī)......476
踢(tī)......476
提(tí)......476
蹄(tí)......476
体(tǐ)......476
剃(tì)......477
涕(tì)......477
替(tì)......477
天(tiān)....477

添(tiān) 478
田(tián) 478
甜(tián) 478
填(tián) 478
挑(tiāo) 479
条(tiáo) 479
调(tiáo) 479
跳(tiào) 480
贴(tiē) 480
铁(tiě) 480
听(tīng) 481
庭(tíng) 481
停(tíng) 481
挺(tǐng) 481
通(tōng) 481
同(tóng) 482
桐(tóng) 484
铜(tóng) 484
酮(tóng) 484
统(tǒng) 484
捅(tǒng) 484
桶(tǒng) 484
筒(tǒng) 485
偷(tōu) 485
头(tóu) 485
投(tóu) 485
透(tòu) 485
凸(tū) 486
秃(tū) 486
突(tū) 486
图(tú) 486
徒(tú) 487
途(tú) 487
涂(tú) 487
土(tǔ) 488
吐(tǔ) 491
菟(tù) 491
湍(tuān) 491
团(tuán) 491
推(tuī) 491
腿(tuǐ) 492
退(tuì) 492
蜕(tuì) 492
褪(tuì) 492
吞(tūn) 492
囤(tún) 492
臀(tún) 492
托(tuō) 492
拖(tuō) 493
脱(tuō) 493
陀(tuó) 494
驼(tuó) 494
妥(tuǒ) 494
椭(tuǒ) 494
拓(tuò) 494
唾(tuò) 494
箨(tuò) 494
挖(wā) 494
洼(wā) 495
瓦(wǎ) 495
歪(wāi) 495
外(wài) 495
弯(wān) 497
丸(wán) 498
完(wán) 498
玩(wán) 498
挽(wǎn) 499
晚(wǎn) 499
椀(wǎn) 499
碗(wǎn) 499
万(wàn) 499
腕(wàn) 499
王(wáng) . . . 499
网(wǎng) . . . 499
往(wǎng) . . . 500
旺(wàng) . . . 500
望(wàng) . . . 500
危(wēi) 500
威(wēi) 500
微(wēi) 500
韦(wéi) 502
为(wéi) 502
违(wéi) 502
围(wéi) 502
桅(wéi) 502
唯(wéi) 502
维(wéi) 502
伪(wěi) 502
尾(wěi) 503
纬(wěi) 503
委(wěi) 503
萎(wěi) 503
卫(wèi) 503
未(wèi) 503
位(wèi) 504
胃(wèi) 504
喂(wèi) 505
温(wēn) 505
瘟(wēn) 506
文(wén) 506
纹(wén) 506
蚊(wén) 506
吻(wěn) 506
紊(wěn) 506
稳(wěn) 506
鹟(wēng) . . . 507
涡(wō) 507
窝(wō) 507
蜗(wō) 507
肟(wò) 507
沃(wò) 507
卧(wò) 507
握(wò) 507
乌(wū) 507
圬(wū) 507
污(wū) 507
呜(wū) 508
屋(wū) 508
无(wú) 508
蜈(wú) 512
五(wǔ) 512
戊(wù) 512
物(wù) 512
误(wù) 513
雾(wù) 513
西(xī) 513
吸(xī) 513
希(xī) 515
析(xī) 515
息(xī) 515
烯(xī) 515
稀(xī) 515
锡(xī) 515
溪(xī) 515
蜥(xī) 516
熄(xī) 516
膝(xī) 516
习(xí) 516
洗(xǐ) 516
铣(xǐ) 516
徙(xǐ) 516
喜(xǐ) 516
系(xì) 517
细(xì) 517
隙(xì) 519
匣(xiá) 519
峡(xiá) 519
狭(xiá) 519
瑕(xiá) 520
下(xià) 520
夏(xià) 521
仙(xiān) 522
先(xiān) 522
纤(xiān) 522
酰(xiān) 523
鲜(xiān) 523
闲(xián) 523
弦(xián) 523
咸(xián) 524
嫌(xián) 524
显(xiǎn) 524
险(xiǎn) 525
藓(xiǎn) 525
县(xiàn) 525
现(xiàn) 525
限(xiàn) 526
线(xiàn) 526
陷(xiàn) 527
腺(xiàn) 527
乡(xiāng) . . . 527
相(xiāng) . . . 527
香(xiāng) . . . 528
箱(xiāng) . . . 529
镶(xiāng) . . . 529
详(xiáng) . . . 529
响(xiǎng) . . . 530
相(xiàng) . . . 530
向(xiàng) . . . 530
项(xiàng) . . . 530
象(xiàng) . . . 530

像(xiàng) . . . 530
橡(xiàng) . . . 531
肖(xiāo) 531
鸮(xiāo) 531
消(xiāo) 531
萧(xiāo) 532
硝(xiāo) 532
销(xiāo) 532
削(xiāo) 532
小(xiǎo) 533
效(xiào) 536
楔(xiē) 536
蝎(xiē) 536
颉(xié) 536
协(xié) 536
胁(xié) 536
斜(xié) 536
谐(xié) 537
携(xié) 537
鞋(xié) 537
缬(xié) 537
写(xiě) 538
泄(xiè) 538
泻(xiè) 538
卸(xiè) 538
榭(xiè) 538
心(xīn) 538
芯(xīn) 538
辛(xīn) 539
锌(xīn) 539
新(xīn) 539
薪(xīn) 540
信(xìn) 540
星(xīng) 540
行(xíng) 540
形(xíng) 540
型(xíng) 541
兴(xìng) 541
杏(xìng) 541
性(xìng) 541
凶(xiōng) . . . 542
胸(xiōng) . . . 542
雄(xióng) . . . 542
熊(xióng) . . . 543
休(xiū) 543
修(xiū) 543
鸺(xiū) 544
髹(xiū) 544
朽(xiǔ) 544
袖(xiù) 544
锈(xiù) 544
嗅(xiù) 544
溴(xiù) 544
须(xū) 544
虚(xū) 544
需(xū) 544
徐(xú) 544
许(xǔ) 544
序(xù) 544
续(xù) 544
絮(xù) 544
蓄(xù) 544
玄(xuán) . . . 545
悬(xuán) . . . 545
旋(xuán) . . . 545
选(xuǎn) . . . 546
镟(xuàn) . . . 547
削(xuē) 547
穴(xué) 547
学(xué) 548
雪(xuě) 548
血(xuè) 548
熏(xūn) 548
薰(xūn) 549
寻(xún) 549
巡(xún) 549
循(xún) 549
训(xùn) 549
讯(xùn) 549
汛(xùn) 549
驯(xùn) 549
蕈(xùn) 549
压(yā) 549
押(yā) 551
鸦(yā) 551
鸭(yā) 551
牙(yá) 551
芽(yá) 551
蚜(yá) 552
崖(yá) 552
哑(yǎ) 552
雅(yǎ) 552
亚(yà) 552
砑(yà) 553
咽(yān) 553
胭(yān) 553
烟(yān) 553
阉(yān) 554
淹(yān) 554
延(yán) 554
严(yán) 554
岩(yán) 554
沿(yán) 554
研(yán) 554
盐(yán) 555
颜(yán) 555
檐(yán) 555
衍(yǎn) 555
掩(yǎn) 555
眼(yǎn) 555
演(yǎn) 556
厌(yàn) 556
砚(yàn) 556
验(yàn) 556
堰(yàn) 556
雁(yàn) 556
燕(yàn) 556
扬(yáng) . . . 556
羊(yáng) . . . 556
阳(yáng) . . . 556
杨(yáng) . . . 557
洋(yáng) . . . 557
仰(yǎng) . . . 557
养(yǎng) . . . 557
氧(yǎng) . . . 557
样(yàng) . . . 557
夭(yāo) 558
腰(yāo) 558
要(yāo) 558
姚(yáo) 558
窑(yáo) 558
摇(yáo) 558
遥(yáo) 558
咬(yǎo) 558
药(yào) 559
要(yào) 559
钥(yào) 559
鹞(yào) 559
椰(yē) 559
冶(yě) 559
野(yě) 559
业(yè) 560
叶(yè) 560
页(yè) 561
曳(yè) 561
夜(yè) 561
液(yè) 561
腋(yè) 562
靥(yè) 562
一(yī) 562
伊(yī) 563
衣(yī) 563
依(yī) 564
仪(yí) 564
夷(yí) 564
宜(yí) 564
移(yí) 564
遗(yí) 565
乙(yǐ) 565
已(yǐ) 566
以(yǐ) 566
蚁(yǐ) 566
倚(yǐ) 566
椅(yǐ) 566
义(yì) 566
刈(yì) 566
议(yì) 566
异(yì) 566
抑(yì) 568
译(yì) 569
易(yì) 569
疫(yì) 569
益(yì) 569
逸(yì) 569
翌(yì) 569
意(yì) 569
溢(yì) 569
缢(yì) 569
翼(yì) 569
因(yīn) 569

阴(yīn).....570
音(yīn).....570
银(yín).....570
引(yǐn).....570
吲(yǐn).....571
饮(yǐn).....571
蚓(yǐn).....571
隐(yǐn).....571
荫(yīn).....571
印(yìn).....571
茚(yìn).....571
英(yīng)....571
婴(yīng)....572
缨(yīng)....572
樱(yīng)....572
鹰(yīng)....572
应(yīng)....572
迎(yíng)....573
荧(yíng)....573
盈(yíng)....573
萤(yíng)....573
营(yíng)....573
楹(yíng)....574
蝇(yíng)....574
颖(yǐng)....574
影(yǐng)....574
瘿(yǐng)....575
映(yìng)....575
硬(yìng)....575
拥(yōng)...575
壅(yōng)...576
永(yǒng)...576
蛹(yǒng)...576
佣(yòng)...577
用(yòng)...577
优(yōu).....577
尤(yóu).....578
由(yóu).....578
油(yóu).....578
疣(yóu).....578
游(yóu).....579
有(yǒu).....580
莠(yǒu).....582
右(yòu).....582
幼(yòu).....582
柚(yòu).....583
诱(yòu).....583
鼬(yòu).....583
迂(yū).....583
淤(yū).....583
余(yú).....584
鱼(yú).....584
娱(yú).....584
渔(yú).....584
隅(yú).....584
逾(yú).....584
榆(yú).....584
虞(yú).....584
与(yǔ).....584
羽(yǔ).....584
雨(yǔ).....584
语(yǔ).....585
玉(yù).....585
郁(yù).....585
育(yù).....585
预(yù).....586
域(yù).....587
欲(yù).....587
阈(yù).....587
御(yù).....587
愈(yù).....587
薁(yù).....587
鸢(yuān)...587
鸳(yuān)...587
元(yuán)...587
芫(yuán)...587
园(yuán)...587
垣(yuán)...588
原(yuán)...588
圆(yuán)...592
缘(yuán)...593
源(yuán)...593
远(yuǎn)...594
约(yuē).....594
月(yuè).....594
阅(yuè).....594
悦(yuè).....594
跃(yuè).....594
越(yuè).....594
晕(yūn)....594
云(yún)....594
匀(yún)....595
芸(yún)....595
耘(yún)....595
允(yǔn)....595
陨(yǔn)....595
孕(yùn)....595
运(yùn)....595
韵(yùn)....596
熨(yùn)....596
扎(zā).....596
杂(zá).....596
灾(zāi).....597
栽(zāi).....597
载(zài).....598
再(zài).....598
在(zài).....599
暂(zàn).....599
錾(zàn).....599
藏(zàng)....599
凿(záo).....599
早(zǎo).....599
枣(zǎo).....599
藻(zǎo).....599
皂(zào).....599
造(zào).....600
噪(zào).....601
责(zé).....601
择(zé).....601
增(zēng)....601
憎(zēng)....602
赠(zèng)....602
轧(zhá).....602
闸(zhá).....602
铡(zhá).....602
栅(zhà).....602
炸(zhà).....602
蚱(zhà).....602
摘(zhāi)....602
宅(zhái)....603
窄(zhǎi)....603
债(zhài)....603
沾(zhān)....603
毡(zhān)....603
蛅(zhān)....603
展(zhǎn)....603
占(zhàn)....603
栈(zhàn)....603
战(zhàn)....603
站(zhàn)....603
绽(zhàn)....603
张(zhāng)..603
章(zhāng)..604
樟(zhāng)..604
蟑(zhāng)..604
涨(zhǎng)..604
掌(zhǎng)..604
丈(zhàng)..604
杖(zhàng)..604
帐(zhàng)..604
账(zhàng)..604
胀(zhàng)..604
障(zhàng)..604
辗(zhǎn)....604
朝(zhāo)....604
招(zhāo)....604
着(zháo)....604
爪(zhǎo)....604
沼(zhǎo)....604
兆(zhào)....605
照(zhào)....605
罩(zhào)....605
遮(zhē).....605
折(zhé).....605
蛰(zhé).....606
摺(zhé).....606
辙(zhé).....606
褶(zhě).....606
蔗(zhè).....606
针(zhēn)....606
侦(zhēn)....607
珍(zhēn)....607
帧(zhēn)....607
真(zhēn)....607
砧(zhēn)....608
榛(zhēn)....608
诊(zhěn)....608
枕(zhěn)....608
阵(zhèn)....608
振(zhèn)....608

震(zhèn)....608
镇(zhèn)....608
争(zhēng)...608
征(zhēng)...608
蒸(zhēng)...608
拯(zhěng)...609
整(zhěng)...609
正(zhèng)...610
证(zhèng)...611
政(zhèng)...611
症(zhèng)...611
之(zhī).....611
支(zhī).....612
只(zhī).....612
汁(zhī).....612
枝(zhī).....612
知(zhī).....613
肢(zhī).....613
织(zhī).....613
栀(zhī).....613
脂(zhī).....613
蜘(zhī).....613
执(zhí).....613
直(zhí).....614
值(zhí).....615
职(zhí).....615
植(zhí).....615
止(zhǐ).....618
纸(zhǐ).....618
指(zhǐ).....619
趾(zhǐ).....619
酯(zhǐ).....619
志(zhì).....619
制(zhì).....619
质(zhì).....620
治(zhì).....621
栉(zhì).....621
致(zhì).....621
秩(zhì).....621
窒(zhì).....621
蛭(zhì).....621
滞(zhì).....621
置(zhì).....622
雉(zhì).....622
稚(zhì).....622
中(zhōng)..622
终(zhōng)..625
钟(zhōng)..626
螽(zhōng)..626
肿(zhǒng)..626
种(zhǒng)..626
踵(zhǒng)..629
种(zhòng)..629
仲(zhòng)..629
众(zhòng)..629
重(zhòng)..629
州(zhōu)....630
周(zhōu)....630
轴(zhóu)....630
肘(zhǒu)....631
帚(zhǒu)....631
绉(zhòu)....631
昼(zhòu)....631
皱(zhòu)....631
骤(zhòu)....631
朱(zhū).....631
侏(zhū).....631
珠(zhū).....631
株(zhū).....632
硃(zhū).....632
蛛(zhū).....632
竹(zhú).....632
逐(zhú).....632
烛(zhú).....632
主(zhǔ).....632
煮(zhǔ).....634
苎(zhù).....634
助(zhù).....634
住(zhù).....634
贮(zhù).....634
注(zhù).....635
柱(zhù).....635
蛀(zhù).....636
铸(zhù).....636
筑(zhù).....636
抓(zhuā)....636
专(zhuān)..636
砖(zhuān)..636
转(zhuǎn)..637
转(zhuàn)..637
桩(zhuāng).638
装(zhuāng).638
壮(zhuàng).639
状(zhuàng).639
撞(zhuàng).639
追(zhuī)....639
椎(zhuī)....639
锥(zhuī)....639
坠(zhuì)....640
准(zhǔn)...640
拙(zhuō)....640
桌(zhuō)....640
着(zhuó)....640
灼(zhuó)....640
浊(zhuó)....640
啄(zhuó)....640
孜(zī)......640
咨(zī)......640
姿(zī)......640
资(zī)......640
滋(zī)......641
髭(zī)......641
子(zǐ)......641
姊(zǐ)......642
籽(zǐ)......642
梓(zǐ)......642
紫(zǐ)......642
自(zì).......642
字(zì).......647
综(zōng)....648
棕(zōng)....648
鬃(zōng)....648
总(zǒng)....648
纵(zòng)....649
走(zǒu).....651
租(zū)......651
足(zú)......651
族(zú)......651
阻(zǔ)......651
组(zǔ)......651
祖(zǔ)......652
钻(zuàn)....652
嘴(zuǐ).....652
最(zuì).....652
左(zuǒ).....654
佐(zuǒ).....654
柞(zuò).....654
作(zuò).....654
坐(zuò).....655
座(zuò).....655

阿贝树脂 Albertol
阿贝折射仪 Abbe refractometer
阿布尼水准仪 Abney level
阿尔本树脂(古塔波树脂) Alban(an)e
阿赫格绷带防腐处理 Ahig process
阿开木油 akee oil
阿拉伯树胶 acacia gum; arabic gum; gum arabic
阿拉伯树胶糖浆 gum syrup
阿拉尔代斯木材防腐法 Allardyce process
阿利定律 Allee's law
阿利氏群聚原则 Allee's principle
阿伦法则 Allen's rule
阿洛波林 alloprene
阿牟离心式圆筛 Apmew (centrifugal) screen
阿姆拉特树胶 amrad gum
阿纳康达防腐处理法 Anaconda process
阿诺德树木园 Arnold arboretum
阿朴莰烯 apobornylene
阿斯木碱 alstonine
阿斯普伦德滤水测定仪 Asplund's freeness-tester
阿斯普伦德热磨机制浆法 Asplund process
阿斯特罗姆剥皮机(链式) Astrom barking machine
阿托品 atropin(e)
阿魏油 asafoetida oil
阿仙药 pale cutch
埃力克森漆膜附着力试验机 Erichsen tester
埃洛石 halloysite
埃默森打浆机 Emerson beater
埃塞俄比亚界 Ethiopian region
埃塞俄比亚树胶 Abyssinian gum
埃氏杯突试验机 Erichsen tester
埃氏发酵管 Emhorn tube
埃斯特拉德防腐法 Estrade process
癌学 carcinology
癌肿 cancer
癌肿的 cancerous
癌肿木 cancer-tree
矮草 short grass
矮干的 low-stemmed
矮干树木 low pole wood
矮高位芽植物 nanophanerophyte
矮灌丛 dwarf shrub
矮桁架 pony truss
矮槲寄生 dwarf mistletoe
矮化 dwarf(ing); stunt
矮化病 dwarf(ing)
矮化的 dwarf(ing); nanous
矮化树 stunted tree
矮化栽培 dwarf culture
矮化砧 dwarf(ing) stock; small stature of stock
矮化作用 dwarf effect
矮胶轮车 dolly
矮林 bosk; bosket; bosquet; chatwood; coppice; coppice forest; coppice stand; coppice with standards; coppice wood; copse; dwarf forest; elfin forest; low forest; scrub (stand); spinny; sprout forest
矮林保残作业法 composite forest system; storied coppice system; coppice with standards system
矮林采伐 cant
矮林地 coppice-land
矮林伐区 coppice coupe
矮林改造 coppice conversion
矮林更新法 regeneration by stool shoots
矮林混农作业 combined system of field crop and coppice; coppice with field crops system
矮林皆伐作业法 clear-cutting coppice method
矮林平茬 coppicing
矮林天然更新 natural regeneration by shoots and suckers
矮林原生苗 maiden
矮林择伐法 coppice selection method
矮林择伐林 coppice selection forest
矮林择伐作业 coppice selection system
矮林作业 coppice method; coppicing; simple coppice system; sprout-system
矮林作业法 coppice system; coppice-forest system; coppice method
矮林作业更新 regeneration under coppice system
矮轮运材车 konaki
矮门 screen door

A

矮乔木限界[线] dwarf tree limit; cripple tree limit
矮曲林 elfin wood; dwarf forest
矮生的 nanous
矮生灌木 suffrutex
矮生树 bushy tree
矮生性 dwarfism
矮石南灌丛 dwarf shrub heath
矮疏林 woodland
矮树 arbuscle; pygmy tree
矮松针油 dwarf pine needle oil
矮态 dwarfism; nanism
矮性砧 dwarf(ing) stock; small stature of stock
矮竹 dwarf bamboo
矮壮剂 growth retardant
矮壮素(调节生长或除草) chlormequat chloride
艾尔顿金字塔 Eltonian pyramid
艾菊油 tanacetum oil
艾伦曲线 Allen's curve
艾氏粉剂 aldrin dust
艾氏剂(杀虫剂) aldrin
艾氏乳剂 aldrin emulsion
艾鼬 fitch; masked polecat
爱尔兰坐犬 Irish setter
隘道 bottleneck; defile
隘口 sag
安哥柯巴树脂素 angocopaloresene
安哥拉树脂(油漆制造) Angola copal
安哥拉紫檀素 angolensin
安古树酊 angostura tincture
安古树酚 angustifoliol
安古树皮油 angostura bark oil
安古树提取物 angostura extract
安乐椅 easy chair; elbow chair; Fauteuil; sleepy hollow chair
安全操作 safe operation
安全侧面(卸车对侧) safe side
安全出口 fire escape
安全存量 safety level
安全措施 safety measure
安全带 life belt; safety strip
安全岛(街道) refuge
安全岛(生物) safety island
安全道(伐木) gateway road
安全工作强度 safe working strength
安全工作压力 safe working pressure
安全工作应力 safe working stress
安全规程 safety code
安全含水量 safe moisture content
安全荷载 permissible load
安全滑板 safety slide
安全密度(使害虫不致猖獗) security density
安全坡度 safety gradient
安全绳 safety line
安全视距 safe sight distance
安全梯 fire escape
安全系数 assurance factor; factor of safety; margin of safety; safety factor
安全系索(滑轮) safety guy
安全限度 margin of safety; safe limit
安全因素 factor of safety
安全应力 safe stress
安全圆锯 safety saw
安全运转 safe handling
安全载荷 safe load
安全着陆方法 safe-landing approach
安全装置 safety device
安山岩 andesite
安锁冒头 lock rail
安妥(杀鼠剂) antu
安息香树胶 gum benzoin
安息香树脂 benzoin resin
安主操作 safe handling
安装 laying; setting
安装角 knife angle
安装索 hay wire; pass line; rigging rope; setting line; straw line
安装索导向滑轮 straw line fairleader
安装索滑轮 straw line block
安装索具 rig
安装索具用绞盘机 rig donkey
安装索卷筒 straw drum
安装索套钩 straw line hook
安装图 installation diagram
安装用短链 pass chain
桉树矮林 mallee
桉树胶素 edudesmin
桉树精油 eucalyptus oil
桉树栲胶 maletto extract; redunca; wandoo extract
桉树林土 malice soil
桉树皮单宁 maletto tannin
桉树烯 eucalyptene

桉树油 eucalyptus oil
桉叶烯 eudesmene
桉油 eucalyptus oil
氨 ammonium
氨化 aminate; ammonification
氨化产物 aminate
氨化肥料 ammoniated fertilizer
氨化水溶紫胶 hydram
氨化物 ammoniate
氨化细菌 ammonifying bacteria
氨基甲酸酯(杀虫剂) carbamate
氨基葡糖 aminoglucose
氨基三唑(除草剂) aminotriazole
氨基树脂胶黏剂 amino resin adhesive
氨基酸 amino acid
氨基纤维素 aminocellulose
氨基脂 aminolipoid
氨水 ammonia liquor; aqua ammonia
氨态氮 ammoniacal nitrogen
鞍接 grafting saddle; saddle grafting
鞍形区 col
鞍形压模(单板加工) saddle
鞍形椅背 saddle back
鞍形栽植(穴底形状) saddle-planting
鞍状壳 ephippium
鞍座 intermediate support; saddle set; saddle
岸 bank; coast; shore
岸壁 quaywall
岸墩 abutment pier; quay
岸礁 fringing reef
岸旁插垛 side jam
岸上带 supralittoral zone
岸上检尺材积 river scale
岸上系排装置 shore hold
岸湾 embayment
按触辊型砂光机 contact wheel sander
按纯折价收益 net discounted revenue
按等造材 buck for grade
按地面位置[计测] at the base
按伐根位置[计测] at the stump
按价 ad valorem
按价关税表 ad valorem tariff
按健全材标准检尺 sound scale
按进出门时间计付工资法 portal-to-portal pay
按立方英尺计付工资 pay-basis by cubic feet
按立木度计算断面积 basal area at stocking
按粒度分级 size
按林班采伐 compartment cutting; compartment felling; felling by compartment
按龄级采伐 stage cut
按木材标准局规定的捡尺 bureau scale
按年轮计算的年龄 chronological age
按钮 button; knob; push button
按钮控制 push-button control
按揉 knead
按属性抽样 sampling for attribute
按统货趸售木材 camp run
按蓄积量抽样 sampling for growing stock volume
按直线英尺计付工资 pay-basis by running feet
按指数调整价格 indexation of price
按重量配料 weighing dosing
胺 amine
胺化水溶紫胶 hydram
胺基防腐剂 amine preservative
胺吸磷(杀虫剂) amiton
案辊 table roll
案上接 bench grafting
暗藏绞链 invisible hinge
暗搭鸠尾榫接 stoppe-lap dovetail joint
暗的 opaque
暗度 opacity
暗度计 opacimeter
暗发芽箱 dark germinator
暗反应 dark reaction
暗沟 blind drain; subdrain; underdrain
暗沟排水 underdrainage
暗河 underground river
暗红壤 dark red soil
暗红色砖红土 dark red latosol
暗呼吸 dark respiration
暗黄色土层 dingy yellow horizon
暗灰色灰黏化土 dark gray gleysolic soil
暗机枪 hammerless gun
暗礁 submerged reef
暗拉手 flush handle

B

暗马牙榫 secret dovetail
暗区 quenched region
暗渠 underdrain
暗色表层 umbric epipedon
暗色黏粒(土壤) drab clay
暗色森林土 dark forest soil
暗色土 fuseous soil
暗色心材 dark heart
暗色针叶树(反射率低) dark conifer
暗生长反应 dark-growth reaction
暗视野 dark field
暗视野显微镜 dark-field microscope
暗适应 dark adaption
暗榫 blind tenon; concealed tenon; dowel; hidden tenon; twist
暗榫眼 blind mortise
暗息 implicit interest
暗销(接合销) dowel pin
暗性发芽 dark germination
暗棕壤 dark brown soil
暗棕色森林土 brown forest dark soil
暗棕色有机质土 dark brown organic soil
肮脏的 masty
凹背锯 hollow-backed saw
凹部 rabbet
凹槽 channel; fillister; flute; groove; rebate
凹槽(木杆道横木中间) saddle
凹槽侧板 notch board
凹槽圆线脚 bead-and-quirk
凹处 recess
凹的 concave; concavus
凹底刨 concave plane
凹雕 diaglyph
凹兜(wā dōu) spin; spur
凹痕 indenture
凹口 indentation
凹口线型 cove moulding
凹面 concave; concavus
凹面体 neiloid
凹面圆锯片 concave saw
凹刨 compass plane; fluting plane
凹曲线体 neiloid
凹饰线脚 bed moulding
凹室 alcove
凹榫 blind mortise
凹凸的 knaggy
凹凸干基 buttress (root); buttressed root
凹凸根干 spin; spur
S形凹凸榫接合 hook rebate
凹凸纹理(机械缺陷) raised grain
凹纹理 dimpled grain
凹陷 hollow
凹陷点(地面或轨道) sag
凹陷接缝 sunken joint
凹形边饰条 scotia
凹形椅 dipped seat
凹形椅座 dropped seat; hollow seat
凹圆线条 flush bead
凹状质壁分离 concave plasmolysis
螯合 chelate
螯合物 chelate substance; chelate
螯合作用 chelation
坳(ào) col
奥地利公式法(收获预定法) Austrian formula
奥尔森型磨蚀计 Olsen-type wearometer
奥卡尔挤压法碎料板制造工艺 Okal-extrusion process
奥卡尔型刨花板 Okal chipboard
奥林匹克国家纪念物及国家公园 Olympic National Monument and National Park
奥斯博恩火情探测仪 Osborne fire finder
奥斯特瓦尔德黏度计 Ostwald viscometer
奥陶巴脂 otoba tallow (butter)
奥陶纪 Ordovician period
澳桉酚 australol
澳柏型加厚 callitrisoid (*or* callitroid) thickening
澳柏型具缘纹孔 callitrisoid (*or* callitroid) bordered pit
澳大利亚界 Australian region
澳大利亚-马来(动物)亚界 Austro-Malayan region
澳洲栗配基 castanogenin
澳洲食蜜鸟 marvel of Peru; four-o'clock
八倍体 octoploid
八基数的 octamerous
八甲磷(杀虫剂) Pestox-III; schradan
八角枫碱 alangine
八角茴香 aniseed

八角茴香油 badian oil; star anise(ed) oil
八角劈材 octagon hewn lumber
八进制的 octal
八进制数位 octal digit
八氢惹烯 octahydro retene
八碳糖 octose
八字腿(家具) splayed foot
巴(压力单位) bar
巴当 badan
巴尔测树仪 Barr dendrometer
巴伐利亚群状渐伐作业(更新期较短) Bavarian group system
巴克表 barkometer; tannometer
巴拉圭茶 mate; paraguay tea
巴拉圭茶碱 yerbine
巴拉塔树胶 balata
巴拉橡胶 para rubber
巴黎绿(杀虫剂) Paris green
巴黎绿防腐剂 Paris green preservative
巴林比重计 Bailing's hydrometer
巴洛克式园林(规则式) Baroque garden
巴拿马角规 Panama angle gauge
巴婆 papaw; pawpaw
巴赛脱犬 basset
巴氏灭菌器 pasteurizer
巴斯德杀菌法 pasteurization
巴特列夫热压机(刨花板连续热压机) Bartrev-press
巴西草原 campo
巴西黑黄檀油 Brazilian rosewood oil
巴西黄樟油 Brazilian sassafras oil
巴西柯巴树脂 Brazilian copal
巴西蜡 Brazil wax
巴西灵 brazilin
巴西玫瑰木油 Brazilian rosewood oil
巴西木素 brazilin
巴西热带雨林次生林 capoeira
巴西疏林草原 campo cerrado
巴西稀树草原 campo sujo
巴西棕榈蜡 carnauba-wax
扒钉 dog; dog iron; double-headed nail; spike
芭蕉蜡 pisang wax
疤痕 scar
疤皮 incrustate
拔齿 wresting
拔出伐根 uprooted stump
拔出力 withdrawal force
拔出阻力 withdrawal resistance
拔根-归堆机 puller-buncher
拔根机 grubber; puller; root dozer; stump extractor; stump-puller
拔节期 elongating stage
拔树机 tree extractor
拔株 regueing
拔桩机 pile drawer; pile extractor
把柄 handle
把工程承包给个人 contract the project out to individuals
把环 ring dog
把手 handle
靶 target
靶板溃疡 target canker
靶子 target
坝 weir
坝顶 crest; weir crest
坝基石块 ballast
坝前防渗层长度 water path
坝前外层覆盖 sheeting
霸鹟科 flycatcher
耙地 harrow; harrowing
耙路机 ripper
耙土机 harrower; rake dozer
白斑(细胞内白色沉积物) floccosoids
白斑病(幼苗主茎受表土灼伤) white spot
白斑腐 white mottled rot
白标签种子 white tag seed
白菖脑 calameone
白菖烯 calamene
白菖油 calamus oil
白菖油萜 calarene
白菖樟脑 calamus camphor
白瓷土 china clay
白度 brightness; degree of whiteness; whiteness
白度测定器 hollow cone
白垩 chalk
白垩沉积物 chalky deposit
白垩的 chalky
白垩化 chalking
白垩纪 Cretaceous period
白飞燕草苷元 leucodelphinidin
白粉病 mildew; powdery mildew

B

白粉菌 powdery mildew
白腐 white rot
白腐菌 white-rot fungus
白干层油 cajeput oil; cajuput oil
白岗岩 alaskite
白桂枝 canella
白厚漆 white paste
白花色苷 leuco-anthocyanin
白花色苷元 leuco-anthocyanidin
白花香茅油 inchi grass oil
白化 albinism; albino; blooming
白化苗 albino
白化现象 albinism
白碱土 white alkali soil
白浆层(漂洗层)(土壤剖面E_a层) albic layer; E_a-layer
白浆土 albic soil
白僵病(菌) muscardine
白蜡 Chinese insect wax; white wax
白蜡虫 pe-la insect; white wax insect
白蜡树酚 Fraxinol
白蜡树苷 Fraxin
白蜡树亭 Fraxetin
白冷杉 Colorado white fir; white fir; white balsam fir
白利比重计 Brix hydrometer
白利糖度 Brix
白令桥 Bering land bridge
白明胶(食用) gelatinum
白木坯 whitewood
白木炭 white charcoal
白柠檬 limette
白柠檬油 lime oil; limette oil
白坯涂漆 open pore coating
白坯涂漆法(不打底色和填孔剂) open-pore finishing
白砒(杀虫剂) Arsenious oxide
白漆苷元 leucofiseltinidin
白千层油 melaleuca oil
白肉桂 canella
白色浆 white paste
白色胶 acid size
白色轮腐 white ring rot
白色磨木浆 white ground-pulp
白色松香胶 free rosin size
白色体 leucoplast
白矢车菊苷元 leucocyanidin
白树油 cajaput oil
白树脂 white resin
白霜 hoar frost
白水 backwater; pulpwater; white water
白水槽 white water chest
白水回收装置 saveall
白松(云冷杉类淡色材树种) whitewood
白檀二烯 amyradiene
白檀灵 amyrolin
白檀酮 amyrenone
白炭 white charcoal
白土层 albic horizon
白锈病 white rust
白液 flesh liquor; white liquor
白蚁 termite; white ant
白蚁巢 termitary; termite hill
白蚁防护罩 house-block cap
白蚁防治 termite control
白蚁塚 termite hill
白蚁通道 termite runway
白蚁危害 termite damage
白蚁窝 termitarium
白蚁稀树草原 termite savanna
白玉兰 white magnolia; white yulan
白云母薄片 isinglass
白云石 dolomite
白云岩 dolomite; dolomitite; dolostone
白樟油 white camphor oil
白珠木烯 gaultherilene
白珠木油 gaultheria oil
白灼病 bark scorch
百标准堆 standard hundred
百分比未出现率 percentage absence
百分比相似性 percentage similarity
百分点 percentage point
百分度的 centigrade
百分率的点 percentage point
40%百分位树高规则 height according to the 40% rule
百分位数 percentile
40%百分位胸径规则 mean DBH according to the 40% rule
百分温度度数 centigrade degree
百份橡胶中的份数 part per hundred parts of rubber
百合池 lily-pool
百合花形花冠 liliaceous corolla
百合园 lily garden

百菌清(杀菌剂) chlorothalonil; daconil
百里香油 thyme oil
百立方英尺(材积计量单位) cunit
百舌鸟 butcher-bird; shrike
百万(10^6) mega-
百万吨 megaton (Mt)
百万分之一 part per million (ppm)
百万克(10^6g) megagram (Mg)
百叶板 luffer board
百叶窗 back-flap; gill; shutter
百叶窗板 louver board
百叶门 slat door
百叶箱 thermometer shelter; thermometer-screen
百叶箱温度表 shelter thermometer
百治磷(杀虫剂) dicrotophos
柏林蓝 Berlin blue
柏木 funereal cypress; Chinese weeping cypress
柏木酮 cedrenone; cedrone
柏木油 cedar leaves oil; cedar oil; cypress oil
柏木沼泽 cypress swamps
柏木浊 cedar (wood) oil
柏型纹孔 cupressoid pit
柏油 asphalt; mineral pitch
柏油胶 asphalt size
柏油烯 cupressene
摆锤冲击试验 pendulum impact test
摆锤冲击试验机 pendulum impact testing machine
摆锤式冲击强度 Charpy impact strength
摆锤式冲击试验机 Charpy (impact) machine; Charpy pendulum machine; simple beam impact machine
摆动 sweep
摆动锤 pendulum hammer
摆动换向齿轮 tumbler
摆动器 oscillator
摆动式上木机 swing charger
摆杆(试验机上) pendulum bar
摆锯 goose saw; pendulum cross cut saw; pendulum saw; swing cut-off saw; swing saw; swinging cross-cut saw
摆压法(防腐) oscillating pressure process
摆压周期(防腐) oscillating pressure cycle
摆转(起重臂) swing(ing)
摆转起重架 guy derrick
摆转索(吊杆) slue line
败坏 deterioration
败育 abortion
败育花粉 abortive pollen
败育卵 abortive egg
拜萨基紫胶 baisacki; bisacki
扳机 sear; trigger
扳机护环 trigger guard
扳机簧 sear spring
扳机拉力 trigger pull
扳手 wrench
班 logging crew; working crew; team
斑(色谱的) spot
斑驳 mottle
斑点 fleck; mottle; speck; spot
斑点病 spot disease
斑点材 speckled wood
斑蛾 leaf skeletonizer moth
斑腐 mottled rot; pecky
斑痕组织 scar tissue
斑块性 patchiness
斑螟亚科昆虫 casebearers
斑纹 brindle
斑岩 porphyry
斑叶植物 variegated plant
斑杂岩 taxite
斑状花纹 mottled figure; roe figure
斑状纹理 mottled grain
搬出距离 extraction distance
搬钩 hook-lever; log jack; peavy; peavie
搬钩工 hookman
搬钩弯钩 peary hook; peavy dog
搬逬费 handling expense
搬运 handle
搬运机械 handling machine
搬运路 extraction rack (road); skid road
搬运作用 transportation
板 general board
板壁 brattice
板壁装饰(室内) cabinet finish
板边欠厚(厚度不足) dubbing

B

板材 board; panel; planking; plate; sheet
板材叉具 mill grapple
板材垛 stack of lumber
板材剪切试验 panel shear test
板材交错堆垛 staggered lumber pile
板材刨平 blanking
板材升降机 lumber elevator
板材制材厂 board (saw) mill
板层岩 flagstone
板尺 board rule
板尺数 footage
板凳 pew
板底滑道 pole-slide with plank-bottom
板垫分离器 separator of board and caul
板堆基柱 pier
板垛 pile of lumber
板方材 lumbering
板根 buttress; buttress root; buttress flare; buttressed root; flange; flare root; root buttress; tree buttress
板滑道 board-chute; plank chute; plank-slide
板簧 flat spring
板基 plate basal
板剪 panel-shears
板块构造 plate tectonics
板块构造学 plate techtonics
板耪 planker
板类制品 panel products
板篱 board fence
板链 plat link chain
板梁桥 plate girder bridge
板料 panel; sheet material
板面 face side
板面不匀(刨花板) show through
板面光泽加工 plate glazing
板面开槽胶合板 grooved plywood
板面量度(面积单位) surface measure
板面罩光 plate glazing
板坯 mat; mattress
板坯成型 mat formation; mat-laying
板坯成型机 mat forming machine
板坯成型输送带 forming belt; mat assembly belt
板坯成型输送机 forming conveyor
板坯放炮(热压鼓泡) blow of mat
板坯横截锯 flying cut-off saw
板坯机 mat former
板坯铺装机 mat forming machine
板坯剔除装置 mat reject station
板坯压缩率 mat compressibility
板坯预装机 mat preloader
板皮 slab
板皮裁边 slab edging
板皮裁边机 slabbing edger
板皮削片机 slabber; slab-chipper
板式家具 panel furniture
板试验 ground board test
板条 lag; lath; stave; wood strip; strip; slat
板条锯 slat saw
板条门 batten door; slat door
板条木管 stave pipe
板条箱 slat crate; crate
板条芯细木工板 laminated board; laminboard; stripboard
板条荫棚 lath-screen
板条圆锯机 stripper
板条运输机 slat conveyor
板条栅 hoading
板芯材结构 lumber-core construction
板芯胶拼机 lumber core edge gluer
板芯拼接机 lumber core composer
板岩 slate
板英尺(体积单位, $=1\text{ft}^2\times1$英寸厚$=\frac{1}{12}\text{ft}^3=0.00236\text{m}^3$) board foot (feet)
板英尺材积表 board-foot volume table
板英尺计量 board scale
板英尺检尺 board-foot scaling; board scale
板英尺-考得转换 board foot-cord conversion
板英尺-立方英尺转换 board foot-cubic foot conversion
板英尺量度 board measure; board-foot measure
板英尺数 footage
板英尺与立方英尺材积比 board foot-cubic foot ratio
板垣 board fence
板院 back yard; lumber yard; service yard; timber yard; timbershed; yard

板院防腐 yard preservative
板院气干 yard seasoning
板院设备 yard equipment
板院贮存材 yard lumber
板栅篱(板桩) sheet-pile
板状板根 plank buttress
板状干基 buttressed root; plank buttress
板状根 palar; plank buttresses root; plank root; tabular root
板状铁结核 tabular ferric concretion
版画 woodcut
半安装的 semi-mounted
半暗机枪 semi-hammerless gun
半抱茎形 semi-amplexicaul
半边接穗法 split-scion graft
半边线脚 measure and a half
半边小斧 half hatchet
半变动成本 semi-variable costs
半变态类 hemimetabola
半不育性 semisterility
半常绿林 semi-evergreen forest
半潮位 half-tide-level
半沉材 bobber
半成材 shop lumber; shop timber
半成品 crude product; half-finished goods; half-finished product; partly finished goods; semi-manufacture
半成品浆 half stuff
半成品浆料 half-stock
半成熟枝扦插 june-struck cutting
半持久型病毒 semi-persistent virus
半椽 half-principal
半存留期 half-life
半搭接 half-lap joint
半担子菌 hemibasidiomycete
半倒吊簧板 semi-bar lock
半倒生的 semi-anatropal; semi-anatropous
半定量浸注法 Lowry process
半端梁 tail piece
半反应期 half-life
半方材 half timber
半方差分析 semivariance analysis
半方形凿齿锯链 semi-chisel chain
半分化种 semispecies
半浮游生物 hemiplankton
半幅单板 half size veneer
半腐层 duff
半腐层湿度表 duff hygrometer
半腐解腐殖质 duff mull
半腐生的 semisaprophytic
半腐生性的 half-saprophyte
半腐生植物 hemisaprophyte
半附生植物 hemiepiphyte
半干法(纤维板工艺) semi dry process; wet-dry method
半干旱的 semiarid
半干旱地区 semi-arid regions
半干旱区 semiarid region
半干性油 semi-drying oil
半干纸浆 semi-dry pulp
半刚质绝缘板 semi-rigid insulation board
半隔离 semi-isolation
半共生 metabiosis
半固定沙丘 semifixed dune
半挂车 fifth-wheel
半灌木 half-shrub; semi-frutex; subshrub; suffrutex; undershrub
半灌木地上芽植物 chamaephyta suffruticosa
半灌木性地上芽植物 suffruticose chamaephytes
半灌木状的 suffrutescent; suffruticose; suffruticulose
半光泽涂饰 semi-gloss finish
半合生的 semiconnate
半合子 hemizygote; merozygote
半河狸尾式(枪柄) semi-beaver tail
半桁架 half truss
半红树林 semi-mangrove
半喉缩 half choke
半化石树脂 semi-fossil resin
半化学浆 semi-chemical pulp
半环孔材 semi-ring porous wood
半荒漠 semi-desert
半灰壤 semipodzol
半活节 partial intergrown knot
半活养生物 hemibiotroph
半机械浆 semi-mechanical pulp
半寄生 half-parasite
半寄生物 hemiparasite; semiparasite; half-parasite
半寄生植物 hemiparasite; semiparasitic plant
半加工材 half-wrought timber

半浆 half stuff; half-stock
半浆料 half-stock
半胶体 hemicolloid
半截站杆 chicot
半径 radius
半径面 edge grain
半鸠尾榫 half dovetail
半鸠尾榫槽接 half-dovetail dado joint
半具缘纹孔对 semibordered pit-pair
半空细胞法(防腐) Lowry process
半控制镶嵌图 semi-controlled mosaic
半露木结构 half-timbering
半露柱 pilaster
半轮裂 half ringshake
半轮生的 hemicyclic; semiverticillate
半螺旋采胶法 half spiralcurve cut
半落叶林 semi-deciduous forest
半履带车(后轮套防滑履带) half-track
半木素 hemilignin
半耐寒的 half hardy
半耐寒性一年生植物 half hardy annuals
半耐久性胶黏剂 semi-durable adhesive
半内生植物 hemiendobenthos
半浓度营养液 half strength nutrient solution
半漂浆 partly bleach pulp
半匍匐植物 hemiepiphyte
半嵌接 halving joint
半嵌鸠尾接 dovetailed halving joint
半鞘翅 hemielytra
半球摄影 hemispherical photography
半融合 meromixis
半乳甘露聚糖 galactomannan
半乳糖苷 galactoside
半乳糖苷酶 galactosidase
半乳糖酶 galactase
半乳糖葡甘露聚糖 galactoglucomannan
半乳糖脂 galactolipid
半散孔材 semi-diffuse porous wood
半沙漠 half-desert
半深海带 bethal zone
半数受限值 median tolerance limit
半数有效量 effective dose 50 (ED50)
半衰期 half-life
半双列杂交 half-diallel crossing
半天然下种 seminature sowing
半天然植被 seminatural vegetation
半同胞(同父或同母) half-sib
半同胞协方差 half-sib covariance
半桶形沙发 tub sofa
半透明的 semi-hyaline; semi-opaque; translucent; translucid
半透性膜 semipermeable membrane
半突变 half-mutation
半突变型 half mutant
半挖填 side-hill
半微量凯氏定氮法 semimicro Kjeldahl method
半下位子房 semi-inferior ovary
半纤维素 half-cellulose; hemicellulose
半纤维素生物降解 hemicellulose biodegradation
半弦纹理 mixed grain
半咸水 brackish water
半咸水草甸 brackish water meadow
半咸水沼泽 brackish water swamp
半显性 semidominance
半休闲地 bastard fallow; half-fallow; occupied fallow
半悬式(集材) high-lead
半悬式钢索集材 high-lead cable logging; high-lead hauling
半悬式集材 high-ball logging; high-lead logging
半悬式集材杆 high-lead spar tree
半悬式集材机 ground skidder; higb-lead skidder
半悬式集材装置 high-lead yarding system
半悬式牵引索 bull line
半悬式牵引索滑轮 high-lead block
半悬式索道集材 skyline-highlead yarding
半悬式主滑轮 high-lead block
半循环湖 meromictic lake
半淹没的 water logged
半燕尾榫 half dovetail
半阳坡 half positive slope; half sunny slope
半叶扦插 divided leaf cutting
半阴坡 half negative slope; half shaded slope
半隐鸠尾榫 half-blind dovetail
半隐芽植物 hemicryptophyte

半隐燕尾榫 half-blind dovetail
半硬枝插穗 semi-hardwood cutting
半硬质纤维板 half-hard (fibre) board; semi-hard board; semi-hardboard
半游离状打浆 semi-fast beating
半鱼骨法 half-herringbone cut
半原木 half-log
半圆材 half balk; half-round; half-round wood
半圆单板 half-round veneer
半圆单板旋切机 half-round veneer lathe
半圆刀具 astragal tool
半圆截薄板 halfround cutting veneer
半圆木料 puncheon
半圆锹 semi-circular spade
半圆式木凿 firmer gouge
半圆饰 bead and reel
半圆筒式剥皮机 semi-drum
半圆线脚 astragal; beading; nosing
半圆镶边嵌板 bead-flush panel
半圆形安乐椅 tub chair
半圆形侧边板 half round profile board
半圆形屋顶 compass roof
半圆形枕木 buck; halfmoon tie; half-round sleeper
半圆旋切 back cut; half-rotary-cutting; half-round cutting; stay-log cutting
半圆旋切单板 half-rotary-cut veneer; half-round veneer
半圆旋切法 half-round; half-round method
半圆旋切机 lathe for eccentric peeling
半圆与纺锤形相间的饰条 spindle-and-bead
半圆枕木 hog-back(ed) sleeper; round-back(ed) sleeper; two-faced tie
半圆柱 half round post
半圆锥锹 semi-conical spade
半缘纹孔 half bordered pit
半缘纹孔对 half bordered pit-pair
半月生殖周期 semilunar reproductive cycle
半月形枕木 halfmoon tie
半载式集材 load bearing
半沼泽土 half bog soil; semiboggy soil
半知菌 deuteromycete; imperfect fungus
半知菌类 fungi imperfecti
半制成品 semi-finished products
半制成商品 semi-manufactured commodity
半制品 semi-products
半致死基因 semi-lethal gene
半致死剂量 medial lethal dose
半窒息 half choke
半重瓣的 semidouble
半子囊菌 hemiascomycete
半自动打捆装置 semi-automatic bundling device
半自动化 semi-automatic
半自动索道跑车 semi-automatic cableway carriage
半自然群落 seminatural community
半自养的 mesotrophic
半族 moiety
伴胞 campanion cell
伴生树种 associated tree species; associated species; associates
伴生种 accompanying species; auxiliary species; companion species
伴细胞 companion cell
伴性(遗传) sex linkage
伴性基因 sex-linked gene
伴性遗传 sex-linked inheritance
伴植 companion planting
拌胶机 blender; glue machine; glue mixer; mixer
拌胶器 glue mixer
拌胶碎料 resin-coated particle
拌土处理法(防腐) puddling treatment
拌种 dressing; seed dressing
瓣 valve
瓣裂 opening by valves
瓣裂的 valvate
瓣片 limb
瓣蹼足 lobed foot
绑紧绳股(切断钢索前) seize
棒 bar; clave (*plur.* clavae); stick
棒材 bar stock
棒节 clave (*plur.* clavae)
棒双价体 rod bivalent

棒条椅背 stick back
棒头 caput
棒形 claviform
棒状的(触角) clavate
棒状晶体 rod-like crystal
棒状体 clave (*plur.* clavae)
棓儿茶素 gallocatechin
棓子 gall nut
棓子单宁 gallotannin
棓子栲胶 gallnuts extract
傍道 sideline
傍管薄壁组织 paratracheal parenchyma
傍管带状薄壁组织 paratracheal banded parenchyma; paratracheal zonate parenchyma
傍管环管束状薄壁组织 paratracheal-vasicentric parenchyma
傍管聚翼状薄壁组织 paratracheal-confluent parenchyma
傍管型 paratracheal
磅秤 platform balance; platform scale
包 pack; package; sack
包被 envelope; peridium
包被结构(种皮等) covering structure
包被物质 encrusting material; entrusting substance
包边 wrap
包尔格莱夫疏伐(择伐式疏伐) Borggreve thinning
包含体(病毒) inclusion
包涵体 inclusion body
包涵物 inclusion
包歼法 containning method
包角(钢索绕过转向滑轮形成的角度) bight angle
包壳弹 jacketed bullet
包埋 embedding
包囊 cyst
包皮 blister; packing
包铁 tree iron; tree plate
包头(捆木索) knob; nubbin
包土栽植 ball planting
包锡 tin
包扎(苗木) burlapping
包扎处理(防腐) bandage treatment
包扎法(防腐) banding method; bandage process; method of bandage
包扎机 strapper
包装 packing
包装材 packing timber
包装材料 package; packing
包装箱 packing box; packing case; casing
包装用单板 packaging veneer
包装纸 baling paper; brown packing paper; wrapping paper
包装纸板 packing board
苞 bract
苞鳞 bract scale
苞片 bract; cover leaf; interfoyles
苞叶 bract leaf; bracteal leaf; subtending leaf
孢瓣 valve
孢粉分类学 palynotaxonomy
孢粉学 palynology
孢梗束 coremium (*plur.* -ia)
孢囊孢子 sporangiospore
孢囊柄 sporangiophore
孢朔(苔藓) capsule
孢子(菌) spore
孢子败育 spore abortion
孢子采集器 spore trap
孢子层 sporulating mat
孢子发芽 spore germination
孢子负荷 spore load
孢子果 sporocarp
孢子花粉素 sporopollenin
孢子角 spore horn
孢子接种体 spore inoculum
孢子囊 sporangium
孢子生殖 spore reproduction
孢子释放 spore release
孢子四胞的 tetradymous
孢子体 diplophyte; sporophyte
孢子同型 isospory
孢子团 spore ball; spore mass
孢子外壁 exospore
孢子形成 sporulation
孢子叶 sporophyll
孢子叶球 strobile; strobilus
孢子印 spore print
孢子雨 spore shower

胞 cyst; loculus
胞壁层 cell wall layer
胞壁雕纹 cell wall sculpturing
胞壁加厚 cell-wall thickening
胞壁间隙 interwall space
胞壁裂缝 cell wall check; spiral cavity
胞壁物质 cell wall substance
胞壁质 dermatoplasm
胞壁组成 cell wall component
胞啶 cytidine
胞苷酸 cytidylic acid
胞间层 intercellular layer; middle lamella; middle layer; true middle lamella
胞间道 interceilular duct; intercellular canal; intercellular passage; interstitial canal
胞间菌丝体 intercellular mycelium
胞间连丝 plasmodesma (*plur.* -mata)
胞间裂开的 loculicidal
胞间腔 intercellular cavity; interstitial cavity
胞间隙 intercellular space; interstitial space
胞间液 intercellular fluid
胞间质 intercellular substance
胞嘧啶 cytosine
胞嘧啶核苷 cytidine
胞嘧啶核苷酸 cytidylic acid
胞囊形成 encystment
胞内酶 cell-bound enzyme; endoenzyme
胞内微管 intracellular canaliculus
胞溶作用 cell lysis
胞束酶 cell-bound enzyme
胞外多糖 extracellar polysaccharide
胞外间隙 extracellular space
胞外酶 extracellular enzyme
胞芽 gemma
胞质分裂 cytokinesis
胞质基因杂合细胞 cytohet
胞质配合 plasmogamy
胞质团 plasmon
雹 hail
雹害 damage by hail; injury by bail
薄板 commercial veneer; lamella (*plur.* -llae/-llas); lamination; plate; slice; thin baord
薄板单位 panel unit
薄板覆面 laminate
薄板条 slat; web
薄壁细胞 parenchyma cell; thin wall cell
薄壁纤维 thin-walled fibre
薄壁组织 parenchyma; soft tissue; spongy tissue
薄壁组织病害 parenchymatous disease
薄壁组织束 parenchyma strand
薄边板(板材缺陷) feather edge board
薄边斜接 feather mitre joint
薄层 layer
薄层色谱法 thin-layer chromatography
薄层土 shallow soil
薄层岩 flagstone
薄荷素油 dementholised peppermint oil
薄荷油 mint oil; peppermint oil
薄荷原油 complete mint oil; crude mint oil; crude peppermint oil
薄锯片 thin-kerf blade
薄膜 film; pellicle; wafer
薄膜胶 film adhesive; film glue
薄膜张力 film force
薄木 veneer
薄木片 splint
薄木条 splint-wood
薄片 chip; film; flake; flatting; lamella (*plur.* -llae/-llas); lamina (*plur.* laminae); layer; slice; stratum (*plur.* strata); wafer
薄片刨花板 flakeboard; waferboard
薄片铺面刨花板 flake-faced particle board
薄铁盘层(土壤剖面Bfe层) placic horizon; Bfe-layer
薄型地下茎 leptomorphic rhizome
薄型锯屑板 thin sawdust board
薄型刨花板 thin wood particleboard
薄叶常绿的 chimonochlorous
薄缘 feather edge
薄缘圆锯 ground-off saw
薄凿 paring chisel

B

薄纸 tissue
饱和层 zone of saturation
饱和差 saturation deficit; vapour pressure deficit
饱和点 saturation point
饱和度 percentage humidity; saturation
饱和红砂壤 eutric arenosol
饱和浸注(防腐) treatment to refusal
饱和空气 saturated air
饱和密度 saturation density
饱和区(地下水) zone of saturation
饱和湿度 saturated humidity
饱和石质土 eutric lithosol
饱和温度 saturated temperature
饱和吸收量 adequate retention
饱和系数 saturation coefficient
饱和压力 saturated pressure
饱和蒸汽 saturated steam; saturated vapour
饱和蒸汽干燥法 seasoning in saturated steam
饱和蒸汽压 saturated vapour pressure
饱满度(种子) filling
饱满种子 filled seed; full seed
宝石紫胶 ruby lac; ruby shellac
保安林 protected forest; protection forest
保本图 break even chart
保残木作业 high forest with reserves (system); high forest with standards (system)
保藏物 preserve
保持区 conservation district
保持时间 retention time
保持水土 conserve water and soil
保持性 retentivity
保持郁闭 maintenance of canopy
保存 preservation
保存带 buffer strip; buffer zone
保存林 demarcated forest; reserved forest
保存性 keeping quality
保兑信用证 confirmed credit
保光性 gloss retention
保护 blocking; conservation; protection
保护层 protective layer
保护带 guard row; protection belt; protection strip
保护地 shelter
保护地带 protective zone
保护法 protective measure
保护法案 conservation law
保护行 guard row
保护候鸟法 migratory bird conservation act
保护剂 protectant
保护经营 conservative management
保护林(树)带 protection belt; protection strip
保护林带 shelterbelt; windbreak
保护贸易 protective trade
保护拟色 pseudosematic colour
保护区 protected area; conservancy; conservation district; district of guard; protective zone; protect-range; ranger district; refuge area; reserve area
保护区管理类型 management categories of protected area
保护色 protecting colouration; protective colour; protective colouration
保护树 nurse-wood; protected tree; shelter-wood; special tree
保护涂层 overcoating
保护物种 protected species
保护性反应 protective reaction
保护性放牧 conservative grazing
保护性拟态 protective mimicry
保护性杀菌剂 protective fungicide
保护性疏伐 conservative thinning
保护性物质 protective substance
保护作物 nurse-crop
保留矮林作业法 reserve coppice method
保留成本 reservation cost
保留带 leave strip
保留地 reserve; reserve cell
保留伐 reserve cutting
保留价格 reservation price
保留利润 retained profit
保留林 reserve-forest; reserve-form stands; reserve-seed forest
保留林作业 system of high forest with standards

保留母树采伐作业法 seed-tree cutting method; seed-tree cutting system
保留母树作业 seed-tree system
保留木 hold over; maintenance of stand; remaining tree; reserving of stand; store; main crop; X-tree; leave tree; reserved tree; residual tree
保留木矮林作业 coppice with reserve
保留木作业 reserve seed method; reserve tree method; reserve-tree system
保留乔林作业 reserve high forest system
保留时间 retention time
保留盈余 retained earnings
保留中林 reserve sprout forest
保留中林作业 reserve sprout system
保留作业 reserve-seed tree method
保色性 colour retention
保墒 hold water in the soil
保湿材料 moisture-holding material
保湿的 moisture-retaining
保湿介质 moisture-holding medium
保水性 water retention
保税货物 bonded goods; goods in bond
保土坝 soil saving dam
保卫细胞 guard cell
保温材料 heat insulator
保温层 thermal insulation
保温性 temperature persistence
保险不足 under-insurance
保险单 insurance policy
保险费 insurance premium; premium
保险杆 bumper
保险凭证 insurance policy
保险人 underwriter
保险丝 fuse
保险装置 protecting device
保香剂 fixative
保续 perseverance; sustainablility
保续采伐量 prescribed yield
保养 maintenance
保养方便性 serviceability
保幼激素 juvenile hormone
保幼激素类 juvenile hormone analong; juvenoid
保幼酮 juvabione
保育 stand tending
保育间伐 opening of the shelter wood
保育目的 tending objective
保育生物学 conservation biology
保育树 nurse tree
保真度 fidelity
保证金 cash deposit as collateral
保证金返还制度 deposit-refund system
保证利得 threshold return
保证书 warranty
报酬 emolument; return
报酬递减律 law of diminishing returns
报酬递增法则 law of increasing returns
报酬率 yield
报废折旧法 retirement method of depreciation
报废资产 abandoned assets
报火时间 report time
报价 bid; offer
抱 brood
抱材车 lumber carrier; lumber container car
抱肩榫 shoulder-clasp joint
抱茎的 clasping
抱茎叶 perfoliate leaf
抱器 harpes
抱室 brood chamber
抱树 brood tree
抱索器 grip
抱握器(卷须) claspers
抱握器(鳞翅目) harpes
鲍恩比率 Bowen's ratio
鲍尔精磨机 Bauer mill; Bauer-refiner
鲍尔氏药剂毒力测定法 Power's method
鲍尔双磨盘精磨机(磨浆) Bauer double disk refiner
鲍尔纤维分离机 Bauer defibrator
鲍威尔浸注法 Powell process; Powellizing process
暴发(林火) blow-up
暴发火 blow up fire
暴发性火 blow-out
暴风(十一级风) violent storm
暴风抗力 wind firmness; windthrow resistance

B

B

暴风雪 blizzard
暴露 exposure
暴露区 exposure plot
暴露试验 exposure test
暴雨 deluge; rain gush; rain gust
暴雨移置(洪水设计) storm transposition
暴长木采伐 felling of wolf trees; liberation cut; liberation felling
暴涨(河流) freshet
爆发式进化 explosive evolution
爆发性病害 explosive disease
爆发性流行病 explosive epidemic
爆裂 burst
爆破 blowing
爆破法 explosion process; Masonite-process
爆破法纤维板 Masonite
爆破法硬质纤维板 Masonite hardboard
爆破工艺 explosion process
爆破枪(爆开大木材) splitting gun
爆破强度 bursting strength
爆燃仪 knockmeter
爆炸 blow back
爆炸拆垛 shoot a jam
爆炸清除伐根 blowing-up stump
爆炸上限 upper explosive limit
爆炸危险 explosion hazard
爆炸楔 explosive wedge
爆震计 knockmeter
刨 plane; planing; shave
刨边机 edger
刨槽 fillister
刨成坯料 blank
刨齿 clearer (tooth); drag
刨床 plane-castes; planer; planing machine
刨刀 plane-iron; rabbet; shaver; shaving knife
刨刀夹角 pitch
刨钩 hook-aroon
刨光 finishing; surfacing
刨光板 dressed board
刨光材 dressed lumber; dressed timber; planed lumber; superfaced lumber; surfaced lumber; surfaced timber
刨光尺寸 finished size
刨光的 surfaced
刨光和装配 dressed and matched
刨花 abatement; chip; flake; scobs; wood chip; shaving
S形刨花 S-type chip
刨花板 chipboard; flakeboard; particle board; shaving board
刨花板板坯 chip cake
刨花板心材 particle board core stock
刨花板心胶合板 particle board core plywood
刨花定向器 flake orienter
刨花干燥机 chip dryer
刨花滑槽 shaving chute
刨花计量 chip metering
刨花模胜 particle moulding
刨花铺装机 chip-spreading machine
刨花切断器(刨光机上) chip breaker
刨机 planning machine
刨裂 planer splitting
刨片 chipping; flake; flaking; flatting; shaving; slice
刨片刀 slicing knife
刨片机 chipper; chipping machine; flake-producing machine; flaker; flake-shaving machine; shaping machine; shaving machine; slicer
刨片机单板刨削机 slicing machine
刨片机进料槽 pocket of shaving machine
刨平 planing
刨平的 surfaced
刨平机 surface planer
刨切单板 sliced veneer; slicing veneer
刨切单板饰面 sliced veneer overlay
刨切的 sliced
刨切机 slicer; slicing machine
刨伤树脂道 traumatic resin duct
刨身(刨刀的木框) plane wood; plane stock
刨丝机 excelsior cutting machine
刨铁 plane-iron
刨削 chipping; gouging; planing; shaping; shaving
刨削车间 planing mill
刨削锯 novelty saw; planer saw
刨削型刨花 cutter-type particle
刨圆削片联合机 round reducer

刨制成材厂 planing mill
刨座 plane stock
曝光 exposure
曝光表 actinometer; exposure meter; sensitometer
曝光不足[的底片] underexposure
曝光计 exposure meter
曝光间隔 exposure interval
曝光试验 exposure test
曝光系数 exposure factor
曝气 aerate
杯式采脂法 cup system
杯形的 calyciform
杯状聚伞花序 cyathium
杯状竖割 cup frilling
杯状竖沟 cup frills
杯状整枝 vase-form training
卑湿的 boggy
北大西洋涛动 north Atlantic oscillation
北方的 boreal
北方干草原的 campestrian
北方混交林 mixed boreal forest
北方偏干性草原地带 campestrian
北方期 Boreal period
北方森林带 boreal forest zone
北方针叶林生物群系 northern coniferous forest biome
北回归线 Tropic of Cancer
北极 north pole
北极草甸 arctic meadow
北极带 arctic zone
北极地衣荒地 arctic lichen barren
北极第三纪植物区系 arcto-tertiary flora
北极垫状草地 arctic mat grassland
北极冻原 arctic tundra
北极风 arctic wind
北极高山 arctic alpine
北极荒原 arctic barren
北极气团 arctic air mass
北极区 arctic region
北极圈 arctic circle
北极生物带 arctic life zone
北极植物 arctic plant
北界 arctogaea; megagaea
北美草原 prairie
北美常绿阔叶灌丛 chaparral
北美黄杉 soft fir
北美金缕梅素 hamamelin
北美鼠李药用树皮 cascara-sagrada
北美苏合香脂 American storax
北欧生态学派 Uppsala school
北坡 north-facing slope
北弯式架空索道 north bend skyline system
北弯式索道集材 north bend yarding
北弯式下坡索道 downhill north bend system
贝 shellfish
贝彻特木材防火处理法 Bachert process
贝茨拟态 Batesian mimicry
贝尔法斯特桁架 Belfast truss
贝尔默漂白机 Bellmer bleacher
贝杰-荣格纤维分离机 Beija-Jung defibrator
贝壳杉胶 kauri gum
贝壳杉柯巴脂 kauri copal
贝壳杉树脂 kauri resin
贝壳杉油 kauri oil
贝类毒素 saxitoxin
贝母云 luminous cloud
贝尼酮 benihione
贝齐纳浮动刀头剥皮机 Bezner floating cutterhead peeler
贝瑟尔满细胞防腐法 Bethel (full-cell) process
贝塔分布 beta distribution
贝形刻痕刀 oyster-knife tooth
备查记录 memorandum entry
备查资料 back-up data
备抵 allowance
备件 spare parts
备忘记录 memorandum entry
备忘录 minute
备用轨道 by-track
备用轮 fifth-wheel
备用清漆 spare varnish
备用森林 forest reserves
备用柱 stand-by column
背[部] dorsum (*plur.* dorsa)
背板 back crossing; back panel; backing board; flitch
背板(胶合板等) back board; back
背板(镜框的) scale board
背板(昆虫) notum (*plur.* nota)
背板单板 back veneer
背冰川的 lee
背部的 dorsal

背侧线 dorsopleural line
背衬 back lining
背衬纸 backing paper
背地弯曲的 apogeoesthetic
背地性 apogeotropism
背风 leeward
背风波 lee wave
背风面 lee side; lee
背风坡 lee slope; leeward slope
背风涡流 lee eddies
背缝[线] dorsal suture
背负式动力喷雾器 motorized knapsack sprayer
背负式活塞喷雾器 knapsack piston sprayer
背负式弥雾喷粉机 knapsack atomizer
背负式喷粉机 knapsack sprayer
背负式喷雾器 backpack pump; knapsack sprayer; manual knapsack sprayer
背负式手动喷雾器 knapsack piston sprayer
背负式自动喷雾器 automatic knapsack sprayer
背腹的 ventri-dorsal
背腹性 dorsiventrality
背割面(割面背面作采脂割面) back face
背合铰链 counter-flap hinge
背角 lip clearance
背景 back ground
背景辐射 background radiation
背景亮度 background luminance
背景适应 background adaptation
背景栽植 background planting
背景噪音 background noise
背景值 base-line
背景种植 background planting
背鳞 elytra
背面 back; backside; dorsum (*plur.* dorsa)
背面(单板) loose face; loose side
背面采脂法(树干割面背面) back cupping
背面的 dorsal
背面支柱 back leg
背面直交单板 back crossing
背坡 backslope
背下横档(家具零部件) back slat lower
背斜 anticline
背斜型[构造] anti form
背血管 dorsal vessel
背倚子叶 incumbent cotyledon
背缘尺 back gauge; back rule
背中线 dorsal line
倍半二倍体 sesquidiploid
倍半香茅烯 sesquicitronellene
倍半氧化物 sesquioxide
倍加作用 diplosis
倍数体 polyploid
倍增器 multiplier
倍增时间(细胞群体) doubling time
倍增效应 multiplicative effect
倍子 plant gall
被 envelope
被保护基团 blocked group
被测品系 tested line
被动感染 passive infection
被动免疫性 passive immunity
被动散布 passive dispersal
被动升温 passive warming
被动式遥感 passive remote sensing
被动土压力 passive earth pressure
被短绒毛 tomentellate
被短柔毛 pubescence
被分析物 analyte
被封含水层 confined aquifer
被覆地 covered land
被害树 damaged tree
被绢毛的 sericeous
被毛的 piliferous
被绵毛的 tomentose
被囊合子 oospore
被片 tepal
被茸毛的 velvety
被绒毛的 tomentose
被软毛的 villose
被微柔毛的 puberulent; puberulous
被压木 oppressed stem (tree); overtopped stem; retarded tree; subordinate; suppressed tree
被叶的 infoliate
被蛹 obtect pupa
被用物 substrate
被遮挡木(角规测树时) hidden tree

被遮阴树 sheltered tree
被子植物 angiosperm
焙烧过度 overcook
焙烧炉 roaster
锛斧 hatchet; hewing axe; mattock; Pulaski
锛机(锛枕木) adzing machine
锛削 adze
锛制枕木 hewing crosstie; hewn tie
锛制中方材 medium hewn square
锛子 adze; shim hoe
本车载挂车回空 humpback
本迪森检验法(检验纤维板表面质量) Bendtsen-method
本底放射性当量强度 background equivalent activity
本底辐射 background radiation
本底谱 background spectrum
本地产的 indigenous
本地的 aboriginal; native
本地浮游生物 autogenetic plankton
本地木材 native wood
本地品种 land variety
本地生植物 indigenous plant
本地植物区系 autochthonous flora
本地种 autochthon; autochthonous species; indigenous species; native species
本金 principal; principal sum
本能 instinct
本色 natural colour
本色涂漆(木材直接涂漆) open pore coating
本体聚合 bulk polymerization
本樟油 hon-sho oil
本质参数 essential parameter
苯 phene; benzene
苯胺(杀虫剂) aniline
苯胺黑 aniline black
苯并呋喃树脂 coumarone resin
苯并咪唑44(杀菌剂) bavistin; carbendazim; carbendazol
苯酚 phenol
苯酚浸渍材 phenolic-impregnated wood
苯酚热固树脂 phenol thermosetting resin
苯基烯丙叉 cinnamal; cinnamilidene
苯甲醚 anisole
苯菌灵 benlate; benomyl
苯醌 quinone
苯来特(杀菌剂) benlate; benomyl
苯亚甲基香豆满酮 benzylidenecoumaranone
崩岗 collapsing hill
崩坏地 slided land
崩积 colluvial
崩积层 colluvium
崩积土 colluvial soil
崩积物 colluvial deposit; colluvium
崩溃 debacle
崩落(地质) avalanche; collapse; scree
崩塌 crushing
绷带处理 bandage treatment
绷带法防腐 banding method; bandage process
绷带法阻止虫道 banding
绷带化 fasciation
绷索 guy; guy cable; guy line; guy rope; side cable; stay
绷索安全角 safe guy angle
绷索钩钉 guy-spike
绷索滑轮 snake block
绷索滑轮系索 stay rope
绷索架杆装车 guy-line loading
绷索架杆装车方式 guy-line system
绷索间距 guy-line spacing
绷索支柱 guy-line post
绷索装车 guy-line loading
绷索装车方式 guy-line system
逼真度 fidelity
鼻梁板(橇脚前面的横板) noseboard
比表面积(面积-重量比) specific surface area; specific surface
比法尔磨浆机 Biffar-mill
比放射性 specific radioactivity
比格尔犬 beagle
比功 specific work
比估计 ratio estimation
比活性 specific activity
比价出卖 sale by successively increased bids
比胶合 specific adhesion
比较成本 comparative cost
比较解剖 comparative anatomy
比较抗腐性 relative durability
比较利益 comparative advantage
比勒法(防腐) Buehler process

比例尺标度 scale
比例分配 proportional allocation
比例极限 proportional limit
比例极限应力 stress at proportional limit
比例税 proportional tax
比流量 specific discharge
比率估计公式 ratio estimator
比强度 specific strength
比热 specific heat
比容(体积-重量比) specific volume
比色法 colourimetric method; colourimetry
比色计 chromometer; colourimeter
比生长速度 specific growth rate
比湿 specific humidity
比松联合干燥机 Bison-combi-dryer
比特(二进位数) bit
比特利希抽样 Bitterlich sampling
比压 specific pressure; unit pressure
比叶面积 specific leaf area
比叶重 specific leaf weight; specific leaf mass
比移值(色谱法) Rf value
比约克曼木素 Bjorkman's lignin
比重 specific gravity
比重保持与股价指数相同 marketing weight
比重瓶 pycnometer
吡啶铜 copper pyridine
吡喃葡萄糖 glucopyranose
彼得堡单位(锯材材积单位,等于4.672m^3) Petersburg standard
彼得格勒单位(锯材材积单位,等于4.672m^3) Petrograd standard
俾斯麦褐染色法 Bismarck brown method
必需氨基酸 essential amino acid
必需元素 essential element
必要器官 essential organ
必要样本数 required sample size
毕特利希可变半径样地 Bitterlich variable radius plot
毕特利希切口 Bilterlich-blade
毕特利希扇形叉 Bitterlich's sector fork
毕特利希速测镜 Bitterlich-relascope
毕特莫尔测仗 Biltmore stick
闭关自守经济 closed economy
闭果 indehiscent fruit
闭合测线 tie
闭合陈化时间(加热加压) closed assembly time
闭合导线 closed traverse
闭合裂缝(胶合板) closed split; bottleneck check
闭合时间 closing time; make time
闭合温度 closing temperature
闭花受精 cleistogamy; close fertilization
闭花受精的 kleistogamous; cleistogamous
闭环摇尺系统 closed-loop setworks
闭孔吸音纤维板 blind holed fibre board
闭模时间(压机) closing time
闭模温度(压机) closing temperature
闭囊壳 cleistocarp; cleistothecium (*plur.* -ia)
闭塞 blocking
闭塞具缘纹孔对 aspirated bordered pit pair
闭塞纹孔 aspirated pit
闭塞纹孔对 aspirated pit-pair
闭塞系统 block system
闭蒴果 cleistocarp
闭锁群体 close population
闭锁装置 blocking device
闭心式整枝 close-centered training
庇护树 nurse tree; shelter growth
庇护所 refuge; shelter
庇护作用 sheltering effect
庇荫 shade
庇荫地 shade area
荜澄茄酚 cubebhol
荜澄茄烯 cadinene
荜澄茄油 cubeb oil; litse cubeba oil
壁 wall
壁板 panel; panelling; sheathing; siding; siding board; wall board; wall panel
壁板划线规 panel gauge
壁板胶合板 plywall
壁板木螺丝 sheathing screw
壁板饰条 wainscoting moulding
壁橱 recess cabinet
壁灯 bracket light
壁骨材 stud(ding)

壁角沙发 cosy corner
壁角食橱 corner cupboard
壁角椅 corner chair
壁脚板 plinth; skirting (board)
壁龛 alcove
壁龛柱 recessed pilaster
壁孔 pit
壁孔突 pit torus; torus (*plur.* tori)
壁炉 fire place
壁炉架 mantel-board
壁炉上的饰架 overmantel
壁台 wall desk
壁柱 pilaster; wall post
避病性 disease-escaping
避车处 refuge
避车道 loop; turnout
避钙植物 calcifuge; calciphobe; calciphobous plant
避光的 photophobic
避碱植物 basifuge
避雷器 arrestor; arrester
避难所 refugium; abri; shelter
避鸟剂 bird repellent
避暑别墅 pavilion; summer-house
避暑胜地 summer resort
避酸的 oxyphobous
避酸植物 oxyphobe
避心下锯法 side cut
避盐种 salt-avoiding species
避阳植物 heliophobe
避阴 shade avoidance; shadephobe; sciophobia
避阴综合征 shade avoidance syndrome
避雨植物 ombrophobe
臂杆回转索(杆式集装机上) running guy
臂间倒位 pericentric inversion
臂内倒位 paracentric inversion
边 lap
边板(板皮) side cut; side cut piece
边板(家具) siding board
边材 alburnum; culled wood; sapwood
边材斑纹 sap streak
边材板 side board
边材变色 colour sapstain; sap stain
边材变色防治 sap stain control; sapstain prevention
边材变色菌 sap stain fungus
边材标记 edge-mark
边材腐朽 external decay; sap rot; sap (wood) rot
边材开裂 sap shake
边材率 percent of sapwood
边材面积 sapwood area
边材木 sap-wood tree; tree with retarded formation of heart-wood; alburnum tree
边撑 side mark
边挡板 fender
边挡木(集材道) sheer skid
边挡木(装卸线) breast log
边对接 edge butt joint
边腐 marginal rot
边号 catch mark; side mark
边号(原木旁边的记号) contramarque
边号(原木主人的记号) bark mark; catch mark
边护木 side log
边花(舌状花, 菊科) ray flower
边火子弹夹 rim-fire cartridge
边几 corner table
边脊 torus (*plur.* tori)
边际采伐 marginal cutting
边际产品 marginal product
边际产值 marginal revenue product
边际成本 marginal cost
边际成本定价法 marginal cost pricing
边际储蓄倾向 marginal propensity to save
边际代换率 marginal rate of substitution
边际法 marginal method
边际分析 marginal analysis
边际购买 marginal purchase
边际价值产品 marginal-value product
边际进口倾向 marginal propensity to import
边际利润 marginal profit
边际利润木材 marginal log
边际取得成本 marginal cost of acquisition
边际生产力 marginal productivity

B

边际生产力均等法则 law of equi-marginal productivity
边际生产力理论 marginal productivity theory
边际生产量 marginal product
边际生产系数 marginal coefficient of production
边际收益 marginal benefit
边际收益产品 marginal revenue product
边际收益率 marginal rate of return
边际消费倾向 marginal propensity to consume
边际效应 marginal effect
边际效用 marginal utility
边际效用递减法则 law of diminishing marginal utility
边际效用均等法则 law of equi-marginal utility
边际需要 margin requirement
边际要素成本 marginal-factor cost
边际支付曲线 marginal outlay curve
边交叉搭接 edge cross lap joint
边角材 scrap wood
边角地带林木 long corner
边角料 offcut timber
边接 edge joint
边节 edge knot
边界 boundary
边界层 boundary layer
边界极限 boundary limit
边界面 boundary surface
边界树 borderline tree; boundary log
边界效应 edge effect
边距 back gauge
边棱 arris
边料 offcut
边裂 edge split
边木 wing log
边刨 plough; plow; rabbet plane; side plane
边皮 flaw-piece
边皮材 slabwood
边坡 side; side slope; highwall
边坡标桩 slope stake
边嵌接 edge rabbet joint
边十字搭接 edge cross lap joint
边榫接 edge rabbet joint
边条 edging; margin strip; rand
边条切碎机 edging grinder
边条纤维回收箱 edge fibre chest
边弯 spring
边弯变形 edgewise distortion
边斜接 edge mitre joint; edge rabbet joint
边选装置 edge sorter
边沿变形 edgewise distortion
边缘 fringe
边缘薄壁组织 marginal parenchyma
边缘采伐 border-cutting
边缘槽刨 border plane
边缘地 marginal land
边缘点火法 edge firing
边缘对接 edge butt joint
边缘发火弹(来复枪子弹) rim-fire
边缘分布 marginal distribution
边缘分生组织 marginal meristem
边缘管胞 marginal tracheid
边缘光滑的 entire
边缘火 edge fire
边缘加厚 border thickening
边缘胶合单板 edge-gluing veneer
边缘角 edge angle
边缘节 margin(al) knot
边缘林 fringe woodland
边缘脉 marginal vein
边缘胚珠 marginal ovule
边缘嵌合体 mericlinal chimaera
边缘群落 edge community; mantle community; marginal community
边缘射线 marginal ray
边缘射线管胞 marginal ray tracheid
边缘射线细胞 marginal ray cell; terminal ray cell
边缘生境 marginal habitat
边缘生物 peribion
边缘胎座式 marginal placentation
边缘涂胶机 edge gluer
边缘位置控制 edge position control
边缘纹孔 marginal pit
边缘效应 border effect; edge effect
边缘修剪锯 edge trim saw
边缘植物 edging plant
边缘种 edge species
边载应力 edge load stress; side load stress
边折弯 edge bend

边枕 slab tie
边装簧板 side plate
编绘 compile
编结的绳环 eye-splice
编结吊索 braided sling
编捆框 pocket
编篱 wattle-fence
编篱作业 fasciae wicker work; wattle-work
编码 code; encoder
编码器 encoder
编目 filing; inventory
编排 booming; raft; rafting
编排场 raft ground
编排工 pond man
编排工出河工 short log cat
编排工人 raftsman
编排工组 swifter gang
编排机 rafting machine
编排绞车 swiftering machine
编排框 crib
编排框架(海排) cradle
编译 compile
编译程序 compiler
编译计算机 compiling computer
编扎(木排) lashing
编栅法 wicker-work
编杖栅栏 wicker-work barrier
编织纹拼板法 basket weave matching
编制 compilation
编制资本预算 capital budgeting
编组线(车辆) side track
蝙蝠蛾 hepialid moth; swifts
蝙蝠葛碱 dauricine
鞭抽 whip
鞭打 lashing
鞭毛(鞭节) flagellum (*plur.* -lla)
鞭子树(擦蹭毗邻木) whip
贬值 depreciation
扁柏酚 chamenol
扁柏素 chamaecin
扁柏烯 chamene
扁斧 adze
扁钩(带舌簧) fid hook
扁化作用 fasciation
扁环节链 plat link chain
扁簧 flat spring
扁甲类 cucujid beetle; flat bark beetle
扁节 spike knot; splay knot
扁平形整枝 flat form training
扁球形的 oblate
扁头钉 sprig
扁突线条 platband
扁叶蜂 webspinning sawfly
扁圆形的 oblate
扁枝烯 phyllocladene
匾额 inscribed tablet
苄氨基嘌呤 benzylaminopurine
苄基香豆满酮 benzylcoumaranone
变成浆末 mash
变成树脂 resinify
变程 range
变定(木材尺寸局部改变) set
变定时间 set time
变动成本 variable cost
变动面积样区法 sampling using variable radius plots
变动权数 changing weight
变动样区法 sampling using variable radius plots
变幅 amplitude
变更 variation
变更基因 modifying gene
变更期森林作业 conversion period
变更因子 modifying factor
变更作业 conversion
变更作业采伐量 conversion cut
变更作业法(变更进行程序) conversion system
变黑降等(因木材中铁与单宁化合) iron degradation
变化 meta~; variation
变化估计 estimating changes
变化探测技术 change detection technique
变换 transformation
变换载荷 alternating load(ing)
变尖削 taper
变量 variate; variable
变量载荷 varying load
变平 flatting
变迁 shift
变色(木材) staining; discolouration; off-colour; stain

B

变色边材 stained sap wood; stained sap
变色材 discoloured wood
变色菌 stain fungus; stainer
变色木材 stained wood; staining wood
变色效应 discolouration effect
B
变色心材 discoloured heart; stained heart (wood)
变渗的 poikilosmotic
变渗透压动物 poikilosmotic animal
变生群丛 faciation
变式 variant
变速齿轮 speed change gear
变速齿轮箱 change speed gear box
变速级数 number of steps of speeds
变速流(河川) variable flow
变速器 gear shift; speed transmission; speed variator
变速箱 gearbox; transmission
变态 abnormality; metamorphosis
变态类 metabola
变态雄蕊 lepal
变态叶 metamorphosed leaf
变位 dislocation
变位酶 mutase
变温 alternating temperature; fluctuating temperature
变温动物 ectotherm; heterothermic animal; poikilotherm; poikilothermal animal
变温性 poikilothermy
变现 realize
变形 anamorphosis; buckling; deformation; deformity; distortion; metamorphosis; shape distortion; warp
变形测定器 extensometer
变形的 deformed
变形曲线 anamorphic curve
变形细胞 amoebocyte
变型 aberration; form(a)
变型学说 phase theory
变性 denature
变性剂 denaturing agent
变性酒精 denatured alcohol
变性木材 transmuted wood
变性土 vertisol
变性纤维素 modified cellulose
变旋光作用 mutarotation
变压器 transformer
变叶病 phyllody
变异 variation; break
变异的 mutant
变异定律 law of variability
变异度 coefficient of variability
变异分子 metagon
变异体 variant
变异系数 coefficient of differentiation; coefficient of variation
变异型体 mutant
变异性 variability
变应力 varying stress
变质 degeneration; deterioration
变质碱土 degraded solonetz
变质景观 metamorphic landscape
变质铝土 degraded allite
变质木片 shiv
变质危害 deterioration hazard
变质岩 metamorphic rock; metamorphosed rock
变质作用 metamorphism
变种 variety
变种间杂交 interbreed
变种生物 mutant
变阻器 rheostat
便道 access road; short cut
便门 access door; postern
便桥 makeshift bridge; temporary bridge
便于作业的用材林 operable timber
遍发侵染 general infection
辨别 mark sensing
辫状河道 braided stream; braided river
标本 specimen
标本树 specimen tree
标称体积 nominal volume
标称压力 nominal pressure
标称直径 nominal diameter
标称重量 nominal weight
标尺 rod; staff; target-rod
标定法(火势) scaling law
标杆 pole; station pole
标高塔尺 target-rod
标号 indicators; marking
标号笔 marking keel

标号规则 marking rule
标号滚轮 marking roll
标记 brand; indicator; label; mark; tag; timber making
标记打印板 marking board
标记读出 mark sensing
标记法 marking method
标记放逐法 mark-and-release method
标记个体 marked individual
标记行为 marking behaviour
标记化合物 label(l)ed compound
标记树木(做记号) spotting; mark
标记原木 marking log
标记原子 tagged atom
标记重捕法 mark-and-recapture method; mark-recapture method; release-recapture method
标价 bid price; list price; posted price
标卖 auction sale
标签 tag
标售 sale by sealed tender
标酸基准 acidimetric standard
标线员 line spotter
标志化石 index fossil; leading fossil
标志基因 marker gene
标志牌 tally board
标志树 bearing tree
标志种 index species
标桩 stake
标准 gauge; standard
标准靶 standard target
标准保持量 typical retention
标准保续采伐面积 yield regulation by area
标准变量 canonical variate
标准材积表 standard volume table
标准材长(16ft) standard log length
标准采伐材积 possibility by volume
标准采伐计算法 yield determination methods
标准采伐量 cut; normal cut; possibility; regular cutting budget; standard cut; standard felling volume; prescribed yield
标准采伐面积 possibility by area; yield regulation by area
标准采伐顺序 normal succession of cutting; normal succession of felling
标准差(标准偏差) root mean square deviation; standard deviation
标准尺寸 normal size
标准大气 standard atmosphere
标准单板旋切机 standard veneer lathe
标准的室外试验法 standardized field-test method
标准等压面气压订正 reduction of pressure to a standard level
标准地 sample area; test plot
标准地法 sample plot method
标准地分区 sub-plot
标准地疏伐 sample thinning
标准地调查 sample-plot survey
标准地调查法 sample-plot survey method
标准堆(针叶树锯材材积单位) standard
标准干围 normal girth
标准规范 criterion (*plur.* -rial)
标准规格 standard specification
标准厚窄板(3英寸×9英寸和12ft) standard deal
标准化 standardization
标准化树高曲线 standardized height curve
标准活节(直径不超过1.5英寸) standard knot
标准菌种(微生物) type culture
标准考得(木材层积单位, 等于128ft^3) apparent cord
标准林分 normal stand
标准令量(25英寸×40英寸的五百张纸重) standard ream
标准密度高程 standard density altitude
标准磨蚀机 standard-abrader
标准木 mean sample tree; model tree; sample tree; standard; type tree; test tree; standard tree
标准木料 stock
标准木年龄 sample-tree age
标准年表 standard chronology
标准年伐量 normal annual yield
标准色 reference coler
标准试验 standard test
标准收获表 normal yield table

B

标准树高曲线 standard height curve
标准条件 standard condition
标准温度压力 normal temperature-pressure
标准误 standard error the mean
标准误差 standard error
标准箱(校正试验机) standard box
标准小区 representative area
标准小试件 small clear specimen
标准小样试验 small clear timber test
标准形率(正形率) normal form quotient
标准形数(正形数) normal form factor
标准旋切机 normal peeling lathe
标准压力 standard pressure
标准硬质纤维板 standard hardboard
标准直径 normal diameter
标准作业方式 standard treatment
飑线 squall line
表阿福豆素 epiafzelechin
表板 face crossing; lace veneer
表板横拼机 face veneer (composer)
表棓儿茶素 epigallocatechin
表荜澄茄酚 epicubehol
表层 deck layer; epipedon; face layer; face ply; overlay; upper horizon
表层腐殖质层(土壤剖面A_h层) surface humus layer; A_h-layer
表层干燥 surface dry
表层刨花 top layer chip
表层皮 skin
表层铺装机 faee spreader; surface spreader
表层侵蚀 sheet erosion
表层纤维应力 extreme fibre stress
表层硬化 case-hardening; case-hardening stress
表儿茶素 epicatechin
表格 table
表观比重 apparent specific gravity
表观弹性模量 apparent elastic modulus
表观光合速率 apparent photosynthetic rate
表观光合作用 apparent photosynthesis
表观密度 apparent density
表观木质素 apparent lignin
表观膨胀 apparent expansion
表观蠕变 apparent creep
表观体积 apparent volume
表观应力 apparent stress
表观重量 apparent weight
表化异构化 epimerization
表流 surface flow
表面 face side
表面变色 surface stain
表面不平度 surface irregularity
表面材积 superficial volume
表面层 skin
表面层次 shell
表面虫孔 surface worm hole
表面处理 superficial treatment; surface treatment
表面处理碎料板 surface-treated particleboard
表面粗糙度 surface roughness
表面单板 face veneer; surface (veneer)
表面腐朽 surface decay
表面干燥 skin dry; surface dry
表面供应量 apparent availability
表面固定性 surface soundness
表面光洁度 surface smoothness
表面活性剂 surface active agent; surfactant
表面积 surface area
表面加工 facing; finishing
表面检修 surface inspection
表面接缝 face joint
表面节子 face knot
表面结构 surface texture
表面结合强度 surface bonding strength
表面结膜(涂层) case-hardening
表面孔隙率 surface porosity
表面量度 superficial measure
表面裂 surface check; surface crack; surface shake; surface split
表面轮廓 surface profile
表面膜 surface film
表面耐沸水性 resistance of surface to boiling water
表面耐高温性 resistance of surface to high temperature
表面耐磨性 resistance of surface to wear

表面耐污性 resistance of surface to stain
表面耐香烟火性 resistance of surface to cigarette burn
表面抛光处理 pollshing of outer skin
表面刨光 dressing; surface dressing
表面膨胀 surface expansion
表面起毛(机械缺陷) surface fuzz(ing)
表面强度(造纸) pick strength
表面缺陷 external defect; open defect; surface blemish; surface defect
表面燃烧 surface combustion
表面渗碳 surface carburization
表面生长 scum; surface growth
表面施胶 surface size(ing); top sizing
表面湿润 wetting of surface
表面涂层 surface coating; top coat
表面涂料 surface coating
表面涂刷 external coating
表面涂装 surface finish(ing)
表面系数 specific surface
表面纤丝化 external fibrillation
表面消毒 surface sterilization
表面消费 apparent consumption
表面修饰 dressing; finish; finishing; surface finish(ing)
表面应力 quadric stress; skin stress
表面硬化(种子等) case-hardening; cementation; surface hardening
表面硬化防止剂 anticrustator
表面油饰 finish coat(ing)
表面张力 surface tension
表面振动器 surface vibrator
表面蒸发 surface evaporation
表面直交单板 face crossing
表面质量 surface quality
表面致密化 surface densification
表面帚化 external fibrillation
表面装饰 coating
表盘式指示器 dial indicator
表皮 cuticula; cuticle; epidermis; integument; periderm; tegument
表皮层 epidermis
表皮侵染 cuticular infection
表皮体 dermatosome
表皮土 rock mantle
表皮下的 infracutaneous; subepidermal
表皮原 dermatogen
表潜层 Ag horizon
表生的 extramatrical
表生动物相 epifauna
表生菌丝体 extramatrical mycelium
表生植物相 epiflora
表施 top dressing
表土 surface soil; top soil; topsoil; uppermost soil
表土层 epipedon
表土生物 epibiont
表土松土机 mulcher
表土中耕 surface cultivation
表现度 expressivity
表现型 phenotype
表现性状 phene
表型 phenotype
表型测定 phenotype test
表型方差 phenotype variance
表型模拟 phenocopy
表型协方差 phenotype covariance
表型选择 phenotypic selection
表型学 phenetics
表型遗传相关 phenogenetic correlation
表型遗传学 phenogenetics
表型值 phenotype value
表岩 mantle rock
表异构物 epimer
裱糊纸 backing paper
别棒 lash pole
别棍 springer
别罗勒烯 alloocimene
别墅 villa
别戊二烯 alloprene
瘪粒 barrenness; blind; empty seed
瘪粒种子 blind seed
玢岩 porphyrite
宾主共栖 xenobiosis
滨海的 littoral
滨海盐土 marine solonchak
滨河园 riverside garden
滨后 back shore
槟榔 areca palm; betelnut palm
濒枯木 dying tree
濒危物种 endangered species; threatened species
冰雹 hail

冰暴 ice storm
冰崩 ice avalanche
冰成的 glacial
冰成岩 glacial rock
冰川 glacier; icing
冰川擦痕 glacial scratch; glacial striae; glacial striation
冰川残遗种 glacial relics
冰川沉积 glacial deposit
冰川沉积平原 apron plain
冰川的 glacial
冰川地貌 glacial landform
冰川后退 glacial retreat
冰川湖 glacial lake
冰川阶地 glacial terrace
冰川孑遗 glacial refuge
冰川裂隙 crevasse of glacier
冰川漂移 glacial drift
冰川瀑布 glacier cascade
冰川侵蚀 erosion of glacier; glacial erosion
冰川三角 glacial delta
冰川世 glacial epoch
冰川退缩 deglaciation; glacial recession
冰川雪崩 glacier-avalanche
冰川植物区系 glacial flora
冰川作用 glaciation; glaciofluvial activity
冰醋酸 glacial acetic acid
冰道 ice road
冰的 glacial
冰冻风化 frost weathering
冰冻腐蚀 freeze etching
冰冻固定 cryofixation
冰冻气候 ice climate
冰冻切片术 freezing microtomy
冰冻蚀刻法 freeze etching
冰冻线 frost line
冰冻作用 frost action
冰害 ice damage
冰河材 glazier's wood
冰河后回弹 post-glacial rebound
冰河期后的森林 post-glacial forest
冰河期前 preglacial period
冰河退却 glacial recession
冰后期 post-glacial period
冰花 flower of ice
冰滑道 ice chute; ice-slide
冰积 icing
冰积层 glacial deposit
冰积物 iceborne sediment; glacial debris; glacial till; moraine; till
冰架 ice shelf
冰搅 cryoturbation
冰窖贮藏 ice-pit storage
冰进期 glacial epoch
冰砾阜 kame
冰砾岩 glacial conglomerate
冰帽 ice cap
冰片 bornyl alcohol
冰期 drifting epoch; glacial age; glacial period; ice age; ice period
冰期残遗种 glacial relicts
冰期的 glacial
冰碛(qì) drift; glacial drift; moraine
冰碛层 drift bed
冰碛沉积 drift deposit
冰碛地形 drift topography
冰碛阜 kame
冰碛湖 drift dammed lake
冰碛阶地 drift terrace
冰碛黏土 drift clay
冰碛物 iceborne sediment; glacial debris; glacial till; moraine; till
冰前平原 apron plain
冰融喀斯特 thermokarst
冰山 iceberg
冰蚀 glaciation
冰蚀高原(北欧等) fjeld
冰霜 glazed frost; hoar frost
冰水沉积平原 outwash plain
冰水的 fluvioglacial
冰水侵蚀 fluvioglacial erosion
冰水作用 glaciofluvial activity
冰丸 sleet
冰雪带 ice-and-snow belt (zone)
冰雪浮游生物 cryoplankton; cryoplanktophyte
冰雪圈 cryosphere
冰雪植物 cryophyta; cryophyte
冰雪植物区系 nival flora
冰乙酸 glacial acetic acid
冰雨 glazed rain; icy shower
冰原 firn; ice field; ice sheet; icing
冰原避难区 glacial refuge
冰原岛峰 nunatak

冰原反气旋 glacial anticyclone
冰缘的 periglacial
冰沼土 tundra soil
冰折 ice-break
冰辙道路 ice rut road
冰柱害 damage by hoar frost
兵蚁 soldier ant
丙氨酸 alanine
丙胺氟磷(杀虫剂) Pestox-15
丙酮酵母 acetone yeast
丙酮油 acetone oil
丙烯酰胺 acrylamide
丙烯酰胺凝胶 acrylamide gel
柄 grip; helve; knob; peduncle; petiole; stalk
蛃型(昆虫) thysanuriform
并发病 complication
并发率 coincidence
并发系数 coefficient of coincidence
并合核 fusion nucleus
并列 juxtaposition
并列维管束 dicollateral bundle
并流 co-current
并排的 jugate
并生副芽 collateral accessory buds
并生木 double-tree
并生维管束(外韧维管束) collateral bundle
并生芽 collateral bud
并生叶 bifoliolate leaf
病斑 lesion; spot
病虫害 damage from pests; pest
病虫害防治 pest control
病虫害根除 eradication of pests
病虫害综合治理 integrated pest management
病毒 virus
病毒病 virosis
病毒病害 virus disease
病毒的 viral
病毒复合体 virus complex
病毒核心 nucleoid
病毒结晶 virus crystal
病毒颗粒 virus particle
病毒粒子 nucleocapsid; virion
病毒农药 viral pesticide
病毒侵染 viral infection
病毒释散 liberation of virus
病毒素 virosin
病毒学 virology
病毒质体 viroplasma
病毒株系 virus strain
病害 disease
病害动力学(病害动态) disease dynamics
病害反应等级 disease reaction scale
病害流行预测 forecast of epiphytotic
病害木材 diseased wood
病害潜力 disease potential
病害调查 survey for disease
病害预报 disease prediction
病害预测 disease forecasting
病害预测模式 disease forecasting model
病害综合症 disease complex
病痕 lesion
病理的 pathologic(al)
病理反应 pathologic reaction
病理流脂 pathological resin
病理轮伐期 pathologic(al) rotation
病理学的 pathologic(al)
病理因素 pathological factor
病粒与虫粒种子 seed carrying pest and insect
病情指数 disease index
病因学 etiology
病原的 pathogen(et)ic; pathogenous
病原寄生物 pathogenic parasite
病原体 causative agent; pathogen(e)
病原物 agent of disease; pathogenic organism
病原细菌 pathogenic bacterium
病原携带者 carrier
病原性 pathogenicity
病原学 aetiology; etiology
病原真菌 pathogenic fungus
病状 sign
拨齿 set tooth; setting; wrest tooth
拨齿机 setting machine
拨齿料 spring set
拨款 appropriation
拨料(锯齿) briar dress; briar dressing; saw set; set; set tooth; setting; spring; tooth set
拨料的 bevelled-dress(ed)
拨料工作台 setting bench
拨料机 saw-setting machine

B

拨料器 saw-wrest; setting gauge; setting iron; setting key
拨轮号锤 marking cog
拨木装置(供单根) feeding single-log mechanism
拨索工 spool tender
波瓣 lobe
波成阶地 built terrace
波动 fluctuation; surging
波动低压 wave depression
波动干燥基准 fluctuating drying schedule
波动加压法 fluctuating-pressure process
波动温度 fluctuating temperature
波动性气旋 wave cyclone
波动压力 fluctuation pressure
波动载荷 fluctuating load
波段 spectral band; spectral interval
波尔(面积单位, $=25.29\mathrm{m}^2$) pole
波尔多液 Bordeaux mixture
波尔兹曼常数 Boltzmann constant
波痕 ripple mark
波莱思电杆防腐法 Poulain process
波浪形饰条 reverse ogee
波美比重计 Baume hydrometer
波谱 spectrum (*plur.* -tra)
波谱鉴别法 spectrometric identification
波谱色谱(法) spectral chromatography
波谱图 spectrogram
波蚀 wave erosion
波斯树脂 galban
波纹 ripple; wavy grain
波纹板 wave board
波纹材 curl
波纹管 bellows
波纹理 curly grain
波纹轧光机 moire calender
波纹纸板 corrugated board
波纹状单板 wavy veneer
波形扣钉 wriggle nail
波形扣件 corrugated fastener
波形跑锯 snake
波形皱缩(早晚材) washboarding; ribbing
波长 wavelength
波状板 wave board
波状的 sinuous
波状防裂钉 crinkle iron
波状花边 cyma reverse
波状花纹 wavy figure
波状锯面 snake
波状嵌条 wave moulding
波状纹理 wavy grain
波状压条 compound layering; serpentine layering
玻璃板 plate glass
玻璃刀 glass knife
玻璃房 glass house
玻璃格条 glazing bar
玻璃刮刀 glass knife
玻璃门 glazed door
玻璃屋顶 glass roof
玻璃纤维 fibreglass
玻璃纸 cellophane; glassine paper
玻璃状胶合接头破坏 glazed joint failure
钵植 pot planting
剥[树]皮 disbark(ing)
剥雕 delamination
剥光 strip
剥含栲胶树皮 fit; fitting
剥壳机 decorticator; dehusker; shucker
剥离 delamination; skip; swatch
剥离器 stripper
剥离强度 delamination strength; peeling strength
剥离试验 peeling test
剥鳞器 scaler
剥落 sealing off
剥落的 spalt
剥落体 incrustate
剥皮 bark; bark removal; barking; debark; debarking; decortication; excoriation; peel; peeling; ross; shelling; skin; skinning; spud; strip
剥皮材 barked wood; peeled wood
剥皮铲 bark scaler; bark spud; barking iron; peeling iron; spud
剥皮车间 rossing-mill
剥皮刀 bark spud; barking iron; spud; tearing tool
剥皮短木段 peeled billet-wood
剥皮反应 peeling(-off) reaction

剥皮斧 peeling axe
剥皮工 barker; peeler; rosser; timber shaver
剥皮机 bark peeler; bark peeling machine; barking machine; debarker; debarking machine; peeler; peeling lathe; peeling machine; rosser; sheller; shucker
剥皮机铣刀头 rosserhead
剥皮锯 barking saw
剥皮林(剥鞣料树皮) tanbark coppice; bark coppice
剥皮器 barker; timber shaver; tree-scriber
剥皮撬棍 peeler bar
剥皮损劣 peeling damage
剥皮损伤 peeling defect
剥皮损失 barking loss
剥皮原木 barked log
剥绒机 linter
剥蚀 abrasion; denudation
剥树皮机 disbarking machine
剥头式剥皮机 barker of rosser-head type
剥制 exfoliate
播幅 sowing width
播前耕耘 preplanting cultivation; presowing cultivation
播撒 seeding
播种 dissemination; drilling; dropping; planting; seed sowing; seeding; sowing
播种板 seed-board
播种板条 seed(ing)-lath; sorting lath
播种草籽 sow grass seeds
播种床 seed-bed
播种行 seed drill
播种机 drill machine; mechanical seeder; planter; seeder; seedling machine; seed-sowing machine; sowing-machine
播种季节 sowing-season
播种角 seed-horn
播种块 seed spot
播种量 rate of seeding; seed rate; seeding rate; seed-quantity; sowing rate
播种苗圃 seeding nursery
播种耙 seed harrow
播种盆 seed-pan
播种期 date of seeding; seeding stage; seeding time; seed-time; sowing time
播种器 sowing-drill
播种浅盆 seed-pan
播种区 sowing division
播种深度 depth of sowing; seed depth; sowing depth
播种实用价(种子) planting value
播种穴 hill
播种造林 reforestation by sowing
播种者 seedsman
播种砖(内放种子的土块) sowing brick
伯的 primary
伯劳鸟 butcher-bird; shrike
伯明翰线规 Birmingham wire gauge
伯内特法 Burnett process
伯内特防腐剂(氯化锌溶液) Burnett's fluid
伯内特注入法 Burnettizing
伯努利方程 Bernoulli equation
驳岸 quaywall
驳运 lighter
泊松比 Poisson's ratio
泊松分布 Poisson distribution
泊位货运量 berth throughput
勃鲁基原木板英尺材积表 Blodgett log rule
铂色紫胶 platina
博尔顿法 Boulton process
博尔顿干燥法 Boulton drying
博物学 natural history
博物学家 naturalist
博奕论 game theory
不变成本 constant cost; fixed cost
不变的 constant
不变景观属性 immutable landscape attribute
不变色边材 bright sap wood; bright sap wood
不变性 invariance
不定风 variable wind
不定根 adventitious root; adventive root
不定花序 adventitious inflorescence
不定胚 adventitious embryo
不定胚生殖 adventitious embryony
不定期 aperiodicity

不定期放牧 off-and-on grazing
不定期型 arhythmic type
不定芽 adventitious bud; aerosor; indefinite bud
不定芽变 accidental sport
不定载荷 varying load
不定枝 adventitious shoot
不定株 adventive
不动产 real estate; real property
不动配子 aplanogamete
不镀锌钢丝绳 ungalvanized wire rope
不对称 asymmetry
不对称法 asymmetric method
不对称分布 asymmetrical distribution
不对称花 unsymmetrical flower
不对称生长 asymmetric growth
不对称弯曲 unsymmetrical bending
不对称吸收谱带 asymmetric absorption band
不对称状态 asymmetric state
不放回(抽样) without replacement
不贯通榫 blind tenon
不混杂种子 pure seed
不健全节 unsound knot
不健全种子 unsound seed
不课税资产 inadmissible assets
不列坦奈斯斑尼犬 Brittany spaniel
不留后备尺寸的 scant size
不密封 blow-by
不耐干燥的种子 desiccation-sensitive seed
不耐寒一年生草花 tender annuals
不耐受的 intolerant
不耐性 intolerance
不耐阴的 intolerant
不耐阴树种 intolerant tree
不配对现象 asynapsis
不翘扭 out-of-wind
不确定性 uncertainty
不渗透层 impermeable bed
不渗透性 impermeability
不适明度测定仪 opacimeter
不适应河 misfit river
不舒适 disamenities
不跳壳枪 non-ejector gun
不透明冰 rime
不透明的 opaque
不透明度 density; opacity
不透明体 opaque
不透明性 opacity
不透气的 airproof
不透气性 imporosity
不透水层 aquifuge; confining stratum; impermeable layer; impervious layer; impervious surface
不透油的 oil-tight
不褪色性 fastness; fastness to light
不易剥落 incrust
不易剥皮 resistance in debarking
不易感受性(对病害) insusceptibility
不予课税 tax exclusion
不育 dysgenesis
不育的 atokous; kakogenic; sterile
不育核 sterile nucleus
不育内稃 sterile palea
不育心皮 paracarpium
不育性 incompatibility; infertility; sterility
不育雄蕊 parastamen; parastemon
不孕性 abortion
不孕性花 sterile flower
不孕性花叶 sterile floral leaf
不孕叶 sterile frond; sterile leaf
不皂化物 unsaponifiables
不致病的 non-pathogenic
卟啉 porphyrin
卟啉原 porphyrinogen
补播(天然林内) reinforcement; enrichment; enrichment seeding; supplementary natural regeneration
补播(造林后) after-culture; aid-to-sowing; beating; beating up; filling; filling-up; infilling; patching up; recruiting; refilling; repairing; supplying
补偿 compensation; counter balance; make good; pay off
补偿薄 balancing sheet
补偿材积 compensating volume
补偿存款金额 compensating balance
补偿贷款 compensatory financing
补偿点 compensation point
补偿法则 law of compensation
补偿贸易 compensation trade
补偿坡度 slope of compensation

补偿期 regulation period
补偿器 compensator; expansion joint
补偿深度 compensation depth
补偿温度 compensation temperature
补偿性捕食 compensatory predation
补偿性财产 compensatory fiscal policy
补偿油缸 compensation cylinder
补充 recruitment
补充伐 subsidiary felling
补充肥料 supplementary fertilizer
补充剂 extender
补充加热管 booster coil; reheat coil
补充木 recruitment
补充生殖型(白蚁) complemental reproductive; neoteinic
补充项目 supplementary item
补充蚁王 substitute king
补充营养 supplemental nutrition
补充主伐 relogging; supplemental final cut
补虫囊 bulb
补给头 feed head
补角变量(单板旋切) diametral pitch
补节工 timber plugger
补进 shortcovering
补块(单板) plug
补块毛糙 rough patch
补片(修补缺陷等) patch; chip plug
补片芽接 patch budding
补漆 touch-up coating
补强剂 fortifier
补色法立体图 anaglyph
补色立体原理 anaglyphic principle
补身精 winterin
补身素 drimenene
补身酮 drimenone
补贴 allowance; subsidy
补正求积仪 compensating planimeter
补植(人工造林) after-culture; aid-to-planting; beating (up); blanking; enrichment; filling (in); filling of blanks; filling-up; infilling; overplanting; patching up; recruiting; refilling; reinforcement (planting); repair planting; repairing; replacement planting; replanting; supplying; comppensatory planting
补植(天然更新) plantation fill in gap; supplementary natural regeneration; reinforcement (planting)
补植木 gap-cover
补助的混交树种 nurse species
补助伐 supplemental cutting
补足 making up
捕虫膏 insect-paste
捕虫沟 insect-trench; trap-ditch; trap-trench
捕获量 catch
捕获物 prey
捕集剂 scavenger
捕浆器 stocksaver
捕捞强度 fishing intensity
捕捞曲线 catch curve
捕鸟网 clapnet; fowling net
捕食 predating
捕食动物 predator
捕食关系 food relationship
捕食假说 predation hypothesis
捕食食物链 predator food chain
捕食压力 predation pressure
捕食者 predation
捕食者-被捕食交互作用 predator-prey interaction
捕食者-被捕食系统 predator-prey system
捕食者-被捕食者间关系 eater-eaten interaction
捕兽夹 iron trap; trap; steel trap
捕捉反应 catching reaction
捕捉-再捕捉法 capture-recapture method
哺乳 lactation
哺乳动物 mammal
不饱和现象 unsaturation
不饱满种子 unsound seed
不产孢的 nonsporous
不产孢子的 asporogenous
不成熟 juvenility
不传导的(对电或热等) opaque
不纯 impurity
不等材长楞堆 random length pile
不等侧面收缩 heterotasithynic

B

不等概率抽样 sampling with varying probability
不等宽门窗边梃 diminished stile
不等斜屋面温室 uneven-span (green) house
不等长度材(造材) random length
不发育的 obsolete
不法经营 log jam
不法正状态 abnormal state
不翻垡的深松土 chisel ploughing
不翻垡耕作 ploughless farming
不分节的 inarticulate
不分类原木(全部原木) woods run
不分离(染色体) non-disjunction
不分离杂种 true-breeding hybrid
不分枝的 eramose
不规则波状纹理 irregular wavy grain
不规则裂缝 random crack
不规则林 irregular stand
不规则扭曲 irregular twist
不规则伞伐作业 irregular shelterwood system
不规则生命期 perturbation lifetime
不规则式 informal style
不规则纹理 irregular grain
不规则演替 anomalous succession
不规则异龄林 irregular all-aged forest; irregular unevenaged stand
不规则圆木 out-of-round log
不规整木堆 jackpot
不含磨木浆白纸 white-wood-free paper
不含磨木浆书写纸 wood-free writing paper
不合尺寸 off-size
不合格 below grade
不合格板坯 miss-spread mat; out-of-standard board
不合格材 non-dimension timber
不合格的 off-grade; out-of-standard; substandard
不合格馏分 slop cut
不合格品 unacceptable product
不合规格材 reject; rejected log
不合规格的纸张 odd sheets
不活动繁殖体 akineton
不结实性 infertility
不均衡分布 unbalanced distribution
不均一性 polydispersity; nonuniformity
不均匀材料 heterogeneous material
不均匀肌理 uneven texture
不均匀结构 uneven grain; uneven texture
不均匀纹理 uneven grain
不均匀性 heterogeneity; inhomogeneity
不开放企业(只雇用某一工会会员单位) closed shop
不开裂果实 indehiscent fruit
不可给态水 unavailable water
不可混用性 incompatibility
不可及林 non-accessible forest
不可及林地 inaccessible forested land
不可及森林 inaccessible forest
不可见光 black light
不可口的 unpalatable
不可逆胶体 irreversible colloid
不可逆扩散 irreversible diffusion
不可逆凝结 irreversible coagulation
不可逆性 irreversibility
不可逆抑制 irreversible inhibition
不可燃的 non-ignitable
不可调型多锯片裁分机 battery edger
不可压缩性 incompressibility
不可遗传性 noninheritedness
不可预知性 unpredictability
不可作业林 non-operable stand
不可作业林地 non-operable forested land
不可作业区 inoperable area
不劳增值 unearned increment
不连贯节 incoherent knot
不连续变异 discontinuous variation; saltation
不连续集材 cold skidding
不连续性 discontinuity
不联会(遗传) asynapsis
不良立木 cull tree; cull wood
不良树种 undesirable species
不毛区 bare area
不毛之地 bare land; sterile land; barren land; barren
不明显的 obsolete
不能漂浮的木材 unfloated timber
不能造林的裸地 denuded unplantable land

不黏尘屑 dust-free
不平衡的 unbalanced
不平衡生长 unbalanced growth
不亲和反应 incompatible reaction
不亲和性 incompatibility
不全变态类(半变态类) heterometabola
不全配子融合 merogamy
不染色的 achromogenic
不稔性 sterility
不溶混性 immiscibility
不溶物 non-soluble matter
不溶性紫檀 insoluble redwood
不熔块 niggerhead
不调和湖泊型 disharmonic lake type
不调和生物区系 disharmonic biota
不同齿的 unequal-toothed
不完全变态 incomplete metamorphosis
不完全变态类 hemimetabola
不完全花 imperfect flower; incomplete flower
不完全花群 incompletae
不完全阶段 imperfect stage
不完全浸注法(防腐) empty-cell process
不完全菌类 fungi imperfecti
不完全燃烧 incomplete combustion
不完全双列杂交 incomplete diallel cross
不完全显性 incomplete dominance; partial dominance
不完全心材 imperfect heart wood
不完全心材树 imperfect heartwood tree
不完全抑制 incomplete inhibition
不完全真菌 imperfect fungus
不完善结晶的外壳 paracrystalline cortex
不完善结晶的纤维素 paracrystalline cellulose
不完善结晶区 paracrystalline region
不完整节 broken knot
不稳定变形 nonsteady deformation
不稳定的 astatic; labile
不稳定腐殖质 unstable humus
不稳定基因 labile gene; unstable gene
不稳定性 instability; lability
不稳定载荷 fluctuating load
不稳定植被 unstable vegetation
不稳态 unsteady state
不相称 disproportionation
不相称河 inadapted river; underfit river; underfit stream
不相容 exclusion
不相容性 incompatibility
不形成孢子的 nonsporing
不需光 light independent
不遗传变异 non-heritable variation
不遗传变异性 non-hereditary variability
不盈不亏 break-even
不匀值 heterogeneity value
不着火的 non-ignitable
不整齐林 irregular stand
不足 hypo- (*eg.* hyposensitivity)
不足额保险 under-insurance
布尔登螺旋管(传压感温) Bourdon tube
布枯叶素 barosmoid
布枯叶脂 barosmin
布朗尼尔特防腐法 Bronnert process
布朗斯天然木素 Braun's native lignin
布朗斯韦克悬浮式干燥机 Bronswerk-dryer
布勒格反射 Bragg reflection
布勒格方向 Bragg direction
布勒格角 Bragg angle
布勒格衍射 Bragg diffraction
布雷恩防腐剂 Brayn preservative
布利登原木材积表(美) Brereton (log) scale
布利尔水银测容计 Breuil mercury volume-meter
布利格洛灌丛 Brigalow scrub
布料器 distributor
布鲁莱斯测高器 Blume-leiss altimeter; Blume-leiss hypsometer
布面椅 upholstered chair
布棚室(栽培设备) cloth house
布氏硬度 Brinell hardness
布纹纸 wove paper
布质背衬(砂带) cloth backing
布置 array; laying-out
步测 pacing
步测法 pace method

步测计 passometer
步道 foot path
步行虫形的幼虫 carabidoid
步行甲虫 ground-beetle
步进处理 stepping treatment
步进计数器 step counter
步枪 rifle
步石 stepping stone
步置法 pace method
步足 periopods
部分安装的 semi-mounted
部分伴性(遗传) partial sex-linkage
部分不育性 partial sterility
部分车间 resaw mill
部分多倍体 segmental polyploid
部分合子 merozygote
部分互交 partial diallel crossing
部分回置抽样 sampling with partial replacement
部分经营合作社 incomplete working co-operative society
部分锯机废材输送机 resaw waste conveyor
部分均衡理论 partial-equilibrium theory
部分连生节 partial adhering knot
部分灭菌 partial sterilization
部分轻度腐朽 partial shallow decay
部分受精 partial fertilization
部分显性 partial dominance
部分性连锁 partial sex-linkage
部分压缩试验 partial compression test
部分严重腐朽 partial deep decay
部分异源多倍体 segmental allopolyploid
部分重复抽样 sampling with partial replacement; sampling with partial replacement
部分周缘嵌合体 mericlinal chimaera
部件 element; package; part; unit
部门 branch
簿木帖面 veneer overlay
擦 chafing
擦后变质(家具涂饰后变色) grayness after rubbing
擦伤 excoriation
擦涂 tamponing
材边 edge; narrow face
材端切余 resection
材段 segment
材堆 pile; stack; wood stack
材堆挡风板 load baffle
材堆顶上风机窑 overhead internal-fan kiln
材堆基础 stack bottom; stack stand
材堆基桁条 foundation bearer
材堆通风口 open space in pile
材堆通气道 chimney in pile
材垛 stack; wood stack
材积 timber volume; volume; volume of wood
材积比方程 volume ratio equation
材积表 scale; tariff; volume table
材积表法 method of volume table
材积测定 determination of volume; estimate survey
材积测定法 volume determination
材积差法 volume difference method
材积超尺(伐区测定材积大于贮木场测定材积) overscale
材积尺 caliper scale
材积-断面积比值 volume-basal area ratio
材积-断面积比值表 volume-basal area ratio table
材积法 method by volume; organization by volume
材积方程 volume equation
材积分配法 volume allotment method; volume-alloting method
材积估测 volume assessment
材积函数 volume function
材积及面积分配法 volume and area allotment method
材积级 volume class
材积计算 volume calculation; volume measurement
材积检查法 volume check
材积净生长量 net volume increase
材积控制法 organization by volume; outturn control; volume control
材积量度 volume measure
材积轮尺 volumetric callipers
材积轮伐期 volume rotation
材积率 percentage of volume; volume per cent
材积平分法 volume frame work
材积曲线 volume curve

材积曲线法 graphic method of valuation
材积生长 growth in volume; increment due to quantity; mass-accretion
材积生长量 volume growth
材积生长率 percentage volume growth; volume increment percent
材积收获 volume production; volume yield; volumetric yield
材积调查 growing stock determination; timber determination
材积调节法 outturn control; volume control; volume regulation
材积形数 timber form factor
材积源生物量 volume-derived biomass
材积增加量 stem volume increment
材积折减(因木材缺陷) dockage
材积指数 volume index
材积指数法 volume exponent method
材棱凹陷 want
材料损伤 material damage
材料指数 material index
材面 face
材面标记 face-mark
材面干燥 skin dry
材面接合 face joint
材面节疤 face knot
材面量度(以面积表示) faee measure
材面模制饰条 struck moulding
材面修整 jointing
材色 wood colour
材质 wood quality
材质因素 quality factor
材种 log assortment; log type; timber assortment; wood assortment; wood sort; assortment; crops; kind of timber
材种表 assortment table
材种别收获表 assortment yield table
材种分等 grading of logs; grading of stems
财产持有费用 carrying charge
财产归属 types of ownership
财产合作社 property co-operative society
财产清册 inventory
财产税 capital levy; property-tax
财务计划 financial plan
财务账目 financial record
财务状况 financial position
财政 finance
财政赤字 financial deficit
财政年度 fiscal year
财政收入 money yield
财政学 finance
裁边 edging
裁边锯 sizing saw
裁边锯下手 edge tailer
裁分圆锯机 edger
裁剪用板 shop-cutting panel
裁切锯 snipper saw
采出 withdrawal
采伐 cutting; harvest felling; cutover; exploitation; extraction; felling; hag; harvesting; logging; timber harvesting
采伐百分率 cutting percent
采伐比 cutting ratio
采伐不足 undercut
采伐簿 record of cutting
采伐成本 felling cost
采伐成熟期 ripe for cutting; ripeness for felling
采伐承包人 felling contractor
采伐处理标准 cutting treatment standard
采伐次序 cutting sequence
采伐带 cutting strip; felled strip; strip; swatch
采伐带排列 alignment of strip
采伐单位 felling unit; show
采伐单元 cutting unit
采伐道 felling-road
采伐登记簿 cutting register
采伐等级 cutting class; felling class
采伐地区交税 taxation of cutover land
采伐点 felling section
采伐点年龄 stump-age
采伐段 allotment; cutting section; parcel
采伐方案 logging plan
采伐方法 exploitation method
采伐方式 cutting system; harvesting system

采伐方向 cutting direction
采伐方针 cutting policy
采伐费用 harvesting cost; felling cost
采伐概算 felling budget
采伐高度 cutting height
采伐工地 logging site
采伐工队 felling crew
采伐工棚 lumbering camp
采伐工长 chopping boss
采伐管理 harvest management
采伐规程 logging rule
采伐规则 cutting rule; logging rule
采伐合同 felling contract
采伐和栽植记录 cutting and planting record
采伐后保留木 stander
采伐机械设备 logging plant
采伐机械司闸 logging shack
采伐基准计算法 yield determination methods
采伐及枯损记录 felling and mortality record
采伐计划 allocation of felling; cutting budget; cutting plan; cutting report; felling budget; felling plan
采伐计划图 harvest plan map; map of harvest operations
采伐记录 cutting record; felling record; record of cutting; scale book
采伐季节 cutting season
采伐迹地 cut area; cutover area; cutover land; cutted land; cutting blank; felling blank; logged(-off) area; logged (over) area; slash area; stump(y) land
采伐迹地更新 cutover restocking
采伐迹地未更新 cutover nonrestocking
采伐间隔期 logging cycle
采伐检索表 felling key
采伐进度 cutting schedule
采伐决定 felling decision
采伐联合机 harvester; tree harvester
采伐量 cutting volume; felling yield; outturn of the felling; yield from felling; harvest volume; removal
采伐量查定 felling volume assessment
采伐量控制 felling control
采伐量确定 determination of cut
采伐量预测 felling yield forecast
采伐列区 cutting series; felling series; stump section; cutting section
采伐列区网络 network of cutting-series
采伐林 cutover forest
采伐林班 cutting compartment
采伐林木 cutting of stand
采伐龄 cutting age
采伐率 cutting rate; cutting ratio; exploitation percent; felling rate
采伐率法 cutting percent method
采伐面 coupe; felling front; felling section
采伐面积 coupe; cut-over; cutting area; felling area; logging area
采伐面形状 form of coupe
采伐木标号 marking trees for felling
采伐木标印 timber making
采伐年表 chronology of cutting
采伐配置 distribution of cut
采伐期 logging period
采伐期间 logging cycle
采伐强度 felling rate; heaviness of felling
采伐趋势 cutting assumption
采伐权 cutting right; timber right
采伐人员 felling crew
采伐日期 cutting date
采伐烧炭材 coal off
采伐设想 cutting assumption
采伐剩余物 brash; debris; logging debris; logging residue; logging slash; waste in wood
采伐剩余物积压(当年未处理) extension slash
采伐剩余物加工剩余物 refuse wood
采伐剩余物利用机械 slash utilization machinery
采伐剩余物切碎机 rotary slasher
采伐时期 felling period
采伐实积 actual cut
采伐收获 felling volume; felling yield
采伐税 severance tax

采伐顺序 cutting procedure; cutting series; cutting succession; felling sequence; felling series; network of cutting-series; order of felling; succession of cutting; succession of felling; chronology of cutting
采伐特许证 timber concession
采伐图 cutting map
采伐系列 cutting series; working section
采伐限度 fell limit
采伐限额 cutting limit; cutting quota; felling quota
采伐限制 logging regulation
采伐小材承包工 short log logger
采伐许可证 clearing permit; felling licence; felling permit
采伐序列 cutting order; felling order; felling series
采伐一览表 cutting budget; cutting list; felling plan
采伐预算 cutting budget
采伐指标 cutting indicator
采伐种类 exploitation method
采伐周期 cutting period; felling cycle
采伐桩木作业 piling show
采伐作业 cutting method; cutting practice; cutting system; show; woods operation
采伐作业计划 logging budget
采伐作业面 logging case; logging show; logging unit
采伐作业区 felling unit; logging chance
采伐作业预算 logging budget
采割 chipping
采割松脂 resin boxing
采集 collection
采集圃 nursery of collection
采集人 food gatherer
采前落果 premature drop
采剩物 lap; lapwood
采剩物处理 slash disposal
采剩物堆 slash pile
采剩物收集机 slash collector
采剩物推叉(装于推土机前) brush rake
采石场 quarry
采石机 rock picker
采食率 grazing rate
采收 harvesting
采收成熟度 detachable maturity
采收费(从伐木到集材诸费用) crop expense
采收机 harvester
采收期 detachable time
采松脂 turpentining
采穗园 cutting orchard; nursery of scion; scion plucking nursery
采条母株 stock plant; stock tree
采销证(林业) forest concession
采运辅助设备 logging accessory
采运工人 logger; lumberer
采运管理处 logging depot
采运机械 logging machinery
采运企业 logging enterprise
采运损伤 extraction damage
采运作业 logging
采运作业损耗 logging loss
采脂 cupping; extraction; resin tapping; rossing; wood tapping
采脂刀 bark hack
采脂斧 cupping axe
采脂割面 resin-blaze
采脂工具 scraping tool
采脂间隔期 periodical tapping
采脂面 face
采脂年 crop year
采脂强度 tapping intensity
采脂树种 resin crop
采种 seed collecting; seed collection
采种带 seed collection zone
采种地段 seed-collection area
S形采种钩 S-hook detacher
采种剪 averruncator
采种林 seed collection stand; seed production stand; seed stand; parent stand
采种母树林 seed production area; seed-collection area
采种区 seed collection zone; seed zone
采种树 seed tree
采种台 harvesting platform
彩绘漆器 polychrome lacquerware
彩色(黑白灰三种颜色除外) chromatic colour

彩色定量 quantification of colour
彩色反转片 colour reversal film
彩色负片 colour negative
彩色改正滤光片 colour correction filter
彩色合成 colour composite
彩色红外片 colour infrared film
彩色灵敏度 colour sensitivity
彩色透明正片 colour transparency
彩色颜料 chromatic pigment
彩色再现 colour reproduction
彩色纸条 dope
参加游憩 recreation participation
参考标桩 recovery
参考情景 reference scenario
参考图像 reference image
参数 parameter
参数反演 parameter retrieval
参数化 parameterization
参数回收模型 parameter recovery model
参数提取 parameter retrieval
参数预估模型 parameter prediction model
参天古树 towering old tree
参杂 inclusion
餐具橱 buffet; sideboard table
餐具柜 cutlery cabinet; kitchen cabinet
餐具架 sideboard table
餐室成套家具 dining (room) suites
残材 chunk; hunk; offal timber; off-cut
残材采伐法 snag-felling law
残材清理 snagging
残材运输车 chunk truck
残茬覆盖 stubble mulch
残差项 error term
残存固有种 relic endemic species
残存林 remaining stand
残存林木 residual stand
残存生态型 relict ecotype
残存特有种 relic endemic species
残存物 survival
残存植物相 relic flora
残存砖红壤 detrital laterite
残毒(农药) residual toxicity
残干 stump
残根采伐法 snag-felling law
残骸 exuviae
残火 slash fire
残积 sedentary
残积层 eluvium
残积的 sedentary
残积土 residual soil
残积物 eluvium; residual material
残料价值 scrap value
残留单优种群落 enclave
残留毒性 residual toxicity
残留木素 residual lignin
残留期 residence time
残留山丘 monadnock
残留物废材 residue
残留性遗传 residual heredity
残留种 survivor
残苗 incomplete plant
残缺染色体 deficient chromosome
残杀威(杀虫剂) Arprocarb
残体(动植物) debris
残屑材 scrap wood; wood loss
残液 raffinate
残遗的 relic; rudimentary
残遗法 relict method
残遗群落 relic coenosinm; relic community
残遗体 relic; relict
残遗土 relic soil
残遗物 relic; relict
残遗小群落(大群落中) enclave
残遗植物 remnants
残遗植物区系 relic flora
残遗植物相 relic flora
残遗种 epibiotic species; relic; relic species; survivor; ancient relicts
残遗种保护区 refugium (*plur.* -ia)
残遗种分布区 relic area
残遗种区 refugium (*plur.* -ia)
残余 residue
残余变定 permanent set
残余变形 offset; permanent deformation; residual deformation
残余产物 residual products
残余的 relic; relict
残余价值 salvage value
残余林木 remaining tree
残余伸长 permanent elongation
残余收缩 residual shrinkage
残余土 relic soil

残余影响 residual impact
残余应办 surplus stress
残余应变 residual strain; unrecovered strain
残余应力 residual stress
残渣 scrap; tailing
残渣浆 tailing pulp
残渣紫胶 refuse lac
残值 junk value
残株 stub; stubble
蚕蛾 silk moth
蚕茧 pod
仓匣进料机 magazine grinder
苍术浸膏 atractylis concrete
舱单 shipped bill
舱顶棚 cabin trunk
舱口 hatch
舱口起重机 hatch crane
藏精器 antherid; antheridium
藏卵器 archegonium
操纵杆 lever
操纵基因 operator gene
操纵价格 administered price
操纵台 control console; control desk
操纵子 operon
操作管路 operating line
操作规程 operating instruction; working-rule
操作时间 hours on stream; operation time; running time
操作性能 performance
操作指令 operating instruction
操作装置 manipulator
操怍 handling
槽 chest; gutter; mesh; pit; pond; spout; sulcus; trough
槽钢 girder
槽浸法(防腐) trough method
槽孔模片 slotted templet
槽孔模片铺设组 laydown
槽口 notch; rabbet
槽口接 fillistered joint; notched joint
槽口雄榫接 notched and cogged joint
槽面板结构 shell construction
槽面的(卷筒) grooved-faced
槽刨 match plane; plough; plow; rabbet; rabbet plane
槽舌接 halving joint; rebated joint
槽舌接合 rabbet joint
槽式链板输送机 trough(ed) chain conveyer
槽式皮带输送机 trough(ed) belt conveyer
槽式施胶 tub sizing
槽榫 chase mortise
槽头螺栓 stove bolt
槽纹钉 corrugated nail
槽纹支承辊 fluted bed roll
槽形屋顶 trough roof
草 fork; grass; herb
草案 projector plan
草白蚁 harvester termite
草被地 grass-covered land
草本矮灌林 field layer
草本层 herb layer; herbaceous layer; herbaceous stratum
草本的 herbaceous
草本的茎 herbaceous stem
草本高位芽植物 phaenerophyta herbaceae
草本群落 herbosa; prata
草本橡胶 grass rubber
草本植被 herbaceous vegetation; herbosa
草本植被区 grassland
草本植物 herb
草本植物群 herbage
草本植物园 herbary
草测 preliminary survey
草叉 fork
草场 range
草场经营 range management
草场利用 range utilization
草场评定 range appraisal
草场调查 range survey
草丛 boscage; boskage; prata; tussock
草地 grass-covered land; grassland; herbland; lawn; ley; sward; turf
草地类型判读 interpretation of grassland types
草地喷灌机 lawn sprinkler
草地生态系统 grassland ecosystem
草地生态学 grassland ecology
草地指示植物 grassland indicator
草甸 meadow; prata
草甸草原 meadow steppe
草甸黑钙土 meadow chernozem

草甸灰化土 meadow podzolic soil
草甸栗钙土 meadow chestnut soil
草甸土 meadow soil
草甸盐碱土 meadow saline soil
草甸盐土 meadow solonchak
草甸沼泽 meadow bog
草甸植物 poophyte
草甸棕壤 meadow burozem
草垫 mattress
草毒死除草剂 allidochlor
草根层 sod
草菇 straw mushroom
草花径 herbaceous walk
草花境 herbaceous border
草花路 herbaceous walk
草浆 straw pulp
草胶 grass rubber
草类冻原 grass tundra
草类石楠群落 grass heath
草镰 grass hook
草蛉 green lacewing; lace wing
草路 grass walk
草螟 sod webworm
草木 greenery
草木犀 clover
草爬子 tick
草皮 sod; turf
草皮谷坊 sodded check dam
草皮泄水道 grass waterway; sodded spillway
草皮栽植 turf planting
草蜱子 tick
草坪 turf
草坪更新 lawn renovation
草坪洒水器 lawn sprinkler
草坪栽培 turf culture
草坪植物 lawn plant
草食的 graminivorous
草食性 phytophagy
草酸 oxalic acid
草酸尿 oxaluria
草炭 turf
草田轮作 rotation of crops and grass
草图纸 detail paper
草图转绘仪 sketchmaster
草屋 lodge
草原 bushveld; grassland; pampa grassland; steppe
草原改良 grassland improvement
草原管理 grassland management
草原灌丛 steppe scrub
草原黑钙土 steppe chernozem
草原气候 steppe climate
草原-森林带 prairie-forest region
草原萎缩 grassland diminishes
草原与草甸生态系统类型自然保护区 nature reserve for steppe and meadow ecosystem
草原植被 grassland vegetation
草毡层(土壤剖面A_0层或O层) sod lyher; sod layer; A_0-layer; O-layer
草沼地 grass moor
草质茎 herbaceous stem
草质茎高位芽植物 phanerophyta herbacea
侧柏 Chinese arbor-vitae; Oriental arbor-vitae
侧柏浸提物 thujoid
侧板(家具等) side panel; side plate; siding
侧板(昆虫骨板) pleura
侧瓣 lateral petal
侧棒(介壳虫) paraphysis
侧背板翅原说 paranotal theory (hypothesis)
侧背片 laterotergite
侧背线 subdorsal line
侧边带式砂光机 edge belt sander
侧边砂光机 edge sander
侧边纤维 lateral fibre
侧边斜接 edge mitre joint; edge rabbet joint
侧表面 side grain surface
侧齿片(锯链) side plate
侧齿片角(锯链) side-plate angle
侧齿刃角 side-edge angle
侧出的 lateral
侧唇舌(昆虫) paraglossa
侧刺 lateral spine
侧单眼 lateral ocelli
侧挡木 floating lateral boom
侧顶木(滑道) top log
侧断面 profile
侧堆 edge piling; edge stacking
侧堆法 edge piling; edge stacking; vertical piling
侧萼 bucca (*plur.* buccae)

侧萼片 lateral sepal
侧方灌溉 lateral irrigation
侧方火 flank of fire
侧方集材(半悬式) diamond lead
侧方截根 lateral cutting; side cutting
侧方收缩 end-contraction
侧方天然下种 lateral natural seeding; marginal seeding
侧方天然下种更新 natural regeneration by marginal seeding
侧方望远镜的经纬仪 theodolite with excentric telescope
侧方下种更新法 regeneration from seed disseminated from outside
侧方迎面火 flanking backfire
侧方遮阴 side shading; lateral shelter
侧方重叠 sidelap
侧分枝式 lateral ramification
侧副卫细胞 lateral subsidiary cell
侧副芽 lateral accessory bud
侧腹片 laterosternite
侧根 branch root; lateral; lateral root; oblique root
侧绠 floating lateral boom
侧沟 side gutter; side ditch; side drain; streak
侧沟角 side gutter angle
侧护木(木板路) side guards
侧积 edge piling
侧迹 lateral trace
侧卷绕(钢索从卷筒上) overlay
侧开式鞍座(索道上) open-sided jack; open jack
侧拉(承载索或吊索) side blocking
侧拦河绠 wingboom
侧连接片(锯链片) side link
侧馏分 side cut
侧脉(血管) lateral vein
侧脉(叶脉) lateral nerve
侧面 side surface
侧面叉车 side fork lift truck
侧面的 lateral
侧面放汽(管路) side relief
侧面火 flank fire; flanking fire; quartering fire
侧面积 lateral surface area
侧面集材 lateral yarding
侧面胶拼单板 edge-gluing veneer
侧面扑火 flank fire suppression; flanking fire suppression
侧面拖集(索道) lateral yarding
侧面悬挂的 side-attached
侧面硬度(径面或弦面) side hardness
侧膜的 parietal
侧木(木排中或车上外侧的原木) wing log
侧渠 lateral canal
侧取馏出物 side cut
侧刃磨损(锯齿) side wear
侧渗 underground runoff
侧生的(旁侧的) lateral
侧生的(雄器) paragynous
侧生分生组织 lateral meristem
侧生迹 lateral trace
侧施 side dressing
侧蚀 lateral corrosion
侧视航空雷达 side looking airborne radar
侧视图 side view
侧丝(真菌) paraphysis
侧条 lateral shoot
侧弯 edge bend
侧位脚档 side stretchers
侧位控制装置 edge position control
侧线 sideline
侧线道路 side track
侧向叠层 lateral lap; lateral overlap
侧向分裂 lateral division
侧向分选装置 edge sorter
侧向抗拉试验 sidewise tension test
侧向抗压强度 sidewise compressive strength
侧向力 lateral force
侧向搂草机 side rake
侧向膨胀 lateral expansion
侧向牵引 bridling (out); side lining
侧向侵蚀 lateral erosion
侧向渗透 side penetration
侧向弯曲 lateral bending
侧向压缩 sidewise compression
侧序 lateral order
侧压 side compression
侧压力 lateral pressure
侧芽 lateral bud
侧曳集材 lateral dragging
侧翼灭火 flanking fire suppression

侧引滑车 flying dutchman
侧枝 lateral; lateral branch; lateral shoot; offshoot; side branches; side shoot
侧枝根 lateral branch-root
侧柱 jamb
测报因子 predictor
测标 measurement mark; measuring mark
测程车 perambulator
测地坐标 geodetic coordinate
测点 point of measurement
测定 measurement; measuring
测定毒性的培养皿法 Petri-dish method of determining toxicity
测定位置 measurement location
测定系统 sensing system
测定要素 growth parameters; increment parameters
测定员 prospector
测定者 tester
测风经纬仪 aerological theodolite
测竿 cruiser's stick
测杆 spindle
测高法 altimetry; Pressler's method of height determination
测高公式 hypsometric formula
测高器 goniometer; height finder; height meter; hypsometer; cathetometer; surveying altimeter; elevation instrument; altimeter
测功器 dynamometer
测荷仪 loadometer
测交 test cross
测交品系 tester strain; tester strain line
测交者 tester
测角镜 angle mirror
测角器 anglemeter; angulometer; demi-circle; goniometer
测井工程车 logging truck
测径尺 caliper; calliper; caliper rule
测径带 diameter band
测径规 caliper gauge
测径器(卡尺) caliper
测径直角尺(轮尺) caliper square
测距仪 perambulator; range finder; telemeter
测力计 dynamometer
测链(66ft或21.1168m) chain; surveyor's chain
测量 measuring; survey; surveying
测量采伐带边界 run line
测量单位 measuring unit; unit of measurement
测量工程 gauging works
测量精度 accuracy of measurement
测量罗盘 surveyor's compass
测量平板仪 surveyor's table
测量树木的手杖 cruiser's stick
测量头 pick-up head
测量误差 measurement error
测量线路 run line
测量学 surveying
测量员 surveyor
测量站 gauging station
测量轴 spindle
测流坝 gauging weir
测流象限仪 stream quadrant
测流堰 measurement weir
测面积法 planimetry
测木尺 lignometer
测皮器 bark measuring instrument
测钎 mark(ing)-pin
测容法 xylometric measurement
测深法 bathymetry
测深仪 depthometer
测湿计 hygrostat
测试链 wrapper chain
测试溶液浓度 checking solution concentration
测树 cruise; estimate survey
测树法 mensuration
测树器 dendrometer
测树学 dendrometry; forest measurement; forest mensuration; mensuration; tree measuration
测树学数据 dendrometric data
测树因子 tree architecture
测树者 culler
测速记录仪 tachograph
测速仪 tachometer
测榫器 thicknesser
测微法(X射线衍射) micromethod
测微计 micrometer; micrometer caliper
测微目镜 micrometer eyepiece
测微偏振计 micro-polarimeter

测线 survey line
测线图方法 alinement chart method
测向计(垂向角方位角) goniometer
测斜器 clinometer
测云器 nepheloscope; nephoscope
测长机构 length measuring mechanism
测长控制盘 length-measuring control dial
测杖 scale-rule; scale-stick
测针 marking-arrow; measuring pin
R-策略 R-strategy
R-策略者 R-strategist
参差变异 fluctuation variation
参差接合 breaking joint
参差羽状的(复叶) interruptedly pinnate
层(层楞中) round
层(单板) ply
层(广义的) course; lay; layer; stratum (*plur.* strata); tier; zone; horizon
层斑岩 planophyre
层别群落 layer society
层次 layering; shades of gray
层叠纸浆 pulp lap
层顶 horizon
层段 epipedon
层化 stratification
层化次生壁 stratified secondary wall
层积 laminate; lamination
层积板 laminated board; laminboard
层积材 cordwood; laminate; laminated wood; stacked timber; stacking wood
层积材积 cord volume; piled volume; stacked volume; superficial volume
层积材价目表 cord-tariff
层积处理 stratification
层积单板 laminated veneer
层积单位 cord volume unit
层积垛 rick
层积拱 laminated arch
层积构件 laminated member
层积归垛 cording
层积计量 ricked measure; stacked measure
层积计算 cord-measure
层积检尺 ricked measure; stacked measure
层积胶合板 laminated plywood
层积结构 laminated structure
层积立方米 raum-meter (rm); stacked cubic meter; cubic meter stacked
层积立方英尺 stacked cubic foot
层积梁 laminated beam
层积量度 stacked measure
层积毛坯料 laminated stock
层积刨花板 laminated particleboard
层积热固塑料 laminated thermosetting plastics
层积容积 stacked volume
层积塑料 laminating plastics
层积碎料板 laminated particleboard
层积纤维板 laminated fibre board
层积芯子细木工板 laminated board; laminboard
层积学说 superposition theory
层积贮藏 stratification
层间剥离强度 interlaminar strength
层间结合力 interlaminar strength
层间流 interflow
层剪 layer-shear
层聚(生态) stratum (*plur.* strata)
层楞 layered pile; open pile; open piling
层离 delamination
层流湍流转变 laminar-turbulent transition
层群 synusia
层析 chromatography
层析谱 chromatogram
层压 laminate; lamination
层压铝纸袋 aluminium-paper laminated bag
层状的 lamellate; storied; storyed
层状电泳[法] zone electrophoresis
层状泥炭 laminated moor
层状黏土 laminated clay
层状排列 storied arrangement; tier like arrangement
层状体 lamina (*plur.* laminae)
层状纹理 lamellation
层状岩石 laminated rock
叉 fork; furca; tine
叉柄 fork arm
叉车 lift truck

叉车司机 fork lift operator
叉齿 prong
叉堆 end racking
叉堆干燥 pole drying
叉杆 fork arm
叉骨 furca
叉积 cross product
叉架 fork
叉架用材 forked wood
叉架桌 trestle table
叉式轮尺 fork caliper
叉式起重车 fork lift truck
叉式倾卸车 dumping fork
叉式升降机 fork lifter
叉式装卸车 fork lift truck
叉式装载机 fork lift loader; fork loader
叉形导木(小型滑道) frog
叉形榫 forked tenon
叉形芽接 prong budding; sprig budding
叉形应力检验片 prong
叉形支座椅 X-shaped chair
叉状菌根 forked mycorhiza
插标(营林标记) at stake
插槽 socket
插床 rooting frame
插垛 jam; jam pier
插垛木 dragger
插垛木堆 centre; center
插腹接 cutting side-grafting; slotted side graft
插干 cutting; firm wood cutting; hardwood cutting; truncheon
插花 floral arrangement; flower arrangement
插肩榫 shoulder-inserted join
插接 cuttage grafting; cutting graftage; cutting grafting; inlaying grafting
插靠接 cutting inarching; inlay approach grafting
插孔楔(伐木) socket wedge
插口(管子的) faucet
插木 slip
插木齿 felling dog
插木造林 planting of slip
插瓶靠接 bottle grafting
插入 cut-in; inlay; insertion
插入端 build-in end
插入件 insert piece
插入锯切法(将导板插入树干或原木) boring cut(ting)
插入物 inlet
插入下锯法 plunge cut
插穗 branch cutting; cutting; cutting wood; big slip; slip
插穗生根 strike
插锁 mortise lock
插条 branch cutting; cutting; cutting wood; big slip; slip
插条(极长的) pole
插条机 cutting planter
插条造林 planting by cuttings
插头 pin; plug
插销 bolt; dowel
插销榫 plug tenon
插削 mortising
插枝遮阴 brush shade
插座 jack; socket
差别反应 differential reaction
差动滑车 chain hoist
差额 balance; margin
差额成本 reduced cost
差角 angle of difference
差热分析-气相色谱-质谱联用法 differential thermal analysis gas chroma-tography-mass spectrography
差示热分析 differential thermal analysis
差示吸收防腐剂 differential adsorption
差速齿轮 differential gear
差向立体异构体 epimer; epimeric form
差向异构 epimerization
差向异构体 epimer; epimeric form
差值年表 residual chronology
茬(硬茎植物) stubble
茬角(弦茬外缘) corner
茶褐色 umber
茶林 tea grove
茶棚 tea-stall
茶室 tea room
茶树种植园 tea plantation
茶亭(广义的) tea-stall
茶亭(中东式) tea-kiosk

茶庭 tea garden
茶油 tea oil
茶园 garden for tea; tea garden; tea plantation
茶桌 tea table; tray top table
茶子油 tea oil
茶座 tea-booth
查定 assessment
查对簿 control-book
查耙任酮 chaparrinone
查氏培养基 Czapek's medium
查找表 look-up table
岔道 turnout
岔节 double knot
岔线(森铁) spur
岔枕 cross sleeper; crossing sleeper; crossing timber; point sleeper; switch tie
拆包机 bale breaker; bale opener
拆垛(搁浅木材) break a jam
拆垛(木材) unstack
拆垛工人 jam cracker
拆垛工组 jam crew
拆垛机 breaker stack; unstacker
拆垛升降机 break-down hoist
拆捆机 bale breaker; bale opener
拆楞 unstack
拆楞推河 break a landing
拆排 dismantling raft
拆卸的 knocked down
拆卸器 puller
拆卸索具 unrigging
拆转式轨道 portable tramway
拆装家具 packaged furniture
拆装式家具 knock-down furniture
柴坝 mattress dam
柴捆 faggot; fasciae
柴捆覆盖工程 covering works with fascine
柴排 mattress
柴束 fasciae
柴束道 fascine work road
柴束护岸 fascine covering
柴油 heavy fuel
柴油集材机 diesel skidder
豺猎期 hunting season
觇板(测量) target
掺合物 blend; admixture
掺水股票 watered stock
掺杂剂 doping agent
掺杂物分离器 rejector
缠绕 twining; wrap
缠绕的 twining; voluble (volubile)
缠绕茎 twining stem
缠绕牵引索于伐根(用于移动绞盘机) burn a stump
缠绕物的一圈 wrap
缠绕植物 twiner; twining plant; voluble plant
蟾背形饰条 toad back moulding
产孢细菌 spore-forming bacteria
产雌孤雌生殖 thelytoky
产单性的 allelogenous
产地 locality; origin
产地试验 provenance trial
产地特性种子 character-seed
产地证明书 certificate of origin
产地种子园 provenance seed orchard
产粉花 pollen flower
产两性单性生殖 amphitoky; deuterotoky
产两性孤雌生殖 deuterotoky
产量 output; yield; out-turn
产量不确定性 yield uncertainty
产量定额 standard performance
产量-呼吸量比 production-respiration ratio (P/R ratio)
产量金字塔 production pyramid
产量密度效果 yield-density effect
产量-生物量比 production-biomass ratio
产量图 yield diagram
产卵 egg deposition; egg-laying; oviposition; spawning
产卵场 spawning ground
产卵后阶段 post-egg stage
产卵洄游 spawning migration
产卵期 oviposition period; spawning-time
产卵器 ovipositor; oopod(eus)
产卵器鞘(蛹) acidotheca
产卵前期 preoviposition period
产率系数 yield coefficient
产膜酵母 film yeast
产囊体 archicarp
产能利用率 capacity utilization rate

产品 commodity; produce
产品差异 product difierentiation
产品等级 product class
产品分收 sharing (of yield)
产品管理 product management
产品检验 produce check
产品设计 product design
产品寿命史 product life history
产品寿命周期 product life cycle
产气细菌 aerogenic bacteria
产权判定 forest settlement
产生生命的 biogenous
产雄单性生殖 androgenetic parthenogenesis; arrhenotoky
产雄孤雌生殖 arrhenotoky
产业 domain
产种[期] seeding
铲 shovel; spade
铲草皮烧火 sod shoveling for fuel
铲叉滑条(装载机) skid tine
铲叉平放位置 level fork position
铲成光面 adzing
铲齿 clearer(tooth)
铲除 eradication
铲除剂(灭树用) eradicant
铲刀缓冲器 blade accumulator cylinder
铲斗(挖土机等) dipper; slip
铲斗挖掘机 bucket excavator
铲斗自由倾斜式推土机 gyrodozer
铲缝栽植法 bar slit method
铲撬检验法(胶合板) knife test
铲式平地机 blade grader
铲形芽接 spade budding
颤动 chatter
菖蒲二烯 acoradiene
常春藤叶油 ivy leaf oil
常雌雄同株 euhermaphrodite
常定风 steady wind
常冻 permafrost
常风 constant wind; regular wind
常规切削 full-round cut(ting)
常规窑干 custom (kiln) drying
常规窑干法 conventional kiln-drying
常见种 common species
常开花的 everblooming
常量 constant
常量沉淀 macroprecipitation
常量成分 macroconstituent
常量分析 macro-analysis
常量元素 macroelement
常绿矮灌木丛 phrygana
常绿草甸 sempervirentiherbosa; sempervirentiprata
常绿的 evergreen; periwinkle; winter-green
常绿防火林 evergreen fire-break
常绿灌丛 evergreen thicket
常绿灌木 evergreen shrub
常绿季节林 evergreen seasonal forest
常绿阔叶林 evergreen broad-leaved forest; laurilignosa
常绿阔叶林带 laurel forest zone
常绿阔叶树 broad-leaved evergreen
常绿林 evergreen silvae; evergreen forest
常绿木本群落 evergreen silvae
常绿扦插 evergreen cutting
常绿群落 evergreen community
常绿下层灌木 evergreen undershrub
常绿硬材林 evergreen hardwood forest
常绿硬叶林 evergreen sclerophyllous forest
常绿针叶林 evergreen coniferous forest; evergreen needle-leaved forest
常绿植丛 aithalium
常绿植物群落 aiphytia
常年伐木作业 year-round logging
常年河流 continuous stream; perennial river
常年花果植物 anabiont
常年林业收益 profit of forestry for annual working
常染色体 autosome; euchromosome
常染色质 euchromatin
常设标志 measurement mark
常湿森林 ever-wet forest
常数 constant
常水位 ordinary water level
常态胶合力试验 dry gluing strength test
常态谱系 normal spectrum
常态侵蚀 normal erosion
常态曲线 normal curve
常温 daily temperature
常温胶合 cold gluing

常温液态杂酚油 cold-application creosote
常现种 constant
常压处理(防腐等) atmospheric pressure process; non-pressure process; non-pressure treatment
常压浸注法 immersion system
常压纤维分离 atmospheric refining
常压纤维分离机 atmospheric refiner
常压油煮调湿法(防腐) boiling-without-vacuum
常压再蒸馏 atmospheric rerun
常异交作物 often cross-pollinated crop
常用支持装置 ordinary back up device
常优种 perdominant
常有丝分裂 eumitosis
常雨灌木群落 pluviifruticeta
常雨林 pluviisilvae
常雨木本群落 pluviilignosa
常雨乔木群落 pluviisilvae
常杂种优势 euheterosis
偿还期 maturity
偿还债务 liquidation
偿债基金 sinking fund
偿债能力 solvency
偿债资金 sinking fund
偿债资金表 sinking fund table
长鼻兵蚁(白蚁) nasute
长柄大木槌(伐木打楔) maul
长柄砍刀 slasher
长柄修枝剪 long-handled pruning shear
长材 length; long timber; long wood; long-length
长材集材法 long log system
长材集运 long logging; long-length logging
长插接 long splice
长插穗 set; sett
长车立柱 extension stake
长翅型 macropterous form
长搭钩 pike pole
长度 length; longitude
长度测量[法] linear measure
长度扫描器(选材) length scanner
长蠹 bostrychid; false powderpost beetle
长方形木排 rectangular raft
长方形栽植 rectangular planting
长匐茎 sarment
长副冠类(水仙等) magni-coronate
长干的 long-boled
长花期 long flowering period
长花柱的 macrostylous
长级 length class
长件挂车 pole trailer
长角果(十字花科植物的果实) silique
长角蒴果 ceratium
长角状的 siliquiform
长截头(原木端头有缺陷的截头) long butt
长径比 slenderness ratio
长距离集材 curve skidding
长距离运材 secondary transportation
长卷锯屑(横截锯产生的) macaroni
长考得(材长大于4ft, 材堆侧面等于32ft^2的层积单位) long cord
长靠椅 settee
长宽比 aspect ratio
长廊 gallery
长廊林 fringing forest; gallery forest
长力眼筛板 slotted plate
长轮伐期 long rotation
长木柱 long wood column
长捻接(钢丝绳) long splice; running splice
长刨 long plane; trying plane
长坡 long slope
长期的 long-term
长期负债 capital liability
长期观点 long run
长期火险性 long term fire danger
长期静应力 long term static stress
长期抗弯试验 long-term bending test
长期拉伸试验(抗拉试验) long term tensile test
长期趋势 secular trend
长期群状作业(渐伐) long-period group system
长期生态研究 Long-Term Ecological Research
长期收入 permanent income
长期消费 permanent consumption
长期演替 permanent succession
长期运转 long run

C

长期载荷 long-time load
长期债务 funded debt
长期阻燃剂 long-term retardant
长日照 long sunshine period
长日照处理 long-day treatment
长日照植物 long day plant
长柔毛的 villose
长沙发(坐卧两用) davenport; lounge
长沙发椅 couch; lounge chair
长伸木(伸出承载梁较长) bunked long
长石砾岩 arkosic conglomerate
长石砂岩 arkose
长石石英岩 arkosite
长寿种子 long-lived seed; macrobiotic seed
长索多支集材方式 Marathon rigging
长条木板 spla; splad
长头的 dolichocephalic
长网案 fourdrinier table
长网部 fourdrinier part; fourdrinier section
长网成型法 fourdrinier-method
长网成型机 fourdrinier board forming machine; fourdrinier machine
长网造纸机 fourdrinier; fourdrinier machine
长细比 slenderness ratio
长镶板 standing panel
长小蠹 ambrosia beetle; ambrosia borer; pinhole borer
长小蠹类 platypodid beetles
长效肥(料) slow-release fertilizer
长效肥料 controlled availability fertilizer; controlled-release fertilizer
长叶蒎烯 longipinene
长叶烯 junipene
长英岩 arkosite
长硬皮壳 incrust
长油度 long oil
长原木 long-cut wood; long-length
长圆盾状的 scutate
长远绿化计划 long-range forestation program
长远趋向 long run
长枕木 long tie
长枝 long shoot
长枝插 long stem cutting
长轴(弹头) long axis
长轴干燥窑 longitudinal shaft kiln
长轴组织 prosenchyma
长柱 long column
长桌 bench table; dischbank
厂内干燥 factory seasoning; shop seasoning
厂务费用 factory burden
场存材 yard lumber
场地 site
场地轨道 yard track
场地排水 ground drain; ground drainage
场地破坏 site destruction
场内移动式起重机 yard(ing) crane
场圃 field
场圃发芽率 field germination
场致电离 field ionization
敞槽法(防腐等) open-tank method; open-tank process; immersion system
敞车 gondola
敞开浸渍法 open infusion method
敞开式真空伏辊 suction couch with open chamber
抄浆机 wet machine
超薄切片 ultrathin section
超薄切片机 ultramicrotome
超倍体 hyperploid
超倍性 hyperploidy
超边界的 super marginal
超测微计 ultramicrometer
超差 oversize
超产 overrun
超出水高 freeboard
超雌性 superfemale
超低量喷雾机 ultra-low-volume sprayer
超低频 super low frequency; ultra-low frequency
超低容量 ultra-low-volume
超顶极植物群落 postclimax
超多倍体 hyperpolyploid
超额材积(工厂检尺量与原木检尺量的差异) overrun
超额尺寸 full cut; full measure; full size
超额出材量 overrun
超额储备金 excess reserve
超额附加税 surtax

超额利润 excess profit; exorbitant profit
超额收益 excess profit
超额准备金 excess reserve
超伐 excessive cutting; overfelling
超分布 over-dispersion
超分子结构 ultra-molecular structure
超高 excess height; excessive height; freeboard; super elevation
超高纯度 ultra-high purity
超高频 super high frequency; ultra-high frequency
超高压浸注法(防腐) ultra-high pressure method
超共轭效应 hyperconjugation; quasi-conjugation
超过尺寸 oversize
超化学计量的 superstoichiometric
超级安全锯链 super guard saw chain
超级细菌 super bacteria
超几何分布 hyper-geometric distribution
超焦距长度 hyper-focal distance
超宽角摄影机 ultra wide-angle camera
超前指标 leading indicator
超前滞后关系 lead-lag relationship
超亲优势 heterobeltiosis
超亲育种 transgression breeding
超全色 hyperpanchromatic
超热的 epithermal
超射频率 ultra radio frequency
超深渊层(海洋) hadopelagic zone
超深渊带 hadal zone
超生物体 supraorganism
超声波 supersonic wave; ultrasonic; ultrasonic wave
超声波的 ultrasonic
超声波检查 ultrasonic inspection
超声波降解 ultrasonic degradation; ultrasonic wave degradation
超声波探伤 ultrasonic inspection
超声波探伤仪 reflectoscope; ultrasonic flaw detector
超声脉冲 ultrasonic pulse
超声频率 supersonic frequency
超声探伤器 supersonic detector
超速(木材串坡) overrunning
超速传动 overdrive transmission
超速离心法 ultracentrifugation
超速离心机 ultracentrifuge
超速离心作用 ultracentrifugation
n维超体积 n-dimensional hypervolume
超透性 hyperpermeability
超微构造 ultimate structure; ultramicroscopic structure; ultramicrostructure; ultrastructure
超微化石 nannofossil
超微技术 ultramicrotechnique
超微结构 fine structure
超微结构薄片 ultrastructural lamella
超微孔 ultramicropore
超微粒子 ultramicroscopic particle; ultramicron
超微量化学 ultramicrochemistry
超微术 ultramicroscopy
超微型浮游生物 ultraplankton
超细粉 micropowder
超显微镜 ultramicroscope
超显性 overdominance
超显性假说(关于杂种劣势) overdominance hypothesis
超显性假说(关于杂种优势) super dominance hypothesis
超限应变 overstrain
超限应力 overstress
超效等位基因 hypermorph; superallele
超性 supersex
超雄性 supermale
超压 excess pressure
超演替顶极 postclimax
超音频率 supersonic frequency
超硬质纤维板 ultra-hardboard
超优势的 predominant
超优势木 predominant tree; superdominant; dominant tree; dominating stem
超有机体 epiorganism; superorganism; supraorganism
超载 overload
超载车 off-highway truck
超载运材 off-highway hauling
超载制动器 load brake
超长 excess length
超长度 overlength

超值商品 merit goods
超种 superspecies
超重 excess weight
超重臂升降 luffing
超重滑轮下降拉索 trip line
晁模油 chaulmoogra oil
巢 nest
巢蛾 ermine moth; small ermine moth
巢路 nesting looks
巢式窑 beehive kiln
巢箱 nesting-box
巢箱法 artificial nesting method
巢域 home range
潮变性土 aquert
潮虫(球鼠妇) sowbug; wood louse
潮带 tidal zone
潮灰土 aquod
潮间带 mediolittoral zone
潮间区 actic
潮间区生物群系 actic association
潮浪 tidal wave
潮老成土 aqult
潮软土 aquoll
潮上带 supralittoral zone; supratidal zone
潮湿 damp
潮湿地 moist land
潮湿面 wetting surface
潮湿森林 moist forest
潮始成土 aquept
潮滩 tidal flat
潮汐 tide
潮汐波 tidal wave
潮汐浸淹地 tidal swamp
潮汐林 tidal forest
潮汐植被 tidal vegetation
潮下带 infralittoral zone; subtidal zone
潮线下群落 subtidal community
潮新成土 aquent
潮氧化土 aquox
潮沼 tidal marsh
车板 sweep
车程表 trochometer
车窗通风器 window ventilator
车床工厂 turnery
车道 carriage way; driveway; lane; lane of traffic; stock driveway; stock route
车底净空 ground clearance
车顶夹 top clamp
车端拦柱 end-rack
车钩 coupler
车行道 carriage way; road tread
车间 mill; plant; workshop
车间成本 shop cost
车间时距 head way
车立柱 bunk hook; car stake; knee; stake; stanchion; stay
车立柱插孔 pocket
车辆 carrier
车辆保有数 vehicle stock
车辆材 cartwrights' wood; timber for vehicle; vehicle stock; wag(g)on plank; wheelwright's wood
车辆干馏釜 car retort
车辆干馏炉 waggon retort
车轮防滑钉 wheel lug
车轮复轨器 humpback
车上测载系统 on-board load-measuring system
车上装车工 head loader
车式甑 car retort
车台 ride
车厢板 side plate
车削材 wood for turnery
车削件构成椅 thrown chair
车辕 thill
车载升降台 cherry-picker
车辙 rut
车轴 axle
车轴草 clover
车轴载荷量 axle load
车桩 knee
掣子(棘轮) dog
尘埃度 dirtness; speckness
尘暴 shaitan
尘层 staubosphere
尘卷 dust devil
尘圈 staubosphere
尘土遮天 clouds of dust
尘屑螺旋输送机 dust screw conveyer
尘屑排出口 dust dispipe
尘旋 shaitan
尘旋中心 dust bowl
沉材 deadhead; dead-wood; sinker; sunken log
沉床花坛 sunken flower bed

沉床园 sunken garden
沉底材 unfloated timber
沉底木 dead-wood; sweep wood
沉淀 accretion; precipitation
沉淀池 pool
沉淀焦油 insoluble tar
沉淀木焦油 precipitated wood tar; settled tar
沉墩 sunk pier
沉耕犁 subsoiler
沉积过程 sedimentary process
沉积景观 sedimentary landscape
沉积体残余 sedimentary relict
沉积物(土壤) deposit; sediment; accretion
沉积物塌滑 sediment slide
沉积学 sedimentology
沉积循环 sedimentary cycle
沉积岩 sedimentary rock
沉积作用 sedimentation; sediment; laydown
沉降 sagging; subsidence
沉降槽 thickener
沉降地 depression
沉降速度 settling velocity
沉降物 fallout
沉降系数 sedimentation coefficient
沉降园 sunken garden
沉胶渣 gad
沉没成本 sunk cost
沉木 down log; sinker; sinker-stock
沉木打捞船 sinker boat
沉木打捞筏 catamaran
沉木打捞排 gun boat; monitor; pantoon
沉入成本 sunk cost
沉沙量 sediment production
沉砂槽 riffler
沉砂盘 sandtrap; gravitator
沉水材 sunk wood; waterlogged wood
沉水草本群落 submersiherbosa
沉水木 deadhead; sunken log
沉水树种 sinker
沉水植物 immersed aquatic plant; submerged (water) plant
沉析纤维 fibrid
沉陷 subsidence
沉香 agarwood; eaglewood; agilawood
沉香呋喃 Agarofuran
沉香油 agar-wood oil; agar oil; agilawood oil; aloewood oil; eaglewood oil
沉堰 ground-weir
沉桩 sunk pier
陈化 ageing
陈化时间 assembly time
陈列柜 showcase; china cabinet; china case
晨昏期 crepuscular period
晨昏迁徙 twilight migration
衬板(胶合板制造等) apron lining; backing board; caul board; retaining board; sheathing; wood slip; head block
衬比剂 contrast agent
衬底纸板 adhesive felt
衬垫 packing; wad; washer
衬面(衬砌、衬祖) lining
衬套 bush; hub
衬条 crosser; stacking strip; strip
衬纸 sheathing paper
衬质 matrix
称重计量 weight scale
称重器 weigher
撑背椅 brace back chair
撑筏者 raftsman
撑架 corbel
撑门(直拼Z形) ledged-and-braced door
撑模器 expander
撑木(房屋) dead shore
撑柱 girder
成氨 ammonification
成本登记 cost keeping
成本分析 cost analysis
成本估计 cost estimate
成本管理 cost control
成本耗损 cost depletion
成本和运费 cost and freight
成本核算程序 costing procedure
成本会计 cost accounting
成本计算 costing
成本记录 cost record
成本加成 mark on; mark up
成本加固定手续费合同 cost-plus-fixed-fee contract
成本价格 cost price

成本价值 cost value
成本控制 cost control
成本利得比率 cost-benefit ratio
成本流动 cost flow
成本收益分析 cost-benefit analysis
成本外摊 external cost
成本项目 cost centre
成本效率 cost effectiveness
成本效率比 cost-effectiveness ratio
成本效益比 benefit-cost ratio(s)
成本效益分析 benefit-cost analysis
成本账目 cost record
成材 lumber; merchantable timber volume; saw lumber; sawed lumber; sawed timber; sawnwood
成材分等工 sorter
成材规格清单 saw bill
成材检尺 mill scale
成材库 timbershed
成材林[的] adolescent
成材乔木 timber tree
成材树阶段 tree stage
成层 layer; stratification
成虫 adult; imago
成虫龄 adult instar
成虫羽化 adult emergence
成串滑道集材 trailing chuting
成串集材 trail skidding
成串集材滑道 trailing slide
成串全拖集材 snake
成串拖集(原木首尾相接) trailing
成簇的 conglomerate
成带现象 zonation
成对的 conjugate
成对染色体 pairing chromosome
成对物 doublet
成对羽状的 conjugate-pinnate
成对掌状的 coujugate-palmate
成分 component; constituent; refinery
成根能力 root-forming ability
成根素 rhizocaline
成根物质 root-forming substance
成功 make good
成果管理 management by result
成行的 serial
成行移植 lining out
成花刺激 floral stimulus
成花激素 florigen; flowering hormone
成花诱导 floral induction; flower induction
成活率 rate of survival; survival percent; survival percentage; survival rate
成浆机 finishing beater
成捆材 bundled stock
成捆堆垛法 package piling; packaged stacking
成捆集材 bundle skidding; package logging
成捆浸注处理 package immersion treatment
成捆造材 bundle bucking
成捆装车 package loading
成捆作业 bulk operation
成粒机 granulator
成列的 serial
成林 established field; established stand; establishment
成林过程 establishment
成林期 establishment period; formation period
成卵作用 ovogenesis
成苗率 plant percent; seedling percent; seedling stand percent; stocking percent; survival percent of seeding; tree percent
成膜剂(成膜物) film former
成膜物质 film-builder
成囊 encystment
成年材 adult wood; mature wood; outer wood
成年阶段 adult phase
成年树 adult tree
成批处理 batch process
成批培养 batch culture
成批生产 batch production; mass production; serial production
成品 end product; finished goods; finished product
成品尺寸 finished size
成品规格 trimmed size
成品宽度 trimmed width
成品紫胶 shellac
成七的 septenate
成熟 maturation; maturity; ripening
成熟材 adult wood; mature timber; mature wood; outer wood; ripe wood
成熟的 mature

成熟度 degree of ripeness; maturity
成熟伐 harvest cutting
成熟分裂 meiosis; meiotic division
成熟工艺 dependable process technology
成熟过程 aging; maturation process
成熟阶段 maturity stage
成熟林 mature crop; mature forest; old growth
成熟林分 mature stand; ripe stand
成熟卵 mature egg
成熟木 adult tree; mature tree; ripe wood tree; mature timber
成熟木质细胞 mature xylary cell
成熟期 date of maturity; date of ripening; maturation period; maturity; maturity phase; maturity stage; adult phase
成熟前期 prematuration period
成熟群丛 mature association
成熟群落 mature community
成熟生境 mature habitat
成熟土壤 mature soil
成熟有丝分裂 meiotic mitosis
成熟指标 maturity index
成熟竹 mature bamboo
成熟紫胶 phunki; rhooki
成束的 fascicled
成套备件 complete spare parts
成套家具 suite; unit furniture
成套交易 package deal
成套设备 complete equipment; complete set of equipment
成套桶板料 shook
成套椅 set of chair
成套转让 package transfer
成图 mapping
成土过程 pedogenic process
成土环境 pedogenic environment
成土特性 soil-building property
成土因素 factor of soil formation
成土作用 pedogenesis
成湾 embayment
成像 imagery
成形 shaping
成形表面砂光机 shaped surface sander
成形车床 shaping lathe
成形机 shaper
成形设备 shaping equipment
成型 moulding
成型板坯 resultant mat; felting
成型浮游生物 morphoplankton
成型辊 forming roll
成型活性炭 pressed active carbon
成型机 former; forming machine; mat former
成型胶合板 formed plywood; moullded plywood
成型接触砂轮 formed contact wheel
成型模 forming die
成型木材 sawn timber
成型铺装机 forming unit
成型铺装台 forming table
成型铺装头 forming head
成型砂光机 moulding sander
成型生产线 forming line
成型铣床 moulding and shaping machine; profiler; too(u)lder
成型性 moldability
成型装置 forming unit
成岩作用 dlagenesis
成叶 adult leaf
成叶素 phyllocaline
成长率 growth percent
成组记录 block record
呈锯齿状的 laciniate
承包 by contract
承包商 entrepreneur
承保人 underwriter
承保限额 underwriting limit
承椽板 plate
承垫车(车组中后面一辆) runner
承接口 birdmouth
承框 template
承力外伸支架 outrigger
承梁板 wall plate
承木 bearer
承水档漂子 fin boom
承索器 tree shoe
承窝 socket
承销人 consignee
承压软管 live hose
承压水层 artesian aquifer; confined aquifer
承载 load bearing
承载抱钩式集材机 bunk jaw skidder

C

承载变形 bearing strain
承载车架式抓钩集材机 grapple-bunk skidder
承载滑轮 depending block
承载回空式索道 return track cable
承载回空式索道集材 Grabinski yarding
承载架空索 loaded skyline cable
承载梁(气动升降车立柱) bunk; air bunk
承载梁(运材车) hercules
承载梁卡杆 bunk jaw
承载能力 bearing capacity; load (carrying) capacity
承载索 carrier cable; main cable; main rope; running skyline; skyline
承载索鞍座 skyline jack; skyline shoe; tree jack
承载索固定式索道 fixed wire cableway
承载索后拉索(安装索) skyline stub
承载索挠度 skyline sag
承载索配重组 counter weight jack
承载索张紧装置 skyline tensioning device
承载索张力 skyline tension
承载索支承滑轮 saddle block
承载索支持索 skyline support line
承载索支架(索道) support skyline
承载索转移 cable change
承载应变 bearing strain
承种样器 seed trap
承重 load bearing
承租人 leaseholder; lessee
城郊园林住宅区 garden suburb
城市大气 urban atmosphere
城市分区(按功能) city zoning
城市风景 civic landscape
城市公园 city park
城市供水 city water supply
城市规划 city planning; urban planning
城市规划法 city Planning law
城市规划区 city planning area
城市花园 city garden
城市化 urbanization
城市环境 urban environment
城市建设 urban construction
城市景观 town scape
城市空地 urban open space
城市空气污染 urban air pollution
城市林业 city forestry; urban forestry
城市绿地 urban open space
城市绿地系统规划 urban green system planning
城市绿化 city planting; urban forestry
城市热岛 urban heat island
城市森林 urban forest; urban forestry
城市设计 urban design
城市生态学 urban ecology
城市园林 city garden; urban forestry
城市园林布局 city garden distribution
城市植被 urban vegetation
城市装饰 urban decoration
乘法器 multiplier
乘积 cross product
乘积法 product method
乘式 multiplier
程序 process
程序编制员 programmer
程序方框图 flow chart
程序估计研究工作 program evaluation research task
程序鉴定技术 program evaluation and review technique
程序控制 program(me) control
程序设计语言 programming language
程序升温器 temperature programmer
程序温度控制器 program temperature controller
程序语言公式转换器 formula translator
程序装置 programmer
程序自动化 automatic programming
橙花苷 aurantiin
橙花基 neryl
橙花素 aurantiol
橙皮 orange peel
橙皮油素 auraptene
橙色紫胶 orange lac

橙酮 aurone; benzylidenecoumaranone
橙叶油 petitgrain oil
橙油 orange oil
秤 scale; weigher
秤杆 scale beam
吃草 browsing
吃刀量 bite
吃蜜性 melliphagy
池 pond; sump
池塘演替 pond succession
池塘园 pool garden
池沼草本群落 aquiherbosa
池沼浮游生物 heleoplankton
迟季的 serotinal
迟季相 serotinal aspect
迟延 lag
迟延剂 retarding agent
迟延显性 delayed dominance
持股公司 holding company
持久极限 endurance limit
持久膨胀 permanent expansion
持久强度 endurance strength
持久型病毒 persistent virus
持久性 permanency; permanence; persistence
持久性肥料 lasting manure
持久性收缩 sustained contraction
持久载荷 load of long duration
持水当量 moisture equivalent
持水力 moisture holding capacity
持水量 moisture holding capacity; water capacity; water-holding capacity; water-storage capacity
持水曲线 moisture retention curve
持水性 water retention
持效性 long-term effectiveness
持续的前引(狩猎) sustained lead
持续锋面雨 prolonged frontal rain
持续农业 sustainable agriculture
持续生产量 sustained yield
持续增长 sustainable growth
持有期 tenure
持有权 tenure
持证者(有执照者) licensee
匙形的 spathulate
匙形腹接 scooped side graft
匙形椅背 spoon back
匙形钻 dowel bit
尺程 foot run; running foot; linear foot
尺寸 dimension; size
尺寸范围 size range
尺寸规格 sizing specification
尺寸精度 accuracy of dimension
尺寸手语规范(制材) code of hand signal
尺寸稳定性 dimensional stabilization
尺寸小于标准 under size
尺度参数 scale parameter
尺度效应 scale effect
尺蛾 canker-worm; looper; spanworm; measuring worm moth
尺杆 measuring stick
尺蠖 canker-worm; looper; spanworm; measuring worm moth
尺码 footage
呎烛光 foot-candle; formative tissue
齿 beard; dens (*plur.* dentes)
齿背 back; back face
齿背角 clearance angle
齿背线 back line
齿仓 chamber; throat
齿仓间隙 depth gauge clearance
齿槽 gullet; throat
齿槽锉 gullet file
齿槽锯 gullet saw
齿槽深度 gullet depth
齿侧修整 side dressing
齿锤 bush hammer
齿顶 top
齿顶斜角 top bevel angle
齿顶修整(锯) stoning; topping; breasting
齿端 teeth point
齿端角 point angle
齿腹 breast
齿高 gullet depth; height of tooth; teeth high
齿跟 heel
齿轨 rack; toothed rail
齿辊 toothed roll
齿喉角 rake angle
齿喉面 cutting surface
齿喉修整器 gummer
齿尖(犁铧等) cusp; teeth point; point

C

齿尖端 tine
齿尖角 sharpness angle; tooth angle
齿节 dens (*plur.* dentes); tooth pitch; tooth spacing
齿距 pitch; space; tooth pitch; tooth spacing
M形齿锯 M-toothed saw
齿菌 hydnaceous fungus; teeth fungus
齿口锯 flooring saw
齿料 saw-wrest
齿轮 gear
齿轮泵 gear-(type) pump
齿轮齿条传动 rack-and-pinion drive
齿轮传动 gear drive; toothed gearing
齿轮传动式摆架 gear bogie
齿轮簧板枪 wheel lock gun
齿轮式摆架 gear bogie
齿轮箱 gearbox
齿轮油泵 gear-(type)oil pump
齿面角 face angle
齿耙 tine rake
齿盘 pan
齿盘磨碎机 teeth disk mill
齿盘式打磨机 toothed disk-mill
齿刨 toothed plane
齿前 breast; front
齿切削量 bit
齿刃角 filing angle
齿深规 raker; raker gauge
齿深规间隙 depth gauge clearance; raker clearance
齿式松土机 tooth scarifier
齿式索轮 gear-type rope wheel
齿式制动 tooth stop
齿室 chamber; gullet; throat
齿室锯屑容量 gullet sawdust capacity
齿室容屑系数 gullet-to-chip area ratio
齿室线 throat line
齿条 gear rack; rack
齿条边接 spline edge joint
齿条锯 rack saw
齿形犁 tine plough
齿形嵌接 indented scarfed joint
齿形装饰 dentil; dentel; denticle
齿型 tooth pattern
齿胸 breast
齿状 dentate
齿状边缘(射线管胞) dentate margin
齿状加厚 dentate thickening; toothed thickening
齿状结构 dentation
齿状射线管胞 dentate ray tracheid
齿状凸出 toothlike projection
齿状销(连结木结构) alligator
齿状运输链 alligator chain
斥力 repulsion
赤潮 red tide; red water
赤道 equator
赤道板(细胞) equatorial plate
赤道低压 equatorial low-pressure
赤道面 equatorial plate
赤道无风带 doldrums
赤道雨林 equatorial rain forest; selva
赤节 red knot
赤经(天文) right ascension
赤霉素 gibberellin
赤色园 red garden
赤松烯 densiflorene
赤糖糊 black strap
赤纬 declination
赤藓糖 erythrose
赤藓型(赤型) erythro form
赤眼蜂 minute egg parasite
赤杨酮 alnusenone
赤云杉油 red spruce oil
赤藻糖 erythrose
赤字 deficit
赤字财政 deficit financing
炽燃火 hotspot
炽燃阶段 glowing time
炽热型燃烧 glow-type combustion
翅 ala (*plur.* alae); pterygium; wing
翅瓣(昆虫) lobulus (*plur.* lobuli); alula (*plur.* -lae)
翅的 alar
翅轭(鳞翅目昆虫) jugum
翅钩列(昆虫) hamulus
翅果 key fruit; pterodium; samara (samera); winged fruit
翅后基叶(鳞翅目) pterygium
翅尖 apex
翅脉 vein
翅脉神经 nerve

翅脉序 venation; wing venation
翅片管 finned pipe
翅室 areae; areolae; cell
翅芽 wing pad
翅展(昆虫) expanse
翅痣 pterostigma; stigma (*plur.* stigmata/stigmas)
翅子树型(瓦状细胞) pterospermum type
冲程 impact height; impact stroke; stroke
冲程次数 number of strockes
冲锤 impact hammer
冲洞 punching
冲沟类型(指溪谷) gully type
冲击 impulse; impulsion
冲击挤压 impact extrusion
冲击剪切强度 impact shearing strength
冲击力 impact force
冲击能 impact energy
冲击强度 impact strength; shock resistance; shock strength
冲击式摆动喷灌机 impulse oscillating sprinkler
冲击式涡轮 impulse wheel
冲击式圆盘打磨机 impact disk mill
冲击试验 impact test
冲击试验机 impact tester
冲击弯曲 impact bending
冲击弯曲强度 impact bending strength
冲击弯曲试验 impact bending test
冲击系数 impact coefficient
冲击型刨片机 turbo-chipper
冲击压强 impact pressure
冲击压头 impact head
冲击应力 impact stress; shock stress
冲击硬度 impact hardness
冲击载荷 impact load(ing); shock load
冲击值 impact value
冲击中心(摆式) centre of percussion
冲积层 alluvial deposit; alluvial horizon; alluvial layer; alluvium
冲积带群落 alluvial community
冲积地层 alluvial formation
冲积河谷 alluvial valley
冲积阶地 alluvial terrace; built terrace
冲积矿床 alluvial deposit
冲积黏土 adobe; alluvial clay
冲积平原 alluvial plain; flood plain
冲积坡 alluvial slope
冲积壤土 fluviatile loam
冲积扇 alluvial apron; alluvial fan; fan
冲积扇阶地 fan terrace
冲积台地 alluvial terrace
冲积滩 alluvial flat
冲积土 alluvial soil; alluvium; fluvisol
冲积土壤 fluviogenic soil
冲积物 alluvial deposit; alluvium; sediment
冲积新成土 fluvent
冲积锥 alluvial cone; debris cone
冲角 angle of attack
冲孔床 puncher
冲量 impulse
冲蚀 erosion
冲蚀度 degree of erosion
冲蚀防治 erosion control
冲蚀力 erosive power
冲蚀面 erosion(al) surface
冲蚀性 erosiveness
冲刷 washing
冲水开关 flush handle
冲洗 flush
冲销 charge off; write-off
冲压工 puncher
冲账 charge off
充气 aerate; aeration
充气阀 gas-filled tube
充气轮胎式压路机 pneumatic tired roller
充气器 aerator
充气箱 plenum chamber
充水 water logging
充水纤维素 water cellulose
充碳酸气喷雾 carbonated mist
充填机 filling machine
充细胞法 full-cell process
充脂材 light wood
充脂木材 resin-soaked wood
虫白蜡 white wax
虫粪 frass

虫害 damage by forest-insects; damage by insects; insect damage
虫害发生预测 forecast of abundance
虫害防治 insect control
虫害蔓延 general infection
虫害木 brood tree
虫害因素 entomological factor
虫黄藻 zooxanthella
虫胶 lac
虫胶腻子 shellac putty
虫胶清漆 lac varnish; shellac varnish
虫胶树脂 lac resin
虫胶油灰 shellac putty
虫孔 worm grub; worm hole; grub hole
虫孔的 worm-eaten
虫口 population
虫蜡 insect wax
虫绿藻 zoochlorella
虫媒 entomophily
虫媒病毒 insect born(e) virus; arthropod-borne virus
虫媒传粉 insect pollination
虫媒的 entomophilous
虫媒花 entomophilous flower
虫媒疾病 insect born(e) disease
虫期 stage
虫漆酶 laccase
虫漆脂 axin
虫生的 entomogenous
虫生腐殖质土(不含蚯蚓和马陆) insect mull
虫生植物 entomophyte
虫生植物的 entomophytous
虫体寄生植物 entomophyte
虫体寄生植物的 entomophytous
虫体内寄生的 entomogenous
虫泄物 worm excretion
虫屑 bore dust; bore-powder; wormdust
虫穴 mole burrow
虫眼 bore hole; borer hole; pinhole; wormhole
虫瘿 cecidium; gall; insect gall
虫蛀的 worm-eaten
虫蛀等级 wormy grade
虫蛀缺陷 boring defect
虫蛀种子 worm-eaten seed
重瓣花 double flower; hose-in-hose flower
重瓣花冠 multiple corolla
重壁层 double wall
重播 reseeding
重捕法 recapture method
重捕获 recapture
重叠 overlap
重叠层(胶合板) lap
重叠度 degree of overlap
重叠峰 overlapped peaks
重叠基因 overlapping gene
重叠区 zone of overlap
重叠区域 overlap area
重叠生态位 overlapping niche
重叠像对 overlapping pair
重叠作用(基因) duplicate effect
重颚钢夹 double jawed trap
重发性状 palingenetic character
重复 duplication
重复抽样 repeated sampling
重复单元 repeating unit
重复断面 double profile
重复基因 duplicate gene; repetitive gene
重复力 repeatability
重复率 repeatability; repetition rate
重复侵染 repeated infection; superinfection
重复性 repeatability
重复选择 repeated selection
重复应力 repeated stress; repetitive stress
重复杂交 recrossing; repeated crossing
重估价盈余 revaluation surplus
重合指数 coincidence index
重寄生 epiparasitism
重寄生的 biparasitic
重寄生物(寄生物的寄生物) hyperparasite
重寄生现象 hyperparasitism; superparasitism; superparasitization
重寄生性 hyperparasitism
重建 reconstitution
重建紫胶 rebulac
重锯齿的 biserrate; duplicato-serrate
重排作用(分子) rearrangement
重三出的 duplicato-ternate

重舌接 double-tongue graft
重生 after-growth
重调节 reconditioning
重纹孔 bordered pit-pair
重现性 reproducibility
重新安排 rearrangement
重新估价 revaluation
重新联合 reassociation
重新绿化(天然的或人工的) revegetation
重新配置 re-allocation; re-allotment
重新评价 reassessment
重新区划 re-allocation; re-allotment
重新造林(荒山荒地) reafforest; reafforestation; reforestation; restocking
重新组合 reassociation
重芽 double budding
重演 recapitulation
重演过程 palingenetie process
重演性变态 palingenetic metamorphosis
重演性状 palingenetic character
重羽状的 duplicato-pinnate
重置成本 replacement cost
重置价值折旧法 depreciation on replacement value
重装 reloading
重装子弹工具 reloading tool
重组 recombination
重组现象 reconstitution
重组子 recon
抽彩式竞争 competive lottery
抽查 spot checking
抽查单位 inventory unit; survey unit
抽出率 extraction rate
抽风 draft; draught
抽风机 exhaust fan
抽片 block cutting
抽气 air exhaust; air-bleed
抽气风扇 extraction fan
抽取 extract
抽水 withdrawal
抽水机 watering machine
抽薹 bolting
抽提法 extraction method
抽提物成分 extraneous component
抽屉 drawer; drawing box
抽屉背板 drawer back
抽屉侧板 drawer side
抽屉导杆 drawer guide
抽屉底板 bottom-board; drawer bottom
抽屉柜 chest of drawers; commode; nest of drawers
抽屉滑槽 drawer slide
抽屉滑道 drawer runner
抽屉滑架 drawer slide
抽屉面板 drawer front
抽屉桌 commode table
抽象群落 abstract community
抽象生长序列 abstract growth series
抽心(伐根劈裂等) splinter pull; stump pull; splinter draw; splinter pulling
抽心下锯 box the heart
抽心下锯法 box heart sawing
抽样 sample; sampling
3P抽样(与预估成比例的概率抽样) three-P sampling; probability proportional to prediction
PPS抽样(与大小成比例的概率抽样) probability proportional to size sampling
P抽样 P-sampling
抽样比 sampling ratio; sample fraction; sampling density; sampling fraction; sampling intensity
抽样单元 sampling unit; unit of sampling
抽样单元组 group of sampling unit
抽样分布 sampling distribution
抽样检查法 sampling inspection
抽样强度 intensity of sampling; sampling intensity
抽样设计 sampling design
抽样调查法 sampling survey method
抽样误差 sampling error
抽株疏伐 spacing thinning
酬金 kickback
稠度 consistency
稠度变量 consistency variable
稠度计 eonsistometer
稠度控制器 consistency controller
稠度曲线 consistency curve
稠度调节器 consistency regulator
稠度系数 consistency coeffcient

稠度值 consistometric value
稠度指数 consistency index
稠化 bodying
稠料器 decker
稠性 consistency
筹码 bargaining counter
筹资 financing
臭败 rancidity
臭虫 bug
臭松油 apinclum; apinol; levomenthone
C
臭腺 scent gland
臭氧层 ozone layer
臭氧洞 ozone hole
臭氧干燥法 ozone seasoning
臭氧化 ozonation
出材端 discharge end; dry end
出材量 mill run; out-turn; outturn of the felling
出材量表 outturn table; timber assortment table
出材率 percentage of cut-turn; percentage of out-turn; recovery; exploitation percent; log scale recovery; out-put
出材损伤 extraction damage
出产量 output
出厂价格 factory price; shop cost
出厂税 factory tax
出轨 derail
出汗(油漆缺陷) sweating
出河工 pond man
出河机 floating conveyer; water-transport elevator
出河平台 loading deck; loading dog; loading jack
出河起重机 quay crane
出河输送机 logway
出河台 live deck
出河斜坡 gangway; log jack; logway; slip
出价 bid
出浆口 stock outlet
出借人 lender
出菌率 plating efiiciency
出孔 exit hole
出口 exit hole; throat
出口材 ship-timber
出口成材 ship lumber
出口气 off-gas
出料端 dry end
出料辊 discharging roll; outfeed roll
出料装置 ouffeed device
出苗 field emergence; seedling emergence
出苗大年(天然下种的) seedling year
出苗后施用 post-emergence application
出苗率 stand establishment percentage
出苗前 pre-emergence
出苗前施用(农药) pre-emergence application
出苗前用量 pre-emergency rate
出纳员 teller
出盘 sale of business
出圃苗 field-run-seedling
出勤时间 attendance time
出射的 emergent
出生率 natality
出售价 realization value; sale value
出土的(子叶) epigaeal; epigeal; epigean; epigeous
出土前中耕 pre-emergence cultivation
出土型发芽 epigeal germination
出现 emergence; occurrence
出血 bleeding
出芽 bud off; budding; chitting
出芽生殖 agmogenesis
出叶 leaf emergence; leaf out
出栽 field planting; field setting; outplanting
出枝 ramify
出租人 lessor
初变形 initial set
初步概算 initial budget estimate
初步流送 predrive
初步推壳(射击动作) primary extraction
初步运材 primary transportation
初步蒸馏 predistillation
初成土 amorphic soil
初次采伐线 first felling-line
初次感染 primary infection
初次寄生虫 protoparasite
初次胶合 primary gluing
初次浸染 primary infection

初次疏伐 first thinning
初沸点 initial boiling point
初腐 dozy
初腐阶段 early stage of decay
初干材重 initial dry weight
初割树脂 virgin dip
初给器 primer
初耕 primary cultivation
初含水率 initial moisture content
初级产品出口国 primary exporting country
初级抽气泵 backing pump
初级抽样单元 primary sampling unit
初级的 primary
初级抵押市场 primary mortgage market
初级飞羽(动物) primaries
初级合作 proto-cooperation
初级结构 primary structure
初级肉食性动物 first level carnivore
初级生产 basic-level production; primary production
初级生产力 primary productivity
初级生产者 primary producer
初级食肉动物 primary carnivore
初级市场 basic-level market
初级消费者 primary consumer
初级性器官 primary sex organ
初加工(截冠、打枝、剥皮、造材、粗锯等) primary conversion
初阶段 initial stage
初距 initial spacing
初龄幼虫 egg-larva; first-instar larva
初滤器 first filter
初磨因素 initial grinding factor
初能(子弹出枪口时的能) muzzle energy
初黏度 tack
初凝 initial set
初期 incipient stage
初期保持量 initial retention
初期材积 initial volume
初期材积总生长 gross growth of initial volume
初期抽空 preliminary vacuum
初期存货 initial inventory
初期的 seminal
初期毒性 initial toxicity
初期腐朽(木材) advance rot; incipient decay; incipient rot; advance decay
初期沟蚀 incipient gullying
初期害虫 primary insect
初期红心腐 firm red heart
初期灰化作用 incipient podzolization
初期菌丝体 primary mycelium
初期扩散指标 initial spread index
初期立木度 density of initial stocking
初期喷蒸处理 preliminary steaming (treatment)
初期气压 initial air pressure; preliminary air
初期蠕变 initial creep; primary creep
初期吸收量 initial absorption
初期蓄积表 initial stock table
初期悬浮干燥 first stage suspension drying
初期资本 initial capital
初生壁 cambial wall; primary lamella; primary wall
初生壁结构 primary wall texture
初生病原物 primary pathogen
初生的 nascent; neonatal
初生分生组织 primary meristem
初生根 primary root
初生木材 primary wood
初生木质部 primary xylem
初生皮 primary bark
初生皮层 primary cortex
初生韧皮部 primary bast; primary phloem
初生射线 primary ray
初生维管束 primary vascular bundle
初生维管组织 primary vascular tissue
初生纹孔场 primary pit-field
初生纹孔缘 initial pit border
初生细胞壁 primary (cell) wall
初生叶 primordial leaf
初生长 primary growth
初生肿瘤 primary tumor
初生主根 root stock
初生组织 primary tissue
初始采伐量 initial harvest
初始角 initial angle

C

初始密度 initial density
初始期 initial stage
初始位置 initial position
初始质壁分离 incipient plasmolysis
初熟 thrifty mature
初霜 first frost; first injurious frost
初速 muzzle velocity
初蜕 amnios
初吸收量 initial uptake
初夏芽接 early summer budding
初型土 amorphic soil
初雪 first snow
初应力 initial stress
初载荷 initial load
初张力 initial tension
初值 initial value
樗鸡 lantern-fly
除草 cleaning; weed clearing; weeding
除草铲 weed(ing) knives
除草锄 draw hoe
除草定(除草剂) bromacil
除草工 weeder
除草机 grass eliminator; stripper; weed puller; weeder; weeding machine
除草剂 chemical weed killer; herbcide; weed killer; weedicide
除草醚(除草剂) nitrofen
除草耙 weed(ing) harrow
除草皮 screefing
除草松土机 weed and loose soil machine
除草油 weed oil
除侧枝 feathering out
除尘器 dust-filter
除虫菊 pyrethrum
除虫菊粉剂 pyrethrum dust
除虫菊杀虫剂 pyrethrum insecticide
除地 exclusion; non-productive forest land
除伐 cleaning; cleaning in small growth; cleaning-cutting; disengagement cutting; release; release cutting; release felling; released thinning
除伐作业 salvage logging
除根 disroot; stubbing
除灌 brush stripping
除灌机 brush cleaner; scrub-clearing machine
除灌剂 bush killer
除节筛 knocker screen; knot screen
除荆机 brush breaker; brush cutter; bush breaker; bush cleaner
除蔓 eradicating of climber
除芒刺机 debeardor
除蘖 tiller cutting
除皮材积 under-bark volume
除皮直径 width in the clear
除去伐根 stumping
除去根出条 suckering
除去果肉 depulping
除沙弯道 sediment trap
除砂沟 sandtrap
除砂盘 gravitator; riffler
除湿干燥[法] drying by dehumidification
除髓打磨机 depithing mill
除髓机 depithing machine
除雾器 demister
除油 degreasing
除莠 weed clearing
除莠剂 chemical weed killer; herbicide; phytocide; weed killer; weedicide
除莠喷雾机 weed sprayer
除莠油 herbicidal oil
除杂装置(木浆处理) knot screen; knotter
除渣器 hydraclone
除枝 delimbing; disbranch; limbing; shaft-cleaning
厨房家具 kitchen furniture
锄 grub-hoe; hoe
锄草 hoeing
锄铲 share
锄刀 tine
锄缝栽植法 grub-hoe-slit-method
雏晶 crystallite
雏鸟 nestling
雏形的 obsolete
雏羽 juvenal plumage
橱 press
橱柜 cabinet
橱柜家具 cabinet-work
橱柜中挺 clap post
橱门拉手 wardrobe handle
储备材积 volume reserve

储备培养 stock culture
储藏薄壁组织 storage parenchyma
储藏物质 reserve substance
储藏组织 reserve tissue
储槽 stock chest
储存期限 storage life
储弹筒 tubular magazine
储气罐 air accumulator
储蓄器 holder
处理 curing; handling
处理槽(防腐) treating tank
处理费用 handling expense
处理罐 treating cylinder
处理过度 overtreating
处理机 processor
处理前准备 prefabrication; preframing
处理性 treatability
处理周期 treatment cycle
处女地 virgin land
处女林 first growth
畜力车 mule cart
畜力集材 animal skidding; bull team logging; dog's work; hand log(ging); hand skid(ding); hand yarding; horse skidding
畜力集材道 horse-road
畜力集材工 hand logger
畜力集材回空道 dray road
畜力耙 oxharrow
畜力拖曳木材(在滑道平缓地段) jigger
畜力运材 shake tongue
畜群 drove
触变防腐剂 thixotropic preservative
触变剂 thixotropic agent
触变性 thixotropism; thixotropy
触变性系统 thixotropic system
触发器 flip-flop
触发装置 trigger
触角 antenna; antennae
触角神经 antennal nerve
触觉器官 tactile organ
触毛 cirrus
触器 tactile organ
触杀剂(杀虫剂) contact agent
触须 cirrus; palpus
触压胶黏剂 contact adhesive
川木香油 elecampane oil
氚化合物 tritiated compound
穿插钢索针 marlin(e) spike
穿刺培养 stab culture; stick-culture
穿过 crossover; traverse
穿过节 passing knot
穿孔 perforation
穿孔板 perforated plate
穿孔处理(应用扩散法) bore-hole treatment
穿孔带 punch tape
穿孔加压注入法 pressure treatment in bored hole
穿孔菌丝 bore hyphae
穿孔卡 punch(ed) card
穿孔卡检索表 punch(ed) card key
穿孔卡片 perforated card
穿孔卡片系统 punch(ed) card system
穿孔膜 perforation plate
穿孔器 puncher
穿孔嵌板型吸音材料 perforated panel type absorber
穿孔索引卡 perforated index card
穿孔硬质纤维板 perforated hardboard
穿孔缘 perforation rim
穿孔纸带 punched tape
穿孔阻力 puncture resistance
穿桥 through-bridge
穿式桁架 through truss
穿梭式集材(一机归堆, 一机集材) shuttle skidding
穿梭式运材 shuttle hauling
穿索孔(捆木) gopher hole; choker-hole
穿索眼钻具 gopher
穿条 lash pole
穿透的 pervious
穿透降水 throughfall
穿透力 bite
穿透曲线 break-through curve
穿透深度 depth of penetration
穿透性 permeability
穿眼纹 transocular stripe
穿衣镜 cheval glass
传病媒介 vector; vector of infection
传播 dispersal; dissemination; propagation; transmission
传播距离(种子) distance of dissemination

传播媒介 vector control
传播时间(电磁波等) travel time
传播式激光(扫描木材检尺) travel(l)ing laser beam
传播体(种子等) diaspore; migrule; disseminule; dispersal unit
传播体类型 disseminule form
传代培养 subculture
传导歪轮 daumen
传导性 conduction
传递 convection; transfer
传递单位 transfer unit
传递点 pass point
传递体 carrier
传递系数 transfer coefficient
传动 drive
传动齿轮 actuating gear; transmission gear
传动链片(突耳) drive lug
传动装置 actuating unit; gear; jack; running gear; transmission
传粉 pollination; fecundation
传粉者 pollinator
传粉作用 fecundation
传感器 sensor; transducer
传感系统 sensing system
传票 voucher
传染 infestation
传染分布 contagious distribution
传染中心 centre of infection; focus of infection
传热 heat transfer; transmission of heat
传热的 diabatic
传热介质 heat transfer mudium
传热面 heat transfer surface
传热系数 coefficient of heat transfer; coefficient of heat transmission; coefficient of thermal transfer; coefficient of thermal transmission; heat transfer coefficient; thermal conductance
传热性 heat conduction
传声性 acoustic(al) conductivity
传声作用 transmission of sound
传输 transmission
传输细胞 transfusion cell
传输组织 transfusion tissue
传输组织束 transfusion strand
传送带 belt
传送辊 transport roll
传送机 transmitter
传送结束讯号 acknowledge (character)
传送链 bull chain
传送装置 conveyer; conveyor
传统成本法 conventional costing
传统花卉 traditional flower
传统家具 conventional furniture; tradition furniture
传统式 traditional style
船板材 strake
船帮材 sheathing
船舶满载吃水线 load line
船舶铺板 skin
船底防污漆 antifouling paint
船骨材 timber
船脊骨 keel
船壳 skin
船壳板 planking
船模试验池 naval tank
船旁交货价格 free alongside ship
船上交货价格 free on board; free on vessel
船式构象 boat conformation
船尾式(弹头) boat-tailed
船用防腐剂 marine preservative
船运 ship
椽胶胶黏剂 rubber-based adhesive
椽梁系板 heel strap
椽式坝 rafter-type splash dam
椽尾梁 hammer beam
椽尾柱 hammer post
椽子 rafter
串(原木) run
串级系统 cascade system
串集 trail
串集作业区 trail setting
串联的绠漂木 string (logs)
串联木材 turn
串联双筒燃油滤清器 dual fliter-agglomerator
串联再锯机铣削装置 in-line resaw moulder system
串联质谱计 tandem mass spectrometer
串联种子清选机 in-line seed cleaner
串联重复 tandem duplication
串坡 coasting; ground sliding; ground-slide

串坡集材 downhill skidding; sidehill logging; sidehill yarding
串坡溜山(集材) jack slip
串形管胞 strand tracheid
串珠嵌条 bead-and-batten
串珠镶边 bead-flush
串珠状端壁 nodular end wall
串珠状加厚 nodular thickening
串珠状缘饰 beading
疮痂病 scab disease
窗格状纹孔 fenestriform pit; window-like pit
窗格状纹孔对 window-like pit pair
窗花格(雕刻) blind tracery
窗槛 windowsill
窗槛滴水槽 check throat
窗框 sash; window frame; window sash
窗框槽刨 sash fillister
窗框底 cill; sill
窗框防腐剂 par-tox
窗框滑轮槽凿 sash pocket chisel
窗框榫凿 sash mortise chisel
窗扇基座 windowsill
窗扇钳夹器 sash cramp
窗扇转动轴 sash centre
窗台 cill
窗台板 window board; window stool
窗台园 window garden
窗梃 stile
窗头板 yoke
窗下座位 window seat
床 bed; couch
床单搁板(橱柜) linen shelf
床高屏 head of bed
床架 bedstead
床脚板 foot of the bed
床踏脚板 bed step
床头 head of bed
床头板 headboard
床头柜 night stand
床头箱 headstock
床柱 bed pillar; bedpost
床作 bedding culture
创办成本 initial cost
创办资本 initial capital
创办资本投资 initial capital investment
创伤 trauma; wound; wounding
创伤胞间道 traumatic intercellular canal; wound intercellular canal
创伤薄壁组织 traumatic parenchyma; wound parenchyma
创伤变色 wound stain
创伤刺激开花 traumatic stimulation of flowering
创伤木栓 wound cork
创伤年轮 traumatic ring
创伤侵填体 traumatic tylosis
创伤射线管胞 traumatic ray tracheid
创伤树胶 wound gum
创伤树胶管 traumatic duct
创伤树脂道 traumatic resin canal; wound resin canal
创伤心材 traumatic heartwood; wound heartwood
创伤形成层 wound cambium
创伤周皮 wound periderm
创伤轴向胞间道 traumatic axial intercellular canal
创伤组织 wound tissue
创始费用 initial outlay
吹出 blow-off
吹风 aeration
吹风除尘 air dusting
吹风处理 fanning treatment
吹风机 air feeder
吹管卸料式削片机 overhead discharge chipper
吹气 blowing
吹蚀 deflation
吹卸管 blow-off pipe
吹卸器 blow-off
吹制玻璃 blown glass
垂度 sag
垂距 vertical interval
垂死木 dying tree
垂索 molly hogan
垂索滑轮 pendant block
垂线点 plumb point
垂向角 vertical angle
垂蛹 suspensi pupa
垂直成层现象 vertical stratification
垂直带逆转 inversion of altitudinal zone
垂直的 vertical
垂直地带性 altitudinal zonality; vertical zonality
垂直点抽样 vertical point sampling
垂直对流 vertical convection

垂直分布 vertical distribution
垂直分布带 altitudinal belt; altitudinal zone; vertical zone
垂直分布替代种 altitudinal vicariad
垂直根 vertical root
垂直构件 vertical member
垂直航空摄影 vertical aerial photograph
垂直环式料仓 vertical ring-magazine
垂直基准面 vertical datum
垂直集材 vertical logging
垂直间距 vertical gap; vertical opening
垂直剪切 vertical shear
垂直结构 vertical structure
垂直锯割 orthogonal cutting
垂直锯木架 log frame
垂直抗病性 vertical resistance
垂直立体高度夸张 vertical exaggeration
垂直流脂割沟(中沟) resin channel
垂直流脂沟(割面上的中沟) gutter
垂直贸易 vertical trade
垂直面 normal surface
垂直喷气干燥机 vert-air-jet impingement dryer
垂直迁移 vertical migration
垂直切割 vertical cutting
垂直射频加热 perpendicular heating; transverse heating
垂直摄像 vertical photography
垂直摄影像片 vertical photograph
垂直树冠 vertical crown
垂直统一管理 vertical integration
垂直线抽样 vertical line sampling
垂直像片草图转绘仪 vertical sketchmaster
垂直于等高线伐倒方式 perpendicular cutting pattern
垂直郁闭 vertical closure
垂直振动 vertical vibration
垂周分裂 anticlinal division
垂周形成层细胞分裂 anticlinal cambial cell division
槌击检验法(检验胶合质量) mallet tapping
槌形扦插 mallet cutting
锤(锤击) hammer
锤击钢印标记 hammer mark
锤击痕 hammer mark
锤角组(鞘翅目昆虫) clavicornia
锤枪 hammered gun
锤筛式打磨机 hammer basket screen mill
锤式剥皮机 hammer type barker
锤式打磨机 hammer mill
锤式粉碎机 hammer crusher; hammer mill; hammer type shredder
锤式破碎机 hammer crusher
锤式碎片机 hammer type shredder
锤式再碎机 hammer hog
锤形触角 capitate antenna
锤型起重机 hammer-head crane
春播 spring seeding
春材 spring wood
春材率 proportion of spring wood
春分点 vernal equinox
春化处理 iarovization; jarovization; vernalization; yarovization
春化阶段 stage of vernalization; thermophase; thermostage
春化作用 iarovization; jarovization; vernalization; yarovization
春季剥皮 sap-peeling
春季采伐 spring felling
春季法正蓄积 spring normal growing stock
春季环流期 spring circulation period; spring overturning
春季相 vernal aspect
春季远足 spring outing
春季植物 spring plant
春雷霉素(春日霉素)(农药) kasugamycin
春夏采伐的杆材 loose bark pole
春汛 freshet
春游 spring outing
春植 spring-planting
纯粹成本 absolute cost
纯粹的 entire
纯度 fineness; purity
纯腐生物 true saprophyte
纯橄榄岩 dunite
纯合的 homozygous
纯合二倍体 homodiploid
纯合体 homozygote
纯合体交配 incross
纯化 brightening

纯化剂 passivating agent
纯寄生物 true parasite
纯剪强度 single shearing strength
纯剪切 pure shear
纯剪应力 pure shear stress
纯净种子 pure seed; clean seed; cleaned seed
纯利率 pure rate of interest
纯利润 pure profit
纯林 pure crop; pure forest; pure stand
纯林培育 monoculture
纯煤焦杂酚油 straight coal-tar creosote
纯挠 simple bending
纯扭转 simple torsion
纯培养 pure culture
纯品系 pure strain
纯切削 clean cut
纯群落 pure community
纯收益 net-income
纯收益最大成熟 maturity of maximum forest net income; maturity of the highest income
纯系 pure line
纯系类型 pure-clonic form
纯系群 biotype
纯纤维素 pure cellulose
纯应变 pure strain
纯应力 pure stress
纯育 breed true
纯育品种 pure-breeding variety
纯原种 foundation seed
纯种 pure bred; pure culture
纯种不变 breed true
莼菜胶 water shield gum
唇 labrum; lip
唇板 apron board; lip
唇瓣 label; labellum (*plur.* labella); lip
唇布 apron
唇基(昆虫) clypeus
唇角 lip angle
唇舌(昆虫) lingula; ligule
唇形变 lipping
唇形的 labiate
唇形花冠 labiate corolla
唇形科的[植物] labiate
醇溶性着色剂(染料) spirit(-based) stain
戳孔植树 peg planting
疵瑕 fault
瓷器柜 china cabinet; china case
瓷釉面硬质纤维板 enamelled hardboard
磁层 magnetosphere
磁带 magnetic tape
磁带操作系统 tape operating system
磁极 magnetic pole
磁力分选机 magnet separator
磁力清选机(种子) magnetic cleaner
磁力探伤器 magnetic defectoscope
磁墨水字符识别 magnetic ink character recognition
磁墨水字符阅读器 magnetic ink character reader
磁盘操作系统 disk operating system
磁偏差 magnetic aberration
磁偏角 declination
磁圈 magnetosphere
磁性门闩 magnetic catch
雌孢子 gynospore
雌核 thelykaryon
雌核发育 gynogenesis
雌花 female flower; fertile flower; pistillate flower
雌花两性花同株的 gynomonoecious
雌花两性花异株的 gynodioecious
雌花无性花同株(菊科) agamogynaecism
雌花雄花两性花同株的 caeno-monoecious
雌花雄花两性花异株的 trioecious
雌鹿 doe
雌配子 oogamete
雌球花 carpellate strobilus; female cone
雌全同株的 gynomonoecious
雌全异株的 gynodioecious
雌蕊 gynoecium; pistil
雌蕊瓣化 pistillody
雌蕊柄 gynophore
雌蕊群 gynoecium
雌蕊托 gynobase; metrophorum
雌蕊先熟 protogyny
雌蕊先熟的 protogynous
雌性先熟 protogyny
雌性先熟的 protogynous

雌雄二形现象 sexual dimorphism
雌雄间体 intersex
雌雄鉴别 sex-discrimination; sexing
雌雄蕊柄 androgynophore; gonophore
雌雄蕊合体的 gynandrous
雌雄蕊熟 homogamy
雌雄榫接合 tenon and mortise joint
雌雄同花序的 androgynous
雌雄同熟 monochogamy; homogamy
雌雄同熟的 monochogamous; homogamous
雌雄同熟花 monochogamous flower
雌雄同穗花序 gynandrous spike
雌雄同体 androgynism; gynandromorph; hermaphrodite; hermaphroditism; monoecy; protogynous hermaphriodism; synoecy
雌雄同体的 ambisexual; androgynous; bisexual; hermaphroditic; monoecious
雌雄同体世代 hermaphroditic generation
雌雄同序 androgyny
雌雄同序的 androgynal
雌雄同株 androgyny; hermaphrodite; hermaphroditism; monoecism
雌雄性比 sex ratio
雌雄异花 dicliny
雌雄异花的 diclinous; unisexual
雌雄异花同株 synoecy
雌雄异花性 diclinism
雌雄异熟 dichogamy; heterogamy
雌雄异熟的 dichogamous
雌雄异熟花 dichogamous flower
雌雄异体 bisexualism; dioecism; dioecy
雌雄异体的 dioecious; heterothallic; unisexual
雌雄异托的 heterocline
雌雄异型 antigeny
雌雄异型遗传 anisogeny
雌雄异株 dioecism; dioecy; heterothallism
雌雄异株的 dioecious; heterothallic
雌雄杂株的 polyoecious
雌雄杂株性 polyoecism

C

次(运材一次) haul
次孢囊 sporangial vesicle
次材 offal timber; secondary timber
次材采伐 recovery logging
次等木材 secondary timber
次级抽样 sub-sampling
次级飞羽 secondaries
次级结构 secondary structure
次级肉食性动物 second level carnivore
次级生产 secondary production
次级生产力 secondary productivity
次级生产者 secondary producer
次级食肉动物 secondary carnivore
次级消费者 secondary consumer
次级效益 secondary benefits
次级硬材 secondary hardwood
次级中心(植物起源) secondary center
次级总体 sub-population
次晶 paracrystal
次林层 substorey
次品 bastard; unacceptable product
次品纸张 culled paper
次期害虫 secondary insect
次生 second growth
次生胞壁 secondary cell wall
次生壁 secondary wall
次生壁层 secondary-wall layer
次生壁内层 inner layer of secondary wall; secondary wall inner layer; S-internal
次生壁外层 outer layer of secondary wall; secondary wall outer layer; S-external
次生壁中层 central layer of secondary wall; secondary wall middle layer
次生代谢物质 secondary metabolites
次生低压 secondary low
次生东质部 secondary xylem
次生冻原 secondary tundra
次生分生组织 secondary meristem
次生根 secondary root
次生构造 secondary structure
次生环境 secondary environment
次生寄生物 secondary parasite
次生加厚 secondary thickening

次生加厚生长 secondary thickening growth
次生空气污染物 derived air pollutant
次生林 second growth; secondary forest; secondary stand; secondary woody growth
次生林评价 valuation of second growth
次生螺纹加厚 secondary spiral thickening
次生泥炭 allochthonous peat
次生皮层 secondary cortex
次生气旋 secondary cyclone
次生群落 secondary community
次生韧皮部 secondary bast; secondary liber; secondary phloem
次生射线 secondary ray
次生生长 secondary growth
次生树皮 secondary bark
次生土 heterochronogenous soil; secondary soil
次生吸收根(半寄生植物的) sinker
次生性休眠 secondary dormancy
次生演替 secondary sere; secondary succession
次生演替系列 subsere
次生幼树 after-growth
次生雨林 secondary rain forest
次生长 second growth; secondary growth
次生肿瘤 secondary tumor
次生周皮 secondary periderm
次生砖红壤 secondary laterite
次生组织 secondary tissus
次显微结构 metastructure
次序 order; sequence
次要产物 secondary product
次要基因 minor gene
次要林道 branch road
次要支道 minor branch road
次要种 minor species; nurse species; secondary species; subordinates pecies; accessory species
次优势木 dominated crop
刺 prickle; spine; acantha (*plur.* -thae); acanthon; aculeus (*plur.* -lei)
刺柏林 juniger forest
刺柏木油 juniper wood oil
刺柏烯 junene; junipene
刺柏油 juniper(us) oil
刺柏籽油 juniper berry oil
刺齿进料辊 spiked roller
刺刀形树(雪折木) bayonet topped
刺蛾幼虫 nettle caterpillar; slug-caterpillar
刺果 lappa
刺果状的 lappaceous
刺槐 black locust; false acacia; locust tree; shipmast locust; yellow locust; bastard acacia
刺槐定 robinetinidin
刺槐苷 acaciin
刺槐黄素 acacetin
刺槐乙素 robinetin
刺槐因 robtein
刺激 stimulus
刺激反应链 stimulus-response chain
刺激剂 stimulant
刺激性气味 pungent odor
刺激因素(与气候有关的) stimuli (climate-related)
刺孔 puncture
刺孔注入法 cobra process
刺毛 chaeta; seta (*plur.* setae)
刺苹婆树胶 karaya gum
刺球花 lappa
刺伤法(橡胶采集) stick-method
刺梧桐树胶 karaya gum
刺吸口器昆虫 sap-sucking insect
刺吸式(口器) piercing type
刺形感器 sensillum chaeticum
刺云实单宁 tara tannin
刺云实栲胶 tara extract
刺状的 aculeate; spinescent
刺状尖棱 beard
枞胺 abietylamine
枞酸基 abietyl
枞酸型酸 abietic-type acid
楤木 dimorphantis; Chinese angelica tree
楤木苷 aralin
楤木烯 araliene
从冰片烯 apobornylene
从动轮 driven wheel
从动轴 driven shaft
从价 ad valorem

从价税 ad valorem tax
从价税率 ad valorem tariff
从量关税 specific tariff
从龙脑烯 apobornylene
从事园艺 gardening
从属种 subordinate species; subordinates pecies
从下口锯出的木片 biscuit
从性遗传 sex-controlled inheritance
丛 mat
丛播 bunch planting; massive seeding
丛晶 crystal conglomerate
丛晶体 cluster crystal
丛林 bosk; bosket; bosquet; brake
丛林火 bushfire
丛毛的 comose
丛生 over-grown; rosette
丛生病 rosette
丛生的 caespitosa; caespitose; cespitose; comate; fasciculate
丛生秆 clump
丛生禾草 bunch grass
丛生禾草草原 bunchgrass steppe
丛生节 cluster knot
丛生萌蘖苗 multiple seedling sprout
丛生枝 deliquescent branching
丛生植物 tuft
丛薮(sǒu) clump
丛枝病 witches' broom; witch's broom
丛枝状吸胞的(菌根真菌) arbuscular
丛植 bunch planting; massive planting; multiple planting
丛状的 fasciculate
丛状根 fasciculated root
丛状混交 mixed by clumps
丛状造林 bunch(ing)
粗剥 shelling
粗铂 platina
粗材(带皮层积材) rough log; rough wood
粗材积 gross scale; gross volume
粗材积增长率 percentage gross volume increase
粗糙的 scabrous
粗糙度 roughness; degree of anchor
粗糙纹理 rough grain
粗糙系数 coefficient of roughness
粗草场放牧 rough grazing
粗测材积 bigness scale
粗齿的 rough serrate
粗出生率 crude birth rate
粗大木质残体 coarse woody debris; mortmass
粗的 bastard; crude
粗端(木材) butt end; large end
粗放放牧地 extensive pasture
粗放耕作 extensive cultivation
粗放经营 extensive management
粗放林业 extensive forestry
粗放牧 rough grazing
粗放农业 extensive agriculture
粗放式管理 extensive management
粗放营林(造林) extensive silviculture
粗放择伐 rough selection felling
粗放作业 extensive treatment
粗榧碱 cephalotaxine
粗浮油 crude tall oil
粗腐殖质 coarse humus; duff; raw humus; mor
粗腐殖质层 duff horizon; duff layer
粗干材 bole; bole wood; trunk
粗根 coarse root
粗耕 coarse ploughing
粗骨土 fragmental soil; skeletal soil; skeleton soil
粗骨性石灰岩土 skeletal limestone soil
粗犷景观 coarse-textured landscape
粗浆 unscreened stock
粗结构 coarse structure; coarse texture; rough construction
粗锯材 rough lumber; undressed timber; unwrought timber
粗锯痕的 rough serrate
粗糠多的 chaffy
粗粒的 coarse-grained
粗粒环境 coarse-grained environment
粗粮 roughage
粗滤 strain
粗滤器 first filter
粗略估测 rough estimating
粗略估计 raw estimate; rough estimate
粗略样地 rough sample plot
粗麻绳 Manila rope
粗面岩 trachyte

粗木醋液 crude pyroligneous acid; crude wood vinegar
粗木方 baulk
粗木工车间 rough mill
粗木精 wood naphtha
粗泥炭 turf
粗刨 fore plane; jack plane
粗皮 coarse bark; corky bark; raw bark; rhytidome; rough bark
粗皮侧耳 oyster cup fungus; oyster mushroom
粗皮山核桃 scalybark hickory; shagbark hickory; thick shellbark hickory; upland hickory
粗群丛表 rough association table
粗砂(径0.2～2.0mm) coarse sand; grit
粗砂土 coarse sand soil
粗砂岩 grit
粗砂质壤土 coarse sandy loam
粗筛 bull screen
粗筛机 coarse screen
粗生长量 gross increment
粗收入 gross income
粗丝期 pachytene
粗饲料 roughage
粗松焦油 crude pine tar
粗松节油 crude turpentine
粗碳酸钾 pearl ash
粗糖紫蜡 wax of kamala
粗填料 coarse aggregate
粗网格 coarse grid
粗网纹 ornate
粗纹的 bastard
粗纹理 coarse grain; open grain
粗细式钢丝绳 filler strand
粗纤丝 macrofibril
粗纤维 coarse fibre; crude fiber
粗线期 pachynema; pachytene; pachytene stage
粗橡胶 coarse para
粗选机 coarse screen
粗削[材] rough hew
粗眼筛 riddle
粗杂材 brush
粗杂材形数 brush form factor
粗枝材 cripple; spray
粗制 rough cut
粗制材 blank; rough sawn timber; rough cut
粗制方材 balk
粗制家具 rustic furniture
粗制蜡 crude wax
粗制木椅 forest chair
粗制品 crude product
粗制椅 rural chair
粗制圆材 rough log
粗制脂 crude tallow
粗质蕊纸 light middle
促成温室 forcing house
促成栽培 forcing culture
促进发芽 accelerating germination; hastening germination
促进剂 accelerant; accelerating agent; accelerator; curing agent
促进萌蘖 forcing of sprouting
促进森林生长经营 rapid forest management
促进生根的 root-promoting
促进生长激素 growth-promoting auxin
促食剂 feeding incitant
促使插穗生根 strike
猝倒病 damping-off
猝倒病菌 damping-off fungus
猝灭剂 quencher
猝灭区 quenched region
醋石 acetate of lime
醋酸 acetic acid
醋酸钙 calcium acetate
簇 fascicle; mat
簇播 bunch planting
簇生 cluster
簇生的 caespitosa; caespitose; cespitose; clustered; comate; conglomerate; fascicled; fascicular; fasciculate
簇生花 fascicle
簇生节 cluster knot
簇生芽孢子 botryoblastospore
簇生叶 fascicle
簇生植物 caespitose plant
簇叶 foliage
簇叶病 rosette
簇植 bunch planting; nest planting
簇状垫状植物草原(高山荒漠) puna
簇状晶 cluster crystal

催干剂 siccative
催化 catalysis; seeding
催化剂 accelerator; activator; catalyst
催芽 accelerating germination; advance germination; chitting; forcing of sprouting; pregermination
催芽床 forcing bed
催育建筑物(温室等) forcing structure
催欲剂 aphrodisiac
脆的 brash
脆点 brittle point
脆度 brashness; brittleness; fragility
D
脆化型胶黏剂 stiff adhesive
脆盘 fragipan
脆磐土 fragipan soil
脆弱木 brittle wood
脆弱性 vulnerability
脆心材 brittle heart
脆性 brashness; brittleness
脆性木材 brittle wood; brash(y) wood
脆性破坏 brash fracture
萃取 extraction
萃取法 extraction method
萃取物 extract; extractive; extraneous substance
萃取用材 extract wood
淬火 quench
淬火器 quencher
翠柏脑 libocedrol
翠柏烯 libocedrene
翠雀定 delphinidin
村有林 hamlet forest
村镇林 town-forest
存板架 veneer storage magazine
存材槽 storage pocket
存材绠 receiving boom
存材廊 storage pocket
存材台 surge deck
存储 memory; storage
存储单元 storage location
存储密度(计算机) packing density
存储器 accumulator; memorizer; memory; storage
存储容量(计算机) storage capacity
存储微程序仪(控制流量) memory microprocessor
存储位置 storage location
存放时间(涂胶单板) open waiting time
存根(财经) stub
存活 survivorship; survival
存活率 survival percent; survival percentage; survival rate
存活曲线 survival curve; survivorship curve
存活者 survivor
存活值 survival value
存货 inventory
存货不足 understock
存货估价调整额 inventory valuation adjustment
存货签 inventory tag
存货周转率 turnover
存量 stocking
存取机理(计算机) access mechanism
存取时间 access time
存入量 load capacity
存在度(法瑞学派) presence degree
搓 laying
搓板 washboard
搓板状 washboarding
搓捻机 buncher
锉 slash
锉齿机 saw doctor
锉刀 file
锉锯 file; filing
锉锯工 filer; saw doctor; saw filer; saw fitter
锉锯工具 saw fitting tool
锉锯机 cutter sharpener; filling machine; gummer; sharpener; sharper
锉锯机械 filing machinery
锉锯夹 filing clamp
锉锯架 file holder
锉锯台 filing clamp
锉平 file
锉屑 filings
错接 stagger joint
错流 cross current
错位铰链门 hingebound door
错误项 error term
搭板 spring board
搭边抽屉 lip drawer
搭叠 shiplap

搭钩 floating hook; gaff; picaroon; raft hook
搭钩伤孔 pickaroon hole
搭挂(伐木) buckwheat; lodge; lodging; snagged; wigwam; hang up
搭挂树 hanging tree; hang-up tree; lodged tree; skybound
搭接 bevelled joint; lap; lap(ped) joint; overlap splice; shiplap joint; splice grafting; whittle grafting
T形搭接 T-lap joint
搭接胶合板 scarfed plywood
搭接强度 lap-joint strength
搭接墙板 drop siding
搭卖 tie in sale
搭桥(料仓) form bridge
达成 close
达尔文适合度 Darwinian fitness
达尔文学说 Darwin theory; Darwinism
达尔文主义 Darwnism
达玛碱不溶树脂(中性树脂) dammar resene
达玛清漆 dammar varnish
达玛树脂 dammar
达玛氧化树脂 dammar resene
达曼斯克斯枪筒(绞筒式枪筒) Damascus barrel
打靶 target shooting
打包 baling
打包机 baler
打包间 bale room
打场 thrashing
打底 layer
打底薄膜 priming foil
打底腻子 clearcole
打底子 precoat
打谷机 thrasher
打夯机 compactor; rammer; tamping machine
打号 marking
打号槌 numbering hammer; number-marking hammer
打号锤 range-hammer
打号器 branding iron; marker
打号印 axe marking; branding; scribe; stamp
打号员 marker
打滑 creep; slip
打火 fire fighting
打火机 lighter
打火拍子 flap
打火扫把 fire beater; fire swatter
打尖 pinching; top removal; topping
打浆 beat; beating; pulping
打浆比压 specific beating pressure
打浆刀 beater bar
打浆度 beating degree; degree of beating
打浆辊 beater roll
打浆过度 over-beating
打浆机 beater; beating machine; macerator
打浆机内施胶 engine sizing
打浆机配料 beater furnish
打浆机施胶 beater sizing
打浆施胶 engine sizing
打浆添加剂 beater additive
打浆系数 beating coefficient
打浆性能 beatability
打浆压力 beating pressure
打窠(kē)(射兔) pot shot; sitting shot
打孔带 perforated tape
打孔钻头 grooved bit
打捆 baling; bunch(ing)
打捆机 buncher; packager; trusser
打捆器 packer
打蜡 cere
打捞沉材 deadhead
打猎季节 open season
打磨机 mill hog
打磨区 milling zone
打腻子机 filling machine
打样子(伐木缺陷) barber chair; tombstone; splinter draw; sloven
打散机 bale breaker
打碎 macerate
打碎机 chopper; mill
打通 opening up
打眼钻 bradawl
打印 chop; marking
打枝 branch cutting; branching; debranch(ing); delimbing; limb; limbing; lopping; prune; pruning; shragging; snedding; trimming
打枝工 knotter; lopper; limber
打枝归堆机 delimber buncher

D

打枝机 debrancher; delimber; limbing machine
打枝截头机 delimber topper
打枝造材工 limber bucker
打枝造材归堆联合机 processor
打枝造材机 bucker-limber; delimber-bucker
打枝造材联合机 delimber and slasher; limber bucker
打枝造财归堆联合机 delimber-bucker-buncher
打枝栅 delimbing gate
打枝桩工 knot bumper; knotter
打桩 pile; piling
打桩锤 monkey
打桩工 spool tender
打桩机 gin; pile driver; piling machine; pole press
打桩排架 pile bent
打桩现场 piling site
打桩栈架 pile trestle
打字纸 manifold paper
大孢子 gynospore; macrospore
大孢子囊 macrosporangium
大孢子叶 carpophyll; macrosporophyll
大孢子植物 carpophyte
大薄片刨花板 large flake board
大比例尺 large scale
大臂杆装车机(装于汽车上) big-stick loader
大材 summer wood
大采伐面 large coupe
大草原(非洲) veld(t)
大草原(南美) llano
大肠菌 coliform
大尺度 macrochore
大翅型 macropterous form
大虫孔 large worm hole
大带锯 log mill
大地测量控制点 geodetic control point
大地测量学 geodesy
大地测量坐标 geodetic coordinate
大地女神假说 Gaia hypothesis
大钉(长于10英寸) spike
大兜 butt swelling; root-swelling
大兜的 barrel-butted; bottle-butted; churn-butted; swell-butted
大豆蛋白松香胶料 prosize
大发生(病虫害) outbreak
大方材 big square; large square
大方料 balk; bau(l)k
大分生孢子 macroconidium
大分子 macromolecule
大分子结构 macromolecular structure
大风(八级) high wind; gale
大风子油 chaulmoogra oil; hydnocarpus oil
大幅面板 giant sized board
大复羽 greater covert
大高位芽植物 macrophanerophyta; megaphanerophyte
大割灌镰 brush scythe
大根 butt root
大根香叶酮 germacrone
大公鹿 stag
大规格材 large dimension timber
大规模的 wide scale; large scale
大规模绿化 intensive afforestation
大规模生产 large-scale production
大规模效应 mass effect
大规模造林运动 large-scale afforestation drive
大规模植树运动 major tree planting drive
大果铁杉 alpine spruce; alpine hemlock; mountain hemlock; Williamson's spruce
大花天竺葵 fancy pelargonium; lady Washington geranium; show pelargonium
大环裂 big cup shake
大寰(地图学) great circle
大黄苷 anthraglucorhein
大黄根脂 aporetin
大黄素 frangula-emodin
大茴香 aniseed
大茴香叉丙酮 anisacetone
大茴香油 star anise(ed) oil
大绘图板 trestle-board
大火 conflagration
大火控制 large fire control
大火团 mass fire
大戟胶树脂 euphorbium
大戟蜡 rhimba wax
大戟树脂 euphorbium

大甲虫 timber inspector
大节 gross knot; large knot
大进化 macroevolution
大径级锯材原木 large sawtimber
大径木 large dimension timber; large timber; sawlog stage
大静脉 main line
大鸠尾锯 carcass-saw
大锯 cleaving-saw
大砍刀 broad sword; matchet
大孔隙(径1000埃米以上) macropore
大块浮游植物 sudd
大冷杉毒蛾 fir tussock moth
大理菊 connnon dahlia; dahlia
大理石 marble
大理石斑腐 marble rot
大理石桌 marble table
大理石桌面 marble top
大理岩 marble
大丽花 connnon dahlia; dahlia; garden dahlia
大梁 girder; summer beam
大量 bulk
大量采集 mass collection
大量繁殖 mass rearing
大量灌溉 heavy irrigation
大量花粉 pollinia
大量渗透 mass infiltration
大量生产 mass production
大量施肥 heavy dressing
大量投入 capital-intensive
大量脱木素作用 bulk delignification
大量营养元素 macronutrient
大量元素 macroelement; major element
大猎物 big game
大林区 division
大林区林务署(美) region office
大林区林长 regional forester
大陆边缘 continental margin
大陆岛 continental island
大陆的 continental
大陆架 continental shelf
大陆架动物区系 shelf fauna
大陆空气 continental air (mass)
大陆块 land mass
大陆裂谷 continental rift
大陆隆起 continental rise
大陆漂移 continental drift; continental migration
大陆漂移假说 contiental drift hypothesis
大陆坡 continental slope
大陆气团 continental air (mass)
大陆性冰川 continental glacier
大陆性的 continental
大陆性气候 continental climate
大陆性森林 continental forest
大麻油精(大麻烯) cannabene
大麻子油 hemp seed oil
大面积皆伐 large area clearcutting; large cleaning; large clear-cutting
大面积伞伐作业 uniform shelterwood system
大苗区 plant-school
大木 breast-summer
大木钉 trunnel
大木蜂 large carpenter bee
大木锯 log saw
大年 on year
大刨 jack plane
大泡状花纹 big blister figure
大配子 macrogamete
大批 bulk
大批生产 lot production
大片刨花 wafer
大片刨花削片机 waferizer
大片森林 large stretches of forests
大片筛 bull screen
大片松林 stretch of pines
大漆 lacquer; lacker
大气 atmosphere
大气窗 atmospheric window
大气电力接收器 sferics receiver
大气动力学法 aerodynamic method
大气反馈过程 atmospheric feedback process
大气浮游生物 air plankton
大气浮游植物 aeroplanktophyte
大气辐射 atmospheric radiation
大气辐射表 pyrgeometer
大气干旱 atmospheric drought
大气候 macroclimate
大气候学 macroclimatology
大气环流 atmospheric circulation
大气浑浊度 atmospheric turbidity
大气扩散 atmospheric diffusion

大气密度 atmospheric density
大气能因 atmospheric agency
大气逆变 atmospheric inversion
大气圈 atmosphere
大气湿度 atmospheric humidity; atmospheric moisture
大气湍流 atmospheric turbulence
大气稳定度 atmospheric stability
大气涡旋 atmospheric vortex
大气污染 air pollution; atmosphere pollution; atmospheric pollution
大气现象(云除外) meteor
大气象学 macrometeorology
大气压力 atmospheric(al) pressure
大气蒸发 atmospheric evaporation
大气作用 atmospheric agency
大乔木 large tree; megaphanerophyte
大染色体 macrochromosome
大生境 major habitat
大生物区系 macrobiota
大生长期 grand period of growth
大书橱 library case; wing bookcase
大树蜂 steel-blue wood-wasp; small blue horntail
大数法则 principle of the mathematics of large numbers
大数据 big data
大苏打 hypo
大条板 large strip
大铁锤 sledge
大厅长椅 hall seat
大头(木材) butt; large end; butt end; bottom end
大头羽裂的(叶) lyrate
大头直径 butt end diameter; diameter at butt end
大突变(染色体) macromutation
大土类 main soil group; main soil type
大腿骨 femur
大西洋酮 atlantone
大纤丝 macrofibril
大小 stature
大小尺寸 format
大小年 biennial bearing
大小组 size class
大型驳船货舱 jumbo hopper
大型底栖生物 macrobenthos
大型浮游生物 macroplankton
大型浮游植物 macrophytoplankton
大型搁栅 heavy joist
大型工程 major works
大型绞盘机 climber
大型木管柱身 trunk
大型沙发 chesterfield
大型生物 macrofauna
大型生物群 macrobiota
大型消费者 macro-consumer
大型叶 macrophyll
大型叶的 macrophyllous
大型植物 macrophyte
大型植物区系 macroflora
大修 capital repair; general overhaul; major overhaul; overhaul; top overhaul; extensive repair
大血管 aorta
大洋区 oceanic province; oceanic region
大样本理论 large sample theory
大样地 large plot
大叶钓樟油 kuromoji oil
大衣柜 cabin trunk
大幼树 large sapling
大圆 great circle
大圆材(胸径8～12英寸) high pole (wood)
大种 macrospecies
大蛀虫(柏木) shingle weaver
呆账 uncollectible account
呆滞年 sluggish year
代表部分程序 representative fraction
代理 commission
代理经营 factoring
代理人 broker
代码 code
代森铵(杀菌剂) amobam
代森锰(杀菌剂) maneb
代森锌(杀菌剂) dithane Z-78; zineb
代收款 collection for others
代替 novation
代谢泵 metabolic pump
代谢产物 metabolic product; metabolite
代谢反馈 metabolic feed back
代谢过程 metabolic process
代谢汇 metabolic sink

代谢活性 metabolic activity
代谢亢进 hypermetabolism
代谢库 metabolic pool
代谢类型 metabolic type
代谢链 metabolic chain
代谢路线 metabolic route
代谢率 metabolic rate
代谢期 metabolic stage
代谢顺序 metabolic sequence
代谢途径 metabolic pathway
代谢型式 metabolic pattern
代谢源 metabolic source
代谢运输 metabolic transport
代谢障碍 dysbolismus
代谢症状 metabolic symptom
代谢支路 metabolic by-pass
代谢阻滞 metabolic block
代谢组织 metabolic tissue
代谢作用 metabolism
代用材 substitute wood
代用物 stand-in
代用枕木 substitute tie
代用资源 replacement resource
带 belt; zone; region
带白霜的 glaucescent
带壁橱卧室 pier bedroom
带病毒的 viruliferous
带病粒 seed carrying pest
带播 band planting; band seeding; belt planting; belt sowing; seeding in strip; sowing in strip; strip seeding
带苍白色的 glaucescent
带槽板 slotted plate
带车立柱的铁路车辆 rack car
带齿扒扛 spiked skid
带齿端钩(装车工具) spiked bell hook
带虫粒 seed carrying pest
带灯塔架拖垃机 lighthouse cat
带电膜 charge membrane
带伐 border-cutting
带伐更新 regeneration by strip-felling
带伐作业 strip system; strip-method
带缝接头 gap joint
带杆轮尺 pole calliper
带隔条堆积 stickered stacking; stripped stacking
带拱架拖拉机 arch cat
带钩吊索 crotch line
带钩阀 gaff
带钩开口滑轮 team snatch block
带钩索套 dog warp
带柜床 bed with storage
带花粉的 pollen-bearing
带化[现象] fasciation
带环铁楔 dog wedge
带夹 band clamp
带锯 band saw; endless saw; log band saw; ribbon saw; strap saw
带锯裁边机 band mill edger
带锯锉锯机 band saw sharpener
带锯机 band mill; band saw mill; endless saw
带锯锯的 band-sawn
带锯拉紧装置 blade straining device
带锯磨锯机 band saw sharpener
带锯跑车 saw-carriage
带锯条 band saw blade
带锯条滚压机 band saw stretcher
带锯制材厂 band mill; band saw mill
带锯制材的 band-sawn
带锯自动磨锯机 automatic band saw sharpener
带卷筒拖拉机 tractor-donkey
带菌体 carrier
带靠山再分锯 line-bar resaw
带棱健全材 square-edge and sound
带利投资 profit-bearing investment
带落钩装置的跑车 slack pulling carriage
带皮板 bark board
带皮材积 volume of over bark; volume outside bark
带皮干围 girth over bark
带皮小干材 barkie
带皮直径 diameter outside bark; diameter over bark; diameter with bark; outside diameter; over-bark diameter
带皮周长 girth over bark
带梢圆材 pole with tops
带梢长材 long log with slender top
带式单板干燥机 band veneer dryer; veneer belt dryer
带式垫网 travel(l)ing belt screen
带式干燥机 band drier; belt dryer

带式计量装置 belt weigher
带式加速运输机 acceleration belt conveyor
带式进料滚筒砂光机 endless bed drum sander
带式均衡器 belt balance
带式砂光机 belt sander; stroke sander
带式输送机 band conveyer; belt conveyer
带式无垫板装板机 belt-type caulles loader
带式预压机 belt precompressor
带式制动器 band brake
带树瘿的 warted
带锁小橱 locker
带屉矮衣柜 lowboy
带通(电子) band pass
带通滤波器 band-pass filter
带土苗 ball-plant
带土球(苗木) bailing
带土球苗 balled seedling; balled stock
带土球树 balled tree
带土球栽植 ball planting; ball(ed) planting
带土栽植 burlap planting
带坨树苗挖掘机 tree balling machine
带线(腐朽材) zone line
带心锯材 boxed-heart (timber)
带薪休假 paid vacation
带形背椅 riband back chair
带形椅背 interlaced chair back
带遥控跑车的双联带锯机 tandem twin bandmill with remote controlled carriage
带叶插穗 leafy cutting
带叶扦插 leafy cutting
带鱼条单板 fishtail
带辕杆单爬犁 loose-tongued sloop
带制动闸的爬犁 brake sled
带状傍管薄壁组织 banded paratracheal parenehyma
带状薄壁组织 banded parenchyma; zonate parenchyma
带状标准地 sample strip
带状冰 ice ribbon; ribbon ice
带状采伐 felling in strip; strip cutting; strip felling; stripping
带状采伐迹地 strip felled area; strip-felling area
带状采伐林 strip-cutting forest
带状采伐母树法 seed tree cutting by the group method
带状采伐作业 strip cutting system; strip felling system
带状测树器 band dendrometer
带状抽样 strip sampling
带状地 strip
带状点火 strip firing
带状伐 cutting in strip
带状伐区 strip felled area
带状分布 zonal distribution
带状耕作 strip cropping
带状更新 regeneration under strip system
带状灌溉 border irrigation
带状花坛 ribbon bed; ribbon flower-bed
带状花纹 ribbon figure; ribbon stripe; ribbon(-stripe) figure; stripe figure
带状环剥 band girdling
带状混交 mixed by strips; mixture by line
带状夹紧装置 band clamp
带状渐伐 strip-shelterwood cutting
带状渐伐法 strip shelterwood method
带状渐伐作业 shelter-wood strip system
带状皆伐 clearcutting in strips; clear-strip felling; strip clearcutting
带状皆伐法 clearstrip method
带状皆伐天然更新法 clearcutting in strips
带状皆伐作业 clear-strip system
带状勘测 strip cruise
带状离管薄壁组织 banded apotracheal parenchyma
带状绿地 green-belt
带状排列管孔 banded pores
带状清理伐 severance cutting; severance felling
带状群状渐伐作业法 strip and group shelter wood system; strip and group system
带状伞伐 shelterwood-strip felling

带状伞伐更新法 strip shelterwood method; strip-stand method
带状伞伐作业 shelter-wood strip system
带状烧除 strip firing; strip burning
带状生长记录器 band dendrometer
带状施药 band application
带状疏伐 line thinning; strip-thinning
带状条播 ribbon sowing
带状调查 strip census; strip cruise; strip survey
带状调查网格法 gridironing method of strip survey
带状微纤丝 ribbon-like microfibril
带状样地 strip plot
带状样地调查 strip-method
带状栽植 band planting; strip planting
带状择伐 strip-selection cutting
带状择伐法 strip selection method
带状择伐作业 selection border cutting method; strip selection method; strip selection system
带状整地 strip soil preparation
带状作业更新林 regenerated forest under strip system
带桌面书橱 bureau bookcase; desk and bookcase
贷方 credit; lender
贷款 call (able) loan
待补税额 deficiency
待采脂林 round timber
待采脂树 round; round tree
待推河木垛 rackheap
待装楞堆 surge deck
怠工 labour slow down
怠速 idle speed
怠速排量 idle discharge
袋 sack
袋虫 bagworm
袋腐 peck; peckiness; pecky dry rot; pocket dry rot; pocket rot
袋模成型 bag moulding
袋式剥皮机 bag barker; Thorne debarker
袋式摩擦剥皮机 pocket debarker
袋形缏 bag boom; sack boom
袋形河缏 round boom
袋形排 bag boom; loose raft; pocket; round-boom type raft; sack boom
袋形排框 round boom
袋形排拖运 towing bag (boom)
袋形拖排 crib log
袋渣胶 kiree
袋状河缏 pocket boom
袋状叶 pocket leaf
袋状栅 bag boom
戴里伦法(木材处理) Drilon process
丹麦式疏伐(综合疏伐) Danish thinning
担孢子 basidiospore; sporidium
担孢子梗 sterigma (*plur.* -mata)
担孢子阶段(锈菌) sporidial stage; sporidial state
担保契约 hypothecation
担架式喷雾机 stretcher mounted sprayer
担子 basidium (*plur.* -ia)
担子层 basidial layer
担子地衣 basidio lichen
担子果 basidiocarp
担子菌纲 Basidiomycetes
单扳机(双筒枪) single trigger
单板 veneer
单板背面 openside; slack face; slack side
单板补节机 patcher; patching machine
单板层积材 laminated veneer lumber
单板带 veneer band; veneer ribbon
单板带锯 band sawing of veneer; veneer bandsaw
单板堆(依切削次序堆叠) flitch
单板堆垛机 veneer stacker
单板垛 veneer flitch; veneer stack
单板垛齐边机 veneer pack jointer
单板分选机 veneer sorter
单板封边(木材横断面) oystering
单板干燥 veneer drying
单板干燥机 veneer drier; veneer dryer
单板割刀 veneer score knife
单板横向胶拼机 veneer edge gluer
单板滑放 wild catting
单板剪裁机 veneer clipper
单板胶带拼缝机 veneer tapping machine

单板紧面(单板接触压尺的一面) tight side (of veneer)
单板锯 veneer cutting saw; veneer saw
单板锯切机 veneer sawing machine
单板卷-松机 veneer reeling and unreeling machine
单板卷筒 veneer reel
单板冷压胶合机 cold veneering press
单板毛方 veneer flitch
单板排锯 veneer gang saw
单板刨切机 veneer slicer
单板拼缝机 veneer composer; veneer stitcher; veneer stitching machine
单板拼缝刨边机 veneer jointer
单板拼接 veneer welding
单板拼接机 veneer composer; veneer splicer
单板饰面碎料板(刨花板) veneer particleboard
单板松面 openside; slack face; slack side
单板贴面刨花板 plyboard
单板挖补机 patching machine; veneer patcher
单板箱盒 veneer chest
单板旋切机 rotary peeling machine; veneer lathe; veneer rotary lathe
单板旋切用刀 veneer knife
单板原木 veneer log
单板原木生产 veneer log production
单板圆锯 circular veneer saw; veneer-circular saw
单板圆木段(旋单板用) veneer bolt
单板铡刀 veneer clipper
单板整边 jointing
单板整张化 veneer unitization
单板贮存架 veneer storage magazine
单板贮存装置 veneer storage system
单板准备工段 veneer preparation area
单孢的(真菌) monospermous
单倍孢子体 haplodiplont
单倍单性生殖 haploid parthenogenesis
单倍核 hemikaryon
单倍配子体无融合生殖 haploid gametophyte apomixis
单倍染色体 haplochromosome
单倍体 haploid; haplont; monoploid
单倍体不育 haplontic sterility
单倍体孤雌生殖 haploid parthenogenesis
单倍体数 haploid number
单倍无配生殖 haploid apogamy
单倍性 monoploidy
单倍性生物 haplont
单倍有丝分裂 haplomitosis
单倍杂种 haplomict
单被的 monochlamydeous
单被花 monochlamydeous flower
单变的 monotropic
单变现象 monotropy
单柄锯 one-handed saw
单采作业 forest management by individual trees
单侧复纹孔导管 unilaterally compound pitting vessel
单侧复纹孔式 unilaterally compound pitting
单侧环管薄壁组织 unilaterally paratracheal parenchyma
单侧聚翼薄壁组织 unilaterally confluent parenchyma
单侧穗状花序 one-sided spike
单侧纹孔 unilateral pit
单侧稀疏环管薄壁组织 unilaterally scanty paratracheal parenchyma
单侧翼状薄壁组织 unilaterally aliform parenchyma
单测法(测量) simple method
单层 monolayer
单层胞 regular
单层采脂法 single-storey tapping
单层的 single layered
单层干燥机 one-deck dryer
单层花瓣的 haplopetalous
单层结构刨花板 single layer particle board; unilayer particleboard
单层结构碎料板 unilayer particleboard
单层连续式滚筒干燥机 continuous one-deck roller-dryer
单层林 one-storey forest; single-storied forest; single-storied stand
单层木排 flat boom
单层压机 single open press

单层液控冷压机 single opening hydraulic cold press
单层原木装载 bunk load
单程集材 single haul
单程蒸馏 simple batch distillation
单程转化率 conversion per pass
单翅的 monopterous
单出数花 monomerous flower
单出蝎尾状聚伞花序 wickel
单处理机 uniprocessor
单穿孔 simple perforation
单穿孔板 simple perforation plate
单穿孔导管 porous vessel
单纯矮林作业 single coppice system
单纯矮林作业法 simple coppice (system)
单纯利用伐 exploitation cutting (felling)
单纯弯曲 simple bending
单纯形法 simplex method
单纯形算法 simplex algorithm
单雌群的(群居膜翅目昆虫) monogynous
单雌蕊的 monogynaecial; monogynian
单雌性 haplometrosis; monogyny; monometrosis
单带锯机 single band-mill
单刀片式伐木剪 single-action shear
单点灭火法 individual station method
单电子键 singlet
单顶极学说 monoclimax theory
单动磨盘机 raffinator
单动盘磨机 single revolving disc refiner
单动三盘磨浆机 triple disc refiner
单独钢丝绳芯 independent wire rope core
单独胶束学说 individual micellar theory
单对等位基因遗传 monomerical inheritance
单饭拼接 veneer splicing
单分散性 monodispersity
单分子层 monolayer
单酚氧化酶 monophenol oxidase
单峰 singlet
单杆起重装置 single stick boom
单干萌生苗 single seedling sprout
单个土体 pedon
单个系列作业法 single periodic block method
单个细胞 individual cell
单个选择 single selection
单根大木装载 jim crow load
单根流送 single out
单根纤维 individual fibre
单根原木进料装置 feeding single-log mechanism
单根原木装载 bunk load
单股环索 molly hogan
单管孔 pore solitary; simple pore; solitary pore
单轨道 monorail
单轨干燥窑 single track kiln
单轨悬挂转运机 monorail transfer
单核 monokaryon
单核的 mononuclear
单核糖体 monosome
单烘缸造纸机 Yankee machine
单花采粉的 monotropic
单环单萜烯 monocyclic monoterpene
单环的 monocyclic
单回路滚筒干燥机 one way-drum dryer
单基火药 single-base powder
单基物 monomer
单基因的 monogenic
单基因抗性 monogenic resistance
单基因性状 monogenic character
单基因杂种优势 single gene heterosis
单级悬浮式干燥机 single-step suspension dryer
单级压缩 single-stage compression
单寄生 monoparasitism; solitary parasitism
单价 unit of cost; unit value; unit-price
单价的 univalent
单价染色体 diad; monovalent chromosome; univalent
单简绞盘机 ground-hog
单交 single cross
单交换 single crossing over
单绞拱 single-hinged arch
单绞盘机一次集材 single yard
单铰拱 one-hinged arch
单脚高架起重机 half-gantry crane

D

单节 single knot
单节扦插(插穗) one-node cutting
单精受精 monospermy
单精受精卵 monospermic egg
单考得(8ft长、2ft宽、4ft高材堆) single cord
单壳机身 monocoque
单孔(孢粉) simple pore
单孔检定 single-boring assay
单跨 single span
单跨楼板 single floor
单跨索道 single-span skyline
单利 simple interest
单利平均生长率 simple average growth rate
单利生长率 simple growth rate
单粒播种 monoseeding; single-seed planting
单连接法 single linkage method
单梁式冲击试验机 simple beam impact machine
单料室立式打磨机(用于蔗茬打磨) one-chamber type vertical shaft mill
单列薄壁组织带 uniseriate parenchyma band
单列射线 linear ray; simple ray; uniseriate ray
单轮的 monocyclic
单轮滑车 gin block
单轮跑车 rider block
单轮雄蕊 haplostemony
单螺形聚伞花序 uniparous helicoid cyme
单缕 monofil(ament)
单面 single-cutting band saw
单面带锯 single band saw
单面的 monofacial
单面封边机 single-sided edge bander
单面斧 single bit
单面胶接(胶层) single glue joint
单面木工压刨床 single side planer
单面刨 single surface planer
单面劈斧 side axe
单面砂光 sanded one side
单面山 cuesta
单面山地形 cuesta landform
单面涂胶 single spread(ing); single surface glue spreading
单面叶 unifacial leaf
单模属 monotypic genus
单木 single tree; single stem; individual tree
单木打枝机 single-stem delimber
单木估价 individual tree valuation; invidual tree valuation
单木竞争指数 individual tree competition index
单木模型 individual tree model; single-tree model
单木作业 forest management by individual trees
单目显微镜 monocular microscope
单宁 tannin
单宁精 tannigen
单宁染色 tan stain
单宁树脂 tannin resin
单宁物质 tanning matter
单宁细胞 tannin cell
单爬犁 lizard; swing dingle
单排流送 raft-section driving
单胚的 monoembryonic
单胚生殖 monoembryony
单配的 monogamous
单配偶制 monogamy
单片存储工艺 monolithic storage technology
单片芽接 veneer budding
单漂流送 drive
单频道 monorail
单歧聚伞花序 incomplete cyme; monochasium; monochasy
单歧式 monochasium
单亲本子代测 one-parent progeny test
单亲的 uniparental
单亲生殖 monogeny
单全同株的 caeno-monoecious
单全异株的 trioecious
单染色体 monosome
单染色体的 monosomic
单染色体组种 monogenomic species
单人动力锯 bucking bar
单人锯 one-handed saw; one-man saw
单刃采种钩 single-hook detacher
单刃带锯 single band saw; single-cutting band saw
单刃斧 single bit; single-bladed axe

单色 monochrom
单色花 self-coloured flower
单色器 monochromator
单色图像 monochrome image
单砂架宽带式砂光机 single head sander
单筛板 simple sieve plate
单扇拉门 single sliding door
单扇摇门 single swinging door
单舌接 simple whip grafting
单生的 solitary
单生仔动物 monotocous animal
单食性 monophage; monophagy; monpohagous
单食性的 monophagous; monotrophic
单食性动物 monophagous animal
单式 simplex
单室的 monolocular
单丝 brin
单索集材 single-wire logging
单索缆车道 single-line inclineway
单胎动物 monotocous animal
单糖 glycose; monosaccharide; monosaccharose; monose; simple sugar
单糖类 monosaccharose; monose
单体 cell; monomer
单体二倍性 monosomic diploidy
单体气球 single-hull balloon
单体生物 monosomic; unitary organism
单体雄蕊 monodelphous stamen
单体雄蕊的 monadelphous; monader
单调(景色) monotony
单萜 monoterpene
单筒单发鸟枪 single barrel single shot gun
单筒连发鸟枪 single barrel repeating shot-gun
单筒望远镜 field scope
单凸面圆锯 bevel saw; swage saw
单位 element; unit
单位安全应力 unit safe stress
单位成本 unit cost
单位冲击强度 specific impact strength
单位负荷 unit load
单位晶格(晶胞) elementary cell
单位矩阵 identity matrix
单位拉伸应力 unit tensile stress
单位流量 specific discharge
单位面积放牧量 stock rate
单位面积林价收入 stumpage revenue per unit area
单位面积年收获量 annual yield by area
单位膜 unit membrane
单位努力捕获量 catch per unit effort
单位切削阻力 specific cutting resistance
单位伸长 unit elongation
单位水文过程线 unit hydrograph
单位体积膨胀 unit volume expansion
单位压力 specific pressure; unit pressure
单位叶速率 unit leaf rate
单位应变 unit strain
单位应力 unit stress
单位元 identity
单位载重 unit load
单纹孔(木材) simple pit; simple poer
单纹孔对 simple pit-pair
单细胞壁 single wall
单细胞变异体 single cell variant
单细胞层 monolayer
单细胞蛋白 single cell protein
单细胞的 single-celled
单细胞培养 single-cell culture
单纤丝 monofil(ament)
单显性组合 simplex
单线集材 one way skidding
单线架空索道 monocable aerial tramway; monorail cableway
单线缆车道 one-way incline; one-way tramway
单线式索道 fixed wire cableway
单线松紧式索道集材 single slackline yarding
单线索道 cable-tramway of single rope; mono cable; single rope cableway; single track-cable; single wire-tramway
单线态 singlet state
单线迎面火 single back fire
单相抽样 mono-phase sampling
单向阀 non-return valve
单向色谱 one-dimensional chromatography

单向循环 one-way circulation
单向应力 pure stress
单向圆盘犁(美) one-way disk plough
单蝎尾状聚伞花序 uniparous scorpioid cyme
单斜的 monoclinic
单斜脊地形 cuesta landform
单斜钠长石 analbite
单心裂(立木) simple heart shake
单心皮雌蕊 monocarpellary pistil; monocarpous pistil
单心皮的 monocarpellary; unicarpellate
单心皮子房 monocarpellary ovary; monocarpous ovary
单形成层的 monocambial
单型属 monotypic genus
单性的 unisexual
单性果实 parthenocarpic fruit
单性合子 azygote
单性结实 parthenocarpy
单性结实的 parthenocarpic(al)
单性卵孢子 aboospore
单性卵裂 parthenogenetic cleavage
单性生殖 agmogenesis; apomixis; parthenogeneic reproduction; parthenogenese; parthenogenesis; parthenogenetic propagation
单性生殖雌虫 parthenogenetic female
单性生殖的 parthenogenetic
单性生殖二倍体 parthenogenetic diploid
单性生殖卵 parthenogenetic egg
单性生殖世代 parthenogenetic generation
单性生殖细胞 parthenogonidium
单性受精生殖 parthenomixis
单性杂交 monohybrid cross
单性杂种 mono-hybrid
单性着果 parthenocarpic fruit-set
单雄群 one-male troop
单雄生殖 androgenesis
单循环病害 simple-cycle disease
单循环的 monocyclic
单芽 simple but
单芽插 eye cutting
单芽插穗 one-eye cutting
单芽靠接 inarching with an eye
单芽扦插 one-eye cutting
单眼 ocellus (*plur.* ocelli)
单叶 simple leaf
单叶的 integrifolious
单叶隙的 unilacunar
单一弹性 unitary elasticity
单一定期施业分区 single periodic block
单一基因杂种优势 one-gene heterosis
单一间隔期的 monocyclic
单一经营 monoculture
单一树种胶合板 homogeneous plywood
单一税 single tax
单一税率 simple tariff
单一斜切法(橡胶采割法) single oblique cut
单一样地抽样 monadical plot sampling
单因素的 monofactorial
单因子差异 one-factor difference
单因子的 monofactorial; monogenic
单因子试验 simple factorial experiment
单因子杂种 mono-hybrid
单优种群丛 consociation
单优种群落 consociation; monodominant community
单优种演替群落 consocies
单元 cell; element; unit
单元单倍体 monohaploid
单元格子 unit cell
单元过程 unit process
单元加热器 unit heater
单元演替顶极 monoclimax
单原木漂子(四周钉有纵向木条) strip boom
单载荷式索道 single-load cableway
单掌状的 unipalmate
单枝靠接 inarching with a branch
单枝压条 simple layering
单种属 autogenus
单种养殖 monocultute
单种栽培 monoculture
单种子的 monospermous
单周期性的 monocyclic
单轴 monopodium
单轴的 monopodial
单轴定向 uniaxial orientation

单轴多锯片圆锯机 slasher; slasher saw; slashing saw
单轴多片圆锯机 single-gang edger
单轴浮雕机 single-spindle carving machine
单轴花序 monopodial inflorescence
单轴拉伸 uniaxial extension
单轴拉伸形变 uniaxial tensile deformation
单轴六轮汽车运材 bob-tailing
单轴木工铣床 single-spindle shaper
单轴汽车运材方式(不经过集材) bob-tail system
单轴生长(树形) monopodial growth; excurrent growth
单轴榫槽机 single-spindle mortiser
单轴型地下茎 leptomorphic rhizome
单轴压缩 uniaxial compression
单轴延伸 uniaxial extension
单株 individual
单株混交 individual-tree mixtue; mixed by individual tree; mixture by single trees; single-tree mixture; tree-by-tree mixture
单株混植 admix; mix by single individuals
单株木样本 single tree sample
单株选择 individual selection; single tree selection
单株移出 single out
单株栽植 planting by single tree
单株择伐 single tree selective cutting; single-tree selection felling
单株择伐法 single tree (selection) method
单株择伐作业 single-tree selection system
单主寄生 monophage
单主寄生的 ametoecious; autoecious; monoxenous
单主寄生现象 autoecism
单主寄生真菌 autoecious fungus
单柱梯 one-legged ladder
单柱屋顶桁架 king post roof truss
单抓钩起吊装置 single-tong system
单转盘式精磨机 single rotating disc refiner
单子叶的 monocotyledonous; monocotylous
单子叶树材 monocotyledonous wood
单子叶树木 monocotyledonous tree
单子叶植物 amphibrya; monocot; monocotyledon
单足的 monopodial
单组份清漆 one-component lacquer
单组染色体 simplex group of chromosomes
胆矾 blue vitriol
胆碱氨甲酰氯 carbamylcholine chloride
胆碱脂酶 choline sterase
胆汁 gall
淡红洋蔷薇 pompon rose
淡季 off-season
淡金水紫胶漆 gold lacquer
淡栗钙土 light chestnut earth
淡色 tint; undertone
淡色表层 ochric epipedon
淡色岩 leucocrate
淡水 fresh water
淡水浮游生物 limnoplankton
淡水生态系统 freshwater ecosystem
淡水生态学 freshwater ecology
淡水生物群落 fresh water community
淡水透镜 freshwater lens
淡水演替 fresh water succession
淡水藻色素 phycochrome
淡水沼泽 pantanal
淡水植物 fresh water plant
淡棕色草原土 light brown steppe soil
弹仓 magazine
弹仓簧 magazine spring
弹串 shot-string
弹夹 clip magazine
弹壳 shell
弹镗 chamber
弹体 elastomer
弹头 bullet
弹头形状 bullet shape
弹头重 bullet weight
弹丸 pellet
弹丸侧飘 lateral drift of shot
弹丸速度 velocity of shot
弹形容器苗栽植 bullet planting
弹沿(来复枪子弹) cartridge rim
弹沿沟 rim grove
弹药库 magazine
弹着点 point of impact

蛋白胨 peptone
蛋白胶 albumen glue; protein glue
蛋白胶黏剂 protein adhesive
蛋白酶 protease; proteinase
蛋白纤维 azelon
蛋白质亚基 capsomer(e)
蛋黄营养的 lecithotrophic
蛋清 albumen
氮代谢 nitrogen metabolism
氮饥饿 nitrogen starvation
氮磷钾肥 azophoska
氮磷酸盐 nitrogen-phosphate
氮气钢瓶 nitrogen cylinder
氮施肥 nitrogen fertilization
氮素代谢 nitrogen metabolism
氮循环 nitrogen cycle
氮氧化合物 nitrogen oxides
氮氧化物(NO_x) nitrogen oxide
当代家具 contemporary furniture
当代显性 ectogeny; metaxenia
当地寄生物 local parasite
当地时间 local time
当地种 endemic species
当归素 angelicin
当归酸 angelic acid
当归酮 angelicone
当归油 angelica oil
当量负荷 equivalent load
当量碱度 equivalent basicity
当量均布荷载 equivalent uniform load
当量孔隙度 equivalent porosity
当量溶液 normal solution
当量酸度 equivalent acidity
当量载重 equivalent load
当年 on year
当年的采剩物 current slash
当年繁殖价值 present reproductive value
当年生长 current growth
当年生枝 annual shoot
当年吸收量 current annual uptake
当年枝 current growth
挡板 apron board; baffle; baffle board; baffle plate; dam board; deflector; flap; hood; siding
挡板汽提塔 baffled stripping column
挡尘板 dust board; dust panel
挡风板 deep bead; verge board
挡火物 fire stop
挡浆板 dam board
挡空气板 shutter
挡块 bunk block; cheese block; chock block; dog
挡块链 bunk chain
挡木 drag; fender; fender skid; rub; sheer log; gap stick
挡木钩 bunk hook
挡木块 monkey block
挡泥板 flipper
挡水堰 overfall
挡土墙 retaining wall
T字形挡土墙 T-shaped retaining wall
挡雾罩 saveall
挡楔 bunk block
挡栅 rake
刀背 back; blade back
刀槽 gullet
刀床 knife beam; knife carriage
刀杆 arbor; knife bar
刀跟 heel
刀耕火种 slash and burn
刀辊型鼓式削片机 cutter spindle chipper
刀痕 gap-cover; knife mark
刀架 blade holder; carriage; cutting carriage; knife block; saddle
刀尖高度(旋刀) knife height; setting height of knife tip
刀尖角 nose angle
刀具 cutter; puncheon
刀口 tine
刀口尺 straightedge
刀梁 knife beam
刀帽 knife cap
刀门 pressure-bar gap
刀面 front; knife face
刀面角 cutting angle; rake angle
刀盘 cutter block; cutter disk; cutter head; disc; head; knife block; knife head
刀片 fly-bar
刀刃 blade; knife edge; lip
刀刃裂隙(单板背面) knife check; lathe check
刀式剥皮机 knife barker; knife debarker

刀式打磨机 cutter mill
刀式切割机 knife(-cutting) machine
刀式碎木机 knife hog
刀头 head; knife head
刀头修整 jointing
刀座 cutter block; cutter head; knife block; knife head
叨齿 cutter
导扳头(油锯) guide bar nose
导板 guide; guide bar; guide plate
导板头 nose
导槽 guide
导承辊 carrier roll
导程 lead
导电复合材料 electrically conductive adhesive
导电率 electric conductivity; electric(al) conductivity
导电涂层 electrically conducting coat
导电性 electric(al) conductivity; electroconductibility; electroconductivity
导电性胶 electrically conductive adhesive; electrically conductive glue
导电性胶黏剂 electrically conductive adhesive
导电性能 electric conductivity; electric(al) conductivity
导管 leader; trachea; vessel; duct
导管病 vascular disease
导管穿孔 vessel perforation
导管的 vascular
导管分子 vessel element; vessel member; vessel segment
导管分子尾部 vessel element extension
导管间纹孔 intervessel pit
导管细菌病害 tracheobacteriosis (*plur.* -ses)
导管真菌病害 tracheomycosis
导管植物 tracheophyta
导管状管胞 vascular tracheid; vesselform tracheid
导轨 guide; guide track; rack; ways
导辊 guide roll; jockey roller; leading roller
导航 navigation
导火线 fuse; tinder
导距(旋切) lead
导块 guide block
导缆器(孔) fairlead; fairleading
导流坝 pier dam
导轮 roller
导热析气计 catharometer
导热系数 coefficient of heat conduction; coefficient of heat conductivity; coefficient of thermal conductivity; heat conductivity; thermal conductivity
导热性 conduction; heat conductivity; temperature conductivity; thermal conductivity
导生岩 derivative rock
导数 differential quotient
导水工程 parallel construction
导水系数 hydroconductibility coefficient
导水性 water conductibility
导水组织 hydrome; water-conducting tissue
导索(使绳索平滑缠绕于卷筒上) spool on
导索工 spool tender
导索滚柱 roller fairlead
导索滑轮 cable guide pulley; guide block; lead block; pulling block
导索金属架 rub
导索器 fairlead; lead
导索器拱架 fairlead arch
导索器旋转部分 barrel
导索直立滚柱 dolly
导网辊 wire (leading) roll
导线 meander line; traverse
导线测量角 meander corner
导线图纸(道路测量) hardshell
导向杆(集材等) guide bar; sampson
导向滑轮 lead; lead block; rider block; runner; snatch block; terrain block; tree block
导向轮 jockey roller
导向木 rib log; sheer log; rub tree
导向漂子 fender boom
导向牵引杆(爬犁或运材车的) tongue
导向曲线 guide curve; guiding curve
导向装置 deflector; distributor
导卸木(滑道终点) whip-poor-will

导游图 tourist map
导脂器 apron; gutter; lip
导纸辊 dancing roll
岛 island
岛屿 islands
岛屿生态学 island ecology
岛屿生物地理学理论 island biogeography theory; theory of island biogeography
捣碎 mash
倒伏 lodging
倒棱(角) chamfer
倒木 blow-down; dead-and-down; fallen tree; fallen trunk; overturned tree
倒木垫板 table
倒木坑 treefall pit
倒木丘 treefall mound
倒树(伐倒、干折、枯倒等树木) down tree
倒树器 forest devil
倒向(伐木) direction of fall; lie
倒向口 cutting throat; wedge cut; felling face
倒向瞄准器 point; sight gun
倒向一致的伐木 directional felling
倒J形曲线 inverse J-shaped curve
倒S形饰条 reverse ogee
倒V形配板 inverted V-match
倒摆式截锯 swing up saw
倒车 gig run; gig-back
倒齿状的 runcinate
倒刺刚毛 barb
倒吊簧板 back action lock
倒丁字形芽接 inverted T-form budding
倒峰 reversal peak
倒刮 back-blading
倒虹吸[管] inverted siphon
倒梨形的 ebpyriform
倒利息 negative interest
倒流 refluence
倒卵球形的 obovoid
倒披针形 oblanceolate
倒三角形 obdeltoid
倒生的 anatropous
倒退 retrogression
倒位 inversion
倒位交换 inversion crossover
倒向羽裂 runcinate
倒楔形栽植 inverted V-planting
倒心形的 obcordiform
倒影 inverted image
倒影镜法 mirage method
倒圆锥形的 obconic
倒转褶皱(地质) overfold; outfold
倒锥形孔洞 inverted cone-shaped opening
到岸价格 cost, insurance and freight
到岸价格加佣金价 cost, insurance, freight and commission
到材点 terminal
到材河埂 terminal boom
到期 maturity
盗寄生现象 cleptoparasitism
盗用 misappropriation
道碴 ballast
道钉 railroad spike; spike; tie bolt
道钉松弛 spike killing
道林纸 super fine paper; white-wood-free paper; wood-free writing paper
道路 pass; road
道路防护林 road protection forest
道路规划 roading
道路交叉点 road junction
道路密度 road intensity; roading
道路曲线 ox bow
道路养护工 grease dauber; greaser
道旁凉亭 wayside pavilion
道旁种植 avenue planting; street planting
道氏滞燃法 Dow process
稻草 rice straw
稻草纸浆 yellow straw pulp
稻壳 paddy hull
得苗率 plant percent
德国可可火药 German cocoa powder
德国疏伐(即下层疏伐) German thinning
德塞蒙德处理法(防腐) Dessemond process
灯蛾 ermine moth
灯蛾类 tiger moth; woolly bear
灯光栽培 light culture
灯黑 lamp black
灯塔 lighthouse
灯台 lampstand

灯心草 bulrush
灯柱 lampstand
灯座 lampstand; torchere
登革热 dengue Fever
等比容线 isostere
等比中项 geometric mean
等臂B染色体 iso-B-chromosome
等臂X染色体 iso-x-chromosome
等臂染色体 isochromosome; metacentric chromosome
等边的 isolateral
等变高风(气象学) isallohypsic wind
等变温线 isallotherm
等变压线 isallobar
等表型线 isophene
等差中项 arithmetic(al) mean
等产量线 isoquants
等成本曲线 isocost curve
等雌蕊的 isogynous
等待时间 lead time
等地温线 geoisotherms
等电pH isoelectric pH
等电点分离 isoelectric fractionation
等电聚焦 isoelectrofocusing
等电位点 isoelectric point
等多度线 isonome
等腐殖质带 iso-humic belt
等高 isometry
等高播种 contour planting
等高草带 contour sod strip
等高带状种植 contour strip cropping
等高耕作 contour cultivation; contour plowing; contour tillage; contouring
等高温的 isotheral
等高温线 isothere
等高线 contour-line
等高线采运方式 contour logging system
等高线地图 contour map
等高线间距 vertical interval
等高线林道 contour (line) road
等高线图 contour (line) map
等高栽植 contour planting
等高整地 field contouring
等高植被线 isophyte
等高种植 contour cropping
等光程放大镜 aplanatic magnifier
等花粉线 isopoll; isopollen line
等花期线(生物) isoanth
等花柱的 isostyled; isostylous
等基因的 isogenic
等基因频率线 isogenes
等级 brand; grade; rank
等级材 graded timber
等级次序 rank order
等级分化 caste differentiation
等级理论 hierarchy theory
等级内增薪 within-grade increment
等级区分 grade separation
等级系数 coefficient of grade
等级相关 rank correlation
等级证明书 grading certificate
等级制 caste system
等季节现象 isophene
等价 parity price
等价交换 equivalent exchange
等价林木(采运成本等于木材售价) marginal tree
等降水量线 equipluve
等角点 isocenter
等径细胞 isodiametric cell
等距 isometry
等距条播 regular drilling
等距栽植 regular planting
等量弯矩 equivalent bending moment
等量吸附热 isosteric heat of adsorption
等量蒸发 equivalent evaporation
等密度线 isostere
等面的 isobilateral; isolateral
等面式 isolaterality
等面叶 isobilateral leaf
等模标本 homeotype; homoeotype
等内材 on-grade
等内材出材量 lumber grade recovery
等内的 sound
等内原木 grading log
等浓度线 isopleth
等气压图 isallobar
等氢离子浓度悬液 isohydric suspension
等日照线 isohel
等深点 isocenter
等深线 depth contour; isobath; isobathyic line

等渗的 isotonic
等渗溶液 isotonic solution
等渗压的 homoiosmotic
等生态域线 isobiochore
等湿度线 isohume
等速带式输送机 constant speed belt conveyor
等速干燥 constant rate of drying
等外板 shop-cutting panel
等外材 back log; below grade; cull; cull log; mill cull; non-dimension timber; off-grade timber; red ink log; reject; rejected log
等外的 off-grade
等外减量(缺陷扣除量) reject; cull
等外减量系数 cull factor
等外锯材 bastard-lumber; cull lumber
等外苗 cull(ed) seedling
等外品 cull
等外树 cull tree
等外原木 submarginal log
等外种子 off-grade seed
等位断裂(染色体或染色线) isoiocus break
等位基因 allele; allelic gene; allelomorph; gene allele
等位基因系列 allelic series
等位基因重组 allelic recombination
等位染色体 allelosomal
等位温 equipotential temperature
等位性 allelism; allelomorphism
等温层 isothermal layer; isothermohyps
等温过程 isothermal process
等温气相色谱法 isothermal gas chromatography
等温图 isothermal map
等温吸附 isothermal sorption
等温线 isotherm; isothermal line
等温线性 isotherm linearity
等温压缩 isothermal compression
等物候线 isophene
等显性的 co-dominant
等响度 equal loudness
等向性材料 isotropic material
等向性木梁 isotropic wood beam
等效负荷 equivalent load
等效孔隙度 equivalent porosity
等效弯曲力矩 equivalent bending moment
等效异位基因 polymeric genes
等胁强 isostress
等形同步分布区 equiformal progressive area
等雪量线 isochion
等压法 isopiestic method
等压力 uniform pressure
等压线 isobarometric line
等压线厂 isobar
等压线色谱法 isobaric chromatography
等压线图 isobaric map
等盐线 isohaline
等应力线 isostress
等优势 co-dominance
等优势木 co-dominant; co-dominant tree
等优势种 co-dominant species
等雨量线 equipluve; isohyet
等云量线 isonephe
等张溶液 isotonic solution
等蒸发量线 isothyme
等值均布负荷 equivalent uniform load
等值曲线 isocost curve
等值线 contour-line; isogram; isopleth
等值线图 isopleth map
等值种 equivalent species
等质环境 isoterra
等周亏格 isoperimetric deficit
等子叶的 isocotylous
凳子 stool
低矮的 low-statured
低背椅 low back chair
低倍检验 macroscopic examination
低车位装车 pit loading
低出叶 cataphyll (*plur.* -lla)
低床 low-bed; lower bed; sunken bed
低床柱床 low post bed
低氮硝化纤维素 collodion cotton; pyroxylin
低档家具 lower-priced furniture
低得率浆 low yield pulp
低等控制点 minor control point
低等植物 lower plant
低地 depression; low-land
低地位沼泽土 fen soil; low moor soil

低地沼泽 lowland moor
低地沼泽群落 hygrophorbium
低地沼泽土 fen soil
低地植物群落 bathyphytia
低地砖红壤 low level laterite
低电压加热 low-voltage heating
低分布 under-dispersion
低风险之森林经营 low-risk forest management
低腐殖质潜育土 low humic gley soil
低功率的 underpowered
低光泽加工(涂饰) low finish
低光泽涂饰 unglazed finish
低河床 lower bed
低河道 lower course
低级针叶材 wrack
低价值木 weed tree
低架 low-bed
低架挂车 low-bed trailer; lowboy trailer
低架宽脚爬犁 mud-boat
低角刨 low-angle plane
低精度削片机 freely cutting machine
低聚物 low polymer
低劣林分 inferior stand
低劣树种 inferior species
低临界温度 lower critical temperature
低轮车 lizard; skidder
低轮集材车 bummer
低密度胶合板 low-density plywood
低密度刨花板 low-density particleboard; low-density wood chipboard
低密度纤维板 lower density fibreboard
低频 low frequency
低频率冲击波 low-frequency shock wave
低倾斜摄影 low oblique photo
低容量喷雾 low volume spray
低容量喷雾处理 low volume spray treatment
低容量喷雾器 low volume sprayer
低渗的 hypotonic
低渗溶液 hypotonic solution
低湿草地 meadow
低双轮车 spool cart
低水位 slack water
低送射 low straight-away shot
低速磨浆机 jordan (mill)
低塔式(塔式亚硫酸钙制造法) low-tower system
低太阳角航空摄影 low-sun angle aerial photography
低碳钢 mild steel
低碳经济 low-carbon economy
低碳生活方式 low-carbon lifestyle
低通滤波器 low-pass filter
低洼地 swale
低位床 low post bed
低位泥炭沼泽 lowland moor
低位湿原 low moor
低位湿原泥炭 low moor peat
低位沼译泥炭 fen peat
低位沼泽 low moor
低位沼泽森林泥炭 low moor wood peat
低温 microtherm
低温泵 cryopump
低温测定法 cold test
低温层积 cold stratification
低温处理 chilling; cold treatment
低温干燥 low-temperature drying
低温固化 cold curing; low-temperature setting
低温固化胶黏剂 low-temperature setting adhesive
低温龟裂 cold cracking
低温恒温器 cryostat
低温计 frigorigraph
低温开裂 cold cracking
低温气候 microthermal climate
低温湿润贮藏 cold moist storage
低温室 cool house
低温温室 cool intermediate house
低温细裂(漆膜表面) cold checking
低温休眠 cold dormancy
低温学 cryogenics
低温硬化 low-temperature setting
低温硬化胶黏剂 low-temperature setting adhesive
低温植物 microtherm; microthermal plant
低温贮藏 low-temperature storage
低息借款 cheap money
低蓄积林分 depleted stand
低血糖症 hypoglycemia
低压冻干法 lyophilization

低压短周期工艺 low pressure short cycle processing
低压计 vacuum-meter
低压杀菌 understerilization
低压贴面装置(工艺) low-pressure laminating system
低压消毒 understerilization
低迎射 approaching low shot
低于估计产量 underrun
低于一车皮载重量的楞堆 less than carload lot
低噪音锯片 low noise sawblade
低真空 coarse vacuum
低枝树 low-branching tree
堤 levee; mound
堤岸 embankment; quay
堤岸侵蚀 bank erosion
堤岸栽植 bank planting
堤坝心壁 core of dam
堤防 dike
堤礁(珊瑚礁的一种) barrier reef
滴测法(松香软化点测定) drop method
滴定 titration
滴灌 dribbling; drip irrigation; drip-watering; trickle irrigation
滴沥误差 drainage error
滴水石 dripstone
滴线(物理) drip-line
狄尔斯-阿德尔反应加成物 Diels-Alder adducts
迪士春(生物切片封固剂) distrene
敌百虫(杀虫剂) chlorofos; dipterex; trichlorfon
敌稗(除草剂) propanil
敌草隆(除草剂) diuron
敌草平(除草剂) dicamba
敌对关系 hostility
敌菌丹(杀菌剂) captafol
敌菌灵(杀菌剂) dyrene
敌克松(杀菌剂) dexon
敌螨普(杀螨剂) arathane
笛形芽接 whistle budding
镝砌 pave
抵减税额 tax credit
抵抗力 hardiness; resistibility
抵抗稳定性 resistant stability
抵销 counter balance; neutralization
抵销性关税 countervailing duty
抵押 hypothecation; mortgage
抵押品 collateral security
抵押债务 mortgage debt
底板(抽屉模等) stool
底板(复合板) back
底板(广义上) bed piece; bed plate; bottom plate; bottom-board; ledger; sole plate
底板(胶合板) ounterveneer
底表动物 epizoan
底表浮游生物 epibenthic plankton
底表生的 epibenthic
底表生物 epibenthos
底部 base unit
底部出料式削片机 bottom discharge chipper
底部加热电缆 electric bottom-beat cable
底部磨损 bottom wear
底部输送带 bottom belt
底部卸车装置 bottom dumper
底材 ground
底槽 base tank
底层 impression
底层地板 subfloor
底层地板用材 subflooring
底层捆木链(装爬犁用) corner bind
底层平面图 ground plan
底层染色 bottom dyeing
底层鱼类 demersal fish
底层原木(装在车上的) bunk log; bunk load
底衬 back lining
底带式进料机(料仓) belt bottom feeder
底刀(机械) bed knife; bedplate bar; counterknife; dead knife; blunt counter knife
底刀板 bed plate
底刀刮涂填孔机 bottom knife coater
底刀盒 bedplate box
底刀座 bed plate
底革 sole leather
底钩 bottom dog
底光 bottom season
底价 appraised stumpage price
底肋(枪筒) bottom rib
底流 underflow
底面负荷 bed load

底面图 ground plan
底面涂布机 bottom coater
底面涂饰 bottom coating
底面直径 basal diameter
底模 bed-die
底木(基础) basal log; sill
底内动物 infauna; endobenthos
底盘高度 ground clearance
底坡 bottom slope
底栖的 benthoal
底栖动物 zoobenthos
底栖动物区系 benthic fauna
底栖群落 bottom community
底栖生物 benthic animal; benthos
底栖生物种群 bottom population
底栖性的 bentho(n)ic
底栖植物 phytobenthos
底栖族群 bottom population
底漆 base coat; first coat of paint; ground coat; precoat; primer; priming paint
底漆腻子 under coat(ing)
底色 bottom colour; undertone
底闩 breech bolt
底图 base; base sheet
底涂 bottom coat
底涂板 prime-coated board
底涂层 base coat; first coat; ground coat
底土 subsoil; undersoil
底网 backing wire
底物 substrate
底岩 bed rock
底用薄膜 priming foil
底着的 innate
底直径 bottom-diameter
底质取样器 bottom sampler
底座(基础) back; basement; foundation; stay
底座(家具) base unit
底座木(水闸或桥梁) mudsill
地 ground
地白蚁 subterranean termite
地板 bottom-board; floor; floor plank; floor plate; floor(ing) board; flooring; planking; timber flooring
地板材 flooring timber
地板-顶板结构 floor-ceiling construction
地板负荷 floor load
地板锯 flooring saw
地板梁 floor beam; joist
地板铺设 timber flooring
地板镶拼 joint flooring
地板障风板 floor baffle
地板支撑 floor strutting
地被物 ground cover; ground flora; soil-covering; surface-cover
地被物采取权 right to litter
地被植物 ground cover; ground cover plant; ground vegetation
地标林 landmark forest
地表被覆 surface ground cover
地表腐殖质层 ectorganic humus layer
地表灌溉 surface irrigation
地表火 grass fire; surface fire
地表径流 surface flow; surface runoff; surface run-off
地表迳流 overland flow
地表流失 surface run-off
地表排水沟 surface drain
地表水 surface water
地表植物 ground flora
地表走茎 surface runner
地槽 geosyncline
地产 acre
地带 belt; terrain; zone
地带区划 zoning
地带性 zonality
地带性顶极 zonal climax
地带性土 zonal soil
地带性预极 primary climax
地带性植被 zonal vegetation
地道 gallery
地点 locality
地动物学 geozoology
地段(森林区划最小单位) lot
地方 terrain
地方21世纪议程 Local Agenda 21
地方病的 endemic
地方材积表 local volume table; one way volume table; site-specific volume table
地方的 local
地方级自然保护区 local nature reserve

D

地方价目表 local tarifs
地方品种 local cultivar; local race; local variety; native variety
地方群丛 local association
地方时 local time
地方时角 local hour angle
地方视时(太阳时) local apparent time
地方收获表 local yield table
地方收获级 local yield class
地方塔里夫材积表 local tarifs
地方太阳时 local solar time
地方特有种 local endemic species
地方性 endemism
地方性磁差 local magnetic attraction
地方性的 endemic
地方性动物病 enzootic; enzootic disease
地方性动物病的 enzootic
地方性密度 endemic density
地方性兽病的 enzootic
地方异速增长方程 site-specific allometric equation
地方种 local species
地核 core
地花蜂 andrena; mining bee
地基梁 foundation beam
地基下层泥土 lower subsoil
地基桩柱 foundation pile
地际(地上1ft, 地下2ft) ground-line
地际处理(防腐) ground-line treatment
地际刻痕 incised groundline
地际面 ground-line area
地价 land value; soil rent value; value of land
地脚螺钉 foot screw
地脚螺栓 anchor bolt; foundation bolt
地窖腐朽 cellar rot
地界林 landmark forest
地景保护 landscape protection
地咖脂 dika fat
地槛 ground sill
地壳 crust; earth crust
地壳变动 diastrophism
地壳均衡说 isostasy
地壳运动 crustal movement
地空侦察系统(防火) detection air-ground system
地块 land parcel; plot
地况 land description
地况记载 destcription of site; site description
地况调查 investigation of locality; land survey; site apparaisal; site apparaisement; site mapping; site survey
地蜡 ceresin wax
地理编码 geocode
地理变型 geographic variant
地理变异 geographic variation; geographical variation
地理变种 geographic(al) variety
地理标志 geographical indicator
地理差型 geocline
地理的替代种 geographical vicariad
地理分布 geographical distribution
地理分布范围 geographic range
地理隔离 geographic isolation
地理隔离种 convivum
地理渐变群 topocline
地理渐变型 geocline
地理空间 geographic space
地理区域 geographic area
地理群 proles
地理群落 geocoenosium
地理生态型 geoecotype
地理小种 geographic race
地理型 geographtype
地理学 geography
地理植物群系 physiographic plant formation
地理族 geographic race
地力 fertility; productibility of soil; productivity of soil; productivity power of soil
地力保持 preservation of fertility
地力级 land-capability class
地利级 accessibility class
地沥青 mineral pitch
地龙 sleeper
地锚 anchor block
地貌 land feature; land form; landform; relief; surface feature
地貌类型判读 interpretation of landform
地貌学 geomorphology
地貌循环 geomorphic cycle

地面 ground
地面导索滑车 ground lead block
地面底点 ground nadir
地面分辨距离 ground resolved distance
地面分辨率 ground resolution
地面钢索集材 ground(-lead) hauling; ground(-line) logging; ground(-line) skidding; ground snigging; ground yarding; ground-lead (cable) logging; ground-line (cable) logging; low-lead skidding
地面滑车 terrain block
地面火 ground fire
地面集材 dragging; ground(-line) logging; ground(-line) skidding; ground hauling; ground snigging; ground yarding; snaking; snigging; surface yarding
地面检查 ground check
地面径流 direct runoff
地面距离 ground distance
地面可燃物 ground fuel
地面控制 ground control
地面控制图 ground control map
地面能见度 ground visibility
地面逆温 surface inversion
地面铺布(气球充气) ground cloth
地面栖的 epigaec
地面生[长]的 epigeal; epigean; epigeous
地面实况 ground truth
地面霜 ground frost
地面水库 surface reservoir
地面索道 incline; inclined cableway; inclined rope railway; surface cableway
地面消毒 ground disinfection
地面校正 ground check
地面信息 ground information
地面压力 ground pressure
地面芽植物 hemicryptophytes (He)
地面芽植物气候 hemicryptophyte climate
地面站 ground station
地面植被层 ground stratum
地面装车工 tong puller
地平经度 azimuth
地平线(圈) horizon
地契 title deed
地堑 rift valley
地球辐射表 pyrgeometer
地球工程 geo-engineering
地球化学循环 geochemical cycle
地球静止气象卫星 geostationary meteorological satellite
地球年代学 geochronology
地球上的 terrestrial
地球同步气象卫星 geostationary meteorological satellite; synchronous meteorological satellite
地球同步卫星 fixed satellite
地球微生物学 geomicrobiology
地球物理航空测量系统 geophysical airborne survey system
地球资源观测卫星 earth resource observation satellite
地球资源技术卫星 earth resource technology satellite (ERTS)
地区 area; area district
地区差别 regional disparities
地区多度 local abundance
地区防火指挥员 fire warden
地区划分 territorial division
地区林业局 county board of forestry
地区群丛 regional association
地区特征种 local character species
地区性变异 local variation
地区植物总称 botany
地圈 geosphere
地热 geothermy
地热的 geothermal
地热能 geothermal power
地上的 epigeal; epigean; epigeous
地上发芽 epigeal germination
地上茎 aerial shoot
地上匍匐枝 surface runner
地上生物量 aboveground biomass
地上性发芽 epigeal germination
地上芽植物 chamaephyte
地上芽植物气候 chamaephyte climate
地上枝 aerial shoot
地生兰 terrestrial orchid
地势 land feature; terrain; topography
地势起伏 relief; terrain relief

地鼠 meadow mouse
地税 land tax
地笋 bugle weed
地毯式铺装成型 carpet laying
地毯状高山草原 carpet-like alpine meadow
地体因素 edaphic factor
地统计学 geostatics
地图 plat
地图点位号 map reference
地图距离 map distance
地图索引 map reference
地图图例系统 legend system
地图坐标系统 map coordinate system
地位 site productivity; yield potential
地位分级 site classification
地位级 grade of locality; locality class; productivity class; quality class; quality of locality; quality of site; site class; site quality class
地位级表 site class table; site-quality table; table of quality of locality; table of site-class
地位级的分等法 classification of site quality
地位因素 site factor
地位指数 productivity index; site indicator
地温颠倒 soil temperature overturn
地温梯度 geothermal gradient
地文学 physiography
地物点 fiducial mark
地峡 isthmian link; isthmus
地下层 subterranean layer
地下成层现象 underground stratification
地下的 hypogaeic; subterranean
地下动物 cryptozoa; subterranean animal
地下工程 underground works
地下灌溉 subirrigation; subsurface irrigation; subterranean irrigation; under(ground) irrigation
地下灌溉(底部灌溉) sub-irrigation
地下含水层 underground reservoir
地下河 underground river
地下火 ground fire
地下截根 undercutting
地下茎 rhizome; stem-tuber; subterranean shoot; subterranean stem; underground rhizome
地下茎插穗 rhizome cutting
地下茎更新 rhizome-wood regeneration
地下茎扦插 rhizome cutting
地下茎鞘 rhizome sheath
地下茎天然更新 natural reproduction by rhizome
地下径流 ground water runoff
地下菌 truffle
地下排水 ground drain; ground drainage; subdrainage
地下器官 subterranean part
地下器官型 radicoid form
地下生的 hypogean; hypogeal; hypogeous
地下室 basement
地下水 ground water; phreatic water; subterranean water; underground water
地下水径流 ground water flow
地下水流量 discharge of ground water
地下水面 ground water level; water table
地下水渗流 seep-off
地下水实地流速 field velocity of ground water
地下水位 ground water table; ground water level; phreatic surface; water table
地下水型水稻土 ground water rice soil
地下水资源 ground water resource
地下性发芽 hypogeal germination
地下芽植物 geocryptophyte; geophyte
地下蚁路 termite runway
地下蚁通道 shelter tube
地向斜 geosyncline
地形 ground state; land feature; landform; orography; surface feature; terrain; topography
地形变化 topographic variation
地形测量单位 topographic unit
地形测量学 topography

地形顶极(群落) physiographic climax
地形改造 topography refotnl
地形渐变群 topocline
地形块 terrain block
地形起伏不平 undulating topography
地形起伏位移 relief displacement
地形倾差 geocline
地形设计 topographical design
地形特征 topographic feature
地形条件 topographical condition
地形图 relief-map; topographical map
地形位移 topographical displacement
地形系列 toposequence
地形线 form line
地形小气候 contour microclimate
地形效应 orographic effect
地形性降水 orographic precipitation
地形学 geomorphology (landform); topography
地形演替顶极 topographic climax
地形因子 orographic factor
地形雨 orographic rain
地形雨量 orographic rainfall
地形阻碍 topographic barrier
地衣 lichen
地衣冻原 lichen tundra
地衣门 Lichenes
地衣型地上芽植物 chamaephyta lichenosa
地衣学 lichenology
地衣沼泽 lichen bog
地衣植物系数 lichen-quotient
地域品种 geographic(al) variety
地域性的 endemic
地域性品种 local race
地罩 ground cloth
地震 earthquake
地植物学 geobotanics; geobotany
地质构造 geologic structure
地质解译 geologic interpretation
地质年表 geologic time scale
地质年代 geologic age
地质判读 geologic interpretation
地质期演替系列 geosere
地质侵蚀 geological erosion
地质图 geological map
地质学 geological science; geology
地质演替 geological succession
地质遗迹 geological formations
地质遗迹类型自然保护区 nature reserve for geological formations
地质作用 geological function
地中海地域 mediterranean region
地中海栎木栓皮 Mediterranean oak cork
地中海气候 Etesian climate; Mediterranean climate
地转风 geostrophic wind
地租 annual equivalent; ground rent; land rent; soil-rent(al)
地租价 land rent value
地租学说 doctrine of soil rent
递电子体 electron carrier
递钩(量圆材周围) timber sword
递耗资产 wasting assets
递减分布 decreasing distribution
递减率 lapse rate
递减税 degressive tax; regressive tax
递减折旧法 diminishing method
递降分解 degradation
递进燃烧药 progressive-burning powder
递推体系 recursive system
递延收益 deferred income
递沿收益 unearned revenue
递增成本 increasing cost
递增修饰因子 plus modifier
第二产品 secondary product
第二次寄生[物] deuteroparasite
第二代 second-generation
第二代上木 standard of the second rotation
第二段原木(在根端材之上) second length; second log
第二割面(采脂作业) yearling face
第二类吸附 secondary adsorption
第三次寄生虫 tritoparasite
第三代 third-generation
第三代上木 standard of the third rotation
第三纪 Tertiary period
第三纪植物 tertiary plant
第三纪植物区系 tertiary flora
第四代上木 standard of the fourth rotation

第四纪 alluvium period; Quaternary period
第四纪植物时代 homostatic period
第一层漆 first coat of paint
第一次减数分裂 initial meiosis
第一粗枝 first thick branch
第一代上木 standard of the first rotation
第一的 primary
第一龄期 first stage
第一周皮 first periderm
蒂特拉塞特木材处理法 Tetraset process
缔约方大会 conference of the parties
缔约方会议(京都议定书) meeting of the Parties (to the Kyoto Protocol)
颠倒 upending
颠倒纹理拼接 reversed matching
颠茄碱 atropin(e)
典范相关 canonical correlation
典型部分 representative fraction
典型抽样 selective sampling
典型腐殖质碳酸盐土 true humus calcareous soil
典型黑色石灰土 eurendzina
典型灰化的 eupodzolic
典型木纤维 true-woodfibre
典型培养 type culture
典型试验 type test
典型相关 canonical correlation
典型相关分析 canonical (correlation) analysis
典型小区 representative area
典型样本 representation sample
典型砖红壤 true laterite
点斑 stigma
点播 seeding in hole; seeding in spot; single-seed planting
点播器 dibber; dibbler
点播穴植 dibbling
点抽样 point sampling
点工 day(-to-day) worker
点估计方程 point estimator
点划线 chain line
点火 ignition; kindling
点火导线 ignition wire
点火枪 flame gun
点火室 igniting-chamber
点火者 firer
点检查 spot checking
点密度 point density
点燃器 transistor pulse ignition
点四分法 point-centered quarter method
点样方分析 point quadrat analysis
点样器 applicator
点源污染 point-source pollution
点阵常数 lattice constant
点种 spot sowing; spot-seeding
点状采伐 spot felling
点状胶合 spot gluing; spotty bond
点状烧除 spot burning
点状涂布 spot coating
点状用火 spot burning
碘染色 iodine staining
碘染色法 iodine staining method
碘染色角 iodine staining angle
碘值 iodine number; iodine value
电场透镜 electric field lens
电池 element
电传打字机 teletype
电磁 electromagnetics; electromagnetism
电磁波 electromagnetic energy
电磁波谱 electromagnetic spectrum
电磁阀 magnetic valve
电磁辐射 electromagnetic radiation
电磁光谱 electromagnetic spectrum
电磁能 electromagnetic energy
电磁谱 electromagnetic spectrum
电磁轫 electromagnetic brake
电磁学 electromagnetics; electromagnetism
电磁闸 electromagnetic brake
电磁振荡器 electromagnetic vibrator
电磁制动[器] electromagnetic brake
电导率 electric conductivity; electric(al) conductivity; electroconductibility; electroconductivity
电灯烘箱 electric-bulb oven
电动搬运车 electric truck
电动测湿计 electric(al) moisture meter
电动叉车 electric truck
电动会计机 electric(al) accounting machine

电动混合器 electric mixer
电动货车 electric truck
电动机 electromotor
电动计算机 electric(al) accounting machine
电动剪刀 electric clipper
电动绞盘机 motor winch unit
电动搅拌器 electric mixer
电动卡车 electric truck
电动捆扎机 electric tying machine
电动链锯 electric powered chain saw
电动生长锥 power-driven increment borer
电动条播机 motor-driven drill
电动土壤加热电缆 electric soil-heating cable
电动摇尺器 electric set-work
电动铡刀机 electric clipper
电动张紧装置 electric tensioner
电风扇 fanner
电杆 telephone pole
电杆材 telegraph-pole wood; wood for telegraph-post
电杆横担木 telegraph pole cross arm
电杆基部浸注罐 tilting cylinder
电杆检查(地面部分腐朽) pole line inspection
电杆木绷带防腐 pole bondaging
电杆线 pole line
电干燥 electric seasoning
电干燥法 electric drying; electric seasoning
电干燥炉 electric dry oven
电功率 electric power
电荷 charge
电荷耦合器件 charge-coupled devices
电烘箱 electric dry oven
电弧率 electric resistivity
电化电位 electrochemical potential
电化学梯度 electrochemical gradient
电话柜 telephone cabinet
电机 electromotor
电集尘器 electric precipitator
电加热系统 electric(al) heating system
电搅拌器 electric mixer
电解 electrolyte
电解膜 electrolyte
电解质 electrolyte
电介质 dielectric
电介质加热 dielectric heating
电镜技术 electron microscopy
电锯 electric powered chain saw; electric saw
电缆 cable
电缆焊接工 jointer
电离 ionization
电离辐射 ionizing radiation
电离子透入法 electrophoresis
电力 electric power
电力干燥 electric drying
电力供热系统 electric(al) heating system
电流计 galvanometer
电膜电位 electric(al) membrane potential
电脑 computer; electronic computer
电能 electric power
电偶 couple
电偶二极管 charge coupled diode
电气除尘器 electric precipitator
电气干燥法 electric seasoning
电气马达 electromotor
电热采暖 electric(al) heating system
电容测湿计 electric-capacity moisture meter
电容率 specific inductivity
电容器用纸 capacitor paper
电容式水分测定计 capactiance moisture meter
电容水分测定计 electric-capacity moisture meter
电伸缩 electrostriction
电渗析 electrodialysis
电势 electric(al) potential
ζ-电势 zeta-potential (ζ-potential)
电视摄像机 television camera
电视探火 television detection
电算机鉴别阔叶树材 computer identification of hardwood species
电缩作用(指溶剂) electrostriction
电梯 elevator
电透析 electrodialysis
电推剪 electric clipper
电位 potential
电位计 potentiometer
电线杆 telegraph pole; telegraph post

电泳 cataphoresis; electrophoresis; ionophoresis
电泳法 electrophoresis
电泳图 electrophoregram; electrophoretogram
电源线 main lead
电晕放电 corona discharge
电载体 electron carrier
电致伸缩 electrostriction
电子操纵相控阵雷达 electronically steerable array radar
电子测角器(垂向角方位角) telegoniometer
电子测湿计 electronic moisture meter
电子测湿仪 electronic moisture detector
电子秤 digital load cell; electronic scale; electronic weighing apparatus
电子传递链 electron transport chain
电子点火 electronic ignition
电子度量 electronic scale
电子对 doublet
电子供体 electron donor
电子光谱 electronic spectroscopy
电子核[磁]双共振 electron nuclear double resonance
电子衡器 electronic scale
电子计量台 electronic weighbridge
电子计算机 electronic computer
电子记账机 electric(al) accounting machine
电子胶合 electronic gluing
电子可控系统雷达 electronically steerable array radar
电子流 electron flow
电子轮尺 recording cal(l)iper(s)
电子密度 electron density
电子能谱学(法) electronic spectroscopy
电子浓度 electron density
电子偶 doublet
电子桥秤 electronic weighbridge
电子扫描雷达 electronically steerable array radar
电子扫描微波辐射仪 electrically scanning microwave radio-meter
电子施主 electron donor
电子湿度计 electronic moisture meter
电子式点火(无触点) electronic ignition
电子受体 electron acceptor
电子束固化 electron beam curing
电子束加速器 electron beam accelerator
电子束探针 electron beam probe
电子数据处理 automatic data processing; electronic data processing
电子数据处理系统 electronic data processing system
电子水分计 electronic moisture meter
电子水分探测器 electronic moisture detector
电子顺磁共振 electron paramagnetic resonance; electron spin resonance
电子顺磁谐振 electron paramagnetic resonance
电子探针 electron beam probe
电子位置传感器 electronic position sensor
电子显微放大器 electron micrograph
电子显微检查法 electron microscopy
电子显微镜 electron microscope; electron microscopy
电子显微术 electron microscopy
电子显微图 electron micrograph
电子显微学 electron microscopy
电子显微照片 electron micrograph
电子旋转共振 electron spin resonance
电子叶 electronic leaf
电子载体 electron carrier
电子钟子计数器 electronic seed counter
电子种子水分仪 electronic seed moisture meter
电子资料处理 electronic data processing
电子资料处理系统 electronic data processing system
电子自旋共振 electron paramagnetic resonance; electron spin resonance
电子自旋谐振 electron spin resonance
电阻 electric resistance; electric resistivity
电阻测湿计 electric-resistance moisture meter
电阻加热 resistance heating

电阻率 electric resistivity
电阻式含水率测定计 resistance-type moisture meter
电阻式水分测定计 electric-resistance moisture meter; resistance-type moisture meter
电阻水分测定计 electric-resistance moisture meter
电阻温度计 electric-resistance thermometer
电阻压力计 electric resistance
玷污 contamination
垫 pad; pedestal
垫板(伐木楔) shim
垫板(广义的) bed piece; caul board; caul plate; floor plate; retaining board; stay plate
垫板呼吸式干燥机 plate dryer
垫板集材 pan skidding
垫板加速运输机 caul speed-up conveyor
垫板条(干燥时用) drying finger
垫板推出机 caul pusher
垫弹塞 wad
垫棍 dolly
垫块(固定作用) heel block; sill; spud
垫块(控制树倒) kicker
垫款 advance
垫木(底层) bearer; bed piece; bedding wood material; bolster; bunk; chock; cill; drop log; dunnage (wood); dwang; piling stick; sill; skid; sole timber; sticker; strip; threshold; wood block
垫木(预装用) spare bunk
垫木(造材时用) dutchman
垫木变色 crosser stain
垫盘 dolly
垫片 spacer
垫圈 burr; washer
垫条(材堆等) stacking strip; crosser
垫铁(集材杆) tree iron; tree plate
垫头规定 margin requirement
垫藓 cushion moss
垫以托板 palletize
垫桩 dolly
垫状植物 cushion plant
淀粉的 amyloid
淀粉胶 starch glue
淀粉胶黏剂 starch adhesive
淀粉粒 starch grain
淀粉酶 amylase; diastase
α-淀粉酶 α-amylase
β-淀粉酶 β-amylase
淀粉溶液 amidin
淀粉水解酶 amylolytic enzyme
淀粉体 amyloid
淀积层(土壤剖面B层) illuvial horizon; illuvial layer; accumulation horizon; illuvium; B-horizon; B-layer
淀积黏粒 illuvial clay
淀积作用 illuviation
淀浆 lodgment of stock
靛黄 indigo yellow
靛玉红 indirubin
凋存的(植物) marcescent
凋落物层(土壤剖面A_0层) A_0-horizon; litter layer
凋落物收集盘 litter trap
凋萎病 blight
貂[皮] marten
雕刻 carving; engraving; incision; sculpture
雕刻材 carving wood; wood for carving
雕刻机 carving machine; sculpturing machine
雕刻木工 sculptor; wood carver
雕刻品 carving; engraving
雕刻漆 incised lacquer
雕刻术 engraving
雕刻条板下横档 frieze rail
雕刻线脚 stuck moulding
雕塑 sculpture
雕纹(木材细胞的形态) sculpturing
雕纹辊 engraved roller
吊车 cable tram; crab; crane; trolley
吊车行走机构 bogie
吊窗 hanging sash
吊窗锤 sash weights
吊床 cot
吊杆(锯木架) horse tree
吊杆(悬吊用) gibbet; lift boom; spreader bar; straight boom; suspension rod
吊杆耳环 ear

D

吊杆滑轮 boom tree block
吊杆式装车 spreader-bar loading
吊杆系索(连接吊钩与吊梁) boom bridle
吊杆柱 king post
吊钩 butt hook; endless tong; hoist hook; loading hook; sling cart; sling dog; sling hook; hell hook
吊钩张开索 opening line
吊挂螺栓 hanger bolt
吊挂门 suspension door
吊行运动 brachiation
吊环 endless tong
吊架 hanger; suspension
吊架挂索 heel strap
吊架螺栓 hanger bolt
吊架装车 tree boom loading
吊截锯 goose saw; overhead saw; overhead trimmer; pendulum cross cut saw; pendulum saw; swing cut-off saw; swing saw; swinging cross-cut saw
吊篮 hanging basket
吊链 chain hoist
吊盘测面求积仪 suspended planimeter
吊炮(树倒后一端着地一端悬空) underbind
吊钳 endless tong
吊桥 suspension bridge
吊索(机械集材等) cable sling; cable strap; hitch; hoist cable; hoist line; sling rope; suspender; suspension guy line; suspension line; suspension rope; tie line; tie-down line; tiedown rope
吊索(气球集材) tag line; tether line
吊索滑轮 side block; squirrel block; suspension block
吊椅(吊送索具工上集材杆) Bosun's chair
吊运 swing(ing)
吊运车 carriage
掉皮枯立木 buckskin
调查 survey
调查报告 survey report
调查表 measurement list
调查表格式 cruising pattern
调查单位 inventory unit; survey unit
调查记录表格 description record form; tally sheet
调查技术设计 cruising design
调查目录 releve
调查数据 cruising data
调查误差 cruising error
调查员 surveyor
调到零处的 zeroed
调度尺 dispatcher's meter
调度工作表 dispatch sheet
调度员 dispatcher
调度指导(防火) dispatching guide
调头 upending
调头装置 upender
调用 call
调转趋性 klinotaxis
调转运动 klinokinesis
跌水 cascade
跌水侵蚀 waterfall erosion
迭代法 iteration
迭接 splice
叠层 overlap; overlapping; overlapping of sheet
叠层梁 laminated beam
叠合法 method of superposition
叠加褶皱 cross fold
叠生的 storied; storyed
叠生分子 storied element
叠生构造 storeyed structure; storied structure
叠生排列 storied arrangement; tier like arrangement
叠生射线 storied ray; stratified ray
叠生形成层 storied cambium; stratified cambium
叠生芽 superposed bud
叠生轴向分子 storied axial element
叠石 rockery making
叠芯(胶合板缺陷) overlap(ping); overlapping of sheet
叠桌(三张套) trio table
碟形翘曲 dish(ing)
蝶蛾幼虫 caterpillar
蝶形花冠 papilionaceous corolla
蝶形花植物 papilionaceous plant
蝶形卷叠式 vexillary
丁坝 groin; groyne
丁苯橡胶 styrene-butadiene rubber

丁基苄基纤维素 butyl benzyl cellulose
丁基纤维素 butyl cellulose
丁香罗勒油 eugenol type basil oil
丁香宁 eugenin
丁形管接头 tee
丁子香酚苄醚 eugenol benzyl ether
丁子香酚型罗勒油 eugenol type basil oil
丁子香花油 clove bud oil
丁子香花油树脂 clove bud oleoresin
丁子香基 eugenyl
丁子香茎油 clove-stems oil
丁子香脑 eugenitol
丁子香宁 eugenin
丁子香亭 eugenitin
丁子香酮 eugenone
丁子香烯 clovene
丁子香叶油 clove leaf oil
丁子香油 clove oil
丁字着生(花药) versatile
酊 tincture
U形钉 staple
钉拔 rigger's bar
钉齿(脱粒滚筒的) peg
钉齿耙 peg-tooth harrow; spike (d-tooth) harrow; spike harrow; straight-tined harrow; tine(d) harrow; tooth harrow; tooth(ed) harrow
钉环 dog
钉接 nailed joint
钉接合 nail(ed) joint
顶坝(第一道坝) top dam
顶板(矿井的) covering timber
顶板(柱顶板) abacus
顶板(最上方) top
顶部量测 top measurement
顶层捆索 top chain
顶插 tip cutting
顶撑 back shore
顶撑条 top strut
顶点 summit; vertex; zenith
顶端 apex
顶端薄壁组织 terminal parenchyma
顶端的 apical
顶端分生组织 apical meristem
顶端嫁接 topwork(ing)
顶端切面 top cut
顶端生长 tip growth
顶端生长点 apical growing point
顶端细胞 apical cell
顶端优势 apical dominance
顶端优势减少 apical reduction
顶端优势取消 apical abolition
顶分泌腺 apocrine gland
顶峰 zenith
顶腐 top rot
顶冠饰条 crown molding
顶果的 acrocarpous
顶极格局假说 climax-pattern hypothesis
顶极格局学说 climax pattern theory
顶极群落 climax community
顶极群落复合体(演替) climax complex
顶极群系 climax formation
顶极森林 climax forest
顶极森林阶段 climax forest stage
顶极适应数 climax adaptation number
顶极植被 climax vegetation
顶极种(演替) climax species
顶件 head
顶浆分泌腺 apocrine gland
顶交 inbred-variety cross; top-cross
顶交测验 top-cross test
顶交亲本 top-cross parent
顶交杂种 strain-linear hybrid
顶接 apical grafting; top grafting
顶径 top diameter
顶枯 top dry
顶枯病 tip blight
顶肋(枪筒) top rib
顶楼 garret
顶蘑菇 button mushroom
顶木架 heel(ing) boom
顶喷 overhead spray
顶喷系统 overhead spray system
顶棚供暖盘管 ceiling coil
顶片角(锯链) top-plate (filing) angle
顶片切削角 top-plate cutting angle
顶刃角 top-plate (filing) angle
顶刃切削角 top-plate cutting angle
顶烧 tip burn
顶梢 leading shoot; top shoot

D

顶梢枯死 dieback
顶梢压条 tip layering
顶稍插穗 terminal cutting
顶生孢子 acrospore
顶生孢子囊 acrosporangium
顶生孢子式的 acrosporous
顶生的 apical; terminal
顶生分生孢子 acroconidium
顶生药 terminal anther
顶芽 end-bud; leading shoot; phyllogen; terminal bud; apical bud
顶支撑 top strut
顶枝 leading shoot; telome
顶质分泌腺 apocrine gland
顶柱 shore
订舱清单 booking list
订合同权 contract authorization
订金 bargain money
订书机 stapling machine
订正处理效应 adjusted treatment effect
订正温度 reduced temperature
钉钉机 nailing machine
钉钩锤 grab driver; grab maul
钉钩工 dogger
定边板 deckle board
定边带 deckle strap
定边滑轮 deckle pully
定边框架 deckle frame
定边装置 deckle
定标 calibration
定标器 scaler
定测 location survey
定尺寸 set
定点 spotting
定额制度 quota system
定额周转金 imprest fund
定幅板 deckle board
定幅装置 deckle
定根 normal root
定购单 buying order
定规格尺寸 sizing
定货 order
定积土 sedentary soil
定极面积仪 polar planimeter
定价 list price; pricing
定价销售 sale at fixed rate; sale at tariff price; sale by royalty
定角测高器 conometer
定金 down payment
定居(植物等) establishment; settlement; ecesis
定居阶段 establishment phase
定距板 peg board
定距疏伐 spacing thinning
定量泵 metering pump
定量猖獗学 quamitative epidemiology
定量地理学 quantitative geography
定量法 empty-cell process; Rueping process
定量辊(涂胶等) doctor roll; metering roll; backward roll
定量环境分类系统 quantitative environmental ranking system
定量流变学 quantitative rheology
定量疏伐 quantitative thinning
定期材积平分法 method of periods by volume; periodic method by volume; volume-period method
定期采伐计划 periodic cutting plan
定期采脂 periodical tapping
定期抵押 term mortgage
定期供应 periodic sale
定期火灾危险 current fire danger
定期间作 periodic intermittent working
定期可伐量 allowable periodic cut
定期枯损 periodic mortality
定期面积材积平分法 method of periods by area and volume combined
定期面积平分法 method of periods by area
定期年采伐量 periodic cut
定期年面积配分法 periodic area allotment method
定期年生长量 periodic annual growth; periodic annual inerement
定期年收获量 periodic annual yield
定期喷蒸 periodic steaming
定期平分法 method of periods
定期平均年间伐量 periodic mean annual intermediate yield
定期平均年生长量 periodic mean annual increment

定期平均生长量 periodic average increment
定期森林资源清查法 periodic forest inventory method
定期生长 periodic growth
定期生长量 periodic increment
定期收获量 periodic yield
定期踏查 periodic cruise
定期调查 multiphase sampling; periodic inventory
定期维修 periodic maintenance
定期样本 periodic sample
定期永续收获 sustained periodic yield
定期游客计数 periodic attendant counts
定期择伐作业 periodic selection system
定期折衷平分法 method of periods by area and volume combined; periodic method by area and volume combined
定期作业 periodic working
定期作业分区 periodic block
定时开闭花 equinoctial flower
定时器 intervalometer; timer
定时运输机 timing conveyor
定速运送器 creeper
定索式集材 fixed-wire logging
定位(测杆等) posting
定位(方向) orientation
定位安装拉索(集材杆) snubbing guy
定位测锤 posting plumb
定位层样条 permanent layer transect
定位点 anchor point
定位分层样带 permanent layer transect
定位规 posting square
定位鉴定 topographical test
定位孔 pilot-hole
定位控制 position control
定位量规 setting gauge
定位器 setworks
定位铅锤 posting plumb
定位渗透试验 on-site penetration test
定位受精 localized fertilization
定位图 location map
定位样地 permanent plot
定位样方 permanent quadrat
定位作用 orientating function
定温过程 isothermal process
定线 pegging; stake
定线测量 location survey
定香剂 fixative
定向 bearing; direction finding; orientation
定向板坯 aligned mat
定向变异 definite variation; determinate variation; directed variation
定向测雨器 vectopluviometer; vector gauge
定向大片刨花板 oriented waferboard
定向度 degree of orientation
定向伐木 directional felling
定向辊涂 direct roll coating
定向华夫刨花板 oriented waferboard
定向角 angle of orientation
定向结构刨花板 oriented structural board
定向进化 orthogenesis
定向刨花板 oriented strand board
定向培育 definite education; directed education; directive breeding
定向喷撒 directed application
定向器 director
定向区 oriented region
定向收缩 directional shrinkage; linear shrinkage
定向树 bearing tree
定向突变 directed mutation
定向吸附 oriented adsorption
定向线 direct line
定向效应 orientation effect
定向选择 directional selection; orthoselection
定向压力 direct(ed) pressure
定向仪 direction finder
定向育种 directive breeding
定向杂交 directed hybridization; judicious mating; purposeful hybridization
定像液 fixing solution
定心(胶合板制造) centr(e)ing
定心机 centr(e)ing machine
定心装置(胶合板制造) centr(e)ing device
定形群体 coenobium

定型行为 stereotypic behaviour
定型颗粒活性炭 pressed active carbon
定型性 constancy
定型杂种 constant hybrid
定芽 normal bud
定影液 fixer solution; fixing solution
定长(造材) measuring off
定植 colonization; field planting; field setting
定植苗 plant(ing) stock; planting stock
定植苗龄 planting stock age
定殖曲线 colonization curve
定制运材卡车 custom-built truck
定子 determinant
丢洛(硬度的标度) Duro
东风带 easterlies
东洋界 oriental region
东印度香叶浊 palmarosa oil
冬孢子 teleutospore; teliospore
冬孢子堆 teleutosorus; telium
冬孢子堆阶段(锈菌) telial stage; telial state
冬虫夏草 entomophyte
冬干梢(小枝、叶等) winter drying
冬候鸟 winter bird; winter resident; winter visitor
冬花园 winter garden
冬季采伐作业 winter show
冬季的 hiemal
冬季发育的 chimonophilous
冬季覆盖 winter mulch(ing)
冬季景象 winter aspect
冬季日灼 winter sunscald
冬季树皮日灼 winter bark-scorch
冬季停滞[期] winter stagnation
冬季相 hiemal aspect; winter aspect
冬季休眠 winter dormancy
冬景园(温室) winter garden
冬裂(树皮) chap
冬绿的 winter-green
冬绿树苷的别名 monotropitin
冬毛 winter coat
冬眠 dormancy; hibernation; winter dormancy
冬青茶 mate
冬青碱 yerbine
冬青素 winterin
冬青烯 winternene
冬青油 wintergreen oil
冬青浊(俗) gaultheria oil
冬霜 winter frost
冬态 winter condition
冬天的 hibernal
冬性一年生植物 winter annual plant
冬芽 winter bud
冬羽 winter plumage
冬贮插穗诱发愈伤组织 winter callusing
动摆测力器 pendulum dynamometer
动臂 lift boom
动臂剪形起重机 shear-legs
动臂起重机 jib-strut crane
动臂式伐木归堆机 cranab
动产 ambulatory chattel
动产抵押 chattel mortgage
动弹性 kinetic elasticity
动电电位 zeta-potential (ζ-potential)
动滑轮 fall block
动力 power
动力闭合(抓钩) power closing
动力播种机 power drill
动力铲运机 motor scraper
动力车(缆车道的) dudley; dudler
动力传达 power transmission
动力打桩锤 power monkey
动力的 dynamic
动力负荷 live load(ing)
动力割灌机 motor cutter
动力工具 power tool
动力供应传感装置 powersupply sensing device
动力刮土机 motor scraper
动力辊 live roll
动力换挡行星轮变速箱 planetary powershift transmission
动力集材 power logging; power skidding
动力锯 power saw
动力螺旋土钻 motor auger
动力黏度 dynamic viscosity
动力黏性系数 dynamic viscosity coefficient
动力喷雾器 power sprayer
动力平地机 motor grader
动力切碎机 power cutter

动力式粉剂喷枪 motorized dust gun
动力索道集材 power(ed) cable logging
动力索道运材 power cable transporting
动力特性 dynamic characteristic
动力条播机 power drill
动力推土机 power dozer
动力消耗量 power consumption
动力性黏着 kinetic adherence
动力学 kinetics
动力油缸 jack
动力载荷 dynamic load
动力支承辊 powered backup roll
动力植树机 motoplanter
动力制动器 power brake
动力中耕机 motor cultivator
动力转向 power steering
动力转向铰接车架 power articulated-frame
动力资源 energy resource
动量 momentum
动脉 aorta
动摩擦 friction of moment; kinetic friction
动能 kinetic energy
动平衡 dynamic balance; inertia balance
动牵引索 moving main line
动索道集材 moving cable yarding
动索系 moving cable system
动态 kinesis
动态抽样 dynamic sampling
动态的 dynamic
动态地位级 dynamic site class
动态分类系统 dynamic classification
动态规划 dynamic programming
动态混合生命表 dynamic-composite life table
动态价格模型 dynamic price model
动态模型 dynamic model
动态平衡 dynamic equilibrium
动态群落 dynamic community
动态蠕变 dynamic creep
动态生命表 dynamic life table
动态生态学 dynamic ecology
动态试验 dynamic test
动态特性 dynamic characteristic
动态稳定 homeorhesis
动态镶嵌体 shifting mosaic
动物(地理)区 faunal region
动物传播 epizoochory
动物传播的 zoochoric
动物传布 synzoochory
动物单位 animal unit
动物地理界 zoogeographical realm
动物地理区 zoogeographical region
动物地理学 zoogeography
动物计数器 zoometer
动物胶 animal adhesive; animal glue; animal size; gelatin
动物恐怖症 zoophobia
动物流行病 epizootic
动物流行病学 epizootiology
动物媒 zoophily
动物内生的 endozoic
动物排泄散布(植物种子散布) endozoochore
动物区系 fauna
动物区系屏障 faunal barrier
动物群落 animal community; zoocoenosis
动物散布 zoochory
动物社会学 animal sociology
动物社群 animal society
动物生境 zootope
动物生态学 animal ecology
动物饲料 animal forage
动物碎屑 offal
动物体上寄生的 epizoic
动物体外散布 epizoochore
动物通道 wildlife corridor
动物危害 tree damage caused by forest fauna
动物习性学 ethology
动物小区系 faunula
动物携带散布(植物种子散布) epizoochore
动物穴 animal furrow
动物园 zoo
动物志 fauna; Historia animalium
动物种集群 faunation
动循环索 moving endless line
动叶片 impeller blade
动应变 dynamic strain
动应力 dynamic stress
动蛹 liberal pupa
动用储金 dissaving

动原体 kinetochore
动载荷 moving loading
动植物胶 natural glue
动植物培育学 thremmatology
动作电位 action potential
冻拔 frost heave; frost heaving; frost lift(ing); lifting by frost; soil-lifting frost; uprooting by frost
冻板道 frozen road
冻剥 frost cleft
冻干 lyophilizing
冻害 chilling injury; freeze; freezing damage; freezing injury; winter injury; winter-killing
冻胶 jelly
冻搅作用 cryoturbation
冻结 congelation; freeze; freezing
冻结干燥法 frozen-dehydration method
冻结温度 freezing temperature
冻结账户 blocked account
冻结真空干燥 lyophilizing
冻棱纹 frost-rib
冻裂 cold checking; cold cracking; frost crack; frost cracking; frost cleft; frost split
冻裂作用 frost action
冻瘤 frost-canker
冻凝温度 congealing temperature
冻融侵蚀 freeze-thaw erosion
冻融作用 freeze-thaw action
冻伤 frostbite
冻伤材 frozen wood
冻伤年轮 frost-ring
冻深线 frost line
冻蚀 frost erosion
冻死 frost killing; winter-kill
冻土 frozen soil
冻土移植法 ice ball method
冻雾 pogonip
冻心材 frost heart (wood)
冻雨 icy shower; sleet
冻原 tundra
冻原草甸 tundra meadow
冻原气候 tundra climate
冻原土 tundra soil
冻胀丘(隆起) frost heaving
冻胀现象 frost heaving
栋梁 ridge beam; roof tree
洞 hole
洞穴 burrow
洞穴的 cavernicolous
洞穴动物 burrowing animal
洞穴环境 spelean environment
洞穴生物群落 cave community
洞穴生物学 speleobiology
洞穴贮藏 nest-hording
兜卸机 unloading gin pole
兜卸绳 drop line
兜转锯切 sawing around
抖动盘 shaking board
抖动筛 shaker bed
陡地缆车道 counterbalance system
陡坡 heavy grading; steep slope
陡坡地 steep land
陡坡减速集材 snub(bing) yarding
陡坡楞台 fly rollway
陡坡爬犁 dry sloop
陡坡剖面 uphill sloping profile
陡坡引擎 steep grade engine
陡峭螺旋 steep helix
斗式输送机 bucket conveyer
斗式提升机 bucket elevator
斗式装载机 loading shovel
斗武起重机 bucket crane
豆粉胶 soybean meal glue
豆荚 legume; pod
豆胶 soybean glue
豆科绿肥 leguminous green manure
豆科植物 legume; leguminous plant
豆科种子 leguminous seed
豆娘类 damsel flies
豆清蛋白 legumelin
豆球蛋白 legumin
都柏林单位(7.65m^3) Dublin standard
毒尘 poisonous dust
毒的 toxic; toxicant
毒蛾 tussock moth
毒饵 poison bait
毒害剂量 toxic dosage
毒剂 toxicant
毒剂保持量(防腐剂吸收量) toxicant retention
毒菌 toxic fungus
毒力 toxicity; virulence
毒力强的 virulent

毒力指数法 poison exponent method
毒毛 poison hair; poison seta
毒囊 poison sac
毒杀 poisoning
毒杀芬(杀虫剂) toxaphene
毒杀环剥 poison girdling
毒杀疏伐 poison thinning
毒死蜱(杀虫剂) chlorpyrifos
毒素 toxin
毒物 toxic compound; toxicant
毒物环割 poison girdling
毒细胞 poison cell
毒腺 poison gland
毒效 toxic value
毒性 toxicity; virulence
毒性测定 toxicity determination
毒性化合物 toxic compound
毒性化学药剂 toxic chemical
毒性极限 toxic limit; toxicity limit
毒性极限值 toxic sill
毒性临界值 toxic limit; toxic threshold
毒性筛选试验 toxicity screening test
毒性试验 toxicity test
毒性危害 toxic hazard
毒性指数 poison exponent
毒蕈 toadstool
毒药散布器 spreader for toxicant
毒质 toxin
毒株系 virulent strain
独寄生 eromoparasitism
独居动物 solitary animal
独居石 monazite
独居型 phase solitaria
独立伐 felling of wolf trees; liberation cut; liberation felling
独立分配 independent assortment
独立经济 autarchy
独立淘汰水平 independent culling level
独立性测验 test of independence
独立账户 separate account
独轮台车 dolly
独轮运材车 box dolly
独木船 canoe; log cabin
独木船浆 canoe paddle
独特并受威胁系统 unique and threatened system
独特景观 unique landscape
独特自然风景区 unique natural area
独占领域 exclusive territory
独资 individual proprietorship
读出 readout
读出信号传感扫描器 sensing scanner
读卡机 card reader
读取时间 access time
犊皮纸 vellum
堵塞 clogging
杜鹃素 farrerol
杜鹃园 rhododendron garden
杜松萜烯 cadinene
杜仲胶 gettama gum; gutta; gutta percha
杜仲收 gutta percha
度 dimension; particle size
度量 scaling; metric
度量单位 unit of measurement
度量性状 metrical character
渡口 river crossing
渡鸟 bird of passage
渡桥 transfer bridge
渡线 crossover
渡越 crossover
镀锡 tin
端壁 end wall; transverse wall
端壁厚度 thickness of end wall
端部搭接 end lap
端部裂缝 end check(ing)
端部缺损 terminal deficiency
端部注入 impregnation from end
端部注入法(防腐) tire tube process; tire-tube method
端部钻孔注入法 end-boring method
端搭叠 endlap
端搭接 end lap; end lap joint
端搭斜接 mitre-with-end-lap joint
端钩(边侧卷钩) end dog; end hook
端钩(装车) salisbury hook
端环裂 end shake
端基 terminal group
端接(木工等) abutting joint; butt matching; butt splice; end joint; end matching; heading joint; butt joint
端梁 end beam
端裂 end check; end check(ing); end split

端裂的 broomed
端轮裂 end shake
端面 end; end surface
端面接头 end joint
端面收缩 end-contraction
端面涂刷(层) end-coating
端面铣削 face milling
端面硬度 end hardness
端面直径 end diameter
端拼 end joint
端翘材 bowed wood
端切面 trimming section
端始种 incipient species
端头 crimp
端头号印 end mark
端头胶合 end gluing
端头截齐机 equalizer
端头拼接 end matching
端头散开的钢丝 porcupine
端头止挡(木材弯曲加工用承压坯上的附件) end stop
端涂防湿层(锯材涂石蜡) end sealed
端铣 end milling; face milling
端铣法 end milling
端向对比试材 end matched specimen
端向切削 end-grain cutting
端向渗透法 end penetration method
端向压力 end pressure
端向注入法 end penetration method
端压 end compression; end pressure
端栅平车 grade destroyer car
端直原木 stick
端柱 end-rack
短把锄 hand pike
短把搭钩(装车) pickaroon
短把刨钩 pulp hook
短边钢索(索道三角形集材带) wasteline
短柄修枝剪 short-handled prunning shear
短柄啄钩 hand pike
短材(按层积计算的) cordwood
短材(原木) log; short end; short-length
短材集运车(带车厢) accumulator forwarder; forwarder
短材集运机(带收集器) accumulator forwarder; forwarding machine
短材框 stillage
短材制材厂 bolter
短尺(锯材不足规格) scant
短翅型 brachypterous form
短粗榫舌 stub tenon
短粗原木 meat block
短吨 short ton
短干木 stunted tree
短钢轨 dutchman
短更新期 short regeneration period
短花柱的 microstylous
短角果(十字花科植物的果实) silicle
短脚衣橱 lowboy
短距离变动集材杆树 jump a tree
短距离集运材 short-haul extraction
短距离运材 primary transportation
短距运材 short haul
短锯材(长度小于6ft) short
短考(材长小于4ft, 垂直剖面不大于32ft^{2}的木垛) short cord
短链 tag chain
短路 short out
短轮伐期 short rotation
短命的 ephemeral
短命植物 ephemeral; ephemerophyte
短命植物荒漠 ephemeral desert
短木段(长1m小头直径7~14cm) billet timber (wood)
短木料 puncheon
短捻接 short splice
短刨(约7英寸长) block plane
短期负债 current liability
短期群状作业 short-period group system
短期失业 frictional unemployment
短期台营 joint venture
短期投资 current investment
短期休耕地 bastard fallow
短期育成林业 forestry on short rotation
短期预测 short-term forecast
短期运转 short run
短鞘翅 brachelytra
短日照 short sunshine period; short-day
短日照处理 short-day treatment
短日照植物 short day plant; shortday plant
短茸毛 down; nap

短绒毛 down
短伸木(稍伸出承载梁) bunked short
短生植物 ephemerophyte
短石细胞 brachysclereid
短时贮存台 live deck
短寿种子 short-lived seed
短索(捆木索与牵引索间) butt rigging
短索(连接吊梁与吊钩) bridle
短索绑扎 short coupled
短索端钩(系滑轮用) becket hook
短头颅的 brachycephalic
短纹 short grain
短纤维 flocks
短纤维纸浆 short stock
短效阻火剂 short-term retardant
短油 short oil
短原木 billet timber (wood); chump; peeler block; short log
短原木段(旋切用) billet
短圆材 block; bolt; yule log
短暂型植被 acheb
短枝 brachyblast; dwarf shoot; short branchlet; short shoot; spur
短枝化小枝 dwarf branchlet
短枝接 twig grafting
短枝芽接 prong budding; scion budding; twig-budding
短轴干燥窑 cross-shaft (dry) kiln
短柱 dolly; puncheon; short column; stub pole
短资 call money
短纵裂 short split
断层 dislocation; fault
断层走向 fault strike
断错 dislocation
断带状花纹 broken stripe figure
断点 break point
断点强度 breaking strength
断缝 breaking joint
断开时间 opening time
断裂 outbreak; rupture; breakage
断裂负荷 load at failure
断裂谷 vale
断裂基因 split gene
断裂极限 limit of rupture
断裂力矩 breaking moment
断裂面 surface of rupture
断裂模数 modulus of rupture
断裂强度 breaking strength; rupture strength
断裂性试验 Breaking-down test
断裂应力 breaking stress; rupture stress
断裂载荷 breaking load
断面 section
断面分析 section analysis
断面积 basal area
断面积表 basal-area table; table of basal areas; table of sectional areas
断面积控制法(计划采伐) basal area control; basal area regulation
断面积平均木 mean basal area tree
断面积平均木树高 height of mean basal area tree
断面积平均木直径 diameter of mean basal area tree
断面积平均直径 quadratic mean diameter
断面积生长 area-increment; growth of cross-sectional area
断面积生长率 area-increment percent; percentage basal area growth
断面积调节法 basal area regulation
断面积系数(角规) basal area factor
断面积中央木树高 median height
断面积中央木直径 median DBH
断面记号 end mark
断面模数 modulus of section
断面图 sectional view
断面直径 basal diameter
断丝(由破损钢丝绳伸出) beard
断条扭转花纹 rope figure
断续器 vibrator
断续雨 rain spell
断氧灭火 smothering
椴苷(椴树叶部) tiliacine
椴树花浸膏 linden concrete
椴树韧皮 bass; lime bast
椴树油 linden oil
煅烧 incineration
煅烧炉 incinerator
锻炼(苗木等) hardening off; hardening
堆板机 log stacker; lumber elevator; lumber jack; mechanical stacker; stacking machine
堆板装置 stack device

D

堆材 parcel
堆藏 mound storage
堆场 piled lot; stacking yard; stock yard
堆存 decking up
堆垛(板材) baling
堆垛(原木等) pack; stacking; stow
堆垛柴捆 fascine work
堆垛衬条 stick; sticker
堆垛成层 tier
堆垛导杆 stacking guide
堆垛定位装置(选材机) stick placer
堆垛工 stacker
堆垛机 mechanical stacker; stacker; stacking machine
堆垛密度(层积) stack(ing) density
堆垛木 stacked wood
堆垛木材 stack wood
堆垛坡度 slope of stack
堆垛提升机 stack elevator
堆垛装置 stack device
堆肥 artificial manure; compost; farmyard manure
堆肥菌 compost bacteria
堆行 windrow
堆积 piling up; rick
堆积变色 bridging blue; pile discolouration
堆积材积 piled volume
堆积材积测量 piled measure
堆积阶地 accumulation terrace
堆积蓝变 piling blue
堆积密度 packing density
堆积缺陷 piling defect; storage defect
堆积容量 stacked content
堆积烧炭法 heap charring
堆积时间 assembly time
堆积台地 accumulational platform
堆积炭化法 kiln-burning
堆积物 accretion
堆积英寻($6ft^3$) piled fathom
堆积制炭法 mailer method; meiler method; meiler process
堆基桁条 foundation bearer
堆烧 pile burn
堆石坝 rock-fill dam; stone weir
堆式烧炭法 charring in heaps
堆土 hilling
堆土压条 mound layering
堆巷(采剩物堆行间地带) stroll
堆植 high-planting
堆装架 stillage
队 logging crew
对瓣的 oppositipetalous
对背信用证 back-to-back letter of credit
对比 correlation
对比度 contrast; degree of contrast
对比法 contrast method
对比率 contrast ratio
对比试验 check experiment; contrast test
对比温度 reduced temperature
对边接 butt edge joint; plain edge joint
对边链片 master link
对丙烯基苯甲醚 anise camphor
对丙烯基茴香醚 anethole
K-对策 K-strategy
r-k对策连续系统 r-k continuum of strategy
对策论 game theory
K-对策者 K-strategist
对称结构 balanced construction; balanced structure
对称木纹拼板法 book matching
对称配板法 balance matching
对称式 symmetrical style
对称性 symmetry
对称轴线 symmetrical axis
对等基金 counterpart funds
对点杆 plumbing bar
对萼的 episepalous
对分检索表 dichotomous key
对缝拼接 plain edge joint
对话式代数语言 conversation algebraic language
对甲氧苯基茴丙酮 anisacetone
对甲氧基苯乙酮 anisketone; anisyl methyl ketone
对角纹理 bastard-sawn grain; diagonal grain
对角线的 diagonal
对接 abutting joint; butt joint; butt splice; end-to-end joint; plain joint
对接板 abutting plate
对接边 abutting edge

对开 bisect
对开材 half balk; split
对开信用状 back-to-back letter of credit
对开枕木 round-back(ed) sleeper; two-faced tie
对抗行为 agonistic behaviour
对抗物 antagonist
对抗作用 antagonistic action
对立 confronting
对立等位基因 oppositional allele
对立基因 oppositional gene
对立形质 allelomorph
对列纹孔式 opposite pitting
对流 convection
对流层 troposphere
对流层顶 tropopause
对流传热 heat transfer by convection
对流式干燥机 convection dryer
对流性气流 convectional current
对流与红外线配合干燥 combination of convection and infrared oven
对流雨 convectional rain
对流柱 convection column
对硫磷 parathion
对剖材 semi-circle cut
对生的 opposite
对生小叶 jugum
对生芽 opposite bud
对生叶 opposite leaves
对生叶的 oppositifolious
对生羽状的 opposite-pinnate
对数变换模型 log-transformed model
对数标尺 log scale
对数材积曲线 logarithmic volume curve
对数分度 log scale
对数级数分布 log series distribution; logarithmic distribution
对数喷雾器 logarithmic sprayer
对数生长期 exponential growth phase; logarithmic growth phase
对数死亡期 logarithmic death phase
对数正态分布 lognormal distribution
对位二氯苯(杀虫剂) paradichlorobenzene
对纹拼板 swing matching
对纹拼合 matching
对纹拼合单板 matched veneer
对向道岔(辙尖) facing point
对烟雾敏感的 smog-sensitive
对异丙基苯酚 australol
对应染色体 homologue
对应物 counterpart
对应性 allelism; allelomorphism
对应支出 counterpart expenditure
对映异构物 enantiomer
对照 check; control
对照材 control piece; matched piece
对照流域 control area
对照品种 check variety
对照情景 base-line
对照区 control area
对照试验 blank test
对照索引 correlation index
对照小区 check plot
对照样地 check plot; control plot
对照值 blank value
对摺的(叶) conduplicate
对摺子叶 conduplicate cotyledon
对中(制材) centr(e)ing
对中剖分锯 centr(e)ing splitting (*or* ripping) saw
对中装置(制材) centr(e)ing device
对重索 counter rope
对转双气流环式精磨机 counter-rotating double-stream ring refiner
对坐桌 partners desk
兑换 conversion
兑现 cash
吨 megagram (Mg); tonn
钝齿 cog
钝化[砂带] dulling
钝化剂 deactivator
钝化作用 deactivation; inactivation
钝角的 obtuse
钝锯齿 crenellate
钝棱 dull edge; wane; wane edge; waney edge
钝棱材 waney lumber; wany lumber
钝头的 blunt-nosed
盾蝽 shield bug
盾片 scutellum
盾片冀 scutel
盾纹面 escutcheon
盾形的 shield shaped
盾形火山 shield volcano

盾形嫁接 sprig grafting
盾形芽接 spade budding
盾状的 peltate; scutiform
盾状嫁接 shield grafting
盾状体 clypeus
盾状芽接 shield-budding
顿挫 abortion
多班制 multishift operation
多抱性 multivoltinism
多抱性昆虫 multlvoltine insect
多倍测图系统 multiplex system
多倍测图仪 multiplex
多倍单倍体 polyhaploid
多倍体 multiploid; polyploid
多倍投影仪 multiplex projector
多倍系统 multiplex system
多倍性 multiploidy; polyploidy
多比例尺立体镜 multiscope
多边形冻原 polygonal tundra
多变量分析 multivariate analysis
多变量直方图 multivariable histogram
多变量总体 multivariable population
多波段彩色增强 multiband colour enhancement
多波段摄影 multiband photography
多波段摄影机 multiband camera
多波段系统 multiband system
多草的 weedy
多侧脉的 venulose
多层存卷架 multi-storey reeling rack
多层带式干燥机 multiple band dryer
多层的 multi-layer; multistratal
多层回转式网带干燥机 multiple-pass conveyor dryer
多层胶合板(五层以上) multi-layered board; multi-ply (plywood)
多层连续式滚筒干燥机 continuous multi-deck roller-dryer
多层林 many storied forest; polylayer forest
多层模压 laminate molding
多层木 delta wood
多层刨花板 multi-layer particle board
多层乔林 many storied high forest
多层乔林作业 many storied high forest system
多层乳剂摄影术 multilayer photography
多层碎料板 multi-layer particle board
多层涂饰 multiple coating
多层吸附 multi-layer adsorption
多层压机 multi-daylight press; multi-open(ing) press; multi-platen press; multiple daylight press
多层压机生产线 multi-opening press-line
多层原木装车 top load
多产的 breedy
多出聚伞花序 multiparous cyme
多出聚伞状分枝式 mutiparous branching
多出数的花 polymerous flower
多雌 polygyny
多次单株选种 multiple single plant selection
多次繁殖生物 iteroparity
多次结实 polycarpic; polycarpicus
多次结实植物 polycarpic plant; popycarpic plant
多次开花 remontant flowering
多次生殖 iteroparous
多次循环干燥窑 recircula ing dry kiln
多次重复授粉法 method of repeated pollination
多刺旱生林(南美) espinal
多刺密灌丛 scariosa
多袋式摩擦剥皮机 multiple-pocket peeler
多刀圆盘削片机 multi-knife disk chipper
多道磁带 multi channel magnetic tape
多地区生长(或居住)的 polydemic
多点引燃 area ignition
多度 abundance
多度频度比 abundance-frequency ratio
多度中心 center of abundance
多段加热干燥法 stepwise heated drying
多对的 multijugate
多对基因杂种 polyhybrid
多分散性 polydispersity

多分枝的 multiramose
多分子的 multimolecular
多酚 polyphenol; polyphenols
多酚氧化酶 polyphenol oxidase
多风的 windy
多负载式索道 multi loaded cableway
多隔膜 multiseptat
多功能团 polyfunctional group
多功能系统 multifunction system
多股钢丝绳 multi-stranded cable
多灌丛的 brushy
多光谱 multispectra
多光谱扫描仪 multispectral scanner
多光谱摄影机 multiband camera
多光谱摄影术 multispectral photography
多光谱卫星影像增强 enhancement of multi-spectral satellite image
多轨式窑 multiple-track kiln
多行起苗机 multiple-row sled tree lifter
多行业公司 multi-industry company
多行原木绠漂 bracket boom
多核蛋白体 polyribesome
多核的 coenocytic
多核配子囊 coenogametangium
多核糖核蛋白体 polysome
多核糖体 polyribesome; polysome
多核细胞 apocyte
多横隔孢子 fragmospore
多花采粉的 polytropic
多花的 multiflorous
多化性 multivoltinism; polyvoltinism
多化性昆虫 multlvoltine insect
多回羽状的 multipinnate
多货位装车场 double header
多基因 multiple gene; polygene
多基因抗病性 polygenic resistance
多基因抗性 multigenic resistance
多基因控制 multigenic control
多基因遗传 polygenic inheritance
多基因杂交 polyallel-cross
多基因杂种 multihybrid
多级谷坊 multiple-drop check dam
多级压缩机 multistage compressor
多季相 multi-seasonal
多寄生 multiparasitism; multiparasitization; poly parasitism; polyparasitism
多价(化学) polyvalency
多价染色体 multivalent chromosome
多交检定 polycross test
多交系 multiple crossing
多角经营 conglomerate
多角体 polyhedron
多角体病毒病 polyhedrosis
多角斜接 polygon mitre joint
多阶抽样 multistage sampling
多阶抽样检验法 multiple sampling inspection plan
多阶资源清查 multistage inventory
多节 knottiness
多节材 full of staybolts; knaggy wood; knotty wood
多节的 bossy; knaggy; knobby; knotty; multinodal; snagged; snaggy
多节干材 knobby-bole
多截面砂光机 variety sander
多晶材料 polycrystalline materials
多精受精 polyspermy
多境的 polytopic
多镜头摄影机 multi-lens camera
多锯机 log frame
多锯片横截锯 multi-saw trimmer
多锯片截锯 slashing saw
多锯片截锯机 trimmer
多锯片锯机 multiple saw
多锯片齐边机 multi-saw edger
多聚酶 polymerase
多菌灵 bavistin; carbendazim (*or* carbendazol)
多孔的 lacunose
多孔菌 polypore fungus; pore fungus
多孔卡检索表 multiple-entry card key
多孔湿球套 porous wet-bulb sleeve
多孔式(内含韧皮部) strychnos type
多孔性 hollowness; porosity
多孔圆筒 perforated cylinder
多孔蒸气管 perforated steam pipe
多跨(索道) multispan
多跨架空索道 multispan skyline

多跨循环索道 multiple continuous circulating cableway
多联带锯机 multiple bandsaw
多量程测量仪表 multimeter
多量效果 manifold effect
多列射线 multiseriate ray; polyseriate ray
多列同形射线 multiseriate homogeneous ray
多列纹孔式 multiseriate pitting
多列异形射线 multiseriate heterogeneous ray
多列杂交 polyallel-cross
多裂的 multifid
多裂片的 plurilobed
多鳞的 squamose
多龄林 all-aged forest; many-aged forest
多轮的 multinodal; pleiocyele; polycyclic
多脉的 nervose
多鬣毛的 erinaceous
多面积样地抽样 palyareal plot sampling
多能拌胶机 universal blender
多能夹紧器 universal clamp
多能锯 variety saw
多能刨片机 universal flaker
多能砂光机 variety sander
多能装配式砂轮砂光机 universal abrasive wheel sander assembly
多年群体 perennial colony
多年生草本 perennial herb
多年生草本植物 renascent herbs
多年生的 perennial
多年生禾草 perennial grass
多年生溃疡 perennial canker
多年生阔叶草本 perennial forb
多年生型 perennial form
多年生真菌 perennial fungus
多年生植物 perennial; perennial plant; perennial plants
多年生子实体 perennial fruit body
多排链条传动 multi-strand chain drive
多排卵 polyovulation
多胚的 polyembryonate; polyembryonic
多胚生殖 polyembryony
多胚现象 polyembryony
多配性的 polygamous
多皮的 corticose
多片圆锯 circular gang-saw (*or* gang-mill)
多品型现象 heteromorphism
多普勒测距系统 Doppler ranging
多普勒定位 Doppler location
多普勒雷达 Doppler radar
多普勒效应 Doppler effect
多歧聚伞花序 pleiochasium
多歧式 pleiochasium
多球果病 polycony
多染色体分析法 multisomic analysis
多肉植物 succulent plant
多乳头的 papillose
多色 multicolour
多色的 polychromatic
多色花卉 multicoloured flower
多生境生物 heterozone organism
多石的 stony
多石地犁 rocky-soil plough
多时期起源 polychronic origin
多时相 multi-temporal
多实的 fruitful
多食性 polyphagous; polyphagy
多室式球磨机 compound ball mill
多室子房 multilocular ovary
多霜地 frost locality
多水高岭土 halloysite
多酸价碱 polyacid base
多缩戊糖 pentosan
多态的 pleomorphic
多态曲线 polymorphic curves
多态现象 polymorphism
多肽 polypeptide
多糖 polyose; polysaccharide
多糖骨架 polysaccharide skeleton
多糖酶 polysaccharase
多体 multimer
多体的 polysomic
多体拟态行为 allelomimetic behaviour
多体生物 polysomic
多体突变型 lata; lata type
多体性 polysomy
多体雄蕊 polyadelphous stamens
多体雄蕊的 polyadelphian; polyadelphous

多通道系统 multi channel system
多头开榫机 multi-head tenoner
多头砂光机 multiple-head sander
多洼的 lacunose
多维超体积生态位 multidimensional hypervolume (niche)
多维空间 hyperspace
多维体积 hypervolume
多维小生境 multidimensional niche
多戊糖 pentosan
多物种成熟生态系统 multi-species mature ecosystem
多雾荒原 fog desert
多系品种 multilineal variety
多系杂交 polycross
多系杂交法 polycross method
多细胞的 multicellular
多线 polyneme
多线染色体 polytene chromosome
多线性 polyteny
多线遗传 multilinear heredity
多线迎面火 multiple backfire
多相抽样 multi-phase sampling
多相取样法 multiphase sampling
多相位 leggy
多相性 heterogeneity; inhomogeneity
多向预测 multiforcasting
多项分布 multinomial distrihution
多小枝的 ramulose; ramulous
多效 manifold effect
多心皮 polygyny
多心皮雌蕊 polycarpellary pistil
多心皮的 polycarpellary
多心皮子房 polycarpellary ovary
多形成层的 polycambial
多形地位指数曲线 polymorphic site curve
多形性 polymorphism
多型进化 polytypic evolution
多型现象 polymorphism
多性杂种 polyhybrid
多雄 polyandry
多雄群 multi-male troop
多雄蕊的 multistaminate; polyandrous
多雄蕊式 polyandry
多雪 snowy
多雪的 nival
多循环湖 polymictic lake
多牙尖榫 tusk tenon
多盐的 polyhaline
多盐类防腐剂 multi-salt preservative
多氧性的 polyoxybiotic
多样的 manifold
多样化 diversification
多样满细胞法处理 miscellaneous full-cell treatment
多样性 diversity
多样性梯度 diversity gradient
多样性指数 diversity index; index of diversity
多叶的 leafy; pleiophyllous
多叶卷叠式 praefoliation; vernation
多叶片通风机 multi-blade fan
多叶隙的 multilacunar
多叶性 leafiness
多伊尔-斯克里布纳原木板积表(估算有缺陷和过熟木材的板材积) Doyle-Scribner log rule
多伊尔原木板积表 Doyle tog rule
多伊尔原木板积表(估算原木板材材积的用表) Doyle log rule
多因子遗传 multiple factor inheritance; multiple inheritance
多用餐桌 universal dining table
多用伐木机 multiprocessing machine; multipurpose timber-harvesting machine
多用家具 utility furniture
多用途经营 multi-use management
多用途林业 multiple-purpose (use) forestry
多用途履带拖拉机 multipurpose crawler tractor
多用途-永续收获法案(美) Multiple Use-Sustained Yield Act
多用椅 multipurpose chair
多用装载机 utility loader
多用桌 multi-use desk
多优势种浮游生物群 polymictic plankton
多雨湖 pluvial age
多雨期 pluvial lake; pluvial period
多元材积表 multiple entry volume table
多元顶极理论 polyclimax theory
多元顶极群落 polyclimax

D

多元分析 multivariate analysis
多元回归 multiple regression
多元碱 polyacid base
多元配置法 multiway-layout method
多元体 polyploid
多圆锥投影 polyconic projection
多云 cloudy
多爪锚 grappling iron
多支管 manifold
多汁果 succulent fruit
多汁性 succulence
多枝的 branchy; ramose
多枝木 grouse ladder; limby tree; squirrel tree
多脂材 fatwood; resinous wood
多脂肪的 adipose
多脂松材 fatwood
多种寄生 multiple parasitism
多种经营农场 diversified farm
多重性 multiplicity
多轴浮雕机 multiple spindle carving machine
多轴卡车 multiple axle vehicle
多轴开榫机 multi-head tenoner
多轴链式榫槽机 multiple chain-saw mortiser
多轴木工钻床 multi-head wood borer; multi-spindle wood borer
多轴榫槽机 multiple hollow-chisel mortising machine
多轴制榫机 end matcher
多轴钻床 multiple borer
多株打枝机(对多株伐倒木) multi-stem delimber
多主寄生物 plurivorous parasite
多主寄生现象 polyphagy
多主脉的 multicostate
多子叶 polycotyl
多子叶的 polycotyledonous; polycotylous
多子叶式 polycotyledony
多子叶植物 polycotyledoae; polycotyledon
多自交系互交 polyallel-cross
多足纲 Myriapoda
多足型 polypod type
夺回 recapture
夺流(河道) piracy
夺流河道 pirate stream
躲避(不利条件) avoidance
垛材积 cord volume
垛材台托板 pallet
垛底 stack bottom
垛计材积 stump scale
剁刀 marking knife
剁斧 chip-axe
舵轮风(英国) helm wind
惰轮 idle pulley; idle wheel; idler; jockey roller
惰性成土因素 passive soil former
惰性腐殖质 inert humus
惰性填料 inert filler
俄寻($7\times7\times7=3430ft^3=9.71m^3$) Russian fathom
鹅 goose
鹅颈式挂车 gooseneck trailer
鹅颈锁榫凿 swan-neck lock mortise chisel
鹅颈弯曲段 swan-neck
鹅颈形或U形物 gooseneck
鹅颈形起重臂 gooseneck boom
鹅卵石 cobble
鹅掌楸碱 liriodendrin; tulipiferine
鹅掌楸苦素 liriodendrin
蛾 moth
额 frons; front
额定飞行速度 rated cruise speed
额定功率 power rating
额定载荷 nominal load; rated load
额外采伐 extra-cutting
额外费用 additional cost
额外股息 extra dividend
额外缴纳资本 additional paid-in capital
额外精核 merocyte
额外染色体 accessory chromosome
额外收益 extraordinary yield
额外填方 extra banking
额外性 additionality
额外应力 extra-stress
额外租金 key money
厄尔尼诺南方涛动 el Niño Southern Oscillation
厄尔尼诺现象 el Niño
恶化 deterioration
恶化指标(干扰) disturbance indicator; deterioration indicator
恶劣(地位级V) poor

恶劣天气 ugly weather
恶性循环 vicious circle
恶性杂草 noxious weed
萼冠同形的 isostomous
萼冠异形的 heterochlamydeous
萼片 calix; calyx; sepal
萼片化 sepalody
萼片状的 sepaloid
萼上的 episepalous
萼檐 limb
萼状苞 exanthium
萼状的 calyciform
萼状总苞 epicalyx
颚式破碎机 alligator
颚爪 fork
颚爪闭度 overarm closure
颚爪开度 overarm opening
颚爪上夹钳 top clamp
鳄嘴裂口(板坯纵向劈裂缺陷) alligatoring
恩格勒氏度 Engler degree
恩格勒氏度数 Engler number
恩格勒氏黏度 Engler degree
恩格勒氏黏度计 Engler viscosimeter
恩格勒氏黏度数 Engler number
恩氏黏度计 Engler viscosimeter
蒽 anthracene
蒽油 anthracene oil; carbolineum
儿茶 cutch; catechu; pale cutch
儿茶单宁 catechotannin
儿茶酚酶 catecholase
儿茶类鞣料 catechol tan
儿茶鞣质 catechotannin
儿茶素 cateehin
儿童公园 children park
耳垂 lobe
耳环支承板 lug support
耳突症 enation
耳羽 ear covert
耳状决明 avaram
耳状体 ear
耳状突起 enation
饵木 trap log; trap tree
二百立方英尺(材积计量单位) unit
二抱性 bivoltinism
二倍单性生殖 ooapogamy
二倍化 diploidization
二倍加一 trisomic diploidy
二倍减一 monosomic diploidy
二倍配子体无融合生殖 diploid gametophyte apomixes
二倍期 diplophase
二倍体 diploid
二倍体植物 diplophyte
二倍性孢子形成 diplospory
二苯醚 diphenyl oxide
二丙烯草胺 allidochlor
二步产物 secondary product
二侧相等的 isobilateral
二层胶合板 two-ply (plywood)
二层林 two-storied
二层林冠 two-storied forest
二叉的 dichotomous
二叉分枝式 dichotomous branching; dichotomy
二叉式检索表 dichotomous key
二叉式体系 dichotomous system
二出聚伞状分枝式 biparous branching
二唇的 bilabiate
二次采伐 recovery logging; second cutting
二次成型 postforming
二次定量加压注入防腐法 double Rueping process
二次分蘖 secondary tiller
二次复型法 two step replica method
二次固化 after-cure
二次关系 quadratic relationship
二次集材(从索道集材终点开始集材) roading; skyline swing; swing(ing); road
二次集材场 swing yard
二次集材道 swing road
二次集材绞盘机 swing donkey
二次加工 secondary conversion
二次加工产品 derivative
二次加工压机 finishing press
二次渐伐 two-stage cutting
二次渐伐法 two-cut shelterwood method
二次胶合 assembly gluing; constructional gluing; secondary gluing
二次萌发 germination by repetition
二次平均偏差 quadratic mean deviation

E

二次平均直径 quadratic mean diameter
二次伞伐法 two-cut shelterwood method
二次休眠 secondary dormancy
二次样品(样本) subsample
二次运材 secondary transportation
二代种子 second generation seed
二单元组 diad
二道底漆 infer-coat; interlayer; intermediate coat
二道锯 secondary saw
二等材 second grade timber
二等肉食性动物 second level carnivore
二叠纪 Permian period
二段矮林 coppice with reserve
二段集材 double yard
二段林 two-storied
二段乔林作业 high forest with reserves (system); high forest with standards (system)
二段乔林作业法 two-storied high forest system
二段式周期 two-stage cycle
二儿茶素 dicatechin
二分法 dichotomy method
二分体 dyad
二分细胞 diad; dyad
二分枝的 bioramous
二合花粉(孢粉) dyad
二化的 bivoltine
二化性 bivoltinism
二回羽叶 pinnule
二回羽状的 binato-pinnate; bipinnate; duplicato-pinnate
二回掌状的 binato-palmate; bipalmate
二基数的 dimerous
二级干燥装置 two-stage drying plant
二级胶合板 type-two plywood
二级结构 secondary structure
二级消费者 secondary consumer
二级悬浮干燥 two-stage suspension dry
二级悬浮式干燥机 two-stage suspension type dryer
二甲苯 dimethyl benzene; xylene
二甲苯酚 xylenol
二甲花翠素 malvidin
二甲基甲酰胺 dimethylformamide
二甲基亚砜 dimethyl sulfoxide
二甲基乙酰胺 dimethylacetylamide
二甲砷酸 cacodylic acid
二价的 bivalent; divalent
二价染色体 bivalent
二碱基的 dibasic
二碱价酸 bibasic acid
二进制编码 binary code
二进制乘法器 binary multiplier
二进制单元 binary cell
二进制的 binary
二进制计数器 binary counter
二进制记数法 binary notation
二进制链 binary chain
二进制码 binary code
二进制数 binary number
二进制数字 binary digit
二进制位 binary digit
二进制装配程序 binary loader
二精入卵 dispermy
二聚戊烯 dipentene
二聚月桂烯 dimyrcene
二类调查 forest management inventory; forest inventory for planning and design; second-category forest inventory
二棱角的 biangulate
二联跑车(轨道式) disconnected trolley
二联跑车(索道式) hati
二列的 biseriate; distichous
二硫化碳(杀虫剂) carbon disulfide
二轮车 cart
二轮运材车 go-devil; logging dolly
二年果 biennial fruit
二年生植物 biennial (plant)
二歧的 dichotomous
二歧聚伞花序 dichasial cyme; dichasium; dichotomous cyme
二歧式 dichasium; dichotomy
二嗪农(杀虫剂) diazinon
二氢吡喃 dihydropyran
二氢丁子香酚 dihydroeugenol
二氢槲皮素 dihydroquercetin
二氢化菲 dihydrophenanthrene
二氢芪 dihydrostilbene
二氢月桂烯 dihydromyrcene

二球悬铃木 European plane; English plane tree; London lane tree
二色花卉(一株或一朵开二色花) bicoloured flower
二色性光栅 dichroic grating
二态分布 bimodal distribution
二态现象 dimorphism
二糖类 bioses; disaccharides
二体雄蕊的 diadelphian
二维相关 two-dimensional correlation
二无色瑟酮定 bileucofisetinidin
二项分布 binomial distribution
二项式系数 binomial coefficient
二硝基苯酚 dinitrophenol
二硝基酚和氟化钠盐剂 PF-salt
二硝散(杀菌剂) nirit
二氧化丁二烯 butadiene dioxide
二氧化碳(CO_2)施肥 carbon dioxide (CO_2) fertilization
二氧化碳补偿点 carbon-dioxide compensation point
二氧化碳当量 CO_2-equivalent; equivalent CO_2 (carbon dioxide)
二氧化碳加浓 carbondioxide enrichment
二氧化碳施肥 carbondioxide fertilizer; CO_2 fertilization
二乙氨乙基纤维素 diethyl-aminoethyl cellulose (DEAE-cellulose)
二乙酰鞣素 tannigen
二异黄樟素 diisosafrole
二元变量正态分布 bivariate normal distribution
二元材积表 general volume table; standard volume table
二元材种出材率表 two-way merchantable volume table
二元的 binary; dibasic; two-entry
二元对称现象 binary symmetry
二元共聚物 biopolymer
二元杂交法 two-way cross
二原型 diarch
二重断面 double profile
二重寄生 diploparasitism
二重嫁接 double grafting; double-work
二重锯齿 biseriate
二重芽接 double budding
发白(涂料) blushing
发病 pathogenesis
发病率 incidence; morbidity
发出的 emergent
发电厂 power plant
发电机组 power plant
发动机 electromotor
发端 initiation
发根 hairy root; root initiation
发根缓慢的 slow-rooting
发根能力 rooting ability; rooting capacity
发光 glaze; luminescence
发光度 luminous emittance
发光器 illuminator
发光器官 luminescent organ
发光生物 luminous organism
发光细菌 luminescence bacteria; photobacterium
发行 emission
发花(漆病) floating
发火点 ignition point
发火机 primer
发火帽 percussion-cap
发火药 primer
发货标记(木材) shipping marks
发货国 country of dispatch
发货人 shipper
发价(外贸) offer
发酵 ferment; fermentation; fermention
发酵层(土壤剖面F层) fermentation horizon; fermentation layer; F-horizon; F-layer
发酵酶(发酵素) ferment
发酵学 zymurgy
发亮的 luciferous
发裂(木材) feather-checking; hair check; hairline crack
发裂(漆膜) crazing
发泡机 frother
发泡剂 foamer; foaming agent; inflating agent
发泡胶合 foam gluing
发泡胶黏剂 foam glue; foamed adhesive
发泡型阻燃涂料 intumescent coating
发情习性显露部分(动物) breeding display area
发情周期 oestrous cycle

发热的 pyrogenic
发热反应 exothermic reaction
发热量 heat output; heating power; thermal value
发色的 chromophoric
发色体 chromogen
发色团 chromophore
发射 emission; emittance
发射场 launching site
发射度 emittance
发射光谱 emission spectrum
发射光谱法 emission spectrography
发射率 emittance
发射谱 emission spectrum
发射谱线 emission line
发射物 effluence
发射线 emission line
发生 evolution; genesis
发生雌雄同体 genetic hermaphrodite
发生炉煤气 generator gas
发生率 incidence
发生期 emergence period
发生器 charcoal-burner; oscillator
发生学 genesis
发生中心 centre of origin
发数(枪弹或箭, 用复数) round
发丝裂缝 feather-checking
发雾(漆膜) milkiness
发香团 odoriphor group; osmophore
发旋光性生物 bioluminescent organism
发芽 break; bud off; chit; emergence; flush; germinate; germination
发芽测定 germinating test
发芽常数 germination consant
发芽成熟 germinating-ripe
发芽迟缓的 slow-germinating
发芽迟滞 delayed germination
发芽床 germinating bed
发芽促进法 method of hastening germination
发芽促进物质 germination promoting substance; germination stimulator
发芽的 gemmiferous; germinant; germinative
发芽盒 germinating-box
发芽孔(花粉) germ pore; pore
发芽力 germinability; germinating (germination) capacity; germinating ability; germinating power; germinating quality; germinative force
发芽力保存期 durability of germination capacity
发芽力试验器 germinator
发芽率 germinating (germination) capacity; germination number; germination percent; germination rate
发芽能力 germinating (germination) capacity
发芽盘 germination tray
发芽器 germinator
发芽前处理 pregermination treatment
发芽曲线 emergence curve
发芽生态学 germination ecology
发芽生殖的 gemmiferous
发芽势 germinating energy; germinating power; germination viability
发芽势期 germination energy period
发芽势指数 germination energy index
发芽试验 germinating test
发芽室 room germinator
发芽数 germination number
发芽速度 germination speed
发芽速率 rate of germination
发芽速率系数 coefficient of velocity in germination
发芽习性 germination (germinating) behaviour
发芽系数 coefficient of germination
发芽箱 germinating-box; germination cabinet; germinator
发芽迅速 fast-germinating
发芽抑制物质 germination inhibiting substance; germination inhibitor
发芽值 germination value
发芽纸 germination blotter
发芽指数 germination index
发芽周期 periodicity of germination
发芽装置 germinating-apparatus
发芽阻抗 germination resistance
发烟器 smoker
发叶 leafing

发育 development
发育变异 developmental variation
发育不充分 underdevelopment
发育不良(胚胎或个体) dysontogenesis
发育不全 abortion; aplasia; hypoplasia; hypoplasty; undergrowth
发育不全的 hypoplastic; obsolete; rudimentary
发育不全胚 rudimentary embryo
发育不全器官 rudimentary organ
发育不全种子 rudimentary seed
发育迟缓 aplasia
发育初期的 brephic
发育单位 developmental unit
发育阶段 developmental stage
发育临界 developmental threshold
发育零点 developmental zero
发育期 development period
发育器官 development organ
发育速率 developmental rate
发育遗传学 developmental genetics; phenogenetics
发育枝 developmental branch
发育周期 developmental cycle
发展 development
伐 lop
伐出费 harvesting costs
伐除 cutover; removal
伐除龄 removal age
伐除森林 disforest
伐除树木(路边或装车场地) slash
伐倒 knocked down
伐倒地点 felling point
伐倒方向 direction of fall; direction of felling
伐倒木 down-timber; fallen tree; fallen trunk; felled timber; felled tree; hag; hewn tree
伐倒木搭挂 swinging in a bind
伐倒木伐区削片 full-tree field chipping
伐倒木集材 full-tree logging; full-tree skidding; whole-tree logging; whole-tree skidding; hewn-tree skidding
伐倒木检尺 scaling felled tree
伐倒木买主 purchaser after felling
伐倒木蓄积资本 capital felling
伐倒木削片 whole-tree chipping
伐倒木削片机 full-tree chipper; whole-tree chipper
伐倒销售(林木伐倒后销售) felled sale
伐倒总方向 lead
伐道影 hatch; swamping
伐点(伐高) felling point
伐根 bole; bole wood; stub; stump
伐根爆破炸药 stumping powder
伐根材 stump wood
伐根残存迹地 stump(y) land
伐根的实际断面积 government square
伐根点现场检尺员 stump scaler
伐根高 cutting height; stump height
伐根更新 reproduction by stump division
伐根横断面 felling section; stump section
伐根解析 stump analysis
伐根空心材 hollow butt
伐根萌芽更新 stump
伐根明子 stump
伐根木片 stump wood chip
伐根年龄 stump-age
伐根膨大的 churn-butted
伐根切碎机 stump cutter; stump crusher
伐根清除 stump extraction
伐根削片机 stump chipper
伐根直径 stump diameter
伐后保留林分 residual crop; residual stand
伐后废材 logging slash
伐后更新[法] after-regeneration; after-reproduction; regeneration after removal of old growth
伐后小块空地 felling gap
伐后造林作业 after-reproduction
伐尽森林 disafforestation
伐开的集材线 swamped line
伐开线 breakthrough
伐木 chop; cutting; falling; fell; felling; throwing; timber felling
伐木臂 feller-boom
伐木场 camp; logging headquarter; logging site; lumber mill

伐木场办公处 depot
伐木场工棚 shanty
伐木场经营者 lumber mill operator
伐木场库棚 logging shed
伐木场主任 logging superintendent
伐木承包者 logging jobber
伐木程序 logging procedure
伐木程序机 tree processor
伐木除枝机 feller-buncher
伐木打枝归堆机 feller-buncher delimber; feller-delimber-buncher
伐木打枝机 cutting-lopping-machine; feller-delimber
伐木打枝造材归堆机 feller-delimber-slasher-buncher
伐木-打枝-造材-归堆机 feller-delimber-bucker-buncher
伐木打枝造材机 feller-limber-bucker; feller-processor
伐木打枝造材集材机 feller-delimber-slasher-forwarder
伐木倒向 felling direction
伐木登记簿 felling register
伐木地平垫 lay
伐木定向器 feller-director
伐木定向仪 timber compass
伐木队 woodcutter's crew
伐木斧 cleaver; felling axe; woodcutter's axe
伐木工 axeman; bush ranger; bushman; chopper; cutter; feller; flathead; log sawyer; logger; lumberman; sawyer; timber cutter; timber jack; tree feller
伐木工队 logging band
伐木工具 choppers' tool
伐木工棚 logging shack; logging camp
伐木工人 lumber jack; wood chopper; woods laborer; woodsman
伐木工头 logging boss
伐木工长 head woodcutter; woods foreman
伐木工组(伐后火烧林地) black gang
伐木工组(伐木) logging band; woods crew
伐木归堆机 feller-buncher; felling bunching machine; tree harvester
伐木归堆联合机 feller-buncher
伐木归堆头 feller-buncher head
伐木规则 felling rule
伐木合同承包人 logging contractor
伐木机 feller; felling machine; felling unit; stumper
伐木集材车 feller-forwarder
伐木集材机 feller-forwarder; feller-skidder
伐木集运材 timber harvest
伐木计划 harvest plan
伐木记数表 cutting list
伐木季节 cutting period; felling season
伐木检尺(测定) measurement of felled timber
伐木剪 logging scissors; tree clipper; tree shears
伐木锯 felling saw; tree feller
伐木控向千斤顶 tree jack
伐木量 fell
伐木裂纹 felling shake
伐木零工 logging jobber
伐木领班 logging boss
伐木留冠干燥法 sour-telling
伐木平垫 bed; table
伐木人 faller
伐木上锯口 felling cut
伐木烧炭 coal off
伐木收益 felling return
伐木踏板 scaffold board; spring board
伐木踏板插口 spring board notch
伐木台架 chopping platform
伐木台账 record of cutting
伐木徒工 buckwheater
伐木下口 felling notch
伐木楔 falling wedge; felling wedge
伐木楔垫板 falling plate
伐木削片机 feller-chipper
伐木业 lumbering
伐木营地 logging camp
伐木造材 extracting
伐木造材程序机 feller-processor
伐木造材费 cost of harvesting
伐木造材工具 wood-cutter's implement
伐木造材工长 bull buck(er)
伐木造材锯 briar
伐木造材损耗 waste in felling and log-making
伐木造材组长 timber inspector
伐木者 feller

伐木证 timber concession
伐木组 timber crew; woodcutter's crew
伐木钻 cutting auger
伐木最小直径 marginal tree
伐木作业 felling operation; logging operation; logging practice; lumber-job
伐期龄 final cutting are; cutting rotation age; exploitation age; felling age; final age; maturity
伐期平均生长量 average yield class; final mean annual increment
伐期收获 final yield; yield of final crop
伐期收入 final income
伐期收益 return from final clearing
伐期直径 final diameter
伐前更新 advance planting; prefelling regeneration
伐前更新法 advance reproduction; pregeneration; preregeneration
伐前更新作业 before reproduction system; method of advance reproduction; preregeneration system
伐前林内幼树 advance growth
伐前选采法 selective cupping
伐前栽植 planting in advance
伐前造林作业 advance-reproduction system; before reproduction system
伐前准备伐(清除障碍物) pre-logging
伐区 cutting block; felling area; cutting area; cutting chance; hag; logging area; chance; coupe; cut area; cut-over; logging unit; setting; show; stage; timber-tract
伐区边角地 odd corner
伐区边缘 front
伐区多种清理 diversified slash disposal
伐区集材 off-road hauling
伐区监工员 logging superintendent
伐区界线 cutting line; felling area line
伐区局部清理 partial slash disposal
伐区清理 clearing of a cutting area; logging clearing; removal of felling area; slash disposal
伐区区划 allotment for felling area
伐区式渐伐作业 shelter-wood compartment system
伐区式皆伐 area-wise felling
伐区式伞伐作业 compartment uniform method; shelter-wood compartment system
伐区调查 coupe inventory
伐区调查记录 coupe inventory record; scale book
伐区形状 cutting form
伐区装车场 timber harvest landing
伐区作业面 side
伐区作业组 side
伐运 timber harvest
伐桩腐朽 stump rot
垡片 furrow slice; upturned sod
阀门 valve
筏 raft
筏道 log pass; logway; rafting channel
筏节 section raft
法定标准 legal standard
法定公路载量 legal highway load
法定股本 authorized (capital) stock
法定货币 legal tender
法定资本 authorized capital
法定最低贮备 legal minimum reserve
法国波特儿犬 French poodle
法国抛光漆 French polish
法国式疏伐 French method of thinning; French thinning
法兰 flange
法兰接合 flange joint
法兰盘 flange
法律实体 legal entity
法呢基 farnesyl-
法人 legal entity; legal person
法向应变 normal strain
法向应力 normal stress
法正材积 normal volume
法正采伐量 normal cut
法正采伐率 normal cutting ratio
法正度 normality
法正伐期龄 normal final age
法正计算上之裸地 theoretical clearfell area
法正立木营积 normal growing stock
法正林 balanced forest; ideal forest; normal forest
法正林分 normal stand

法正林模式 ideal forest model; normal forest model
法正龄级 normal age-class
法正龄级分布 normal (ideal) age classdistribution
法正龄级分配 normal age-class distribution
法正龄级配置 normal age-class arrangement
法正龄阶序列 normal series of age gradations
法正轮伐期 normal rotation
法正年伐量 normal annual yield
法正生长量 normal increment
法正生长率 normal growth rate
法正收获表 normal yield table
法正收获量模式收获量 normal yield
法正同龄林 normal even-aged forest
法正完满立木度林分 normal-fully stocked stand
法正蓄积 normal growing stock; normal stock; normal stocking
法正蓄积本利价 capital rental value of normal growing stock
法正蓄积法 method of regulating yield (by comparing actua with normal crops); normal stock method; normal volume method; volume and increment control method
法正蓄积价 cost value of normal growing stock; value of normal growing stock
法正异龄林 balanced unevenanged stands; normal uneven-aged forest
法正择伐林 balanced selection stand
法正择伐作业 ideal selection system
法正主伐龄 normal final age
法正作业级森林租金价 rentavalue of forest of normal working section
珐琅 enamel
帆布挡板 canvas retainer plate
帆布贴面胶合板 canvas covered plywood
番红花 saffron
番荔枝碱 anonaceine
番木鳖苷 loganin
番木瓜 papaya; melon tree
番木瓜碱 carpaine
番樱桃酚苄醚 eugenol benzyl ether
番樱桃酚型罗勒油 eugenol type basil oil
番樱桃基 eugenyl
番樱桃脑 eugenitol
番樱桃宁 eugenin
番樱桃亭 eugenitin
番樱桃酮 eugenone
蕃椒油 capsicin
翻板 flap
翻板器 board turner
翻地 ploughing
翻地的垡片 furrow slice
翻斗车 dump car; skip
翻滚摘挂 rolling off
翻楞机 nigger
翻路机 scarifier
翻埋 covering
翻门(家具零部件) flap
翻木机(制材) log rotater; log turner
翻木器 nigger
翻木上料装置 loading flipper
翻松机(道路) soil pulverizer
翻椅(戏院等用) tip-up seat
翻转 upset
翻转的草皮层 upturned sod
翻转辊 tucking roll
翻转器 flipper; turner
翻转装置 tilter; upender
凡士林 liquid petrolatum
矾鞣 alum tanning; aluminium tanning
矾土 alumina
繁茂度 thrift
繁茂性 thriftiness
繁育系统 breeding system
繁殖 breed; breeding; multiplication; procreation; propagation
繁殖成本 reproductive cost
繁殖成效 breeding success (rate)
繁殖地 breeding area; breeding place
繁殖法 reproduction method
繁殖幅度 breeding range
繁殖洄游 breeding migration
繁殖季节 breeding season
繁殖阶段 reproductive phase
繁殖框架 propagating case
繁殖力 fertility; reproduction power; reproductive capacity

繁殖率 breeding efficiency; breeding rate; propagation coefficient; reproductive rate
繁殖能力 reproduction capacity
繁殖努力 reproductive effort
繁殖器官 propagative organ
繁殖潜力 breeding potential; reproduction potential
繁殖曲线 reproduction curve
繁殖群 breeding herd; deme
繁殖生理学 reproduction physiology
繁殖时期 reproductive phase
繁殖室 propagation house
繁殖体 propagule; propagulum (*plur.* -la)
繁殖温室 propagation house
繁殖系数 propagation coefficient
繁殖型 migrule form; migrule type
繁殖芽 brood bud; reproductive bud
繁殖种群 breeding population
反补贴税 anti-subsidy duty; countervailing duty
反捕食者行为 antipredator response
反茬 setting back
反差 contrast; definition
反铲式挖掘机 backhoe
反铲式挖掘-装载机 backhoe loader
反铲式装载机 backer loader
反常 abnormality
反常波 anomalous wave
反冲(齿尖或刀刃的) kickback; kick
反冲洗 backwash
反刍 ruminate
反刍动物[的] ruminant
反萃取 back-extraction
反弹(树倒时) jump
反弹效应 rebound effect
反方位角 back azimuth
反分化 dedifferentiation
反复冲击试验机 alternating impact machine
反复加压法 alternating pressure method
反复嫁接 cross-grafting; regrafting
反光立体镜 mirror stereoscope
反光速测镜 mirror relascope
反回归系数 inverse regression coefficient
反击式水轮机 reaction wheel
反季风气候 Etesian climate
反接 inversed grafting
反接重复 reverse duplication
反锯口 undercut
反卷的 evolute
反砍口(伐木) Humboldt notch
反馈 feedback
反馈环 feedback loop
反馈控制 feedback control
反馈抑制 feedback inhibition
反立体 inverted stereo
反流 regurgitation
反密度制约 inversely density dependent
反密码 anticode
反密码子 anticodon
反捻(钢丝绳) regular lay
反喷(热磨) blow back
反坡梯田 gradoni
反气旋 anticyclone
反翘 spring; springing
反倾销税 countervailing duty
反乳化 demulsify
反射板 baffle board
反射比 reflectance
反射角 angle of reflection
反射率 albedo; reflectance; reflectivity
反射密度计 reflection densitometer
反射投影转绘仪(反射投影器) reflecting projector
反射效应 albedo-effect
反渗透 reverse osmosis
反式 anti form
反束光导摄像管 return beam vidicon
反推 inversion; retrieval
反胃 regurgitation
反相色谱法 reversed phase chromatography; reversion phase chromatography
反向 transoid
反向铲 backhoe
反向承载索(气球集材) inverted skyline
反向跟踪 back tracking
反向河弯 pothook
反向连锁 backward linkage
反向螺旋 reverse acting spiral
反向平行的 antiparallel

F

反向曲线 S-curve
反向蠕变 reverse creep
反向应变 reverse strain
反向应力 reversed stress
反向转移 antiport
反硝化细菌 denitrifying bacteria
反硝化作用 denitrification
反效等位基因 antimorph
反效用 disutility
反协同剂 antagonist
反协同效应 antagonistic effect
反信风 antitrade (wind)
反絮凝作用 deflocculation
反压电性 electrostriction
反演 inversion; retrieval
反应 reaction; response
反应产物 resultant
反应时间(由见猎物至发射的时间) personal lag
反应速率 reaction rate
反应性 responsivity
反应性防火剂 reactive fire retardant
反应性胶黏剂 reaction sensitive adhesive
反应滞时 reaction time lag
反映性物种 reactive species
反载体 holdback carrier
反照效应 albedo-effect
反周期行动 countercyclical action
反转录酶 reverse transcriptase
反转透明正片 reversal transparency
反足核 antipodal nucleus
反足细胞 antipodal cell
反作用锁 back action lock
返程 kickback
返黄 brightness reversion
返回 return
返回冲程 back stroke
返回行程 back stroke
返回捆木索的短索 bunching hand line
返回迁徙 remigration
返扩散 back diffusion
返祖突变 degressive mutation
返祖现象 atavism
返祖遗传 reversion
泛北极植物区 holarctic floral kingdom
泛顶极 panclimax
泛滥 divagation; flowage
泛滥地 washland
泛滥地森林 forest on land liable to inundation
泛滥盆地 flood basin
泛滥平原 inundated plain
泛滥盐土 flooded solonchak soil
泛配 pangamy
泛频峰 overtone
泛区森林 forest on land liable to inundation
泛热带的 pantropic
泛热带区 pantropical
泛热带植物 pantropical plant
泛色乳剂 panchromatic emulsion
泛生假说 hypothesis of pangenesis
泛生粒 pangen
泛音 overtone
泛域的 azonal
泛域群落 azonal community
泛域群系 azonal formation
泛域土 azonal soil
泛子(生命力最小单位) pangen
范德华力 Van der Waals force
范围 range; territory; zone
方(面积单位, $=100\text{ft}^2$) square
方边凿 firmer chisel
方材 square beam; square log; squared (squaring) log; squared stuff; square-edged timber; square-sawn; cant; square; hewn square
方槽刨 square rebate plane
方差 variance
方差比 variance ratio
方差分量 variance component
方差分析 analysis of variance; variance analysis
方差检验 variance test
方差齐性 homoscedasticity
方法(工作体系或模式) regime (*eg.* thinning regime)
方法(具体的) process; sort; instrument
方沸石 analcime; analcite
方沸岩 analcimite
方格门 square panelled door
方格图案(木材拼花) chequer
方格网 square grid
方格镶板 square framed panel
方角柜 square-corner cabinet

方解石 calcareous spar; calcite
方块地板 flooring block
方框图 block diagram
方棱的 square-edged; square-jointed
方棱木 dice-square log
方木块胶合法 blocking
方式(行动或行为) regime (*eg.* thinning regime); mode
方栓边接 spline edge joint
方头 square head
方位 bearing; direction of slope
方位-高度(天体) azimuth-elevation
方位角 azimuth; azimuthal angle
方位指示器 direction indicator
方向 heading
方形板 quadrate plate
方形刀头(刀座) square cutterhead
方形堆积 square type piling
方形垛 box pile
方形射线细胞 square ray cell
方形细胞 square cell
方形线脚刀头(刀座) square molding head
方形芽接 square budding
方形样地 square plot
方形凿齿(链锯上) chisel tooth; chisel
方形凿齿锯 chisel tooth saw
方形凿齿锯链 chisel saw chain
方原木 square log
方枕(木) squared tie
芳基-甘油-β-芳基醚结构 aryl-glycerol-β-aryl ether structure
芳香 fragrance (material)
芳香烃代谢 aromatic hydrocarbon metabolism
芳香团 enomosphore
芳香物 fragrant substance
芳香植物 aromatic plant
芳油 shiu oil
芳樟油 ho oil; shiu oil
防白蚁垫片 termite shield; termite-proof course
防白蚁刨花板 termite-proof particleboard
防变色处理 anti-stain treatment
防变色化学药剂 anti-stain chemical
防波堤 breakwater; groin; groyne; jetty; pier
防潮 hydrofuge
防潮的 moisture-proof; moisture-resistant
防潮防霉漆 damp-proof fungicidal paint
防潮胶黏剂 moisture proof adhesive
防潮林 tidal prevention forest; tide water control forest; tidebreak; tide-water-prevention forest
防潮漆 damp-proof paint
防潮涂料 damp-proof paint; moisture-proof(ing) coating
防尘板 dust board; dust guard; dust guard plate; dust panel
防尘盖 dust guard
防尘密封 dust seal
防尘面罩 dust mask
防尘罩 dust guard; dust hoop
防尘装置 dust proofing
防冲铺砌 apron
防虫胶合板 borer-proof plywood
防虫纸 insect repellent paper
防虫蛀胶合板 antiborer plywood
防臭木油 verbena oil
防冻 antifreezing
防冻剂 antifreeze
防反弹罩 kickback guard
防风 windbreaking
防风带 shelter-belt
防风根烯 bisabolene
防风固沙林 sand-shifting control forest
防风林 breakwind; wind breaking forest; wind breaks; windbreak forest; windbreak trees
防风林带 shelterbelt; wind belt; windbreak; wind-break
防风林缘 wind-mantle
防风障 breakwind; wind-break
防风竹林 bamboo brake
防腐 antisepsis; preservation
防腐绷带 preservative bandage
防腐材 preservation timber; preserving timber
防腐处理材 treated timber; treated wood
防腐的 antiseptic; aseptic
防腐法 antisepsis; antiseptic process; treating process
防腐工厂 treating plant

防腐合剂 preservative combination
防腐化学药品 preservation chemicals
防腐机理 preservation mechanism
防腐剂 antiseptic; aseptic; preservative; preserving agent
防腐剂持久性 preservative permanence
防腐剂毒性 toxicity of preservative
防腐剂分类 preservative classification
防腐剂含量梯度 gradient preservative
防腐剂挥发性 preservative volatility
防腐剂回出量 kickback of preservative
防腐剂检验法 preservative test method
防腐剂降解 degradation of preservative
防腐剂解毒 preservative detoxification
防腐剂渗透 preservative penetration
防腐剂始效值 toxic sill
防腐剂梯度 preservative gradient
防腐剂停滞 preservative retention
防腐剂选择 preservative selection
防腐剂药筒 preservative cartridge
防腐剂有效性 preservative reliability
防腐剂注入钻孔法 preservative in bored hole
防腐胶合板 preserved plywood
防腐经济 preservation economics
防腐蚀 corrosion prevention
防腐蚀剂 spoilage inhibitor
防腐涂层 ground
防腐涂料 anticorrosive paint
防腐效应 antiseptic effect
防腐药剂 antiseptic chemical
防腐油 dead oil; preservative oil
防腐枕木 treated tie
防腐纸 preservative paper
防腐注射器 springer-presser
防拱系统 arch breaker system
防滚块 chock block
防滚楔 cheese block
防寒处理 winter protection
防洪 flood control
防洪堤 levee
防洪工程 flood control project
防洪林 flood control forest; flood damage prevention forest; protection forest for flood hazard
防洪水库 flood control reservoir
防洪闸门 flood gate
防护板 balk board
防护成熟 protection maturity
防护剂 repellent
防护梁 protective beam
防护林 protection forest; protective forest; shelter-forest; shelter-wood
防护林带 protective belt; shelter-belt
防护林带栽植 belt planting
防护林地 protection forest land
防护林体系 protection forest system
防护林业 protection forestry
防护木(集材道旁) fender skid
防护区 guard plot
防护套 lag
防护衣 protective clothing
防护栅 grill(e) guard
防护罩 hood
防护装置 protecting device
防花粉林 pollen prevention forest
防滑链 antiskid chain
防滑轮爪 strake; wheel girdle; wheel lug
防滑性能 antiskid capability
防滑转前驱动轴 no-spin front axle
防火 fire prevention; fire protection
防火材料 fire proofing material
防火策略 fire control tactics
防火层 fire-proofing
防火处理 fire proofing treatment; fire-proofing
防火带 fire belt; fire break; fire break belt; fire guard; firebreak; fire-trace
防火地区 fire protecting zone
防火队 prevention guard; prevention patrolman
防火方案 fire control plan
防火封禁季节 close(d) fire-season
防火隔离线 fire-lane
防火隔墙 fire stop
防火沟 fire trench; gutter trench; trench

防火构造 fire protecting construction
防火规划 fire control plan
防火季节 closed season
防火剂 fire proofing agent; fire proofing chemical; fire retardant
防火监督人 line boss
防火监护员 fire boss
防火胶合板 fireproof plywood; fire-resistant plywood
防火犁 fire plough; fire plow
防火瞭望塔 fire tower
防火瞭望台 fire watch tower
防火林 combat forest fires; fire break forest; fire break; fire-prevention forest
防火面积 fire area
防火母树(保护幼树) fire insurance (seed) tree
防火木材 fireproofed wood; fireproofing wood
防火气象站 fire-weather station
防火区 fire district
防火人员 fire fighting crew
防火设施 fire control improvement
防火树 fire prevention tree; fire protection tree
防火树带 fire break tree belt; fire-mantle
防火塔 fire tower
防火图表集 fire atlas
防火涂料 fire retardant coating; fire retardant paint
防火网罩 spark arrester-screen
防火位置 anchor point
防火线(林间) control line; fire break; fire line; fire protection break; cut-off
防火线施工设备 fire line construction equipment
防火性 fire resistance
防火油漆 fire retardant paint
防火员 fire guard
防火站 fire control station
防火障 fire barrier; fire break
防火植物带 living fire break; living firebreak; living fuelbreak
防火指挥员 fire warden
防火准备措施 fire presuppression
防火总图 master fire operations map
防火组织 fire control organization
防迹漆(不生碳黑迹) anti-tracking varnish
防溅器 saveall
防酵剂 antiferment
防卡槽板(起重滑车上的) flying dutchman
防浪板(船舶) weather boarding; surf-board
S形防裂扒钉 S-iron
C形防裂钉 C-iron
防裂钩扒钉 anti-checking iron
防裂夹具 anti-checking clamp
防裂器 cleat
防裂铁钉 anti-split iron
防裂铁箍 iron sheet bandage
防裂涂料 anti-checking coating
防裂之形铁(Z形) Z-iron
防霉漆 fungicidal paint
防磨板(削片机等) wear plate; bed plate
防磨装置 shoe
防沫剂 anti foaming agent
防黏剂 releasing agent
防鸟剂 bird repellent
防泡剂 anti foaming agent
防青变浸渍处理 anti-blue stain dip
防沙林 forest for protection against soil denudation; protection forest for shifting sand prevention; sand defence forest; sand protecting plantation; sand protection green; sand break; sandbreak forest
防沙丘 protective dune
防沙设施 sand protection facility
防沙栅栏 fence for heaping sand
防沙障 sand control hedge
防湿 hydrofuge
防湿处理 moisture proofing
防湿的 moisture-proof
防蚀剂 anticorrosive agent; corrosion preventive
防蚀造林 erosion control planting
防霜 frost prevention; frost protection
防水材料 water proofing material
防水剂 sizing agent; water repellency agent; water repeller; water-repellent

F

防水胶 waterproof glue; water-repellent size
防水胶合层 waterproof glueline
防水胶黏剂 waterproof adhesive
防水效率 moisture excluding efficiency
防水性防腐剂 water-repellent preservative
防水堰 coffer dam
防缩处理 shrinkage retarding treatment
防脱环 clamp ring
防弯压辊 back-up roller
防弯装置 back roll assembly
防卫的领域 defended territory
防污漆 antifouling coating; antifouling paint
防污染 pollution abatement
防雾林 fog prevention forest
防雾林带 fog prevention forest zone
防楔(铁路) crest
防斜装置(装卸桥) anti-skew protection
防锈剂 anticorrosive agent; rust preventative; rust preventing agent
防锈漆 antirust paint; rust resisting paint; rust-inhibiting paint
防锈涂料 antirust paint
防锈油漆 anticorrosive paint
防雪崩栅 fence for protection against avalanche
防雪林 protection forest against snow; snowbreak forest
防雪屏障 snowbreak
防雪设备 snow shield
防雪树带 snow catch
防烟林 fume-protection forest; smoke protection forest
防蚁垫片 ant-proof course
防蚁剂 termi-repellent
防蚁黏带 barrier ant tape
防蚁罩 ant-proof course; termite shield; termite-proof course
防油的 oil proof
防御反射 protective reflex
防御行为 defense behaviour
防灾林 disaster prevention forest
防藻漆 antifouling paint
防噪声的 noise-shielded
防震手柄(油锯) devibrated handle bars
防止变色 stain control
防止插垛 snib
防止风沙 curb the sand
防止侵蚀林 protection forest for erosion control
防止沙化 halt desertification
防止水土流失 prevent water loss and soil erosion
防止土壤流失 halt soil erosion slow down
防治 control
防治白蚁 termite-proof
防治密度下限 control threshold
防皱缩处理 collapse treatment
防蛀箱 mothproof chest
房 loculus
房屋建筑 building construction
仿鳄鱼皮纸 alligator imitation paper
仿古加工 antique finishing
仿皮纸 vellum
仿日本局纸 imitation Japanese vellum
仿生学 bionics
仿乌木(木材染色后代替乌木) ebonize
仿形车床 copying lathe
仿形机床 copying machine
仿型刨床 copying planer
仿羊皮纸 imitation parchment
仿真器(程序) emulator
仿制材 imitative wood
仿制纹理 graining
纺锤根(大叶南洋杉实生苗的) fusiform radical
纺锤体 fusiform body; spindle
纺锤形薄壁细胞 fusiform (wood) parenchyma cell; substitute fibre
纺锤形杆背椅 spindle back chair
纺锤形射线 fusiform ray; lenticular ray
纺锤形细胞 fusiform cell
纺锤形形成层原始细胞 fusiform (cambial) initial
纺锤形原始细胞 fusiform (cambial) initial; fusiform initial (cell)
纺锤状 cambiform; spindle-like shape
放出 exhaust

放出的 effluent
放出钒索 pay out
放出钢索 pay off
放大 gain
放大记录仪 amplifying recorder
放大镜 magnifier
放大镜检查 lens examination
放大镜鉴定 hand-lens identification
放大率(望远镜的) magnifying power
放大器 amplifier
放壶架 urn stand
放回(抽样) with replacement
放火 incendiary
放火者 firer
放浆 blowdown; blowing
放浆管 blow pipe
放浆热回收系统 blow down heat recovery system
放浆上网 discharge onto the wire
放卷 unreeling
放卷机 unreeling machine
放料 blowdown; blowing
放料上网 discharge onto the wire
放料闸门 tap
放流堰 horizontal boom
放牧 grazing; pasture
放牧草甸 grazing meadow
放牧草食动物 grazing herbivore
放牧地 grazing ground; pasture; pasture field; pasture land; pasture range; range land
放牧地价 grazing land value
放牧地指示植物 grazing indicator
放牧动物 grazing animal
放牧法 grazing system
放牧分配区 grazing allotment
放牧环境 grazing habit
放牧极限 grazing capacity
放牧季节 grazing season
放牧林 grazed woodlot; grazing forest
放牧偏好 grazing preference
放牧期 grazing period
放牧强度 grazing capacity
放牧权 common of pasture
放牧容量 grazing capacity
放牧食物链 grazing food chain
放牧线 grazing line
放牧作业 grazing system
放能反应 exergonic reaction
放排舵工 steerman
放排工 boom poke; rafter
放排钩杆 rafting rod
放气(热压机等) air drainage; air-bleed; degassing; breathing
放气阀 bleed valve
放热反应 exothermic reaction; thermo-positive reaction
放热面 radiating surface
放射 emission; emittance
放射的 radial
放射光谱 emission spectrum
放射化分析 activation analysis
放射密度测定 radiodensitometry
放射生态学 radiation ecology; radioecology
放射线 radioactive ray
放射型的 actinomorphic
放射性 radioactivity
放射性标记的 radiolabelled
放射性尘埃 radiation dust
放射性沉降物 radioactive fallout
放射性坏死 radionecrosis
放射性磷 radioactive phosphorus
放射性年代测定法 radioactive dating
放射性示踪剂 radioactive tracer
放射性同位素 radioactive isotope; radioisotope
放射性同位素法 radioisotope method
放射性同位素检测器 radioisotope detector
放射性污染 radioactive pollution
放射性污染物 radiation pollutant
放射性指示剂 radioactive indicator
放射遗传学 radiation genetics
放射自显影 radioautograph; radioautography
放射自显影法 autoradiography
放射自显影图 radioautograph
放树杆 timber rake
放索 spool off; unwinding
放线菌 actinomycete
放线菌颉颃体 actinomycete-antagonist
放线菌噬菌体 actinophage
放线菌素 actinomycin
放线菌酮 actidione; cycloheximide

放压 release of pressure
放养 stocking
放养度 stocking; stocking rate
放养-射杀比(野兽) release-kill ratio
放液 tapping
放鹰捕猎 falconry
放油孔 exhaust outlet
放淤 warp
放淤土 warp soil
放在托板上 palletize
飞刀 fly(ing) knife
飞刀辊 beater roll
飞地 enclave
飞点扫描器 flying spot scanner
飞过射击 pass-shooting
飞行高度 flight altitude; flight height; flying height
飞行高度层 flight level
飞行极限(鸟一次飞行最大距离) flight limit
飞行路径 flight path
飞行路线 flight line
飞行式起重机(装于直升飞机上) flying crane
飞行水平面 flight level
飞火 fire rapidly spreading; jump(ing) fire; spot fire; spotting fire
飞火扩散 breakover
飞机播种 aerial planting; aerial seeding; aerial sowing; air drilling; air seeding; air-sowing
飞机防治 aero control
飞机观测 airplane observation
飞机喷粉 air dusting
飞机喷雾 aerial spraying; aero spraying; air-spraying
飞机喷雾机 aircraft sprayer
飞机喷雾器 aerial sprayer
飞机洒射(消毒) aerial spraying
飞机上交货价格 free on plane
飞机施肥 aerial dressing
飞机探测 aircraft sounding
飞机投水弹 water bombing
飞机用材 aeroplane timber; airplane lumber
飞机种子撒播饥 helicopter-mounted broadcaster
飞溅带 spray zone
飞溅润滑 splash-lubrication
飞轮马力(有效功率) flywheel horsepower
飞散极限 flight limit
飞沙 drift sand
飞砂 blow sand
飞艇 aerostat
飞艇集材 blimp logging
飞翔行为 flight behavior
飞翔距离 flight distance
飞檐 cornice
飞燕草色素 delphinidin
飞羽 flight feather
飞云(雨雾雪的) scud
飞枝(树倒时) sailer
非(计算机) negate
非保留林区 unreserved forest land
非必要元素 non-essential element
非闭合循环 open cycle
非标准尺寸 oversize
非标准设备 nonstandard equipment
非标准长度 random length
非病原的 non-pathogenic
非彩色 achromatic colour
非彩色的 achromatic
非层状堆积 unstratified drift
非持久性病毒 nonpersistent virus
非抽样误差 non-sampling error
非导体 non-conductor
非等位基因 non-allelic gene
非地带性的 azonal
非地带性顶极 azonal climax
非地带性植被 azonal vegetation
非地居性白蚁 non-subterranean termite
非典型 off-type
非点源污染 non-point-source pollution
非叠生分生组织 non-stratified meristem
非叠生形成层 unstoried cambium
非对映异构物 diastereoisomer
非多倍体种 nonpolysomatic species
非法采伐 illegal cutting
非法正采伐量 abnormal cutting amount
非法正伐期龄 abnormal final age
非法正林 abnormal forest
非法正作业级 abnormal working section

F

非放牧林地 ungrazed woodlot
非分泌性细胞间隙 non-secretory intercellular space
非附件一国家(缔约方) Non-Annex I Country (Party)
非干性油 non-drying oil
非刚性路 nonrigid road
非高峰负荷期间 off-peak period
非公路运材 off-highway hauling
非公路载重车 off-highway truck
非共生的 non-symbiotic
非共生固氮 non-symbiotic nitrogen fixation
非光敏种子 non-light-sensitive seed
非光照面的 non-luminated
非规定采伐(经营计划外) unclassed felling; unregulated felling
非规定尺寸 off-size
非规则采伐 irregular cutting
非规整立木度 irregular stocking
非过敏性抗性 nonhypersensitive resistance
非禾本杂草 forb
非划线支票 uncrossed cheque
非化石燃料能源 energy derived from non-fossil-fuel source
非环苯甲基芳基醚结构 non-cyclic benzyl aryl ether structure
非挥发物质 non-volatile matter
非回交亲本 non-recurrent parent
非混交林 unmixed stand
非机械物理性质 non-mechanical physical property
非基因变异 non-genic variation
非极性胶黏剂 non-polar adhesive
非极性胶黏体 non-polar adherent
非极性溶剂 non-polar solvent
非季节性苗圃 off-season nursery
非寄生的 non-parasitic
非寄生性病害 non-parasitic disease
非寄主作物 nonhost crop
非加温样地为对照 paired with unwarmed plots
非尖削的 non-tapering
非降解性污染物 non-degradable pollutant
非结晶区 noncrystalline region
非结晶碳 agraphitic carbon
非经济疏伐 uneconomic thinning
非晶部分 amorphous portion
非晶基质 noncrystalline matrix
非晶区 amorphous area; amorphous region
非晶态纤维素 amorphous cellulose
非竞争性抑制 noncompetitive inhibition
非绝热的 diabatic
非均匀抽样 nonuniform sampling
非均质材料 heterogeneous material
非均质土壤 heterogeneous soil
非孔隙度 imporosity
非控速索道 noncontrol cableway
非劳力增值 unearned increment
非利润盈余 paid-in surplus
非连续法 discrete method
非连续分布区 disjunctive area
非连续性生态位 niche particulate
非林地 non-forest area; non-forest land; non-forest soil; non-forestry land; non-productive forest land
非淋洗处理 non-leaching treatment
非隆起纹理染色 non-grain-raising stain
非轮回亲本 non-recurrent parent
非密度制约 density-independent; density-independent controls
非密度制约死亡[率] density independent mortality
非密度制约因子 density-independent factor
非木本植物 non-woody plant
非木材人造板 non-wood based panel
非木质化薄壁组织 unlignified parenchyma
非木质素化合物 nonlignin compound
非目的树 undesirable tree
非目的树种 undesirable species
非耐用品 nondurable goods
非挠性 inflexibility
非内稳态生物 non-homeostatic organism
非偶联呼吸 uncoupled respiration
非配子生殖 agamic reproduction
非平衡理论 non-equilibrium theory
非破坏试验 non-destructive test(ing)
非破损应力 non-destructive stress
非齐次性总体 heterogeneous population

非气候性成土 aclimatic soil formation
非侵染性病害 non-infectious disease; non-infective disease
非侵袭性菌系 non-aggressive strain
非亲缘交配 outcrossing
非染色体的 achromosomal
非染色原生质 achromatoplasm
非染色质 achromatin
非染色质纺锤体 achromatic spindle
非染色质丝 achromatic fiber
非染色质物质 karyoplasm
非染色质象 achromatic figure
非鞣质 non-tan; non-tannin; non-tanning substance
非商品原木 unmerchantable log
非商业性疏伐 non-commercial thinning; thinning-to-waste
非商业用材林地 non-commercial forest land
非伸缩性供应 unresponsive supply
非生产地 non-productive forest land
非生产季节 off-season
非生产林 unproductive stand
非生产林地 unproductive forest land
非生产性财产 unproductive property
非生配体 agamont
非生物成分 obiotic component
非生物成因的 abiogenic
非生物的 abiotic
非生物防治 abiotic control
非生物浮聚物 abioseston
非生物环境 abiotic environment
非生物环境条件 non-living environmental condition
非生物物质 abiotic substances
非生物性浮游物 tripton
非生物悬浮物 abioseston
非生物学的 abiological
非生物因子 abiotic factor
非生物源的 abiogenic
非生物源有机分子 abiogenic organic molecule
非生物组分 abiotic component
非生源论 abiogeny
非石灰性土 non-calcareous soil
非石灰性棕色土 noncalcic brown soil
非石墨碳 agraphitic carbon
非市场价格 non-market value
非市场影响 non-market impact
非适应性特征 non-adaptive character
非数字计算机 non-digital computer
非双折射层 noubirefringent layer
非随机抽样 non-random sampling
非随机分布 non-random distribution
非随机原因 assignable cause
非特化法则 law of non-specialized
非调转趋性 tropotaxis
非同化系统 non-assimilation system
非同源染色体 non-homologous chromosome
非纤维成分 non-fibrous component
非纤维素 non-cellulose
非纤维素多糖 non-cellulosic polysaccharide
非纤维索组分 non-cellulosic constituent
非纤维物质 non-fibrous matter
非纤维型 non-fibrous type
非显晶岩 aphanite
非线性 non-linearity
非相干噪声 incoherent noise; salt-and-pepper effect
非休眠的 non-dormant
非选择性滤色镜 non-selective filter
非循环光合磷酸化 non-cyclic photophosphorylation
非压缩板 non-compressed (fibre) board
非一致性削度方程 noncompatible taper equation
非遗传变异 non-hereditary variation
非遗传性因素 non-heritable agency
非营林费 non-silvicultural cost
非原木要素 non-log factor
非原质体 apoplast
非匀流 non-uniform flow
非杂种繁殖 noninterbreeding
非整倍体 aneuploid; dysploid
非正常采伐利用 extraordinary felling; extraordinary harvest
非正常林 abnormal forest
非正常损失 abnormal spoilage
非正常型 off-type
非正交设计 non-orthogonal design
非政府组织 Non-Governmental Organization
非重叠三联体(遗传密码) non-overlapping triplet

非重叠生态位 non-overlapping niche
非洲马钱子碱 akazgine
非洲檀香油 African sandalwood oil
非洲香脂油 African copaiba balsam oil
非专化反应 nonspecific reaction
菲岛桐蜡 wax of kamala
菲斯特防腐法 Pfister process
绯红园(植物以淡红、粉红、玫瑰色等为主) pink garden
鲱骨法 herring-bone method
鲱骨形割面(采脂) herringbone face
鲱骨状花纹 herring-bone figure
肥大 hypertrophy
肥大根端 bell-shapled butt
肥大生长 diamemter growth
肥力 fertility
肥力梯度 fertility gradient
肥料 fertilizer; manure
肥料保证成分 fertilizer guarantee
肥料比例 fertilizer ratio
肥料颗粒 pellet
肥料配方 fertilizer formula
肥料配合比例 fertilizer ratio
肥料配合式 fertilizer formula
肥料撒施机 fertilizer spreader
肥料树 soil improving tree
肥料注入机 fertilizer injector
肥田植物 soiling plant
肥田作物 cover crop; soiling crop
肥土 fat soil
肥土营养的 eutrophic
肥土植物 eutrophyte
肥沃草甸 fertile meadow
肥沃的 fertile; fruitful; nutrient rich
肥沃黑色石灰土 eurendzina
肥沃牧场 fertile pasture
肥沃生态系统 fertile ecosystem
榧烯 torreyene
榧子油 kaya oil
废板坯回收料箱 reject mat hopper
废材 culled wood; mill waste; offal timber; refuse; residual; splinter; spoil(age); wastage; waste; waste in wood; waste stuff; waste wood; wastewood; wood loss; wood residue
废材处理 slash disposal
废材堆 brush pile; slash pile
废材粉碎机 refuse grinder
废材计量螺旋输送器 wood-waste metering screw
废材加工车间 slash mill
废材利用裁分机 salvage edger
废材利用价值 salvage value
废材燃烧炉 refuse burner; teppee (tipi) burner; waste burner; wigwam burner
废材输送机 refuse conveyer
废材输送线 reclaim line
废的 exhausted
废地放牧 waste range
废梁 crooked log
废料 abatement; outthrows; residual
废料锅炉(烧树皮锯屑) disposal system
废料价值 junk value
废料排出口 trash gate
废料输送机 waste conveyor
废馏分 slop cut
废苗率 cull percentage
废木料 dunnage; wood refuse; wood waste
废皮革 leather waste
废品(种苗等) discard
废品抛出器 rejector
废气 effluent; exhaust; off-gas
废气净化 exhaust purification
废气气流干燥机 waste-gas stream dryer
废气脱硫 exhaust gas desulfurization
废弃道路 dead-line road
废弃农田 abandoned field
废汽 exhaust steam; waste steam
废汽蛇管 waste steam coil
废热回收 heat reclaim
废梢头木 wasted top
废石 spoil
废水 effluent; sewage; slops
废水槽 waste water chest
废水处理 effluent disposal; wastewater treatment
废水箱 waste water chest
废物燃烧 debris burning
废液 black liquor; effluent; waste liquor; slop
废液回收 waste liquor recovery
废油 slops

F

废蒸气 bled steam
沸点 boiling point
沸石 zeolite
沸腾 ebullition
沸腾温度 boiling temperature
沸油干燥[法] drying by boiling in oily liquid
费比恩原木板英尺材积表 Fabian log rule
费尔德群落 veld(t)
费用 charge; commission; due; fee
费用概算 cost estimate
分(角度) minute of angle
分(时间) minute
分板机 board-caul separator
分板器 separator of board and caul
分板楔(防夹锯黏连等) board splitter; outfeed guide; riving knife; spreader wheel
分板装置 board separating device
分贝 decibel
分辨单元 resolution cell
分辨力 resolution; resolving power
分辨率 definition; resolution; resolving power
分辨能力 resolution capacity
分别透性膜 differentially permeable membrane
分蘖 tillering
分布 distribution
χ^2分布 chi-square distribution
t分布 t-distribution
J形分布 J-shaped distribution
分布边缘地区 distribution fringe; fringe area
分布带 zonation
分布格局 distribution pattern
分布刮板带 distributing scraper band
分布刮板链 distributing scraper chain
分布函数 distribution function
分布盘 distributing disk
分布区 range; distribution range; distribution area
分布区不重叠的 allopatric
分布区重叠的 sympatric
分布型 distribution type
分布序列 distribution series
分布学 chorology
分布于世界各地的 cosmopolitan (species)
分布载荷 distributed load
分布障碍 distributional barrier
分布指数 index of dispersion
分布中心 center of distribution
分布周缘 distribution fringe
分步成本法 process costing
分步生产 process production
分层 delamination; layering; stratification
分层矮林作业法 storied coppice method
分层剥离 flake; flatting
分层抽样(分类抽样) stratified sampling
分层堆垛 tier
分层耕作 layer ploughing
分层乔林 storeyed high forest; storied high forest
分层射线 stratified ray
分层生长 growth in storeys
分层随机抽样法 stratified random sampling
分层随机样本 stratified random sample
分层型树冠 stratified crown
分层样木 stratified sample
分层刈割法 stratified clip method; stratified clip technique
分层植物群落 stratified community
分层装车 layer
分层总体 stratified population
分叉 divarication
分叉吊钩 crotch grab
分叉钢索的短索 diamond line
分叉节 branch(ed) knot
分叉木 fork; forked tree
分叉牵引索 crotch chain; crotch line
分叉生长 forked growth
分叉索横滚装车 crotch line loading
分叉原木 pronghorn
分尘机 dust separator
分成五部的 quinquepartite
分等产品出材率 sorted outturn of production
分等规则 grading regulation
分等级 classification
分段多项式 segmented polynomials

分段法 discrete method
分段分布 segmented distribution
分段赶羊流送工 river guardian
分段合成 salvage
分段集材 double yard; tandem yarding
分段牵引机组(长距离集材) battery
分段式出料辊 sectional outfeed roll
分段式断屑器 sectional chip breaker
分段式进料辊 sectional infeed roll
分段式气控托板 sectional air plate base
分段式橡胶进料辊 rubber sectional infeed roll
分段水解 grade hydrolysis
分段武托板 sectional plate
分段武纵向弹性压带器 sectioned longitudinal pad
分段蒸煮 stage cooking
分隔薄壁细胞 septate parenchyma cell
分隔的(植物) chambered
分隔的伐区 segregated coupe
分隔的小伐区 small segregated coupes
分隔木纤维 septate (wood) fibre
分隔髓部 chambered pith
分隔纤维 septate fibre
分隔纤维管胞 septate fibre-tracheid
分隔育种法 breeding by isolation
分根 root division
分根更新 reproduction by root cutting
分根造林 planting by rootcuttings; root planting
分工 division of labour; division of labour
分光光度测定法 spectrophotometric method; spectrophotometry
分光光度计 spectrometer
分光光度计法 spectrophotometriy
分光光度学 spectrophotometry
分光镜 spectroscope
分光仪 spectrometer; spectroscope
分果 schizocarp
分果瓣 mericarp
分果核 pyrene
分红制 profit sharing
分洪 water spreading
分户账 ledger
分化 differentiation
分化等位基因 divergent allele
分化期 idiophase
分化系数 coefficient of differentiation
分化中心 differentiation center
分级 dressing; fractionation; grading; sorting
分级标准木法 class mean sample tree method
分级规范 grading rule
分级机 classifier; sizer
分级台 grading table
分级者 grader
分角法 angle method
分解 decay; decomposing
分解代谢 catabolism; destructive metabolism
分解者 decomposer
分解蒸馏 destructive distillation
分解作用 decomposition
分界点(卷筒上每两层钢索间) crossover
分开 part; partition
分块更新 compartment under regeneration
分类 assorting; assortment; classification; selection; sort; sorting; taxa
分类表 assortment table
分类场 sorting plant
分类处理 diversified disposal
分类单位 taxon (*plur.* taxa)
分类绠 sorting boom
分类工 sorter
分类号印 classification mark
分类卡片索引 card sorting key
分类器 grader; sorter
分类前 pre-stratification
分类群 taxon (*plur.* taxa)
分类台 assorting table; sorting table
分类系统 system of classification
分类学 systematics; taxonomy
分类学的 systematic
分类学者 systematist
分类账 ledger
分离 part; scission; segregation; separation
分离槽 separatory tank

分离的 liber
分离法则 law of segregation
分离花柱 meristyle
分离机 eliminator
分离剂(防止胶料沾黏) parting agent
分离块 explant
分离培养(病原物) isolation
分离培养基 isolation medium
分离筛 shaker bed
分离室 separating chamber
分离税 severance tax
分离液 macerating fluid
分离中央胎座式 free central placentation
分立光谱 discrete spectrum
分粒 classification
分粒器 sizer
分量 component
分料辊 distributing roller
分料器 diverter
分料装置 diverting system
分裂 disrupt; fission; scission
分裂的 lobate; lobed
分裂间期 interphase
分裂选择 disruptive selection
分裂子 oidlum (*plur.* -ia)
分裂组织 meristem
分裂作用 lysis
分流 diffluence
分馏(同位素等) fraction; separation; fractionation
分路 by-pass
分泌胞间隙 secretory intercellular space
分泌素 secretin
分泌物 excreta; secretion
分泌细胞 excreting cell; secretory cell
分泌组织 secretory tissue
分泌作用 exudation; secretion
分娩季节 delivery season
分沫器 catch-all
分蘖 tiller
分蘖性 bushiness
分蘖造林 planting by sucker; planting by tillers; planting by vegetative propagation
分凝 segregation
分配 assortment; contribution
分配拨款 allotment
分配层析 partition chromatography
分配迟延 distributed lag
分配行为 sharing behaviour
分配器 dispatcher; distributor
分配色谱法 partition chromatography
分配数量单位 assigned amount unit
分配系数 distribution coefficient; partition coefficient
分配液 partition liquid
分配原理 principle of allocation
分配柱 partition column
分批采伐 fall in rounds
分批称量秤 batch scale
分批法 batch process
分批混合器 batch mixer
分批培养 batch culture
分批蒸馏器 batch still
分批装运 partial shipment
分期 period; subperiod
分期采伐 partial cutting; periodic cutting
分期采伐面积 periodic cutting-area
分期伐区 periodic coupe
分期法 allotment method
分期付款 installment
分期付款信贷 installment credit
分期还本付息 installment and interest charge
分期区划采伐作业 allotment logging
分区 block; zonation
分区法 zoning law
分区轮伐法 allocated cutting method
分区取样 zoning
分群 swarm
分散度 degree of dispersion
分散过程 dispersive process
分散剂 dispersion agent
分散竞争 diffuse competition
分散体 dispersion
分散性土壤 dispersed soil
分散贮藏(动物) scatter hording
分散资金 proliferation of funds
分散作用 dispersion
分生孢子 conidiospore; pycnidiospore
分生孢子层 conidial layer

分生孢子梗 conidiophore
分生孢子梗束 synnema (*plur.* -ata)
分生孢子盘 acervulus(*plur.* -li)
分生孢子器 pycnidium (*plur.* -dia)
分生孢子座 sporodochium (*plur.* -ia)
分生节孢子 meristem arthrospore
分生区 meristematic zone
分生细胞 meristematic cell
分生组织 generating tissue; generative tissue; meristem; meristematic tissue; nascent tissue
分生组织带 meristematic zone
分生组织培养 meristem culture
分室薄壁细胞 chambered parenchyma
分室薄壁组织 chambered parenchyma
分室干燥窑 apartment kiln; compartment kiln
分室含晶细胞 chambered crystalliferous cell
分水界 topographic(al) divide; water divide; water-parting; watershed; watershed divide
分水蛉 drainage divide
分水岭 divide; topographic(al) divide; water divide; water-parting; watershed; watershed divide
分水闸 side regulator
分丝 devillicate
分摊 contribution
分摊型竞争 scramble type of competition
分析测定 assay determination
分析程序 assay procedure
分析纯度 analytical purity
分析方法 assay procedure
分析机(计算机) analytical engine
分析仪 analyzer; analyser
分线箱 feed(er) box
分形几何 fractal geometry
分形维数 fractal dimension
分选 separation
分选后的原木仓位 log bin
分选机 classifier
分压梯度 partial pressure gradient
分液槽 knockout tank
分液器 knockout
分支 branch
分支公司 affiliate
分支进化 cladogenesis
分枝(从主轴上分出) ramification; ramify; branching; divarication; offshoot; frotrude
T形分枝 T-branch
分枝程度 degree of branching
分枝的 ramose; branching
分枝角度 branch angle
分枝式 ramification
分枝特性 branching habit
分枝纹孔 ramiform pit
分枝纹孔式 ramiform pitting
分枝习性 branching habit
分枝系统 branching system
分枝性 limbiness
分殖 separation
分趾蹄形支脚 pad foot
分株 division; offset; offshoot; plant division
分子孢子 conidium (*plur.* -ia)
分子病 molecular disease
分子结构学 molecular architecture
分子进化 molecular evolution
分子扩散 molecular diffusion
分子筛 molecular sieve
分子生态学 molecular ecology
分子生物学 molecular biology
分子团 micell(e); micella(*plur.* -llae)
分子吸着 molecular sorption
分子遗传学 molecular genetics
分子杂交 molecular hybridization
芬兰抛物线式轮尺 Finnish parabolic caliper
酚 phenol
焚风 foehn (wind)
焚烧采伐剩余物 slash burn
粉斑螟蛾 almond moth; dried fig moth
粉孢子 aleuriospore; oidlum (*plur.* -ia)
粉笔记号 swede level
粉蠹 dry wood beetle; powder-pest beetle; powder-post beetle; powder-post borer
粉蠹危害 powder-post damage
粉蠹危害缺陷 powder-post defect
粉化 efflorescence

F

粉剂 dust
粉蚧 mealybug
粉粒 silt
粉螨 acarid mite
粉霉病 mildew
粉砂 silt
粉砂黏壤土 silty clay loam
粉砂壤土 silty loam
粉虱 white fly; mealy-wing
粉虱形 aleyrodiform
粉碎 size degradation
粉碎机 cracker; disintegrater; disintegrator; hog(ger); hogging machine; pulverizer
粉原植物 pollen plant
粉状活性炭 powdered activated carbon
粉状胶 powdery glue
粉状胶黏剂 powdered adhesive
粉状树脂 powder resin
粉状雪崩 avalanche of loose snow
粉状滞火剂 powdered fire retardant
粪便 faeces; feces
粪肥 manure
粪肥池 manure receiver
粪花粉学(研究排泄物中花粉) copropalynology
粪化石 coprolite
粪尿 night soil
粪生植被 coprophilous vegetation
粪液 liquid manure
丰产年 mast year
丰度 abundance; richness
丰度比 abundance ratio
丰富 abundance
丰花月季 floribunda rose
丰满的 non-tapering
丰满度 fullness
丰年 abundant year
丰收 bumper crop
风暴潮 storm surge
风暴信号 cone
风播 anemochory
风播的 aelophilous
风播植物 anemochore; anemosporae
风车 fanner
风成[作用] eolation
风成地形 aeolian landform
风成堆积 aeolian accumulation
风成盆地 adierstein; aeolian basin
风成土 aeolian soil
风成岩 aeolian rock; aeolianite; eolian rock
风虫媒 anemoentomophily
风吹落果 windfall
风大的 windy
风挡 air damper; damper
风倒 blowdown; windblow; windthrow
风倒木 blowdown; blowdown tree; wind fallen tree; windblow; windfall; windthrow
风倒区 blowdown; windblow; windfall
风道 gallery
风动绞车 air hoist
风洞试验 tunnel test
风干 air seasoning; air-dry; seasoning
风干材 air-dried wood; air-dry wood
风干的 air-dried
风干失重 air-drying loss
风干重 air dry weight
风谷 wind gap; wind gorge
风刮的 windswept
风管 duct
风光 scenery
风害 gale damage; storm damage; wind blast; wind damage
风害废材 wind slash
风害迹地 wind slash
风荷载 wind loading
风化 effloresce; efflorescence; seasoning; weathering
风化变色 weather stain
风化材 weathered wood
风化残积物 weathering residue
风化层 mantle rock; rock mantle; weathered layer
风化带 belt of weathering
风化面 erosion(al) surface
风化试验 weathering test
风化物 weathered material
风化岩石 mantle rock
风积层 eluvium
风积土 aeolian soil
风积物 aeolian deposit; aeolian sediment; eolian deposit
风积作用 eolian deposition

风级 degree of wind
风景 landscape; scenery
风景爱好 landscape preference
风景保护林 scenic beauty protection forest
风景保留区 landscape reservation
风景发展 landscape development
风景分析 landscape analysis; scenic analyses
风景感觉 perception of landscape
风景工程 landscape engineering
风景观点 aesthetic point
风景管理 landscape management
风景规划 landscape planning
风景化城市 landscaped city
风景画 landscape
风景经营 landscape management
风景控制点 landscape control point
风景林 aesthetic forest; amenity forest; scenic beauty forest; scenic forest
风景林业 aesthetic forestry
风景美 scenic beauty
风景评价 appreciation of scenery; landscape judging
风景区 scenic area; scenic spot
风景设计 landscape design
风景设计师 landscape architect
风景式(园林) landscape style
风景树木 landscape tree and shrub
风景素质 scenic quality
风景型 landscape type
风景印象 perception of landscape
风景质量 scenic quality
风口 wind gap
风棱石 ventifact
风力堆积 aeolian accumulation
风力分选 wind sifting
风力分选机 wind sifter
风力记录仪 anemograph
风力灭火机 pneumatic extinguisher
风力削片输送机 chipper-blower
风力移动 blowout
风力自记曲线 anemogram; anemograph
风裂(立木) wind-crack; wind shake
风流 wind current
风玫瑰 wind rose
风媒 anemophily
风媒传粉 anemophilous pollination; wind-pollination
风媒的 aelophilous
风媒花 anemophilous flower; wind flower
风媒授粉 anemogamy; anemophilous pollination
风媒植物 anemophic plant; anemophilae; anemophilous plant; wind-pollinated plant
风门 screen door; throttle
风门杆(油锯) choke lever
风频率 wind frequency
风沙中心 dust bowl
风筛清选器 air-screen cleaner
风扇 fan
风扇控速器 fan brake
风扇式干燥窑 fan type kiln
风扇式鼓风机 fan blower
风蚀 deflation; eolian erosion; erosion by wind; weathering; wind erosion
风蚀变色 weather stain
风蚀材 weathered wood
风蚀程度 degree of wind erosion
风蚀地区 blowout
风蚀沟 wind gully
风蚀坑 blowout
风蚀裂纹 weather shake
风蚀强度 intensity of wind erosion
风蚀仪 weatherometer
风送施肥机 fertilizer blower
风送式喷雾机 air blast sprayer; air-carrier sprayer; air-flow sprayer; blower sprayer
风速 wind velocity
风速计 anemogram; anemograph; anemometer
风速图 anemogram; anemograph
风土的 endemic
风歪(因风倾斜) windlean
风弯(因风弯曲) wind bend
风涡 wind eddy
风险补贴 risk premium
风险负担 risk-taking
风险控制 control risk
风险因素 factor of risk
风险资本 risk capital; venture capital
风箱 bellows; blower

风箱树苷 cephalanthin
风向 wind direction
风向袋 cone
风向的 windward
风向风速仪 aerovane; anemovane
风向频率图 wind rose
风斜木 windlean
风信子 Dutch hyacinth; common hyacinth
风信子浸膏 hyacinth concrete
风选 fanning
风选机 fanner
风压 wind pressure
风压计 anemometer
风眼 air hole; ventilation hole
风摇 windrock(ing)
风雨板 weather board
风载荷 wind load
风障 windscreen
风折 wind-break; wind-breakage
风折木 stub; ram pike
风折枝 windfall
风致林 scenic beauty forest
风致美 scenic beauty
风致素质 scenic quality
风阻力 wind firmness; windthrow resistance
枫香树脂 liquidambar resinoid
枫香烯 liquidene
封闭层 closing layer
封闭的火灾季节 close(d) fire-season
封闭底漆 wash coat
封闭法 containning method
封闭横木(原木分类通廊入口处) gap stick
封闭季节 closed season
封闭剂 sealant
封闭节 blind knot
封闭膜 closing membrane
封闭木节 knotting
封闭腻子 under seal
封闭式钢丝绳 seal pattern wire rope
封闭式钢索 sealed rope
封闭式公司(私营公司) closed corporation
封闭式循环 close cycle
封闭视景[线] closed vista
封闭系统 closed system
封闭效应 blocking effect
封闭性溃疡 closed canker
封闭循环 closed circulation
封闭蒸汽 closed steaming
封边 edge seal(ing); hiding edge
封边薄木封边单板 edge-veneer
封边机 edge banding machine; edge bonding machine
封边木料 banding; lipping; railing
封边芯板 banded core
封顶链 top chain
封顶木(装车) binding log; peaker; sky piece
封端 blocking
封禁的 closed
封禁林 timber reserve
封口套 closed socket
封圃 sealing
封入 lute in
封山育林 closed forest; closing the land for reforestation
封檐板 verge board; weather board
疯长 overgrowth
峰度 kurtosis
峰谷法(表面粗糙度测定) peak to valley method
峰口 peak day
峰值 crest; peak value
峰值发芽率 peak germination percent
葑基 fenchyl
锋后雾 post-frontal fog
锋面低压 wave depression
蜂翅花纹 bee's wing figure
蜂房 comb
蜂蜡 beeswax; cera
蜂媒花 melittophilae
蜂式(幼虫) apoid
蜂团 cluster
蜂王质 queen substance
蜂窝板 cellular board
蜂窝腐朽 honeycomb rot
蜂窝结构 cellular construction; cellular structure; cellular texture
蜂窝结构材料 cellular honeycomb material
蜂窝裂 hollow homing; honeycomb check; honeycombing
蜂窝裂的 hollow-horned

F

蜂窝式真空伏辊 suction couch with cellular perforations
蜂窝状白腐 white pocket rot
蜂窝状采伐带 cell-shaped strip
蜂窝状堆积 honeycomb stacking
蜂窝状褐腐 brown pocket rot
蜂窝状结构 honeycomb structure
蜂窝状心层 honeycomb core
冯曼特公式 Von Mantel's formula
缝 suture
缝拼机 stitching machine
缝隙 slit
缝隙坝 slit dam
缝隙式真空伏辊 suction couch with cellular grating
缝楔 board splitter; spreader wheel
缝植 planting in notch; planting in slit
T形缝植 T-notching
缝植法 planting in notches
缝植锹 notching spade
缝制 making up
凤仙花 touch-me-not; garden balsam
佛蒙特州原木板材积表(美) Vermont log rule
佛青 ultramarine
佛手酚 bergaptol
佛手油 bergamot oil
佛焰苞 spathe
佛焰苞片 proper valves
佛焰花序 spadix
佛焰花序的 spadiceous
夫琅和费谱线 Fraunhofer line
夫琅和费谱线鉴别器 Fraunhofer line discriminator
夫牛 longhorn beetle
呋喃丹虫螨威(杀虫剂) carbofuran
呋喃甲基腺嘌呤(一种细胞分裂素) kinetin
呋喃甲酰 furoyl
呋喃树脂 furan(e) resin
呋喃树脂胶黏剂 furan(e) resin adhesive
呋喃糖 furanose
呋喃亚甲基 furfurylidene
跗节(昆虫) tarsus (*plur.* tarsi)
稃状的 paleace; paleaceous
孵 brood
孵化 eclosion; hatching; incubation
孵化率 hatchability
孵化期 incubation period
孵化所 hatchery
孵化因子 hatching factor
孵卵器 incubator
敷料机 applicator
敷面料 surface dressing
敷设 laying
敷设缏漂 hang stick
敷设线路 laying-out
敷药膏 salve
敷着作用 apposition
弗卢里辛防腐法 Fluorising process
伏地火 creeping fire
伏辊 couch press; couch roll
伏河 sinking creek
伏流 subterranean river
伏条 slip
扶垛 counterfort
扶手 balustrade; elbow; hand rail; rail
扶手(椅子的) arm rail; arm rest; bolster arm
扶手椅 arm chair; Fauteuil
服务地点差价调整数 post adjustment
服务地点等级 post classification
氟化锌二次浸注法 two-movement treatments with zinc chloride
氟乐灵(除草剂) trifluralin
氟橡胶 fluoro gum
氟乙酰胺(杀虫剂) Fluoroacetamide
浮标 buoy; drogue; float gauge; floater; floating mark
浮标站 buoy
浮沉材 bobber
浮点标度计算 floating-scale calculation
浮雕 emboss; embossing; relief; relier
浮雕机 embosser; embossing machine
浮雕胶合板 embossed plywood
浮雕性能 embossability
浮雕装饰 relief ornament
浮吊 derrick barge
浮动测标 floating mark
浮动挡栅 floating-boom
浮动轮胎(不压入地面的宽轮胎) floatation tire
浮动状态 float(ing) position

浮杆(承压构件) floater
浮根 floating root
浮绠 floating-boom
浮力 buoyancy; buoyant force
浮力原理 buoyancy theory
浮膜 pellicle; scum
浮木钢索河绠 cable boom with floats
浮木装夹法 floating log set
浮桥 pontoon; pontoon bridge
浮石 pumice
浮石凝灰岩 trass
浮式超重机 floating crane
浮式传送机 floating conveyer
浮式输送机 floating conveyer; floating transporter
浮水材 floater
浮水横档材 lateral floating boom
浮水树种 floater
浮水性试验 floatation test
浮桶式疏水器 bucket steam trap; open bucket type steam trap
浮桶式阻汽排水器 open bucket type steam trap
浮筒水尺 float gauge
浮凸 emboss
浮土 regolith
浮性 buoyancy
浮性卵 pelagic egg
浮选 flotation
浮选促进剂 collector
浮选试验 floatation test; flotation test
浮叶水生植物 floating-leaved aquatic plant
浮叶植被 floating-leaved vegetation
浮叶植物 floating leaved plant; floating-leaf water plant
浮油沥青 tall oil pitch
浮油衍生物 tall oil derivative
浮游的 planktonic
浮游动物 zooplankter; zooplankton
浮游生物 plankon; plankter
浮游生物带 emersion zone
浮游生物连续取样器 continuous plankton recorder
浮游生物网 plankton net
浮游微生物 plankton
浮游幼体 larval plankton
浮游植物 phyplankton
浮游植物极盛繁殖 phytoplankton bloom
浮游植物群落 phytoplankton
浮渣 scum; skimming
浮栅(流送材的) floating rake
浮质 skimming
浮子验潮仪 float gauge
符号代数语言 symbolic algebra
符号识别读出器 mark-sense character reader
符合 coincidence
符合系数 coefficient of coincidence
匐枝植物 runner plant
幅度 range
辐合 convergence
辐刨(双柄刨) spokeshave
辐射 emission; emittance; radiate; radiation
γ-辐射 gamma radiation
X光辐射 X-radiation
辐射不透明的 radiopaque
辐射测量学 actinometry
辐射场 radiation field
辐射传热 heat transfer by radiation
辐射对称 radial symmetry
辐射对称的 actinomorphic; radiosymmetric
辐射功率 radiant power
辐射管孔材 wood with radial band
辐射计 radiometer
辐射剂量 radiation dose
辐射距离 radial distance
辐射绝育 radiation sterilization
辐射抗性 radioresistance
辐射孔材 radial porous wood
辐射亮度(光源发光量) radiance; radiancy
辐射裂 radial check; radial shake
辐射率 radiance
辐射脉 radiate veined; radiate veins
辐射脉的 diverginervius
辐射灭菌 radiation sterilization
辐射敏感度 radiosensitivity
辐射能 radiant energy; radiation
辐射平衡 radiative balance
辐射强度 intensity of radiation; radiant intensity; radiation intensity
辐射强迫 radiative forcing

辐射强迫情景 radiative forcing scenario
辐射热测定器 balometer
辐射三角测量 radial triangulation
辐射-散射仪 radiometer-scattermeter
辐射霜 radiation frost
辐射体 radiator
辐射透明的 radioparent
辐射突变型 radiomutant
辐射维管束 radial bundle; radial vascular bundle
辐射线 radial line; radiation
辐射线固化 radiation curing
辐射线交会图 radial line plot
辐射线转绘成图 radial line plotting
辐射线转绘仪 radial line plotter
辐射选种 radioselection
辐射诱导畸变 radiation-induced aberration
辐射育种 radiation breeding
辐射源 radiation source
辐射照度(受光量) irradiance; irradiancy
辐射状的 radial; radiate
辐射状排列 radial arrangement; radiation
辐条 spoke
辐条木块 spoke stock
辐向的 radial
辐照 irradiation
福尔根氏反应 Feulgen reaction
福利经济学 welfare economics
福利商品 merit goods
福美联(杀菌剂) thiuram
福美砷(杀菌剂) asomate
福美双(杀菌剂) arasan
福美锌(杀菌剂) zerlate
福木色素 fucugetine
福斯特曼测高器 Faustmann's hypsometer
抚育 care; nursing; stand tending; tending
抚育采伐 tending cutting (felling); tending thinning
抚育措施 cultural practice
抚育伐 disengagement cutting; improvement cutting; improvement felling; non-commercial thinning; released thinning
抚育伐进度表 improvement cutting schedule
抚育费 tending cost
抚育复壮 reproduction by tending treatment
抚育改良 improvement
抚育更新 reproduction by tending treatment
抚育计划 tending plan
抚育间伐 intermediate cutting
抚育间隔期 thinning interval
抚育目标 tending objective
抚育期 tending period
抚育疏伐 thinning for tending
抚育幼苗 lay out young trees
抚育择伐 improvement-selection cutting
抚育制度 thinning regime
抚育作业 cultural treatment; cultural work; tending operation
抚育作业周期 cycle of tending operations
斧背 head
斧柄 axe handle; helve
斧伐法 felling with axe
斧口 cutting leak; cutting throat; felling face; notch
斧面 face; side
斧刃 bit
斧刃断面 bit profile
斧刃规 axe gauge
斧身 blade
斧式扣块枪管 chopper lump tubes
斧式犁 chopper plough
斧头镢 axe mattock
俯伏的 ventricumbent
俯角 angle of depression; depression angle
俯卧的 pronate
辅椽 auxiliary rafter
辅加能量 energy subsidy
辅件 auxiliary
辅酵素 co-ferment
辅锯箱 mitre box
辅酶 coenzyme; cozymase
辅酶A coenzyme A
辅因子 cofactor
辅增塑剂 co-plasticizer; plasticizer extender
辅助绷索 gill guy

F

辅助槽 auxiliary tank
辅助畜力(驻于陡坡段) snatch team
辅助的 aceessary; accessory; adjuvant
辅助电动机 pilot motor
辅助伐(主伐后) assistance cutting; subsidiary felling
辅助罐 auxiliary tank
辅助光照 supplementary illumination
辅助河绠 haywire boom
辅助记录 subsidiary record
辅助绞盘机 auxiliary winch; swing donkey
辅助卷筒 outboard drum; swing drum; utility drum
辅助能 auxiliary energy
辅助扭矩支承辊 auxiliary torque backup roll
辅助跑车 auxiliary carriage
辅助漂子 stay boom
辅助去节器 back knotter
辅助溶剂 auxiliary solvent
辅助色素 accessory pigment
辅助设备 accessory equipment; aceessary; accessory; additional equipment; ancillary equipment; auxiliary; auxiliary equipment
辅助生殖 supplementary reproduction
辅助授粉 accessory pollination; complementary pollination; subsidiary pollination; supplementary pollination
辅助树(装设集材杆的) dummy tree
辅助树种 accessory species
辅助索(集材索道) hay wire; line of attachment; tag line; grass line
辅助投资 ancillary investment
辅助效益 ancillary benefits
辅助因素 cofactor
辅助装置 accessory equipment; booster
辅助资源 ancillary resource
辅助作业 assessory system; jackpot
辅助作业法(特指二层乔林) accessory system
辅佐树种 accessory species
腐败 decomposition; petrefaction; putrefaction; pythogenesis
腐草土 grass mold
腐化 pythogenesis
腐坏 taint
腐坏种粒 decayed seed
腐烂 decomposing; perish; rottenness
腐烂的枯枝落叶层 vegetable refuse; vegetable remains
腐泥 gyttja; sapropel
腐泥煤 sapropelite
腐泥土 gyttja soil
腐泥岩 sapropelite; sapropelith
腐肉 offal
腐生 pythogenesis; saprophytism
腐生的 saprobic; saprophilous; saprophytic
腐生方式 saprophytic form
腐生菌 saprogen; saprophytic fungus
腐生链 saprophytic chain
腐生生物 saprotroph(s)
腐生生物群落 saprophyte-community
腐生食物链 detrital food chain
腐生物 saprobiont; saprophage
腐生营养 saprophytic nutrition
腐生植物 saprophyte
腐食性生物 saprotrophs
腐蚀 corrosion; erosion
腐蚀剂 corrosivene; etchant
腐蚀试验 corrosion testing
腐蚀抑制剂 corrosion inhibitor
腐蚀作用 corrosive effect
腐熟 decomposing
腐熟腐殖质 mull
腐熟厩肥 decomposed dung; decomposed manure
腐心 punk heart
腐心木 pumped tree
腐朽 decay; doat; dote; doty; mouldering; punk; rottenness
腐朽必需条件 decay requirement
腐朽材 punky wood; rotten timber; rotten wood
腐朽程度分类 decay hazard classification
腐朽传播 spread of decay

腐朽节(腐朽部分超过$\frac{1}{3}$) decayed knot
腐朽进级期 advanced stage of decay
腐朽菌 decay fungus
腐朽控制 decay control
腐朽类型 decay type
腐朽蔓延 spread of decay
腐朽木 decayed timber; doty wood; punk wood; doted tree
腐朽囊 decay pocket
腐朽器 rot
腐朽试验 decay test
腐朽条纹 rotten streak
腐朽危害 decay hazard
腐殖化有机质 humified organic matter
腐殖化作用 humification
腐殖积土 cumulose soil
腐殖煤 humulith
腐殖水铝英石土 humic allophane soil
腐殖酸 humic acid
腐殖酸盐 humate
腐殖土 humus soil; muck
腐殖物质 humic substance
腐殖岩 humulith
腐殖质 humus
α-腐殖质 alpha humus
腐殖质层(土壤剖面A层) humus horizon; humus-layer; mold-cover; A-layer
腐殖质层湿度码 duff moisture code
腐殖质淀积层(土壤剖面B_n层) sombric horizon; B_n-layer
腐殖质钙质假潜育土 humus-calcic pseudogleyed soil
腐殖质硅铝土 humic siallite
腐殖质灰壤 humus podzol
腐殖质灰土 humod
腐殖质碱 humic alkali soil
腐殖质老成土 humult
腐殖质潜育土 humic gley soil
腐殖质碳酸盐土 humic carbonated soil; humocarbonate soil
腐殖质铁盘 humic iron pan
腐殖质土 leaf mould
腐殖质氧化土 humox
腐殖质营养学说 humus theory
腐殖质植物 sathrophyta
腐殖质植物群落 sathrophytia
腐殖质砖红壤 humic latosol
腐质 detritus
腐质食物链 detritus food chain
父本 pollen parent; pollen supplier
父本供粉系 P-line
父本类型 paternal form
父系继嗣 patriliny
父性遗传 paternal inheritance
付款 disbursement
付款交单 document against payment
付款人 payer
付讫支票 cancelled check
付清 pay off
付运费的货物 payload
负二项分布 negative binominal distribution
负反馈 negative feedback
负号林分 minus stand
负号树 minus tree
负荷 load
负荷高峰 peak load
负荷量 carring capacity
负横向地性 negative diageotropism
负积温 negative cumulated temperature
负密度制约因子 negative density-dependent factor
负屈曲 negative curvature
负染色法(超薄切片技术) negative staining
负税计划 negative-tax plan
负投资 disinvestment; negative investment
负弯曲 negative curvature
负吸附 negative adsorption
负吸收 negative adsorption
负相关 negative correlation
负向地性 negative geotropism
负向光性 negative phototropism
负向性 negative tropism
负斜向地性 negative plagiogeotropism
负杂种优势 negative heterosis
负载 burden
负载偏移量 loaded deflection
负载系数 load factor
负债 liability
负债净值比率 debt-to-net-worth ratio

负直向地性 negative orthogeotropism
附壁植物 walling plant
附表 accompanying table
附带成本 incidental cost
附单 coupon
附果 accessory fruit
附果萼 systellophytum
附加费 premium
附加峰 additional peak
附加股息 extra dividend
附加林带 supplementary windbreak
附加税 surtax
附加体 episome
附加填方 extra banking
附加系 addition line
附加险 accessory risk
附加演替系列 adsere
附加应力 extra-stress
附加账户 adjunct account
附件 aceessary; accessory; fixings
附件二国家 annex II country
附件一国家 annex I country
附聚[物] agglomerate
附容器栽植法(用塑料管等) bullet planting
附生的 adnascent
附生动物 epizoan
附生兰 epiphytic orchid
附生生物 epibiont; epiorganism
附生植物 epiphyte
附生植物的 epiphytic
附生植物群落 epiphyton
附生植物商数 epiphyte-quotient
附饰物 applique
附属成本 associated costs; incidental cost
附属担保品 collateral; collateral security
附属的 aceessary; accessory; collateral
附属机构 auxiliary body
附属林分 secondary crop
附属丝 appendage
附属作业 auxiliary activities (works)
附条件的销售 conditional sale
附物纹孔 cribriform pit; vestured pit
附息票据 interest-bearing note
附有条件的合同 tying contract
附着胞 appressorium
附着的 scandent
附着根 adhesive root
附着力 adhesion; cohesion
附着力测定仪(漆膜) adherometer; Adohero gauge
附着器 appressorium
附着植物 epiphyte
附肢(昆虫) appendage
附著应力 bond stress
赴火场时间 speed of attack
复[年]轮 double (annual) ring
复比例尺模片 multiple scale templet
复变态 hypermetamorphosis
复测法 repeating method
复测经纬仪 repeating theodolite
复层混交林 stratified mixed stand
复层林 compound storied forest; irregular forest; irregular plantation; many storied forest; multi-storied forest; multi-storied stand
复层乔林 multi-storeyed high forest
复尺 check scaling
复尺员 check scaler
复穿孔 multiple perforation
复穿孔机 reproducer
复雌蕊 compound pistil
复等位基因 multiple allele; multiple allelomorph
复等位基因系列 multiple allelic series
复顶枝 multiple leaders
复二倍体 allotetraploid; amphidiploid
复二歧聚伞花序 compound dichasium
复二歧式 compound dichasium
复发燃烧 after flaming
复分解 metathesis
复浮子 double float
复耕(耙地、碎土、松土、平整、镇压等) over tillage; secondary ploughing; secondary tillage
复共振(共鸣) multiple resonance
复管孔 multiple pore; pore multiple
复果 multiple fruit
复合板 composite board; composition board; composite

panel; particle board core plywood; plyboard; sandwich composites
复合板材 board
复合胞间层 compound middle lamella
复合背衬(砂带) combination backing
复合材积表 compound tariff
复合材料 composite
复合成材 composite lumber
复合断层崖 composite fault scarp
复合肥料 compound fertilizer; mixed fertilizer
复合钢材 bimetal
复合胶合板 composite plywood
复合结晶 co-crystallization
复合梁 composite beam
复合芒状髓 compound (medullary) ray
复合磨料 abrasive compound
复合木 composite wood
复合刨花板 composite particle board
复合频率分布 compound frequency distribution
复合碎料板 composite particle board
复合体 complex; compound body
复合纤维素 cellulosan; conjugated cellulose
复合因素 composite factor
复合应力 compound stress
复合优势比 multiplied dominance ratio
复合油 compound oil
复合杂交 composite crossing
复合增长年率 compound annual rate of growth
复合症状 syndrome
复合子房 compound ovary
复合组织 complex tissue; compound tissue
复花序 compound inflorescence
复滑轮 tackle; tackle block
复活作用 revivification
复寄生 superparasitization
复寄生现象 superparasitism
复检 check scaling
复交杂种 multiple hybrid
复节 knot cluster
复聚伞花序 compound cyme
复卷机 rewinder
复利 compound interest
复利增长率 compound growth rate
复裂 compound shake
复年轮 multiple (annual) ring; multiple (growth) ring
复年轮重叠 double of multiple annual ring
复配法 method of double crossing
复燃 reburn
复染色体 multiple chromosome
复伞房花序 compound corymb
复伞形花序 tompound umbel; universal umbel
复筛板 compound sieve-plate
复梢 double heart-wood; double summit
复舌接 complex whip grafting
复射线 compound (medullary) ray
复生植物群落 saprophyte-community
复式簿记 double-entry bookkeeping
复式齿 combination teeth
复式齿锯(刨齿间隔锯齿) peg-raker saw
复式齿轮绞盘机 compound geared donkey
复式带锯机 multiple bandsaw; twin band-saw
复式带锯制材厂 twin band (saw) mill
复式滑轮 block tackle; heel block; purchase block; tightening block
复式火山 composite volcano
复式记载体系 multiple entry system
复式接头 manifold
复式结构 duplex
复式截锯 double cut-off saw
复式收获表 multiple-yield table
复式压缩 two-stage compression
复式杂交 multiple crossing
复式张紧滑轮 heel block; multisheave heel tackle
复苏 anabiosis
复穗状花序 compound spike; spiculate spike
复头状花序 compound head; many headed
复网状生长假说 multinet growth hypothesis
复网状学说 multinet theory

复显微镜 compound microscope
复相对因子作用 multiple allelism
复相关 multiple correlation
复向斜 synclinorium
复消色差透镜 apochromatic lens
复斜接 compound miter joint
复斜纹理 double cross grain
复写原纸 carbon tissue
复写纸 duplicating paper
复心材 double summit
复新者 revitalizer
复型 replica
复型法 replica method
复型技术 replication technique
复性染色体 multiple sex chromosome
复循环病害 multipie-cycle disease
复芽 multiple bud
复眼 compound eye
复叶 compound leaf
复叶的一片 leaflet
复印纸 manifold paper
复杂性理论 complexity theory
复杂杂交 composite crossing
复制 duplication; reproduction
复制差错 copy error
复制机 reproducer
复制类型 replicative form
复制酶 duplicase; replicase
复制品 replica
复壮 rejuvenation
复壮现象 rejuvenescence
复总状花序 panicle
副薄壁组织 paraplectenchyma
副本 doublet
副产品 accessory product; by-product; coproduct; minor produce; minor product; residual products
副产物 by-product; off products
副产物使用权 servitude to by-product
副椽 auxiliary rafter
副次抽样 sub-sampling
副次样地 sub-plot; cell
副道(苗圃) secondary road
副低压 secondary low
副底板 sub-back
副淀粉 paramylum
副萼 accessory calyx; calycule; epicalyx
副伐 subsidiary felling
副防风林带 supplementary windbreak
副果 accessory fruit
副花瓣 parapetalum
副花冠 corona; crown; paracorolla
副极地 subpolar
副极地气候 subarctic climate
副集材杆 back spar; tail spar; tail tree
副集材杆索具 back rigging
副价 secondary valence
副茎叶体的 epicormic
副林班线 accessory frame; minor ride
副林带 auxiliary forest belt
副林道 auxiliary road; branch forest road; branch road; by-pass road; by-road; lateral road; subsidiary road
副林木 dominated crop
副模标本 paratype
副木 secondary species
副片岩 paraschist
副起重臂 outer boom
副气旋 secondary cyclone
副染色体 accessory chromosome
副梢 lateral shoot; side shoot
副生物圈 parabiosphere
副树种 nurse species; subordinates pecies
副髓磷脂 paramyelin
副卫细胞 accessory cell
副腺 accessory gland
副小枝 secondary branchlet
副形成层 accessory cambium
副型 paratype
副芽 accessory bud; secondary bud
副枝 accessory branch; ramulus (*plur.* ramuli)
副作用 side effect
赋税亏损结转 tax-loss carryforward
赋税亏损退算 tax-loss carryback
赋税优惠期 tax holidays
傅立叶分析 Fourier analysis
傅立叶转换 fourier transform

富硅高岭土 anauxite
富积层 enrichment horizon
富集 enrichment
富集层 enrichment horizon
富集培养 enrichment culture
富集系数 concentration factor
富集指数 index of concentration
富铝风化作用 allitic weathering
富铝土 allite
富养植物 eutrophyte
富营养 eutrophy
富营养的 eutrophic
富营养化 eutrophication
富于生殖力的 breedy
腹板 sternum (*plur.* sterna); web
腹柄(昆虫) petiole
腹部 abdomen
腹担子 endobasidium (*plur.* -ia)
腹缝线 ventral suture
腹杆 web member
腹环接 side rind graft
腹接 side grafting
腹菌 gasteromycetes
腹面的 ventral
腹片 lobule
腹神经节 abdominal ganglion
腹足 abdominal leg; false leg; proleg; uropod
缚缢 wiring
覆翅(前翅) tegmen (*plur.* tegmina)
覆盖 covering; mulching; pave; roofing
覆盖部件 covering element
覆盖层 blanket; mulch; pave
覆盖度 coverage
覆盖工程 covering works
覆盖面积 coverage
覆盖器 coverer
覆盖物 mulch
覆盖着 be covered with
覆盖植物 cover plant
覆盖作物 cover crop
覆面 overlay
覆土 covering; earthing
覆土机 sand spreader
覆土耙 covering harrow
覆土器 coverer
覆土深度 covering depth
覆瓦状的 imbricate
覆网案辊 wire-covered table roll
伽玛分布 gamma distribution
伽玛辐射 gamma radiation
伽玛射线 gamma-ray
改变 modification
改变位置 translocation
改道 road changing
改进伐 improvement cutting; improvement felling
改进集团育种法 modified mass method
改进谱系育种法 modified pedigree system
改进疏伐 improvement-thinning; preliminary thinning
改锯(制材) resawing
改锯裁分锯 reman edger
改良木 improved wood
改良皮下接 modified rind graft
改良品种 improved variety
改良期 regulation period
改良土壤树种 soil improving tree species
改良式喉缩 modified choke
改良式落差防腐法 prescap process
改良栽植(林分改造) improvement planting
改良种子园 improvement seed orchard
改排 raft remaking; rerafting
改算面积 equalized area; equi-productive area
改性 modification
改性木 improved wood
改性木材 modified wood
改性松香 modified rosin
改性纤维素 modified cellulose
改制锯 reman saw
钙层土 pedocal
钙成土 calomorphic soil
钙化松香 limed rosin
钙化作用 calcification
钙积层(土壤剖面BCa层) calcic horizon; BCa-layer; caliche
钙镁磷肥 fused calcium-magnesium phosphate; serpentine-fused phosphate
钙黏土 calcium clay
钙生植物 calcicole; calcicolous

钙铁榴石 andosol
钙土恒有群丛 lime-constant association
钙土植物 calcicole; calcicole plant; calciphyte; lime plant
钙藻类 calcareous algae
钙长石 anorthite
钙质 calcareous; calcic
钙质泥岩 calcilutite
钙质砂岩 calcareous sandstone
钙质铁结核(沼泽土中) blansand
钙质土 calcium soil
盖 housing; opercule
盖板(防护用) sheathing
盖板(家具零部件) deck
盖度 percent cover; cover degree; cover(age); fractional cover
盖度级 cover class
盖果 pyxidium; pyxis
盖帽(枪柄上) cap
盖片 lap
盖屋顶 roofing
概测报告 reconnaissance report
概估 raw estimate; rough estimate
概率 probability; stochastics
概率抽样 probability sampling
概率单位 probit
概率密度 probability-density
概率密度函数 probability density function
概率与单元大小成比例抽样 sampling with probability proportional to size; sampling proportional to size; proportional to size sampling
概率值 probit
概貌图 physiographic map
概念模型 conceptual model
概算 budget estimate; budget proposal; rough estimating
甘草 glycyrrhiza
甘菊蓝 azulene
甘露 manna
甘松油 spikenard oil
甘酞树脂 glyptal resin
甘油冻胶 glycerin jelly
甘油明胶 glycerin gelatine
甘油树胶 glycerin gum
苷化 glycosidation
柑 king orange
柑果 hesperidium
柑橘粉虱 citrus white fly
柑橘业 citriculture
杆(长度单位, =5.029m) pole
杆材 low pole; pole crop; pole stage; pole stock; round timber; small pole
杆材林 crop of small poles; pole-wood; small pole forest
杆材木 pole size timber; pole
杆材劈裂 split stuff
杆材期森林 pole stage forest
杆材枕木 pole tie
杆顶绷索(集材) top guy
杆根横木(电杆) underbrace
杆滑车 leverblock
杆滑道 pole chute
杆柱材 round wood
干板坯 dry mat
干包环割法 dry-packing (method)
干标本 exsiccata
干材计数 dry tally
干材制材厂(无贮木池) dry (saw) mill
干藏 dry storage
干草原 steppe
干草原灌丛 steppe scrub
干草原和半荒漠 steppe and semidesertу
干草原森林 steppe forest
干冲积地 dry wash
干冲积物 dry wash
干电池 unit cell
干端 discharge end; dry end
干法(纤维板工艺) dry-process; dry method
干法气流铺装成型机 dry fiberfelter
干防腐盐净吸收量 net dry salt retention
干风 desiccating wind
干腐 dry rot
干割面(无脂割面) dry face
干谷 dry valley; oued; arroyo
干果 dry fruit
干果斑螟 almond moth; dried fig moth
干旱 drought; drouth
干旱程度 degree of drought
干旱带 arid zone

干旱的 arid
干旱地区 arid area
干旱锻炼 drought hardening
干旱防护林 protection forest for drought control
干旱高原(南非) karoo
干旱季节 drought; drouth
干旱码 dry code
干旱气候 dry climate
干旱区 arid region
干旱区小河 arroyo
干旱土 xerosol
干旱土壤 arid soil
干旱指标 drought index
干旱指数 aridity index
干旱棕色森林土 brown dry forest soil
干河床 blind creek; dry wash
干涸湿地 dried-out peatland
干黑钙土 dry chernozem
干红辣椒 paprika
干花(干燥后) dried flower
干滑道 dry chute; dry flume; dry slide
干环压强度 dry ring crush
干荒漠 dry desert
干荒漠群落 siccideserta
干荒盆地 playa
干荒植物 chersophyte
干灰化土 dry podzolic soil
干季 dry season
干季烧除 late burning
干剂围树处理法 dry-packing (method)
干酱型乳化防腐剂 bodied mayonnaise-type emulsion
干胶合 dry gluing
干胶膜 glue film
干胶片 wafer
干接合 dried joint
干节 dry knot
干枯面(采脂) dead face
干酪素 casein
干冷锋 dry cold front
干料 siccative
干料仓 dry bin
干裂 crack; drought crack; drying split; seasoning check; seasoning crack; split; weather shake; air-crack
干馏 destructive distillation; dry distillation; pyrogenic distillation
干馏材 acid wood
干馏釜 retort
干馏炉 oven for dry distillation
干馏物 pyrolyzate
干毛布 dry felt
干敏种子 desiccation-sensitive seed
干母(蚜虫) fundatrix
干木白蚁 dry-wood termite
干木材贮存 dry wood storage
干木片 dry chip
干木片料仓 dry chip bin
干盘离合器 dry type plate clutch
干刨花 dry chip
干刨花料仓 dry chip bin
干喷漆间 dry spray booth
干砌石坝 loose stone dam; dry-laid masonry dam
干砌石谷坊 mortarless stone check dam
干球 dry bulb
干球温度 dry-bulb temperature
干球温度计 dry-bulb thermometer
干扰 interference
干扰发射机 jammer
干扰竞争 contest competition; interference competition
干扰滤波器 interference filter
干扰模式 disturbance regime
干扰素 interferon
干扰因子 disturbance agent
干扰状况 disturbance regime
干热 dry heat
干热淋溶土 xeralf
干热灭菌 hot air sterilization
干热灭菌器 sterilizing oven
干肉粉(肥料) ammonite
干梢 fore-stock; upper log; upper stem
干涉 interference
干涉滤光片 interference filter
干湿球湿度表 wet and dry bulb hygrometer
干湿球湿度计 dry-and-wet bulb hygrometer; psychrometer
干湿球温度差 wet bulb depression
干湿球温度计 catathermometer; wet and dry-bulb thermometer

干湿球自记湿度表 wet and dry-bulb recording hygrometer
干式滚筒剥皮机 dry-type rotary drum barker
干式宽带砂光机 dry type wide belt sander
干式转鼓剥皮机 dry-type rotary drum barker
干缩 air shrinkage; dry shrinkage; drying shrinkage; shrinkage; shrinkage on drying
干缩比 shrinkage ratio
干缩量 shrinkage mass
干缩裂缝 desiccation fissure; shrinkage crack
干缩率 percentage shrinkage; shrinkage in percent
干缩系数 coefficient of shrinkage
干缩应变 shrinkage strain
干缩应力 shrinkage stress
干态胶合强度试验 dry gluing strength test
干透的 bone-dry
干物质含量 dry matter content
干物质生产[量] dry matter production
干性-抽空处理槽 dri-vac tank
干性落叶林 dry deciduous forest
干性油 drying oil
干血粉 dried blood
干椰子仁 desiccated coconut
干燥 desiccation; seasoning
干燥不均 uneven drying
干燥部 dry (*or* drier) part
干燥材 seasoned wood; seasoned timber; dry wood
干燥材抗拉强度 dry tensile strength
干燥程序自动控制记录器 kilnboy
干燥的 siccative
干燥底子 siccative base
干燥动力学 dynamics of drying
干燥度 aridity; dryness; dryness index
干燥工 drier; dryer
干燥过程 dry run
干燥环裂 seasoning shake
干燥机 drier; dryer
干燥机出料端 dryer (*or* drier) end
干燥机出料装置 dryer (*or* drier) unloader
干燥机废气 dryer emission
干燥机理 drying mechanism
干燥基准 drying schedule; seasoning schedule
干燥基准模板 cam
干燥季焚烧采剩物 one match burn
干燥剂 desiccant
干燥架 drying bed
干燥间 drying room
干燥降等 seasoning degrade
干燥胶黏性 tack dry
干燥阶段 drying period
干燥介质 drying medium
干燥力 drying power
干燥炉 dry kiln; kiln
干燥轮裂 seasoning shake
干燥率 index of aridity
干燥黏着 dry tack
干燥棚 drying shed
干燥劈裂 seasoning split
干燥器 dehumidifier; desiccator; drier; dryer; drying oven
干燥强度 dry strength
干燥曲线 dry(ing) curve
干燥缺陷 drying defect; seasoning deflect
干燥热风 chinook
干燥杀菌 dry sterilization
干燥时间 drying time
干燥时间自动记录仪 drying time automatic recorder
干燥时期 drying period
干燥势 drying power
干燥室 dry kiln; kiln
干燥室褐变 kiln brown stam
干燥疏林 miombo
干燥速度 drying rate; drying velocity; rate of drying
干燥速率 drying rate
干燥条件 drying condition
干燥温度 drying temperature
干燥箱 loft dryer; oven
干燥阳坡 dry sunny slope
干燥窑 dry kiln; kiln; seasoning kiln
干燥窑车轨道 kiln track
干燥窑装材过程 kiln charge run
干燥窑装载量 kiln charge

干燥因子 drying element
干燥应力 drying stress
干燥应力变定 drying stress set
干燥指数 dryness index; index of aridity
干燥周期 drying cycle; drying period
干燥状态 drying condition; drying regime
干枝(枝条干枯) fore-stock
干重 dry weight
干状断裂长度 dry breaking length
干状剪切(单板剪切) dry clip
杆菌 bacillus (*plur.* -lli); bacterium (*plur.* -ia); rod-shaped bacteria
杆菌素 bacillin
杆菌状 bacilliform
杆式挂车 pole trailer
杆式角规 stick-type angle gauge
杆式链锯 pole chain saw
杆式喷头系统 vertical boom-nozzle system
杆式喷雾机 boom sprayer
杆式四轮车 pole wag(g)on
杆式圆锯 pole circular saw
杆式运材挂车 pole carriage
杆形二价染色体 rod bivalent
杆形染色体 rod chromosome
杆状病毒 rhabdovirus
杆状的 bacillar; baciilaris; baculiform
杆状菌 bacillus
杆状菌的 bacillar; baciilaris
秆 cane
秆柄(草本) culm stalk
秆柄(竹类) rhizocaul
秆基 culm base
赶出猎 hunt by chasing or beating
赶楞工 decker
赶套子工 chain tender
赶羊流送 bill to mature; drive; loose driving; loose floating; river driving; stream driving
赶羊流送工 log driver
赶羊流送扫尾 mop up stray
感病反应 susceptible reaction
感病体 suscept
感病性 susceptibility
感光板灵敏度 speed of photographic plate
感光计 actinometer; sensitometer
感光阶梯光楔 sensitometric step wedge
感光细胞 photoreceptor cell
感光性 photonasty
感光纸 bromide paper
感觉刚毛 sensillum chaeticum
感觉毛 sensillum trichodeum
感觉器官 sense organ
感觉温度 sensible temperature
感染 taint
感染材 infected timber
感染剂量 infective dose
感染焦点 focus of infection
感染型 infection type
感热筒 thermobulb
感热性 thermonasty
感色的 colour sensitivity
感受器 receptor
感受性 receptivity; sensitivity
感水性 hydronasty
感温阶段 thermophase; thermostage
感温球 thermobulb
感性运动 nastic movement
感夜性 nyctinasty
感夜运动 nyctinastic movement
感应 induction
感应毛 sensitive hair
感应遗传 telegony
感震性 seismonasty
橄榄钙长岩 allivalite
橄榄石 olivline
橄榄石熔成磷肥 olivine-fused phosphate
橄榄树 varnish tree
橄榄油 olive oil
橄长岩 allivalite
干材(主干) log; long wood; bodywood
干材积式 tree volume equation; tree volume function
干材面积指数 bole area index
干材期 bole stage
干腐 stem rot; trunk rot
干花(老茎生花) trunciflory
干基 butt
干基材 butt cut; butt length; butt log; root-swelling
干基段 butt block

干基肥大 swell
干基腐朽 butt rot; stump rot
干基割伤 basal wounding
干基含水率 dry basis moisture content
干基环剥 basal stem girdle
干基膨大的 barrel-butted; bottle-butted; churn-butted; swell-butted
干基树皮处理(药剂带涂布) basal bark treatment
干基原木 bottom log
干基直径 butt diameter; diameter at butt end
干距 spacing (between trees); tree spacing
干空(空心木) centre hole
干流 master river; master stream
干生花的(茎干生花) cauliflorous
干围 girth
干围级 girth class
干围界限 girth limit
干围卷尺 girth tape
干围率 girth quotient
干围生长量 girth increment
干线 line; main line; main road; trunk
干线道路 primary road
干线林道 arterial forest road; skeleton forest road
干形级 form class
干形级材积表 form class volume table
干形圆满 full-bodied
干形指数 form index
干朽菌 tear fungus; dry rot fungus
干支线交会点 hooking-on point
干轴 stem axis
冈崎碎片 Okazaki fragments
冈崎碎片冈崎片段 Okazaki piece
刚度 stiffness
刚果玷巴(树脂) Congo copal
刚劲因数 stiffness factor
刚毛 bristle; chaeta; seta (*plur.* setae)
刚毛排列法 chaetotaxy
刚毛状的 bristly; setaceous
刚石[粉] emery
刚性 inflexibility; rigidity
刚性副河绠 stiff-leg
刚性河绠 rigid boom
刚性结构 rigid structure
刚性绝缘板 rigid insulation board
刚性模量 rigidity modulus
刚性因子 stiffness factor
刚玉[粉] emery
刚玉磨料 aluminium oxide mineral
肛门 anus
纲 class
纲索磨痕 rope print
钢带 steel belt
钢垫片 steel shim
钢箍 loop
钢管家具 tubular furniture
钢轨 rail
钢夹 steel trap
钢架杆(集材) steel tower
钢架杆绞盘机绷索 walking guy
钢筋混凝土围桩木防护法 Koetitze pile armour process
钢锯 hack-saw
钢连杆 steel link
钢木结构 wood-steel construction
钢盘(集材) pan
钢盘集材 pan skidding
钢钎 jumper
钢琴材 wood for pianoforte
钢丝锯 fret saw; web saw
钢丝捆扎 wiring
钢丝绒 steel wool
钢丝绳 cable; wire rope
钢丝绳承座 wire rope slipper
钢丝绳垂度 wire rope sag
钢丝绳打绞 tangle wire rope
钢丝绳反捻(丝与股捻向相反) ordinary lay
钢丝绳股 wire strand
钢丝绳股芯 independent wire strand core
钢丝绳过卷(卷筒上) overwind
钢丝绳挠度 wire rope sag
钢丝绳圈(临时连接钢丝绳) molle
钢丝绳套管 wire rope thimble
钢丝绳套节 wire rope socket
钢丝绳套筒 wire rope sleeve; wire rope socket
钢丝绳芯 core
钢丝绳油压张力计 wire rope tensiometer

钢丝索 wire strand
钢丝网坝 woren wire dam
U形钢丝芯撑 jammer
钢索 cable; line; wire rope
钢索安全固定器(卷筒上) breakaway cable anchor
钢索鞍座(索道) cable saddle
钢索插接器 hercules
钢索穿过滑轮 reeving
钢索定位辊 line retainer
钢索断线探测器 defectoscope
钢索挂钩桥 cable-suspension bridge
钢索活节 line swivel
钢索集材 cable extraction; cable logging; cable yarding; simple wire log-slide
钢索集材装置 cable yarding system
钢索集材作业 cable work
钢索集运材 cable log haul
钢索夹子 wire rope clip
钢索铰接 line swivel
钢索纠结 crimp
钢索空载挠度 empty cable sag
钢索联结 wire rope rigging
钢索临时绳圈 wild eye
钢索扭结 tangle wire rope
钢索牵引 roping
钢索容量 line capacity
钢索上伸出来的断丝 jagger
钢索松弛度(允许向一旁拉动) runout
钢索踏板(集材机旁) cable step
钢索套管 rigging socket
钢索套环 cable strap; rigging socket
钢索调头(平衡磨损) change ends
钢索跳出滑轮槽 foul
钢索涂油器 cable wiper
钢索外层钢丝 crown wire
钢索下放(从卷筒) run
钢索引出端 leading end
钢索装配 wire rope rigging
钢枕 steel tie
钢支架 steel yoke
钢支座 bridle iron
钢纸 vulcanized fibre
钢制集材杆 high spar
杠杆 lever; leverage
杠杆臂 lever arm
杠杆式拉力滑车 leverblock
杠杆式同时闭合装置 beam simultaneous closing device
杠杆作用 lever-action; leverage
杠钩(抬木头用) swede hook; lug hook
高薄荷酮 homomenthone
高背椅 wing chair
高背长靠椅 settle
高倍接物镜 high power objective
高边 high side
高表樟脑 homoepicamphor
高丙体六六六 lindane
高草 high grass; takakusa; top grass
高草本层 tall field layer; tall herbaceon layer; tall herbaceous layer
高草草原(北美) tall-grass prairie
高草群落 altherbosa (altoherbosa); altoherbiprata; altoherbosa (altherbosa)
高草植物 altherbosa (altoherbosa); altoherbosa (altherbosa)
高层大气 upper atmosphere
高层空气 upper air
高层云 altostratus
高差量测器 height gauge
高产草 high-yield grass
高产率 high yield
高产色谱法 high productive chromatography
高产种 high-flelding seed
高潮滩 high water bed
高成分肥料 high analysis fertilizer
高程 elevation
高程测量 altitude surveying; height measurement
高程基准 datum level; vertical datum
高床 raised bed
高磁场 upfield
高次谐波 higher harmonic wave
高档家具 high-priced furniture
高得率 high yield
高得率浆 high yield pulp
高的 alto*sim*
高等菌类 higher-fungus
高低缝搭接 rebate
高低缝接合 ship-lap joint

G

高低速双转盘水力碎浆机 Hi-lo pulper
高低位镶嵌沼泽 aapamoor; Aapa moor
高地林 upland forest
高地泥炭土 upland peat soil
高地砖红壤 high level laterite
高度 altitude; alto～; stature
高度补偿涡轮增压器(集材机上) altitude-compensating turbocharger
高度测定器 height finder
高度传染 hyperinfection
高度地方性的 hyperendemic
高度菌根营养的 highly mycotrophic
高度量测 height-measure
高度曲线 hypsometric curve
高多孔板 high porous board
高伐根材积折减 stump deduct
高分辨力红外线辐射仪 high resolution infrared radiometer
高分辨力接触印象像片 print of high resolution
高分辨率 high resolution
高分辨率红外线辐射探测仪 high resolution infrared radiation sounder
高分子 macromolecule
高风化氧化土 acrox
高割面(采脂) pulling face
高拱式跨车 straddle-carrier
高光谱 hyperspectrum
高光谱的 hyperspectral
高光泽纸 glazed paper
高海松烯 homopimanthrene
高含水材 wet timber; wetwood
高寒带 subnival belt
高寒森林 cold forest
高寒土 subnival soil
高环 epipodium
高积云 altocumulus
高级材 high value assortments
高级单板 high-grade veneer
高级废水处理 advanced wastewater treatment
高级国有林 superior national forest
高级锯材(无节无缺陷年轮致密) ladder stock
高级线性规划系统 advanced linear programming system
高级遗传学 advanced genetics
高级侦察卫星 advanced reconnaissance satellite
高价材 high value assortments
高架梁式齐边机 overhead beam edger
高架门 overhead door; up-and-over door
高架内扇窑 overhead internal-fan kiln
高架坡道 elevated ramp
高架起重机 overhead crane
高架式装车 overhead loading system
高架索道 tower cable way; trestle cable
高架移动式起重机 overhead travelling crane
高架荫棚 high cover
高间隙拖拉机 high clearance tractor
高间隙挖掘机 high clearance digging tractor
高脚柜 highboy; tallboy
高脚烛台(装饰用) torchere
高接 top grafting
高截冠 high topping
高聚合物 high polymer
高聚糖 polysaccharide
高聚物 high polymer
高峻地形 high relief
高空 upper air
高空架索工 high climber
高空降水 precipitation aloft
高空暖锋 upper warm front
高空气候学 aeroclimatology
高空微扰 upper perturbation
高跨运材车 straddle truck
高立木度 dense stocking
高龄 senility
高岭黏土 kaolinton
高岭石 kaolinite
高岭石化 kaolinization
高岭土 china clay; kaolin; kaolinite
高岭土化作用 kaolinization
高轮车 high wheels
高轮集材车 logging-wheels
高轮拖车(轮轴拖挂木材) slip-tongue cart; high-wheeler
高轮运材车 high wheels; katydid; stump jumper; sulky
高锰酸钾值 K value

高锰酸盐值 permanganate number
高密度胶合板 high density plywood
高密度刨花板 extra heavy particleboard; high-density particleboard
高密度数字磁带 high density digital tape
高密度碎料板 extra heavy particleboard
高能化合物 energy-rich compound
高能键 energy-rich bond
高能喷射 high-energy jet
高黏度黏合剂 trowel adhesive
高黏度燃料 heavy fuel
高浓单动盘磨机 chemi-finer
高浓度浆泵 high density stock pump
高频 high frequency; radio frequency; radio frequency
高频滴定 high frequency titration
高频电介质干燥法 high frequency (dielectric) drying
高频干燥 electric drying; high-frequency drying; radio-frequency drying
高频干燥法 high frequency (dielectric) drying
高频功率 high frequency power
高频固化 radio-frequency curing
高频加热 high frequency heating; radio-frequency heating
高频胶合 high frequency gluing; radio-frequency gluing
高频率超声波 high-frequency ultrasonic wave
高频摄影像片 high oblique photo
高频硬化 high frequency hardening
高频预热 radio-frequency preheat
高频真空干燥法 high frequency vacuum drying
高频振动 high frequency vibration
高频装置 high frequency equipment
高剖面(采脂) high face
高气压 anticyclone
高萜烯 homoretene
高山矮曲林 elfin forest; elfin wood; krummhol(t)z
高山矮曲树 elfin tree; pygmy tree
高山残遗种 alpine relict
高山草甸 alpine meadow
高山草甸区 alpine meadow region
高山草甸土 alpine meadow soil
高山草甸植物群落 mesophorbium
高山草皮 alpine sward
高山草原土 alpine steppe soil
高山带 alpine belt; alpine region; alpine zone
高山地衣 alpine lichen
高山动物区系 alpine fauna
高山冻原 alpine tundra
高山腐殖质土 alpine humus soil
高山寒土植物 psychrophyte
高山花园 alpine garden
高山家畜 domesticated alpine animal
高山景观 alpine landscape
高山森林 high mountain forest
高山石质荒漠 alpine stony desert
高山苔藓 alpine moss
高山苔原 upland tundra
高山土 alpine soil
高山性的 alpine
高山驯化 altitude acclimatization
高山植被 alpine vegetation
高山植垫 alpine mat
高山植物 acrophyta; alpine plant
高山植物区系 alpine flora
高山植物群落 acrophytia
高山植物园 alpine garden
高山种 alpine species
高渗的 hypertonic
高渗溶液 hypertonic solution
高生长 elongation of tree stem; longitudinal growth; upward growth
高湿处理 high-humidity treatment
高疏密度 dense stocking
高水位 high water bed
高斯理论 Coase's theory
高斯-牛顿 Gauss-Newton
高斯原理 Gause's axiom; Gause's principle
高松油基甲酮 homoterpenylmethylketone
高送射 high straight-away shot
高速搅拌机 high-speed mixing machine

高速精浆机 hydrafiner
高碳单糖 higher-monosaccharide
高通滤波器 high-pass filter
高位泥炭 high moor peat
高位酸 high moor
高位酸沼地 high peat-bog
高位压条 aerial layering; air layering
高位芽 phanero~; phaner~
高位芽植物 phaenerophyte; phanerophytes (Ph)
高位沼泽 high bog; high moor; raised bog
高温 megatemperature
高温层积处理 warm stratification
高温处理 high-temperature processing
高温分解 pyrogenic decomposition; pyrolysis
高温干燥 high-temperature drying
高温固化 hot-setting
高温固化胶 high-temperature setting glue
高温固化胶黏剂 hot setter
高温计 pyrometer
高温胶黏剂 high-temperature adhesive
高温灭菌法 high-temperature pasteurization
高温期 hypsithermal interval (period)
高温气候 megatherm climate; megathermal climate
高温休眠 high-temperature dormancy
高温窑干 high-temperature kiln drying
高温蒸馏 pyrogenic distillation
高温植物 macrotherm; megatherm; megatherm plant
高温贮藏 warm storage
高狭温的 polythermal
高性能液相色谱 high performance liquid chromatography
高修枝 high pruning
高压 circumposition; pot layerage; stool layering
高压侧 high side
高压法 high-pressure process
高压锅 pressure cooker; pressure sterilizer
高压胶合板 high density plywood; superpressed plywood
高压灭菌 autoclaving
高压灭菌器 autoclave
高压喷漆 airless spray(ing)
高压喷枪 airless sprayer
高压气相色谱 high pressure gas chromatography
高压条法 marcottage; marcotting layerage
高压贴面工艺 high-pressure laminating system
高压液相色谱 high pressure liquid chromatography
高压蒸汽消毒器 high pressure steam sterilizer
高压质谱分析 high pressure mass spectrometry
高压贮胶罐 storage pressure vessel
高压贮气罐 high-pressure air receiver
高盐度水 hypersaline water
高盐培养基 Murashige and Skoog high salt medium MS
高隐棚(狩猎) high stand
高迎射 approaching high shot
高游离松香胶 high free rosin size
高原(台地) plateau
高原生境 upland site
高原适应 altitude acclimatization
高栈台装车 pit loading
高樟脑 homocamphor
高沼地 moorland
高枝剪 averruncator; lopper; lopping shears; tree pruner
高枝锯 pole saw
高种群适度 high-population fitness
高紫檀素 homoptercarpin
睾丸 testis; testis spermatogonium
睾丸雌性化 testicular feminization
睾丸管(昆虫) testicular follicle
睾丸囊 testicular envelop
镐 grub-hoe
镐缝栽植法 mattock-slit method
镐式斧 Pulaski tool
戈壁 gobi

戈德施米特气压表 Goldschmidt barometer
戈登坐犬 Gordon setter
疙瘩(树瘤等) lump
哥本哈根发芽台 Copenhagen table
哥德堡层积单位(180层积ft^3, 120实积ft^3) Gothenburg standard
鸽 dove
鸽子 pigeon
搁板 shelf
搁板木砖 shelf nog
搁材架 tree rack
搁插 shelf hole hardware
搁脚凳 stool
搁浅 strand
搁浅材(流送) neap; neap(ed) timber; dragger
搁浅木材 dry rear
搁扎工 tier
搁栅 joist
搁置 laydown
割草机 clipper; lawn-mower; mower; right of grass-cutting; weeding machine
割插 cleft cutting; split cutting
割刀 bark hack; spur knife
割沟(采脂) scrape; streaking
割灌刀 bush knife; matchet; slasher
割灌机 brush and weedcutter; brush breaker; brush-eating device; bush breaker; bush cleaner; bush cutter; scrub-clearing machine; scrub cutter; scrubber
割灌锯 brush cutting saw; brush saw
割灌木机 brush cutter
割灌弯刀 brush hook
割浆法 tapping system
割接器 cleft-grafting chisel
割靠接 cleft inarching
割捆机 knotter
割裂靠接 inarching by cleaving
割面(采脂等) face; tapping face; blaze
割面高峰(割面肩) shoulders
割面间隔带 bark bar
割面肩(鱼刺式采脂割沟的尖顶部) shoulders
割晒机 swather
割脂(割胶) tapping
割脂法 tapping system
割脂伤斑 pitch defect
革菌 leader fungus; thelephoraceous fungus
革兰氏染色法 Gram's stain; Gram-stain
革兰氏染色剂 Gram-stain
革兰氏染色阳性 Gram-positive
革兰氏染色阴性 Gram-negative
革质 leather-like
阁楼 garret; loft
格板门扇 panelled door
格构细工 trelliswork
格局 pattern
格楞 package pile
格罗森保允许采伐量公式(收获预定) allowable cut formula-Grosenbaugh
格洛格氏定律 Gloger's rule
格筛 riddle
格式 format; pattern
格调 aspect
格网 grid
格网状水系 trellis drainage pattern
格网坐标 grid co(-)ordinate
格韦克加压注入吸收法 Gewecke pressure and suction method
格栅 grill(e); grating
格栅坝 horizontal grilled dam
格栅式起重臂 lattice boom
格状沙丘 latticed dune
格子窗 grill(e)
格子单位 unit cell
格子花样 lattice-work
格子门 hatch
格子墙 lattice fence
格子图 trellis diagram
格子纹饰 fret; greek fret; labyrinth
格子细工 lattice-work; trelliswork
格子椅背 lattice back
格子状栅栏 lattice fence
隔板 apron lining; backing board; baffle plate; balk board; boarding; brattice; compartment; orifice; partition; screen
隔板回空线 return line for caul plate
隔胞 cystidium (*plur.* -ia)
隔带采伐 felling in alternate strips; staggered setting

隔行疏伐 line thinning; row thinning; strip-thinning
隔火带 fire line
隔离 isolation; quarantine
隔离层薄膜 barrier film
隔离层涂刷 insulating coating
隔离带 barrier; buffer zone; isolation strip
隔离防火法 fire containment
隔离群 conivium; isolate
隔离说 isolation theory
隔离型 seclusion type
隔离育种法 breeding by separation
隔离种 insular species
隔膜 diaphragm; septum (*plur.* -ta); partition wall
隔膜泵喷雾机 diaphragm sprayer
隔年结果 biennial bearing
隔年结实 alternate (year) bearing; biennial bearing; fruit-bearing in alternate years
隔年连续作业 intermittent sustained working
隔年生产量 intermediate yield
隔年收获 intermittent yield
隔年择伐 intermittent selective cutting
隔年作业 intermittent managemmt; intermittent working; intermittent yield system
隔年作业林业收益 profit of forestry for intermittent working
隔片 parting slip
隔坡梯田 alternation of slope and terrace
隔热门 heat isolated door
隔热幕 heat curtain door
隔热效应 thermoinsulation effect
隔热纸 sheathing paper
隔声板 sound proof board; sound-insulating board
隔水层 confining stratum; impervious surface
隔丝 paraphysis
隔条 crosser; fillet; parting bead; spacer; spacing sticker; stacking strip; sticker; strip
隔条变色 sticker stain
隔条对准 sticker alignment
隔条放置机 stick placer
隔条痕迹 sticker markings
隔条间距 sticker spacing
隔条木变色 crosser stain
隔芯板(层压板) barrier sheet
隔芯薄膜 barrier film
隔芯层 barrier sheet
隔音板 baffle board; sound proof board; sound-insulating board
隔音材料 acoustic insulating material; deadening
隔音纤维板 isolation fibre board
隔音装置 sound arrester
隔育群落 gamodeme
葛缕烯酮 carvenone
葛缕子油 caraway oil
个别计量销售 sale-lot of separate pieces
个别生境 individual site
个人所有权 individual proprietorship
个体 individual; unit
个体变异 individual variation
个体的 individual
个体发生 individual ontogenesis; ontogeny
个体发生说 ontogenetic theory
个体发育 ontogeny
个体发育说 ontogenetic theory
个体基因型 idiotype
个体密度 individual density
个体膜 unit membrane
个体生境 individual habitat
个体生态学 autecology; autoecology; idioecology; individual ecology
个体生物学 autobiology; idiobiology
个体数 number of individuals
个体所在地 individual habitat
个体选择 individual selection
个体遗传型 idiotype
个体增长率 individual growth rate
个体转化率 individual converse rate
各向同性 isotropy
各向同性材料 isotropic material
各向同性层 isotropic layer
各向同性的 isotropic
各向同性纤维 isotropic fibre
各向同性现象 isotropism
各向异性 anisotropism; anisotropy
铬绿 chromoxide green

铬皮粉 chromed hide powder; chrominated hide powder
铬鞣革 chrome tanned leather
铬天青S试剂 chrome azurol S reagent
铬铜砷合剂 ascuerdalith
给电子体 electron donor
给光量 light supply
给料 feeding
给料斗 feed hopper; hopper
给料辊道 feed table
L形根(J形根, 窝根栽植所致) L-root
根癌病 crown gall
根拔 root up; uproot; uprooting
根白腐 annosum root rot
根部 rhizine
根部冻害 root frozen
根部堆土 hilling
根部扩大 root flare
根部膨大 root-swelling
根部伤害 root injury
根部习居菌 root inhabiting fungus
根部窒息 root suffocation
根材 root timber; root wood
根材单板 stump veneer
根材积 root volume
根插[穗] root cutting
根缠绕 root wrapping
根出的 radicicolous
根出苗 ratoon
根出条 root sprout; root sucker; sucker; water sprout
根出叶 root leaf
根除 eradication; uproot
根除剂 eradicant
根床接 bench grafting on root
根刺 root thorn
根的 radicular
根端 butt; root apex; butt end; bottom end
根端处理 butt treatment
根端花纹 butt figure
根端膨大的 barrel-butted
根端跳起(倒树时) kick; kickback
根端修齐机 butt-end reducer
根端修切(原木) recutting
根端原木 first log
根端整形机 butt-end reducer
根端直径 base diameter; butt end diameter; diameter at foot
根段 butt
根段材(跳过采脂面的) jump-butt
根段单板 butt veneer
根段原木 butt cut; butt length; butt log; meat block
根腐 butt rot; stump rot
根腐病 root rot
根腐材 squaw log
根冠 calyptra; cap; pileorhiza; piles-rhiza; root cap
根冠比 ratio of root to shoot; root-shoot ratio
根际 rhizosphere
根际材垛 stump wood
根际材积 stump scale
根际直径 bottom-diameter; diameter at foot
根寄生 root parasitism
根间泥土(根拔后所带) root wad
根接 graft on root; root grafting
根结线虫 root-knot eel-worm; root-knot nematode
根结指数 root-knot index
根茎 rhizocaul; root-shoot; rootstock; stem-tuber
根茎比 ratio of root to shoot; root:shoot ratio
根茎插 root-and-shoot cutting
根茎地下芽植物 geophyta rhizomatosa; rhizome geophyte
根茎过渡区 root-stem transition zone
根茎芽 rhizome bud
根茎状的 rhizomoid
根颈 collar; corona; neck; rhizome neck; root collar; root crown
根颈接 crown grafting
根颈直径 collar diameter
根颈蛀虫 collar borer
根径立木材积表 stump scale
根靠接 root grafting by approach
根瘤 nodule; root cancer; root nodule; root tubercle; tubercle
根瘤病 knot-root disease; root knot
根瘤菌 legume bacteria; nodule bacteria; root nodule bacteria; root tubercle bacteria
根瘤菌剂 nitrogin

G

根瘤菌属 rhizobium (*plur.* -ia)
根瘤细菌 nodule bacteria
根毛 root hair
根毛层 piliferous layer
根密度 root density
根面(土壤) rhizoplane
根蘖 ratoon; root sprout; root sucker; root-shoot; stool shoot; stump shoot; stump sprout
根蘖更新 regeneration from root suckers; reproduction by sucker
根攀缘植物 root climber
根盘 root plate
根区 root zone
根群 root mass
根梢两用抓具 butt and top grapple
根生的 radical
根条 root
根托 rhizophore
根外追肥 foliage spray
根围 rhizosphere
根围作用(土壤中生物之间) rhizosphere effect
根膝(植物) knee
根系 radication; root system
根系层 root zone
根系分泌物 root exudate
根系盘绕 root spiralling
根系特性 rooting habit
根系修剪 root pruning
根型培育器(育苗容器名) Rootrainer
根序 rhizotaxis
根压 root pressure
根原基 root primodium
根原细胞 root initial
根源 source
根砧 understock
根枝接 scion grafting on root
根质量 root mass
根株 stock; stump
根株材 stump wood
根株残存迹地 stump-land
根株花纹 stump wood figure
根株萌 reproductive power from stool
根株萌蘖 stool sprout
根株萌条 stool shoot
根株木纹 stump wood figure
根株形数 rootstock form factor
根桩 parent stock; rootstock
根状的 radiciform; rhizomorphoid
根状茎 rhizome
根状茎的 rhizomatic
根状茎植物 rhizome plant
根状菌索 rhizomorph
跟单汇票 documentary bill
跟面 heel
跟踪程序 trace routine
庚糖 heptose
耕层 plough layer; tilth
耕地 cropland; cultivated land; plough land
耕地红线 arable land minimum; red line of farmland area; warning limit of arable land; minimum threshold of farmland area
耕翻 ploughing
耕垦 cultivation
耕宽 tilling width
耕深 tilling depth
耕作 farming; tilth
耕作层(土壤剖面A_p层) agric horizon; arable layer; top soil
耕作冲积土壤 anthropogenic-alluvial soil
耕作法 farming
耕作防治 cultural control
耕作机具 tiller
耕作熟化层(土壤剖面) agric horizon
耕作土壤 cultivated soil
耕作土壤学 edaphology
耕作制度 cropping system
埂 ridge
埂植 ridge-planting; step planting
绠场 booming ground; holding area; holding ground
绠存木 boom
绠链 boom chain
绠漂 raft stick
绠漂横连木 swifter
绠漂连接木 lap
绠漂链 toggle chain
绠漂木 boom stick; set of stick; stick
绠漂木链钩 booming dog
绠漂木柱塞 boom pin
绠漂钻 boom auger
绠绳 boom cable
绠围木材 crib; crib log

绠围排 boom raft
梗 stalk; stem
梗节 pedicel
梗塞 bottleneck
更换率 turnover rate
更生性杀虫剂 eradicant insecticide
更新(天然或人工) regeneration; rejuvenation; renewal; renovation; reproduction; restocking
更新成本 replacement cost
更新成熟 regeneration maturity
更新带 belt under regeneration
更新地块 compartment under regeneration
更新伐 progressive shelterwood felling; regeneration cutting; regeneration felling; regeneratory cutting; reproduction cutting; reproduction felling
更新法 method of regeneration; reproduction method
更新方向 direction of regeneration
更新后作业 system of after regeneration
更新级 regeneration class; reproducting class
更新间隔期(从下种伐至后伐) regeneration interval
更新林地 regenerated forest land
更新轮伐期 silvicultural rotation
更新率 reproduction rate
更新面 regeneration coupe
更新面积 regeneration area
更新目标(目的) regeneration target
更新能力 reproduction capacity
更新期 regeneration period; reproducting period
更新潜力 reproduction potential
更新区 regeneration block; regeneration coupe
更新世 Pleistocene
更新速度 reproduction speed
更新调查 reproduction survey
更新修剪 renewal pruning
更新样地 reproduction plot
更新营养 regenerative nutrition
更新幼林 restocking
更新障碍 regeneration barrier
更新者 revitalizer
更新作业法 system of regeneration
工厂 mill; plant
工厂废水 plant effluent
工厂负荷 factory burden
工厂管理 factory management
工厂机械安装 millwork
工厂间接生产费用分配率 factory overhead rate
工厂交货[价] ex factory
工厂绿地 factory garden
工厂绿化 factory planting
工厂门市部 manufacture's sales branches office
工厂设备 full-scale plant
工厂庭园 factory garden
工厂停工损失 factory shutdown loss
工厂预制件(如门窗等细木工部件) millwork
工厂园林 factory garden
工厂装置 full-scale plant
工场交货价 ex factory
工程地质及勘察地质图 geological map
工程化木制品 engineered wood products
工程木材产品 engineered wood product
工程效益 project benefits
工地 camp site
工段(作业地段) working unit; working section
工蜂 spado; worker
工件 piece of work; workpiece
工具 instrument
工具变钝 dulling
工具柄 helve
工具材 implement stock; wood for tool and implements
工具箱 instrument case
工棚(伐木工) bungalow; camp
工人 worker
工舍(木排上) ark
工时 man-hour
工时测定 work measurement
工时研究 time study
工效分析 performance analysis
工薪 payroll
工薪税 payroll tax

工序内部传送 intraprocess movement
工序之间传送 interprocess movement
工业安全 industrial safety
工业安全卫生法 Industrial Safety and Health Law
工业废料 industrial waste
工业革命 industrial revolution
工业固氮 industrial nitrogen fixation
工业黑化现象 industrial melanism
工业化 industrialization
工业化社会 industrial(ized) society
工业林 industrial forest
工业林企业 industrial forestry enterprise
工业森林经营 industrial forest management
工业设备 commercial plant
工业污水 industrial sewage
工业性林业 industrial forestry
工业用材 commercial timber; industrial cut stock; industrial wood
工业用革 industrial leather
工业用木片 mill-size chip
工业用漆 industrial finish
工蚁 spado; worker
工艺 process
工艺采伐龄 technical cutting age
工艺成熟 technical maturity
工艺程序图 process chart
工艺伐期龄 technical final age
工艺规程 technical schedule
工艺过程 technological process
工艺流程图 process chart; process flow diagram
工艺轮伐期 technical rotation
工艺刨花 engineered particle
工艺品 workmanship
工艺树种 technological tree
工艺特性 technological aspect
工艺性湿斑 manufacture damp
工艺性质 technological property
工资 wage
工资单 payroll
工资负担 payload
工字钢 Z-iron
工字木梁 I-joist
工字形生芽 I-shaped budding
工作单 job card
工作负荷 live load(ing); operating load
工作负载 working load
工作岗位责任制 organization of worker's position
工作记录 log
工作量 yield
工作强度 working strength
工作台 bench; dischbank; table; workbench; working bench; working table
工作台夹钳 bench screw
工作台升降装置 table lifting device
工作项目 job
工作项目分析 job analysis
工作压力 operating pressure; working pressure
工作因素 working factor
工作应力 working stress
工作站等官署名称 forest name
工作制度 working system
工作质量 workmanship
工作组 working group
弓把手 handle
弓背梁 camber-beam
弓锯 bow-saw; turning saw
弓弯 camber; longitudinal distortion
弓弦式桁架 bowstring truss
弓形材 bowed wood
弓形顶 bow top
弓形堵漏夹 C-clamp
弓形拱 segmental arch
弓形桁梁 bowstring truss
弓形锯 coping saw; hack-saw
弓形锯导板 bowguide
弓形拉手 bow handle
弓形链锯 rotary chain-saw
弓形翘 camber
弓形翘曲 longitudinal warping
弓形退水闸门 segmental waste gate
弓形弯曲 bowing
弓形椅背 bow back
弓形正面(橱柜) sweep front; swell front; swept front
弓状结构 arch
公差 allowance; tolerance
公差等级 quality of tolerance
公差范围 margin tolerance

公称尺寸 nominal size; specified dimension
公称体积 nominal volume
公称压力 nominal pressure
公称直径 nominal diameter
公称重量 nominal weight
公定容重 basic volumetric weight
公共设施 communal facility
公害 environmental disruption; public hazard; public nuisance
公害控制 nuisance control
公路 high-road; high-way
公路干线 arterial highway
公路排水沟 highway ditch
公路树 highway tree
公路运材 on-highway hauling
公麋 stag
公墓 cemetery
公平价格 just price
公平交易 arm's length transaction
公平市价 arm's length pricing
公顷株数 number of trees per hectare
公式变换 formula transformation
公式法 formula method
公式翻译 formula translation
公式翻译程序 formula translator
公式转换 formula translation
公司工会 company union
公司间利润 intercompany profits
公司内标准 company standard
公司性营业 incorporated business
公司章程 by-law
公司注册手续 incorporation procedure
公鸭 drake
公益林 public welfare forest
公用设施 convenience
公用事业公司 public utility company
公有地 communal land
公有林 common forest; communal forest; community forest; forest for community; forest under common ownership; public corporation-forest; public forest; yield sharing forest
公有林主 public forest owner
公有土地 crown land
公园 park; public park
公园大道 boulevard; parkway
公园道路 park way
公园管理 management of park; park administration; park management
公园规划 park planning
公园苗圃 park nursery
公园设计 park planning
公园土木工程 park engineering
公园吸引力 park attractiveness
公园系统 park system
公众露营 public camp ground
功力器 booster
功率不足的 underpowered
功率计 dynamometer
功率损耗系数 power loss factor
功率消耗量 power consumption
功率因素 power factor
功能反应 function response; functional response
功能抗性 functional resistance
功能群 functional group; functional guild
功能设计 functional design
功能生态位 functional niche
攻角 angle of attack
宫庭 court
宫苑 palace park
供电车 power wagon
供肥模式 fertilizer formula
供给的 feeding
供给装置 feedway
供浆量 stock volume
供料 feeding; furnishes
供料槽 supply chute
供料台 log dump; log storage deck; rollway
供气 air supply
供气管道 air supply duct; air supply line
供气装置 air feeder
供热 heat input
供热管道 heat supply pipeline
供水槽 feeder
供体 donor
供应槽 charging tank
供应的 feeding
供应全部食物 mass provisioning
供应资金 finance

G

拱 arch
拱顶 crown; dome; tree canopy
拱顶结构 dome-like structure
拱顶内侧 soffit
拱度 camber
拱腹 soffit
拱架 arch
拱架吊钩 arch hook
拱架集材 arch logging
拱架牵引索(拖拉机) arch line
拱架拖拉机 tomcat
拱廊 arch
拱门 arch; arched door
拱桥 arch bridge
拱形壁 overarching wall
拱形洞 grotto (*plur.* grotto(e)s)
拱形集材车 timber bob
拱形水坝 arch dam
拱形下横档椅 arched stretcher
拱形枕木 crowntree
拱轴集材车 arched axle logging wheel
拱柱 counterfort
拱座 abutment
珙桐 handkerchief tree; dove tree
共变数分析 analysis of covariance
共存 co-existence
共存属 binding genera
共存系数 coincidence index
共轭的 conjugate
共沸干燥法 azeotropic drying
共寄生 symparasitism
共价晶体 homopolar crystal
共建种 coedificator
共交种 cenospecies
共接枝 cograft
共聚物 copolymer
共聚作用 copolymerization; copolymerization
共鸣板 sound(ing) board
共鸣材 resonance wood
共栖 commensalisms; joint occurrence; mutualism
共栖的 commensal
共栖关系 symbiotic relationship
共栖现象 sympatric phenomena
共生 commensalisms
共生的 commensal
共生腐生植物 symbiotic saprophyte
共生菌 fungal symbiont
共生蓝藻 syncyanose
共生生物 symbiont
共生体 symbiont; association
共生同氮作用 symbiotic nitrogen fixation
共生细菌 symbiotic bacteria
共生现象 consortism; symbiosis
共生效益 co-benefit
共生藻类 symbiotic algae
共生真菌 symbiotic fungus
共生滋养的 symbiotrophic
共双倍体 syndiploidy
共同保险 coinsurance
共同海损 general average
共同基金 mumal fund
共同利益 co-benefits
共同受精 joint fertilization
共同提取 coextract
共同析晶 co-crystallization
共同作用 coactions
共显性 co-dominance
共享行为 sharing behaviour
共享摄食场所 common feeding ground
共有的自然资源 shared natural resources
共有林 shared forest
共振 resonance
共振板 sound(ing) board
共振材 resonance wood
共振法 resonance method
共振壳 resonant shell
共振频率 resonant frequency
共质体 symplasm; symplast
共质移动 symplasmic movement
共轴 neutral axis
共主寄生 symparasitism
勾尺 yard stick
沟 channel; ditch; gully; gutter; sulcus; trench; trough
U形沟(侵蚀) U-gully
沟播 furrow drilling; furrow planting; trench-sowing
沟播机 lister-planter
沟槽(加工的) groove
沟槽(木材缺陷) flute
沟道防护林 gully erosion control forest

沟灌 ditch irrigation; furrow irrigation
沟壑 gully; ravine
沟壑密度 gully density
沟横埂 balk
沟胫天牛 lamiid
沟流 channeling
沟垄耕作 furrow and ridge tillage
沟渠 canal
沟蚀 gully erosion
U形沟蚀 U-shaped gullying
V形沟蚀 V-shaped gulling
沟头防护 gully head protection
沟头侵蚀 head erosion; headward erosion
沟植 ditch planting; furrow planting
沟状侵蚀 gully erosion
沟状栽植 trench planting
钩 claw; grab; hook
U形钩 clevis
钩尺 hook scale
钩齿(生长锥探杆) hook teeth
钩齿粗木锯 whip saw
钩齿锯 gullet saw
钩刺状的 lappaceous
钩钉 dog
钩竿(采收果实等) nuthook; picaroon
钩杆 floating hook; hook-lever; peavy (peavie)
钩骨 unciform
钩环 shackle
钩介幼虫 glochidium
钩紧 hook
钩口 throat
钩链式辕柯 hook catch tongue
钩毛 glochid; glochidium
钩索杆 rifting rod
钩索铁杆(捆木) canary
钩头道钉 dog spike
钩形突 hamulus
钩眼 picaroon hole; pickaroon hole
钩爪(运材车横梁端) deer foot
钩住 hook
钩状的 unciform
钩状毛 barb
钩状刃口 hook
篝火 campfire
枸橼叶油 citron leaf oil
枸橼油 cedrat oil; citron oil
构材 member
构架 carcass; frame; frame-work; truss (frame)
构架材 carcassing timber; framing timber
构架梁 frame girder
构架式栈桥 framed trestle
构件 member; module
构件胶合 assembly gluing; constructional gluing
构件生物 modular organism
构象 conformation
构型 configuration; conformation
构造 making up
构造湖 tectonic lake
构造盆地 tectonic basin
购股证 warrant
购买力 buying capacity
购买力评价 purchasing power parity
购买者 purchaser
购制 charge account
估测 visual estimation
估测法 appraisal method
估测器 presumascope
估测与市场价 assessed versus market value
估定价值 appraised value
估定评价 assessed valuation
估计标准误差 standard error of estimate
估计价值 assessed value
估计量 estimator
估计平均寿命 estimated average life
估计式(器) estimator
估计寿命 life expectancy
估计者 estimator
估价 appraisal; assessment; costing
估价条款 valuation clause
估价员 assessor
估税员 tax assessor
估算利息 imputed interest
估算收入 imputed income
估算资本流动 imputed capital flow
估算资本值 imputed capital value
估值 estimator
咕巴 copaene
咕咕声(浅水鸭类鸣调) quack lingo
孤雌生殖 apomixis; female parthenogenesis; gynogenesis;

G

parthenogeneic reproduction; parthenogenesis; parthenogenetic propagation
孤雌生殖雌虫 parthenogenetic female
孤雌生殖的 parthenogenetic
孤雌生殖世代 parthenogenetic generation
孤雌受精生殖 parthenomixis
孤存林 forest outlier
孤岛 island
孤立的 solitary
孤立的丘 butte
孤立花坛 flower plot
孤立林分 single stand; isolated stant
孤立木 isolated tree; open grown tree
孤立栽植 isolated planting
孤生的 precocial
孤尾梢 whorl-less
孤雄生殖 male parthenogenesis; patrogenesis
孤植树 isolated tree; specimen tree
蓇葖 follicle
蓇葖的 follicular
蓇葖果 follicle
酤芭油清漆 copal oil varnish
箍(桶) hoop(ing)
箍缠 wiring
古北界 Palaearctic region; Palearctic realm
古北区 Palaearctic province; Palearctic realm
古病理学 palaeopathology
古彩涂饰 antique finishing
古代侵蚀 ancient erosion
古代演替系列 pterosere
古代园林 ancient garden
古地理学 paleogeography
古地中海 Mesogean Sea
古典建筑 classic architecture
古典式 classic style
古典型 classical style
古典园林 classical garden
古革尔法(热冷处理法) Gugel process
古迹 historic relic; historic site
古柯碱 cocaine
古昆虫学 palaeoentomology
古老特有分布 epibiotic endemism
古老特有性 epibiotic endemism
古老特有种 epibiotic endemic species
古老植物 epibiotic plant
古鸟亚纲 Archaeornithes
古蓬香脂 galban
古气候学 paleoclimatology
古侵蚀面 fossil erosion surface
古群落分布学 paleosynchorology
古群落生态学 paleosynecology
古热带的 palaeotropical
古热带间断分布 paleotropical disjunction
古热带植物区 paleotropical region; paleotropical floral kingdom
古生代 Palaeozoic era
古生茎叶植物 palaeocormophyta
古生态学 palaeoecology; paleoecology
古生物地理学 palaeobiogeography
古生物群落 paleobiocoenosis
古生物学 palaeontology; palaeobiology
古生物遗迹 ancient organisms remains
古生物遗迹类型自然保护区 nature reserve for ancient organisms remains
古生植物学 palaeobotany
古塔胶 gutta; gutta percha
古特有种 palaeo-endemic species
古土壤 fossil soil; pal(a)eosol; paleosoil
古土壤学 paleopedology
古温 paleotemperature
古细菌 archaeal
古演替顶极 eoclimax
古遗传学 palaeogenetics
古遗迹学 paleoichnology
古芸素 gurjuresene
古芸烯 gurjunene
古芸烯酮 gurjunene ketone
古芸香 gurgan; gurjun
古芸香脂(龙脑香树干所含) gurjun balsam
古芸香脂油 gurjun balsam oil
古植物 paleophyte
古植物分布学 synchronology
古植物学 palaeobotany; palaeophytology; phytopaleontology

古种子 aged seed
谷 vale
谷氨酰胺 glutamine
谷草 cereal straw
谷蛋白 glutelin; gluten
谷底 valley bottom
谷地 valley
谷地森林 ravine forest
谷坊 check dam
谷坊群 multiple-drop check dam
谷风 valley wind
谷壳 chaff
谷粒 kernel
谷粒颖果 grain
谷硫磷(杀虫剂) azinphos methyl
谷仁乐生(杀菌剂) granosan
谷穗 ear
谷值 valley value
股本 capital stock
股东权益 stockholders' equity
股份公司 joint stock company
股份固定投资公司 closed-end investment company
股份申请 application for share
股骨 femur
股金总额 capital stock
股利 dividend
股权公司 holding company
股权资本 equity capital
股息 dividend
股息收益 dividend yield
骨骼 skeleton
骨骼菌丝 skeletal hyphae
骨骼物质 skeletal substance
骨骼纤维素 skeletal cellulose
骨架 skeleton; structural framework
骨架物质 framework substance
骨胶 bone glue
骨幕 tentorium
骨湃油 copal oil
骨髓腔 medullary cavity
骨炭 animal black; animal charcoal; bone char
骨突 apophysis (*plur.* -hyses)
骨质的 osseous
鼓 drum; barrel
鼓包(胶合板缺陷, 树干上的凸起) bump
鼓风 air-blast; air-blowing
鼓风干燥机 fan dryer
鼓风干燥器 blast drier
鼓风干燥窑 fan blower kiln
鼓风机 air blower; blast blower; blower; fan blower
鼓风净种器 seed blower
鼓风式干燥窑 blower (dry) kiln
鼓风系统 air blast system
鼓膜 tympanum
鼓泡(锯身隆起) hump
鼓泡(液体) blister
鼓泡检测仪 blister detector
鼓起嵌板 elevated panel
鼓丘 drumlin
鼓式剥皮机 barker drum; barking drum; drum barker; drum debarker; drum-peeler
鼓式剥皮机装置 drum debarker system
鼓式刀辊削片机 circular saw chipper
鼓式干燥机 drum dryer
鼓式砂光机 drum sander
鼓式筛 drum screen
鼓式削片机 cutter cylinder chipper; drum chipper
鼓式制动器 drum brake
鼓室 tympanum
鼓形枕木 pole tie
固持作用 immobilization
固氮根瘤 nitrogen-fixing root nodule
固氮菌 azotobacter; nitrogen-fixing bacteria
固氮菌肥料 azotobacterin
固氮菌剂 azotogen
固氮酶 azotase; nitrogenase
固氮细菌 azotobacteria
固氮作用 azotification; nitrogen fixation
固定 stay
固定保护区 fixed forest; fixed reserve
固定绷索钩钉 guy-spike
固定标准地 fixed-size plot
固定测产笼 permanent cage
固定成本 constant cost; fixed cost
固定程序计算机 fixed program computer
固定床 fixed bed
固定存贮器 read-only storage
固定刀 bed knife

固定刀式刨床 fixed knife planer
固定刀头齐边机 fixed head jointer
固定的林业工人 constant forest-labourer
固定费用 fixed cost
固定负债 capital liability
固定干燥窑 stationary drying kiln
固定钢索重力集材 single-wire logging
固定根 anchor root
固定更新区 fixed regeneration block
固定横挡栅 fixed lateral boom
固定剂 fixative
固定寄生 stationary parasitism
固定价格 fixed price
固定架杆式集装机 stationary jammer
固定锯卡 fixed guide
固定缆索起重机 stationary cable crane
固定劳务 determinate service
固定梁 fixed beam
固定螺母 retaining nut
固定螺栓 retaining bolt
固定面积样地抽样 fixed-area plot sampling
固定面积样区法 fixed area sampling
固定苗圃 fixed nursery; permanent nursery
固定磨环 fixed stone ring
固定年伐量 fixed yearly cut
固定年龄分布 stationary age distribution
固定沙匠 fixed dune
固定沙丘 stabilized dune
固定设备 capital equipment
固定式承载索 standing line
固定式绞盘机 stationary winch
固定式金属探测器 stationary metal detector
固定式铺装机 stationary spreader
固定式起重机 stationary crane
固定式起重架 rigid boom
固定式牵引索(缆索起重机的) stationary traction line
固定式索道 standing skyline; tight(ening) skyline
固定式索道集材 standing skyline yarding
固定水解器 stationary digester
固定索 anchor cable; tie-down line; tiedown rope; tie-up line
固定炭窑 standing charcoal-kiln
固定投入 fixed input
固定卧式干燥机 fixed horizontal dryer
固定相 fixed phase; immobile phase
固定需求曲线 stationary demand curve
固定压尺 fixed nosebar
固定样地 fixed-size plot; growth plot; permanent growth plot; permanent sample area
固定液 fixing solution; immobile liquid
固定预备林 fixed forest; fixed reserve; fixing reserve; standing reserve
固定载荷 fixed load; quiescent load
固定侦察点(防火) detection fixed point
固定蒸煮器 stationary digester
固定支索 guy line
固定制材厂 stationary (saw) mill
固定贮存器 read-only storage
固定装卸桥 stationary loading bridge
固定装置(承载索) anchorage
固定资本 fixed capital
固定资产 capital goods; fixed assets; fixed capital
固定作业区 permanent periodic block
固化 cure; curing; set
固化不足(胶接) undercure
固化剂 curing agent
固化期 setting-up period
固化时间 curing time; setting time
固化树脂 solidified resin
固化速率 curing rate
固化温度 setting temperature
固链板 chain plate
固黏性 firmo-viscosity
固坡设施 grade stabilization structure
固沙 sand fixation; sand solidification
固沙草类 sand-binding grasses
固沙造林 dune fixation afforestation

固沙植物 sand binder; sand binding plant
固态逻辑技术 solid logic technology
固碳 carbon sequestration; sequestration
固体 solid
固体板圆锯 solid-plate circular saw
固体肥料 solid fertilizer
固体废弃物 solid waste
固体含量 solid content
固体径流 sediment discharge; sediment runoff
固体径流量 sediment yield
固体酒曲(日本) solid koji
固体培养 solid culture
固体培养基 solid medium
固体燃料 solid fuel
固体树脂 solid resin
固有的抗性 preformed resistance
固有缺陷 inherent vice
固有释放机制 innate releasing mechanism
固有树种 endemic species
固有休眠 innately dormant
固有应力 inherent stress
固有栽培植物 apophyte
固有种 endemic species; local species
固着的 sessile
固着动物 sessile animal
固着器 holdfast
固着生物 periphyte; sessile organisms
固着水生植物 hydrophyta adnata
故障 breakdown; failure; fault; jam; malfunction
故障时间 down time
雇佣劳动力 employed labour force
雇员伤害保险费 compensation of employee
刮 adzing; shave
刮板 doctor blade; sweep
刮板输送机 drag conveyer; flight conveyor; rake conveyer; scraper conveyor
刮除器 sweeper
刮刀(采脂用) hack
刮刀(各种) adze; doctor; jigger; scraper; scraping tool; shaver; shaving knife; spatula
刮刀(双柄) draw knife; drawshave
刮光 scraping
刮光机 scraper; scraping machine; scrapper
刮痕 scratch
刮痕硬度 scratch(ing) hardness
刮具 stripper
刮路机 scraper
刮毛槎齿 sliver tooth
刮面(刮皮面) bark shaving face
刮刨 scraper plane
刮皮 rossing; scrape; scribe
刮皮刀(采脂用) scrape iron
刮皮刀(清理病斑等) bark blazer; bark shaver; shave
刮皮工(采脂) bark blazer
刮皮面 bark shaver face
刮皮器(采脂) rossing tool
刮皮器(各种) bark scraper; scratcher; scribe
刮平 shaving
刮平辊 scalping roll
刮砂板 straightedge
刮杓 burning-in knife
刮饰 screeding
刮涂 blade coating; knife coating
刮涂机 knife coater
刮削 scraping
刮脂 dipping
刮脂刀 dip iron
刮脂工 dipper
刮脂钩 scraper
刮走 blow-off
寡基因 oligogene
寡基因抗性 oligogenic resistance
寡基因性状 oligogenic character
寡食性 oligophagy
寡食性的 oligophagous
寡污水的 oligosaprobic
寡盐的 oligohaline
寡养植物 oligotrophic plant
挂车 trail car; trailer
挂车式割灌机(剪枝机) trailed bush cutter
挂锤击碎机 hinged hammer pulverizer
挂耳(伐木) side cutting; side notching; splint cut; dutchman; pull

挂耳伐木 corner cutting
挂耳伐木法 cornering
挂钩 butt hook; hook; hookup
挂钩工 hookman; tong choker; tonger
挂钩牵引力 hitching force
挂钩调节索 tongline
挂号 marking
挂号码牌(固定样地树木标号) tag; tagging
挂料(锅炉) hanging-up
挂瓦条 gauge lath
挂衣棍 clothes-rail
挂鹰嘴工 carriage rider; dogger
拐点 inflexion point; point of inflexion
怪峰 ghost peak
怪峰效应 ghosting effect
关闭 close; locking
关键刺激 key stimulus
关键区域 key area
关键因素 key factor
关键因子分析 key factor analysis
关键种 key species; keystone species
关节式起重臂(吊杆) Knuckle-boom
关联系数 association coefficient; coefficient of association
关税 customs duty
关税表 tariff
关税及贸易总协定 general agreement on tariff and trade
关税率 tariff
观测窗 observation window
观测的评权 weighting of observation
观测方向 direction of view
观测分量 observatoinal component
观测剖面 observed profile
观测设备(照片) viewing equipment
观测室 lookout cabin; lookout cupola; observatory; tower cupola
观测收缩率 observed shrinkage
观测站 observation station
观察孔 eyelet; inspection hole
观察孔盖 access plate
观察站 lookout station
观光 sightseeing; tourism
观果树木 ornamental fruit trees and shrubs
观花灌木 flowering shrub
观赏价值 aesthetic value
观赏树木 landscape woody plant; ornamental trees and shrubs
观赏树种 ornamental species
观赏图样 ornamental design
观赏园艺 ornamental horticulture
观赏植物 decorative plant; landscape plant
观赏植物的 ornamental
观赏植物育种 breeding ornamental plant
观赏竹类 ornamental bamboos
观叶植物 foliage plant
棺材 coffin
冠 corona; crown
冠层分析仪 canopy analizer
冠幅 crown breadth; crown diameter; crown width
冠腐病 crown rot
冠根比 crown root ratio; stem-root ratio; top-root ratio
冠接 crown grafting
冠径 crown diameter
冠径比(树冠直径与胸径之比) crown diameter ratio
冠淋 crown wash
冠毛(菊科瘦果) aigret(te); pappus
冠毛状的 pappiform
冠生雄蕊 epipetalous stamen
冠形 crown form
冠形质离 cap plasmolysis
冠芽 crown bud
冠芽接 crown budding; whittle grafting
冠檐(花) corolla limb; limb
冠长树高比 crown length ratio
管板 tube sheet; tube-plate
管胞 tracheid
管胞壁 tracheid wall
管胞测定 tracheidometry
管胞的 tracheidal
管胞腔 tracheid lumina
管道 conduit
管道工程 pipework
管道模型理论 pipe model theory
管道清除 blowout
管道运输(木片) pipeline transport
管腐原木 piped log
管核 tube nucleus

管间的 intervascnlar
管间纹孔 intertrachear(y) pitting
管间纹孔式 intervascular pitting
管接头 sleeve
管茎 calamus
管孔 pore
管孔带 pore zone
管孔带状排列 zonate pore
管孔火焰状排列 pore flame-like arrangement
管孔链 pore chain
管孔排列 pore arrangement
管孔排列式样 pore pattern arrangement
管孔配列图型 pore pattern
管孔群 grouped pores
管孔团 pore cluster
管孔线 vessel line
管理 regulation
管理伐木场 run camp
管理费用 administration cost; handling expense; overhead cost
管理机构 organization of administration
管理基金 management fund
管理计划 business planning
管理价格 administered price
管理能力 management ability
管理人员 managing staff
管理信息系统 management information system
管理职能(业务) management function
管路坡度线 grade-line
管胚 bobbin square
管栖动物 tubicole; tubicolous animal
管塞 plug
管式挂车 pole trailer
管式喷胶机 tube sprayer
管式轴流通风机 lube-axial fan
管束 tube-bundle
管束式干燥机 tube bundle dryer
管筒材 bobbin wood
管线 manifold
管形的 vasiform
管形芽接 tubular budding; whistle budding
管状部分 tracheal portion
管状的 siphonaceous; vasiform
管状分子 tracheary element
管状腐朽 pipe(d) (rot)
管状孔(昆虫) tubuli
管状桥 tubular bridge
管状锈子器 roestelium
管状组织 tracheary tissue
管子明接 open joint
管子清洁器 go-devil
贯彻指射法 pointing-out method
贯穿脉 solidnerve
贯穿深度 depth of penetration
贯穿叶 perfoliate leaf
贯通环裂 through shake
贯通节 passing knot
贯通鸠尾榫 through dovetail
贯通裂缝 through check
贯通轮裂 through shake
贯通榫 through tenon
惯性(惯量) inertia
惯性矩 moment of inertia
惯性平衡 inertia balance
惯用名 trivial name
灌丛 bush; scrub; scrub (stand); thicket
灌丛稠密度 thickness of brush
灌丛带 shrubland
灌丛地 brushland
灌丛荒地 bush fallow
灌丛火 brush fire
灌丛火灾 bush fire
灌丛林 scrub forest
灌丛牧场 brush pasture
灌丛清除机 brushmaster
灌丛清除削片机 biomass harvester
灌丛清理 brush disposal
灌丛群落 scrub community
灌丛沙障 bush hedge
灌丛消灭剂 brushkiller
灌丛型 jungle form
灌溉 irrigation
灌溉泵 irrigation pump
灌溉喷射系统 irrigation injector system
灌溉渠 feed ditch; irrigation channel
灌溉网 irrigating net
灌后中耕 post-irrigation cultivation
灌木 arboret; arbuscle; brake; bush; chatwood; scrub; shrub
灌木层 shrub layer

灌木铲除机 scrub slasher
灌木丛 boscage; boskage; brush; brushwood; scrub wood
灌木丛阶段 thicket stage
灌木丛生的 copsy
灌木丛生地区 bushveld
灌木带 shrub zone
灌木的 shrub
灌木冻原 fruticous fundra
灌木干草原 shrub savanna
灌木花境 shrub border
灌木荒漠 fruticous desert; shrub desert
灌木阶段 brush-stage; shrub stage
灌木犁 brush (land) plough; brush breaker plough; scrub plough; swamp plough
灌木林 brush-field; scrub (stand); shrubbery; spinny
灌木林地 shrub land
灌木搂耙 scrub rake
灌木绿篱 bush hedge
灌木茂盛的 lingy
灌木切碎机 scrub pulverizer; scrub slasher
灌木清理 scrub clearance
灌木清理组 brush crew
灌木群 shrubbery mass
灌木群落 fruticeta
灌木树种 shrub species
灌木性的 frutescens
灌木沼泽犁 brush-and-bog plough; heath plough
灌木植被 shrubby vegetation
灌木植物 suffruticosa plant
灌木状的 frutescens; shrubby
灌木状地衣 fruticose lichen
灌木状生长的 arbuscular
灌油桥台 loading bridge
灌淤土 anthropogenic-alluvial soil
罐门(防腐) cylinder door
光 light
光暗循环 light-dark cycle
光斑(林内) sunfleck
光饱和 light saturation
光饱和点 light saturation point
光补偿点 light compensation point
光材 finish
光蝉 lantern-fly
光刺激 photostimulation
光催化剂 photocatalyst
光弹性的 photoelastic
光弹性试验 photoelastic test
光导摄像管 vidicon
光电倍增管 multiplier phototube; photomultiplier tube
光电倍增器 photomultiplier
光电比色法 photoelectric colourimetry
光电池 photoelectric cell; sensitive cell
光电分析天平 photoelectric analytical balance
光电管 photoelectric cell; sensitive cell
光电光泽计 photoelectric glossmeter
光电输入机 optical reader
光电效应 photo-electric effect
光电直读光谱仪 photo-electric direct reading spectrometer
光电浊度计 photoelectric turbidometer
光电阻 light resistance
光动态 photokinesis
光度测定法 photometric titration
光度计 lightmeter; photometer
光度学 photometry
光钝化作用 photoinactivation
光反射 light reflection
光反应 light reaction; photoreaction
光辐射强度 luminous emittance
光感效应种子 photoblastic seed
光感应 photoperception; photoreception
光固化剂 photocuring agent
光固化涂料 photo-cured coating
光还原作用 photoreduction
光合薄壁组织 photosynthetic parenchyma
光合本领 photosynthetic capacity
光合比率 photosynthetic ratio
光合部分 photosynthetic part
光合产量 photosynthetic yield
光合产物 photosynthate
光合光量子通量密度 photosynthetic photon flux density
光合面积 photosynthetic area
光合能力 photosynthetic ability

光合器官 photosynthetic organ
光合商 photosynthetic quotient
光合生产力 photosynthetic productivity
光合系统 photosynthetic system
光合细菌 photosynthetic bacteria
光合效率 photosynthetic efficiency
光合有效波长 photosynthetic active wave length
光合有效辐射(400~700nm) photosynthetically active radiation (PAR)
光合作用 photosynthesis
光合作用率 photosynthetic rate
光合作用面 photosynthetic surface
光合作用细胞 photosynthesing cell
光呼吸 light respiration; photorespiration
光滑雌蕊 leiogynus
光滑刮面(采脂) smooth face
光滑果子的 leiocarpus
光滑菌落 S-colony
光化层 chemosphere
光化大气 photochemical atmosphere
光化复合物 photochemical complex
光化过程 photochemical process
光化吸收 actinic absorption
光化性 actinism
光化学 photochemistry
光化学的 photochemical
光化学反应 photochemical reaction
光化学空气污染物 photochemical air pollutant
光化学污染 photochemical pollution
光化学烟雾复合物 photochemical smog complex
光化学诱导 photochemical induction
光化学作用 photochemical reaction
光化烟雾 photochemical smog
光化作用 photochemical process
光化作用的 photochemical
光活化 photoactivation
光机扫描器 optical-mechanical scanner
光间歇 light break
光降解 light degradation
光洁度 finish
光洁木板 clean cut
光截获 light interception
光解作用 photolysis
光卷筒 bare-drum
光量测定 actinometry
光量子通量密度 photon flux density
光路辐射通量 pathradiance
光密度法 densitometry
光密度计 densitometer
光面 gloss
光面板 clear face board
光灭活作用 photoinactivation
光敏感性 photosensitivity
光敏剂 photosensitizer; sensitizer
光敏色素(植物) phytochrome
光敏种子 light-sensitive seed
光敏作用 photosensitization
光能营养生物 phototroph
光皮 prima; smooth bark
光票 clean bill
光谱 spectrum (*plur.* -tra)
光谱背景 spectral background
光谱测定法 spectrometry
光谱带 spectral band
光谱定量分析 quantitative spectrochemical analysis
光谱定性分析 qualitative spectrochemical analysis
光谱反射 spectral reflectance
光谱分析 spectral analysis
光谱激发 excitation of spectra
光谱间隔 spectral interval
光谱曲线 spectral curve
光谱色谱(法) spectral chromatography
光谱特征 spectral signature
光谱投影仪 spectrum projector
光谱图 spectrogram
光谱仪 spectrograph
光期钝感植物 day neutral plant
光漆 lac varnish
光强 light intensity
光散射 light scattering
光伤害 light injury
光渗 irradiation
光生物学 photobiology
光生长反应 ljght growth reaction
光生长抑制作用 Light-growth inhibition

光适应 light adaptation
光束 beam
光台效率 photosynthetic effect
光通量 light flux
光通量密度 light flux density
光同化 photoassimilation
光透射 transmission of light
光秃的 calvescent
光秃海边 bare seashore
光系统II photosystem II
光线的 radial
光楔 optical wedge
光形态发生(建成) photomorphogenesis
光学测树仪 optical dendrometer
光学处理 optical processing
光学的 optic
光学轮尺(测定树干上部直径) optical calliper
光学密度 optical density
光学树脂胶 Chinese balsam
光学纤维束 fibre bundle
光学显微镜 light microscope
光艳斑点 slick spot
光氧化作用 photooxidation
光抑制种子 light-inhibited seed
光抑制作用 photoinhibition
光因子 light factor
光影划线(制材) shadow line
光诱导 light inducing
光诱致褪色 light-induced bleaching
光源 light source
光泽 brightness; brilliance; glazing; lustre; luster
光泽不匀 uneven brightness
光泽剂 gloss agent
光泽减少 gloss reduction
光泽油 gloss oil
光栅光谱仪 grating spectrograph
光照 illumination
光照充足的 sunny
光照锻炼 light hardening
光照固化的 irradiation-cured
光照阶段 photophase; photostage
光照延续时间 light duration
光质 light quality
光周期 photoperiod; photoperiodism
光周期反应 photoperiodic response
光周期现象 photoperiodicity; photoperiodism
光周期效应 photoperiodic effect
光周期性 photoperiodicity
光周期诱导 photoperiodic induction
光自动氧化 photoautoxidation
光自养的 photoautotrophic
广地带性的 euryzonous
广分布性的 eurychoric
广幅的 eurytopic
广幅营养性的(生境) eutrophic
广告费 advertising cost
广光的 euryphotic
广藿香油 patchouli oil
广肩小蜂 eurytomid; jointworm; seed chalcid
广角 wide angle
广居的 polydemic
广卵形的 broad-ovate
广木香烯 costene
广谱杀菌剂 broad spectrum fungicide
广谱性杀虫剂 broad spectrum insecticide
广谱性系 broad spectrum
广栖的 euryecious
广栖性的 euryoecious
广深的 eurybathic
广生境的 amphiecious; euryoecic; euryoecious
广生态幅度 wide ecological amplitude
广湿的 euryhygric
广湿性的 euryhydric
广食的 euryphagous
广食物性 euryphagic
广食性 euryphagy
广食性的 euryphagic
广视场 wide-field-of-view
广适的 euryokous
广适性 euryoky
广适性植物 euryvalent
广适应性的 eurytopic
广适种 euryspecies; eurytopic species; indifferent species
广适种植物 indifferent plant
广水性的 euryhydric
广酸碱性的 euryhadric

广温生物 eurythermal; eurythermic
广温性的 eurythermal; eurythermic
广效性系 broad spectrum
广压的 eurybaric
广延区 extension
广盐分 euryhaline
广盐性 euryhaline; holeuryhaline
广盐性的 euryhaline; euryhalinous
广氧性的 euryoxybiotic
广义法正林 generalized normal forest
广域分布的 eurychoric
广域性的 eurychoric; euryzonal
广状伞伐更新法 uniform shelterwood system
归巢本领 orientation
归成一排 windrow
归堆 bunch(ing); line log; pile; preyarding; rick
归堆叉架 accumulator clamp
归堆焚烧(采伐剩余物) piling-and-burning
归堆工 buncher
归堆机 buncher; preyarder
归垛 stack; stacking
归垛工 log stacker
归垛机 piler
归化 naturalization
归化植物 naturalized plant
归化种 alien; naturalized species
归楞(伐区) decking up; piling; skid up; stack; stockpile
归楞(推河楞场) bank; banking
归楞扒杠 piling jack
归楞工 decker
归楞机 deck loader; log stacker
归冷楞 cold deck
归密楞 close piling
归纳 induction
归纳成本法 absorption costing
归纳法 inductive method
归先突变 degressive mutation
归因 attribution
龟裂 alligatoring; chap
龟裂冻原 polygonal tundra
龟裂树皮 rimose bark
龟裂土 takir; takyr
龟叶甲 tortoise beetle
规程 schedule
规定尺寸 bare size; dead size; regular size
规定的最低保持量 specific minimum retention
规定强度 specified strength
规定收获量 yield prescription
规定用火 prescribed fire; prescribed burning
规范限度 specification limit
规格 dimension; gauge; size
规格板材 stock board
规格标准 specification standard
规格材 dimension board; dimension stock; dimension-lumber; on-grade; stock lumber
规格化 standardization
规格木瓦 dimension shingle
规格外品 outthrows
规划论 planning theory
规划远景 planning horizon
规律(行为) regime (*eg.* precipitation regime)
规律式(园林) formal style
规律式园林 formal garden
规模 size
规模成比收益 return to scale
规模经济 economy of scale
规则(工作制度) regime (*eg.* thinning regime)
规则分布 regular distribution
规则林 regular plantation
规则式 regular style; formal style
规则样方法 regular quadrat method
规章 by-law; regulation
规章措施 regulatory measure
硅胶 silica gel
硅孔雀石 liparite
硅酸钠胶黏剂 sodium silicate adhesive
硅酸盐 silicate
硅酮树脂 silicone resin
硅藻 diatom
硅藻软泥 diatom ooze
硅藻土 diatomaceous earth; diatomaceous silica; diatomite; infusorial earth; kieselguhr
硅质黏粒 siliceous clay
硅质砂岩 siliceous sandstone
硅质细胞 silice cell

硅质岩 acidite
硅质硬盘 duripan
轨道 cart track; orbit; orbital; race; rail; rail track; track
轨道拨正 lining
轨道参数 orbital parameter
轨道式路面 strip road
轨道手车 track barrow
轨道效应 race track effect
轨道运材 log tramway system
轨道周期 orbital period
轨底 flange
轨管机 tube mill
轨迹 locus; trajectory
WRE轨迹 WRE profile
轨迹点连线 drip-line
轨间绝缘 gauge insulation
轨距 gauge; gauge of track
轨距规线 gauge line
轨距余宽 slack
轨枕 crosstie
柜 box; chamber; chest; press
柜旁板 case side
柜台(餐馆) bar
柜台式饮食台 buffet
贵重材 high value assortments
贵重材料 precious material
桂花浸膏 osmanthus concrete
桂金合欢定 guibortacacidin
桂皮油 cinnamon bark oil
桂芯 cassia scrapped
桂樱油 cherry laurel oil
桂竹素 cheirolin
跪式(射击) kneeling
辊(gŭn) barrel
辊道 rollway
辊式输送机 live roller conveyer
辊式卸料头 roller discharge head
辊筒式砂光机 cylinder grinder
辊涂机 roll coater; roller coater
辊涂压花法 design roller coating
辊压机 roll press
辊压刨花板工艺 calender particleboard process
辊子 roller
滚板刀 bedplate bar
滚材 rolling; throwing
滚材垫木 skid pole
滚材工 decker
滚存价值 carrying value
滚刀式粉碎机 knife grinder
滚动负荷 roiling load
滚动进料带锯 roller bandsaw
滚动摩擦系数 coefficient of rolling friction
滚动台锯 flat-top saw; rolling-table saw
滚动装车 moving loading; roiling load
滚花辊 design roller
滚楞(装车) tail-down; rolling
滚木 ball hooting
滚木工 ballhooter; tailer-in
滚式除灌机 roll-type bush breaker
滚水坝 overflow dam; rolling dam; spillway dam
滚筒 reel; roll; roller; rotor; tumbler
滚筒剥皮机 drum barker; drum debarker; drum-peeler; rotary barker
滚筒胶压 roll laminating
滚筒绞盘机 drum winch
滚筒砂光机 drum sanding
滚筒式拌胶机 cylinder gluing machine
滚筒式单板干燥机 roller conveyer veneer drier; roller veneer drier; veneer roller dryer
滚筒式干燥机 roller dryer; rotary drom kiln
滚筒式抛光机 drum polishing machine
滚筒式砂光机 drum sander
滚筒式涂胶法 roller glue laying method
滚筒式涂胶机 roller gluing machine
滚筒涂胶机 roll spreader; roller spreader
滚压机 roller press
滚压剪切 rolling shear
滚压剪切试验 rolling shear test
滚辗 tumbling
滚珠轴承 ball bearing
滚柱导向轮(导板头) roller nose
滚柱压尺 roller nosebar
棍 stick
棍子 truncheon
锅顶料仓 digester silo
锅垢 scale

锅炉 boiler
锅炉间 boiler plant
锅炉马力 boiler horsepower
锅炉装置 boiler plant
国产木材 domestic timber; native wood
国道 country-road
国花 national flower
国际地球观测年 International Geographical Year
国际景观生态学会 International Association of Landscape Ecology
国际能源机构(IEA) international Energy Agency
国际鸟类保护协会 International Council for Bird Preservation
国际生态学会 International Association for Ecology
国际生物科学联合会 International Union of Biological Sciences
国际生物学计划 International Biological Programme
国际水文发展十年计划 International Hydrologic Decade
国际原木板积表 international log rule
国际原木英尺材积表 international (log) rule
国际栽培植物命名法规 International Code of Nomenclature for Cultivated Plants
国际植物命名法规 International Code of Botanical Nomenclature
国际制土壤质地等级 international textural grade
国际种子检验规程 international rules for seed testing
国际种子验证 international seed analysis certificate
国际自然及自然资源保护同盟 International Union for the Conservation of Nature and Nature Resources
国家安全规程 state safety code
国家保护林 state forest in the custody
国家风景步道(小径) national scenic trails
国家公园 country park; national park
国家河川风景区 national scenic river ways
国家湖滨风景区 national lakeshores
国家火险等级系统 national fire danger rating system
国家级自然保护区 national nature reserve
国家纪念公园 national memorial park
国家军事公园 national military park
国家科学研究专用地 national scientitle reserves
国家历史公园 national historic park
国家绿化委员会 National Forestation Commission
国家木材制造业协会 national lumber manufactures association
国家森林保护委员会 national forest reservation commission
国家森林审核会议 national forest board of review
国家森林游憩调查 national forest recreation survey
国家森林游憩资源审查 national forest recreation resource review
国家森林咨询委员会 national forest advisory council
国家森林资源连续清查 national forest inventory; first-category forest inventory
国家首都公园 national capital park
国家天然公园 national park
国家野牛保护区 national bison range
国家野生动物保护系统 national wildlife refuge system
国家银行 national bank
国家游憩区 national recreation area
国家造林 national afforestation
国家战场纪念公园 national battlefield park
国家沼泽公园 everglades national park
国家征用权 eminent domain
国家重点保护物种 wildlife under special state protection
国家资源计划会议 national resources planning board
国家资源委员会 national resources committee
国家自然保护区 national nature reserve
国家总收益 gross national product
国立公园 national park
国民净产值 net national product
国民生产总值 gross national product
国内检疫 inland quarantine; internal quarantine

国内净产值 net domestic product
国内生产总值 gross domestic product
国内市场 home market
国内消费税 excise tax
国树 national tree
国营采伐 state-run logging
国营林场 state owned forest farm
国营农场 state farm
国有林 crown forest; federal forest; national forest; state forest
国有林保存 retention of state forest
国有林代办机构 state forest agencies
国有林地 national forest land
国有林经济 state forest economy
国有林经营 working of state-forest
国有林区 national forest area
国有水道 national waterway
果瓣柄 carpophore
果孢子 carpospore
果柄 carpopodium
果顶 blossom end
果冻 jelly
果梗 pedicel
果核 stone
果荚 hull
果酱 jam
果胶 pectic material; pectin(e)
果胶基质 pectic matrix
果胶酶 pectase; pectinase
果胶酸 pectic acid
果胶糖 pectose
果胶纤维素 pecto-cellulose
果胶质 pectin substance
果接 fruit grafting
果颈 carpopodium
果聚糖 fructosan; levan
果壳 hull
果棱(伞形科) jugum
果篱 fruit hedge
果鳞 cone scale
果爿 carp (*or* carpel); hystrella
果皮 carpodermis; hull; pericarp; peridium; rind; seed vessel; shell
果皮的 pericarpial; pericarpic
果肉 pulp; sarcocarp
果肉分离机 depulper
果实 fruit
果实采集机 picker
果实采取权 right to pannage
果实抖落器 shaker; tree knocker
果实分类学 carpology
果实接 fruit grafting
果实调制 fruit-processing
果实形成期 fruit-formation period
果实直感 ectogeny; metaxenia
果树 fruit tree
果树树冠喷雾器 top fruit sprayer
果糖 laevoglucose; laevulose
果托 hypocarpium
果序 infructescence
果园 orchard
果园化 oreharding
果园喷雾机 orchard spray
果枝 fruit branch; spur
果子狸 gem-faced civet; masked civet
裹铁绝缘桩 ironbark dropper
过成熟材期 over-mature timber stage
过程 course; process
过大立木度 overstocking
过当量的 superstoichiometric
过低种群密度 undercrowding
过度 overcrowding
过度捕捞 over-fishing
过度采伐 over-exploitation
过度发育 over development
过度放牧 over grazing
过度灌溉 heavy irrigation
过度开发利用 overexploitation
过度冷却 supercooling
过度利用 overuse
过度喷涂 overspray
过度生长 hypertrophy; over-grown; overgrowth
过度生长的 hypertrophic
过度特化 over-specialization
过度调湿处理 overconditioning
过度通风 overdraft
过渡 transition
过渡材 intermediate wood
过渡层 intermediate layer; transition lamella
过渡带 intermediate belt; transition zone; transition(al) zone
过渡阶段 transition phase

过渡孔 transitional pore
过渡类型 intermediate form; passage form
过渡区 transition region; transition zone
过渡区群落 ecotonal community
过渡曲线 transition curve
过渡蠕变 transition-creep
过渡微生物 transient microbe
过渡性泥炭土 transitional moor soil
过渡性泥炭沼泽 transitional moor
过渡装置(输送机用) transit device
过渡状态 transition state
过渡组织 transition tissue
过伐 excessive cutting; excessive felling; excessive tree felling; over cut; overcutting; over-exploitation; overfell; overfelling; overworking
过伐林 hag
过繁茂 overluxnriant growth
过肥的 eutrophic
过干单板 over-dry veneer
过江藤油 lippia oil
过境税 transit duty
过梁 trimming joist
过量蓄积(超过法正蓄积量) overstock
过滤 filtration; percolate; sift; sifting
过滤池 leaching field
过滤介质 filter medium
过滤空气 filtrated air
过滤器 filter; strainer
过滤性病毒 filtrable virus
过密 overcrowded; overstock
过密的 overstocked
过敏的 hypersensitive
过敏反应 allergic reaction
过敏抗性 hypersensitive resistance
过敏性 allergy; hypersensibility; hypersensitivity
过敏性危机 allergic crisis
过敏原 allergen
过桥式重型木工车床 heavy patternmaker's gap lathe
过去成本 past costs
过热 superheat
过热蒸汽 superheated steam; superheated vapor
过热蒸汽干燥 superheated steam drying; superheated steam seasoning; superheated vapor drying
过热蒸汽干燥法 drying with superheated steam
过热蒸汽干燥窑 superheated steam kiln
过热蒸汽快速干燥窑 high-velocity superheated steam dry kiln
过剩 overflow
过剩采伐 surplus felling
过剩生产能力 surplus capacity
过失误差 error of mistake
过熟(林木) declining
过熟材 overmature wood; overstand wood; overyeared wood
过熟林 overmature forest
过熟林分 overmature stand
过熟木 old standard
过水断面积(河流或渠道) discharge area
过水路面 ford; watersplash
过酸 peracid
过小木片 undersized chip
过压缩(轧压时的) overdraft
过氧化氢酶 catalase
过氧化物酶 peroxidase
过氧化物酶体 peroxisome
过载安全阀 overload relief valve
过载恒温器 overload thermostat
过闸下驶 lock down
过长 overlength
过账 posting of account
过重 excess weight
过重超重 overweight
过煮 overcook
哈笛网 Hartig net
哈蒂氏材积平分法 Hartig's method
哈蒂氏网(胞间菌丝网) Hartignet
哈加测高器 Hasa altimeter
哈斯门盐类混合注入法 Hasserman process
哈特湿度计 Hart moisture gauge
哈提曼杆材端部注入法 Hartman process
蛤蚌毒素 saxitoxin
蛤蜊蕈 oyster cup fungus; oyster mushroom
胲(hǎi) hydroxylamine

海岸(森)林 coastal forest
海岸沉积 beach deposit; coastal deposit
海岸的 coastal; littoral
海岸防风林 coastal windbreak
海岸防护林 protective coast-forest
海岸固沙 coastal sand dune fixation
海岸湖 strand lake
海岸林 littoral forest; maritime forest
海岸侵蚀 beach erosion
海岸群落 coastal community
海岸沙地造林 coastal sand dune afforestation
海岸沙丘 coastal dune; coast dune; littoral dune
海岸线 coast line
海岸效应 coastal effect
海岸原生林 virgin strand forest
海岸植被 maritime vegetation
海岸植物 beach plant
海岸桩木 marine pile
海拔 altitude; elevation
海拔分布带 altitudinal belt
海拔高度替代种 altitudinal vicariad
海拔气候适应 altitudinal acclima(tiza)tion
海拔适应 altitudinal adaptation
海拔仪(测高仪) elevation instrument; altimeter
海滨 beach strand
海滨公园 sea-side park
海滨泥地 mudflat
海滨植被 strand vegetation
海滨植物 shore plant
海冰 sea ice
海波 hypo
海产养殖 mariculture
海潮防护林 protection forest against tide water damage
海成腐殖质 marine humus
海成黏土 marine clay
海成岩 neptunic rock
海成盐土 marine solonchak
海虫 gribble; sphaeroma
海胆灵 echinuline
海堤 seawall
海底的 benthal; benthic
海底等深线 bathymetric contour
海底气候 submarine climate
海底植物群 benthic flora
海地油 pennyroyal oil
海防林 shelter belts to protect sea coasts
海茴香烯 crithmene
海角球根(花卉) cape bulb
海浸 transgression
海里 nautical mile
海龙卷 water spout
海绵皮层 spongy cortex
海绵砂光机 sponge sander
海绵湿磨 sponge sanding
海绵土 mellow earth
海绵橡胶 foam rubber
海绵质 spongioplasm
海绵质髓 porous (spongy) pith; spongy pith
海绵质叶 spongophyll
海绵状白腐 white spongy rot
海绵状腐朽 spongy rot
海绵状褐腐 brown spongy rot
海绵组织 spongy tissue
海面群落 pelagium
海鸟 seabird
海平面升降变化 eustatic sea-level change
海平面长期变化 sea level secular change
海漆 blind-your-eye-tree; blinding tree
海茄冬式 Avicennia type
海山 sea-mount
海上浮标 marine buoy
海上漂散木 sea drift
海深测量法 bathymetry
海生菌 marine fungus
海生钻木动物 marine borer
海蚀平原 marine plain
海蚀作用 marine erosion
海松二烯 pimaradiene
海松树脂 galipot (resin)
海松酸型树脂酸 pimaric type acid
海滩 foreshore
海滩冰 beach ice
海滩土 marine clay
海田梁 bunk
海豚头形支脚 dolphin foot
海豚形铰链 dolphin hinge
海湾 embayment; estuary; inlet

海湾冰 bay ice
海湾形采伐 bay cutting; bay felling
海相沉积 marine deposit
海啸 tidal wave
海洋 marine; ocean
海洋孢粉学 marine palynology
海洋传输带 ocean conveyor belt
海洋带状分布 marine zonation
海洋岛屿 oceanic island
海洋的 oceanic
海洋度 oceanity
海洋浮游生物 thalassoplankton
海洋和海岸生态系统类型自然保护区 nature reserve for ocean and seacoast ecosystem
海洋洄游 oceanodromous migration
海洋木排 ocean raft; ocean-going raft; cigar raft
海洋盆地 ocean basin
海洋平排 wheeler raft
海洋气候 oceanic climate
海洋软泥 marine ooze
海洋生态系统 marine ecosystem; marine system
海洋生态学 marine ecology
海洋生态演替 marine succession
海洋卫星 SEASAT(sea satellite)
海洋细菌 marine bacteria
海洋性气候 marine climate
海洋性森林 maritime forest
海洋学 oceanology
海运木排 cigar raft
海藻 marine algae
海藻酸 alginic acid
海藻植被 marine (algae) vegetation
害虫 insect pest
害虫大发生 outbreak of insect
害虫管理 pest management
害虫综合防治 integrated control of insect pests
害鸟 injurious bird
害兽 injurious beast
含氮量 nitrogen content
含酚杀虫油 phinatas oil
含晶薄壁组织 crystalliferous parenchyma
含晶分子 crystalliferous element
含晶厚壁组织 crystal sclerenchyma
含晶木薄壁组织 crystalliferous xylary parenchyma
含晶室 crystal locule
含晶细胞 crystal cell; crystalliferous cell
含晶异细胞 crystal idoblast
含钠熔渣 soda slag
含沙量 sediment charge; silt concentration
含沙率 sediment charge
含砷杂酚油(防腐剂) arsenical creosote
含树脂的 resiniferous
含水层 water bearing layer
含水量 moisture content; water content
含水量测定 moisture (content) determination
含水量测定仪 moisture meter
含水量分布试片 moisture distribution section
含水量临界值 threshold moisture level
含水量试片 moisture (content) section
含水量阈 threshold moisture level
含水率 moisture content
含水率测定 determination of moisture content
含水试片绝干重 oven-dry weight of moisture section
含水油 wet oil
含松脂树根 pitch stump
含髓锯材 boxed-heart (timber)
含碳气溶胶 carbonaceous aerosol
含碳土 carbonaceous soil
含羞草碱 mimosine
含盐的 saline
含盐度 salineness
含盐量 salinity
含盐土壤残值 salty soil
含油率 oil length
含油树脂 oleoresin
含脂材 lightwood; resinous wood; turpentine-timber
含脂节 resinified knot
含脂纤维素 adipo-cellulose
焓 enthalpy; heat content; total heat
寒带 arctic zone; frigid belt; frigid zone
寒带的 boreal
寒带林带 cold forest zone
寒带林区 frigid forest-region

寒害 cold damage; cold injury; winter-killing
寒荒漠群落 frigorideserta
寒极 cold pole
寒冷的 hiemal
寒冷指数 coldness index
寒流 cold current; cold snap
寒漠 cold desert; fell field
寒土植物 psychrophyte
寒温带混交林 mixed boreal forest
寒温带针叶林 boreal forest; taiga
寒武纪 Cambrian period
罕见景观 unique landscape
汉兹利克公式 Hanzlik's formula
汉兹利克氏公式 Hanzlik formula
汗腺 sweat gland
旱成土 aridisol
旱风 desiccating wind
旱谷 arroyo
旱害 damage by drought; drought damage
旱轮(年轮) drought ring
旱坡 xerocline
旱期休眠 drought dormancy
旱生 xero-
旱生的 xeric
旱生动物 xerocole
旱生林 espinal
旱生群落 xeric community
旱生树种 xerophilous tree species
旱生演替 xerarch succession; xeric succession
旱生演替系列 xerosere
旱生植被 xeromorphic vegetation
旱生植物 xerophyte
旱生植物的 xerophytic
旱生植物群落 xerophytia
旱性结构 xeromorphism
旱性结构的 xeromorphic
旱灾 drought; drouth
焊接 jointing; solder
焊料 solder
焊枪 burner
夯 rammer
夯具 compactor
夯式压路机 tamping roller
行 tier
行播 drill; line seeding; line sowing; row seeding; seeding in line; sowing in line; sowing in row
行堆 windrow
行堆间栽植(采伐残余物堆间) windrow planting
行间除草机 in-row weeder
行间除草小耙 drill-harrow
行间的 inter-row
行间伐 row thinning
行间施肥 side dressing
行间中耕 hoeing; inter-row cultivation; length(wise) cultivation
行间中耕器 drill-hoe
行距 between-row space; distance between rows; planting width; row spacing
行列 rank
行列式 determinant
行内的 in-row
行情调查机构 market information institution
行业协会 trade association
行植 drill planting; line-planting; row planting
行状混交 line mixture
行状配置 strip spacing
行状小区 linear plot
行状栽植 planting in row
绗缝花纹(类似绗棉被缝线) quilted figure
航测林分蓄积量表 aerial stand volume table
航测树木材积表 aerial (tree) volume table
航测制图 aerial mapping
航带 flight strip
航点 waypoint
航高 flight altitude; flight height; flying height
航迹 flight path
航空材积表 aerial volume table
航空测量 aerial photogrammetry; air survey
航空高差记录器 airborne profile recorder
航空护林 aerial forest fire protection
航空胶合板 aircraft plywood; airplane plywood; microplex
航空胶片 aerial film

H

航空曝光指数 aerial exposure index
航空倾斜摄影像片 aerial oblique photograph
航空摄影测量 aerial (photo) survey; aerial photogrammetry
航空摄影测量学 aerocartography
航空摄影机 aerial camera
航空调查 aerial surveying; flying survey
航空卫星 aerosat
航空像片比例尺 scale of aerial photograph; aerial photograph scale
航空像片测树 aerial photo mensuration
航空像片分层 stratification in aerial photograph
航空像片类型 aerial photograph type
航空像片略图 air photo mosaic
航空像片索引 aerial photograph index
航空像片镶嵌图 air photo mosaic
航空用材 aircraft stock
航空照片 aerial photograph
航空照相图 photo map; photographic map
航拍 aerial photography
航偏 drift
航偏角 angle of drift
航片覆盖区 aerial photo coverage
航片判读 aerial photo interpretation
航摄带 flight strip
航摄失真中心 isocenter
航摄像片转绘仪 aero sketchmaster
航天学 astronautics
航线 flight line; lane; track
航线图 flight map
航向 course; heading
航向重叠 end lap
蒿柳 basket willow; common osier; osier willow
豪猪 porcupine
好眼力 master eye
好碱植物 alkaline plant
好气分解 aerobic decomposition
好气固氮细菌 aerobic nitrogen-fixing bacteria
好气培养 aerobic culture
好气生活 aerobiosis
好气微生物 aerobe
好气细菌 aerobic bacteria
好氧生物 aerobiont
好营养的 eutrophic
号锤 branding hammer; log stamp; marking hammer; marking iron; stamp ax(e); stamping hammer
号码牌(树木标号用) number tag
号印 brand; branding iron
号印斧 stamp ax(e)
耗减 depletion
耗减百分比 percentage depletion
耗减优惠(扣除额) depletion allowance
耗尽土壤 exhausted soil
耗能 energy consumption
耗汽量 steam consumption
耗热量 heat consumption; thermal losses
耗散结构 dissipative structure
耗散结构理论 dissipative theory
耗散热 burn-off
耗水量 water consumption
耗损 detrition; drain
耗损的 waste
耗损率 attrition rate
耗氧量 oxygen consumption
禾本科的 gramineous
禾本型地上芽植物 chamaephyta gramminidea
禾本植物学 agrostology
禾草类 graminoid
禾草类指数 agrostological index
禾草稀噬草原 grass savanna
禾草学 agrostology
禾谷植物 cereal grass
禾木胶 acaroid gum; acaroid resin
合板(胶合板) plywood
合瓣的 gamopetalous; menopetalous; sympetalous
合瓣花冠 sympetalous corolla
合胞体 symplasm; symplast
合被的 synphyllous
合并(机构或数据等) incorporation; pool
合并资产负债表 amalgamated balance sheet
合巢群聚 synoecium
合成单宁 synthetic tannin; synthetic tanning agent

H

合成风 resultant wind
合成胶 synthetic glue
合成胶料 synthetic size
合成孔径雷达 synthetic aperture radar
合成孔径天线(雷达) synthetic aperture antenna (radar)
合成培养基 synthetic medium
合成鞣剂 syntan; synthetic tannin; synthetic tanning agent
合成树脂 synthetic resin
合成树脂胶黏剂 synthetic resin adhesive
合成天线 synthetic antenna
合成纤维 synthetic fibre
合成像片 composite photograph
合成橡胶 elastomer; synthetic rubber
合成橡胶胶黏剂 synthetic rubber adhesive
合成樟脑 synthetic camphor
合成指数 composite index number
合点(植物) chalaza
合叠 ply
合萼的 gamosepalous; monosepalous
合法避税 tax avoidance
合法的林业 regulated forest management
合法命名 legitimate name
合法属 legitimate genus
合格的 certified
合格浆料 accepted stock
合格木片 accepted chip
合格试验 qualification test
合格证 certification
合股机 buncher
合核细胞 syncaryocyte
合伙关系 partnership
合伙经营 pool
合接 fit grafting; splice grafting
合理报酬率 possible rates of return
合理化 rationalization
合理交配 judicious mating
合理造材 buck for grade
合力 resultant; resultant of forces
合排 joining rafts
合蕊的 gynandrian
合蕊柱 gynostemium
合生 coalesce
合生花冠 gamomery
合生群落 concrete community
合生维管束 conjoint bundle
合生纹孔口 coalescent pit aperture; fused aperture
合生子房的 gamogastrous
合体共生 conjunctive symbiosis
合同 instrument
合同销售 license sales
合线 zygotene
合线期 zygotene stage
合心皮果 syncarp; syncarpous fruit
合营企业 partnership
合约价 exercise price
合轴孢子 sympodulospores
合轴分枝的 sympodial branching
合资企业 joint venture
合子 cytula; zygote
合子不育性 zygotic sterility
合子核 zygotic nucleus
合子减数分裂 zygotic meiosis
合子前隔离 prezygotic mechanism
合作 cooperation
合作剂 synergist
合作林管理法案 cooperative forest management act
合作社 cooperative
合作狩猎 cooperative hunting
河岸林 riverain forest; riverine forest
河岸园 riverside garden
河岸植树 plant trees on river banks
河边集材场 bank
河边植被 riverside vegetation
河冰的 fluvioglacial
河成沉积 fluviatile deposit
河成的 fluvial
河成阶地 fluvial terrace
河川 river
河川等级 stream rating
河川径流 stream flow
河川流送 stream driving
河川频数(单位面积河流数) stream frequency
河川群落 stream community
河川外贮木池(制材厂) off-stream storage pond
河川学 potamology
河川贮木池 on-stream storage pond

河床 stream bed; water course
河床沉积 channel fill deposit
河床型 river bed type
河道 lade
河道出口 outfall
河道侵蚀 stream channel erosion
河道取直 channel straightening
河道网 hydrographic net
河道贮留水量 channel storage
河绠 boom; storage boom
河绠冰害防护木 ice guard
河绠工 boom poke
河绠使用费 boomage
河谷 vale
河间地 interfluve
河口 debouch; embouchure; estuarine; estuary; influx; lade; river mouth or estuary
河口浮游生物 estuarine plankton
河口湾 estuary
河狸尾式(枪柄) beaver tail
河流沉积 fluviatile deposit
河流冲击物 riverwash
河流的 fluvial; potamic
河流浮游生物 potamoplankton
河流干涸 rivers die out
河流阶地 fluvial terrace; river terrace
河流侵蚀 fluvial erosion
河流取直 cut-off
河流群落 potamium
河流学 rheology
河流综合开发 multiple purpose river development
河漫滩 alluvial flat; flood land; flood plain; washland
河漫滩草甸 flood plain meadow
河漫滩森林 floodplain forest
河面温度 river (surface) temperature
河鸟 dipper
河水剥蚀作用 pluviofluvial denudation
河滩地 overflow land; flood land
河湾 embayment; ox bow
河系 river system; water system
荷尔蒙杀树剂 hormone silvicides
荷花 Chinese water-lily; lotus
荷兰球根 Dutch bulb
荷兰式打浆机 Hollander (beater)
荷兰式打浆机飞刀辊 Hollander roll
荷兰园林 Dutch garden
荷兰原木板英尺材积表 Holland log rule
核 kernel; nucleus (*plur.* nuclei); putamen
核磁共振 nuclear magnetic resonance
核磁共振谱测定法 NMR spectrometry
核蛋白 nucleoprotein
核定概算 approved estimate
核定股本 authorized (capital) stock
核定投资额 capitalized cost
核定预算 approved budget
核定资本值 capitalized cost; capitalized value
核断裂 karyoclasis
核对 check
核对表 checklist
核对簿 check book
核对器 verifier
核对清查法 check method; control method; examination method; method of control
核对清查体系 check system
核对调查 check cruise
核纺锤体 nuclear spindle
核分裂 caryokinesis; karyokinesis
核苷 nucleoside; ribo(nucleo)side
核苷酶 nucleosidase
核苷酸 nucleotide; ribo(nucleo)tide
核苷酸酶 nucleotidase
核果 drupe; kernel fruit; stone fruit
核果肉 naucum (*plur.* nauca)
核果状果实 pyrenocarp
核基因 nuclear gene
核浆 karyoplasm
核接合 karyapsis
核菌 Pyrenomycete
核菌果 pyrenocarp
核壳 putamen
核壳体 nucleocapsid
核膜 karyolemma; nuclear membrane
核内的有丝分裂 endomitosis
核内多倍性 endopolyploidy
核内复制 endoreduplication
核内核分裂 endomitosis

核内有丝分裂 endomitosis; mesomitosis
核内再复制 endoreduplication
核配(同性配子之互相融合) karyogamy
核仁 nuclcolus (*plur.* -li)
核仁组成中心 nucleolar organizer
核融合 karyogamy
核实 verification
核素 nuclein
核酸 nucleic acid
核酸酶 nuclease
核酸周期 nucleic acid cycle
核算方法 accounting approach; accounting method
核糖核 ribonuclease (Rnase/RNAse)
核糖核酸 ribosomal RNA
核糖体 ribosomal RNA; ribosome
核糖体RNA ribosomal RNA
核桃球蛋白 juglansin
核桃油 walnut oil
核酮糖 ribulose
核心地区 focal region
核心家庭 nuclear family
核心降水区 precipitation cell
核心区 core area; core zone; central area
核型 caryotype; karyotype; nuclear type
核型多角体病毒 nuclear-polyhedrosis virus
核型图 caryogram; karyogram
核验 verification
核遗传学 caryogenetics
核再复制 endoreduplication
核质 karyoplasm
核质不亲合性(不协调) nuclear-cytoplasmic incompatibility
核质相互作用假说 nucleo-cytoplasmic interaction hypothesis
核状结构 nutty structure
盒 case
盒形梁 box beam
翮(hè) calamus
赫尔克里士滴测法 Hercules drop method
褐斑 foxiness
褐变 brown stain
褐醋石 brown calcium acetate
褐腐 brown rot; carbonizing rot; dry rot
褐腐菌 brown-rot fungus
褐色胶 brown size
褐色磨木浆 steamed groundwood
褐色木浆 brown wood pulp
褐色土 cinnamon soil
褐色萎缩 brown atrophy
褐色乙酸钙 brown calcium acetate
褐土 cinnamon soil; drab soil
鹤嘴锄 mattock; pick mattok; pick-haue
壑 clove; sink
黑瘤肿 black knot
黑白红外胶片 black and white infrared film
黑白瓶法 light and dark bottle method
黑斑病 black spots
黑斑纹 roe
黑潮 kuroshio
黑达马树脂 black dammar
黑貂 sable
黑方解石 anthraconite
黑粉病 smut
黑粉菌 smut-fungus
黑腐病 black rot
黑腐酸 humic acid
黑钙土 chernozem; tshernosem
黑根腐病 black root rot
黑化(有机质渗入表层矿质土) melanization; melanism
黑火药 black gun-powder
黑碱土 black alkali soil; black alkaline soil
黑碱液 black lye
黑椒酚 chavicol
黑椒素 chavicine
黑结[病] black knot
黑荆树栲胶 acacia extract; black wattle extract
黑胫病 black leg
黑淋溶土 altalf
黑榴石 melanite
黑棉土 black cotton soil; regur
黑木耳 black fungus; Jew's ear; Judas's ear
黑木炭 black wood charcoal
黑木蚁 black carpenter ant
黑软土 altoll

黑色草炭土 black turf soil
黑色裂纹(夹皮导致的木材缺陷) black check; black streak
黑色湿草原土 black prairie soil
黑色石灰土 rendzina
黑色石灰性冲积土 borovina; borowina
黑山公式(收获调整法之一) Black Hills formula
黑松烯 thunbergene
黑穗病 smut
黑穗菌 smut-fungus
黑炭 black charcoal
黑碳 black carbon
黑体(全部吸收辐射能) blackbody
黑体发射 black-body emission
黑体辐射 black-body radiation
黑土 black earth; black soil
黑线病(嫁接不亲和) black line
黑箱模型 black-box model
黑心病 black heart
黑心材 black heart
黑杨潮虫(棉鼠妇) sowbug; wood louse
黑液 black liquor
黑云母 biotite
痕 mark; scar
痕迹 trace
痕量级 trace level
痕量元素 trace element
亨德哈根公式 Hundeshagen's formula
恒等式 identity
恒定 constancy
恒定的 constant
恒风 steady wind
恒化器 chemostat
恒渗透压动物 homoiosmotic animal
恒湿器 humidistat; hygrostat
恒水植物 homeohydric plant; homoiohydric plant
恒态 steady state
恒温的 homoiothermic; idiothermal
恒温动物 endotherm; homeotherm; homeothermal animal; homoiotherm; homoiothermal animal; homothermal animal
恒温器 thermostat
恒温箱 incubator
恒温性 homoiothermy
恒向趋性 menotaxis
恒续林 continuous cover forest; continuous forest
恒续林施业 continuous cover forest management system
恒续林思想 idea of continuous forest
恒雪带 nival belt
恒有度 constancy
恒有种 constant; constant species
恒浊器 turbidostat
桁 girder
桁构梁 trussed beam
桁架 girder; truss (frame)
桁架臂架 lattice boom
桁架构件 truss member
桁架节点 panel point
桁架中柱 crown post; king post
桁条 purline; stringer piece
桁下横木(地龙) lower cord
鸻(héng) plover
横摆装车 gunning
横臂 crossbar
横槽 dado
横槽搭口接 dado-and-rabbet joint
横槽舌接 dado-and-lip joint
横槽榫接 dado-and-tenon joint
横撑 bolster; flying-shore; wale
横撑臂分叉索 sail-guy spreader line
横撑臂支索 sail guy
横撑链(两侧壁之间的支撑件) spreader chain
横担木 bunk; cross arm; cross piece; jib(arm)
横档 capsill; cross member; rail; wale; top rail
横档木 arm-tie
横的 transverse
横垫木(装车时使木材滚向中心) dutchman
横动喷气驱尘装置 air cleaner traverse
横渡河流 river crossing
横断工事(河川的) transverse construction
横断裂 cross break
横断面 cross section; transverse section; traverse surface
横断面测量尺 cross-section rod
横断面尺寸 cross dimension
横断面积 cross-sectional area

横堆法 cross piling; cross stacking; cross-piled loading; crosswise piling
横杆 beam
横隔[膜] transverse septum
横绠 lateral boom
横沟 transverse trench
横贯线 intersecting line
横滚式装车 parbuckling
横滚装车 parbuckle
横滚装车设备 slap happy loading rig
横滚装车索具 parbuckle
横过的 transverse
横过路面的明水沟 Irish bridge
横河绠 transverse boom
横积炭窑 horizontal kiln
横截 cross cutting; transverse cut
横截的 trimmed
横截端面 crosscut end
横截法 transect
横截后备长度 allowance for trimming
横截锯 briar; buck saw; cross cutter; cross-cut saw; cross-cutting saw; cut off saw; docker; docking saw; swede fiddle
横截面 cross section
横截圆锯机 trimmer
横锯 cross cutting
横口斧 adze; shim hoe
横梁 bolster; cross rail; crossbar; cross-beam; floor beam
横梁木 span piece
横梁式运材车 bolster truck
横列射线细胞 procumbent ray cell
横列型(气孔) diacytic type
横裂 cross break; transverse dehiscence; transverse shake
横林班线 cross-ride
横流风机 transverse-flow fan
横流式干燥窑 cross current type dry kiln
横眉 breast-summer
横面切片 transection
横磨 cross grinding
横木 ledger; rail; traverse
横木杆道 corduroy (road); skidding road; skipper road
横木滑道 dragging rack; dragging track
横木条 stretcher
横拼机 cross feed splicer
横坡 transverse grade
横剖面 cross section
横切 cross-sectioning
横切面 cross-sectional surface; end surface; end-grain section; transection; transverse section; transverse surface
横切片 transverse section
横伸梯臂支索 sail guy; sailing line
横生的 amphitropous; amphitropic; hemitropal; hemitropous
横式挤压机 couch press
横式起重机 swing-jack
横式气道 horizontal flue
横闩(木材结构中穿孔板) bar
横条 trabecula (*plur.* -lae)
横凸榫 cross-tongue
横弯 edge bend
横弯材 cupped wood
横纹导管 trabecular vessel
横纹断裂负荷 perpendicular to grain load at failure
横纹剪切 shear perpendicular to the grain
横纹胶合板 cross grained plywood
横纹抗剪 vertical shear
横纹抗剪强度 shear strength perpendicular to grain
横纹抗拉强度 tensile strength perpendicular to grain
横纹抗拉试验 tension test perpendicular to grain
横纹抗压强度 compression strength perpendicular to grain
横纹拉力 tension perpendicular to the grain
横纹刨 block plane
横纹强度 strength perpendicular to the grain
横纹切削 end-grain cutting
横卧细胞 procumbent cell
横线支票 cross check
横向 transverse
横向变形 lateral deformation
横向变形系数 Poisson's ratio
横向槽 cross slot

H

横向成型铺装头 cross direction forming head
横向搭接 cross lap joint
横向弹性 transverse elasticity
横向弹性压带器 transversal pad
横向地性 diageotropism
横向吊装 gunning
横向定向结构刨花板 oriented structural board-cross direction
横向分裂 transverse division
横向共振 transverse resonant vibration
横向滚柱 line up roll
横向滚装 crosshaul(ing); parbuckling
横向集材 cross yarding
横向力 lateral force
横向排水[沟] cross drain
横向膨胀 lateral expansion
横向气流冲击式粉碎机 cross-stream impact mill
横向翘曲 cupping; transverse warping
横向切削 transverse-longitudinal cutting
横向侵蚀 lateral corrosion; lateral erosion
横向热膨胀 transverse thermal expansion
横向沙丘 transverse dune
横向输送机 cross conveyor; horizontal conveyor; transverse conveyor
横向树脂道 horizontal resin canal
横向弯曲 lateral bending
横向性的 diatropic
横向循环干燥窑 cross current type dry kiln
横向应变 lateral strain
横向应力 lateral stress; transverse stress
横向载荷 transverse loading
横向造材 sidewise bucking
横向振动 lateral vibration; transverse vibration
横向中耕 cross(wise) cultivation
横向重叠 side lap
横向装材 crosshaul(ing)
横向装车机 log elevator
横向装载 cross-piled loading
横压力 lateral pressure
横轧辊 cross roll
横折 cross break
横褶皱 cross fold
横枕连接件 traverser-link
横轴干燥窑 cross-shaft (dry) kiln
横装材堆 cross piling; cross stacking; cross-piled loading; crosswise piling
横装短材齐端 load aligning
烘房 drying room
烘干 oven-dry
烘干部 dry (*or* drier) part
烘干材 oven-dried wood; oven-dry wood
烘干的搪瓷器 stoving enamel
烘干法 oven-drying method
烘干漆 baking varnish
烘干重 oven-dried weight; oven-dry weight
烘干状态 oven-dried condition
烘缸毛布 dry felt
烘炉 warming-up
烘漆 baking finish; stoving enamel
烘箱 drying oven; oven
哄抬价格 price pushing
红变 red stain
红变菌 red staining fungus; rough staining fungus
红顶病 red-top
红豆树碱 ormosine
红豆素 jequiritin; abrin
红豆酸 abric acid
红豆因 abrin; jequiritin
红腐 red rot
红根 red root
红根栲胶 cherokee rose extract; red root extract
红厚壳树脂 tacamahac
红花罗勒油 sweet basil oil
红花素 carthamin
红环腐 red ring rot
红金鸡纳皮 red cinchona bark
红橘素 tangeritin
红橘油 tangerine oil
红利 bonus; dividend
红利股息 bonus dividend
红栗钙土 reddish chestnut soil
红林蚁 red wood ant; red forest ant
红没药烯 bisabolene

红莓子蜡 cranberry wax
红黏土 adamic earth
红皮书 red data book
红皮松类 red pine
红铅 minium; red lead
红曲红素 monascorubrin
红曲黄素 ankaflavin
红曲霉 monascus
红壤 krasnozem; red earth; red loam; red soil
红壤化 lateri(ti)zation
红色名录 red list
红色漠境土 red desert soil
红色黏土层 argillaceous red bed
红色湿草原土 red prairie soil
红色石灰土 limestone red loam; terra rossa
红色天然染料 hypernic
红肾圆盾蚧 agave scale; California red scale
红树林 mangrove; red mangrove; mangrove forest
红树林沼泽地 mangrove swamp
红树林沼泽土 mangrove soil
红树皮单宁 mangrove tannin
红树皮栲胶 mangrove extract
红树皮鞣质 mangrove tannin
红树群系 mangrove formation
红土 laterite; red earth; red soil
红土化作用 laterization
红外辐射 infrared radiation
红外光谱法 infrared spectrometry
红外光区 infrared spectrum
红外加热 heating by infrared rays
红外胶片 infrared film
红外扫描器 infrared scanner
红外探火仪 fire scan; infrared fire detector
红外吸收 infrared absorption
红外吸收光谱 infrared absorption spectum
红外吸收光谱法 infrared absorption spectroscopy
红外线 infrared radiation; infrared ray; ultrared ray
红外线的 infrared (IR); ultrared
红外线分光辐射仪 infrared radiometer spectrometer
红外线辐射计 infrared radiometer
红外线感光照相 infrared sensing photographic
红外线干燥 ultrared drying
红外线干燥法 infrared drying; red-ray drying
红外线加热 heating by infrared rays
红外线频率 infrared frequency
红外线扫描器 infrared line scanner
红外线探测系统 infrared detection system
红外线探火仪 infrared scanner
红外线透射 infrared transmission
红外照片 infrared photography
红心腐 red heart
红玄武土 bole
红雪(水文) red snow
红叶螨 red-spider
红园 red garden
红樟油 brown camphor oil
红蜘蛛 red-spider
红紫胶素 erythrolaccin
红棕钙土 reddish brown (steppe) soil
红棕色森林土 reddish brown forest soil
红棕色砖红壤性土 reddish brown lateritic soil
宏单元 macroelement
宏观的 macroscopic
宏观分析 macro-analysis
宏观构造 gross structure; macroscopic structure; macrostructure
宏观检查 macrography
宏观检验 macroscopic examination
宏观解剖特征 gross anatomical feature
宏观进化 macroevolution
宏观特征 gross feature; macroscopic feature
宏观症状 macroscopic symptom
宏观组织 gross structure; macrostructure
虹彩云 iridescent cloud
虹吸管 siphon
虹吸管雨量器 siphon raingauge
虹吸器 crane
洪泛平原 flood land; flood plain
洪峰 flood peak; peak flow
洪峰阶段 crest stage
洪峰流量 flood peak discharge; flood peak rate; peak discharge

洪峰流速 flood peak rate
洪峰纵剖面 crest profile
洪积层 deluvium
洪积世 Pleistocene
洪积台地 diluvial upland
洪积土壤 diluvial soil
洪积物 proluvium
洪流 flood current; torrent
洪流控制工程 torrent control works
洪水 deluge; flood; inundation
洪水标记 flood mark
洪水冲蚀 flood erosion
洪水赶羊天数(平均) stream rating
洪水痕迹 flood mark
洪水季节 flood season
洪水控制 flood control
洪水流量 flood discharge
洪水期 flood period; flood season
洪水期河床 high water bed
洪水侵蚀 flood erosion; pluvial erosion
洪水位 flood height; flood level; flood stage
洪水灾害 flood damage
喉凸 palate
猴 monkey
后(昆虫) queen
后把 stock
后备长度 bucking allowance; overlength; overtrim
后背板(昆虫) post notum
后绷索 back guy
后冰期沉积 late glacial deposit
后柄 stock
后侧片 epimeron (*plur.* -era)
后肠 posterior intestine
后成岩 deuterogene
后处理 after treatment
后床腿 back leg
后代 advanced generation; offspring; posterity; progeny
后代测验 progeny test; progeny trial
后代的 filial
后代回归律 law of filial regression
后代杂交 advanced generation cross
后导向板 rear guide shoe
后顶极(群落) post climax
后端固定索 tail hold
后端自卸车 end dumper; tail dumper
后发性枯梢 postemergence tipburn
后伐 clearfell; post-logging; removal cutting; secondary felling
后方交会(测量) resection
后杆(集材工具) end mast
后固化 postcure
后横档 back rail
后滑 kickback
后继共生 metabiosis
后价 end value; future value; prolonged value
后焦距 back focus
后角 clearance angle; lip clearance
后进先出法 last in first out
后扣块 rear lump
后拉角 tieback angle
后拉锚索 tail hold
后拉索 bridle; tieback line
后拉索锚桩(伐根) tieback stump
后瞄准器 rear sight
后排气 final vacuum
后坡 backslope
后期(细胞分裂) anaphase
后期抽空 final vacuum
后期腐朽 advanced decay
后期空气压 final air press
后期流送 post drive
后期气压 final air press
后期疏伐 advanced thinning; later thinning; old thinning
后期杂草防除 late weed control
后期蒸汽浴 final steam bath
后生成本 after cost
后生厚角组织 metacollenchyma
后生木质部 deutoxylem; metaxylem
后生韧皮部 metaphloem
后生树 after-growth; regrowth
后生形成层 postcambial
后生形成层阶段 postcambial stage
后生异粉性 metaxenia
后食 adult feeding
后适应 post-adaptation
后手柄(油锯) rear handle
后熟 late-maturity
后熟处理 after-ripening treatment

后熟型休眠 after-ripening type of dormancy
后熟种子 after-ripening seed
后熟作用 after-ripening
后水解 posthydrolysis
后台(计算机部件) back ground
后台处理 background processing
后膛枪 breech loader
后天井 back yard
后天性状 acquired character
后庭 back yard
后头(昆虫) occiput
后退演替 regressive succession
后向反射 backreflection
后向散射 backscattering
后效 after-effect
后斜角 back bevel angle
后斜面 back bevel
后胸 metathorax
后胸背板 metanotum
后胸侧板 metapleuron
后延企业联合 forward integration
后裔 descendance; offspring
后硬化 postcure
后院 back yard
后沼泽 back swamp
后植核酸 metagon
后轴驱动 pusher drive
后装架 rear mount
后足 metapedes
后坐(伐木时) setting back
后坐垫 recoil pad
后坐力 recoil
厚板 thick board; plank board; plank
厚薄加工 caliper finish
厚背锯 back saw
厚壁(细胞) thick wall
厚壁含晶细胞 thick-walled crystal cell
厚壁细胞 sclerenchymatous cell
厚壁纤维 thick-walled fibre
厚壁组织 selerenchyma
厚层泥岩 argillite
厚层土壤 deep soil
厚度规 gap gauge; thickness gauge
厚度选择器 thickness selector
厚革 heavy leather
厚果皮的 pachycarpous
厚角组织 collenchyma
厚角组织细胞 collenchymatous cell
厚壳桂碱 cryptocarine
厚块 slab
厚膜(词头) chlamydo-
厚膜孢子 chlamydospore
厚木板英尺材积表 board-foot log rule
厚木片 coarse splinter
厚膨胀 thickness swelling
厚片 slab
厚漆 paste paint
厚镶板 solid panel
厚芯板结构 lumber-core construction
厚芯胶合板 wood core plywood
厚叶的 dasyphyllous; pachyphyllous
厚硬纸板 cardboard
厚窄板 deal
候鸟 bird of passage; migrant; migratory bird; transient; migrant bird
候鸟类 migratory birds
候鸟迁徙 bird migration
候鸟移动期 time of migration (bird)
呼笛(诱捕鸟兽) calls
呼拉锯(双锯片截锯机) hula saw
呼吸分离 adsorption stripping
呼吸高峰(种子成熟时) climacteric
呼吸高峰前期 preclimacteric rise
呼吸根 aerating root; breathing root; pneumatophore; pneumatophoric root; respiratory root
呼吸计 respirometer
呼吸孔 breathing pore; stoma (*plur.* stomata/stomas)
呼吸链 respiratory chain
呼吸强度 respiratory rate
呼吸商 respiratory quotient
呼吸生物量比 respiration-biomass ratio (R/B ratio)
呼吸式单板干燥机 breathing veneer drier
呼吸式干燥机 platen drier; breather drier
呼吸水流 respiratory current
呼吸速率 breathing rate; respiratory rate

H

呼吸消耗 respiration loss
呼吸跃变 climacteric
呼吸作用 respiration
狐尾锯 drag saw; dragsaw; fox-tail saw
狐尾梢 foxtail
狐尾榫 fox wedged tenon
弧 segmental arch
弧口薄凿 paring gouge
弧口中凿 middle gouge
弧裂 arc shake; burst check; cup shake
弧形端接法 serpentine end matching process
弧形桁架 Belfast truss
弧形泄水闸门 segmental waste gate
弧状(脉) campylodromous
弧状菌 bracket fungus
胡薄荷油 penner oil
胡伯尔求积式 Huber's formula
胡椒油 black pepper oil; pepper oil
胡萝卜素 carotin
胡萝卜素酶 carotinase
胡敏-富里酸 humo-fulvic acid
胡敏素 humin
胡敏酸肥料 humic fertilizer
胡荽子油 coriander oil
胡桃(核桃) corcassian walnut; common walnut; European walnut; Persian walnut
胡桃根皮浸膏 juglandin
胡桃醌 juglone
胡桃钳 carpenter's pincers
胡桃素 juglandin; juglansin
胡桃酮 juglone
胡桃油 nut oil
壶穴侵蚀作用 evorsion
葫芦状的 lageniform
湖泊的 lacustrine
湖泊类型 lake type
湖泊效应 lake effect
湖沉积 lacustrine deposit
湖成的 lacustrine
湖成阶地 lake terrace
湖成泥炭 limnic peat
湖底沉积物 gyttja
湖底的 pythmic
湖底腐殖质土 gyttja soil
湖底生物 geobenthos
湖积白垩粗质土 lake chalk raw soil
湖积黄土 lake loess
湖积土壤 lacustrine soil
湖上层 epilimnion
湖下层 hypolimnion
湖沼带生物群落 communities in the limnetic zone
湖沼学 limnology
槲果 acorn; glans
槲寄生 giant mistletoe; mistletoe
槲树皮栲胶 oak bark extract
槲树皮鞣法 oak bark tannage
蝴蝶桌 butterfly table
糊粉 aleurone
糊剂 magma; paste
糊精 amylin; British gum; gummeline
糊状树脂 paste resin
虎斑 brindle
虎斑花纹 silver figure
虎斑纹 tiger grain
虎斑纹单板贴面桌 tiger table
虎斑纹理 silver-grain
虎蛾类 tiger moth; woolly bear
虎克定律 Hooke's law
虎皮楠碱 daphniphylline
虎鼬 tiger weasel; marble polecat
琥珀 amber; fossil resin
琥珀色 amber
互变异数分析 analysis of covariance
互补基因 complementary gene
互补碱基 complementary base
互补碱基序列 complementary base sequence
互补色 complementary colour
互补试验 complementary assay
互补纹孔 complementary pit
互补因子 complementary factor
互补作用 complementary action
互搭 probit
互搭板壁 drop siding
互搭鸠尾榫 lap dovetail
互换 interchange
互换性 interchangeability
互惠共生 mutualism; mutualistic symbiosis
互交 intercross; reciprocal crosses
互交可孕的 interfertile
互利共生 endoecism

互利群聚 essential aggregation
互连 interlock
互列纹孔式 alternate pitting
互助公司 mutual company
互作变量 interaction variance
户外嫁接 outdoor grafting
户外游憩 outdoor recreation
户外游憩价值 outdoor recreational value
户外游憩模式 outdoor recreation model
户外游憩资源可利用性 outdoor recreation resource availability
护岸 bank protection; breast-wall
护岸绠 stiff-leg
护岸林 stream bank protection forest
护岸漂子 side boom
护坝 counter-dam
护板 apron; protective shield; sheeting
护被 chlamydia
护壁板 dado; wainscot
护壁板顶木条 dado rail
护道 berm
护堤(保护城堡) mound
护堤壁 retaining wall
护盖 protecting cover
护根接 nurse-root grafting
护轨 anti-derailing rail
护轨夹 yoke
护林 protect forests
护林鸟 tree-protecting bird
护林协会 log patrol
护林员 forest guard; forest ranger; prevention patrolman; warden
护林员制度 district-ranger system
护林制度 forest-guard system
护路边石 guard stone
护路沟 highway ditch
护面板 board liner
护木(滑道) prevent log
护木鸟 tree-protecting bird
护目镜 goggle
护牧林 pasture protection forest
护墙 breast-wall; counterfort
护墙板 clapboard
护石 guard stone
护套 sheath
护田林 shelter belts of protect farmland
护铁(爬犁脚) chain plate
护铁(双铁刨) back iron
护檐板 clapboard
护罩 protecting cover
戽(hù) clamshell
戽式链轮 bucket wheel
瓠果 pepo
花 bloom; flower
花柏酚 hinokiol
花柏酚亭 chanootin
花瓣 petal
花瓣上着生的 epipetalous
花瓣维管束 petal bundle
花瓣与萼片互生 alternisepalous
花瓣状的 petaloid
花瓣状花萼 petaloid calyx
花被 floral envelope; perianth; perigone; perigonium
花被瓣 perianth lobe; perianth segment
花被的 perianthial
花被卷叠式 aestivation; praefloration; prefloration
花被卷叠武 estivation
花被片 tepal
花边 moulding
花边刨 cornice plane
花钵 flower pot
花部退化变态 antholysis
花彩孔材 festoon porous wood
花彩型管孔排列 pore flgured arrangement
花彩状 festoon-like
花侧柏酚 cuparophenol
花侧柏酮 cuparone
花侧柏烯 cuparene
花程式 floral formula
花蝽 flower bug
花丛 anthemia; anthemy
花丛花坛 cluster flower-bed
花丛花坛花卉 cluster bedding plant
花簇 anthemia; anthemy; blossom cluster; flower cluster
花的 florulent; floscular; flosculous
花端 blossom end
花萼 calyx
花萼变花瓣(瓣化) calycanthemy

花发端 floral initiation; flower initiation
花房 green house
花粉 pollen
花粉病 pollinosis
花粉不育性 pollen sterility
花粉铳 pollen gun
花粉催取室 extractor
花粉代替物 pollen substitute
花粉带 pollen zone
花粉分类学 palynotaxonomy
花粉分析(古气候学) pollen analysis
花粉盖 stopples
花粉供给者 pollen supplier
花粉管 pollen tube; spermary
花粉化石 fossil pollen
花粉浸出物 pollen extract
花粉块 pollinia; pollinium
花粉粒 microdiode; pollen grain
花粉粒有丝分裂 pollen mitosis
花粉蒙导 pollen mentor
花粉母细胞 pollen mother cell
花粉囊 microdiodange; pollen sac
花粉培育 pollen (grain) culture
花粉喷射器 pollen gun
花粉剖面 pollen profile
花粉谱 pollen spectrum
花粉亲本 pollen parent
花粉室 pollen chamber
花粉素 pollenin
花粉图式 pollen diagram
花粉团 pollinium
花粉小块 massulae
花粉携带者 pollen carrier
花粉学 palynology
花粉症 pollinosis
花粉植株 pollen plant
花粉制取器 extractor
花蜂科 anthophorids
花盖 lid; perigone
花岗闪长岩 granodiorite
花岗岩 granite
花岗岩砂粒 granite sand
花格 grill(e)
花格结构 trellis
花梗 pedicel; peduncle; scape
花梗鞣素 pedunculagin
花冠 corol; corolla
花冠喉 corolla throat
花冠上着生的 epipetalous
花冠上着生的雄蕊 epipetalous stamen
花灌木 flowering shrub
花痕 leafy scale
花候 florescence
花环 wreath
花黄色素 anthoxanthin
花卉布置 floral arrangement; flower arrangement
花卉防腐剂 floral preservative
花卉品种分类学 systematic floriculture
花卉瓶饰 flower vase
花卉企业 florist industry
花卉色图 flower colour chart
花卉业 florist industry
花卉园 flower garden
花卉展览 flower show
花卉中心 flower centre
花卉装饰 flower decoration
花卉装饰设计 floral design
花架 flower trellis; jardiniere
花椒 wild pepper
花径 herbaceous walk
花境 herbaceous border; flower border
花篮 flower basket
花蕾(单花) Alabastrum
花蕾脱落 flower abscission
花篱 flower hedge
花粮(切花保养用) bloom-food
花路 herbaceous walk
花轮减少 oligotaxy
花落 blossom fall
花蜜腺 floral nectar
花面狸 gem-faced civet; masked civet
花木(有花树) flowering tree
花盆 flower pot
花盆过大 over-potting
花盆架 jardiniere
花瓶 flower vase; jardiniere; vase
花圃 flower bed
花期 anthesis; blooming date; blooming period; flowering period
花期隔离 anthesis isolation
花期调整 anthesis regulation

花旗松 Columbian pine; Douglas fir; douglas-tree; Oregon pine; soft fir
花旗松油(俗) Douglas fir oil
花青定 cyanidin
花青素 anthocyanidin; cyanidin
花青素-3-葡糖苷 cyanidine-3-glucoside
花青素苷 anthocyanin
花楸属 sorb; scorb
花蕊同长 homogony
花蕊异长 heterogony
花蕊异长的 heterogonous
花色素 anthocyan; anthocyanidin; cyanidin
花色素苷 anthocyanin; erythrophyll
花色素苷载体 anthocyanophore
花色素酶 anthocyanase
花束 bouquet
花丝 filament; nema
花丝柱 synema
花穗 tassel
花坛 flower bed
花坛植物 bedding plant
花坛种植 bedding
花葶 scape
花桶 flower tub
花托 cephalophorum; receptacle; torus (*plur.* tori)
花纹 figure; figuring
花纹材 figured wood; veined wood
花纹单板 figured veneer
花纹伐根材 figured stumpwood
花纹干基材(根端材) figured butt
花纹木材 figured wood
花香吸取法 enfleurage
花谢 blossom fall
花心 eye of flower
花型 flower form; flower type
花序 inflorescence
花序芽 floral bud
花序轴 rachis (*plur.* raches); rhachis
花芽 blossom bud; floral bud; flower bud; reproductive bud
花芽发生 floral initiation
花芽接 fruit bud grafting
花檐 cornice
花眼 eye of flower
花样 pattern
花药 anther; semet
花药的药爿 theca (*plur.* thecae)
花药培养 anther culture
花药室 anther cell
花药叶化 antherophylly
花药异常 antheromania
花叶病 mosaic; mosaic discase
花叶病毒 mosaic virus
花蝇 root maggot fly
花园 floretum; flower garden
花园城市 garden city
花园路 parkway
花园小径 garden path
花园洋房 garden house
花园中心 garden centre
花原基 flower initial; flower primordium (*plur.* -ia)
花原始体 flower initial
花展 flower show
花枝 flower branch; flowering wood
花轴 floral axis; mamelon
花柱 style
花柱的 stylar
花柱异长 ditopogamy; heterostyly
划行器 marking gauge
划痕硬度试验 scratch-test
划手 oar
划线 laying-out
划线刀 marking gauge; marking knife
划线法(测定施胶度) stroke method
划线工具 scribe
划线规 marking gauge
划线盘 thicknesser; thicknessing machine
划线培养 streak culture
划线平台 level(l)ing block
划线器 marker
划线下锯 pattern line sawing
划线员(造材) marker
划线支票 cross check
划子 canoe
华橙花油 Chinese neroli oil
华夫刨花板 wafer board; waferboard
华氏温度 Fahrenheit (temperature) scale
华氏温度计 Fahrenheit thermometer
铧犁 mouldboard plough

铧式犁 mouldboard plough; share plough
滑把式鸟枪 slide-action shotgun
滑把式枪 pump gun
滑斑 slick spot
滑槽榫接 pulley mortise
滑车 pulley; sleigh
滑车环索 blockstrop
滑车轮 block sheave
滑车系梁 brow skid; lead log
滑车系索 rigging rope
滑锤 sliding weight
滑道 chute; ramp; shoot; slide; sliding track; slip; ways
滑道边挡木 breast-work log
滑道侧木 side log
滑道挡木 chute stick; glancer
滑道底木(圆木) base log
滑道会合点 frog
滑道集材 ball hooting; chuting
滑道集材工 ballhooter
滑道集运材 sliding
滑道减速齿钉 grab
滑道索引集材 trailing chuting
滑道下侧挡木 glancer
滑道制动齿 gooseneck
滑道主木 top log chute
滑动 creep; sliding; slippage
滑动承受面 shoot
滑动方向 glide direction
滑动接合 slip joint
滑动卡尺 slide caliper
滑动量 slippage
滑动面 slide plane; slip plane
滑动摩擦系数 coefficient of sliding friction
滑动平均 moving average
滑动切片机 sliding microtome
滑动套管(捆木索) shaw socket
滑动载荷 sliding weight
滑端胶合 slip joint
滑放(木材) sliding out
滑杠 skid frame; skid log
滑钩(捆木索) sliding hook
滑轨 rack
滑过生长 gliding growth; sliding growth
滑行 coasting; running
滑行坡度 coasting grade
滑架 skid frame
滑剪动作(液压剪) slicing action
滑脚 runner
滑脚链 runner chain
滑轮 block; pulley (wheel)
滑轮槽 pulley groove; throat
滑轮壳 block shell
滑轮系环 swivel yoke
滑轮系索 block strap; blockstrop; strop
滑轮组 block and tackle; block sheave; block tackle; block-and-a-half; tackle; tackle block; teagle
滑门阀 gate valve
滑坡 apron; coasting; land slide; landslip; slip
滑坡侵蚀 slip erosion
滑橇 stone boat; swing dingle
滑橇绞盘机 sledgewinch
滑撬 joe-log
滑石 agalite; talcum
滑石处理 talc(k)ing
滑套(捆木索) bell
滑脱环 slip hook
滑移层 glide lamella
滑移面 glide mirror; glide plane; slip plane
滑移平面 glide mirror; glide plane
滑移线 slip line
滑油枪 grease gun
滑榆酸 ulmic acid
滑走 sliding
滑走木 sledge runner
划分 measuring off
划割锯(与主锯机配套的先导锯) scribing saw
划记号 scratch
划界 delimitation; demarcation
划温冷却器 pistorius condenser
华山松烯 armanene
化成白云石 dolomitization
化肥撒肥机 fertilizer distributor
化感物质 allelochemical
化感作用 allelopathy
化能合成 chemosynthesis
化能无机营养生物 chemolithotroph

化能营养的 chemotrophic
化能自养菌 chemolithotroph
化能自养生物 chemoautotroph
化脓 maturation
化石 fossil
化石带 fossil zone
化石二氧化碳排放 fossil CO_2 (carbon dioxide) emissions
化石根 radicite
化石化作用 fossilization
化石侵蚀面 fossil erosion surface
化石燃料 fossil fuel
化石生物群落 fossil biocoenosis
化石树脂 fossil resin
化石岩相 ichnofacies
化石种 fossil species
化性(一年世代数) voltinism
化学变色 chemical stain(ing)
化学剥皮 chemical barking; chemical debarking
化学不孕剂 chemosterilant; sterilant
化学沉积 chemical sediment
化学除草 chemical weed control; chemical weeding
化学处理 chemical treatment
化学发光 chemiluminescence
化学防卫 chemical defense
化学防治 chemical control
化学防治法 chemical control method
化学分类学 chemotaxonomy
化学风化(作用) chemical weathering
化学腐蚀 chemical attack
化学改变 chemical modification
化学感觉器官 chemosensory organ
化学感受器 chemoreceptor
化学干燥 chemical drying; chemical seasoning
化学合成的 chemosynthetic
化学褐变 chemical brown-stain
化学化石 chemical fossil
化学环割 banding
化学机械浆 chemimechanical pulp
化学机械制浆 chemical-mechanical pulping; chemi-mechanical pulping
化学加工用材 chemical wood
化学降解 chemical degradation
化学进化 chemical evolution
化学绝育 chemical sterilization
化学绝育剂 chemosterilant
化学疗法 chemotherapeutics; chemotherapy
化学灭火 chemical in fire suppression
化学灭菌 chemical sterilization
化学磨木浆 chemical-ground pulp; chemi-ground wood pulp
化学木浆 chemical (wood) pulp; chemical wood-pulp
化学侵蚀 chemical attack; chemical erosion
化学热机械制浆 chemi-thermo-mechanical pulping
化学热磨法制浆 chemi-thermo-mechanical puping
化学生态学 chemical ecology
化学势 chemical potential
化学受体 chemoreceptor
化学疏伐 chemical thinning; poison thinning
化学透光伐 chemical cleaning
化学弯曲 chemical bending
化学位移 chemical shift
化学吸附 chemisorption
化学纤维 man-made fibre
化学向性 chemotaxis
化学修枝 chemical pruning
化学需氧量 chemical oxygen demand
化学永久定型 chemical permanent set
化学诱变 chemomorphosis
化学诱变育种 chemical induced breeding
化学纸浆 chemical pulp; dissolving pulp
化学治疗 chemotherapy
化学治疗剂 chemotherapeutant
化学治疗学 chemotherapeutics
化学贮藏 chemical storage
化学组成 chemical composition
化蛹 pupation
化油器 carburett; carburetter
化妆品原料 cosmetic raw material
化妆椅 shaving chair
化妆桌 shaving stand; shaving table
画栏 picture-field
画廊 gallery
桦多孔菌 razor-strop fungus; birch conk

桦木焦油 birch tar oil
桦木脑 betulin(ol)
桦木炭黑 birch black
桦木油 birch oil; sweet birch oil
桦皮焦油 birch bark tar
桦树皮 birch bark
桦芽油 birch bud oil
怀特氏培养基 White's medium
怀孕的 gravid
槐碱 sophorine
槐树豆胶 carob gum
踝 malleolus
坏死 necrosis
坏死斑 necrotic lesion
坏死病 necrotic disease
坏死的 necrotic
坏死反应 necrogenous reaction
坏死陷斑 necrotic depression
坏死组织 necrotic tissue
坏账 uncollectible account
还林 reforestation
还原卟啉 porphyrinogen
还原酶 reductase
还原糖 reducing sugar
还原性末端基 reducing end group
还原者 reducer
还债 pay out
环 eye; kink; race
U形环 shackle; yoke; catch shackle
环斑病(病毒病症状) ring spot
环保 environmental protection
” 环保人士 environmentalist
” 环剥 girdling; ringing
环剥法 ringing (method)
环剥压条法 layerage by girdling
环刀 cutting ring
环刀取样器 foil sampler
环的 cyclic
环钉 ring dog
环幅式地板 mushroom floor
环腐 ring rot
环割 band girdling; girdling; ring-girdling; ringing; deadning
V形环割 notch-girdling; notching
环庚三烯酚酮 tropolone
环钩 fit hook; log dog; log grab
环管薄壁组织 vasicentric parenchyma
环管的 vasicentric
环管管胞 vasicentric tracheid
环管密型薄壁组织 vasicentric abundant parenchyma
环管稀疏型薄壁组织 vasicentric scanty parenchyma
环痕孢子 annellospore
环化生橡胶 cyclocaoutchouc
环礁 atoll; atoll reefs
环接 eye-splice; ring-joint
环接系数 linkage coefficient
环节 chain
环节动物 annelid
环境 environment; locality
环境白皮书 environmental white paper
环境保护 environmental protection; protection of nature
环境保护论 environmentalism
环境保育 environmental conservation
环境变积 environmental covariance
环境变异 environmental variation
环境财产 environmental goods
环境差异 environmental deviation
环境承载力 environmental capacity
环境恶化 environment deterioration; environmental degradation; worsening of environment
环境反应 biapocrisis
环境废物 sanitary waste
环境负荷量 environmental capacity
环境复合体 environmental complex
环境工程 environmental engineering
环境公害 environmental pollution
环境激素 environment hormone
环境监测 environment monitoring
环境教育 environmental education
环境勘测卫星 environment survey satellite
环境抗力 environmental resistance
环境科学 environmental science
环境离差 environmental deviation
环境伦理学 environmental ethics
环境论 environmentalism
环境绿化 environmental greening
环境密度 environmental density
环境敏感性 environmental sensitivity
环境偏差 environmental deviation
环境评价 environmental assessment

H

环境破坏 environmental disruption
环境气候 environmental climate
环境取向 environmental preference
环境容量 environmental capacity
环境设想(园林) environmental consideration
环境社会学 environmental sociology
环境生理学 environmental physiology
环境生态学 environmental ecology
环境生物学 environmental biology
环境适度 fitness of environment
环境梯度 environmental gradient
环境条件 environmental condition
环境调查卫星 environment survey satellite
环境退化 environmental degradation
环境危机 environmental crisis
环境温度 ambient temperature
环境问题 environmental problem
环境污染 environmental pollution
环境污染水平 environmental degradation
环境无害技术 environmentally sound technologies
环境小气候 environmental climate
环境协方差 environmental covariance
环境信号 environment signal
环境压力 ambient pressure
环境研究 environmental consideration
环境要素 environmental factor
环境宜人场所 oasis (*plur.* oases)
环境异质性 environmental heterogeneity
环境因素 environmental agency; environmental factor
环境因子 environmental factor
环境影响评价 environmental impact assessment
环境影响因素 environmental factor
环境园艺学 environmental horticulture
环境灾害 environmental disaster
环境噪声 environmental noise
环境指标 environmental indicator
环境质量 environmental quality
环境资源斑块 environmental resource patch
环境综合体 environment complex; environmental complex
环境阻力 environmental resistance
环孔材 ring porous wood; ring-pored wood
环连 linkage
环裂 cup shake; ring crack; ring shake; round shake; shake; shell shake; wind shake
环裂根端木 shaky butt
环流 circulation
环内键 cyclic bond
环切 girdling; ring-girdling; ringing
环球法 ball and ring method
环染色单体 loop chromatid
环绕的 circumferential
环生的 verticillate
环生体 verticil
环式剥皮机 barker of rotary-ring type; ring (rotary) barker
环式磨木机 ring grinder; ring type grinder
环式转刀打枝机 ring-type delimber
环树坑(树干周围积雪融化形成) tree well
环双价体 ring bivalent
环索(连接安装索等) becket; loop; molly hogan
环索活动接头 live end of loop
环纹(病毒病症状) ring line
环纹导管 annular vessel; ringed vessel
环纹管胞 ringed tracheid
环纹加厚 ringed thickening
环纹细胞 annular cell
环线(森铁或轻轨道的岔线) run-around
环形二价染色体 ring bivalent
环形伐 circular cutting
环形干燥机 festoon dryer
环形构造业法 curving linear structure
环形锯 log band saw
环形联合 circular integration
环形喷嘴 annular nozzle
环形染色单体 ring chromatid
环形染色体 ring chromosome
环形钻 annular bit
环型 cell
环型辊式破碎机 ring roll mill
环压强度 ring crush compression resistance

环压试验 ring crush test
环氧化作用 epoxidation
环氧树脂胶黏剂 epoxy resin adhesive
环志 bird-banding; banding
环状剥离 concentric exfoliation
环状剥皮 bark ringing; girdling; ring-bark(ing); ring-stripping
环状剥皮机 ring debarker
环状的 cyclic
环状皆伐 ring-formed clear-cutting
环状排列 ring arrangement
环状破坏 ring failure
环状染色体 circular chromosome
环状条丝褐腐 brown ring stringy rot
环状芽接 annular budding; ring budding
环状择伐作业 ring shaped selection system
缓冲板 baffle
缓冲槽 knockout tank
缓冲衬层 breaker
缓冲存贮部件 buffer memory unit
缓冲带 buffer strip; buffer zone; protection belt; protection strip
缓冲固化剂 buffered hardener
缓冲基因 buffering gene
缓冲剂 buffer
缓冲楞堆 surge deck
缓冲料仓 buffering bin
缓冲木 fender; fender skid
缓冲器 buffer; bumper; dashpot; slack; vibroshock
缓冲区 buffer zone; easement zone
缓冲溶液 buffer solution
缓冲生物 buffer
缓冲塔 tempering tower
缓冲涂面层 buffer coat
缓冲效应 buffer effect
缓冲硬化剂 buffered hardener
缓冲栅 wolfbrake
缓冲种 buffer species
缓冲作用 buffer action
缓弹性变形 delayed elastic deformation
缓发显性 delayed dominance
缓和干燥(法) moderate drying
缓和距离 transition distance
缓和曲线 transition curve
缓和物质 appeasement substance
缓流(因插垛引起的) slack water
缓苗 transplanting shock
缓坡 gentle slope; glacis
缓蚀剂 corrosion inhibitor
缓释肥料 controlled-release fertilizer
缓释杀草剂 controlled-release herbicide
缓煮法 slow cooking process
幻灯片 diapositive
幻日 parhelion
幻月 paraselene
换钵 repotting
换茬 rotation
换档 gear shift
换能器 transducer
换盆 repotting
换算表(图) conversion chart
换算系数 expansion factor
换线路 change road
换线索 guinea line
换向开关 on-off switch
换羽 moult
荒地 badland; desolation; left-over area; wasteland; wild land; wilderness (area); wildland
荒废地 devastated land; waste land
荒废区域 blighted area
荒火 wildfire
荒漠 desert
荒漠的 eremium
荒漠灌丛 desert scrub
荒漠化 desertization
荒漠群落 eremium; eremus
荒漠生的 eremophilous
荒漠生态系统类型自然保护区 nature reserve for desert ecosystem
荒漠土壤 desert soil
荒漠植被 desert vegetation
荒漠植物 eremophyte
荒山 bare hill
荒山造林(史上无林) afforestation
荒山造林与森林恢复 afforestation and reforestation
荒芜地 barren (land)
荒溪 devastated torrent
荒溪侵蚀 torrent erosion
荒溪治理 torrent control
荒野游憩 wilderness recreation
荒原 fellfield

荒原保护区 wilderness area
皇室林 imperial forest
黄柏 phellodendron
黄柏酮 obakunone; obakurone
黄变(阔叶材的一种真菌变色) yellow stain
黄标签种子 yellow tag seed
黄草灵(除草剂) asulam
黄粉(鞣花类鞣质分解物) bloom
黄蜂 wasp
黄蒿油 caraway oil
黄胡蜂 yellow jacket; hornet
黄胡蜂类 potter wasps; hornets
黄花园 yellow garden
黄化 chlorosis; etiolation; yellowing; yellows
黄化法 etiolation method
黄化现象 aetiolation; eitolation phenomenon; etiolation; xanthochroism
黄化效应 etiolation effect
黄化型病毒 yellow virus
黄化植物 aurea
黄昏活动的 crepuscular
黄兰叶油 champaca leaf oil
黄兰油 champaca oil
黄连碱 coptisine
黄料浆池 brown stock chest
黄栌心材 fuste
黄麻杆 jute stalk
黄木心材 fustic wood
黄芪胶 tragacanth (gum)
黄芪属 poison-vetch; milkvetch
黄壤 yellow earth; yellow soil; zheltozem
黄肉楠碱 actinodaphnine
黄色灰化土 yellow podzolic soil
黄色鼠李苷 xanthorhamnin
黄杉毒蛾 fir tussock moth
黄蓍树胶 gum tragacanth
黄石公园森林保护地 Yellow Park forest reservation
黄石国家公园 Yellowstone National Park
黄熟 yellow maturity
黄鼠 gopher
黄鼠狼 kolinsky; yellow weasel; weasel
黄素 flavin(e)
黄素蛋白 flavoprotein
黄素氧化还原蛋白 flavodoxin
黄檀素 dalbergin
黄体素 lutein
黄酮 anthoxanthin; flavon(e)
黄酮类单宁 flavonoid tannin
黄酮类化合物 flavonoid
黄酮体单宁 flavonoid tannin
黄土 loess
黄土沉积 loess deposit
黄土类的 loessal
黄烷醇 flavanol
黄鼬 kolinsky yellow weasel
黄雨 sulphur rain
黄樟素 shikimol
黄樟油 yellow camphor oil; sassafras (wood) oil
黄棕壤 yellow-brown earth
黄棕色砖红壤性土壤 yellowish brown lateritic soil
黄棕森林土 yellow-brown forest soil
蝗虫 grasshopper; locust
磺化栲胶 sulfonated extract
磺化器 sulfonator
磺化作用 sulfonation
簧板(机枪) lock
灰白色的 glaucous
灰斑 grey stain
灰醋石 gray acetate of lime; gray lime acetate
灰度 shades of gray
灰度标 gray scale
灰度等级(遥感图像) step wedge
灰度级 gray scale
灰分 ash
灰腐殖酸 gray humic acid
灰钙土 serozem; sierozem
灰黑变色 grayish-black stain
灰化 podzolization
灰化层 podzolic horizon
灰化的 podzolic; spodic
灰化淀积层 spodic horizon
灰化腐殖质层(土壤剖面E_s层或A_2层) spodic humus layer; E_s-layer; A_2-layer
灰化红黄壤 red yellow podzolic soil
灰化黄壤 yellow podzolic soil
灰化土 gley podxol; podsol; podsolic soil; podzol; podzolic soil
灰化作用 podzolization

灰黄霉素 griseofulvin
灰烬 ember
灰绿曲霉素 echinuline
灰霉 grey mould
灰霉素 grisein
灰漠土 gray desert soil; grey desert soil
灰泥天花板 plaster ceiling
灰泥岩 calcilutite
灰黏层 gley horizon
灰黏化 gleization
灰黏化过程 gley process; glei process
灰黏化作用 gleization
灰壤 gley podxol; podzol; podsol
灰壤化 podsolization
灰色冲积土 gray warp soil
灰色胡敏酸 gray humic acid
灰色热带黏土 grey tropical clay
灰色森林土 gray forest soil; grey forest soil; grey wooded soil; phaeozem
灰色铁质土 grey ferruginous soil
灰色土 ashen-grey soil
灰色脱碱土 gray solodic soil
灰条 lath
灰土 gray soil; grey soil; orthod; spodosol
灰像法 spodography; microincineration
灰心材(木材缺陷) gray heart
灰翳 gray shade
灰棕荒漠土 gray-brown desert soil
灰棕壤 gray-brown forest soll
灰棕色灰化土 brown illimerized soil; brown podzolic soil; grey-brown podzolic soil
灰棕色森林土 gray-brown forest soll
挥发成分 volatile content
挥发物 volatile matter
挥发物含量 volatile content
挥发性瓷漆 lacquer enamel; spirit enamelling
挥发性漆 lacquer; lacker
挥发性溶剂 volatile solvent
挥发油 benzine; volatile oil
恢复 recovery; restoration
恢复基因 restoring gene
恢复生态学 restoration ecology
恢复稳定性 resilient stability
辉绿岩 diabase; ophite
辉长岩 gabbro
回采率 extraction rate
回采期 period of return
回潮 damping; moisture regain; rewetting
回车场 turn-around; turning bay; turning place
回车线 turn-around
回程(广义的) return
回程(拖拉机集材) outhaul trip
回程货 return cargo
回出[量] kickback
回弹 kickback; spring-back
回弹距离(伐木) throwing
回弹性 resilience; resiliency
回动刮板 reverse scraper
回动刮板链 reverse scraper chain
回返对流 return convection
回复变异 reversion
回复蠕变 recovery creep
回复体 revertant
回复突变 back mutation; reverse mutation
回复突变型 back mutant; revertant
回复值 recovery value
回复子 revertant
回购价 buy-back price
回归 regression
回归参数 regression parameter
回归方程 regression equation
回归分析 regression analysis
回归估计 regression estimation
回归估计方程 regression estimator
回归年 cutting cycle; harvesting cycle; cutting interval; cutting rotation; felling interval; return interval; return period; turn
回归系数 regression coefficient
回归直线 regression-line
回火 tempering
回浆(打浆机) carry-over of stock; overfall
回交 backcross
回交能育 paragenesis
回交亲本 backcross parent
回交育种 backcross breeding
回交育种法 backcross method of breeding

H

回交杂种 backcross hybrid
回交种 comeback
回截 heading back
回空(牵引索) outhaul; haulback
回空道(畜力集材等) go back road; short road; drag road
回空滑轮 head trip block; outhaul block
回空链 return chain
回空索 back line; haulback; haulback line; haul-out line; messenger line; outhaul line; pull back (line); receding line; rehaul line; return line; tail line; tail rope; trip line
回空索侧滑轮 haulback side block
回空索导向轮 haulback lead block
回空索滑轮 haulback block; receding block; side block; squirrel block; tail block
回空索急剧转向点 corner
回空索卷筒 haulback drum
回空索尾端滑轮 haulback tail block
回空转向线段 wasteline
回扣 kickback; rebate
回馈 feedback
回馈控制 feedback control
回流 back flow
回流作用 backwash effect
回路 loop
回气道 return air duct
回色 brightness reversion; colour reversion
回升水位 back-up water level
回生 anabiosis
回声测探 echo sounding
回声定位 echolocation
回湿性 moisture regain
回收槽(造纸) accumulator
回收浆料 reclaimed stock; recovered stock
回收率 recovery; return rate
回收凝汽阀 return trap
回收期 pay off period
回收线路 reclaim line
回收液 released liquor
回水 backwater; return water
回水管 return line
回送 loop back
回缩 recovery; retraction
回填(土壤) backfill
回填机 backfiller
回跳肖氏硬度 Shore hardness
回跳硬度计 scleroscope
回头棒子(树倒时) sailer
回头伐 re-logging; subsidiary felling
回吐物 pellet
回洗 backwash
回旋净种机 clipper cleaner
回音测深仪 fathometer
回油槽 oil-recovery tank
回油真空处理 recovery vacuum
回运货物 return cargo
回转 slewing
回转半径(车辆) turning radius
回转半径(旋转) radius of gyration
回转刀架旋切机 turret veneer lathe
回转阀 rotary valve
回转环 rotating ring
回转角 slewing angle
回转接头 swivel joint
回转扭矩 slewing torque
回转器 rotator
回转筛 rotating screen
回转式成型系统 merry-go-round system
回转式分配盘 trough carousel
回转式剪板机 rotary clipper
回转式进料机 rotary feeder
回转式喷气干燥机 rotary nozzle dryer
回转式起重机 swing crane
回转式热压机 rotary hot press
回转式蒸球(蒸煮) rotary ball digester
回转应变 rotational strain
茴香 anisum
茴香基 anisyl
茴香基丙酮 anisyl acetone
茴香基甲酮 anisketone; anisyl methyl ketone
茴香醚 anisole
茴香脑 anethole
茴香素 anisoxide
茴香油 anise oil; fennel oil
茴香樟脑 anise camphor
茴香子 aniseed
茴香子油 aniseed oil

毁林 deforest; forest destruction; forest devastation
毁林者 forest destroyer
毁馏 destructive distillation
毁灭 extirpation
毁灭林地的 stand-leveling
毁损 fault
毁约 delinquency
汇 sink
汇编 round up
汇合的 confluent
汇流 affluent
汇流盆地 confluence basin
汇票 bill of exchange
汇票费 charges for remittance
汇票通知书 advice of drawing
汇入 afflux
汇水 charges for remittance
汇总 compilation
会车道 turnout
会费 due
会聚 convergence
会议桌 conference table
荟萃分析 meta-analysis
绘图板 drawing board
绘图平板 drawing-table
绘图椅 draft chair
喙(昆虫) rostrum
喙(鸟兽) beak; proboscis
喙侧叶(部分鞘翅目昆虫) pterygium
喙裂接 gape grafting
喙状的 rostrate
昏厥 asphyxy
婚(姻)色 nuptial colouration
婚飞 nuptial flight
婚羽 nuptial plumage
婚赠 nuptial gift
浑浊度 opacity
混播 intermingle; mingle; mixed seeding
混草北美草原 mixed prairie
混草造林 reforestation with grass culture; reforestation with pasture culture
混肥播种 combine drilling; ferti-seeding
混合 admixture; blend; knead
混合比 mixing ratio
混合槽 mixing tank
混合层 mixed layer
混合层积 compound stratification
混合成本制度 hybrid cost system
混合齿 combination teeth
混合二甲苯 xylol
混合法 bulk system
混合肥料 compost
混合腐朽 mixed rot
混合花境 mixed border
混合花序 mixed inflorescence
混合货车(运材) mixing truck
混合嫁接 mixed grafting
混合胶 mix(ed) glue
混合胶黏剂 mixed adhesive
混合接 graft with cutting
混合经济 mixed economy
混合林 highly structured forests; mixed woodland
混合络合物 mixed (ligand) complex
混合谱系法 mass pedigree method
混合气调节螺钉(油锯) mixture screw
混合侵染 mixed infection
混合授粉 mixed pollination
混合树种胶合板 mixed plywood
混合涂布法 intermix application
混合纹理 mixed grain
混合物 blender
混合像素 mixed pixel
混合选择 mass selection
混合芽 mixed bud
混合育种 mass breeding
混合育种法 bulk method of breeding; mass method of breeding
混合植物群落 mixed plant community
混交方法 pattern of mixture
混交分布 mixture distribution
混交类型 type of mixture
混交林 highly structured forests; mingled forest; mixed crop; mixed forest; mixed stand; mixed wood; mixed woodland
混流 interflow
混流风机 mixed-flow fan
混流涡轮 mixed flow turbine
混牧林 grazing forest
混凝土 concrete
混凝土坝 concrete dam

H

混凝土板 concrete slab
混凝土轨枕 concrete tie
混凝土模板 board liner; concrete form; concrete shuttering
混凝土匣子(电杆防腐) concrete casing
混农林 forest combined with agriculture
混农林牧系统 agrosilvopastoral system
混农林系统 agroforestry system
混农林业 agri-silviculture; agroforestry; agro-forestry
混农人工林 taungya plantation
混农作业 silviculture taungya plantation
混农作业法 taungya method; taungya system
混溶性 compatibility
混生节 intergrown knot
混生林 mixed woodland
混生群 mixed flock
混生树种 associated species
混系种 multilineal variety
混响室 reverberant room
混淆 mixture; species mixture
混优种浮游生物群 pantomictic plankton
混杂遗传性 mixed inheritance
混杂种子 accessory seed
混植 intermingle; mingle; mixed planting
混种 mistus; mixtus
混种群 mixed-species population
混浊度 cloudiness
混作 mixed cropping
活板 flap
活的可燃物 green fuel
活地被物 living ground cover
活地被物层 ground vegetation
活动坝 movable dam
活动百叶窗 venetian blind
活动百叶帘板条 venetian blind slat
活动扳手 monkey; monkey wrench
活动半径 radius of action
活动层(活性层)(冻土区) active layer
活动范围(动物) home range
活动范围(踪迹) orbit
活动房屋 prefabricated house
活动画景 panorama
活动节律 activity rhythm
活动锯 cordwood saw
活动面 active surface
活动平台 stack
活动桥 movable bridge
活动梯度 activity gradient
活动性适应 active adaptation
活动芽 active bud
活动样方 movable quadrat
活动载荷 roiling load
活动资产 active assets
活动资金 working capital
活短枝 twig
活盖导向滑轮 auto-snatch block
活根 green root
活化 activation
活化分析 activation analysis
活化硅胶 activated silica gel
活化剂 activator
活化炉(用于活性炭) activating oven
活化能 activation energy; energy of activation
活化途径 activation pathway
活火 open fire
活节 adhering knot; link lock; live knot; red knot; tight knot
活劳动 live labour
活力 activity; light
活力变异 vigour modification
活力等级 vigour rating
活力分级 vigour classification
活力级 vigour class
活力评定 vigour evaluation
活力指数 vital index
活力种子 viable seed
活立木 live wood; living tree; green timber
活媒鸟 live decoy
活期贷借 call (able) loan
活期贷款市场 call money market
活期借款 call money
活塞 piston
活塞泵 piston-type pump
活塞冲程 piston displacement
活塞阀 bucket valve
活塞杆 connecting rod
活塞排量 piston displacement
活生作诱饵用的动物 live decoy

活食者 biophage
活树防腐法 standing tree method
活树冠率 live crown ratio
活树冠面积 live crown surface
活套 kink
活体检定 bioassay
活体内的 in vivo
活体外 in vitro
活体营养 biotroph
活物寄生菌 parasitic fungi
活细胞 living cell
活细胞培养法 biocytoculture
活性 activity
活性白土 atlapulgite
活性成分 active component; active constituent; active ingredient
活性防火剂 reactive fire retardant
活性腐殖质 active humus
活性剂 activator
活性酵素 active ferment
活性酸度 active acidity
活性炭 activated carbon; activated char(coal); active carbon; active char(coal)
活性炭纤维 activated carbon fibre
活性污泥 activated sludge; active sludge
活性污泥法 activated sludge method; activated sludge process
活性物质 active substance
活性中心 active centre
活血丹叶油 ivy leaf oil
活页绞链 drop-leaf hinge
活页直角接 rule joint
活页肘桌 elbow table
活页桌 drop-leaf table; falling table
活跃区 active region
活载荷 live load(ing)
活枝 green branch; live branch
活枝修剪 live pruning; green pruning
活质体 energid
活蛀虫 live borer
火疤 fire scar
火暴 blow up fire; fire storm
火柴材 match block; wood for match sticks and cases
火柴材段 match wood bolt
火柴杆坯料 match block
火柴梗 match splint; match stick; splint
火柴梗材 wood for match sticks and cases
火柴梗用材 splint-wood
火柴簧板枪(狩) match lock gun
火柴用板 match plank
火柴用材 match stock
火柴用原木 match-wood log
火场 fire area
火场隔离 contain fire
火场型 burn pattern
火场侦察员 line scout
火车上交货价格 free on rail
火成片麻岩 orthogneiss
火成石英岩 orthoquartzite
火成岩 eruptive rock; igneous rock
火成演替 pyrrhic succession; pyrogenic succession
火道 quirk
火风暴 fire storm
火锋 flame front
火夫 black gang
火管理 fire management
火管试验(木材在管中燃烧的失重试验) fire-tube test
火行为 fire behavior
火环[伤害] fire ring
火迹 fire scar
火胶的 collodion
火胶棉 collodion cotton
火界 pyrosphere
火警演习 fire drill
火口原 atrio
火力干燥窑 fire kiln
火烈度 fire severity
火蔓延 fire propagation
火蔓延速度 rate of fire spread
火棉 pyroxylin
火棉胶 collodion
火器 firearm
火前阻火 front fire
火枪 firearm
火情 fire behaviour
火情分析 fire analysis
火情判断 fire appraisal
火情扫描 fire scan
火三角 firetriangle
火山湖 crater lake; volcanic lake

火山荒漠 volcanic desert
火山灰 volcanic ash
火山灰年代学 tephrochronology
火山灰土 volcanic ash soil
火山口 crater
火山砾 lapillus (*plur.* -lli)
火山链 volcanic chain
火山泥石流 volcanic mud flow
火山喷出物 volcanic product
火山喷发岩 extrusive rock
火山砂 volcanic sand
火山土 volcanic soil
火山岩 volcanic rock
火伤 fire wound
火伤暗疤 hidden fire scar
火伤疤痕 fire scar
火伤轮 fire ring
火伤木 fire injury tree
火烧材(防腐) fire killed timber
火烧的 pyrrhic
火烧顶极 pyroclimax; fire climax
火烧顶极群落 fire climax community
火烧洞(立木基部的) goosepen
V形火烧法 chev(e)ron burn
火烧耕作法 taungya system
火烧迹地 burn; burned area
火烧控制线 fire-trace
火烧木 salvage timber; fire damaged
火烧强度 fire intensity
火烧演替顶极 fire climax
火生态学 fire ecology
火绳枪 match lock gun
火石簧板枪 flight lock gun
火势 fire behaviour; fire behavior
火势测定 fire calibration
火势控制 control of a fire
火势扩张速度仪 rate-of-spread meter
火势蔓延 fire spread
火室 fire box; fire chamber; fire place
火速计算尺 rate-of-spread meter
火头 fire head; head; head of fire
火尾 fire rear; rear
火险 fire danger
火险报道板 danger board; fire-danger board
火险表 danger table
火险尺度 fire danger scale
火险等级 fire danger rating
火险等级系统 fire danger rating system
火险度 fire hazard
火险度指标 fire hazard index
火险负荷指标 fire load index
火险估测仪 danger meter
火险观测站 fire-danger measurement station
火险级 danger class; fire-damage class
火险计算表 fire danger scale; fire-danger table
火险计算尺 danger meter; fire danger meter
火险频度 fire frequency
火险期 fire danger season
火险区 division; fire danger division
火险天气 fire danger weather; fire weather
火险天气预报 fire weather forecast
火险天气指标 fire weather index
火险图 fire danger map; fire risk map
火险性 fire risk
火险因子 risk factor
火险预警仪 fire danger meter
火险指标 danger index; fire danger index
火险指标等级 degree of risk index
火险指数计数器 burning index meter
火线营部 line camp
火旋风 fire whirl; fire whirlwind
火焰除漆法 flame cleaning
火焰穿透试验 flame-penetration test
火焰传播 flame propagation; flame spread
火焰传播速度 rate of flame spread
火焰的线性蔓延 linear flame propagation
火焰电离检测器 flame ionization detector
火焰反应 flame reaction
火焰光度计 flame photometer
火焰光谱法 flame spectrometry
火焰孔材 flaming porous wood
火焰蔓延 flame propagation; flame spread; spread of flame
火焰蔓延速度 rate of fire travel
火焰清洁法 flame cleaning
火焰状 flaming

火焰状管孔排列 flame-like pore arrangement
火焰状花纹 flame figure; flamy figure
火焰状排列 flame-like arrangement
火药 gun powder
火药棉 gun cotton
火疫病 fire blight
火翼 fire flank; flank fire; flank of fire
火因种类 fire cause class
火缘 fire edge; fire margin
火源 fire origin; ignition source; fire brand
火灾保险公司 fire office
火灾采伐剩余物 fire-slash
火灾等级(按火灾范围) fire size class
火灾定位图 fire-finder map
火灾定位仪 fire-finder
火灾发生图 fire occurrence map
火灾发现 fire discovery
火灾范围图 fire occurrence map
火灾后检查余烬 feeling for fire
火灾季节 fire season
火灾间隔期 fire return interval
火灾监测 fire detection
火灾警报器 fire alarm
火灾警报箱 fire-alarmbox
火灾警报指示器 fire-alarm indicator
火灾控制系统 deluge fire control system
火灾控制组织 fire control organization
火灾类型 fire type
火灾模拟器 fire simulator
火灾频率(火灾次数/年) fire concentration
火灾评估 fire appraisal
火灾强度 fire intensity
火灾三要素(温度, 氧气和燃烧物) firetriangle
火灾探测 fire detection
火灾图 fire occurrence map
火灾危险度 fire hazard
火灾危险度指标 fire hazard index
火灾危险级 fire danger class
火灾危险计算尺法 fire-danger meter method
火灾危险性 fire risk
火灾巡逻[队] fire patrol
火灾因子 pyric factor
火灾预防 fire prevention
火灾预防计划 fire-prevention plan; fire-prevention project
火灾侦查仪(林区防火) fire detective device
火阵旋风(火的涡旋) fire devil
或然率 law of probability
货币工资 nominal wage
货币价 present value
货币收获 financial value; monetary value
货币收获表 money yield table
货车集场 collecting yard
货车上交货价格 free on lorry; free on truck
货到付款 cash on delivery
货架 goods shelf
货物 commodity
货物抵押贷款 goods credit
货物税 excise tax
获得抗病性 acquired resistance
获得免疫性 acquired immunity
获得物(用复数) gaining
获得性状 acquired character
获得性状遗传 inheritance of acquired character
获利率 yield
获利能力 profitability
获许可的人 licensee
霍普斯吨 Hoppus ton
霍普斯量度法 Hoppus measure
霍普斯片英尺 super-foot Hoppus
霍普斯英尺 Hoppus foot
霍普斯原木英尺材积表 Hoppus Rule
霍普斯圆材检尺法 Hoppus string measure
霍沃思式 Haworth formula
击穿点 yield point
击发阻铁 sear
击碎器 beater
机变授粉 allautogamia
机铲 mechanical digger
机车场 locomotive yard
机车车辆 rolling stock
机车司机 glucogallin
机床身 bed piece
机动采种升降梯 power ladder for tree
机动车运材 motor truck hauling

机动刮皮器(采脂) motorized rossing tool
机动化游憩 motorized recreation
机动喷粉机 power duster
机动性 maneuverability
机动闸 power brake
机动中耕机 power cultivator
机构 mechanism
机构用地 institutional land
机会 opportunity
机会成本 opportunity cost
机架 frame; frame-work; husk; rack
机理 mechanism
机面 action face
机能障碍 dysfunction
机器代码 machine code
机器待命时间 machine waiting time
机器工时 machine hour
机器可读信息 machine sensible information
机器空耗时间 machine idle-time
机器语言 machine language
机器运转时间 machine running-time
机器指令 machine instruction
机身 action body
机闩 action bolt
机碎板皮 hogged slab
机体 basement
机头座 headstock
机外施胶 external sizing; off-machine sizing
机误 chance error
机误方差 error variance
机械搬运设备 mechanical handling equipment
机械剥皮 mechanical barking; mechanical peeling
机械布置图 machinery layout
机械擦伤 mechanical scarification
机械操纵设备 mechanical handling equipment
机械操作码 machine operation code
机械沉积 mechanical sediment
机械抽样(等距抽样) mechanical sampling
机械处理 mechanical treatment
机械打桩 mechanical piling
机械法制浆 mechanical pulping
机械分级 machine grading
机械分析 mechanical analysis
机械分选 machine separation
机械分样器 mechanical sampler
机械附着力 mechanical adhesion
机械杠杆式锯条张紧装置 mechanical lever strain system
机械归楞 mechanical piling
机械化采伐 mechanical logging
机械环剥 mechanical girdling
机械加工 machining
机械加工精度 machining precision
机械降解 mechanical degradation
机械胶合 mechanical adhesion
机械锯 machine saw
机械抗性 mechanical resistance
机械磨焦 machine bum
机械磨损 mechanical wear
机械木浆 ground pulp; ground (wood) pulp
机械黏着力 mechanical adhesion
机械平地 machine grading
机械强度 mechanical strength
机械缺陷 machining defect; mechanical defect
机械式抖动机 mechanical shaker
机械手 autohand; manipulator
机械疏伐 mechanical thinning
机械损伤 mechanical damage
机械调节 mechanical conditioning
机械效率 mechanical efficiency
机械性能 mechanical property
机械性损伤 mechanical injury
机械应变 mechanical strain
机械应力 mechanical stress
机械振动 mechanical vibration
机械纸浆 mechanical pulp
机械装配工 millwright
机械组织 mechanical tissue
机遇法则 law of chance
机载辐射温度计 airborne radiometer thermometer
机罩 hood
机制 mechanism
机制纸 machine-made paper
机组 element; unit
机钻 machine drill
肌构(木材构造) texture
肌胃 gizzard; muscular stomach
矶松根碱 baycurine

鸡骨常山碱 alstonine
鸡冠状的 cristate
鸡冠状突起的 crested
鸡心环(编结索眼用) thimble
唧筒式鸟枪 slide-action shotgun
唧筒式枪 pump gun
积材 pile of wood
积材费 pay for stacking
积分热 integral heat
积累 build-up
积木式砂光机 built-up type sander
积水坝 splash dam
积体线路 monolithic storage technology
积温 accumulated temperature; cumulated temperature; heat sum; temperature sum
积雪 snow pack
积雪持续时间 snow cover duration
积雪带 nival belt
积雪地带采伐作业 winter show
积雪梯度 snow accumulation gradient
积雪线 snow cover line
积雨云 cumulonimbus cloud
积云 cumulus cloud
积云状雨云 nimbus-cumuliformis
积状云 cumuliform cloud
姬蜂 ichneumon fly
基 pedestal
基板附器(昆虫) paraphysis
基本标准 primary standard
基本参数 essential parameter
基本操作系统 basic operating system
基本单位 base unit
基本的 primary
基本的飞机高度 cardinal altitude
基本分生组织 fundamental meristem; ground meristem
基本干缩率 basic shrinkage
基本耕作 primary cultivation
基本工资 original wages
基本核型 basikaryotype
基本径流 base flow
基本流量 base flow; base run-off
基本密度 basic density
基本培养基 basal medium; minimal medium
基本器官 essential organ
基本群落 concrete community
基本容重 basic volumetric weight
基本收缩 basic shrinkage
基本图 base sheet; special map
基本纤丝 elementary fibril
基本应力 basic stress (value)
基本种 elementary species; fundamental species
基本种群 base population
基本种子 basic seed
基本组织 fundamental tissue; ground tissue
基部(树干或树枝的) base
基部切面 basal cut
基部受精 basigamy
基材 substrate
基层 basement
基出脉的 basal-nerved
基础 foundation; ground work; substructure; ground
基础代谢 basal metabolism; basic metabolism
基础代谢率 basal metabolic rate; basic metabolic rate
基础结构 infrastructure
基础梁 foundation beam
基础螺丝 foot screw
基础生态位 foundational niche; fundamental niche
基础土壤学 pedology
基础植物 foundation plant
基础种植 foundation planting
基带(植被) basal zone
基底 substrate
基底膜 basement membrane
基底木 mudsill
基底片麻岩 fundamental gneiss
基底胎座 basal placenta
基底胎座式 basal placentation
基底岩 fundamental rock
基底种群 base population
基点 fiducial point
基点标志树(测量) witness tree
基点制(经济) basing-point system
基段插穗 basal cutting
基肥 basal dressing; basal fertilizer; basic manure; ground fertilizer
基峰 base peak
基腹片 basisternum

基盖度 basal cover(age)
基干人员清册(防火) masterlist
基根系统 stump root system
基花 basiflory
基价 base price
基建费 capital expenditure
基脚 footing
基节(足) coxa
基节缝(足) coxa suture
基金 base capital; capital stock; principal
基金的本金 corpus
基金准备 funded reserve
基径(树干基部) collar diameter; basal diameter; diameter at base
基流 base current; standing current
基面方向(摄影) direction of tilt
基诺树胶 gum kino; kino
基配 basigamy
基期 base period
基群丛(法瑞学派) sociation
基色 primary colour
基生花 basiflory
基生花柱 basilar style
基生叶 radical leaf
基数(染色体) basic number
基态 ground state
基体 basal body; matrix (*plur.* matrixes/matrices)
基调(园林等) main key; key-note
基调树(园林布景) dominant tree
基团 radical
基线 base; base-line; basic line; ground-line
基线法 method of intersection
基线航高比 base-height ratio
基线漂移 baseline drift
基线数据 baseline data
基线位移 baseline shift
基线校正 baseline correction
基线噪声 baseline noise
基线长 base length
基性岩 basic rock
基岩 bed rock; parent bed rock; underlying rock
基耶达法 Kjeldahl method
基因 gene
基因变异 genovariation
基因表现度 expressivity
基因带 genophore
基因的主效应 primary effects of gene
基因点突变 point mutation
基因多效性 pleiotropy; pleiotropism
基因分离 gene segregation
基因剂量 gene dosage
基因库 gene pool
基因块 gene block
基因粒 genomere
基因量子理论 quantum theory of gene
基因流 gene flow
基因内变化 intragenic change
基因漂变 genetic drift
基因频率 gene frequency
基因平衡 genic balance
基因冗余性 gene redundancy
基因数量 gene dosage
基因体 genosome
基因突变 gene mutation
基因外区 extron
基因外突变 extragenetic mutation
基因位点 gene locus
基因系统 gene system
基因线(原核类) genonema
基因相加效应 additive genetic effect
基因型 genotype
基因型的 genotypic
基因型方差 genotypic variance
基因型-环境互作 genotype-environment interaction
基因型-环境相关 genotype-environment correlation
基因型-环境相互作用 genotype-environment interaction
基因型频率 genotypic frequency
基因型选择 genotypic selection
基因型值 genotypic value
基因序列 gene order; genetic code
基因障阻 gene block
基因直线排列法 linear order of genes
基因重复 gene redundancy
基因重组 gene recombination; genetic recombination
基因重组价 cost of gene recombination
基因重组现象 gene recombination
基因转换 gene conversion

基因转移 transgenosis
基因资源 genetic resources
基因综合体 gene complex
基因组 gene block; genom(e)
基因作用 gene action
基原纤丝 elementary fibril
基质 host; matrix; matrix substance; medium (*plur.* -ia); substratum (*plur.* -ta)
基柱 pillar
基桩 foundation pile
基准 benchmark
基准材面 face cut; slab cut
基准点 fiducial mark; fiducial point
基准高度 base height
基准面 base face; base level; datum level
基准年 base year
基准年龄 key age; reference age
基准期间 base period
基准曲线 guiding curve
基准线 baseline
基座 holder; pedestal; footing
畸变 aberration; distortion; malformation
畸态 monostrosity
畸形 abnormality; deformation; deformity; malformation; monostrosity; monster
畸形病 distorting disease
畸形的 deformed
畸形构造的 ataxinomic
畸形苗 deformed seedling
畸形木疙瘩 blind conk
畸形树 wild timber; misshapen tree
畸形树干 badly formed stem
畸形学 teratology
畸形有丝分裂 mitosis aberrant
畸形原木 misshapen log
稽核簿 control of operation; management record; work tracking
稽核法 examination method; methods of control
激动素 kinetin; kinin
激发因子 evocator
激光 laser
激光测距仪 laser range finder
激光测树仪 laser dendrometer
激光导管 laser guide
激光雷达 laser radar
激光器 laser
激光切割 laser cutting
激光扫描仪 laser scanner
激光束探测仪 laser beam detector
激光探测和测距装置 laser detection and ranging
激光图像处理扫描仪 laser image processing scanner
激化温度 thermal activating point
激活 activation
激活物 activator
激酶 kinase
激素 hormone
激素生成 hormonogenesis
激肽 kinin
几丁质 chitin
几率比例取样法 probability proportional to size (pps); proportional sampling
几率律 law of probability
几全缘的 subentire
吉布森方形海排 Gibson raft
吉丁 jewel beetle
吉丁虫 buprestid-beetle; flat headed borer; flathead (wood) borer; metallic wood borer
吉拉德树木材积计算法 Girard's rule of thumb for tree volume
吉拉德形级 Girard form class
吉拉德形率 Girard form quotient
吉纳紫檀 kino
吉枝索 geijerin
吉枝烯 geijerene
级 caste; class
级别 classification
级距 class interval
级数 number of stages
级位 level
极不利价格 penal rate
极低频 ultra-low frequency; very low frequency
极地带 polar zone
极地附近的 circumpolar
极地湖 polar lake
极地荒漠 arctic desert
极地气候 polar climate
极地森林界线 arctic timber line
极度的 faulty

极端天气事件 extreme weather event
极端值 extreme value
极峰 polar front
极高分辨辐射计 very high resolution radiometer
极高频 extremely high frequency; very high frequency
极惯性矩 polar moment of inertia
极光 aurora
极核 polar nuclei
极化 polarization
极化滤波器 polarizing filter
极密林 very dense forest
极谱扫描器 polarographic scanner
极强度疏伐 drastic thinning
极圈 polar circle
极弱度疏伐(树冠级疏伐) A-grade thinning
极生鞭毛 polar flagellum
极盛地质期 hemera
极湿地 wet soil
极限承载量 ultimate bearing capacity
极限抗拉强度 ultimate tensile strength
极限抗弯强度 ultimate bending strength
极限抗压强度 ultimate compression strength
极限亮度 ultimate brightness
极限坡度(陡坡缆车道) choking limit
极限强度 ultimate strength
极限位置 extreme position
极限压力 extreme pressure
极限应力 ultimate stress
极限载荷 ultimate load
极性 polarity
极性分子 polar molecule
极性化合物 polar compound
极性胶合体 polar adherent
极性胶黏剂 polar adhesive
极性溶剂 polar solvent
极性运动 polar movement
极压 extreme pressure
极移 polar wandering
极值 extreme value
极坐标法 polar method
即将成年的 sub-adult
即期付现 prompt cash payment
即期票据 demand note
急尖的 acute
急剧干燥 intensive drying
急流 chute; torrent; white water driving
急流带 rapid zone
急流河川 torrent(ial) river
急流区 rapid zone; riffles zone
急流群落 swift-water community
急坡 steep slope
急射 snapshot
急弯 elbow; sharp corner
急性毒性 acute toxicity
急性中毒 acute intoxication
急性中毒剂量 acute dose
急性子 touch-me-not; garden balsam
急骤干燥法 flash drying
疾风 high wind
棘 spine
棘轮 ratchet wheel
棘轮制动器 brake ratchet wheel
集材(广义集材) log assemblage; assemblage; assembling wood; bringing in log; extracting; haul; logging; minor transportation; primary transportation; log removal
集材(索道集材) yard; yarding
集材(拖曳集材) skid; skidding; twitching
集材场 brow; collecting place; collecting yard; depot; forest depot; upper landing; wood yard; wood depot
集材场检尺材积 road scale
集材场交货 delivery at landing
集材场清理工 sapling chopper
集材车 lifting trailer; prehauler
集材承索辊(免与地面摩擦) road roller
集材大钩 swamp hook
集材带(索道) road-line
集材带宽度 yarding span
集材导木 guide tree
集材到装车场 drag(ging) in
集材道(常规的) cart track; drag road; dragging track; logging road; lumber road; skid(ding) trail; skidding road; skidway; snig

track; snig(ging) trail; stock driveway; stock route; travois road
集材道(爬犁的) dray road
集材道边导木 breast-work log
集材道伐开面积 hatch
集材道间距 yarding span
集材道清道工 run cutter
集材道设置 trail setting
集材吊杆 stiff boom
集材方式 yarding system
集材费 cost of hauling
集材杆 mast; pole; spar pole; swing tree
A形集材杆 A-frame spar
集材杆底部辅助索 breast line
集材杆吊索 bracket guy line
集材杆护板 spar cap
集材杆帽 rigging cap; spar cap
集材杆树 mast-tree; spar tree
集材杆索具装置 tree rig
集材杆支脚 spar foot
集材钢杆 steel spar; yarding tower
集材钢塔架 yarding tower
集材工 skidder
集材工队长 hook tender
集材工长 log getter
集材拱架 integral arch; lifting trailer; logging arch; sulky; yarder
集材钩 bull hook
集材钩钉 dragging dog; dragging pin
集材归堆机 skidder-piler
集材滑车(可转动) yarder
集材滑道 road-slide
集材滑轮 bouse block; yarding block
集材机 log getter; skidder
集材机集材 yarder cable way
集材机械 skidding machinery
集材机移动滑轮 construction block
集材机移动曳索 construction line
集材架 pallet
集材架杆 head spar; spar
集材架秆 head spar tree
集材绞盘机 dudley; dudler; skidding winch; yarder; yarding donkey; yarding wheels
集材距离 extraction distance; skidding reach
集材卡钩 logging tongs
集材卡阻 hang up
集材连接索 tongline
集材链 snig chain
集材领班 cat hooker
集材路 extraction rack (road); skid road
集材帽 baptist cone; dragging cap; skidding cone; cap
集材爬犁 dray; go-devil; logging sledge; skidding sled; sled; sloop; travois; wood boat; yarding sled
集材期 removing time
集材牵引索 yarding line
集材钳 skidding scissors
集材橇 skidding sled; wooden sled; yarding sled
集材撬 logging sledge; wood boat
集材区 cutting chance; show
集材时间 removing time
集材树 swing tree
集材索 logging line; skidder line; skidding line; skidding rope; snaking line
集材索道跑车 yard carriage
集材趟数 skidding trip
集材托盘 skid(ding) pan; skidding dish; skidding shoe; snigging pan
集材拖板 dragging shoe; dragging slip
集材拖杆(畜力) spreader
集材拖拉机 cat; logging tractor; running cat; skid cat; timber tractor; tractor skidder
集材线 extraction rack (road); skid road
集材线路 corridor; road-line
集材小道 alleyway; extraction lane; run; skidding lane; snaking trail; yarding trail
集材小爬犁 drag-lizard
集材遭 driveway
集材爪钩 skidding scissors
集材柱 mast
集材抓钩 logging grapple
集材装车机 jammer; loader-tractor; skid loader; skidder-loader; yard(ing) crane
集材装车联合机 yarder-loader

集材锥形套筒 skidding cone
集材阻力 skidding resistance
集材阻力系数 skidding coefficient
集材作业 chance; timber collecting
集材作业区 logging case; logging show; logging unit
集草器 buckrake
集尘 dust collection
集尘管道 dust collecting duct
集尘器 dust collector
集成 integration
集成单板(胶合板) integrated veneer
集成岩 composite rock
集堆机 buckrake
集肤效应 skin effect
集合花坛 grouping bed
集合群落 metacommunity; meta-community
集合种群 metapopulation
集局型 settlement pattern
集聚地 settlement
集块岩 agglomerate
集拢 assemblage
集群 assembly; clan; colonization; colony
集群分布 clumped distribution; clustered distribution
集群分布的 clumped
集群分布种群 contagious population
集群开花 gregarious flowering
集群效应 group effect
集群植物 settlement
集水 catchment
集水沟 collecting channel; cut-off drain
集水排水沟 collecting drain
集水盆地 catchment basin
集水区 water catchment; watershed
集水区保护 watershed protection
集水区经营 watershed management
集水区域 catchment area; watershed area
集体林 collectively owned forest
集体林场 collective tree farm
集团谱系育种法 mass pedigree method
集团选种 bulk selection
集团育种法 mass method of breeding
集团子代测定 bulk progeny test
集约放牧 dense pasture
集约经营 conservative management; intensive management
集约经营林 intensive managed forest
集约林业 intensive forestry
集约森林经营 intensive forest management
集约育林 intensive silviculture
集约栽培 intensive culture
集约择伐 intensive selection felling
集运(木材) extraction
集运材 remove; log removal
集运材车 bunk
集运材成本 hauling cost
集运材道尽头 dead-end of logging road
集运材起重机 logging crane; logging hoist
集运材线路 lead
集运点 landing; ramp
集植 massive planting
集中 concentration
集中放牧 close herding
集中灭火法 individual station method
集中应力 concentrated stress
集中载荷 concentrated load(ing); point load
集中贮藏(动物行为) ladder-hording
集装架 load cradle
集装箱 container
集装箱运货 unitized shipment
籍板材 packing case wood
几何定心 geometrical centr(e)ing
几何分布 geometric distribution
几何平均数 geometric mean
几何式(园林) geometrical style
几何式花园 geometric style garden
几何体 geometric solid
几何原理 geometrical principle
几何增长 geometric growth
己聚糖 hexosan
己酸 caproic acid
己糖 hexose
己糖激酶 hexokiuase
挤出(钉子) springing
挤出机 extruder
挤奶(从母体中分离子体同位素) milking

挤水辊 squeeze roll
挤涂机 extrusion coater
挤压 bear; extrusion; squeeze-out
挤压成型 extrusion moulding
挤压导火索 crimp
挤压法(刨花板制造) extrusion method; extrusion-pressing
挤压钢模 extrusion die
挤压辊 nip roll
挤压机 extruder; extrusion press
挤压力 extrusion pressure
挤压模 extrusion die
挤压碎料板 extruded particle board
挤压系统 extrusion system
挤制模型 extrusion die
脊 carina (*plur.* carinae); lophos; raphe; ridge
脊的 raphal
脊梁 ridge beam; ridge-piece; roof tree
脊椎的 vertebral
计尘器 dust counter; konimeter
计划、规划和预算系统 planning, programming and budgetary system
计划报废 planned obsolescence
计划采伐量 prescribed yield
计划采伐期 cutting period
计划采伐效益 allowable cut effect
计划概要 general plan
计划利润 project benefits
计划利用量(牧草及饲料植物) allowable use
计划评审法 program evaluation and review technique
计划坡度 designed slope
计划期间 forest management period; planning period
计划烧除(废材) controlled prescribed burning
计划手册 management instructions; management prescriptions; management requlations
计划外采伐 unbudgeted cutting
计划用火 prescribed fire
计划增长数 program growth
计价 pricing
计件采伐 bushel
计件产品 piece product
计件工合同 piece-work agreement
计件工人 piece worker
计件工资 pay-basis by the piece; piece rate; piece wage; price wage; price wages
计件工资制 piece wage rate; piece-rate system
计件工作 piece-work
计量 account; accounting
计量泵 metering pump
计量槽(木材防腐) charging tank
计量方法 accounting approach; accounting method
计量分配箱 equalizing vat
计量罐 working tank
计量检验装置 check weigher
计量经济模式 econometric modelling
计量经济学 econometrics
计量料仓 dosing bin; dosing silo
计量料斗 weigh hopper
计量器(混凝土送料) batcher
计量输送带 dosing band
计量箱 dosage bunker; weigh hopper
计日工 day(-to-day) worker
计日工资 daily wage
计入期 crediting period
计时工资 time wage
计时工资率 time wage rate
计时器 keyer; time device
计示硬度 durometer-hardness
计数板 counting board
计数器 counter; counting device
计数销售 sale-lot of numbers of pieces
计数样方 count quadrat method
计税基数 tax base
计税基准 tax base; taxable base
计算尺 slide rule
计算机 computer; electronic computer
计算机程序 computer program
计算机辅助分析 computer assisted analysis
计算机辅助设计 computer aided design
计算机兼容磁带 computer compatible tape
计算机模拟分析 computer simulation analysis

计算机模拟输入 computer analog input
计算机线性程序设计模型 computerized linear programing projection model
计算期 calculation-period
计算器 calculator; computer
计算图表 nomograph
计量装置 metering device
记号 mark
记录本 level book
记录放大装置 recorder magnification setting
记录器记数器 register
记录器响应讯号 recorder response
记码员 checker
记时员 time officer
记账单位 management planning unit
纪念公园 memorial park
纪念巨广场 monumental square
技术 technology
技术标准 quality specification
技术规程 technical schedule
技术规范 technical specification
技术或性能标准 technology or performance standard
技术经济指标 technical economic index
技术潜力 technological potential
技术说明书 technical specification
技术转让 technology transfer
忌光种子 negatively photoblastic (seed)
季风 monsoon
季节[性]变化 seasonal change
季节层片 seasonal synusia
季节隔离 seasonal isolation
季节洄游 seasonal migration
季节节律 seasonal rhythm
季节频率 seasonal frequency
季节迁移 seasonal migration
季节替代种 seasonal vicariad
季节相 seasonal aspect
季节性变异 seasonal variation
季节性波动 seasonal fluctuation
季节性的 seasonal
季节性放牧 seasonal grazing
季节性分布 seasonal distribution; seasonal prevalence
季节性浮游生物 meroplankton
季节性降雨 seasonal reinfall
季节性景色 seasonal aspect
季节性迁移放牧 transhumance
季节性生长量 seasonal increment
季节性消长 seasonal prevalence
季节性演替 seasonal succession
季节性沼泽灌丛 seasonal swamp thicket
季节性沼泽森林 seasonal swamp forest
季节性状貌 seasonal aspect
季节周期 seasonal periodicity; seasonal rhythm
季戊炸药 penta
季相(物候) aspection; seasonal aspect; aspect
季相林 seasonal forest
季雨林 monsoon forest; monsoon rain forest
剂量(以每单位面积重量或容积计) dose
剂量效应 dosage effect
剂量效应关系 dose-response relationship
迹 mark
迹地法 cut area method
迹地更新 reforestation
迹隙 trace gap
既得利益 vested interest
既往成本 past costs
继承性 hereditability
继电器 relay
继电器装置 relay set
继动器 relay
继发性感染 secondary infection
继发性休眠 secondary dormancy
继生枝 lammas shoot
寄存(计算机) load
寄存能力(计算机) load capacity
寄存器 register
寄存信息段 loader module
寄生 parasitism
寄生虫学 parasitology
寄生的 parasitic
寄生蜂 chalcid-fly; parasitic hymenoptera; parasitic wasp
寄生根 parasitic root
寄生过程 parasitic process
寄生昆虫 insect parasite

寄生潜力 parasitic potential
寄生生物 parasites
寄生树上的 dendrophilous
寄生维管植物 parasite vascular plants
寄生物 parasite
寄生物抗性 parasite resistance
寄生物上寄生的 biparasitic
寄生物-宿主相互作用 parasite-host interaction
寄生物学 parasitology
寄生物样的 parasitoid
寄生现象 parasitism
寄生性昆虫 parasitic insect
寄生性食物链 parasite food chain
寄生性双翅类 parasitic diptera
寄生性种子植物 parasitic seed plant
寄生蝇 larva-fly
寄生幼虫 larva
寄生植物 parasitic plant
寄食生物 inquiline
寄食现象 inquilinism
寄售 sale on commission
寄托 bailment
寄销 consign
寄销品 goods-out on consignment
寄宿 lodging
寄主 host
寄主范围 host range
寄主寄生物复合体 host-parasite complex
寄主寄生物互作 host-parasite interaction
寄主寄生物间关系 host-parasite relationship
寄主寄生物组合 host-parasite combination
寄主抗性 host resistance
寄主密度 host density
寄主适合性 host suitability
寄主树 primary host tree
寄主选择性 host-selection
寄主植物 host plant
寄主专化病原物 host-specific pathogen
寄主专一性 host specificity
蓟马 thrips
系船浮标 mooring buoy
系船索 bridle
系船柱 mooring post
系紧 anchor; tie
系紧螺栓 tie bolt
系链(固定滑轮) rigging chain
系梁 binder; tie beam
系留索(气球等) tether line; tiedown rope
系排桩 dolphin
系索(导向滑轮等) lasher; strap; tie line; dutchman
系索钩 line hook
系索连接板 rigging plate
系柱 hitching post
加氨 ammonification
加班 overtime
加倍染色体 doubling chromosome
加倍余额递减法 double-declining-balance
加标(提高标价) mark on
加标签 tagging
加成的 additive
加成聚合作用 polyaddition
加德纳式干燥机 Gardner dryer
加德纳颜色 Gardner colour
加法电路 adder
加法器 adder
加工(木材加工修饰) conversion; curving
加工报酬 conversion return
加工材 finished stock; surfaced timber
加工产品 converted product
加工厂 millwork plant; conversion plant
加工厚度 working thickness
加工锯材 manufacturing lumber
加工宽度 working width
加工木材 converted timber
加工能力 work(ing) ability
加工缺陷 mismanufacture
加工剩余物 off-cut
加工性能 working quality
加工用材 factory timber; industrial lumber
加工用锯材 factory lumber
加固 reinforcement
加合物 adduct
加和的 additive
加厚 bodying

加厚程序 thickening proceed
加环标 ringing
加碱水解 alkaline hydrolysis
加劲钢丝 bracing wire
加里格群落(地中海常绿矮灌丛) Garrigue
加力漂子 stay boom
加料 charge; feed
加料斗 charging hopper; feed hopper
加料漏斗 hopper
加料器 feeder; filler
加拿大标准打浆度 Canadian Standard Freeness
加拿大冷杉香脂 Canada balsam
加拿大树胶 fir balsam
加拿大庭园 Canadian garden
加拿大香脂 Canada balsam
加拿大园林 Canadian garden
加拿大原木板英尺材积表 Canadian log rule
加强件 reinforcement
加强筋 buttress; gill
加权平均盘存法 weighted-average inventory method
加权平均数 weighted average; weighted mean
加热 heating-up; warming-up
加热层 zone of heating
加热导管 healing duct
加热度日 heating degree days
加热管 heating coil
加热滚筒 heated roll; heating drum
加热胶合 hot gluing
加热器 heater; heating coil; heating unit
加热区 zone of heating
加热杀菌法 pasteurization
加热时间 heating period; heating time
加热速度 heating rate
加热温度 heating temperature
加热雾化室 heated nebulization chamber
加热元件 heating element
加热周期 heating cycle; heating period
加入 cut-in
加入量 additive
加色法 additive colour process
加色观察器 additive viewer
加色合成 additive colour formation; additive colour mixture
加砂浆膏围干法(防腐) sand collar method
加速风化 accelerated weathering
加速干燥法 accelerated air-drying
加速老化 accelerated ageing
加速器 accelerator; oscillator
加速侵蚀 accelerated erosion
加速蠕变 accelerated creep
加速森林成长 speed the growth of the forests
加速生长 accelerated growth
加速生长量 accelerated increment
加速试验 accelerated test
加速输送带 accelerating belt
加速输送机 acceleration conveyor
加速死亡期 accelerating death phase
加速折旧 accelerated depreciation
加酸漂白 sour bleach
加酸器 acid feeder
加条纹割沟 streak
加系(染色体) addition line
加楔 wedging
加楔伐 wedge felling
加性方差 additive variance
加性基因 additive gene
加性基因效应 additive genetic effect
加性遗传方差 additive genetic variance
加性因素 additive factor
加压 compression; pressing; upset
加压部分 press part; press section
加压冲击处理 pressure-stroke process
加压冲击周期 pressure-stroke cycle
加压处理 pressure treatment
加压罐 pressure tank
加压过程 pressure process
加压阶段 pressure period
加压时间 clamp(ing) time; pressing time
加压温度 pressing temperature
加压纤维板 compressed fibre board
加压纤维分离 pressurized refining
加压纤维分离机 pressurized refiner
加压真空交替法 alternating pressure process
加压周期 pressing cycle; pressing period

加压周期阶段 pressure cycle stage
加压周期自动定时仪 press cycle timer
加氧作用 oxygenation
加腋嵌槽接 haunch rabbet joint
加腋榫 haunched tenon
加腋榫槽接 haunch mortise and tenon joint
加油车 fuel(l)er
加油工 grease dauber; greaser
加油器 fuel(l)er
加载速率 rate of loading
加植 filling in; filling of blanks
加州胡椒油 schinus oil
加桩 plus stake; spiling
家蚕 silk worm; mulberry silk worm
家畜 livestock
家具 furniture
家具板 furniture board
家具背板 furniture back
家具部件 furniture components
家具材 wood for furniture
家具覆盖饰物 upholstery
家具工业 furniture industry
家具规格材 furniture dimension stock
家具胶合板 furniture plywood
家具模型 mock-up furniture
家具木工 cabinetmaker
家具窃蠹 common furniture beetle; anobium beetle; furniture beetle
家具清漆 furniture varnish
家具商 upholsterer
家具业 upholstery
家具用材 cabinetmaker's wood
家具用小五金 cabinet hardware
家具制造 cabinetmaking
家具蛀虫 furniture beetle
家菌(建材腐朽菌) domestic fungus
家庭苗圃 home nursery
家庭园林 home garden
家庭园艺 home gardening
家系 family; kin
家系内选择 within family selection
家系图 family tree
家系选择 family selection
家系遗传力 family heritability
家系优势 family dominant
家用林 home-use forest
夹 holder
夹板 fish plate; pallet; splice bar
夹板接[头] fished joint
夹板式锯卡子 sandwich guide
夹层锅 jacketed kettle
夹层梁 sandwich beam
夹缝钉 dowel
夹角接 lock mitre joint
夹紧 chuck
夹紧装置 clamp; clamping device; gripping device
夹具 clamp iron; gripping device
C形夹具 C-clamp
夹锯 bind; binding; pinch; saw pinching; side bind; timber bind
夹皮 bark pocket; bark seam; enclosed pocket; inbark; ingrown bark; trapped bark
夹钳 clamp; iron jaw; tongs
夹钳-承载梁式集材机 clam-bunk skidder
夹钳链锯(伐木) grapple saw
夹钳式索轮 clamp-type rope wheel
夹钳松开装置 clamp release
夹生 under cook
夹生浆 undercooked pulp
夹生木片 under cooked chip
夹式储弹器 clip magazine
夹兽者 trapper
夹送辊 pinch roll
夹套蒸馏锅 jacketed still
夹心板 sandwich board; sandwich composites; sandwich panel
夹心板结构 sandwich construction
夹心梁 sandwich beam
夹杂水 entrained water
夹杂物 foreign matters; impurity; inert matter
夹钻(生长锥) corer locking; corer sticking
茄替胶 ghatti gum
荚 shale
荚果 cod; legume; pod
荚膜 capsule
荚膜多糖 capsular polysaccharide
颊 bucca (*plur.* buccae); gena
颊囊 cheek pouch
蛱蝶科 Nymphalidae
甲板 deck plank; decking

甲板材 ship-plank
甲拌磷(3911)(杀虫剂) phorate
甲虫 beetle
甲酚树脂胶 cresol resin adhesive
甲氟磷(杀虫剂) Pestox-14
甲基托布津(杀菌剂) topsin-M
甲壳类动物 shellfish
甲壳钻孔虫 crustacean borer
甲克动物学 carcinology
甲腊 formazan
甲酸的 formic
甲烷 methane
甲氧苯基 anisyl
甲氧苯基甲基酮 anisyl methyl ketone
甲氧基 methoxyl group
甲氧基苯 anisole
甲瓒 formazan
钾泵 potassium pump
钾硝 saltpeter
钾硝石 kali saltpeter
假薄壁组织 pseudoparenchyma
假背斜 pseudoanticline
假边材 false sapwood
假彩色 pseudo colour
假彩色图像 false colour image
假潮(海洋) seiche
假顶生的 pseudoterminal
假顶芽 pseudoterminal bud
假定向演化 pseudo-orthogenesis
假定蓄积量 assumed growing stock
假多倍性 agmatoploidy
假多效性 false pleiotropy
假二歧分枝 false dichotomy
假二歧式 false dichotomy
假峰 ghost peak
假峰效应 ghosting effect
假附生植物 pseudo-epiphyte
假根 rhizine; rhizoid; rizoid
假根茎 rhizomoid
假果 accessory fruit; pseudo-carp
假红紫素 pseudopurpurin
假虎刺苷 carissin
假花瓣 petalode
假花冠 pseudocorolla
假花纹 faux swirl
假灰化层 albic horizon
假计成本 imputed cost
假菌核 pseudosclerotium
假立体模型 pseudo scopic stereo model
假粒状结构 pseudo granular structure
假面状花冠 personate corolla
假年轮 false annual (growth) ring; false annual ring; false ring; pseudoring
假配合 pseudogamy
假皮 imitation leather
假潜育 pseudogleyzation
假山 artificial hill; rockery
假山园 rockery
假射线 false ray
假受精 pseudogamy
假死 anabiosis; asphyxy
假死态 thanatosis
假显性 pseudodominance
假想圆柱 ideal cylinner
假向斜 pseudosycline
假像图谱 deceptive spectrun
假心材 false duramen; false heartwood; false-heart
假形成层 pseudocambium (*plur.* -bia)
假性紫罗兰酮 citrylidene acetone
假羊皮纸 vegetable parchment
假叶 phylloclade; phyllode
假叶烯 phyllocladene
假杂种 false hybrid
假噪声 pseudo-noise
假直向演化 pseudo-orthogenesis
假植 heel(ing)-in; heelin; sheugh(ing); temporary planting
假指示 false point
假种皮 aril
假子座 pseudostroma
槚如酚 anacardol
槚如坚果 cashew nut
价格比 price relative
价格变动条款 escalator clause
价格变异 price variation
价格不确定性 price uncertainty
价格幅度 pirce range
价格机制 price mechanism
价格极限 price limit
价格-收益比率 price-earnings ratio
价格调整条款 escalator clause

价格形成 price formation
价格引导 price wise
价格增长 price increment
价格增长率 price growth percent; price increment percent
价格支持计划 price-support program
价格指数 price index
价格指数化 indexation of price
价值 value
价值方程式 equation of value
价值平分法 value frame work
价值增长 price increment; value increment
价值增长率 price growth percent; price increment percent; value increment percent
驾驶仪 pilot
架 holder; rack
架撑 shelf supports
架底短柱 ground prop
架顶构件 head
架杆 gin pole; guide tree; straight boom
A形架杆 A-frame
架杆绷索 snap line
A形架杆集材绞盘机 A-frame yarder
架杆起重机 pole derrick
A形架杆起重机 A-frame derrick
A形架杆装载机 A-frame loader
架空槽道 suspension chute
架空单轨道 aerial monorail-tramway
架空电线 pole line
架空轨道 hanging rail
架空式摆动喷灌机 overhead oscillating sprinkler
架空式集材 aerial skidding
架空索 aerial line; overhead cable
架空索道 aerial cableway; aerial ropeway; aerial tramway; aerial wireropeway; running skyline; skyline; skyroad; suspended skyline; track cable; track line
架空索道集材 aerial skidder; aerial skidding; log tramway system; overhead cableway skidding; overhead skidding; overhead yarding
架空线集材 yarder cable way
架空线路 aerial line
架空线式 yarder cable way
架木 baulk
架设 set-up
架设索道 rig
架设索道用滑轮 passblock
架线滑轮(索道) rigger's block; rigging block
架线集材 cable extraction
架线卷筒 rigger's drum
架轩兜卸机 frame unloader
架子 shelf
嫁接 graft; grafting; imp; ingraft; ingraftment
V形嫁接 notch graft
嫁接不亲和性 graft incompatibility
嫁接法 graftage
嫁接接合部 graft union
嫁接框匣 grafting case
嫁接苗 grafting plant
嫁接嵌合体 grafting chimaera
嫁接亲和力 grafting affinity
嫁接匣 grafting case
嫁接杂种 burdo; chim(a)era; graft hybrid
尖 cusp
尖齿(锯) lance tooth
尖顶承载梁 spear edge bunk
尖端细的 tapering
尖镐 pick mattok
尖角(刀具) knife chamber angle
尖角节 knot horn
尖节 horn knot; mule ear; mule-ear knot; slash knot; spike knot; splay knot
尖枯 tip burn
尖梢 high taper
尖舌蜂 acutilingues
尖舌类 acutilingues
尖头锯 fox-tail saw
尖头式(弹头) spitzer
尖突 cusp
尖弯头 beak
尖削 fall-off; taper; taperness
尖削的 tapering
尖削度 declining; fall-off; high taper; taper; taper coefficient; tapering grade; taperingness; taperness; rise
尖削度值 taper value

尖削锯材 tapering lumber
尖削率 taper rate
尖削四方支腿 spade foot
尖削系数 taper coefficient
尖削原木 tapered log
尖削原木下锯法 taper sawing
尖削指数 taper coefficient
尖叶番泻树 Alexandria senna
坚固的 solid
坚固红心腐 firm red heart
坚固性 fastness
坚果 mast; nut (*plur.* nuces)
坚木 quebracho
坚韧材 tough wood
坚韧纤维 stiff fibre
坚实度 compactness
坚实节 coherent knot; firm knot; fixed knot; light knot
坚硬红壤 indurated red earth
坚硬黏土 gumbo
坚硬细胞 sclereid
坚硬性 stiffness
间冰期 interglacial period
间齿 raker (tooth); clear teeth
间道 by-pass road
间渡 intergradation
间距 space; spacing
间生 paragenesis
间性体 intersex
间质 mesenchyma; mesenchyme; stroma (*plur.* -ata)
间柱材 stud(ding)
肩挂式喷雾机 shoulder sprayer
肩角 humeral angle
监测[器] monitor
监查护林员 prevention guard
监督 supervisor
监督区 inspection district
监督系统 lookout system
监控程序 monitoring program
监视器 monitor
兼农林业 farm forestry
兼任经理 interlocking directorate
兼容信息系统 compatible information system
兼容性 compatibility
兼性腐生物 facultative saprophyte
兼性孤雌生殖 facultative parthenogenesis
兼性好氧生物 facultative aerobe
兼性寄生菌 facultative parasite
兼性寄生物 facultative parasite; hemiparasite
兼性盐生植物 facultative halophyte
兼性厌氧菌 facultative anaerobe; facultative anaerobic bacteria
兼性厌氧生物 facultative anaerobe
兼性自养生物 facultative autotroph
拣掉道材(木滑道上) sack the slide
茧 cocoon
茧蜂 braconidfly; supplementary ichneumon fly
捡尺 inspection
检波 rectifying
检测器 detector
检测组人员 measuring staff
检查 inspection; overhaul
检查表 checklist
检查法 check system; control method; examination method
检查过的 zeroed
检查孔 inspection hole
检查孔盖板 inspection plate
检查口盖板 access plate
检查哨 check(ing) point
检查用铰链板 check-up flap
检查余烬 feeling for fire
检查站 check(ing) point; check(ing) station
检潮仪 tide gauge
检尺 measuring; scale; scale measure; scaling; scaling of felled timber
检尺簿 measurement list
检尺法 method of measurement
检尺分级仪(木材) proof grader
检尺复查员 check scaler
检尺杆 cheat stick; thief stick
检尺规则 inspection rule; scaling rule
检尺计数器 tally counter
检尺记录表 tally sheet
检尺记录牌 tally board
检尺径 scaling diameter
检尺盘盈 overscale
检尺清单 scale bill
检尺台 scaling ramp

检尺野账 scale bill
检尺员 cheater; grader; inspector; scaler; tally man; measurer
检尺站 scaling station
检尺长 length class
检尺杖 scale-rule; scale-stick; scaling stick; thief stick
检定种子 certified seed
检漏器 leak detector
检卤漏灯 halide lamp
检偏振器 polarization analyzer
检票员 teller
检石机 rock picker
检视门(窗) inspection door; small door
检索 retriever
检索表 identification key; key
检索分类卡 key sort cards
检索卡片 key card
检修标准 inspection standard
检修孔 access hole
检修门 access door
χ^2检验 chi-square test
t检验 t-test
检验标准 inspection standard
检验输送带 inspection conveyor
检验样地 test plot
检验员 checker; inspector; verifier; controller
检验证明书 inspection certificate
检样器 sampler
检疫 quarantine; quarantine inspection
检疫法规 quarantine law
检疫规章 quarantine regulation
检疫员 quarantine inspector
检疫证书 bill of health
减产效应 depressor effect
减毒 attenuation
减毒株 avirulent strain
减反率法 Gentan probability method
减耗资产 wasting assets
减活化剂 deactivator
减活化作用 deactivation
减排 mitigation; mitigation of carbon emission; emmision reduction
减排单位 emission reduction unit
减排能力 mitigative capacity
减轻 abatement
减色成色 subtractive colour formation
减色法 subtractive colour process
减少投资 negative investment
减少因采伐和退化造成的排放 reducing emissions from deforestation and forest degradation (REDD)
减少种(者) decreaser
减湿 dehumidification
减湿干燥[法] drying by dehumidification
减数分裂 meiosis; meiotic division; reduction division
减数分裂分离比偏移 meiotic drive
减数分裂价 cost of meiosis
减数分裂驱动 meiotic drive
减数分裂失常 meiotic disturbance
减税 abatement; tax abatement
减速齿轮 reduction gear
减速干燥期 falling rate period
减速链(缠于下滑原木上) bull chain
减速期 falling rate period
减温率 lapse rate
减压处理 pressure reduction treatment
减压阀 pressure reducing valve
减压干燥法 vacuum seasoning
减压干燥器 vacuum drier
减压油煮法(防腐) boiling under vacuum; Boulton process
减压周期 vacuum cycle
减震 quench; vibes-less
减震垫式油锯 cushion power saw
减震器 absorber; damper; dashpot; quencher; vibration damper; vibroshock
剪板机 clipper; guillotine; guillotine shears; plate shears; veneer clipper
剪板试验 plate shear test
剪草机 weeding machine
剪草坪 mowing
剪草器 lawn-mower
剪刀撑 bridging; double bridging
剪断 scission
剪根伐木机 root shear feller
剪力面积 shear area
剪力图 shearing force diagram

剪裂 shear fracture
剪切 clipping; shear; trimming
剪切弹性模量 modulus of shearing elasticity; shear elasticity; shear modulus
剪切工具 clipper
剪切机 shear; snippers
剪切拉伸强度 shear tensile strength
剪切力 shear; shear force
剪切破坏 shear failure
剪切强度 shear(ing) strength
剪切区域 shear zone
剪切试件 shear-test block
剪切试验 shear test
剪切梯度 shear gradient
剪切系数 coefficient of shear
剪式伐木装置 tree shearing device
剪式桁架 scissors truss
剪应变 shear strain; shearing strain
剪应力 shear(ing) stress
剪枝 lop; lopping; lopping of branches; lopping off
简便经纬仪 simple theodolite
简并 degeneracy
简并密码 degenerate code
简单矮林作业 simple coppice system
简单茎花 simple cauliflory
简单逻辑斯蒂曲线 simple logistic curve
简单随机抽样 simple random sampling
简单塔里夫材积表 simple tariff
简单遗传 simple inheritance
简单异速生长 simple allometric growth
简单易位 simple translocation
简单应变 simple strain
简单应力 simple stress
简卡压痕试验 Janka indentation test
简卡硬度 Janka hardness
简卡硬度试验 Janka-test
简图 diagrammatic(al) view
简易柴捆梯田 simple terracing with fascine
简易谷坊 simple check dam
简易架空索道 simple aerial cableway
简易经营计划 preliminary plan
简易拦沙坝 simple check dam
简易芒属草埂梯田 simple terracing with miscanthus
简易求积仪 hatchet planimeter
简易梯田工程 simple terracing works
简易温床(无加热设备的) cold frame
简支梁 free beam; freely supported beam; simple beam
碱斑 alkaline patche
碱不溶树脂 resene
碱抽出物 alkali extractive
碱处理 alkaline treatment
碱度 alkalinity; basicity
碱法半化学浆 alkaline semichemical pulp
碱法浆 alkali pulp; alkaline pulp
碱法蒸煮 alkaline cooking
碱法纸浆 alkaline pulp
碱钙岩系 alkali-lime series
碱耗 alkali consumption
碱化(作用) solonization; mercerization
碱化度 causticity
碱化灰钙土 alkaline s(i)erozem
碱化土 alkaline soil
碱化液 green liquor
碱化棕钙土 mallisol; solonized brown soil
碱回收 alkali recovery; soda recovery
碱基 base
碱基饱和 base saturation
碱基饱和度 base saturation degree
碱基对 base pair
碱基配对 base pairing
碱基取代 base substitution
碱基顺序 base sequence
碱基替换 base substitution
碱基值 base status
碱解法 alkaline hydrolysis
碱氯法(制浆) Pomolio-Celdecor process
碱木质素 alkali lignin
碱熔 alkali fusion
碱石灰 soda-lime
碱土 alkali soil; solonetz; solonetz soil
碱土植物 basicole
碱性 alkalinity; basicity
碱性草甸土 alkaline meadow soil
碱性湖 alkali lake
碱性胶 alkaline glue

碱性品红 fuchsin
碱性染料 basic dye
碱性溶离 alkaline solvolysis
碱性水解 alkali hydrolysis
碱性土 alkaline soil
碱性亚硫酸盐法 alkali-sulfite process
碱性岩类 alkaline rocks
碱性植物 alkaliplant
碱性纸浆 alkaline pulp
碱纸浆 soda pulp
间壁 mid-board; partition
间断[生长]轮 discontinuous (growth) ring
间断波状纹 interrupted wavy grain
间断层 discontinuity layer
间断分布 discontinuous distribution; disjunctive distribution
间断分布区 disjunctive area
间断花序 interrupted inflorescence
间断开沟植树机械 intermittent furrow planter
间断琴背纹 interrupted fiddle-back
间断叶序 discontinuous phyllotaxy
间发性 periodicity
间伐 intermediate cutting; thinning out; intermediate
间伐方式(类型) thinning method; type of thinning
间伐收获 intermediate felling yield; intermediate yield; thining yield
间伐收入 financial returns from thinnings
间隔 compartment; inter space; skip; interval
间隔带 inter space
间隔垫木 spacing sticker
间隔法 distance method
间隔基 spacer
间隔距离数 spacing figure
间隔炉 compartment furnace; compartment kiln
间隔期 plastochrone; return interval
间隔期指数 plastochrone index
间隔收获 intermittent yield
间隔团 spacer
间隔窑 compartment furnace; compartment kiln
间隔择伐 intermittent selective cutting
间隔作业 intermittent yield system
间混作 intercropping
间接成本 indirect cost; overhead cost
间接分裂 caryokinesis
间接复型法 indirect replica (method)
间接估计 indirect estimate
间接加热 indirect heating
间接减数分裂 indirect reduction division
间接劳动 indirect labour
间接灭火 indirect fire suppression
间接灭火法 indirect method of suppression
间接评定法 method of indirect evaluation
间接扑灭法 method of indirect suppression
间接汽蒸 indirect steaming
间接税 hidden tax
间接细胞分裂 indirect (cell) division
间接效应 indirect effect
间接选择 indirect selection
间接营养体分裂 indirect vegetative division
间接蒸煮工艺 indirect cooking process
间竭河 blind creek
间苗 gap; gapping; plant thinning; singling; thinning; thinning out
间苗机 gapping machine; plant thinning machine; seedbed thinner; single machine
间苗器 singler
间喷温泉 geyser
间期 intermediate stage
间位薄壁组织 metatracheal parenchyma
间位宽带状薄壁组织 metatracheal wide parenchyma
间位星散状薄壁组织 metatracheal diffused parenchyma
间位窄带状薄壁组织 metatracheal narrow parenchyma
间隙(广义) gap; inter space
间隙(芯板) tunnel
间隙尺寸 gap size
间隙的 interstitial
间隙动物 interstitial animal
间隙动物区系 interstitial fauna

间隙规 gap gauge
间歇法 batch process
间歇干燥装置 part time drying system
间歇河 intermittent river; intermittent stream
间歇湖 intermittent lake
间歇灭菌 discontinuous sterilization
间歇喷泉 geyser
间歇喷雾 intermittent mist
间歇喷蒸 intermittent steaming; periodic steaming
间歇式拌胶机 batch mixer
间歇式干燥机 plate dryer
间歇式进料系统 intermittently working feed system
间歇式量秤 batch scale
间歇式磨木机 intermittent type grinder
间歇式圆筒剥皮机 intermittent barking drum
间歇式蒸煮器 batch digester
间歇性常年作业 complete intermittent working
间歇性经营 intermittent managemmt
间歇性择伐(回归期多为10年以下) intermittent selection
间歇蒸馏器 batch still
间歇作业 forest under intermittent management; intermittent production
间歇作业窑 intermittent kiln
间型木 intermediate tree
间种的作物 catch crop
间作 catch crop; companion cropping; intercropping; intermediate culture; interplanting
间作作物 companion crop
件长堆(木件单位) rick
饯椽(四坡屋顶的) hip rafter
建立者效应 founder effect
建立者原则 founder principle
建群种 edificator; constructive species
建设期 building phase
建议 offer
建筑板材 building board
建筑材 building log; building lumber; building timber; strength timber; structural timber; structural wood
建筑材料 structural material
建筑材卧孔菌 building-rot fungus; building poria
建筑结构家具 architectural furniture
建筑结构涂饰 architectural finish
建筑绝缘板 structural insulation board
建筑砌块 construction block
建筑色彩效果 architectural colour effect
建筑设计 architectural design
建筑声学 architectural acoustics
建筑式园林 architectural style garden
建筑纤维板 fibre building board
建筑学 architecture
建筑用板 structural panel
建筑用材 construction timber; crop of formed tree
建筑造型 mould-making
建筑柱型 order of architecture
剑状的 ensate
健康的 disease-free
健康树 healthy tree
健康证明书 bill of health
健全材 sound wood
健全材面锯切法 sound cutting
健全的 sound
健全节 sound knot
健全缺陷(无腐朽的缺陷) sound defect
健全无疵的 sound on-grade
健全心材 sound heart-wood
渐变 anamorphosis; gradation
渐变结构刨花板 graduated-particle board; size-graded particleboard
渐变结构碎料板 graduated-particle board
渐变密度 graded density
渐变密度刨花板 graded-density particleboard
渐变群 cline
渐变群变异 clinal variation
渐变态类 paurometabola
渐不发育性状(逐代) obsolete character

渐伐 shelterwood cutting; shelter-wood cutting; shelter-wood felling; stage cut
渐伐更新 shelterwood regeneration
渐伐更新作业 system of successive regeneration fellings
渐伐林 shelter-wood forest
渐伐择伐作业 shelter-wood selection system
渐伐作业 preregeneration system; shelter-wood method; shelter-wood system
渐伐作业法 method of successive thinning
渐伐作业更新 regeneration by compartments
渐伐作业更新林 regenerated forest under the system of successive regeneration fellings
渐行伞伐 progressive shelterwood felling
渐尖的 acuminate
渐进法(灭火) progressive method
渐进改良 progressive improvement
渐进化 progressive evolution
渐进性坏死(细胞) necrobiosis
渐近线 asymptote
渐渗杂交 introgression; introgressive hybridization
渐新世 Oligocene
腱鞘囊肿 ganglion
溅击侵蚀 splash erosion
溅散(涂胶辊上胶) throwing
溅湿地 flush
溅涂 spattering
鉴别 differentiation; identification
鉴别寄主 differential host
鉴别卡 identification card
鉴别染色 heterochromaty
鉴别着色 differential colouration
鉴别种 diagnostic species
鉴定 diagnosis; identification; valuation
鉴定层 diagnostic horizon
鉴定峰 diagnostic peak
鉴定卡 identification card
鉴定试验 design test; identity test
鉴界 boundary correction; rectification of boundaries
键 bond; key; link
键合 link
键合相色谱法 bonded-phase chromatography
键接 keyed joint
键孔锯 keyhole saw
键控穿孔机 keypunch
键能 bond energy; bond strength
键扭转 bond twisting
键弯曲 bond bending
键映射表 key map
箭靶垫 butt
箭毒木毒汁 antiar
箭头状的 sagittal
箭形椅背 arrow back
江河的 potamic
江河分水区 interfluve
江口 river mouth or estuary
江心洲 centre-field
将搁浅木滚入水中 dry roll
将树拉倒 pull; pulling
姜黄 curcuma; turmeric
姜黄素 curcumin
姜黄酮 turmerone
姜黄油 curcuma oil
姜烯 zingiberene
姜油 ginger oil
浆 mash; plasma
浆板 pulp board; sheet pulp
浆板离解机 bale pulper
浆泵 stock pump; stuff pump
浆池 pulp chest
浆幅 pulp web
浆缸底刀 bedplate of beater
浆膏 paste
浆膏密封处理 slurry seal process
浆疙瘩 drag spot
浆果 bacca; berry
浆果皮 sarcotesta
浆果状的 baccate
浆果状毬果 galbulus
浆块 clump; stuff knot
浆料 fibre stuff; fibrestock; pulp stock; pulp stuff; stuff
浆料流送系统 flow approach
浆料流速 stock velocity
浆料上网 approach flow of stock
浆料贮仓 pulp bin
浆浓度 pulp consistency

浆片 glumelle; lodicule
浆粕 pulp
浆砌石谷坊 mortar stone check dam
浆团 clump
浆液 serous fluid; serum
僵顶病 stolbor
僵化 inactivation
讲价 bargaining
奖金 bonus; bounty; gaining; premium
奖励费 incentive payment
桨 oar; paddle
桨船 oar
桨片 paleola
桨式搅拌机 paddle-type blender
桨式搅拌器 paddle type agitator
桨叶 blade
桨叶搅拌器 blade agitator
降等 degradation; degrade; deterioration; downgrade
降低成本 reduced cost
降低火险度 fire hazard reduction; fuel reduction
降低角(雷达遥感) depression angle
降低土地壤湿度 reduce soil moisture
降低账面价值 write-down
降菲烯 horphenathrene
降海繁殖的 catadromous
降河产卵鱼 catadromous fish
降河的 catadromous
降解 degradation
降落式选材槽 drop pocket
降落式选材装置 drop sorter
降水 precipitation
降水机制 precipitation mechanism
降水积效指数 build-up index
降水截流[量] interception of precipitation
降水量 amount of precipitation
降水性云 precipitus
降水有效指数 precipitation effectiveness index
降水蒸发比 precipitation-evaporation ratio; precipitation-effectiveness ratio
降水蒸发指数 precipitation evaporation index
降雪[量] snowfall
降雪线 snow precipitation line
降压 release of pressure
降压期 pressure reducing period
降压自动定时仪 decompression timer
降雨 rainfall
降雨变率 rainfall variability
降雨过剩 rainfall excess
降雨可靠率 rainfall reliability
降雨类型 ombrotype; rainfall type
降雨量 rain height; rainfall
降雨频率 rainfall frequency
降雨期间 rainfall duration
降雨强度 rainfall intensity
降雨侵蚀 rain wash; rainfall erosion
降雨时数 rainfall hours
降雨特性 rainfall characteristic
降雨型 rainfall regime
降雨-蒸发量比 P/E ratio
交边运动 alternating motion
交变变形 alternating deformation
交变冲击试验机 alternating impact machine
交变负荷 alternating load(ing)
交变抗弯试验 alternating bending test
交变拉伸应力 alternating tension stress
交变强制横向循环 alternating forced cross circulation
交变位移 alternating displacement
交变压力 alternating pressure
交变应力 alternating stress; reversed stress
交变应力试验机 alternating testing machine
交哺现象(生态) trophallaxis
交叉 chiasmata; cross; crossover; resection
交叉保护 cross protection
交叉壁孔 crossed pit
交叉播种 cross sowing
交叉测定法 cross-measurement
交叉差异 chiasmata deviation
交叉场(主要用于松柏类木材) cross field
交叉场纹孔式 cross field pit(ting)
交叉道岔 double crossover
交叉点 intersection point
交叉点桩 intersection point mark
交叉堆垛 transverse piling
交叉堆垛法 cross stacking

交叉法 method of intersection
交叉反应 cross-reaction
交叉分类 crossed classification
交叉拱 groin; groyne
交叉归楞 transverse piling
交叉绗纹(绗缝花纹) quilted figure
交叉回收方法 cross-recovery method
交叉接种 cross inoculation
交叉结构 criss-crossed structure
交叉靠架堆板法 end racking
交叉楞 cross pile
交叉侵染 cross infection
交叉射击 cross-shooting
交叉纹孔口 crossing aperture
交叉线 intersecting line
交叉效应 interaction effect
交叉形纹孔对 X figure pit-pair
交叉遗传 criss-cross inheritance
交叉中耕 cross(wise) cultivation
交叉重叠法 juxtaposition
交错采伐 patch logging
交错层 cross stratification
交错层理 cross bedding
交错齿 zigzag tooth
交错堆垛(法) staggered stacking
交错归楞 staggered piling
交错环状剥皮 crossed girdle
交错结构 interwoven texture
交错刻痕 staggered incision
交错区群落 ecotonal community
交错群落 ecotone
交错纹理 alternating spiral grain; interlocked grain; interwoven grain
交代岩 metasomatite
交点 node
交付木材(在集材场) put in
交割延期费 back wardation
交互带状采伐 felling in alternate strips
交互带状皆伐 alternate felling; clearcutting in alternate strip; clear-strip felling; coulisse felling; screen felling
交互带状皆伐更新 alternate felling; clear-strip felling; coulisse felling; screen felling
交互带状皆伐天然更新法 clearcutting in alternate strip
交互带状皆伐作业法 alternate clear strip system; alternate clearcutting method
交互带状作业法 alternate strip system
交互对生的 decussate
交互加压法 alternating pressure process
交互嫁接 reciprocal grafting
交互抗性 cross resistance
交互三出的 tricussate
交互因子 reciprocal factor
交互作用方差 interaction variance
交互作用离差 interaction deviation
交互作用项 interaction term
交换 crossing-over; interchange
交换常数 exchange constant
交换价值 exchange-value
交换量 exchange capacity
交换性酸度 exchange acidity
交换性盐基 exchange base; exchangeable base
交换性阳离子 exchangeable cation
交换抑制基因 crossover suppressor
交换值 cross-over value
交换子 recon
交会摄影测量 intersection photogrammetry
交货付款 cash on delivery
交货价格 delivered price
交货期 lead time
交键 cross linkage
交界处 edge
交界桥 transfer bridge
交联 cross link; cross linkage
交联剂 cross-linking agent
交联键 cross link
交流 interflow
交配 copulation; mate; mating; pairing
交配成本 cost of mating
交配行为 copulatory behaviour; mating behaviour
交配季节 mating season
交配囊 bursa copulatrix
交配前行为 premating behaviour
交配群 mating continuum
交配室 copulation chamber; nuptial chamber

交配体系 mating system
交配同类群 gamodeme
交配型(低等生物) mating type
交配制度 mating system
交让木碱 daphniphylline
交替采伐的伐区 staggered setting
交替生育(结实) alternate (year) bearing
交尾 copulation; mating
交椅 cross leg chair
交易 trade; transactions
交易场 pit; trade ring
交易所 bourse; market house; market place
交织 interfelting
郊区风光 rural scenery
郊区住宅 villa
郊外 outskirt
郊游 hike
浇冰道 icing; sprinkling
浇冰道水车 ice wag(g)on
浇水 watering
浇水量 sprinkling norm
浇涂 flow coating
浇涂法 flowing method
浇涂设备 flow coater
胶 glue
胶斑 glue stain; gum spot; gum vein; size speck
胶版印刷纸 super fine paper
胶槽 glue container
胶层 adhesive layer; bond line; glue film; glue line
胶层比 coreline ratio
胶层滑移 slippage
胶层加热 glue-line heating
胶层黏度 glue line viscosity
胶层强度 glue line strength
胶带 adhesive tape; gum(med) tape; tape; tape adhesive
胶带拼缝机 taping machine
胶固着度 size fastness
胶罐 glue pot
胶合 adhere; adhesion; bond; cementing; glue; gluing
胶合板 laminated wood; plywood; veneer
V形槽胶合板 V-grooved plywood
胶合板板坯 unbonded plywood assembly
胶合板材 lathe log; peeler; peeler log; peeler block; veneer log
胶合板材等级 peeler grade
胶合板地板 plywood flooring
胶合板垫板(热压) plywood caul
胶合板干燥室 dry kiln for plywood
胶合板工业 plywood industry
胶合板箱 veneer chest
胶合板芯板 plywood core veneer
胶合板修整 repairing of plywood
胶合板选材员 peeler picker
胶合板凿离试验 chisel test of plywood
胶合板中板 plywood core veneer
胶合剥离试验 glue line cleavage test; peel adhesive test
胶合层积板 glued lamination board
胶合层积材 glue lam; glulam
胶合层积构件 glued lamination member
胶合层积木构件 glue laminated wood member
胶合层楔开试验 glue line cleavage test
胶合分层 delamination
胶合剂 adhesive; paster
胶合间隙 open joint
胶合剪切试验 glue shear test
胶合力 adhesive force
胶合力计 adhesiometer
胶合力破坏 adhesion failure
胶合黏合叠合 splice
胶合破坏 adhesive failure
胶合强度 adhesive strength; glue strength; glue-joint strength
胶合特性常数 characteristic adhesion constant
胶合体破坏 adherend failure
胶合系数 joint factor
胶合性 adhesivity
胶合性质 glueability
胶合压力 gluing pressure
胶合应力 bond stress
胶合毡状纸 adhesive felt
胶合作用 gluing
胶化剂 gelatinizer
胶化纤维素 amyloid
胶浆 mucilage

胶接 glue joint; glued joint; gumming
胶接破坏 glue joint failure
胶接器 paster
胶接强度 joint strength
胶接榫 glued joint
胶结 cure; glue bond
胶结层 cementation zone
胶结层脱壳(砂带) shelling
胶结带 cementation zone
胶结料 binder
胶结指数 cementation index
胶结作用 cementation
胶粒 micell(e); micella(*plur.* -llae)
胶料 glue; size; sizing
胶料溶剂 adhesive solvent
胶料污染 glue stain
胶料堰 pond
胶棉 collodion
胶棉丝 collodion silk
胶膜 film adhesive; film glue; glue film; skin; tego (glue) film
胶膜纸 film adhesive; resin film
胶木 ebonite
胶囊浮游生物 kalloplankton; kollaplankton
胶泥 mastic
胶黏 gluing
胶黏层 bond line
胶黏带 paster; splice
胶黏的 adhesive
胶黏剂 adhesive; agglutinant; binder; binding agent; bonding agent; cement; cementing agent; mastic; mastic gum; sticking agent; tackifier; tackiness agent
胶黏剂分散液 adhesive dispersion
胶黏剂含量 binder content
胶黏剂漆 adhesive lacquer
胶黏破坏 adhesion failure
胶黏强度 adhesive strength; bonding strength
胶黏体 adherent
胶黏性 adhesiveness; adhesivity; stickability; stickness; tack; tackiness
胶黏纸 adhesive paper; bond paper
胶黏质量 bond quality
胶凝 gelatinize
胶凝点 gel(ling) point
胶凝剂 gelatinizer
胶凝时间 gelation time
胶凝作用 gelatination
胶皮管 spray hose
胶皮糖香树脂 liquidambar resinoid
胶片 film
胶拼 glue joint; splicing
胶拼机 glue jointer
胶拼纵锯机 glue-joint-ripsaw
胶溶 peptize
胶乳 latex (*plur.* latices)
胶乳管 latex tube
胶乳器 laticifer
胶束 micell(e); micella(*plur.* -llae); micellar strand; supermolecule
胶束假说 micellar hypothesis
胶束间反应 intermicellar reaction
胶束间溶胀 intermicellar swelling
胶束间隙 intermicellar space; micellar cavity
胶束结构 micellar structure
胶束索 micellar string
胶束网状学说 micellar network theory
胶水 mucilage; viscose glue
胶丝 collodion silk
胶素 pectic body; pectin
胶缩 syneresis
胶体(态) colloid
胶体脱水收缩作用 syneresis
胶体状态 gel state
胶团 glue ball
胶压 gluing pressure
胶与填料胶结(砂磨材料) glue and filler bond
胶羽 floc
胶原 collagen
胶着力试验 adhesion test
胶纸 film glue
胶质 colloid; galactane
胶质层 gelatinous layer; G-layer
胶质卵囊(线虫的) gelatinous matrix
胶质强度 jelly strength
胶质物 jelly
胶质纤维 gelatinous fibre
胶质纤维管胞 gelatinous fibre tracheid

胶质形成作用 gum formation
胶珠 glue ball
焦茶 umber
焦点 focus point
焦点面 focal plane
焦化 char
焦化法(木材防腐) carbonization
焦化反应 pyrogenic reaction
焦距 focal distance; focal length
焦炉焦油 coke-oven tar
焦炉煤焦油 coke-oven coal tar
焦平面 focal plane
焦梢[树] spike-top
焦糖色试验 caramel test
焦油 oil tar; tar; tar oil
焦油工厂 tar-works
焦油碱 tar-base
焦油沥青 tar oil pitch
焦油木材防腐剂 carbolineum
焦油酸 tar acid
焦油型防腐剂 tar-oil type preservative
焦油状残渣 tarry residue
礁盖 opercule
礁湖 lagoon
礁滩 reef flat
嚼烂状的 ruminate
角斑病(叶部) angular leaf spot
角部锁紧 lock corner
角槽刨 rebate plane
角撑 corner brace; knee brac(ing)
角尺 angle board; carpenter's square; square; try square
角刺浮游生物 chaetoplankton
角蛋白 keratin
角蛋白酶 keratinase
角豆树胶 carob gum
角度计 goniometer; pantometer
角度离散 angular dispersion
角度量具 angle block
角度偏转 angular deflection
角钢 angle bar
角规 angle gauge
角规测量 angle measurement
角规常数 basal area factor; gauge constant
角规抽样 Bitterlich sampling; plotless sampling; point sampling; variable radius plot sampling
角规抽样调查 angle gauge sampling
角规计数抽样 angle count sampling
角规计数法 angle count method
角规计数调查 angle count cruising
角规计数样本 angle count sample
角规计数样地 angle count plot
角规切口 Bilterlich-blade
角规求和法 angle-summation method
角化作用 cutinization
角接 angular joint
角砾岩 breccia
角鹿 Caribou; reindeer
角木块 angle block
角刨 angle plane
角鲨烯 squalene
角闪片岩 hornblende schist
角铁 angle bar
角页岩 hornfels
角藻属 ceratium
角支柱 angle post
角质 cutin
角质层 cuticle
角质层形成 cuticularization
角质层蒸腾 cuticular transpiration
角质的 corneous; horny
角质化 cuticularization
角质纤维素 cutocellulose
角柱 angle post
角状节 horn knot; horny knot; mule ear; mule-ear knot; slash knot; spike knot; splay knot
角锥形孔洞 pyramid-shaped opening
绞车 winch; winder
绞车横杆十字块 cross piece
绞棍 springer; twister
绞集 fairleading; winching
绞集原木 line log
绞接 splice
绞排平台(排节) headwork
绞盘 capstan; cathead; gin; gypsy; gypsy spool; reel; spool
绞盘车 winch truck
绞盘机 donkey; ground skidder; hoister; log hoist; teagle; tugger; winch

J

绞盘机集材 donkey skidding; drum skidding; high lead yarding
绞盘机集材范围 yard
绞盘机集材杆 yarder mast
绞盘机卷筒 winch drum
绞盘机卷轴 winch spool
绞盘机爬犁 skid
绞盘机牵引力 winch pull
绞盘机牵引索 winch line
绞盘机手 stolon
绞盘机调向 change ends
绞盘机座横木 head block
绞盘机座木 sled
绞盘卷轴 capstan spool
绞杀植物 strangler
绞筒式枪筒 twisted barrel
绞状漂白紫胶 hank shellac
铰接 hinge
铰接点 pivot point
铰接端 hinged end
铰接滑轮 knuckle sheave
铰接式铲斗装载机 articulated loading shovel
铰接式车架 articulated frame
铰接式车立柱 hinged stake
铰接式起重臂 hinge-type jib boom
铰接式松土机 articulated ripper
铰接式原木装载机 articulated log loader
铰接式支柱 hinge-type tower
铰接武起重臂 articulated boom
铰接转向集材机 articulated frame steering skidder
铰链 hinge; swivel joint; pivot
铰链安置板 hinge mounting plate
铰链接合 hinged joint
铰链门梃 hanging stile
铰链五金件 hinge hardware
铰链销 hinge pin
铰支承 hinged bearing
铰支架杆机 jib-strut crane
矫平机(板材) level(l)ing machine
脚蹬 cripple
脚凳 foot board; footstool
脚架 leg
脚轮矮床 truckle bed; trundle bed
脚手架 cripple; framing scaffold; scaffold; staging
脚踏电门 foot switch
脚踏截锯机 under cut trimmer saw
脚踏开关 foot pedal; foot switch
脚注 footer
搅拌 knead; paddle
搅拌机 blender; mixer
搅拌器 agitator; beater; impeller
搅拌箱 mixing tank
搅拌轴 agitator shaft
搅打 beating
缴款通知 demand note
缴款通知书 call
较差法 difference method
较劣材面 worse face
教区森林(教堂森林) parish forest
教堂长椅 pew
教学林 experiment forest; demonstration forest; instruction forest; training forest; university forest; college forest
窖藏 pit storage
酵解 glycolysis
酵母抽提物 yeast extract
酵母发酵作用 yeast fermentation
酵母浸膏 yeast extract
酵素 zymase
酵素游离木素 enzymatically liberated lignin
校核 check
校平垫块 level(l)ing block
校验仪表 correcting apparatus
校正板 ali(g)ning board
校正材积 adjusted volume
校正辊 guide roll
校正基因 suppressor
校正量规 setting gauge
校准 calibration
阶地 terrace
阶段 phase
阶段多态现象 phase polymorphism
阶段发育 phasic development
阶式梯田 bench terrace; graded bench terrace
阶梯等位基因 step allelomorph(ism)
阶梯斜接 stepped scarf joint
阶梯形采伐 step felling
阶田耕作 terracing
阶状装车场 split-level landing

J

皆伐 clean cutting; clean felling; clean logging; clear cutting; clear felling; clear-cut logging; clearfell; complete cutting; complete exploitation (felling); cut-over
皆伐矮林作业 coppice system by clearcutting
皆伐地 cut over land; cut-over area
皆伐更新 regeneration under clear cutting system
皆伐迹地 clearcut area; cleared area; clearfell area
皆伐林 clear cutting forest
皆伐萌芽林作业 sprout system by clear cutting
皆伐面积 clearcut area; clearfell area; clear-felled area
皆伐区 clear-cut area
皆伐设计 clear cutting design
皆伐天然下种更新法 clearcutting method with natural regeneration by seeds
皆伐作业(法) clear cutting; clear cutting method; clear cutting operation; clear cutting system; clear-felling system
皆伐作业法 clean-cutting method; clear felling method
皆伐作业天然更新 natural regeneration by clearcutting method
接触表面黏结破坏 cohesion failure
接触除莠剂 contact herbicide
接触传染 contact transmission
接触传热 heat transfer by contact
接触干燥 touch dry
接触辊 contact roll
接触剂 accelerator
接触加热 contact heating
接触胶黏剂 contact sticker
接触角 contact angle
接触孔 touch hole
接触面 interface
接触面积 contact area
接触杀草剂 contact herbicide
接触杀虫剂 contact insecticide
接触式干燥机 contact dryer
接触式砂光机 abrasive planer-contact sander; contact sander
接触温度计 contact thermometer
接触性毒剂 contact poison
接触性化学物质 contact chemicals
接触印相照片 contact print
接地板 ground plate
接地压力 ground pressure
接缝 joint
接缝角 splice angle
接缝口 seam
接缝刨 jointer
接骨木花浸膏 elder flower concrete
接合(连接) concrescent; conjugation; joint; swivel joint
接合(胚) zygosis
L形接合 L-joint
接合孢子 zygospore
接合薄壁组织 conjugate parenchyma
接合部 graft union
接合处 layup
接合点 joint
接合菌 zygomycete
接合面 joint
接合器 jointer
接合染色体 zygosome
接合受精 joint fertilization
接合体 amphiont; conjugant
接合稳固期 joint aging time
接合物 joiner
接合线 seam
接合型 mating type
接合养护期 joint conditioning time
接合子 conjugon
接胶箱 catch up box
接锯机 brazer
接口 interface
接蜡 grafting wax
接料工 tailer
接木 grafting; imping; oculation
接收缏 receiving boom
接穗 imp; ingraftment; scion; scion wood
接穗生根 scion-rooting
接穗效应 scion effect
接穗砧木的交互作用 scion-stock interaction
接替 relay
接通时间 make time
接头 layup
接头期 joint aging time

接线 wiring
接芽母树 budwood
接枝 ingraft
接枝反应 graft reaction
接枝共聚 graft copolymerization
接枝共聚物 graft copolymer
接枝作用 graft reaction
接种 inoculate; inoculation
接种环 inoculating loop; transfer loop
接种架 catching frame
接种期 seeding stage
接种体 inoculum
接种体来源 source of inoculum
接种物 inoculum
接种罩 inoculating hood
接种者 inoculator
接种针 transfer needle
揭起 pick
街道花园 street garden
街道绿地 street green area
街道园 roadside garden
街道园林 street garden
街道种植 avenue planting; street planting
街坊公园 neighborhood park
街坊花园 neighbourhood garden
结果 fruit setting
结果多的 breedy
结果能力 bearing capacity
结果树 bearing tree
结果枝 bearing branch; fertile branch; fruit-bearing shoot
结实 bearing; fructification; fruit bearing; fruit setting
结实的 solid
结实丰年 full seed year; seed year
结实率 fruit bearing percentage
结实年 fruiting year
结实年龄 age of seed-bearing; bearing age; seed-bearing age
结实平年 half seed year; medium seed year
结实习性 bearing habit; fruit character
结实小年 partial seed year
结实性 fruitfulness
结实周期 periodicity of seed bearing
结一次果的 monocarpic; monocarpous
结籽 seed setting
结籽率 seed setting percentage
孑遗 relic; relict
孑遗生态型 relic ecotype
孑遗植物 relic plant
孑遗种 relic; relic species; relict
节 bump; gnarl; knag; node; segment
节疤群 knot cluster
节孢子 arthrospore
节丛 knot cluster
节的 nodal
节点 panel point
节段互换 segmental interchange
节腐 knot rot
节隔膜 nodal diaphragm
节荚 loment
节荚状的 lomentaceous
节间 internode
节间清理 clear internode
节间伸长 internodal elongation
节间生长 intercalary growth
节间生长点 intercalary growth point
节间纹孔 intersegment pit
节间长度 panel length
节距 pitch
节锯(锯制薄木) segment(ed) saw
节孔 branch hole; knot hole
节孔钻床 knot hole boring machine
节理(地质) joint
节理平面 jointing plane
节流[阀] throttle
节流口 orifice
节流闸 side regulator
节瘤 knot
节律 phonological; rhythm
节律生长 rhythmic growth
节律失常 arhythmicity
节律同步 rhythmic timing
节律性 rhythmicity
节囊 condyle
节气门 air damper
节群 knot group
节省劳力 labour saving
节心 knotty core
节圆直径 pitch circle diameter

J

节直径 knotty diameter
节制因素(经济) abstinence element
节制闸 controlling gate
节状壁(端壁节状加厚) transverse wall nodular
节奏 rhythm
节奏性 rhythmicity
杰奥基安茨形点法 Geoorkiantz form point method
杰弗里离析法 Jeffrey's maceration method
洁净的 clean
结 node
结疤 cicatrization; cicatrize
结构 fabric; structural framework; texture; tissue
A-C结构(土壤剖面) A-C soil
B-C结构(土壤剖面) B-C soil
结构材 strength timber; structural timber; structural wood
结构材料 structural material
结构单板 structural veneer
结构改变 structural change
结构隔离 constitution isolation
结构工程 structural work
结构横梁 structural jointer
结构基因 structural gene
结构胶黏剂 structural adhesive
结构器官 structural component
结构色 structural colour
结构特征 structural feature
结构体 ped
结构用板 structural panel
结构杂种 structural hybrid
结垢 incrustation
结合 associate; conjugate
结合点谱 junction point spectrum
结合力 binding power
结合料 binder
结合能 bond energy
结合鞣质 fixed tan
结合水 bond water; bound moisture; bound water; combined water; imbibed water; tie water
结合体 coenobium
结合种 binding species
结合组织 conjunctive tissue
结合作用 amphigamy
结核 tubercle
结核结(节) tubercle
结核结构(地质) concretionary texture
结核体 concretion
结核砖红壤 concretionary laterite
结节 nodule; protuberance
结节材 nodal wood
结节的 nodal
结节状的 tubercular
结晶 crystallize
结晶度 crystallinity; degree of crystallinity
结晶化 crystallization
结晶间润胀 intercrystalline swelling
结晶链 crystals in chain
结晶面 crystal plane
结晶倾向 crystallization tendency
结晶区 crystalline region; crystallized region
结晶区内润胀 intracrystalline swelling
结晶趋势 crystallization tendency
结晶细胞 crystalogeneous cell
结晶性 crystallinity
结晶中心 crystalline core
结壳 incrust; incrustation
结壳土壤 crust soil
结壳物质(木素与细胞膜结合物) encrusting material; entrusting substance
结块(树疙瘩等) lump
结块紫胶 blocked shellac
结皮粗腐殖质 crust mor
结欠(尚欠余额) balance due
结石(木材中) stone
结束标志 end mark
结凇 rimming
结算 balance
结算表 balance sheet
结网虫 leaf-tier (tyer)
结形花边 knotting
结扎 ligation
结账 clearing; close
结账日 account day; balance sheet date
结种子的 seed-bearing
结转 carry-over
结转价值 carrying value
捷径 short cut

截 lop
截边后规格 trimmed size
截边后宽度 trimmed width
截除缺陷(选材时) jump cut
截掉 cut-off
截掉根朽部分 jump-butt
截顶凹面体 frustum of neiloid
截顶抛物线体 frustum of paraboloid; truncated paraboloid
截顶体 frustum
截顶圆锥体 truncated cone
截顶锥体形数 frustum form factor
截端 end cutting; end trimming
截端材 equalize
截短 cutting back; docking; heading back; lopping
截断 lopping; resection; trim
截断材 breakdown timber
截断工 cross cutter
截断和散铺(采伐剩余物) lopping and scattering
截方头(将原木一端截成方形) jump-butt
截干林 stem-shoot forest
截根机 root cutter; rootprunner
截根苗 ratoon; truncated plant
截冠 beheading; dehorning; heading off; topping
截技工 lopper
截锯机 cutting saw; trimming saw
截离根出条 suckering
截留 interception; intercept
截流雨量 interception storage
截面 crossover; section
截面积 size
截面密度(人造板) profile density
截面模量 section modulus
截齐 equalize; equalizing; trimming
截梢 top-lopping; topping
截梢苗 truncated plant
截水地埂 drainage terrace
截水沟 cut-off drain; intercepting ditch; trap drain
截头 butt; butt off; docking; end; end cutting; end trimming; end-butt(ing); pollarding; trimming
截头锯 butting saw; cut off saw; docker; docking saw; trimmer
截头木 pollard; pollarded tree
截头锥体 frustum
截土坝 soil saving dam
截尾正态分布 truncated normal distribution
截芽 bud pruning; debudding; disbudding
截液器 catch-all
截圆锯 compass saw
截枝 lopping; lopping of branches; lopping off
截枝矮林 coppice with pollard
截枝刀 hatchet
截枝斧 lopping-axe
截枝更新 lopping regeneration; lopping reproduction; regeneration by stem-shoots
截枝林 lopping forest
截枝萌芽 pollard shoot
截枝树 lopping tree
截枝用具 lopping tool
截枝作业 lopping system
截止日期 cut off date
截阻保险 intercepting safety
截阻保险截阻块 mlercepting safety stop
截阻保险扣杆 intercepting safety gear
解除伐 isolation; release; release cutting; release felling; relief cutting
解除抑制 derepression
解冻 debacle; thaw
解冻风 aperwind
解毒法 disintoxication
解毒剂 antidote
解毒素 toxolysin
解毒药 antidote
解毒作用 disintoxication
解放伐 disengagement cutting; liberation cutting; liberation felling; relief cutting
解离木材 macerated wood
解磨机 disintegrater; disintegrator
解偶联 uncouple
解偶联剂 uncoupler
解剖 anatomy
解剖构造 anatomical structure

J

解剖区域 anatomical area
解剖特征 anatomical feature
解剖学 anatomy
解湿作用 desorption of moisture
解释机(程序) interpreter
解索 chase
解索工 block tender; chaser; unhooker
解体 knocked down
解酮作用 ketolysis
解吸 adsorption stripping; logging site
解吸等温现象 desorption isotherm
解吸能 energy of desorption
解吸器计算机 desorber
解吸作用 desorption
解析航空三角测量 analytical aerial triangulation
解析模型 analytical model
解絮(种子) depulping
解絮凝 deflocculation
解旋 unwinding
解约 extinguish
介电常数 dielectric constant; specific inductivity
介电常数测湿计 dielectric constant moisture meter
介电强度 dielectric strength
介电系数 dielectric coefficient
介电性质 dielectric property
介壳 scale
介壳虫 scale-insect
介体 medium (*plur.* -ia); vector
介质 medium (*plur.* -ia)
介质损耗 dielectric loss
介质温度 ambient temperature
芥子油 mustard oil
界(动植物分布) realm
界标 beacon; monument
界标设置 boundary settlement
界面 interface
界面表面张力 interfacial surface tension
界面化学反应 topochemical reaction
界石 boundary stone
界限林地 marginal land
界限以下用地 submarginal land
界线 line; mark
界线说明 boundary description
蚧 scale-insect
借贷 borrowing
借贷资本 borrowing capital; loan capital
借方(借记) debit
借景 scene borrowing
借款 borrowing
借向 dutchman
借向伐木 pull felling
借向楔 dutchman
借项 debit
今鸟亚纲 Neornithes
金箔 leaf gold
金刚砂 emery; garnet; silicon carbide (mineral)
金刚砂布 emery cloth; emery fillet
金刚砂带 emery belt; emery fillet; emery tape
金刚砂卷带 emery tape
金刚砂轮 emery wheel
金刚砂纸 emery paper
金刚钻刀 diamond knife
金龟子 chafer; cockchafer; grass grub; scarab; Scarab beetle
金合欢 acacia
金合欢基 farnesyl-
金合欢荚果 bablah; babool; babul
金合欢胶 acacia gum; acacin
金合欢宁 acacinin
金合欢素 acacetin
金合欢稀树草原 acacia savanna
金合欢油 acacia oil
金鸡纳定 cinchonidine
金鸡纳宁 cinchonine
金鸡纳树皮 cinchona
金鸡纳辛 cinchonicine
金橘(四季橘) calamondin orange
金缕梅单宁 hamamelitannin
金霉素 chlortetracycline
金钱收获 financial value; monetary value
金钱松 coin larch; golden larch
金钱总收益最高之轮伐期 maximum cash flow rotation; rotation of maximum cash flow
金雀花碱 sophorine
金融 finance
金融业务 financing
金属板滑道 tryon chute
金属薄板 sheet-metal

金属箔 metal foil
金属箔贴面 metal foil overlay
金属导槽 metal channel
金属垫板 metal caul
金属浮筒河绠 steel pontoon boom
金属构件 hardware
金属合叶 metal hinge
金属化胶合板 metallized plywood
金属黄素蛋白 metalloflavoprotein
金属家具 metal furniture
金属夹 metal clip
金属角撑 metal brace
金属铰链 metal hinge
金属浸注木材 metal-impregnated wood; metallized wood; metal-treated wood
金属卤化物灯 halide lamp
金属酶 metalloenzyme
金属配件 metal fitting
金属喷涂(镀) metal shadowing
金属软管 flexible metallic conduit; flexible metallic tubing
金属丝网筛 wire screen
金属探测器 metal detector
金属贴面胶合板 armor-plywood; metal-faced plywood; plymax; plymetal
金属网带式输送机 wire mesh belt conveyor
金属网栅 woven-wire fence
金属温度计 metal thermometer
金丝桃科 Johns wurtfamilyr-hypericum famiy)
金叶树苷 monesin
金字塔形树冠 pyramidal crown
津贴 allowance
筋 string
紧板铁马(扣紧楼地板的马钉) floor cramp
紧度 closeness
紧固 fastening
紧固套 adapter sleeve
紧急采伐 emergency cutting
紧急刹车 emergency brake
紧急反应 alarm reaction
紧急迎面火 emergency backfire
紧急制动[器] emergency brake
紧链杆 jim binder
紧链器 chain dancer; twister
紧密的 solid
紧密度 compactness
紧密节 tight knot
紧密树冠 close canopy
紧实脆盘层(土壤剖面B_x层) compact fragipan horizon; B_x-layer
紧实土 compacted soil
紧缩方案 austerity program
紧索杆 load binder
紧索环钩 fit hook
紧索器 binder; cableway tension device; line tensioner; rope-tightener
紧索器细索 cinch line; cinch rope
紧索式索道 tight line; tight(ening) skyline
紧索式索道集材 tight (sky) line yarding
紧索式索迫 standing line
紧索式装车 tight-line loading
紧索越障(半悬式集材时) tightlining
紧索运行(集材) running tightline
紧索装置 line tensioner
紧线器 turnbuckle
紧胀 turgor
紧胀度 turgidity
紧胀压 turgor pressure
锦葵苷 malvin
锦葵花素 malvidin
尽头 end
进材端 charge end; green end
进刀 bite
进刀角 knife angle; knife pitch; knife setting angle
进度图(作业) progress map
进给 feed
进给速度 feed speed
进给速率 feed rate
进行性干扰 progressive disturbance
进化 anagenesis; evolution
进化分枝图 cladogram
进化控制素 bion
进化力 evolution force
进化论 theory of evolution
进化生态学 evolution ecology; evolutionary ecology
进化生物学 evolutionary biology
进化时间 evolutionary time

进化树 cladogram
进化速度 evolution rate
进化延滞 evolutionary retardation
进货员 purchaser
进浆口 stock inlet
进界材积生长 ingrowth volume
进界木 recruitment
进界生长 ingrowth
进口材 imported timber
进口压力 inlet pressure
进料 feed; feeding; incoming charge; incoming stock
进料槽 feed chute
进料斗 feeding funnel
进料费用 incoming charge
进料管 feeder pipe
进料辊 feed roll; feed roller; infeed roll
进料滚筒 feed roll; feed roller
进料机构 feedworks
进料口 feed head
进料链 feeding chain
进料链式输送机 feeding chain conveyor
进料漏斗 feeding funnel
进料盘 feed table
进料器悬索 feeder suspension cable
进料输送带 feeding belt
进料输送机 feed conveyor
进料速度 feed speed; input speed
进料速率 feed rate
进料台 feed table; feedbed
进料调节杆 feeding adjusting lever
进料头 feed head
进料箱 feed hopper; feed(er) box; head box
进料贮仓 feeding bin
进料装置 feed attachment; feeder; feeding apparatus; feeding device
进路 approach
进气道 air inlet flue
进气端 feed head
进气管 air feeder; air inlet flue
进气口 air inlet port; inlet
进气系统 gas handling system
进食 food intake
进水沟 feed ditch
进水口 inlet
进水口工程(水滑道) headwork
进水闸 head gate
进位(计算机) carry-over
进蓄出水曲线 inflow-storage-discharge curve
进样 specimen handling
进样口 injection port
进油调节针(油锯) inlet needle
进展突变 progressive mutation
进展演替 progressive succession
近侧的 proximal
近地点 perigee
近顶生的 subapical
近端的 proximal
近端着丝的(染色体) acrocentric
近灌木状的 frutescent
近海岸的 inshore; submaritime
近红外辐射 near-infrared radiation
近红外线 near infrared (ray)
近基的 proximal
近尖端的 subapical
近渐尖的 acuminose; subacuminate
近交 inbreeding
近交衰退 inbreeding depression
近交系 inbred line
近交系数 inbreeding coefficient
近零坍度 near-zero slump
近苗中耕 close cultivation
近乔木的 subarborescent
近亲 cognation
近亲繁殖 close breeding; inbreeding
近亲繁殖程度 amount of inbreeding
近亲交配 consanguineous mating; incest breeding; sib-mating
近亲受精 close fertilization
近亲授粉 close pollination
近亲杂交 closed crossing
近亲植物 close relative plant
近群种 coenospecies
近日点 perihelion
近似计算法 approximate calculation method
近似真实生长序列 approximated real growth series
近似值 approximate value
近熟林 near-mature forest
近天然经营林分 naturally managed forest; near to nature forest
近透明的 diaphan(o)us

近无毛的 calvescent
近缘授粉 sib-pollination
近缘植物 kindred plant
近轴傍管薄壁组织 adaxial paratracheal parenchyma
近轴薄壁组织 adaxial parenchyma
近轴的 proximal
近轴种子 proximal seed
近自然经营之林分 naturally managed forest; near to nature forest
劲度 stiffness
浸沉木 snag
浸膏 concrete; extract
浸膏油 absolute oil; concrete oil
浸碱处理 mercerization
浸胶木材 prewood
浸焦油 resinify
浸解 maceration
浸解液 macerating fluid
浸溃 dip; imbibition; impregnation; steep
浸溃法 soaking method
浸溃纸贴面(树脂) paper-face overlay
浸离液 macerating fluid
浸没深度 immersion depth
浸泡池 soaking pit
浸漆纸 varnished paper
浸染 steep(ing); tincture
浸软 macerate
浸润 infiltration
浸石灰木材 limed wood
浸蚀剂 etchant
浸水 water soak
浸水区 emersion zone
浸水堰 ground-weir
浸提 extracting; extraction; leach
浸提法 extraction method
浸提罐 extractor
浸提罐组 battery of extractor; extraction battery
浸提率 extraction rate
浸提器组 battery of extractor
浸提试验 leaching test
浸提松节油 steam distilled wood turpentine; wood turpentine
浸提松香 steam distilled rosin; wood rosin
浸提物 extract; extractive; extraneous substance
浸提液 leaching liquor
浸透 percolate; water logging
浸透水的 water logged
浸透水木材 waterlogged wood
浸涂 dip-coating; dipping
浸涂设备 flood coater
浸析木灰 leached wood ashes
浸脂木材 prewood
浸脂压缩木 compreg; compregnated wood
浸种 soaking; steep; steep(ing)
浸煮器 digester
浸注 impregnation
浸注处理 impregnating process
浸注器 impregnator
浸注外层 impregnated shell
浸注性 accessibility; impregnatability
浸渍 dipping; maceration; soaking; steep(ing)
浸渍槽 dipping tank; steeping tank
浸渍处理 immersion treatment; dipping; impregnation; infusion method; dip-method; dipping treatment process
浸渍锅 infusion kettle
浸渍机 impregnator
浸渍剂 impregnant
浸渍木 impreg
浸渍木材 impregnated timber
浸渍器 macerator
浸渍试验 immersion test
浸渍液 impregnation liquid
浸渍纸 impregnated paper
禁闭技术和规范 lock-in technologies and practice
禁伐 exploitation ban; preservation order
禁伐防护林 protection forest under prohibition of cutting
禁伐林 ban forest; protected forest; cutting prohibited forest; protection forest; timber reserve
禁伐林区 emergency district
禁伐令 exploitation ban; preservation order
禁猎 protection of game
禁猎地 preserve
禁猎季节 close(d) game season

禁猎林 timber reserve
禁猎鸟类 non-game birds
禁猎期 close time; closed season; fence month
禁猎区 sanctuary; game preserve; game refuge; game cover; game park; refuge; refuge area; reserve area; wildlife reserve
禁令 embargo
禁牧林地 ungrazed woodlot
禁牧区 exclosure
禁渔期 closed season
禁运 embargo
禁止荒废 prohibition of devastation
禁止皆伐 clear fell ban; prohibition of clear cutting
禁止开发 exploitation ban; preservation order
禁止开垦 prohibition of exploitation
茎 shaft; stem; stirps (*plur.* stirpes); shoot
茎材纤维 stem fibre
茎端结实的 acrocarpous
茎腐 stem rot
茎秆切碎机 haulmiser
茎干绞缠(结扎) stem strangulation
茎根比 shoot-root ratio; stem-root ratio
茎基 caudex
茎流(树干表面径流) stem flow
茎生的 caulicole
茎生花的 cauliflorous
茎维管束 cauline bundle
茎叶体 cormus
茎轴 shoot
京都机制 Kyoto Mechanism
京都议定书 Kyoto Protocol
经典遗传学 classical genetics
经度 longitude
经费 appropriation; appropriations reserve
经过点 pass point
经过检验的 certified
经过时间 cycle time
经纪费 brokerage charge
经纪人 broker; jobber
经济成熟 economical maturity; financial maturity
经济成熟度 economical maturity
经济单位 economic unit
经济的防火理论 economic fire-control theory; least-cost fire-control theory
经济的利用龄 exploitable age
经济低迷 economic downturn
经济抵制 economic sanction
经济地租 economical rent
经济伐期 economic maturity
经济分析 economical consideration
经济干围级限界 economic girth limit
经济集约度 economic intensity
经济计量学 econometrics
经济价值少的树种 nurse species
经济间伐收获 financial returns from thinnings
经济间距(集材运输) economic space
经济径级限界 economic diameter limit
经济考虑 economical consideration
经济利益 economical consideration
经济林 economic forest; non-timber product forest
经济龄 economic age
经济轮伐期 commercial rotation; economic rotation; financial rotation
经济轮伐期理论 theory of financial rotation
经济密度 economic density
经济年限 economic life
经济平衡 economic balance
经济起飞 economic take-off
经济潜力 economic potential
经济生活 economic life
经济失调 economic ailments
经济收获 financial yield
经济受损水平 commercial timber; tree of economic value
经济树木 tree of economic value
经济外部性 economic externality
经济效果 economic effect
经济效率 economic efficiency
经济效益 economic effect
经济蓄积 economic volume
经济阈值 economic threshold
经济择伐 economic selection cutting; economical selection cutting
经济择伐法 economical selection method; selective method
经济植物 economic plant
经济制裁 economic sanction

经济转型(EITs) economies in transition
经济租金 economical rent
经理(广义) working
经理(狭义) management
经理单位 management unit
经理方法 method of management; system of management
经理集约度 management intensity
经理计划检订 revision of management plan
经理类型 management type
经理目标 management goal
经理期 management-period; working-plan period
经理期间 working period
经理期内第一分期(前五年) first working period (5 years)
经理调查 management survey
经售处 sale depot
经纬距 coordinate; latitude and departure
经纬距差 difference of co-ordinate
经纬距法 coordinates method
经纬仪 theodolite; transit
经纬仪测量 theodolite surveying
经验收获表 empirical yield table
经营(广义) management
经营(开发利用) exploitation
经营安全性 low-risk forest management
经营簿 control of operation; work tracking
经营采伐法 silvicultural cutting method
经营单位 management unit
经营法规 management prescription
经营法则 working-rule
经营方案 management plan
经营方案检订 renewal of working plan
经营方案修订 management plan revision
经营方法 method of management
经营方针 management policy
经营费用 carrying charge
经营附属地 submarginal land
经营纲要 note regarding future treatment
经营规定(指令) management instructions; management prescriptions; management requlations
经营规划 working scheme
经营集约度 intensity of management; management intensity
经营计划 business planning; working-plan
经营计划编订 working-plan report
经营计划单位 management planning unit
经营计划的执行 plan of operations
经营计划检订 working plan revision; working-plan control
经营计划面积 working-plan area
经营计划手册 management instructions; management prescriptions; management requlations
经营计划说明书 working-plan document
经营计划修订 working-plan renewal
经营记录 management record
经营记账单位 management planning unit
经营技术监督 technical supervision of management
经营类型 management class; management type; working group
经营林 forest under management; managed forest
经营率 working-percentage
经营面积 area under management
经营目标 management goal; management objective
经营目标类型 type of management objective
经营目的 management objective
经营能力 management ability
经营期 management-period; working-plan period
经营潜在等级 management potential class
经营区 area under management
经营区划 management subdivision
经营史 history book
经营狩猎场 shooting preserve

J

经营图 management map; working-map
经营蓄积清查 management-volume inventory
经营原则 management prescription
经营择伐 intensive selection cutting; management selection cutting
经营周期 period
经营组织 managerial organization
荆树栲胶 mimosa extract
荆树皮浸膏 black wattle extract
荆树皮栲胶 mimosa (bark) extract; wattle extract
荆条耢 brush drag
晶胞 unit cell
晶簇 crystal druse; druse
晶格 crystal grating
晶格常数 lattice constant
晶格结构 lattice structure
晶核 crystalline core
晶化(木材防腐) blooming
晶库 crystal repository
晶粒 crystal grain; crystallite
晶粒尺寸 grain size
晶粒取向 grain orientation
晶区 crystalline area
晶砂 crystal dust; crystal sand
晶体 crystal; crystalloid
晶体材料 crystalline material
晶体单胞 crystal unit cell
晶体的 crystalloid
晶体取向 crystal orientation
晶体物质 crystalline material
晶体贮藏所 crystal repository
晶轴 crystallographic axis
晶锥 crystalline cone
精胞 antherid; antheridium
精巢 testis
精度 accuracy; precision
精度等级 accuracy rating
精核 male nuclei
精华 elite; flower
精加工 finishing
精炼厂 refinery
精炼的 refined
精炼茴油 rectified aniseed oil
精量播种 precision drilling
精馏釜 rectifying still
精滤器 fine filter
精密播种 precision drilling
精密播种机 precision seeder
精密辊涂机 precision roll coater
精密刨床 super surface
精密调整 fine setting
精磨机 mill; precision grinder; refiner
精磨木浆 refiner ground pulp
精确度 accuracy rating
精确窑干 precision kiln drying
精砂光机 final sander
精神食粮 manna
精算师 actuary
精提用材 extract wood
精细胞 sperm cell
精选 pick; refining
精选树 elite tree
精选种子 elite seed
精研机 refiner
精英 elite
精英树 champion tree; etlite tree
精油 essence; essential oil; volatile oil
精油分离瓶 florence flask
精油树脂 oleoresin
精原细胞 testis spermatogonium
精整模 forming die
精制 milling
精制虫胶 button lac
精制的 refined
精制机 refiner
精制焦油 prepared tar
精制磨木浆 refiner ground pulp
精制木松香 refined wood rosin
精制松节油 refined turpentine
精制松香 refined rosin
精制脂肪 corpa
精制紫胶(经脱蜡漂白) refined shellac
精子细胞 androcyte
精子学 spermatology; spermology
井字垛 pen
颈 cervix; neck
颈腐病 crown rot; neck rot
颈卵器 archegonium
颈片(蜂) collar
颈深(单板拼缝机或补洞机的) throat depth
颈缩 neck-down

景观 landscape
景观凹地 landscape depression
景观保护 landscape protection
景观保护区 landscape reservation; protected landscape
景观变化 aspection; scenery variation
景观单元 landscape unit
景观地势 landscape grain
景观多样性指数 landscape diversity index
景观发展 landscape development
景观分析 landscape analysis
景观丰富度指数 landscape richness index
景观风格 landscape style
景观格局 landscape pattern
景观工程 landscape engineering
景观规划 landscape planning
景观价值 amenity value
景观结构 landscape structure
景观经营 landscape management
景观均匀度指数 landscape evenness index
景观开发 landscape development
景观控制点 landscape control point
景观评价 landscape judging
景观破碎化 landscape fragmentation
景观设计评价 landscape design judging
景观生态学 landscape ecology
景观镶嵌体 landscape mosaic
景观形态 landscape style
景观型 landscape type
景观要素 landscape element
景观优势 physiognomic dominance
景色 prospect
景深(摄影) depth of field
景天庚酮糖 sedoheptulose; sedoheptose
景天酸代谢 crassulacean acid metabolism
景天酸代谢植物 CAM plant
景物图 pictorial drawing
景致 landscape
警报声 alarm call
警报外激素 alarm pheromone
警报物质 alarm substance
警告标志 warning signal
警告色 warning colour; warning colouration
警号 warning signal
警戒反应 alarm reaction
警戒拟态 Batesian mimicry
警戒色 warning colour; advertising colour; aposematic colouration; warning colouration
警戒态 aposematism
警戒信息素 alarm pheromone
警戒作用 aposematism
径高曲线 diameter-height curve
径厚比 radius-thickness ratio
径级 diameter class; diameter grade
径级标准木 diameter class-specific sample tree
径级法 diameter-classes method
径级幅度 class width; width of diameter class
径级界限 diameter class limit; diameter limit
径级收获调整 yield regulation by diameter classes
径级择伐 diameter limit selection cutting; diameter-limit cutting; dimension cutting; extensive selection cutting
径阶 diameter grade; diameter class
径阶分布模型 size-class distribution model
径阶距 diameter-grade interval
径阶整化 diameter classification
径节(齿轮) diametral pitch
径截面的 rift sawed
径锯成材 rift sawed lumber
径锯面 quarter sawn grain; quarter section; quarter surface
径锯切[面] radial cut
径锯纹理 rift grain
径列 radial alignment; radial arrangement; radial row
径列分子 radial element
径列复管孔 multiple radial pore; radial (pore) multiple; radial multiple
径列管孔材 radial porous wood
径列条 trabecula (*plur.* -lae)
径列纹孔式 radial pitting
径裂 cross shake; radial check; radial shake; simple check; star check

径流 flow-off; runoff; run-off; surface flow
径流的 run-off
径流分析 runoff analysis
径流率 runoff ratio
径流式风机 radial fan
径流系数 coefficient of runoff; discharge coefficient; drainage ratio; runoff coefficient
径脉的(虫) radial
径面材 rift sawed lumber; vertical-grain lumber
径面花纹 comb-grained; quarter-sawn figure
径面锯材 quarter-sawed lumber; quarter-sawn stock
径面纹理 edge grain; felt grain; quarter sawn grain; rift grain; vertical grain
径面旋切 quarter-rotary-cutting
径切 rift sliced
径切板 edge-grain lumber; quarter-sawed lumber; quarter-sawn stock; vertical-grain lumber
径切表面 edge-grain surface
径切材 radial wood
径切的 edge-sawn; quarter-sawed; quarter-sawn
径切花纹 quarter-cut figure
径切面 centre-sawed; edge grain; radial plane; radial section
径向胞间道 horizontal intercellular canal; radial intercellular canal; transverse intercellular canal
径向薄壁组织 radial parenchyma
径向薄壁组织分子 transverse parenchymatous element
径向壁 radial wall
径向齿前角 radial rack angle
径向冲击硬度 radial impact hardness
径向的 radial
径向对称 radial symmetry
径向分裂 radial division
径向分子 horizontal element; transverse element
径向干缩 radial shrinkage
径向管胞 cross tracheid
径向管孔排列 radial pore arrangement
径向环流 meridional cell
径向间隙 radial clearance
径向角变量(镟切) automatic pitch; diametral pitch
径向锯材 edge-grain lumber
径向锯解 quarter sawing; quarter-cut; quartered; radial sawing; radial-sawn; rift sawed
径向扭曲 radial twist
径向膨胀 radial swelling
径向皮厚 radial bark thickness
径向润胀 radial swelling
径向渗透 radial penetration
径向生长 radial growth
径向生长量 radial increment
径向收缩 radial shrinkage
径向树脂道 horizontal resin canal; radial resin canal; transverse resin canal
径向位移 radial displacement
径向细胞 transverse cell
径向下锯的 centre-sawed
径向下锯法 center sawn
径向压力 radial pressure
径向叶片 radial blade
径向翼片 radial wing
径向-纵列分裂 radial-longitudinal division
径向纵切面 radial longitudinal section
净保持量 net retention
净剥皮 clean bark
净材 clear; finished stock
净材积(扣除缺陷) net scale; net volume
净尺 net scale
净尺寸 finished size
净初级产物 net primary production
净初级生产力 net primary production; net primary productivity
净初级生产量 net primary production
净得率 net yield
净第一性生产力 net primary productivity
净第一性生产量 net primary production
净度 analytical purity; percentage of purity; purity
净度纯度 degree of purity

J

净度分析台 purity board
净辐射 net radiation
净干长 clean length
净光合 net photosynthesis
净光合速率 net photosynthetic rate
净化 eradication; scouring
净化剂 scavenger
净化器 cleaning cartridge; eliminator
净降水量 net precipitation; throughfall
净截流量 net interception
净空 freeboard
净空高度 head way; headroom
净利 net profits
净利润 retained profit
净面板 clear face board
净面材 clear cutting board
净漂的 floated bright
净生产量 net production
净生产率 net production rate
净生态系统生产量 net ecosystem production
净生长 net growth
净生殖率 net reproduction rate; net reproductive rate
净收入 net proceeds; net revenue
净体积吸收量(防腐) net volumetric absorption
净同化 net assimilation
净同化速率 net assimilation rate
净同化作用 net assimilation
净投资 net investment
净吸收量 net absorption
净相关 partial correlation
净油 absolute oil; concrete oil
净载重量 payload
净增率 percentage net growth
净增长量 net increase; net-income
净增值率 net reproduction rate; net reproductive rate
净值 net worth
净种 cleaning of seed; cleaning-up of seed; seed-cleaning
净种器 seed cleaner; seed separator
净种筛 seed riddle
净重 net weight
胫骨 tibia
胫管式土面穿入器(土壤气体薰蒸装置) ground-penetrating shank implement
胫节 tibia
胫节距(昆虫) tibial spur
胫肢节(昆虫) carpopodium
痉挛状质壁分离 cramped plasmolysis
竞标 competitive bidding
竞争 competition
竞争出价 competitive bidding
竞争共存 competitive coexistence
竞争互斥原理 principle of competitive exclusion
竞争假说 competition hypothesis
竞争密度效果的互反方程 reciprocal equation of common-density effect
竞争密度效应 competition density effect
竞争密度效应幂方程 power equation of common density effect
竞争木政策 competing tree
竞争能力 competitive ability; competitive capacity
竞争排斥原理 competition exclusion principle; competitive exclusion principle
竞争排除 competitive exclusion
竞争曲线 competition curve
竞争群丛 competitive association
竞争替代 competitive displacement
竞争系数 coefficient of competition; competition coefficient; competitive coefficient
竞争相互作用 competitive interaction
竞争性抑制 competitive inhibition
竞争压力 competition pressure
竞争因素 competition factor
竞争优势 competitive advantage
竞争者 competitor
竞争指数 competition index
竞争种组 competitive association
静靶 still target
静点 quescent point
静电滤尘器 electric precipitator
静电喷枪 electrostatic spray gun
静电喷涂 electrostatic coating

静电喷雾呛 electrostatic spray gun
静电手提式喷枪 electrostatic hand gun
静电涂装(覆) electrostatic coating
静电雾化器 electrostatic nebulizer
静荷载 dead load(ing)
静力弹性常数 static elastic constant
静力弹性系数 static modulus of elasticity
静力矩 static moment; static torsion
静力拉伸持久载荷 static endurance load in tension
静力平衡 static equilibrium
静力试验 static test
静力弯曲 static bending
静力弯曲破坏 stat bending failure
静力弯曲强度 static bending strength
静力弯曲试验 static bending test
静力弯曲试验机 static bending machine
静力学 statics
静力硬度 static hardness
静摩擦 friction of rest
静配子 aplanogamete
静曲应变能 strain energy of bending
静水 slack water
静水的 lenetic; lenific
静水浮游生物 stagnoplankton
静水生态系统 lentic ecosystem
静水位 hydrostatic level
静水压力 hydrostatic pressure
静态的 static
静态地位级 static site class
静态刚度 static rigidity
静态经济 static economy
静态生命表 static life table
静压 static pressure
静压剪(伐木工具) hydrostatic clipper
静压缩 static compression
静应力 static stress
静运材滚轮(非电动的) dead rollers
静载荷 quiescent load; statical load
静载应力 dead load stress
静止 quiescence
静止代谢 resting metabolism
静止的 static
静止点 rest point
静止锋 stationary front
静止火 stationary fire
静止角 angle of repose
静止期 rest period; resting stage
静止气旋 stationary cyclone
静止卫星 fixed satellite
静止状态 quiescence
静滞地下水 perched ground water
静滞水位 perched water table
静重 dead weight
静驻需求曲线 stationary demand curve
静转矩 static torsion
境边花坛 border; flower border
境界 prospect
境界标 boundary-mark
境界补正 boundary correction; rectification of boundaries
境界簿 boundary-register
境界查定 delimitation; demarcation
境界纠正 boundary correction; rectification of boundaries
境界描述 boundary description
境界树 border tree; borderline tree; boundary tree; edge tree
境界调查 metes and bounds survey
境界图 boundary-map
境界线 boundary line
镜框式线脚 bolection moulding
镜面 flooring
镜面反射率 specular reflectance
镜面光泽 specular gloss
镜面精加工 mirror finish
镜面抛光 mirror polish
镜筒式瞄准器 scope sight
纠缠 foul
纠正 rectifying
纠正点 rectified point
纠正仪 rectifier
鸠尾槽 dovetail groove; dovetail slot
鸠尾搭接 dovetail lap joint
鸠尾角 dovetail angle
鸠尾接合 dovetail joint
鸠尾锯 dovetail saw
鸠尾开榫机 dovetailer; dovetailing machine
鸠尾榫 dovetail; dovetail tenon
鸠尾套接 dovetailed housing joint
鸠尾铣刀 dovetail cutter
鸠尾钻 dovetail bit

九倍体 enneaploid; nonuploid
九倍性 enneaploidy
久效磷(杀虫剂) monocrotophos
酒吧椅 bar chair
酒化酶 zymase
酒精水准器 spirit-level
酒椰 raffia
酒椰蜡 raffia wax; raphia wax
酒椰纤维 raffia
酒桌 grog table
旧火烧迹地 old burns
旧石器 palaeolith
旧石器时代 palaeolithic age
救护货物 salvage
救护绳 tail line
救火安全网 fire net
救火车 fire engine
救火道 fire trail
救火演习 fire drill
救生圈 life saving ring
救生设备 life saving equipment
厩肥 barnyard manure; farmyard manure; stable manure; yard manure
厩肥堆 mixen
厩肥坑 stercorary
厩肥液 stable water
厩肥汁 liquid manure; manure juice
厩粪 stable dung
厩液 liquid manure
就地保护 in-situ conservation
就眠运功 nyctinastic movement
就业成本效益分析 employmem benefit cost analysis
就业状况 work status
居间分生组织 intercalary meristem
居间生长 intercalary growth
居留地 settlement
居民区绿化 residential region planting
居住区公园 residential district park
居住区绿地 residential district green area
居住小区 neighbourhood unit
居住小区花园 community park
居住于森林中的 silvicolous
居住者 occupant
掬网(猎取野禽) scoop-net
局播 partial seeding; partial sowing
局部病变(斑) local lesion
局部采伐 partial cutting
局部处理 partial disposal
局部磁引力 local magnetic attraction
局部的 local
局部多度 local abundance
局部分布区产地试验 limited range provenance trial
局部分裂 merokinesis
局部感染 local infection
局部更新 regeneration in group
局部化学反应 topochemical reaction
局部坏死 local necrosis
局部坏死斑 local necrotic lesion
局部坏死病斑 necrotic local lesion
局部加热 spot heating
局部胶合 spot gluing
局部免疫 local immunity
局部侵染 local infection; localized infection
局部融合 meromixis
局部森林调查 partial forest survey
局部生境 partial habitat
局部施肥 partial fertilization
局部性种源 local seed source
局部用火(烧出隔离带) patch burning
局部照射 local irradiation
局部整地 partial soil preparation
局部症状 local symptom
局部种群 local population
局地气候 local climate
局限型不亲和性 localized incompatibility
菊胺 chrysanthemine
菊粉 alantin; inulin
菊粉酶 inulase
菊石 ammonite
橘皮皱纹 orange peel
橘子单宁 Chinese tannin
橘子油 mandarin oil
咀嚼 mastication
咀嚼式(口器) mandibulate type
咀嚼式口器(昆虫) chewing mouthpart
矩尺 carpenter's square
矩形侧边空心板 square profile board
矩形断面 rectangular profile
矩形集材区 rectangular setting

矩形木排 rectangular raft
矩形芽接 rectangular patch budding
矩形样地 rectangular plot
矩阵 matrix
举重台 lifting platform
巨变 macromutation; sport
巨变性因子 catastrophic factor
巨大态 gigantism
巨漂砾 erratic block
巨杉 giant redwood; giant sequoia
巨噬细胞 macrophages
巨兽 big game
巨树 giant tree
巨形角砾岩 megabreccia
巨形细胞 giant cell
巨型(巨体) gigas
巨型浮游生物 megaloplankton
巨型染色体 giant chromosome
巨型生物 megafauna
巨型性 gigantism
巨型叶 megaphyll
拒食剂 anti-feedant
拒食作用 anti-feeding effect
拒受点(防腐) refusal point; virtual refusal
具白粉的 pruinate; pruinous
具白霜的 glaucous
具斑点的 macular; maculose; notate
具备花 perfect flower
具边缘的 marginate
具翅轭的(鳞翅目昆虫) jugate
具翅花梗的 pterygopous
具刺的 acanthaceous; erinaceous; spinescent
具大鳞的 scutate
具单一雌蕊的 monogynous
具单一雄蕊的 monandreous
具倒刺的 barbed
具点的 punctate
具端着丝点的 telocentric
具短角的 siliculose
具短茎的 acauline
具短毛的 barbellate
具盾片的 scutate
具二回羽叶的 pinnulate
具二锯齿的 biserrate
具二室的 bilocular
具二叶的 bilobate
具佛焰苞的 spathaceous
具复出掌状脉的 palinactinodromous
具副萼的 caliculate; calycled; calyculate
具副花冠的 crowned
具刚毛的 bristly; setaceous
具隔的 septiferous
具隔膜的 septate
具根的 radicated; radiciferous
具沟的 channeled
具钩毛的 glochidiate
具冠的 crowned
具果霜的 pruinate; pruinous
具花梗的 pedicellate
具花托的 thalamifloral; thalamiflorous
具花柱的 stylate
具假隔膜的 septulate
具坚果的 nuciferous
具浆果的 bacciferous; cocciferous
具桨片的 paleolate
具较小叶的 meiophyllous
具节荚的 lomentaceous
具茎的 caulescent
具九雌蕊的 enneagynian
具九雄蕊的 enneandrous
具卷须的 capreolate; tendrillous
具菌褶的 lamellate
具颗粒的 granulose
具蜡的 sebiferous
具肋的 costate
具两萼片的 disepalous
具两果爿的 dicarpellary
具两心皮的 dicarpellary
具两叶的 diphyllous
具两种头状花序的 heterocephalous
具裂片的 lobose
具鳞根出条 turion
具鳞片的 squamate
具流苏的 fringed
具龙骨瓣的 keeled
具龙骨状突起的 carinate; keeled
具毛的 piliferous; trichiferous
具毛果实 comospore
具绵毛叶的 dasyphyllous
具绵状毛的 lanate; lanose
具内稃的具鳞毛的 paleoliferous
具皮刺的 aculeate

具皮孔的 lenticellate
具脐状突起的 umbonate
具气囊 saccate
具鞘的 vaginate
具球果的 coniferous
具缺刻的 incised; nicked
具髯毛的 barbate
具茸毛的 nappy
具柔毛的 pubescent
具葇荑花序的 julaceous; juliform
具乳头的 papillate; papillose
具乳头状小突起的 papillose
具三肋的 tricostate
具三脉的 trinervate; trinerved
具三主脉的 tricostate
具十萼片的 decasepalous
具十花瓣的 decapetalous
具十叶的 decaphyllous
具十种子的 decaspermial
具十字花的 cruciferous
具疏绵状毛的 lanuginose
具四瓣的 quadrivalvate; quadrivalvular
具四室的 tetradymous
具四小叶的 quadrifoliolate
具四雄蕊的 tetrandrous
具穗状花序的 spiculate
具繸的 fringed
具同形孢子的 homosporic
具托苞的 paleace; paleaceous; paleoliferous
具托叶的 stipulate
具托叶鞘的 greaved
具微圆齿的 subcrenate
具五活瓣的 quinquevalvate
具五裂爿的 quinquevalvate
具细刺的 spiculate
具细锯齿的 ciliato-dentate
具细孔的 puncticulate
具细脉的 nervulose
具细脉纹的 venulose
具纤毛的 ciliate
具显脉的 nervose
具腺的 glandular
具腺点的 glandular punctate
具腺毛的 glandular pubescent
具小点的 puncticulate
具小梗 pedicellate
具小锯齿的 serrulate
具小裂片的 lobulate
具小条裂的 lacinulate
具小叶的 lobulate
具小羽片的 pinnulate
具小圆齿的 crenellate
具新月痕的 lunated
具雄蕊的 staminiferous
具须的 barbate
具芽的 gemmiferous
具叶柄的 petioled
具一小叶的 unifoliolate
具一叶的 unifoliate
具硬壳的 incrustate
具油脂的 sebiferous
具有颊片的枪柄 Monte Carlo stock
具羽状脉的 feather veined
具羽状三小叶的 pinnately trifoliolate
具圆齿的 crenate
具圆锥花序的 paniculate
具缘毛的 ciliate
具缘纹孔闭塞 aspiration of bordered pit
具缘纹孔对 bordered pit-pair
具掌状脉 actinodrome
具掌状三小叶的 palmately trifoliolate
具摺的 plicate
具枝隙的 cladosiphonic
具指状脉的 digitinervate
具中间着丝粒的 metacentric
具中脉的 costate
具重圆齿的 bicrenate
具总苞的 enveloping; involucrate; involucred
剧毒性 acute toxicity
距(虫) spur
距离 spacing; range
距离法 distance method
距离效应 distance effect
距离有关的 distance-related
距平值 anomaly; departure; deviation from average
飓风(十二级以上) hurricane
锯 mill; saw; slash
锯板紧密堆垛 close stracking
锯背 back; blade back
锯比子(锯机上的) linebar
锯柄 saw handle

J

锯材 lumber; lumbering; saw log; saw lumber; saw timber; sawed lumber; sawed timber; sawn product; sawn timber; sawnwood
锯材板芯结构 lumber-core construction
锯材材积表 lumber scale
锯材材积核定 lumber volume determination
锯材叉架(铲) lumber fork
锯材产量(单位时间) cut
锯材尺寸检量规 lumber gauge
锯材出材率 lumber recovery
锯材等级 lumber class; lumber grade; sawlog class
锯材短原木 saw log short
锯材堆垛机 lumber elevator; lumber stacker
锯材垛 pile of lumber
锯材分级规则 sawnweod grading rule
锯材干燥 lumber drying
锯材干燥窑 lumber drying kiln; lumber kiln
锯材价格 sawnwood price
锯材价格差额 lumber value differential
锯材检尺 lumber scale
锯材区分 sort
锯材树(胸径12～20英寸) standard
锯材台账 lumber tally; mill tally
锯材消耗量 lumber consumption
锯材心裂 boxed heart check
锯材芯板胶合板 lumber core plywood
锯材原木 saw log (sawlog)
锯材原木部分 saw log portion
锯材原木等级 mill grade; saw log grade
锯材运输 lumber transportation
锯材贮存室 lumber storage room
锯成方材的 square cut
锯成棱边 edge; edging
锯齿 saw-teeth; serrature; tooth
M形锯齿(刨齿) M-tooth
锯齿夹角 fleam
锯齿紧固器 holder shank
锯齿料 tooth setting
锯齿偏角(锯齿口和锯条面所形成的角) fleam
锯齿斜度 bevel
锯齿形的 serrate; zigzag
锯齿形风选机 zigzag wind sifter
锯齿修整 wrest of saw; saw set
锯齿修整器 saw set
锯齿一侧(横截锯) breast
锯齿整平 jointing
锯齿状年轮 indented ring
锯锉 saw doctor
锯导扳 saw guide
锯导板 bar; saw bar
锯断[法] sawing-off
锯方 square cut
锯蜂 sawfly
锯缝 curf; cutting leak; kerf; saw cut; saw draft; saw kerf
锯干[燥]裁联合加工法 saw, dry and rip process
锯规 gauge
锯痕 saw cut
锯后尺寸 off-saw size
锯机(大型) sawmill
锯机副手 off-bearer; saw tailer
锯机台架 bench
锯架 breast bench; saw buck
锯截 cutting; trimming
锯截不足 undercut
锯解 breakdown; cut; sawing
锯解过程 sawing procedure
锯具 cutter
锯卡 saw vice
锯开 saw down
锯口 curf; kerf; saw draft; saw kerf
锯口嫁接 saw kerf graft
锯框 gate; husk; sash; saw frame
锯链 saw chain
锯链锉 chain file
锯链锉磨机 chain grinder; saw chain filing machine
锯链顶片角 filing angle
锯链加油器(加润滑油) chain oiler
锯链连接片 link
锯链刃磨机 saw-sharper
锯链突耳(限制锯切深度) lug
锯链圆形凿齿 chipper
锯料 set
锯路 curf; saw kerf; saw line
锯木厂 lumber mill; sawmill
锯木工 log sawyer; sawyer

锯木架 horse tree; saw block; sawhorse
锯木三脚架 jack
锯木支架 trestle for sawing
锯钮 box
锯片 saw blade
锯片调整 saw adjustment
锯片凸出高度 saw-plate projection; saw-blade projection
锯剖 saw buck; sawing jack
锯剖图 sawing pattern
锯切分级法 cutting system
锯切毛尺寸的 sawn full
锯切坯料 cutting blank
锯去板皮的原木 slabbed log
锯去背板 slab; slabbing
锯去木端号印 dehorn
锯入 cut-in
锯上口(伐木) severing cut
锯身 cutter bar
锯台 breast bench; buck; saw bench; saw buck; sawing jack
锯条 blade; saw blade
锯条固定夹 saw vice
锯条辊压机 sawblade stretcher; stretcher
锯条焊接器 saw brazing clamp
锯条夹 blade holder
锯条压辊 stretcher roll
锯条张紧装置 sawblade strain system
锯下口 belly cut
锯线 saw line
锯楔 saw wedge
锯屑 coom; sawdust; scobs
锯屑机 excelsior cutting machine
锯屑聚合制品 sawdust polymer
锯屑与废材输送机 sawdust and log waste conveyor
锯削联合机 chip-N-saw
锯罩(安全罩) saw guard
锯制单板 sawed veneer; sawn veneer
锯制方材 sawn square
锯制桶板 sawn stave
锯轴 mandrel; saw arbor; saw mandral; saw shaft; saw spindle
锯组 battery of saws
聚苯并咪唑 polybenzimidazole
聚苯乙烯 polystyrene
聚变核 fusion nucleus
聚丙烯 polypropylene
聚丙烯腈 polyacrylonitrile
聚丙烯腈纤维 poiyacrylonitrile fiber
聚丙烯酰胺 polyacrylamide
聚点识别 cluster identification
聚合薄壁组织 aggregate parenchyma; confluent parenchyma
聚合杯状聚伞花序 incyathescence
聚合程度不大的油 short oil
聚合度 degree of polymerization
聚合度分布性 polydispersity
聚合改良法 convergent improvement
聚合果 aggregate fruits; coenocarpium; conocarpium
聚合核果 drupetum
聚合基因型 aggregate genotype
聚合酶 polymerase
聚合射线 aggregate ray
聚合树脂细胞 aggregate resin cell
聚合松香 polymerized rosin
聚合体 polymer
聚合物 polymer; polymeride
聚合现象 polymerism
聚合种 cenospecies
聚合作用 polymerization
聚花果 collective fruit; multiple fruit
聚花月季 floribunda rose
聚黄酮体 polyflavonoid
聚集 aggregation; granulation
聚集地 gathering ground
聚集点 focus point
聚集分布 aggregated distribution
聚集态 aggregation
聚集信息素 gregarization pheromone
聚集指数 aggregation index
聚类分析 cluster analysis
聚落 settlement
聚落类型 settlement pattern
聚氯乙烯 polyvinyl chloride
聚氯乙烯(PVC)帖面 PVC film overlay
聚木糖 polyxylose
聚偏二氯乙烯 polyvinylidene chloride
聚羟甲基脲 poly methylolurea
聚群反应 aggregative response

聚群指数 aggregation index
聚伞花序 cyme; tuft
聚伞花序的 cymose
聚伞圆锥花序 racemose cyme; thyrse
聚伞圆锥花序的 thyrsiferous
聚伞状伞形花序 cymose umbel
聚伞总状花序 botry-cymose
聚生的 aggregate; gregarious
聚水坑 sump
聚水区 gathering ground
聚糖 glycan; polyose; polysaccharide
聚糖酶 polysaccharase
聚糖体化合物 polysaccharoid
聚体雄蕊 fraternity
聚戊糖 pentosan
聚心皮果 aggregate fruits; baccausus; etaerio
聚乙烯 polyethylene
聚乙烯薄膜 polyethylene film
聚乙烯容器 polyethylene container
聚乙烯树脂胶黏剂 polyvinyl adhesive
聚翼薄壁组织 aliform-confluent parenchyma; confluent parenchyma
捐款 contribution
捐赠成本 endowment cost
卷板-松板机 veneer reeling and unreeling machine
卷板系统 reeling system
卷边 dog eared
卷尺 measuring-tape; tape
卷尺围量法 tap measure
卷顶病 curly top
卷动装置 rolling device
卷蛾 tortricid
卷蛾幼虫 budworm
卷簧机 winder
卷积 conyolution
卷荚 cochlea
卷巾测定 roll test; rolled towel test
卷口 wire edge
卷帘 shutter
卷门 flexible door; roll front
卷曲 crimp; curl
卷曲的 convolute
卷曲花纹 curly figure
卷曲木丝 curl
卷曲刨花 short curly chip
卷曲纹理 curl; curly grain
卷绕 coiling; reeling
卷绕钢索 spool
卷绕机 winder
卷索 reeling; winding
卷索滚筒架 roller frame for reel
卷索轴(筒) cable reel
卷筒 cylinder; drum; gypsy; spool; bobbin
卷筒机 reeling machine
卷筒牵引力 linepull
卷筒绕绳长度 spool length
卷筒绕绳直径 spool diameter
卷筒容绳隔板(压紧钢索) drum narrower
卷筒容绳量(绞盘机等) drum capacity; rope (storage) capacity
卷尾状扶手端 monkey tail
卷线机 reel
卷须 cirrhus (*plur.* cirri); clavicle; cyrrhus; tendril
卷须植物 tendril climber
卷须状的 cirriform; pampiniform
卷扬机 hoist; winder
卷叶病 leaf roll
卷叶虫 leaf-roller; needle roller
卷叶蛾 leaf-roller; tortricid
卷叶昆虫 leaf-rolling insect
卷云 cirrus
卷轴 gypsy spool; reel; spool
卷轴材 spool wood
卷轴式绞盘机 gypsy yarder
绢云母 sericite
绢质的 sericeous
眷群 harem
孓遗特有的 epibiotic
孓遗种 epibiotics
决定模型 deterministic model
决定系数 coefficient of determination
决定因素 determinant
决算 actual budget
决算表 final statement
绝顶 zenith
绝对保值(险) absolute cover
绝对报酬 absolute return
绝对比重 true specific gravity
绝对成本 absolute cost
绝对程序设计 absolute programming

绝对大气压 absolute atmosphere
绝对方位 absolute orientation
绝对高度 absolute altitude
绝对含水率 absolute moisture content
绝对林地 absolute forest land; absolute forest soil
绝对螺钉握持力 absolute screw-holding power
绝对密度 absolute density
绝对黏度 absolute viscosity
绝对平面 dead level
绝对生殖值 absolute reproductive value
绝对湿度 absolute humidity; absolute moisture content
绝对视差 absolute parallax
绝对收益 absolute return
绝对腾贵生长 absolute price increment
绝对同龄林 absolute even-aged stand
绝对温标 absolute temperature scale; Kelvin temperature scale
绝对温度 absolute temperature; Kelvin temperature
绝对握钉力 absolute screw-holding power
绝对误差 absolute error
绝对形率 absolute form quotient
绝对形数 absolute form factor
绝对休闲地 bare fallow
绝对削度 absolute taper
绝对压力 absolute pressure
绝对应变值 absolute strain value
绝对值 absolute value
绝对值计算机 absolute value computer
绝对重量 absolute weight
绝对最大剪力 absolute maximum shearing force
绝对最大坡度 absolute maximum gradient
绝对最大弯曲力矩 absolute maximum bending moment
绝对最低致死温度 absolute minimum fatal temperature
绝对最高致死温度 absolute maximum fatal temperature
绝干 oven-dry
绝干材 absolutely dry wood; oven-dried wood; oven-dry wood
绝干材比重 specific gravity in over dry
绝干的 bone-dry
绝干密度 oven-dry density
绝干重 absolute dry weight; bone dry weight; oven-dried weight
绝干状态 absolutely dried condition; oven-dried condition
绝热板 thermal insulation board
绝热材料 thermal insulating material
绝热门 heat isolated door
绝热体 insulator
绝热性质 thermal insulating property
绝热蒸发 adiabatic(al) evaporation
绝热值 thermal insulating value
绝热纸 sheathing paper
绝育 sterilization
绝缘板 felt board; insulating board; insulation board
绝缘材料 insulator; non-conducting material
绝缘层 insulating layer
绝缘电极 insulated electrode
绝缘介质 isolation medium
绝缘刨花板(碎料板) insulating-type particleboard
绝缘涂层 insulating coating
绝缘纤维板 insulating fibre board
绝缘性能 insulation potential
绝缘性质 insulating property
绝缘纸 condenser paper; insulating paper
绝缘子 insulator
绝种 die-off
掘除伐根 rooting out of stump
掘伐根 extraction of stump; stump; stump out
掘根 grub; removal of stump; stub out; stump extraction
掘根齿 root bit
掘根锄 grub-hoe
掘根伐 grub felling
掘根斧 dipping axe; grubbing axe
掘根机 grub puller; grubber; root dozer; root grubber; root puller; stumper
掘根犁 grub-breaker

掘根器 forest devil; grubbing-instrument
掘坑 pitting
掘树机 tree puller
掘穴作用 excavation
蕨类石南灌丛 bracken heath
蕨类温室 fern house
蕨类植物 fern; pteridophyta
蕨叶病 fern leaf
爵床科的 acanthaceous
攫食动物 predator
军刀形支腿 sabre leg
军旗 vexillum
均变论 uniformitarianism
均差 mean deviation
均二羟甲脲 dimethylol urea
均方差 error of mean square; mean square deviation; mean square of the deviation
均方差比 variance-mean ratio
均方根 root-mean-square
均方根差 root mean square deviation
均方偏差 mean square deviation
均方误差 error of mean square
均方值 mean square value
均衡 equilibrium
均衡的连年永续作业 equalized annual sustained working
均衡辊 equalizing roller
均衡陆地移动 isostatic land movement
均衡配料箱 equalizing vat
均衡异龄林 balanced unevenaged stands
均衡择伐林 balanced selection stand
均化 level(l)ing
均化剂 level(l)ing agent
均化器 homogenizer; homogenizing mill
均积 mean product
均聚物 homopolymer
均裂 homolysis
均染(均涂) level(l)ing
均相水解 homogeneous hydrolysis
均一抗性 uniform resistance
均一性 identity
均匀变量载荷 uniformly varying load
均匀材色 even colour
均匀抽样 uniform sampling
均匀度 equitability; evenness
均匀分布 regular distribution; uniform distribution
均匀分布的 uniform
均匀分布载荷 uniformly distributed load
均匀负荷 even load
均匀覆盖 homogeneous cover
均匀混和 intimate mixing
均匀肌理 even texture; uniform texture
均匀剪切 homogeneous shear
均匀渐伐 uniform shelterwood cutting
均匀渐伐作业 uniform method; uniform system
均匀结构 even texture; smooth texture; uniform texture
均匀结构材 uniform-textured wood
均匀林分 uniform crop; uniform stand
均匀强度 uniform strength
均匀乔木 uniform high forest
均匀伞伐作业 uniform method; uniform system
均匀色调 even shade
均匀水解 homogeneous hydrolysis
均匀纹理 even grain; uniform grain
均匀性 homogeneity; homogeneousness
均匀压力 uniform pressure
均匀颜色 even colour
均匀应力 homogeneous stress
均匀载荷 even load; uniform load
均匀质地 even texture
均质的 homogeneous
均质林分 uniform stand
均质木梁 homogeneous wood beam
均质刨花板 homogeneous particle board
均质髓 homogeneous pith
均质碎料板 homogeneous particle board
均质土壤 homogeneous soil
均质纤维板 homogeneous fibre board
均质性 isotropy
菌柄 stipe
菌腐 fungal decay
菌盖 cap; pileus

菌根 bacteriorhiza; mycorhiza; mycorrhiza; mycorrhizae
菌根共生体 mycorrhizal association
菌根群丛 mycorrhyzal association
菌根树 mycorrhizal tree
菌根营养 mycotrophy
菌根真菌 mycorrhizal fungi; mycorrhizal fungus
菌根植物 mycorrhizal plant
菌管 tubuli
菌害 fungus damage
菌害木 foxty-tree
菌核 sclerotium
菌环 annulus; ring
菌块 mushroom spawn
菌类 fungus (*plur.* -gi)
菌类传播 fungus spread
菌类腐朽 fungal decay
菌类蔓延 fungus spread
菌轮 ring
菌落 coenobium; colony
L型菌落 L-type colony
菌落局变 saltation
菌绿素 bacteriochlorin
菌膜 pellicle
菌幕 veil
菌内 context
菌盘核病 sclerotiniose
菌圃(白蚁) fungus garden
菌褥 mycelial felt; mycelial pad
菌伞 pileus
菌扇 mycelial fan
菌丝 hypha (*plur.* hyphae)
菌丝层 subiculum (*plur.* -la)
菌丝垫 mycelial mat; mycelium mat
菌丝孔 hyphal hole
菌丝融合 fusion of hyphae
菌丝噬菌体 mycelial phage
菌丝束 hyphal strand; mycelial strand; strand
菌丝体 mycelium (*plur.* -lia); spawn
菌丝体的 mycelial
菌丝砖(栽培蘑菇用) spawn
菌丝状体 fungus filament
菌髓 medulla (*plur.* -llae); trama (*plur.* -ae)
菌索 strand
菌台 lawn
菌体 thallus
菌托 volva
菌瘿 cecidium; fungous gall; fungus gall; mycocecidium
菌原体 mycoplasma
菌藻植物 thallophyte
菌藻植物门 thallophyta
菌褶 gill; lamella (*plur.* -llae/-llas)
菌质体 mycoplasma
菌株 strain
菌紫素 bacteriopurpurin
喀斯特(主要为石灰岩) karst
喀斯特地貌 karst landscape
卡巴呋喃 carbofuran
卡伯值 Kappa number
卡车运材 truck hauling; truck logging; trucking
卡尺 caliper (calliper)
卡德浸注防腐法 Card process (treatment)
卡尔群落(疏林) carr
卡尔文循环 Calvin cycle
卡尔再碎机 Carr's disintegrator
卡方测验 Chi-square (*or* χ^2) test
卡方分布 chi-square distribution; χ^2 distribution
卡方检验 chi-square test; χ^2 test
卡钩 grip; tongs
卡钩扎眼 dog hole
卡可基酸(杀虫剂) cacodylic acid
卡拉烯 calamene
卡藜油 cascarilla oil
卡鲁荒漠 karroo
卡米尔连续蒸煮器 Kamyr digester
卡米尔磨浆机 Kamyr grinder
卡姆比奥剥皮机 Cambio-barker
卡脑巴棕蜡 carnauba-wax
卡盘 chuck
卡片穿孔装置 card punching unit
卡片代码 caxd code
卡片检索表 card key
卡片整理机 card collator
卡片纸 carton paper
卡片纸板 cardboard
卡特基紫胶 Katki
卡廷加群落 caatinga
卡爪 chuck

卡值 caloric power; caloric value; heating power
卡轴 spindle
卡阻 hang up; lodging
开办费 capital account; initial cost; organization costs
开闭杆 trip timber
开采税 severance tax
开槽 gaining
开槽刀 grooving cutter
开槽刀头 dado head
开槽机 groover; grooving machine
开槽锯 drunken saw; grooving saw; wobble saw
开槽口 notching
开槽刨 grooving plane
开槽榫 chase-mortising
开齿机 gulleting machine; gummer; saw shaper
开齿器 saw punch
开发 development; opening up
开发方式 exploitation method
开发价值 exploitability; exploitable value; exploitation value
开发林道 woods highway
开发区 developed land
开反锯口 undercut
开方 radication
开放(芽或花蕾) break
开放传粉 open pollination
开放道路 break out
开放脉序 open venation
开放培养 open culture
开放系统 open system
开放系统循环 circulation of open system
开放循环 open circulation
开放贮藏 open storage
开缝接头 open joint
开工率 operating rate
开沟 ditching; furrowing; streaking; trenching
开沟播种 listing
开沟播种机 lister-planter
开沟铲 ditching blade
开沟机 ditch cutter; ditcher; ditching machine; drain digger
开沟犁 ditch plough; furrower; middle breaker plough; trenching plough
开沟裂 middle buster
开沟器 furrower; share; trencher; opener
开沟压条 trench layering
开沟移植 trench transplanting
开沟移植两用板 combined trenching and transplant board
开沟栽植 listing
开沟中耕器 lister cultivator
开沟种植机 lister plougher
开花 anthesis; effloresce; efflorescence; flower
开花弹 mushroom bullet
开花的 abloom; blooming
开花地 abloom
开花激素 flowering hormone
开花期 blooming date; blooming stage; efflorescence; florescence; flowering period
开花起初 flower initiation
开花受精 chasmogamy
开花诱导 flower induction
开花植物 flower
开荒犁 breaker plough; breaking-up plough; newground plough; prairie breaker; prairiebuster plough; reclamation plough
开角 angular aperture
开垦 forest clearing; open up; pioneering; reclamation
开孔移植 dibble transplanting
开口 hatch
开口滑车(一侧夹板可开闭) openside block
开口滑轮 knock-down block; logging block; snatch block
开口集材滑车 spring link yarder
开口接合 open joint
开口跑车 openside trolley
开口榫槽 open mortise; slot mortise
开口榫槽接合 open mortise-and-tenon joint
开口索套 open socket
开口系环(滑轮) snatch yoke
开跨接 cocking; cogging caulking
开扩性溃疡 open canker

开阔单车道 open single lane
开阔群丛 open association
开阔群落 open community
开阔阳坡 open sunny slopes
开阔植被 open vegetation
开裂 crack; cracking; dehiscence; fracture
开裂材 cracked wood; shaky wood
开路循环 open cycle
开启时间 opening time
开嵌板槽 trenching
开始(生长发育等) onset (*eg.* dormancy onset); commence
开始干枯 declining
开始生长 commence growth; commencing of growth; growth onset
开氏温标 absolute temperature scale; Kelvin temperature scale
开氏温度 Kelvin temperature
开式陈化时间 open assembly time; open waiting time; open time
开式存放时间 open time
开式循环 open cycle
开榫 tenoning
开榫槽 dado
开榫机 tenoner
开榫锯 tenon saw
开拓地 polder
开拓种 colonizing species
开下茬 set-up
开下口 undercut
开下口工 undercutter
开斜槽 oblique notching
开穴移植 hole transplanting
开穴栽植法 dug-hole methods
开叶期 leaf developing period
开腋榫 haunching
开云玫瑰木油 Cayenne linalor oil; Cayenne rosewood oil
开栽植沟 trenching
开展 divarication
开展的圆锥花序 open panicle
揩涂 tamponing
凯氏定氮法 Kjeldahl method
凯氏法桩木外壳 Koetitze pile armour
凯氏桩木防护法 Koetitze pile armour process
勘测 reconnaissance
勘测队 cruising party
勘查抽样 exploratory sampling
勘察报告 reconnaissance report
勘界 boundary settlement
勘探员 prospector
看车工 head loader
看守人 warden
坎伯兰木材处理法 Cumberland method
坎达蜡(一种蜂蜡) ghedda wax
坎普电柱围护法 Camp process
砍 adzing; lop; slash
砍柴人 woodman
砍除杂灌 slashing
砍刀 bush knife; cleaver; hatchet; machete
砍道工 axeman
砍道影 brush out; chunk out
砍道影工 run cutter
砍伐 cutting off; felling with axe; hew
砍伐森林 deforest; deforestation
砍伐适龄 exploitation age; felling age
砍伐者 hewer
砍斧 broad axe; chopping axe; scoring ax
砍灌刀 machete
砍号 axe marking; blaze; hag; hammer mark
砍号斧 marking axe
砍号工 bark blazer
砍口(原木端头) crotching; chop
砍劈 hack
砍皮 blaze
砍平(枕木承轨部分) adzing
砍缺口(在木杆上) saddle a skid
砍树标(浅削树皮) blaze
砍下口 fit(ting); round up
砍斜口(伐木) snipe
砍削 hack
砍削枕木 hack tie
砍原木端头企口工 corner man
砍枝 chop
砍枝工 chopper
砍枝桩 dehorn
康杜克斯磨浆机 Condux mill
康格法(木材冷热浸渍防腐) Conger method

康乐室 recreation room
康斯坦丁原木板英尺材积表 Constantine log rule
康斯坦丁原木丈量法 Constantine measure
糠 chaff
糠秕状的 scurfy
糠酰 furoyl
抗拔力 withdrawal resistance
抗白蚁 resistant against termite
抗白蚁性 termite resistance
抗爆剂 antidetonant; antidetonator
抗病的 disease-resistant
抗病性 disease-resistance; disease-resistant
抗补体 antialexin
抗扯力 tearing resistance
抗扯强度 tearing resistance
抗冲击性能 shock resistance
抗虫性 insect resistance
抗臭氧植被 ozone-resistant vegetation
抗戳穿力 puncture resistance
抗代谢物质 antimetabolite
抗冬性 winter resistance
抗冻性 freezing resistance; frost resistance
抗毒素 antitoxin; toxolysin
抗毒性质 antitoxic property
抗风的 windfirm
抗风化化学药剂 anti-weathering chemical
抗风力 wind firmness; windthrow resistance
抗风树 windfirm tree
抗风性 wind resistance
抗风种 windfirm species
抗辐射效应 antiradiation effect
抗腐蚀剂 corrosion inhibitor
抗腐性 decay proof; decay resistance; resistance to decay
抗腐朽 rot-resistant
抗腐组织 decay-resistant tissue
抗寒的 winter hardy
抗寒锻炼 cold hardening
抗寒力 cold tolerance; frost-resisting power
抗寒性 cold resistance; winter hardiness; winter resistance
抗寒作物 hardy crop
抗旱保墒耕作 storing water tillage
抗旱性 drought resistance; dry resistance
抗划痕性 marresistance
抗划伤性 scratch resistance
抗坏血酸 ascorbic acid
抗坏血酸氧化酶 ascorbic acid oxidase
抗剪刚度 shear stiffness
抗剪角 angle of shearing resistance
抗剪力 shear resistance
抗剪强度 shear(ing) strength
抗剪效率系数 shearing effect figure
抗碱纤维素 alkali resisting cellulose
抗降解性 degradation resistance
抗焦(剂) antiscorch(ing)
抗菌的 antibiotic
抗菌活性 antibiotic activity
抗菌谱 antibiogram; antibiotic spectrum
抗菌素 antibiotic
抗菌物 antibiont
抗抗毒素 antiantitoxin
抗抗体 antiantibody
抗拉弹性 tensile elasticity
抗拉剪切试验 tension shear test
抗拉强度 tenacity; tensile strength
抗涝性 water-logging resistance
抗流变 flow resistance
抗酶 antienzyme
抗霉性 mould resistant
抗磨力 wearing resistance
抗磨性 abrasion resistance; abrasive resistance; antifriction property
抗墨性能 ink resistance
抗逆的 resistant
抗逆性 resistance
抗黏附性 block resistance
抗凝固物质(抗凝剂) anticoagulant
抗扭刚度 torsional rigidity; torsional stiffness
抗扭剪切强度 torsion-shear strength
抗扭剪切试验 torsion shear test
抗扭强度 torsional strength
抗扭性 torsion property
抗劈力 cleavage resistance
抗疲劳性 fatigue resistance
抗氰呼吸 cyanide resistant respiration
抗屈强度 yield strength
抗热性 heat resistance

抗蠕变力 creep resistance
抗蠕变强度 creep strength
抗生[作用] antibiosis
抗生的 antibiotic
抗生素 antibiotic
抗生素微生物 antibiotic microbe
抗生物 antibiont
抗生物质 antibiotic substance
抗生长激素 antigrowth hormone
抗盛开剂 anti-blooming agent
抗湿 slushing
抗蚀墙 counter-dam
抗受精素 antifertilizin
抗水性 water resistance
抗缩处理 anti-shrinkage treatment
抗缩效率 antishrink(age) efficiency
抗踏性 resistance to trampling
抗体 antibody
抗蜕皮素 anti-ecdysone
抗弯刚度 bending rigidity; bending stiffness; flexural rigidity
抗弯强度 bending strength; buckling strength; cross-breaking strength; flexural strength; flexure strength; transverse strength
抗弯屈服点 bending yield point
抗弯阻力 buckling resistance
抗性 degree of hardiness; hardiness; resistance; resistibility; resistivity
抗性品种 resistant variety
抗性适应 resistance adaptation
抗性育种 resistance breeding
抗性预测 resistance prediction
抗性砧木 resistant stock
抗血清 antiserum
抗压材 brace
抗压构件 strut
抗压极限 crushing limit
抗压极限强度 maximum crushing strength
抗压强度 compression strength; compressive strength; crushing strength
抗压试验 compression test
抗压试验机 compression testing machine
抗烟力 smoke resistance
抗烟性 fume resistance; smoke resistance
抗淹性 water-logging resistance
抗盐性 saline resistance; salt resistance
抗氧化剂 antioxidant
抗油性试验 grease resistance test
抗诱变剂 antimutagen
抗原 antigen
抗原性 antigenicity
抗原作用 antigenic action
抗胀缩处理 anti-movement treatment
抗折强度 folding strength
抗真菌的 antifungal; antimycotic
抗蒸腾剂 anti-transpirant
抗植物生长素 antiauxin
抗皱性 crease resistance
抗紫外聚乙烯 ultraviolet-ray-resisting polyethylene
考得(木材层积单位: 1个标准考得等于4×4×8=128ft^{3}或3.624m^3) cord
考得量度(以考得为计算单位) cord-measure
考得数(层积材) cordage
栲胶 tannin extract; tanning extract; vegetable tannin extract
栲里松脂 kauri gum
栲树脂 kauri resin
烤漆 baking finish
靠板 fence; linebar
靠背(椅) seat-back
靠背椅 back chair
靠边原木(装车) mush log
靠窗桌(桌面宽度等于窗宽) pier table
靠接 ablactation; approach grafting; grafting by approach; inarching; inarching grafting
靠墙桌 side table
靠山(木工机器上) fence; linebar
苛性(苛化度) causticity
苛性碱液 causticlye
苛性钠法 soda process
柯巴油(同上) copal oil
柯巴油清漆(同上) copal oil varnish
柯巴脂 copal
柯本气候分区 Koppen's climatic province
柯赫氏法则 Koch's postulates

柯赫氏蒸气灭菌器 Koch's steam-sterilizer
科 family
科布施胶度测定法 Cobb tester
科尔曼法(木材蒸汽干燥处理) Collmann method
科技兴林 promote forestry with science and technology
科间嫁接 interfamily grafting
科内尔泥炭矿肥营养土 Cornell "Peat-Lite" Mix
科氏模压成型工艺 Collipress
科学认识水平 level of scientific understanding
颗粒 grain; particle
颗粒层增殖 granulosis
颗粒尺寸 grain size
颗粒大小 particle size
颗粒定向 grain orientation
颗粒度 granularity
颗粒肥料 granulated fertilizer
颗粒活性炭 granular activated carbon
颗粒剂 granule
颗粒密度 block density; particle density
颗粒体病毒 granulosis virus
颗粒体病毒病 granulosis
颗粒物 particulate material
颗粒状的 granulose
颗粒紫胶 seed lac
髁(kē) condyle
壳 shale; shell; theca (*plur.* thecae)
壳多糖 chitin
壳粒 capsomer(e)
壳糖胺 chitosamine
可保风险 insurable risk
可被生物腐蚀的 bioerodible
可编程序计算器 programmable calculator
可变半径样地抽样 variable radius plot sampling
可变半径样区法 sampling using variable radius plots
可变产价值 realization value
可变成本 alternative cost; variable cost
可变概率 varying probability
可变概率抽样 variable-probability-sampling
可变火险性 variable (fire) danger
可变焦距立体镜 zoom stereoscope
可变焦距转绘仪 zoom transfer scope
可变密度全林分模型 variable-density growth and yield model
可变密度收获表 variable density yield table
可变现价值 realization value
可变性 changeability; variability
可变样地 variable plot
可变样地法 variable plot method
可变样地调查 variable plot cruising
可变异的 variate
可捕量 catchability
可捕猎的鸟 game bird
可捕性 catchability
可采储量 reserves
可采伐材 harvestable timber
可采伐的 exploitable
可采伐的立木 operable timber
可采伐径级 exploitable size
可采伐利用龄 exploitable age
可采伐林分 exploitable stand
可采期 detachable time
可操纵性 maneuverability
可侧方拖集的架空索道 skyline crane
可测径级 measurable size
可拆装家具配件 knock-down fittings
可拆装椅座 loose seat
可承受窗口方法 tolerable-windows approach
可持续发展 sustainable development
可持续利用 sustainable use
可持续林业 sustainable forestry
可持续森林经营 sustainable forest management
可传让的 heritable
可电离基因 ionogen
可伐量 allowable cut; allowable cut amount
可伐林木 loggable timber
可耕层 arable layer
可耕的 arable
可耕地 arable land; tillable acres
可更新资源 renewable resource
可供狩猎的种类 game species
可行性研究 feasibility study
可混性 compatibility
可及等级 accessibility class
可及林 accessible forest

可及林地 accessible forested land
可及性 accessibility
可继承的 heritable
可加工性 machinability
可加性 additivity
可见冠幅(航摄) visible crown diameter
可见光辐射 visible (light) radiation
可见光和红外线辐射计 visual and infrared radiometer
可见光谱 visible spectrum
可见裂缝 visible crack
可见树高(航摄) visible tree height
可见突变 visible mutation
可见症状 visible symptom
可降解的污染物 degradable pollutant
可降式吊钩 lowerable load hook
可交配性 cross(ing) ability
可交配性模式 crossability pattern
可结晶性 crystallizability
可结实 fruitful
可居住部分(世界范围) ecumene
可开发的 exploitable
可开钩环 knockout shackle
可靠性 reliability
可靠最低估计 reliable minimum estimate
可可碱 theobromine
可可蜡 cacao-tallow
可可油 cocoa butter
可可脂 cacao-butter; cacao-fat; cocoa butter
可扩充语言 extensible language
可乐茶素 cola
可利用材 merchantable timber volume
可利用材积生长量 merchantable timber increment
可利用的 exploitable; merchantable
可利用径级 exploitable diameter
可利用梢头 utilized top
可利用最小梢头木 merch top
可利用干围级 exploitable girth
可利用饲料面积 forage acre
可挠性 flexing
可能破坏面 potential rupture surface
可能收获木 potential crop tree
可能再流走的搁浅木 floating rear
可能最大降水量 probable maximum precipitation
可逆辊涂机 reverse roll coater
可逆横向循环 reverse cross circulation
可逆强制横向循环 alternating forced cross circulation
可逆系统 reversible system
可逆循环 reversible circulation
可逆循环干燥窑 reversible circulation kiln
可逆抑制 reversible inhibition
可漂白程度 bleachability
可屈性 pliability
可取消更新区 revocable regeneration block
可燃气体 fuel gas
可燃物 fuel
可燃物测湿棒 fuel stick
可燃物储量 fuel loading
可燃物含水量 fuel moisture content
可燃物类型 fuel type
可燃物类型分类 fuel-type classification
可燃物危险度 fuel hazard
可燃性 combustibility; flammability; ignitability; ignition probability; inflammability
可燃性生物岩(有机岩) caustobiolith
可染色的 stainable
可染性 dyeability
可溶解有机物 dissolved organic matter
可熔型胶带 fusible tape
可入肺颗粒物 particulate matter less than 2.5 μm in diameter (PM2.5)
可烧除地 allowable burn area
可渗透种子 permeable seed
可生化降解污染物 biodegradable pollutant
可湿性 wettability
可湿性粉剂 wettable powder
可视角 visibility angle
可收回价值 recovery value
可售材积 sal(e)able volume
可售材长度 commercial height
可售高度(树高) sal(e)able height
可售苗 sal(e)able plant
可塑伸展性 plastic extensibility

可塑性 compliance; moldability; plasticity
可塑性指数 plasticity index
可溯成本 traceable cost
可提取物 extractable material
可调型裁分机 selective edger
可调整平价 sliding parity
可涂布时间 spreadable life
可吸入颗粒物 particulate matter of 2.5–10 μm in diameter (PM10)
可稀释度 dilutability
可销售材积 merchantable volume
可销售的 marketable; sal(e)able
可信界限 fiducial limits
可信区间 fiducial interval
可信样品 authentic sample
可压缩性 compressibility
可遗传性 hereditability
可用程度 serviceability
可用宽度 available width
可用时间 usable life; working life
可用水分 available water
可用性 availability
可用余额 available balance
可用资源 available resource
可育性种子 viable seed
可孕性 receptivity
可运用性 maneuverability
可杂交群(种内) commiscuum
可再生的 fertile
可再生能源 renewable
可再生资源 renewable resource; renewable resources
可再生资源回收 renewable resource recycling
可造林裸地 denuded plantable land
可折叠的 accordion
可折椅 folding chair
可植苗 plantable seedling
可转换的 fertile
可转让信用证 assignable credit
可作业地 operable land
可作业林 operable stand
可作业区 operable area
可坐小圆桌 chair table
克贝纤维素 Cross and Bevan cellulose
克菌丹(杀菌剂) captan
克拉夫特分级法 Kraft's tree classification
克拉松木素 Klason lignin; sulphuric acid lignin
克劳斯纳测高器 klausner hypsometer
克雷布斯循环 Krebs cycle
克里斯登测高器 Christen hypsometer
克里希测高器 Konig's hypsometer
克龙(百万年单位) cron
克隆生长 clonal growth
克罗西特木业公司(美) Crossett Lumber Company
克罗西特实验林(美) Crossett Experimental Forest
克氏瓶(烧瓶的一种) Kolle flask
克特(虫胶稠度单位, 磅(虫胶)/加仑(酒精)) cut
刻串珠线脚器 bead router
刻点 puncture
刻度盘 dial gauge; dial; scale; gradual dial
刻度曲线(列线图) graduating curve
刻度线 graduated reticle
刻号刀 marking knife
刻号工具 scratcher
刻痕 incising; indentation; notch; score
V形刻痕 V-shaped notch
刻痕齿 incising tooth
刻锯齿者 jagger
刻面 facet
客栖 synoecy
客厅套装家具 drawing room suites
客厅坐椅 parlour chair
客土 soil dressing
客重 volumetric weight
课税公平 tax equity
课税减免 tax credit; tax deduction
课税年度 taxable year
课税值 taxable base
课税准则 canons of taxation
垦复 improvement of stand condition; improvement of site condition; improvement of soil condition
垦荒 land reclamation
啃牧 browse; browsing
啃牧线 browse line; browsing level

坑 pit
坑壁填木 lagging
坑道 gallery; mine
坑道系统 gallery pattern system
坑腐 pitted sap rot
坑锯 pit saw; whip saw
坑锯法 whip sawing
坑木 mine-prop; minig timber; picket; pit wood; post pillar; stull
坑丘地形(根拔时形成) pit-and-mound topography
坑烧法 pit kiln process
坑式烧炭法 charcoal-making in pit; charring in pit; charring-making in pit
坑窑(烧炭) pit kiln
坑植 pit planting
空秕的 exhausted
空瘪粒(仅具种翅的) wing without seed
空播 sow tree seeds by plane; sow tree seeds from aircraft
空车牵引索 spotting line
空车停车线 tail track
空档 neutral position
空洞节 pith knot
空稃 sterile lemma
空盒气压计 aneroid; aneroid barograph; aneroid barometer
空花现象(无雌雄蕊) cenanthy
空架运材车 mutibo
空间尺度 spatial scale
空间次序 spatial order
空间分辨率 spatial resolution
空间分布 spatial distribution
空间格局 spatial pattern
空间和时间尺度 spatial and temporal scale
空间环境监测器 space environment monitor
空间模拟 space simulation
空间群 spacer
空间生态学 space ecology
空间梯度 spatial gradient
空间胁迫 spatial stress
空间艺术 spatial art
空间异质性 spatial heterogenity; special heterogeneity
空间直观模型 spatial explicit model
空间秩序 space ordering
空间自相关分析 spatial autocorrelation analysis; spatial autorelation analysis
空间坐标 space coordinate
空降救火 aerial smoke jumping
空降灭火员 smoke jumper
空降扑火队 helitack crew
空卷筒(未绕绳的) bare-drum
空壳 exuviae
空孔节 hollow knot
空粒种子 empty seed; hollow seed
空泡 vacuole
空栖白蚁 aerial termite
空气尘度计 konimeter
空气传播 aerial transmission
空气传播病害 airborne disease
空气传染 airborne infection
空气干燥 airing
空气过滤器 air filter
空气环流 air circulation
空气环流系统 air-circulation system
空气加热器 air heater
空气减压阀 air reducing valve
空气净化器 air purifier
空气净化设备 air-purification equipment
空气净化系统 air-purification system
空气-空气热交换器 air-to-air heat exchanger
空气冷却系统 air cooling system
空气量 air volume
空气流量 air supply
空气流速 air velocity
空气滤清器 air cleaner; air filter
空气喷口 air jet
空气喷嘴 blow pipe
空气容积 air volume
空气容量 air capacity
空气扫除机 sir sweeper
空气射流 air jet
空气生物学 aerobiology
空气湿度 air humidity
空气调节器 air governor; air register; air regulator
空气污染 aerial contamination; air pollution
空气污染观测站 air pollution observation station

空气污染潜势 air pollution potential
空气污染物 air contaminant; air pollutant
空气修根(育苗技术) air pruning
空气蓄压器(蓄能器) air accumulator
空气悬浮 air suspension
空气循环 air circulation
空气压力表 air-gauge
空气压缩机 air compression machine; air compressor
空气油类分离器 air oil separator
空气预热 air preheating
空气预热器 air preheater
空气制动器 air brake; air damper
空气置换 replacement of air
空气阻力 air resistance
空气阻塞(蒸汽管道) air binding
空腔 void volume
空试(射击) dry practice
空调 air-conditioning
空调器 air-conditioning unit
空调室 air-conditioned channel
空调装置 air-conditioning unit
空头(金融) bear
空细胞法防腐 open cell process; empty-cell process; Rueping process
空细胞浸渍 empty-cell impregnation
空想 maggot
空心坝 hollow dam
空心板 tubular board
空心秆(竹秆禾秆等) culm
空心胶合板 hollow plywood
空心结构 hollow-core construction
空心木 centre hole
空心刨花板 hollow-core particleboard; tubular particle board
空心髓 excavated pith
空心碎料板 hollow-core particleboard; tubular particle board
空心细木工板 hollow-core board
空心原木 piped log
空心凿 hollow chisel
空心轴 sleeve
空心钻头(生长锥) hollow cutting bit
空颖 sterile glume
空载 no load
空载承载索 unloaded skyline cable
空载挠度(钢索) unloaded deflection
空载跑车 unloaded carriage
空指令(计算机) skip
空中传播 air-borne
空中垂直摄影像片 aerial vertical photograph
空中花园 hanging garden
空中基线 air base
空中集材(气球或直升机) aerial logging; air lift logging; vertical logging
空中集材绞盘机 aerial skidder
空中勘测 aerial reconnaissance
空中可燃物(树冠) aerial fuels
空中灭火指挥 air attack boss
空中漂浮生物 aeroplankton
空中三角测量 aerial photo triangulation; aerial triangulation
空中摄影站 air station; camera station
空中索道 cableway
空中探测 aerial detection
空中巡逻 aerial patrol
空中压条 ablactation; air layering; circumposition; marcottage; pot layerage; stool layering
空中压条法 marcotting layerage
空中压条高压法 Chinese layering
空中压枝 aerial layering
空中样地测量 aerial plot measurement
空中种群 aerial population
空转轮 idle pulley; idle wheel; idler
空转消耗 idling consumption
孔 bore; eye; foramen (*plur.* -mina); hole; pitting; trema (*plur.* tremata)
孔出孢子 porospore
孔茨处理法(木材防腐) Kuntz process
孔道尺寸 pore size
孔度计 porosimeter
孔盖 opercule
孔环 halo; pore ring; porous circle
孔径 bore; pore diameter; pore size
孔径分布 pore size distribution
孔径角 angular aperture
孔口(机械) orifice
孔口(子囊果或分生孢子器的) ostiole

孔口拦沙坝 sediment storage dam with hole
孔裂蒴果 pore capsule
孔率计 porosimeter
孔容分布 pore volume distribution
孔容积 pore volume
孔隙 pore space
孔隙度 porosity
孔隙度分析 lacunarity analysis
孔隙率 porosity
孔隙容积分布 pore volume distribution
孔隙水压 pore water pressure
孔眼 eyelet
孔缘(孢粉) labrum
孔罩 escutcheon
孔状白腐 white pocket rot
孔状伐 gap cutting; gap felling; patch method
孔状伐作业 gap cutting system
孔状腐朽 cavity forming rot
孔状皆伐作业 gap clear-cutting system; patch clear-cutting system
空白地 blank
空白试验 blank test
空地 open-space; patch; vacancy
空格状堆积 honeycomb stacking
空林地 unstocked area
空位 vacancy
空隙 cellule; open-space; void
空隙量 void content
空隙容积 void volume
空闲地 unoccupied land
控股公司 conglomerate; holding company
控件 control piece
控湿系统 humidity control system
控速架空索道 control skyline
控速索道集材 gravity cable logging
控速装置(钢索) snubbing device
控制 check; control
控制扳 steering flap
控制板 panel
控制变量 control-variate
控制部件 control unit
控制传粉的后代检验 control(led)-pollination progeny test
控制点 control point
控制点位图 radial line plot
控制点镶嵌图 controlled mosaic
控制法 control method
控制翻板 control flap
控制杆 trip timber
控制航带 control strip
控制火烧 controlled burning
控制基因 controlling gene
控制经济 command-directed economy
控制论 cybernetics
控制片 control strip
控制器 controller; keyer
控制燃烧 controlled burn
控制烧除 controlled prescribed burning
控制时间 control time
控制台 control console; control desk
控制温包 control bulb
控制污染 control pollution
控制压力 operating pressure
控制用火 control(led) burning
控制杂交 purposeful hybridization
口 bucca (*plur.* buccae)
口部 trophi
口疮 canker
口附肢 buccal appendage
口紧(拒带镪修锯不当) tire
口蘑 tricholoma
口器 buccal appendage; trophi
口器的 trophic
口器易足现象(口器足) proboscypedia
口腔 mouth cavity; oral cavity
口穴 excavation
口针(工具) spear
口针(昆虫) stylet
口针携带病毒 styler-borne virus
口锥 cone
叩解 milling
叩解度 beating degree
叩头虫 click-beetle
叩头虫幼虫 wire worm
扣减 abatement
扣缴税款 withholding
扣接合 lock joint
扣链 bitch chain
扣留 detain

K

扣闩 action bolt
扣闩槽 slot for action bolt
扣榫 coak
扣押权 lien
寇文离心式圆筛 Cowan screen
刳补锯 dowel saw; plug saw
枯斑 necrosis; necrotic spot
枯草层 litter of weed
枯草隆(除草剂) chloroxuron
枯倒木 dead-and-down tree; down dead wood; down-timber; fallen dead wood; lying dead tree
枯顶的 stagheaded
枯而不落的(植物) marcescent
枯伐 ringing
枯立木 dead standing tree; deadhead; dead-wood; dry wood; dying tree; ram pike; standing dead tree; stem dryness; snag
枯立木采伐 dry wood cutting; snagging
枯立木推杆(装于拖拉机前) snag pusher
枯茗油 cumin oil
枯梢 dead-top; dried top; top dry; top drying; topkill
枯梢病 die-back; stagheaded
枯梢的 dry-topped; stagheaded
枯梢木 candelabra-topped tree; staghead; spike-top; dry topped tree
枯熟期 dead stage; dead-ripe stage
枯水河 oued
枯水流量 base run-off
枯死材 dead timber
枯死木 dead and dying tree; dead trees; deadwood; drying tree
枯死木采伐 dying tree cutting
枯损 mortality
枯损模型 mortality model
枯萎 scorch; perish
枯萎病 wilt disease
枯折木 snag
枯枝 dead arm; dead branch
枯枝材 dry fallen wood
枯枝层 branch litter
枯枝度 cleanness of bole
枯枝落叶 litter
枯枝落叶层(土壤剖面A_0层或O层) litter layer; dead soil covering; duff; ground-litter; leaf-cover; leaf-litter; A_0-layer; O-layer
枯枝落叶层分解 litter decomposition
枯枝落叶层温度 litter temperature
枯枝修枝 dry pruning
苦艾油 absinthe oil; wormwood oil
苦橙 bigarade orange
苦楝根碱 azedarine
苦楝根皮 amargosa bark; azadarichta
苦楝树皮 amargosa; azadarach
苦楝素 mersosin
苦木素 quassin
苦木素类 quassinoids
苦配巴香脂 copaiba balsam
苦味素 bitter principle
苦味质 bitter substance
苦杏仁苷酶 amygdalase
苦杏仁油 bitter almond oil
苦杏素 amaroid
库存 inventory; on hand; stock
库存充足的 well stocked
库尔巴盐剂 Kulba salt
库房 storage premises
库斯米紫胶 kusmi lac; koosmie
库塘系统 fish pond-dike system
酷暑 scorching
夸莱斯法(木材防腐) Quarles process
跨车 timber carrier
跨档 span-rail
跨度 span
跨接片 jumper
跨界保护区 transboundary protected areas
跨距 space; span
跨距中央 mid-span
跨式装卸车 straddle-carrier
跨装车组 sett; solid content
会计程序 accounting procedure
会计核算 accounting
会计经济业务 accounting transaction
会计科目 account; accounting item
会计年度 fiscal year
会计凭证 accounting voucher
会计学 accounting
会计业务 accounting procedure
会计长 controller

会计账目 accounting record; financial record
会计制度 accounting system
块播 patch-sowing; sowing in patch; sowing in spot
块根 root tuber
块金方差 nugget variance
块茎 tuber
块菌 truffle
块密度 block density
块体运动 mass movement
块闸 block brake
块植 block planting
块状斑点花纹 block mottle figure
块状剥皮 patch bark; scorch; spot bark
块状播种 plat sowing; plot sowing
块状采伐 cutting in block; felling by group; felling for group system; patch logging
块状采伐母树法 seed tree cutting by the group method
块状除草皮 scalping
块状点火 area ignition
块状伐 cutting by group method
块状腐朽 cubical rot
块状更新 regeneration in group
块状褐腐 brown cubical rot
块状混交 group mixture; mixture by group; mixture in group
块状混交林 mixed forest in groups
块状渐伐 group-shelter wood cutting
块状皆伐 block cutting; clearcutting in patches
块状皆伐天然更新法 clearcutting in patches
块状皆伐作业 patch clear-cutting system
块状结构 blocky structure; massive structure
块状菌落 massive colony
块状灭火法 area suppression method
块状群体 massive colony
块状烧除 patch burning
块状松土机 patch-scarifier
块状雪崩 avalanche of massive snow
块状栽植 group planting; patch planting
块状择伐 cutting by group selection method
块状整地 soil microsite preparation; spot soil preparation
快干漆 fast-drying lacquer
快火 rapid fire
快门 shutter
快凝水泥 accelerated cement
快熟 thrifty mature
快速拌胶机 high-speed mixing machine; short-retention(-time) blender
快速闭合 quick make
快速剥皮机 rapid-peeler
快速发芽试验 accelerated germination test
快速返回 quick return
快速高分辨质谱分析 real-time high resolution mass spectro metry
快速工艺 fast technology
快速固化 expediting setting
快速关闭门 quick-locking door
快速管式干燥机 flash tube dryer
快速横截锯 flying cut-off saw
快速回行 quick return
快速移动 quick travel
快速蒸煮 high speed cooking
快艇 yacht
宽波段 wide range
宽带锯 wide band-saw
宽带锯条 wide band-saw blade
宽带砂光机 wide-belt sanding
宽带式砂光机 wide belt sander; wide belt sanding machine
宽度 width; cross
宽埂梯田 sloping terrace
宽行条播 broad line seeding; sowing in broad drill
宽脚榫 stump tenon
宽口手斧 broad hatchet
宽口圆锯 drunken saw
宽量程 wide range
宽年轮的 broad-ringed; broad-zoned; coarse-grown; fast grown; open-grown; wide-ringed
宽刃斧 block bill; broad axe
宽射线 broad ray
宽条板芯细木工板(条板宽大于30mm) battenboard
宽叶的 platyphyllous
宽叶香蒲 bulrush

宽硬叶植物 broad-sclerophyll plant
宽窄条相间壁板 bar and batten paneling
狂风 gustiness
狂火 fire storm
狂燃火 conflagration fire
矿床 deposit
矿道顶木 lagging
矿化细胞法 mineralized-cell process
矿井 pit
矿泉疗法 balneotherapy
矿物质 mineral; mineral matter
矿物质循环 mineral cycling
矿物组成分 mineral-constituent
矿用材 mine timber; mining timber; pit prop; pit wood
矿用枕木 mine tie
矿质变色 mineral stain
矿质肥料 mineral manure
矿质化作用 mineralization
矿质灰分 mineral ash
矿质缺乏 mineral deficiency
矿质条纹 mineral streak
矿质土壤 mineral soil
矿质营养 mineral nutrition
矿质元素 mineral element
矿柱 jamb; mine-prop; pit prop; pit wood; prop; stull
矿柱材 colliery timber; pillarwood
矿柱采伐作业 short log show
框 mullion
框板 deckle board
框带 deckle; deckle strap
框格 sash
框构门 framed and braced door
框架 frame; frame-work; framing; set
框架加工 framing
框架家具 stick furniture
框架锯 buhl saw
框架楼板 framed floor
框架式拆垛机 crib unstacker
框架式压机 frame press
框架外部构件 outside lining
框锯 frame saw; gang saw; gate saw; log frame-saw; sash saw; span saw; stock-gang; web saw
框锯车间 frame (saw) mill
框锯加工的 frame-sawn
框锯磨锯机 frame saw sharpener
框锯制材厂 frame (saw) mill
框式成型法 deckle box-method
框式家具 frame-type furniture
框式结构门 framed door
框式排锯 sash gang-saw
框条(门窗) sash bar
亏本出售 distress selling
亏绌 deficiency
亏损 deficit; loss
盔瓣 mitra
盔状的 cassideous
魁北克原木板英尺材积表 Quebec log rule
馈电箱 feed(er) box
馈给 feed
馈给压头 feed head
溃陷 collapse; crimp
溃疡 canker
溃疡变色 canker stain
昆虫病原体的 entomopathogenic
昆虫传播 insect transmission
昆虫纲 class insecta
昆虫寄生 entomophagous parasite
昆虫寄生病 entomosis
昆虫媒介 insect vector
昆虫排泄物 frass
昆虫区系 insect fauna
昆虫群落 insect society
昆虫身上的 entomogenous
昆虫体寄生 entomoparasitism
昆虫体寄生的 entomogenous
昆虫相 insect fauna
昆虫学 entomology
昆虫学家 entomologist
昆虫志 insect fauna
髡枝 pollarding
醌类 quinone
捆 bundle; pack; package
捆把机 looper
捆包机 strapper
捆材 fastening
捆槽 dee
捆车链 binding chain; wrapper chain
捆车器 load binder
捆带 binding strap
捆吊索 sling
捆挂工 block tender; choker setter; stump jumper
捆挂工长 head rigger; hook tender

K

捆挂索吊梁 spreader
捆挂组长 chain tender
捆紧链 binding chain
捆链(集材用) sling chain; toggle chain
捆链钩 toggle hook
捆链联锁器 jim binder
捆链压木 binding log
捆木 choke; set bead; set choker
捆木工 choker man; choker rester; choker setter; set knob; tonger
捆木链 binding chain; butt chain; load binder; tag
捆木器件 binder
捆木索 chocker; choker; load binder; necktie; rigging choker; sling rope; strop
捆木索包头 ferrule
捆木索扁钩 bear claw
捆木索钩 choker hook; line hook
捆木索挂钩 bull hook
捆木索滑钩 running hook
捆木索滑套 choker hook; hook; yarding bell
捆木索集材机 cable skidder; choker skidder
捆木索结 timber hitch
捆木索拉杆 choker gun
捆木索连接钩 slip grab hook
捆木索套节 bead
捆木索套筒 choker socket
捆绳打捆装置 twine-tying device
捆索 binder; binding strap; lashing; strap
捆索夹头 shackle
捆条(成捆装车用) strap; with
捆扎 bunch(ing); bind; binding; lashing; tie
捆扎工 toggler
捆扎装置 tier
捆枝机 branch bundler
捆枝作业 fascine work
捆装(按木捆装车) bundle load; sling load
捆装车 sling cart
困河材 neap; neap(ed) timber
扩程器 multiplier
扩基 root buttress
扩基上部干围 girth above buttress
扩孔器 nicker
扩孔钻头 expanding bit
扩散 diffusion
扩散度 diffusibility
扩散法 diffusion process
扩散剂 spreading agent
扩散率 diffusivity
扩散势 diffhsion potential
扩散速度 diffusion velocity; rate of spread
扩散速率 diffusivity ratio
扩散梯度 diffusion gradient
扩散洗涤 diffuser washing
扩散系数 difflusivity; dlffusion coefficient
扩散现象 diffusion phenomena
扩散性 difflusivity
扩散压亏缺 diffusion pressure deficite
扩散压力 diflusion pressure
扩散指数 diffusion index
扩散阻力 diffusion resistanee
扩延树冠火(先于地面火) running crown fire
扩展性溃疡 diffuse canker
扩展因子 expansion factor
扩张 extension
扩张模压(树脂浸渍压缩木成型法) expansion moulding
扩张器 expander; extender
扩张式离合器 expanding type clutch
蛞蝓 limacid; slug
蛞蝓型幼虫 slug
阔剑 broad sword
阔叶材 deciduous wood; hard wood
阔叶材层积单位(75层积ft^3, 50实积ft^3) lead square
阔叶草本干草原 forb steppe
阔叶草本植被 broad-leaved herb
阔叶草本植物 forb
阔叶常绿灌木群落 laurifruticeta
阔叶常绿林群落 aiphyllium
阔叶常绿木本群落 laurilignosa
阔叶常绿乔木群落 laurisilvae
阔叶的 broad leaved; broad leafed; latifoliate; leafy
阔叶林 broad-leaved forest; broad-leaved woodland; broad-leaf forest; broad-leafed

forest; hardwood forest; leaf wood; angiospermous forest
阔叶树 broad-leaved tree; broad-leaf tree; deciduous wood; foliaged tree; leaf tree
阔叶树材 hardwood; pored wood; porous wood
阔叶树插穗 hardwood cutting
阔叶树害虫 pests of broad-leaf trees
阔叶树林 hardwood forest
阔叶树群系 hardwood formation
廓羽 contour feather
垃圾分类 refuse sorting
拉 draft; draught
拉办释放 release of tension
拉出钢索 overhaul
拉德(辐射剂量单位) rad
拉丁方 Latin-square
拉杆 lever; linkage; queen bolt; tension member; tie
拉割(采脂) pulling
拉机柄 bolt-handle
拉紧钢索 bracing wire; tight line
X形拉紧器 X-stretcher
拉紧索锚桩 walking anchor
拉捆木索工具(从原木穿过) duke
拉缆 guy cable
拉力 tensile force; tension
拉力计 tensiometer
拉力试验 tensile test
拉力试验机 tensile testing machine
拉力调节悬挂装置 draft control hitch
拉裂 tension break
拉马克氏学说 Lmnarckism
拉马克式遗传 Lamarckian inheritance
拉马克效应 Lamarckian effect
拉曼光谱学 Raman spectroscopy
拉门(横向移动) sliding door
拉尼娜 la Niña
拉诺群落 llano
拉片工(印度) bhilwaya
拉平控制机构 flush control
拉切式伐木剪 draw shear
拉伸变定 tension set
拉伸破坏 tension failure
拉伸性 stretchability
拉伸应变 extensional strain; tensile strain; tension strain
拉伸指数 tension index
拉绳 stay
拉手 knob
拉树(倒向与倾斜方向相反时) tree pulling
拉丝 webbing
拉索 guy cable; tag line
拉应变 tension strain
拉应力 tensile stress
拉长 elongation
拉桌(桌面可拉长) draw table; extending table
喇叭形根端 butt flare
腊叶标本 exsiccata; herbarium
蜡梅花浸膏 chimonanthus fragrans concrete
蜡螟 wax moth
蜡膜 cere
蜡熟 wax(en) maturity
蜡腺 wax gland
蜡质的 ceraceous
蜡质塑料绘图纸 waxed plastic sheet
蜡状的 ceraceous
辣椒红素 capsanthin
辣椒胶 capsicin
辣椒玉红素 capsorubin
来复枪 rifle
来复枪瞄准器 rifle sight
来复枪皮带 rifle sling
来复枪枪柄 rifle stock
来复枪射击姿势 rifle shooting position
来复枪子弹筒 rifle cartridge
来复线 rifling
来回 turn
来料 furnishes
来年芽 statoblast
来源 origin; source
莱恩伍德刨花板制造工艺(卧式挤压工艺) Lanewood particleboard process
莱姆病 lyme disease
莱氏集材装车机 pine logger (Lidgerwood)
莱特氏效应 Wright effect
赖氨酸 lysine
赖百当胶 labdanum(gum)
赖百当油 labdanum oil
兰科植物 orchids

L

兰类温室 orchid house
兰姆氏干馏釜 Lambiotte retort
兰室 orchid house
拦板 stake
拦材格栅 grating for stopping floating wood
拦河坝 barrage; lasher
拦河绠 catch boom
拦河收漂场 transverse holding ground
拦洪 flood retention
拦洪容量 flood absorption capacity
拦截漂子 catch boom
拦截逃逸材河绠 spider web
拦木架 catching trestle
拦沙坝 check dam; debris barrier; detention dam; sediment storage dam; debris dam
拦沙谷坊 debris dam
拦沙坑 sediment trap
拦水 impounding
拦水坝 check dam
拦水沟埂 retaining ditch and embankment
拦网式河绠 fence holding boom
拦阻漂子 holding boom
栏杆 balustrade; hand rail; rail; railing
栏杆小柱顶 abacus
栏杆摇床 crib
栏杆柱 baluster
栏杆柱椅背 baluster back; banister back
蓝斑 blue stain
蓝变 blue sapstain; blue stain; blueing
蓝变材 blue lumber; blue stained wood
蓝变菌 blue sapstain fungus (*plur.* -gi); blue stain fungus
蓝丹油 frankincense oil
蓝矾 blue vitriol
蓝领工人(体力劳动者) blue-collar workers
蓝绿藻 blue-green algae
蓝图 blueprint; master plan
蓝樟油 blue camphor oil
篮栽植物(花卉) basket plant
篮植 basket planting
缆车 cable car
缆车道 cable tram; cable-railroad; drum-system cable railway; incline; inclined cableway; inclined rope railway; log hoist; rope railway
缆车道卷扬系统 drum-system of inclineway
缆车道运材 inclined haulage
缆车钢索 carrier cable
缆绠 cable boom with floats
缆索 hawser
缆索归楞法 log handler
缆索滑道 cable chuting
缆索起重 cable lift
缆索起重机 cable crane; cable lift
缆索运输机 cable conveyor
缆索支架 cable holder
榄仁树油 talisai oil
榄香 elemi
榄香树脂 elemi resin
榄香素 elemicin
榄香烯 elemene
榄香烯酮 elemenone
榄香油 elemi oil
榄香脂 elemi
榄香脂精油 elemi oil
滥伐 destructive cutting; destructive lumbering; indiscriminate cutting; misappropriation; overworking; rapine; reckless cutting; ruinous-exploitation; cut-off and get-out; cut-out and get-off; deforestation; forest denudation
滥计成本 inflated cost
廊柱 porch column
廊柱试验 veranda(h) post test
廊状栽植 corridor planting
朗伯正形投影 Lambert conformal projection
朗凯尔系数(木材构造) Runkel coefficient
朗克尔频度定律 Raunkiaer's law of frequency
朗克尔生活型系统 Raunkiaer's system of life-form
浪溅带 splash zone
浪蚀 wave erosion
浪蚀台地 abrasion platform
莨菪苷 scopoline
捞木筏 pontoon

捞木平底船 pond-dozer
捞网 dredge
劳丹脂 labdanum(gum)
劳动费用 labour cost
劳动力 labour force
劳动力的垂直流动性 vertical labour mobility
劳动力周转率 labour turnover
劳动能力 labour capacity
劳动强度 intensity of labour
劳动日 labour day
劳动生产率 labour productivity; productivity of labour
劳动生产率最高成熟 maturity of largest labor productivity
劳动生理学 labour physiology
劳动时间 working time
劳动效率 efficiency of labor
劳动效率差异 efficiency variance
劳动协议 labour agreement
劳动组织 organization of labour
劳恩凯尔频度谱 Raunkiaer's frequency spectrum
劳恩凯尔生活型分类 Raunkiaer's life-form classification
劳工保险 labourer-insurance
劳瑞灵 laureline
劳务 services
劳资纠纷 labour trouble
老成土 ultisol
老鹳草油 geranium oil
老化 ageing
老化过程 ageing process
老化时间 aging time
老化条件 ageing condition
老化质量 aging quality
老茎开花现象 cauliflory
老狼木 wolf tree
老林 old crop
老林分 old height crop; old stand
老林木 old timber
老林土 taiga soil
老龄大径木 veteran
老龄林 pole size timber
老迈年高 senility
老年期(树木) veteran stage
老年土 aged soil; senile soil
老农用林 old field stand
老球根(花卉) back bulb
老熟幼虫 full grown larva
老树 old timber
老树期 old timber stage
老枝插 truncheon
老株 old individual
老竹 old bamboo
烙画 pyrograph
烙画术 pyrography
烙绘饰面 poker-picture
烙焦法 scaring
烙铁 branding iron
烙印(在动物皮肤作标记) branding
涝 waterlogging
耢 drag; planker
酪氨酸 tyrosine
酪氨酸酶 tyrosinase
酪蛋白 casein
酪蛋白纤维 casein fibre
酪朊 casein
酪朊豆胶 casein-soybean glue
酪朊胶 casein glue
酪朊纤维 casein fibre
酪素糊浆 casein-paste
酪素胶 casein glue; casein-paste
酪素胶黏剂 casein adhesive; casein cement
乐果(杀虫剂) dimethoate; rogor
乐器材 wood for musical instruments
勒刀 spur; spur knife
勒克司计 luxmeter
雷暴 thunderstorm
K波段雷达 K-band radar
L波段雷达 L-band radar
X波段雷达 X-band radar
雷达阴影(盲区) radar shadow
雷达自动导航系统 radar automatic navigation
雷电火险 lightning risk
雷电伤害材 lightning-damaged tree
雷电探测系统 lightning detection system
雷电形干裂 lightning shake
雷杜克斯胶合工艺 Redux-process
雷汞(作火帽用) mercury fulminate
雷管 percussion-cap
雷击火 lightning fire; lightning-caused fire
雷蒙德研磨机 Raymond mill

雷姆生带锯 Ramson
雷诺数 Reynolds number
雷雨云 thundercloud
雷阵雨 thunder shower
蕾期可孕性 bud-fertility
蕾期授粉法 bud pollination
累高法(立木材积测定) height accumulation
累积赌注 jackpot
累积扩散指数 cumulative diffusion index
累积数量折扣 cumulative quantity discount
累积损失 aggregate loss
累积误差 accumulated error; aggregate error; cumulative error
累积选择法 cumulative selection
累积叶量 integrated leaf amount
累积影响 aggregate impact
累积雨量计 accumulative raingauge
累积指数 cumulative number
累计材积检尺表 cumulative volume tally sheet
累计成本法 cost-plus method
累计穿孔机 accumulated total punch
累计分布曲线 cumulative frequency curve
累计干旱指数 build-up index
累计耗减 accumulated depletion
累计亏损 accumulated deficit; accumulated loss
累计生长曲线 cumulative growth curve
累计摊销 accumulated amortization
累计折旧 accumulated depreciation
累加器 accumulator
累加因子 additive factor
累进税 progressive tax
累退税 regressive tax
肋 rib
肋部 flank
肋材(造船) ship knee
肋梁 ribbon stripe
肋形盘管 rib-type coil
泪柏二萜 dacrene
泪柏木油 huon pine wood oil
泪柏烯 dacrydene
泪柏型纹孔 dacrydioid pit
泪菌 tear fungus; dry rot fungus
类别 classification; sort; category; guild
类病毒 viroid
类垫状植物 mat-forming plant
类儿茶素 acacatechin
类腐殖物质 humic-like substance
类核 nucleoid
类胡萝卜素 carotenoid
类基因 genoid
类集抽样 cluster sampling
类晶体 paracrystal
类菌原体 mycoplasma like organism
类立克次氏体 Rickettsia(e); Rickettsia(e) like organism
类立克次氏细菌 Rickettsia(e) like bacteria
类囊体 thylakoid
类凝胶 gellike
类染色体 chromosomoid
类水溶剂 waterlike solvent
类蜥蜴爬行动物 lizard
类纤维 fibrid
类型(广义) form(a); mould; type
类型(行为模式) regime (*eg.* temperature regime)
类型图 type map
类脂化合物 lipid(e); lipoid; lipin
棱 arris; rib
棱角 edge angle
棱镜 prism
棱镜测谱仪 prism spectrograph
棱镜角规 optical wedge; prism relascope
棱镜经纬仪 theodolite with prism
棱镜轮尺 prism calliper
棱镜速测镜 prism relascope
棱镜校正 calibration of prism
棱镜校准 prism calibration
棱形木片 prismatic wood chip
棱缘节 arris knot; corner knot
棱柱状结构 prismatic structure
楞场 bank; banking ground; dump; log storage deck; ramp; ramp site; stock yard; timber yard; yard
楞场蓝变(木材缺陷) bridging blue; piling blue
楞场装车工 deck loader
楞地 piled lot; piling site
楞顶盖(单斜面) pitched roof

楞堆 deck; log stack; lot; pile; piled lot; stockpile; timber deck
楞堆边缘木 lead log
楞堆侧面 side
楞堆垫木 skid timber
楞堆方向(前面) facing of pile
楞垛 deck; stack
楞垛层积单位(108ft^3) stack
楞间通道 decking aisle
楞间通路 alley
楞区 log pile area; piling site; stacking area; stacking yard; stock yard
楞台 mill deck
楞台截锯 deck saw
楞台原木检尺员 deck scaler
楞台原木制动器 deck stop
楞腿 brow skid; piling stick; skid
楞腿木 skid timber
楞腿下垫木 head block
楞位 decking aisle
冷白荧光灯 cool-white fluorescent lamp
冷白荧光管 cool-white fluorescent tube
冷变定胶黏剂 cold setting adhesive
冷藏 cold storage
冷藏间 locker
冷藏库 cold storage; refrigerated storage
冷藏箱 congealer
冷层积 cold stratification
冷床 cold frame
冷冻 chilling
冷冻干燥 freeze drying; lyophilize
冷冻过的种子 chilled seed
冷冻剂 cryogen
冷冻脱水 lyophilization
冷度指数 coldness index
冷对流 return convection
冷锋面 cataphalanx
冷固化胶黏剂 cold setting adhesive
冷固性胶 cold setting glue
冷固性胶黏剂 cold setting adhesive
冷光 luminescence
冷挤压 impact extrusion
冷碱纸浆 cold-soda pulp
冷胶合 cold gluing
冷胶合木块 cold-glued block
冷接点 cold junction
冷浸处理 steeping treatment
冷浸法 cold-soaking method
冷楞 cold deck
冷楞装车 cold deck loading
冷磨浆 cold ground pulp
冷磨木浆 cold groundwood pulp
冷凝 condensate; condensation
冷凝弹丸 chilled shot
冷凝干燥窑 condenser (dry) kiln; condensing (dry) kiln
冷凝胶合 chilled (glue) joint
冷凝热 heat of condensation
冷凝水 condensed water
冷凝水返回 condensate feed back
冷凝水集结槽 condensate collecting vat
冷凝物 condensate
冷凝液 condensate; phlegma
冷却池 quencher
冷却度日 cooling degree days
冷却翻板机 board cooler
冷却管道 cooling duct
冷却率温度计 catathermometer
冷却器 congealer
冷却式干燥窑 condenser (dry) kiln; condensing (dry) kiln
冷却水 cooling water
冷却水管 cooling water pipe
冷却塔 cooling tower
冷却液 liquid coolant
冷却装置 cooling device
冷热浸渍法 open-tank process
冷溶浸提 cold-soluble extract
冷溶栲胶 cold-soluble extract
冷杉胶 abieninic balsam; fir balsam
冷杉球果油 fir cones oil
冷杉树脂 fir resin
冷杉香胶 abies balsam; Chinese balsam
冷杉油 fir (needle) oil
冷试法 cold test
冷适应 cold adaptation
冷室 cool house
冷水浸种 cold soak(ing)
冷水性世界种 cold water cosmopolitan
冷水种 cold-water species
冷吸法 enfleurage

冷压 cold press; cold pressing
冷压机 cold press
冷压胶合 cold gluing
冷硬化 cold hardening
冷硬化剂 cold hardener
冷用杂酚油 cold-application creosote
冷浴处理 cold bath treatment
冷浴法(防腐) cold-soaking method
冷预压 cold-prepressed
狸 badger
离岸的 offshore
离岸价格 free on board
离岸流 offshore current
离瓣花类 polypetalae
离瓣式 polypetaly
离材堆空气 leaving air
离层 absciss layer; abscission layer; delaminating; separation layer
离差 deviation
离巢雏鸟 fledgling
离巢性 nidifugity
离巢幼龄动物群 creche
离地间隙 ground clearance
离萼的 chorisepalous
离萼片的 polysepalous
离伐 isolation; isolation cutting; severance felling; severance cutting
离缝(胶合板缺陷) open glue joint; gap
离缝组合柱 spaced column
离管薄壁组织 apotracheal parenchyma
离管带状薄壁组织 apotracheal banded parenchya
离管星散聚合薄壁组织 apotracheal diffuse-in-aggregate parenchyma
离合鼓 clutch drum
离合器操纵机构 clutch lever mechanism
离合器弹簧 clutch spring
离解机 deflaker
离散变量 discrete variable
离散的 discrete
离散度 dispersion
离散分析 analysis of variance
离散过程 dispersive process
离散世代 discrete generation
离散随机变量 discrete random variable
离散相关 discrete correlation
离散型模型 discrete model
离散游 discrete spectrum
离色 bleeding
离生雌蕊 liberate pistil
离生的 liber; schizogenetic; schizogenous
离生心皮 liberate carpel
离生雄蕊 adelphia (*plur.* -phiae)
离生雄蕊的 adelphic; adelphous
离生子房 liberate ovary
离室药 distractile anther
离体 in vitro
离体胚 excised embryo
离体胚测定 excised embryo test
离体胚培养 culture of excised embryos
离体培养 explantation
离析 macerate
离析木材 macerated wood
离析作用 maceration
离心管 centrifugal tube
离心花序 centrifugal inflorescence
离心机 centrifuge
离心净浆机 centrifugal cleaner
离心力 centrifugal effort
离心力干燥[法] centrifugal seasoning
离心力干燥法 centrifugal (force) drying
离心木质部 centrifugal xylem
离心喷胶法 centrifugal spraying method
离心喷射 centrifugal jet
离心皮 apocarpy
离心皮雌蕊 apocarpous pistil
离心生长 excentric growth
离心式风机 centrifugal fan
离心式鼓风机 centrifugal blower
离心式喷胶机 centrifugal glue sprayer
离心式通风机 paddle fan
离心式雾化头 centrifugal atomizing head
离心脱水机 centrifugal dehydrator
离心作用 centrifugation
离蛹 free pupa; liberal pupa
离真概率 probability of discrepancy

L

离中趋势 dispersion
离子半径 ionic radius
离子泵 ion pump
离子对抗作用 ion antagonism
离子化 ionization
离子活动性 ionic mobility
离子活度 ionic activity
离子交换 ion exchange
离子交换层析 ion-exchange chromatography
离子交换反应 ion-exchange reaction
离子交换纤维素色谱法 ion exchange cellulose chromatography
离子流 ionic flow
离子平衡 ionic balance
离子强度 ionic strength
离子取代作用 ion substitution
离子水化作用 ion hydration
离子碎化反映 ionic fragmentation reaction
离子梯度 ionic gradient
离子调节 ionic regulation
离子载体 ionophore
梨果 pome
梨果宿萼 eye
梨莓树 limette
梨莓油(调味) lime oil; limette oil
犁 plough; plow
犁把 ploughstaff; ploughtail
犁壁 breast
犁柄 ploughstaff; ploughtail
犁刀 coulter
犁底层 plough pan
犁沟播种 lister planting; sowing in furrow
犁沟底 sole
犁沟移植 furrow transplanting
犁沟栽植 lister planting
犁过的土地 plough; plow
犁铧 ploughshare; share
犁路 plough line
犁体 body
犁头式拌胶机 ploughshare mixer
犁托 frog; ploughshoe
犁辕 beam
犁柱 leg
篱笆 fence; wattle; wattle-fence
篱架式整枝 espalier
篱列 hedgerow
篱柱 fence post
李拔克防腐法 Lebacq process
李比希木材腐朽理论 Liebig theory of decay
李比希最低限定律(木材防腐) Liebig's law of minimum
李奇修枝工具 Rich pruning tool
李属斑纹 plum mottle
李属团状花纹 plum-pudding design
里程表 trochometer
里程计 pedometer
里克求积式 Riecker's lbrmula
里网 backing wire
浬(海里, 等于1852m) nautical mile
理财成熟 economical maturity
理财收获(林业上的) internal rate of return; money yield
理论上之无立木地 theoretical clearfell area
理想含水状态 ideal moisture condition
理想立木度 normal stocking
理想林 ideal forest
理想林模式 ideal forest model; normal forest model
理想流体 ideal fluid
理想剖面 ideal profile; idealized profile
理想蓄积 ideal growing stock; ideal standing volume; normal stocking
理想择伐作业 ideal selection system
力偶 couple
力形材堆 box pile
力学的不平衡 mechanical nonequilibrium
力学效率 mechanical efficiency
力学性质 mechanical property; strength properties
力学阻力 mechanical resistance
历史生物地理学 historical biogeography
历史园林 historical garden
历史植物地理学 historical plant geography
立标(标示位点或栽植点) staking out; pegging
立标杆架线路 poling
立带式砂光机 vertical head sander
立底 action face
立地 locality; site

立地尺度(垂直异质性) topological dimension (vertical heterogeneity)
立地级估计 site class estimation
立地类型 site type; site-type class
立地类型分类 site type classification
立地类型图 site map
立地描述 destcription of site; site description
立地评价 site evaluation
立地强度 stability of stand
立地生产力 site quality
立地-树种组台 site-species combination
立地条件记载 description of locality
立地条件确定 site determination
立地调查 site apparaisal; site apparaisement; site mapping; site survey
立地图 site type map
立地稳定性 stability of stand
立地因子 locality factor
立地指数 site index
立地指数表 site index table
立地指数级 site index class
立地指数曲线 site index curve
立地质量 site quality
立方单位量度(材积) cubic measure
立方吨(锯材为25~50ft^3, 圆材为25~50霍普斯英尺) cubic ton
立方米 stere
立方容积 cubical content
立方英尺材积表 cubic foot volume table
立方英尺检尺 cubic-foot scaling
立方英寻(材积单位) fathom
立管 downpipe
立即采伐 quick liquidation
立枯 standing death
立枯病 damping-off
立枯法 girdle
立木 standing timber; standing tree; stumpage; timber; body
立木不足林分 subnormal-under stocked stand
立木材积 stand volume; standing volume; tree volume
立木材积表 stand volume table; tree volume table
立木侧倾 side lean
立木成本 stumpage
立木成本价 cost value of standing timber
立木出卖 sale of standing trees
立木出售 sale of standing timber; sale on the stump; stump sale
立木登记簿 stock book
立木等级 tree grades
立木地 stocked area; stocked land
立木度 degree of stocking; density of stocking; stocking; stocking percent; crop density
立木度不足 deficient stocking; under stocking; understocked
立木度适中的 well stocked
立木度完满的 full-stocked
立木度稀疏 poor stocked
立木防腐 treating living tree
立木腐朽 decay of living tree
立木估价 valuation of stumpage
立木估价中的利润比 profit ratio in stumpage appraisal
立木过密林分 overstocked stand; abnormal overstocked stand
立木环状剥皮干燥法 girdle
立木火灾保险 standing timber fire insurance
立木季节含水率 seasonal water contents of living tree
立木价 forest charge; stand value; standing timber value; value of a stand; value of stand; value standing
立木价的未来增长 future increase in stumpage price
立木价格 standing timber price; stumpage; stumpage price
立木检尺 tree scale
立木买主 purchaser of standing tree
立木密度 stem density; stock density; stocking density
立木评价 appraisal of stumpage
立木山价 stumpage price; stumpage value
立木数 number of stems
立木调查 timber cruise
立木图解 diagram of tree
立木完全出卖 sale by area
立木位置图 DBH distribution
立木稀疏 reduction in stock

立木下倾(向下坡倾斜) head lean
立木销售 stand sale; stumpage sale
立木销售市场 stumpage market
立木蓄积 growing stock; standing stock; stocking; volume of standing timber
立木蓄积调整 growing stock regulation
立木蓄积一览表 compartment stand and stock table
立木蓄积资本 capital growing stock
立木样方区 stocked quadrat
立木原价 stumpage charge
立木株数 number of stems
立木注药防腐法 capping method
立式(射击) standing
立式车床 turning-lathe
立式成形铣床 spindle moulder
立式带锯 vertical band saw
立式釜 vertical retort
立式挤压机 vertical extruder
立式挤压刨花板 Okal chipboard
立式绞车 capstan; capstan winch
立式绞盘 capstan winch
立式绞盘机 capstan
立式开榫机 vertical spindle tenoner
立式刨切单板机 vertical veneer slicer
立式刨切机 vertical slicer
立式砂光机 spindle sander
立式镗床 vertical boring machine
立式圆盘刨片机 vertical disk type flaker
立式圆盘削片机 vertical disk type chipper
立式贮料仓 vertical storage bin
立式钻床 vertical boring machine
立体测摄 stereoscopic measurement
立体测图仪 stereoscopic plotter
立体测图仪器 stereo plotting instrument
立体测微仪 stereomicrometer
立体的 stereoscopic
立体地图 relief-map
立体风景画 stereoscopic view
立体覆盖 stereoscopic coverage
立体观测 stereoscopy
立体观察 stereoscopic vision
立体观察练习 sausage exercise
立体交叉 grade separation
立体景观 stereoscopic view
立体镜 stereoscope
立体镜基线 stereoscopic base
立体镜效应 stereoscopic effect
立体可变焦距转绘仪 stereo zoom transfer scope
立体量测仪 stereometer
立体门 flush door
立体三角测量 stereotriangulation
立体调查 stereoscopic examination
立体显微镜 stereoscopic microscope; stereomicroscope
立体像对 stereo (scopic) pair
立体像片 stereoscope picture
立体像片对立体摄影测量 stereopair
立体效应 stereoscopic effect
立体影像 stereoscope imagery
立体影像转换器 stereo image alternator
立体照片法 stereogram method
立铣 end milling
立约出卖 sale by private contract
立竹 living bamboo
立柱机 pole press
利比希最低量法则 Liebig's law of the minimum
利链菌素 streptolydigin
利率 interest rate; rate of interest
利润 margin
利润分成 profit sharing
利润及风险边界 margin for profit and risk
利润率 profit margin; return; return rate
利润面 profit area
利润压缩 profit squeeze
利润总额 gross profit
利他行为 altruism
利益相关者 stakeholder
利益值 benefit value
利用 exploitation; utilisation; utilization
利用材积 economic volume
利用材积生长量 merchantable timber increment
利用材积收获表 assortment yield table
利用度 exploitability
利用价 exploitable value; exploitation value
利用价值 utilization value

利用竞争 exploitive competition; scramble competition
利用率 exploitation percent; harvest percent; use percent; using percentage; utilization percent; yield; yield percent
利用率法 rational method; utilization percent method; yield determination using harvest percent
利用期 utilization period
利用期间 working period
利用强度 intensity of utilization
利用系数 utilization coefficient
利用许可证 utilization permit
利用种类 exploitation method
沥滤 leach
沥滤场 leaching field
沥滤器 leach
沥滤试验 leaching test
沥滤液 leaching liquor
沥青 asphalt; bitumen; pitch
沥青处理(防腐) asphaltic process
沥青膏 bituminous cement
沥青灰岩 anthraconite
沥青混合剂 bituminous compound
沥青胶黏剂 bituminous cement
沥青焦炭 pitch coke
沥青浸注绝缘板 bitumen-impregnated insulating board
沥青玛碲脂 bituminous cement
沥青黏合绝缘板 bitumen-bonded insulating board
沥青水泥 bituminous cement
沥青纸板 bituminous board
例行维修 routine maintenance
例外条款 escape clause
栎林 oakery
栎木 oak wood
栎木栲胶 oak (wood) extract
栎木拼花地板 oak parquet
栎树矮林 oak coppice-wood
栎树剥皮林 oak-bark coppice
栎树虫瘿 oak apple
栎树丛 oak grove
栎树皮栲胶 oak bark extract
栎树皮鞣法 oak bark tannage
栎树群丛(拉丁) quercetum (*plur.* -ta)
栎型射线 oak-type ray
栗酚苷 chesnatin
栗钙土 castanozem; chestnut earth; chestnut soil
栗壳变色 iron-tannate stain
栗木单宁 chestnut bark tannin
栗木栲胶 chestnut extract
栗木鞣质 chestnut bark tannin
栗木素 castalin
栗木蛀虫 chestnut pole borer
栗球蛋白 castonin
栗色的 castaneous; spadiceous
栗疫病 chestnut blight
砾 pebble
砾漠 serir
砾石 gravel
砾石荒漠群落 rupideserta; saxideserta
砾石群落 petrium
砾土 lithosol; litho-soil
砾岩 conglomerate
砾质壤土 gravelly loam
砾质砂壤土 gravelly sandy loam
砾质土 chisley soil; fragmental soil; gravel ground; gravel soil; gravelly soil; rubby soil
砾质岩 psephite; psephyte; rudite; rudyte
粒度 grain; grain size; granularity; granule size; grit; particle size
粒度比 fineness ratio
粒度分析 grading analysis
粒粪 pellet
粒化 graining; granulation
粒化作用 granulation
粒级 granule size
粒雪 firn
粒渣紫胶 molamma
粒重 grain weight
粒状结构 granular structure
粒状栲胶 granular (tannin-)extract
粒状片麻岩 granular gneiss
粒状闪岩 granular amphibolite
粒状石灰岩 granular limestone
粒子 corpuscle
粒子大小 particle size
连测点 tie point

连带产品 joint product
连带需要 joint demand
连萼瘦果 cypsela; epiachene
连杆 connecting link; connecting rod; reach
连杆机构 connecting rod gear
连杆台车 Seattle car
连杆拖车 runner
连根拔 root up; chunk out
连根伐 grub felling; root up; stub out; stump out; pull felling
连根挖除 stubbing
连贯节 coherent knot; firm knot; fixed knot; light knot
连贯性 coherence
连旱指数 build-up index
连旱作用 build-up
连枷式刀片 swinging blade
连接 coherence
连接板(连接牵引索) section plate; shaw
连接薄壁细胞 conjugate parenchyma cell
连接边缘 jojnt flange
连接编辑程序 linker
连接度 connectivity
连接耳板(绞盘机座的) sled plate
连接法兰 jojnt flange
连接杆 beam
连接钩 butt hook
连接滑轮 monkey block
D形连接环 dee
连接件(家具) connecting fittings
连接件(接头) connector; linker
连接链(拖拉机集材等) butt chain; tag chain
连接酶 ligase
连接片(锯链) tie strap
连接器 coupler; hook
连接索(各种) connecting line; connecting rope; tag; tag line
连接索(牵引索和回空索之间) butt rigging
连接索(牵引索与捆索之间) skidding extension rope
连接套 connecting sleeve
连接套管 adapter sleeve
连接套筒 joint sleeve
连接细胞 conjugate cell
连接销 pivot
连接轴 connecting shaft
连接装置 bridle
连接组织 conjunctive tissue
连结杆 tie rod
连结件 tie
连结索 stub line
连年材积生长量 current annual volume growth
连年采伐 annual cutting; annual felling
连年的 current annual
连年价格增长 current annual value increment
连年价格增长率 current annual value increment percent
连年经营收益 annual supervisory return
连年利率 current annual rate of interest
连年商品材收益 annual stumpage return
连年生长 current annual growth
连年生长量 current annual growth; current annual increment
连年收获 current annual yield; current yield
连年收获作业 current yield system
连年收利率 current annual forest percent; indicating percent
连年正常估损量 normal annual mortality
连年作业 annual working
连橇 twin-sled; two sleigh
连上一个碳原子的 primary
连生节 adhering knot
连锁 linkage
连锁编辑 linkage editor
连锁关系 linkage relationship
连锁群(基因) linkage group
连锁商店 chain store
连锁图 linkage map
连锁系数 linkage coefficient
连锁银行 chain banking
连锁与交叉法则 law of linkage and crossover
连锁折扣 chain discount
连锁指数 chain index
连体生物 parabiont
连续变量 continuous variable

连续变形微胞 continuous deformed micelle
连续变异 continuous variable; continuous variation
连续波 continuous wave
连续采伐 successive cutting; successive felling
连续采伐带 successive strip
连续抽样 sampling at successive occasion
连续带式成型系统 continuous forming belt system
连续带式热压机 continuous band hot press
连续带状采伐 felling in progressive strips
连续带状皆伐 clearcutting in progressive strip
连续带状皆伐式渐伐作业 progressive clear-strip system
连续带状皆伐天然更新法 clearcutting in progressive strip
连续带状摄影机 continuous strip camera
连续单板带 continuous ribbon; flow of veneer
连续的 serial
连续动作式计算机 linkage computer
连续放牧(允许全年放牧) all-year grazing; continuous grazing; year-long grazing
连续分布区 continuous areal
连续干燥法 full time drying system
连续更新 successive regeneration
连续更新采伐 successive regeneration cutting; successive regeneration felling
连续更新作业 successive regeneration system
连续供料剥皮机 continuous feed debarker
连续观察样方 continuous quadrat
连续及平均年生长量 current and mean annual growth
连续集材 hot skidding
连续胶合 continuous gluing; progressive gluing
连续进料 continuous feeding
连续决策 sequential decision making
连续开沟植树机 continuous furrow planter
连续开花 perpetual bloom; perpetual flowering
连续开花的 everblooming
连续勘察 continuous cruise
连续林木覆盖 continuous tree cover
连续流动干燥机 continuous-flow drier
连续灭菌 continuous sterilization
连续盘存法 continuous inventory method
连续培养 continuous culture; open culture
连续喷涂防腐法 deluging
连续喷雾 continuous mist
连续清查 successive inventory; sequential inventory
连续区 continuum
连续区指数 continuum index
连续曲线式索道 continuous curved cableway
连续群 continuum; linkage group
连续群落 continuous community
连续热磨机 caterpillar grinder
连续森林调查 continuous forest inventory
连续森林调查法 continuous forest inventory with permanent samples
连续上胶 continuous gluing
连续生产线 in-line system
连续施胶 progressive gluing
连续世代 successive generation
连续式拌胶机 continuous (type) mixer
连续式打浆机 continuous beater
连续式单板干燥机 conveyer (veneer) dryer
连续式干燥窑 continuous (dry) kiln
连续式进料系统 continuously working feed system
连续式纤维板干燥机 continuous fibreboard dryer
连续式压板干燥机 progressive plate dryer
连续式压机 continuous press
连续式预压机 continuous prepress
连续式圆筒剥皮机 continuous barking drum
连续式蒸煮器 continuous digester
连续送料 friction feeding

连续随机变量 continuous random variable
连续踏查 continuous cruise
连续稳定性群落 continuously stable community
连续循环式索道 continuous circulating cableway
连续压条 continuous layering
连续载荷 continuous loading
连续资源清查体系 continuous inventory system
连续作业干燥窑 progressive (drying)kiln
连着 cohesion
连作 continuous cropping; repeated tillage
帘式淋涂 curtain coating
莲 Chinese water-lily; lotus
莲藕 lotus root
莲座叶丛 rosette
莲座叶植物 rosette plant
莲座状 rosette
涟纹 ripple mark
联邦及各州联合防火系统 federal-state fire control system
联邦林 federal forest
联邦林务局 federal forest service
联邦税收法 federal revenue law
联邦种子条例(美) Federal Seed Act
联苯胺染色试验 benzidine stain test
联动计算机 linkage computer
联动式集材机 interlocking skidder
联动式卷筒 interlocking drum
联动装置 interlocking gear
联拱廊 arcade
联合(企业机构等) associate; association; conjugate; federation; pool; integration
联合大企业 conglomerate
联合割灌机 bush combine
联合国气候变化框架公约 United Nations Framework Convention on Climate Change
联合机臂杆伐木半径 tree boom cut radios
联合经营 joint venture
联合流送(木主不同) union drive
联合履行 joint implementation
联合群丛 alliance association
联合式裁边机 combination edger
联合销售 combination sale
联会消失 desynapsis
联机操作 on line operation
联机信息处理 on-line message processing
联接 anastomosis
联结杆(车辆或爬犁的) gooseneck
联结架 headstock
联结孔 eye hitch
联结梁 binding-beam
联结门 braced door
联结器 coupling
联生的 adnate
联生叶的 symphyllous
联锁 interlock
联锁棒 lock bar
联锁董事会 interlocking directorate
联锁机构 interlock arrangement
联锁式卷筒(索道) interlocking drum
联锁装置 interlock arrangement
联系种 binding species
联线操作 on line operation
联线实时处理器 on-line real time processor
联营商店 chain store
联运单据 combined transport document
联运提单 combined transport bill of lading
联轴节 coupler; coupling
联珠线脚(转角处的) return bead
廉价拍卖 loss leader
廉价抛售 distress selling
镰刀式打枝机 cradle type delimber
镰形的 falcate
镰形聚伞花序 sickle-like cyme
镰状的 falcate
镰状聚伞花序 drepanium; sickle cyme
脸盖(蜻蜓若虫) mask
脸面 face side
脸盆架 washstand
链板法测定纵断面(索道) chain board profile
链板式输送机 apron conveyor
链板输送机 flight conveyor
链传动 chain drive; sprocket-and-chain drive

链锤式打枝装置 flail type delimbing device
链刀式打枝机 knifebelts limbing device
链斗挖掘机 continuous bucket excavator
链钩(挂于链环) grab hook
链钩环 grab link; grab ring
链环 link
链击式剥皮机 chain-flail barker
链击式打枝机 chain-flail delimber
链接编辑程序 linkage editor
链节 link; segment
链锯 chain saw; endless chain saw
链锯导板 chain bar; chain blade; chain guide; cutter bar; rail; sword
链锯护罩 saw protector
链锯前端 stinger
链锯切齿片 cutting plate
链锯式伐木头 chain saw felling head
链捆集材 chaining
链轮 chain pulley; chain sprocket; sprocket wheel
链轮式导板头(油锯) sprocket nose
链霉溶菌素 streptolydigin
链霉素 streptomycin
链片 link
链片脱落(受敲击) link peened
链片踵部磨损 heel wear
链式剥皮器(机) chain peeler
链式打眼机 chain-saw mortising machine
链式打印机 chain printer
链式多层选材装置 chain tray sorter
链式开榫机 chain mortiser
链式连续磨木机 continuous chain grinder
链式磨木机 caterpillar grinder; chain grinder
链式手锯 chain hand saw
链式输送机 chain conveyor; chain flight conveyer; chain shifter; chain transfer; moving-slat conveyor
链式榫槽机 chain-saw mortising machine
链式压机 chain press
链式原木输送机 log chain conveyor
链式装载机 chain loader
链条 chain
链条集材 chain skidding
链条输送机 jack chain
链武剥皮机 chain type barker
链线 chain line
链终止密码子 chain-terminating codon
链状分子 chain molecule
链状细胞 chainlike cell
楝树碱 azedarine
良姜酮 alpinone
良种 high-quality seed
良种繁育 elite breeding; stock breeding
凉亭 alcove; arbour; kiosk; pavilion; pergola; summer-house
梁 beam; square beam
梁材 balk-timber; stringer
梁端垫块 template
梁木 balk
梁托 corbel
椋鸟 starling
粮食危机 food insecurity
两边带屉写字台 knee-hole desk
两边刨光材 dressed two edges
两侧对称 bilateral symmetry; lateral symmetry
两侧对称的 zygomorphic; zygomorphous
两端处理 end-for-end treatment
两端卡木跑车 end dogging carriage
两段环接绷索 split guy
两段集材 cold skidding
两伐期矮林 coppice of two rotations
两个轮伐期的矮林作业 two-rotation coppice (system)
两级压缩 two-stage compression
两阶抽样 two-stage sampling
两裂片的 bilobate
两轮同数的(雄蕊) isadelphous
两面光(人造板) smoothed two sides; good two sides
两面光硬质纤维板 both surface smooth hardboard; hard-board
两面刨光材 dressed two sides; surfaced two sides
两面气孔的(上下表皮皆有气孔) amphistomatic
两面下锯法 two-faced sawing

L

两面一边刨光材 dressed two sides, one edge; surfaced two sides, one edge
两年生的 biennial
两年生植物 biennial plant
两栖的 amphibious
两栖动物 amphibian
两栖昆虫 amphibiotica
两栖植物 amphibious plant; amphiphyte
两蕊柄 gynandrophore
两室的 bilocular
两属的 bigeneric
两丝期 amphitene stage
两头沉写字台 knee-hole desk
两头钉 dowel pin
两头鹤嘴锄 double-pointed hoe
两系杂种 bilinear hybrid
两性雌蚜 amphigonic female
两性单宁 amphoteric tannin
两性的 ambisexual; amphiprotic; amphoteric; bisexual; digenetic; hermaphroditic
两性个体 amphigous individual
两性花 hermaphrodite flower; monoclinic; monoclinous flower
两性花的 monoclinous
两性胶体 ampholytoid; amphoteric colloid
两性离子 amphion
两性群 bisexual group
两性融合 amphimixis
两性生殖 amphigenesis; amphigony; syngamy
两性生殖的 amphigenetic
两性生殖族群 amphimictic population
两性世代 hermaphroditic generation
两性同序的 digamous
两性异形 sex dimorphism
两眼间距 inter-ocular distance
两用的 dual-function
两用斧(防火) Pulaski tool
亮度 brightness; brilliance; luminosity
亮度迭加性 additivity of luminance
亮度范围(摄影) brightness range
亮度温度 brightness temperature
亮光涂饰 gloss finish
亮色针叶树(反射率高) light conifer
亮油 gloss oil
量测 measurement
量测法 method of measurement
量程 range
量尺 measuring; measuring for bucking; measuring off; measuring stick
量尺及标号(造材前) fit(ting)
量度 measurement
量纲 dimension
量规(计) gauge
量角仪 angulometer
量热器 calorimeter
量水堰 gauging weir; measurement weir
量筒 measuring cylinder
量雪器 nivometer
量雨器 precipitation gauge
量云器 nephometer
量载规 loading gauge
量子 quantum
量子产量 quantum yield
量子生物学 quantum biology
量子式进化 quantum evolution
量子效率 quantum efficiency
量子需要量 quantum requirement
晾干 air-dry; airing
疗养胜地 health resort
瞭望点(防火) lookout point
瞭望护林员 lookout fireman
瞭望室 lookout cabin; lookout cupola; lookout room
瞭望死角 blind area
瞭望塔 lookout tower; tower cupola
瞭望台 lookout observatory
瞭望台联络图 seen-area map; visibility map; visible-area map
瞭望调度 lookout dispatcher
瞭望系统 lookout system
瞭望员 lookout; towerman
瞭望站 lookout; lookout house; lookout station
料板子 setting key
料仓 bin; magazine
料仓壁 bin wall
料仓出料装置 silo outfeed device
料仓式裁分机 pocket edger
料斗 bin; gondola; magazine

撂荒 abandonment
撂荒地 abandoned land; black fallow
撂荒地造林 field abandoned stocking
撂荒灌丛 bush fallow
列表抽样 list sampling
列当 broomrape
列联表 contingency table
列宁格勒单位(锯材材积计量单位, 等于4.672m^3) Leningrad standard
列线图 alignment (alinement) chart; nomograph
列线图法 alinement chart method
列线图解法 nomography
列照小区 control plot
列植 line-planting; planting in row
列柱 colonnade
劣等材 wrack
劣地 badland
劣苗 cull(ed) seedling
劣生(有害遗传特性) cacogenenesis
劣生的 dysgenic; kakogenic
劣势地位 subordination
劣势木 dominated stem; subordinate tree; inferior tree
劣性 pessimum
劣质林分 minus stand
劣质树 minus tree
劣种 rogue
烈火 running fire
烈性噬菌体 virulent phage
猎场 chase
猎房 hunting cabin
猎狗 hound
猎狐犬 foxhound
猎獾狗 basset
猎获 bag
猎获量 bag; game cropping
猎具 hunting tackles; hunting-instrument
猎鹿犬 deerhound
猎枪 fowling piece; shot-gun
猎区 hunting-field; hunting-ground
猎区主 game-preserver
猎犬 sport(ing)-dog
猎舍 hunting-box
猎兔犬 beagle
猎物 game animal; prey; quarry
猎物被捕量方程 disc equation
猎物飞走速率 speed of quarry
猎物栖息处 game habitat
猎物隐蔽 game cover; game park
猎物隐蔽处 game preserve
猎鹰 falcon
猎鹰训练 falconry
猎用步枪 sporter; sporting rifle
猎用枪支 firearm
裂 cleft
裂变 fission
裂成五份的 quinquefid
裂度 divergent angle of check
裂断长 fracture length
裂缝 crack; fissure; score; seam
裂缝状树皮 fissured bark
裂构 degradation
裂谷 rift valley
裂果 dehiscent fruit
裂化 cracking
裂环 split ring
裂节 checked knot; open knot
裂解(质谱) fission
裂开 rupture
裂开散播 ballochore
裂开散布 bolochory
裂口 slash
裂榄树脂 chibou
裂片(果) valve
裂片(叶片) lobe; lobus
裂片状的 lobate
裂区设计 split-design
裂区试验 split-plot experiment
裂溶生 schizo-lysigenous
裂溶生创伤树脂道 schizo-lysigenous traumatic canal
裂溶生树脂道 schizo-lysigenous duct
裂溶生细胞 schizo-lysigenous cell
裂生[作用] schizogenesis
裂生胞间腔 schizogenous intercellular cavity
裂生胞间隙 schizogenous intercellular space
裂生创伤树脂道 schizogenous traumatic canal
裂生的 schizogenetic; schizogenous
裂生树脂道 schizogenous canal; schizogenous duct; schizogenous resin canal
裂生细胞 schizogenous cell

L

裂生子叶 schizocotyledon
裂纹 check; craze; flaw
裂纹方向 direction of check
裂隙 crack; crevice
裂隙状单纹孔 slit-like simple pit
裂隙状具缘纹孔 slit-like bordered pit
裂叶(病毒病害症状) ragged leaf
裂殖菌 schizomycete
裂殖体 agamont
裂殖植物 schizophyte
鬣山(鬣 liè) cuesta
邻苯二酚酶 catecholase
邻苯基苯酚 ortho-phenylphenol
邻二氯苯(防虫防蚁剂) ortho-dichlorobenzene
邻接分析 adjacency analysis
邻接皆伐带 adjoining cleared strip
邻接木 adjoining wood
邻接效应 effect of neighbours
邻近分离 adjacent segregation
邻近流送河道的林区 front
邻里公园 neighborhood park
邻式关系 neighborhood
邻位效应 ortho-effect
邻域性物种 pjarapatric speciation
林班(森林区划) compartment; division; forest compartment
林班登记簿 compartment history
林班记录 compartment history; compartment register
林班间小道 section line
林班界 compartment boundary
林班界线 ride
林班境界线 section line
林班区划 division into compartment
林班图 compartment map
林班线 compartment line; cut-through; lane
林班小道 lane
林班株数 compartment stand and stock table
林班作业 stand-method
林班作业法 compartment system
林层 storey
林产盗窃 theft of forest-produce
林产化工产品 silvichemicals
林产化学 forest chemistry
林产化学产品 forest chemical product
林产化学加工 chemical processing of forest products
林产品 crops; forest product; forest production
林产物 forest commodities; forest produce; forest products
林产物利用 utilisation; utilization
林产业 wood processing industry
林场 forest farm; wood-run
林场经营档案 management archives of forest farm
林带 forest belt
林丹(农用杀虫剂) lindane
林道 forest road; haul road; lumber road; wood road
林道规划 road layout
林道计划 forest road contruction plan; roading plan
林道间距 road spacing
林道密度 forest road density
林道旁造材归堆机 strip-road processor
林道网 forest road net; forest road network; forest road system; road network
林道系统 forest road system; road network
林道选线 selection of route
林道运材 on-road hauling
林道支线 branchline forest road
林德曼比率 Lindeman's ratio
林德曼效率 Lindeman's efficiency
林地(林业用地) forestry land; land for forestry
林地(有林地) timber-land; wooded land; forested land; forest land
林地百分率 forest percent
林地采草权 removal sod right
林地常居动物 patowle
林地纯收益最大的轮伐期 highest soil rent rotation
林地登录 stand register
林地地表 forest floor
林地动物 patobiont
林地改良 forest amelioration
林地规划 forest subdivision
林地毁灭的 stand-replacing
林地价 soil expectation value
林地价值 cost value of forest soil
林地勘测 cruise

林地可利用限度 forest land capability
林地利用分类 forest land-use class
林地利用规划 forest land-use planning
林地利用图 forest land-use maping
林地面积 wooded area
林地偶居动物 patoxene
林地评价 valuation of forest land
林地期望价最大的轮伐期 highest soil expectation value rotation
林地清理 ground clearance
林地清理机 clearing machine; land clearing machine
林地清理机械 site preparation machinery
林地区划网(森林区划线总称) network of accessory frame
林地生产力 productive capacity on area
林地剩余物 forest residue
林地施肥 forest fertilization
林地税 forest-land tax
林地土壤改良 forest land amelioration
林地租约 timber-land lease
林分 crop; forest crop; forest stand; growth; land parcel; parcell; stand
林分边界 edge of a stand; stand edge
林分边缘保护 edge protection
林分表 stand table; stock table
林分表法 stand table method
林分表推算法 stand-table projection
林分表预估法 stand-table projection
林分材种出材量测定 stand merchantable volume measurement
林分测高法 height measures of stand
林分成长龄级 stand development class
林分出材率 outturn of stand
林分垂直结构 vertical stand structure
林分等级 stand-class
林分断面积调查仪 relascop
林分断面图 cross-section of a stand; stand profile
林分发育 development of a stand
林分法正配置 normal arrangement of stand; normal distribution of stand
林分分类 stand classification
林分抚育 care of stands; stand tending
林分改造 stand conversion; stand improvement
林分高度 stand height
林分更新方程 stand renewal equation
林分管理 management on a stand basis; stand management
林分航空材积表 stand aerial volume table
林分混交 mixed by stands
林分级 stand size class
林分记录簿 crop description; stand assessment
林分记载 stand description
林分记载检索分类卡 stand description key sort card
林分价值 stand value; value of a stand
林分建造 crop establishment; establishment (of a stand); stand establishment
林分结构 stand structure
林分结构表 stand table
林分界 stand boundary; sub-compartment boundary
林分经济法 management by compartment
林分经营 management on a stand basis; stand management
林分经营法 compartment management method; individualizing method; management by compartment
林分经营计划 stand management planning
林分类型 stand type
林分类型图 type map
林分历史 history of stand
林分龄级组成 stand age class organization
林分密度 crop density; density of crop; density of forest stand; density of stocking; stand density
林分密度级 stand density class
林分密度控制图 stand density control diagram
林分密度特征曲线 characteristic curve of stand density

林分密度指数 stand density index
林分面积龄 stand area-age
林分面积调查仪 relascop
林分描述 stand description
林分年龄 age of stand; crop age; stand age
林分平均高 average height of stand
林分平均年疏伐度 average annual stand depletion
林分评价 stand valuation
林分剖面 cross-section of a stand; stand profile
林分期望价 stand expectation value
林分起源 origin of stand; stand origin
林分区 area
林分生物量 stand biomass
林分生育期(阶段) development of a stand
林分生长 growth of stand; stand growth
林分生长指标 stand-grow index
林分收获模型 stand yield model
林分疏开 opening-up of a stand
林分树高分布 stand height distribution
林分树高结构 stand height structure
林分数据 stand data
林分水平结构 horizontal stand structure
林分速测镜 Bitterlich-relascope
林分特性估测 estimation of stand characteristics; stand estimation
林分特征 stand characteristics
林分调查 stand survey
林分调查因子 stand description factor; stand description factors
林分图 stand map
林分稳定性 stand stability
林分信息 stand information
林分形数 forest form factor; stand form factor
林分形状高 form height
林分胸高断面积 stand basal area
林分蓄积 stand volume
林分蓄积分类 growing stock volume classification
林分蓄积级 stand size class
林分蓄积量表 stand volume table
林分蓄积量公式 stand volume equation
林分蓄积增长量 stand volume increment
林分演进 stand development
林分直径分布 stand diameter distribution
林分直径结构 stand diameter structure
林分质量 quality of stand; stand quality
林分株数调查 enumeration
林分资源清查 inventory of stand; stand inventory
林分组成 constitution; species composition; stand composition; stand structure
林分作业 stand-method
林副产品 forest by-product; minor forest produce
林覆盖率 percent forest cover
林耕法 taungya; taungya method; taungya system
林管局 district administration
林管区(若干营林区组成) district of forest-ranger; district of forest-warden; district
林管区主任 ranger
林冠 canopy; forest canopy; leaf conopy
林冠层 canopy class; storey
林冠覆盖 canopy cover
林冠覆盖级 canopy cover class
林冠覆盖面积 area under (crown) cover; area under canopy
林冠级 canopy class
林冠截留 canopy interception
林冠面 canopy surface
林冠内光线 inside-light of canopy
林冠披覆 cover
林冠疏开 opening of canopy; opening-out of leaf canopy
林冠疏透度 crown openess
林冠树木 canopy tree
林冠下清除 lifting the canopy; lifting the canopy base
林冠郁闭 canopy closure; closed canopy
林化产品 forest chemical product
林火 forest fire
林火等级 forest fire size class
林火定位图 fire plotting map
林火防护 forest fire protection

林火控制 forest fire control
林火蔓延 fire spread
林火蔓延距离 run
林火破坏力 forest fire destructive power
林火扑灭 forest fire suppression
林火探测 forest fire detection
林火通信系统 forest fire communication system
林火巡护[员] fire patrol
林火预报 forest fire prediction
林火预测预报 forest fire prognosis and prediction
林火种类 kinds offorest fire
林价 cost value of forest; forest value; value of growing stock; value of stand
林价比 stumpage ratio
林价边界 stumpage margin
林价成本 stumpage cost
林价公式 stumpage formula
林价评估 stumpage appraisal
林价算法 forest valuation
林间车道 ride
林间空地 glade
林间苗圃 bush nursery; field nursery; nursery in forest
林间通道 breakthrough; cut-through
林间幼树 interplant
林间栽植 inter-planting
林间增播(包括天然下种) interseeding; intersowing; sod seeding
林警 forest ranger
林警规程 rules of forest police
林况 forest condition; stand condition
林况公式 formula of forest condition
林况记载 description of forest; forest description
林况类别图 condition class map
林况调查 investigation of stand-conditions
林况图 stand condition map
林利 forest rent(al)
林利最大伐期龄 final age of the maximum forest rent
林龄 age of stand
林龄空间 age-class space
林龄面积法 age-area system
林龄向量 age-class vector
林龄转移概率 age-class-transition probability
林木 body; crop; forest-tree; growth; stand; standing timber; timber; wood
林木不齐 unevenness of stand
林木材积测定员 land looker
林木材积式 tree volume equation
林木采销证 timber lease; timber licence
林木成年期 sexual maturity of trees
林木出售 timber sale
林木出售投标记录 timber sale bidding record
林木打号 timber marking
林木单株选种 individual tree-selection
林木的郁闭 closing of standing-crop
林木费用价 cost value of stand
林木分布 DBH distribution
林木分布不均 unevenness of stand
林木抚育 improvement of forest; tending of woods
林木改换价 stumpage conversion value
林木改良 forest tree improvement; improvement of forest
林木更新成熟期 sexual maturity of trees
林木估价 timber appraisal
林木估价员 land looker
林木价 stand value; value of a stand
林木价格 stumpage value
林木界限 timber limit
林木勘测 timber acquisition; timber cruise
林木可及性 accessibility of timber
林木绿化率 tree cover; woody-plant cover
林木茂盛 woodiness
林木密度 thickness of stand; tree density
林木苗圃 forest-tree nursery
林木评价 stand valuation
林木期望价 stand expectation value
林木清单文件 tree list file
林木山价 stock-value
林木生长预测系统 tree growth projection system
林木收获 timber yield

林木收利(林木蓄积资本的利息) rental of stumpage
林木收入 rent of growing stock
林木调查 timber inventory
林木调查成本 timber cruising cost
林木调查员 timber cruiser
林木投资 timber investment
林木现存生物量 standing crop
林木像片直接量测 timber direct photomeasurement
林木销售 stumpage sale
林木信息 tree information
林木形数 timber form factor
林木蓄积 standing crop; store; timber growing stock
林木选择 selection of tree
林木营养 forest tree nutrition
林木育种 forest tree breeding
林木种子 forest tree seed
林木种子园 tree seed orchard
林木株数 stem-number
林木资本 stock capital
林木租金 rent of growing stock
林木租约 timber lease
林牧地 forest range
林牧结合制度 silvi-pastoral system
林奈种 Linnean species
林内 forest interior
林内草地 grassplot
林内放牧 forest grazing; forest pasture; grazing in forest; wood pasture
林内光线 inside-light of woods
林内集材道 wood (skid) trail
林内降雨量 rainfall under tree crown
林内居民 woodman
林内潜居民 forest squatter
林内擅居民(擅自在林内居住) forest encroacher
林内移民 forest colony
林内雨 canopy drip
林内运材 off-road hauling
林鸟 forest bird
林栖动物 arboreal animal
林丘 holt
林区 forest district; forest range; forest region; timber-tract
林区大道(供开发用) woods highway
林区工舍 woods cabin
林区公路 forest highway
林区实验室 forest laboratory
林区实验所 field laboratory
林区署 district (forest-)office
林区长 district forester
林区制材 mountainous sawmilling
林权 forest ownership; forest right
林权裁决员 forest settlement officer
林权制度改革 forest right system reform
林套 forest mantle
林外运材道路 exterior logging road
林网 network of forest
林位 stand quality
林务管理员 woodman
林务局 Bureau of Forestry; forest service
林务区长(林管区技术主任) regarder
林务署 Bureau of Forestry; forest supervisor's headquarter
林务署长 forest supervisor
林务员 forest officer; forester; woodward
林隙树种(林窗更新的树种) gap-phase species
林隙置换(林窗树种更替) gap-phase replacement
林下层 floor layer; floor stratum; lower storey of forest
林下栽植 underplanting; understorey management
林下植被 floor vegetation
林下植物 floor plant; undergrowth
林线(水平或垂直分布界限) forest limit; forest line; timber line; forest line
林相 forest form; form of crop
林相图 forest map; stand map; stock map
林相制图 timber type mapping
林型 cover type; crop type; forest type; stand type
林型图 cover type map; stock map; type map
林型学 forest typeology; forest typology
林学 forestry
林学会 forest association; forestry association
林学特性 silvics

林学习性 silvical habit
林学系 Department of Forestry
林学原理 silvics; silvicultural principle
林学院 forest college
林野 wildland
林业 forestry
林业播种机 seed drill for forestry
林业部 Ministry of Forestry
林业部门 forestry sector
林业财政收益 financial returns of forestry
林业测量 forest survey
林业成本 forestry cost
林业措施 forest practice
林业法规 forestry regulation
林业辅助用地 support land; auxiliary land
林业工人 forest-labourer; woods laborer; woodsman
林业工长 woods foreman
林业工作者 forester
林业合作社 forest co-operative society
林业机械 forestry machinery
林业基金 forestry fund
林业计划体系 forestry planning system
林业计划委员会 forestry program committee
林业计算学 forest mathematics
林业经济 forest economy
林业经济学 economies of forestry; forest economics; forestry economics
林业局总体设计 overall project of forest enterprises
林业科学 forest science
林业劳动力 labour in forestry
林业利率 rate of forestry interest
林业利益 profit of forestry
林业企业 forest enterprise
林业区划 division of forestry
林业杀草剂 forestry herbicide
林业生产力 productivity of forestry
林业生产要素 productive factor of forestry
林业史 forestry history; history of forestry
林业试验场 forest experimental station
林业税制 forestry taxation system
林业所得税 forest-income tax
林业推广 forest extension
林业推广服务 forest extension service
林业推广机构(人员) forestry extension agent
林业委员会 forestry commission
林业信息检索 forest information retrieval
林业学校 forest school
林业要素 element of forestry
林业用地(简称林地) forestry land; land for forestry
林业用地评价 valuation of land for forestry
林业政策 forest policy
林业职员 forest-staff
林业植树法(区别于园艺栽植) forestry planting
林业资本 forestry capital
林衣 forest mantle
林荫大道 boulevard; parkway
林荫大道系统 parkway system
林荫道 avenue; mall
林荫散步道 Alameda
林用点火枪 forestry torch
林用犁 forestry plough
林用龙门起重机 forestry gantry crane
林用喷雾机 tree-base sprayer
林用装卸桥 forestry overhead travelling crane
林园 forest homestead
林园地势 landscape grain
林缘 border; border of forest; edge of a stand; edge of forest; forest border; forest edge; forest fringe; forest margin; stand edge
林缘保护 edge protection
林缘带(防风) edge
林缘木 marginal tree; tree of forest border; border tree; edge tree
林缘群落 marginal community
林苑 park
林宅 forest homestead
林政学 forest policy
林中草地 grass glade

林中村 forest hamlet; forest village
林中废材 woods cull
林中空地 blank; clearance; clearing of a cutting area; fall place; forest glade; gap; glade; opening; slash
林中空地植被 gap-cover
林中小屋 woods cabin
林种 category of forest; forest category
林主协会 forest owner's association
林租 forest rent(al)
林租学说 doctrine of forest rent
临界点 critical point
临界风速 critical wind velocity
临界高度 critical height; critical altitude
临界高度抽样 critical height sampling
临界光周期 critical photoperiod
临界含水量 critical moisture content
临界流速 critical velocity
临界摩擦速度 critical friction velocity
临界深度 critical depth
临界试验 critical experiment
临界水平 critical level
临界松弛时间 critical relaxation time
临界途径法 critical path method
临界途径分析 critical-path analysis
临界拖曳力 critical tractive force
临界弯曲应力 critical buckling stress
临界温度 critical temperature
临界压力 critical pressure
临界压缩比 critical compression ratio
临界夜长 critical night length
临界载荷 critical load
临界值 critical value; threshold value
临界昼长 critical day-length
临时标准地 temporary sample plot
临时采伐 accidental cutting; incidental felling; extraordinary felling; extraordinary harvest
临时道路 shoofly
临时费用 non-recurring expenses
临时工 jobber
临时河绠 haywire boom; logan; pokelogan
临时混交 temporary mixture
临时寄生 temporary parasitism
临时捆木 hot choking
临时楞 hot deck
临时楞堆 hot pile
临时苗圃 field nursery; flying nursing; shifting nursery; temporary nursery; wandering nursery
临时桥 makeshift bridge; temporary bridge
临时收益 accident yield
临时水道 swale
临时小营地 line camp
临时样地 temporary sample plot
临时样方(长方形) temporary quadrat
临时营地 fly camp
临时用捆木索 hot choker
临时作业点 setting; set-up
临危种 endangered species
淋滤 eluviation
淋滤带 leaching zone
淋滤酌 eluviation
淋溶 eluviation; leach
淋溶层(土壤剖面E层) eluvial horizon; eluvial layer; eluvium; leached layer; leaching zone; E-horizon; E-layer
淋溶黑钙土 leached chernozem
淋溶水采集装置 lysimeter
淋溶水成土 eluvial-hydromorpbic soil
淋溶土 alfisol; eluvial soil; luvisol
淋溶作用 eluviation; leaching
淋失 eluviation
淋涂机 curtain coater
淋洗假潜育土 lessive pseudogley
淋洗侵蚀 leaching erosion
淋洗土 lessive
淋洗作用 lessivation
淋余土 pedalfer
淋浴装置 shower bath
磷胺(杀虫剂) phosphamidon
磷化氢(熏蒸剂) phosphite
磷化锌(杀鼠剂) zinc phosphide
磷酸化作用 phosphorylation
磷酸戊糖途径 pentose phosphate pathway
磷循环 phosphorus cycle

L

磷脂 phosphatide
磷质砂岩 phosphatic sandstone
磷质页岩 phosphatic shale
鳞盾(球果鳞片) apophysis (*plur.* -hyses)
鳞茎 bulb
鳞茎地下芽植物 geophyta bulbosa
鳞茎皮 tunic
鳞落 scale-off
鳞毛 palea; palet; pale
鳞片 cataphyll (*plur.* -lla); interfoyles; lodicule; squama; scale
鳞片体 lepidosome
鳞片叶 scaly leaf
鳞片状的 scaly; squamaceous; squamiform; squamose
鳞片状侵蚀 squamose erosion
鳞脐 umbo (*plur.* umbones)
鳞芽 scaly bud
鳞状的 scaly; squamate
鳞状树皮 scale bark; scaly bark
鳞状叶 phyllade
檩椽材 purlin and rafter
檩条 purline
檩条屋架 purlin roof
灵活采伐量 flexibility in cut
灵活机制 flexibility mechanism
灵活性 elasticity; maneuverability
灵敏度 response; sensibility
灵敏度分析 sensitivity analysis
灵敏高度表 statoscope
灵猩(一种猎犬) borzoi
陵园 cemetery
菱形变形 diamonding
菱形晶体 rhomboidal crystal
菱形拼接 diamond matching
菱形纹配板 diamond matching
零点 zero point
零点校核 check of zero
零度等温线 zero isotherm
零合基因 nullozygous gene
零和游戏 zero-sum game
零件 detail; detail part; element; part
零界采伐 zero margin logging
零界采伐周期 zero margin cutting cycle
零界林木 zero margin tree
零界林业 zero margin forestry
零买 buy at retail
零式 nulliplex
零售记账 charge account
零售价格 retail price
零售苗圃 retail nursery
零售贮木场 retail yard
零下温度 subfreezing temperature; subzero temperature
零星交易 odd lot
零增长线 zero net growth isoline
龄(昆虫) instar
龄级 age class; age-grade
龄级比例轮伐面积 proportionate area by age class
龄级表 age-class table
龄级法 age class method; compartment system
龄级法正分布 normal distribution of age classes
龄级分布 age calss distribution; age-class distribution; distribution of age-classes
龄级结构 age calss distribution
龄级配置 distribution of age-classes
龄级期 age-class interval; age-class period
龄级顺序 gradation of age classes
龄级系数 age-class factor
龄级周转 turnover
龄阶 age gradation; age grade
龄期(虫) stadium
龄组 age group
岭 ridge
领土 territory
领先指标 leading indicator
领域 area; domain; realm; territory
领域分布 parapatry
领域行为 territorial behavior
领域性 territoriality
另类投资 alternative investment
令(1令纸500张) ream
溜槽 spout
溜放式平底跑车 skyscooter
溜质包裹弹头 lubaloy-coated bullet
溜质包裹弹丸 lubaloy-coated shot
留材场(临时截留流送木材) capturer

留茬 sloven; tombstone
留茬伐根(伐木缺陷) tombstone stump
留充木 filler
留床 non-transplanting
留存木 main crop after thinning
留后备长度 snipe
留空堆积 open stacking
留兰香油 spearmint oil
留母树带状采伐法 seed-tree method of strip cutting
留母树法 reserve-seed tree method
留母树群天然下种更新法 group seed tree method
留母树天然更新 natural regeneratiov by seed tree method
留母树主伐 harvest cut with seed trees
留母树作业法 seed-tree method
留鸟 permanent bird; resident bird
留土的(子叶) hypogean; hypogeal; hypogeous
留土型发芽 hypogeal germination
留土子叶 hypogaeous cotyledon
留弦 bridge; crest; hinge wood; holding wood; key; leave; tipping edge
留有后备尺寸 full cut; full measure; full size
留有余量尺寸 overcut size
留置权 lien
流 flowage
流变行为 rheological behaviour
流变图形 flow figure
流变性质 rheological proterty
流变学 rheology
流产 abortion
流程表 flow sheet
流程示意图 flow diagram
流程图 flow chart; flow diagram; flow sheet
流出(黏合剂) sagging
流出(水等) effluence; efflux; outflow; outpouring; run-off
流出的 effluent
流出量 outflow
流出物 effluence; effluent; efflux
流出液 effluent
流迭闸坝 sluice dam
流动 flow; flowage
流动比率 current ratio
流动变形 flow figure
流动层 fluidized bed
流动产资 circulating asset
流动点 yield point
流动负债 current liability
流动干燥窑 floating dryer
流动基金 current fund
流动曲线 flow curve
流动沙丘 active dune; migratory dune; mobile dune; wandering dune
流动施膜 flow coating
流动投资 current investment
流动涂覆 flow coating
流动相 mobile phase
流动性 fluidity; liquidity
流动性范围 yield limit
流动应力 flow stress
流动制材厂 mobile (saw) mill; transportable (saw) mill
流动资本 circulating capital; floating assets; floating capital; fluid capital; liquid capital
流动资产 accrued asset; current asset; floating assets; immediate assets; liquid asset; working assets
流动资金 circulating asset
流动资金周转 turnover of circulating capital
流动阻力 flow resistance; fluid resistance
流度 fluidity
流度计 fluid(i)meter
流挂 run; sag; sagging
流行 outbreak
流行病 epidemic
流行病学 epidemiology; quamitative epidemiology
流行的 epidemic
流化 fluidization
流化床 fluidized bed
流化床干燥机 fluidized bed dryer
流化作用 fluidization
流浆箱 flow box; head box
流胶 flow of gum; gummosis
流胶反应 gummous reaction
流胶现象 gummosis
流控技术 fluidics

流况曲线 flow duration curve
流量 discharge; efflux; flow; stream discharge; stream flow
流量测定 discharge measurement; flow measurement
流量计 flow meter
流量控制 flow control
流量历时曲线 flow duration curve
流量率 specific discharge
流量曲线 discharge curve
流量调节 flow control
流量图 flow chart
流量系数 discharge coefficient
流量站 discharge site
流料箱 flow box
流明(发光度单位) lumen (*plur.* lumina)
流入 afflux; influx
流入液 influent
流沙 drift sand; quick sand; shifting sand
流沙地(荒地) drifting sand waste
流沙固定 fixation of shifting sand
流沙荒漠 arvideserta
流沙荒漠群落 mobilideserta
流沙控制林 shift sand control forest
流沙沙堆 sand-drift
流失 spillage
流失木 flood trash
流失木材 long-shore drift
流石滩 scree; talus
流石滩植被 talus vegetation
流水 lotic water
流水槽 sluice
流水的 eotic
流水楞 hot deck
流水群落 running water community
流水生态系统 lotic ecosystem
流水石 flowstone
流水式采运(采伐后直接运到制材厂) hot logging
流水式二次集材 swing hot
流水式集材 hot skidding
流水式集材转载场(不堆存) hot deck
流水台阶 cascade
流水系群落 running water community
流水植物 rheophyte
流水装车台(随到随装) hot skidway
流水作业 tact system
流送 float; floating; white water driving
流送材 floater; floating wood
流送工 cherry-picker; log driver; rafter
流送工队长 river boss
流送工人 driver
流送河道整治 floating channel realignment
流送距离 water lead
流送木材(向贮木池或缏场) water logs; drive
流送木道 floating log path
流送扫尾 after floating; floating rear; pick the rear; rear; sack; sack the rear; sweep the rear
流送扫尾工人 sacker
流送设施 floating facility
流送水道 floatable waterway; floating-channel; rafting channel
流送水闸 log lock
流送损耗 waste in floating
流送线路 channel; drift-road; driving road
流送用钩杆 floating hook
流送用小船 drive "chaland"
流送闸坝 flood dam
流送中断 hang up
流送中未变色木材 floated bright
流送周年降水量(结冰霜至翌年流送完毕) water year
流苏 fringe; tassel
流速测量 flow measurement
流速或流量 streamflow
流速计 current meter; hydrodynamometer
流体泵 fluid pump
流体静力学 hydrostatics
流体静压 hydrostatic pressure
流体力学 hydromechanics
流体阻力 fluid resistance
流通面积 flow area
流通税 turnover tax
流通蒸汽 live steam
流涂 flow coating
流涂法 flowing method
流涂设备 flow coater
流唾液多涎 salivation

流纹岩 liparite; rhyolite
流线 streamline
流线型的 stream-lined
流星 meteor
流性曲线 flow curve
流岩 flowstone
流域 basin; catchment; drainage area; drainage basin; river-basin; water catchment; watershed
流域变量 watershed variable
流域分界线 drainage divide
流域管理 watershed management
流域界线 drainage line
流域漏水 watershed leakage
流域面积 watershed area
流域图 drainage map
流域吸收量 basin recharge
流域下垫面类型判读 interpretation of bedding surface types of river basin
流域蓄水 basin storage
流脂 resin flow; resin flux; resinosis
流阻 flow resistance
硫代硫酸钠(漂白后脱氯用) antichlor
硫代木质素 thiolignin
硫化 cure; curing
硫化度 sulphidity
硫化期 curing time
硫化物生物群落 sulphide community
硫化纤维 vulcanized fibre
硫化雨 sulphur rain
硫黄细菌 sulphur bacteria
硫黄炸药 aerolite; aerolith
硫脲 thiourea
硫酸木素 sulphuric acid lignin
硫酸钠硫酸铜混合液(杀虫杀菌剂) soda bordeaux
硫酸铜溶液处理法 Margary's process
硫酸烟碱 nicotine sulphate
硫酸盐半化学纸浆 kraft semi chemical pulp
硫酸盐法 sulphate process
硫酸盐还原菌 sulfate-reducing bacteria
硫酸盐黑液 sulfate (cooking) liquor; sulphate (cooking) liquor
硫酸盐化作用 sulfation
硫酸盐浆 sulphate pulp
硫酸盐松节油 sulphate turpentine
硫酸盐蒸煮 sulphate cook
硫酸盐蒸煮液 sulfate (cooking) liquor; sulphate (cooking) liquor
硫酸盐纸浆 kraft pulp
硫酸盐制浆法 kraft process
硫细菌 sulfur bacteria
硫辛酸 lipoic acid
硫循环 sulfur cycle
硫氧还蛋白 thioredoxin
硫质喷气孔 solfatara
馏分 cut fraction; fraction
馏分收集器 fraction collector
S形榴口边(装饰) ogee sticking
榴莲型(瓦状细胞) Durio type
瘤(木材缺陷) bump; gall; protuberance; wart; bluge
瘤层(细胞壁) warty layer
瘤节 excrescence burl
瘤胃(反刍动物的第一胃) rumen
瘤锈病 gall rust
瘤状的 tubercular
瘤状花纹 burr
瘤状结构(细胞壁) wart-like structure
瘤状突起 sphaeroblast
瘤状异常生长 excrescence burl
柳谷坊 willow pile check dam
柳前黄菱沫蝉 willow spittle bug
柳杉酚 cryptojaponol; sugiol
柳杉酚基 sugoyl
柳杉树脂酚 sugiresinol
柳杉酮 cryptomerone
柳杉烯酮 cryptomerione
柳属 osier; willow
柳属(灌木柳类) sallon tree; willow
柳树皮栲胶 willow extract
柳条 wicker
柳条坝 mattress dam
柳条编的 wicker
柳条椅 basket chair; wanded chair; wicker chair
柳条制品 wicker
柳锈病 willow-rust
柳织蛾 sallow flat body Depressaria conterminella Zeller
六倍体 hexaploid(6n); sextuploid
六分仪 sixtant
六氟化硫(SF_6) sulfur hexafluoride

六根原木材(构成1000板英尺) six-log timber
六行移植机 six-row transplant machine
六基数的 hexamerous
六价的 sexivalent
六价染色体 sexivalent
六角形芯层 hexagon cell core
六列的 sexfarious
六六六(杀虫剂) agrocide; benzene hexachloride
六羟基联苯 hexahydroxy diphenyl
六氢惹烯 hexahydroretene
六十度(60%)喉缩 sixty percent choke
六雄蕊的 hexandrian
六亚甲基四胺 Urotropine
六株木样本 six-tree sample
鹨(liù) pipit
龙胆苷 gentiin
龙骨 backbone; keel
龙骨瓣(花) carina (*plur.* carinae); kneel
龙骨垫木 keel block
龙骨墩顶垫 cap block
龙骨状的 carinate
龙骨状突起 carina (*plur.* carinae)
龙卷 spout
龙卷风 tornado
龙卷漏斗柱 trunk
龙门架 gantry
龙门起重机 transfer gantry
龙脑 bornyl alcohol
龙脑基 bornyl
龙脑香素 gurjuresene
龙脑香烯 gurjunene
龙脑香烯酮 gurjunene ketone
龙脑香油 Borneo camphor oil
龙脑香脂 gurjun balsam
龙脑香脂油 gurjun balsam oil
龙头 faucet; tap
龙涎香 ambergris
龙涎香酊 ambergris tincture
龙血白质 dracoalban
龙血树脂 dragon's blood; sanguis draconis
隆根 buttress flare; flange
隆起 elevation; hump; rise
隆起线 raised line
隆起线型 crown molding
隆肿 elevation
拢排场 rafter harbor
垄 ridge; bed
垄背播种 seeding on ridge; sowing on ridge
垄背栽植 ridge-planting
垄播 ridge drilling
垄侧栽植 side planting
垄耕 ridge plowing
垄沟 drill
垄植 step planting
垄作 hill culture; ridge culture
垄作中耕机 go-devil
楼板梁 floor beam
楼地板搁栅 floor joist
楼房 storied building
楼基盖板 string board
楼面板 floor plank; floor(ing) board
楼梯 staircase
楼梯扶手 stair rail
楼梯扶手处中柱 newel
楼梯扶手处中柱接头 newel joint
楼梯扶手处中柱装饰 newel cap
楼梯踏步梁 carriage
楼梯斜梁 stringer piece
楼梯斜梁侧板 string board
蝼蛄类 mole crickets
髅状突起 condyle
搂根耙 rootrake
镂花锯 jig saw; jigger
镂空 hollowed out carving
镂铣 routing
镂铣机 router (machine)
漏播地 blank spot; gap
漏窗 leaking window
漏斗 filler
漏斗喂料机 hopper feeds
漏斗状的 funnel-form
漏耕 balk
漏耕地 balk; blank spot; gap
漏刨 hit-and-miss; skip-in-dressing; skip-in-planing; skip-in-surfacing
漏气 blow-by; blowing; blowout
漏气试验 air leakage test
漏失量 leakage
漏税 tax evasion
漏涂 void
漏脂 pitching

露白(发芽) radical emergence
露底(油漆缺陷) cratering; crystalline bloom
露角 corner defect
露面榫 bare faced tenon
卢埃林坐犬 Lewellyn setter
芦苇沼泽 reed swamp
芦席纹配板法 basket weave matching
炉 fire place
炉床(炉底) bed of furnace
炉干 oven-dry
炉干材比重 specific gravity in over dry
炉煤气 generator gas
炉热干燥窑 furnace-heated kiln
卤灯 halide lamp
卤化 halogenation
卤化物灯 halide lamp
卤化物防火剂 halogenated fire retardant
卤化银 silver halide
卤化阻燃剂 halogenated fire retardant
卤化作用 halogenation
鲁特送风机 Roo't s blower
陆半球 continental hemisphere
陆背斜 anteclise
陆槽 syneclise
陆沉带 geosyncline
陆地上的 terrestrial
陆地生态系统 terrestrial ecosystem
陆地生态学 terrestrial ecology
陆地生物 geobenthos
陆地卫星 Landsat
陆地卫星影像 landsat imagery
陆地贮材 land storage
陆地贮木 storing wood in land
陆地贮木场 land-depot
陆风 land breeze
陆封种 land-locked species
陆界 geosphere; lithosphere
陆栖的 terrestrial
陆桥 land bridge
陆桥说 continental bridge hypothesis; land bridge hypothesis
陆上猎物 upland game
陆生草本群落 terriherbosa
陆生的 terrestrial; terricolous
陆生动物 terrestrial animal
陆生植物 terrestrial plant
陆线 landing line
陆性率 continentality
陆旋风 terrestrial whirl wind
录波器 oscillograph
录音磁带 tape
鹿 deer
鹿斑花纹 roe figure
鹿斑条纹 roe stripe
鹿场 deer yard
鹿害 deer damage
鹿角的叉枝 knag
鹿角状进料输送机 buckhorn feed conveyor
鹿角状运输链(原木) buckhorn chain
鹿脚(斧柄形状) fawn foot
鹿脚斧柄(斧柄末端) deer foot
鹿磨(鹿对树皮的磨损) fraying
鹿皮 buckskin
鹿肉 venison
鹿伤害 tree damage caused by forest fauna
鹿苑 deer yard
路边楞堆 roadside stockpile
路程表 hodometer
路堤 embankment
路拱坡度 crown slope
路基 base; bed; formation; ground work
路基石块 ballast
路基硬实层 hardcore
路肩 berm; shoulder
路肩排水沟 let; water-bar
路径点 waypoint
路枯马木苷 lucumin
路面过水沟 open-top culvert
路面排水沟 let; water-bar
路旁带 roadside strip
路旁绿地 roadside greenspace
路旁园 roadside garden
路旁植物 aletophyte
路堑 cut
路线 course; line; race
路线设计 location
路线调查 route survey
路型 road pattern
路牙 curbstone

路源沉积[物] terrigenous sediment
路障 road block
露 dew
露地花卉 outdoor flower
露地嫁接 field grafting
露地苗圃 nursery in field
露点 dew point
露点调节 dew point control
露点温度 dew-point temperature
露台 terrace
露天干燥 open air seasoning
露天慢腐(木材) eremacausis
露天台级 perron
露头岩层 outcrop
露尾甲 sap beetle
露营场地 camp ground
露营地 camp site
孪生薄壁细胞 disjunctive parenchyma cell
孪生薄壁组织 disjunctive parenchyma
孪生管胞 conjugate tracheid; disjunctive tracheid
孪生细胞 disjunctive cell
孪生种定律 law of geminate species
銮基尼紫胶(印度紫胶品系) rangeeni lac
卵 egg; ovum (*plur.* ova)
卵孢子 oospore
卵巢 ovary
卵巢管 ovariole
卵睾 ovotestis
卵核 oosphere nucleus
卵核分裂 ookinesis
卵寄生蜂 egg-parasite wasp
卵寄生者 egg parasite
卵精巢 ovotestis
卵菌 oomycetes
卵壳 chorion
卵孔 micropyle
卵块 egg batch; egg mass
卵块发育(无核) merogony
卵裂 cleavage
卵膜 ootypus
卵母细胞 oocyte
卵囊 cocoon; oocyst
卵片发育 merogony
卵鞘 egg-case
卵球 oosphere
卵球形的 ovoid
卵生 oviparity; oviparous reproduction
卵石海岸 shingle beach
卵石器 pebble tool
卵石沙漠 serir
卵室 ooecium
卵胎生 ovoviviparity
卵细胞 egg cell; ootid; ovum (*plur.* ova)
卵形节 oval knot
卵形饰 ovolo
卵原论者 ovist
卵原细胞 oogonium
卵圆形的 ovate; ovoid
卵状小体 ovule
卵状小体的 ovular
卵子 oosporen
卵子发生 ovogenesis
乱纹 wild grain
乱纹理 random grain; wild grain
掠夺式采伐 exploitation cutting (felling)
掠夺式经营 exploitation management; high grading
略具花梗的 subpedunculate
略具尖头的 subapiculate
伦琴射线 Röntgen ray
伦琴射线图 X-ray diagram
伦琴射线照片 Röntgenogram
伦琴射线照相术 Röntgenography
轮尺 caliper; calliper; caliper scale; diameter gauge; slide caliper
轮尺刻度 caliper scale
轮尺量测 caliper measure
轮尺杖 caliper-stick
轮伐 circular cutting
轮伐龄 rotation age
轮伐期 circulation period; cutting cycle; cutting interval; cutting rotation; exploitation age; felling age; felling cycle; felling interval; felling rotation; period of return; regulatory rotation age; rotation; rotation length; rotation of cuttings
轮伐期的财政收获 money yield of financial rotation

轮伐期决定 rotation determination
轮伐期选择 rotation choice
轮伐区 periodic coupe
轮辐 spoke
轮腐 ring rot
轮毂 hub; nave
轮毂用材 hub block
轮箍 tire
轮换刀(削片用) turnknife
轮换寄主 alternate host; alternative host
轮换拖车运材 shuttle hauling
轮回 palingenesis
轮回定时器 cyclic timer
轮回公式 recurrent formula
轮回亲本 recurrent parent
轮回选择 recurrent selection
轮机人员 black gang
轮界薄壁组织 marginal parenchyma; terminal parenchyma
轮界层 terminal layer
轮距 track
轮垦(毁林) roving agriculture; shifting agriculture; shifting cultivation
轮廓 silhouette
轮廓尺寸 overall size
轮廓砂光机 contour sander
轮廓线 form line; visible line
轮列的 cyclic
轮裂 cup shake; ring crack; ring shake; round shake; shake; shell shake
轮裂心材 ring shaken heart
轮牧 rotational grazing
轮伞 verticillaster
轮生 whorl; whorle
轮生的 verticillate
轮生节 whorl knot
轮生体 verticil; whorl
轮生叶 verticillate leaf
轮式刮土机 wheel scraper
轮式铧犁 sulky plough
轮式集材机 wheel skidder
轮式中耕机 wheeled cultivator
轮胎 tire
轮胎缓冲器(索道) rolling stock
轮纹病 ring spot
轮压机 calender
轮缘材 felloe wood
轮藻植物 charophyta
轮枝 whorl branch
轮转计 hodometer; trochometer
轮状的 verticillate
轮状聚伞花序 verticillaster
轮状破坏 ring failure
轮状星散薄壁组织 diffuse-zonate parenchyma
轮作 crop rotation; rotation
轮作牧场 rotation pasture
轮作制 rotation system
罗得(木材材积计算单位) load
罗汉柏二烯 dolabradiene
罗汉柏酚 dolabrinol
罗汉柏木油 hiba wood oil
罗汉柏素 dolabrine
罗汉柏叶油 hiba leaf oil
罗汉松烯 podocarptorene
罗汉松型纹孔 podocarpoid pit
罗勒烯 ocimene
罗勒烯酮 ocimenone
罗勒油 basil oil
罗盘 compass
罗盘测手 compass man
罗盘方位 compass bearing
罗盘经纬仪 theodolite with compass
罗盘仪测量 compass surveying
罗素效应 Russell effect
逻辑单元号 logical unit number
逻辑斯谛方程 logistic equation
逻辑斯谛曲线 logistic curve (Sigmoid curve)
逻辑斯谛增长 logistic growth
逻辑语言 logical language
逻辑指令 logical instruction
螺钉 bolt; fid
螺钉接合 screw joint
螺环 volution
螺栓 bolt
螺栓孔处理器 bolt-hole treater
螺栓连接 bolting
螺栓位移 bolt displacement
螺栓支承强度 bolt-bearing strength
螺栓支承应力 bolt-bearing stress
螺丝扣 turnbuckle
螺丝起子 screw driver
螺纹导管 spiral duct; spiral vessel

螺纹管胞 spiral tracheid
螺纹加厚 spiral thickening
螺纹接头 nipple
螺纹裂缝 cell wall check; helical cavity
螺纹条沟 spiral striation
螺纹支腿 whorl foot
螺线角 helix angle
螺形绷带处理 banding treatment
螺形聚伞花序 helicoid cyme
螺旋导承辊 spiral bedroll
螺旋方向 direction of helix; direction of spiral
α-螺旋构象 alpha-helix conformation
螺旋构型 helics
螺旋钻 auger bit
螺旋辊 screw roller; worm roller
螺旋辊涂胶机 ribbon spreading machine
螺旋滚筒 worm roller
螺旋挤压机 screw extrusion press
螺旋桨(直升飞机) rotor
螺旋桨式风扇 propeller fan
螺旋角 helix angle
螺旋结构(染色体或DNA等) helics; spiral texture
螺旋进料 screw feeding
螺旋进料器 auger; screw feeder
螺旋裂 spiral checking; spiral crack; spiral cavity
螺旋面 helical surface
螺旋千斤顶(起重器) screw jack
螺旋切割法(橡胶采割法之一) spiral-curve cut
螺旋生长 twisted growth
螺旋式伐根切碎机 helical stump cutter
螺旋式分配输送机 distribution screw
螺旋式磨木机 screw grinder
螺旋输送机 screw conveyer; spiral conveyor
螺旋榫钉 spiral dowel
螺旋纹理 corkscrew-growth; spiral grain
螺旋线 helix
螺旋线(Z形的) Z-helix
S形螺旋线 S-helix
螺旋形 volution
螺旋形刨花 spill
螺旋形物 wreath
螺旋型割法 spiral system cut
螺旋压机 screw press
螺旋压榨 screw-press process
螺旋叶序 spiral phyllotax
螺旋运输机 helical conveyer
螺旋支承辊 spiral bedroll
螺旋柱 wreathed column
螺旋状管孔排列 heliciform pore arrangement
螺旋状聚伞花序 helicoid cyme
螺旋状排列 spiral arrangement
螺旋钻 auger
螺旋钻探 auger boring
螺原体 spiroplasma
螺状聚伞花序 bostrix; bostrychoid cyme; false raceme
螺钻 auger borer
裸的 nude
裸地 bare area; bare ground; denuded land
裸地年租 annual rent for bare land
裸地评价 appraisal of bare land
裸根 open root
裸根苗 bareroot seedling; bare-rooted seedling; naked-rooted plant
裸根栽植 bare-root(ed) planting; naked root(ed) planting; open-root(ed) planting; open-root(ing) planting
裸花叶 sterile floral leaf
裸机 bare machine
裸茎的 nudicaulous
裸蕨类植物 psilopsida
裸蕨亚门 psilopsida
裸蕨植物 psilophyte
裸离子 bare ion
裸露 denudation
裸露的 naked
裸露地 bare land; unoccupied land; unstocked area
裸露纹孔膜(不具结壳物质的) unencrusted membrane
裸山 bare hill
裸锈子器 caeoma
裸芽 naked bud
裸岩 bare rock; exposed rock; rock outcrop

裸蛹 free pupa; liberal pupa
裸子植物 gymnosperm
洛维邦德色调计 Lovibond tintometer
络合的 complex
络合点 hapto
络合物(络合基) complex
络合物形成 ligation
落差 head
落差损失 loss of head
落差注入法 capping method
落刀 low ring roll
落地式车床 face lathe
落地式砂光机 floor-type belt sander
落钩跑车 sharkey carriage
落钩索 slack puller; slack puller line; slack pulling line
落钩索滑轮 slack pulling block
落钩索卷筒 slack pulling drum
落果 shedding
落花 defloration
落料辊 doffing roll
落皮层 rhytidome
落球冲击试验 falling ball impact test
落叶 deblade; leaf drop; leaf-fall; shed leaf; leaf abscission
落叶病 leaf cast; needle cast
落叶层截留损失量 litter interception loss
落叶的 defoliate; leaf-losing
落叶杜鹃 azalea
落叶灌木 deciduous shrub
落叶剂 defoliant
落叶阔叶林 deciduous broad-leaved forest
落叶林 deciduous forest
落叶木本群落 deciduilignosa
落叶期 abscission period
落叶树 deciduous tree
落叶树苗 deciduous seedling
落叶树种 deciduous species
落叶松癌肿病 larch canker
落叶松栲胶 larch (bark) extract
落叶松溃疡 larch canker
落叶松松节油 larch turpentine; Venice turpentine
落叶松松脂 larch turpentine
落叶松叶锈病 larch-needle-rust
落叶松叶油 larch leaf oil
落叶针叶林 deciduous coniferous forest
落羽杉烯 taxodine
落针病 needle cast
吕宾法 Rueping process
吕宾罐(防腐) Rueping cylinder
吕宾空细胞法循环周期(防腐) Rueping empty-cell cycle
吕特格尔法 Rütgers process
旅馆园林 hotel garden
旅行社 travel agency
旅行游乐 travel for pleasure
旅行者 traveller
旅鸟 passing migrant; traveler bird
旅游 recreation; sightseeing; tourism; travel for pleasure
旅游承载力 tourism carrying capacity
旅游环境容量 tourist environmental capacity
旅游林 forest attraction
旅游区 resort area
旅游胜地 sightseeing resort
旅游业 tourist trade
旅游营地 travel(l)er's camp
铝矾土性砖红壤 bauxite-laterites
铝罐 aluminium can
铝牌(树木标号用) aluminium tag
铝漆 aluminium paint
铝铁土 allite
铝涂料 aluminium paint
铝氧 alumina
铝英石 allophane
铝质土 allite
履带 caterpillar; track
履带板 track pad; track plate; track shoe
履带板突棱 lug
履带板抓地齿 grauser
履带片 shoe
履带片主销 master pin
履带牵引装置 crawler
履带式伐木归堆机 crawler feller-buncher
履带式刮土机 track scraper
履带式起重机 crawler crane
履带式拖拉机 crawler; track-laying tractor
履带式装载机 track loader
履带拖拉机 caterpillar tractor

率 index; rate
绿变 green-stain
绿带 green-belt
绿地 green area; green space; greenland
绿地覆盖率 green cover percentage
绿地率 open space percentage
绿雕塑 topiary
绿雕园 topiary garden
绿度指数 normalized difference vegetation index (NDVI)
绿肥 green manure
绿肥作物 green manure crop
绿腐 green-rot
绿高岭石 nontronite
绿痕(木器涂饰中的缺陷) greenish cast
绿化 afforestation; greening; treeplanting; revegetation
绿化标准 afforestation standard
绿化成绩 achievement in afforestation
绿化带 belt of trees
绿化工程 landscape engineering
绿化面积 green coverage; green area
绿化苗木 forest seedling
绿化区域(面积) afforested areas
绿化沙地 afforest desert
绿化委员会 greening commission
绿化先进单位 advanced unit in afforestation
绿化先进工作者 advanced workers in afforestation
绿化运动 afforestation drive; greening campaign
绿化祖国 cover the country with trees; green the country
绿篱 hedge; hedgerow; live fence
绿篱剪 hedge shears
绿篱培育 hedging
绿篱修剪机 hedge cutter; hedgemaker; hedger trimmer
绿篱修整 hedging
绿帘岩 epidosite
绿霉[病] green mold
绿泥 dregs
绿色 green
绿色GDP green gross domestic product
绿色草木 greenery
绿色革命 green revolution
绿色建筑 green building
绿色就业 green jobs
绿色食品 green food; pollution-free food; natural food
绿色消费 green consumer
绿色组织 chlorenchyma
绿心硬木 greenheart
绿修 green pruning; live pruning
绿雪 green snow
绿盐 green salt
绿盐剂 ascuerdalith
绿叶植物 greens
绿液 green liquor
绿枝扦插 green-wood cutting
绿洲 oasis; oasis (*plur.* oases)
葎草烯 humulene
氯 chlorinity
氯草灵(除草剂) chlorbufam
氯丹(杀虫剂) chlordane
氯丁二烯橡胶 chloroprene rubber
氯丁橡胶 chloroprene rubber; duprene
氯丁橡胶胶黏剂 neoprene adhesive
氯氟碳化物(CFCs) chlorofluorocarbon
氯化法(制浆) Pomolio-Celdecor process
氯化汞浸注法 kyanizing process; kyan process
氯化汞冷浸防腐处理 kyanising
氯化苦(熏蒸剂) acquinite; chloropicrin; nitrochloroform
氯化牻牛儿基 geranyl chloride
氯化木素 chlorinated lignin
氯化松香 chlorinated rosin
氯化物印相纸 chloride paper
氯化香叶基 geranyl chloride
氯化橡胶 alloprene
氯化橡胶胶黏剂 chlorinated rubber adhesive
氯化锌 butter of zinc
氯化锌单宁法 zinc tannin process
氯化锌法 chlorzinc process; zinc chloride process
氯化锌浸渍法 Burnettizing
氯化锌-杂酚油浸注法 zinc creosote process
氯化锌注入法 Burnett process

氯化血红素 hemin
氯价 chlorine number
氯硫磷(杀虫剂) chlorthion
滤波器 eliminator; filter
滤尘器 dust-filter
滤出物 siftings
滤光 filter; filtration
滤光色片 filter
滤光作用 filtration
滤膜 filterable membrane
滤清 filtration
滤声器 acoustic filter
滤食 filter feeding; filter food
滤食动物 suspension feeder
滤食生物 filter feeder
滤水度 freeness
滤水秒 defibrator second
滤水性 drainability
滤网 sieve
滤液 filtrate
滤渣 residue
滤渣胶 kunhi
滤纸 filter
滤质 filter medium
麻花钻 auger bit
麻花钻头 twist bit
麻黄式穿孔 ephedroid perforation
麻黄式穿孔板 ephedroid perforation plate
麻黄式导管穿孔 ephedroid vessel perforation
麻芯 fibre core
马鞍菌 saddle fungus
马鞭草烯 verbenene
马鞭草油 verbena oil
马勃 puffball
马达 electromotor
马道 bridle path
马兜铃属 dutchman spipe; birthwort
马尔柯夫类模型 Markovian model
马尔柯夫链 Markov chain
马尔莱夫演替模型 Margalef's model of succession
马尔萨斯参数 Malthusian parameter
马尔萨斯折棒模型 MacArthur's brocken stick model
马尔文日照计 Marvin sunshine recorder
马基群落(地中海夏旱灌木群落) maquis; macchia
马基思群落 macchia; maquis
马克样板(可调节的样板) Maco template
马拉滑道 drag chute
马拉硫磷(马拉赛昂) malathion
马来化作用 maleation
马来松香 maleated rosin
马莲垛(纵横叠置的木支座) cribwork; open piling
马莲垛基森铁 piling railroad
马莲垛路基森铁岔线 piling track
马陆 millipede
马尼拉柯巴 almaciga
马尼拉麻绳 Manila rope
马钱子 nuxvomica poisonnut; snake wood
马钱子苷 loganin
马钱子型(内含韧皮部) strychnos type
马桑碱 coriarine
马氏管(昆虫) Malpighian tube
马松奈脱法(纤维分离) Masonite-process
马松奈脱纤维板 Masonite
马松奈脱纤维爆破筒 Masonite-fibre gun
马松奈脱硬质纤维板 Masonite hardboard
马套子集材 dog's work
马蹄形餐桌 wine table
马蹄形防裂扒钉 beegle iron
马蹄形支腿 horse-hoofs leg
马尾松 horsetail pine; Chinese red pine; Masson's pine
马醉苷 asebotin
玛迪树胶 gum mastic
码(3ft或91.44cm) yard
码尺 yard stick
码垛堆积 palletize
码头 quay
码头费 wharfage
码头交货 at the dock
码头交货价 ex dock; ex pier; ex quay
码头交货价格 free on quay
码头起重机 quay crane
码头设备 wharfage
码头完税价 ex quay
码头蛀虫 wharf borer
蚂蝗钉 dog iron; double-headed nail

M

蚂蚁(英国方言, 古语) emmet
蚂蚁引诱剂 attractant of ant
埋板 ground plate
埋藏 burying storage
埋藏腐殖质层(土壤剖面A_b层) hide humus layer; A_b-layer; crypto-mull
埋藏细腐殖质层 crypto-mull
埋干造林 planting by layerings
埋生的 innate
埋条 level cutting
埋条造林 planting by layerings
埋葬群 taphocoenosis
埋葬学 taphonomy
埋植(苗木) embedding; heel(ing)-in
埋置 embedding
霾 haze
买价 bid price
买卖价 exchange-value
买主 vendee
买主垄断 monopsony
迈耶定期补偿材积年伐量计算公式 amortization formula-Meyer
迈耶尔防腐法 Mayerl process
麦草畏 dicamba
麦谷蛋白 glutenin
麦加香脂 mecca balsam
麦精 malt extract
麦克法兰单线松紧式索道集材 Mcfarlane slackline yarding
麦克吉弗特双臂集材机(两侧集材) MeGiffert skidder
麦克莱恩横伸梯臂装车 McLean boom loading
麦克劳德真空规 McLeod vacuum gauge
麦鞘形椅背 wheatsheaf back
麦氏真空计 McLeod vacuum gauge
麦芽 malt
麦芽法 malt process
麦芽膏 malt extract
麦芽浆 mash
麦芽琼脂(培养基) malt agar
麦芽琼脂培养法 malt-agar method
麦芽琼脂培养试验 malt-agar culture test
麦芽三糖 maltotriose
麦芽糖 malt; maltobiose; maltose
麦芽糖化酶 β-amylase
麦芽糖酶 maltase
麦芽提取液(麦芽汁) malt extract
麦芽汁培养基 malt extract medium
卖后租回 sale and leaseback
卖空 bear; short sale; shortcovering
卖空市场 bear market
卖青山 stumpage sale
卖主 vendor
脉冲 impulse; pulse
脉冲编码装置 moder
脉冲代码 pulse code
脉冲发生器 pulse generator
脉冲回音法 pulse-echo method
脉冲开关 impulse switch
脉冲宽度 pulse width
脉冲长度 pulse length
脉动 pulse
脉动喷烟机 pulse jet fog machine
脉动压力 fluctuation pressure
脉动载荷 fluctuating load
脉管的 vascular
脉管状的 vasiform
脉序 nervation; neuration; venation
脉肢目昆虫 lace wing
蛮石 niggerhead
满岸水位(满槽水位) bankfull stage
满浸注法 full immersion method
满细胞法 full-cell process
满细胞浸渍 full-cell impregnation
螨 mite; plant mite
蔓生 rambling
蔓生植物 rambler
蔓性月季 climbing rose
蔓延 creep; infestation; overgrowth
蔓延成分 spread component
蔓延地上芽植物 chamaephyta velantia
蔓延度 contagion
蔓延火 creeping fire
蔓延时间(火焰的) sweep time
蔓延速率 rate of spread
蔓延速率计算器 rate-of-spread meter
蔓越橘蜡 cranberry wax
蔓状的 pampiniform
漫地水流 overland flow
漫反射 diffuse reflection
漫灌 basin irrigation; flood(ing) irrigation; flooding; over irrigation

漫洪(漫流) sheet flood
漫射 diffusion
漫射光 diffuse light
漫水坝 basin dam
漫水路 Irish bridge
漫滩沉积 flood-plain accumulation
漫滩阶地 flood plain terrace; flood-plain bench
漫滩沼泽 back marsh; flood-plain marsh
慢火 slow fire
慢火焰 slow flame
慢慢地移动 creep
慢燃物(原木等大径可燃物) heavy fuel
慢射 slow fire
慢生长的 close-grown; slow-grown
慢速拌胶机 long-retention-time blender
慢性毒力 chronic toxicity
慢性氧化 eremacausis
慢性中毒剂量 chronic dose
芒 awn; barba
芒柄花素 onocerin; onocol
芒刺状的 lappaceous
芒果胶 mango gum
芒塞尔彩色系统 Munsell colour system
芒状(触角) aristiform
杧果 common mango; mango
盲肠 caecum
盲沟 blind drain
盲节 blind knot
盲囊 caecum
盲区 blind area
盲树(不开花结果) blind tree
盲纹孔 blind pit
牻牛儿素 geraniin
牻牛儿烯 geranene; geraniolene
猫脸纹(树干伤疤未全愈合) cat face
猫头鹰 owl
猫眼节 cat eye knot
T形毛 T-shaped hair
毛板裁边 slab edging
毛板下锯 saw alive
毛板下锯法 saw through and through; sawing alive; through-and-through sawing
毛边材 unedged lumber
毛边的 unedged
毛冰(飞机上) rime
毛布面 felt side
毛材积(未扣除缺陷) gross scale; gross volume; bigness scale
毛材积增长率 percentage gross volume increase
毛尺寸(不扣除木材缺陷部分) bare measure; bare size; full cut; full measure; full size; full scale; overcut size
毛虫 caterpillar
毛疵(单板上的毛糙) patch
毛刺 burr; fuzzy grain; grainlifting; heavy irrigation; woolly grain
毛刺沟痕 chip bruising; chip mark
毛刺砂光 scuffing
毛地板 carcass-flooring; subfloor
毛地黄 common foxglove; foxglove
毛地黄毒[素]糖 digitoxose
毛地黄苷 digitalin
毛地黄黄酮 luteolin
毛发湿度表 hair hygrometer
毛发湿度计 hair hygrograph
毛方 cant; flitch
毛方材 slabbed cant
毛方带锯 flitch band saw
毛方送料装置 cant flipper
毛纲草酚 eriodictyol
毛管边缘 capillary fringe
毛管间孔隙 infracapillary space
毛管孔隙度 capillary porosity
毛管联系破裂含水量 moisture of the capillary bond disruption
毛管凝聚 capillary condensation
毛管上限 capillary fringe
毛管势 capillary potential
毛管性 capillarity
毛巾架 towel rack
毛孔 pore
毛利率 profit margin
毛利润 gross profit
毛瘤(昆虫) verruca (*plur.* -ae)
毛楼板 counter-floor
毛面板方材 rough timber
毛生长 gross growth
毛生长量 gross increment
毛石 quarry stone

毛收获最多的轮伐期 highest gross yield rotation
毛松香 barras; scrape; solidified oleoresin
毛损 gross loss
毛吸收量 gross absorption; gross uptake
毛细持水量 capillary moisture capacity
毛细的毛发状的 capillary
毛细根 fibrous root
毛细管 capillary; capillary pipe; capillary tube
毛细管冷凝 capillary condensation
毛细管力 capillary force
毛细管黏度计 capillary viscosimeter
毛细管凝结水 capillary condensed water
毛细管气相色谱法 capillary gas chromatography
毛细管渗滤作用 capillary percolation
毛细管渗透 capillary penetration
毛细管水 capillary water
毛细管水分容量 capillary moisture capacity
毛细管瞬态水 capillary-transient water
毛细管弯液面 capillary meniscus
毛细管引力 capillary attraction
毛细管状态 capillary form
毛细管酌带 capillary fringe
毛细管作用 capillarity; capillary action
毛橡胶 caoutchouc
毛形感器 sensillum trichodeum
毛序 chaetotaxy
毛依莱反应 Maeule reaction
毛毡花坛 carpet (flower) bed; carpet bed
毛毡花坛栽植 carpet bedding
毛毡花坛植物 carpet herb
毛毡式栽植 carpet bedding
毛毡转向辊 return felt roll
毛重 apparent weight; gross weight
毛状纹 fuzz grain; woolly grain
毛状纹理(机械缺陷) fuzz grain
矛尖形的 lanceolate
锚 anchor
锚定 anchor; anchorage
锚定板 Jonson's form point
锚定螺栓 anchor bolt
锚定位置 anchorage
锚杆 anchor rod; rock bolt
锚根 buttress flare
锚具 anchorage
锚块 anchor block
锚链 anchor line; hawser
锚木 anchor log
锚栓 cathead
锚索 anchor cable
锚桩 anchor log; deadman; deadman anchor
锚桩伐根 anchor stump
铆合 seam
冒口 feed head
冒泡 effervescence
贸易 trade
贸易差额 balance of trade; visible balance
贸易效应 trade effects
帽盖状饰线(细木工) cap moulding
帽柱碱 mitragynine
帽状体 calyptra
帽状质壁分离 cap plasmolysis
玫瑰草油 palmarosa oil
玫瑰红 rose-bengal
玫瑰油 rose oil
眉斑(昆虫) supercilium
眉叉(鹿角) brow tine
眉条(木材构造) bars of sanio; rim of sanio
梅尔比率 Meyer ratio
梅尔降水系数 Meyer's coefficient of precipitation
梅苷 antimellin
梅花鹿 spotted deer; sika deer
梅花式的 quincuncial; quincunxial
梅花形栽植 quincunx planting
梅里特测高器 Merritt hypsometer
梅里亚姆生命带 Merriam's life zone
梅雨 plum rains; bai-u
媒焦杂酚油 coal-tar creosote
媒介 medium (*plur.* -ia)
媒鸟 decoy
媒质 medium (*plur.* -ia)
楣 summer beam
煤焦油 coal tar
煤焦油沥青 coal tar pitch
煤气发生炉 gas generator

煤气房煤焦油 gas-house tar
煤气焦油 coal gas tar
煤污病 sooty mould
煤烟 coom
煤油石蜡松香混合剂 petroleum-paraflin rosin mixture
煤砖 briquet(te)
酶 enzyme
酶促氧化(酶性氧化或酶催氧化) enzymatic oxidation
酶催化 enzyme catalysis
酶蛋白 apoenzyme; zymoprotein
酶活性 enzyme activity
酶解 enzymolysis
酶解作用 enzymolysis; zymolysis
酶联免疫吸附测定 enzyme-linked-immuno-sorbent-assay
酶缺乏 azymia
酶朊 zymoprotein
酶树脂 enzyme resin
酶学 enzymology
酶游离木素 enzymatically liberated lignin
酶原 proenzyme
霉变范围 incidence of molding
霉菌 mould; mould fungus
霉菌变色 mould-stain
霉菌糖化法 amylo process
每车每趟装载量 scale bill
每分钟卡片数 card per minute
每分钟卡片张数 cards per minute
每毫米线对数(影象) line per millimeter
每轮伐期采伐一次 one cut per rotation
每木估价 estimating by tree
每木检尺 tally
每木检尺记录簿 scale book
每木调查 complete enumeration; enumeration survey; tally
每木调查表 measurement list
每年结实 annual bearing
每年一轮的(针叶树枝条) uninodal
每年盈利 annual earning
每千板英尺固定费用 costs fixed per m.b.f.
美吨(907.2kg或0.9072t(公吨)) short ton
美国白蛾 American white moth
美国成材标准 American lumber standard
美国大山核桃 shellbark hickory (big shellbar hickory)
美国公地调查法 U. S. public land survey system
美国矩形测量法 U. S. rectangular surveying system
美国垦务局 Bureau of Reclamation
美国试验材料标准 American standard of testing materials
美国试验材料标准编号 ASTM designation
美国试验材料学会 American Society for Testing Materials
美国试验材料学会标准 ASTM standard
美国试验材料学会技术规范 ASTM specification
美饰涂层 decorative coating
美术纸 art paper
美学功能 esthetic function
美学观点 aesthetic point
美学价值 aesthetic value
美学评价 appraisal of aesthetic value
美学特性 aesthetic property
美叶桉树胶 marri kino
美洲茶碱 ceanothine
美洲乳香 American frankincense
美洲土荆芥油 American wormseed oil
镁灰岩 dunstone
闷火 smouldering fire
闷木 doty wood
闷热 close
门(生物分类) division; phylum
门把手 hand bar
门板 door skin
门边梁 square head
门边梃 door cheeks
门衬板 door lining
门窗钩 cabin hook
门窗框木砖 framed grounds
门窗框条 sill course
门窗框线条板 architrave
门窗直棂 mullion
门底脚螺栓 foot bolt
门架 upright
门槛 cill; door sill; sill; threshold
门槛嵌缝条 door strip
门框 door case; door frame

M

门廊 galilee; porch; portico
门碰珠 ball catch
门嵌板 door panel
门式起重机 gantry; gantry crane; log handling portal crane
门梃 stile
门头线墩子 architrave block
门托架(窑) door carrier
门心板 panelling
门眼镜框 frieze panel
门中梃 munting
门柱 door post; gate strut
门座式撑架 portal bracing
虻 tabanid
萌发 germinate; germination
萌发的 germinant
萌发管 germ tube
萌发孔 trema (*plur.* tremata)
萌蘖 sprout; sprouting
萌蘖根株 stool
萌蘖更新 sprout regeneration; sprout reproduction; reproduction by stool shoot
萌蘖更新法 sprout method
萌蘖林 sprout forest; sprout land; sprout stand
萌蘖林作业 sprout-system
萌蘖树 sprouting tree
萌蘖枝 imp
萌生灌丛 sprout clump
萌生林 coppice; coppice forest; copse
萌生林地 coppice-land
萌生苗 seedling sprout
萌生树 coppice grown stock
萌条 coppice shoot; coppice sprout; sap shoot; stump plant; sucker
萌芽 sprouting
萌芽繁殖体 blastochore
萌芽更新 sprout regeneration; sprout reproduction; copse regeneration; regeneration by sprout; regeneration from sprouts; regeneration under sprout method; reproduction by sprout
萌芽更新法 sprout method
萌芽力 power of sprouting; power of sprouting form stools
萌芽林 coppice with standards; coppice wood; sprout forest; sprout land; sprout stand
萌芽林皆伐作业 clearcutting sprout system
萌芽林作业 sprout-system
萌芽天然更新 natural regeneration by shoots; natural regeneration by sprouts
蒙特卡洛枪柄 Monte Carlo stock
蒙特利尔议定书 Montreal Protocol
蒙脱石 montmorillonite
猛烈雷暴 rattler
猛禽 bird of prey; raptor
锰肥 manganese fertilizer
锰土 wad
孟德尔定律 Mendel's law
孟德尔法则 Mendel's principle
孟德尔群体 Mendelian population
孟德尔式比率 Mendelian ratio
孟德尔式群体 Mendel's population
孟德尔式性状 Mendelian character
孟德尔式遗传 Mendelian inheritance
孟德尔式杂种 Mendelian hybrid
孟德尔遗传学 Mendelian genetics
弥尔烯 mirene
弥缝剂 sealer
弥散 diffusion
弥雾机 atomizer
弥雾器 mist blower
迷迭香油 rosemary oil
迷宫(园林) labyrinth; maze
迷鸟 straggler bird
迷向(防虫措施) confusant
醚抽出物 ether extractive
米彻利希盆栽试验 Mischerlich's pot experiment
米彻利希植物生长律 Mischerlich's law of plant growth
米凯利斯常数 Michaelis constant
米勒拟态 Müllerian mimicry
米丘林遗传学 Michurin genetics
米制 metric system
觅食 foraging
觅食行为 food-seeking behaviour; foraging behaviour
泌乳 lactation
泌液 bleeding
泌脂薄壁组织 epithelial parenchyma

泌脂流 resin flow; resin flux
泌脂漏(异常分泌) resin bleeding; resinosis
泌脂细胞 epithelial cell
泌脂细胞层 epithelial layer; epithelium
密闭的 leak-free
密闭管道 closed conduit
密闭林分 closed stand
密闭系统循环 circulation of closed system
密闭蒸煮 closed steaming
密播 close planting; dense sowing; heavy seeding
密触型胶黏剂 close-contact adhesive
密丛林 thicket
密簇聚伞花序 fascicle
密度 degree of density; density
密度比 density ratio
密度测定术 densitometry
密度分割 density slicing
密度分级刨花板 graded-density particleboard
密度级 density class
密度计 densimeter; densitometer; densometer
密度控制反应 density-governing reaction
密度控制因子 density-governing factor
密度梯度 density gradient
密度梯度离心 density-gradient centrifugation
密度效应 density effect
密度依赖学说 density-dependent theory
密度仪(测郁闭度) densimeter
密度指数 density index
密度制约 density-dependent; density-dependent controls
密度制约死亡率 density dependent mortality
密度制约因子 density-dependent factor
密堆 close stacking; solid piling; solid type piling
密堆法 bulk stacking
密垛法 block stacking; bulk stacking
密封 packing; sealing; under seal
密封带式运输机 enclosed belt conveyor
密封的 leak-free; vacuum-tight
密封件 packing
密封圈 lock ring
密封容器 sealed container
密封贮藏 hermetical storage; sealed storage
密缝接合 hooked joint
密冠树 close-crowned tree
密花石栎 tan oak; tan-bark oak
密集 cluster; compression
密集林木 crowded trees
密集群体 close population
密集树冠 compact crown
密集纹孔式 closed pitting
密集小林 wood
密集叶的 dasyphyllous
密聚花 aggregate flowers
密楞 solid piling
密林 dense stand; close stand; dense crop; dense forest; full-stocked stand; fully stocked woods; overstocked forest; thick forest
密林抚育 tending of thickets
密码(编码) code
密码(病毒分类) cryptogram
密码翻译器 code converter
密码子 codon
密气匣 stuffing box
密三糖 raffinose
密伞花序 fascicle
密伞圆锥花序 panicled thyrsoid cyme
密生林 closed forest
密生群丛 closed association
密实 densify
密实堆积法 tight stacking
密实归楞 solid piling
密实归楞的 block-piled; close-plied; dead-piled
密实化 densification
密实楞 bulk pile
密实木材 densified wood
密实心层 densified core
密丝组织 plectenchyma
密植 close planting; closely spaced planting; compact planting;

M

condensed planting; dense planting; overstocked planting
幂等矩阵 idempotent matrix
幂数曲线 exponential curve
蜜蜂学 apiology
蜜环菌 honey mushroom; honey fungus
蜜环菌根腐病 Armillaria root rot
蜜露 honeydew
蜜腺 nectary
蜜腺盘 epipodium
蜜源植物 bee plant; honey plant; Nectar plant
蜜汁 honeydew
绵毛 tomentum
绵蚜 woolly aphid
绵状的 lanate; lanose
绵状毛 erion
棉短绒 cotton linters; linter
棉花壳 cotton seed hull
棉花色素 gossypetin
棉绒 lint
棉绒纸浆 linter pulp
棉纤维 lint
棉子皮亭 gossypetin
棉子糖 raffinose
免除条款 escape clause
免费苗木 free planting stock
免耕法 no tillage
免税 tax exemption
免税单 bill of sufferance
免税期 tax holidays
免疫的 immune
免疫反应 immune response
免疫化学 immunochemistry
免疫性 immunity
免疫学 immunology
免疫血清 immune serum
免疫作用 immunization
面板(家具零部件) top
面板(胶合板) face
面板(平板) deck; face plate; face veneer; facing
面层 facing; surface layer
面层符号 face-mark
面朝下的 ventricumbent
面积 area
面积表 area table
面积簿 area book; area register
面积-材积检查 area-volume check
面积材积控制法 area and volume method combined
面积材积平分法 area and volume period method
面积采伐法 cut area method
面积测定 area determination
面积抽样误差 area sampling error
面积法 area-method; method by area; organization by area
面积分配法 area allotment method
面积负荷 area load
面积估计量 area estimate
面积归一化 area normalization method
面积划分 area division
面积计 planimeter
面积计算器(求积仪) area calculator
面积考得(材长均为1ft的一个短考) face cord
面积控制法 area control method; area control system; area regulation method; organization by area
面积量测 area measurement
面积平分法 area frame work; area period method
面积式采伐 area-wise felling
面积调整法 area regulation
面积效应 area effect
面积需求计算公式 acreage requirement formula
面积周长比 area-to-perimeter ratio
面筋 gluten
面具 mask
面盘 facial disk
面漆 overcoating; top coat
面漆下涂层(底漆+封闭底漆+中间涂层) under coat(ing)
面蚀 surface erosion
面向机器的语言 machine oriented language
面向计算机的语言 computer oriented language
面向信息的语言 information-oriented language
苗 shoot
苗床 nursery bed; seed plot; seedbed; seed-bed
苗床成型 seedbed formation

M

苗床除草机 seedbed weeder
苗床框架 seedbed frame
苗床形成器 bed former; bed shaper
苗床熏蒸 seedbed fumigation
苗床支柱 seedbed stake
苗根沾泥浆的 root-puddled
苗垄 nursery row
苗木(圃育) nursery stock; planting stock
苗木(天然) plant; stock
苗木包装 seedling packing
苗木钵(泥炭和木浆所制) Jiffy pot
苗木铲根 wrenching
苗木打包机 seedling packing machine
苗木费 cost of planting stock
苗木分级 seedling sorting
苗木供应 plants supply
苗木盒 plant box
苗木密度 bed density; stocking
苗木培育 nursery-stock growing; stock production
苗木商 nurseryman
苗木调查 nursery inventory
苗木箱 plant box
苗木修剪 stripling
苗木选择 nursery selection
苗木移植机 seedling transplanter
苗木贮藏 seedling storage
苗圃 nursery; nursery garden; plant-school; seed plot; tree nursery
苗圃病害 nursery disease
苗圃播种机 nursery planter
苗圃地 nursery land
苗圃地膜覆盖机 nursery tarp roller
苗圃工作者 nurseryman
苗圃设计 nursery planning
苗圃土壤 nursery soil
苗圃学 nursery gardening
苗圃筑床机 nursery bed former
苗圃作业 nursery operation; nursery practice
苗期 seedling stage
苗梢木质化 hardened off
描绘 tracing
描金漆器 gold painted lacquer ware
描述统计学 descriptive statistics
描图纸 detail paper; detail tracing paper; tracing paper
描准倒向 gunning
瞄准点 point of aim
瞄准器 fore sight; hairline
瞄准树倒方向 swing(ing)
瞄准星 bead sight
灭茬机 parer
灭虫碱(杀虫剂) anabasin
灭害威(杀虫剂) aminocarb
灭火 attack a fire; fire suppression; inhibition; suppression action
灭火策略 fire strategy
灭火法 suppression method
灭火方案 tactical plan
灭火飞机 air tanker
灭火机 fire engine
灭火力量 strength of attack
灭火沫 fire foam
灭火器 fire annihilator; fire beater; fire extinguisher; fire swatter; quencher
灭火器材贮藏所 fire-tool cache
灭火系统 fire suppression system
灭火装置 fire suppression system
灭绝 extinction
灭绝率 extinction rate
灭绝曲线 extinction curve
灭绝限界(恢复种群的最少个体数) extinction threshold
灭菌 sterilization
灭菌丹(杀菌剂) phaltan
灭菌检验 steriling test
灭菌器 sterilizer
灭鼠 deratization
灭瘟素 blasticidin
灭蚁灵(杀虫剂) mirex
灭幼脲(杀虫剂) diflubenzuron
灭幼脲(抑制昆虫脱皮) dimilin
篾缆 bamboo cable
民族植物学 ethnobotany
敏感性 receptivity; sensibility; sensitiveness; sensitivity
敏感元件 sensor
敏感指数 sensitive index
敏感种 sensitive species
名称 nomenclature
名胜 famous site; tourist interest
名胜古迹 famous historical site

名胜区 famous historical site
名义尺寸 nominal measure; nominal size; specified dimension
名义工资 nominal wage
名义价格 nominal price
名义密度 nominal density
名义上的 nominal
明槽 open channel
明度 luminosity
明矾 alum
明矾土 alum earth
明沟 open ditch; open-top culvert; surface channel; water furrow
明沟排水 open (ditch) drainage
明火 open fire
明机枪 hammered gun
明胶 gelatin
明铰链 back-flap hinge
明利 explicit interest
明亮剂 brightener
明脉症(病毒病害) vein-clearing
明排水沟 surface drain
明渠 open canal; open channel; open ditch
明视野 bright field
明榫 bare faced tenon
明晰度 definition
明细表 measurement list
明显缺陷 open defect
明显租金 explicit rent
明线(用于种子) luminous line; light line
明子 fatwood; light wood; lightwood; pitch stump; resinous wood
明子材 squaw log; stump wood
明子干馏 resinous wood distillation
明子松节油 stump turpentine
鸣管 syrinx
鸣鸟 song-bird
鸣腔 tympanum
鸣禽 soundbird
螟蛾 snout moth
命令 command
命名法 nomenclature
命名模式 nomenclature type
摹图 tracing
模 form work
模糊化加工 mat finish
模糊聚类 fuzzy clustering
模量 modulus
模拟 analogue
模拟场地 simulator enclosure
模拟的 model
模拟电路 simulator
模拟方法 simulation
模拟花纹 faux swirl
模拟计算机 analog computer; analogue computer
模拟计算装置 analogue computing device
模拟量 analogue
模拟模型 simulation model
模拟器 emulator; simulator
模拟实验 model experiment
模拟-数字 analog-digital
模拟-数字记录器 analog-digital recorder
模拟-数字-模拟转换器 analogy-digital-analog converter
模拟-数字转换器 analog-digital converter
模拟-数字综合转换器 analog-digital integrating translator
模拟调查数据更新 simulated inventory data updating
模拟物 stand-in
模拟物质 model substance
模拟系统 model system
模拟语言 simulation language
模拟装置 simulator
模式(广义的) pattern
模式(行动或行为) regime (*eg.* disturbance regime)
模式标本 type; type specimen
模式产地 type locality
模式分类 pattern classification
模式空细胞处理(防腐) normal empty-cell treatment
模式立木蓄积 normal growing stock
模式林 model forest; normal forest
模式龄级分配 normal age-class distribution
模式龄阶序列 normal series of age gradations
模式区别 pattern discrimination
模式生长量 normal increment
模式识别 pattern recognition
模式属 type genus

M

模式体系 model hierarchy
模式有丝分裂 eumitosis
模式种 type species
模数 modulus
模型 matrix; mould; mould(ed) timber; model; template
模型工车床 pattern maker's lathe
模型化合物 model compound
模型区域 model area
模型庭园 model garden
模压 moulding
模压胶合板 formed plywood
膜 hymen; membrane; pellicle
膜被 tunic
膜的 membranous
膜电势 electric(al) membrane potential; membrane potential
膜电位 membrane potential
膜电阻 membrane resistance
膜盒 bellows
膜片 diaphragm
膜状的 membranous
膜阻力 membrane resistance
摩(基因交换单位) morgan
摩擦 friction
摩擦传动 friction drive
摩擦传动装置 friction-gearing
摩擦翻木机 friction nigger
摩擦分级机(用于种子) friction separator
摩擦风 antitriptic wind
摩擦复位器 friction receder
摩擦杆 friction lever
摩擦角 sliding angle
摩擦进料 friction feeding
摩擦离合器 friction
摩擦力 force of friction; friction; frictional force
摩擦轮 cathead
摩擦强度 frictional strength
摩擦式剥皮机 friction debarker
摩擦速度 friction velocity
摩擦性失业 frictional unemployment
摩擦装置 friction device
摩擦阻力 friction resistance
摩尔比例 mole fraction
摩尔氏反应 Maule's reaction
摩尔氏木质素测定法 Maule's test for lignin
磨 mill
磨边机 edger
磨床 grinder
磨刀 sharpening
磨刀机 cutter sharpener; knife grinder; knife grinding machine; sharpener
磨刀夹 blade holder
磨光 abrade; abrasive planing; finish; polishing
磨光工作 millwork
磨光机 abrader; glazing machine; sander; sanding machine
磨光漆 rubbing varnish
磨光试验 abrading test
磨光室 sanding station
磨耗 abrasion; attrition
磨耗面 abrasive surface; wearing surface
磨耗试验 abrasion test; attrition test
磨耗试验机 abrasion testing machine; wear testing machine
磨合 break-in
磨具 sharper
磨锯 saw sharpening
磨锯齿 saw gumming
磨锯机 saw sharpener; sawblade sharpener; sharping machine
磨锯机锯条轨道 saw carriage track
磨锯链机 chain grinder
磨粒 abrasive particle
磨粒号数 grit number
磨粒密布(砂带) closed coat
磨料 abradant; abrasive; abrasive material; abrasive substance; grinding material; grit
磨料颗粒 abrasive grain
磨料粒度 grit size
磨料刨床 abrasive planer
磨料硬度 abrasive hardness
磨轮 abrasive disk; abrasive wheel; sand disk
磨石 abrasive stick; grind stone
磨石变钝 glaze
磨蚀 abrade; abrasion
磨蚀的 abradant; abrasive
磨蚀剂 abradant; abrasive
磨蚀试验机 abrader
磨蚀性 abradability

磨损 abrade; abrasion; abrasive wear; chafing; detrition; excoriation; wear; wear-and-tear
磨损程度 degree of wear
磨损处 excoriation
磨损零件 wear parts
磨损路 worn-out road
磨损面 wearing surface
磨损试验 attrition test
磨损性 abradability
磨损硬度 abrasion hardness
磨损原木端头锯下的木段 lily pad
磨条 abrasive stick
磨头 mill-end
磨细 size degradation
磨屑嵌填(砂带) loading
磨削 ablation; abrasive planing
磨削滚筒 sanding cylinder
磨削加工 abrasive machining
磨削面 grinding face; grinding surface
磨削压力 grinding pressure
磨研角 filing angle
磨制木浆 defibrination
蘑菇 mushroom
蘑菇菌丝体 mushroom spawn
蘑菇圈 fairy ring
蘑菇栽培室 mushroom hot house
抹头 heading back
抹芽 debudding; disbudding
没食子 gall nut; nutgall
没食子单宁 gallotannin
没食子儿茶素 gallocatechin
没收性赋税 confiscatory taxation
没药树脂 budelium
没药烯 heerabolene
磨浆机 attrition mill; refiner
磨浆石 pulp stone
磨浆压力 grinding pressure
磨木比压 specific pressure of grinding
磨木机 grinder
磨木机贮屑袋 grinder pocket
磨木浆 ground pulp; ground(wood) pulp; mechanical pulp
磨木浆用木芯 groundwood core
磨木木素 milled wood lignin
磨木木索 Bjorkman's lignin
磨木木质素 milled wood lignin
磨木纸浆 ground wood pulp
磨盘 abrasive disk; grinding disk; grinding pan
磨盘间隙指示器 disk clearance indicator
磨室 grinding chamber
磨碎 milling
磨碎的 ground
磨碎机 attrition mill
末道漆 overcoating
末端 end
末端搭接 end lap
末端卷须 solitary terminal
末端小叶 solitary terminal
末端氧化 terminal oxidation
末端氧化酶 terminal oxidase
末端载荷 end load(ing)
末端枝条 solitary terminal
末端直径 small-end diameter; top (end) diameter
末端钻孔注药法 end-boring method
末口直径 small-end diameter; top (end) diameter
末批赶羊木材 rear
末期腐朽 final decay
茉莉属 Jasmine Jasminum
沫蝉 cuckoo-spit insect
莫尔圆 Mohr's circle
墨卡托投影 Mercator projection
墨汁病 ink disease
模板 pattern; template
模板工程 form work
模板纸 template paper
模槽压痕 impression
模锻工 puncher
模壳 form work
模刻板 mould(ed) timber
模塑胶合板 plymold
模子 former
母(昆虫) queen
母本类型 maternal form
母本区 mother block
母本性状 maternal character
母本植物 maternal plant
母峰 parent peak
母根 parent root
母公司 parent company

母核 mother nucleus; mother nuclide
母坑道 mother gallery
母群 parent stock
母树 maternal plant; mother tree; seed bearer; seed tree
母树林 elite stand; parent stand; normal seed stand; reserve-seed forest; seed stand; selected crop; selected stand
母树亲本树 parent tree
母树选择 seed tree selection
母体 precursor; stem
母体发芽(植物) viviparity
母体性决定 maternal sex determination
母系半同胞 maternal half-sib
母系后代 maternal progeny
母系选种法 maternal line selection
母系遗传 maternal inheritance; matrilinear inheritance
母细胞 female cell; matricyte; mother cell
母性伪杂种 metromorphic hybrid
母性影响 maternal influence
母岩 native rock; parent bed rock; parent rock
母岩层(土壤剖面D层或R层) rock stratum; D-layer; R-layer
母岩接触层(土壤剖面) lithic contact
母液 mother liquor
母质 matrix; parent material
母质层(土壤剖面C层) parent material horizon; C-layer; C-horizon
母株 maternal plant; stock
母竹 mother bamboo; mother culm
牡荆碱 castine
木矮凳(垫脚) cricket
木坝 log dam; timber dam; wooden dam; wooden splash dam
木白蚁 kalotermite
木板 board; wooden plate
木板坝 slab dam
木板道 plank road; stringer road
木板谷坊 slab check dam
木板路导轨 sheer rail
木板桥 plank-floored bridge
木板印刷术 xylography
木版画 wood engraving
木薄壁细胞 wood parenchyma cell
木薄壁组织 wood parenchyma; xylem parenchyma
木薄壁组织束 wood parenchyma strand
木本的 woody
木本攀缘高位芽植物 phanerophyta scandentia
木本群落 lignosa; wood; woods
木本藤蔓植物 woody lianas
木本显花植物型 arborescent phanerogamic type
木本植被 lignosa
木本植物 ligneous plant; wood plant; woody plant
木本植物细胞 plant woody cell; woody plant cell
木表面(靠树皮的材面) bark side
木材 lignum; summer wood; timber; wood; lumber
木材氨塑化处理 plasticizing wood with ammonia
木材白坯封闭涂装 sealing bare wood
木材薄片 ply
木材保存 preservation of wood; wood preservation
木材保存法 powell process Powell; preservation process of wood
木材保管 log preservation and management; wood protection in storage
木材保管员 cheater
木材保护 wood protection
木材本色填孔料 natural wood filler
木材比重 specific weight of wood
木材比重计 xylometer
木材编码 lumber code
木材变色 timber stain; wood stain; wood staining
木材变色真菌 wood-staining fungus
木材变质 wood deterioration
木材标本 wood sample
木材标准 log standard; timber and lumber standard; timber and lumber standards; timber standard
木材标准化 timber and lumber standardization
木材材积 wood volume
木材采伐权 common of estovers
木材采运 log; logging off; timber harvest

M

木材测容仪 xylometer
木材层积立方米 cubic meter of piled wood
木材叉铲 log-lumber fork
木材叉车 log fork-lift truck
木材叉架 log-lumber fork
木材插垛(流送) rackheap
木材产量 timber yield
木材产品 wood product
木材超微结构 wood ultra-structure
木材承包商 wood contractor
木材承销商 woods contractor
木材炽热 glowing of wood
木材出河 hauling up logs from water
木材出口 timber export
木材出口贸易 timber export trade
木材传送装置 log hauler
木材船运 wood barging
木材打印 timber marking
木材代号 lumber code
木材单宁 wood tannin
木材盗伐 timber trespass
木材的选择吸收 selective absorption of wood
木材等级 timber grade
木材淀粉 wood starch
木材雕刻 wood carving
木材堆场 stacking stall
木材堆存台 timber dock
木材垛 stack of timber
木材防腐 wood preservation
木材防腐厂 timber treating plant
木材防腐处理 wood-preserving process
木材防腐剂 wood preservative; murolineum
木材分堆 timber lot
木材分类 timber assortment; wood assortment
木材分选场 break-down landing
木材分子 wood element
木材腐朽 timber decay; wood decay
木材干腐 dry-rot of wood
木材干馏 destructive distillation of wood; dry distillation of wood; wood destructive distillation; wood distillation
木材干缩 wood shrinkage
木材干燥 timber drying; timber seasoning; wood seasoning; wood drying
木材干燥机 wood drier
木材干燥机理 mechanism of timber drying
木材干燥器 wood drier
木材干燥窑 lumber dry kiln; lumber kiln
木材港口 timber port
木材高聚糖 wood polyose
木材各向异性 anisotropy of wood
木材工业 timber industry; wood industry; wood processing industry
木材工艺学 wood technology
木材供给 wood procurement
木材供应 timber supply
木材供应地 timbershed
木材构架估算、存档、报价计算机系统 computer truss estimating, filing and quoting system
木材构造 structure of wood; wood structure
木材购入 purchase of timber; purchase of wood
木材关税 duty on wood
木材归垛层 course
木材规格 timber specification
木材过坝 transfer logs over dam
木材害虫 timber pest; timber worm
木材花纹 figure in wood; wood figure
木材化石 xyltile
木材化学 wood chemistry
木材灰泥板 wood-gypsum board
木材混凝土复合梁 composite wood-concrete beam
木材火烘干燥法 drying of wood by direct fire
木材机械保护法 mechanical protection of wood
木材肌理 woody texture
木材集中 assembling wood
木材加工 wood conversion; wood processing; wood working
木材加工机床 wood working machine
木材加工业 wood processing industry
木材加固件 timber fastener
木材价格 timber price; wood price

木材检尺 timber scaling
木材检尺明细表 scale ticket
木材检尺评等 quality cruise
木材检验 timber inspection
木材检验局 lumber inspection bureau
木材检验所 timber inspection bureau
木材检验员 timber inspector
木材鉴别电算机化 computerized wood identification
木材降等 dockage; wood degradation
木材交易 timber exchange; timber transaction
木材交易税 timber transaction
木材结构 woody texture
木材解剖 wood anatomy
木材解剖术 xylotomy
木材解剖学 wood anatomy
木材进口定额 timber import quota
木材进口贸易 timber import trade
木材浸渍 wood impregnation
木材锯截 wood cutting
木材科学 timber science
木材库 timber magazine
木材库存台账 log ledger
木材库棚 timbershed
木材力学 timber mechanics
木材沥青 wood pitch
木材连接器 timber connector
木材淋洗 leaching of wood
木材磷光现象 phosphorescence of wood
木材流送 floatage; wood floatage
木材陆运 log land transportation
木材滤取法 leaching of wood
木材码头 timber port
木材买主 timber purchaser
木材贸易 lumber trade; timber trade
木材密度 wood density
木材明细表 timber specification
木材内含物 wood inclusion
木材膨胀 wood swelling
木材评等员 timber grader
木材破坏 wood destroy
木材破坏率 wood failure rate
木材起重输送机械 log handling and conveying machinery
木材气化 wood gasification
木材气味 odour of wood
木材强度 strength of wood
木材切片术 xylotomy
木材切削 wood cutting; wood machining
木材氰乙基化作用 cyanoethylation of wood
木材屈挠性 pliability of wood
木材趋势研究 timber trend study
木材权 timber right
木材缺乏 timber famine
木材燃料 wood fuel
木材燃烧炉 wood burning furnace
木材染料 wood dye
木材染色 wood staining
木材染色剂 wood stain
木材热解 heat decomposition of wood; wood pyrolysis
木材热软化 thermal softening of wood
木材热学性质 wood thermal property
木材认领 timber claim
木材鞣质 wood tannin
木材蠕变 creep of wood
木材入口(水上选材场) gap
木材润胀 wood swelling
木材三切面 three sections of wood
木材色素 xylochrome
木材商 lumber merchant; lumberman
木材升降机 timber elevator
木材生产 timber production; wood production
木材生产机械 logging machinery
木材生物降解 wood biodegradation
木材剩余发光 residual luminescence in wood
木材石膏板 wood-gypsum board
木材识别 wood identification
木材实积立方米 cubic meter of dense timber; cubic meter of trunk wood
木材实质比重 specific gravity of wood substance
木材实质密度 density of wood substance
木材使用权 wood right
木材市场 log market; lumber market; timber market; wood market

木材市价 ruling price timber
木材试样 wood sample
木材收获 timber crop; timber harvest; timber harvesting
木材收缩 wood shrinkage
木材树种 wood species
木材水解 wood hydrolysis; wood saccharification
木材水运 log transportation by water; wood transport by water
木材松脱(集材时) sideswipe
木材塑化 wood plasticization
木材塑料 wood plastic
木材炭化 carbonization of wood; wood carbonization
木材探伤器 woodchecker
木材糖化 wood saccharification
木材糖浆 wood molasses
木材提升机 timber elevator
木材填孔 wood filling
木材填孔剂 wood sealer
木材填孔料 wood filling
木材调拨 log allotment
木材停泊场 timber dock
木材拖挂装置 timber hitch
木材拖运 timber haulage
木材弯曲 wood bending
木材纹理 wood grain
木材稳定性 wood stabilization
木材无屑切削 chipless wood-cutting
木材物理学 timber physics; wood physics
木材物质(多糖类和木素) wood substance
木材物质比重 specific gravity of wood substance
木材习居菌 wood-inhabiting fungus
木材消费 wood consumption
木材消耗 timber consumption
木材销售 timber selling
木材销售计划 timber sale program
木材卸车 timber unloading
木材性质 wood property
木材需求 demand for wood
木材学 timber science; wood science
木材烟熏和燃气干燥法 drying of wood in smoke and firing gas
木材验号员 lumber checker
木材样品 wood sample
木材油漆 wood painting
木材预干太阳灶 solar lumber pre-dryer
木材寓居真菌 wood-inhabiting fungus
木材运费 timber haulage
木材运输 timber transport; wood transport(ation)
木材债券 timber bonds
木材债务 timber loans
木材着色 wood painting
木材正常使用年限 normal timber life
木材支拨 log allotment
木材直流电性质 direct-current property of wood
木材质量 wood quality
木材主要等级 main assortment; main category
木材贮存 log storage
木材蛀虫 timber borer
木材蛀孔害虫 wood boring insect
木材转载装置 log transfer
木材装车 log loading
木材装船 timber shipment
木材装卸 log handling
木材装卸工 log loader
木材装卸桥 log handling portal crane
木材装运 timber shipment
木材装运记录 shipping tally
木材装载机 log loader
木材状的 ligniform
木材资本 wood capital
木材资源 timber resources; wood resource
木材综合利用 wood comprehensive utilization
木材阻燃 fire-retarding of wood
木材阻燃处理 wood preservation against fire
木材阻燃剂 wood fire retardant
木材组织 wood organization; wood tissue
木材组织学 histology of wood; timber histology; wood histology
木车床 pole lathe
木杵插 cutting held by splinter
木船 wooden-ship
木槌 beetle; mallet
木醋酸 wood vinegar

木醋液 pyrolignucs acid
木地板 wood floor
木雕工 wood carver
木雕工艺 wood carving
木钉 knag; peg; tack; tre(e)nail; trunnel; wooden pin; wooden plug
木钉孔钻机 dowel hole boring machine
木钉生产机 dowelling machine
木豆 Congo pea; pigeon pea; cajan
木蠹蛾 carpenter moth
木段 cut log; section; wood block
木段定心 centering
木段定心装置 block centering device
木段汽蒸 block steaming
木段扫描定心 scanning block centring
木段水煮 block hot-water treating
木段蒸汽处理 block steaming
木段装机 log charging
木堆(由一定数量构成) timber lot; lot
木多糖 wood polyose
木垛试验(阻燃) crib test
木蛾 wood moth
木耳 Jew's ear
木筏 log-raft; wood raft
木筏流送作业 rafting operation
木房(用原木制的) block house
木粉 sander-dust; sanding dust; wood flour; wood powder
木粉填料 wood meal
木蜂 carpenter bee
木浮标 deadhead
木腐菌 conk; wood decay fungi; wood-decaying fungus; wood-destroying fungus; wood-rotting fungus
木杆(汽车)轨道 pole chute
木杆坝 rafter-type splash dam
V形木杆槽道 pole road
木杆道 stringer road
木杆道润滑脂 skid grease
木杆轨道 pole road; pole slide; pole tram-road
木杆铺道 skid up
木杆水坝 post-type dam
木杆水栅 pole retard curtain
木工 carpenter; wood worker; wood working
木工扁凿 firmer chisel
木工车床 wood lathe; wood-turning lathe
木工车间 joinery; woodshop
木工成形车床 wood shaping lathe
木工尺 carpenter's square
木工粗刨 carpenter's jack (trying) plane
木工多刀车床 wood multi-cut lathe
木工仿形车床 wood copying lathe
木工工艺 woodcraft
木工工作台挡头木 bench hook
木工机械 wood processing machine; wood working machinery
木工技术 woodcraft
木工夹 hand screw clamp
木工角尺 framing square
木工锯 carpenter's wood saw
木工刨 sizer
木工平刨 surfacer
木工钳 carpenter's pincers
木工水平尺 carpenter's level
木工台挡头 bench stop
木工铣床 milling machine
木工旋臂钻床 radial wood borer
木工业 carpentry
木工用材 carpenter's wood
木工用弧口凿 firmer gouge
木工凿 carpentry tongue
木工折尺 carpenter's rule
木工钻床 wood borer
木工钻孔车床 wood boring lathe
木工作业 timberwork
木拱 wood arch
木构件 wooden member
木管 wood pipe; wooden pipe
木管乐器 wood-wind
木管用材 pipe stock
木轨道 tramway
木轨路(轨条断面矩形) wooden stringer road
木桁架 timber truss
木桁架桥 wooden truss bridge
木滑道 log slip; timber chute; timber slide; wooden slide
木滑轮 wooden block
木化 lignification; lignifying; lignify

木化石 dendrolite
木荒 wooden-famine
木灰 wood ash
木屐 wooden clogs
木集材杆 wood mast; wooden spar; wooden tree
木甲虫 wood beetle
木架建筑 timber-framed building
木架结构 wood frame construction
木间的 interxylary
木间韧皮部 interxylary phloem
木间栓皮 interxylary cork
木建筑 wood construction; wooden building
木姜子油 litse cubeba oil
木浆 mechanical pulp; wood pulp
木浆浮油 liquid rosin; tall oil; tallol
木浆浮油松香 tall oil rosin
木焦油 wood tar; wood tar oil
木焦油沥青 wood tar pitch
木焦油提制的防腐剂 woodllin
木焦杂酚油 wood tar creosote
木接合 timber connection
木节 gnarl; knar(l)
木节封闭 knot sealing
木节封闭漆 knotting varnish
木结构 timber construction; timberwork; wood construction; wood structure
木截头 offcut timber
木聚糖 xylan
木聚糖酶 xylanase
木刻 wood engraving; woodcut; xylography
木刻工作者 wood engraver
木刻术 wood engraving
木口面 edge grain
木块 block; wood block; wooden block
木块地板 block floor
木块路面 wood block pavement
木块拼花地板 wood block flooring
木块试验(菌类培养试验或耐腐试验) wood block test
木框架 wood frame
木框架机械 timbcr-framing machinery
木框架建筑 wood frame construction
木捆 bundled stock
木捆保全率 bundle-survival ratio
木捆吊钩 piling hook
木捆流送 bundle floating
木捆排运 bundle rafting
木捆装车 sling load
木兰类 magnolia
木梁 wood beam; wooden beam
木料 stuff; timber; tree; wood
木料接合器 timber connector
木馏油 creosote; wood creosote
木瘤 bunion; burr; gnarl; knar(l)
木笼 crib; cribwork; gabion
木笼坝 crib dam; pier dam
木笼桥 crib bridge
木笼桥台 logged-crib abutment
木笼闸 crib dam
木螺钉 wood screw
木螺钻 auger bit
木麻黄栲胶 she-oak extract
木麻黄素 casurin
木马道 kimma sled road
木毛(包装填塞物) wood wool; excelsior
木煤气 wood gas
木棉 cotton tree; wood cotton-tree; kapok
木棉漆 pyroxyline lacquer
木模 lagging; template; wood form
木模工 pattern maker
木模工车床 pattern maker's lathe
木排 boom; log-raft; raft; wood raft
木排边扎木 side stick
木排横压木 swifter
木排流送规程 craft raft clause
木排路 corduroy (road)
木排拖船 crab
木刨花 wood shavings
木刨压板 chip cap
木盆 tub
木片 chip; cleaver; spill; wood chip
V形木片 wedge
木片仓 chip loft
木片打包机 chip packing machine
木片打碎机 chip breaker
木片分配管 chip distribution pipe
木片分配器 chip distributor
木片风送机 chip blower
木片黑煮 burning of chip
木片集装箱 chip container

木片接触式干燥机 chip contact dryer
木片进料螺旋 chip feeding screw
木片料仓 chip bin
木片磨木浆 supergroundwood
木片抛散辊 chip-throwing roll
木片刨片机 chip flaker
木片破碎机 chip crusher
木片筛选机 chip screen
木片筛选装置 chip thickness screening slicing system
木片输送机 chip conveyor
木片水洗 chip washing
木片水洗机 chip washer
木片拖车 chip van
木片研磨机 chip grinder
木片再碎机 chip disintegrater; rechipper
木片蒸煮 chip steaming
木片贮仓 chip silo
木片贮存器 chip container
木片装锅器 chip distributor
木片装料机 chip filler
木铺装 mat formation
木器 carpentry; tree ware; wood ware; wooden article; woodenware
木签 tally
木桥墩 bent
木琴 xylophone
木塞 plug; wood plug
木塞切刀 plug cutter
木商 wood merchant; wood-dealer
木射线 wood ray; xylary ray; xylem ray
木射线细胞 xylary ray cell
木生的 xylophilic
木虱 gall aphid; psylla
木薯粉 cassava starch; tapioca flour
木薯胶 cassava glue
木薯橡胶 ceara rubber plant; manihot caoutchouc
木栓 cork; peg; phellem; suber; cork oak; tre(e)nail; wooden pin; wooden plug
木栓板 cork board
木栓层 phellem layer
木栓化 suberification
木栓离层 absciss phelloid; abscissphelloid
木栓形成层 cork cambium; phellogen
木栓毡 cork-carpet
木栓脂 suberin
木栓制品 cork product
木栓质 suberin
木栓质的 corky
木栓组织 cork tissue
木丝 excelsior; wood wool
木丝板 excelsior board; excelsior plate; wood-wool board; wood-cement board; wood-wool slab
木丝机 excelsior cutting machine; excelsior-shredding machine
木丝刨切 wood wool machining
木丝刨制机 wood-wool machine
木松节油 steam distilled wood turpentine; steam-distilled (wood) turpentine; wood turpentine
木松香 steam distilled rosin; steam-distilled (wood) resin; wood rosin
木松脂[制品] wood naval stores
木素蛋白复合体 ligno-protein complex
木素蛋白体(化合物) ligno-proteinate(s)
木素蛋白质络合物 ligno-protein complex
木素分解 lignolysis
木素分解酶 hadromase
木素腐殖质 ligno-humus
木素前体 lignin precursors
木素-碳水化合物复合体 lignin-carbohydrate complex
木素-纤维素 ligno-cellulose
木素原体 lignin precursor
木塑复合材 wood plastic composite
木塔 wooden tower
木踏板绳梯 jack ladder
木炭 char; charcoal; vegetable carbon; vegetable charcoal; wood char (coal); wood coal
木炭规格 charcoal standard
木炭炉 charcoal-burner
木糖 wood sugar
木糖苷 xyloside
木糖胶酶 xylanase

木条 cleat
木条固定 cleat
木条胶接缝加热 strip heating
木条胶拼机 lumber core composer
木条芯细木工板(木条芯宽为7~30mm) blockboard
木条椅背 lath back
木条阴帘 lath-screen
木酮糖 xyloketose; xylulose
木桶制作 cooperage
木拖鞋 wooden clogs
木瓦 shake; shide; shingle; tile of trend
木瓦材 shake bolt
木瓦材栅框(水上拖运) shingle crib
木瓦短圆材 shingle bolt
木瓦工人 shingle weaver
木纹 vein
木纹交错拼合 mismatching; miss matching
木纹漆 graining paint
木纹刷 graining brush
木纹研光机 moire calender
木纹印刷 grain printing; wood grain printing
木纹印刷机 grain printer
木纹纸 grained paper; veined paper
木屋 log cabin
木犀草素 luteolin
木犀浸膏 osmanthus concrete
木细胞 xylary cell
木纤维 wood fibre; xylogen
木纤维素 wood cellulose
木香根油 costus root oil
木箱 coffin
木销 coak
木楔 plug; wooden wedge
木屑 sawdust; wood chip; wood residue; wood shavings; cleaver
木屑模压制品 sawdust molding product
木芯 core; veneer core
木芯防弯装置 core bracer assembly
木芯胶合板 wood core plywood
木芯子 centre; center
木型 wood form
木堰 timber dam
木堰堤 wooden dam
木蚁 carpenter ant
木油 wood oil
木杂酚油 wood creosote
木栅栏 post-and-paling
木枕 blocking
木制的 ligneous
木制家具 wooden furniture
木制品 tree ware; wood; wood ware; wooden article; wooden product; woodenware; woodwork
木制水槽 wooden tank
木制纸 wood paper
木质 lignin; wood; woody part; xylogen
木质部 hadromestome; xylem
木质部传导作用 xylem translocation
木质部发生 xylogenesis
木质部份 woody part
木质部母细胞 xylem mother cell
木质部内的 intraxylary
木质部束 xylem bundle
木质部外的 extraxylary
木质部细胞 xylary cell
木质部子细胞 xylem daughter cell
木质的 ligneous; ligniferous; lignous; woody
木质构件 wooden member
木质灌木 ligneous shrub
木质光敏素 xylochrome
木质果 xylodium
木质果的 xylocarpous
木质化 hardened off; lignification; lignify
木质化的 lignified
木质化阶段 phase of lignification
木质化细胞壁 lignified cell wall
木质化纤维 lignified fiber
木质化组织 lignified tissue
木质化作用 lignificationlignifying
木质林产品 harvested wood product
木质刨花板 wood chipboard; wood particleboard; wood shaving board
木质人造板 wood-based panel
木质人造板材料 wood-base panel material
木质素 lignin
木质素分解菌 lignin-decomposing fungus
木质素骨骼 lignin skeleton

木质素结构单元 lignin building unit
木质素结构基团 lignin building stone
木质素酶 ligninase
木质素母体 lignin precursors
木质素溶解菌 lignin-dissolving fungus
木质素生物降解作用 lignin biodegradation
木质素塑料 lignin plastic
木质碎粒(碎料) wood particle
木质碎料板 wood chipboard
木质碎料饭 wood particleboard
木质填料 wood filler
木质细胞 xylem cell; xylematic element
木质纤维 xylem fibre
木质纤维板 wood fibre board
木质纤维壁板 sundeala
木质纤维素 ligno-cellulose
木质纤维填充剂 lignocellulosic filler
木质与树皮分离 wood-bark separation
木质制成的 xyloid
木质组织 ligneous tissue; woody tissue
木柱 wood column; wooden column
木蛀虫 wood moth
木砖 wood block; wood brick; ground
木桩 picket; spiling; timber pile
木桩试验(防腐) stake trial
木状的 ligniform
目(分类单位) order
目标 objective; target
目标地区 target area
目标管理 management by objectives
目标函数 objective function
目标和时间进程 targets and time table
目标经营法 management by objectives
目标林 objective forest; target forest
目标射击 target shooting
目标蓄积 target growing stock
目标直径(理想直径) target diameter
目测 ocular measurement; visual estimation
目测法 estimation by eye; ocular estimate
目的程序 object program
目的论 teleology
目的取样法 purposive sampling
目的树 desirable tree
目高直径 diameter at height of the eye
目检 visual inspection
目降水量 daily precipitation
目镜(显微镜) ocular
目录 schedule
目视估计立木材积 ocular volume estimation; volume estimation by sight
目视解译 visual interpretation
苜蓿 alfalfa
牧草 forage crop; forage plant; herbage; pasturage; pasture; rank weed
牧草茂盛的 rank
牧场 meadow; pasturage; pasture; rangeland
牧场改良 range improvement
牧场管理 pasture management
牧场火灾 range fire
牧场经理 range manager; range officer
牧场条件 range condition
牧地 ley
牧豆树胶 mesquite gum
牧食生物 grazer
幕骨(昆虫) tentorium
幕式淋涂 curtain coating
日柳杉烯 kiganene
钠长石 albite
钠质土 sodic soil
奈曼分布 Neyman's distribution
耐病性 disease-enduring
耐擦伤性 marresistance
耐擦性 erasibility
耐藏力 storability
耐藏潜力 storage potential
耐冲击材料 high-impact material
耐虫性 insect resistance
耐船蛆性 teredo resistance
耐醇性 alcohol resistance
耐冻裂性 cold-crack resistance
耐冻性 freezing resistance; frost hardiness
耐风化胶合板 exterior plywood

耐风力 wind firmness; windthrow resistance
耐风性 wind resistance
耐辐射的 radioresistance
耐腐性 decay durability; decay proof; decay resistance
耐干燥种子 disiccation-tolerant seed
耐光的 light-resistant
耐光度 fastness to light
耐光性 fastness to light; light fastness; light permanency; light resistance
耐寒的 hardy
耐寒的多年生植物 hardy perennials
耐寒力 cold endurance; frost-resisting power
耐寒树种 winter hardy tree species
耐寒性 cold endurance; cold hardiness; cold tolerance; frost hardiness
耐寒性的 frost-hardy
耐寒性一年生植物 hardy annuals
耐寒宿根植物 hardy perennials
耐旱性 drought hardiness; drought resistance; drought tolerance
耐火材料 fire proofing material
耐火处理 fire resisting finish
耐火的 flame-proof
耐火极限 limit of fire resistance
耐火剂 fire proofing agent; fire proofing chemical
耐火胶合板 fireproof plywood; fire-resistant plywood
耐火木材 refractory timber
耐火黏土 fireclay
耐火树种 fire resistant tree
耐火涂料 fire resisting finish
耐火纤维板 flame-retardant fibre board
耐火性 fire resistance; flame resistance
耐火性试验 fire-resistance test
耐火植物 pyrophyte
耐碱性 alkali resistance
耐久 wear
耐久材 durable wood
耐久强度 endurance strength
耐久试验 durability test
耐久树种 durable species
耐久限度 endurance strength
耐久性 durability; fastness; persistence
耐久性试验 endurance test
耐量 tolerance
耐淋洗能力 leach resistance
耐磨 wear
耐磨强度 abrasive resistance
耐磨清漆 rubbing varnish
耐磨试验 abrasion test
耐磨试验机 abrasion testing machine
耐磨性 abrasion resistance; abrasive resistance; friction resistance; wearability
耐磨硬度 abrasion hardness; abrasive hardness
耐疲劳性 fatigue durability
耐破度 bursting strength
耐破度试验器 bursting tester
耐破强度 pop strength
耐气候性 weatherability
耐热性 heat resistance; heat tolerance
耐热性胶黏剂 thermoresistance adhesive
耐热植物 thermophyte
耐揉性 crumpling resistance
耐晒性 light permanency
耐湿性 humidity resistance
耐湿性涂料 moisture-resisting paint
耐湿纸 wet strength paper
耐受度 tolerance degree
耐受限度 limit of tolerance
耐水胶 water-resistance glue
耐水胶合板 waterproof plywood
耐水黏合剂 water-resistance adhesive
耐水性 water resistance
耐酸处理 acid proofing treatment
耐酸的 acid-tolerant; aciduric
耐酸涂料 acid-proof coating
耐酸细菌 acid fast bacteria
耐酸性 acid fastness; acid resistance
耐酸性染色 acid fast stain
耐酸植物 acid plant
耐酸砖 acid-resisting tile
耐踏性 resistance to trampling
耐铜性菌类 copper-tolerant fungi
耐性 degree of hardiness; hardiness; tolerance
耐性定律 law of tolerance

耐性生态学 toleration ecology
耐朽性 decay durability
耐压的 barotolerant
耐烟雾性 fume resistance
耐烟植物 fume-tolerant plant
耐淹性 flood tolerance
耐盐的 salt-tolerant
耐盐性 salinity tolerance; salt endurance; salt tolerance
耐盐植物 salt-tolerant plant
耐盐种 salt enduring species
耐氧的 aerotolerant
耐药性菌 tolerant fungi
耐药真菌 resistant fungi
耐蚁剂 termi-repellent
耐阴的 shade tolerant; shade-bearing; shade-enduring
耐阴度 shade-endurance; shade tolerance
耐阴类 shade bearing guild
耐阴树 shade-bearer; shade-enduring tree; shade-kind; shade-tolerating tree; tolerant-tree; shade-bearing tree
耐阴树种 shade-tolerant tree species
耐阴下木 tolerant understorey
耐阴性 tolerance of shade; shade tolorance
耐阴性理论 theory of tolerance
耐阴植物 shade-enduring plant; shade-tolerant species
耐阴中生树 tolerant mesophytic tree
耐阴种 shade-tolerant species; shade-beare; tolerant species
耐用材 durable wood
耐用树种 durable species
耐油的 oil proof
耐油度 grease proofness
耐油性试验 grease resistance test
耐油纸 pergamyn paper
耐折叠强度 creasing strength
耐折度 folding number
耐折度测定仪 folding tester
耐折强度 folding endurance; folding strength
耐皱性 creasability; crease resistance
耐贮性 keeping quality
萘乙酸 naphthalene acetic acid
南北贸易 vertical trade
南方涛动 southern Oscillation
南方针叶林带 Austroriparian life zone
南非干燥高原 karroo
南回归线 Tropic of Capricorn
南极 south pole
南极带 antarctic zone
南极的 antarctic
南极荒漠 antarctic desert
南极区 antarctic region
南极圈 antarctic circle
南极生物带 antarctic life zone
南界 notogaea
南坡 south-facing slope
南蛇藤碱 celastrine
南弯式索道(下坡)集材 south bend yarding
南五味子油 kadsura oil
南洋杉树林 pinheiro
南洋杉型纹孔 araucaroid pit
难发芽的种子 refractory seed
难干燥材 refractory wood
难加工的难处理的 refractory
难加工木材 refractory timber
难浸注材 impermeable wood; resistant wood
难控性(火) resistance to control
难燃性 fire retardant
难以诱导萌发的(种子) refractory
楠木油 machilus oil; nanmu oil
楠叶油 machilus leaf oil
囊 sac
囊柄 suspensor
囊层基 hypothecium (*plur.* -ia)
囊地鼠 gopher
囊果 cystocarp
囊果皮 pericarp
囊托(蕨类等) apophysis (*plur.* -hyses); receptacle
囊脱石 nontronite
囊形突(昆虫) saccus
囊枝状菌根 vesicular-arbuscular mycorrhiza
囊状 scrotiform
囊状的 saccate
囊状栅 bag boom
挠度 buckling; deflection; sag
挠度计 deflectometer

挠曲 bending; bending flexure; deflection
挠曲刚度 flexural rigidity
挠曲强度 flexural strength; modulus of rupture
挠曲系数 flexibility coefficient
挠曲性 plasticity
挠曲应变 flexural strain
挠性 flexibility
挠性管 flexible pipe
挠性胶合板 flexible plywood
挠性联轴器 flexible coupling
挠性软管 flexible hose
挠性塑料垫板 flexible plastic caul
挠折模量 modulus of rupture
脑皮层 cortex (*plur.* -ces)
内凹 cove
内孢 endospore
内孢子 endogenous spore; endospore
内胞浆 endoplasm
内薄层 inner lamella
内壁(孢子及花粉) entine; intine
内标物(化学) internal standard
内标准(光学) internal standard
内表面(细胞的) inner surface; inside surface
内禀增长力 inner capacity increase
内禀增长率 intrinsic rate of increase
内禀自然增长率 intrinsic rate of natural increase
内部 interior
内部报酬率 internal rate of return
内部边材变色 interior sap stain
内部变率 internal variability
内部变色 interior stain(ing)
内部叠层 inner overlap
内部腐朽 inside rot
内部干裂 interior check; internal check
内部共生 intestinal symbiont
内部归一化 internal normalization; internal standard normalization
内部红斑 inside red spot
内部环裂 internal shake
内部寄生物 endoparasite
内部检疫 internal quarantine
内部接线图 cut-away view
内部节律 internal rhythm
内部扩散 inner diffusion
内部蓝变 interior blue stain
内部排水 internal drainage
内部缺陷 inside defect; internal defect
内部审计 internal audit
内部施胶 internal sizing
内部收益率 internal rate of return
内部通信系统 intercommunications system
内部纤维化 internal fibrillation
内部种 interior species
内部帚化 internal fibrillation
内材面(向髓心的一面) inner surface of wood
内侧傍管薄壁组织 adaxial paratracheal parenchyma
内侧薄壁组织 adaxial parenchyma
内层 inner layer
内插地方材积表 interpolation for local volume table
内成岩 endogenetic rock
内翅类(昆虫) endopterygote
内唇(昆虫) epipharynx; sublabrum
内存 memory
内导杀虫剂 systemic insecticide
内地水域 inland water
内动力土壤 endodynamic soil
内动力型土壤 endodynamomorphic soil
内多倍性 endopolyploidy
内翻 plough in
内分布型 internal distribution pattern
内稃(禾本科) palea; palet; pale; glumelle
内腐 hollow heart; internal decay
内果皮 endocarp
内含边材 double sap wood; false sapwood; included sapwood; inside sapwood; internal sapwood
内含成本 imputed cost
内含节 enclosed knot
内含韧皮部 included phloem; interxylary phloem
内含体 inclusion body
内含纹孔口 included (pit) aperture
内含物 inclusion
内含子 introne
内弧表面砂光机 scroll sander
内环 inner ring

N

内环境 inner-environment
内环维管束 inner circle of bundles
内混型喷枪 internal mixing gun
内击锤枪 hammerless gun
内寄生 endoparasitism; internal parasitism
内寄生虫 endoparasite
内寄生的 endophytic; endotrop(h)ic
内寄生菌 endophyte
内寄生物 endoparasite
内夹皮 closed bark-pocket
内胶结强度 internal bond strength
内酵素 endoenzyme
内景 interior
内径 inner diameter; inside diameter
内径规 internal gauge; plug gauge
内聚 coherence
内聚的 cohesive
内聚力 coherence
内聚力破坏 cohesion failure
内聚力学说 cohesion theory
内聚性 cohesion
内聚性物种 cohesive species
内聚作用 cohesive action
内卷 involution
内菌根 endotrop(h)ic mycorrhiza
内卡 internal gauge
内莰烯 endocamphene
内拉力 interior tension
内涝 inland inundation
内冷凝管干燥窑 internal condensing-coil kiln
内裂 inner checking
内流盆地 endorheic basin
内陆部分 interior
内陆排水水系 inland drainage; interior drainage
内陆沙 inland-sand
内陆湿地和水域生态系统类型自然保护区 nature reserve for inland wetland and water area ecosystem
内陆水系 inland drainage; interior drainage
内轮对瓣的 proteropetalous
内毛管水 inner capillary water
内酶 endoenzyme
内面 inner surface; inside surface
内面校正 internal adjustment
内摩擦 internal friction
内胚层质 endoplasm
内皮 endodermis; hypophloedes; inner bark; under bark; underbark
内皮层 endodermis
内蹼 inner web
内倾水平梯田 gradoni
内曲 incurvation
内曲花药 incumbent anther
内曲胚根 incumbent radicle
内燃机 internal combustion engine
内热点 endothermic point
内热釜 internal gas-heated retort
内韧的 endophloic
内韧皮部 endophloem
内渗 endosmosis
内生孢子 endogenous spore; endospore
内生担子 endobasidium (*plur.* -ia)
内生的 endogenous
内生分生孢子 endoconidium (*plur.* -ia)
内生菌 endophyte
内生菌根 endomycorrhiza; endotrophic mycorrhiza
内生内皮层 inner endodermis
内生韧皮部 internal phloem
内生生物 endobiont
内生生物群集 endobiose
内生芽孢 endogenous spore
内生岩 endogenetic rock
内生真菌 endophyte
内生植物 endophyte
内生植物的 endophytic
内食的 endophagous; entophagous
内始式 endarch
内始式木质部 endarch xylem
内始式原生木质部 endarch protoxylem
内斯托斯控速式集材 Nestos snubbing yarding
内填生长 intussusception growth
内填生长学说 intussusception theory
内填作用 intussusception
内贴板 lining
内外翻垡法(轮流内翻垡和外翻垡) balk-ploughing

内外生菌根 ectendotrophic mycorrhiza
内维型磨蚀机 Navy abrader
内维型磨蚀计 Navy-type wearometer
内温性 endothermy
内稳定过程 homeostatic process
内稳态 homeostasis
内稳态生物 homeostatic organism
内吸的 systemic
内吸磷(杀虫剂) systox; demeton
内吸杀虫剂 systemic insecticide
内吸杀菌剂 systemic fungicide
内陷气孔 sunken stomata
内芯 inner core
内休眠 internal dormancy
内循环干燥窑 internal circulating kiln
内因 intrinsic factor
内因动态演替 endodynamic succession
内因性的 endogenous
内因性迁移 endogenous migration
内因演替 autogenic succession
内应变 internal strain
内应力 internal stress
内原生质 endoplasm
内原型原生木质部 centrifugal protoxylem
内圆线刨 concave plane
内源的 endogenous
内源节律 endogenous rhythm
内源性的 endogenous
内在环境 internal environment
内在优势 intrinsic superiority
内在周期 inner periodicity
内脏 viscera
内增塑作用 internal plasticization
内酯 lactone
内酯酶 lactonase
内质 endoplasm; endosarc
内质网 endoplasmic reticulum
内种皮 endotesta; inner coat
内重复 endoreduplication
内轴 inner spindle
内珠被 inner integument; internal integument
内转 involution
内装风机干燥窑 internal-fan kiln
嫩插条 wand
嫩化剂(肉类) tenderizer
嫩节 epicormic knot
嫩芽 chit; imp
嫩叶 leaflet; young leaf
嫩叶的 therophyllous
嫩枝 burgeon; epicormic branch; imp; shoot
嫩枝插 shoot cutting
嫩枝插穗 softwood (stem) cutting; summerwood cutting
嫩枝扦插 softwood (stem) cutting
嫩枝叶 browse
能传病毒的 viruliferous
能发芽的 fertile; germinable
能耗 energy consumption
能见度 visibility
能见度计 visibility meter
能见距离 visibility distance
能见区地图 seen-area map; visibility map; visible-area map
能垒 energy barrier
能力建设 capacity building
能量变化 energy change
能量变换 energy transformation
能量传输 energy transfer
能量代谢 energy metabolism
能量耗损 energy dissipation
能量金字塔 pyramid of energy
能量库 energy sink
能量垒 energy barrier
能量流动 energy flow
能量平衡 energy balance
能量散逸 energy dissipation
能量势垒 energy barrier
能量释放因素 energy release component
能量收支 energy budget
能量守恒定律 law of conservation of energy
能量输入 energy input
能量位垒 energy barrier
能量消耗 energy consumption; energy dissipation
能量消散 energy dissipation
能量转换 energy transformation
能量转换者 energy transformer
能流 energy flow; energy flux
能流分析 energy analysis

能流结构 energy flow structure
能流路径 energy flow path
能配群 comparium
能稔花 fertile flower
能散X射线分析 energy-dispersive X-ray analysis
能手 cattyman
能通量 energy flow
能育的 fertile
能育花粉 fertile pollen
能育性 fertility
能育枝 fertile branch
能源 energy resource; energy source
能源材 energy wood
能源服务 energy service
能源林 energy forest; energy plantation
能源强度 energy intensity
能源税 energy tax
能源消耗 energy consumption
能源植物 energy crop
能源转化 energy conversion
能源资源 energy resource
能值 emergy
尼古丁 nicotine
尼龙刷辊 nylon brush roll
泥岸 muddy shores
泥鸽射击 skeet
泥鸽射击喉缩 skeet choke
泥灰土 marl soil; marly soil
泥灰岩 marl
泥浆 sludge
泥坑 mud hole
泥砾 boulder clay
泥流 earth flow; earthflow; mud flow; mudflow; solifluction
泥流砾岩 cenuglomerate
泥煤蜡 peat wax
泥内生的 endopelic
泥盆纪 Devonian period
泥沙径流 sediment runoff
泥沙输移 silt transportation
泥沙输移比 sediment delivery ratio
泥石崩坍 detritus avalanche
泥石沉积 detrital deposit
泥石流 debris flow; mud-rock flow; mud-stone flow
泥石流侵蚀 debris flow erosion
泥炭 peat; peet
泥炭表层 histic epipedon
泥炭层(土壤剖面H层) peat horizon; H-horizon; H-layer
泥炭地 peatland
泥炭火 peat bog fire
泥炭焦油 peat tar
泥炭块(育苗基底材料) peat block; peat cube
泥炭黏土 peaty clay
泥炭潜育土 peaty gley soil
泥炭苔 peat moss
泥炭土 peat moss; peat soil; peaty soil
泥炭藓 sphagnum moss
泥炭藓型地上芽植物 chamaephyta sphagnoides
泥炭藓沼泽 Sphagnum bog; Sphagnum moor
泥炭营养钵(营养砖) peat plant block
泥炭沼泽 bog; moor; muskeg; ombrotrophic bog; peat bog
泥屑灰岩 calcilutite
泥沼状态 bogginess
泥质板岩 argillaceous slate; argillite
泥质胶结构 argillaceous cement
泥质石灰岩 argillaceous limestone
泥质岩 argillite; pelite
泥质岩相 argillaceous facies
拟保幼激素 juvabione; paper factor
拟表型 phenocopy
拟柄状的 subpedunculate
拟侧丝 pseudoparaphysis
拟等位基因 pseudo-alleles
拟古式(园林或建筑) classic style
拟合度 goodness of fit
拟横向分裂 pseudotransverse division
拟横向形成层细胞分裂 pseudotransverse cambial cell division
拟黄土 loess-like soil
拟寄生 parasitoidism
拟寄生的 parasitoid
拟警戒色 pseudoaposematic
拟连锁 pseudo-linkage
拟木栓细胞 phelloid cell
拟侵填体 tylosoid; thylosoid
拟染色体 chromosomoid
拟散孔 foraminoid

拟态 mimesis; mimic; mimicry
拟态基因 mimic gene
拟蜕皮激素 ecdysoid
拟网状 ornate
拟显性 pseudodominance
拟脂 lipid(e); lipoid
逆表层硬化 case tension
逆表面硬化 reverse casehardening; reverse(d) casehardening
逆差 deficit
逆风 opposing wind
逆风切 upshear
逆风用火 back burn
逆行 retrogression
逆行分化 regressive differerentiation
逆行进化 evolutionary reversion
逆行气流 back draft
逆行循环 reverse circulation
逆行循环喷气式干燥窑 reverse circulation jet blower type kiln
逆行演化 regressive evolution
逆行演替 regressive succession; retrogression of succession; retrogressive succession
逆境 stress
逆境测定 stress test
逆境发芽 stress germination
逆锯(从下向上锯) undercutter
逆锯口(从下面开的锯口) undercut
逆锯造材(由下向上锯截) underbucking; undercutting
逆利息 negative interest
逆联动装置 backward linkage
逆流 refluence
逆流式干燥窑 counter current type (dry) kiln
逆密度制约因子 inverse density-dependent factor
逆平行的 antiparallel
逆坡 adverse grade
逆坡集材 uphill skidding; uphill yarding
逆切变 upshear
逆水层 inverse stratification
逆温 inversion; temperature inversion; thermal inversion
逆纹 bristled grain
逆铣 conventional milling; up milling
逆向反捻索 regular lay rope
逆向河 anticonsequent stream
逆向锯齿 retroserrate
逆向空气环流 reversing air circulation
逆向扭转 antidromal torsion
逆向气流 counter current air
逆向气流干燥窑 counter current type (dry) kiln
逆向渗透 reverse osmosis
逆芽接 reversed budding
逆掩断层 overthrust
逆增 inversion
腻斑 grease spot
腻子 mastic; sealant; wood putty; wood sealer
年变幅 annual amplitude
年表 chronology
年产量 annual output
年次数字总和逐次比例折归法 sum-of-the-year's-digits depreciation
年次相关 year-to-year correlation
年代序列 chronosequence
年代学 chronology
年等温线 annual isotherm
年度检修 annual overhaul
年度预算编制法 annual budget formulation directive
年度作业计划 annual operation plan
年伐量 annual cut
年伐率 annual cutting percent
年伐面积 annual coupe; annual cutting area; annual felling area
年工作日 annual working days
年际变化 inter annual change; annual change; annual variation
年际的 inter annual
年降水量 annual precipitation; precipitation per annum
年较差 annual range
年节律 circannian rhythm; circannual rhythm
年结实量 annual bearing
年金计算公式 annuity formulae
年金折旧法 depreciation-annuity method
年径流量 annual runoff
年利率 annual interest rate
年龄比率 age ratio
年龄测定 age determination

年龄分布 age distribution
年龄级(动物出生年) year class
年龄结构 age structure
年龄金字塔 age pyramid
年龄面积假说 age and area hypothesis
年龄组成 age composition
年流量 annual discharge; annual flow
年轮 annual (growth) ring; annual ring; ring; yearly ring
年轮边界 ring boundary
年轮测宽仪 dendrochronograph
年轮层 annual (growth) layer
年轮带 annual zone
年轮分析 annual ring analysis
年轮分析仪 tree ring measurement system; tree ring measuring device; tree ring analyzer; tree ring positioning table; tree ring positiometer
年轮计数 ring count
年轮界 annual ring boundary
年轮均匀 even grain
年轮宽度 annual ring breadth; annual ring width; ring width
年轮宽度指数 ring width index
年轮宽松的 open-grown; wide-ringed
年轮密度 annual ring density
年轮年代学 dendrochronology
年轮气候学 dendro-climatology; tree-ring climatology
年轮取样器 core extractor
年轮生态学 dendroecology
年轮条 increment core; tree-ring core
年轮学 dendrochronology
年轮样品 core sample
年毛利 annual gross earning
年平均林利率 mean annual forest percent
年平均增长率 average annual growth rate
年侵蚀量 annual erosion
年生产量 annual production
年生长 annual increment; growth (annual) increment
年生长层 annual (growth) layer
年生长量 annual growth; annual increment
年收获 yearly yield
年收获量 annual yield
年收入 yearly receipts
年收入率 yield
年收支 annual budget
年损失量 annual loss
年温较差 annual range of temperature
年增长 annual growth
年支出费用 annual charge
年直接流出率 coefficient of annual direct runoff
年最大流量 annual flood
黏板 fiber sticking
黏板岩 clay slate
黏程 plastic range
黏稠度 sliminess
黏稠性质的 heavy bodied
黏弹计 viscoelastometer
黏弹特性 viscoelastic behavior; viscous-elastic behaviour
黏弹体 visco-elastic body
黏弹性 viscoelasticity
黏弹性流动 visco-elastic flow
黏点 sticky point
黏点试验仪 sticky point tester
黏度 consistency; degree of viscosity
黏度杯 flow cup
黏度单位 viscosity unit
黏度计 fluid(i)meter; viscosimeter
黏多糖 mucopolysaccharide
黏附 adhere; adhesion
黏附计 adherometer; adhesiometer
黏附体破坏 adherend failure
黏附作用 sticking
黏合 bind; binding; bond; cementing; cohesion
黏合的 cohesive
黏合计 adherometer
黏合剂 binder; binding material; cement; emulsifier
黏合期 joint aging time
黏合强度 bond strength
黏合清漆 adhesive varnish
黏合体 bond
黏合状态 tacky state
黏合阻力 block resistance
黏化层(土壤剖面B_t层) argillic horizon; B_t-layer

N

黏化的 argillic
黏胶 glut; mucilage; viscoid; viscose glue
黏胶法 viscose process
黏胶丝 viscose silk
黏胶纤维(黏胶液) viscose
黏接 cementing
黏结土 cohesive soil
黏结作用 cohesive action
黏锯 gum up
黏均分子量 viscosity-average molecular weight
黏菌 myxomycete; slime-fungus; slime-mold
黏粒 clay
黏粒含量 clay content
黏粒淋失 clay impoverishing
黏盘 claypan
黏盘土 terras soil
黏磐 clay pan
黏磐黑钙土 clay pan chernozem
黏磐土 planosol
黏壤土 clay loam
黏态 plastic state
黏土 clay; clay soil; slime
黏土矿物 clay mineral
黏土硬层 clay pan
黏网 sticking
黏性 plasticity; stick; viscosity
黏性持续度 tenacious consistency
黏性胶体 viscoloid
黏性流 viscous flow
黏性末端 cohesive end
黏性凝胶 viscogel
黏性浓度 tenacious consistency
黏性期 tack range
黏性土壤 cohesive soil
黏液 mucilage; mucosa; slime
黏液道 mucilage canal; mucilage duct
黏液化(浸水木材表面黏液) obstruction by slime
黏液流溢 slime-flux
黏液细胞 mucilage cell
黏着 blocking; cling
黏着的 adherent; adhesive
黏着剂 adhesive; sticker
黏着力 adhesion; adhesive force
黏着强度 tack strength
黏着性 stickability; stickness
黏着诱捕法 sticky-trap method
黏着诱捕器 sticky trap
黏质灰壤 clayed podzol
黏质砂岩 clay sandstone
黏质渗出物 slimy exudate
黏质细胞 mucilage cell
黏质纤维 mucilaginous fibre
黏质纤维素 muco-cellulose
黏质页岩 clay shale
黏质棕色壤土 clay brown loam
黏滞度 viscosity
黏滞流动 plastic flow; viscous flow
黏滞性 tenacity
黏重土壤 heavy soil
黏重质地 heavy texture
黏状打浆 shiny beating; wet beating
黏状浆 slow stock; wet stock
黏状叩解 slow beating
捻 laying
捻接 splice
碾碎 pulverization
碾压剥皮 compression debarking
碾压剥皮机 compression debarker
念珠形(昆虫触角) moniliform
念珠状射线 beaded ray
酿酶 zymase
酿造 brew
酿造锅 brewing kettle
酿造学 zymurgy
鸟巢菌 birds' nest fungus; nidulariaceous fungus
鸟粪 guano
鸟苷 guanosine
鸟喙形线脚 qurik moulding
鸟瞰 bird's-eye view
鸟瞰图 birds-eye view
鸟类保护 protection of birds
鸟类保护区 bird sanctuary
鸟类观察家 birdwatcher
鸟类区系 avifauna
鸟类学 Ornithology
鸟类学家 ornithologist
鸟媒 ornithophily
鸟媒花 ornithophilous flower; ornithophily
鸟媒性 ornithophily

鸟媒植物 ornithophilous plant
鸟嘌呤核苷 guanosine
鸟枪 fowling piece; shot-gun
鸟枪独弹 shot-gun-slug
鸟枪猎枪 scatter gun
鸟枪子弹 shell; shot-gun shell
鸟舍 aviary
鸟兽保护区 sanctuary
鸟眼花纹 bird's eye
鸟眼纹 bird's eye grain
鸟眼纹理 bird's eye figure; bird's eye grain
鸟眼纹木材 birds' eye wood
鸟浴池(园林装饰小品) bird-bath
鸟啄痕 bird peck
鸟足状的 pedate
尿苷 uridine
尿苷酸 uridylic acid
尿嘧啶核苷 uridine
尿素 urea
尿素树脂 urea resin
尿素树脂处理材 urea-resin treated wood
尿素树脂胶黏剂 carbamide resin adhesive; urea-resin adhesive
脲 urea
脲尿素 carbamide
捏和 knead; kneading
捏和机 kneader
啮齿动物忌避剂 rodent repellent
啮齿类毒杀剂 rodenticide
啮害 gnawed
啮合 joggle; mesh
啮合榫接头 joggle joint
啮接 bridle joint
啮蚀状的 eroded
镊合状的(花被卷叠代) valvate
镊子 forceps
柠黄质 citroxanthin
柠檬桉精油 eucalyptus citriodora oil
柠檬桉油 citriodora oil; eucalyptus citriodora oil
柠檬叉丙酮 citrylidene acetone
柠檬虫胶 lemon lac
柠檬精油 lemon oil
柠檬苦素 limonin
柠檬素 citrin
柠檬酸循环 citric acid cycle; Krebs cycle
柠檬萜烯 citrine
柠檬烯 cinene
柠檬油 lemon oil
柠檬紫胶 lemon lac
凝冻时间 jelly time
凝固 set; syneresis
凝固点 set(ing) point
凝固时间 set time
凝灰岩 tuff
凝集剂 agglutinant
凝集素 agglutinin
凝集作用 agglutination
凝胶 gel; gelatin
凝胶稠度 gel consistency
凝胶分散作用 gel peptization
凝胶过滤色谱 gel filtration chromatography
凝胶化 gelatination; gelatinization; gelation
凝胶胶溶作用 gel peptization
凝胶能力 geling power
凝胶排阻色谱 gel exclusion chromatography
凝胶强度 gel strength
凝胶色谱法 gel chromatography
凝胶渗透色谱 gel permeation chromatography
凝胶体 gelatin
凝胶纤维 gelatinous fibre
凝胶状态 gel state
凝胶作用 gelatinization; gelation
凝结 agglomerate; condensation; precipitate
凝结热 heat of condensation
凝结时间 setting time
凝结水 condensed water
凝汽阀 steam trap; trap
凝乳 curd
凝缩 condensation
凝缩单宁 condensed tannin
凝缩类栲胶 condensed tannin extract
凝缩率 condensing factor
凝缩鞣质 condensed tannin
拧茬 pull; pull out; pull around
拧紧力 screw-holding power
牛蒡(根) lappa
牛蒡油 bardane oil
牛顿求积式 Newton's formula

牛轭 oxbow
牛轭湖 oxbow lake
牛放牧日 cow-day
牛角瓜苷 calotropin
牛力集材 ox-skidding
牛马道 horse-road
牛马套子集材 bull team logging
牛排菌 liver fungus; beef steak fungus
牛皮纸 brown packing paper; kraft
牛皮纸浆 kraft pulp
牛皮纸浆制造法 kraft process
牛头刨床 shaper
牛月单位 cow-month
牛樟油 sho-gyu oil
牛脂(做皂烛等) tallow
扭秤 torsion balance
扭结 kink
扭矩 leverage; torque moment; torsional moment; twisting moment
扭矩扳手 torque wrench
扭矩计 torsionmeter
扭力 torsional force; torsional load; turning force
扭力计 torsionmeter
扭力黏度计 torsion viscosimeter
扭力天平 torsion balance
扭曲 distortion; twist; twisting
扭曲方向 direction of twisting
扭曲强度 torsional strength tester
扭曲树干 twisting stem
扭曲形扶手 wreathed hand-rail
扭曲状态 twisted state
扭应变 torsional strain
扭应力 torsional stress
扭枝压条 layerage by twisting
扭转 twist; wrench
扭转常数 torsion constant
扭转共振 torsional resonant vibration
扭转极限 torsional limit
扭转角 twist(ing) angle
扭转力偶 twisting couple
扭转模量 torsion modulus
扭转式针阀芯 wiggle wire
扭转弯曲 torsional bending
扭转纹 corkscrew-growth; torsion-fibre; twisted grain
扭转纤维 twisted fibre
扭转载荷 torsional load
扭转振动 torsional vibration
钮状紫胶 button lac
农场防风林 farmstead windbreak
农场经营者 agriculturist
农场林 farm forest; farm woodlot
农场林地 farm woodland
农场林合作社 farm forest cooperative
农场林业 farm forestry
农场林主 farm forest owner
农场森林 farm woods
农场周围防风林 farmstead windbreak
农地指示生物 agricultural indicator
农家肥 farmyard manure
农家林 farm forest; farm woodland; farm woods
农具用材 timber for agricultural implement
农林地质学 agroforestrial geology
农林复合生态系统 integral agroforestry ecosystem
农林合作法案 cooperation farm forestry act
农林间作 combination of forest and farm crop; taungya plantation
农林业 agroforestry
农牧轮作 rotation pasture
农田 farmland
农田防护林 farm shelterbelt; farmland shelter-belt
农田防护林带 farm shelterbelt
农田生态系统 farmland ecosystem
农学 agronomy
农药 agricultural chemical; agricultural chemical pesticide; agromedicine; pesticide
农药的 agromedical
农药污染 pesticide contamination
农业(学)家 agriculturist
农业经济区位 agricultural economical zoning
农业气象预报 agrometeorological forecast
农业生态工程 agricultural eco-engineering
农业生态系统 agroecosystem
农业生态学 agricultural ecology; agroecology

农业土壤学 edaphology
农业资源 agricultural resources
农业自然区划判造 interpretation of agricultural regionalization
农业自然条件判读 interpretation of agricultural natural conditions
农艺品种 agronomic variety
农艺土壤学 edaphology
农用材 wood for agricultural purpose
农用抗菌素 agricultural antibiotic
农用木材 agricultural timber
浓白水 enriched water
浓差陡度 concentration gradient
浓度 concentration; consistency
浓度梯度 concentration gradient
浓度调节 regulation of consistence
浓浆 slush pulp
浓胶 concentrated extract
浓色 overtone
浓缩 concentration; enrichment; thickening
浓缩肥料 high analysis fertilizer
浓缩粉 dense powder
浓缩机 thickener; thickner; decker
浓缩机浆槽 thickener vat
浓缩浆池 thickening chest
浓缩浸提物 concentrated extract
浓缩系数 concentration factor
浓液浸蘸 concentrated-solution-dip
浓荫采伐 dark-felling
耨锄 draw hoe
奴尔斯焦油 Knoweles tar
努里法 Lowry process
努氏硬度试验 Knoop-test
暖藏 warm storage
暖层积处理 warm stratification
暖房 green house; greenery
暖锋 warm front
暖锋面 anaphalanx; warm front surface
暖管 warming-up
暖流 warm current
暖色 warm colour
暖水种 warm-water species
暖温带 temperate warm region; temperate warm zone; warm-temperate district; warm-temperate zone
暖温带落叶林 warm temperate deciduous forest
暖温带雨林 warm temperate rain forest
暖温带植物 warm temperate plant
疟疾 malaria
挪威云杉 common spruce; Norway spruce; European spruce
诺布尔发芽皿 Noble germinating dish
诺尔德海姆防腐法(空细胞法之一) Nordheim process
诺模图 nomograph
噢(wō) benzylidenecoumaranone
噢哢 aurone
欧夹竹桃油 oleander oil
欧石南群落 ericifruticeta; heath
欧石南型木本群落 ericilignosa
欧石楠丛生荒野 heathland
欧亚板块 Euro-Asia plate
欧洲式木块菌类培养试验 European wood-block test
偶差 chance error
偶氮苯 azobenzene
偶发心材树 tree with faculatatively coloured heartwood
偶发异交 unexpected outcrossing
偶发杂交 vicinism
偶合剂 coupling agent
偶合染色体 mixochromosome
偶极子 doublet
偶见的 accidental
偶见种 accidental species; casual species; incidental species; occasional species
偶来泉的 crenoxenous
偶来污水的 saproxenous
偶联反应 coupled action
偶联呼吸 coupled respiration
偶联木(造材时未完全锯断) bucker's coupling; coupled log
偶然变异 accidental variation
偶然单性生殖 parthenogenesis facultative
偶然浮游生物 tychoplankton
偶然停车 accidental shut down
偶然误差 accidental error; random error
偶然相关 contingency correlation
偶然芽变 accidental sport
偶然杂交 occasional crossing

偶深性 bathoxenous
偶适地性生物 geoxene
偶数多倍体 evenpolyploid
偶数羽状的(复叶) abruptly pinnate; interruptedly pinnate; even pinnate; paripinnate
偶数羽状复叶 paripinnate leaf
偶线 zygotene
偶线期 zygotene stage
偶遇植物 stranger plant
偶遇种 stranger; stranger species
耦合 couple
耦合器 coupler
藕 lotus root
爬虫 crawler; creeper
爬虫学 herpetology
爬杆 pole switch
爬杠 skid; skid log
爬杠装车 send-up
爬行 crawling
爬犁 catamaran; joe-log; jumper; scoot; sledge; sleigh; stone boat
爬犁道 kimma sled road; sledge road; travois road
爬犁滑脚 moccasin
爬犁滑木(集材) alligator
爬犁集材工 sled tender
爬犁集运材 sledging
爬犁脚两侧护铁 bridle iron
爬犁连接杆(成V形) crotch tongue
爬犁列车 trailer
爬犁起动 break out
爬犁翘头 nosing
爬犁铁配件 sled iron
爬犁腿 runner; sled runner; sledge runner
爬犁腿前端 nose
爬犁腿前端护铁 sled shoe
爬犁腿制动链 runner chain
爬犁腿制动爪 runner dog
爬犁运材 sled hauling
爬犁装卸工 chain tender; sled tender
爬坡能力 gradient capability
爬山者 climber
爬树刺钉 spur
爬树刺靴 tree climbing spur
爬树截梢工 climber; high climber
爬树器 climbing-iron
爬树绳 climbing rope
耙 harrow; rake
耙齿 clear teeth; peg; rake finger; raker (tooth)
帕马桑防水剂 Permasan
帕施克圆筒剥皮机 Paschke (drum-peeler)
拍卖 auction; auction sale; sale by auction
拍卖价格 upset price
拍卖市场 auction market
排 tier
排斥 exclusion
排出阀 bleed valve
排出管 blow-off pipe
排出物标准 effluent standard
排出压力 delivery pressure
排出装置 distributor
排除 exclusion
排除法 removal census; removal method
排除卡阻木 break out
排钉 raft nail
排队论 queuing theory; waiting line theory
排放 blow-off; discharge; emissions
排放标准 effluent standard
排放的 effluent
排放量减少单位 emissions reduction unit
排放贸易 emissions trading
排放配额 emissions quota
排放清单 emission inventory
排放情景 emissions scenario
排放税 emissions tax
排放速度 evolution rate
排放许可 emissions permit
排风机 exhaust fan
排绠 rafting boom
排管式釜 calandria pan
排灌犁沟 water furrow
排架 framed trestle
排节 brail; crib; raft section; section raft
排节流送 raft-section driving
排节撬棒 boom pin
排锯 frame saw; gang saw; gate saw; log gang saw; log gang saw; sash saw
排锯工 gang sawyer
排锯制材厂 gang(saw)mill
排锯装料工 cant setter

排框 rafting boom; standing boom
排捆 raft bundle
排涝沟 surface drain
排梁木(木排横木) lash pole; stringer piece; stringer; lock down; set of stick
排梁索 swifter line
排料头(刨花) pass-out head
排料针辊 discharge spiked roller
S形排料装置 S-discharging device
排列 orientation; taxis
排卵 ovulation
排泥孔 mud hole
排气 air drainage; air exhaust; degassing; exhaust; ventilation
排气道 air outlet flue; flue
排气阀 blow-off valve
排气风扇 exhaust fan; extraction fan
排气管 air outlet flue; exhaust pipe; nozzle
排气孔 exhaust port
排气口 air exhaust; exhaust outlet; exhaust port; vent
排气口出口截面 outlet
排气噪音 exhaust noise
排气装置 air exhaust; exhaust; ventilator
排遣 egestion
排绕子 swifter line
排沙量 silt discharge
排上不浸水木材 dry-floated timber
排上运的(木材在水面以上) floated bright
排水 catchment; drain; drainage; water drainage
排水槽 drain tank; drainage tray
排水池 drain tank
排水出路 discharge outlet
排水干渠 arterial drainage
排水沟 drain; drainage; leader; offtake; outlet drain; turnout; water drain
排水沟挖掘机 land-drainer
排水管 drain pipe
排水孔 drainage hole
排水口 exhaust outlet; outfall; outlet; water spout
排水良好的 well-drained
排水量 withdrawal
排水流量 drain flux
排水面积 drainage area
排水器 hydathode
排水区 drainage basin
排水区域 drainage area
排水渠 conduit; offtake
排水设备 drainage
排水时间 drainage time
排水土沟 soil drainage
排水系统 drainage; drainage system
排水细胞 hydathodal cell
排头 crab
排头木 head stick
排污标准 effluent standard
排泄 excretion
排泄阀 blow-off valve
排泄口 discharge outlet
排泄物 egesta; excreta; faeces; feces
排泄细胞 excreting cell
排序 ordination
排烟脱硫 exhaust gas desulfurization
排用环钉 rafting dog
排用小滑轮 rafting block
排运 raft; rafting
排运材 raft-wood
排运工人 raftsman
排运水库 rafting reservoir
排运作业 rafting operation
排障木 rub
排障器(装于集材机前) stinger
排种 drilling; dropping
排种器 seed metering device
排桩 bent; pile bent
排字工人 setter
排阻色谱 exclusion chromatography
排钻 gang borer
排钻采石法(采矿) line drilling
牌价 posted price
牌桌 basset table
派克木碱 parkine
派生地下水 allochthonic ground water
派生效益 secondary benefits
派生需求 derived demand
潘迪亚连续蒸煮器(管式) Pandia continuous digester
攀附的 scansorial
攀禽类(啄木鸟等) climber
攀树工具 climbing tool

攀树脚镫 climbing-iron
攀缘的 climbing; scandent
攀缘附生植物 scandent epiphyte
攀缘灌木 climbing shrub
攀缘茎 climbing stem
攀缘性地上芽植物 chamaephyta scandentia
攀缘植物 climber; climbing plant; scandent plant; scantentes; runner; vine
盘 tier; tray
盘存 inventory
盘存标签 inventory tag
盘点存货 stocktaking
盘菌 cup fungus; discomycetes
盘菌类 discomycetes
盘磨 teeth disk mill
盘曲 coiling
盘曲的 crumpled
盘山道 winding mountain path
盘式剥皮机 disk barker
盘式打磨机 disk mill
盘式粉碎机 Wiley mill
盘式风机 disk fan
盘式精磨机 disk refiner
盘式静电喷涂器 disk type electrostatic sprayer
盘式刨片机 disk planer
盘式砂光机 disk sander
盘式削片机 disk-type chipper
盘头 round up; snap; snipe; snout
盘旋 coiling
盘旋根(在容器中形成) spiralling root
盘旋线 zigzag
盘状花托 cotyloid receptacle; discopodium
盘状屋顶 slab roof
磐层 pan
蟠根植物 leaner
蟠桃 saucer peach; flat peach
判别分析 discriminatory analysis
判别函数 discriminant function
判别路径法 critical path method
判别路径分析 critical-path analysis
判读辅助手段 interpretation aids
判读检索表 interpretation key
判读立体像片 interpretation stereograms
判读模片 interpretation template
判读统计分析 interpretation statistical analysis
判读仪(员) interpreter
判据 criterion (*plur.* -rial)
庞然大物 jumbo
旁板(机械或家具零件) side; side panel
旁锯法 cornering
旁锯浩 corner cutting
旁拉绷索 gill guy
旁拉索 dutchman
旁路 by-pass
旁蚀 lateral erosion
旁通管 by-pass
旁向重叠 lateral lap; lateral overlap; side lap
旁引(侧拉钢索, 绕过障碍) spring
旁引索 dutchman line
抛光 brightening; brush polishing; glaze; gloss; gloss finish; polishing; satin finishing
抛光板 caul board
抛光垫板(热压机) polished caul plate; polished press plate
抛光粉 abrasive grain
抛光机 buffing machine; glazing machine
抛光轮 buffing wheel; polishing wheel
抛光清漆 polishing varnish
抛光刷 burnish brush
抛光压板(热压机) polished press plate
抛捆器 sheaf discharge
抛木杆 kicker
抛木机 log ejecter; log kicker; scrambler
抛木器 flipper
抛木装置 kicker system
抛石[护岸] riprap
抛物面 paraboloid
抛物线 parabola; parabolic curve
抛掷辊 throwing roll
抛掷式铺装头 throwing roll spreader head
抛掷式气流分选 throw sifting
抛掷输送带 throwing belt

P

跑车(地面的) bicycle; carriage; carrier
跑车(索道的) hati
跑车车上操作工 carriage rider
跑车带锯 band carriage
跑车搁凳 knee
跑车轨道 carriage way
跑车轨道清扫器 carriage track cleaning device
跑车进料(制材) carriage feed
跑车落钩装置 squirrel carriage slack puller
跑车退避装置 carriage offset
跑车小带锯 pony rig bandsaw and carriage
跑车止动器 carriage stop
跑车座架 head block
跑锅 surging
跑锯 overcutting
跑锯材 miscut lumber; miss cut lumber
跑偏控制 edge position control
泡沸石 analcime; analcite
泡林藤单宁 guarana tannin
泡沫 effervescence
泡沫发生器 foamer
泡沫剂 foaming agent
泡沫胶 foam glue
泡沫胶合 foam gluing
泡沫胶黏剂 foamer; foamed adhesive
泡沫结构 foam-like structure
泡沫灭火剂 fire foam
泡沫塑料 polyfoam
泡沫橡胶 foam rubber
泡沫芯层 foam core
泡囊 vesicle
泡胀 swelling
泡状花纹 blister figure
泡状结构 foam-like structure
泡状菌根 vesicular mycorrhiza
泡状纹 blister
泡状纹理 blister grain
泡状锈子器 peridermium
炮眼 gopher hole
炮眼钎子(挖伐根) bull stick
疱(泡囊) blister
疱斑(病毒病症状) puckered
疱病 blister
疱锈病 blister rust
疱状突起 blister figure
疱状疫病 blister blight
胚 corcule; embryo; embryon; plantlet
胚比 embryo ratio
胚柄 suspensor
胚层 blastoderm
胚成熟 embryo ripening
胚的延长部分(木麻黄属) caecum
胚动 blastokinesis
胚根 corcule; embryonic root; radicel (radicle)
胚根出现 radical emergence
胚根的 radicular
胚核 embryo nucleus
胚嫁接 embryo grafting
胚孔 foramen (*plur.* -mina)
胚鳞 scutellum
胚莽 germinal bud
胚囊 embryo sac
胚囊管 secondary embryo-cord
胚盘 blastoderm
胚培养 embryo culture
胚腔 embryo cavity
胚乳 albumen; endosperm
胚乳的 endospermous
胚熟 embryo ripening
胚胎 embryo; embryon
胚胎败育 embryo abortion
胚胎发育 embryonic development
胚胎腐殖质层 embryonic humus-layer
胚胎阶段 embryo phase
胚胎培养 embryo culture
胚胎期 embryo phase
胚胎期发育不良 dysembryoplasia
胚胎腔 embryo cavity
胚胎植入 embryonic implantation
胚细胞 elementary cell
胚休眠 embryo dormancy
胚芽 corcule; embryo (*or* embroyonic) bud; embryo; embryon; gemmule; germ; pangen; plantule; plumule
胚芽的 plumular
胚芽鞘 coleoptile
胚移植 embryonic implantation
胚轴 embryonic axis; plumule axis

P

胚珠 ovule
胚珠败育 ovule abortion
胚珠的 ovular
胚珠托(银杏) collar
胚状体 embryo; embryon; embryoid
陪氏培养皿 Petri dish
培垄 ridge plowing
培土 covering; earthing; earthing up; hilling; hipping; molding-up; ridging
培土机 banker
培土器 molder; ridger
培养基 culture; culture medium; medium (*plur.* -ia); Murashige and Skoog medium MS; nutrient medium; substrate; substratum (*plur.* -ta)
培养基试验 cultures test
培养介体 culture medium
培养皿 culture dish; double dish
培养皿琼脂法 Petri-dish-agar method
培养皿试验 Petri dish test
培养木菌的木片 feeder strip
培养箱 incubator
培养液 culture solution
培育 breed; cultivation; culture; incubation; raising
培育成本 cultural cost
培育措施(不含整地修枝疏伐) cultural operations
培育费用 cultural expense
培育期 incubation period
培育设施 growing facility
培育树苗 grow saplings
培育作业的采伐记录 cutting record of cultural operation
赔偿 make good
赔偿保险 compensation of employee
配板 assembly; layup; matching
配板失当 mismatching; miss matching
配电盘 panel board
配对 associate; association; pairing
配对部分 mate
配对抽样 matched sampling
配对染色体 pairing chromosome
配对物 counterpart
配方 formulation
配合 couple
配合力 combining ability
配合面 matching surface
配价的 coordinate
配碱槽 alkali make-up tank
配件 spare parts
配胶 glue preparation
配胶机 glue dissolver
配胶室 glue preparing station
配酒桌 grog table; mixing table
配聚作用 olation
配料 blend; compounding
配料比率 composition of furnish
配料仓 dosing bin; dosing silo
配料车间 furnishing department
配料输送带 dosing band
配料箱 dosage bunker
配偶 mate
配糖物 glycoside
配体 ligand
配位的 coordinate
配位体 ligand
配伍性 compatibility
配重 counterweight; rolling stock
配重索 counter rope
配重索滑轮 squirrel block
配重支架 counterbearing
配子 gamete; gametes
配子败育 gamete abortion
配子核 gametic nucleus
配子减数分裂 meiosis gametic
配子囊 gametangium
配子囊柄 gametophore
配子配合 syngamy
配子染色体数 gametic number
配子数 gametic number
配子体 gametophyte
配子体异型 anisogamonty
配子同型 isogamy
配子托 gametophore
配子小型 merogamy
配子异型 anisogametismanisogamety; anisogamy; heterogamety; heterogamy
喷(木材层积单位, 等于$\frac{1}{5}$考得) pen
喷布 sparge
喷程 throw

P

喷出口 spout
喷出器 blow-off
喷出岩 eruptive rock; extrusive rock
喷镀 shadowing
喷发岩 eruptive rock
喷放 blowing; blowout
喷放池 blow pit
喷放阀 blow valve
喷放管 blow-off pipe
喷放损失 blow-out loss
喷放压力 blow-out pressure
喷粉 dusting
喷管 spray hose
喷灌 overhead irrigation; spray irrigation
喷灌机 irrigator; watering machine
喷灌器 sprinkler apparatus
喷灌装置 rainer
喷壶 sprinkling can; watering pot
喷溅性 overspray
喷浆法 gunite method
喷胶机 glue sprayer
喷胶枪 glue gun
喷淋 sprinkling
喷漆 lacquer; lacker; paint coating; spray painting
喷漆器(涂漆用) paint gun; sprayer
喷漆枪(作记号用) marking gun
喷漆嘴 paint nozzle
喷气 air jet
喷气除尘装置 jet air cleaner; jet cleaning device
喷气管 jet tube
喷气管式干燥机 jet-tube dryer
喷气管束烘干机 jet turbine bundle dryer
喷气孔(火山) fumarole
喷气式单板烘干机 jet veneer dryer
喷气式干燥机 air jet dryer; jet dryer
喷气式干燥器 jet dryer
喷气式干燥窑 jet dry kiln
喷汽干燥窑 steam jet blower kiln
喷汽旋刀 steam-injection knife
喷枪 spray gun
喷枪拉丝(漆未雾化) combwebbing
喷泉 fountain
喷泉井 artesian well
喷洒 spray
喷洒器 sprinkler
喷砂木雕 wood carving by sand blast
喷射 emission; jet; sparge; spraying; spurt
喷射法 jet method
喷射鼓风机 jet blower
喷射角 jet angle
喷射气流 jet-stream
喷射器 jet; knockout
喷射器枪 ejector gun
喷射送风机 jet blower
喷射隧道 spray tunnel
喷射注入 jet impregnation
喷束 spray fan
喷水 water spreading
喷水催芽 water spray seeding
喷水防腐 decay control by water spray
喷水干燥窑(调湿) water spray (dry) kiln
喷水控制腐朽 decay control by water spray
喷水设备 sprinkler
喷水嘴 water spout
喷嚏木 sneezewood; nieshout
喷头 spray head; sprayer
喷涂 shadowing; spray; spray coating
喷涂处理(木材防腐) spray(ing) treatment
喷涂法 gunite method
喷涂机 spraying coater
喷涂技术 shadowing technique
喷涂外逸 overspray
喷雾 mist; sparge; spray; spray dust; spray mist
喷雾床 mist bed
喷雾法 spraying
喷雾繁殖 mist propagation
喷雾繁殖设备 mist propagating installation
喷雾干燥 spray dry
喷雾干燥器 spray dryer
喷雾机 mist sprayer; spraying machine
喷雾飘移 spray drift
喷雾器 atomizer; pulverizer; spray producer; sprayer; vapourizer
喷雾式施胶机 spray glue applicator

P

喷雾损失(飞逸) spray loss
喷雾头 mist nozzle
喷雾装置 mist system
喷烟机 aerosol sprayer; fog(ging) machine; fogger
喷烟喷雾两用机 fogging and spraying machine
喷药保持力 spray retention
喷药除草 spray cultivation
喷蒸处理(解除应力) steaming treatment; reconditioning
喷蒸管 steaming pipe
喷蒸汽 steaming
喷蒸真空处理(防腐) steaming and vacuum treatment
喷嘴 jet; jet nozzle; muzzle; nipple; nozzle; orifice; pulverizer; spray nozzle
喷嘴式流浆箱(网前箱) nozzle headbox
喷嘴式堰板 nozzle type slice
喷嘴速度系数 nozzle velocity coefficient
喷嘴调节 nozzle control
喷嘴系数 nozzle coefficient
盆钵移植 pot transplanting
盆地 basin
盆花 potted plant
盆景 bonsai; miniature landscape; penjing
盆栽 pot culture
盆栽花卉 potted plant; potting plant
盆栽试验 pot experiment
盆栽营养土 potting mixture
盆栽植物 potted plant; potting plant
棚板 shelf-board
棚干备料 shed drying stock
棚内堆积 shed piling
棚内干燥法 shed seasoning
棚屋 wigwam
硼-氟-铬-砷防腐剂 BFCA salt
篷式天窗 awning window
膨大(树干基部) butt swelling
膨大细胞 enlarged cell
膨缩应力 swelling compressive stress
膨压 turgor; turgor pressure
膨胀 bulking; dilatation; inflate; swell; swelling; turgescence; turgescency
膨胀比(弦向或径向) swelling ratio
膨胀钉 expansion bar
膨胀度 degree of swelling
膨胀法 bulking
膨胀缝 dilatation joint
膨胀缝隙 dilatation fissure; expansion fissure
膨胀剂 bulking agen; swelling agent
膨胀节 expansion joint
膨胀裂缝 expansion fissure
膨胀率 percentage swelling
膨胀漆 intumescent paint
膨胀热 heat of swelling; swelling heat
膨胀涂层 intumescent coating
膨胀系数 coefficient of expansion
膨胀效应 bulking effect
膨胀型潜溶剂 swelling type cosolvent
膨胀性黏土(黏粒) expansive clay
膨胀压 extensive pressure
膨胀压力 swelling pressure
膨胀应力 swelling stress
膨胀浴处理(空细胞法) expansion-bath treatment
碰伤 gouging
批发价格 wholesale price
批发苗圃 wholesale nursery
批发商 jobber; wholesaler
批发市场 wholesale market
批发物价指数 wholesale price index
批号 batch number
批件(销售单位) parcel
批量操作 bulk operation
批量处理 batch process
批量生产 batch production; lot production; volume production
批量种子 seed lot
坯材 bare wood
坯料 blank
披钯炭 palladium charcoal
披针形的 lanceolate
劈板斧 froe
劈材 cleft wood; split timber
劈材工 splitter
劈材工具 splitting tool
劈材机 log splitter; splitter
劈茬(伐木时) chair
劈柴 split fire-wood

劈柴机 cordwood splitter
劈刀 cleaver; riving knife
劈方 hew
劈斧 axe wedge; clearing axe; cleaving-axe; riving hammer
劈根机 stump splitter
劈痕 kerf
劈接 cleft grafting
劈开 cleavage; riving
劈开材 crevice
劈开的 spalt; split
劈开的短木段 split billet
劈开的薪材 split fire-wood
劈开面 hewn surface
劈理 cleavage
劈裂 cleavage; splinter pulling; split
劈裂材 splitwood
劈裂插 split cutting
劈裂强度 cleavage strength
劈裂楔 splitting wedge
劈裂性 cleavability; fissility
劈裂枕木 split tie
劈马蹄插穗 heel cutting
劈马蹄更新 reproduction by sprout with stumpbark
劈马蹄扦插 heel cutting
劈面 split face
劈篾机 bamboo splitting machine
劈木工 river
劈木工具 river
劈木机 splitting machine
劈木楔 splitting wedge
劈损 slabbed
劈楔 axe wedge; froe
劈制 riving
劈制材 cleave timber; cleft timber
劈制大方材 large hewn square
劈制的薪材 billet
劈制木材 hewn timber
劈制小方材 small hewn square
劈制枕木 hack tie
皮 cortex (*plur.* -ces); hide
皮包节(含树皮或周围包被树脂的死节) encased knot
皮鞭 lasher
皮层 cortex (*plur.* -ces); skin
皮层维管束 cortical bundle
皮层细胞 cortical cell
皮层形成作用 certification
皮层原 periblem
皮刺 prickle
皮带变速装置 belt speeder
皮带传动 belt drive
皮带轮 belt pulley
皮粉 hide powder
皮粉分析法 hide powder method
皮肤 tegument
皮革家具 leather furniture
皮号(原木上的号印) side mark
皮号(原木削去一块树皮作标识) watermark
皮厚减量 bark deduction
皮环节 bark-ringed knot
皮胶 boiled hide; hide glue; skin glue
皮焦 bark blister; bark necrosis; bark scorch
皮接 bark grafting
皮靠接 rind approach graft
皮孔 lenticel; lenticelle
皮孔侵染 lenticel infection
皮孔蒸腾 lenticular transpiration
皮枯 bark dieback
皮枯斑 bark blister; bark necrosis; bark scorch
皮棉 lint
皮囊 bark seam; inbark; ingrown bark
皮伤 bark injury
皮下的 hypodermal
皮下腹接 side bark-grafting
皮下接 bark grafting
皮下靠接 inarching by bark incision; rind inarching
皮下组织 subcutis
皮质 cortex (*plur.* -ces)
枇杷浸膏 concrete of neflier
疲劳变形 fatigue deformation
疲劳断裂 fatigue failure; fatigue fracture
疲劳骨折 fatigue fracture
疲劳极限 endurance limit; fatigue limit; limit of fatigue
疲劳耐久性 fatigue durability
疲劳破坏 fatigue failure
疲劳强度 endurance strength; fatigue strength
疲劳失效 fatigue failure

疲劳试验 durability test; endurance test; fatigue test
疲劳损伤 fatigue damage
疲劳限界 fatigue limit
疲劳性质 fatigue behavior
疲劳应力 fatigue stress
啤酒 brew
蜱(草爬子) tick
蜱螨目 Acarina
蜱螨学 acarology
偏摆锯 drunken saw; wobble saw
偏差 biased error; declination; deflection; departure; deviation; drift
偏差角 angle of deviation
偏度 skewness
偏父的 patroclinous
偏父遗传 patroclinous inheritance
偏冠树 lopsided tree
偏光计 polarimeter
偏光力 chirality
偏光显微镜 micropolariscope
偏害共栖 amensalism
偏航 crab; yaw
偏航角 yaw
偏回归 partial regression
偏角(钢索的) fleet angle
偏角(广义的) declination; deflection; edge angle
偏角法 method of deflection angle
偏离中心的(生长锥入钻方向偏离) off-center
偏母遗传 matroclinous inheritance
偏母杂种 matroclinous hybrid
偏僻伐区 backwoods
偏倾 tip
偏上性 epinasty
偏途[演替]顶极 disclimax
偏途顶极 disturbance climax; disclimax
偏途顶极群落 plagioclimax
偏途演替系列 plagiosere
偏位胚 excentric embryo
偏相关 partial correlation
偏向滑轮 deflecting block
偏向角 angle of deviation
偏向木 sway wood
偏斜 drift
偏斜板 deflect board
偏斜的 oblique
偏心 eccentricity; wandering heart
偏心力 eccentric force
偏心率 eccentricity
偏心磨损 eccentric wear
偏心年轮 eccentric ring
偏心胚胎 excentric embryo
偏心驱动装置 eccentric drive
偏心生长 eccentric growth
偏心髓 eccentric pith
偏心调整 eccentric adjustment
偏心下锯法 cutting off centre; offset sawing
偏心应力 eccentric stress
偏心载荷 eccentric load(ing)
偏心振动筛 gyratory screen
偏性遗传 genesclinic
偏性杂种 heterodynamic hydrid
偏移角 angle of difference
偏倚 bias
偏载 side heavy
偏振 polarization
偏振计 polarimeter
偏振滤光镜 polarizing filter
偏振转(光学的) optical rotation
偏转 yaw
偏转滑轮 deflecting block
偏转器 deflector
偏转式喷头 deflection nozzle
胼胝 tylosis (*plur.* -se)
胼胝体 callus
胼胝质 callose
片 filter
片材 sheet
片弹簧 flat spring
片沸石 heulandite
片胶 flake shellac
片接 plate budding; veneer grafting
片锯 web saw
片利共生 karpose
片流 sheet flood
片麻岩 gneiss
片切单板 slicing veneer
片蚀 sheet erosion
片英尺(板材计量单位) super ficial foot
片纸浆 sheet pulp
片状冰片 borneol flake
片状剥落的 exfoliate

P

片状剥皮 scale
片状材(木材缺陷) flaky wood; scaly wood
片状的 lamellate
片状结构 platy structure
片状黏土 laminated clay
片状闪电 luminous cloud
片状树皮 flake bark
片状髓心 lamellated pith
片状细胞(薄) tabular cell
片状芽接 plate budding
片紫胶 flake shellac; shellac
漂(木材计量单位, 等于18罗得) float
漂变[遗传] drift
漂浮 floating; flotation
漂浮草甸 floating meadow
漂浮测定 flotation test
漂浮岛屿 floating island
漂浮能力 floatability
漂浮期 drift epoch
漂浮生物 neuston
漂浮水生植物 hydrophyta natantia
漂浮性 floatability
漂浮植物 floating plant
漂浮植物群落 floating plant community
漂积物 drift
漂节(构成漂子的单节木材) boom stick
漂块 erratic; erratic block
漂砾 drift boulder; erratic; erratic block
漂流 drift
漂流搁浅木 waif (wood)
漂流生物 drifting organism
漂木道 floating log path; logway
漂鸟 wandering bird
漂前浆池 brown stock chest
漂散材 flood trash
漂散的无号印原木 stray
漂散木(木材流送) driftwood; erratic
漂散木搜集工 comber
漂砂矿床 alluvial deposit
漂石 erratic block
漂移 drift
漂移的 erratic
漂子 boom
漂子固定位置浮标 boom buoy
漂子固定物 boom stay
瓢虫 lady bird; lady-beetle
漂白 bleach
漂白层 albic horizon; bleached layer
漂白虫胶 bleached lac
漂白度 degree of bleaching
漂白干紫胶 bone-dry shellac
漂白过程 bleaching process
漂白灰化作用 bleaching podzolisation
漂白浆 bleach pulp; bleached pulp
漂白率 bleachability
漂白木材 bleaching wood
漂白黏土 bleaching clay
漂白软土 alboll
漂白森林土 bleached forest soil
漂白土 blcached earth; bleached mineral soil; bleached soils; china clay; fuller's earth
漂白紫胶 bleached lac; white shellac
漂洗矿质土 bleached mineral soil
漂洗淋溶棕色森林土 bleached brown forest soil
漂洗森林土 bleached forest soil
漂洗土 blcached earth
票据到期 bill to fall due; bill to mature
票据到期日 bill to mature
票面价值 face value; par value
嘌呤 purine
撇渣 skimming
拼缝 edge joint
拼缝机(单板) splicer
拼缝间隙 gap joint
拼缝胶黏剂 gap-filling adhesive; gap-filling glue
拼缝硬化剂 gap-filling hardener
拼缝纸条 tape
拼合接头 splice joint
拼合宽度 aggregate width
拼合梁 joggle beam
拼花 inlay
拼花板 panel board
拼花边缘 parquet border
拼花地板 block floor; parquet; parquet floor
拼花地板料 parquetry stave
拼花地板条 parquet block; parquetry stave
拼花面 mosaic surface

拼花木制品 tarsia
拼花贴面 oystering
拼花镶嵌 intarsia
拼花小木条 parqnet strip
拼接 joint; mate; splice; splice grafting
拼接单板 jointed veneer
拼接宽度(砂带) splice width
拼接砂带 segment belt
拼接梯 sectional ladder
拼接修整 jointing
拼色 colour matching
拼桌(可拼成大桌的一套小桌) range table
贫瘠的 arid; nutrient poor; skeletal
贫瘠地 infertile land; sterile land
贫瘠林地 infertile forest land
贫瘠土壤 impoverished soil; sterile soil
贫养的(生境) dystrophic; oligotrophic
贫氧酸沼地 high moor; high peat-bog
贫营养 oligotrophy
贫营养湖 oligotrophic lake
贫营养水 oligotrophic water
频带扩展 band spread
频度 frequency
频度级 frequency class
频度律 frequency law
频度曲线 frequency curve
频度图谱 frequency spectrum
频度系数 coefficient of frequency
频发突变 recurrent mutation
频率 frequency
频率百分比 frequency percentage
频率表 frequency table
频率分布 frequency distribution
频率特征曲线 frequency-response curve
频率图 frequency diagram
频率系数 coefficient of frequency
频率响应曲线 frequency-response curve
频数图 frequency diagram
频压法 alternating pressure process
品等 grade
品等区分 assorting
品红(碱性) magenta; fuchsin
品位 purity
品系(遗传) line; strain
品系繁育 line breeding
品系间杂交 line cross
品系间杂种 interstock hybrid
品系试验 strain test
品质鉴定 valuation
品种 assortment; brand; breed; race; variety
品种纯度 purity of variety
品种复壮 renovation of variety
品种更新 renovation of variety
品种间杂交 interbreed; intervarietal crossing; mongrelize
品种间自由异花授粉 intervarietal free cross-pollination
品种间自由杂交 intervarietal free crossing
品种内异系交配 incross
品种特征 varietal characteristic
品种退化 varietal deterioration
品种问杂交现象 mongrelism
品种性 varietalness
品种性状 varietal character
品种资源 varietal resource
平板 plate
平板车 fubo; platform truck
平板辐射仪 flat plate radiometer
平板干燥机 plate-press dryer
平板干燥室 panel dry kiln
平板挂车 flat bed; low-bed trailer
平板接合 board joint
平板卡车 flat bed truck
平板料 slab
平板培养计数法 plate count
平板筛浆机 flat strainer
平板系统 tablet system
平板仪 plane table
平板仪测量 plane table surveying
平板振动筛 flat vibrating screen
平舱费 trimming charge
平茬 stump
平茬苗 root-and-shoot cutting; stump plant; stump transplant
平车 butt carriage; platform truck
平车道 tramway
平齿器 jointer
平底 level bottom
平底驳船 landing barge

平底槽刨 dado plane
平底齿仓 flat-bottom gullet
平底船 gondola; logging barge
平底船式河绠 barge boom
平底谷地 flat-bottomed valley
平底轨 flat-bottom rail
平底排 baker raft
平底烧瓶 florence flask
平底鱼尾板 angle bar
平底圆形栏杆扶手 mop-stick hand-rail
平底座 flat bed
平地(伐木) make bed
平地机 blade machine; grader; land leveling machine; land scraper; level(l)er
平地器 drag
平垫 make bed
平顶驳船 flat-topped barge
平顶峰 flat top peak
平顶搁栅 ceiling joist
平顶山 mesa
平堆 flat piling; flat stacking; flat type piling; open piling
平堆法(板材水平堆积) flat piling; flat stacking
平方关系 quadratic relationship
平方平均 quadratic mean
平方英尺 surface foot
平房 bungalow
平分 bisect
平分法 allotment method; frame work method
平伏的 procumbent
平根植物的 liorhizal
平光饰面 dead finish
平行伐木 falling to the lead
平行进化 parallel evolution; parallelism
平行脉 parallel veins
平行脉的 parallelodromous
平行脉叶 parallelinervate
平行群落 parallel community
平行水系(型) parallax drainage pattern
平行下锯(造材) saw alive
平行下锯法(原木造材) saw through and through
平和的 gentle
平衡 balance; compensation; counter balance; equilibrium
平衡棒(昆虫) halter
平衡薄 balancing sheet
平衡薄板 compensator; compensator sheet; equalizing layer
平衡表 balance sheet
平衡补偿薄板 equalizing layer
平衡层 compensator sheet; equalizing layer
平衡处理 equalization treatment; equalizing
平衡锤 counterweight
平衡多态现象 balanced polymorphism
平衡范式 equilibrium paradigm
平衡含水量 equilibrium moisture content
平衡和瞬变气候实验 equilibrium and transient climate experiment
平衡横截锯 balance cut-off saw
平衡活塞式锯 balance piston saw
平衡假说 balanced hypothesis
平衡结构 balanced construction; balanced structure
平衡结构胶合板 plywood of balanced construction
平衡截断锯 balance cut-off saw
平衡块 balance weight; counter balance
平衡理论 equilibrium theory
平衡链 tail chain
平衡料仓 in-process-storage-unit
平衡密度 equilibrium density
平衡模数 equilibrium modulus
平衡木(装于卡钩上) slack puller
平衡配板法 balance matching
平衡坡度 equilibrium slope
平衡期 regulation period
平衡器 halter
平衡强度 equilibrium strength
平衡群体 equilibrium population
平衡湿量 equilibrium moisture content
平衡石 statolith
平衡式窗扇 balanced sash
平衡图 phase diagram
平衡型 balanced type
平衡选择 balancing selection
平衡压力 balance pressure

平衡营养溶液 balanced nutrient solution
平衡致死基因 balanced lethal gene
平衡重 ballast
平衡重量 balance weight
平衡重熏(吊钩) rolling stock
平衡装载 right on the pin
平衡纵断面(河道) profile of equilibrium
平滑材面 smooth grain
平滑的 laevigate
平滑度 smoothness
平滑结构 smooth texture
平滑面凸面(纤维板的) smooth side convex
平化 flatting
平缓的 gentle
平缓地面 gentle ground
平缓螺旋[线] flat helix
平价 parity price
平接 butt splice; end-to-end joint; plain joint
平截(苗根) undercutting
平截头棱锥体公式 prismoidal formula
平截头体 frustum
平局分析 break even analysis
平均半径 average radius
平均比 average ratio
平均比率 mean of ratio
平均标准木法 mean tree method; method of mean sample tree
平均材积表 average volume table
平均材积树直径 diameter of the tree with the mean volume
平均成本 average cost
平均乘积 mean product
平均出厂品质界限 average outgoing quality limit
平均断面积 average basal area
平均断面积求积式 Simalian's formula
平均发芽时间 mean length of incubation time
平均费用 average cost
平均孵化期 mean length of incubation time
平均干围 mean girth
平均高 mean height
平均高度 average height
平均冠幅 crown mean breadth; crown(mean)width
平均冠径 crown mean breadth
平均海平面 mean sea level
平均海平面温度订正 reduction of temperature to mean sealevel
平均含水量 average moisture content
平均进风量 average intake
平均净生产力 mean of annual net productivity
平均可变成本 average variable cost
平均立木度 average stocking
平均林分高 average stand height
平均林分年龄 average stand age
平均林龄 average age
平均流速公式 mean velocity formula
平均面积 mean area
平均木 average stem; average tree; mean sample tree; mean tree
平均木法 method of mean tree
平均年生长量 mean annual increment
平均年收获量 mean annual yield
平均年吸收量 mean annual uptake
平均偏差 average deviation; mean deviation
平均频度 average frequency
平均气候 average climate
平均区域火灾危险 average regional fire danger
平均日发芽率 mean daily germination
平均生长 average growth
平均生长量 average accretion; average increment; mean accretion; mean increment
平均生长木 mean growth tree
平均世代时间 mean generation time
平均寿命 average life
平均树高(洛业) Lorey's mean height
平均误差 mean error
平均吸入量 average intake
平均纤维长度框架 average fiber length grit
平均显性度 average dominance
平均信息含量 average information content
平均形级 average form class
平均胸径 average diameter at breast height

P

平均样本 mean sample
平均拥挤(度) mean crowding
平均拥挤度-平均密度法 mean crowding-mean density method (m*-m method
平均优势度 average dominance
平均优势木高 mean dominant height; mean predominant height
平均有效压力 mean effective pressure
平均折旧法 depreciation-straight-line method
平均直径 average diameter; mean diameter
平均值 average; mean value
平均值标准差 standard error the mean
平均总生长量 mean gross increment
平口刀具 flat nose tool
平楞 open piling
平流层 stratosphere
平流辐射霜 advection radiation frost
平流霜 advective frost
平流雾 advection fog
平流作用 advection
平路 grade; grading
平路机 grade shovel; rutter
平门 jib door
平面(平板) table; plane
平面测量学 plane-surveying
平面车床 face lathe
平面光栅摄谱仪 plane grating spectrograph
平面门 flush door
平面模压工艺(浮雕) flat molding process
平面抛光 flat finish
平面刨床 surface planer
平面培养 plate-culture
平面偏光镜 plane polarizer
平面砂磨 flat sanding
平面图 plat
平面镶板门 flush panelled door
平面斜嵌接 plane scarf joint
平皿 Petri dish
平皿接种率 plating efficiency
平皿培养 plate-culture
平年 even year; normal year
平排 flat boom; flat raft; plane type raft
平排侧绠 floating lateral boom
平刨(木工工具) planer; trying plane
平拼接 slab joint
平铺的 procumbent
平切面 side grain
平三角堆积法 cribbing
平筛 flat screen; flat strainer
平筛分选系统 flat screen classifying system
平时密度 endemic density
平闩锁 dead lock
平送射 straight-away shot
平台 flat bed
平台基准面 face plate
平台式干燥机 platform dryer
平台园 terrace
平坦 level
平坦的 flat; plain
平庭(日本式庭园) flat garden
平头搬钩 swing dog
平头接合 square joint
平头木螺丝 flathead woodscrew
平头式(弹头) flat-nosed
平头手把 flush handle
平土 grading
平土机 grader; scraper
平推动力除草器 vibro-hoe
平尾 even tail
平纹横镶板 lying panel
平纹镶板 lay panel
平卧的 prostrate
平卧茎 prostrate stem
平镶板 flush panel
平斜接 flat mitre joint
平压(刨花板工艺) flatpressing; platen press
平压板(刨花板) flat-pressed board
平压机 platen press
平压碎料板(刨花板等) flat-platen-pressed particle board; platen-pressed particle board
平迎射 approaching direct shot
平原 dead level; plain
平原绿化工程 afforestation project in the flat-land areas
平圆方角接合 bead-butt and square
平圆接合 bead-butt
平整 grade; out-of-wind

平整地 grade
平整地面 laying
平整耙 smoothing harrow
平整田地 smooth(ing)
平直嵌板 plane panel
平直正面 straight front
平周分裂 periclinal division
平周形成层细胞分裂 periclinal cambial cell division
平作 level culture
评等 grading
评等标准 grading rule
评等站 grading station
评定 assessment
评定立木价 appraised stumpage price
评工法 job evaluation
评估报告 assessment report
评估步骤 appraisal procedure
评价 appreciation; assessment; valuation
评价单元 assessment unit
评审员 assessor
评税 tax assessment
评税人 tax assessor
苹果 common apple; apple tree
苹婆胶 hog gum
凭单付款 cash against document
凭证 voucher
凭证标本 voucher specimen
屏蔽 screening
屏蔽种植 screen planting
屏风 screen door
屏幕原木扫描器 curtain log scanner
屏障 barrier; screen
屏障地区 sheltered ground
屏障效应 barrier effect
屏障栽植 screen planting
瓶梗孢子 phialospore
瓶接 bottle grafting
瓶颈现象 bottleneck
瓶颈效应 botfleneck effect
瓶饰 vase
瓶形栏杆小柱 vase baluster
瓶形椅背板 vase splat
坡道 ramp
坡底 slope base; slope bottom
坡地 hilly land
坡地坍塌 slope failure
坡地栽培 hill culture
坡地植树机 steep slope tree planter
坡度 grade; gradient; slope
坡度改正 slope correction
坡度极限 grade limit
坡度角 angle of slope
坡度量测 slope measurement
坡度稳定结构 grade stabilization structure
坡度线 grade-line
坡度转折点 sharp pitch
坡度阻力 grade resistance
坡度阻力指数(坡度角的正弦) slope index
坡积土 clinosol; slope wash soil
坡积物 deluvium; slope wash; talus
坡降 decline
坡面渗流 effluent seepage
坡式梯田 sloping terrace
坡向 aspect; direction of slope
破产 insolvency
破除休眠 breaking of dormancy
破除休眠措施 dormancy-breaking measure
破坏 decay; failure
破坏伐 indiscriminate cutting
破坏荷载 load at failure
破坏荷重功 work to ultimate load
破坏极限 limit of rupture
破坏强度 breaking strength
破坏性试验 Breaking-down test; destructive test
破坏应力 breaking stress
破坏载荷 breaking load
破壳机 rejuvenator
破裂 break; burst; outbreak; rupture
破泡剂 foam breaker
破生脂沟 lysigenous duct
破碎 fragmentation; hogging
破碎板浆 cracker pulp
破碎机 breaker; crusher; Raymond mill
破碎纸浆 cracker pulp
破碎紫胶用的石磨 chatki
破损面 surface of rupture
破云器(播撒干冰用) cloudbuster
剖刀 doctor
剖分(原木) secondary breakdown
剖分带锯 band resaw

P

剖分锯 slab saw; slabber; splitting saw
剖开材 breakdown timber
剖料 breakdown; primary breakdown
剖料锯 breakdown saw; head saw; headrig; log saw; rift saw
剖面 profile; section
剖面法 bisect method
剖面图 profile diagram; cross section
剖面样条 bisect
剖切的种子 excised seed
剖视图 cut-away view; sectional view
扑火 attack a fire; fire fighting; fire suppression
扑火队员 fire fighter
扑火外站 stand-by crew
扑火指导员 line boss
扑火总指挥 fire boss
扑灭 extinguish; suppression action
扑灭林火 extinguish
铺板 boarding; decking; planking
铺草路 grass walk
铺草皮 sod; turfing
铺道碴 ballast
铺道横木 cross skid
铺道木杆 skid
铺地草本植物 carpet herb
铺地木材 wood paving
铺地木块 flooring block
铺地设计 paving design
铺地植物 mattae
铺垫 laying
铺垫层 bed
铺盖布机(苗圃) tarp-layer
铺盖布装置 tarp-laying device
铺路工事 pavement
铺路横木 puncheon
铺路木 wood paving
铺路木块 paving block; road block; wood paving block
铺路石 flagstone
铺面 pave
铺木 wood block
铺木杆道 skidding
铺木路面 wood pavement
铺砌石块 pitching
铺条机 swather
铺展剂 spreader
铺纸台 paper laydown station
铺装 spread
铺装辊 spreading roll
铺装机 felting machine; spreader; spreading machine
铺装角 spreading-angle
铺装失误板坯 miss-spread mat
铺装头 spreader head
铺装长度 spread-length
匍匐的 procumbent; prostrate
匍匐地上芽植物 chamaephyta reptantia
匍匐灌木 scandent shrub
匍匐茎 creeping stem; prostrate stem; runner; stole; stolon
匍匐茎的 diageic
匍匐生根的 reptant; running
匍匐性植物 groundling; prostrate plant; stoloniferous plant
匍匐枝 creeper; runner; stolon
匍匐植物 creeper; creeping plant; mattae; trailer; trailing plant
菩提花浸膏 linden concrete
菩提树 peepul; bodhi; peepul tree
葡甘聚糖 glucomannan
葡聚糖 dextran(e); glucan; glucosan
葡聚糖酶 glucanase
葡聚糖凝胶 dextran gel
葡糖倍苷 glucogallin
葡糖二酸 glucaric acid
葡糖苷 glucoside
葡糖苷酶 glucosidase
葡糖苷酸 glucuronide
葡糖化酶 glucase
葡糖没食子苷 glucogallin
葡糖酶 dextrase
葡萄 grape; wine grape
葡萄糖 dextrose; glucosum; glycose
葡萄糖胺 chitosamine
葡萄糖甘露聚糖 glucomannan
葡萄糖酐 anhydroglucose
葡萄糖基移转作用 transglycosylation
葡萄牙栎 Lusitanian oak; Portuguese oak
葡萄状的 racemose
葡萄状花纹(木材) vinaceous figure
蒲福风级 Beaufort (wind) scale

蒲式耳(谷物容量单位) bushel
蒲桃皮苷 antimellin
蒲苇(南美) pampas grass; pampa
蒲苇草原(南美) pampa; pampas grassland
脯氨酸 proline
圃捕 stalking
普遍的 pandemic
普查地段 census tract
普第式双闩 Purdey double bolt
普兰克常数 Planck constant(b)
普兰克定律 Planck's law
普雷斯勒材积法 Pressler's method of volume
普雷斯勒生长率 Pressler's growth percent
普雷斯勒生长锥 Pressler's (increment) borer
普列利群落 prairie
普鲁士蓝 Berlin blue; Prussian blue
普纳群落 puna
普适方程 generic equation
普通测量学 ordinary surveying
普通插穗 straight cutting
普通铲 ordinary spade
普通椽子 common rafter
普通搁栅 common joist
普通轨枕 regular tie
普通合接 ordinary splice grafting
普通嫁接 common grafting
普通胶合板 normal plywood; plywood for general use
普通鸠尾榫(木工) common dovetail
普通卡车 line truck
普通开环摇尺系统 conventional open-loop setwork
普通梁 beam with simply supported ends
普通磨木法(磨石旋转与木纤维方向成直角) cross grinding
普通配合力 general combining ability
普通漂白紫胶 regular bleached lac
普通舌接 simple whip grafting
普通疏伐 universal thinning
普通形数 common form factor
普通旋切机 normal peeling lathe
普通压条 simple layering
普通原木 grading log
普通枝插 simple cutting
普通组织学 general histology
谱 spectrum (*plur.* -tra)
谱带拖尾 band tailing
谱尾 tailing
谱系 lineage; pedigree
谱系法 pedigree method; pedigree system
谱系繁殖 pedigree culture
谱系化验法 pedigree trial method
谱系选择 family selection; pedigree selection
谱系选种 pedigree selection
蹼趾 webbed toe
瀑布 chute; fallen; water fall
瀑布风景 waterfall landscape
七倍体 heptaploid; septuploid
七出的 septenate
七氯(杀虫剂) heptachlor
七位字节 septet
七弦琴形背椅 lyre back chair
七叶苷 aesculin
七月份收获的久树紫胶 Jethuri
七重峰 heptet; septet
栖花类(食花甲虫) anthobian
栖居荒漠的 eremophilous
栖居沙漠的 eremophilous
栖菌动物 mycetocole
栖木 roost
栖木白蚁 wood-dwelling termite
栖息地 habitat
栖息地管理 habitat managemen
栖息地恢复 habitat restoration
栖息地选择 habitat selection
栖息所 roost
栖蚁巢动物 myrmecocole
桤木酮 alnusenone
期 phase
期货合同 froward contract
期货价格 forward price
期货交易 forward business
期货市场 future market
期末存货 closing inventory; ending inventory; final inventory
期末库存 ending inventory
期票 promissory note
期望 expectation
期望法正结构 desired normal structure

期望法正蓄积 desired normal volume
期望回归线 expected regression line
期望使用年限 expected life-period
期望树 promising tree
期望值 expectation; expectation value; expected value
期望值模型 expected-value-model
期中数据更新 midcycle data updating
欺骗色 deceiving colouration
漆(日语) urushi
漆(涂漆) coating; lacquer
漆病 lifting
漆层 impression
漆酵素 gummase
漆蜡 lacquer wax; urushi tallow; urushi wax
漆酶 laccase
漆膜 paint film
漆膜形成 varnish formation
漆膜增厚(局部) lap
漆器 lacquer ware
漆器材 plain wood of lacquerware
漆器用材 plain-wood
漆树 lacquer tree; varnish tree
漆树漆 che shu lacquer; rhus lacquer
漆树酸 rhusinic acid
漆素 urushin
漆叶单宁 sumac(h) tannin
漆叶栲胶 sumach extract
漆叶鞣质 sumac(h) tannin
漆油 urushoil
漆脂 urushi tallow
齐边 trim
齐边的 tumbled-in
齐边机 edger
齐边-胶合-裁截联合加工法 edge, glue and rip process
齐边锯 edge trim saw
齐边锯机副手 strip catcher
齐端 square the break; square-end
齐端截锯机 equalizer
齐端装置 load squeezer
齐根器 packer
齐苗 established field; established stand; establishment
齐明镜 aplanat
齐平端头 even-end
齐头 butt off; end trimming; end-butt(ing); even-end
齐头板 bumping plate
齐头的 trimmed
齐头锯 butting saw
齐性总体 homogeneous population
奇倍体 anisoploid; anisopolyploid
奇多倍体 anisopolyploid
奇数羽状的(复叶) imparipinnate; odd-pinnate; unequally pinnate
奇数羽状复叶 odd pinnately compound leaf
奇数长度材 odd lengths; odd-number lengths
歧管 manifold
歧化[作用] dismutation
歧化松香 disproportionated rosin
歧化作用 disproportionation
歧散生长 deliquescent growth
脐 Hilum; hylum; hylus; navel
脐带 umbilical cord
畦 bed; border; trench
畦播 drill seeding
畦灌 border irrigation
畦植 drill planting
崎岖地面 short ground
骑接 saddle grafting
棋盘式点火 area grid ignition
棋盘形播种 checkered seeding
棋盘形植树 checkcrow
蛴螬(金龟子幼虫) white grub; grub
蛴螬孔 grub hole
蛴螬形的 scarabaeoid
旗瓣 vexillum
旗腹姬蜂 ensign-fly
旗舰种 flagship species
旗形树 flag-shaped tree
鳍形漂子 fin boom
企口槽榫接 rabbet and dado joint
企口地板 joint flooring
企口接 grooved and tongued joint
企口接合 match(ed) joint; rabbet joint
企口开槽刨 match plane
企口连接 tongue(d) and groove(d) joint
企口镶板 match board(ing); matching

企业 enterprise
企业倒闭 business failure
企业管理 business management
企业家 entrepreneur
企业林 commercial forest; industrial forest
企业森林经营 industrial forest management
企业实体 business entity
企业形成期 formative period of enterprise
企业性林业 industrial forestry
杞柳苷 salipurposide
启动区 promoter
启动子 promoter
起步肥(定植用) booster solution; starter solution
起测径级 measurable class
起测直径 minimum DBH (by inventory)
起吊滑轮 pick-up block
起动杆 starting bar
起动注油系统 priming system
起动阻力 resistance of start
起伏地 undulating ground
起伏地形 rugged terrain
起伏平原 rolling plain
起伏式调整 rolling readjustment
起拱 hanging-up
起钩钉锤 dog knocker
起关键作用 play a vital role
起货单 landing account
起脊式装车 peaker
起脊装车法 ridging loading method
起壳(单板) flake
起鳞 delamination
起垄犁 ridger
起落式截锯机 drop trimmer
起码利润 threshold return
起毛 fluff; fuzz; grainlifting
起毛刺 grain raising
起毛纤维 fuzz fibre
起毛现象 fluffiness
起锚索 anchor line
起苗 lifting; lifting of seedlings
起苗机 lifter; mechanical digger; plant lifter; seedling harvester; seedling lifter; tree digger
起苗犁 plant lift plough
起苗器 tree lifter
起苗作业 lifting operation
起泡 blowmark; ebullition; effervescence
起泡剂 frother
起泡胶黏剂(有过量气泡) foamy adhesive
起始 initiation
起始价值 starting value
起始期 inception phase
起始区 initiator
起始因子 initiator
起霜(表面发白，漆病) frosting; bloom; blooming
起霜油(松香干馏产物) bloom oil
起纹 graining
起卸机 elevator
起源 genesis; origin
起源中心 center of dispersal; centre of origin
起重 hoisting
起重臂 boom; crane; gibbet; jib(arm); lift boom
起重臂摆转导向滑轮 boom swing lead block
起重臂摆转拉索 boom swing line
起重臂变幅 luffing
起重臂吊索 boom strap; bracket guy line; nose guy
起重臂固定吊索 boom hold-up strap
起重臂固定索 inspector
起重臂回空索 boom haul back line
起重臂回转索 swing line
起重臂回转索滑轮 squirrel line swing block
起重臂配重 squirrel
起重臂小滑轮 space block
起重车 crane truck
起重钩 grab hook; grapple hook
起重滑轮 fall block; gin block; hoisting block; hoisting tackle; loading block; rigging block
起重滑轮组 purchase block
起重机 crane; elevator; hoist; hoister; lift; lifter; liftor
起重机承载车 crane carrier
起重机挺杆 crossbar
起重机小车 monkey
起重卷筒 hoisting drum

起重链 jack chain
起重能力 lifting capacity
起重器 jack
起重索 hoist cable; hoist line; hoisting rope; lifting line; main cable; main lead line; main line; main rope; operating cable; tongline
起重索吊杆 spreader
起重索滑轮 main-line block
起重臂油缸 boom cylinder
起重桅杆 gin pole
起重悬臂索 boom-suspension guy
起重注油器 primer
起重作业 lifting operation
起皱 shrink
起皱纹的 crumpled
气胞 gas vacuole
气成的 pneumatogenic
气成岩 atmolith; atmospheric rock
气窗 window ventilator
气刀刮涂机 air knife coater
气垫 air support; air suspension
气动 air drive
气动薄膜阀 air diaphragm valve
气动剥皮机 air debarker
气动测微计 air-gauge
气动带锯整形机 band saw air-operated shaper
气动翻转装置 air-operated puffer
气动分离设备 pneumatic separation arrangement
气动辊式砂磨机 pneumatic drum sander
气动恒温器 air-operated thermostat
气动夹具 air clamp; pneumatic clamps
气动剪板机 air clipper; air-operated clipper
气动卡钩(装车用) earwig
气动卡木鹰嘴 air dog; air-operated dog
气动控制 aerodynamic control; pneumatic control
气动量规 air-gauge
气动马达 air motor
气动起重机 air hoist
气动湿度控制器 air-operated humidity controller
气动输送 pneumatic conveying
气动送风机 pneumatic conveyor fan
气动调节 pneumatic control
气动温度控制器 air-operated temperature controller
气动系统 pneumatic system
气动液压泵 air-powered pump
气动铡板机 air clipper; air-operated clipper
气动自动控制 air-operated automatic control
气动自计调节器 air-operated recorder regulator
气阀 air valve
气封 air seal
气干 air seasoning; air-dry; natural seasoning
气干板院 drying yard; seasoning yard
气干板院卫生 sanitation in seasoning yard
气干材 air-dried wood; air-dry lumber; air-dry wood; air-seasoned timber; natural seasoned timber; natural seasoned wood; seasoned timber; seasoned wood
气干材比重 specific gravity in air dry
气干材变色 yard stain
气干场 drying yard; seasoning yard
气干的 air-dried; air-dry; air-seasoned
气干法 air-drying
气干含水率 air-dry moisture content
气干锯材 air-dried lumber
气干密度 air-dry (wood) density
气干失重 air-drying loss
气干条件 air-dried condition; air-dry condition; air-seasoning condition
气干状态 air-dried condition; air-dry condition; air-seasoning condition
气缸 cylinder
气根 air root
气固色谱 gas-solid chromatography
气管 trachea
气候 climate; clime
气候变动 climatic fluctuation
气候变化 climate change; climatic change

气候变化框架公约 framework Convention on Climate Change
气候变异 climate variability
气候变种 climatic variety
气候波动 climatic vacillation
气候带 climatic zone; zonation of climate
气候顶极 climate climax
气候恶劣 awful weather
气候反馈 climate feedback
气候敏感性 climate sensitivity
气候模式(体系) climate model (hierarchy)
气候迁移 climatic migration
气候情景 climate scenario
气候融资 climate financing
气候生活型 climatic life form
气候实验室 climatizer
气候适宜期 climatic optimum
气候谈判 climate negotiation
气候图 clim(at)ograph
气候图解 climagraph; climatograph
气候稳定性 climatic stability
气候系统 climate system
气候型 climatype
气候性成土作用 climatic soil formation
气候性的 climatogenic
气候性演替 climatic succession
气候序列 climosequence
气候学派 climate school
气候雪线 climatic snowline
气候循环 climatic cycle
气候驯化 acclimatization
气候严酷 harsh climate
气候演替顶极 climatic climax
气候演替顶极群落 climatic climax community
气候演替系列 clisere
气候因素 climatic factor
气候预报 climatological forecast
气候预测 climate prediction
气候韵律 climatic rhythm
气候植物群系 climatic plant formation
气候指标 climatic indicator
气候资料 climatic material
气候宗 climatic race
气候最适度 climatic optimum
气化 gasification
气界 aerosphere
气孔(动植) spiracle; spiracula; stoma; pore
气孔(植物) stomata; pore; stoma
气孔的 stomatal
气孔关闭 stomatal closure
气孔计 porometer
气孔侵染 stomatal infection
气孔运动 stomatal movement
气孔蒸腾 stomatal transpiration
气孔阻力 stomatal resistance
气控分段式托板 pneumatic sectional plate base
气控滚筒式砂光机 pneumatic drum sander
气控压力 pneumatic control pressure
气力传动 pneumatic drive
气力输送 pneumatic conveying
气力输送管 pneumatic tube
气力输送机 pneumatic conveyor
气流 air current; air flow; air stream; air-blast; draft; draught; wind current
气流表 air-meter
气流分选 air classification; air separation; pneumatic separation
气流分选机 air separator
气流分选铺装机 air grading spreader
气流干燥 pneumatic conveying dry
气流干燥机 airflow dryer; pneumatic conveying dryer; pneumatic dryer; stream dryer
气流喷雾机 air-flow sprayer
气流铺模 air-felting
气流铺装 air-felting
气流铺装板坯 air-felting sheet
气流铺装机 wind spreader
气流铺装室 wind-spreader chamber
气流筛选机 air-screen cleaner
气流速度 air velocity; velocity of air
气流循环分选器 air circular classifier
气路 channel
气门(动物) spiracle; spiracula; stigma; stoma
气门(机械) air valve; air-drain
气门(昆虫) aeriductus
气门侧沟(介壳虫) canella
气门的 stomatal
气密的 vacuum-tight

气密封 air tight seal
气密接合(接头) air tight joint
气幕 air curtain
气囊 air bladder; air sac; saccus; wing
气囊砂光机 air drum sander
气泡 gas vacuole
气泡痕纸病 blowmark
气泡水准仪 air bubble level; air level
气球 aerostat
V形气球 Vee-balloon
气球集材 balloon logging; high-flying logging
气球集材机 aero-yarder
气球拖曳 balloon drag
气圈 aerosphere
气热干燥窑 gas fired kiln
气溶胶 aerosol; aerosols
气溶胶芳香产品 aerosol fragrance product
气溶胶去臭剂 aerosol deodorant
气溶胶效应 aerosol effect
气生变态 aeromorphosis
气生层片 aerosynusia
气生根 aerial root; pneumatophore; pneumatophoric root; respiratory root
气生菌丝 aerial hypha(e)
气生菌丝体 aerial mycelium
气生同型 aerosynusia
气生植物 aerial plant; aerophyte
气生种群 aerial population
气栓 air embolism
气提吸附 adsorption stripping
气体处理 gas treatment
气体处理系统 gas handling system
气体导电能力 gas conductivity
气体的计量 gasometry
气体发生器 gas generator
气体交换系数 gas exchange quotient
气体净化 gas sweetening; gas treatment
气体释放 gas evolution
气体污染物 gaseous pollutants
气体循环 gaseous cycle
气体逸出 gas evolution
气体滞留 gas hold-up
气调贮藏 controlled-atmosphere storage
气团 air mass
气味浓度 odorousness
气吸式种子分离机 seed aspirator
气相色谱 gas chromatography
气相色谱质谱 gas chromatography-mass spectrography
气相色谱质谱计算机 gas chromatography-mass spectrography computer
气相色谱质谱联用仪 gas chromatograph-mass spectrometer
气象符号 meteorological symbol
气象卫星 meteorology satellite; weather satellite
气象现象 meteorological phenomena
气象学 meteorology
气象要素 meteorological element
气旋 cyclone
气旋降水 cyclonic precipitation
气旋性雨 cyclonic rain
气压 air pressure
气压表 barometer
气压测高表 pressure altimeter
气压称重装置 air scale system
气压传动装置 shuttle
气压计 air pressure gauge; air-gauge; barograph; barometer
气压排泄管 barometric pipe
气压喷射法 pneumatic injection
气压温度计 barothermograph
气压温湿计 barothermohygrometer
气压形势 baric topography
气压自记曲线 barogram
气液层析法 gas-liquid chromatography
气液分配色谱法 gas-liquid partition chromatography
气液色层分离法 gas-liquid chromatography
气液色谱 gas-liquid chromatography; gas-liquid partition chromatography
气液色谱法 gas-liquid chromatography
气源 air supply
气闸 air brake; damper; ventilating damper

气针孔 gas pin
弃耕 abandonment
弃耕地 abandoned land; old field
弃土 spoil
汽车道用材 motorway timber
汽车库 garage
汽车列车运材 truck-trailer log hauling
汽车起重机 cranemobile
汽车前轮外倾 camber
汽车上交货价格 free on car
汽车运材 log transportation by truck; motor truck hauling
汽车装卸机 mobile loader
汽车装载限量指示链(悬于承载梁上) tattle tale
汽化 vapourization
汽化计 atmidometer; atmometer
汽化器 carburettor; carburetter; vapourizer
汽化热 heat of vaporization; vapourization heat
汽热淬冷处理(防腐) steam-and-quench treatment
汽射法 steam blow; vapour injection process
汽提 logging site
汽提塔 stripper
汽相 vapour phase
汽压 vapo(n)r pressure
汽压表 steam gauge
汽压平衡法 vapo(n)r pressure equilibrium method
汽油 benzine
汽油动力锯 wade; wader
汽油集材机 gasoline skidder
汽油酒精混合燃料 gasohol
汽障 vapour barrier
汽蒸池 steaming pit
汽蒸处理 steaming; steaming treatment
汽蒸期 steaming period
汽蒸时间 steaming time
契据 title deed
契约 covenant; indenturc
契约利息 explicit interest
槭蜡 maple-wax
槭蜜 maple honey
槭树汁 maple syrup
槭糖 maple sugar
槭糖浆 maple cream; maple syrup
槭叶单宁 acer tannin
槭汁酸 aceric acid
器孢子 pycnidiospore
器官发生 organogenesis
器官培养 organ culture
器官形成 organgenesis
卡木(家具零部件) block; corner block
卡木齿(链锯) tree dog; felling dog
卡木钩 board dog; log dog; log grab; dogger jaw
卡木钩爪 log dogging jaw
卡木鹰嘴 board dog
卡木桩 knee
卡木装置(半圆旋切) stay log
卡钳 calliper
卡索 skybound
千板英尺量度 thousand-foot board measure
千吨 gigagram
千分尺 micrometer; micrometer caliper
千分英亩($\frac{1}{1000}$英亩为研究单位, $=6.6\text{ft}^2$) milacre
千斤顶 jack
千粒重 thousand grain weight; 1000-seed weight
千枚岩 phyllite
千年计划 millenium plan
千年评估 millenium assessment
千万亿(10^{15}) petra-
千万亿克(10^{15}g)(→十亿吨) petragram (Pg)
千兆 giga-
千足虫 millipede
扦插 cleft cutting; cutting
Y形扦插 Y-cutting
扦插材料 cutting material
扦插床 cutting bed; rooting bed
扦插法 cuttage
扦插繁殖 cutting propagation
扦插基质 cutting medium; rooting medium
扦插介质 cutting medium
扦插苗 cultivated cutting
扦插区 cutting plot

Q

扦插台 rooting frame
扦插造林 afforestation by cuttings
迁出 emigration
迁出者 emigrant
迁飞路线 flyway
迁入 immigration
迁入率 immigration rate
迁入曲线 immigration curve
迁入者 immigrant
迁徙 migration
迁徙鸣禽 migratory songbird
迁移 migration
迁移单位 transfer unit
迁移分水岭 shifting divide
迁移机制 migration mechanism
迁移适应 modification for migration
迁移速率 migration rate
迁移植物 migrant
牵牛花素 petunidin
牵引 draft; draught; drag; pull; pulling; tow
牵引伐木 pull felling
牵引方向 lead
V形牵引方向 V-lead
牵引杆 drawbar; reach
牵引根 contractile root
牵引钩 drawbar; line hook
牵引钩集材 drawbar skidding
牵引钩牵引力 drawbar pull
牵引环 eye hitch
牵引机械 hauling equipment
牵引绞盘机 towing winch
牵引力 propulsive effort
牵引量 hauling capacity
牵引流 tractive current
牵引能力 hauling capacity
牵引爬犁 drag sled
牵引起重索 haul-in line
牵引式平地机 pull-grader
牵引式旋转割灌机 pull-type rotary cutter
牵引索(半悬式) snaking line
牵引索(各种) haul ahead line; haul line; haul(age) cable; haul-in line; hawser; inhaul line; pulling line; pulling-in line; running line; skidder line; skidding line; skidding rope; traction cable; traction line; traction rope
牵引索(维森索道) snubbing line
牵引索(主控) lead line; main cable; main lead; main lead line; main line; main rope; operating cable; operating line; operating rope
牵引索导向轮(索道跑车上) depending sheave
牵引索滑轮 bull block; main lead block; main-line block
牵引索卷筒 skidding line drum
牵引索速度 line speed
牵引索索具 mainline rigging
牵引索托轮 operating line support
牵引索止动器 grab
牵直管道 uptake
铅 lead
铅笔描图 pencil tracing
铅淬火高强度钢丝 plow steel wire
铅丹 minium; red lead
铅漆 white paste
铅砂 bird shot
铅直对流 vertical convection
铅直风速表 vertical anemometer
铅直风速仪 vertical anemoscope
签发信用证 credit
签证 certification
签证种子 certified seed
前把闩钩 bolt loop
前坝 counter-dam
前扳机 fore trigger
前绷索(防反倾) snap guy
前侧片(昆虫) episternum
前肠 fore-gut
前车架 front frame
前翅 anterior wing; fore wing; primaries
前抽空 initial vacuum
前出叶 prophyll
前处理 pretreatment
前代 previous generation
前导向块 front guide shoe
前顶极 perclimax
前顶极群落 preclimax; preclimax community
前定变量 predetermined variable
前端面 front face
前端升降机 front-end lift

前端装载机 front loader; front-end lift; front-end loader
前端自卸车 front-end dumper
前盾片(昆虫骨板) prescutum
前伐 advance felling; advance-cutting; beginning of regeneration-cutting; cutting in advance; pre-logging
前峰 leading peak
前腹片(昆虫骨板) presternum
前割面(采脂) front face
前更新伐 preregeneration cleaning
前光阑 front stop
前寒武纪 Precambrian (period)
前横档 front rail
前后重叠 end lap
前胡宁 imperatorin
前级 backing stage
前级泵 backing pump
前级真空泵 forepump
前角 rake angle
前进式干燥室 progressive drying-kiln
前进式干燥窑 continuous (dry) kiln; progressive (drying) kiln
前进突变 forward mutation; progressive mutation
前景 foreground; perspective; prospect
前景调查 anticipation survey
前颏(昆虫) prementum
前馈控制 feedforward control
前廊(教堂的) galilee
前利用 intermediate
前面(伐木的倒向侧) front
前面(位于前头的面) front face
前排气 initial vacuum; preliminary vacuum
前期(细胞分裂) prophase
前期降水 antecedent precipitation
前期经理 former management
前倾面 rake
前染色体 prochromosome
前生杆材 pre-existing pole
前生树 advance growth; advance reproduction; pre-existing tree; shelter growth; volunteer growth
前手柄(油锯) front handle
前台处理(优先处理) foreground processing
前滩 foreshore
前镗枪 muzzle loader
前体 precursor
前庭种植 forecourt planting
前维生素 provitamin
前位 anteposition
前胃 proventriculus
前斜角 front bevel angle
前胸 prothorax
前胸背板(昆虫骨板) pronotum
前胸侧板(昆虫骨板) propleuron
前悬挂式割灌机 front-mounted scrubclearing equipment
前沿 leading end
前沿分析 frontal analysis
前演替顶极 preclimax
前叶轴 epipodium
前诱变剂 promutagen
前原卟啉 protoporphyrinogen
前缘脉(昆虫) costal vein
前缘木 head log; top log
前震(地震) foreshock
前制点 front stop
前置切杆式动力中耕机 front-mounted cutter bars power cuitlvator
前轴 epipodium; steering axle
前轴驱动 puller drive
前作 preceding crop
钳齿 set tooth; wrest tooth; wresting
钳子 forceps
潜蛾 ribbed cocoon maker
潜伏 delitescence; incubation
潜伏的变异性 potential variability
潜伏根原基 latent root primordium
潜伏期 incubation period; latency; latent period
潜伏期带菌者 incubating carrier
潜伏性病毒 latent virus
潜伏性侵染 latent infection
潜伏学习(动物) latent learning
潜伏芽 latent bud; resting-bud
潜伏遗传 latent heredity
潜伏状态 latency
潜行猎 stalking
潜火 hangover fire

潜近猎 hunt by stalking
潜力 potential
潜猎 poaching
潜流 subterranean river; underflow; underrun
潜燃火 hangover fire
潜热 latent heat
潜溶剂 cosolvent
潜水 phreatic water
潜水灰化作用 ground water podzolization
潜水灰壤 ground water podsol soil; ground water podzol
潜水面 ground water level; phreatic surface
潜水土 ground water soil
潜水位 ground water table
潜水植物 phreatophyte
潜水砖红壤 ground water laterite
潜水砖红壤性土 ground water lateritic soil
潜酸性 potential acidity
潜象 latent image
潜鸭 diving duck; pochard
潜芽 embryo (*or* embroyonic) bud
潜叶虫 leaf miner
潜叶蛾 leaf miner; leafminer fly
潜蝇 leafminer fly
潜影 latent image
潜育 gleization
潜育层(土壤剖面G层) glei horizon; gley; gley horizon; G-horizon; G-layer
潜育高草原土 gley-prairie soil
潜育过程 gley process; glei process
潜育化腐殖质层(土壤剖面A_g层) gleying humus layer; A_g-layer; A_g-horizon
潜育化森林土 gleyed forest soil
潜育灰壤 gley-podzol
潜育灰色森林土 glei forest grey soil
潜育森林土 gleyed forest soil
潜育水稻土 ground water rice soil
潜育土 gley soil; gleyed soil; gleysol
潜育土壤 ground water soil
潜育状灰壤 gley-like podzol
潜育状水稻土 gley-like rice soil
潜育棕色冲积土 glei alluvial brown soil
潜育作用 gleization
潜在变形 potential deformation
潜在的 potential
潜在顶极(群落) potential climax
潜在断裂 potential break
潜在肥力 inherent fertility; potential fertility
潜在分布区 potential area; potential distribution area
潜在光合能力 potential photosynthetic capacity
潜在害虫 potential pest
潜在生产量 potential production
潜在生产线 potential production line
潜在生物相 potential biota
潜在生长量 potential growth
潜在食物 potential food
潜在寿命 potential longevity
潜在蓄积 potential growing stock
潜在盐土 hidden solonchak
潜在影响 potential impact
潜在增长 potential increase
潜在蒸发散 potential evapotranspiration
潜在自然景观 potential natural landscape
潜在自然森林社会(植群) potential natural forest community
潜在自然植被 potential natural vegetation
潜在自然植被图 map of potential natural vegetation
潜在自然植生 potential natural forest community
潜在自然最终群落 potential natural terminal community
浅层对流 shallow convection
浅抽屉 tray
浅的 shallow
浅翻 shallow tillage
浅根的 shallow-rooted
浅根性 shallow rootedness
浅根植物 shallow rooted plant
浅耕 loosen; shallow ploughing; shallow tillage
浅耕机 loosener; parer
浅耕犁 paring plough
浅耕灭茬 skim ploughing
浅海带 sublittoral zone

浅裂 surface check; surface crack
浅裂的 lobate; lobed
浅裂片 lobe
浅绿色 light green
浅色 tint
浅色松香 pale rosin
浅色紫胶 blonde
浅水处 shallow
浅水生长的 epigeal
浅松土 soil wounding
浅榫眼 stub mortise
浅滩 shoal
浅滩浅水的 shallow
浅箱(育苗用) flat
浅休眠 shallow dormancy
浅沼泽 pocosin
欠载运行 underrun
茜草 madder
茜草色 madder
嵌板 panel; panel board
嵌板背椅 pan back chair; panel back chair; wainscot chair
嵌板框架结构 panel-and-frame construction
嵌壁家具 build-in (fitted) furniture; fitted furniture; wall furniture
嵌补 patching
嵌补片 insert
嵌槽 spline
嵌槽边接 rabbet edge joint
嵌槽斜接 mitre-with ribbet joint; offset mitre joint
嵌齿 inserted tooth
嵌齿轨道 cogged rail
嵌齿锯 inserted tooth saw
嵌齿圆锯 inserted tooth circular-saw
嵌带 intercalation
嵌腹接 side grafting by inlaying
嵌固件 fixings
嵌合体 chim(a)era
嵌花 applique
嵌花胶合板 mosaic plywood
嵌环 thimble
嵌环式滑轮 thimble block
嵌接 rabbet; rebated joint; table
嵌接合 dowelling joint
嵌金刚石的圆锯 diamond impregnated circular saw
嵌锯 pad saw
嵌靠接 inarching by inlaying
嵌框门 sash door
嵌木 spacer
嵌木细工 marquetry
嵌木芽接 chip budding
嵌入 insertion; intercalation; lute in
嵌入腹接 side inlaying
嵌入靠接 inarching by inlaying
嵌入木片 spline
嵌套式回归 nested regression
嵌条 mould
歉年(产量) fail-year; off year; quarter-mast
歉年(结实) no-seed year
枪 gun
枪柄式植苗锥 pistol-grip dibble
枪机 action
枪口 muzzle
枪口绷带(瞄准用) muzzle bandage
枪口速率 muzzle velocity
枪口炸震 muzzle blast
枪口支托 muzzle rest
枪时(打猎强度单位) gun-hour
枪闩 breech bolt
枪筒(管) barrel
枪筒喉缩 choke
枪筒口径 bore
枪筒落槽(装入枪炳槽内) bedding
枪筒瞄准 bore sighting
枪筒尾部 breech
枪托 gunstock
枪托板 butt plate
枪支 firearm
枪注法(木材防腐) gun-injection process; gun-injection
腔(孔洞或内室) cavity; cavitas
腔(昆虫) lumen (*plur.* lumina)
腔隙的 lacunar
强毒株 virulent strain
强度 intensity
强度采伐 heavily cut
强度采伐区 heavily cutover area
强度伐 heavily cut
强度灰化土 intensely podzolized soil
强度曲线 intensity curve
强度试验 strength test

强度疏伐(D度疏伐) heavy thinning; severity thinning; strong thinning; D-grade thinning
强度疏伐林 heavily thinned stand
强度系数 strength factor
强度性质 strength properties
强度修枝 heavy pruning
强度因数 strength factor
强度-重量比 strength-weight ratio
强度轴 intensity axis
强风 fresh gale
强风化黏磐 nitosol
强光抑制作用 light inhibition
强化层积材 densified laminated wood
强化干燥 intensive drying
强化胶合板 metallized plywood; reinforced plywood
强化木材 metallized wood
强化松香 fortified rosin
强化松香胶料 fortified rosin size
强化塑料 reinforeed plastic
强化砧 invigorating rootstock
强聚合油 long oil
强力磨削(砂磨) heavy duty grinding
强烈疏伐 drastic thinning
强淋溶土 acrisol
强流电子束加速器 electron beam accelerator
强韧的 coriaceous
强势木生产力 productivity index
强酸性氧化土 acrox
强心酚 cardol
强修剪的 hard-pruned
强阳性树 strong-light demander
强制剥离法(胶合板胶黏力试验) forcible separation method
强制采伐 forced cutting
强制发芽 forced germination
强制进料 forced feed
强制气干 accelerated air-drying; fan drying; forced-air drying
强制烧除 forced burning
强制通风 forced draft
强制通风干燥窑 forced draught kiln
强制喂料 forced feed
强制休眠 imposed dormancy
强制循环干燥窑 fan type kiln; forced circulation kiln; forced draught kiln
强制重复循环干燥窑 forced recirculating dry kiln
强制自花授粉 enforced self-pollination
强重比 stiffness to weight; weight-strength ratio
墙板 wall board
墙壁园 wall garden
墙壁纸板 plaster board
墙基植物 foundation plant
墙基种植 foundation planting
墙间系梁 couple-close
墙角桌 corner table
墙园 wall garden
蔷薇与带状线脚(饰条) rose and ribbon moulding
蔷薇园 rosary
强迫 external forcing
强迫共栖 synechthry
强迫休眠 enforced dormancy; imposed dormancy
强迫杂交 compulsory crossing
抢夺型竞争 scramble-type competition
抢救 salvage
羟胺 hydroxylamine
羟化酶 hydroxylase
羟化作用 hydroxylation
羟基 hydroxy
羟基白藜芦酚 hydroxyresveratrol
羟基络合物 hydroxy complex
羟基芪 hydroxystibene
羟基血红素 haematin
羟甲基化 methylolation
羟喹啉 hydroxyquinoline
羟联 olation
锹 notching spade; spade
锹甲 stag beetle
锹植(以锹开缝) bar planting
敲打 hammer
敲击检查 acoustic(al) inspection
敲碎 batter
敲挞机 thrasher
橇 sledge; sleigh
橇板摇椅 slipper rocker
橇道 sledge road

橇运 sled hauling; sledging
乔灌木水藓沼泽 Muskey
乔林混农作业 combined cropping system; combined system of field crop and high forest
乔林皆伐作业 clear-cutting high-forest system
乔林伞伐作业 high forest compartment system
乔林作业 high forest system; seedling forest system
乔木 arbor; high tree; tall tree; crop of formed tree; tree
乔木层 tree layer; tree stratum
乔木带 tree zone
乔木的 arboreal
乔木林 high forest; seed forest; seedling-forest; timber forest; tree forest
乔木树种 arbor species; arboreal species; tall tree species
乔木幼树期 brush-stage
乔木植物 magaphanerophytes
乔木状的 arbor form; arboreous; arborescent
桥 bridge
桥墩 abutment; pier
桥桁 saddler's wood
桥接 bridge grafting; bridging
桥接法 bridge method
桥梁路面 floor system (of bridge)
桥梁用材 timber for bridge
桥面 deck plank
桥面系统 floor system (of bridge)
桥式起重机 bridge crane; loading bridge; overhead crane; overhead travelling crane
桥台 abutment pier
樵夫 firewood cutter; woodsman
壳斗 acorn cup; cup; cupula; mast; semi-capsula
壳内层(种子) endotesta
壳圈 geosphere
峭壁 cliff; crag
峭度 kurtosis
翘曲 sweep; warp; warping; cup
翘曲材 warped timber
翘头 nose
翘弯 crosswise concaving
撬棒 rigger's bar
鞘 sheath; theca (*plur.* thecae)
鞘翅(昆虫) elytra
鞘蛾 casebearers
鞘状细胞 sheath cell
切 trim
切变线 shear line
切齿 cutter tooth
切齿侧刃 side cutting edge
切齿侧刃角 side cutting edge angle
切齿链片 cutter link
切齿修整 jointing cutting teeth
切除 ablation; cut-off; resection
切除枝节 bumping knot
切段 dissection
切断 crossover; cut-off; scission
切方头 butt
切缝栽植 notch planting; slit planting
T形切缝栽植 angle planting
切腹接 side cleft graft
切割 chipping; incision
V形切割法 V-shaped incision
切割钢索的榔头 soft hammer
切割面 cutting surface
切割器 knife bar
切割挖根机 cutter-lifter
切根虫 cutworm
切根机 root cutter; rootprunner
切根起苗机 cutter-lifter
切根器 root breaker
U形切根器 U-blade cutter
切花保鲜剂 floral preservative
切花用花卉 cut flower
切肩刨 shoulder plane
切接 common grafting; cut-grafting; proper crown grafting; veneer grafting
切开 incision
切开检验(种子) cut(ting) test
切靠接 inarching by veneering
切口 notch; rabbet; slash
切力线 shear line
切面 splay mould; tangential plane
切片 scction; slab; slice; sliced sheet
切片操作 sectioning
切片刀 slicing knife
切片法 sectioning
切片机 microtome

切片取样 section sampling
切片染色 section staining
切片术 microtomy
切去断裂部分(造材) break cut
切入 cut-in
切入量 bite
切伤压条法 layerage by notching
切梢机 plant-top removing machine; topper
切梢器 topper
切树腿(发目前清理根部周围) laying in
切碎机 chopper; shredder
切条 big slip
切条机 cutting cutter
切线测径杖 tangent stick
切向壁 tangential wall
切屑 turnings
切削(切屑) chipping; cut; cutting
切削刀片 cutting tip
切削法 cutting system
切削方向 cutting direction
切削工具 puncheon
切削功率消耗 cutting consumption
切削沟纹 chipped grain
切削规 cutting gauge
切削轨迹 cutting trajectory
切削痕 cutter mark
切削加工 working
切削夹具 cutting jig
切削角 cutting angle
切削力 cutting force; force of cut
切削量 cutting output
切削刃 cutting edge; cutting lip; lip
切削刃位移 cutting edge displacement
切削速度 cutting speed; cutting velocity
切削型碎料 cutter-type particle
切削性 machinability
切削圆(刀尖旋转轨迹) cutting circle
切削振动 shudder
切削阻力 cutting resistance
切枝杂交 cutting cross
窃蠹 drug store beetle
侵害 disoperation
侵染部位 infected site
侵染材 affected timber
侵染的 infective
侵染点 infection court
侵染过程 infection process
侵染介体 vector of infection
侵染量 amount of infection
侵染密度 density of infection
侵染频率 frequency of infection
侵染期 infective stage
侵染丝 infection thread
侵染途径 infection gate
侵染物 infectious organism
侵染现象 infection
侵染型 infection type
侵染性 infectivity
侵染性病害 infectious disease
侵染性菌丝 infection hypha
侵染循环 cycle of infection
侵染中心 infection center; infective center
侵扰 infestation
侵入(演替等) intrusion; invasion; penetration; encroachment
侵入菌丝 penetration hypha
侵入期 infective stage; invasion stage
侵入生长 interposition growth; intrusive growth
侵入体 intrusion
侵入岩 intrusion; irruptive rock
侵入种 invader; invader species
侵蚀 corrosion; erosion; scouring
侵蚀地 eroded land
侵蚀地形 erosion(al) landform
侵蚀度 erodibility; erosiveness
侵蚀防治 erosion control
侵蚀风景 destructional landscape
侵蚀高原 erosion plateau
V形侵蚀沟 V-gully
侵蚀谷 erosion(al) valley
侵蚀湖 erosion(al) lake
侵蚀基面 base of erosion
侵蚀基准面 base level of erosion; erosion basis
侵蚀阶地 erosion terrace
侵蚀控制 erosion control
侵蚀面 erosion(al) surface
侵蚀模数 erosion modulus
侵蚀能力 erosive power
侵蚀盆地 erosion(al) basin; scooped basin

侵蚀平原 erosion plain
侵蚀水 invading water
侵蚀土 eroded soil
侵蚀性 erosiveness
侵蚀循环 cycle of erosion; erosion cycle
侵蚀崖 erosion(al) scarp
侵蚀因素 factor of erosion
侵蚀周期 cycle of erosion
侵蚀作用 abration; erosion
侵填体 tylosis; tylosis (*plur.* -se)
侵填体雏形层 tylosis forming layer
侵袭 infestation
侵袭力 aggressivity
侵袭性菌系 aggressive strain
侵占型 aggressive type
侵占植株 aggressive individuals
侵占种 aggressive species
侵殖种 colonizing species
亲本 parent; progenitor
亲本材料 parent material
亲本对 parental pair
亲本树 mother tree
亲本投资 parental investment
亲本细胞壁 parent cell wall
亲代养育 parental care
亲电试剂 electrophilic reagent
亲电体 electrophile
亲电子剂 electrophile; electrophilic reagent
亲和标记 affinity labeling
亲和层析 affinity chromatography
亲和的 compatible
亲和反应 compatible reaction
亲和嫁接 congenial graft
亲和力 affinity
亲和色谱法 affinity chromatography
亲和性 compatibility
亲核加成 nucleophilic addition
亲核取代 nucleophilic substitution
亲核试剂 nucleophilic reagent
亲核物质 nucleophile
亲核性 nucleophilicity
亲实体性 stereotropism
亲属关系 kinship
亲属选择 kin selection
亲水的 hydrophilic; hydrophilous; lyophilic
亲水胶 hydrophilic gum
亲水胶体 hydrophile; hydrophilic colloid
亲水物 hydrophile
亲水性纤维 hydrophilic fibre
亲液的 lyophilic
亲液胶体 lyophilic colloid
亲裔交配 parent-offspring mating
亲缘 affinis; affinity
亲缘交配 related crossing
亲缘特征类似 affinity
亲缘种 sibling species
亲脂的 lipophilic
亲子间关系 offspring-parent relationship
亲子间回归 offspring-parent regression
亲子相关 parent-offspring correlation
芹子烯 selinene
琴背椅 fiddleback chair
禽鸟园 aviary
青变 blue sapstain; blue stain; blueing
青变材 blue lumber; blue stained wood
青变菌 blue sapstain fungus (*plur.* -gi); blue stain fungus
青春激素 juvenile hormone
青蜂 cuckoo wasp; ruby-tailed wasp
青腐 blue-rot; green-rot
青冈材 oak wood
青蒿酮 artemisia ketone
青蒿油 artemisia oil
青壳纸 vulcanized fibre
青绿饲料 green fodder
青霉病 green mold
青苗 young crop
青年河 adolescent river
青年期 juvenescent phase
青少年的 juvenile
青苔 lichen
青香茅油 inchi grass oil
青液 green liquor
青榆烯 lacinilene
青枝绿叶(装饰用) greenery
青苎麻 green ramie
轻便剥皮机 portable peeler
轻便动力牵引机 motor mule

轻便动力挖坑机 portable power hole digger
轻便构架 balloon framing
轻便轨道 portable tramway
轻便集材爬犁 crotch
轻便绞盘机 smack winch
轻便锯 portable saw
轻便刨床 portable planing machine
轻便牵引车 mule cart
轻便式木材劈裂机 portable log splitter
轻便索道 light skyline; wire tramway
轻便铁轨道 light railway
轻便贮水箱 portable tank
轻便桌 occasional table
轻度放牧 under grazing
轻度火烧 light burn
轻度林火 light forest fire
轻度淋溶 weakly leached
轻轨道 tramway
轻轨运材道(临时) tram
轻磨 touch sanding
轻木 balsa wood; balsa; lightwood
轻木焦油 light tar
轻木油 lightwood oil
轻黏土 light clay
轻侵染 subinfection
轻砂 touch sanding
轻烧(清除可燃物) light burning
轻施胶 half-sized
轻松土 light soil
轻微变色 light discolouration; light stain; slight stain
轻微滑刨 slight skip
轻微缺陷 slight imperfection
轻微着色 tint
轻小种子 light seed
轻型集材车 light forwarding vehicle
轻型可燃物 light fuel
轻型索道 light cableway
轻型无线电遥控绞盘机 portable radio-controlled winch
轻型运输车 pickup
轻型装载机 cherry-picker
轻亚黏土 sand loam
轻油 light oil
轻油溶剂处理(防腐) light solvent treatment
轻质胶合板 low-density plywood
轻质壤土 light loam
轻质碎料板 light particleboard
轻质土 light soil
轻中量级 light middle
氢氟碳化物 hydrofluorocarbons
氢供体 hydrogen donor
氢化枞胺 hydroabietylamine
氢化黄樟丁香酚 hydrosafroeugenol
氢化裂解 hydrocracking
氢化裂解作用 hydrogenolytic cleavage
氢化木质素 hydrolignin
氢化器 hydrogenator
氢化松香 hydrogenated rosin
氢化松香胺 hydroabietylamine
氢化作用 hydrogenation
氢火焰电离检测器 hydrogen flame ionization detector
氢键 hydrogen bond
氢解作用 destructive hydrogenation; hydrogenolysis
氢离子活性 hydrogen ion activity
氢离子浓度 hydrogen ion concentration
氢受体 hydrogen acceptor
氢焰离子检测器 flame ionization detector
氢氧磷灰石 hydroxyapatite
氢质黏土 hydrogen clay
倾覆力矩 overturning moment
倾角 angle of inclination; dip; dip angle; inclination angle; rake
倾角计 clinometer
倾盆大雨 rattler
倾群 cline
倾塌 sagging
倾销 dumping
倾斜 decline; dip; slope; steep; tilt; leaning
倾斜材堆 pitched pile
倾斜度 batter; grade; gradient; slope
倾斜堆积 pitched pile; slope pile
倾斜方向 direction of tilt
倾斜活板指示器 bevel indicator flap
倾斜角 angle of obliquity; rake angle
倾斜节理 cutter
倾斜面 inclined plane

倾斜木 cripple; lopsided tree; leaning tree
倾斜木纹 short grain
倾斜坡度 dip slope
倾斜切削 inclinel cutting
倾斜摄影 oblique photograph
倾斜式带锯机 slant band saw; tilted bandmill
倾斜式输送机 tilt down conveyor
倾斜式圆锯机 tilted circular saw
倾斜树干 leaning stem
倾斜位移(航空摄影) tilt displacement
倾卸 dumping
倾卸承载梁 dumping bunk
倾卸装置 tilter; tipper
倾轴式多能圆锯 tilting arbor variety saw
清查控制样地 control plot
清除 eradication; liquidation; removal
清除场地(残留物) chunk
清除场地(灌木丛等) swamp
清除伐 liquidation cutting; refining
清除伐根 stump pull
清除共生 cleaning symbiosis
清除灌木 brush out
清除森林枯枝落叶层 removal of forest litter
清除用火 fuel-reduction burning
清单 inventory; list
清耕法 clean cultivation
清灌斧 clearing axe
清烘漆 baking varnish
清洁发展机制 clean development mechanism
清洁共生 cleaning symbiosis
清洁剂(清洁器) cleaner
清理 clearance; clearing; liquidation; remove
清理部分采伐剩余物法 partial slash disposal
清理采伐场地 logging clearing
清理齿 clearer(tooth)
清理伐(后伐) final clearing; disengagement cutting
清理过的伐区 cleared area
清理火场 mopping up
清理价值 liquidation value
清理线路 swamping
清理作业 cleaning-up operation
清漆 varnish; varnish paint
清漆底子 varnish base
清漆涂层 varnish coating
清漆用树脂 varnish resin
清扫机 cleaning-machine
清扫器 sweeper
清水流送的 floated bright
清水性的 katharobic
清算 clearing
清晰度 definition
清洗机 cleaning-machine
清洗液 leaner
清选 dressing
清液 serum
清油 boiled oil
清枝 shaft-cleaning
清装列车(装运单一材种) unit train; solid train
蜻蜓 dragon fly; dalmon fly
情报 message
情节 storyline
情景 scenarios
晴空湍流 clear air turbulence
晴空雨 serein
晴朗 clear
晴色土 ando soil
晴雨通车道 all-weather road
氰不敏感呼吸 cyanide insensitive respiration
氰敏感呼吸 cyanide sensitive respiration
氰乙基纤维素 cyanoethyl cellulose
穹棱 groin
穹棱屋顶 groined roof
琼森形点 Jonson's form point
琼脂 agar; agar-agar
琼脂木块法 agar-wood block method
琼脂木块腐朽试验 agar-block decay test
琼脂木块培养试验 agar-block test
琼脂培养基 agar medium
琼脂平面(培养) agar plate
琼脂平面培养法 agar plate method
琼脂深层培养 agar deep culture
琼脂斜面培养 agar slant
丘 hummock
丘陵 hill
丘陵地 hilly land; undulating ground

丘陵地带 foothill
丘陵区 hilly zone
丘氏集材钢球 Chuball
丘泽 aapamoor; Aapa moor
丘植 hill planting; mound planting
丘状沙 dune-sand
秋播 autumn seeding; autumn sowing; fall planting; fall seeding
秋材 autumn wood
秋材率 percent of autumn wood
秋翻地 autumn ploughing
秋耕 autumn ploughing; fall-plough
秋海棠色素 carajurin
秋花坛 autumn flower-bed
秋季法正蓄积法 autumn normal growing method
秋季景观 autumnal aspect
秋季相 autumn circulation; autumnal aspect
秋季循环 autumn circulation; autumnal circulation
秋色 autumnal aspect
秋梢 autumn growth
秋霜 autumn frost
秋芽(秋枝) lammas shoot
秋植 autumn planting; fall planting
蚯蚓细腐殖质 earthworm mull
求爱行为 presenting
求反(计算机) negate
求积法 determination of volume
求积式 cubing formula; formula for determining the volume
求积仪 planimeter
求偶(期) courtship
求偶表态 courtship display
求偶行为 courtship behaviour; courtship behavior
求偶喂食 courtship feeding
求算面积截距法 transect for area determination
球掣 ball catch
球蛋白 globulin
球根 corm
球根栽植器 bulb planter
球根盆 bulb pan
球根牵牛苷 orizabin
球根牵牛树脂 jalap
球根植物 bulbous plant; tuberous plant
球果 strobile; strobilus
球果苞片 cone-bract
球果不裂(因干燥处理不当) case-hardening
球果采集机 cone harvesting machine
球果采摘机 cone picking machine
球果迟熟 cone serotiny
球果干燥室 cone kiln
球果类树木(泛指针叶树) cone bearer; cone bearing tree
球果破碎机 cone breaker
球果球花 cone
球果收集机 cone collector
球果脱粒 extracting seeds from cone
球果脱粒机 cone shaker
球果锈病 cone rust
球果震落机 cone shaker
球果直播 cone-sowing
球果植物 coniferophyte
球花 strobile
球茎 corm; solid bulb
球茎地下芽植物 bulbous geophyte
球菌 coccus
球磨机 ball mill pulverizer
球形的 sphaeroid; spherical
球形蒸煮器 spherical boiler
球芽 propago; sphaeroblast
球蚜 gall aphid
球状膨润 sphelrical swelling
区 district; township (corner); zone
区别 differentiation
区别种 differential species
区带电泳[法] zone electrophoresis
区段 section
区分 fractionation
区分段(测树) section
区分段测量 sectionwise measurement
区分段法(测树) sectional method
区分嵌合体 sectorial chim(a)era
区分求材积法 sectional measurement by means of volume
区分求积 volune measurement by sections
区分求积法 sectional measurement
区分周缘嵌合体 sectorial-periclinal chim(a)era
区公园 district park
区划 zoning

区划法 zoning law
区划副线 accessory frame; minor ride; secondary compartment line
区划更新区 division into regeneration areas
区划轮伐法 division into annual coupes
区划线 compartment line; ride
区划主线 main compartment line; main frame; major ride
区间系统 block system
区内碎部 interior detail
区系动物地理学 faunal zoogeography
区系省 faunal province
区系组成 floristic composition
区县林 county forest
区域材积表 regional volume table
区域地理学 regional geography
区域定位 zone location
区域分类 area classification
区域公园 regional park
区域划定 regional assignment
区域化 compartmentalization; compartmentation
区域计划 regional planning
区域生物地理学 regional biogeography
区域时间 zone time
区域试验 regional test
区域性群丛 regional association
区域重新开发规划 area redevelopment program
区组 block
区组纪录 block record
驱虫剂 repellent
驱虫纸 insect repellent paper
驱动 drive
驱动滚筒 live roll
驱动滚筒输送机 live roller conveyer
驱动力 drive force; driving effort
驱动链轮 driving sprocket
驱动链片(锯链) drive link
驱动轮 driver; driving wheel
驱动面(锯链) drive flank
驱动排料滚筒 live discharge roll
驱动压辊 powered backup roll
驱动轴 drive shaft
驱动装置 actuating device; actuating gear
驱赶 trailing
驱猎 beat
驱鸟剂 bird repellent
驱鸟指示 flush point
驱气 degassing
驱轴 pivot
屈服点 yield limit; yield point
屈服极限 yield limit
屈服强度 yield strength
屈服应变 yield strain
屈服应力 yield stress
蛆(无足幼虫) maggot
蛆室 maggot chamber
躯干 truncus
趋触性 stereokinesis; stereotaxis; thigmotaxis; thixotaxis
趋触性的 thigmotactic
趋触性的反应 thigmotropic reaction
趋地性 geotaxis
趋风性 anemotaxis
趋光节律 phototactic rhythm
趋光性 phototaxis; phototaxy; phototropism
趋光性的 phototactic
趋光源性 topophototaxia
趋化性 chemotaxis
趋激性 telotaxis
趋流性 rheotaxis
趋气性 aerotaxis
趋渗性 osmotaxis
趋声性 phonotaxis
趋实体性 stereotaxis
趋势面分析 trend surface analysis
趋同 convergence; homoplasy
趋同波动 convergent oscillation
趋同进化 convergent evolution
趋同适应 convergent adaptation
趋位性 topotaxis
趋温性 thermotaxis
趋性 taxis; tropism
趋氧性 aerotaxis
趋异 divergence
趋异波动 divergent oscillation
趋异适应 divergent adaptation; radiation adaptation
鸲(qú) robin
渠 channel; ditch
渠道比降 longitudinal slope of canal

渠道沉积 channel fill deposit
渠道淤积 channel sedimentation
渠首 headwork
渠首闸门 head gate
蠼螋 earwig
曲背锯片 sway sawback
曲臂式装车机 dog leg crane
曲柄孤口凿 bent gouge
曲柄锯 cranking saw
曲柄摇钻 brace
曲槽刨 compass rebate plane; router plane
曲尺 try square
曲垂 bight
曲干材 compass-timber
曲根材 sabre butt
曲颈甑 retort
曲静载荷 static load(ing)
曲流 meander
曲率 buckling
曲率半径 radius of curvature
曲率中心 centre of curvature
曲霉 aspergillus
曲霉病 aspergillosis
曲面板结构 shell construction
曲面胶合板 curved plywood
曲面面积 surface area
曲面刨 compass plane
曲面刨床 bending planer
曲线 wow
S形曲线 S-curve; sigmoid curve
曲线坝 curved dam
曲线板 mould
曲线带伐作 curve strip-cutting system
曲线的 curvilinear
曲线横断面 curved cross-section
曲线集材 curve skidding
曲线锯 compass saw; scroll saw
曲线量测计 curvimeter
曲线求积法 graphic method of valuation; volume-curve method
曲线设定法 curve setting
S形曲线饰条 ogee
曲线索道 crooked cable way
曲线调和化 harmonization of curve
曲线阻力 resistance due to curvature
曲型线条 falling mould
曲折 tortuosity
曲折带状采伐 zigzag felling
曲折河岸 indentation
曲折型起重臂 gooseneck boom
曲枝 bending
取材时间 access time
取代度 degree of substitution
取代行为 displacement activity; displacement behaviour
取得成本 acquisition cost
取花粉 extraction
取景器 viewfinder
取食的 trophic
取食共生 trophobiosis
取食生态位 feeding niche
取食者 consumer
取向 orientation
取向区 oriented region
取向效应 orientation effect
取芽树 budwood
取样 sample; sampling
取样法 sampling methods
取样率 sampling rate
取样器 probe; sample collector; sampler
取样匙(生长锥) extractor
取样数 sample number; size of a sample
取样调查法 sampling methods
去氨基 deamination
去翅(种子) dewing; removal of wing
去翅机 dewinger
去除方法 removal census; removal method
去春化作用 devernalization
去顶 beheading; decapitation
去核的 enucleated
去荚 hulling
去尖 decapitation
去角动物(牛、羊、鹿等) pollard
去节环(剥皮机) de-knotter ring
去节机 knocker screen; knot borer; knot catcher; knotter
去节器 knot borer
去壳 shelling
去壳机(种子) seed huller
去矿质水 demineralized water
去离子水 deionized water
去离子作用 deionization
去劣 rogue

去劣种子园 rogued seed orchard
去零分布 zerotruncated distribution
去木质素 delignification
去皮 barking; peeling; shelling; stripping
去皮材 barked wood; peeled wood
去皮材积 under-bark volume; volume inside bark
去皮的 bark-free
去皮杆 dressed pole
去皮干围 girth under bark
去皮机 decorticator
去皮器 dehusker; tree-scriber
去皮研磨器 debarking abrader
去皮原木 barked log; buckskin; peeled log
去皮直径 diameter inside bark; diameter under bark; diameter without bark; in-bark DBH; inside diameter
去皮种子 decoated seed
去皮周长 girth under bark
去色 logging site
去湿 dehumidification
去湿作用 desiccation
去势(动物) castration; emasculation
去髓短木段 billet
去碳酸基 decarboxylation
去心板材 centre board
去芯厚板 centre plank
去雄(植物) castration; emasculation
去芽 debudding; disbudding
去叶 defoliation
去杂(样品处理) scalping
去杂(种子处理) conditioning
去枝 debranch(ing); top-lopping
去脂(毛布) scouring
去脂材 bled timber
去籽机 macerator
趣味园艺 amateur gardening
圈 race; zone
圈肥 yard manure
圈纹园 knot(ted) garden
圈椅 easy chair
权利金 royalty
权益 equity
权益补偿 commutation of rights
权益资本 equity capital
权重 weighting
权重分摊 weight apportionment
全板门 flush door
全北界植物区系 holarctic floral kingdom; holarctis
全北区(昆虫) Holarctic Region
全变态昆虫 holometabolic insect
全变态类 holometabola
全剥皮 clean bark
全部成本法 absorption costing
全部的 entire
全部破坏 complete failure
全部树冠覆盖 full crown cover
全部填装(沙发) stuff-over
全部限额 overall limit
全成本定价 full-cost pricing
全尺寸 full scale
全磁迹 full track
全担子 holobasidium
全单板胶合板 all-veneer plywood; veneer plywood
全动物营养性的 holozic
全对称(结晶) pantomorphism
全伐林班 completed compartment
全伐区立木出售 sale of the whole lot
全分布区产地试验 wide range provenance trial
全氟化碳(PFCs) perfluorocarbons
全干(木材干燥) dry through
全干材(绝干材) absolutely dry wood
全干材作业(全部树干) full-length logging
全干重(绝干重) absolute dry weight
全干状态(绝干状态) absolutely dried condition
全高 total height
全挂车 full trailer
全光 full sun; full light; full sun light
全轨 full track
全国劳动关系法 national labour relations act
全国绿化运动 national forestation drive
全国土地分类 state-wide land classification
全国义务植树活动 national compulsory tree-planting activities
全国植树节 national tree-planting festival
全合子(遗传) holozygote

Q

全黑位置(偏振光显微镜) blackout position
全回流 infinite reflux
全胶合板构造 all-veneer construction
全金属活动温室 all-metal prefabricated greenhouse
全浸没法 full immersion method
全景 full view; panorama; whole scene
全景廓线图 panoramic profile map
全鸠尾榫槽 fuil dovetail dado
全垦 complete cultivation; full cultivation; general cultivation
全裂片 segment
全林分生长模型 whole stand model
全林分推算法 total stand projection
全林渐伐作业 shelterwood uniform method; shelter-wood uniform system; uniform method; uniform system
全林利用 full-forest utilization
全林伞伐作业 compartment uniform method; shelterwood uniform method; shelter-wood uniform system; uniform method; uniform system
全龄的 all-aged
全龄林 all-aged forest
全龄林分 all-aged stand
全螺旋型切削法 full spiral cut
全履带[车] full track
全酶 holoenzyme
全面保险 all risk insurance
全面播种 full seeding
全面拆修 general overhaul
全面点火 area ignition
全面耕翻 complete cultivation; full cultivation
全面耕作 general cultivation
全面经济 economy-wide
全面控制 comprehensive control; over-all control
全面清查 complete enumeration
全面烧荒 broadcast burning
全面施用 over-all application
全面调查 complete enumeration
全面消减 across-the-board-cut
全面整地 full cultivation; overall soil preparation
全面作业 stand-method
全能枪 all-purpose gun
全能自动制图编制装置 universal automatic map compilation equipment
全年放牧地 yearlong range
全年生芽 perennating bud
全盘控制 over-all control
全蹼足 totipalmate foot
全氢化松香 perhydrogenated rosin
全球变化 global change
全球变暖 global warming
全球表面温度 global surface temperature
全球大气研究计划 Global Atmospheric Research Program
全球定位系统 global position system
全球环境监测系统 Global environmental monitoring system
全球生态学 global ecology
全球增温潜势 global Warming Potential
全人字形切法 full-herringbone cut
全日照 full sun
全色片 panchromatic
全色乳剂 panchromatic emulsion
全上方荫蔽 full overhead shade
全烧 clean burn
全烧伐带 clear-burning strip
全身的 systemic
全深海的 holopelagic
全视图 full view
全树采运(带主根) whole-tree harvesting
全树干材(全部树干) full-length tree
全树高 total tree height
全树集材 full-tree logging; full-tree skidding; whole-tree logging; whole-tree skidding
全树利用(地上部分) complete-tree utilization; whole-tree utilization; full-tree utilization
全树皮碎料板 all bark particle board
全树削片 whole-tree chipping
全树削片机 complete tree utilizer; full-tree chipper; whole-tree chipper
全损(保险) total loss
全套工具 solid content
全体(整体)计划 intergrated planning

Q

全天候 all weather
全天候道路 all-weather road
全天候道路集运作业 winter show
全天候基础 all-weather foundation
全天候抓钩装载机 all-weather grapple loader
全同胞 full sib
全图 panorama
全拖车 full trailer
全拖集材 surface yarding
全拖式 ground(-line) logging
全拖式钢素集材 ground-lead(cable) logging
全拖式钢索集材 ground(-line) logging; ground(-line) skidding; ground snigging; ground yarding; ground-cable logging
全拖式集材 ground(-line) skidding; ground snigging; ground yarding; snaking; snigging; bob-tailing
全拖式集材方式 bob-tail system
全吸收法(木材防腐) full-cell process
全纤维素 holocellulose
全纤维素制浆法 holopulping process
全效率 total efficiency
全心材的 all-heart
全新世 Holocene
全型 holotype
全雄群 all-male group
全旋转式起重机 full circle crane
全循环湖 holomictic lake
全叶插穗 entire leaf cutting
全叶扦插 entire leaf cutting
全隐鸠尾榫 blind dovetail
全隐鸠尾榫接 housed dovetail joint
全郁闭 continuous canopy
全域 population
全圆切削 full-round cut(ting)
全缘[的] entire
全缘叶的 integrifolious
全载集材(尤指短材) forwarding
全载集材机 forwarding machine
全载式集材机 forwarder
全站仪 total station
全轴驱动 all-axle drive
全株 whole tree
全转式起重机 full circle crane
全装载集材 prehaul; prehauling
全装载集材机 prehauler
全自动调温温室 completely automated heating and cooling greenhouse
泉源 fountainhead; springhead
拳卷的 circinal; circinate
犬齿 canine tooth
犬火灾 conflagration
缺尺 scant
缺对二倍性 nullisomic diploidy
缺对染色体的 nullisomic
缺对染色体生物 nullisomic
缺额材积(原木检尺量与工厂检量之差额) underrun
缺乏 deficiency
缺肥症状 deficiency symptom
缺钙 acalcerosis
缺钙症 acalcicosis
缺胶 starved failure; starved joint
缺胶接合 starve joint
缺刻 incision
缺刻状的 eroded; gnawed
缺口 gap; incision; indentation; recess
缺口大小 gap size
缺口式瞄准器 open sight
缺口堰 notch weir
缺棱 wane; wane edge; waney edge
缺棱材 wany lumber
缺绿病 chlorosis
缺锰症 manganese deficiency
缺区 missing plot
缺少影响 moisture stress
缺失(不足或没有) deficiency; depletion
缺失(遗传) deletion
缺失纯合体 deficiency homozygote
缺失型突变体(变型体) depletion mutant
缺失杂合体 deficiency heterozygote
缺失重现染色体 deficient-duplicated chromosome
缺陷 blemish; defect; failure; fault; flaw; imperfection
缺陷材 defected wood; defective wood
缺陷分级法 defect grading; defect system
缺陷检查仪 defectoscope

Q

缺陷扣除 deduction for defect; defect deduction; allowance for defect; allowance for fault
缺陷显示器 defectoscope
缺陷因素 cull factor
缺陷原木 defective log
缺陷长度 defective length
缺陷折减系数 cull factor
缺氧的 anoxic
缺氧湖 oxygen deficient lake
缺氧环境 anoxic condition
缺氧培养 anaerobic culture
缺雨 lack of rainfall
缺雨区 anhyetism
缺雨性 anhyetism
缺株点 blank
雀丝 woolly grain
确定变异性 definite variability
确定采伐量的周期调查法 recurrent inventory system in cut determination
确限度 exclusiveness; fidelity
确限种 exclusive species
裙板 apron board
裙板线脚 apron lining
裙布 apron
群 group; group of trees; swarm
群丛 association; coenobium
群丛表 association table
群丛单位理论 association unit theory
群丛复合体(植物) association complex
群丛隔离 association segregate
群丛个体 association individual
群丛林 association forest
群丛片段(断) association fragment
群丛温度 association temperature
群丛型(植物) association type
群丛组 association group
群飞 swarming
群集(鱼) schooling
群集的 gregarious
群集度(法瑞学派) sociability
群集度(广义的) gregariousness; stocking
群集调节 community regulation
群集稳定性 community stability
群集形态学 mass morphology
群集指数 aggradation
群集总生产率 community gross prodution rate
群寄生 gregarious parasitism
群间关系 intergroup relation
群间过渡 intergradation
群间选择 interdeme selection
群节 group knot
群居(社会性昆虫) communal
群居的 gregarious
群居动物 social animal
群居寄生 social parasitism
群居距离 social distance
群居昆虫 gregarious insect
群居习性 gregarious habit; social habit
群居相 gregarious phase
群居型 phase gregaria
群居性动物 gregarious animal
群居性寄生昆虫 gregarious parasitoid
群居性昆虫 colonial insect
群聚求偶 communal courtship
群聚指数 aggradation
群落 coenosis; coenosium; communal; community
群落表 community table
群落表面 community surface
群落动态 community dynamics
群落动态学 syndynamic; syndynamics
群落发生 syngenesis
群落发生学 genetic synecology
群落发生演替 syngenetic succession
群落分布学 synchorology
群落分类学 syntaxonomy
群落复合体 community complex
群落过渡带 ecotone
群落过渡区的 ecotonal
群落函数 community function
群落呼吸 community respiration
群落环 circle of vegetation; community ring
群落机能 community function
群落渐变群 coenocline
群落交错 alterne
群落交错区 ecotone; zono ecotone
群落结构 community structure

群落净生产力 net community productivity
群落局变 saltation
群落类型 community type
群落连续体 community continuum
群落配置 community matrix
群落平衡 community equilibrium
群落生活 coenobiosis
群落生境 biotope; communal habitat
群落生理学 synphysiology
群落生态群 coenocline
群落生态系统 community ecosystem
群落生态学 community ecology; synecology
群落生物量 community biomass
群落生物学 coenobiology
群落属(法瑞学派) alliance
群落调查记录 releve
群落调节 community regulation
群落外貌分类 physiognomic classification of community
群落稳定性 community stability
群落系数 coefficient of community
群落系统发生学 synphylogeny
群落系统分类学 synsystematics
群落相似系数 coefficient of community similarity
群落镶嵌 community mosaic
群落学 coenology
群落演替 community succession; syngenesis
群落遗传学 syngenetics
群落综合性 community integration
群落组成 community composition
群落组织 community organization
群目(法瑞学派) order
群青 ultramarine
群生性 gregariousness
群属(法瑞学派) alliance association
群体 assembly; colony
群体繁殖 colonial breeding
群体间基因交换 introgressive hybridization
群体连续体 continuum
群体生态学 synecology
群体狩猎 group hunting
群体调节 group regulation
群体选择 group selection
群体压力 population pressure
群体遗传学 population genetics
群团(属性特征相同的一群) guild
群系 formation
群系纲 class formation
群系团 group of formation
群系组 formation group; group formation
群相(法瑞学派) facies; faciation
群型种 cenospecies
群植 massive planting
群种统计学 demography
群众运动 mass movement
群桩 dolphin
群状(孔状)伞伐 group shelterwood
群状采伐 felling by group; felling for group system; group cutting; group felling
群状抽样 cluster sampling
群状渐伐 group shelterwood cutting; group shelterwood felling; shelterwood felling by groups
群状渐伐作业 group cutting method; group system; shelter-wood group system
群状皆伐作业 group clearcutting method
群状母树作业天然更新 natural regeneration by group seed tree method
群状配置 group spacing
群状伞伐 group shelterwood cutting; group shelterwood felling; shelterwood felling by groups
群状伞伐作业 group shelterwood system; shelterwood group system; shelter-wood group system
群状天然下种更新法 group shelterwood
群状栽植 block planting
群状择伐 group selection felling; group-selection cutting
群状择伐作业 group selection management; group selection system; group-selection (system)
群状择伐作业法 group selection method; group-selection (system)
群状作业 group system

群状作业更新 regeneration by groups; regeneration under group system
群状作业更新林 regenerated forest under group system
群状作业天然更新 group system of natural regeneration
髯毛 barba
燃材 firewood
燃点 burning point; ignition point; kindling point
燃耗分析 burnup analysis
燃力 burning-power; heating power
燃料 fuel
燃料层 fuel bed
燃料堆 fuel bed
燃料负荷指标 fire load index
燃料值 fuel value
燃料砖(棒球) briquet(te)
燃料转换 fuel switching
燃烧 catching fire; combustion; inflammation
燃烧边缘 burning edge
燃烧弹 incendiary
燃烧地段 burning block
燃烧点 point of ignition
燃烧堆 burning pile
燃烧过程 combustion process
燃烧阶段 burning period; combustion phases
燃烧期限 burning period
燃烧强度 burning intensity
燃烧区 burning block
燃烧失重率 rate of lostmass after fire
燃烧室 burner; fire box; fire chamber; fire place
燃烧速度 rate of combustion
燃烧温度 ignition temperature; kindling temperature
燃烧系统 combustion system
燃烧要素 component of combustion
燃烧值 fuel value
燃烧指标等级 burning index rating
燃烧指标系统 burning index system
燃烧指数 burning index
燃烧指数率 burning index rating
燃油计量膜片 fuel metering diaphragm
燃油吸气系统 fuel and induction system
染菌 contamination
染料 dyestuff
染料木(可提取染料) dyewood
染绿素 chlorophoria
染色 stain; tincture; tintage; tinting; staining
染色差异 differential colouration
染色单体 chromatid
染色单体随机分离 random chromatid segregation
染色的 tinct
染色反应 staining reaction
染色剂 stain
染色力 dyeing capacity
染色粒 chromomere
染色木材 staining wood
染色素 chromosin
染色体 chromosome
S染色体 S-chromosome
T染色体 T-chromosome
染色体臂 chromosome arm
染色体不育 chromosome sterility
染色体断裂 chromosomal breakage
染色体断片 chromosome fragment
染色体分离 chromosomal disjunction
染色体构形 chromosome configuration
染色体固结 chromosomal nondisjunction
染色体基数 chromosome basic number
染色体基质 matrix chromosome
染色体畸变 chromosomal aberration; chromosome aberration
染色体间畸变 heterosomal aberration
染色体结构杂种 structural hybrid
染色体结合 synapsis
染色体局部交换 segmental interchange
染色体联会 chromosome pairing; synapsis; syndesis
染色体内的 intrachromosomal
染色体内畸变 homosomal aberration
染色体配对 chromosome pairing; syndesis; synapsing chromosomes
染色体桥 chromosome bridge
染色体群 chromosome complex
染色体数 chromosome number

染色体数目改变 numerical chromosome change
染色体四分子 chromosome tetrad
染色体随机分离 random chromosome segregation
染色体损伤 chromosomal damage
染色体突变 chromosome mutation
染色体图的制作 mapping of chromosome
染色体位点 chromosomal loci
染色体纤丝 chromosomal fibrilla
染色体显带 chromosome banding
染色体小片 minute
染色体形态变异 amphiplasty
染色体型(品种) chromosomal race
染色体学说 chromosome theory
染色体遗传学说 chromosome theory of heredity; chromosome theory of inheritance
染色体重叠 chromosome repeat
染色体重排 chromosomal rearrangement
染色体重组 chromosome recombination
染色体族 chromosomal race
染色体组 chromosome set; complex; genom(e); set of chromosome
染色体组型 caryotype; idiogram; karyotype
染色体组型模式图 karyogram
染色物质 colouring matter
染色线 chromonema
染色线测丝 anastomosis
染色性 chromaticity
染色休机制 chromosomal mechanism
染色者 tinter
染色质 chromatin
染色质核蛋白 chromonucleoprotein
染色质组 skein; spireme
染色质排出(细胞核中) achromatic
壤黏土 loam clay
壤土 loam; loamy soil
壤质 loamy texture
壤质粗砂土 loamy coarse sand
壤质砂土 loamy sand
壤质上 loamy soil
壤质细砂土 loamy fine sand
让价 concessional rate
让与 transfer
扰动剖面 disturbed profile
绕过伐根集材 sideline
绕过障碍(木排) saddle bag
绕圈的 round
绕射 diffraction
绕线器 winder
绕向(钢丝绳) lay
绕子(捆扎用的枝条) withe
热板式干燥机 iron drier
热变定 heat set; hot-set
热变形 heat deformation; heat distortion
热波段 thermal band
热不稳定性 thermal instability
热成土 thermogenic soil
热处理 heat treatment; tempering; thermal treatment
热处理板 heat-treated board
热处理材 heat-treated wood
热处理木材 heat-treated wood; staybwood
热处理室 tempering chamber
热处理压缩木 staypak
热处理硬质纤维板 tempered hardboard
热穿透 heat penetration
热传导 heat conduction; thermal conductance
热传导性 heat transmissibility
热传递 transmission of heat
热脆性 hot shortness
热代谢作用 thermometabolism
热带 torrid zone
热带材 tropical timber; tropical wood
热带丛林 jungle
热带的 tropical
热带反气旋 tropical anticyclone
热带腐殖质黑黏土 grumosol; grumosolic soil
热带钙层土 tropical pedocal
热带高山植物 oreithalion tropicum
热带高压 tropical anticyclone
热带海岸林 tropical strand forest
热带旱季落叶的 tropophytic
热带旱生林 thorn forest
热带黑土 melanite
热带红壤 tropical red loam
热带红壤土 tropical red soil
热带红土 tropical red earth
热带荒漠 hot desert

R

热带灰色黏土 gray tropical clay
热带棘林 thorn forest
热带渐伐作业 tropical shelterwood system
热带阔叶林 tropical hardwoods
热带林 tropical forest
热带林带 tropical forest belt; tropical forest region; tropical forest zone
热带林区 tropical forest region
热带落叶林 tropical deciduous forest
热带美洲雨季 invierno
热带气旋 tropical cyclone
热带区 tropical zone
热带伞伐作业 tropical shelterwood system
热带始成土 tropept
热带外地区 extratropical belt; extratropical zone
热带无风带 tropical calm zone
热带稀树草原植被型 dendropoion
热带稀树大草原 tree veld
热带雨林 selva; tropical rain forest
热带雨林气候 megathermal climate; tropic rain forest climate
热带植丛 jungle
热带植物 tropical plant
热导池检测器 thermal conductivity (cell) detector
热导计 catharometer
热导率 temperature conductivity; thermal conductivity
热岛 heat island
热电联产 co-generation
热电偶 electric thermo-couple; thermocouple; thermoelectric couple
热电偶高温计 thermoelectric pyrometer
热电偶温度计 thermocouple thermometer
热电偶自记温度计 thermocouple thermograph
热电性能 electric conductivity
热定型 heat set
热对流 heat convection
热法磨木浆 hot-ground wood pulp
热分解 thermal decomposition; thermolysis
热分解作用 thermal decomposition
热风干燥 hot air seasoning
热风器 air heater
热辐射 heat radiation; radiation; thermal radiation
热功率 heat output
热固化 hot-set; hot-setting
热固化剂 hot hardener
热固剂 hot setter
热固胶黏剂 hot-setting adhesive
热固木 staybwood
热固性 thermosetting
热固性合成树脂 thermosetting synthetic resin
热固性胶黏剂 thermosetting adhesive
热固性聚合物 thermosetting polymer
热固性清漆 thermosetting varnish
热固性树脂 heat-setting resin; thermosetting resin
热固性树脂胶黏剂 thermosetting resin adhesive
热固性塑料 thermosetting plastic
热固性增塑剂 thermosetting plasticizer
热惯性 thermal inertia
热辊压 hot-rolling
热害 heat lesion
热含量 enthalpy; heat content; thermal content
热函 enthalpy; total heat
热红外 thermal infrared
热后成型 postforming
热回收 heat reclaim; heat recovery
热降 heat drop
热降解 heat degradation; thermal degradation
热交换器 heat exchanger; heating coil
热接点 hot junction
热解 pyrogenic decomposition; pyrolysis
热解的 pyrogenic; pyrolytic
热解反应 pyrolytic reaction
热解气相色谱 pyrolysis gas chromatography
热解气相色谱法 thermal cracking gas chromatography
热解物 pyrolyzate

热解重量分析 thermogravimetric analysis
热介质 heating medium
热进-冷出周期 hot-cold circle
热进-热出周期 hot-hot circle
热静力式疏水器 thermostatic type steam trap
热静力式阻汽排水器 thermostatic type steam trap
热绝缘 thermal insulation
热绝缘材料 heat insulator
热空气干燥 heated-air seasoning
热空气干燥法 heated-air seasoning; hot air drying
热空气干燥炉 air-oven
热空气干燥窑 hot air drying kiln
热空气供气道 supply hot air duct
热空气烘箱 air-oven
热扩散 thermodiffusion
热扩散率 thermal diffusivity
热冷槽法(防腐) hot and cold bath method; hot and cold bath process; boiling and cooling process
热冷敞槽处理 dual-bath treatment; hot-and-cold open-bath treatment
热冷敞槽法 hot and cold open tank process
热冷浴槽法 method of hot and cold bath
热冷浴法 boiling and cooling process
热力-机械制浆 thermal-mechanical pulping
热力效率 thermodynamic efficiency
热力学 thermodynamics
热量 quantity of heat
热量测定 calorimetric measurement; calorimetry
热量单位 thermal unit
热量计 calorimeter
热量平衡 heat budget
热量输送 heat transfer
热量转移 heat transfer
热疗法 heat therapy
热流 heat flux
热滤法 hot filtration process
热煤焦油涂刷 hot coal-tar coating
热敏电阻 thermo-variable resistor
热敏电阻器 thermistor
热敏探测仪 heat-sensitive equipment
热磨 hot grind(ing)
热磨机 defibrator; defibrater
热磨机法 defibrator method
热磨机械浆 thermomechanical pulp
热磨秒 defibrator second
热能 thermal energy
热黏度 thermoviscosity
热黏度计 thermoviscosimeter
热喷清漆 hot-spray lacquer
热膨胀系数 thermal coefficient of expansion; thermal expansion coefficient
热膨胀性 thermal expansion
热平板式干燥机 hot-plate dryer
热平衡 heat balance; thermal equilibrium
热谱图 thermogram
热气流管道 hot gas duct
热气预热干燥机 hot air predryer
热侵蚀 thermal erosion
热容量 capacity of heat; heat capacity; thermal capacity
热溶胶 thermosol
热熔胶黏剂 hot melt adhesive
热扫描仪 thermal scanner
热杀(灼死但未焦化) heat kill
热渗入 heat penetration
热输出 heat output
热输入 heat input
热水池 warm water bath
热水处理 hot-water treatment
热水加热板式干燥机 hot-water platen dryer
热水浸种 hot soaking
热水蛇管 hot-water coil
热塑性 thermoplasticity
热塑性胶黏剂 thermoplastic adhesive
热塑性树脂 thermoplastic resin
热塑性树脂胶黏剂 thermoplastic resin adhesive
热塑性塑料 thermoplast; thermoplastic plastic
热塑性塑料薄膜 thermoplastic foil
热塑性纤维 thermoplastic fibre
热损失 heat loss; thermal losses
热态堆积 hot stack
热调节器 thermoregulator
热通量 heat flux

热同性材料 thermosetting material
热同性树脂 heat convertible resin
热稳定性 heat stabilization; thermostability
热污染 heat pollution; thermal pollution
热雾化室 heated nebulization chamber
热吸收 heat absorption
热线流速计 hot-wire anemometer
热效应 heating effect
热效指数 thermal-efficiency index
热休眠 thermodormancy
热循环 heating cycle
热压 extrusion; hot pressing; hot press
热压板 heating platen; hot plate
热压板温度 press platen temperature
热压釜 autoclave
热压机 heat press; hot plate press; hot press
热压器 cooker
热压杀菌器 pressure cooker; pressure sterilizer
热烟道气体 hot flue gas
热液浸渍 hot dip
热应变 thermal strain
热应力 thermal stress
热硬化 thermosetting
热硬化剂 hot hardener
热油干燥 oil drying
热油干燥法 oil seasoning
热浴法处理 hot bath treatment
热辗轧 hot-rolling
热涨漆 intumescent paint
热值 caloric power; caloric value; fuel value; heat output; heat value; heating power; thermal value
热治疗 heat therapy
热致死时间 thermal death time
热重量分析-气相色谱-质谱联用法 thermal gravity-gas chromatograph-mass spectroscopy
热周期 thermoperiodicity
热阻系数 thermal resistivity
人粪尿 human excrement
人工剥皮 hand barking
人工剥皮器 hand barker; hand peeler
人工播种 artificial seeding
人工操作式干窑燥 manually operated kiln
人工草地 artificial grassland
人工成本 labour cost
人工池塘 artificial pond
人工传染 artificial infection
人工促进更新 artificial measures promoting regeneration
人工催熟 artificial ripening
人工单性生殖 artificial parthenogenesis
人工的 man-made
人工洞窟(造园设施) grotto (*plur.* grotto(e)s)
人工繁殖 artificial propagation
人工防火线 fire guard
人工防治 artificial control
人工放牧 artificial stocking
人工辅助能 artificial auxiliary energy
人工富养化 artificial eutrophication
人工干燥 artificial drying; artificial seasoning
人工干燥积材法 piling lumber for kiln drying
人工更新 artificial reforestation; artificial regeneration; artificial reproduction; man-made reforestation
人工更新混农作业 artificial regeneration with the aid of field crop
人工海岸沙丘 artificial coast sand dune
人工海水 artificial sea water
人工湖 artificial lake
人工花粉 pollen substitute
人工荒山造林 artificial afforestation
人工混交林 mixture plantation
人工加速老化 accelerated ageing
人工降水 artificial precipitation
人工降雨 artificial rainfall
人工降雨器 sprinkler
人工降雨装置 artificial rain device; rainer
人工接种 artificial inoculation
人工进料 hand-feed
人工进料带锯 hand feed band-saw

人工进料刨床 hand (feed) planer
人工进料窄带砂光机 hand block sander
人工景观 cultivated landscape
人工控制传粉 controlled pollination
人工老化 tempering
人工林 artificial crop; artificial forest; artificial plantation; man-made forest; plantation; planted forest
人工林分 artificial stand
人工林分改造 artificial conversion
人工林经营 plantation management
人工林林缘 man-made forest edge
人工林收获 yield of plantation
人工螺旋纹理 artificial spiral grain
人工磨料 manufactured abrasive
人工牧场 man-made pastrre
人工培养 artificial culture
人工培养物内 in vitro
人工气候室 climatic chamber; phytotron(e); psychrometric room
人工强制授粉 artificial compulsory pollination
人工区分 artificial division
人工区划 artificial division
人工区划法 artificial division method
人工群落 artificial community
人工群种 artificial population
人工上料 hand charging
人工生态系统 man-made ecosystem
人工生长箱 growth cabinet; growth chamber
人工授粉 artificial pollination
人工水槽 artificial water tank
人工特征判读 interpretation of man made
人工添加试验 artificial enrichment ex-periment
人工通风 artificial ventilation
人工通风干燥窑 artificial ventilating dry kiln
人工心材 artificial heartwood
人工信息流 artificial information flow
人工修枝 artificial pruning
人工选择 artificial selection
人工穴装夹 artificial hole set
人工幼林 young plantation
人工幼林抚育 care of plantations
人工幼龄林 pre-thicket
人工诱变 artificial induced mutation
人工鱼礁 artificial reef
人工杂交 artificial hybrid; man-made crossing
人工栽植 artificial planting
人工造林 artificial afforestation; artificial forestation
人工增益调整 manual gain control
人工整枝 artificial pruning
人工直接调控 artificial direct regulation
人工竹林 bamboo plantation
人工贮藏 artificial storage
人行道 path; side walk; walkway
人均森林面积占有量 average forest area per capita
人均增长率 per capita growth rate
人口普查 census; census(ing)
人口统计 vital statistics
人口统计学 demography; human demography
人类聚集地 human settlement
人类生态学 human ecology
人类系统 human system
人力采伐 hand log(ging)
人力采运工 hand logger
人力操作绞盘机 manual cable winch
人力除草 hand weeding
人力集材 hand log(ging); hand prehauling; hand skid(ding); hand yarding; hand-bag (hand-bank); manual skidding
人力集材工 hand banker
人力拖排 hand-bag; hand-bank
人力镇压器 hand yoller
人力装车 hand loading
人力资本 human capital
人力资源 human capital; human resource
人为变更作业 artificial conversion
人为冲积土壤 anthropogenic-alluvial soil
人为传播体 brotochore
人为的 anthropogenic
人为干扰 human disturbance
人为火发生指标 man-caused fire occurrence index
人为排放 anthropogenic emission

人为侵蚀 anthropogenic erosion
人为散布 anthropochore
人为险(火险) man-caused risk
人为演替 anthropogenic succession
人为因素 anthropic factor; human factor
人为因子 anthropogenic factor
人为引入植物 anthropophyte
人为植被 anthropogenic vegetation
人文旅游资源 cultural tourism resource
人心果胶 sapote gum
人与生物圈计划 Man and the Biosphere Programme
人造板 artificial board; wood-based panel
人造板单位 panel unit
人造板工业 board industry
人造板机械 wood-based panel manufacturing machinery
人造草场 cultivated meadow
人造的 man-made
人造革 artificial leather; imitation leather
人造环境 man-made environment
人造磨石 artificial grinding stone; synthetic stone; artificial pulpstone
人造气候 man-made climate
人造茜草染料 madder
人造群落 man-made community
人造石 synthetic stone
人造树脂 artificial resin
人造丝绳 rayon cord
人造纤维 artificial fibre; man-made fibre
人造纤维帘子线 rayon cord
人造幼龄林 pre-thicket
人造樟脑 artificial camphor
人字架杆 A-frame
人字起重架 shear-legs; sheer legs; shear; sheer
人字屋顶 gable roof
人字形 fishbone system; herring bone system
人字形伐倒 herringbone felling
人字形伐倒法 fishbone system; herring bone system
人字形拼配花纹 V-matched figure
人字形屋顶 V-roof
仁(果实) kernel
仁果 kernel fruit; pome
稔性 fertility
刃 edge
刃磨次数 sharpening number
刃磨角 grinding angle; sharpness angle
认购价格 exercise price
认股书 application for share
认股证书 warrant
认可资本 authorized capital
任意点火 unregulated fire
任意铺装(碎料) random forming
任意长度 random length
任意长度材 odd lengths
任意值 arbitrary value
韧度 tenacity; toughness
韧化脆性单板 toughening fragile veneer
韧皮 bast
韧皮薄壁细胞 phloem parenehyma cell
韧皮薄壁组织 bast parenchyma; phloem parenchyma
韧皮部 inner bark; liber; phloem
韧皮部母细胞 phloem mother cell
韧皮部石细胞 phloem sclereid
韧皮层 bast zone
韧皮射线 bast ray; phloem ray
韧皮束 phloem bundle
韧皮纤维 bast fiber; bast fibre; phloem fiber; phloem fibre; stem fibre
韧型木纤维 libriform wood fibre
韧型纤维 libriform fibre
韧性 tenacity; toughness
韧性胶 tough glue
妊娠的 gravid
日本扁柏 hinoki cedar; Japanese cypress; obtuse ground cypress
日本扁柏酚 hinokiol
日本扁柏酸 hinokiic acid
日本扁柏酮 hinoki ketone
日本扁柏油 hinoki oil
日本大漆-A jap-A-lac
日本和平友好绿化协会 Japanese Peace and Friendship Forest
日本金松油 parasol pine oil

日本金针松油(高野槙油) Koyamaki oil; kinsho oil
日本栗宁 cretanin
日本栗亭 crenatin
日本漆 Japanese lacquer; varnish tree
日本漆蜡 Japan tallow; Japan wax
日本石(dàn)(120板英尺) koku
日本樱花 someiyoshine; Yoshino cherry; Tokyo cherry; Yoshino-zakura
日变化 daily variation
日变异显着 diurnal inequality
日补偿点 daily compensation point
日产量 daily output; daily production capacity
日常保修 routine maintenance
日常费用 running cost
日常维修 maintenance; running repair
日度 day degree
日光 daylight
日光灯 fluorescent lamp
日光干燥 solar drying
日活动规律 daily activity rhythm
日积温 daily cumulative temperature
日较差 daily range
日节律 daily rhythm
日金松定 sciadin
日金松酮 sciadinone
日流量 daily flow
日平均流量 daily mean discharge
日射 insolation; solarization
日射表 solarimeter
日生产能力 daily production capacity
日温动物 heliotherm
日消长 daily succession
日演替 daily succession
日益适宜温度带 zone of increasingly favorable temperature
日允许摄入量 acceptable daily intake
日照长度 length of day
日照中性植物 day neutral plant
日周期 daily periodicity
日灼 sun burn; sunscald
日灼病 sun-scald; sunscorch
茸角 antler
茸毛 pappus; pubescence
绒毛 fuzz; nap
绒毛被 tomentum
绒屑 flocks
容材量 storage capacity
容差临界值 margin tolerance
容错计算机 fault-tolerant computer
容积百分率 volume per cent
容积比 volume ratio
容积度 voluminosity
容积计 volumeter
容积流量仪 bulkmeter
容积配合比 composition rate by volume
容积热容量 volumetric heat capacity
容积重 volumetric weight
容积重量 weight by volume
容量 volumetric capacity
容量计 volumeter
容器单板 container veneer
容器化育苗系统 containerized nursery system
容器苗 container seedling; container-grown seedling
容器苗圃 container nursery
容器苗栽植机 pot seedling setting machine
容器苗栽植器 planting tube
容器培育 container-growing
容器育苗装播机 filling and sowing equipment for containerset
容器栽植 container planting
容器制作机 containerset making machine
容绳量(电缆) cable capacity
容水量 water capacity
容索量 line capacity
容隙 tolerance
容限 tolerance
容许采伐单位 allowable cutting unit
容许承载量 allowable bearing capacity
容许承载能力 allowable load capacity
容许荷载 permissible load
容许量 allowance
容许猎杀比 permitted hunter-kill ratio
容许强度 allowable strength

容许缺陷 allowable defect; permissible defect
容许收益 permissible revenue
容许误差 allowable error; allowance
容许误差限 limit of allowable error
容许压力 allowable pressure
容许应力 allowable stress; permissible stress
容许载荷 allowable load
容许载荷挠度 allowable loaded deflection
容许载重量 allowable load capacity
容许值 allowed value
容许阻力 allowable resistance
容重 bulk density; bulk specific gravity; volume weight
溶胞作用 lysis
溶度计 lysimeter
溶杆菌素 bacilysin
溶化分离法 liquidation
溶剂 thinner
溶剂分解(作用) solvolysis
溶剂干燥(法) solvent drying
溶剂干燥材 solvent dried wood
溶剂化(作用) solvation
溶剂回收 solvent recovery
溶剂活性胶 solvent-activated adhesive
溶剂前沿 solvent front
溶剂型胶黏剂 solvent adhesive
溶胶 sol
溶接 cementing
溶解产物 lysate
溶解度 solubility
溶解焦油 soluble tar; dissolved tar; soluble wood tar
溶解酵素 lysozyme
溶解热 enthalpy
溶解树胶 dissolved gum
溶解形成的 lysigenetic; lysigenous
溶解氧 dissolved oxygen
溶解油 dissolving oil
溶解纸浆 dissolving pulp
溶解作用 lysis
溶菌斑法 plaque assay
溶菌产物 lysate
溶菌反应 lytic response
溶菌酶 lysozyme
溶菌素 bacteriolysin
溶菌作用 bacteriolysis; lysis
溶离 milking; solvolysis
溶酶体 lysosome
溶生创伤树脂道 lysigenous traumatic canal
溶生的 lysigenic
溶生细胞 lysigenous cell
溶生性的 lysigenetic; lysigenous
溶生脂沟 lysigenous duct
溶失量 loss by solution
溶蚀 corrosion
溶血性的 haemolytic
溶岩洞风景 stalactite cave landscape
溶液培养 hydroponics; solution culture; water culture
溶胀 swelling
溶胀度 degree of swelling
溶脂素 lipolysin
榕树蜡 gondang wax; java wax; kondang wax
熔成磷肥 fused tricalcium phosphate
熔点 melting point; point of fusion
熔化 fuse
熔胶炉(印度) bhatta
熔融磷酸三钙 fused tricalcium phosphate
熔融物 melt
熔岩 lava
熔岩刀片 lava flybar
熔岩流 lava flow
融冻层 active layer
融合核 fusion nucleus
融合遗传 blending inheritance
融化 melt; thaw
融雪期 snowmelt duration
融雪侵蚀 snow melt erosion
融资资本 capital financing
冗余度 redundancy
柔光 silky-sheen
柔光机 satining machine
柔光涂饰 semi-gloss finish
柔量 compliance
柔流 flowage
柔毛 pappus
柔曲性 flexibility
柔韧的 pliable
柔韧弯曲面 flexible bent surface
柔韧系数 flexibility coefficient; pliability coefficient

柔韧性 pliability
柔软度 softness
柔软韧皮部 soft bast
柔软性 softness
柔性 compliance; flexibility
柔性河绠 limber boom
柔性栲胶 sweet extract
柔荑花序 ament; catkin
柔荑花序植物 amentaceous plant
柔质夹心板 flexible coreboard
柔组织细胞 parenchymatous cells
揉搓涂层 squeegee coat
揉曲性 pliability
揉皱的(叶) puckered
揉作规模 scale of operation
鞣革 tan; tannage; tanning of leather
鞣花单宁 ellagitannin
鞣剂 tanning agent
鞣料 tanning material; tanstuff
鞣料粉碎机 tan mill
鞣料浸出器 tanning materials extractor
鞣料浸膏 tannin extract; tanning extract
鞣料磨 tan mill
鞣酸 tannin
鞣透度 degree of tanning
鞣液比重计 barkometer; tannometer
鞣制系数 degree of tanning
鞣质细胞 tannin cell
肉垂(禽鸟) wattle
肉豆蔻 muskatel
肉豆蔻蜡 nutmeg tallow
肉豆蔻油 nutmeg oil
肉豆蔻脂 nutmeg butter
肉冠 comb
肉桂 cinnamon
肉桂基 cinnamyl
肉桂皮 cinnamon bark
肉桂皮油 cassia (bark) oil; cassia oil; Chinese cinnamon oil
肉桂酰胺 cinnamic amide
肉桂叶粉 cassia leaf powder
肉桂叶油 cinnamon leaves oil
肉桂油 cinnamon oil
肉桂汁 cinnamon water
肉果 fleshy fruit; sarcocarp
肉茎高位芽植物 phaenerophyta succulenta
肉茎植物 stem-succulent
肉食甲[的] adephagid
肉食昆虫 predacious insect; predator
肉食性昆虫 carnivor(o)us insect
肉食植物 carnivorus plant
肉穗花序 spadix
肉穗花序的 spadiceous
肉穗花序状的 spadicose
肉眼 unaided eye
肉眼检查 lens examination; macrography; macroscopy; visual inspection
肉眼检验 macroscopic examination
肉眼可见的 macroscopic
肉眼识别检索表 lens key
肉叶地上芽植物 chamaephyta succulenta; succulent chamaephyte
肉叶植物 leaf succulent
肉质的 carnose; fleshy
肉质根 fleshy root
肉质果 fleshy fruit
肉质性 succulence
肉质植物 succulent; succulent plant
肉质种子 fleshy seed
肉重 flesh weight
如刺猬的 erinaceous
如掌状 palmately
蠕变 creep
蠕变单位(1000h内0.1%的变形) creep unit
蠕变断裂 creep rupture
蠕变断裂试验 creep rupture test
蠕变回复 creep recovery
蠕变极限 creep limit
蠕变流动 creep flow
蠕变曲线 creep curve
蠕变柔量 creep compliance
蠕变-时间曲线 creep-time curve
蠕变试验 creep experiment; creep test
蠕变松弛 creep relaxation
蠕变速度 creep speed
蠕变速率 creep rate
蠕变阻力 creep resistance
蠕虫状的 vermiform
蠕陶土 anauxite
乳白蜡(一种氢化植物蜡) opalwax
乳的 lacteal

R

乳管系 lacticiferous system; laticiferous system
乳光 opalescence
乳化剂 emulsifier; emulsion
乳化浓缩液 emulsion concentrate
乳化器 emulsifier
乳剂 emulsion
乳胶 emulsion
乳胶特性曲线 emulsion characteristic curve
乳酪素胶 casein-paste
乳色 opalescence
乳熟 milky maturity
乳熟期 milk-ripe stage
乳酸发酵 lactic fermentation
乳酸菌酶 lactacidase
乳糖 lactose
乳糖酶 lactase
乳头状的 papillar(y); papillate
乳头状突起 papilla (*plur.* papillae)
乳头状突起的 papillar(y)
乳香 gum mastic; mastic
乳香黄连木树胶 mastic gum
乳香胶 mastic gum
乳香树脂 boswellic resin; olibanum resin; olibanum resinoid
乳香树脂素 olibanoresene
乳香树脂制剂 olibanum resinoid
乳香油 frankincense oil; olibanum oil
乳香脂 mastix
乳液 emulsion
乳液防腐剂 emulsion preservative
乳油(杀虫剂) oil miscible concentrate
乳汁 latex (*plur.* latices)
乳汁导管 lacticiferous vessel
乳汁道 latex canal; latex duct; latex trace
乳汁管 lacteal; lactiferons duct; lactiferous canal; lactiferous tube; latex canal; latex tube; laticifer; laticiferous tube
乳汁迹 latex canal; latex trace
乳汁糖 lactose
乳汁细胞 latex cell; laticifer
乳状的 lacteal
乳状液 emulsion
乳浊 milkiness
乳浊液 emulsion
入口 eye; inlet
入梅(入霉) setting-in of bai-u
入片刨花板 flake board
入侵刺 infection peg
入侵物种 invasive species
入射光 incident light
入射光束 incident beam
入射角 angle of incidence; incident angle
入射粒子分布 incident-particle distribution
入射能量 incident energy
入射强度 incident intensity
入射倾角 incidence
入射线 incident ray
入钻方向(生长锥取样) boring direction
软X射线 soft x-ray
软X射线机 soft x-ray machine
软百叶帘 venetian blind
软材 softwood
软材板 softwood board
软材焦油 softwood tar
软材焦油沥青 softwood tar pitch
软材林 softwood forest
软材木炭 softwood charcoal
软材干馏 softwood distillation
软材纤维 softwood fibre
软垫椅 upholstered chair
软吊排 loose raft
软腐 soft rot; spongy rot
软腐病 wet rot
软腐耐药性 soft rot tolerance
软腐真菌 soft-rot fungus
软腐殖质 mild humus
软钢 mild steel
软钢丝绳 soft wire-rope
软管 flexible conduit; flexible pipe
软化点 softening point
软化点滴测法 drop softening-point method
软化剂(软化器) softener
软化与防水油脂 dubbing
软件工程 software engineering
软浆蒸煮 soft cook
软蚧 soft scale
软块 mash
软阔叶树材 soft deciduous wood
软阔叶树林 light hardwoods
软毛 pubescence

软木 cork
软木板 cork board
软木化 suberification
软木填料 cork filler
软木脂(木拴层中的脂肪) suberin
软木纸 cork sheet
软木制品 cork product
软木砖 cork block
软泥 ooze
软拟胶骨架 soft gel-like matrix
软埝 broad-base terrace
软片 film; pellicle
软片暗盒 magazine
软片片基 film base
软漂子 limber boom; rolling boom
软树皮 soft bark
软树脂 barras; malengket
软松类 soft pine
软体动物学 malacology
软体蛀虫 molluscan borer
软土 mollisol
软橡胶叶片泵 flexible rubber impeller pump
软心(材) soft heart
软枝插 soft stem cutting
软枝嫁接 softwood grafting
软纸浆 soft pulp
软质(纤维)板 softboard
软质滚筒(表层覆盖柔性材料) soft roll
软质纤维板 felt board; insulating fibre board; insulation board; non-compressed (fibre) board; soft fibre board
蕊柱 column
芮木泪柏树脂 rimu resin
芮木泪柏烯 rimuene
芮木烯 totarene
锐端分子 prosenchymatous element
锐端细胞 prosenehymatous cell
锐端细胞组织 prosenchyma
锐角的 acutungular
锐角转角 sharp corner
锐棱的 acutungular
锐裂的 incised
锐缘铁板(踏板前端)(伐木) dog
瑞典林分高度 Swedish top height
瑞利散射 Rayleigh scattering
瑞士择伐林 Switzerland selection forest
瑞香油 daphne oil
闰年 leap year
润滑 lubricate; lubrication
润滑工 grease dauber; greaser
润滑剂 lubricant
润滑器具 greaser
润滑油泵 gun
润滑油值 oil number
润滑作用 lubrication
润湿 damping
润湿度 degree of wetness
润湿热 heat of wetting
润湿水 wet water
润湿系数 coefficient of wetness
润胀 swelling
润胀度 degree of swelling
润胀法 bulking
润胀剂 bulking agen; swelling agent
润胀效应 bulking effect
润胀压力 swelling pressure
润胀应力 swelling stress
若虫 nymph
若虫龄 nymphal instar
若尔当种 Jordanon
若叶 leaflet
弱度疏伐(B度疏伐) light thinning; B-grade thinning
弱光层 disphotic zone; dysphotic zone
弱灰化土 weakly podzolic soil
弱碱性土 weakly alkaline soil
弱能见度 poor visibility
弱水河 inadapted river; misfit river; underfit river; underfit stream
弱优势木 weak dominant
洒水 sprinkling; water dropping; watering
洒水爬犁(洒水车) sprinkler (sled)
洒水器 watering pot
撒播 broadcast seeding; broadcast sowing; surface seeding; surface sowing
撒播机 surface drill; seed broadcaster
撒播施肥机 broadcast seeder fertilizer
撒肥机 fertilizer distributor; fertilizer spreader; manure spreader
撒肥卡车 spreader truck

S

撒土机 sand spreader
撒药机 applicator
萨布香脂 cebur balsam; tagulaway
萨拉橡胶 ceara rubber plant; manihot caoutchouc
萨列阿法则 Sanio's law
萨瑟林开口滑轮(集材拖拉机后部) Sutherlin side block
萨王纳草原 grass savanna
塞尔尺(锯材体积单位, $8.5ft^3$或$0.71m^3$) serch
塞缝片 spline
塞角 block; corner block
塞角木块 corner block
塞块 chock
塞伦防腐法 Cellon process
塞莫戴恩模压刨花板成型工艺 Thermodyn process
塞片 chip plug; patch
塞条 shim
塞缘(纹孔塞) margo
塞钻(生长锥) borer clogging

S

鳃叶触角 lamellate antenna
鳃状管 aeriductus
赛果(杀虫杀螨剂) amidithion
赛璐玢 cellophane
三棓酸 trigallic acid
三北防护林 Three-North protection forest
三倍体 triploid
三边切纸机 trilateral paper trimming machine
三层结构 three layer
三层结构碎料板(刨花板等) three-layer particleboard
三层木片筛 triple deck chip screen
三出的 ternately compound
三出复叶 ternate compound leaf
三出复叶的 ternately compound; ternately trifoliolate
三出脉的 trinervious
三出数的花 tri-merous flower
三出羽状的 ternate pinnate
三次捕获法 triple catch method
三次分裂 tertiary split
三次渐伐法 three-cut shelterwood method
三次伞伐法 three-cut shelterwood method
三叠纪 Triassic period
三段加压法 three stage pressing cycle
三对基因杂种 trihybrid
三分的 triparted; tripartite
三辊铺装头 three roll spreader head
三回路滚筒式干燥机 three-way drum dryer
三回三出叶的 triternate
三回羽状的 tripennate; tripinnate
三回羽状复出的 tripinnately compound
三回掌状的 tripalmate
三回掌状复出的 tripalmately compound
三级胶合板 type-three plywood
三级结构 tertiary structure
三级脉 tertiary vein
三级消费者 tertiary consumer
三甲基硅醚 trimethylsilyl ethers
三甲基硅醚化作用 trimethylsilylation
三甲基葡萄糖 trimethyl glucose
三价的 trivalent
三价染色体 trivalent
三尖杉碱 cephalotaxine
三交 three-way cross
三交杂种 triple hybrid
三角测点 triangulation point
三角测量 triangulation
三角齿 peg tooth
三角齿锯 peg saw; peg-tooth saw
三角点 triangulation point; triangulation station
三角法测高器 trigonometrical hypsometer
三角高程测量 trigonometrical leveling
三角横档 arris rail
三角滑轮 saddle block
三角栏杆 arris rail
三角气球 delta-wing balloon
三角区分法 triangular division method
三角屋顶 gable roof
三角屋架 collar roof
三角形播种 chequered seeding
三角形侧边空心刨花板 triangular profile board
三角形齿 lance tooth

三角形顶 cuneus
三角形堆垛法 crib stacking
三角形锯齿 scriber teeth
三角形留弦伐木法 corner cutting
三角形栽植 planting in triangle; triangle-planting; triangular planting
三角凿 burr
三角洲 delta
三角洲沉积 delta deposit
三角洲洪积扇 deltaic fan
三角桌 knee table
三脚架 tripod
三脚起重机 gin
三脚桌 cricket table; tripod table
三阶抽样 three-stage sampling
三进制的 ternary
三九一一(杀虫剂) thimet
三聚氰胺树脂 malamine resin
三卡钩跑车 three kneed carriage
三类调查 forest operational inventory
三棱镜罗盘仪 prismatic compass
三联带锯机 triple band saw
三联密码 triplet code
三联体(遗传密码) triplet
三联再剖锯机 tri-band resaw
三氯硝基甲烷 chloropicrin
三面木工刨床 three side planing and moulding machine
三面下锯法 three faced sawing
三名制(动植物拉丁学名) trinomial system
三木漂子(断面呈三角形) three-log boom
三歧聚伞花序 trichasium
三歧式 trichasium
三强雄蕊的 tridynamous
三切面 three planes of section
三砂架宽带式砂光机 three heads wide belt sander
三深裂的 triparted; tripartite
三生壁 tertiary lamella
三生细胞壁 tertiary (cell) wall
三十碳六烯 squalene
三羧酸循环 Krebs cycle; tricarboxylic acid cycle
三台绞盘机分段集材 triple haul
三糖 trisaccharide
三体(染色体) trisome
三体二倍体 trisomic diploidy
三体生物(2n+1) trisomic
三体雄蕊的 triadelphous
三萜类 triterpenes
三萜烯化合物 triterpenoid
三通管 tee
三筒绞盘机 triple drum
三筒绞盘机索道集材 three-drum skyline yarding system
三弯腿 three-bend leg
三维空间外观 three-dimensional appearance
三维空间外貌 three dimensional-look
三维应力 stress in three dimension
三系杂交 three-way cross
三相分布 three phase distribution
三相容积比 distribution of three phases
三斜面端头(原木) three lick snipe
三性花同株 agamogynomonoecism
三性同株 coenomonoecia
三雄蕊的 triander
三序(昆虫趾钩) triordinal
三叶草 clover
三叶轮生 tricussate
三叶隙的 trilacunar
三乙氨基乙基纤维素 TEAE-cellulose(triethyl aminoethyl cellulose)
三乙基氨基乙基纤维素 triethylaminoethyl cellulose
“三有”动物 animal species with high values for wellfare, economy and science
三元材积表 form class volume table; three-way volume table
三元的 ternary; three-entry
三原型 triarch
三源杂交 three-way cross
三掌出的 tridigitate
三重寄生 triploparasitism
三重立体 stereo triplicate
三重立体组 stereo triplet
二重蠕变 tertiary crccp
三重线态 triplet state
三轴应力 stress in triaxial
三子树油 bagillumbang oil
三足梯 tripod ladder

S

伞 fringe
伞伐 shelter-wood cutting; shelter-wood felling
伞伐矮林 shelterwood coppice
伞伐法 shelter-wood method
伞伐更新 regeneration under shelterwood; regeneration under shelterwood system; shelterwood regeneration
伞伐更新法 shelter-wood cutting; shelter-wood felling
伞伐林 shelter-wood forest
伞伐择伐作业 shelter-wood selection system
伞伐作业 compartment system; felling for shelter-wood system; over cover system; seed-tree method; shelter-wood method; shelter-wood system; stand-method
伞伐作业法 shelterwood method; shelterwood system
伞伐作业更新 regeneration by compartments
伞伐作业更新林 regenerated forest under shelterwood system
伞伐作业天然更新 natural regeneration by shelterwood method
伞房花序 corymb
伞房花序的 corymbous
伞房花序形 corymbose
伞房状聚伞花序 corymbose cyme
伞护种 umbrella species
伞菌 agaric (fungus); cap-fungus; mushroom; parasol mushroom
伞形花序 umbel
伞形花序枝 ray
伞形聚伞花序 umbellate cyme
伞形酮 umbelliferone
伞型屋顶 umbrella roof
散材堆积 stacking in drift
散弹枪 scatter gun
散点图 scattergram; scatterogram
散堆 bulk pile; wild pile
散堆材积 loose volume
散股(钢索) strand
散孔 foramen (*plur.* -mina)
散孔材 diffuse porous wood
散孔式 foraminate type
散孔式穿孔 foraminate perforation
散孔式穿孔板 foraminate perforation plate
散孔式内含韧皮部 foraminate included phloem
散粒状结构 loose granular structure
散乱的片石铺砌 crazy paving
散落 dispersal
散片紫胶 free shellac
散漂 loose driving; loose floating
散漂材 loose floating
散漂原木(流送时) floatage
散射 diffusion; scattering
散射光 diffused light; scattered light
散生 irregular stocking; rambling
散生混交 mixture by isolated plants
散生混交林 scattered mixed forest
散生节 distributed knot; scattered knot
散生林地 irregularly stocked open woodtract; scarcely forested land
散生母树作业天然更新 natural regeneration by scattered seed tree method
散生射线 diffuse ray
散生树 scattered tree
散生植被 open vegetation
散生中柱 atactostele
散生竹类 running bamboos
散穗花序 panicle; panicled spike
散植 scattered planting
散装物输送和存放系统 bulk system
散装种批 bulk lot of seeds
散状定期更新区 scattered periodic block
散状伐 scattered felling
散状留母树采伐法 scattered seed-tree method
散布 dispersal
散布孢子 diaspore
散布力 vagility
散布图 scatter diagram
散布系数 coefficient of dispersion
散步中心 center of dispersal
散出 efflux
散热管 heating coil
散热片 gill; louver board; web
散热器 radiator
散热作用 thermolysis

散烧法(采伐剩余物) broadcast burning
桑 mulberry
桑白盾蚧 mulberry scale; dadap shield scale
桑蚕 mulberry silk worm; silk worm; silk moth
桑橙素 maclurin
桑给巴尔柯巴(树脂) Zanzibar copal
桑基图(理论热平衡图) Saukey diagram
桑寄生 parasite scurrula
桑毛虫 yellow-tail moth
桑树 weeping mulberry; white mulberry
桑酮 morindone
扫除 sweep
扫粉器 powder sweeper
扫描电子显微镜 scanning electron microscope
扫描电子显微镜图 scanning electron micrograph
扫描辐射计 scanning radiometer
扫描器 scanner
扫描影像 scanner imagery
扫雪机 snow plough
色斑 eye; spot
色彩 hue
色彩的联想性(园林) colour association
色彩透视法 colour perspective
色彩效果(园林) colour effect
色差 chromatic aberration
色带 zone
色度 Chroma; chromaticity
色度计 colourimeter
色度图 chromaticity diagram
色度学 colourimetry
色辉计 tintometer
色料 colourant; colouring matter
色料扩散(表面涂饰) bleeding
色浓度 saturation
色品 Chroma
色谱(词头) chromato-
色谱板(薄层) chromatoplate
色谱法 chromatography
色谱法的 chromatographic
色谱法分离 chromatograph
色谱分离 chromatographic fractionation
色谱分析的 chromatographic
色谱级 chromatographic grade
色谱图 chromatogram
色谱学 chromatographia
色谱仪 chromatograph
色谱仪的 chromatographic
色谱柱 chromatographic column
色漆 coloured paint
色区 zone
色散 dispersion
色树脂 chromoresin
色素 colourant; pigment
色素花纹 pigment figure
色素基因 chromogene
色素适应 chromatic adaptation
色素细胞 chromatophore
色调 hue; tincture; tint; tone
色调过深 overtone
色调计 tintometer
色调强度 tinting strength
色酮 chromone
色脱犬 setter
色温度 colour temperature
色线 streak
色原 chromogen
色泽 colouration; tinct
色重团 pathochrome
森林 forest; wooded land
森林保护 forest conservation; forest protection
森林保护法 forest conservation law
森林保护法案 forest protection act
森林保护机构 organization of forest-protection
森林保护机械 forest protection machinery
森林保护区 forest range; forest reserve area; forest-beat
森林保护群众组织 civilian conservation corps
森林保险 forest insurance
森林病害 forest disease
森林病理学 forest pathology
森林步道 forest nature trail; forest walk
森林部分(区别于农田等的林地) forest sector

S

森林簿 crop description; stand assessment; stand record
森林簿记 forest bookkeeping
森林财产 forest estate; forest property
森林财务机构 organization of forest cash-office
森林财政学 forest finance
森林采伐 forest cut; forest harvesting
森林采伐机械 felling machinery
森林采伐联合机 forest harvesting combine
森林采运 forest engineering
森林采运工程 logging engineering
森林采运机械化 logging mechanization
森林采运机械系统 logging machine system
森林草原 forest steppe
森林草原土 phaeozem
森林测量 forest surveying
森林产权判定 forest settlement
森林产业 forest estate; forest property
森林产业成本价 cost value of forest property
森林成本资金 forest-cost-capital
森林成熟 forest maturity
森林抽样调查 forest inventory by using sampling method
森林储备 forest store
森林储备基金 forest reserve funds
森林纯利价值 forest-rental value
森林纯利理论 forest rent theory; theory of forest rent(al)
森林纯收获 forest net yield
森林纯收获量 removal
森林纯收益 forest rent (al); forest returns
森林纯收益最大的轮伐期 highest forest rent rotation
森林纯收益最高之轮伐期 rotation of maximum income; rotation of the highest net income
森林村 forest hamlet; forest village
森林大火 large forest fire
森林带 forest zone; woodland belt
森林地被物(死活) forest floor
森林地带 forest region
森林地籍簿 forest cadastral register; forest cadastre
森林地名 forest name
森林地域 forest area
森林佃户 forest squatter
森林凋落物 forest litter
森林顶极 forest climax
森林动物 forest animals
森林动物学 forest zoology
森林动植物 forest fauna and flora
森林冻原 forest-tundra
森林多种利用 forest multiple use
森林发育 forest development
森林法 code forestier; forest act; forest law
森林法令 forest ordinance
森林防火公约 forest fire protection compact
森林费用 forest charge
森林分布判读 interpretation of forest distribution
森林分布图 forest distribution map
森林分类 forest classification
森林分区署 divisional forest-office
森林风景 forest landscape
森林风景管理 forest landscape management
森林抚育 care of woods; tending of crop; tending of woods
森林抚育机械 forest cultivation machinery
森林腐殖质 forest humus
森林覆被 forest cover; forest-clad
森林覆盖率 forest coverage; forest cover percent; forest percent; pcercent of forest cover; percentage of forest cover
森林改良土壤 amelioration with protection forest
森林更新 forest regeneration; forest renewal
森林更新计划 reforestation program
森林更新调查 forest regeneration survey
森林更新问题 reforestation problem
森林更新作业 reforestation work
森林工程[学] forest engineering
森林工业 forest industry; timber industry
森林工业协会 industrial forestry association

森林工艺学 forest technology
森林工作手册 forest service manual
森林公益机能 social benefits of forests
森林公园 forest garden; forest park; park forest; woodland park
森林功能 forest land use; forest use; forset function
森林功能图 forest functions map
森林估测 forest cruise
森林估价 forest assessment; forest valuation
森林关税 forest tariffs
森林管理 forest administration; forest husbandry
森林管理学 forest administration
森林管理员 forest officer
森林规划 forest planning
森林规整体系 system of forest regulation
森林轨道(轻铁轨或木轨) forest-tramway
森林航空调查 aerial forest survey; forest aerial survey
森林和牧场可更新资源规划法案(美) Forest and Rangeland Renewable Resources Planning Act
森林化学 forest chemistry
森林环境管理 forest environmental management
森林荒废 forest devastation
森林恢复 reforestation
森林火险等级 forest fire danger rating
森林火险计算尺 forest fire danger meter
森林火灾 forest fire; stand fire
森林火灾保险 forest fire insurance
森林火灾法 forest fire law
森林火灾气象学 forest fire meteorology
森林火灾危险度 forest fire danger rating
森林火灾危险率 risk of forest fire
森林火灾预防 forest fire prevention
森林火灾侦察 detection of forest fire
森林机能 forest land use; forest use; forset function
森林机能分类 forest land-use class
森林稽查 forest inspectorate
森林极限温度 forest limit temperature
森林集水区试验 forest-watershed experiment
森林计算学 forest mathematics
森林技师 forest experiment officer; forest master
森林价 forest assets; forest value; value of forest
森林间接效用 indirect utility of forest
森林监督局 forest inspection bureau
森林检查员 forest examiner
森林较利学 forest statics
森林结构 constitution; forest structure
森林界限 forest boundary; forest limit
森林经济 forest economy
森林经济目标 forest economy goal
森林经济拓荒期 frontier period of forest economy
森林经理 forest organization; forest regulation
森林经理表 forest management table
森林经理单位 forest management unit
森林经理调查 forest management inventory; forest inventory for planning and design; second-category forest inventory
森林经理复查 revision of forest management plan; revision of working plan
森林经理规程 forest management rules
森林经理机构 organization of forest management
森林经理计划 forest management plan
森林经理检定 revision of the management
森林经理局 Bureau for the Revision of Working Plans
森林经理目标 forest management purpose
森林经理图 management map
森林经理学 forest management; forest organization
森林经营 forest management
森林经营标准 forest management scale

森林经营方案修订 revision of forest management plan; revision of working plan
森林经营风险回避(低危险) low-risk forest management
森林经营规模 forest management scale
森林经营合作方案 cooperative forest management program
森林经营计划 forest mnagement plan
森林经营计划期 forest management period; planning period
森林经营局 forest management office
森林经营目标(目的) forest management objective
森林经营调查报告 forest mnagement plan
森林经营细部计划 stand management planning
森林经营学 forest management
森林经营中期计划 forest management planning
森林景观 forest landscape
森林景观经营 forest landscape management
森林警察 forest police
森林开发 forest exploitation
森林勘测员 cruiser
森林可及度 accessibility of forest
森林可及性 forest accessibility
森林课税 forest taxation
森林矿物质循环 mineral cycling in forest
森林昆虫 forest insect
森林昆虫学 forest entomology
森林类型制图 timber type mapping
森林理事会 forest council
森林立地 forest site
森林立地类型 forest site type
森林立木度 forest stocking
森林利用 forest exploitation; forest land use; forest use; forest utilization; forset function
森林率 percent forest cover
森林美学 forest esthetics
森林密度 forest density
森林苗圃 forest nursery
森林灭火预防措施 forest fire presuppression
森林脑炎 forest encephalitis; forest spring encephalitis; lyme disease; Russian spring-summer encephalitis; tick-borne encephalitis
森林能量平衡 energy equilibrium in forest
森林泥炭 forest peat
森林年租 annual rent for forest
森林培育 forest culture; silviculture
森林评价 forest evaluation
森林评价法(学) forest appraisal
森林破坏 devastation of forest; forest destruction
森林普查 extensive forest survey
森林期望价 expectation value of forest; forest expectation value
森林企业 forest enterprise
森林气候 forest climate; stand climate
森林气象观测站 forest weather station
森林气象学 forest meteorology
森林清查 forest survey
森林清查方法 inventory method
森林清理 forest liquidation
森林区 silva; sylva
森林区分计划 scheme of division
森林区划 area division; division into compartment; division of forest area; forest compartment; forest district; forest division; forest subdivision; subdivision forest
森林燃料 forest fuel
森林燃烧性 forest inflammability
森林社会 forest community
森林社会效益(利益) social benefits of forests
森林摄影测量 forest photogrammetry
森林生产 forest production
森林生产力 forest prductivity; forest productivity
森林生产物 forest production
森林生产资金 forest-cost-capital
森林生态体系 forest community
森林生态系统 forest ecosystem
森林生态系统类型自然保护区 nature reserve for forest ecosystem
森林生态学 forest ecology

森林生物量 forest biomass
森林生物生态空间调查 forest biotope mapping
森林生物学 forest biology
森林生物质 forest biomass
森林生长及收获学 forest growth and yield science
森林生长区域 forest growth region
森林生长与收获学 growth and forest yield science
森林施肥 forest fertilization
森林施业地 forest area under management
森林施业方案(计划) forest working plan
森林施业调整 forest management regulation
森林实务 forest practice
森林使用权 forest easement; forest privilege; forest right; forest servitude; right of forest use
森林事业年度 forest year
森林事业体 forest enterprise
森林试验场 forest experiment station
森林收获 forest crop; forest yield
森林收获模拟模型 forest harvesting simulation model
森林收益性 forest profitability; profitability of forestry
森林收益学 forest statics
森林数学 forest mathematics
森林衰退 forest degradation
森林水文学 forest hydrology
森林税 forest tax
森林税收 forest taxation
森林税制 forest taxation
森林死地被物(枯枝落叶层) forest litter; litter; vegetable refuse; vegetable remains
森林所有权 forest property
森林所有权分割 forest fragmentation; fragmentation of forest holdings
森林踏查 forest cruise; forest inspectorate
森林苔藓泥炭 forest moss peat
森林特许权 forest concession
森林条例 forestry act
森林调查 forest description; forest inventory; forest survey
森林调查簿 crop description; forest inventory sheet; stand assessment
森林调查方法 inventory method
森林调查目标 forest inventory objective
森林调查设计 forest inventory design
森林调查员 cruiser
森林调查专业人员 forest inventory specialist
森林调节法 system of forest regulation
森林调整 forest regulation
森林调整技术 forest regulatory technique
森林铁路 forest railroad; forest railway; logging railroad (railway); logging-track
森林统计学 forest statistics
森林突发灾害 forest catastrophe
森林图 forest map; forest plot
森林土地价 cost value of forest soil
森林土壤 forest soil
森林土壤改良 forest reclamation
森林土壤生产力 forest soil productivity
森林土壤学 forest soil science
森林退化 forest deterioration
森林外貌 physiognomy of forest
森林卫生 forest hygiene; forest sanitation
森林卫生学 forest hygienic
森林无性更新 vegetative forest regeneration
森林物质循环 material cycling in forest
森林现行法规 forest practice rule
森林限界温度 forest critical temperature
森林线(海拔高限界) tree line; wood limit
森林消防车 forest fire-fighting caravan
森林小气候 forest microclimate
森林小区 forest plot
森林小区测绘 forest biotope mapping
森林小屋 lodge
森林效益 forest effect

森林信用 forest credit
森林刑法 forest criminal law
森林型 forest type
森林蓄积 forest reserve
森林蓄积价 growing stock value
森林巡逻 forest inspection
森林巡视 forest-warden
森林演替 forest succession
森林业务 forest business
森林遗产税 forest inheritance tax
森林遗传学 forest genetics
森林影响 forest influence
森林永续利用 sustained yield of forest
森林永续收获 sustained yield of forest
森林游乐林 recreational forest
森林游憩 forest recreation
森林游憩林 recreational forest
森林游憩设施 forest recreation facility
森林娱乐 forest recreation
森林预算 forest budget
森林杂草 forest-weeds
森林灾害 forest calamity
森林宅园 forest homestead
森林占有率 percent forest cover
森林沼泽 forest swamp
森林职员 forest-staff
森林植被 forest vegetation
森林植物 forest plant
森林植物带 forest region
森林植物学 forest botany; tree botany
森林指率 forest percent
森林指示植物 forest indicator
森林志 silva; sylva
森林制图 forest mapping
森林质量 quality of standing-crop
森林致死遮阴度 forest fatal shade
森林种子更新 forest regeneration from seeds
森林主 forest owner
森林主产物 forest staple product; principal forest product
森林贮备区 forest reserve area
森林转移稳定性 forest transition stability
森林资本 forest capital
森林资产 forest assets
森林资源 forest reserve; forest resource; forest resources
森林资源代码 forest resource code
森林资源档案 forest resource archives; forest resource record
森林资源分布图 forest resource distribution map
森林资源规划设计调查 forest management inventory; forest inventory for planning and design; second-category forest inventory
森林资源连续清查 continuous forest inventory; sequential forest inventory
森林资源明细表 synopsis of the forest resource
森林资源评价 forest resources appraisal
森林资源评价系统 forest resources evaluation system
森林资源数据库 forest resource data base
森林资源信息系统 forest resource information system
森林资源周期清查法 recurrent (forest) inventory method
森林综合利用 integrated forest utilization
森林棕壤 brown forest soil
森林总监督 general supervision of the forest
森林租金 forest rent (al); forest returns
森林组成 forest composition
森林组织 forest organization
森林作业法 silvicultural system
森林作业记录 record on forest operation
森铁台车 bogie; detached truck; disconnected truck; forestry railroad car; logging truck; rattler
森铁运材 forest railroad transportation
森铁支线尽头 stub track
杀病毒的 viricidal; virucidal
杀病毒剂 virucide
杀草剂 herbicide
杀草快(除草剂) diquat
杀草强(除草剂) amitrole; aminotriazole
杀虫粉 insect powder

杀虫剂 biocide; insecticide
杀虫剂抗性 insecticide resistance
杀虫剂停用[天敌]法 insecticidal check method
杀虫脒(杀虫剂) chlordimeform
杀虫漆 insecticide paint
杀虫畏(杀虫剂) tetrachlorvinphos
杀虫烟雾(薰烟) insecticidal smoke
杀虫油 insecticidal oil
杀花粉基因 pollen killer gene
杀花粉剂 pollen killer
杀菌处理 sterilizing treatment
杀菌灯 germicidal lamp
杀菌防腐剂 fungicidal preservative
杀菌剂 bactericide; fungicide; germicide; germifuge; microbicide
杀菌素 bactericidin
杀菌涂料 fungicidal paint
杀菌效力 fungicidal effectiveness
杀菌效力指数 index of fungicidal effectiveness
杀菌效率 fungicidal efficiency
杀卵剂 ovicide
杀螨剂 miticide
杀螨特(杀螨剂) aramite
杀螟硫磷 fenitrothion
杀螟松(杀虫剂) fenitrothion
杀鸟剂 avicide
杀软体动物剂 molluscicide
杀生的 biocidal
杀生物剂 biocide
杀势障木 bushy tree
杀鼠剂 rodenticide
杀鼠灵(杀鼠剂) warfarin
杀微生物的 microbicidal
杀线虫剂 nematocide
杀蚜剂 aphidicide
杀真菌剂 mycocide
沙 sand
沙岸的 psammolittoral
沙岸生物 amnicolous; psammolittoral organism
沙巴拉群落 chaparral
沙暴 sand storm
沙表的 epipsammic
沙藏 stratification
沙草 sand binder
沙床埋接 bench grafting
沙地 dene
沙地柏油 savin(e) oil
沙地生物 psammon
沙地园 sand garden
沙地植被 sand vegetation
沙地植物 prammophyte
沙发包布 upholstery
沙发床 day bed
沙发桌 davenport table
沙柑 madarin orange; king orange
沙岗 sand hill
沙荒土壤 desert soil
沙间生物 mesopsammon
沙粒 sand
沙面动物 epipsammon
沙漠 desert; sand-barren
沙漠花园 desert garden
沙漠化 desertification
沙漠气候 desert climate
沙漠群落 eremium; eremus
沙漠园 desert garden
沙漠植被 desert vegetation
沙漠植物 eremophyte
沙内生物 endopsammon
沙栖的 psammophilous
沙栖生物 psammon
沙栖性的 mesopsammic
沙丘 dune; dune-sand; sand dune; sand hill
沙丘固定 dune fixation
沙丘链 chain of sand dunes
沙丘群落 thinium
沙丘群落的 thinic
沙丘推进 advance of sand dunes; desert advancing
沙丘未熟土 sand dune regosol
沙丘植被 sand dune vegetation
沙丘植物 dune plant; sand dune plant
沙生的 sabulose
沙生演替系列 psammosere
沙生植被 psammophytic vegetation
沙生植物 psammophyte
沙滩群落 sand-beach community
沙滩生物 amnicolous
沙土草地 sand grass
沙蚤 sand flea
沙障 checkerboard protection
沙针苷 osyrit(r)in
沙针油 osyris oil

S

沙洲 bar; sand bar; sandbar; shoal
沙洲砾石 bar gravel
沙嘴 sand spit
纱管 bobbin spool; spool
纱管材 bobbin wood
纱管坯 spool bar; spool blank
纱管用材 spool bar; spool blank; spool wood
砂边机 edge belt sander; edge sander
砂布 sand cloth; abrasive cloth; emery cloth; emery fillet
砂带(磨光用) abrasive band; abrasive belt; coated abrasive belt; emery belt; emery fillet; emery tape; sand belt
砂带底涂 make coat
砂带垫压块 sanding pad
砂带规格 abrasive belt size
砂带张紧辊 abrasive belt tension roll
砂粉 sander-dust; sanding dust
砂光 abrasive planing; sanding; surfacing
砂光机 grinder; sander; sanding machine; abrasive planer
砂光刨 abrasive planer
砂辊 sanding cylinder
砂基培养 sand culture
砂砾 granule; gravel
砂砾地生的 sabulicole
砂砾路 gravel path
砂砾培养 gravel culture
砂粒 sand
砂轮 abrasive disk; abrasive wheel; buffing wheel; emery wheel; grind stone; sand disk
砂轮机 knife grinding machine
砂轮片 grinding disk
砂轮修整器 abrasive dresser
砂磨 sanding
砂磨室 sanding station
砂磨刷 abrasive brush
砂磨压力 sanding pressure
砂囊 gizzard
砂壤土 sandy loam; sandy soil
砂碎屑岩 arenite; arenyte
砂土 arenosol; sand; sand soil
砂土流 sand run
砂土植被 sandy soil vegetation
砂新成土 psamment
砂削 sanding
砂岩 arenite; arenyte; sandstone
砂杂酚油包扎法(防腐) sand-creosote-collar method
砂纸 abrasive paper; emery paper; sand paper
砂质沉积物 arenaceous sediment
砂质粗骨土 psammitic regosol
砂质土 sandy soil
砂质岩 arenaceous rock
莎纶纤维 Saran fabric
筛 screen; sieve; sifting
筛板 sieve disk; sieve plate
筛板塔 sieve-plate column; sieve-plate tower
筛场 sieve field
筛出物 siftings
筛分 screen fractionation; sieve; sift; sifting; sizing
筛分法 sieving method
筛分试验 sieve test
筛分析 mesh analysis; screen analysis; screen classification; size test
筛分子(包括筛管和筛胞) sieve element
筛管 sieve tube
筛管分子 sieve-tube element
筛环式打磨机 sieve mill
筛浆机 pulp cleaner; pulp screen
筛孔 mesh; screen mesh; sieve mesh; sieve pore; slotted hole
筛磨区 sieve area; sieve field; sieving zone
筛目 screen mesh; sieve mesh
筛上 oversize
筛筒 perforated cylinder
筛网 grit; screen mesh
筛析 sieve analysis
筛析色谱 exclusion chromatography
筛细胞 sieve cell
筛下料 under size
筛下木片 undersized chip
筛选 bolting; screen; screening; sieve; sieve analysis; sieve selection; sifting
筛选机 screen; sieve machine; sizer
筛选说 sieve selection hypothesis
筛选损失 screening loss
筛眼 orifice; screen mesh; sieve mesh

筛域 sieve area
筛状薄壁组织 cribral parenchyma
筛状纹孔 cribriform pit
筛状纹孔式 sieve pitting
晒斑日烧病 sun burn
晒干 insolation
晒裂 sun-crack; sun-eheck
晒伤 sun burn; sun-scald; sunscorch
山凹 cove
山崩 debacle; landslide; mass movement
山扁豆素 cassiamin
山扁豆叶蜡 senna leaf wax
山苍子油 litse citrata oil; litse cubeba oil; litsea citrata oil; litsea cubeba oil; cubeba oil
山苍子脂 undung
山茶花(日本) Japanese camellia; camellia flower
山茶油(日本茶油) camellia oil
山茶皂苷元 camelliagenin
山场售价 at stump
山地 mountain
山地草甸 upland meadow
山地草甸土 mountain meadow soil
山地带 montane belt; montane zone
山地黑钙土 mountain chernozem
山地灰化土 mountain podzolic soil
山地栗钙土 mountain chestnut soil
山地牧场 mountain pasture
山地泥炭 mountain peat
山地气候 mountain climate
山地森林 montane forest; mountain forest
山地生境 upland site
山地野生动物保护区 mountains wildlife refuge
山地雨林 montane rain forest
山地沼泽 montane bog
山地针叶林 montane coniferous forest
山地植物 montane flora
山地制材 mountainous sawmilling
山靛碱 mercurialine
山靛灵 mercurialine
山顶缆车下放机 headwork
山豆根碱 dauricine
山风 mountain wind
山腹路 mid-slope read
山岗 hummock
山谷 valley
山谷风 anabatic wind
山核桃油 pecan oil
山洪漫蚀 torrential flood erosion
山洪排导工程 torrential flood drainage works
山脊 ridgetop
山脊口 col
山尖 summit
山间盆地 intermontane basin
山间平原 intermontane plain
山涧 ravine
山椒鸟 minivet
山脚 base
山榄果胶 sapote gum
山林 wold
山林巡查 woodward
山麓冲积平原(北美洲西南部) bajada; bahada; piedmont alluvial plain
山麓冲积扇 bajada
山麓地带 piedmont zone
山麓小丘 foothill
山脉 range
山毛榉坚果 beech nut
山帽云 cap cloud
山民锯木厂 woodpecker mill
山坡 hill side; hillslope; incline; side-hill
山坡被覆工程 hillside covering works
山坡编枝栅栏工程 hillside wicker works
山坡防护林 slope protection forest
山坡工程 hillside works
山坡沟渠工程 hillside channel works
山坡截流沟 drainage ditch on slope
山坡侵蚀防治 hillside erosion control
山坡石方工程 hillside stone masonry
山坡水土保持工程 technical measures of soil and water conservation on slope
山坡梯田工程 hillside terracing
山坡下方 downhill (side)
山丘 massif
山区的 montane
山雀 tit
山上锯木厂 peckerwood mill
山上楞场 forest depot; landing; mountain timber yard; upper landing
山上劈制的方材 hewed mountain square

山上贮木场 mountain timber yard
山水盆景 potted miniature landscape
山水纹理 landscape grain
山水园 mountain and water garden
山桃 wild peach
山头风 helm wind
山杏 wild apricot
山岳 orography
山岳俱乐部 sierra club
山岳学 orography
山岳指示种 montane indicater species
山岳志 orography
杉类树种 fir
杉木型纹孔 taxodioid pit
杉木型纹孔对 taxodioid pit pair
杉木油 san-mou oil
珊瑚 coral
珊瑚礁 coral reef; reef coral
珊瑚礁白化 coral bleaching
珊瑚石灰岩土 coral limestone soil
煽式干燥室 fan blower kiln
闪点 flash(ing) point; point of flammability
闪电 lightning
闪电害 damage by lighting strike; damage by lightning
闪化辉绿岩 ophite
闪急蒸发器 flash evaporator
闪石 amphibole
闪烁器 scintillator
闪长岩 diorite
闪蒸器 flash evaporator
扇 booster
扇骨形脉的 flabellinerved
扇积砾 fanglomerate
扇锯 segment(ed) saw
扇砾岩 fanglomerate
扇形沉积[层] fan deposit
扇形带集材 fan-shaped yarding
扇形的 flabellate
扇形地 fan
扇形二分枝式 flabellate dichotomy
扇形聚伞花序 fan
扇形连接器(位于机尾) fan tail
扇形嵌合体 sectorial chim(a)era
扇形拖集 fantail towing
扇形整枝 fan training
扇状聚伞花序 chipidium; rhipidium
扇状整枝 fan training
善变生殖 versatile reproduction
伤疤节 scar (branch) knot
伤擦机(种皮处理) scarifier
伤腐 wound-rot
伤割插穗 wounding cutting
伤害 damage
伤害促进反应 wounding improved response
伤害症状 injury symptom
伤口敷料 wound dressing
伤口寄生物 wound parasite
伤口侵染 wound infection
伤口侵入微生物 wound-invading organism
伤口愈合 occlusion of wounds
伤流 bleeding
伤瘤 wound tumour
伤皮愈合 rind-gall
伤蚀(种皮) scarification
伤纹 score
伤枝节 scar (branch) knot
商(数学) quotient
商标 trade mark
商标名称 brand name
商定期限 agreed period
商盘 windowsill
商品 commodity
商品材 commercial timber; merchantable log; merchantable timber; timber
商品材材积 merchantable volume
商品材高 merchantable height
商品材规格 merchantability specification
商品材积 merchantable volume
商品材径级界限 merchantable diameter limit
商品材林地 commercial timberland
商品材面积 commercial acreage
商品材年龄 merchantable age
商品材形数 merchantable form factor
商品材长 merchantable length
商品材长度 commercial height; merchantable height
商品材直径 merchantable diameter
商品差额 merchandise balance

商品单板 commercial veneer
商品干材 merchantable bole
商品规格材 commercial-sized timber
商品花卉栽培 commercial floriculture
商品架 goods shelf
商品林病害 merchantable forest disease
商品目录价格 catologue price
商品梢头木 merch top
商情预测 business forecasting
商业经营 business management
商业区 commercial district
商业性皆伐 commercial clear cutting
商业性林地 commercial forest land
商业性林业 commercial forestry
商业性疏伐 commercial thinning; extraction thinning
商用规格 market specification
商誉 good will
熵(热力学函数) entropy
赏金 bounty
赏鸟 bird watching (birding)
上钵 potting
上部的 upper
上部结构 superstructure
上侧片 epipleurite
上层 epipelagic zone; upper storey; uppermost layer
上层浮游生物 autopelagic plankton; epiplankton
上层抚育 thinning from above; thinning in dominant; thinning in upper storey
上层间伐 upper storey thinning
上层平流层 upper stratosphere
上层疏伐 crop-tree thinning; crown pruning; crown thinning; French method of thinning; French thinning; high pruning; high thinning; thinning from above; thinning in dominant; thinning in upper storey; upper storey thinning
上层树层 upper tree layer
上层树高 height of dominant trees
上层树冠高 upper crown height
上层树冠级 main crown class
上层树冠长度 upper crown length
上层鱼类 pelagic fish
上层郁闭 cover-overheaded
上茬(伐木锯口) back cut
上承夹钳式集材机 bunk jaw skidder
上承桥 deck-bridge
上冲断层 overthrust
上唇(昆虫) labrum; epichil; epichile
上唇板 top lip (of slice)
上等材 clean timber; clear; high-grade stock; prime log; upper stock
上等结构 superstructure
上等锯材 clear lumber
上等紫胶(商品等级) fine shellac
上底 precoat
上颚(昆虫的) mandible
上方保护 overhead protection
上方庇荫 overhead cover
上方灌溉 overhead irrigation
上方光线 over light
上方喷洒 overhead sprinkling
上方施肥 top dressing
上方下种天然更新 natural regeneration under shelterwood
上方郁闭 cover-overheaded
上方遮阴 overhead shading
上伏辊 upper couch roll
上腹板 episternum
上供料式削片机 top feed chipper
上光蜡 polishing wax
上行色谱法 ascending chromatography
上颌骨 maxilla
上滑面 upslide surface
上夹锯 top bind
上胶量 spread
上胶器 sizer
上界 upper bound
上锯口(伐倾斜木时) high side; back cut
上卷筒 coiling
上料装置 charger
上流水面(堰堤) back face
上毛毡 top felt; upper felt
上楣(柱) cornice
上楣头 capsill
上面 epi-
上面光线 overhead-light

S

上木 high cover; overgrowth; overstorey; overstorey tree; tree of overstorey; tree of upper storey; upper growth; upper storey
上木定心装置 feeding and centering unit
上木伐 felling of wolf trees; liberation cut; liberation felling
上木高度 height of dominant trees
上木装置(胶合板制造) charger; block charger; X-Y charger
上内唇 epipharynx
上排料式削片机 top discharge chipper
上胚轴 epicotyl
上胚轴休眠 epicotyl dormancy
上盆 potting
上皮 epidermis; epithelium
上皮薄壁组织 epithelial parenchyma
上皮层 epithelial layer; epithelium
上皮细胞 epithelial cell; epithelium
上皮细胞层 epithelial layer; epithelium
上皮组织 epithelium
上坡 adverse grade
上坡材 uphill timber
上坡风 anabatic wind
上坡集材 uphill skidding; uphill yarding
上坡位 uphill side
上漆 lacquering
上气道 uptake
上清漆 varnishing
上绕(钢索从卷筒上面卷绕) overwind
上砂式宽带砂光机 top side wide belt sander
上射式水车 overshot wheel
上升割面(采脂) ascending face
上升谷风 upvalley wind
上升坡 ascending grade
上升射击 ascending shot
上升式风力分选机 rising wind sifter
上升水 ascending water
上生萼 epjgynous calyx
上市体重 marketing weight
上树脚镫 tree bicycle; tree climbing iron
上树脚扣 tree bicycle
上树梯 tree climbing ladder
上树脂胶 resin sizing
上凸下凹的波状花边 cyma reverse
上涂料 top sizing; washing
上位(遗传) epistasis; epistasy
上位的(花瓣上着生) epipetalous
上位的(花被及雄蕊) epigynous
上位萼 epjgynous calyx
上位花 epigynous flower
上位花的 epicarpanthous; epicarpous; epicarpius
上位花盘 epigynous disc
上位基因 epistatic gene
上位离差 epistatic deviation; interaction deviation
上位偏差 epistatic deviation
上位式(花被或雄蕊等) epigyny
上位细胞 epistatic gene
上位效应 epistatic effect
上位性基因 epistatic gene
上位性效应 epistatic effect
上位子房 superior ovary
上位子房周位花 perigynous with superior ovary
上下锯机 top-and-bottom saw
上下开关的窗扇 window sash
上弦 top chord
上现蜃景(海市蜃楼) looming
上限 upper limit
上新世 Pliocene
上胸骨 episternum
上压榨 top press; upper press
上压榨辊 top press roll; upper press roll
上芽接 up-budding
上咽 epipharynx
上一层 upper storey
上涌 upwelling
上游水面 up-stream face
上毡垫圈 upper felt
上涨(河水) rise
上足 epipodium
烧材 fuel wood
烧除 burning off; burning out; clean burn
烧除易燃物 fuel-reduction burning
烧光 burning off; burning out
烧化 burn-off

烧画 pyrograph
烧绘(木材烙花) poker-picture
烧碱液 white liquor
烧焦 char; charring
烧焦喷涂处理(防腐)
charring-and-spraying treatment
烧结剂 agglutinant
烧尽试验 burn-out test
烧垦 slash and burn agriculture
烧木材锅炉 wood boiler
烧瓶形的 lageniform
烧伤 burn
烧伤木 burnt-wood
烧失量 ignition loss; loss-on-ignition
烧炭 charcoal-burning; charring
烧炭工 charcoalman
烧炭业者 charcoal-burner
烧窑 kiln
梢变 shoot mutation
梢材 stem topwood; top log; top
wood; upper log; upper stem
梢插 tip cutting
梢端 small end; top end
梢端腐朽 top rot
梢端原木 top log
梢端直径 diameter at small end;
diameter at top end
梢枯病 dieback; top dry
梢料覆盖层 fascine covering
梢杀 high taper
梢头 small end
梢头材 head log; tip
梢头木 top; top stem; top wood;
topwood
梢头直径 small-end diameter; top
diameter
勺轮 bucket wheel
芍药花青素 peonidin
芍药色素苷 peonin
芍药素 paeonidin
杓 dipper
杓形钻 dowel bit
少基数的 oligomerous
少量成分 minor component
少木的 manoxylic
少年期 juvenile period
少雄蕊的 oligandrous;
oligostemonous
少循环湖 oligomictic lake
少叶的 oligophyllous
少圆孔的 oligoforate
少种子的 oligospermous
少壮河 adolescent river
奢侈吸收 luxury absorption
赊 charge account
赊欠账户 book account
赊售 charge sale
赊销 charge sale; sales on credit
舌(昆虫) palate; hypopharynx
舌槽企口板 match board(ing)
舌搭接 tongued-lap joint
舌接 tongue grafting; whip-grafting
舌靠接 tongued approach grafting
舌片(菊科) ligulate
舌榫 slip feather
舌下颊囊 sublingual pouch
舌形小花 strap
舌状导管分子延伸 ligulate
vessel-element extension
舌状的 ligular; ligulate
舌状腹接 side tongue grafting
舌状花 ligulate flower; ray flower
舌状花冠 ligulate corolla
舌状靠接 inarching by tongueing
舌状片 ligule
舌状铁套(搬钩前端) toe ring
舌状小花 tongue
蛇床烯 selinene
蛇根木苷 ophioxylin
蛇菰蜡 balanophora
蛇菊糖苷 stevioside
蛇蛉 snakefly
蛇绿岩 ophiolite
蛇皮管 flexible conduit; flexible pipe
蛇纹石 ophiolite
蛇纹岩 serpentine; serpentinite
蛇蜥媒植物 saurochore
蛇形丘(地质) esker; eskar
蛇形沙丘 barchan
蛇形纹(锯口不平的切面) snake
蛇嘴法(防腐) cobra process
设备 gear; plant; unit
设备利用率 capacity utilization rate
设备容量 plant capacity
设计洪水 design flood
设计坡度 designed slope

设计图 blueprint
社会成本 social cost
社会等级 social hierarchy
社会公益 social benefit
社会价值 community value
社会经济潜力 socio-economic potential
社会经济学 socio-economics
社会经济状态 socioeconomic status
社会-经济-自然复合系统 social-economic-natural complex ecosystem
社会生态学 social ecology
社会性动物 social animal
社会性昆虫 social insect
社会性易化 social facilitation
社会总劳动量 total social labour
社交旅游 social travel
社区 community
社区共管 community co-management
社区森林 community forest
社群 social group; society
社群大小 group size
社群行为 social behaviour
社群化 socialization
社群结构 social structure
社群距离 social distance
社群联结 social bond
社群生物学 socio-biology
社群图 sociogram
社群性群集 social aggregation
社群优势 social dominance
社团林 association forest; corporation forest
射程 range
射出的 emergent
射击 gunning
射击场 shooting range
射击横向偏斜 lateral drift of shot
射击立场 standing place
射击者 shooter
射箭方位 standing
射流 efflux; jet
射流剥皮机 stream barker
射流技术 fluidics
射频 radio frequency
射频含水率测定器 radio-frequency moisture meter
射频加热 heating by radio frequency
射频水分测定计 radio-frequency moisture meter
射束 beam
射线 ray
α-射线 alpha ray
γ-射线 gamma-ray
X射线 X-ray
射线斑纹 ray fleck
射线薄壁细胞 ray parenchyma cell
射线薄壁组织 ray parenchyma
X射线衬比法 X-ray contrast method
射线导管间的 ray-vessel
射线导管间纹孔式 ray-vessel pitting
射线迭生 ray storied
X射线分光计 X-ray spectrometer
X射线分析 X-ray analysis
X射线分析法 X-raying
射线管胞 cross tracheid; ray tracheid
射线管胞交叉场 ray tracheid cross field
射线间隔 ray spacing
射线交叉场 ray crossing
X射线胶片 X-ray film
射线接触区 ray contact area
X射线结构 X-ray structure
γ-射线控制系统 gamma-ray-control-system
射线宽度类型 type of ray-width
射线母细胞 ray mother cell
X射线谱 X-ray spectrum
X射线软片 X-ray film
射线筛管 ray sieve tube
射线上皮细胞 ray epithelial cell
X射线摄谱仪 X-ray spectrograph
射线摄影分析 radiography analysis
X射线摄影术 X-radiography
X射线探测 X-ray detection of defect
X射线探伤法 X-ray defectoscopy
射线体积 ray volume
X射线透视过的 X-rayed
X射线图 X-ray diagram
X射线图式 X-rayogram
X射线图型 X ray pattern
射线细胞 ray cell
X射线显微术 X ray microscopy
X射线显微照相术 microradiography
X射线衍射 diffraction of X-ray; X-ray diffraction

X射线衍射图谱 X-ray diffraction pattern
X射线荧光分析 X-ray fluorescence analysis
射线原始细胞 ray initial; ray mother cell
射线照片 radiogram; shadow picture; shadowgraph; skiagram; skiagraph; skotograph
X射线照片 Röntgenogram; X-radiograph
X射线照相 X ray photograph
X射线照相术 Röntgenography
射线周边细胞 ray epithelial cell
射线组成类型 type of ray-component
射心 centre of impact
涉禽 shorebird (wader)
摄谱仪 spectrograph
摄取 ingestion
摄食 food intake
摄食场所 feeding place
摄食地点 feeding site
摄食高度 grazing height
摄食行为 feeding behaviour
摄食量 food consumption
摄食率 grazing rate
摄食水流 feeding current
摄氏温度 Celsius degree
摄氏温度标 Celsius' thermometric scale
摄氏温度度数 centigrade degree
摄氏温度计的 centigrade
摄像管 camera tube
摄影测量 photogrammetry; photographic surveying
摄影测量调查 photogrammetric cruising
摄影测量学 photogrammetry
摄影初感 threshold value
摄影感光板 photographic plate
摄影机方向 camera orientation
摄影基线 photo base; photographic base
摄影经纬仪 phototheodolite
摄影曝光 photographic exposure
麝香 musk
麝香蓍油 iva oil; musk yarrow oil
申报价值 declared value
申请用地 acquisition of land
伸出部 extension
伸幅(起重臂) reach
伸价条款 escalator clause
伸锯(给予锯片预应力) tension
伸缩臂(装卸机) telescopic boom
伸缩缝(钢轨) dilatation joint
伸缩杆 telescopic pole
伸缩管 slip pipe
伸缩夹盘 retractable chuck
伸缩接缝 expansion joint
伸缩接合 slip joint
伸缩接头 expansion joint
伸缩卡头 retractable chuck
伸缩配件 extension fitting
伸缩式集材杆 telescoping spar
伸缩式起重臂 snorkel boom; telescoping jib
伸缩梯 extension ladder
伸缩性 elasticity
伸悬臂梁 outrigger
伸延 extrusion
伸展 extension
伸展蛋白(伸展素) extensin
伸长 extension
伸长百分率 elongation percentage
伸长变定 elongation set
伸长计 extensometer
伸长率 tensile stretch
伸长期 elongating stage; elongation period
伸长生长 elongation; growth in elongation; extension growth
伸长速率 elongation rate
伸长应变 elongation strain
伸长张量 elongation tensor
伸长主枝 extension leader
砷糖液沸煮法 Powell process; Powellizing process
砷盐(防腐) arsenic salt
深槽饰 quirk
深层(海洋) bathypelagic zone
深层渗透 deep percolation
深冲蚀 deep erosion
深虫孔 deep worm hole
深度废水处理 advanced wastewater treatment
深度腐朽(木材) advanced decay

深度腐朽阶段 advanced stage of decay
深度规 depth gauge
深度计 depthometer
深翻 deep ploughing; deep tillage
深根的 deep-rooted
深根性 deep-rootedness
深根植物 deep rooted plant
深耕 deep ploughing; deep plowing; deep tillage
深耕铲 pan-breaker
深耕犁 heavy duty plough; subsoil plough
深耕松土 subsoiling
深耕中耕机 grubber
深谷 ravine
深海带 abyssal zone; bathyal zone
深海底的 bathybic
深海底栖生物 abyssal benthos
深海潜水器 bathyscaph
深海区 abyss
深海声波散射层 deep scattering layer
深海细菌 deep-sea bacteria
深海性 abyssal
深海鱼 deep-sea fish
深海渔业 deepsea fishery
深框锯 buhl saw
深蓝色 ultramarine
深裂 deep shake
深裂的 cleft
深裂纹 heavy torn grain
深切曲流 incised meander
深色假心材 dark false heartwood
深色细胞内含物 dark cell content; dark cell inclusion
深色砖红壤化土壤 euchrozems
深水[暗层]浮游生物 skotoplankton
深水带生物群落 communities in the profundal zone
深水腐泥土 deep water muck
深水温度测量仪 bathythermograph
深水形成 deepwater formation
深松犁 chisel plough
深位柱状碱土 deep-columnar solonetz
深温仪 bathythermograph
深休眠 deep dormancy
深渊 abyss
深渊层 abyssopelagic zone
深渊带 habal zone
深栽 deep planting
深栽钻孔机 deep planting auger
深直根系 deep taproot system
神殿园 temple garden
神经的 neural
神经毒素 neurotoxin
神经行为学 ethology
神经节(节) ganglion
神经元 neuron(e)
审核单位 control unit
审核员 checker
审记员 comptroller; controller
审美功能 esthetic function
审美观 aesthetic point
肾形的 kidney shaped; Kidney-form; nephroid
渗出 effluent seepage; exudation; percolate; seepage; seep-off; sweating
渗出流(地下水) effluent seepage
渗出物 exudate
渗出性 leachability
渗出液 diffusate; exudate
渗胶 bleeding; bleed-through; glue penetration; strikethrough
渗量计 lysimeter
渗流 effluent seepage; seepage flow; vadose
渗流泉 vadose spring
渗流水 vadose water
渗漏 blow-by; effluent seepage; filtration; seepage
渗漏率 seepage rate
渗漏区 seepage area
渗滤 diffusion; percolate; percolation
渗滤比 percolation ratio
渗滤法 percolation method
渗滤率 percolation rate; percolation ratio
渗滤器 percolator
渗滤速度 percolation rate
渗入 exosmosis; infiltration
渗入能力 infiltration capacity
渗入速率 infiltration rate
渗碳剂 carburizer
渗碳作用 carburization

S

渗透 impregnation; osmosis; penetration
渗透测量计 infiltrometer
渗透层 permeable layer
渗透处理 osmose treatment; permo-treatment
渗透的 pervious
渗透法 osmose process
渗透功 osmotic work
渗透机理 mechanism of penetration
渗透剂 penetrant
渗透力 penetrating ability; penetrating power
渗透率 infiltration rate; percolation rate
渗透浓度 osmotic concentration
渗透深度 depth of penetration
渗透势 osmotic potential
渗透水 osmotic water; seepage water
渗透水分 permeant water
渗透速度 infiltration velocity
渗透调节 osmoregulation
渗透物 osmotica
渗透系数 coefficient of permeability; osmotic coefficient
渗透性 accessibility; penetrability; permeability
渗透性防腐剂 osmose preservative
渗透压 osmotic pressure
渗透压力测定法 osmometery
渗透压梯度 osmotic pressure gradient
渗透盐剂(木材防腐剂) osmosar
渗透作用 infiltration
渗析 dialysis; osmosis
渗压计 osmometer
渗养者 osmotroph(s)
渗育层 precogenic horizon
渗征塑剂防腐洗 osmoplastic method
椹果 sorose
升程 lift
升华 vapourization without melting
升华干燥 lyophilization
升降舵 elevator
升降滑轮 fall block
升降滑轮集材(增力式索道) falling block yarding
升降机 elevator; hoister; lift; lifter
X形升降机 X-lifter
升降架 crane
升降锯 rise-and-fall saw; rising and falling saw
升降螺丝杆 elevation screw
升降门 lift gate
升降平台 elevator
升降式吊桥 lift bridge
升降式闸门 lift gate
升降丝杆 elevation screw
升降台 lifting platform; lifting table
升降梯 elevator
升降圆截锯 jump saw
升降运送车 lift truck
升力 lift
升力线 lifting line
升麻素(升麻脂) macrotin
升麻脂 macrotin
升坡 elevation
升起 rise
升水 premium
升温 heating-up; warming-up
升温致冷全自动温室 completely automated heating and cooling greenhouse
升压器 booster
升值 revaluation
生材 fresh wood; green; green lumber; green timber; green wood
生材比重 specific gravity in green
生材含水率 green moisture content
生材计数 green-chain tally
生材密度 green density
生材台账 green-chain tally; mill tally
生材重 green weight
生草 tussock
生草丛冻原 tussock tundra
生草灰化土 derno podsolic soil
生草泥炭沼泽 grass moor
生草森林土 soddy forest soil
生草土 sod
生产 forest production; yield
生产参数 production parameter
生产成本 production cost
生产定额 production quota
生产工人 production worker
生产过剩 over production
生产函数转向点 point of inflexion
生产环境 production context

生产计划 planning of production; production planning; program of production
生产技术财务计划 technical productive financial plan
生产结构 productive structure
生产结构图 productive structure diagram
生产经营会计 activity accounting
生产扩大 expansion of production
生产力 productiveness; productivity; yield capacity; yield-power
生产力金字塔 pyramid of productivity
生产力相等的林地 equi-productive area
生产力指数 productivity index
生产量 production; throughput capacity; turnout
生产量估 assessment of yield-capacity
生产路线 production path
生产率 output; production rate; rate of production
生产率条款 productivity clause
生产目标 production target
生产能力 throughput capacity
生产期 production time; productive time
生产潜力 site productivity; yield potential
生产生态学 production ecology
生产时间 production time; productive time
生产水平 yield level
生产税 yield tax
生产效率 production efficiency
生产性财产出售 sales of productive property
生产性森林 production forest
生产性作业时间 production time; productive time
生产要素 element of production; factor of production
生产要素费用 factor cost
生产者 producer
生产者议价实力 bargaining strength of producer
生产蒸散关系 production evapotranspiration
生产资料 capital goods
生产组织 organization of production
生成物 resultant
生存 survival
生存比 survival ratio
生存环境 living environment
生存经济 subsistence economy
生存竞争 struggle for existence
生存率 survival rate
生存年代不重叠的 allochronic
生存潜能 survival potential
生存水平 subsistence level
生存性 viability
生根 radication; rooting
生根带 rooting zone
生根激素 rhizocaline
生根介质 rooting medium
生根率 rooting percentage
生根能力 root-forming ability; rooting ability; rooting capacity
生根物质 root-forming substance
生化处理 biochemical treatment
生化工程 biochemical engineering
生化过程 biological process
生化燃料电池 biochemical fuel cell
生化染色试验 biochemical staining test
生化衰退 biochemical deterioration
生化相克 allelopathy
生化需氧量 biochemical oxygen demand; biological oxygen demand
生化遗传学 biochemical genetics
生荒地 virgin land
生活场所 life arena
生活方式 living pattern
生活废物 sanitary waste
生活费用 maintenance
生活力 viability; vital force; vitality
生活力缺失 abiosis
生活力染色(检验法) viability stain
生活期 life stage
生活期限 life span
生活强度 vitality
生活设施 convenience
生活史 life cycle; life history
生活史对策 life history strategy
生活史学(昆虫) bionomics
生活习性 life habit

生活型 growth habit; life form; lift form; vegetative form
生活型级 life-form class
生活型谱 biological spectrum; life form spectrum
生活型优势 life-form dominance
生活样式 living pattern
生活周期 life cycle; life history
生机论 vitalism
生计畜牧 subsistence herding
生计活动 subsistence activity
生胶层 green glue-line
生节 sound knot
生精管(胚胎) testicular follicle
生境 habitat; oikos; site
生境多样性 habitat diversity
生境分离 habitat segregation
生境隔离 habitat isolation
生境孤岛 habitat island
生境管理 habitat management
生境恢复 habitat rstoration
生境类群 habitat group
生境类型分类 habitat type classification
生境偏爱 habitat preference
生境片段化 habitat fragmentation
生境群(组) habitat group
生境生态位 habitat niche
生境小区 biotope
生境型 habitat form
生境选择 habitat selection
生境因素 habitat factor
生境破碎化 habitat fragmentation
生境专有性 despotism
生境状况 condition of habitat
生蜡 crude Japan tallow
生篱 hedge
生理成熟 physical maturity; physiological maturity
生理出生率 physiological natality
生理伐期龄 physiological maturity
生理分类学 physiological taxonomy
生理辐射 photosynthetically active radiation (PAR)
生理干旱 physiological drought
生理干燥 physiological drying; physiological dryness
生理干燥法 sour-telling
生理隔离 physiological isolation
生理化学 chemophysiology; physiological chemistry
生理碱性肥料 physiological basic fertilizer
生理节律 circadian rhythm; physiological rhythm
生理节奏的节律 circadian rhythm
生理抗性 physiological resistance
生理可塑性 physiological plasticity
生理零点 physiological zero
生理轮伐期 biological rotation; physical rotation
生理年龄 physical age
生理燃烧 physiological combustion
生理生活史 physiological life history
生理生态学 physiological ecology
生理失调 physiological disorder
生理寿命 physiological longevity
生理树脂 physiological resin
生理衰退 physiological deterioration
生理水分 physiological moisture; physiological water
生理酸性肥料 physiological acid fertilizer
生理同步 physiological timing
生理小种 biological strain; physiological race
生理性病害 physiological disease
生理性虫害 physiological insect-pest
生理休眠 physiological dormancy
生理学 physiology
生理学的 physiological
生理循环 circulation
生理遗传学 physiological genetics
生理钟 physiological timing
生理种 physiologic species; physiological species
生理宗(生态) biological race
生瘤(立木皮囊) cat face
生命表 life table
生命产生的 biogenous
生命次序 order of life
生命带 life zone
生命单元 bion
生命地带 biochore
生命起源 origin of life
生命网 web of life
生命维持系统 life-support system
生命系统 life system

S

生皮 hide
生漆 che shu lacquer; Chinese lacquer; raw lacquer
生气剂 inflating agent
生氰糖苷 cyanogenetic glucoside
生氰植物 cyanogenous plant
生热反应 pyrogenic reaction
生乳 lactation
生色团 chromophore
生石灰 burned lime; quick lime
生水 raw water
生死比率 birth-death ratio
生态安全 ecological safety
生态背景 ecological setting
生态变异 ecocline; ecological variation
生态变种反应 ecophene
生态表现型 ecophene
生态表型 ecophenotype
生态补偿 ecological compensation
生态补偿机制 ecological compensation mechanism
生态层 ecosphere
生态持续性 ecological sustainability
生态冲击 ecological backlash
生态出生率 ecological natality
生态等价 ecological equivalence
生态等值种 ecological equivalent; ecological equivalent (species)
生态地理分歧 ecogeographical divergence
生态地理学 ecological geography
生态地区 ecoregion
生态地区适应 ecotopic adaptation
生态顶极 ecological climax
生态定位 ecological location
生态动物地理学 ecological zoogeography; ecozoogeography
生态度假村 Ecolodge
生态对策 bionomic strategy
生态多度计算 ecological bonitation
生态多态现象 ecological polymorphism; physiological polymorphism
生态发生学 ecological embryology
生态分布 ecological distribution
生态分歧 ecological divergence
生态幅 ecological amplitude; ecological valence; ecological valency
生态复合体 ecological complex
生态隔离 ecological isolation
生态工程 ecological engineering
生态光照单位 ecological light unit
生态过渡带 ecotone
生态红线 red line of ecological protection; ecological red line
生态化合物 eco-chemical
生态技术 ecological technique
生态价 ecological valence; ecological valency
生态价值 ecological value
生态渐变群 ecocline
生态交错带假说 ecotone hypothesis
生态交错区 ecotone
生态结构 ecological structure
生态金字塔 ecological pyramid
生态进化 ecogenesis
生态经济区位 eco-economical zoning
生态力 ecological force
生态临界 ecological threshold
生态零值 ecological zero
生态龄 ecological age
生态旅游 ecotourism; ecological tourism
生态密度 ecological density
生态耐性 ecological tolerance
生态内稳定 ecological homeostasis
生态能量学 ecoenergetics; ecological energetics
生态农业 ecological agriculture
生态农业模式 ecological agriculture model (pattern)
生态浓集度 ecological concentration
生态胚胎学 ecological embryology
生态品种 ecological race
生态平衡 ecological balance; ecological equilibrium; eubiosis
生态平衡表 ecological balance sheet
生态气候 ecoclimate
生态倾差 ecocline
生态倾群 ecocline
生态区 ecotope
生态区适应 ecotopic adaptation
生态区位 ecological location
生态趋异性 ecological divergence

生态圈 ecosphere
生态群 cline; ecological group
生态入侵 ecological invasion
生态设计 ecological setting
生态生理学 ecological physiology; ecophysiology
生态生物地理学 ecological biogeography
生态生长效率 ecological growth efficiency
生态时间 ecological time
生态史 ecogenesis; ecological history
生态适应型 ecophene
生态释放 ecological release
生态寿命 ecological longevity
生态死亡率 ecological mortality
生态梯度 ecological gradient
生态替代种 ecological vicariad
生态条件 ecological condition
生态调查法 ecological survey method
生态调节 ecoregulation
生态同类群 ecodeme
生态同源种 ecological homologue
生态统计学 biodemography
生态危机 ecological crisis
生态位 ecological niche; niche
生态位分离 niche separation
生态位幅度 niche breadth
生态位宽度 niche width
生态位理论 niche theory
生态位漂移 niche shift
生态位维(度) niche dimension
生态位压缩 niche compression
生态位重叠 niche overlap
生态文明 cecological civilization (eco-civilization); conservation culture
生态稳定性 ecological stability
生态习性 silvical habit
生态系 ecosystem
生态系列 ecological series
生态系列法 ecological serial method
生态系生产力 productivity of ecosystem
生态系统 biogeocoenosis; ecological system; ecosystem; holocoen
生态系统的 holocoenotic
生态系统的发育 ecosystem development
生态系统服务 ecosystem services
生态系统功能 ecosystem function; ecosystem functioning
生态系统管理 ecosystem management
生态系统价值评估 ecosystem value assessment
生态系统建模 ecosystem modeling
生态系统健康 ecosystem health
生态系统界面 ecosystem interface
生态系统净交换量 net ecosystem exchange
生态系统净生产力 net ecosystem productivity
生态系统类型 ecosystem type
生态系统力能学 ecosystem energetic
生态系统模拟 ecosystem modeling
生态系统生态学 ecosystem ecology
生态系统完整性 ecosystem integrity
生态系统稳定性 ecosystem stability
生态系统预测 ecosystem prediction
生态相同种 ecological equivalent (species)
生态小种 ecological race
生态效率 ecological efficiency
生态效益 ecological effect; ecological benefit
生态形态学 ecological morphology
生态型 climatype; ecological type; ecotype
生态型变异 ecotypic variation
生态型分化 ecotypic differentiation
生态性能 ecological performance
生态性物种形成 ecological speciation
生态序列数 ecological sequence number
生态学 ecology; oecology; bionomics
生态学(脊椎动物) ethology
生态学范式 ecological paradigm
生态学过程 ecological process
生态学家 ecologist
生态循环 ecocycling
生态亚系统 ecological subsystem
生态研究网络 ecosysterm research nNetwork
生态演替 ecological succession

生态遗传学 ecogenetics; ecological genetics
生态意义 ecological significance
生态因子 ecological factor
生态因子替代 replacement of ecological factors
生态影响评价 ecological impact assessment
生态优势 ecological dominance
生态优势种 ecological dominant
生态域 biochore
生态阈值 ecological threshold
生态灾害 environmental disaster
生态灾难 eco-catastrophe; environmental disaster
生态植物群落系列 ecologic-phytocoenostic series
生态指示种 ecological indicator
生态种 biological species; ecospecies
生态种发生 ecogenesis
生态种分化 ecological speciation
生态重叠 ecological overlap
生态周转 ecological turnover
生态锥体 ecological pyramid
生态族 ecological race
生态最适度 ecological optimum
生统遗传学 biometrical genetics
生土 immature soil
生物 biont
生物安全 biosafety
生物半衰期 biological half-life
生物变异度 biological variability
生物材料 biologicals
生物残留群 liptocoenosis
生物测定 bioassay; biological assay; biotest
生物测量 bio-measurement
生物测量学 biometry
生物层积学 biostratonomy
生物差异 biological variability
生物柴油 biodiesel
生物产品 biologics
生物常数 biological constant
生物触媒 biocatalyst
生物次序 order of living beings
生物催化剂 biocatalyst
生物带 biozone; life belt; life zone
生物的 biotic
生物的环境调节 biological conditioning
生物地层学 biostratigraphy
生物地理化学循环 biogeochemical cycle
生物地理界 biogeographic realm
生物地理气候区 biogeoclimatic zone
生物地理区 biogeographic region; biotic province
生物地理群落 biogeocoenosis; geobiocoenosis
生物地理省 biogeographic province
生物地理学 biogeography; biological geography; chorology
生物地理障碍 biogeographic barrier
生物地球化学 biogeochemistry
生物地球化学循环 biogecochemical cycle; biogeochemical circulation
生物电 bioelectricity
生物电源 biogalvanic source
生物顶极群落 biotic climax
生物动力农业 biodynamic agriculture
生物动力学 biodynamics
生物多样性 biodiversity
生物多样性保护区域 biodiversity conservation region
生物多样性关键地区 critical regions of biodiversity
生物多样性监测 biodiversity monitoring
生物多样性经济价值 economic value of biodiversity
生物多样性热点地区 biodiversity hotspots
生物发光 bioluminescence
生物发生律 biogenetic law
生物反馈控制 biofeedback control
生物防治 biological control; biotic control
生物放大 biological magnification; biomagnification
生物分布学 chorology
生物分解 biological breakdown; biolysis
生物分类学 biosystematry
生物分子 biomolecule
生物风化 biological weathering

生物辅助能 biological energy subsidies
生物腐蚀 biodeterioration
生物富集 biological enrichment
生物隔离 bioisolation; biological isolation
生物工程学 bioengineering
生物固氮 biological nitrogen fixation
生物光度计 biophotometer
生物还原作用 bio-reduction
生物合成 biosynthesis
生物互助 biosocial facilitation
生物化学 biochemistry
生物化学进化 biochemical evolution
生物环境调节技术 biotronics
生物活素 bios
生物活性 biological activity
生物机能结构学 biostatics
生物机制 biomechanism
生物积累作用 biological accumulation
生物计 biometer
生物计法 biometer method
生物计量法 biometric method
生物剂 biotic agent
生物间的 interbiotic
生物监测 biological monitoring
生物碱 alkaloid
生物降解 biodegradation
生物节律 biological rhythm; biorhythm
生物结石 biolite
生物经济学 bioeconomics
生物净化 biological cleaning
生物静力学 biostatics
生物酒精 bioethanol
生物聚合物 biopolymer
生物科学 bio-science
生物可用率 bioavailability
生物控制剂 biocide
生物控制论 biocybernetics
生物控制系统 biocontrol system
生物矿物 biolite
生物扩张力 biotic pressure
生物廊道 biological corridor
生物利用率 bioavailability
生物量 biomass
生物量产量 biomass yield
生物量方法 biomass method
生物量呼吸量比 biomass-respiration ratio
生物量换算因子 biomass expansion factor
生物量金字塔 pyramid of biomass
生物量密度 biomass density
生物量生产力 biomass productivity
生物量碳 vegetative carbon
生物量鲜重 green biomass
生物量原料 biomass feedstock
生物量增量 biomass increment
生物模型 living model
生物内生的 endobiotic
生物能降解的 biodegradable
生物能量效率 biological energy efficiency
生物能疗法 bioenergetics
生物能学 bioenergetics
生物拟态 biomimesis
生物年代学 biochronology
生物农药 biotic pesticide
生物农业 biological agriculture
生物浓缩 bioconcentration; biological concentration
生物品系 biological strain
生物平衡 biological equilibrium; biotic balance; biotic equilibrium
生物起源 biogenesis
生物气候 bioclimate
生物气候带 bioclimatic zone
生物气候定律 bioclimatic law
生物气候分带 bioclimatic zonation
生物气候室 biotron
生物气候学 bioclimatics; bioclimatology
生物气候学的 bioclimatic
生物气象学 biometeorology
生物潜能 biotic potential
生物侵害 biological attack
生物区 biochore; biotic region; zono biome
生物区系 biota
生物圈 biosphere; vivosphere
生物圈保护区 biosphere reserve
生物群 biota
生物群个-生境小区系统 biota-biotop system
生物群居促进作用 biosocial facilitation

生物群落 biocoenose; biocoenosis; biome; biotic community
生物群落学 biocoenology; biocoenotics
生物群区 biome
生物群-生物小区系统 biota-biotop system
生物群系 biotic formation
生物群系型 biome type
生物燃料 biofuel
生物人口统计学 biodemography
生物入侵 biological invasion
生物色素 biochrome
生物社会学 biosociology; coenobiology
生物生产 biological production
生物生产系统 biological production system
生物生态地理学 ecobiogeography
生物生态学 bioecology
生物势(繁殖和生存) biotic potential
生物适合性 biocompatibility
生物适应性 biological fitness
生物数学 biomathematics
生物衰老 biological decay
生物体 organism
生物统计法 biometric procedure
生物统计学 biometrics; biometry; biostatistics
生物温度 biotemperature
生物污染指数 biological index of pollution
生物物理学 biophysics
生物雾 biofog
生物系统 biological system; biosystem; ecosystem
生物系统分类学 biosystematics
生物系统效率 efficiency of biological system
生物系统学 biosystematry
生物相 biofacies; biota
生物相容性 biocompatibility
生物相图 biofacies map
生物小区 biotope
生物小循环 biological micro-cycle
生物效率 biological efficiency
生物效能 biological activity
生物效应 biological effect
生物型 biological type; bion; biotype
生物性演替顶极 biotic climax
生物悬浮物 bioseston
生物学 biology
生物学过程 biological process
生物学化 biologization
生物学混杂 biological admixture
生物学零度 biological zero
生物学时间 biological time
生物学试验 biological test
生物学致死温度 biological zero point
生物学种 biological species
生物学最低温度 biological minimum temperature
生物压力 biotic pressure
生物岩 biolite; biolith
生物演替 biotic succession
生物氧化 biological oxidation; bio-oxidation
生物遥测术 biotelemetry
生物因素 biotic factor
生物因子 biological factor; biotic agency
生物元素 biological element
生物原型 living prototype
生物指数 biotic index
生物制剂 biologicals
生物制品 biologicals; biologics
生物质 biomass
生物质发电厂 biomass power plant
生物质锅炉 biomass boiler
生物质能源 bioenergy
生物质收获机 biomass harvester
生物钟 biochronometer; biological clock; living clock
生物钟学 chronobiology
生物种群 biological population
生物周期 biorhythm
生物周期现象 biological periodism
生物资源 biological resources; living resources
生物宗 biological race
生物总量的 bio-massy
生物族 biological race
生物组分 biotic component
生橡胶 caoutchouc; crude rubber; elastic gum; raw rubber
生效硬化剂 aggressive hardener
生芽的 gemmiferous
生瘿的 cecidogenous

S

生于地上的 epigaeal; epigeal; epigean; epigeous
生于动物体内的 endozoic
生于高含盐水中的 euhalabous
生于根部的 radicicolous
生于基物表面的 extramatrical
生于湿地的 mesophilus
生育 procreation
生育地描述 destcription of site; site description
生育地生产力 site productivity; yield potential
生育地调查 site apparaisal; site apparaisement; site mapping; site survey
生育地图 site type map
生育地整治 clearing of the felling area; site clearing
生育空间 growing space
生育控制 birth control
生育力 fecundity
生育力表 fecundity schedule
生育型 growth form
生源论 biogenesis
生源体 biophore
生长 growth
S形生长 sigmoid growth
生长表 growth table; increment table
生长参数 growth parameters; increment parameters
生长测定 growth test
生长层 growth layer
生长大周期 grand period of growth
生长带 growth zone
生长点 growing point
生长伐(近轮伐期前的强度疏伐) accretion cutting; increment cutting; increment thinning; increment felling
生长方程 growth equation
生长分析 growth analysis
生长估测的林分投影法 stand projection in growth estimation
生长过快 overgrowth
生长函数 growth function
生长活力 vigour
生长激素 auxin; growth hormone; natural growth hormone
生长激素抑制素 somatostatin
生长级 crown class
生长计 auxanometer
生长记录器 auxograph
生长季 vegetation season
生长季嫁接 growing season grafting
生长季节 growing season
生长季长度 growing season length
生长加速 growth acceleration; increase in growth; increase in growth rate
生长空间 growing space; growth space
生长空间比(树冠高-树高比等) growing-space ratio
生长量 growth; increment
生长量法 growth method
生长量数据 growth data
生长量研究 increment study
生长量与采伐量之比 growth-cut ratio
生长量增加 increase in growth rate
生长轮 growth ring
生长轮界线 growth-ring boundary
生长率 growth percent; growth percentage; growth percentage; growth rate; increment coefficient; increment percent; rate of growth; rate of increment
生长模型 growth model
生长逆境 growth stress
生长期 growing period; growth period; vegetation period
生长潜力 vigour
生长强度 intensity of growth
生长区 growth region; growth zone; region of growth
生长区域 growth region
生长曲线 growth curve; increment curve
S形生长曲线 sigmoid growth curve
生长缺陷 defect in growth
生长试验 growth test
生长疏伐 accretion thinning
生长素 bios
生长素载体 carrier for auxin
生长速率 growth percent; growth rate
生长速率函数 growth-rate function
生长条纹 growth band

生长调节剂 growth regulator
生长图谱法 auxanographic method
生长习性 growth habit; habit of growth
生长系数 coefficient of growth; increment coefficient
生长限定基质 growth-limiting substrate
生长相关 growth correlation
生长相关作用 correlation of growth
生长效率 growth efficiency
生长效应 growth effect
生长胁迫 growth stress
生长型 growth form; growth type
生长序列 growth series
生长研究 growth study
生长抑制剂 growth retardant
生长因子 growth factor
生长应力 growth stress
生长于砾土的 glareous
生长于叶面的 epigeous
生长预测 growth prediction; growth prognosis; growth projection
生长在地上的 epigeal
生长指数 index of growth
生长中断 growth interruption
生长周期 growth cycle; vegetative cycle
生长锥 increment borer; accretion borer; increment gauge
生长锥取样匙 increment borer extractor
生长锥手柄 borer handle
生长锥探杆 core extractor; increment borer extractor
生长赘瘤的 snagged; snaggy
生枝打枝 green pruning
生枝烧除 live burning
生殖 genesis; propagation
生殖成效 reproduction effect
生殖格局 reproductive pattern
生殖隔离 misogamy
生殖根 stolon
生殖行为 breeding behaviour; reproductive behaviour
生殖核 generative nucleus
生殖洄游 breeding migration
生殖力表 fertility table
生殖率 birth rate; natality; reproduction rate
生殖母细胞 auxocyte; gonocyte
生殖器官 sexual organ
生殖潜力 reproductive potential
生殖腔 genital chamber
生殖切离 reproductive isolation
生殖托 receptacle
生殖系统 genitalia; reproductive system
生殖细胞 generator cell; germ cell
生殖腺 gonad
生殖芽体 gonophore
生殖障碍 dysgenesis
生殖值 reproductive value
生殖质 germplasm
生殖滞时 reproductive time lag
生殖周期 reproductive cycle
生殖组织 generative tissue
声波发射器 pinger
声传导 transmission of sound
声导率 acoustic(al) conductivity
声混响 reverberation of sound
声混响时间 reverberation time of sonud
声级计 noise meter; sound meter; sound-level meter
声检 sound
声纳 sonar
声能 acoustic(al) energy
声频法 audio method
声强 intensity of sound
声强计 acoustimeter
声桥 sound bridge
声透射 sound transmission
声响反射系数 acoustic reflection coefficient
声学性质 acoustical property
声压级 sound pressure level
声源定位 sound location
声障 sound barrier
声阻 acoustic resistance
声阻抗 acoustic impedance
牲畜道 driveway; stock driveway; stock route
绳测(木材检尺) string measurement
绳股 strand
绳环 eye-splice; roll splice
绳结 swifter knot
绳结园 knot(ted) garden

绳卡 shackle
绳圈 bight
绳索 cordage
绳索缠绕 snake
绳索减速(绳索缠于树上) snub check
绳索减速制动 snub
绳索节距 rope pitch
绳索捆扎 roping
绳套 becket
绳梯 rope ladder
绳状花纹 rope figure
省立公园 provincial park
省钱的 cost-effective
省有林 provincial forest
圣草酚 eriodictyol
圣诞树 Christmas tree
盛行风 cardinal wind; prevailing wind
盛行西风带 prevailing westerlies
盛燃时间 burning period
盛夏季相 estival aspect
剩余保持量 residual retention
剩余财产 equity
剩余储备金 excess reserve
剩余弹性 residual elasticity
剩余的 waste
剩余繁殖价值 residual reproductive value
剩余方差 residual variance
剩余价值 residual (value); surplus value
剩余精核 merocyte
剩余林价 residual stumpage price
剩余物(采伐后) wastage; slash
剩余物收集机 residue harvesting machine
剩余应力 surplus stress
剩余阻力 residual resistance
剩值 residual (value)
尸体群 necrocoenosis; thanatocoenosis
尸学 necrology
失光(漆病) gloss reduction; loss of gloss
失灵 failure; malfunction
失水 anabiosis
失效控制线 lost line
失真 anamorphosis; distortion
施莱特火焰蔓延试验 Schlyter test
施肥 dressing; fertilization; fertilizer; fertilizer application; fertilizing; manuring
施肥播种[联合作业] ferti-seeding
施肥风机 fertilizer blower
施肥灌溉 fertilizer irrigation
施肥机 applicator; fertilizer apparatus; fertilizer applicator; fertilizer distributor; fertilizing machinery
施肥技术 fertilizer practice
施肥器 fertilizer applicator
施肥区 area of application
施肥装置 fertilizer apparatus
施粪肥 dunging
施工能力 work(ing) ability
施工设计 work design
施基肥 basal dressing
施胶 glue blending; size; sizing
施胶槽 sizing tub; sizing vat
施胶辊 applicator roll; sizing roll
施胶机 glue applicator; gummer
施胶剂 sizing agent
施胶强度 size fastness
施胶桶 sizing tub; sizing vat
施胶箱 sizing box
施胶性能 sizability
施料辊 applicator roll
施奈德公式 Schneider's formula
施皮格尔速测镜 Spiegel relascop
施药区 area of application
施业 working
施业案 management plan; working-plan
施业案编定 preparation of working plan
施业案编制员 working plans officer
施业案管理区 working plans circle
施业案面积 working-plan area
施业案修订 management plan revision
施业案总监 working plans conservator
施业簿 management record; management-book
施业单位 working area
施业分区 district; forest district; territorial division; working section

施业规划 working scheme
施业级 working figure
施业计划 operable plan; operating plan; plan of management
施业计划检订 working plan revision
施业林 managed forest
施业目的 management objective
施业期 utilization period; working-plan period
施业区 block; circle; division; forest circle; territorial circle; working circle; working cycle; working unit; working area
施业区划 working division
施业区主任 divisional officcr
施业指南 operable guide
施业总方案(规划) general working-plan
施业总计划 general plan of working
施用绿肥 green manuring
施釉 glazing
湿板坯 wet lap; wet mat; wet sheet
湿材 fresh wood; green; green lumber; green timber; green wood; moist wood; wet timber
湿材场地 green lumber yard
湿材堆场 green lumber storage
湿材分类台 green sorting deck
湿材含水率 green moisture content
湿材料 wet stock
湿材选材台 green table
湿材运输链 green chain
湿材重 green weight
湿藏 wet storage
湿草甸 moist meadow
湿草原 grass fen
湿草原群落 hygrophorbium
湿草原土 brunizem; plansol
湿打浆 wet beating
湿单板 green veneer
湿地 damp ground; marsh; quagmire; wetland
湿地制图 wetland mapping
湿度 humidity; moisture
湿度表 hygrometer
湿度范围 moisture range
湿度计 hygrograph; hygrometer; hygroscope; moisture (content) meter; moisture register
湿度计算器 psychrometric calculator
湿度控制器 humidity controller
湿度链(土壤) moisture catena
湿度曲线 hygrogram
湿度势 moisture potential
湿度梯度 moisture gradient
湿度调节器 humidity controller
湿度图(表) humidity chart; hygrogram
湿度仪 hygrograph
湿度指示器 atmidoscope
湿端(干燥窑) charge end; green end
湿法 wet process
湿法胶合板 wet-glued plywood
湿法胶合的 wet-cemented; wet-glued
湿法铺装 wet-felting
湿法砂光 wet sanding
湿法硬质纤维板 wet-method hardboard
湿法装饰 wet finishing
湿腐 damp rot; water rot; wet rot
湿复型法 wet replica method
湿害 wet damage
湿旱生植物 tropophyte
湿灰化土 wet podzolic soil
湿基含水率 wet basis moisture content
湿胶层 moist glue line
湿胶合 wet gluing
湿绝热[线] saturation adiabat
湿空气 moist air; moisture-laden air
湿空气比容 humid volume
湿料仓 wet bin
湿木白蚁类 damp-wood termites; moist-wood termites
湿木蛀虫 wetwood borer
湿气排除能力 moisture excluding efficiency
湿强度 wet strength
湿强度纸 wet strength paper
湿球 wet bulb
湿球位温 wet-bulb potential temperature
湿球温度 wet-bulb temperature
湿球温度计 wet-bulb thermometer
湿热 moist heat

湿热灭菌法 moist heat sterilization
湿润 damp; imbibition
湿润比容 humid volume
湿润的 hygric
湿润低温处理 moist-chilling
湿润低温预处理 moist prechilling
湿润范围 wetted perimeter
湿润剂 wetter; wetting agent
湿润角 wetting angle
湿润拉伸 wetting tension
湿润气候 humid climate
湿润器 humidifier; wetter
湿润曲线 wetting curve
湿润热 wetting heat
湿润伸长 damping stretch
湿润土壤 moist soil
湿润指数 index of moisture
湿润周界 wetted perimeter
湿润棕色石灰壤土 humid limestone brown loam
湿润作用 humidification
湿烧除(新鲜采伐剩余物) live burning
湿生 hygro-
湿生的 hygric
湿生动物 hygrocole
湿生树种 hygrophilous tree species
湿生植物 hygrophyte
湿式[精]砂光机 final water sander
湿式剥皮机 waterpour debarker
湿式宽带砂光机 wet type wide belt sander
湿式喷粉机 dew duster
湿土 moist soil
湿稳性 hygro-stability
湿物料 wet stock
湿心材(活树中形成) water heart; water soak; water-core; wet heart; wetwood
湿压法 wet pressing
湿原 moor
湿纸幅强度 wet web strength
湿制程 wet process
湿重 fresh weight; wet weight
湿重含水率 wet basis moisture content
湿重木材(白冷杉类) sap fir
蓍花油 iva oil
十倍体 decaploid
十二倍体 dodecaploid
十花柱的 decagynian; decagynous
十进制板英尺材积表 decimal log rule; decimal rule
十进制乘法器 decimal multiplier
十进制数字 decimal digit
十心皮的 decagynian; decagynous
十一倍体 hendecaploid
十一雌蕊的 endecagynian; endecagynous
十一雄蕊的 endecandrous
十亿(10^9) giga-
十亿吨(10^9t) gigaton (Gt); petragram (Pg)
十亿克(10^9g) gigagram
十字镐 hack
十字架(测量横断面) cross-staff
十字交叉归楞 cross piling
十字交叉形河绠(断面) Johnson boom
十字楞 staggered piling
十字锹 pickaxe; pick-haue
十字球 mattock
十字人行道 cross walk
十字丝 cross-hair reticule
十字形 cruciform
十字形芽接 cross-shaped budding
石 stone
石坝 masonry dam; stone splash-dam; stone weir
石白粉(矽酸镁的白色填料) agalite
石板 slate
石板坝 slab dam
石板谷坊 slab check dam
石标 boundary stone
石床园 morainc garden
石灯笼 stone lantern
石膏 gypsum
石膏模型 plaster figure
石膏像 plaster figure
石膏岩 gypsum rock
石戈壁植被 rock pavement vegetation
石构建筑 stone construction
石磙打浆机 stone roll beater
石化森林 petrified forest
石化作用 petrification
石灰化 chalking
石灰结核 caliche; lime concretion

S

石灰坑 dolina
石灰硫黄合剂 lime sulphur (solution)
石灰泥炭 limnic peat
石灰熔合分析法 lime ignition method; lime-fusion method
石灰熔融法 lime-fusion method
石灰石 limestone
石灰松香 limed rosin
石灰性母质(黄土起源的) ergeron
石灰性黏土 calcium clay
石灰性土壤 calcareous soil; limy soil
石灰岩 limestone
石灰岩红色土 limestone red loam
石灰岩性植物群落 calcipetrile
石灰硬磐 lime hardpan
石灰质 calcic
石灰质的 calcareous
石灰质土 calcareous soil
石家具 stone furniture
石锯 rock saw
石坑 quarry
石块 quarry stone
石块耙集机 rock rake
石蜡胶 paraffin size
石蜡浸注材 paramn-impregnated wood
石蜡切片 paraffin section
石蜡乳液 paraffin emulsion
石蜡油 paraffin(ic) oil
石蜡油干燥法 drying in paraffin(ic) oil
石蜡纸杯 paraffined paper cup
石栗仁油 lumbang oil
石林风景 rockery and forest landscape
石榴单宁 pomegranate tannin
石榴皮 pomegranate peel
石榴色紫胶 garnet lac
石榴石(磨料) garnet
石笼 crib; gabion; quay
石笼坝 pier dam
石笼河绠(水栅) gabion boom
石卵 pebble
石棉 asbestos; asbestus; cotton asbestos
石棉板 asbestos board; asbestos panel; asbestos plate
石棉薄板 asbestos sheet
石棉壁板 asbestos wallboard
石棉粉 asbestine
石棉夹心胶合板 asbestos-veneer plywood
石棉绒 asbestos fibre
石棉水泥板 asbestos cement plate; asbestos cement sheeting
石棉纤维 asbestos fibre
石棉纤维板 asbestos fibreboard
石棉毡 asbestos blanket
石棉纸板 asbestos cardboard
石面的 epilithic
石面生物 epilithon
石漠 rock desert
石墨 mineral black
石南灌丛灰壤 heath podzol
石南灌丛泥炭 heath peat
石内生的 endolithic
石内植物 endolithophyte
石蕊[试剂] litmus
石生演替系列 lithosere
石生植物 chomophyte; lithophyte; petrophyte; rock plant; rupicolous plant
石生植物的 lithophytic
石塑 sculptured stone
石炭纪 Carboniferous period
石炭酸 phenol
石细胞 sclereid; sclerotic cell; stone cell
石隙植物 chasmophyte
石窑 stone kiln
石英斑岩 quartz porphyry
石英粗面岩 quartz trachyte
石英岩 quartzite
石油发动机 petroleum engine; petroleum moter
石油乳剂(煤油乳状液) kerosene emulsion; petroleum emulsion
石油软化剂 oil extender
石质的 rocky; stony
石质地 stony land
石质冻原 rocky tundra
石质荒漠群落 rupideserta
石质沙漠 hammada
石质土 chisley soil; lithosol; litho-soil; stony soil
石质棕色钙质硅铝土 litho-morphic brownish calsiallitic soil
石竹园 dianthus garden

S

时价 current ruling price; ruling price
时间表 schedule
时间尺度 temporal scale; time scale
时间度标 timescale
时间隔离 allochronic isolation; temporal isolation
时间基准表 time schedule
时间间隔 time-lag
时间间隔计 intervalometer
时间间隔遥感技术 technique of time-lapse remote sensing
时间鉴别分析 time discriminate analysis
时间量程 timescale
时间序列 time series
时间序列分析 time series analysis
时间障碍 time barrier
时间秩序 time ordering
时间滞后 time-lag
时间周期控制器 time cycle controller
时空变化 spatiotemporal variability
时空的 spatio-temporal
时期 stage
时速控制标准 hour-control standard
时序分析 time series analysis
识别 identification; recognize
识别号 identification number
识别检索 identification key
识别卡 identification card
识别种 differential species
实测 main survey
实测干缩率 observed shrinkage
实齿 solid tooth
实齿锯 solid tooth saw
实齿圆锯 solid tooth circular saw
实得 yield
实得工资 take-home pay
实地测量 field measurement
实地调查 field survey
实堆 close piling
实堆法 block stacking
实堆积 solid piling
实干(漆膜) dry to handle
实行照查(稽核) control of prescriptions
实积(木材) solid volume; solid content; solid cubic content
实积的 block-piled; bulked down; close-plied; dead-piled; solid-piled
实积立方米 festmeter(fm); fm (festmeter); solid meter
实积量度 true volume measure
实积量法 solid measure
实积形数 reducing factor
实际材积 actual volume
实际材积检尺 round measure; roundwood equivalent
实际采伐量 actual cut
实际产量 actual yield
实际成本 actual cost
实际出生率 realized natality
实际发芽率 practical germination percent
实际放牧度 actual use; actual utilization; actual incidence
实际供给量 actual supply
实际立木价格 actual stumpage price
实际利率 effective (interest) rate
实际流量 actual discharge
实际密度 actual density
实际面积 actual area
实际年龄 chronological age
实际排量 positive displacement
实际容量 actual volume; practical capacity
实际生产能力 practical capacity
实际生态位 realized niche
实际使用 actual use; actual utilization
实际收获 actual cut; actual yield
实际收获表 actual yield table; real yield-table
实际收入 real income
实际死亡率 realized mortality
实际误差 actual error
实际效率 actual efficiency
实际蓄积量 actual growing-stock
实际应力 actual stress
实际余额效应论 real-balance-effect theory
实际蒸散 actual evapotranspiration
实际装载能力 operating capacity
实际资产 tangible assets
实楞 close pile; solid pile
实木产品 solid wood products
实木家具 solid wood furniture

实木胶合材 solid glued-up stock
实片英尺 super-foot true
实生矮林 seedling coppice
实生林 high forest; seed forest; seedling crop; seedling stand
实生苗 seedling; seedling-plant; tree seedling; wild seedling
实生苗龄 seedling age
实生树 seedling
实生植株 seedling-plant
实生种子园 seedling seed orchard
实施计划 action program
实施记录簿 execution record
实时 real time
实收款项 proceeds
实体 solid
实体短木段 solid billet
实体堆积的 block-piled; close-plied; dead-piled; solid-piled
实体改性木材 solid modified wood
实体基线 stereoscopic base
实体木材 solid wood
实体提升式闸门 solid lift gate
实物单位 physical units
实物股息 dividend in kind
实物市场 physical market
实物税 tax in kind
实物资本 real capital
实物资产 physical assets
实习林 experiment forest; demonstration forest; instruction forest; training forest; university forest; college forest
实心窗格 blind tracery
实心的 solid
实心地板 solid floor
实心髓 solid pith
实心子弹 solid bullet
实验单元 experimental unit
实验林 demonstration forest; institutional forest; instruction forest; training forest; university forest
实验林场 forest experiment station
实验流域 experimental basin
实验区(保护区区划) experiment zone
实验生态学 experimental ecology
实验室打浆机 laboratory beater
实验室发芽率 laboratory germination percentage
实验室腐朽试验 laboratory decay test
实验室搅拌器 laboratory beater
实验小区 experimental plot
实验形数 experimental form factor
实验植被 experimental vegetation
实验种群 experimental population
实业模拟博弈 simulation games for business
实用防火装置 practical fire precaution
实用家具 utility furniture
实证经济学 positive economics
实质 essence
实质性休眠 constitutive dormancy
实重 absolute weight
拾物犬 retriever
食草的 herbivorous
食草动物 grazing animal; herbivore; herbivorous animal
食草性生物 herbivore
食虫的 entomophagous; entomophilous; insectivorous
食虫寄生 entomophagous parasite
食虫昆虫 entophagous insect
食虫性 entomophagous; entomophagy
食虫植物 carnivore; carnivorous plant; insectivorous plant
食橱 cupboard; kitchen cabinet
食橱桌 livery board
食道 oesophafus
食道神经节 oesophageal ganglion
食底泥动物 deposit feeder
食底栖生物者 benthos feeder
食动物的 zoophagous
食饵诱捕法 baiting method
食粪的 saprophagous
食粪性 coprophagy
食浮游生物的 planktotrophic
食浮游生物者 planktophile
食腐的 rypophagous; saprophagous
食腐动物 saprophagous animal; scavenger
食腐生物 saprophagous organism
食腐性 saprophagy
食腐者 saprovore

食腐殖质的生物 humivore
食腐质者 detritivore
食管上神经节 supraoesophageal ganglion
食管下神经节 suboesophageal ganglion
食果的 frugivorous
食果动物 frugivorous animal
食果性 fruit-eating
食具柜 dresser
食菌的 fungivorous; mycetophagous
食菌性 mycetophagy
食客 trophobiont
食昆虫的 entomophagous
食料植物 food plant
食蜜的 melliphagous
食木的 lignivorous; xylophagous
食泥动物 ilyotrophe
食品风味 food flavor
食品添加剂 food additive
食品香料 food flavoring
食品香味 food flavor
食肉的 carnivor(o)us; sarcophagous; zoophagous
食肉动物 carnivore; sarcophaga
食肉马 bird of prey
食肉生物 carnivore
食肉性 sarcophagy
食肉植物 carnivorous plant
食尸 necrophagy
食尸的 necrophag(o)us
食死地被物动物 litter feeder
食碎屑者 detritivore
食土的 geophagous
食土者 geovore
食微粒性 microphagous
食微生物动物 microbivore
食微小生物的 microphagous
食物的 trophic
食物基地 food base
食物链 food chain
食物链结构 food chain structure
食物生态位 food niche
食物网 food web
食物循环 food cycle
食屑生物 detritus feeder
食性 feeding habit; food habit
食性层次 trophic level
食血的 haemophagous
食叶昆虫 defoliator
食用成熟度 edible maturity
食用酵母 food-yeast
食用菌 edible fungus
食用香料 flavour; flavorant
食鱼动物 ichthyovorous animal; piscivore
食藻动物 algivore
食真菌动物 fungivore
食植动物 herbivorous animal; phytophaga; phytophagous animal
食植物的 phytophagous
食植物动物 plant eater
蚀刻剂 etchant
土壤流失量 soil loss amount
史迹园林 historical garden
史前归化植物 prehistoric naturalized plant
史前期栽培植物 archaeophyte
史前时代 prehistoric age
史前植物 prehistoric plant
矢尖蚧 yanon scale; arrowhead scale
矢量数据 vector data
矢状的 sagittal
使坚固 bind
使昆虫生病的 entomopathogenic
使失效 neutralization
使用[地]点 end-use point
使用安全 user safety
使用费(付给林地所有者) royalty
使用负载 working load
使用记录 service record
使用年限 serviceable life; work life
使用年限幅度 span of useful life
使用期 service life; service time
使用期试验 end-use test
使用期限 life span
使用权境界 boundary of servitude
使用试验 end-use test; service test; service trial
使用寿命 active life; usable life; work life; working life
使用税 use tax
使用特性 operational performance
始爆器 primer
始成土(美国土壤分类) inceptisol
始发站台 platform starting

始新世 Eocene
示波法 oscillography
示波器 oscillograph; oscillometer
示范林 demonstration forest
示范区 demonstration plot
示构式 rational formula
示性式 rational formula
示意图 schematic drawing; sketch map
示踪 tracking
示踪弹 tracer shot
示踪分析 tracer analysis
示踪同位素 isotopic tracer
示踪物 isotopic tracer; tracer
示踪原子 tagged atom
世代 generation
世代促进法 rapid generation scheme
世代交替 alloiobiogenesis; alternation of generations; digenesis; heterogony; metagenesis; metagenetic cycle
世代交替中的有性世代 gamobium
世代时间 generation time
世代重叠 overlapping generations
世界(广布)种 cosmopolitan (species)
世界时 universal time
世界野生动物基金会 World Wildlife Fund
世界种 cosmopolite; cosmopolitic species
世界种植种 cosmopolitan plant
世袭林 entailed forest
世袭林财产 entailed forest property
市场 market house; market place
市场呆滞 heavy market
市场的 marketable
市场的分散 dispersion in marketing
市场份额 market share
市场概况 market profile
市场工资 market wage
市场供应量 apparent availability
市场供应能力 marketing efficiency
市场汇率 market rate
市场激励机制 market-based incentive
市场价格 market price
市场经济 market economy
市场利率 market rate
市场潜力 market potential
市场渗透 market penetration
市场影响 market impact; marketing weight
市场占有率 market share
市场障碍 market barrier
市价 current ruling price; market value; ruling price
市面价值 market value
市内公园 city park
市区[面积] urban area
市有林 municipal forest
势 potential
势能 potential energy
事故 emergence
事故保险 accident insurance
事故停车 accidental shut down
事业 enterprise
事业创办者 entrepreneur
事业区 circle; forest circle; territorial circle
事业区林长 conservator
饰变 modification; modificational variation
饰面 facing
饰面烙纹 poker-picture
饰面压机 finishing press
饰皿(园林) ornamental vessel
饰条(家具) projection
试材 test(ing) piece
试车 break-in; run
试错学习 trial and error learning
试管内 in vitro
试管培养 tube culture
试行标准 tentative standard
试件 test(ing) piece
试区 experimental site
试销 approval sale; memorandum sale
试验板 sample board
试验场 experimental area
试验程序 test(ing) procedure
试验地 experimental site; study plot
试验点 field plot
试验方法 test(ing) procedure
试验分类学 experimental taxonomy
试验工厂 pilot plant
试验火 test fire
试验机 testing machine; torsional strength tester

试验机测量头 testing machine head
试验林 demonstration forest; instruction forest; pilot forest; test plantation; training forest; university forest
试验轮伐 test rotation
试验区 experimental area; experimental site
试验设计 experiment design
试验设计法 method of experimental design
试验生态系统(小型) microcosm; microcosmos
试验台 test bay
试验台试验 bench run
试验误差 experlmental error
试验小区 experimental plot
试验杂交 test cross
试验站 experiment-station
试验作业区 parcel
试样 sample
试用枪 try-gun
试运转 break-in
视差 parallax
视差测高法 parallax method
视差测高方程式 parallax height equation
视差测微器 parallax micrometer
视差杆(视差尺) parallax bar
视差较 difference parallax; parallax difference
视差水系(型) parallax drainage pattern
视差楔 parallax wedge
视场 angle of view; field of view
视场角 angle of field
视程 visual range
视高 apparent height
视角 angle of view; angle of vision; visual angle; viewing angle
视觉的 optic
视觉敏锐度 visual acuity
视觉器官 visual organ
视界 visual field
视景弱点 visual vulnerability
视距 apparent distance; visibility distance
视距标尺 stadia rod
视距测定法 tacheometry
视距测量 stadia surveying; tachymeter-survey
视距测树仪 range finder dendrometer
视距经纬仪 tachymeter-transit
视距丝 stadia hairs; stadia wire
视频 video frequency
视频磁带录像机 video tape recorder
视频数字转换器 video digitizer
视体积 apparent volume
视线 line of vision
视野 field of view; visual field
视野辖域 scope
视重量 apparent weight
柿涩 crude persimmon tannin
柿涩酚 shibuol
柿油 persimmon oil
适冰雪的 cryophilous
适氮植被 nitrophilous vegetation
适氮植物 nitrophile; nitrophilous plant; nitrophyte
适当采伐 optimum cut
适地适树 right tree on right site; matching species with the site
适度放牧 conservative grazing
适度放牧量 proper stocking
适度干燥(法) moderate drying
适度利用系数 proper use factor
适度疏伐 moderate thinning
适度疏开林冠 moderate opening of canopy
适腐的 saprophilous; saprophytic
适腐性 saprophile
适钙植物 calciphile; calciphilous plant
适耕性 ploughing profitability
适硅植物 silicicolous plant
适寒植物 hekistotherm
适旱变态的 xeroplastic
适旱的 xeric; xerophil(e); xerophilous
适合度 fitness
适荒漠的 eremophilous
适季植物 tropophyte
适碱的 basiphil(ous)
适碱植被 basophilous vegetation
适碱植物 alkaline plant
适流性植物 rheophilic vegetation

S

适配温度(热压时温度与压力、时间的上升配合) maturing temperature
适泉的 crenophilous
适沙性的 psammophilic
适沙植物 psammophile
适酸的 oxyphilic; oxyphilous
适酸植物 oxylophyte; oxyphyte; oxyphile
适温的 mesothermophilous; thermophilous
适温生物 thermophilous organism
适压的 barophilic
适压细菌 barophilic bacteria
适盐生物 halobion
适盐植物 halophile; halophyte
适宜采伐量 optimum cut
适宜采伐期 optimum cut period
适宜含水率 suitable moisture content
适宜湿度 preferential wetting
适宜种 preferent
适蚁植物 ant plant; myrmecophilous plant
适阴植物 sciophile
适印性 printability
适应 accommodation; adaptation
适应变异性 adequate variability
适应补偿 adaptive compensation
适应不当 maladaptation
适应反应 adaptation reaction
适应分化 adaptive differentiation
适应辐射 adaptive radiation
适应复合体 adaptive complex
适应高峰 adaptive peak
适应规范 adaptation norm
适应合宜性 adaptive fitness
适应环境(驯化) acclimation
适应火的 fire-adapted
适应阶段 adaptive phase
适应进化 adaptive evolution
适应类型 adaptive form
适应酶 adaptive enzyme
适应耐性 adaptation tolerance
适应能力 adaptability; adaptive capacity; adaptive faculty
适应气候 climatize
适应迁移 adaptive migration
适应色 adaptive colouration
适应特性 adaptive character
适应型 ecad; epharmone
适应性 adaptability; adaptivity; fitness
适应性变异 adaptive variation
适应性成本 adaptation cost
适应性评估 adaptation assessment
适应性趋同 adaptive convergence
适应性试验 compatibility testing
适应性状发生 adaptiogenesis
适应选择 adaptive selection
适应值 adaptive value
适应综合征 general adaptation syndrome
适用期 pot life; shelf life; usable life; working life
适用性 adaptivity
适用于机器的语言 machine oriented language
适于耕种的 arable
适于栖息的 insessorial
适雨植物 ombrophile; ombrophyte
适者生存 survival of the fittest
适中立木度 adequate stocking; optimum stocking
适中疏密度 adequate stocking
室(箱体房间等) chamber; compartment
室(子房或花药等) locule
室背开裂 loculicidal dehiscence
室背开裂的 loculicidal
室间开裂的 septicidal
室内播种 indoor seeding
室内花卉装饰 indoor flower decoration
室内花园 house garden; room garden
室内家具 indoor furniture
室内嫁接 indoor grafting
室内接 bench grafting
室内绿化 indoor planting
室内气干 indoor seasoning
室内气候 indoor weather; kryptoclimate
室内人造气候学 conditional climatology
室内庭园 indoor garden
室内小气候 cryptoclimate
室内用胶 interior glue

室内用胶合板 interior plywood; plywood for internal use
室内用清漆 interior varnish
室内园林 indoor garden
室内园林装饰 interior landscape
室内植物 house plant; indoor plant
室内装饰 interior decoration
室内装饰板 interior finish board
室内装饰品 upholstery
室内装饰商 upholsterer
室内装饰业 upholstery
室内装修 interior trim
室外家具 outdoor furniture
室外嫁接 field grafting
室外抗白蚁试验 termite field trail
室外气干 open air seasoning; outdoor seasoning
室外试验 exposure test
室外梯级 perron
室外用胶合板 exterior plywood; plywood for external use; weatherproof plywood
室温 ambient temperature; room temperature
室温固化胶黏剂 room temperature setting adhesive
室轴开裂的 septifragal
释放剂 releasing agent
释放因子 releaser
嗜光的 photophilous
嗜寒生物 psychrophilic organism
嗜火性菌类 fireplace fungus
嗜流性植被 rheophilic vegetation
嗜热的 thermophilic
嗜热生物 thermophile
嗜酸的 acidophilous
嗜温生物 thermophilic organism
嗜旋光性浮游生物 panteplankton
嗜雪植物 chianophile
嗜压的 barophilic
嗜盐微生物 halophile
嗜阴植物 shade-requiring plant
噬菌体 bacteriophage; phage
噬菌体的 bacteriophagic
噬菌细胞 phagocyte
噬菌现象 bacteriophagia
噬菌性的 bacteriophagic
噬菌性溶解 bacteriophagic lysis
噬菌作用 phagocytosis
噬真菌体 mycophage
螫针 sting
收兑 call
收额 crop
收费 charge
收费表 tariff
收割年 crop year
收购土地 acquisition of land
收回(对森林特许权) extinguish
收回成本 cost-recovering
收回液 released liquor
收回溢价(贴水) call premium
收获 crop; produce; yield
收获保险 harvesting insurance
收获表 yield table
收获表样地(编表用) yield(-table sample) plot
收获查定 assessment of yield; yield estimation
收获成熟度 harvest maturity
收获次序 temporal order
收获地域(地点)顺序 spatial order
收获法 harvest method
收获费用 harvesting costs
收获公式 yield formula
收获规整(正) cutting control
收获函数 yield function
收获函数估计值 yield function estimate
收获级 yield class
收获价 yield-value
收获控制或管制 cutting control
收获量 harvest volume
收获量测定法 determination of yield
收获量估计 yield estimate
收获量调节 yield regulation
收获率 harvest percent; using percentage; utilization percent; yield percent
收获模型 yield model
收获前喷药 preharvest spray
收获曲线 yield curve
收获时序 temporal order
收获树 final crop tree; future tree
收获水平 yield level
收获税 yield tax
收获损失 harvest loss; harvest waste

收获调整 budget regulation; yield regulation
收获调整法 method of yield regulation; methods of yield regulation
收获预测 yield forecast; yield prediction
收获预定 yield determination
收获周期 harvesting cycle
收集采伐剩余物 chunk up
收集剂(器) collector
收集者 gatherer
收紧排链 cross link a chain
收紧索 cinch line; cinch rope
收卷钢索 pull
收利率(林分) rate earned
收敛 convergence
收敛剂 astringent
收漂工 pick-up man
收漂工程 booming engineering; holding project
收漂及水上选材费用 boomage
收漂区 holding area; holding ground
收漂作业 holding work
收容埂(木材分类后) pocket boom; holding; holding boom
收入 proceeds; revenue
收入分配 income distribution
收入价 income value
收入轮伐期 income rotation
收入速度 income velocity
收入周期 income cycle
收入-资本比 income-capital ratio
收缩 cissing; contraction; shrink; shrinkage
收缩比 shrinkage ratio
收缩的 astringent
收缩根 contractile root
收缩开裂 shrinkage crack
收缩量 shrinkage mass
收缩率 percentage shrinkage; shrinkage in percent
收缩系数 coefficient of contraction
收缩应变 shrinkage strain
收缩应力 shrinkage stress
收索 winding
收炭率 yield of charcoal
收益 gaining; return; revenue
收益表 preceding table
收益递减 diminishing returns
收益递减规律 law of diminishing returns
收益矩阵 pay off matrix
收益循环 revenue recycling
收益支出营业支出 revenue expenditure
收支表 charge-and-discharge statement
收支两抵 break-even
收脂 dipping
收脂工 dipper
手扳葫芦 leverblock
手扳千斤顶 hercules
手柄 hand lever; spoke
手播器 hand seeder
手铲 dibble
手抄纸 handsheet
手持动力锯除伐 motor-manual cleaning
手持反射水准器 hand reflected level
手持开沟器 hand trencher
手持肘钉机 portable stapling machine
手传动 hand-power
手电钻 hand drill
手动 hand-power
手动杠杆 hand lever
手动夹钳 hand clamp
手动剪板机 manual clippers
手动绞盘机 hand-operated winch
手动螺杆 handscrew
手动螺旋夹 hand screw clamp
手动喷雾器 hand sprayer
手动起重器 handscrew
手段 instrument
手扶园艺拖拉机 autogardener
手斧 hand axe; hatchet
手工剥皮 manual debarking
手工采集 manual collecting
手工操作带锯整形机 band saw hand-operated shaper
手工除草 hand weeding
手工伐木二人小组 hand set
手工具 hand tool
手工刻痕器 hand incisor
手锯 arm saw; hand-saw
手磨机 hand mill
手刨 hand plane

手枪式把手(步枪与鸟枪柄) pistol grip
手砂线(家具生产) hand sander line
手提电刨 electric hand plane
手提钉钉机 portable nailing machine
手提式打眼机 portable mortising machine
手提式电动镂铣机 portable electric router
手提式静电喷枪 electrostatic hand gun
手提式磨刀器 portable knife sharpener
手提式喷雾机 portable sprayer
手提式喷雾器 portable spraying unit
手推播种车 seed barrow
手推车 bogie; buggy; hand cart; perambulator; push car; trolley
手推车挂车(集材) cart
手推除草铲 push hoe
手推倾卸车 tipping cart
手推碎木机 hand-feed grinder
手推中耕器 push hoe
手型分子 chiral molecule
手性 chirarity
手续 process
手续费 fee; service charge
手压块砂光机 hand block sander
手压式单滚筒砂光机 open drum sander
手摇干湿表 sling psychrometer
手摇绞车 hand winch; hand-operated winch
手摇绞盘机 manual cable winch
手摇切片机 rotary microtome
手摇钻 hand drill; twist gimlet
手用狭条锯 hand jig
手征性 chirality
手制动器 hand brake
手钻 gimlet; hand drill
守车(森铁) hack
首次采伐 first cutting
首次稳定相 initial stational phase
首付 down payment
首木(串集时) lead log
首木卡钩 head(log) grab
寿命 duration of life; life span; lifespan; lifetime; longevity
受保护树 preserved tree; preserved wood
受光伐 increment cutting; interlucation; liberation cutting; liberation felling; light cutting; light felling; opening of the shelter wood; overhead release felling; removal felling; removal cutting
受光伐作业 open wood system; open-stand system
受光器 photoreceptor
受光生长 increment due to increased light; light increment; open-stand increment
受光生长量 light increment
受光疏伐 opening of the shelter wood
受光体 photoreceptor
受害材 affected timber; attacked timber; infected timber
受害树 damaged tree
受剪面积 shear area
受精 fecundation; fertilization; impregnation; receptivity
受精的 gamic
受精卵 cytula; oosperm; oospore; zygote
受精卵核 cytococcus
受精囊 seminal receptacle; spermatheca
受精丝 trichogyne
受精作用 amphigamy; fertilization
受款人 beneficiary; payee
受拉材(木材缺陷) tension wood
受拉构件 tension member
受力外板 stressed skin panel
受料工 tailer
受纳器 receptor
受器 acceptor
受热变形 thermal strain
受水面积 catchment area
受体 acceptor; receptor
受土壤条件影响的 edaphic
受压材(木材缺陷) compression wood
受压构件 compression member
受压面积 compression area
受压木 subordinate
受益人 beneficiary
受益值 benefit value

受孕 fecundation; impregnation
受载荷的承载索 skyline loaded
受折磨的 lacerate
受脂器(松脂采集) cup; tin
受脂器安装工 tin nailer
狩猎 game; hunting; shooting
狩猎场 game area; game land; hunting-field; hunting-ground
狩猎动物 game animal
狩猎动物保护者 game protector
狩猎动物管理员 game manager
狩猎法 game law
狩猎管理 game management
狩猎纪念物 trophy
狩猎家 sportsman
狩猎警察(监督员) game warden
狩猎牧场 game pasture
狩猎期 open season
狩猎权 hunting-right
狩猎营区 hunting camp
狩猎苑 game park
狩猎者 hunter; huntsman
狩猎证 game-license; hunting licence; shooting license; licence
狩猎执照 game-licen; hunting-license
授粉[作用] pollination
授粉树 pollenizer; pollinizer
授粉者 pollina(tor)
授精 impregnation
授权银行 authorized bank
授乳 lactation
售后成本 after cost
售后租回 sale and leaseback
售货收入款 sales proceeds
兽害 tree damage caused by forest fauna
兽媒 zoophily
兽炭 animal black; animal charcoal
兽炭黑 animal black; animal charcoal
兽穴 lodge
兽疫学 epizootiology
瘦果 achene; capsella
瘦黏土 lean clay
瘦土 exhausted soil; infertile soil
瘦长的 leggy
书橱 bookcase
书籍 liber
书架 book rack
书皮纸板 millboard
书箱 bookcase
书写扶手椅 tablet(-arm) chair
书页式拼板法 book matching
书桌 secretaire; secretary; writing desk
枢纽 headwork
枢轴 heel; pin
梳 comb
梳齿花纹 comb-grained
梳齿接 combed joint; corner locked joint
梳齿形应力检验片 prong
梳辊 rake roller
梳理 grooming; preening
梳膜(鸟眼) pecten
梳取 stripping
梳形背椅 comb back chair
梳摘 stripping
梳妆台 chamber table; toilet table
梳妆椅 toilet chair
梳状刀 comb knife
梳状构造(地质) comb structure
舒尔茨离析法 Schultze's maceration method
舒适性资源 amenity resources
舒适指数(对环境舒适感的指标) comfort index
疏播 thin sowing
疏布(磨粒占背衬面积60%~70%) open coat
疏堆 open stacking
疏伐 thinning
A度疏伐 A-grade thinning
C度疏伐 C-grade thinning
E度疏伐 E-grade thinning
疏伐措施 thinning treatment
疏伐打枝造材机 thinning processor
疏伐等级 thinning grade
疏伐度 thinning rate
疏伐法 thinning method; thinning system; type of thinning
疏伐方式 thinning regime
疏伐后的林分 stand after thinning
疏伐基准表 thinning schedule; thinning table
疏伐计划 thinning plan

疏伐间隔期 thinning cycle; thinning interval
疏伐绞盘机 thinning winch
疏伐经济学 economics of thinning
疏伐类型 thinning type
疏伐量 thinning weight
疏伐率 rate of thinning; thinning rate
疏伐频度 thinning frequency
疏伐期 thinning period
疏伐强度 degree of thinning; intensity of thinning; thinning intensity; thinning weight
疏伐收获 thining yield; yield of thinning
疏伐收入 financial returns from thinnings
疏伐收益 return from thinning
疏伐体系 thinning regime
疏伐系统 thinning system
疏伐型 thinning method; type of thinning
疏伐序列 thinning series
疏伐预测 thinning forecast
疏伐周期 thinning cycle; thinning periodicity
疏伐作业 thinning system
疏伐作业过程表 thinning schedule
疏冠树 open-crowned tree
疏果 thinning fruit
疏果剂 fruit thinning agent
疏解(纤维) fluff
疏解机 fluffer; slusher
疏开(林冠) opening up; light open; light; open out
疏开的 roomy
疏开伐 severance felling
疏林 open crop; open forest; open stand; parkland; under stocking; sparse stand
疏林地 open forest land; thin stocked land; sparse stand
疏林干草原 steppe savanna
疏密度 degree of stocking; crop density
疏密适中的 well stocked
疏绵状毛的 lanuginose
疏苗 singling
疏苗机 gapping machine; seedbed thinner
疏苗机铲 thinner blade
疏苗器 singler
疏木的 manoxylic
疏散辊 picked roller
疏上胶层 open coat
疏水的 hydrophobe; hydrophobic
疏水胶体 hydrophobe; hydrophobic colloid
疏水器 steam trap; trap
疏水效应 hydrophobic effect
疏丝组织(真菌) prosenchyma
疏松母质层 regolith
疏松土壤 loosening of the soil
疏透度 openess
疏涂层 open coat
疏液的 lyophobic
疏液胶体 lyophobe sol; lyophobic colloid
疏郁闭的 undercrowded
疏枝 lighting of tree
疏枝度 cleanness of bole
疏脂的 lipophobic
疏植 thin planting
输出(计算机) output
输出端引出线 outlet
输出功率 delivered power
输出量 output; retirement area; work output
输出软管 delivery hose
输出许可证申请书 application for export permit
输出压力 delivery pressure
输导束(用于苔藓植物) conducting bundle
输导细胞 transfusion cell
输导组织 conductive tissue; transfusion tissue
输导组织束 transfusion strand
输浆管道 stock line
输卵管 oviduct
输入 load
输入程序 loader
输入模块 loader module
输入器 loader
输入-输出控制系统 input-output control system
输入转换器 input translator

输沙量 sediment discharge; sediment load; sediment yield; silt discharge
输沙率 sediment flux
输送(液)管 spout
输送带 apron
输送垫板 transport caul
输送管 duct
输送管道 suction pipe line
输送机 conveyer; conveyor
输送链 transport chain
输送链钩 bunk
输送装置 feedway
输纸辊 feed roller
输种管 drop tube
赎回溢价 call premium
熟材 imperfect heart wood
熟材树 imperfect heartwood tree; tree with imperfect heartwood
熟地 fermented soil
熟化 curing
熟化剂 curing agent
熟化期 curing time
熟荒地 abandoned land
熟练伐木工 catty logger
熟练流送工(尤指能在急流中拆垛的) white water man
熟练流送工人 log watch
熟土 fermented soil; mellow earth
熟油 boiled oil
属 genus (*plur.* genera)
属间嫁接 intergeneric grafting
属间杂交 bigeneric cross; genus cross
属间杂种 bigener; bigeneric
属模标本 genotype
属全模标本 genosyntype
属完模标本 genoholotype
属性 attribute
属性抽样 attribute sampling
属选模标本 genolectotype
鼠洞式孔 gopher hole
鼠害 damage from rats
鼠尾草酚 salviol
鼠尾草烯 salvene
薯蓣 Chinese yam
薯蓣粉 tapioca flour
曙红 eosin
束 fascicle; pack; package; strand
束的 fascicular
束缚 bind
束缚水 bound moisture; bound water; combined water
束间木质部 interfascicular xylem
束间区域 interfascicular region
束间韧皮部 interfascicular phloem
束间射线 interfascicular ray
束间髓线 interfascicular medullary ray
束间形成层 interfascicular cambium
束生的 fascicular
束水坝 pier dam; wing dam
束水堤 side pier
束丝 coremium (*plur.* -ia); synnema (*plur.* -ata)
束外形成层 extra-fascicular cambium
束腰板 frieze panel
束中木质部 fascicular xylem
束中形成层 fascicular cambium
束状薄壁组织 strand parenchyma
束状的 fasciculate
束状管胞 strand tracheid
树疤 burl; burr; knag; knob; lump
树病学 dendropathology; tree pathology
树层 tree layer
树杈 fork
树杈单板 crotch veneer
树杈状花纹 crotch figure; crotch swirl
树茬 crest
树槎 stump
树丛 boscage; boskage; bosk; bosket; bosquet; clump; grove
树倒方向 lay; lie
树倒位置 lay
树的 arboreal
树顶 apex; treetop
树顶端 tree tip
树蜂 horntail; wood wasp; horn tail
树干 shaft; stem; tree stem; trunk
树干材 stem timber; stemwood
树干材积 stem volume
树干断面积 cross-sectional area
树干伐点 felling point
树干分布 DBH distribution

S

树干附生苔藓 bark inhabiting bryophyte
树干火 stem-fire; fire in stem
树干基部直径 base diameter
树干级 stem class; stem quality class
树干级木 stem class
树干解析 stem analysis; tree analysis
树干理论 tree-stem theory
树干萌芽 stem sprout
树干剖面模型 stem profile models
树干倾斜 setm inclination; stem lean
树干曲线 stem curve
树干曲线方程 stem curve equation
树干上部环切(放炸药) crease
树干上部直径 upper-stem diameter
树干上着花 cauliflory
树干体积 tree stem volume
树干形成材 stem-formed wood
树干形数 bole form factor; stem form factor
树干形质 stem quality
树干形状 stem form
树干圆盘(查定树木生长量) disk; disc
树干长 stem length; timber length
树干直径 stem diameter
树干纵剖面 stem profile
树高 height of tree; stature; tree height
树高测定 height determination; height measurement; hypsometrical measurement
树高测杆 tree height measuring pole
树高抽样法 height sampling methods
树高分级 height classification
树高级 height class; height grade
树高级立木材积表 height class volume table
树高类型 type of tree height
树高曲线 height curve
树高曲线拟合 fitting height curves
树高生长 growth in elongation; growth in height; growth in length; height growth
树高生长量 height accretion; height increment; longitudinal increment
树高生长曲线 height age curve; height development curve; height growth curve
树高-直径曲线 height-diameter curve
树高指数 height index
树挂 air hoar; rime
树冠 crown; tree canopy; tree crown
树冠被覆 cover
树冠材 crown-formed wood
树冠测定器 crownmeter
树冠层 canopy; crown canopy; crown layer; overstorey; roof-canopy
树冠尺寸 crown dimension
树冠大小 crown dimension; crown size
树冠覆被度 crown coverage
树冠覆盖 canopy; crown cover; crown spread
树冠覆盖密度 density of crown cover
树冠高 crown height
树冠构造 crown structrue
树冠厚度 crown depth
树冠火 canopy fire; crown fire; crown(ing) fire
树冠级 crown class; tree crown class
树冠计数 crown count
树冠结构 crown structure; crown texture
树冠截留 crown interception
树冠截留损失量 crown interception loss
树冠竞争 crown competition
树冠竞争因子 crown competition factor
树冠枯萎 topkill
树冠轮廓 canopy silhouette
树冠率 crown percent (ratio); crown ratio
树冠密度 crown density
树冠面积 crown surface
树冠面积指数(立木冠长与冠径的乘积) crown-area index
树冠内光线 inside-light of tree crown
树冠平均直径 crown mean diameter
树冠烧焦 crown scorch
树冠生物量 aerial biomass
树冠疏枝 setting free the crown
树冠特征 crown characteristics
树冠投影 crown projection
树冠投影面积 crown projection area
树冠投影图 crown projection diagram
树冠投影制图 crown mapping

树冠狭小木 small-crowned tree
树冠形数 brush form factor
树冠形态 crown architecture
树冠修枝 crown pruning
树冠长 crown length
树冠直径测量楔 crown diameter wedge
树号 tree number
树基花纹 butt figure
树间中耕 inter-tree cultivation
树胶 gum; wood gum
树胶斑 gum spot; gum streak
树胶道 gum duct
树胶管 gum canal
树胶脉纹 gum vein
树胶囊 gum check; gum pocket
树胶腔 gum cavity
树胶糖浆 gum syrup
树胶条纹 gum streak
树胶脂 gum resin
树节 knob; knot
树径采脂法(按树径决定采脂剖面) diameter limit cupping
树径记录仪 dendrograph
树蕨 tree fern
树篱 espalier; hedge
树林 wood
树林带 woodland belt
树龄 tree age
树瘤材 burr wood; knaggy wood
树瘤状花纹 burl figure
树瘤状纹理 burl grain
树轮 tree annual ring
树木 tree; arbor
树木保护 protection of trees
树木材积 total tree volume
树木材积测量 tree volume measure
树木侧冲(伐木时) sideswipe
树木打号 identification number
树木打印(采伐标示) tree marking
树木倒向控制器 tree controller
树木分类 tree classification
树木粉碎机 tree crasher; tree crusher
树木腐朽 trunk rot
树木个体间竞争 inter-tree competition
树木航空材积表 tree aerial volume table
树木护罩喷雾器 tree-guard sprayer
树木基因库 tree (gene) bank
树木级 tree class
树木价格增长 growth of tree value
树木界限 forest boundary; tree limit
树木昆虫学 dendro-entomology
树木密度 density of trees
树木年代学 dendrochronology
树木年轮分析 tree ring analysis
树木年轮历 tree ring calendar
树木生理学 tree physiology
树木生长 arboreal growth
树木生长标准地 tree increment plot
树木生长自记图 dendrograph
树木提取物 tree extractive; wood extractive
树木挖掘机 tree bailing machine
树木外科术 tree surgery
树木线 tree line
树木像片材积表 tree photo volume table
树木形数 tree form factor
树木形态 tree architecture
树木修补 tree surgery
树木修复 tree repair
树木学 dendrology
树木学家 dendrologist
树木养护 tree maintenance
树木移植机 tree spade; tree transplanting machine
树木育种 tree breeding
树木园 arboretum; breeding arboretum; tree (gene) bank
树木栽培 arboriculture
树木栽培工作者 arboriculturist
树木栽培学 arboriculture
树木造型 topiary
树木造型艺术 topiary art
树木整形 tree work
树木重量表 tree weight table
树木注射 tree injection
树木状的 dendroid(al)
树棚 espalier
树皮 bark; cortex (*plur.* -ces); rind
树皮百分率 bark percent(age)
树皮斑点 bark speck
树皮采收权 right of bark
树皮测定器 bark gauge
树皮的 corticose

树皮反转 bark inversion
树皮粉碎机 bark grinding mill; bark hog
树皮规(测皮厚) bark gauge
树皮厚度 bark thickness
树皮厚度测定器 bark-thickness measurer
树皮厚度计 bark gauge
树皮浸提物 bark extract
树皮扣除 allowance for bark; bark deduction
树皮扣除率 allowance for bark; bark allowance
树皮率 percent of bark
树皮囊 bark pocket; trapped bark
树皮疱 bark blister
树皮破碎机 bark breaking machine; bark crusher
树皮切断机 bark cutter
树皮切碎机 bark shredder
树皮球(燃料) bark pellet
树皮鞣料 bark tanner; tanbark; tanning bark
树皮生长 bark growth
树皮生长量 bark increment
树皮苔藓 corticolous bryophyte
树皮条剥 bark stripping
树皮条剥机 bark stripping machine
树皮调整系数 bark adjustment factor
树皮脱离 bark slip(ping)
树皮系数 bark factor
树皮蛀虫 bark borer; bark miner
树皮灼伤 bark scorching
树栖的 arboreal; dendricola; dendrocola
树栖生境 arboreal habitat
树栖适应 arboreal adaptation
树墙 espalier
树杈 crotch
树群 group; group of trees
树上附生植物 arboreal epiphyte; epiphyta arboricola
树梢 apex; tip; top; treetop
树梢高 top height
树势 tree vigour
树图 diagram of tree
树下栽植 understorey management
树下种植 underplanting
树心劈裂 heart split
树形 shape of tree; tree form
树形不良林分 scrub (stand)
树形高 tree form height; volume-basal area ratio
树形高生长 form height increment
树形生长 form increment
树型级 stem quality class; tree class
树桠 limb
树叶饲料 leaf-fodder
树液 sap; tree sap
树液抽出干燥法 seasoning by leaching of wood
树液流动 circulation of sap
树液流动期 sap-season
树液酸蚀 sap acid-etching
树液置换处理(防腐) sap displacement (treatment)
树艺学 arboriculture
树瘿 wart
树枝(粗大的树枝) bough
树枝石 dendrolite
树枝状的 arborescent
树枝状排水系统 dendritic drainage
树枝状排水型 dendritic drainage pattern
树枝状水系 dendritic drainage
树脂 pitch; resin
树脂斑 resin gall; rosin speck; scab pitch pocket; scab resin pocket
树脂斑纹 pitch streak; resin streak
树脂板 resin plate
树脂包 resin bulb
树脂薄膜 resin film
树脂采集器 resin-tapper
树脂道 resin canal; resin duct; resin passage
树脂的 resinous; resinaceous
树脂多的 resinous; resinaceous
树脂覆盖率 resin coverage
树脂改性木材 resin modified wood
树脂管 pitch tube; resin tube
树脂管胞 resinous tracheid
树脂含量 resin content
树脂化 resinify
树脂化作用 resinification
树脂痂 scab pitch pocket; scab resin pocket
树脂胶合板 resin-bonded plywood
树脂胶黏剂 resin adhesive

树脂浸渍板坯 resinated mat
树脂浸渍压缩木 compreg; compregnated wood
树脂浸渍纸 resin-impregnated paper
树脂孔 pitch hole
树脂枯竭材(采脂) worked-out timber; tapped-out timber
树脂溃疡 pitch canker
树脂扩散率 resin spread rate
树脂裂 pitch shake
树脂裂缝 resin seam; resin shake
树脂瘤 resinous cancer
树脂囊 pitch pocket; resin cyst; resin gall; resin pocket
树脂黏锯 gum up
树脂漆 resin varnish
树脂腔 resin cavity
树脂缺陷 resin defect
树脂鞣 resin tanning
树脂鞣法 resin tannage
树脂伤害 resinous wound
树脂伤痕 pitch seam
树脂渗透 resin soak(ing); resinosis
树脂酸钙 limed rosin; calciumresinate
树脂填充 resin impregnation
树脂条痕 pitch seam; pitch streak
树脂条纹 resin streak
树脂系统 resiniferous system
树脂细胞 resin cell
树脂腺 resin gland
树脂瘿 resin gall
树脂皂 resin soap
树脂障碍(制浆时) pitch trouble
树脂转移(碎料间) resin transfer
树脂紫胶 gum-lac
树种 pecies; species of trees; tree species
树种变更 species conversion
树种变换(更) change of species
树种代码 species code
树种更替 change of species
树种混合率调节 control of mixture
树种混淆 mixture; species mixture
树种交替学说 theory of alternation of species
树种识别(鉴定或判读) species identification; tree species identification
树种选择 choice of tree species
树种组 species group
树种组成 species composition
树桩 stub; stub snag
树桩标号 stump mark
树桩盆景 miniature potted-tree; potted miniature tree
树桩挖掘机 stub puller; stump-puller
树状的 dendriform
树状图 tree
树状物 tree
树状植物 pfanerophytes
竖板机 upender
竖堆 end piling; end-way piling; endwise piling
竖堆法 end piling; end stacking; end-way piling; endwise piling
竖堆烧炭(法) stand heap charing
竖沟环割 frill girdling
竖沟切口 frill cut
竖井(坑) shaft
竖框梃 stile
竖立电杆 standing pole
竖木 shore
竖琴状的 lyrate
竖曲线 vertical curve
竖式蒸馏罐 vertical retort
竖线锯 jig saw
数度开花 perpetual flowering
数据 data
数据变换 data transfer
数据采集 data logging; data collecting; data acquisition
数据采集器 data logger
数据处理 data handling; data processing
数据处理辅助系统 data processing subsystem
数据处理机 data processing machine; datatron
数据处理设备 data processing equipment
数据处理中心 data processing centre
数据传送 data transfer
数据存储 data storage
数据简化 data reduction

数据库 data bank; data library
数据收集系统 data collection system
数据下载 data downloading
数据压缩 data compression
数据中心库 central data bank
数据转换 transformation of data
数均分子量 number-average molecular weight
数均聚合度 number-average degree of polymerization
数控进料装置 digital feed system
数理经济学 mathematical economics
数理气候 mathematical climate; solar climate
数理统计 mathematical statistics
数理统计学的样本分析 statistical analysis of inventory data
数量成熟期 quantitative maturity
数量分类学 numerical taxonomy
数量化 quantizing
数量级 order of magnitude
数量金字塔 pyramid of number
数量抗性 quantitative resistance
数量生态学 quantitative ecology
数量性状 quantitative character
数量遗传 quantitative inheritance
数量遗传学 quantitative genetics
数码盘式按厚度进料机构 dual thickness feed system
数式法 formula method; method of regulating yield (by comparing actua with normal crops)
数式平分法 formula method of periods by area and volume combine
数式原木板英尺材积表 mathematical log rule
数突变(染色体) numerical mutation
数学模型 mathematical model
数学生态学 mathematical ecology
数值反应 numerical response
数值分析 numerical analysis
数字表面模型 digital surface model (DSM)
数字传真系统 digital facsimile
数字磁带记录仪 digital tape recorder
数字地形模型 digital terrain mould
数字电子计算机 digital computer
数字高程模型 digital elevation model (DEM)
数字化 digitization
数字轮宽测定仪(年轮仪) digital positiometer
数字图像 digital image
数字图像处理 digital image processing
数字显示计算器 digital display counter
数字显示器 digital indicator
数字响应[值] digital response
数字转换器 digitizer
数组 array
刷光 brush polishing
刷辊(砂光机) clean brush; brush roll
刷漆 brushing lacquer
刷式去翅机 brush-type dewinger
刷涂机 brush spreader
刷型砂光机 abrasive brush
衰败 degeneration
衰变 ageing; decay
衰变产物 decay product
衰变期 decay period
衰变网图 decay scheme
衰化时间 aging time
衰减(图片褪色) fade
衰减(物理量) attenuation
衰减率 attenuance
衰减效应 depletion effect
衰减指数 attenuation index
衰竭 depletion
衰竭的 prostrate
衰竭色素 wear-and-tear pigment
衰竭土 senile soil
衰老 aging; doat; senesce; senescence
衰落区 blighted area
衰弱木 back log
衰退 depression; deterioration; recession; slump; decline
衰退变异性 fading variability
衰退病 decline disease
摔抛轮 stripping wheel
甩车卷筒(绞盘机上) car spotter
甩车索 snubbing line; spotting line
闩 bolt; lock
闩柄 bolt-handle

闩锁式连发鸟枪 bolt action repeating shotgun
栓 dowel
栓化 suberization
栓化纤维素 adipo-cellulose
栓化作用 suberification; suberization
栓孔锯 keyhole saw
栓内层 phelloderm
栓皮 cork; phellem; phellum; suber
栓皮粉 ground cork
栓塞锯 dowel saw; plug saw
栓质层 suberinlamella (*plur.* -llae)
双 binary
双凹面圆锯 hollow-ground saw
双扳机(鸟枪) double trigger
双瓣卧式闸门 bear-trap gate
双孢体 dyad
双倍体 amphiploid
双倍体生物 diploid organism
双壁犁 double mould board plow
双壁雪犁 V-plough
双边齿锯 double teeth saw
双边对称 bilateral symmetry
双边筒管 bobbin spool
双变量常态分配 bivariate normal distribution
双柄剥皮刀 timber shave
双柄锯 two-handled saw
双拨齿锯 double set saw; double swage saw
双波状纹 double wavy grain
双槽圆线脚 double quirk-head
双侧喷雾机 double-sided sprayer
双层采割法(采脂) two-storey tapping
双层床 bunk bed
双层干燥机 two-deck dryer
双层割面(采脂) two-storey face
双层柜 chest-on-chest
双层滚筒干燥机 double pass rotary dryer
双层膜 double membrane
双层输送机 two-deck conveyor
双层装板架 two level board pick-up
双叉树 double-topped tree
双车道 two-lane traffic way
双车立柱 double dray
双车头列车 double header
双翅的 bialatus
双翅果 double samara
双带锯[组]制材厂 double band (saw) mill
双带蛀虫 two lined borer
双单体 double monosomic
双刀架旋切机 turret veneer lathe
双刀刨 double plane iron
双刀片伐木剪 double-action shear
双道集材 two-way skidding
双电子键 doublet
双对流混合湖 dimictic lake
双盾形芽接 double-shield budding
双多倍体 amphipolyploid
双二倍体 amphidiploid
双二倍体组合 amphidiploid combination
双二元体 amphidiploid
双分子膜泡 cellule
双酚A(中间体) bisphenol A
双峰 doublet
双峰分布 bimodal distribution; bipolar distribution
双幅门 double margined door
双杆划线规 double bar gauge
双干材 double trunk
双干树 doub bole
双工组制材厂(两台主锯机) double mill
双蓇葖的 bifollicular
双鼓轮刨片机 knife ring flaker
双管式干燥机 double-dryer
双轨剪切试验 two-rail shear test
双辊铺装头 double roll dissolution head
双辊柱压尺 double roller bar
双滚筒涂胶机 double roller machine
双行播种 double-row planting; twin drilling
双行条播 double-row drilling
双行栽植 double-row planting
双行植树机 two-row planter
双核仁 amphinucleolus
双铧犁 two-furrow plough
双环的 dicyclic
双机牵引(上坡) double the hill
双基火药 double-base powder
双剂型和三剂型合剂 duplex and triplex mixture
双尖的 bicuspid
双尖装车钩 deer foot hook
双剪强度 double shearing strength

双渐尖的 biacuminate
双交 two-way cross
双交换 double crossing over
双交杂种 double hybrid
双绞拱 two-hinged arch
双节橇 double dray
双金属材料 bimetal
双鸠尾榫键 double dovetail key
双锯口伐木 match sawing
双锯面对开材 crowntree
双锯片裁分锯 double block edger
双聚伞花序 dicyme
双卷筒绞盘集材拖拉机 double-winch tractor
双卡轴 telescopic spindle
双卡轴旋切机 double spindle lathe; dual spindle lathe; peeling lathe with telescopic spindle
双开式旋转铰链 helical hinge
双口喷枪 two-component spray gun
双框格吊窗 double hung sash window
双扩散法 double-diffusion process
双联带锯机 tandem twin bandmill; twin band-saw
双联带锯制材厂 twin band (saw) mill
双联卷筒 tandem drum
双联控制温包 dual control bulb
双联鞔杆(畜力集材) stretcher
双联双刃带锯机 double cut twin band-mill
双梁爬犁 double-rack sled
双料室立式打磨机 two-chamber type vertical shaft mill
双列射线 biserial ray
双列杂交 diallel cross
双列栽植 two-row planting
双轮滑轮 saddle block
双轮集材车 bogie; bummer; drag
双轮索道跑车 double block
双轮运材车 lumber buggy
双轮运材拖车 drag cart
双马鞔绳镶木 double-tree
双萌发孔 ditreme
双面裁边机 double sizer
双面黏带 double-back tape
双面刨床 double surface planer; double sides planer; two-faced planing machine
双面砂光 sanded two sides
双面上胶 double sizing
双面凸镜形纹孔口 lenticular pit aperture
双面涂布机 two-side coater
双面涂胶 double spread(ing)
双面修整 side dressing
双名法 binomial nomenclature
双名制 binomial system
双木漂子 two-log boom
双目显微镜 binocular microscope
双年轮 double (annual) ring; double ring
双爬犁 double dray; jumbo; twin-sled; two sleigh; wag(g)on sled
双排锯制材厂 twin-frame (saw) mill
双蒎烯 dipinene
双盘精磨机 double revolving disc refiner; two disk refiner
双皮厚 double bark thickness
双片横截锯 double saw; two-saw trimmer
双片截锯 double cut-off saw
双起重索装车 duplex loading system
双气流式打磨机 double stream mill
双橇 double sledge
双切齿锯 peg-raker saw
双亲的后代检验 two-parent progeny test
双亲行为 parental behaviour
双亲后裔 biparental progeny
双亲遗传 amphilepsis
双亲中值 midparent
双氢刺槐亭 dihydrorobinetin
双曲拱桥 double-curvature arch bridge
双曲面 hyperboloid
双曲线 hyperbola
双曲形家具腿 French leg
双人锯 double-handed saw; two-man saw
双人课桌 double desk
双人椅 love seat
双人长片锯 pit saw; whip saw
双刃采种镰 double-hook detacher
双刃带锯 double-cutting band saw; double-cutting bands
双刃刀 double edge knife
双刃斧(刃口垂直) Pulaski

双刃再分带锯 double edge cutting band resaw
双韧维管束 bicollateral bundle
双三体生物 double trisomic
双舌槽接合 double match joint
双舌接 double-tongue graft
双嗜性的 amphophilic
双瘦果 achenodium
双竖沟环割 double-frill girdling
双髓心 double pith
双索导索器 double deck fairleader
双索式 two way system
双索抓钩 two-cord grapple; two-line grapple
双索装车[绞盘]机 duplex loader
双体气球 twin-hull balloon
双体双壁犁 two-bottom mouldboard plough
双萜 diterpene
双筒绞盘机 tandem drum
双筒鸟枪 double barrel shotgun
双筒望远镜 binocular
双头钉 double-headed nail
双头钢轨 chair rail
双头开榫机 double end tenoner; double end tenoning machine
双头宽带砂光机 two heads wide belt sander
双头垄断(两家卖主垄断) duopoly
双头喷枪 double headed spray gun
双头圆锯 twin circular saw
双凸面圆锯 double-conical saw; splitting saw; taper-ground saw
双凸腔(纹孔) biconvex cavity
双凸榫 double tenons
双腿爬犁 double sled; dry sloop
双拖拉机放树(用钢索拉倒小树) siamese cat
双维管束的 diploxylic
双涡卷形扶手 scroll-over arm
双戊烯 dipentene
双戊烯树脂 dipentene resin
双显性组合 duplex
双线架空索道 double-rope cableway; double-wire tramway; bicable aerial tramway
双线架孔索道 bicable system
双线缆车道 double-line incline-way
双线期 diplonema; diplotene stage
双线松紧式索道集材 double slackline yarding; duplex flyer yarding
双线索道 bicable; bicableway; double line system; two-way cableway
双线索道大滑轮 equalizing block
双线往复式索道 double standing line
双线性方程 bi-linear equation
双线循环索道 double endless cableway; Potlatch tram
双线循环索系 double endless system
双向风向标(水平及垂直两个方向) bivane; bi-directional vane
双向风信器 bi-directional vane
双向犁 alternate plough; reversible plough; two-way plough
双向木楔法 counter wedging
双向圆盘犁(英) one-way disk plough
双向纸色谱法 two-dimensional paper chromatography
双像摄影测量 double-image photogrammetry
双心 double heart; double pith
双心材 double heart-wood
双心皮的 bicarpellary
双悬果 cremocarp
双压齿锯 double set saw; double swage saw
双眼绠漂 rider
双眼射击 binocular shooting
双眼视觉 binocular vision
双摇门 double-acting door
双叶折叠腿圆桌 double gate leg table
双椅背 double back
双翼除草铲 duck foot
双因素法 two-factor method; two-way method
双因子杂合子 dihybrid
双因子杂交 diallel cross
双因子杂种 dihybrid
双因子直线方程 bi-linear equation
双雨季 birainy
双圆锯[组]制材厂 double circular (saw) mill
双圆锯裁边机 double edger; twin edger; two-saw edger
双圆网机 double cylinder machine
双圆线脚 double-bead

双杂合子 double heterozygote
双杂交 double cross
双爪链钩(用短链连接的) double-crotch grab
双爪抓钩 box hooks
双折面桌 butterfly table
双折射 birefringence; double refraction
双着丝点染色体 dicentric chromosome
双真空法 double vacuum process
双真空法循环周期 double vacuum cycle
双真空木材防腐处理法 vac-vac process
双支柱带锯机 double-column bandmill
双制材跑车 double carriage
双中心分布 bicentric distribution
双重泊松分布 double Poisson distribution
双重抽样 double sampling
双重抽样技术 double sampling technique
双重点抽样 double point sampling
双重红利 double dividend
双重回归抽样 double sampling with regression; regression double sampling
双重嫁接 double grafting; double-work
双重交配法 method of double crossing
双重课税 double taxation
双重连续扩散处理 double-diffusion treatment
双重谱线 doublet
双重侵染 dual infection
双重受精 double fertilization
双重休眠 double dormancy
双重增力索(通过一个动滑轮) double line
双周期的 dicyclic
双轴多锯片圆锯 double arbor gang
双轴木工铣床 double spindle shaper
双轴齐边圆排锯 double gang edger
双轴驱动 live tandem
双轴压缩 biaxial compression
双轴应力 biaxial stress; stress in biaxial
双轴圆锯的上锯片 overhead saw
双轴再锯机 double arbor resaw
双柱桁架 queen post truss; queen-post
双柱架杆 queen-post
双柱梁(上撑) queen post girder
双柱式 queen-post
双柱梯 two-legged ladder
双柱屋架(上撑) queen post roof truss
双子叶的 dicotyledonous; dicotylous
双子叶树木 dicotyledonous tree; dicotyledonous wood
双子叶植物 dicotyledon; dicotyls
双组分喷枪 two-component spray gun
双作用的 dual-function
霜 frost
霜带 frosty-zone
霜冻 frost; frost action
霜冻伤疤 frost-rib
霜冻穴 frost hollow
霜害 damage by frost; frost damage; frostbite; frost-injury; injury by frost
霜降 first frost
霜举 frost heave; frost heaving; frost lift(ing); soil-lifting frost
霜孔 frost hollow
霜裂 frost crack; frost split
霜瘤 frost-canker
霜霉病 downy mildew; false mildew
霜棚 frost-shed
霜蚀 frost erosion
霜凇 air hoar; hard rime
霜洼 frost hole; frost locality; frost pocket; frost pool; frost spot; frost hollow
霜线 frost level; frost line
霜穴 frost hole; frost locality; frost pocket; frost pool; frost spot; frost hollow
霜肿 frost-rib
水岸林 waterside forest; water bank forest; riparian forest
水坝 dam
水白色的 water-white
水斑 water stain
水饱和亏缺 water saturation deficit
水泵组 pump crew
水边低沙丘群落 foredune community
水边贮木场 timber bow
水表薄膜 surface film

水表层浮游生物 epineuston
水玻璃胶黏剂 water glass adhesive
水玻璃漆(涂料) water glass paint
水钵(日本式园林点缀物) water basin
水布植物 hydrochore
水材化学变色 wood chemical stain
水槽箱 flume box
水草地 meadow
水草甸 aquiprata
水层区 pelagic division
水层生物 pelagos
水插(繁殖法) water cutting
水产养殖 aquaculture; aquiculture
水菖蒲烯 calarene
水成磷矿石 sedimentary phosphlate
水成论 neptunian theory; neptunism
水成片岩 paraschist
水成土 hydromorphic soil
水成岩 aqueous rock; hydatogen rock; hydatogenous rock; hydrogenic rock; hydrolith; neptunic rock; sedimentary rock
水程 water lead
水池 pool
水存法 water storage
水弹 water bombing
水道 canal; lade; race; water course
水道流送 ship
水稻土 paddy soil; rice soil
水底的 benthal; benthic
水底动物(区系) bottom fauna
水底群落 bottom community
水底植物 benthophyte; submerged plant
水底植物区系 bottom flora
水底植物群落 phytobenthos
水泛地 overflow land
水分保持 water conservation
水分保持力 moisture holding capacity
水分测定计 moisture (content) meter
水分测定仪 moisture detector
水分电测计 electric(al) moisture meter
水分交换 water exchange
水分亏缺 water deficit
水分利用效率 water use efficiency
水分平衡 hydrologic balance; hydrologic budget; hydrological balance; water balance; water budget; water equilibrium
水分平衡方程 hydrologic equation
水分平衡图解 water balance diagram
水分驱动力(木材内部) moisture driving force
水分梯度 gradation of moisture; moisture gradient
水分吸力 moisture suction
水分循环 hydrologic cycle; hydrological cycle; water cycle
水分移动(在木材内部) moisture movement
水分移动通路(木材内部) moisture travel
水分应力 moisture stress
水分张力 moisture tension
水分状况 hydrologic regime; moisture regime
水封管 barometric pipe
水工材 wood for hydraulic works
水工用材 construction timber for water-work
水垢 water scale
水管 conduit
水管敷设法 hose-lay
水罐车 piggy back tank
水罐直升机 helitank
水合度 hydrature
水合莰烯 camphene hydrate
水合酶 hydratase
水合氢离子 hydronium
水合数 hydration number
水合物 hydrate
水合茚三酮反应 ninhydrin reaction
水合作用 hydration
水华 water bloom
水滑道 flume; log sluice; sluice; water slide; wet chute; wet flume; wet slide
水滑道槽 flume box
水滑道底木 backbone
水滑道运材 log sluice; sluice
水化变质 hydrochemical metamorphism
水化度 degree of hydration
水化腐殖质砖红壤 hydrol humic latosol

水化精磨机 hydrafiner
水化酶 hydrase
水化纤维素 hydrated cellulose
水化纤维素II cellulose hydrate II
水化纤维素I cellulose hydrate I
水化氧化物 hydrous oxide
水化云母 hydromicas; hydrous micas
水化作用 hydration
水混浊度 water turbidity
水际园 aquar garden
水浆研磨机 magazine grinder
水窖 water cellar
水结碎石路面 water-bound macadam
水解 hydrolyze
水解产物 hydrolysate; hydrolyzate
水解常数 hydrolysis constant
水解催化剂 hydrolyst
水解单宁 hydrolysable tannin; hydrolyzable tannins
水解反应锅 hydrolysis reactor
水解降解 hydrolytic degradation
水解类栲胶 hydrolysable tannin extract
水解酶 hydrolase
水解木质素 hydrolytic lignin
水解器 hydrolyzer
水解鞣质 hydrolysable tannin; hydrolyzable tannins
水解酸度 hydrolytic acidity
水解纤维素 hydrocellulose
水解质 hydrolyte
水解紫胶 hydrolyzed shellac
水解作用 hydrolysis; hydrolytic action
水界 hydrosphere
水浸 submersion; water logging; waterlogging
水浸原木变色 sour log stain
水景 water course
水景园 water garden(ing)
水均衡 hydrologic budget
水库 impounded body; reservoir
水库防护林 reservoir protection forest
水蜡虫 mealybug
水涝地 water logged land
水力板皮剥皮机 hydraulic slab debarker
水力半径 hydraulic radius
水力剥皮 hydraulic debarking; jet barking
水力剥皮机 hydraulic barker; hydraulic debarker; hydraulic debarking machine; hydraulic peeler; jet barker; stream barker
水力测功机 hydraulic dynamometer
水力环式剥皮机 hydraulic ring barker
水力计 hydrodynamometer
水力劈木机 hydraulic splitter
水力切削 liquid cutting
水力去皮机 hydraulic debarking machine
水力散浆机 hydrapulper
水力碎浆机 aquapulper
水力碎浆设备 hydrabrusher
水力梯度 hydraulic gradient
水力调节器 hydraulic regulator
水力学 hydraulics
水利工程 hydraulic engineering
水利工程建筑物 hydrotechnical construction
水利技术土壤改良 hydrotechnical amelioration
水利枢纽 hydro-junction
水利土壤改良 hydro-melioration
水量提取 water withdrawal
水龙带 spray hose
水陆联运运送补给 portage
水陆两用集材机 amphibian
水陆运费 freight and cartage
水铝英石 allophane
水媒传播 hydrochory
水媒的 hydrophilous
水煤气焦油 water-gas tar
水煤气焦杂酚油 water-gas-tar creosote
水门 sluice
水面浮标 surface-float
水面漂浮生物 pleuston
水螟 casebearers
水膜试验 water-film test
水磨 liquid honing
水磨土 lemnian earth
水母 jellyfish
水囊 water sac
水泥 cement

水泥木丝板(建筑) cement-bond building panel; wood wool cement board
水泥刨花板 cement particle board; cement-bond building panel; cement-bonded particleboard
水泥碎料板 cement-bonded particleboard
水鸟 natatorial bird; water bird; waterfowl
水凝胶 hydrogel
水培床 soilless bed
水培法 water culture
水平 horizon; level
水平带整地 terracing
水平地带性 horizontal zonality
水平点抽样 horizontal point sampling
水平分布 horizontal distribution
水平分枝 horizontal branching
水平分子 horizontal element
水平根 horizontal root
水平埂 contour ridge
水平沟 contour trench; horizontal ditch; level trench
水平管束式干燥机 horizontal tube-bundle dryer
水平辊式剖分带锯 band resaw with horizontal roll
水平混交 horizontal mixture
水平间距(镟切机) horizontal gap; horizontal opening
水平剪切 horizontal shear(ing)
水平阶 level bench
水平截流 horizontal interception
水平进料削片机 horizontal feed chipper
水平抗性 horizontal resistance
水平拉杆 tie beam
水平犁沟 contour furrow
水平路堑 level cutting
水平面 level; water-level
水平能见度 horizontal visibility
水平起重臂 horizontal boom
水平气流 horizontal air flow
水平迁移 horizontal migration
水平式贮料仓 horizontal storage bin
水平输送机 horizontal conveyor
水平台 level(l)ing bench
水平梯田 bench terrace; graded bench terrace
水平土壤地带性 horizontal soil zonality
水平纹孔口 flattened pit orifice
水平线抽样 horizontal line sampling
水平压块 level(l)ing block
水平压力 horizontal pressure
水平样线法(直线样品) line-plot; line-plot survey
水平郁闭 horizontal closure
水平转臂 gill poke
水栖的 aquatic
水栖生物 nekton
水汽 aqueous vapour
水汽凝结物 precipitate
水禽 waterfowl
水禽保护区 waterfowl refuge
水青冈坚果 beech nut
水青冈碱 fagine
水青冈林带 fagetum
水圈 hydrosphere
水泉群落 crenium
水染剂 water stain
水溶胶 hydrosol
水溶式制浆法 hydrotropic pulping
水溶性防腐剂 aqueous preservative; waterborne(-type) preservative
水溶性腐殖质 water humus
水溶性化学药剂 waterborne chemical
水溶性润胀剂 aqueous bulking agent
水溶性盐类 waterborne salt
水溶助长剂 hydrotropic solvent
水乳化漆 oil bound water paint
水砂纸 wet abrasive paper
水上编排场 raft pocket
水上分类廊 raft pocket
水上交材 in the water
水上码头 pontoon
水上起重机 derrick barge; floating crane
水上拖运 towage
水上选材 sluice
水上选材廊 corridor
水上原木检尺 water scale
水上贮木场 river depot
水上贮木区 pocket
水上作业场 boomage
水深 stage

水深计 fathometer
水生草本群落 aquiherbosa; aquiprata
水生大型植被 aquatic macrophyte vegetation
水生的 aquatic; hydric; hydrophilous
水生动物 aquatic animal
水生动物疫病控制 aquatic animal pest control
水生附着生物 periphyte
水生根 water root
水生昆虫 aquatic insect
水生群落 aquatic community
水生生态系统 aquatic ecosystem
水生生物学 hydrobiology
水生苔藓 aquatic mosses
水生演替 hydrarch succession
水生演替系列 hydrosera; hydrosere
水生植物 aquatic plant; aquatic vegetation; hydrophyte; water plant
水声测位仪 sonar
水湿地 swamp-land
水湿地的 marshy
水蚀 erosion by water; water erosion
水蚀程度 degree of water erosion
水蚀强度 intensity of water erosion
水势 hydraulic potential; water potential
水松型纹孔 glyptostroboid pit
水苏糖 stachyose
水损失 water loss
水套加热烘箱 water-jacket oven
水梯 water ladder
水体循环 overturn circulation
水头 head
水头势 hydraulic potential
水头损失 loss of head
水土保持 soil and water conservation; soil protection and water regulation
水土保持措施 soil and water conservation measures
水土保持耕作措施 soil and water conservation tillage measures
水土保持工程 soil and water conservation engineering
水土保持工作条例 Act of soil and water conservation
水土保持规划 soil and water conservation planning
水土保持林 soil and water conservation forest
水土保持林草措施 forest-grass measures for soil and water conservation
水土保持乔林 high-forest with soil protecting wood
水土保持区划 soil and water conservation regionalization
水土保持效益 soil and water conservation benefit
水土保持作物 soil conserving crop
水土流失 soil and water losses; soil erosion
水团 water mass(es)
水湾 inlet
水网 hydrographic net
水位 river stage; stage; water-level
水位变动 fluctuation of water table
水位标志 watermark
水位表 river gauge
水位差 hydraulic gradient
水位持续曲线 stage-duration curve
水位计 water-level recorder
水位升降 fluctuation of water table
水位图 water table map
水位箱 head box
水温制图 water temperature mapping
水文测量工程 gauging works
水文测量站 gauging station
水文地层单元 hydrostratigraphic unit
水文地质学 hydrogeology
水文分水界 hydrological divide
水文观测站 hydrometric station
水文过程曲线 hydrograph
水文年度 water year
水文气象学 hydrometeorology
水文条件 hydrographical condition; hydrological condition
水文统计学 hydrological statistics
水文图 hydrograph
水文学 hydrography; hydrology
水文站 hydrographic station; hydrological station
水文状况 hydrological condition
水文状况估算 hydrologic budget
水纹(菌类引起的变色斑) watermark

水纹病 watermark disease
水污染 water pollution
水污染控制 water-pollution control
水雾繁殖 mist propagation
水稀释涂料 waterthinnable paint
水螅虫 hydroid
水洗机 water washer
水洗损失 washing loss
水系 channel net; hydrographic net; river system; water system
水系判读 interpretation of river system
水下光度计 submarine photometer
水下支墩 sunk pier
水藓泥炭 sphagnum peat
水线 dry line
水腺 water gland
水箱 reservoir
水胁迫 water stress
水锌矿 zinc bloom
水性漆 water based paint; water paint
水性清漆 water-varnish
水性染料 water based stain
水性填料 water filler
水性涂料 water based paint; water paint
水性着色剂 water stain
水悬浮液 water suspension
水选法 water test
水压骨骼 hydrostatic skeleton
水压试验 water test
水堰 weir
水杨皮质激素 salicortin
水银蓝油膏 blue ointment
水银水准仪 mercurial level
水印 watermark
水印痕 dandy mark
水域 hydrological basin; hydrosphere
水域的 aquatic
水域影响带 water-influence zone
水源 springhead; water source
水源层 aquifer
水源防护林 shelter belts to protect waterhead; water source protection forest
水源涵养林 head water conservation forest; protection forest for water conservation; water conservation forest; water reservoir forest
水栽法 hydroponics
水闸 dam; lock; sluice gate
水枝 coppice shoot; coppice sprout; water sprout
水质 water quality
水质控制法案 water quality control act
水致发白(漆病) water blush(ing)
水中呼吸器 scuba
水中检尺材积(木材推河或出河材积) river scale
水中气候 hydroclimate
水中气候图 hydroclimograph
水中煮沸法(防腐) boiling-in-water method
水中贮材 storing wood in water
水中贮木 ponding; storing wood in water
水煮 cooking; hot-water treatment
水准标尺 level(l)ing (rod-)staff
水准测量 level(l)ing
水准测量员 level(l)er
水准基点 benchmark
水准调节管 level(l)ing tube
水准仪 level; level(l)er; level(l)ing-instrument; telescope level; water-level; Y-level
水资源 water resource
水资源保护 water conservation
水资源应用 water resources application
水渍 water stain
水渍粗腐殖质 sog mor
水族馆 aquarium
税单 return
税额评定 tax appraisal
税负 tax burden
税-交叉效应 tax-interaction effect
税金 tax
税金负担 tax burden
税款 due
税率 tax rate
税收罚款 tax penalties
税收管理 tax administration
睡莲池 lily-pool; water-lily pool
睡莲叶子 lily pad
睡椅 couch
顺磁共振 electron spin resonance

S

顺磁位移 paramagnetic shift
顺从性 amenability
顺从姿态 submissive posture
顺读游标 direct-vernier
顺反异构体 cis-trans-isomer
顺反子 cistron
顺风的 downwind
顺风火 head fire
顺河绠 longitudinal boom
顺河绠收漂场 parallel holding ground
顺锯 climb sawing
顺裂纹理 tip grain
顺流下漂原木 running log
顺木(将木材顺成与装车线平行) square around
顺捻(钢丝绳) albert lay; lang-lay
顺捻钢丝绳 uniform lay wire rope
顺坡 downward grade; favorable grade
顺坡集材 downhill skidding
顺坡烧除 strip burning
顺绕钢丝绳 lang lay rope
顺时针螺旋生长(木材缺陷) clockwise spiral growth
顺时针旋转 clockwise rotation
顺水的 downstream
顺弯 bowing; longitudinal distortion; spring
顺纹 with the grain
顺纹层积材 straight laminated wood
顺纹剪切 shear along the grain; shear parallel to the grain
顺纹胶合板 long grained plywood
顺纹锯切 saw with the grain
顺纹抗剪强度 shear strength parallel to grain
顺纹抗拉强度 endwise tensile strength; tensile strength parallel to grain
顺纹抗拉试验 endwise tension test
顺纹抗压 end compression; endwise compression
顺纹抗压强度 compression strength parallel to grain; endwise tensile strength
顺纹抗压试验 endwise tension test
顺纹拉力 tension parallel to grain
顺纹压缩法 longitudinal compression process
顺纹应力 parallel-to-grain stress
顺铣 climb cutting; climb milling; down milling
顺向水系 consequent drainage
顺序 sequence
顺序采伐 successive cutting; successive felling
顺序法则 sequence rule
顺序配板法 slide matching
顺序相关 order correlation
瞬变气候响应 transient climate response
瞬变现象 transient
瞬间 minute
瞬间干燥 wink-dry
瞬间畸变 transient distortion
瞬间流率 transient flow
瞬间毛细管 transient capillary
瞬间视场 instantaneous field of view
瞬时出生率 instantaneous birth rate
瞬时干燥法 flash drying
瞬时蠕变 transient creep
瞬时死亡率 instantaneous death rate; instantaneous mortality rate
瞬时增长率 instantaneous growth rate; instantaneous rate of increase
说明书 detailed account
蒴苞 perianth
蒴柄 chaeta; seta (*plur.* setae)
蒴盖 opercule
蒴果 capsule; theca (*plur.* thecae)
蒴果状的 capsular
蒴托 apophysis (*plur.* -hyses)
司机 driver
司凯脱 skeet
司炉 black gang
司旗员 flagman
丝氨酸(栓质) serine
丝光化(丝光处理) mercerization
丝光纤维素 mercerized cellulose
丝绢橡胶树 lagos silkrubber; kickxia rubber
丝色的 bombycinous
丝镇气压计 Naudet's barometer
丝质的 bombycinous
丝状白腐 white stringy rot

丝状的 filiform; sericeous
丝状腐朽 stringy rot
丝状菌 filamentous fungus; thread fungus
丝状体 filament
丝状纤维 filamentary fibril
丝状原纤维 filamentary fibril
丝状真菌 filamentous fungus
私人成本 private cost
私人庭园 private landscape
私人游乐 private recreation
私有林 private forest
私有林森林经理 private forest management
私有林业 private forestry
私有林业补助金 private forestry grants
私有林主 private forest owner; private timber manager
斯班尼犬 spaniel
斯波尔丁原木板积表 Spaulding log rule
斯伯林格斯班尼犬 springer spaniel
斯凯脱喉缩 skeet choke
斯克里布纳原木板积表 Scribner log rule
斯克里布纳原木材积表(板英尺单位) Scribner log scale; Scribner rule
斯里兰卡桂皮油 Ceylon cinnamon bark oil
斯里兰卡香茅 Ceylon citronella grass
斯马林公式(求立木材积) Smalian's formula
斯马林区分求积法 sectional measurement by means of Smalian's formula
斯塔福德干馏法 Stafford process
斯特劳德测树仪 Stroud dendrometer
撕裂 tearing; tension break
撕裂刀具 tearing tool
撕裂度 tear strength
撕裂强度 tearing strength
撕裂强度试验机 tearing tester
撕裂速率 tear speed
撕裂纹(加工缺陷) torn grain; pick-up grain
撕裂状的 lacerate
死尺寸 dead size
死打浆(浆料叩解过度) dead beating
死地被物 dead soil covering
死地被物层 litter layer; O-layer
死地被物输入量 litter supply
死浮游生物 necroplankton
死后 post-mortum
死湖 dead lake
死节 black knot; dead knot; encased knot; non-adhering knot
死结 dead lock
死枯可燃物 dead fuel
死木寄生菌 saprogen; saprophyte
死木生物 saproxylobios
死梢 dead centre
死树皮 dead bark
死水 stagnant water
死体营养性病原体 necrotrophic pathogen
死体营养性的 necrotrophic
死土层 dead horizon
死亡 mortality; abiosis; die-off
死亡量 mortality
死亡率 death rate; mortality rate
死亡率曲线 mortality curve
死亡木 dead tree
死线公路 dead-line road
死养生物 perthotroph
死种子 blind seed
似变形虫的 amoeboid
似共轭效应 quasi-conjugation
似灌木状的性质或状态 frutescence
似黑色石灰土 borovina; borowina
似乎不相关回归 seemingly unrelated regression
似爵床科的 acanthaceous
似木质的 xyloid
似披针形的 lanceate
似侵填体 tylosoid; thylosoid
似亲群体 autocolony
四倍的 quadruple
四倍体 tetraploid; tetraplont
四边形 quadrangle
四不像 Caribou; reindeer
四车道 four-lane traffic way
四出数的花 quadrimerous flower
四碘四氯荧光素 rose-bengal
四方形栽植 square planting
四分孢子 quartet; tetrad; tetraspore
四分的 quartered
四分法 quartering

四分锯 quarter saw
四分体 quadrant; tetrad
四分位 interquartile
四分圆饰条 quarter round
四分枕木 quarter tie
四分之一跨度载荷 quarter point loading
四分周法 quarter-girth method
四分周干围 quarter girth
四分周检尺法 quarter-girth measure
四分周卷尺 quarter-girth tape
四氟化铀 green salt
四合花粉 tetrad
四核子基 quartet
四回羽状的 quadripinnate
四活塞计量泵 four-piston metering pump
四级结构 quaternary structure
四季开花 perpetual bloom; perpetual flowering
四季开花的 everblooming
四季园 seasonal garden
四价的 quadrivalent; tetravalent
四价染色体 quadrivalent; tetravalent
四价体 quadrivalent
四碱价酸 tetrabasic acid
四开材 quarter stuff; quarter timber
四开的(圆材沿两直径成90°锯开) quartered
四开断面 quarter section
四开枕木 quarter tie
四棱正边板 square edged board
四楞规正的方材 die-square
四联带锯 quad band saw
四联单刃带锯机 single cut quad band-mill
四联体(染色体) tetrad
四联体(遗传密码) quadruplets
四链环 tour paws
四裂爿的 quadrivalvate; quadrivalvular
四零四九(杀虫剂) malathion
四轮车 mule cart; wag(g)on
四轮车运材 wag(g)on; wag(g)on hauling
四轮运材车 log-wagon
四氯化碳(熏蒸剂) carbon tetrachloride
四氯异苯腈 daconil
四面锯切的 square-sawn
四面木工刨床 four sides planing and moulding machine; four-sided matchboarding machine; matcher; planer and matcher
四面刨 four side moulder; matcher; planer and matcher; sizer
四面刨床 four side moulder; four sides planing and moulding machine; four-sided matchboarding machine
四面刨光材 dressed four sides; surfaced four sides
四面刨削机床 four cutter
四面下锯 four-faced sawing
四旁植树 four-side tree planting
四片对拼法 tour-piece butt matching
四破木料 edge-grain lumber
四剖分锯解 quarter-circle cut; quarter-cut
四浅裂的 quadrilobate
四腔快速拌胶机 four-chamber short-retention-time blender
四强的 tetradynamous
四强雄蕊 tetradynamous stamen
四强雄蕊的 tetradynamous
四切齿一组的截锯 four cutter
四式型 quadruplex
四室的 quadrilocular
四糖 tetrasaccharide
四体生物 tetrasomic
四萜类 tetraterpenes
四系杂交 four-way cross
四显性组合 quadruplex
四线期交换 four-strand crossing over
四小叶的 quadrinate
四溴荧光素 eosin
四氧化三铅 minium; red lead
四叶的 quadrifoliate
四元酸 tetrabasic acid; tetraprotic acid
四原型 tetrarch
四爪链钩 double-coupler; tour paws
四至(图幅边界) four boundaries
四重的 quadruple
四重峰 quartet
四重线 quartet
四足梯 stepladder
四唑染色法 tetrazolium staining

寺庙林 shrine forest; temple forest
寺庙园林 monastery garden; temple garden
饲草生产 hay production
饲草系数 forage factor
饲草总消费量 range utilization
饲料 fodder; forage
饲料地食料地 food patch
饲料果的(栎槲等树) masty
饲料酵母 feed yeast; fodder yeast
饲料颗粒 pellet
饲料树木 fodder tree
饲料添加剂 feed supplement
饲料运送器 feed conveyor
饲料植物 fodder plant; forage plant
饲料作物 fodder crop; forage crop
饲养者 feeder
饲用酵母 feed yeast; fodder yeast
饲用植物 forage plant
松柏科木材 coniferous wood
松柏类植物 coniferophyte; coniferous plant
松柏树(泛指) coniferous tree
松柏园 pinetum
松材线虫 pine wood nematode
松弛(锯口) loose
松弛模量 relaxation modulus
松弛平衡度 relaxation balance
松弛曲线 relaxation curve
松弛试验 relaxation test
松弛速度 relaxation velocity
松弛旋切单板 loose-cut veneer
松垂钢索 slack
松动木板(板路或桥面上木板翘起) gill poke
松干蚧 Matsumura pine scale
松根干馏 softwood distillation
松根油 pine root oil
松果 pine nut
松簧时间(从扣动板机至子弹发射的时间) lock time
松鸡 grouse
松浆油 liquid rosin
松焦油 pine tar; softwood tar; stockholm tar
松焦油沥青 pine tar pitch; softwood tar pitch
松节油 spirit of turpentine; turpentine; turpentine oil; turps
松节油代用品 turpentine substitute
松紧螺旋扣 turnbuckle
松紧式(索道集材) fall block system
松紧式索道 live skyline; slack line system; slackline system
松紧式索道集材 flyer slackline yarding; slacking; slackline yarding; slacklining
松紧式索道集材绞盘机 slacker
松卷装置(单板) unreeling system
松林 pinery
松林采伐作业 short log show
松林地区(美国) short log country
松林区伐木工人 short log logger
松露 truffle
松落针病 needle shedding; pine-leaf-cast
松毛虫 pine caterpillars; pine-moth (*Dendrolimus pini* L.)
松密度 bulk; bulk density
松面 loose face; loose side
松木 pine wood
松木焦油 pine wood tar
松木油 pine wood oil
松木杂酚油 pine-wood creosote
松皮材 loose bark pole
松软单板 loose veneer
松软节 unsound knot
松软土 malice soil
松软心材 spongy heart
松散机 thrasher
松散轮 loosening wheel
松散纹理 loosened grain
松散性 bulking property
松砂机 fluffer
松杉目 Coniferales
松疏节 loose knot
松属素 pinocembrin
松鼠 squirrel
松树 pine
松树的 pinic
松树象鼻虫 pine-weevil
松树杨梅色素 pinomyricetin
松树油 pine-tree oil
松突圆蚧 pine greedy scale

松土 loosen; pulverization; scarification
松土齿 tine
松土机 ripper; scarifier; soil pulverizer
松土犁 mellow-soil plough
松土器 loosener; rejuvenator; scarifier
松线式集材机 slacker
松香 abietyl; colophany; colophony; rosin; colophonium
松香胺 abietylamine; rosin amine
松香斑 rosin speck; rosin spot
松香的 rosiny
松香甘油酯 ester gum
松香基 rosinyl
松香胶料 rosin size
松香结晶现象 rosin crystallization
松香腈 rosin nitrile
松香沥青 rosin pitch
松香清漆 gloss oil; rosin varnish
松香色级 colour grades of rosin; grade of rosin
松香树脂 rosin resin
松香素 abietine
松香酸 rosin acid
松香亭 abietin
松香盐 rosin salt
松香颜色等级 grade of rosin
松香衍生物 rosin derivative
松香油 rosin oil; rosinol
松香皂 rosin soap; rosinate soap; tall oil soap
松香脂 abietin
松小蠹虫 pine beetle (*Myelophilus piniperda* L.)
松型纹孔 pinoid pit
松型纹孔对 pinoid pit pair
松锈病 pine-cluster cups
松烟 carbon black; pine-black; pine-soot
松杨梅精 pinomyricetin
松叶锈病 pine-needle rust
松叶油 pine leaf oil
松夜蛾 pine beauty (*Noctu piniperda*)
松油 pine oil
松油防腐剂 Termiteol
松油精 dipentene
γ-松油烯 crithmene
松油皂 tall oil soap
松藻虫 back swimmer
松针油 fir (needle) oil; pine leaf oil; pine needle oil
松脂 galipot (resin); gum of pine; oleoresin; pine gum; pine oleoresin; turpentine
松脂采集 tap for resin
松脂产量 resin crop
松脂林 turpentine-orchard
松脂破碎机 gum crusher
松脂气味 resinous odor
松脂制品 gum navel stores; naval stores
松装体积 apparent volume
松子 pine nut
松子油 pine-seed oil
淞雾 rime fog
送材车 log carriage; lumber carrier; support carriage
送风 air supply; blowout
送风阀 blow valve
送风干燥窑 blower (dry) kiln
送风机 fan
送风室 plenum chamber
送货费用 delivery expense
送料 feeding
送料台 deck plank
送射 straight forward shot
送水管 feeder
搜集剂(螯合金属) gathering agent
搜索 retriever
搜索常数 quest constant
搜索像 searching image
搜索沿岸滞留材 sack the banks
苏氨酸 threonine
苏打法 soda process
苏打石灰水 Burgundy mixture; soda bordeaux
苏打土 sodic soil
苏丹香茅油 naa oil
苏方荚 algarobilla
苏方荚单宁 algarobilla tannin
苏方荚栲胶 algarobilla extract
苏方荚素 algarobin
苏合树烯 liquidene
苏合香 styrax
苏合香树脂 Levant storax

S

苏合香素 styracin
苏合香油 storax oil
苏木酚 sappanin
苏木精 haematoxylin
苏木栲胶 hemartine extract
苏木素 haematoxylin
苏糖 threose
苏铁属 cycad cycas
酥溶性清漆 spirit varnish
俗名 trivial name
素材 rough timber; untreated timber
素坯 biscuit
素质 predisposition
速测镜 relascope
速成胶 accelerated gum
速成培育 accelerating culture
速动资产 quick assets
速度表 speed meter
速度计 speed meter; velometer
速度记录图 tachograph
速高比 velocity-height ratio
速行火 running fire
速率基因 rate gene
速燃火药 quick-burning powder
速射 rapid fire; snapshot
速生材 fast growing wood
速生的 fast grown
速生丰产林 fast-growing and high-yield plantation; fast-growing and high-yielding forest
速生林作业 free growth management
速生木材 fast growing wood
速生树 fast growing tree; rapidly growing tree
速生树种 fast growing species
速胀钵(育苗容器) Jiffy pot
速煮法(亚硫酸木浆蒸煮) quick cook(ing) process
嗉囊 crop; gizzard; gorge
塑变 flow
塑度计 plastometer
塑合木 wood plastic composite; wood-polymer
塑化 plastify
塑化剂 plasticizer; plasticizing agent
塑化木 plastic wood
塑化木材 plasficizing wood
塑化纸板 papreg
塑料 plastomer
塑料板坯 hide
塑料薄膜 plastic film
塑料钵(育苗用) polypot; plastic pot
塑料的塑料 plastic
塑料防裂夹头 plastic antichecking clamp
塑料粉末喷涂 plastispray
塑料家具 plastic furniture
塑料跑道 plastic track
塑料棚 plastic covered greenhouse; plastic house
塑料容器 polyethylene container
塑料溶胶 plastisol
塑料贴面 plastic overlay
塑料贴面胶合板 plastic faced plywood
塑料温室 plastic covered greenhouse; polyethylene greenhouse
塑料制品 plastic
塑入 lute in
塑性 plastic property
塑性变形 creep; plastic deformation
塑性变形范围 plastic range
塑性材料 plastic material
塑性测定法 plastometry
塑性常数 plastometer constant
塑性的 plastic
塑性高分子物质 plastomer
塑性极限 plastic limit
塑性计 plastometer
塑性流动 plastic flow
塑性流动极限 plastic flow limit
塑性流限 limit of plastic flow
塑性黏土 plastic clay
塑性破裂 plastic crack
塑性收缩 plastic shrinkage
塑性体 plastomer
塑性弯曲 plastic bending
塑性下限 lower plastic limit
塑性限界 plastic limit
塑性学 plasticity
塑性应变 plastic strain
塑性状态 plastic state
溯河产卵洄游 anadromous fish
溯源侵蚀 head erosion; headward erosion

宿瓣蒴果 repletum
宿被瘦果 diclesium; dyclesium
宿存病原物 persistent pathogen
宿存萼 persistent calyx
宿萼瘦果 scleranthium
宿萼蒴果 diplotegia; dyplotegia
宿根植物 perennial plants
酸败[度] rancidity
酸不溶木素 acid-insoluble lignin
酸橙 bigarade orange; seville orange; sour orange
酸橙花油 bigarade oil
酸当量 acid equivalent
酸度 acidity
酸法蒸煮 acid cooking; aid cooking
酸化 acidation; acidification
酸化剂 acidulant
酸碱度 pH-value
酸碱平衡 acid-base balance; acid-base equilibrium
酸胶基 acidoid
酸溶木素 acid-soluble lignin
酸食性的 acidotrophic
酸蚀 acid scarification
酸污染 acid smut
酸腺 acid gland
酸性 acidity
酸性沉降物 acid deposition
酸性腐泥 moder
酸性腐殖化作用 acid humification
酸性腐殖质 acid humus; sour humus
酸性高山草甸土 acid alpine meadow soil
酸性固化剂 acid hardener
酸性胶 acid size
酸性胶体 acidoid
酸性橘红 acid orange
酸性硫酸盐土 mangrove soil
酸性木材 acid wood
酸性黏土 acid clay
酸性培养基 acid medium
酸性泉 acid spring
酸性染料 matching stain
酸性熔岩 acidic lava
酸性生境指示植物 acidophilous indicator plant
酸性树脂 acidic resin
酸性松香胶 acid rosin size
酸性土 sour soil
酸性土壤 acid soil
酸性亚硫酸法 bisulfite process
酸性亚硫酸盐半化学浆 acid sulfite semichemical pulp
酸性亚硫酸盐法 acid sulfite process
酸性岩 acidic rock; acidite
酸性演替系列 oxarch; oxysere
酸性硬化剂 acid hardener
酸性棕壤 acid brown soil
酸性棕色森林土 acid brown forest soil
酸雨 acid precipitation; acid rain
酸枣苷元 jujubogenin
酸枣仁皂苷 jujuboside
酸沼 acid bog; bog; muskeg
酸值 acid number; acid value
酸中毒 acidosis
算法 algorithm
算法语言 algorithmic language; algorithmic oriented language
算术平均 arithmetic mean; arithmetic(al) mean
算术平均标准木 arithmetical mean sample tree
算术平均标准木调查法 method of the arithmetical mean sample tree
算术平均树高 arithmetic mean height
算术平均直径 arithmetic(al) mean diameter
算术指令 arithmetic instruction
算数平均直径 arithmetic mean diameter
随车起重车 crane car
随到随装 hot (deck) loading
随动装置 follow-up; follow-up device
随堆随烧(处理采伐剩余物) progressive burning; swamper burning
随机变动 random fluctuation
随机变化 random variation
随机变量 random variable; stochastic variable
随机抽样 random sampling
随机次级样本 random subsample
随机存储器 random access memory
随机存取 random access
随机存取分类程序 random access sort
随机的 random
随机对法 random pairs method

随机分布 random dispersion; random distribution
随机过程 stochastic process
随机交配 random mating
随机交配单位 panmictic unit
随机交配群 deme
随机交配群体 panmictic population; panmixia (panmixis)
随机离差 deviate at random
随机模型 stochastic model
随机配种 random mating
随机拼板 random assembly; random matching
随机起点 random start
随机区组 randomized block
随机失配 random mismatching
随机调查 random cruising
随机误差 random error
随机性 randomness
随机性偏差 deviation from randomness
随机样本 random sample
随机育种单位 random breeding unit
随即匹配 random matching
随集随运 hot (deck) loading
随手式(射击) off hand
随体(染色体) satellite; trabant
随体区 SAT-zone
随意性 randomness
随遇的 indifferent
随遇种 ubiquitist
髓 pitch
髓斑 medullary blemish; medullary spot; pith fleck
髓部的 medullary
髓节 pith knot
髓腔 hollow pith; medullary cavity; pith cavity
髓鞘 pith sheath
髓射线 medullary ray; pith ray
髓体丧失(种间杂交) amphiplasty
髓外部分 external medulla; external pith
髓维管束 medullary bundle
髓线斑 pith ray fleck
髓心 heart centre; medulla (*plur.* -llae); pitch; pith
髓心板 heart board
髓心材 core wood; pith wood
髓心的 medullary
髓心厚饭 heart plank
髓心裂 pithshaken
髓心偏离 wandering heart
髓心区 boxed pith
岁入 yearly receipts
碎部测量 survey in detail
碎部点 detail point
碎单板剪板机 salvage clipper
碎浆 pulping
碎浆机 deflaker; pulper
碎届 scrap
碎砾岩 psephite; psephyte; rudite; rudyte
碎料 outthrows; particle
碎料板 chipboard; flakeboard; particle board; shaving board
碎料板带 shoot
碎料板心材 particle board core stock
碎料称量台 furnish weighing station
碎料分离器 particle separator
碎料塞 chip plug
碎料再碎设备 secondary particle reduction unit
碎裂纹 chipped grain
碎木机 hammer hog; hog(ger); hogging machine
碎木机(粉碎机) hogger
碎木块 chunk; hunk
碎皮机 bark hog
碎片 debris; fraction
碎片屑 splinter
碎石 scree
碎石坝 loose rock dam
碎石道 macadam road
碎石堆 talus
碎石谷坊 loose rock check dam
碎石坡 scree; scree slope
碎石铺道法 macadam pavement; macadamizing
碎食物链 detrial food chain
碎食性生物 detritivore
碎树皮(制堆肥) shredded bark
碎土机 clod crusher; pulverizer
碎土耙 pulverizer harrow
碎屑 detritus
碎屑沉积 detrital deposit
碎屑沉积物 mechanical sediment

S

碎屑食物链 detritus food chain
碎屑压痕(在材面上) chip bruising; chip mark
碎屑岩 clasolite; conglomerate; detrital rock; fragmental rock
碎屑砖红壤 detrital laterite
碎屑状的 clastic
碎纸机 deflaker; kneader
隧道 tunnel
隧道干燥窑 tunnel kiln
隧道净空 tunnel clearance
穗行选种 ear-to-row selection
穗状的 spicate
穗状花序 spike
穗状聚伞花序 spicate cyme
损耗 depletion; shrinkage; wastage; wear; wear-and-tear
损耗抵偿 depletion allowance
损耗角正切 loss tangent
损耗率 waste rate
损耗因数 loss tangent
损坏商品 damaged merchandise
损伤 blemish; damage
损伤材 damaged wood
损伤节 damaged knot
损伤木 broken timber
损失 loss
损益计算书 profit and loss statement
损益两平点 break even point
损益平衡表 break even chart
隼(sǔn) falcon
榫(sǔn) dovetail
榫槽 gain
榫槽接 grooved and tongued joint
榫槽接合 tenon and slot mortise; tongue and groove joint; tongue(d) and groove(d) joint
榫齿 mortise chisel
榫钉 dowel
榫钉边接合 dowel edge joint
榫钉嵌装机 dowel inserting machine
榫接 joint; mortise joint; rabbet; rebate; table
榫接合 feather joint
榫孔 mortise; mortice
榫舌 tongue
榫头 rabbet
榫销接合 tenon dowel joint
榫腋脚 haunch
莎草酮 cyperone
莎草油 cyperus oil
梭子 shuttle
蓑蛾 bag-worm moth; casebearers
羧化酶 carboxylase
羧化阻力 carboxylation resistance
羧基生物素 carboxybiotin
羧激酶 carboxykinase
羧甲基纤维素 carboxymethyl cellulose; triethyl aminoethyl cellulose
羧酸 carboxylic acid
缩边 corner defect
缩差率(弯曲材的凸、凹面长度差) upset
缩尺 reduced scale
缩度计 ompressometer
缩放仪 pantograph
缩合状态(热硬化树脂) B-stage
缩聚(作用) polycondensation
缩裂 crack
缩入 intussusception
缩微照相术 microphotography
缩叶病 leaf curl
所得税 tax on income
所得税分配 income tax allocation
所得值 income value
所有权 claim
所有权境界 boundary of property
所有权凭证 title deed
所有权型 types of ownership
索 strand; cable; rope; wire
索道 cable tram; cable tramway; cableway; ropeway; wire ropeway
索道安装组长 chain tender
索道大滑轮 equalizing block
索道吊车 load pendulum
索道吊钩 skyhook
索道滑轮 logging block
索道集材 cable logging; cable skidding; high-ball logging; skyline logging; skyline skidding; skyline system; skyline yarding; skyline-crane yarding
索道集材(架空) suspended cable yarding
索道集材带 skyline road; skyline strip

S

索道集材范围 strip
索道集材方式 cable system; cable yarding system; skidder yarding system
索道集材工组长 rigger
索道集材机 flying machine; skyline skidder
索道集材区 setting
索道集运材 cable hauling; cable log haul; cartier cable logging
索道绞盘机 cableway winch; donkey snubber; skyline winch; skyline yarder
索道控速装置 aerial snubbing system
索道跨距 reach
索道跑车 bicycle; buggy; cableway carriage; flyer carriage; skycar; skycrane; skyflyer
索道跑车(可侧方拖集) skyline-crane carriage
索道驱动轮 ropeway driving wheel
索道始发站 driving station
索道尾端锚桩(伐根) skyline tail stump
索道硬木滑轮 skyline sheave
索道运材 cableway hauling
索道张紧点 tension station
索道张紧装置 cableway tension device
索道中间支架 skyline support
索道抓具 cableway bucket
索端活钩 slip hook
索恩式剥皮机 Thorne debarker
索饵洄游 feeding migration
索格-卡普防腐法(端部注入法) Saug-Kappe process
索环 strap socket
索节 funic(u)le; funiculus
索结 swifter knot
索具 cordage; rig; rigging; tackle
索具安装 rigging
索具安装工 rigger
索具安装工长 head rigger
索具安装工组 bull gang
索具部件 rigging assembly
索具纠缠 foul
索具纠结 hang up
索具配套 rigging assembly
索具硬件(金属零件) rigging hardware
索轮 rope wheel
索赔 claim
索式抓具 cable grapple
索套 lashing; roll splice; strap socket
索套(捆木材) noose
索梯 jack ladder
索系 cable system
索眼 eye-splice
索引 index; indicators
索引图 key map
索引镶嵌 index mosaic
索状管胞 septate tracheid
索状结构 strand-like structure
锁 lock
锁定承载梁 Seattle bunk
锁定链环 bitch link
锁峰(核磁) lock
锁杆 lock bar
锁钩(安装索具) bitch hook
锁环 lock ring
锁接 lock joint
锁节 link lock
锁紧 locking
锁紧器 load binder
锁紧楔块 lock wedge
锁紧圆榫 lock dowel
锁扣装置 locker
锁斜接 lock mitre joint
锁眼盖 escutcheon
锁状连合 clamp connection
他感作用物 allelopathic substance
塌方 ground avalanche; landslide; landslip
塌棱(砂磨效果) dubbing
塌陷 collapse
塔 pagoda
塔尺 level(l)ing (rod-)staff; staff
塔拉栲胶 tara extract
塔里夫材积表 Tarif(f) table
塔式集材杆 tower spar
塔式集材机 tower-yarder
塔式起重机 derrick; hammer-head crane
塔形树冠 pyramidal crown
塔状生长(树形) monopodial growth; excurrent growth
踏板 foot board; foot path; paddle

踏板(伐木) staging board
踏板插孔(伐木) board hole
踏板端头齿钉(伐木) board iron
踏板端头止钉(伐木) gem-faced civet; masked civet
踏查 cruise; reconnaissance
踏查率 cruising percentage
踏脚石 stepping stone
胎生(动物) viviparity
胎生(植物) vivipary
胎生植物 vivipary plant
胎座 placenta; spermaphore
台板胶合板 base plywood
台背斜 anteclise
台车 hati; mutibo
台车撬杆 car mover
台秤 platform balance; platform scale
台地 bench terrace; mesa; platform; tableland; terrace
台地园 terrace garden
台风 typhoon
台架(机器的) working table; gantry
台架运转试验 bench run
台脚 pedestal
台锯 bench saw
台面 bottom bed; decking; stand top
台面可倾式多能圆锯 universal circular saw with tilting table
台刨 bench plane
台刨把手 toat
台桥 deck-bridge
台生的 coalescent
台时 machine hour
台式车床 bench lathe
台式带锯 table band-saw
台式电子计算机 desk-size electronic computer
台式铣床 bench shaper
台式再分带锯 table band resaw
台式振动分级机 oscillating table separator
台湾扁柏叶油 Formosau hinoki leaf oil
台湾暖流 kuroshio
台向斜 syneclise
台榭 terraced building
抬钩 lug hook
苔类植物 liverwort
苔藓层 moss layer
苔藓类阶段 moss stage
苔藓林 mossy forest
苔藓林带 mossy forest zone
苔藓群落学 bryocoenology
苔藓森林 moss(y) forest
苔藓系数 moss-quotient
苔藓植物 bryophyte
苔原带 tundra zone
薹草 sedge (*Carex*)
太平洋板块 Pacific plate
太平洋冷杉 red silver fir
太平洋木材检查所 Pacific Lumber Inspection Bureau
太阳常数 solar constant
太阳防护剂 sunscreening agent
太阳辐射 solar radiation
太阳辐射曲线 solar radiation curve
太阳高度角 angle of sun
太阳黑子 sunspot
太阳活动 solar activity
太阳能 solar energy; sun power
太阳能干燥 solar-energy drying
太阳能焦耳 solar emjoules
太阳能值转换率 solar transformity
太阳同步 sun synchronous
态变 phase variation
钛白 titanium white
泰伯试验机 Taber-apparatus
泰国大风子油 lukabro oil
泰加林 boreal coniferous forest; boreal forest; taiga
泰加林土壤 taiga soil
泰勒氏索道 Tyler system
泰勒氏索道集材 Tyler yarding
泰勒式循环索系 endless Tyler system
泰勒雪夫粗糙度检查仪 Tallysurf measurement
泰罗斯-N号卫星 TIROS-N
坍方 avalanche
坍坡 landcreep
摊还 amortization
摊派的计划费用 assessed program cost
摊晒机(工) tedder
摊数 assessment
摊销 amortization
滩 bank
滩岸 beach strand

弹果 sling-fruit
弹簧秤 spring balance
弹簧床垫 spring mattress
弹簧夹紧器 spring clamp
弹簧式张力计 spring balance
弹簧锁钩 latch hook
弹簧支柱伐木(弹簧连于锯端) spring pole falling
弹力 elastic force; elasticity; resilience; resiliency
弹裂蒴果 regma; rhegma
弹黏塑性体 elastic-viscoplastic body
弹黏体系 elastic-viscous system
弹器 furca
弹塑性 elastic-plastic property; elastoplasticity
弹塑性变形 plastoelastic deformation
弹塑性应变 elastic-plastic strain
弹性 elasticity; resilience
弹性变形 elasticity; temporary set
弹性材料 elastomer
弹性采伐量 flexibility in cut
弹性断裂 elastic failure
弹性复原 elastic recovery
弹性高分子物质 elastomer
弹性辊 spring roll
弹性后效 elastic after effect; elastic consequence; elastic hysteresis
弹性回复 spring-back
弹性回能 elastic resilience
弹性极限 elastic limit; limit of elasticity
弹性极限应力 stress at elasticity limit
弹性价格 flexible price
弹性胶 elastic gum
弹性介质 elastic medium
弹性进料滚筒 elastic feed roll
弹性力 elastic force
弹性模量 elastic modulus; modulus of elasticity; resilient modulus
弹性木材 rubbery wood
弹性能量 elastic energy
弹性疲劳 elastic fatigue
弹性破坏 elastic failure
弹性气控压带器(宽带式砂光机上) flexible contact platen
弹性强度 elastic strength
弹性失效 elastic failure
弹性树胶 elastic gum
弹性饲料辊 elastic feed roll
弹性体 elastomer
弹性天然橡胶 elastic caoutchouc
弹性位能 elastic potential energy
弹性系数 coefficient of elasticity
弹性橡胶 elastic caoutchouc; elastic rubber
弹性形变 elastic deformation
弹性需求 elastic demand
弹性压带器 flexible contact platen
弹性应变 elastic strain
弹性应变能 elastic strain energy
弹性应力 elastic stress
弹性余效 after effect of elasticity
弹性馀纹 elastic after effect
弹性振动 elastic vibration
弹性滞后 elastic hysteresis
弹性阻力 elastic resistance
坛 bed
坛状花冠 urceolate corolla
檀香木 white sandalwood; yellow sandalwood
檀香脑 santal camphor
檀香油 sandalwood oil; sanders oil; santal oil
炭 char; charcoal
炭材 charcoal-wood
炭粉 charcoal dust
炭化 charring
炭化木材 charred wood
炭化喷油处理(防腐) furnos treatment
炭化喷油法(防腐) furnos process
炭化速率 charring rate
炭化长度 char length
炭阱 gharcoal trap
炭疽病 anthracnose
炭窑 charcoal kiln; charcoal oven; country kiln; earth kiln
炭窑制炭法 kiln-burning
炭砖 briquet(te); carbon brick
探测单元(热感) detector element
探测和归因 detection and attribution
探测器 detector; finder
探测设备 sensor
探测伪装胶片 camouflage detection film
探测元件 detector element

探访 inspection
探火热像仪 thermal imaging system for fire detecting
探空观测 radiosonde observation
探浅球 bathysphere
探热仪 heat probe
探伤仪 defectoscope; flaw-detector
探深球 bathysphere
探针 probe; seed trier
碳-14测定年龄 carbon-14 dating
碳储量 crbon storage; carbon stock
碳氮比 carbon-nitrogen ratio
碳氮化合 carbon-nitriding
碳抵消 carbon offset
碳固定 carbon sequestration
碳还原循环 carbon reduction cycle
碳黑 carbon black
碳化法 carbonization
碳化硅磨料 silicon carbide (mineral)
碳化钨 tungsten-carbide
碳化作用 carbonization
碳汇 carbon sink
碳汇林 forest for carbon sequestration
碳汇造林 afforestation for carbon sequestration
碳积累 carbon gain
碳交易市场 carbon trading market
碳库 carbon pool; carbon reservoir
碳排放 carbon efflux; carbon emission; carbon emmission
碳配额 carbon allowance
碳三途径植物 C_3 plant
碳水化合物 carbohydrate; carbonhydrate
碳税 carbon tax
碳四途径植物 C_4 plant
碳素相纸 carbon tissue
碳酸钙 calcium carbonate; lime carbonate
碳酸化作用 carbonation
碳酸钠波尔多液 Burgundy mixture
碳酸盐岩 carbonate rock
碳损失 carbon loss
碳通量 carbon flux
碳同化(作用) carbon assimilation
碳鎓离子 carbonium ion
碳钨镶齿圆锯 tungsten carbide tip saw
碳吸收 carbon sequestration
碳纤维 carbon fiber
碳纤维复合材料 carbon-fibre composite
碳酰胆碱(药) carbachol
碳酰二胺 carbamide
碳泄漏 carbon leakage
碳信用 carbon credit
碳蓄积 carbon storage
碳循环 carbon cycle
碳源 carbon source
碳增加 carbon gain
碳债务 carbon debt
碳质页岩 carbonaceous shale
碳中性 carbon neutral
碳追踪器 carbon tracker
碳足迹 carbon footprint; carbon consumption
羰化反应 oxo reaction; oxonation
搪瓷 enamel
搪瓷面硬质板 enamelled hardboard
搪瓷漆 lacquer enamel
搪桥(伐倒木中间悬空) in the swing; swinging in a bind
搪桥(料仓) hanging-up
镗锯法 boring cut(ting)
糖蛋白 glucoprotein; glycoprotein
糖苷 glucoside; glycoside
糖苷键 glucosidic bond
糖苷酶 glycosidase
糖苷配基 aglycon(e)
糖化[作用] saccharification
糖化酶 β-amylase
糖化物 saccharide
糖基化作用 glycosylation
糖浆 molasses
糖胶树胶(制口香糖用树胶) chicle
糖酵解途径 glycolytic pathway
糖解酶 glycolytic enzyme; glycolytic ferment
糖精 saccharin(e); saccharol
糖类 carbohydrate; saccharide
糖量测定法 saccharimetry
糖量计(比重法) saccharometer
糖量计(旋光法) saccharimeter
糖酶 carbohydrase
糖蜜 molasses
糖朊 glucoprotein; glycoprotein
糖树脂 glucoresin

糖原(元) glycogen
糖原酵解 glycolysis
糖原酶 glycogenase
糖脂 glucolipid(e); glycolipid; glycolipin
螳螂 mantid; praying mantis
躺椅 lounge
烫号铁 marking iron
趟(运材一次) run; haul
逃避 escape
逃避机制 escape mechanism
逃避运动 avoiding movement
逃税 tax evasion
逃亡蜂群 absconded swarm; absconding swarm
桃胶 peach gum
桃拓罗汉松 totara tree
桃拓酮 totarolone
桃拓烯 totarene
桃汛 freshet
桃檐 cornice
陶钵 clay pot
陶土 kaolin
淘汰 cull; reject; rogue
淘汰分析 siltanalysis
淘析 elution
淘析分析 elution analysis
讨价能力 bargaining power
套(外壳) case; socket
套(轴承) stirps (*plur.* stirpes)
套层蒸馏器 jacketed still
套层注射器 jacketed syringe
套袋注入法 impregnation from end
套叠桌 nested table
套杆 loop bar
套钩(拖拉机后连接捆木索套节) basket hook
套管 slip pipe; thimble
套管法 tire tube process; tire-tube method
套管冷却 jacket cooling
套管楔 socket wedge
套冠花 hose-in-hose flower
套锅 jacketed kettle
套环 burr; thimble; toggle
套环钩(绠漂链连接杆的) toggle hook
套节 knob; nubbin
套节钢箍 ferrule
套捆木索 set bead; set choker
套轮胎外履带 over tire track
套膜 envelope
套式手钻 shell gimlet
套索 sling; sling rope
套索(装车用) noose
套筒 sleeve
套筒式滑轮 thimble block
套筒式取样器 sleeve-type trier
套筒形花 hose-in-hose flower
套筒指轴 telescopic spindle
套准装置 alignment device
特别规定的灌木林 shrubland specified as forest; shrubland assigned as forest; ; administratively assigned shubland
特别收获 unordinary yield
特别危险地段 area of special risk
特产的 endemic
特大火灾 project fire
特定出生率 specific natality
特定年龄存活率 age specific survival rate
特定年龄生命表 age specific life table
特定年龄死亡率 age specific mortality
特定生境 individual site
特定生境的 site-specific
特定时间生命表 time-specific life table
特定死亡率 specific mortality
特福力(茄科宁除草剂) treflan
特赋 special assessment
特厚装垫(沙发) overstuffing
特化进化 gerontomorphosis
特惠税率 preferential tariff
特级硬质纤维板 super-hardboard
特立中央胎座式 free central placentation
特留树 special tree
特柔型起重钢丝绳 extra-pliable hoisting rope
特色 feature
特殊的 individual
特殊动力效应 specific dynamic action
特殊反应 specific reaction
特殊砍伐 special cutting

特殊密化硬质纤维板 special densified hardboard
特殊配合力 specific combining ability
特殊纹理 specified grain
特殊性 specificity
特殊营养的 idiotrophic
特殊植被地理学 special vegetation geography
特税 special assessment
特提斯海 Tethys Sea
特性 character; mark
特性曲线 characteristic curve
特性释放 character release
特许采伐权 lumbering concession
特许权税 excise tax
特许证 patent
特许证书 letters patent
特异抗病性 differential resistance
特用经济树种 economic trees of special use
特用树木 tree of special use
特用树种 technological tree
特优母树林 elite stand; selected crop; selected stand
特优母株 elite plant
特优树 champion tree; etlite tree
特有的 endemic
特有分布 endemism
特有树种 endemic species
特有现象 endemism
特有性 endemism
特有植物 endemic plant
特有种 endemic species
特征 attribute; character; feature
特征波谱 characteristic spectrum
特征尺度 characteristic scale
特征动物 characteristic animal
特征光谱 characteristic spectrum
特征函数 characteristic function
特征明显的变种 distinct variety
特征释放 character release
特征提取技术 feature extraction technique
特征替代 character displacement
特征吸收峰 characteristic absorption peak
特征种 characteristic species
特种伐木 special cutting
特种干燥法 special drying method
特种胶合板 special plywood
特种纤维板 special fiberboard
特种用途林 forest for special use
誊写纸 duplicating paper
藤 rattan
藤本 vine; voluble shrub
藤本的 lianoid
藤本高位芽植物 phaenerophyta scandentia
藤本型 lianoid form
藤本植物 liana; liane
藤架 pergola
藤孔状的 caney
藤器 rattanware
藤纤维 vine fibre
藤椅 basket chair
藤制家具 rattan furniture
藤制品 rattanware
剔出 cull
剔出材 reject; rejected log
剔除 charge off
剔除材(造材时) wood cull
剔除纸 culled paper
梯臂回转索配重滑轮(装车) squirrel suspension block
梯臂式装车 hayrack (boom) loading
梯凳 stepladder
梯度 gradient
梯度薄层色谱 gradient thin-layer chromatography
梯度分析 gradient analysis
梯度功能 gradient capability
梯度假说 gradient hypothesis
梯级 spoke
梯级地理倾差 graded topocline
梯级台地 graded terrace
梯级样板 height board
梯田 terrace; terraced field
梯田埂造林 terrace ridge afforestation
梯纹导管 scalariform duct; scalariform vessel
梯纹的 scalariform
梯纹管胞 scalariform tracheid
梯纹加厚 scalariform thickening
梯纹细胞 scalariform cell
梯形堆积 stepped stacking
梯形接合 scalariform conjugation

T

梯形开沟器 ladder ditcher
梯形纹孔排列 echelon pore arrangement
梯形堰 trapezoidal weir
梯形椅背 slat back
梯状薄壁组织 scalariform parenchyma
梯状穿孔(木材构造) ladder-like perforation; scalariform perforation
梯状穿孔板 latticed perforation plate; scalariform perforation plate
梯状穿孔导管 scalariform vessel
梯状的 scalariform
梯状加厚 scalariform thickening
梯状具缘纹孔 scalariform bordered pit
梯状纹孔式 scalariform pitting
梯子横木 stave
锑白 antimony oxide
踢脚板 baseboard; skirting (board); washboard
踢木器 log kicker
提纯 purification; scouring
提高标价 mark up
提高账面价值 write up
提供资本 capital financing
提供资金 financing
提货单 bill of lading; document against acceptance
提款 withdrawal
提款通知书 advice of drawing
提前 advance
提前期 lead time
提琴背纹 fiddleback; fiddleback figure; fiddleback grain; fiddleback mottle
提琴形背椅 fiddleback chair
提琴用树干基材 fiddle butt
提取 extract; extracting; milking
提取器组 extraction battery
提取松节油 turpentining
提取速度 extraction rate
提取物 extractive; extraneous substance
提取液 extract
提升 elevation; lifting
提升机 lifting transporter
提升量 lifting capacity
提升设备 liftor
提索杆 cable holder
提余液 raffinate
蹄 ungula
体被 integument
体壁 body wall; integument
体表共生 epoekie
体表寄生的 epizoic
体表寄生菌 epiphyte
体表生的 epizoi(ti)c
体干 soma
体积 bulk
体积比 volume ratio
体积传导率 bulk conductivity
体积弹性 bulk elasticity; volume elasticity
体积弹性模量 bulk modulus; volume modulus of elasticity
体积电阻率 specific volume resistance
体积含水量 volumetric water content
体积回弹 voluminal resilience
体积混合比 volume mixing ratio
体积计 volumeter
体积克分子浓度 molar concentration
体积氯度 chlorosity
体积黏弹性 bulk viscoelasticity; volume viscoelasticity
体积膨胀 volume expansion; volumetric swelling
体积膨胀性 volume expansivity
体积收缩 volume shrinkage; volumetric shrinkage
体积收缩量 volume shrinkage mass
体积稳定处理 bulk treatment
体积稳定性 dimensional stability
体积形变 volume deformation
体积应变 volume strain; volumetric strain
体积应力 bulk stress
体节(动物) somite; segment
体壳 sleeve
体内(寄)生的 endobiotic
体内的 endosomatic
体内照射 internal exposure
体内钟 internal clock
体腔 body cavity; coelome
体躯 truncus
体染色体 euchromosome

体视显微镜 stereoscopic microscope; stereomicroscope
体外寄生 ectoparasitism; epiparasitism; epizoite; epizootic
体外生的 epibiotic
体外适应 exoadaptation
体外营养的 ectotrophic
体系(工作方法) regime (*eg.* management regime)
体系的 systemic
体系突变 systemic mutation
体细胞 soma; somatic cell; vegetable cell
体细胞单性生殖 ooapogamy
体细胞的 somatic
体细胞分裂前期 somatic prophase
体细胞染色体加倍 somatic doubling
体细胞染色体减数 somatic reduction
体细胞染色体交换 somatic crossing over
体细胞染色体联会 somatic synapsis
体细胞染色体配对 somatic pairing
体细胞突变 somatic mutation
体细胞有丝分裂 somatic mitosis
体细胞杂交 somatic (cell) hybridization
体细胞杂种 somatic cell hybrid
体型 body size; somatotype
体育公园 sports park
体长 body length
体质 soma; somatoplasm
体质形成 somatogensis
剃 shave
剃齿机 shaving machine
涕灭威(氨基甲酸醋类农药) aldicarb
替代(猎物)种 buffer species
替代成本 opportunity cost
替代发展道路 alternative development path
替代猎物 alternative prey
替代林 substitution forest
替代能源 alternative energy
替代群落 vicarious community
替代物 proxy
替代现象 vicarism
替代植被 substitution vegetation
替代种 vicariad; vicarious species
替换 conversion
替换成本 alternative cost
替换值 replacement value
替续器 relay
天蚕蛾 giant silkworm moth
天车 overhead travelling crane
天敌 natural enemy
天底点 nadir point
天顶 zenith
天顶距 zenith distance
天冬酰胺 asparagines
天蛾 hawk moth; sphinx moth
天蛾幼虫 horn worm
天沟侧板 cant board
天花板 ceiling; ceiling board
天花板挡板 ceiling baffle (board)
天花板托梁 ceiling joist
天际线 skyline
天空光 skylight
天空漫射辐射 diffuse sky radiation
天门冬氨 asparagin
天牛 capricorn-beetle; cerambycid-beetle; long-horned beetle; round-headed borer; ruuudheaded wood borer; sawyer
天牛幼虫 round-headed borer
天牛幼虫型 cerambycoid
天气判断 weather evaluation
天气图 synoptic chart; weather map
天桥 scaffold
天然本底 natural background
天然播种 self-sowing
天然草原 natural grassland
天然成分 natural ingredient
天然单宁 natural tannin; vegetable tannin
天然的 crude
天然堤 levee
天然防腐 natural preservation
天然肥力 inherent fertility
天然丰度 natural abundance
天然风景资源 natural scenery resource
天然干燥 natural drying
天然干燥法 air-drying
天然高聚物 natural high polymer
天然更新 natural regeneration; natural reproduction; natural afforestation; natural formation of wood; self-reproducing; volunteer
天然更新树 volunteer growth

T

天然更新择伐作业 selection system of natural regeneration
天然公园 landscape park; natural park
天然海岸林 natural strand forest
天然互交 open intercrossing
天然纪念物 natural monument
天然架杆 guide tree
天然胶 natural glue
天然景观 natural landscape
天然抗性 natural resistance
天然林 indigenous forest; natural forest; natural stand; wild wood
天然林发育过程(阶段) development phases of (natural) forests
天然林分 wild stand
天然林演替相(阶段) development phases of (natural) forests
天然落枝 self pruning
天然免疫 natural immunity
天然磨石 natural pulpstone; natural sandstone
天然木素 Braun's native lignin; native lignin
天然牧场 natural pasture
天然耐久性 natural durability
天然耐久性试验 natural durability test
天然平衡坡度 equilibrium slope
天然漆 lacquer; lacker
天然区研究 research natural area
天然缺陷 natural defect
天然染料 natural dye
天然鞣质 natural tannin; vegetable tannin
天然色 natural colour
天然食用香料 natural flavour
天然树脂 natural resin
天然树脂清漆 natural resin varnish
天然水 connate water
天然松香 unmodified rosin
天然脱枝 self pruning
天然污染 natural pollution
天然下种 natural seeding; self-sown seed
天然下种更新 natural regeneration by seeds; natural regeneration by seed-shedding; natural regeneration by self-sown seeds
天然纤维 native fiber; native fibre
天然纤维素 native cellulose
天然纤维素纤维 native cellulose fibre
天然橡胶 native rubber
天然修枝 self pruning
天然压条 natural layering
天然颜料 natural pigment
天然幼树 seeding-growth
天然杂种 natural hybrid; vicinism
天然组分 natural ingredient
天文测量法 astronomical method
天文气候 mathematical climate; solar climate
天线 antenna
添加剂 additive
添加系 addition line
田埂 balk
田间 field
田间表现(定植苗等) field performance
田间测定 field test
田间成活率 field survival
田间成苗值 field planting value
田间持水量 field capacity; field moisture capacity; field water capacity
田间的 campestral
田间抗性 field resistance
田间设计 field design
田间试验 field experiment; field test
田间水分当量 field moisture equivalent
田间水分亏缺 field moisture deficiency
田间最大持水量 field maximum moisture capacity
田间作业 field-work
田鼠 meadow mouse
田野的 campestral
田野风光 rustic scenery
田园都市 garden city
田园风景 rural scenery
甜茶苷 stevioside
甜橙油 sweet orange oil; sweet orange peel oil
甜桦油 sweet birch oil
甜罗勒油 sweet basil oil
甜性栲胶 sweet extract
填补处理 presizing
填补木节 knotting
填充 filling; pack; packing

填充床 bed of packing; packed bed
填充剂 filler
填充料 stuff
填充毛细管柱 packed capillary column
填充式钢丝绳 filler strand
填充塔 packed column
填充物 filling; matrix substance
填充柱 packed column
填方 fill
填缝 caulking; filling
填缝胶黏剂 gap-filling adhesive; gap-filling glue
填缝硬化剂 gap-filling hardener
填高 banking
填力 embankment
填料 filler; filling; loading; packing; padding; sizing
填料函 gland; stuffing box
填料器 sizer
填塞匣 stuffing box
填土 embankment
填土动物穴 crotovina
填土机 backfiller
填土涨方 swell
填隙片 shim
挑口板 gutter board
挑选 pick
条 bar
条斑(病毒病症状) stripe
条斑(花纹) streak
条斑状白腐 white mottled rot
条斑状褐腐 brown mottled rot
条板 batten; louver board; ribbon board; riffler
条板芯细木工板 lumber-core board
条播 drill planting; drill seeding; drill sowing; drilling; line drilling; line seeding; line sowing; row seeding; seeding in line; strip sowing
条播锄 drill-hoe
条播机 drill machine; driller; scuffle drill
条播器 drill seeder; seed driller
条蜂科 anthophorids
条根比 shoot-root ratio
条痕 streak
条件 chance
条件常数 conditional constant
条件反射 conditional reflex; conditional response; conditioned reflex
条件气候学 conditional climatology
条件直线复回归 conditional multiple regression
条件直线回归 conditional linear regression
条例 regulation
条裂的 laciniate
条纹 streak; stria (*plur.* striae); striation; stripe
条纹单板 striped veneer
条纹加工 rep finish; repp finish
条纹长石 perthite
条形叶 linear leaf
条状剥皮 barking in strip
条状剥皮机 bark stripping machine
条状加热 strip heating
条状节 knot cut lengthwise; knot horn
条状结构 streaky structure
条状碎片 sliver
条状型刨花板 strandboard
条状增厚 barlike thickening
调风器 air register; air regulator
调合漆 prepared paint
调和湖沼型 harmonic lake type
调和平均数 harmonic mean
调和漆 mixed paint; ready mixed paint
调和曲线 harmonized curve
调和型生物相 harmonic biota
调和中项 harmonic mean
调浆箱 stuff box
调胶 glue preparation
调胶机 glue dissolver; glue mixer
调胶圆筒 glue drum
调节(对温度、湿度、光照等) conditioning; accommodation; regulation
调节处理 conditioning (treatment)
调节基因 regulator gene
调节链 bunk chain
调节漂(流送) catpiece
调节气候 regulate climate
调节器 automatic humidity controller; controller; governor

调节装置 actuating device; adjusting device; pilot
调料 flavour
调配防腐剂 mixing preservative
调频 frequency modulation
调平 level(l)ing
调平生长 balanced growth
调漆油 paint naphtha
调情 mash
调色 colour matching; toning
调色剂 toner
调梢装置 taper set
调湿干燥窑 humidity regulated dry kiln
调湿调温处理 conditioning (treatment)
调湿调温处理室 conditioning chamber
调速器 governor
调味 seasoning
调温凝气阀 thermostatic trap
调温器 temperature controller
调向摘挂(摘挂搭挂树) rolling off
调整 compensation
调整采伐量 regulation of felling budget
调整定位时间 setting time
调整范围 adjusting range; setting range
调整基因 modulator
调整精度 accuracy of adjustment
调整林龄级 age classes in regulated forest
调整期 regulatory period
调整时间 adjustment time
调整条款 escalator clause
跳出滑轮槽 foul
跳刀 chatter
跳动 jump
跳过 skip
跳甲 flea beetle; halticid beetle
跳接 jump cut
跳接线 jumper
跳壳机头 ejector tumbler
跳壳枪(自动) ejector gun
跳伞灭火员 smoke jumper
跳移(跳动) saltation
跳跃 leapfrogging
跳跃进位 skip
贴板 sheeting
贴地生长的 epigeal
贴面 facing; laminate
贴面板 laminated board; overlay
贴面单板 overlaid veneer
贴面胶合板 faced plywood
贴面砂光机 over laying sander
贴面设备 laminated surfacing equipment
贴面纤维板 overlaid fibreboard
贴皮芽接 patch budding
贴生的 adnate
贴水 agio; premium
贴现 discount
贴现边际成本 discounted marginal cost
贴现边际收入 discounted marginal revenue
贴现材积 discounted volume
贴现公司 discount houses
贴现价值 discounted value
贴现净收入 present net worth
贴现率 discount rate
贴现投资 discounted investment
贴现系数 discount factor
贴现银行 discount houses
铁道枕木 railway sleeper
铁-腐殖质硬磐 iron-humus ortstein
铁钩 grappling iron
S形铁钩 pothook
铁管界标 iron-pipe monument
铁黑 iron oxide black
铁红 iron oxide red
铁黄 iron oxide yellow
铁夹 beegle iron; iron clamp; iron jaw
铁结核 iron concretion
铁结核层 marrum
铁路 rail
铁路敞车 gondola
铁路高架装车机 decker; decker loader
铁路机车起重机 locomotive crane
铁路运材 railway logging
铁路终点交货 at-railhead
铁路终点停车场(待编组) yard
铁铝土 ferrallitic soil; pedalfer
铁铝性黏土 sesquioxidic clay
铁铝氧化物淀积层(土壤剖面B_{in}层或B_{ox}层) iron and aluminum

oxide deposition layer; Bin-layer; Box-layer
铁铝质土 ferallitic soil; ferrallitic soil
铁模 swage
铁牌(树木标号用) iorn tag
铁磐 ortsrein
铁器时代 iron age
铁锹 shovel
铁杉栲胶 hemlock extract
铁杉木酚素 conidendrin; tsugaresinol
铁杉树皮栲胶 hemlock extract
铁杉树脂素 tsugaresinol
铁杉油 hemlock spruce oil
铁杉脂素 conidendrin
铁丝测径计 wire gauge
铁丝蛇笼(水土保持工程用) wire cylinder
铁丝网围栏 wire fence
铁条 iron strap
铁细菌 iron bacteria
铁线子胶 balata
铁氧化还原蛋白 ferredoxin
铁硬盘 iron hardpan
铁赭土 paint rock
铁砧 anvil
铁质硅铝土 ferric-siallitic soil
铁质灰壤 iron podzol
铁质结壳土壤 iron crust soil
铁质矿泉 chalybeate spring
铁质黏土 iron clay
铁质砖红壤性土 ferruginous lateritic soil; iron lateritic soil
铁桩 iron stake
听觉临界 threshold of hearing
听觉器官 auditory organ
听觉信号 acoustic signal
听觉阈 threshold of hearing
庭园 courtyard; garden; home garden
庭园陈设 garden furniture
庭园雕塑 garden sculpture
庭园雕像园林雕塑艺术 garden statuary
庭园建筑 garden architecture
庭园空间 garden room
庭园设计 garden design
庭园设计方案 garden lay-out
庭园沿边花坛 border
庭园装饰物 garden ornament
庭院 court; courtyard
庭院露台 garden terrace
停泊处 anchorage
停产 laying off
停车视距 non-passing sight distance
停车支腿(半拖车) parking leg
停点 rest point
停工 laying off
停工的采伐作业区 starvation show
停工期 shut-down period
停工时间 down time
停工损失 factory shutdown loss
停机时间 machine down-time; machine idle-time
停排港 rafter harbor
停业 discontinued operation
停育型蚜 sistens type
停滞 delay; stagnation
停滞阶段 lag phase
停滞空气 stagnant air
停滞木(水滑道曲线处) binding
停滞期 lag phase; stagnation period
挺水植物 emerged plant; emergent
挺水植物带 zone of emergent vegetation
通便 def(a)ecation
通草纸 rice paper
通常竞争 conventional competition
通道(广义) channel; corridor; lane; passage
通道(林区通道) access road
通道细胞 passage cell
通风 aeration; airing; draft; draught; ventilation
通风处理 fanning treatment
通风道 ventilation tunnel; air-drain
通风干湿表 aspirated psychrometer; ventilated psychrometer
通风干燥 aeration-drying
通风干燥窑 ventilating dry kiln
通风管 blow pipe
通风机 fanner
通风降温 aeration-cooling
通风井 air pit
通风孔 flue; vent
通风口 ventilation hole
通风炉 wind furnace

通风温度表 aspirated thermometer; ventilated thermometer
通风闸 ventilating damper
通风装置 vent; ventilating device; ventilator
通钩 gaff
通过动物体内的 endozoic
通过速度 throughput speed
通行权 way leave
通行许可证 right-of-way grant
通径分析 path analysis
通径图 path diagram
通径系数 path coefficient
通廊(水上选材) alley
通量调整 flux adjustment
通路 approach
通气 aeration
通气层(土壤表层) zone of aeration
通气道 admission passage
通气根 aerating root
通气管 air-drain; chimney
通气孔 air hole; ventilating pit
通气口(昆虫) ventilation hole
通气软管 air hose
通气性 aeration; air permeability
通气组织 aerenchyma
通汽软管 steam hose
通信 communication
通信技术卫星 communication technology satellite
通信控制台 communication console
通信卫星 communication satellite
通用材积表 general volume table
通用材料试验机 universal testing machine
通用的 versatile
通用方程 generic equation
通用机床 general purpose machine
通用计算机 general purpose computer
通用计算机语言 universal computer oriented language
通用解算机 general problem solver computer (GPS)
通用商业语言 common business oriented language
通用塔里夫材积表 general Tarifs
通用土壤流失方程 universal soil loss equation
通用拖拉机 universal tractor
通用问题解算机 general problem solver computer (GPS)
通用语言 all-purpose language; general purpose language
通用原木材积表 universal log rule
通用自动计算机 universal automatic computer
通知退约 withdrawal by notice
同胞 sib; sibling
同胞交配 adelphogamy; sib-mating
同胞种 sibling species
同倍体 homoploid
同步 synchronization; synchronous
同步的 synchronic; synchronous
同步培养 synchronous culture
同步调节 synchronization regulation
同步卫星 synchronous satellite
同步系数 synchronization factor
同成岩作用 syndiagenesis
同等条件区 equiconditional area
同等位基因 iso-allele
同等性 homology
同地演替系列 consere; cosere
同电子排列体 isostere
同分异构物 isomer
同工酶 isoenzyme; isozyme
同合子 autozygote
同核的 homokaryotic
同花柱的 isogynous
同化 assimilate
同化产物 assimilation product; assimilation substance
同化的 assimilative; assimilatory
同化呼吸比 assimilation-respiration ratio
同化率 assimilation ratio
同化商 assimilation quotient
同化室 assimilation chamber
同化数量 assimilation number
同化系统 assimilation system
同化箱 assimilation box; assimilation chamber
同化效率 assimilation efficiency; efficiency of assimilation
同化组织 assimilating tissue; assimilatory tissue
同化作用 assimilation
同环二烯 homoannular diene
同基因的 isogenetic

同极的 homopolar
同境群落 aerosynusia
同聚反应 homopolymerization
同类分子 congeneric element
同龄的 even-aged
同龄林 even-aged crop; even-aged forest; even-aged group stands; forest consisting of even-aged stand
同龄林分 even-aged stand
同龄林作业 even-aged system
同龄乔林 even-aged high forest
同龄群生命表 cohort life table
同龄人工混交林 even-aged artificial mixed forest
同龄型 even-aged type
同模标本 homotype
同模式标本 homoeotype; isotype
同配生殖 homogamy; isogamy
同期 synchronization
同亲个体间交配 brother-sister mating
同群生命表 ohort life table
同生群 cohort; guild
同生演替系列 consere; cosere
同时闭合装置 simultaneous closing units; simultaneously closing device
同时扩散 simultaneous diffusion
同属种 congeneric species
同素异形体 allotrope
同态的 isomorphous
同体嫁接 autoplastic graft
同体移植 autograft
同位素 isotope
同位素标记 tagging
同位素计(表) isotope gauge
同位素示踪 tagging
同位素示踪法 isotope tracer
同位素稀释法 isotope dilution method
同位素指示剂 isotopic tracer
同位素追踪剂 isotope tracer
同位性 allelism; allelomorphism
同窝(单指后代) brood
同系的 homologous
同系交配 endogamy
同系聚合物 polymeric homologue
同向捻钢丝绳 uniform lay wire rope
同向气流干燥窑 parallel current type kiln
同效基因 mimic gene
同心层 concentric lamina
同心的 concentric
同心环 concentric ring
同心连接 concentric jointing
同心式(内含韧皮部) Avicennia type
同心式构造 concentric structure
同心式管孔排列 concentric pore arrangement
同心式内含韧皮部 concentric included phloem
同心式维管束 concentric vasculer bundle
同心条纹 concentric stria
同心圆样地 concentric sample plot
同形孢子 homospore; isosporic
同形的 isomorphous
同形林 uniform stand
同形配子 isogamete
同形射线 homogeneous ray
同形种 cryptic species
同型 homeotype; homotype
同型的 homogeneous; homozygous
同型纺锤体 homotypic spindle
同型分裂 homotypic division
同型分子 congeneric element
同型花被的 homochlamydeous
同型花传粉 legitimate pollination
同型花组合 legitimate combination
同型接合体 homozygote
同型木 homoxylous wood
同型配子 homogamete
同型配子的 homogametic
同型配子结合 isogamy
同型同境群落 synusia
同型细胞的 homocellular
同型细胞射线 homocellular ray
同型小种 biotype
同业公会 trade association
同业联合 horizontal integration
同一性 identity
同域分布 sympatry
同域物种形成 sympatric speciation
同域性的 sympatric
同域亚种 sympatric subspecies
同域种 sympatric species
同源 cognation

同源单倍体 autohaploid
同源的 homologous; isogenetic
同源多倍体 autopolyploid
同源基因 isogenes
同源联会 autosynapsis; autosyndesis
同源染色单体 homologous chromatids
同源染色体 chromosome homologous; homologous chromosome; homologue
同源十倍体 autodecaploid
同源四倍体 autotetraploid
同源体起源 autoploid origin
同源系列(群落) homologous series
同源性 homology
同源异倍体 autoheteroploid
同源异形 homoeosis
同源异源多倍体 autoallopolyploid
同源原生质 autoplasm
同源种 congeneric species
同质单倍体 autohaploid
同质的 homogeneous
同质结合 homologous association
同质林 uniform stand
同质性 homogeneity
同质性系数 coefficient of homogeneity
同质异晶 allomorphism
同质异能素 isomer
同质种群 homogeneous population
同种的 conspecific; homogeneous
同种寄生 autoparasitism; adelphoparasitism
同种嫁接 homeoplastic graft; homoplastic graft
同种相食 cannibalism
同种小集团演替系列 family sere
同种异体的 allogenic
同种异型酶 allozyme
同种植物 kindred plant
同株嫁接 autoplastic graft
同株异花受精 geitonogamy
同住的 commensal
同宗配合 homothallism
同宗配合的 homothallic
桐树 tung-tree
桐油 China wood oil; Chinese (wood) oil; lumbang oil; mu oil; tung oil; wood oil
桐油-亚麻子油清漆 China wood-linseed varnish
铜铵法人造丝 cuprammonium rayon
铜铵木素 cuoxam lignin
铜铵溶液 cuprammonium solution; cupriammonium solution; Schweitzeros reagent
铜铵丝 cuprammonium silk
铜包弹丸 copper-coated shot
铜化加铬砷酸锌(防腐剂) copperized boliden salt; copperized chromated zinc arsenate
铜碱液 copper-soap
铜壳 brass shell
铜绿色(家具) patina
铜帽 percussion-cap
铜索传送装置 rope transfer
铜网张紧辊 wire stretch roll
铜网转向辊 wire return roll
铜污(枪筒) metal-fouling
铜锌氨酚防腐剂(溶液) Aczol; Ac-zol
酮分解 ketolysis
统保 all risk insurance
统材[的] common(s)
统计计算 statistical calculation
统计决策 statistical decision
统计员 actuary; statistician
统块(枪扣机) through lump
统售价格 flat rate
统一材积表 unified volume table
统一横轴墨卡托投影 universal transverse Mercator (UTM)
统一收费率 flat rate
统一预算 unified budget
捅钩 hook-aroon; pike pole
桶(容器) tub
桶(石油计量单位, 等于50加仑) barrel
桶(松香计量单位, 等于520磅) drum
桶板 clapboard; lag; wooden stave
桶板材 cooperage bolt; cooperage stock; cooper's-wood; hoop wood; stave-wood; stove bolt; tight cooperage stock
桶板短圆材 stave bolt
桶板锯 bilge saw
桶板条 stave
桶材 tubbing
桶底[板] barrel head
桶端[板] barrel head

桶盖桶底材 heading
桶浸渍法(防腐) barrel method
桶笼用材 coop and cask timber
桶形锯(锯桶板材的弯曲表面) barrel saw; cylinder saw
桶形沙发 tub sofa
筒 barrel
筒仓 silo
筒蠹 ambrosia beetle; ambrosia borer; pinhole borer; ship-timber beetle
筒式磨机 tube mill
筒支梁 beam with simply supported ends
筒状弹仓 tubular magazine
偷盗木材 timber trespass
偷猎 poaching
偷猎者 poacher
偷税 tax evasion
偷渔 poaching
偷渔者 poacher
头道锯 head saw; headrig
头等肉食性动物 first level carnivore
头顶 vertex
头段运输 minor transportation
头盖的 cpicranial
头火 head fire
头架 headstock
头奖 jackpot
头壳 head capsule
头木林 pollarding-forest
头木林更新 regeneration under pollarding method
头木林作业法 branch copic method
头木作业 headwood system; pollard method; pollard(ing) system; pollarding
头木作业更新林 regenerated forest under pollarding method
头细胞[藻] capitulum
头胸部 cephalothorax
头指数 cephalic index
头状花序 anthodium; capitulum; cephalanthium
头状聚伞花序 cymose inflorescence
头状体 capitulum
投标 bid
投标程序 bidding process
投标价格 tender price
投标决策 bidding decision
投入-产出分析 input-output analysis
投入-产出结构 input-output framework
投入资本 venture capital
投影 projection
投影面积 project area; projection area
投影描绘器 camera lucida
投影体积 biovolume
投影图 projection drawing
投资 capitalization
投资保证 investment guarantee
投资回收额 recovery of invested capital
投资集约度 intensity of investment
投资倾向 propensity to invest
投资收益率 rate of return on invested capital
投资预算支出 below-the-line expenditure
投资周转额 turnover
投资资本 invested capital
透翅蛾 clearwing moth
透顶涵洞 open-top culvert
透风堆积 open stacking
透光层 euphotic stratum; photic zone
透光带 euphotic zone
透光伐 interlucation; lighting of tree
透光率 light transmission percent; transmittance
透胶 bleeding; bleed-through; glue penetration; strikethrough
透景伐 vista cutting
透景线 vista
透镜射线 lenticular ray
透裂 through check
透流坝 overflow dam
透明的 diaphan(o)us; hyaline
透明度 pellucidity; transparency
透明度测定法 transparency test
透明胶带 adhesive glassine tape
透明片 transparency
透明性 pellucidity
透明正片 diapositive; transparence positive
透明纸 glassine paper
透平风机 turbo fan
透气度 air resistance

透气度测定仪 densometer
透气性 air permeability; air-conductivity; gas conductivity; gas permeability; permeability
透热厚度 heat penetration
透射 transmission
透射比 transmissivity; transmittance
透射率 transmissivity
透射密度计 transmission densitometer
透射式电子显微镜 transmission electron microscope
透射式电子显微镜图 transmission electron micrograph
透射系数 transmissivity
透声性 acoustic(al) permeability
透视 perspective
透视孔(步枪瞄准器) peep hole
透视图 perspective; perspective drawing; perspective view
透视线 vista
透水的 pervious
透水汽的 moisture transmitting
透水性 water permeability
透榫 hollowed out tenon; through tenon
透雾滤光器 haze-penetrating filter
透析 dialysis
透性膜 permeable membrane
透性系数 permeability coefficient
透性障碍 permeability barrier
透支 overdraft
凸凹形橱柜顶(中央高起两侧凹下) serpentine top
凸凹形橱柜面(正面中央凸出, 两侧凹入) serpentine front
凸出边缘 cock-bead
凸出部(树干鼓包等) bump
凸轮 cam
凸面 crown
凸面圆锯 single-conical saw
凸起 buttress; emboss; lump
凸式线脚 bolection moulding
凸榫 tenon
凸尾 graduated tail
凸线脚 risen moulding
凸线装饰 reeding
凸镶板 raised panel
凸圆线脚 bead
凸圆线脚刨 bead-plane
凸缘 flange; rand
凸缘饰边桌 galleried table
凸缘线脚接缝的 bead-jointed
凸状质壁分离 convex plasmolysis
秃的 bald
秃积雨云 cumulonimbus calvus cloud
秃山 bald mountain; bare mountain
秃山披绿装 clothe bare mountains in verdure
突变 heterogenesis; heterogeny; idiovariation; mutation; saltation; sport
突变变种 emergency sport
突变基因 mutant gene; mutator gene
突变率 mutation rate
突变体 lata
突变型体 mutant
突变压力 mutation pressure
突变载荷 mutation load
突变子 muton
突出层 emergent layer
突出的 emergent
突出木 emergent; emergent tree
突出位置(园林中) foreground
突出物 bulge
突发(大火) conflagration
突发(花叶) flush
突发生态种 abrupt ecospecies
突加载荷 shock load; suddenly applied load
突片 lug
突起 enation; tubercle; umbo (*plur.* umbones)
突起处(集材道) hump
突起生长 enation
突起物 protuberance
突然关机 accidental shut down
突生进化 emergent evolution
突缘 nosing
突跃 break; jump
图案基本花纹 motif
图单位 map unit
图号 identification number
图画衬板 picture-backing board
图画台纸 picture-backing board

图解法 drawing diagram; graphic method
图解估计法 graphical estimation
图解内插法 graphic interpolation
图解式生命表 diagrammatic life table
图解样方 chart quadrat
图解原木板材积表 diagram log rule
图距单位(染色体) map unit
图距离(染色体) map distance
图片自动传输 automatic picture transmission
图示 diagrammatic(al) view
图式栏 picture-field
图像 imagery
图像编码 image encoding
图像表示 image representation
图像处理 image processing
图像分割 image segmentation
图像复原 image restoration
图像描绘 image description
图像信息数字器系统 image data digitizer system
图像压缩 image compression
图像移动补偿器 image-motion compensator
图像增强 image enhancement
图像转换 image transformation
图像转换法 conversion system
图形 figure
图形分类 pattern classification
图形级 pattern class
图形区别 pattern discrimination
图形识别 pattern recognition
图形学 cartography
图形演示 graphical presentation
徒步旅行 hike
徒步原野旅游者 wilderness hiker
徒手画曲线 free hand curve
徒手绘制 freehand fitting
徒手切片 free-hand section
徒手曲线 free hand curve
徒长 overgrowth
徒长的 epicormic; leggy
徒长节 cpicormic knot
徒长枝 epicormic; epicormic branch; turion; water sprout
途行时间(从救火队出发到现场) travel time
途行时间图(救火队) travel-time map
涂白 whitewash
涂布辊 coating roll
涂布树脂(树脂胶结层上) resin over resin bond; resin over glue bond
涂层 coating; washing
涂底漆 primer coating
涂敷浆膏 plastering
涂覆层 overlay
涂光 seasoning
涂胶 gelatinize; glue spread(ing); gumming; spread
涂胶单板 blanket
涂胶辊 spreading roll
涂胶机 glue applicator; glue coater; glue spreader; mechanical glue spreader; mechanical spreader; spreader; spreading machine
涂胶纸 paster
涂焦油 resinify
涂蜡 cere
涂料 coating; padding; paint; washing
涂料封底板 prime-coated board
涂料脱落处 skip
涂料堰 pond
涂料用树脂 coating resin
涂面 lap
涂膜 paint film
涂抹培养 smearing culture
涂末道漆 finishing
涂漆 coating; japanning; lacquering; paint; paint coating; painting; varnish; varnish coating; varnishing
涂漆钉板法(防海生钻孔动物) paint-and-batten method
涂漆特性 painting characteristic
涂饰 finish
涂饰木节 knotting
涂饰器 applicator
涂饰性能 coatability; paintability
涂饰纸 varnished paper
涂刷 brush
涂刷处理(防腐) brush treatment
涂油 gumming
涂油灰 slushing
涂装 coating; painting

T

土 nitosol
土坝 earth dam
土被 soil mantle
土崩 earth fall; ground avalanche; landsliding; landslip
土表结皮 earth crust; soil crust
土层 horizon of soil; soil horizon; soil layer; solum
土场 loading area
土锄 soil pick
土道爬犁 ground sledge
土地 land; soi; ground; acre
土地本利价 capital rental value of soil
土地测绘 site apparaisal; site apparaisement; site mapping; site survey
土地测量 land survey
土地纯收益说 theory of soil rent
土地纯收益最高之轮伐期 rotation of maximum soil rent
土地纯益论 theory of soil rent
土地登记[簿] land register
土地登记处 land registry office
土地等级区划 land class zoning
土地多样性 land variety
土地法 land law
土地分配 land parcelling
土地分配法 land parcelling legislation
土地负荷力 land carrying capacity
土地覆被图 land cover map
土地改良 land amelioration; land improvement; land reclamation
土地耗损 exhaustion of soil
土地价 soil value; value of soil
土地价格 value in land
土地价值 cost value of soil; land value
土地经营 land management
土地可用度分级 land-use capability classification
土地可用度调查 land-use capability survey
土地可用率 land-use capability
土地垦殖计划 land settlement project
土地类型 land-use type
土地类型判读 interpretation of land types
土地利用 land use
土地利用变化 land-use change
土地利用成本 land-use cost
土地利用分类 land-use classification
土地利用规划 land-use planning
土地利用计划委员会 land-use planning committee
土地利用模型 land-use model
土地利用图 land-use map; land utilization map
土地利用现状判读 interpretation of the present state of land use
土地利用型 land-use pattern; land-use type; land-use classification
土地利用政策 land-use policy
土地面积(以英亩计算) acreage
土地描述 destcription of site; site description
土地年金 annual equivalent; soil rent; soil rental
土地平整 land leveling
土地评价 land appraisal
土地期望价 land expectation value; soil expectation value
土地倾斜度 slope of soil
土地区分 land parcel
土地区划 land parcelling
土地生产力 productibility of soil; productivity power of soil
土地使用权 way leave
土地税 land tax
土地所有权(制) land ownership
土地台账图 cadastral map; cadastre; property map
土地特征判读 interpretation of land features
土地调查 land survey
土地调查局 land survey office
土地图 land map
土地休耕补贴制 land retirement
土地增值税 increment tax on land value
土地占有制 land tenure
土地征用权 eminent domain
土地指示 edaphic indicator
土地资本 land capital
土地资源 land resource
土地资源评价判读 interpretation of land resources appraisal

土地租金 rent of soil
土地租金价 soil rent value
土豆-葡萄琼脂培养基 potato-dextrose agar
土堆 hill; mound
土墩 mound
土耳其棓子单宁 Turkish gallotannin
土耳其大黄 Turkey rhubarb
土方工程 dirt work; earth work; ground work
土方计算 calculation of cutting and filling
土方少的筑路 scratch grading
土纲 soil class; soil order
土工作业 earth work
土滑 landslide
土滑道 dirt chute; dirt slide; dry flume; dry slide; earth chute; earth slide; ground chute; ground-slide
土荆芥油 ambrosia oil; chenopodium oil
土居生物 edaphon
土菌消毒(杀菌剂) tachigaren
土科 soil family
土坑(观察土壤剖面用) soil pit
土块 clod
土类 soil great group; soil type
土粒 soil fraction
土链 catena; soil catena
土流 earth flow
土路 dirt path; dirt road; earth road; strip road; unpaved road
土路(无路面的) unmetalled road; unsurfaced road
土埋试验(木材耐腐) graveyard test
土幂 soil mantle
土蜜树叶蜡 wax of bridelia leaf
土面穿入器(土壤气体薰蒸设备) ground-penetrating implement
土木工用材 construction timber for earth work
土木香粉 alantin; inulin
土木香油 elecampane oil
土木用材 earth and underground working timber
土内腐殖质 endo-humus
土内腐殖质层 endorganic humus layer
土内水流 interflow
土磐 pan
土栖白蚁 soil-inhabiting termite; subterranean termite
土栖昆虫 soil insect
土撬 ground sledge
土丘 mound
土壤保持 soil conservation
土壤保护 soil protection
土壤冰柱 ice columns in ground
土壤测定 soil test
土壤层的 edaphic
土壤冲蚀 soil erosion
土壤稠性 soil consistence
土壤处理 soil treatment
土壤传播的真菌 soil borne fungus
土壤传带病害 soil-borne disease
土壤大型动物区系 soil macrofauna
土壤带 soil zone
土壤的 edaphic; pedonic
土壤的生产力 site productivity; yield potential
土壤的植物群落安定相 edaphic climax community
土壤地层学 soil stratigraphy
土壤地带 soil zone
土壤地带性 soil zonality
土壤地理学 soil geography
土壤顶极 edaphic climax
土壤顶极群落 edaphic climax community; pedoclimax
土壤动物 soil animal; soil fauna
土壤动物区系 soil fauna
土壤发生 pedogenesis; soil genesis
土壤发生过程 pedogenic process
土壤发生学分类 genetical classification of soils
土壤反应 soil reaction
土壤肥力 soil fertility
土壤分类 soil classification
土壤分析 soil analysis
土壤风蚀 soil drifting
土壤复壮 soil renewal
土壤覆盖 soil mulch
土壤改良 melioration; soil amelioration; soil amendment; soil improvement
土壤改良剂 soil amendment
土壤改良树 soil improving tree
土壤概测 reconnaissanee soil survey
土壤干旱 soil drought

土壤耕作 soil cultivation
土壤管理 soil management
土壤耗损 exhaustion of soil
土壤呼吸 soil respiration
土壤寄居菌 soil invader
土壤加热电缆 electric soil-heating cable
土壤碱度 soil alkalinity
土壤胶体 soil colloid
土壤接种物 soil inoculant
土壤结持度 soil consistence; soil consistency
土壤结构 soil constitution; soil structure
土壤结合力 binding nature of soil
土壤经营 soil management
土壤抗蚀性 soil anti-erodibility
土壤颗粒 soil particle
土壤颗粒分组 soil separate
土壤空气 soil air
土壤孔隙 pore of soil; soil pore; soil pore space
土壤孔隙度 porosity of soil; soil porosity
土壤矿物 soil mineral
土壤昆虫 soil insect
土壤类型判读 interpretation of soil groups
土壤粒组 soil separate
土壤流失 soil loss
土壤流失性 erodibility
土壤灭菌 soil sterilization; soil sterization
土壤木块试验(木材防腐) soil block test
土壤内排水 internal soil drainage
土壤内蒸发 inner evaporation of soil
土壤排水 soil drainage
土壤剖面 soil profile
土壤气候 soil climate
土壤切片制备 soil split preparation
土壤侵蚀 land retirement; soil erosion
土壤侵蚀类型 soil erosion type
土壤侵蚀判读 interpretation of soil erosion
土壤区划 soil regionalization
土壤圈的 edaphic
土壤壤抗冲性 soil anti-scouribility
土壤溶液 soil solution
土壤杀虫剂 soil insecticide
土壤渗透度 soil permeability
土壤生产力 soil productivity
土壤生态型 edaphic ecotype
土壤生态学 edaphology
土壤生物 edaphon(e); geobiont; soil biota
土壤生物学 soil biology
土壤湿度 soil humidity; soil moisture
土壤试验 soil test
土壤适耕性 soil tilth
土壤水(分) soil water
土壤水分常数 soil moisture constant
土壤水分带 belt of soil water
土壤水分特征曲线 soil moisture characteristic curve
土壤松化 loosening of the soil
土壤酸度 soil acidity
土壤酸性 soil sourness
土壤体系 soil system
土壤调查 land appraisal; soil survey
土壤通气性 soil aeration
土壤透水性 soil permeability
土壤图 soil map
土壤团粒 soil aggregate
土壤退化 land retirement
土壤微动物区系 soil microfauna
土壤微气候 soil microclimate
土壤微生物 edaphon; phytoedaphon; soil microbe; soil microorganism
土壤微生物群 edaphon
土壤微生物学 soil microbiology
土壤温度 soil temperature
土壤无脊椎动物 soil invertebrate
土壤物质 soil material
土壤习居菌 soil inhabiting fungus
土壤习居生物 soil inhabitant
土壤系统 soil system
土壤消毒 soil disinfection; soil sterilization; soil sterization
土壤消毒剂 soil disinfectant; soil sterilant
土壤形成 soil formation; soil former
土壤形成过程 soil formation process; soil forming process
土壤形成因素 soil forming factor
土壤形态[学] soil morphology
土壤性极峰相群集 edaphic climax community
土壤悬液 soil suspension

土壤学 edaphology; pedology; soil science
土壤熏蒸 soil fumigation
土壤压实 soil compactness
土壤亚纲 soil suborder
土壤亚类 soil subgroup
土壤盐渍度 soil salinity
土壤颜色 soil colour
土壤演替顶极群落 edaphic climax community
土壤养分流失 loss of soil nutrient
土壤抑菌作用 soil mycostasis
土壤因素 edaphic factor
土壤营养 soil mixture
土壤硬化 hardening of the soil
土壤有机质 soil organic matter
土壤有效深度 available depth of soil
土壤再生 soil renewal
土壤整段标本 soil monolith
土壤植被关系 soil-vegetation relationship
土壤指标 edaphic indicator
土壤指示剂 edaphic indicator; soil indicator
土壤指示植物 edaphic indicator
土壤质地 soil texture
土壤中型动物区系 soil mesofauna
土壤终极群落 edaphic climax community
土壤注射系统 soil injector system
土壤自然结构体 ped
土壤总含水量 holard
土壤组成 soil constitution
土壤组合 soil association
土壤最小单位 pedon
土砂崩坍防护林 protection forest for land slide prevention
土砂道 mud-road
土砂路 unpaved road
土生的 autochthonous
土生土长的 edaphic
土体 bulk soil
土体层 solum
土系 soil series
土相 soil phase
土型 soil type
土压 earth pressure
土压计 earth pressure cell
土窑 earth kiln
土质 quality of soil
土种 soil species
土柱 soil core
土著的 aboriginal; endemic; indigenous
土著民族 indigenous peoples
土著性 endemism
土砖 adobe
土族 soil family
土钻 earth auger; soil auger
吐鲁香脂 tolu balsam
吐弃 regurgitation
吐水 guttation
吐水毛 hydathodal hair
吐水细胞 hydathodal cell
吐水作用 guttation
吐芽 radical emergence
吐渣 scum
菟丝子 dodder
湍动 turbulence
湍流 turbulence; turbulent flow
湍流层 turbosphere
湍流喷胶 turbulent jet
团集聚伞花序 glomerule
团聚 agglomerate
团聚体 aggregate
团聚性土壤 aggregated soil
团聚作用 aggregation; granulation
团块(大块) crumb
团块(木浆) chunk
团粒 crumb; granule
团粒化泥炭 granulated peat
团粒结构 crumb structure
团粒状粗腐殖质 granular mor
团粒状团聚体 granular aggregate
团流(细胞质中细胞器的流动) mass flow
团伞花 flower cluster
团伞花序 glomerule
团体林 corporation forest
团状伐 gap cutting; gap felling
团状伐作业 gap cutting system
团状择伐体系 pleater system
推板器 board ejector; diverter
推板装置 diverting system
推车杆 jill poke
推出板 headboard
推弹环 magazine follower
推定交货 constructive delivery

推动 impulse
推杆(伐木) killig
推杆(木材集堆) push stick; push(ing) pole; timber rake
推杆(卸材时) kicker
推杆夹 bar clamp
推广费用 sales promotion expense
推河(原木) break a rollway; dump; launching; rolling
推河场 banking ground; launching site; log dump; pokelogan
推河河湾 logan
推河机 log dozer
推河楞场 break-down landing
推河流送 watering
推河斜坡 rollway
推进式干燥窑 tunnel kiln
推力滚珠轴承 ball thrust bearing
推木船 pushing-boat
推木杆 punch bar; pusher bar
推木工(装车) tailer-in
推木器(伐木时) kicker system; log kicker
推木器(制材时) nigger
推木拖拉机 pusher cat
推木小船 water bronco; dozer boat
推石机 rock blade
推式伐木剪 guillotine shears
推手板(保护门上的漆器) finger plate
推树杆(伐木) sampson
推树机 tree dozer; tree eater
推树器 stinger; tree-pusher
推头(伐木用) timber tipper
推土铲 blade; bulldozer blade
推土铲托架 blade bracket
推土机 blade grader; bulldozer; dozer; loading shovel
推土机刀片 bulldozer blade
推土机刮板 blade
推土整辙机 dozer rotter
腿节 femur
退避装置 offset
退耕还林 reversion of agricultural lands to forest; concede the land to forestry
退行进化 retrogressive evolution
退化 degeneracy; degeneration; involution; regress
退化的 dysgenic; rudimentary
退化根 reduced root
退化黑钙土 degraded chernozem
退化碱土 degraded solonetz; soloti soil
退化林 degraded forest
退化铝土 degraded allite
退化特征 degenerative character
退化土 degraded soil
退化型 involution form
退化性状 degenerative character
退化雄蕊 staminode
退化演替 retrogressive succession
退化种 regression species
退化组织 vestigial tissue
退回 return
退菌特(杀菌剂) tuzet
退溶胀作用 deswelling
退隐处所 retirement area
蜕(昆虫) pellicle
蜕壳 exuviation
蜕皮 ecdysis; exuviae; exuviation; moult; moulting
蜕皮激素 ecdysone
褪绿 chlorosis
褪绿病 aetiolation
褪色 discolouration; fade
褪色材 discoloured wood
褪色非标准色 off-colour
褪色效应 discolouration effect
吞并遗传性 absorbing inheritance
吞并杂交 absorptive crossing
吞肩(凸榫两翼肩臂凸出部) relish
囤场 enclosure
臀板(介壳虫) pygidium
臀角 anal angle
臀栉 anal comb
托板升降装置(宽带砂光机) base lifting device
托苞(菊科) palea; palet; pale
托臂梁 hammer beam
托布津(杀菌剂) topsin
托弹簧 magazine spring
托花的 thalamifloral; thalamiflorous
托架 bearer; bracket; chair; lug support; tray
托架带式模框成型系统 tray belt moulding system

托架带式装板机(无垫板) tray belt press loader
托架灯 bracket light
托梁 bearer; joist; saddler's wood
托盘 pallet; pan
托盘集材 pan logging
托盘天平 counter balance
托墙梁 breast-summer
托收票据 bill for collection
托索器 cable holder
托叶 interfoyles; peraphyllum; stipule
托叶痕 stipule scar
托叶鞘 ochrea
托叶芽鳞 stipular scale
托枕 bearer
托座(墙上伸出的) bracket
拖 drag
拖板 planker
拖板耙 sweeper harrow
拖驳 tow boat
拖车 trail car; trailer
拖车连杆 stinger
拖车式运材汽车 tandem truck
拖出索 slack pulling line
拖船 tow; tow boat; tug boat
拖船筏 tug raft
拖杆(畜力集材) single tree
拖钩 towing hook
拖挂集材车 cat bummer
拖集木 pulling
拖拉机铲土车 tractor shovel
拖拉机队队长 cat hooker
拖拉机集材 tractor hauling; tractor logging; tractor skidding
拖拉机集材(无拱架) bob
拖拉机集材牵引索 bull line
拖拉机集材索 tractor-cable
拖拉机库 garage
拖拉机牵引爬犁 tractor sledge
拖拉机生产数据 tractor production data
拖缆 towing cable
拖链 drag line; drag(ging) chain
拖链槽(原木端头砍口) dee
拖链除灌 chaining; pulling
拖链孔(在集材原木端头钻的孔) drag hole
拖轮 tug; tugger
拖排 tow; tug raft
拖排绞盘机 towing winch
拖欠年金 annuity in arrears
拖欠债务 delinquency
拖索除灌木 cabling
拖尾 tailing
拖曳 tow; towage
拖曳板 drag-lizard
拖曳锯 dragsaw; log cut-off saw; oscillating saw
拖曳索 drag line
拖运(水上) sack; towage
拖运费 towage
拖运水道 tow-way
脱氨基酶 deaminase
脱层 delaminating
脱层现象 delamination
脱翅 dealation
脱氮细菌 denitrifying bacteria
脱分化 dedifferentiation
脱辅基酶蛋白 apoenzyme
脱附能 energy of desorption
脱附器 desorber
脱钙作用 decalcification
脱钩器 release
脱钩索 throw line
脱轨[器] derail
脱碱化 solod
脱碱化土壤 solodic soil
脱碱土 solod; solod soil; soloth soil; soloti soil
脱碱作用 solodization
脱胶 degumming
脱胶苎麻 degummed ramie
脱焦油 detar
脱壳 hulling
脱壳机(种子) seed huller; huller
脱矿质水 demineralized water
脱蜡 dewaxing
脱蜡漂白紫胶 refined bleached lac
脱蜡脱色紫胶片 dewaxed decolourized shellac
脱蜡紫胶 dewaxed shellac
脱离子水 deionized water
脱粒 extraction; seed extraction; seed husking; thrashing
脱粒厂 extractory
脱粒干燥室 seed kiln
脱粒滚筒 extraction drum

脱粒机 huller; seed extractor; sheller; thrasher
脱粒器 extractor
脱粒室 extractor; seed extractory
脱粒筒 extraction drum
脱硫作用 desulfurization
脱落 scale-off; shedding
脱落的 caducous
脱落节 falling knot; hollow knot; loose knot
脱落素 abscisin
脱落酸 abscisic acid
脱落物 exuviation
脱氯剂 antichlor
脱镁叶绿素 pheophytin
脱模(塑物) knockout
脱模剂 mould release agent; parting agent; release agent
脱墨 deinking
脱木质素 delignification
脱皮激素 molting hormone
脱皮液 molting fluid
脱漆剂 paint remover
脱气 degassing
脱潜育化 degleyfication
脱羟基作用 dehydroxylation
脱氢 dehydrogenation
脱氢酶 dehydrogenase
脱氢松香 dehydrogenated rosin
脱鞣 detanning
脱乳 demulsify
脱色剂 decolourant; decolourizer
脱色栲胶 decolourlzed (tannin) extract
脱色力 decolourizing ability
脱色炭 decolourizing carbon; decolourizing char(coal)
脱色紫胶片 decolourized shellac; decolourized shellac
脱色作用 decolouration; depigmentation
脱湿 desertion of moisture
脱水 dehydration; desiccation; dewatering
脱水成型机 dewatering machine
脱水二儿茶素 anhydrodicatechin; dehydrodicatechin
脱水机 decker; thickner
脱水基质 anhydrobase
脱水剂干燥法 drying with absorptive chemical
脱水酶 dehydrase; dehydratase
脱水作用 slushing
脱羧酶 decarboxylase
脱盐作用 desalinization
脱氧红紫胶素 deoxyerthrolaccin
脱氧檀香灵 deoxysantalin
脱叶 defoliation
脱脂 degreasing; scouring
脱脂木片 spent turpentine chip
脱种机 extractor
陀螺状龙脑香 gurgan; gurjun; mai yang
驼峰(森铁) hump
妥尔油 liquid rosin; tall oil; tallol
椭圆节 oval knot
椭圆粒子 ellipsoid particle
椭圆轮式计量仪 oval wheel type meter
椭圆形生长 elliptic growth
拓扑学 topology
唾腺 salivary gland
唾腺染色体 salivary gland chromosome
唾液分泌过多 salivation
唾液管 salivary duct
箨(tuò) vagina
箨耳 sheath auricle
箨叶(片) sheath blade
挖补 stop
挖补腻子 putty retouch
挖除伐根 uprooting
挖方 cut; excavation
挖根机 extractor; root extractor; root grubber; rootrake; uprooter; uprooting machine
挖根机掘齿 rooter bit
挖根犁 root plough; root plow
挖根螺旋钻 stump auger
挖根器 rooter
挖沟机 ditch cutter; trench cutting machine; trencher
挖沟器 ditcher
挖掘铲 digger blade
挖掘机 digger; excavator; raiser; shovel
挖掘机工作半径 shovel reach
挖掘犁 undercutter

挖掘装载机 loader-digger
挖坑机 digging machine; earth auger; earth boring machine; hole digger; planting hole machine
挖坑螺旋钻 digging auger
挖孔器 dibber
挖苗机 tree digger; tree lifter
挖苗犁 grubber
挖泥机(船) dredge
挖树机 tree mover
挖土 excavation
挖土机 banker; excavator; ground-hog
挖穴机 dibbler; digging machine; earth boring machine; hole borer
挖穴器 dibble board; hole borer
挖穴式植树机 automatic injection planter
洼地 depression; low-land
瓦格纳带状伞伐法 Wagner's border cutting
瓦格纳带状伞伐作业法 Wagner's "Blendersaumschlag" system; Wagner's method
瓦解 disintegration
瓦块式制动器 block brake
瓦楞 flute
瓦楞板材料 corrugated sheet material
瓦楞钉 corrugated nail
瓦楞夹心原纸 corrugating medium
瓦楞胶合板 corrugated plywood
瓦楞面 corrugated surface
瓦楞纹(材面) washboarding
瓦楞形影响 washboarding effect
瓦楞纸板 corrugated board
A形瓦楞纸板 A-flute
瓦楞皱缩(早晚材收缩不均) ribbing
瓦利施胶度测定仪 Valley size tester
瓦弯材 cupped wood
瓦状细胞 tile cell
歪曲 distortion
歪长石 analbite; anorthoclase; anorthose; anorthoclasite
外苞[对向]叶 subtending leaf
外被 tegument
外壁 ectotheca; exospore; peridium
外壁内层 nexine
外表层 ectoderm
外表皮 exocuticule
外部(散孔材) outer part
外部冻裂 outside frost cleft
外部腐朽 external decay
外部环境 external condition
外部冷凝干燥窑 external condensing-coil kiln
外部缺陷 exterior defect; external defect
外部施胶 external sizing
外部收益 external benefit
外部物质 extraneous material
外部性 externality
外部因素 extrinsic factor
外部应力 external stress
外部装修 exterior finish
外材面(锯材) outer surface; outside surface
外侧傍管薄壁组织 abaxial paratracheal parenchyma
外侧薄壁组织 abaxial parenchyma
外侧卷筒(绞盘机) side spool
外侧尾羽 outer rectrix (*plur.* -trices)
外侧原木 mush log
外层(细胞) ectoderm; ectotheca
外层(细胞壁) outer layer
外层腐朽 shell rot
外层含水率 shell moisture content
外插(数学) extrapolation
外差法 heterodyne
外成土 ectodynamomorphic soil; ektodynamomorphic soil
外代谢产物 extra-metabolites
外担菌属 exobasidium
外地种 exotic
外动力土 ektodynamic soil
外动力土壤 ectodynamic soil; ektodynamic soil
外动力形成的 ectodynamomorphic
外动力型土 ektodynamomorphic soil
外动力型土壤 ectodynamomorphic soil; ektodynamomorphic soil
外动性土壤 ectodynamic soil
外动性演替 ectodynamic succession
外毒素 exotoxin
外颚叶(昆虫) galea
外翻 plough out
外分泌物 ectocrine substance
外佛焰苞 outer spathe

外稃 lemma
外敷层 overcoating
外附生生物 epicole
外骨骼 exoskeleton
外果壳 husk
外果皮 ectocarp; epicarp; exocarp
外函 ectotheca
外弧口凿 outside gouge
外花被 outer perianth; outer perigone
外激素 ecto-hormone; exohormone; pheromone
外寄生 ectoparasite; ectoparasitism; epiparasitism; epizoic; external parasitism
外寄生虫的 ectoparasitic
外寄生的 ectoparasitic; ectotrophic
外寄生物 ectoparasite
外寄生性 ectoparasitism
外加力 applied force
外加应力 applied stress
外加转矩 external torque
外夹皮 exposed bark-pocket; outer bark pocket
外界环境 external environment
外界条件 external condition
外界压力 ambient pressure
外界因素 external factor
外景(园林) exterior view
外景(自然) outskirt
外径 external diameter
外径规 gap gauge
外卷叶 revolute leaf
外卡轴 outer spindle
外壳 hood; housing; peel
外空生物学 exobiology
外口式的 ectognathous
外廓材积 superficial volume
外来病害 introduced disease
外来材 exotic timber; exotic wood
外来的 adventive; alien; allochthonous; exotic
外来的冰碛覆盖物 nappe
外来花粉 foreign pollen
外来泥炭 allochthonous peat
外来泉局的 crenoxenous
外来染色体 alien chromosome
外来树 exotic tree
外来物种 alien species; exotic species
外来因素 external factor
外来植物 adventitious plant; alien plant; exotic plant; foreign plant
外来植物区系 allochthonous flora
外来种 adventitious species; alien; alien species; allochthonous species; exotic species
外力 external force
外裂片 outer lobe
外鳞片 surfoyl
外流 efflux
外露木材(沙发等) show wood
外轮对瓣的(雄蕊群) obdiplostemonous
外轮对萼的 proterosepalous
外貌(群落) physiognomy
外门 exterior door
外面 epi-
外膜(病毒) envelope
外派津贴 assignment allowance
外胚被 outer indusium
外胚层 ectoderm
外胚乳 perisperm
外胚叶 ectoderm
外皮 ectoderm; exodermis; outbark; outside bark; over bark; peel; periderm; rhytidome; skin
外皮层 exodermis
外蹼 outer web
外鞘 ectotheca
外热釜 externally fired retort
外韧维管束 collateral bundle; collateral vascular bundle
外扇干燥窑 external-fan kiln
外伤 trauma; wound
外伸伐根 overhung stump
外伸支腿(装卸机) outrigger leg
外渗 exosmosis
外生孢子 exospore
外生代谢作用的 exogenous
外生担子 exobasidium
外生的 ectotrophic; epibiotic; exogenous; exotrophic
外生菌根 ectomycorrhiza; ectotrophic mycorrhiza; exomycorrhiza; exotrophic mycorrhiza
外生菌根菌 ectomycorrhizal fungi
外生菌根真菌 ectomycorrhiza

外生韧皮部 external phloem
外生现象 ectogeny
外生殖器 genitalia
外生中柱鞘 outer pericycle
外食的 ectophagous
外始式 exarch
外始式原生木质部 exarch protoxylem
外视图 exterior view
外树皮 outer bark
外涂层 external coating
外推法 extrapolation
外围保护地带 outside buffer zone
外显成本分析 explicit analysis of cost
外显率 penetrance
外显子(遗传) exons; extron
外向 ectad
外向通量 efflux
外消旋作用 racemation; racemization
外形尺寸 external dimension; overall size
外形线 visible line
外循环 extrinsic cycle
外循环干燥窑 external circulating kiln
外咽片 basilaire; gula
外延 extrapolation
外延生长率曲线 extrapolating growth percent curve
外养生物 ectotroph
外业记录簿 field-note
外业设计 field design
外业数据 field data
外叶 outer lobe
外因 external factor; extrinsic factor
外因的 allogenic
外因性迁移 exogenous migration
外因性休眠 exogenous dormancy
外因演替 allogenic succession
外因周期 extrinsic cycle
外颖 lower glume; outer glume
外应力 external stress
外用清漆 exterior varnish
外用罩面漆 exterior finish
外原型原生木质部 centripetal protoxylem
外圆角[刨] bull-nose
外缘(翅) outer margin
外缘集材距离(由伐区边界到装车场) external yarding distance
外源的 exogenous
外源地下水 allochthonic ground water
外源浮游生物 allelochemical(s); allogen(et)ic plankton
外源河 exotic stream
外源性 exogenous
外源休眠 exogenous dormancy
外源因素 extrinsic factor
外在环境 external environment
外在优势 extrinsic superiority
外展纹孔口 extended (pit) aperture
外展作用 accretion
外植 explantation
外植体(组织培养) explant
外种皮 exotesta; outer coat
外周神经系统 peripheral nervous system
外珠被 external integument; outer integument; primine
外驻扑火队 stand-by crew
外装风扇干燥窑 external-fan kiln
外子座 ectostroma
弯板机 bender
弯背椅 hogarth chair
弯刀(修灌用) bush knife
弯刀(修枝用) billhook; matchet
弯道集材 snake
弯管 angle pipe; siphon
弯管接头 deflection hood
弯管器 bender
弯脚钉合 clench nailing
弯脚器 cal(l)iper
弯矩 moment of flexure
弯矩图 bending moment diagram
弯曲 bending; bending flexure; bent; buckling; crook(ing); curve; incurvation; kink; sweep; sweeping; tortuosity; warp; wow
弯曲半径 bending radius
弯曲变形 bending deflection; buckling deformation; flexural displacement
弯曲材 saddle; curved timber
弯曲弹性模量 modulus of elasticity in bending
弯曲的 sinuous

W

弯曲多节材 kneed timber; kneewood
弯曲干基 pistol butt
弯曲机 bender
弯曲极限 buckling limit
弯曲胶合板 bent plywood
弯曲劲度 bending stiffness
弯曲力矩 bending moment
弯曲临界载荷 buckling load
弯曲木(人工) bent wood
弯曲木(自然形成) crooked stem; bending-wood; compass-timber; crook; cruck; knee; sweeper; sweeping tree; wow
弯曲木定心下锯 sweep sawing
弯曲木段 bent log
弯曲木家具 bendwood furniture; bentwood furniture
弯曲挠度 bending deflection
弯曲疲劳试验 fatigue-bending test
弯曲破坏 bending failure; flexural failure
弯曲伸缩率 upset
弯曲试验 angular test; bending test; flexural test
弯曲树 crooked tree
弯曲位移 flexural displacement
弯曲应变 bending strain; buckling strain; flexural strain
弯曲应力 bending stress; buckling stress; flexure stress; stress in bending
弯曲应力系数 buckling stress coefficient
弯曲原木 bent log; crooked log
弯曲折断载荷 buckling load
弯曲振动 bending vibration; flexural vibration
弯曲竹秆 curved culm
弯缺 sinus
弯入 embayment
弯头 elbow
弯液面 meniscus
弯状带伐 bay cutting; bay felling; cove strip-cutting
丸化种子 pelletted seed
完顶体 full cone
完寄生 micanoparasitism
完了行为 consummatory act
完满度(树干) non-taperness
完满立木度 full stocking; fully stocking percent
完满立木度的同龄林 fully stocked even-aged stand
完满立木度林 fully stocked woods
完满立木度林分 full-stocked stand
完满树干 full-bole; full-boled stem
完模标本 holotype
完全变态 complete metamorphosis
完全剥皮 complete barking
完全的 entire
完全肥料 complete fertilizer
完全花 complete flower; monoclinic; perfect flower
完全花的 monoclinous
完全寄生物 holoparasite
完全浆料 perfect stuff
完全经营合作社 complete working co-operative society
完全均衡分离 complete equational segregation
完全密度依赖因子 perfectly density-dependent factor
完全苗 complete plant material
完全耐水胶合板 type-one plywood
完全培养基 complete medium
完全燃烧 perfect combustion
完全渗透 solid penetration
完全双列杂交 complete diallel cross
完全调整林 fully regulated forest
完全显性 complete dominance
完全小花 perfect floret
完全性抑制 complete inhibition
完全叶 complete leaf; perfect leaf
完全抑制点(防腐剂致死浓度) total inhibition point
完全营养溶液 full nutrient solution
完全郁闭 complete canopy
完全致死基因 absolute lethal gene
完熟 dead maturity; full maturity
完熟种子 fully ripened seed
完税价格 dutiable value
完整虫孔 sound wormy hole
完整林 intact forest
完整苗 complete plant material
玩忽职守 malpractice
玩具材 wood for children's toy

玩具用材 timber used in children's toy
挽杆 whiffle-tree; whipple-tree
挽绳杆(畜力集材) swingle bar; swingle-tree; single tree
晚材 autumn wood; late wood; summer wood
晚材带 late wood zone; summer wood zone; zone of late wood
晚材导管分子 late wood vessel element; summer wood vessel segment
晚材率 late wood percentage; percent of late wood; percent of summer wood; proportion of late wood; proportion of summer wood
晚材区 zone of late wood
晚成期的(需要母鸟照顾一段时期的) altricial
晚放叶品系 late-flushing strain
晚放叶组 late-flushing groups
晚开裂的(叶花果) serotinous
晚皮(树皮) hard bark
晚期树皮 late bark
晚熟的(果实) serotinous
晚熟性 late-maturity
晚霜 last injurious frost; late frost
晚夏的 serotinal; serotinal
晚夏景色 serotinal aspect
椀刺 tirnao
椀子 valonia
碗橱 cupboard
万能工具磨床 universal tool grinder
万能胶 epoxy resin adhesive
万能木工机床 universal woodworking machine
万能试验机 universal test machine; universal testing machine
万能削片机 multipurpose tree chipper
万能圆锯 radial (arm) saw; radial-arm saw
万能钻孔机 universal drill
万向导索器 bull's eye fairleader; universal lead
万亿(10^{12}) tera-
万亿克(10^{12}g) teragram (Tg)
万用表 multimeter
万有引力 gravitation
腕锁(脊椎) jugum
王草因 imperatorin
王室林 crown forest; crown timber land; crown wood
王室林地 crown land
网案 forming table
网案架 wire frame
网部 wire part
网采浮游生物 net-plankton
网蝽科 lace bug
网带干燥机 wire belt dryer
网带式单板干燥机 belt veneer dryer; chain-belt conveyer veneer dryer; conveyer (veneer) dryer
网带式干燥机 mesh dryer; mesh-belt drier
网带输送机 mesh-belt conveyor
网点板 dot grid
网点抽样 dot grid sample
网格线 grid line
网痕 wire mark
网痕面(纤维板) screen-back
网架组织 trabecula (*plur.* -lae)
网结 anastomosis
网孔 aperture
网络分析 network analysis
网面 wire side
网目 aperture; mesh
网前箱 breast box; flow box; head box
网前箱浆料液面 pond
网球场 court
网速 wire speed
网纹 webbing
网纹导管 reticulate vessel
网纹管胞 reticulate tracheid
网印 wire mark
网印痕凸面(纤维板成型模压) wire side convex
网状薄壁组织 reticulate parenchyma
网状穿孔 reticulate perforation
网状穿孔板 reticulate perforation plate
网状格局 network pattern
网状河道 braided stream
网状河绠 grating for stopping floating wood
网状加厚 reticulate thickening
网状结构 network structure

网状裂隙土 fissure network soil
网状脉 reticulate vein
网状脉序 netted venation
网状石质土 net-like stone soil
网状顺河绠 parallel boom of net type
网状-梯状穿孔板 reticulate scalariform perforation plate
往返 turn
往返迁移 return migration
往复带式运输机 shuttle belt conveyor
往复横截锯 dragsaw; log cut-off saw
往复锯 oscillating saw; reciprocal saw; reciprocating (blade) saw
往复式打榫眼凿 oscillating mortising chisel
往复式框锯 deal flame saw
往复式铺装线 reversible forming line
往复式榫槽机 oscillating-bit mortiser
往复式索道 jig-back cableway; return track cable; returning cableway; to-and-fro cableway
往复式索道系统 jig-back system
往复系统 to-and-fro system
往复运动 alternating motion
往来账户 book account
旺盛生长木 thriving tree
望板 boarding; sheathing; lagging
望点 Pressler reference point
望高 pressler reference height
望高法 Pressler method
望火员 towerman
望台望楼 outskirt
望远镜 telescope
望远镜水准仪 telescope level
望远速测镜 tele-relascope
危害分级(等级) hazard rating
危害现象 infestation
危险的人 masty
危险度 risk factor
危险断枝(挂在树上) widow maker
危险水位 flood stage
危险线(洪水水位) danger line
威巴(农药) velpar
威百亩(杀菌剂) sodium methyl dithiocarbamate; Vapam
威布尔分布 Weibull distribution
威利粉碎机 Wiley mill
威廉斯双层振动筛 Williams 2-deck vibrating screen
威尼斯松节油 Venice turpentine
威吓色 threat(en)ing colouration
威胁行为 threat behaviour
微胞 supermolecule
微胞束 micellar strand
微薄层 micro lamella
微薄木 micro-veneer
微波 microwave
微波传感器 microwave sensor
微波干燥法 microwave drying
微波光谱仪 microwave spectrometer
微波谱法 microwave spectroscopy
微波探测器 microwave sounder
微层填覆工艺 micro-sealing process
微尘 dust
微沉淀试验 microprecipitation test
微成层现象 microstratification
微程序设计 microprogramming
微处理机 microprocessor
微刺 spinule
微地表火 light surface fire
微地形 micro topography
微地形鼓起(如倒木或木桩等) elevated microsite
微电解池 micro cell
微动气压计 statoscope
微分[法] differentiation
微分方法 differential method
微分膨胀 differential swelling
微根管 minirhizotron tube
微共生物 microsymbiont
微观构造 microscopic structure
微观构造破坏 microscopic structural fracture
微观检验 microscopic examination
微观结构 microstructure
微观特征 microscopic feature
微观症状 microscopic symptom
微光周期性 crepuscular periodicity
微环境 micro-environment
微极化池 micro cell
微嫁接 micrografting
微尖的 acutate
微结构单位 fabric unit
微进化 microevolution
微晶 crystallite; micell(e); micella(*plur.* -llae); micro crystal

微晶结构 microcrystalline structure; microlitic structure
微晶纤维素 microcrystalline cellulose
微晶纤维素酶 avicelase
微菌落 microcolony
微孔(纹孔膜上) pit membrane pore; pore; micropore; microporous
微粒 corpuscle; fleck; particulate matter
微粒体 microsome
微量 trace
微量音调计 microtonometer
微量营养元素 micronutrient; minor nutrient element
微量营养元素缺乏症 micronutrient deficiency
微量元素 microelement; micro-nutrient element; minor element; trace element
微量注射器 microsyringe
微裂 crazing
微毛细管 microcapillary
微磨刀刃 jointing
微黏度计 microvisco(si)meter
微凝集试验 microprecipitation test
微扭计 torsion micrometer
微起伏 microrelief
微气候 microclimate
微气象学 micrometeorology
微缺的 nicked
微染 tint
微蠕变 micro-creep
微伤 scratch
微商 differential quotient
微生境 microsite; niche
微生境优势种 dominule
微生态系统 microecosystem
微生物 germ; microbe; microorganism
微生物防治 microbial control
微生物分解 microbial decomposition
微生物分析 microbiological analysis
微生物降解 microbiological degradation
微生物区系 microbiota
微生物群落 microbiota
微生物杀虫剂 microbial pesticide
微生物生态学 microbial ecology
微生物学 microbiology
微生物学测定法 microbiological assay
微生物遗传学 microbial genetics
微丝 fibril
微丝晶 fibrillar crystal
微酸的 acescent
微体繁殖 micro-propagation
微体化石 microfossil
微调发动机 vernier
微团 micell(e); micella(*plur.* -llae)
微细管道 microtubule
微细裂缝 microscopic check
微细纤维 microfibril
微细纤维质索 fine cellulosic strand
微细遗传学 microgenetics
微纤丝 microfibril
微纤丝参量 microfibril parameter
微纤丝方向图像 microfibril(lar) orientation pattern
微纤丝分布 microfibril distribution
微纤丝间质 intermicrofibrillar substance
微纤丝角 microfibril(lar) angle
微纤丝螺旋排列方向 helical orientation of microfibril
微纤丝束 microfibril strand
微纤丝网系 microfibril(lar) network
微纤丝轴 microfibril axis
微纤丝走向 microfibril(lar) orientation
微纤丝组织 microfibril(lar) organization
微咸水草甸 brackish water meadow
微咸水沼泽 brackish water swamp
微小浮游生物 microplankton
微效多基因 polygene
微效基因 minor gene
微效基因抗病性 minor gene resistance
微斜面宽度(旋刀) joint width
微形叶 leptophyll; nanophyll
微型比色槽 micro cell
微型底栖生物 microbenthos
微型浮游生物 nannoplankton
微型计算机 microcomputer
微型游泳生物 micronekton
微型园艺 miniature gardening
微型真菌(包括半知菌类及子囊菌) micro-fungus
微型植物 microphyte

微型植物群落(区系) microflora
微需氧细菌 microaerophile
微芽接 micro budding
微亚种 microsubspecies
微演替系列 serule
微液槽 micro cell
微应变 microstrain
微疣 wart
微雨量器 ombrometer
微圆齿状的 subcrenate
微阵雨 sprinkle
韦尔豪斯防腐法 Wellhouse process
韦斯测高器 Weise's hypsometer
为害分生组织的昆虫 meristem insects
违反合同 breach of contract
违犯林业规章 forest offence
违犯森林警规 contravention of forest-police
违章作业 malpractice
围场 holding paddock; paddock
围尺 diameter tape
围地 exclosure
围封 enclosure
围干处理 banding treatment
围隔 enclosure
围埂分块灌溉 border irrigation
围绠 rafting boom
T形围绠 keel towing boom
围级择伐 girth-limit cutting
围剿火势 contain a fire
围控(控制于防火线内) contain a fire; corral a fire
围控火势 corral a fire
围控时间 corral time
围栏 enclosure; fence; railing
围栏封育 enclosure
围篱桩 fence post
围梁 girth
围猎 enclosing of game; hunt units
围生殖孔腺(介壳虫) circumgenital gland
围食膜 peritrophic membrane
围食膜的 peritrophic
围限 minimum girth
围眼的 circumocular
围堰 coffer dam
围蛹 coarctate pupa; puparium
围脏窦 perivisceral sinus
围长(树干) girth; periphery
桅杆 mast; spar
桅杆材 mast timber
桅杆起重机 mast crane
桅杆清漆 spar varnish
唯美主义 aestheticism
维 dimension
维度分带 latitudinal zonation
维恩位移定律 Wien's displacement law
维管的 vascular
维管分生组织 vascular meristem
维管骨架 vascular skeleton
维管管胞 vascular tracheid
维管射线 vascular ray
维管束 bundle; conducting bundle; fibro-vascular bundle; vascular bundle; vascular cylinder
维管束的 fascicular
维管束痕 bundle scar
维管束间的 intervascnlar
维管束鞘 bundle sheath; vascular bundle sheath
维管束系统 fascicular system
维管束原 provascular strand
维管束原组织 provascuar tissue
维管束植物 vascular plant
维管系 vascular system
维管形成层 vascular cambium
维管植物 tracheophyta
维管柱 vascular cylinder
维管组织 fibro-vascular tissue; vascular tissue
维护 handling
维森索系 Wyssen system
维生素 vitamin
维生素C ascorbic acid
维生素P citrin
维生素原 provitamin
维氏捆木索钩 Wirkkala hook
维氏松紧式索道集材 Wirkkala (slackline) yarding
维氏索道集材法(双绞盘机松紧式) Wirkkala system
维氏硬度试验 Vickers (hardness) test
维修工作量 maintenance load
维修用漆 spare varnish
伪不变特征 pseudo-invariant feature
伪彩色 pseudo colour

伪菌根 pseudo mycorrhiza
伪年轮 double ring; false ring; fictitious ring
伪髓线 false ray
伪心材 false heartwood; false-heart
伪形率 false form quotient
伪形率(胸高形率) breast-height form quotient; artificial form quotient; false form quotient
伪形数(胸高形数) breast-height form factor; artificial form factor; chest-height form factor; common form factor; form factor at breast height; false form factor
伪杂种 false hybrid
伪着性植物 pseudoepiphytes
伪装 camouflage
伪装识别 false identification
伪装探测胶片 camouflage detection film
伪足 false leg
尾 cauda
尾板 pygidium
尾材 tailing
尾端滑轮 tail block
尾端物件 tail piece
尾端原木(成串集材) trail log; trailer
尾峰 end peak
尾钩 tail hook
尾集材杆 back spar; tail spar; tail tree
尾集材杆鞍座(承载索) tail jack
尾集材杆索具 back rigging
尾节 pygidium
尾流 backwash
尾馏分 last runnings
尾片 cauda; stylus (*plur.* styli)
尾气 off-gas
尾筛 back knotter
尾上复羽 upper tail covert
尾绳(矿) tail rope
尾水河段 tailwater
尾下复羽 lower tail covert
尾须 circus; stylus (*plur.* styli)
尾旋 backwash
尾翼截面(风向标) airfoil section
尾羽 tail feather
尾渣浆 tailing pulp
尾脂腺 oil gland; uropygial gland
尾制动链 tail chain
尾制动索钩 tail grip
尾柱 end mast; tail spar; tail tree
尾柱集材 end mast
尾状管 aeriductus
尾锥形的 boat-tailed
尾足(昆虫) uropod; anal leg
纬度 latitude
纬线 abscissa
委托(把　委托给) consign; bailment
委托购买证 authority to purchase
委托林 consigned forest
委托书 proxy statement
委托销售 sale on commission
萎黄矮化病 chlorotic dwarf
萎蔫 wilting
萎蔫病 wilt disease
萎蔫点 wilting point
萎蔫率 wilting percent
萎蔫湿度 wilting point
萎蔫系数 wilting coefficient
萎缩 atrophy
萎锈灵(杀菌剂) carboxin
卫矛苷 euonymin; euonymus
卫生伐 sanitation cutting; sanitation felling
卫生害虫 sanitary insect
卫生设备 sanitary equipment
卫星 satellite
卫星红外光谱仪 satellite infrared spectrometer
卫星数据处理系统 satellite data processing system
卫星探火 fire detection by satellite remote sensing
卫星影像 satellite imagery
卫星资源清查 satellite inventory
未饱和空气 non-saturated air
未剥皮材 unbarked wood
未剥皮原木 rough timber
未采伐边角林地 corner
未成林造林地 young afforested land
未成年 immaturity
未成商品材前的疏伐 precommercial thinning
未成熟 immature; immaturity
未成熟材 immature wood
未成熟的 sub-adult

未成熟林 immature stand; immature wood
未成熟期 immature stage
未成熟土 immature soil
未成熟心材 imperfect heart wood
未成熟心材树 imperfect heartwood tree
未成熟轴向管胞 immature longitudinal tracheid
未成熟轴向木质部细胞 immature longitudinal xylary cell
未充分漂白的 under bleached
未充分硬化胶层 moist glue line
未出现率 percentage of absence
未处理板材 untreated board
未处理材 untreated timber
未发育 immaturity
未发育的 rudimentary
未伐小块林木 pocket
未分配利润税 undistributed profit tax
未分配盈余 retained earnings
未风化矿质土壤 unweathered mineral soil
未改性松香 unmodified rosin
未干材 green lumber
未干燥材 unseasoned wood
未干燥材比重 specific gravity in green
未固化胶层 green glue-line
未加工材 unhewn timber
未加紧悬索桥 unstiffened suspension bridge
未交定货的积压 backlog of unfilled orders
未交股本 uncalled capital
未经营林 unmanaged stand
未开发林 unexploited forest
未开发林地 inaccessible forested land
未开发林区 backwoods
未开发区 under-developed area
未开发森林 inaccessible forest
未开垦的 uncultivated
未来成本 future cost
未立木地 unstocked area
未利用材 unutilized wood
未刨成材 rough lumber
未刨光材 undressed timber; unwro(ugh)t timber
未刨光成材 rough timber
未刨光厚板 rough plank
未漂白纸浆 unbleached pulp
未漂的(木浆) green
未切边尺寸 untrimmed size
未鞣透的 under tanned
未伸到屋脊的主椽 half-principal
未施业林 unmanaged forest
未收资本 uncalled capital
未熟(种子) incomplete ripeness
未熟林分 immature stand
未熟落果 premature drop
未熟紫胶(紫胶虫涌散前采收) ari
未涂饰(木材) in the white
未涂油漆木材 whitewood
未消耗成本 cost residue
未用余额 unexpended balance
未郁闭 free growing; free-to-grow
未造林地 nonforested land
未长成的 exhausted
未蒸解物(制浆) knot
未支配款额 uncommitted amount
未支配余额 unencumbered balance
未植点 blank
位(比特) bit
位标 locant
位次 locant
位错 dislocation
位点(基因的) locus
位能 potential energy
位向 aspect
位移 displacement; movement
位于叶腋下的 infra-axillary
位蒸发散 potential evapotranspiration
位置 locality; location
位置参数 location parameter
位置拟等位基因 position pseudoalleles
位置拟等位性 position pseudoallelism
位置图(森林) location map
位置效应(树体) topophysis; position effect
位组 byte
胃 ventriculus
胃毒剂 stomach agent
胃毒杀虫剂 stomach insecticide
胃盲囊 gastric caeca

喂料斗 feed hopper
喂料口 spout
喂入端 feed head
喂入箱 feed(er) box
喂送装置 feeding device
喂养 feeding
温包 thermobulb
温差电偶 thermocouple; thermoelectric couple
温床 forcing bed; forcing house; frame; hotbed
温床花卉 frame culture flower
温带 extratropical belt; extratropical zone; temperate zone
温带阔叶林 temperate hardwood
温带林 temperate forest
温带林区 temperate forest region; temperate forest zone
温带气候 mesotherm climate
温带雨林 temperate rain forest
温带植物 mesotherm plant
温度传导率 thermometric conductivity
温度垂直递减率 lapse rate to temperature
温度动态 temperature regime
温度反应系数 Q_{10} value
温度分散 temperature dispersion
温度极限 temperature limit
温度计 thermometer
温度记录器 thermograph; thermometrograph
温度距平 temperature anomaly
温度控制器 temperature controller
温度落叶林 temperate deciduous forest
温度逆增 temperature inversion
温度曲线 temperature profile
温度适应 thermal adaptation
温度梯度 temperature gradient
温度调节 homo(io)thermism
温度调节器 program temperature controller; temperature regulator; thermoregulator
温度系数 temperature coefficient
温度效应 temperature effect
温度循环 temperature cycle
温度异常 temperature anomaly
温度雨量图 hydrotherm figure; hythergraph
温度噪声系数 noise temperature ratio
温度状况 temperature regime
温度自记曲线 thermogram
温度纵剖面辐射计 vertical temperature profile radiometer
温感应生物 thermonasty
温克勒法 Winkler's method
温敏胶黏剂 temperature sensitive adhesive
温暖干燥期 xerothermic interval (stage)
温暖期 hypsithermal interval (period)
温暖指数 warmth index
温切斯特(美独弹牌名) Winchester
温泉 hot spring
温泉场 hot spring ground
温泉生物 hot-spring organism
温泉游乐园 hot spring ground
温热作用 hyther
温湿计 hygrothermograph
温湿曲线 hydrotherm figure
温湿图 hytherograph
温湿自记器 hygrothermograph
温室 conservatory; forcing house; glass house; green house; greenery; greenhouse; hot-house; stovehouse
温室管理 greenhouse management
温室花卉 greenhouse flower; greenhouse plant
温室加热 greenhouse heating
温室气体 greenhouse gas
温室群 greenhouse cluster
温室效应 greenhouse effect
温室芽接 greenhouse budding
温室栽培 greenhouse culture
温室植台 bench in greenhouse
温室植物 greenhouse plant
温室中心工作室 greenhouse headhouse
温室种植台 bench in greenhouse
温水浸种 hot soaking
温效指数 temperature-efficiency index
温性腐殖质 mild humus
温盐环流 thermohaline circulation

温雨图 hythergraph
温育 incubation
温跃层 thermocline
温周期 thermoperiod
温周期现象 thermoperiodicity; thermoperiodism; thertnoperiodism
瘟疫 plague
文化多样性 culture diversity
文化公园 cultural park
文化景观 cultural landscape
文件(计算机) file
文件号码 file number
文件区(计算机) file area
文件依据 documentary basis
纹孔 pit
纹孔闭塞 aspiration; pit aspiration
纹孔场 pit field
纹孔穿孔 pitted perforation
纹孔次生壁 pitted secondary wall
纹孔导管 pitted vessel
纹孔道 pit canal
纹孔对 pit-pair
X形纹孔对 X figure pit-pair
纹孔分子 pitted element
纹孔管胞 pitted tracheid
纹孔环 annulus; pit annulus
纹孔口 aperture; pit aperture; pit orifice
纹孔类型 type of pit
纹孔膜 pit membrane
纹孔内口 inner pit aperture
纹孔腔 pit cavity
纹孔区 pit area; pit zone
纹孔塞 pit torus; torus; torus (*plur.* tori)
纹孔式 pitting
纹孔室 pit chamber
纹孔外口 outer aperture; outer pit aperture
纹孔细胞 pitted cell
纹孔缘 pit border
纹理 grain; pattern; vein
纹理方向 grain direction
纹理角度 grain angle
纹理结构(影像判读) texture
纹理排列方向 grain orientation
纹理偏差 deviation of grain
纹理破坏 grain rupture
纹理特性 grain character
纹理斜度 slope of grain
纹理纸 grained paper
纹泥 varved clay
纹盘机 crab
蚁酸的 formic
吻合术 anastomosis
吻针(线虫) spear
紊流 non-uniform flow; turbulent flow
紊流扩散 turbulent diffusion
稳定材 stabilizing wood
稳定度 stability
稳定反应 stabilizing reaction; stopping reaction
稳定风 constant wind
稳定河槽 stable channel
稳定化 stabilization
稳定恢复力 stability-resilience
稳定极限环 stable limit cycle
稳定剂 stabilizer
稳定阶段 steady-state phase
稳定进料 steady feed
稳定流 permanent flow
稳定年龄分布 stable age distribution
稳定品种 distinct variety
稳定平衡 stable equilibrium
稳定坡度 equilibrium slope
稳定气团 stable air (mass)
稳定器 stabilizer
稳定情景 stabilization scenario
稳定群落 permanent community
稳定蠕变 stationary creep
稳定时期 stationary (growth) phase
稳定同位素 stable isotope
稳定性 constancy; stability
稳定性因子 fixity factor
稳定选择 stabilizing selection
稳定载荷 steady load
稳定振动 stationary motion
稳定植被 stable vegetation
稳固腐熟腐殖质 firm mull
稳进地表火 creeping surface fire
稳进树冠火(与地面火并进) dependent crown fire
稳流 permanent flow
稳态 steady state

鹟 flycatcher
涡度相关 eddy covariance
涡度相关法 eddy correlation method
涡卷花纹 swirl crotch figure
涡卷饰 scroll
涡卷纹 swirl crotch
涡卷形三角楣饰 scrolled pediment
涡卷形饰条 scroll moulding
涡卷形镶嵌 scrolled marquetry; seaweed marquetry
涡卷形椅扶手 rollover arm
涡卷形支腿 scroll foot; scroll leg
涡卷形装饰 scroll ornament
涡流 eddy; vortex
涡流风扇 turbo fan
涡流浆料除渣器 vortex action pulp cleaner
涡流净浆器 vortex action pulp cleaner; vortex cleaner
涡流侵蚀作用 evorsion
涡轮干燥器 turbo-dryer
涡轮式刨片机 turbo-chipper
涡轮增压器 turbocharger
涡纹 swirl; whirl
涡旋 eddy
窝工时间 down time
窝浆 lodgment of stock
窝卵数 clutch size
窝眼筒清选机(种子) indent cylinder seperator
窝仔数 litter size
蜗牛媒介花 malacophilous flower
肟基 hydroxyimino
沃伦链式磨木机 Warren chain grinder
沃土 fat soil
沃扎利特模压刨花板成型工艺 Werzalit process
沃洲 oasis (*plur.* oases)
卧木 ground sill; ledger
卧牛石 niggerhead
卧式(射击) prone (position)
卧式带锯 horizontal band-saw
卧式单板排锯 horizontal veneer gangsaw
卧式多层木片平筛 horizontal multiple-deck chip screen
卧式多轴木工钻床 horizontal multi-spindle wood borer
卧式釜 horizontal retort
卧式干馏釜 lying retort
卧式挤压机 horizontal extruder
卧式进料器 horizontal feeder
卧式开榫机 horizontal spindle tenoner
卧式木工钻床 horizontal wood borer
卧式排锯 horizontal gang saw
卧式排钻 horizontal gang borer
卧式刨切单板机 horizontal veneer slicer
卧式刨切机 horizontal slicer
卧式劈木机 horizontal splitter
卧式燃烧炉试验 horizontal furnace test
卧式榫眼机 horizontal mortiser
卧式炭窑 lying charcoal kiln; lying meiler
卧式推料离心机 horizontal push-type centrifuge
卧式圆盘刨片机 horizontal disk type flaker
卧式圆盘削片机 horizontal disk type chipper
卧式再分带锯 horizontal band resaw
卧式凿孔机 horizontal mortiser
卧式贮料仓 horizontal storage bin
卧室套装家具 bedroom suites
卧桩 deadman; deadman anchor
握柄 grab handle
握钉力 nail-holding ability; nail-holding capacity; nail-holding power; screw-holding power
握钉力试验 nail test
握力 holding power
握索器 clip
乌桕油 Chinese tallow; stillingia oil; stillingia tallow
乌瞰图 aerial view
乌鲁斯索道集材 Urus logging
乌仑代栲胶 urunday extract
乌洛托品(防腐剂) Urotropine
乌兽禁猎区 sanctuary
乌索酸 ursolic acid
乌头 aconite
乌头碱 aconitine
乌鸦 crow
乌贼 squids
圬工坝 masonry dam
污斑 stain
污点 speck; stain

污泥 muck; mud; sludge
污泥处理 sludge treatment
污泥生的 epipelic
污染 impurity; pollution
污染大气 polluted atmosphere
污染负荷量 pollution load
污染监测 monitoring pollution
污染空气 polluted atmosphere
污染区 contaminated zone
污染水 polluted water
污染物 contaminant; contaminate; pollutant
污染物质 polluter
污染源 pollution source; source of pollution
污染者 polluter
污染指数 pollution index
污水 sewage; sewerage; slops
污水泵 sewage pump
污水池 sump
污水处理厂 sewage farm
污水浮游生物 saproplankton
污水灌溉 under(ground) irrigation
污水生的 saprobic; saprobiotic
污水生物 saprobia; saprobiont
污水生物的 saprobiotic
污水污染 sewage pollution
污着生物 fouling organism
污浊负荷量 pollution load
呜呜声(潜水鸭类的叫声) purr lingo
屋顶 roof tree; roofing
M形屋顶 trough roof
屋顶板 shingle
屋顶材料 roofing
屋顶拐角试验 roof corner test
屋顶花园 roof garden; rooftop garden
屋顶架 frame of roof
屋顶式闸门 bear-trap gate
屋脊 ridge roof
屋架 roof truss
屋面板 roof boarding; roof sheathing; roof slab; roofing board
屋面绝缘板 roof insulation board
屋缘 couple-close
无孢子的 asporous
无孢子生殖 apospory
无孢子形成 apospory
无被的 achlamydeous; naked
无被花 naked flower
无边有肩的(子弹) rimless shouldered
无变态类 ametabola
无柄的 sessile
无柄叶 sessile leaf
无病的 disease-free
无病毒的 virus-free
无病毒苗 virus-free stock
无病合格证 disease-free certification
无病原体的 pathogen-free
无病砧木 pathogen-free stock
无草休闲地 bare fallow
无层次树冠 nonstratified crown
无层次树形 nonstratified form
无车立柱装车法 loading without car stake
无尘 dust-free
无翅的 apterous; wingless
无翅型 aptera(e); apterous form
无穿孔管胞 imperforate tracheary cell
无疵材 clean
无疵的 clear
无疵锯材 clear lumber
无疵木板 clean deal
无疵木材 clean-cut timber; clear wood
无疵小材试验 small clear timber test
无疵小试样 small clear specimen
无雌及完全花同株 agamogynomonoecism
无雌蕊的 agynous
无大小年 even year
无带胶拼机 tapeless splicer; splicer
无氮浸出物 nitrogen free extract
无垫板成型生产线 caulless forming line
无垫板加压系统 caulless pressing system
无垫板系统 caulless system
无垫板预压机 caulless prepress
无垫板装板机 caulless loader
无垫木堆积 bulk pile
无顶树 pollard; pollarded tree
无定期踏查 non-periodic cruise
无定形部分 amorphous portion
无定形的 amorphous

无定形区 amorphous area; amorphous region
无定形碳 agraphitic carbon
无动力索道 gravity skyline system
无毒性的 avirulent
无毒株 avirulent strain
无端链锯 endless chain saw
无粉管植物 asiphonogamae
无风(零级风) calm
无风险利息 risk-frec interest
无风险率 risk-free rate
无缝地板 jointless floor
无缝管 seamless pipe
无扶手直背椅 side chair
无腹柄的 sessile
无覆盖播种 clean cultivation
无隔孢子 amerospore
无隔担子 holobasidium
无隔膜的 coenocytic
无隔纤维 unseptate fibre
无梗的 acaulescent
无功消耗 idling consumption
无股钢丝绳 non-spinning rope
无光 flatting
无光带 aphotic zone
无光的 mat
无光漆 lusterless paint
无光清漆 flatting varnish
无光饰面 dead finish
无规聚合物 atactic polymer
无规立构聚丙烯 atactic polypropylene
无规立构聚合物 atactic polymer
无规则的 random
无号印木材 free log
无号原木 prize log
无核的 acaryotic; anucleate; enucleated; seedless
无核突变 anucleate mutation
无核细胞 akaryote; anucleate cell; cytode
无核状态 seedlessness
无后代 infertility
无后的 atokous
无花年 even year
无滑轮钢索集材 straight haul
无患子 Chinese soapberry; soap nut tree
无灰干重 ash-free dry weight
无灰重量 ash-free weight
无悔机会 no-regrets opportunity
无悔潜力 no-regrets potential
无悔选择 no-regret option
无悔政策 no-regrets policy
无机肥料 inorganic fertilizer; mineral manure
无机胶黏剂 inorganic adhesive
无机胶体 inorganic colloid
无机营养 lithotrophy
无机阻燃剂 inorganic fire retardant
无级变速 infinitely variable speed
无极细胞 apolar cell
无脊椎动物 invertebrata
无计划经营 management without plan
无监督分类 unsupervised classification
无浆布 unsized fabric
无胶干法工艺 glueless dry process
无铰拱架 hingeless arch
无节 knot-free
无节材 clean; clean wood; clear
无节的 inarticulate
无节干材 clean-bole; clear-bole
无节良材 knot free lumber
无节木材 clean timber
无节树干 clear trunk
无节制采伐 unclassed felling; uncontrolled cutting
无茎的 acaulescent; acauline
无菌的 aseptic; sterile
无菌过滤 aseptic filtration
无菌空气 filtrated air
无菌培养 aseptic culture; axenic culture; nonsymbiotic culture
无菌培育 aseptic culture
无菌室 clean room
无菌箱 germ-free box; sterile board; sterilized box
无卡盘单板旋切机 chuckless veneer lathe
无抗性的 nonresistant
无靠背睡榻 ottoman
无空气喷漆枪 airless sprayer
无空气喷枪 airless spray gun
无空气喷涂 airless spray
无孔材 non-pored wood (timber); non-porous wood (timber)
无孔性 imporosity

无控制林火 uncontrolled fire
无控制镶嵌图 uncontrolled mosaic
无框架结构 non-frame construction
无犁耕作 ploughless farming
无力偿付 insolvency
无力浮游生物 akineton
无立地木 unstocked area
无立木地 unstocked blank
无立木林地 non-stocked forest land; unstocked forest land; unstocked stand
无利润疏伐 uneconomic thinning
无裂纹单板 check-free veneer
无林的 woodless
无林地 nonforested land; non-stocked land; treeless land
无林区 non-forested area
无鳞片的 elepidote; esquamate
无硫制浆 non-sulfur pulping
无路地区 roadless area
无马蹄插穗 non-heel cutting
无黏流体 ideal fluid
无胚的 ablastous
无胚乳的(种子) exalbuminous; exendospermous
无胚乳种子 exalbuminous seed; non-endospermic seed
无配繁殖 anogamic propagation
无配生殖 agmogenesis
无配生殖的 agamogenetic
无配生殖体 apomict
无配生殖种 agameon
无配系群 agamic complex
无配种 agamospecies
无配子生殖 agamic reproduction; agamogony; apogamy
无劈裂伐木机 no-shatter feller
无皮边材 bright sap(wood)
无皮层的 ecorticate
无皮鳞茎 naked bulb; scaly bulb
无偏估计 unbias(s)ed estimate
无气门式的 apneustic
无气喷涂 airless spray(ing)
无驱动滚筒输送机 dead roller conveyor
无缺株的人工林 fully-stocked plantation
无人操纵的 pilotless
无人机 drone; unmanned aircraft; unmanned aerial vehicle; unmanned plane
无溶剂环氧清漆 solventless epoxy varnish
无溶剂清漆 solventless varnish
无融个体 agamont
无融合兼雌两性 agamogynomonoecy
无融合兼两性 agamomonoecy
无融合结籽 agamospermy
无融合生殖 apomixes; apomixia
无融合体 apomict
无融生殖种 agamospecies
无蕊花 asexual flower; neuter flower; neutral flower
无色刺槐亭定 leucorobinetinidin
无色翠雀定 leucodelphinidin
无色的 hyaline
无色非瑟酮定 leucofiseltinidin
无色盖波亭定 leucoguibourtinidin
无色花青素 leucocyanidin
无色花色苷 leuco-anthocyanin
无色花色素 leuco-anthocyanidin
无色染剂 leuco base stain
无色天竺葵色素 leucopelargonidin
无伸缩性价格 nonflexible price
无生活力的 novriable
无生命杂质(种子净度) inert matter
无生源说 abiogenesis
无声齿轮 silent gear
无声进料 silent feeding
无树皮的 ecorticate
无霜带 green-belt; verdant zone
无霜季 frost-free season
无霜期 frost-free period; frost-free season; frostless season; non-frost period
无水醋酸 glacial acetic acid
无水酒精 absolute alcohol
无水孔隙 free pore space
无水亚黏土草原 karoo
无丝分裂 amitosis
无髓心板 side cut; side cut piece
无提取物木材 extraction-free wood
无土苗床 soilless bed
无土栽培 soilless culture
无托架系统装板机 trayless-system loader
无托叶的 estipulate

无维管木质部 hadrome
无细胞抽出物 cell-free extract
无细胞系统 cell-free system
无显性组合 nulliplex
无限的 infinite
无限公司 company of unlimited liability
无限花序 indefinite inflorescence
无限群体 infinite population
无限润胀 unlimited swelling
无限伞房花序 indefinite corymb
无线电报 radiogram
无线电定向和测距 radio directing and ranging
无线电高空测候器 radiosonde
无线电探空测风 rawinsonde
无线电探空仪 radiosonde
无线电遥测 radio telemetry
无线电追踪 radio tracking
无向性 isotropy
无效播种 dead sowing
无效等位基因 amorph
无效孔隙度 inactive porosity
无效水 non-available water; unavailable water
无效水分(固着于土粒) echard
无屑切削 chipless cutting
无形损失 non-physical loss
无形资产 immaterial assets; intangible assets
无性白蚁 sterile caste
无性孢子 asexual spore
无性的 neutral; sexless
无性繁殖 asexual multiplication; asexual propagation; asexual reproduction; vegetative propagation; vegetative reproduction
无性繁殖体 vegetative propagule
无性繁殖系 clone
无性分离 vegetative segregation
无性分裂 indirect vegetative division
无性工蚁 sterile caste
无性后代 vegetative progeny
无性花 asexual flower; neuter flower; neutral flower; vegetative flower
无性假多倍体 agamo-pseudopolyploidy
无性阶段 imperfect stage
无性接近法 method of vegetative approachment
无性扩展 vegetative spread
无性两性花同株 agamohermaphroditism
无性胚 vegetative embryo
无性期 imperfect stage
无性生殖 asexual reproduction
无性生殖的 agamogenetic
无性生殖细胞 asexual germ cell
无性生殖种 agameon
无性生殖周期 asexual cycle
无性世代 agamobium; asexual generation
无性系 pure-clonic form
无性系保存 clonal preservation
无性系变异体 clonal variant
无性系测定 clonal test
无性系的 clonal
无性系繁殖 clone propagation
无性系分株 ramet
无性系母株 ortet
无性系群 agamic complex
无性系试验 clohal test; clonal trial
无性系试验场地 clone trial site
无性系选择 clonal selection
无性系种子园 clonal seed orchard
无性杂交 asexual hybridization; vegetative hybrid; vegetative hybridization
无性种 agamospecies
无休眠的 non-dormant
无须鸢尾(花类) beardless iris
无压盖圆锯 collarless circular saw
无芽的 ablastous
无氧处理 anoxic treatment
无氧呼吸 anaerobic respiration
无样地抽样 plotless sampling
无样地调查 plotless cruising
无叶的 aphyllous; defoliate; leafless
无叶的树木 bald tree
无叶片的 elaminate
无液气地计 aneroid barograph
无液气压计 aneroid; aneroid barometer
无益共生 hamabiosis
无意义突变 nonsense mutation
无翼的 apterous
无影像片 shadowless photograph

无应力木 stressfree wood
无用的 exhausted; waste
无源遥感 passive remote sensing
无辕橇 mud-boat
无载体胶黏剂 unsupported adhesive
无着丝点(染色体) acentrics
无着丝点倒位 acentric inversion
无着丝点断片 acentric fragment
无着丝点环(染色体) acentric ring
无着丝点染色体 acentric chromosome
无枝的 clean; clean-stemmed; clear; eramose
无枝干材 bodywood; clean-bole
无枝性 branchlessness
无枝圆木 pruned log
无中心粒 acentrics
无踵插穗 non-heel cutting
无主材 waif (wood)
无子叶植物 acotyledon
无籽的 seedless
无籽状态 seedlessness
无足类 apode
无组织的 amorphous
蜈蚣 centipede
五(希腊字头) penta
五半裂的 quinquefid
五棓子蝇 gall fly
五倍体 pentaploid
五倍子 Chinese gall; Chinese gallnut; gall; gall apple; gall nut; knopper gall; nutgall
五倍子虫 Chinese gall aphid
五倍子单宁 Chinese gallotannin; Chinese tannin
五倍子蜂 gall wasp
五出叶的 pentafoliate
五雌蕊的 pentagynous
五点形的 quincuncial; quincunxial
五斗橱 commode
五斗柜 chest of drawers
五对的 quinquejugate
五分周法 fifth girth method
五花草甸 lawn
五环糖 furanose
五价体 quinquevalent
五角棱镜轮尺 penta prism caliper
五金 hardware
五联带锯 quint band saw
五轮花 pentacyclic flower
五轮裂的 pentacyclic
五氯苯酚(杀菌剂或除草剂) pentachiorophenol; chlorophenasic acid
五氯酚(除草剂) penehlorol; penta
五氯酚钠 sodium penta chloro phenate
五氯酚钠-钠-硼砂合剂 noxtane
五氯酚铜盐 copper pentachlorophenate
五氯硝基苯(杀菌剂) pentachloronitrobenzene
五腔的 quinquelocular; quinqueloculate
五深裂的 quinquepartite
五室的 quinquelocular; quinqueloculate
五室瘦果 pentachaenium; pentakenium; polachena; polackena
五水合硫代亚硫酸钠 hypo
五水合硫酸铜 blue vitriol
五体雄蕊 pentadelphous stamen
五小叶的 quinate
五叶的 quinquefoliolate
五原型 pentarch
五针松 five-leaved pine; five-needle pine
五指状的 pentadactylous
五中脉的 quinquecostate
五株栽植(配置四角及中心) superposed square planting
五株正方形栽植 quincunx planting
五子叶的 pentacotyl
戊聚糖 pentosan
戊糖 pentose
戊糖苷 pentoside
物候观测 phenological observation
物候规避 phenology avoidance
物候历 phenological calendar
物候谱 phenological spectrum
物候期 phenological phase; phenophase
物候相 phenological phase
物候镶嵌 phenology mosaic
物候学 phenology
物候学隔离 phenological isolation

物候演替 phenological succession
物价指数 price index
物镜 objective; objective lens
物镜测微尺 object micrometer
物理单位 physical units
物理风化 physical weathering
物理干燥 physical dryness
物理气候 physical climate
物理生态学 physical ecology
物理吸附 secondary adsorption
物理现象 physical phenomena
物理性质 physical property
物理休眠 physical dormancy
物理因子 physical factor
物理阻力 physical resistance
物料平衡[表] material balance
物料通过量 throughput capacity
物质收获 yield in material
物质收获表 material yield table; yield-table in material
物质收支 material budget
物质循环 matter cycle
物种 species
物种多样性 diversity of species
物种多样性指数 index of species diversity; species diversity index
物种丰富度 species richness
物种红皮书 red data book
物种均匀度 species evenness
物种密度 species density
物种内聚性 species cohesion
物种平衡 species equilibrium
物种栖息地保护区 habitat management area
物种起源 origin of species
物种起源学说 theory of origin of species
物种受威胁等级 categories of threatened species
物种维系 species cohesion
物种星座图 species constellation
物种形成 speciation
物资周转 turnover of material resource
误差变量 error variance
误差常态曲线 normal curve of error
误差传播 propagation of error
误差方差 error variance
误差扩散 error propagation
误差列线图 error nomogram
误差率 percentage error
误差偏差 error variance
误差项 error term
误伐 accidental cutting; miscut(ting)
雾 fog; mist
雾带 fog belt
雾滴 fog drip; fog droplet
雾化 atomization; nebulization; pulverization
雾化度(胶滴) degree of atomization
雾化器 atomizer
雾林 fog forest
雾霾 haze, smog
雾室 fog chamber
雾凇 rime
雾消(日消雾) burn-off
雾翳滤光片(翳 yì) haze catting filter
雾锥角 spray-cone angle
雾浊 blushing
西柏烯 cembrene
西风 westerly wind
西谷米(西谷椰子树干所含淀粉) sago
西拉橡胶 ceara rubber
西力生(杀菌剂) ceresin
西马林求积式 Simalian's formula
西玛津(除莠剂) simazine
西蒙得木蜡 jojoba wax
西维因(杀虫剂) carbaryl; Sevin
吸潮 absorption of moisture
吸尘机 dirt-catcher; dust separator
吸尘器 collector; dust accumulator; dust arrestor; dust catcher
吸尘罩 dust collecting hood
吸虫器 suction catcher; suction trap
吸风 draft; draught
吸风式圆筒干燥机 suction drum drier
吸附[作用] adsorption
吸附层析法 adsorption chromatography
吸附床 adsorption bed
吸附等温线 adsorption isotherm
吸附剂 absorbent; sorbing agent
吸附能力 adsorbability
吸附器官 holdfast

吸附曲线 adsorption curve
吸附热 heat of adsorption
吸附色谱法 adsorption chromatography
吸附水 adsorbed water; adsorptive water; bond water
吸附水层 adsorbed water layer
吸附性 adsorbability
吸附选择性 selectivity of adsorption
吸附柱 adsorption column
吸根(寄生植物的) haustorium (*plur.* -ia); sucker
吸光测定法 absorptiometry
吸光测定计 absorptiometer
吸光度 absorbance; absorbancy
吸光分析法 absorptiometry
吸光剂 absorptiometer
吸光指数 absorbance index
吸力堆垛机 suck stacker
吸力运输机 suction carriage
吸磷系数 phosphorus absorption coefficient
吸墨性 ink absorption
吸墨性能 ink receptivity
吸能反应 endergonic reaction
吸气 aerate
吸气缸 suction cylinder
吸气管 air suction pipe; suction duct
吸气口 air suction port
吸器 haustorium (*plur.* -ia); sucker
吸热点 endothermic point
吸热反应 endothermic reaction; heat-absorbing reaction; thermo-negative reaction
吸入管 suction duct
吸入式喷枪 suction feed type gun
吸入水流 inhalant current; inhalant stream
吸声 sound absorption
吸声板 acoustic(al) board
吸声材料 acoustic(al) material
吸声瓦 acoustic tile
吸声系数 acoustic absorption coefficient (factor); acoustic absorptivity; coefficient of sound absorption; sound absorbing coefficient
吸声纤维板 acoustic fibre board; acoustic(al) fibreboard
吸湿 absorption of moisture; imbibition; moisture regain
吸湿极限 hygroscopic limit
吸湿剂 hygroscopic agent
吸湿式水分测定计 hygrometric moisture meter; hygroscopic moisture meter
吸湿水 hygroscopic water
吸湿水分 hygroscopic moisture
吸湿物质 hygroscopic substance
吸湿系数 hygroscopic coefficient
吸湿性 hygroscopic nature; hygroscopic property; hygroscopicity; moisture regain
吸湿性材料 hygroscopic material
吸湿性处理剂 hygroscopic treating agent
吸湿性能 hygroscopic behavior
吸湿盐剂结块 caking of hygroscopic salt
吸湿滞后 sorption hysteresis
吸收 absorption; imbibition
吸收比 absorption ratio
吸收带 absorption band
吸收度试验 mounting test
吸收复合体(土壤) adsorption complex
吸收根 feeder root; feeding root
吸收光谱 absorption spectrum; apsorption spectrum
吸收机制 mechanism of absorption
吸收计 absorptiometer
吸收剂 absorbent; absorber
吸收-解吸动力学 absorption-desorption kinetics
吸收量 uptake; loading
吸收率 absorbance; absorptance; absorptive power; absorptivity
吸收能 absorbed energy
吸收能力 absorbing capacity; absorption ability; absorptive power; absorptivity
吸收谱带 absorption band
吸收期 absorption period
吸收器 absorber
吸收器官 foraging component
吸收区 absorption zone
吸收曲线 absorption curve
吸收湿度计 absorption hygrometer

吸收水 absorbed water; absorption water; imbibed water
吸收速率 absorption rate
吸收系数 absorption coefficient
吸收性 absorptivity
吸收性能 absorbing quality
吸收杂交 absorptive crossing
吸收指数 absorbance index
吸水辊 suction roll
吸水力 suction force
吸水量 soakage
吸水率 water absorption
吸水能力 water-absorbing capacity
吸水膨胀系数 hydroexpansivity
吸水箱 suction box
吸水箱盖板 suction box cover
吸水性 water absorption
吸水压 suction pressure
吸血的 haemophagous
吸血性昆虫 blood feeder
吸音 sound absorption
吸音板 peg board
吸引范围 ducking area
吸引力 attracting force
吸油量 oil number
吸胀作用 imbibition
吸着(作用) sorption
吸着等温线 sorption isotherm
吸着剂 sorbing agent
吸着式单板堆垛机 veneer suction stacker
吸着水 adsorptive water; bond water; hygroscopic water; sorption water
吸着性质 sorptive nature
吸着滞后 sorption hysteresis
吸种机 vacuum seed harvester
希弗尔公式 Schiffel's formula
希腊回纹饰 greek fret
析出(造纸) accretion
析离 isolation
析因实验 factorial experiment
析皂 graining
息票 coupon
烯丙树脂 allyl resin
稀表面性浮游生物 spanipelagic plankton
稀播 thin sowing
稀堆(每层用垫木隔开) open piling
稀堆法 stick stacking
稀垛 open stacking
稀垛法 stick stacking
稀灌丛干草原 shrub steppe savanna
稀浆 slurry
稀释计数 dilution count
稀释剂 diluent; liquifier; thinner
稀释平板计数 dilution plate count
稀释效应 dilution effect
稀释营养溶液 attenuated nutrient solution
稀释终点 dilution end point
稀疏 thinning
稀疏傍管型 scanty paratracheal
稀疏处理 rarefaction
稀疏电子密度部分(初生壁) less electron dense portion
稀疏灌丛 open shrubland
稀疏环管薄壁组织 scanty paratracheal parenchyma; sparsely paratracheal parenchyma
稀疏阶段 thinning phase
稀疏林冠 open canopy
稀疏群丛 open association
稀疏群落 open community
稀疏树冠 free-setting (of) the crown
稀疏植被 open vegetation; sparse vegetation
稀树草原 parkland; savanna(h)
稀树草原林(热带) savanna forest; savanna woodland
稀树干草原 tree savanna
稀树沙漠 arboreous desert
稀液浸渍 dilute-solution-soaking
稀有 rare
稀有碱基 minor base
稀有种 rare species
稀有自然风景区 unique natural area
稀淤泥 muskeg
稀植 sparse planting; thin planting; wide planting
锡 tin
溪沟 torrential valley
溪谷(广义的) dale
溪谷(有林地带) dene
溪河底栖的 bentho-potamous
溪流 stream
溪流群落 namatium

蜥蜴 lizard
熄灭 quench
膝状材 knee
膝状弯曲的 geniculate; geniculatus
习得性行为 learned behaviour
习性 habit
洗出移液管 wash out pipette
洗涤层 wash coat
洗涤剂 detergent
洗涤器刮板 rip
洗鼓 wash drum
洗浆 pulp washing
洗浆池 pulp washer
洗净器 washer
洗漆剂 paint scrubber
洗色桶 colour-washing tank
洗沙机 sand-washing machine
洗提 elution
洗提的 elutropic
洗提液 eluant; eluent
洗脱 elution
洗脱剂 eluant; eluent
洗脱液 effluent; eluant; eluent
洗选 washing
洗选机 washer
洗液 washing
洗衣板 laundry board; washboard
洗种器 seed washer
铣槽刀 grooving cutter
铣槽机 slot mortiser
铣槽锯 grooving saw
铣槽组合刀锯 dado-head saw
铣成边接 milled edge joint
铣床 shaper
铣刀头 cutter block; cutter head; shaper head
铣削 mill; milling
铣削材 milled wood
铣削式剥皮机 rosser head debarker
铣削型剥皮机 milling head debarker
铣削型多刀剥皮机 multiple cutterhead peeler
徙动常数 migration constant
喜白蚁的 termitophilous
喜冰雪的 cryophilous
喜虫的 entomophilous
喜氮植物 nitrophile
喜冬的 chimonophilous
喜腐生的 saprophilous; saprophytic
喜腐生物 saprophile
喜钙的 calciphilous; chalk-loving
喜钙植物 calcipete; calciphile; calciphilous plant
喜高温真菌 pyrophilous fungus
喜光的 photophilous; heliophilous; light-demanding
喜光类 light demanding guild
喜光树 light-demanding tree
喜光树种 intolerant tree species; light-demander
喜光树种纯林 pure intolerant stand
喜光性 shade intolarance; sciophobia; heliophilia
喜光种子 light-favored seed
喜旱的 xerophil(e); xerophilous
喜好(干湿阴阳等) demand; require; philia
喜好的(干湿阴阳等) philous; philic; loving; demanding; requiring; ~phile
喜花的 anthophilous
喜碱的 basiphil(ous)
喜碱植物 alkaline plant; basiphile
喜冷的 cryophilous
喜泉的 crenophilous
喜沙的 ammophilous; psammophilous
喜沙性的 psammophilic
喜沙植物 psammophile
喜深水的 bathophilous
喜湿的 hygrophilous
喜湿性的 hygrophil(e)
喜树的 dendrophilous
喜树碱 camptothecin
喜酸的 oxyphilic; oxyphilous
喜酸植被 acidophilous vegetation
喜酸植物 oxylophyte; oxyphyte; oxyphile
喜土的 geophile
喜温的 mesothermophilous; thermophilic; thermophilous
喜温生物 thermophilous organism
喜温树种 thermophilous tree species
喜温植物 thermophyte
喜旋光性浮游生物 phaoplankton
喜盐生物 halobion

喜盐植物 halophile; halophyte
喜阳植物 heliophile; heliophyte; intolerant species; sun plant; light-demander; light-demanding species
喜氧细菌 aerobic bacteria
喜蚁植物 ant plant
喜阴的 shade-requiring; shade-demanding; shade loving; sciophilous; sciophilic; heliophobous
喜阴植物 heliophobe; sciophile; shade species; sciad; shade-loving plant; sciophilous species; heliophobous species
喜雨的 ombrophilous
喜雨植物 ombrophile; ombrophyte
系杆 tie rod
系列(设备) line
系列图 serial map
系列相关系数 serial correlation coefficient
系列育种 line breeding
系谱树 family tree; genealogical tree
系谱图 family tree
系谱选择 linear selection
系数 coefficient
系统 system
系统抽样(机械抽样) systematic sampling
系统带状抽样 systematic strip sampling
系统的 genealogical; systematic; systemic
系统动物学 systematic zoology
系统发生系列 phylogenetic series
系统发育 historical development; phylogeny; phytogeny; systematic development
系统方法 system approach
系统分布 systematic distribution
系统分析 system analysis
系统管理 system administration
系统论 system theory
系统模型 system model
系统胚胎学 systematic embryology
系统侵染 systemic infection
系统侵染病 systemic disease
系统设计 system design
系统生态学 system ecology
系统疏伐 systematic thinning
系统树(演化关系) family tree; genealogical tree; systematic tree
系统突变 systemic mutation
系统误差 systematic error
系统学 systematics
系统样本 systematic sample
系统样地抽样 systematic plot sampling
系统优化 system optimization
系统育种 line breeding
系统栽培 pedigree culture
系统植物社会学 systematic plant sociology
系统植物学 systematic botany
系统装入程序 system loader
细 particle size
细班 subplot; sub-plot; sub-subcompartment
细胞 cell; corpuscle
细胞板 cell plate
细胞壁 cell wall
细胞壁加厚阶段 phase of cell-wall thickening
细胞壁加厚学说 theory of cell-wall thickening
细胞壁显微构造 microscopic cell-wall structure
细胞变态 cytomorphosis
细胞的 cellular
细胞分部分离 cell fractionation
细胞分化 cell differentiation; cytodifferentiation
细胞分级分离 cell fractionation
细胞分类学 cytotaxonomy
细胞分离酶 cell separating enzyme
细胞分裂 cell division; segmentation
细胞分裂素 auxetic; cytokinin; kinin
细胞分子 cellular element
细胞核 cytoblast; nucleus (*plur.* nuclei)
细胞核学 karyology
细胞核直接分裂 karyostenosis
细胞化学染色 cytochemical staining
细胞坏死 cellular necrosis
细胞减生的 hypoplastic
细胞减生现象 hypoplasia; hypoplasty

细胞角隅 cell corner
细胞结构 cell structure
细胞空隙率 cell space ratio
细胞块悬浊液 cell-dump suspension
细胞类型 cell pattern
细胞膜 cell membrane
细胞内的 intracellular
细胞内含物 cell content; cell inclusion
细胞内寄生的 endotrop(h)ic
细胞器 cell organ; cell organell; organoid
细胞嵌合体 cytochimera
细胞腔 cell cavity; cell lumen; cell space; lumen (*plur.* lumina)
细胞溶解 cell lysis
细胞溶素 lysin
细胞融合 cell fusion
细胞色素 cytochrome
细胞色素过氧化酶 cytochrome peroxidase
细胞色素还原酶 cytochrome reductase
细胞色素氧化酶 cytochrome oxidase
细胞生理学 cellular physiology
细胞生态学 cytoecology
细胞生物学 cytobiology
细胞体 cyton
细胞外多糖 extracellar polysaccharide
细胞外酶 extracellular enzyme
细胞形成 cytomorphosis
细胞形态学 cellular morphology; cyto-morphology
细胞型(胚乳) cellular type; cytotype
细胞学 cytology
细胞盐化防腐法 mineralized-cell process
细胞液 cell sap; sap
细胞遗传 cell heredity
细胞遗传学 cell genetics; cytogenetics
细胞异固缩 heteropycnosis
细胞增大 cell enlargemet
细胞增大性生长 auxesis
细胞增生 hyperplasia
细胞质 cytoplasm
细胞质基因 cytoplasmic gene; genoid; plasmagene
细胞质基因组 plasmon
细胞质雄性不育 cytoplasmic male sterile
细胞质遗传 cytoplasmic heredity; cytoplasmic inheritance; extra-nuclear inheritance
细胞致死 cell lethality
细胞转化 cell transformation
细胞状结构 cellular structure
细胞自动机 cellular automata
细胞组织 cellular tissue
细波花纹 fiddleback; fiddleback figure; fiddleback grain; fiddleback mottle
细部点 detail point
细部转绘 transfer of detail
细齿木锯 half-rip saw
细度 degree of fineness; fineness
细缝 hairline
细腐殖质 mull
细干卷(昆虫) cirrus
细根 fine root; rootlet
细管灌溉 spaghetti tube watering
细环裂 fine shake
细混土腐殖质 fine mull
细尖的 acuminulate
细浆 screened pulp; screened stock
细节 detail; pin knot
细节丛 pin-knot group
细结构 close texture; fine texture
细结构材 fine-textured wood
细晶岩 haplite
细菌 bacterium (*plur.* -ia); germ
L型细菌 L-form bacteria
细菌病 bacteriosis
细菌病毒 bacterial virus
细菌病害 bacterial disease
细菌毒素 bacterial toxin; bacteriotoxin
细菌分解作用 bacterial decomposition
细菌分类学 systematic bacteriology
细菌活化剂 bacteria-activator
细菌镜检法 bacterioscopy
细菌菌根 bacterorrhiza
细菌绿素 bacterial chlorin
细菌杀虫剂 bacterial insecticide
细菌素 bacteriocin(e)
细菌污染 bacteria contamination
细菌无性繁殖法 bacterial cloning

细菌纤毛 cilium (*plur.* -ia)
细菌性穿孔病 bacterial shot-hole
细菌学 bacteriology
细菌叶绿素 bacteriochlorophyll
细菌遗传学 bacterial genetics
细菌溢脓 bacterial ooze
细菌营养的 bacteriotrophic
细菌状的 bacteriform; bacteroid
细孔 puncture
细孔分布 pore size distribution
细孔状的 punctate
细粒 under size
细粒的 fine-grained
细粒环境 fine-grained environment
细粒状土壤 fine-grained soil
细料铺面刨花板 flake-faced particle board
细裂 hair crack
细裂片 segment
细裂纹 hair check; hairline crack
细滤器 fine filter
细毛 nap
细木杆排水沟 pole drain
细木工 cabinetmaker; cabinetmaking; joinery; woodwork
细木工板 block board; lumber core plywood
细木工刮刀 cabinet scraper
细木工锯 dimension saw
细木工料 small timber
细木工人 joiner
细木工手锯 pad saw
细木工用材 cabinet wood; cabinetmaker's wood
细木工凿 Jojner's chisel
细木工制品 joinery
细木工作业 cabinet-work
细木片 sliver
细目筛网 fine mesh wire
细黏土 fine clay
细鸟眼花纹 fine bird's eye (figure)
细刨 trying plane
细刨花 excelsior; wood wool
细劈裂 hair split
细砂(径0.25~0.1mm) fine sand
细砂带 fine-grit belt
细砂壤土 fine sandy loam
细筛 lawn
细绳 string
细竖锯 jig saw
细碎料 fine particle
细土 fine earth
细团块 fine crumb
细网格 fine grid
细微构造 minute structure
细微结构 fine structure
细微特征 minute feature
细纹材 fine-grained wood
细纹理 close grain; close texture; fine grain
细纹木材 close-grained wood; fine-grained wood
细纤维 fibril; fine fibre
细纤维化 fibrillation
细线 string
细线期(细胞分裂中) leptotene
细小可燃物 fine fuels
细小可燃物湿度码 fine fuel moisture code
细屑粒状 fine crumb
细叶脉 veinlet
细长八面体 elongated octahedron
细致结构 fine texture
细致纹理 fine grain; smooth grain
隙缝 blank
隙间部分 interstitial fraction
隙栖生生物群 pholadobiose
隙望室 observatory
匣式簧板 box lock
峡 clove; defile
峡谷 canyon; defile; gorge; notch; ravine
峡谷风 canyon wind; ravine wind
峡谷跑车(吊钩可落至深谷) canyon carriage
峡门 embouchure
峡湾 fjord; flord
狭凹槽 quirk
狭板 narrow board; planchette
狭槽刨 quirk router
狭槽榫接 slot mortise
狭带性的 stenozonal
狭低温的 oligothermal
狭缝 slit
狭缝喷嘴 slit nozzle
狭幅的 stenotopic
狭光的 stenophotic

狭口 bottleneck
狭木条 fillet
狭年轮的 narrow-zoned
狭栖的 stenoecic
狭栖性 stenoky
狭栖性的 stenoecious
狭湿性的 stenohydric
狭食性 stenophagy
狭食性的 stenophagic; stenophagous
狭适性的 stenoecic; stenoecious
狭水性的 stenohydric
狭条单板 fishtail
狭温性 stenothermal
狭温植物 stenothermic plant
狭盐性的 stenohaline
狭氧性的 stenooxybiotic
狭样区 transect
狭叶胡椒油 matico oil
狭叶态 stenophyllism
狭域性 stenoky
狭长餐桌 refectory table
狭长花床 border
狭长桌 davenport table
瑕疵 blemish
下凹嵌板 sunk panel
下凹饰条 sunk moulding
下凹桌面 sunk top
下半径圆 lower branch
下部夹锯(造材时) underbind; bottombind
下部渐深海底带 infrabathyal zone
下部结构 substructure
下层 underlayer
下层浮游生物 hypoplankton
下层抚育 thinning from below; thinning in lower storey
下层灌丛 underbrush
下层林 understorey
下层林丛 underwood
下层木 understorey; lower storey; subordinate tree
下层漂游生物 hyponeuston
下层疏伐 low thinning; ordinary thinning; thinning from below; thinning in lower storey
下层疏伐法 low-thinning method
下层树冠高 lower crown height
下层树冠长 lower crown length
下层树种 understorey (tree) species
下层土 undersoil
下层枝 lower branch
下层枝叶层火 duff fire
下层植被 lower storey
下层植生 understorey
下层植物 understorey
下茬(伐木) face; felling face; lek; undercut; gullet
下沉 subsidence
下沉逆温 subsidence (temperature) inversion
下沉气流 down draft
下承板梁桥 through plate girde bridge
下承桁架 through truss
下承桥 through-bridge
下池 ponding
下吹风 canyon wind
下垂 lop; sag
下唇(昆虫) labium (*plur.* labia)
下唇板 bottom lip (of slice)
下唇瓣(花冠) labium (*plur.* labia)
下唇花瓣 palate
下担子 hypobasidium
下垫层 underlayer
下垫面 underlying surface
下颚(昆虫) maxilla
下放重车的绞盘机 donkey snubber
下风(背风) lee; downwind
下伏辊 lower couch roll
下伏岩石 underlying rock
下斧口(伐木) cutting nick; wedge cut
下行色谱法 descending chromatography
下行纸分配色谱 descending paper partition chromatography
下颌骨 mandible
下横档 underbrace
下滑式斜索(索道) clothes line
下降风 canyon wind; katabatic wind
下降割面(采脂割面向下扩展) descending face
下降管道 downpipe
下降花序 descending inflorescence
下降坡度 descending grade; downward grade
下降射击 descending shot
下脚材 mill cull

下脚料 offcut; off-cut; offcut timber; offout
下锯(伐木) sawing; cutting
下锯点 cutting point
下锯法 sawing procedure
下锯口 bird's mouth
下锯木工(通常站在地下土坑内) pit man
下锯手 off-bearer
下锯图 sawing pattern
下卷式 underwind
下口(伐木) birdmouth; dip; face; felling face; felling notch; gullet; lek; set-up; throat; undercut; Humboldt notch; nick
下口内缘 key
下口平面 bed
下口式的 hypognathous
下拉伐根(用作锚桩) sucker-stump
下拉滑轮 sucker-down block
下料 laying off
下落时间 down time
下落式选材机 let-down drop sorter
下毛纸(毡布) bottom felt
下面 hypo-
下木 lower storey; lower storey of forest; tree of understorey; underbrush; undergrowth; underplant; understorey; underwood
下木层 understorey
下木栽植 understorey management
下木作业 underplant-operation
下胚轴 hypocotyls
下皮 hypodermis
下皮层 subcutis
下皮的(植物) hypodermal
下坡 downgrade; downhill (side); downward grade; favorable grade
下坡材 downhill timber
下坡集材 downhill skidding
下坡索道集材 downhill cable logging
下倾(枪柄底板) pitch down
下绕 underwind
下热压板 bettom hot platen
下砂式宽带砂光机 under side wide belt sander
下伸根(主根以外的) sinker
下水 launching
下水道污物 sewerage
下网 lower wire
下位的 hypostatic
下位萼 inferior calyx
下位花 hypogynous flower
下位基因 hypostatic gene
下位式 hypogyny
下位瘦果 hypoachene
下位子房 inferior ovary
下位子房周位花 perigynous with inferior ovary
下弦杆 lower chord
下限直径 diameter limit
下陷气孔 cryptopore
下向锯齿 retroserrate
下压辊 bottom press roll; lower press roll
下压式热压机 down-stroke hotpress
下芽接 down budding
下咽(昆虫) hypopharynx
下延的 decurrent
下曳气流 down draft
下刈 slashing
下游 lower course; lower reach
下游的 downstream
下种 seed release; seed shedding
下种槽 seeding trough
下种地(天然) seed-bed
下种伐 dark-felling; reserve cutting; seed(ing) cutting; seed(ing) felling
下种管 seed-horn
下种萌芽混交林 mixed seeding and sprout forest
下种萌芽林 composite coppice
下种树 bearing tree
夏孢子 summer-spore; ured(i)ospore; urediniospore
夏孢子堆 uredinium; uredium; uredosorus
夏孢子堆阶段(锈菌) uredial stage; uredial state
夏材 summer wood
夏伐材剥皮 sap-peeling
夏候鸟 summer bird; summer resident; summer visitor
夏季的 aestival; estival
夏季花园 summer garden
夏季停滞期 summer stagnation period

夏季相 aestival aspect
夏季休闲 clean summer fallow
夏季一年生植物 aestival annual plant; summer annual (plant)
夏季硬枝扦插 june-struck cutting
夏季作业伐木场 summer show
夏蜡梅碱 calycanthine
夏绿灌木群落 aestatifruticeta
夏绿林 aestilignosa; summer green forest
夏绿落叶林 summer-green deciduous forest
夏绿木本群落 aestilignosa
夏绿乔木群落 aestatisilvae
夏绿硬叶草本群落 aestiduriherbosa
夏绿硬叶林 aestidurilignosa; estidurilignosa
夏绿硬叶木本群落 aestidurilignosa
夏眠 summer dormancy
夏眠(生态) aestivation; estivation; summer dormancy
夏芽 aestival bud
夏叶的 therophyllous
夏羽 summer plumage
夏蛰 aestivation; estivation
夏滞育 summer diapause
仙人掌胶 cholla gum
仙人掌素 cactin
仙人掌温室 cactus house
仙食菌 ambrosia fungus
先成根原细胞 preformed root initial
先成河 antecedent stream
先出叶 prophyll
先导阀(液压) pilot
先导孔 pilot-hole
先导压辊(旋切机) pilot operated back-up roll
先端 apex
先锋类群 pioneer guild; pioneer species
先锋企业 pioneer enterprise
先锋群落 initiative community; pioneer community
先锋树种 pioneer; pioneer tree species
先锋植物 pioneer; pioneer plant
先锋种 pioneer species
先父遗传 telegony
先进先出 first in first out
先菌丝 promycelium (*plur.* -lia)
先期开花 proanthesis
先期流送 predrive
先期天然更新 advance growth
先期演替系列 eosere
先驱调查 pilot inventory; pilot survey
先天的 autochthonous; innate
先天行为 innate behaviour
先天萎缩 aplasia
先天性代谢病 inborn error of metabolism
先天性行为 instinctive behaviour
先天性休眠 innate dormancy
先天与后天 nature and nurture
先于田野的 campestral
先质 precursor
先祖 progenitor
纤匐枝 sarment
纤丝 fibril; fibrilla (*plur.* -llae); pirn
纤丝层 fibril layer
纤丝定向 orientation of fibril
纤丝化 fibrillation
纤丝角 fibril angle
纤丝排列 orientation of fibril
纤丝图式 fibrillar pattern
纤丝走向 fibrillar orientation
纤维 fiber; fibre
纤维板 fiber board; fibreboard; pressed-fibre board
纤维板芯胶合板 fibreboard core plywood
纤维饱和点 fiber saturation point; fibre saturation point
纤维背衬(砂带) fibre backing
纤维钵 fibre pot
纤维丛 fibre bundle
纤维蛋白 fibrin
纤维蛋白酶 plasmase
纤维定向 fibre orientation
纤维二糖酶 cellobiase
纤维分级 fiber classification; fibre classification
纤维分解 celluloslysis
纤维分离 defibering; defibration
纤维分离机 attrition mill; defibering machine; defiberizer; defibrator; defibrater; disintegrater; disintegrator; fluffer
纤维分离器 fibre separator

纤维分离温度 defibration temperature
纤维根 fibrous root
纤维固定 set in fiberlayer
纤维管胞 fibre tracheid
纤维化 fiberize
纤维回收 fibre recovery
纤维间接合 inter fibre bonding
纤维浆槽 stock chest
纤维结构 fibre texture
纤维精磨 fiber refining
纤维净化处理 fibre cleaning
纤维聚糖 cellulosan
纤维块(育苗容器) fibre block
纤维离析 defibrination; fiberization; defibration
纤维帽 fibre cap
纤维膜 fibre membrane
纤维泥炭 fibrous peat
纤维黏合 fibre bonding
纤维扭转树 tree with torse fibre
纤维排列 fibre orientation
纤维刨花板 fiber particleboard
纤维配比 fibre furnish
纤维皮层 fibrocortex
纤维铺装头 fibre spreader head
纤维强度 fibre strength
纤维鞘 fibre sheath
纤维朊酶 plasmase
纤维筛分 fibre fractionation
纤维绳芯 fibre core
纤维梳散机 shredder
纤维束 fiber bundle; fibre bundle
纤维水悬浮液 fibre-water-suspension
纤维丝 filament
纤维素 cellulose
α-纤维素 alpha-cellulose; α-cellulose
β-纤维素 beta cellulose; β-cellulose
γ-纤维素 γ-cellulose
纤维素分解菌 cellulose-dissolving fungus
纤维素分子链 cellulose molecular chain
纤维素分子排列 orientation of cellulose molecule
纤维素构成体 cellulose matrix
纤维素降解 cellulose degradation
纤维素胶黏剂 cellulose adhesive
纤维素类交换剂 cellulosic exchanger
纤维素链 cellulose chain
纤维素酶制备 cellulase preparation
纤维素酶制剂 cellulase preparation
纤维素醚类 cellulose ethers
纤维素漆 cellulose lacquer
纤维素三糖 cellotriose
纤维素生物降解 cellulose biodegradation
纤维素束 cellulose bundle
纤维素索 cellulosic strand
纤维素纤丝 cellulose fibril
纤维素纤维 cellulose base fibre; cellulose fiber; cellulose fibre
纤维素衍生物 cellulose derivative
纤维素乙醚 ethyl cellulose
纤维素酯类 cellulose esters
纤维细胞 fibrous cell
纤维形态 morphology of fibre
纤维悬浮液 fibre suspension
纤维亚麻 fibre flax
纤维应力 fibre stress
纤维原料 fibre stuff; fibrestock
纤维照影机 fibrograph
纤维制品 fibre; fiber
纤维质层 cellulosic layer
纤维质地 fibre texture
纤维质泥炭 fibrous peat
纤维质纤维 cellulosic fibre
纤维状导管分子 fibriform vessel element; fibriform vessel member
纤维状的 filiform
纤维状管胞 fibre tracheid; fibrous tracheid
纤维状细胞 fibrous cell
酰胺 amide
酰胺化作用 amidation
酰胺酶 amidase
鲜贝壳杉脂 bush kauri
鲜色夹皮 bright bark pocket
鲜重 fresh mass; fresh weight; green weight; wet weight
闲暇活动 leisure activity
闲置时间 lead time
闲置资金 idle fund
弦茬(伐木时未锯断部分) beard; key; sloven; bridge
弦尺(测量桶板材单位) chord-foot
弦径向收缩比 ratio of tangential to radial shrinkage
弦锯 crown-cut
弦锯板 plain-sawed lumber

弦锯材 bastard-lumber; flat-sawn timber
弦锯面 flat grain face; flat-sawn surface; plain grain face
弦锯面的 plain-sawed
弦锯纹理 bastard-sawn grain; flat-sawn grain
弦面 side grain surface; tangential plane
弦面板 side board
弦面材 flat grain lumber; slash-grain lumber
弦面花纹 plain cut figure; slash figure; slash grain
弦面花纹裂 slash figure check
弦面切片 tangential section
弦面纹理 bastard grain; slash figure; slash grain
弦切 flat cut
弦切材 plain-sawed lumber
弦切花纹 plain cut figure
弦切面 bastard face; flat grain face; plain grain face; tangential face; tangential section; tangential surface
弦切纹理 flat (cut) grain; flat-sawn grain
弦向 tangential direction
弦向板 tangential plank
弦向薄壁组织 tangential parenchyma
弦向分裂 tangential division
弦向干缩 tangential shrinkage
弦向锯切 back cut; back-sawing; tangential cut
弦向锯切的 back-sawn; bastard-sawn; crown-cut; flat-sawed; flat-sawn
弦向孔材(管孔弦列) tangential porous wood
弦向排列 tangential arrangement
弦向膨胀 tangential swelling
弦向切割 flat cut
弦向收缩 tangential shrinkage
弦向纹孔式 tangential pitting
弦向细胞壁 tangential wall
弦向下锯的 plain-sawed
弦向下锯法 flat sawing; plain sawing
弦向应力 tangential stress
弦向载荷 tangential loading
弦向纵切面 tangential longitudinal section
咸度 salinity
咸水湖 lagoon
嫌恶(干湿阴阳等) phobe; phobia; phobi; fuge; avoidance
嫌恶的(干湿阴阳等) phobous; phobic; fugous; avoiding
嫌钙植物 calcifuge; calcifuge plant; calcifugous plant; calciphobe; calciphobous plant
嫌光性浮游生物 knephoplankton
嫌碱植物 basifuge
嫌气微生物 anaerophytobiont
嫌气细菌 anaerobic bacteria; anerobic bacteria
嫌气性果胶分解细菌 anaerobic pectin decomposing bacteria
嫌气性纤维素分解细菌 anaerobic cellulose-decomposing bacteria
嫌深性 bathoxenous
嫌水液体 hydrophobic liquid
嫌水液体干燥法 drying in hydrophobic liquid
嫌酸的 oxyphobous; acidophobous
嫌酸植物 oxyphobe
嫌雪植物 chianophobe
嫌压的 barophobic
嫌阳的 heliophobous
嫌阳植物 heliophobe
嫌蚁的 myrmecophobic; myrmecophobous
嫌雨植物 ombrophobe; phreatophyte
显光合作用 apparent photosynthesis
显花植物 aerogams; anthophyta; carpophyte; phanerogam; phenogam
显露 emergence
显热 sensible heat
显色反应 colour reaction
显示 exposure
显示器 indicator; scope
显微操作 micromanipulation
显微操作技术 micro-manipulative technique
显微操作术 micrurgical technique
显微测树仪 microdendrometer
显微分光光度测定法 microspectrophotometry
显微分光镜 microspectroscope

显微构造 microscopic structure
显微光度计 microphotometer
显微灰化 microincineration
显微绘图 micrography
显微技术 microtechnique
显微结构 microstructure
显微解剖 micro-dissection
显微解剖器 micro-dissection apparatus
显微解剖枝术 micro-dissection technique
显微镜 microscope
显微镜分辨率 resolution of microscope
显微镜分析 microscopic(al) analysis
显微镜观察 microscopic observation
显微镜检查 microscopic examination
显微镜描图器 camera lucida
显微镜视野 field of microscope
显微切片 microtomed section
显微伸长测定仪 micro-extensometer
显微术 microscopy
显微特征 microscopic feature
显微投影器 microprojection apparatus
显微形态学 microscopic morphology
显微学 micrology
显微遗传学 microgenetics
显微映象器 microprojection apparatus
显微照相机 microscope camera
显微照相设备 microphotographic apparatus
显微照相术 micrography
显像(层析法) development; developing
显像纸 developing paper
显心材树种 obligatorily coloured heartwood tree
显性 dominance
显性定律 law of dominance
显性度 degree of dominance
显性方差 dominant variance
显性基因 dominant gene
显性假说(杂种优势) dominance hypothesis
显性突变型 dominant mutant
显性性状 dominant character
显性修饰因子 dominigene
显性杂合体 goneoclin
显性致死因子 dominant lethal
显影 development
显影不足 underdevelopment
显影剂 developer
显影液 developer solution
显域分布 zonal distribution
显域土 zonal soil
显域植被 zonal vegetation
显着水平 level of significance
显著差异 significant deviation
显著度 conspicuousness; significance
显著特征 dominant feature
显著性 significance
显著性测验 significance test
显著性检验 test of significance
险峭的山脊 arete
藓类 moss
藓类冻原 moss-tundra
藓类林 mossy forest
藓类沼泽 moss-moor
县林业局 county board of forestry
县营林员 county ranger
县有林 prefectural forest
现场处理 field treatment
现场鉴别 field identification
现场勘测 field reconnaissance
现场踏查 field reconnaissance
现存 on hand
现存量 standing crop biomass
现存林 existing forest
现存植被 extant vegetation
现存植被图 real vegetation map
现代家具 contemporary furniture; modern furniture
现代建筑 modern architecture
现代侵蚀 recent erosion
现代园林 modern garden
现地 in situ
现付成本 current outlay cost
现行价值 current value
现行市价 going market price
现货价格 spot price
现货升水 back wardation
现货市场 physical market
现金 cash; hard cash
现金存款 primary deposit
现金购买 cash purchase
现金净流量 net cash flow
现金流量贴现 discounted cash flow

现金债券 cash bond
现进 cash purchase
现净值 present net worth
现款 cash
现钱 quick assets
现时指令 current order
现实立地 actual site
现实连年总生长量 gross current annual increment
现实林 actual forest; real forest
现实毛材积生长量 current gross volume growth
现实生长量 current accretion; current increment
现实收获表 empirical yield table
现实蓄积 real growing-stock
现实遗传力 realized heritability
现实植被 actual vegetation
现实植被图 actual vegetation map
现在蓄积 present yield
现值 current value; present value
现值成本 present value cost
限定基金 restricted fund
限定流速 limiting water velocity
限度 mark
限量采伐(年采伐量少于年生长量) undercut
限量与交易(碳排放与碳汇) cap-and-trade
限外区 exclosure
限性遗传 one-sided inheritance
限制法 Rueping process
限制坡度 absolute maximum gradient; ruling gradient
限制使用林 limited use forest
限制水位 limiting water level
限制相似性 limiting similarity
限制性贷款 tied loan
限制性内切酶 restriction enzyme
限制因子 limiting factor
限制砧 restrictive rootstock
线(果实的) suture
线虫 Eelworm; nema; nematode
线虫病 nematodiasis
线虫学 nematology
线虫瘿 nematode gall
线抽样 line sampling
线弹性 linear elasticity
线规 wire gauge
线角 moulding
线脚 mould; moulding
线脚钉夹钳 brad setter
线脚端头 stop moulding
线脚刨 banding plane
线脚刨床 moulding plane
线脚切割器 moulding cutter
线锯 coping saw; fret saw; hand jig; narrow band-saw
线框 wire frame
线粒体 mitochondrion
线路(集材道, 运材道, 钢索道) road
线路图 road pattern
线膨胀 linear expansion
线膨胀系数 linear expansivity
线圈 bobbin; coil; winding
线润胀 linear expansion
线上信息处理 on-line message processing
线速度 line speed; linear velocity
线条机 moulding machine
线条刨 cornice plane; too(u)lder
线筒 bobbin spool
线团 coil
线纹(病毒病症状) line pattern
线系进化 phyletic evolution
线相交抽样 line intersect sampling
线形的 filiform
线形纹孔 linear pit
线型模式 linear model
线型刨床 moulding plane
线型设计 linear programming
线性 linearity
线性变换应变仪 linear transformer strain gauge
线性变位传感器(压机) linear variable displacement transducer
线性参数 linear parameter
线性抽样 linear sampling
线性抽样样地 linear (sample) plot
线性分子 linear molecule
线性规划 linear programming
线性回归 linear regression
线性回归估计 linear regression estimate
线性量测 linear measurement
线性收缩 directional shrinkage; linear shrinkage

线性特征判读 interpretation of linear features
线性因子 linear factor
线性应变 linear strain
线性预应力 linear prestressing
线性阵列 linear array
线叶(病毒病症状) line leaf
线重力 linear gravity
线轴 reel; spool
线状腐朽 thread rot
线状磨屑(整个刃口成一条线落下) wire edge
线状侵蚀 linear erosion
线状调查 linear survey
线状样地 linear plot
线状样地法 line-plot method
线状样地调查 line-plot survey
陷阱 pitfall; pit-fall trap; trap-hole
陷笼(捕兽) box trap
陷穴 mud hole
陷穴侵蚀 sinking hole erosion
腺 gland
腺鳞 glandular scale
腺毛 glandular hair
腺嘌呤核苷 adenosin(e)
腺体 gland
腺胃 proventriculus
腺制诱剂 gland lure
乡村车削家具 country turned furniture
乡村风景 rustic scenery
乡村林 village forest
乡村林业 rural forestry
乡土的 autochthonous; indigenous
乡土动植物 aboriginal
乡土树种 endemic species; indigenous tree species; native species; native tree; native tree species
乡土植物 indigenous plant
乡土种 autochthon; endemic species; indigenous species
相称 equilibrium
相斥 repulsion
相斥系 repulsion series
相对保留时间 relative retention time
相对变异 relative variance
相对方位角 relative azimuth
相对盖度 relative cover
相对光照度 percentage illumination; percentage irradiance
相对光照强度 relative light intensity
相对海平面 relative sea level
相对含水率 relative moisture content
相对孔径 relative aperture
相对密度 relative density
相对拧紧力 relative screw-holding power
相对频度 relative frequency
相对燃烧性指标 relative flammability index
相对热平衡 relative heat balance
相对涩度 relative astringency
相对生长法 relative growth method
相对生长量 relative growth
相对生长速率 relative growth rate
相对生长系数 relative growth coefficient
相对生殖值 relative reproductive value
相对湿度 relative humidity
相对湿度百分率 percentage humidity; percentage relative humidity
相对收敛值 relative astringency
相对收入假设 relative income hypothesis
相对疏伐强 relative thinning intensity
相对同龄林 relative even-aged stand
相对位置 anteposition
相对握钉力 relative screw-holding power
相对性状 allelic pair; contrasting character
相对需光量 relative light requirement
相对削度 relative taper
相对优势度 relative dominance
相对有利条件 comparative advantage
相对雨量系数 relative pluviometric coefficient
相对蒸腾 relative transpiration
相对直径法 method of relative diameter
相多态现象 phase polymorphism
相干图像雷达 coherent image radar
相干性 coherence
相干噪声 coherent noise

相关分析 correlogram analysis
相关索引 correlation index
相关系数 coefficient of correlation; correlation-coefficient
相关性状 correlation character
相关性状群 pleiade
相关指数 correlation index
相关作用 correlation
相害性竞争 disoperative competition
相互掺混 interblend
相互动态 co-dynamic
相互干扰常数 mutual interference constant
相互交错样本 interpenetrating samples
相互适应 coadaptation
相互授粉 mutual pollination
相互易位 reciprocal translocation
相互作用 coactions; interact; interaction
相互作用者 coactor
相克生物 antibiont
相克因子 oppositional factor
相邻峰 adjacent peak
相邻染色体 adjacent chromosome
相容曲线 consistency curve
相容性 consistency
相似度 degree of similarity; similarity
相似规则 rule of similarity
相似系数 coefficient of similarity
相似性矩阵 similarity matrix
相似性指数 index of similarity
相思豆毒素 abrin; jequiritin
相思树胶 Australian gum
相异 disconcordance
相异的 alien
相异系数 coefficient of difference
相引组 coupling series
香柏一类树种的总称 arbor-vitae
香柏油 thuja oil
香槟油 champagne
香草隆(除草剂) cyperon
香草酸 vanillic acid
香椿木油 calantas wood oil
香椿脑 toonalol
香椿属 toona Toona
香椿素 toonin
香豆冉酮 coumaranone
香豆素 coumarin
香度计 osmometer
香附酮 cyperone
香附油 cyperus oil
香膏 pomade; resinoid
香根草油 oil of cuscus
香根油 cuscus oil; vetiver oil
香花园 perfume garden
香精 essence; perfume
香精油 essential oil
香苦木油 cascarilla oil
香辣味 spicy acrid taste
香兰酮 vanilione
香料 essence; fragrance (material); odoriferous material; perfume; spice
香鳞 scent scales
香茅基 citronellyl
香茅烯 citronellene
香茅油 citronella oil; pamorusa oil
香柠檬酚 bergaptol
香柠檬亭 bergamottin
香柠檬烯 bergamotene
香柠檬叶油 bergamot leaf oil; bergamot petitgain oil
香柠檬油 bergamot oil
香农多样性指数 Shannon's index of diversity
香农-威纳指数 Shannon-Weiner index
香农-韦尔方程 Shannon-Weaver function
香气 essence; fragrance (material)
香芹艾菊酮 carvotanacetone
香芹蓝烯 carvomenthene
香芹孟酮 carvomenthone
香芹鞣酮 carvotanacetone
香芹烯 carvene; corvestrene
香芹烯酮 carvenone
香树素 aromadendrin
香树酮 aromadendrone
香树脂 resinoid
香水 perfume
香水油 ylang-ylang oil
香桃木油 myrtle oil
香调 aspect
香豌豆苷 oroboside
香味 flavour; fragrance (material)
香叶烯 geranene; geraniolene
香叶油 geranium oil

香橼叶油 citron leaf oil; citron petitgrain oil
香橼油 cedrat oil; citron oil
香韵 aspect
香脂 balsam; styrax; pomade
香柱病 choke disease
箱 box; case; chamber; chest; housing
箱板 box board
箱板材 box lumber; box shook timber; box timber; case wood; packing timber; wrack
箱凳(座部为箱的凳) box stool
箱盒用材 short end
箱式窗框 cased frame
箱式法(采脂) box-system
箱式干燥器 loft dryer
箱式干燥窑 box kiln
箱式磨木机 magazine grinder
箱锁 box lock
箱体成型法 collipress process
箱匣地木 body
箱匣鸩尾榫 box dovetail
箱形采脂法 box-system
箱形断面起重臂(装载机的) box-section loader arm
箱形堆垛 box(-end) stacking
箱形堆积 box stacking
箱形涵洞 box culvert
箱形冷凝器 box-condenser
箱形梁 box beam
箱形榫 box tenon
箱型百叶窗 boxing shutter
箱型框架 boxed frame
箱型桩 box pile
箱植 box planting
镶板(门或墙) boarding; panel board; panelling; fielded panel
镶板胶合板 plypanel
镶板门 panel(ed) door
镶边(装饰) edging
镶边平面门 margined flush door
镶边手锯 back saw; bead saw
镶边植物 edging plant
镶玻璃框条 glazing bar
镶玻璃饰条 glazing bead
镶齿 inserted tooth; tipped-tooth
镶齿的(圆锯) tipped
镶齿锯齿刃 bit
镶齿圆锯 rim saw
镶端板条 end strip
镶盖 veneer
镶椽接合 mortise and tenon joint
镶合腹接 side veneer-grafting
镶合靠接 inarching by veneering
镶尖榫接头 tusk and tenon joint
镶接 hatched grafting; veneer grafting
镶接木板 board-and-brace work
镶铝装饰胶合板 plymax
镶木地板 parquet; parquet floor; parquetry
镶木工艺 wood-mosaic
镶木涂装(板条镶木板涂装) strip coating
镶木细工 parquetry
镶木制品 tarsia
镶片芽接 flute budding
镶片圆锯 rift saw; wing saw
镶嵌 inlay; mosaic
镶嵌板条 parqnet strip
镶嵌边缘 parquet border
镶嵌工艺品 inlay
镶嵌花坛植物 mosaic bedding plant
镶嵌接头 housed joint
镶嵌结构 mosaic
镶嵌木块 parquet block
镶嵌式栽植 mosaiculture
镶嵌索引图 index mosaic
镶嵌条 fillet
镶嵌条饰 cover fillet
镶嵌图(航片) mosaic
镶嵌土壤 mosaic soil
镶嵌物 insert
镶嵌细工 inlaid work; marquetry
镶嵌细木工 tarsia; wooden mosaic; wood-mosaic
镶嵌显性 mosaic dominance
镶嵌型刀具 inserted tool
镶嵌杂种 mosaic hybrid
镶嵌栽培 mosaiculture
镶嵌植被 mosaic vegetation
镶嵌作业 inlaid work
镶榫接合 mortice and tenon joint
镶芽接 veneer budding
详细经营方案 detail plan
详细木材资料 detailed timber data

详细设计 design in detail
响度 loudness
响应 response
响应时间 response time
响应性 responsivity
相 phase
相变 phase variation
相衬显微镜 phase-contrast microscope
相位偏移 phase deviation
相位图 phase diagram
向岸流 onshore current
向触性 haptotropism; thigmotropism; thixotropism
向低温性 cryotropism
向地性 geotropism
向顶的 acropetal; apical
向风 wind exposure; windward
向风面 windward side
向风性 anemotropism
向光的 heliotropic
向光性 eutropic; heliotropism; phototropism
向化性 chemotropism
向基的(孢子着生方式) basipetal
向碱性 alkaliotropism
向冷性 cryotropism
向量 vector
向内生长[物] ingrowth
向气性 aerotropism
向器官性 organotropy
向前伐倒 pull down front
向热性 thermotropism
向日的 heliotropic
向日性 heliotropism
向日性的 eutropic
向洒水箱充水斗架 water ladder
向上鼓风切割机 blowing-up cutter
向上调整 upward adjustment
向上直射 perpendicular shot
向渗性 osmotropism
向声性 phonotropism
向实体性 stereotropism
向水性 hydrotropism
向天空的 sky-ward facing
向外 ectad
向温性 thermotropism
向斜 syncline
向心花序 centripetal inflorescence
向心木质部 centripetal xylem
向性 tropism
向性运动 tropic movement
向阳的 sunny
向氧性 aerotropism
向雨性 ombrotropism
向自性 autotropism
项目边界 project boundary
象鼻(水滑道终点) elephant
象鼻虫 collar borer; weevil
象甲 snout beetle; weevil
象细胞的 cellular
象限分线规 quadrant dividers
象限角 bearing; quadrant angle
象限锯 quarter saw
象限内伐木(绞盘机集材时) fall a quarter
象形机器(如铣床或开槽机等) elephant
像点 image point
像幅 format
像革质的 coriaceous
像片比例尺换算 photo scale reciprocal
像片材积 photo volume
像片材积表 photo volume table
像片材积估测 photo volume estimate
像片测定 photo measurement
像片成图辐射仪 photo mapping radiometer
像片对 pair of photographs
像片反差 contrast on photograph
像片勾绘 delineation on photograph
像片林木勘察 photo timber cruise
像片判读 photo interpretation
像片判读技术 technique for photo-interpretation
像片判读检索表 photo-interpretation key
像片判读员 interpreter for photograph; photointerpreter
像片色调 photo tone
像片位移 displacement on photograph
像片样地 photo plot
像片阴影测定法 shadow method in photo

像片中心 photo center
像片重叠 overlap of photograph
像片资源清查 photo controlled inventory
像片坐标 photograph coordinate
像元(素) pixel
橡胶采割 tap for resin
橡胶袋压模 rubber bag moulding
橡胶辊 rubber roll
橡胶基质 isoprene
橡胶浆 latex (*plur.* latices)
橡胶接触辊 rubber contact roll
橡胶乳液 rubber latex
橡胶软化剂 rubber softener
橡胶树 bamboo grows; para rubber tree; para caoutchouc tree; seringa
橡胶树脂 gum-lac
橡胶园 rubber estate; rubber plantation
橡胶植物 rubber bearimg plant; rubber plant; rubber producing plant; rubber yielding plant
橡精 datiscetin
橡木 oakwood
橡皮垫脚 rubber cushion glide
橡皮刮涂 squeegee coat
橡皮头移液管 teat pipette
橡实 acorn
橡实播种器 acorn planter
橡椀(橡碗) valonia; acorn cup
橡椀栲胶 valonea extract; valonia extract
橡椀鞣质 valonia tannin
橡椀酸 valoneaic acid
橡椀氧杂蒽酮 valoneaxanthone
橡叶纹(病毒病症状) oak line pattern
橡子仁 acorn nut
肖楠酚 heyderiol
肖楠脑 libocedrol
肖楠烯 libocedrene
肖氏打浆度 degree Schopper-Riegler; Schopper-Riegler grade; Schopper-Riegler beating degree; Schopper-Riegler-Mahlgrad
肖氏打浆度测定仪 Schopper-Riegler beating degree tester
肖氏叩解度 Schopper-Riegler-Mahlgrad
肖氏硬度 Shaw hardness
鸮(xiāo) owl
消除 abatement
消除表面硬化(木材) relief of casehardening; remedy of casehardening
消除反应 elimination reaction
消除器 eliminator
消毒 sterilization
消毒剂 disinfectant; sterilant
消毒器 sterilizer
消毒杀菌 disinfection
消毒种子 disease-treated seed
消防 fire fighting; fire protection
消防部门 fire department
消防车 fire engine
消防措施 fire protection
消防队 fire brigade; fire fighting crew
消防队员 fire fighter; fireman
消防队长 fire master
消防管理 fire management
消防署全体人员 fire department
消防栓 hydrant
消防水带 fire belt
消防水桶 fire bucket
消防演习 fire drill
消防营地 fly camp
消防员 fire extinguisher
消费不足 underconsumption
消费单位 spending unit
消费价格指数 consumer price index
消费倾向 propensity to consume
消费信贷 consumer credit
消费信贷公司 consumer finance company
消费性资产 wasting assets
消费需求 consumer demand
消费者 consumer
消费者主权 consumer sovereignty
消光 flatting
消光度 absorbancy
消光剂 achromatic colours; flatting agent
消光涂饰 mat finish
消光系数 extinction coefficient; light extinction coefficient
消光整理 mat finish
消光装饰 flatting finish
消耗 drain

消耗不足 underconsumption
消耗期 exhaustion stage
消化效率 digestive efficiency
消化性 digestibility
消泡剂 defoamer agent; defrother; foam breaker; foam killer
消泡器 defrother; foam breaker; foam killer
消球差放大镜 aplanatic magnifier
消球差镜 aplanat
消去反应 elimination reaction
消融(冰雪) ablation
消弱系数 extinction coefficient
消色 compensation
消色期 achromic period
消色颜料 achromatic colours
消声漆 anti-noise paint
消声器 acoustic filter; buffer; deadener; deafener; sound damper; sound eliminator
消声装置 sound damping device
消失点 vanishing point
消退型 regressive type
消闲活动 leisure activity
消音胶合板 sound-deadening plywood
消音器 muffler; quencher
萧条 depression
硝铵 aerolite; aerolith
硝化(纤维人造)丝 pyroxyline silk
硝化甘油 nitroglycerine
硝化呼吸 nitrate respiration
硝化木素 nitrolignin
硝化细菌 nitrifier; nitrifying bacteria
硝化纤维 gun cotton; nitro-cellulose
硝化纤维漆 nitro-lacquer
硝化纤维素 nitro-cellulose
硝化纤维素乳液 nitrocellulose emulsion
硝化作用 nitration; nitrification
硝基瓷漆 lacquer enamel
硝基氯仿 nitrochloroform
硝基漆 nitrocellulose lacquer; nitro-lacquer; pyroxyline lacquer
硝基纤维漆 lacquer; lacker; nitrocellulose lacquer
硝石 saltpeter
硝石层 caliche
硝酸根离子 nitrate ions
硝酸钾 kali saltpeter
硝酸细菌 nitric bacteria
硝酸盐 nitrate
硝酸盐细菌 nitrate bacteria
硝酸盐植物 nitrate plant
销 pin
销钉 dowel bar
销钉接合 dowelling joint
销货清单 account sales
销口 lock joint
销路说明 merchantability specification
销售的重要性 marketing weight
销售堆 sale lot
销售效率 marketing efficiency
销售折扣 sales discount
销子 fid; pin
削边 feather edge
削边机 chipping edger; edge chipper
削皮 skin; skinning
削皮标志 bark hack
削皮工 spudder
削皮器 bark blazer; spudder
削片 chipping
削片材 chip wood
削片裁边机 edge chipper
削片吹送机 chipper-blower
削片废材处理 slash disposal of chipping
削片机 chipharvestor; chipper; chipping machine; flake-producing machine; flaker; flake-shaving machine; shaving machine
削片机出料口 chipper spout
削片机刀盘 chipper disc
削片碾磨机 chipper-mill
削片刨方机 chipper canter
削片齐边机 chipper-edger
削片去皮法 chip barking method
削片压痕 bruising
削片研磨机 chip grinder
削片制材联合机 chip-N-saw
削片制材主锯机 chipping headrig
削片制方机 canter; chipper canter; chipping canter
削平钻孔机(枕木) adzing and boring machine
削凿刀 paring chisel

削枕机 adzing machine
小(型)孢子囊 sporangiole
小班 lot; stand; subcompartment; sub-compartment
小班界 stand boundary; sub-compartment boundary
小班经理法 subcompartment management method
小班经营法 compartment management method; individualizing method; subcompartment management method
小班调查 in-place inventory; subcompartment investigation
小半径转角 sharp corner
小苞片 bracteole; bractlet
小孢子 microspore; sporidium
小冰期 little ice age
小冰山 bergy bit
小柄 sterigma (*plur.* -mata)
小波分析 wavelet analysis
小材(直径7cm以下) small wood
小潮 neap tide
小车 bogie; pallet
小尺度 fine scale
小齿轮传动绞盘机 simplex donkey
小翅的 micropterous
小虫 bug
小虫孔 needle hole; pin-hole; shot hole
小抽屉柜 bachelor's chest
小船(木材流送中运绞盘机等设备) alligator
小船肋骨 canoe rib
小蠹 ambrosia borer
小刺 spinule
小带锯 pony band-saw
小带锯机 pony rig bandmill
小带锯制材厂 pony rig bandmill
小单优种群丛 consociule
小道(行走用) alley; by-path; foot path
小道(集材用) runway
小凳子 tabouret
小地产 minifundium
小地段 plot
小地形 microrelief; minor feature
小豆蔻油 cardamon oil
小蠹 ambrosia beetle; bark beetle; pinhole borer
小蠹虫 engraver beetle
小盾片 escutcheon
小方材 batten; butten; rail; scantling; small square; baby square
小费 key money
小分生孢子 microconidium
小蜂 chalcid wasp
小蜂媒花 micromelittophilae
小幅板 small board
小幅航空摄影像片 small-format aerial photograph photo
小复查 intermediate revision
小副冠类(水仙等) pavi-coronate
小杆材 rod; wood strip
小杆材期 thicket stage
小干果 coccus
小港 creek
小高位芽植物 microphanerophyte
小搁栅 sleeper
小根 radicel; radicle; rootlet
小梗 sterigma (*plur.* -mata); trichidium
小共生体 microsymbiont
小沟侵蚀 rill erosion
小钩 hamulus; hooklet
小谷 dale; dell
小冠巴西棕蜡 ouricury wax
小灌木 semishrub; semi-shrub; shrublet; subshrub; undershrub
小灌木状的 fruticulose
小规格材 small-dimension stock
小规模林业 forestry on small scale; petty forestry
小规则式 irregular style
小果 fruitlet
小河 rill
小河床 minor (river) bed
小河绠 crib
小核果 drupel; drupelet; ossiculus
小花 floret
小花的 floscular; flosculous
小花樟酚 microl
小花轴 rachilla
小滑车 jigger
小环境 micro-environment
小环裂 slight shake

X

小茴香油 fennel oil
小茴香子油 fennel seed oil
小机车 dolly
小集材道 rack; rack way
小集中(短距离集材) bank; bunch(ing); preyarding; fairleading
小集中绞盘机 preyarder
小尖刺 stimulus
小坚果 nutlet; pyrene
小茧蜂 supplementary ichneumon fly
小绞车 twister
小进化 microevolution
小精管 testicular follicle
小径 alley
小径材 slender-branched; small timber
小径木(单株木) submarginal log; undersized log; undersized tree
小径木(林子) pole crop; pole stage
小酒桌 wine table
小锯材 small scantling
小聚伞花序 cymelet; cymule
小刻线刀 scratch
小孔 eyelet; foramen (*plur.* -mina)
小跨距桁架桥 pony truss bridge
小块 tablet
小块灌溉 basin irrigation
小块茎 tubercle
小块林地 wood lot
小块状采伐 small gap felling
小块状更新造林 cell-type reforestation
小粒弹 bird shot
小梁侧封板 apron lining
小猎兽(如兔等) ground game
小裂片 lobelet; lobule
小林地 holt
小鳞茎 bulblet; clove
小鳞片 ramentum
小令(每令480张纸) short ream
小流域综合治理 small watershed management
小瘤 tubercle
小龙卷 water spout
小路 by-pass road; by-road; path
小轮裂 slight shake
小脉 veinlet
小猫 kitten
小门 hatch
小面 facet
小面积集团森林经营 group management
小苗 plantlet
小苗移植 pricking out
小木段 peeler block
小木片 mill-size chip
小牧场 enclosure; holding paddock; paddock
小泥铲 trowel
小年 off year; partial mast
小鸟 hatchling
小农场 minifundium
小爬犁 bob
小排 float
小牌桌 ombre table
小配子 microgamete
小批量生产 short run
小片 tablet
小片林 wood; woodland
小平板 tablet
小平面 facet
小瀑布 cascade
小栖息地 microhabitat
小鳍的 micropterous
小气候 microclimate
小气候学 microclimatology
小气象学 micrometeorology
小丘 hillock; pingo
小球果 conelet
小球茎 cormlet
小区(地形图的方格) quadrangle
小区(事业分区以下) sector
小区(样地) plot
小群 group
小群落 microcommunities; minor community
小染色体 microchromosome
小伞形花序 umbellule
小山 hill
小商品原木 peewee
小生境 biotope; ecological niche; microhabitat; niche
小石子 scree
小食橱 livery cupboard
小室(子房或花药之室) loculus
小手锄 planting hammer
小手锯 chest saw

小书桌(翻板式) davenport
小树 arboret
小树林 grove; shaw
小树枝(直径小于0.5cm) twig
小树脂斑纹(针叶材) small pitch streak
小穗 spicula; spikelet
小穗轴 rachilla
小穗状花序 spikelet
小特厚板 scantling
小体 corpuscle
小体配合(原生动物) merogamy
小头 small end; small(er) end; top end
小头钉 sprig
小头直径 diameter at small end; small-end diameter
小土铲 trowel
小团状花纹 small blister figure; small pomele figure
小托叶 stipel(la)
小湾 creek
小五金 hardware
小溪 dell; rill
小细胞 cellule
小细长裂片 lacinule
小峡谷 cove
小纤维 fibrilla (*plur.* -llae)
小纤维束(化学木粕中纤维尚未充分离解) sheaves; shives
小线虫 Eelworm
小形叶 microphyll
小形叶的 microphyllous
小型采运设备 one-donkey show
小型带锯 pony rig
小型带锯制材厂 pony band mill
小型单筒绞盘机 log haul(-up)
小型底栖生物 meiobenthos
小型电打印机 mini electro printer
小型伐区 one-donkey show
小型鼓式剥皮机 pocket debarker
小型烘缸 baby dryer
小型化 miniaturization
小型计算机 mini-computer
小型链锯 mini-size chain saw
小型轮式集材机 wheel-mounted small skidder
小型磨浆机 pocket grinder
小型染色体 minisome
小型森铁机车 dinkey
小型生物 microfauna
小型拖拉机 dinkey; kitten; midget tractor
小型挖土机 gopher
小型消费者 microconsumer
小型运材挂车 pup trailer
小型造纸机 sheet machine
小修 minor overhaul; running repair
小芽 gemmule
小眼面(昆虫) facet
小演替 micro-succession
小演替系列 microsere; serule
小样本理论 small sample theory; theory of small sampling
小样方 sampling plot; small quadrat; sub-plot
小样区 small quadrat
小叶 foliole; leafit; leaflet; lobule; lobulus (*plur.* lobuli)
小叶柄 petiolule
小叶病 little leaf disease
具2～5小叶的 bipentaphyllus
小叶迹 leaflet trace
小叶突 lobelet
小夜班 swing shift
小异染色质的 microheterochromatic
小蝇媒花 micromyiophilae
小幼树 small sapling
小鱼 fingerling
小羽片 pinnule
小雨 light rain
小原木(小头直径11英寸或28cm以下) batten
小圆材 round-billet (wood); small pole
小圆锯 burr
小圆木 rod
小圆凿 quick gouge
小针刺 stylet
小阵雨小阵风 scud
小支柱材 small pit-prop
小枝 branchlet; ramellus; ramulus (*plur.* ramuli); sprig
小植物 plantlet
小制材厂(主锯为小带锯或小圆锯) pony mill
小种 microspecies; minor species; race

小种非专化抗性 race-nonspecific resistance
小种圈 circle of microspecies
小种专化抗性 race-specific resistance
小烛树蜡 candelila wax
小柱材 low pole
小锥 bradawl
小总苞 involucret
小总状花序 racemule
效价 potency
效率 coefficient of efficiency; efficiency; output; performance
效能差异 efficiency variance
效益-成本法 benefit-cost procedure
效益整合 co-benefits
楔边板 feather edge board
楔刀 froe
楔钉榫接 cogged-scarf joint
楔角(锯链切齿等) bevel angle; grinding angle; knife chamber angle; lip angle; sharpness angle; tooth angle; wedge angle; face angle
楔接 wedge grafting; wedge joint
楔面齿接 indented scarfed joint
楔片 cuneus; key
楔入 wedge
楔形板 taper plate
楔形的 sphenoidal
楔形垫块(打入钢轨下) shim
楔形钉钩 dog wedge
楔形缝 wedge-shaped slit
楔形夹紧套 wedge socket
楔形劈裂 wedge split
楔形切片法 wedge shaped sectioning
楔形伞伐(作业) shelterwood wedge system
楔形伞伐作业法 wedge system; wedge-shaped shelter-wood system
楔形榫 wheelwright's tenon
楔形套筒 wedge socket
楔形栽植 wedge planting
楔状渐伐作业 shelter-wood wedge system; wedge system; wedge-shaped shelter-wood system
楔状伞伐作业 shelter-wood wedge system
楔子 cleat
蝎 caterpillar
蝎尾状的 scorpioid; scorpioidal
蝎尾状聚伞花序 cincinnal cyme; cincinnus; scorpioid eyme
颉颃 antagonism
颉颃菌 antagonistic organism
颉颃效应 antagonistic effect
颉颃作用 antagonism; antagonistic action
协变量 covariant
协定年度 agreement year
协方差 covariance
协方差分析 analysis of covariance
协方差矩阵 covariance matrix
协合剂 synergist
协合作用 synergism
协调曲线 harmonization of curve
协调生长 harmonious growth
协同进化 coevolution (co-evolution)
协同生长 symplastic growth
协同适应系统 coadapted system
协同相互作用 cooperative interaction
协同效应 co-benefits; synergistic effect
协作 cooperation
协作的公方机构 cooperating public agency
协作的永续经营单位 cooperating sustained yield unit
胁迫休眠 enforced dormancy
斜边 bevel edge
斜槽 chute
斜槽刨 skew rebate plane
斜槽取直 chute cutoff
斜层理 oblique bedding
斜插 oblique cutting
斜铲式推土机 angledozer
斜衬木 brail
斜撑 arm-tie; knee brac(ing)
斜撑爬杠(在承载梁上) bunk
斜撑扦起重架 stiff-leg
斜齿 oblique tooth
斜挡水栅 oblique boom
斜刀切根机 oblique blade root pruner
斜道 ramp
斜的 oblique

斜堤 glacis
斜度 banking
斜堆 slope piling
斜堆法 inclined piling; inclined stacking
斜对接 bevelled halving
斜垛 slope piling
斜垛法 inclined stacking
斜方晶体 rhomboidal crystal
斜根 oblique root
斜拱 oblique arch
斜构材件 diagonal member
斜辊 cross roll
斜行狭圆锯 piercing saw
斜河绠 oblique boom
斜滑台 skid ramp
斜架 tilter
斜角 angle of inclination; bevel; bevel angle
斜角尺 mitre square
斜角夹紧器 mitre clamp
斜角接 mitre
斜角接头 franking
斜角企口按 rabbet-mitre joint
斜角切削 inclined cutting
斜角射击 quartering shot
斜角锁紧 lock mitre
斜接 bevelled joint; hopper joint; juxtaposition; oblique joint; scarf; scarfjoint; mitre
斜接(45°角) mitre
斜接规 mitre gauge
斜截 mitre cutting
斜晶束 raphides
斜锯盒 mitre box
斜锯架 mitre block
斜锯切 bevel cutting
斜跨山头集材 quartering to the tree
斜跨铁路的推卸杆 sheer arm
斜梁隅角槽 cut-and-mitred-string
斜列管孔 oblique pore arrangement
斜面(广义) angular surface; batter; bevel; cant; dip section
斜面(指下口) scarf
斜面板 cant board
斜面方向 direction of slope
斜面加工(接头) scarf
斜面接 scarf joint
斜面接合 mitre joint
斜面锯 bevel saw
斜面培养 slant culture
斜面琼脂 slant agar
斜劈接 oblique cleft grafting; oblique wedge grafting
斜拼合 tongue mitre
斜平面 angular plane surface
斜坡 decline
斜坡冲刷 slope wash
斜坡度 incline
斜剖面 oblique profile
斜切 mitre cutting; oblique cutting
斜切锯 mitre saw
斜筛过滤器 sloping screen filter
斜榫 bevelled tenon; oblique tenon
斜榫接 tongue mitre
斜台 tilter
斜纹 angle grain
斜纹材 diagonal wood
斜纹的 diagonal
斜纹胶合板 diagonal plywood
斜纹理 cross grain; incline grain; sway grain
斜纹木材 cross-grained wood
斜纹刨切单板 endy veneer
斜屋顶 pitched roof
斜向地性 plagiogeotropism
斜向集材 diamond lead
斜向剪切 oblique shear
斜向性 plagiotropism
斜楔 dip cleat
斜形狭圆锯 compass saw
斜倚垛 pole piling
斜运动 klinokinesis
斜长辉长石 anorthositic gabbro
斜长岩 anorthosite; plagioclase rock
斜植 oblique planting
斜置木杆(木杆道转弯处) slew skid
谐振峰 harmonic peak; resonance peak
谐振共鸣 resonance
谐振曲线峰值 resonance peak
携播 phoresy
鞋带菌 honey mushroom; honey fungus; shoestring fungus
缬氨酸 valine
缬草油 valerian oil

写字扶手(椅) writing arm
写字扶手椅 writing-chair
写字台 secretaire; secretary; writing desk; writing table
泄出 run-off
泄洪 flood discharge
泄料池 blow pit
泄流管道 outlet conduit
泄漏 leakage; spillage
泄漏量 spillage
泄水坝 flush weir
泄水处 offtake
泄水闸 sluice
泄油口(阀) drain port
泻出 outpouring
泻湖 lagoon
泻溜 earth debris flow
泻药 jalap
卸板机(热压机械) unloading cage; unloader
卸板机传动辊 unloader live-roller
卸板机液压泵 unloader hydraulic pump
卸板器 diverter
卸材 dumping
卸材场 dumping ramp; log storage deck
卸材地 dumping ground
卸材工 unloader
卸材机 offloader; unloader
卸材流送 launching
卸材器 cant flipper
卸材斜木(水滑道) snub
卸材斜木(推木用) pole switch; kicker
卸材斜台 dumping ramp
卸车 off-loading; unloading
卸车场 dump
卸车工 dumper
卸车机 unloader
卸车撬杆 sweep
卸车台 unloading deck; unloading platform
卸车推杆 hercules
卸车支柱 unloading gin pole
卸货费用 discharging expense
卸货装置 discharging device
卸捆器 sheaf discharge
卸料车 dumper
卸料池 dump chest
卸料机 offloader
卸木 log haul(-up)
卸液罐 unloading tank
卸载 debark; off-loading; spot; spotting; unloading
卸载阀 unloader valve
卸载偏导装置 unloading deflector
卸载台 off-load table
卸载性能 off-load capability
榭 kiosk
心部木材 core
心材 duramen; heartwood; truewood
心材变色 heart stain
心材腐朽 heart rot; heartwood rot
心材开裂 boxed heart check
心材轮裂 heart ring-shake
心材内含物 heart substance
心材树 heartwood tree; tree with coloured heartwood
心材形成 heartwood formation
心茬(抽心后留在伐根上的树茬) stump pull
心腐 centre rot; heart rot; inside rot; spongy heart; stem rot
心腐材 punky wood
心裂(树木) heart shake; pith check; rift crack; heart check
心皮 carp; carpel; carpophyll; hystrella
心皮柄 carpophore; thecaphore
心皮的 carpellary
心皮化 pistillody
心皮胶合(蜂窝结构) fillet
心皮胶合强度(蜂窝结构) fillet strength
心皮群 sorema
心皮维管束 carpellary bundle
心室 heart chamber
心土 subsoil
心土层 subhorizon
心土层松土 subsoiling
心土犁 subsoiler
心土下层 lower subsoil
心轴 arbor; mandrel; spindle
心轴车削 spindle turning
芯板(胶合板) center; cross band veneer; cross band; cross banding; crossing veneer; crossing; core
芯板材 coreboard

芯板带(层) coreline
芯板料 core stock
芯板木条 core strip
芯板拼缝机 core veneer splicer
芯层 core layer
芯层单板 core veneer
芯层密度 core density
芯层刨花 middle layer chip
芯层铺装机 core spreader
芯层碎料搅拌机 core blender
芯饭拼接机 core composer
辛可尼定 cinchonidine
辛可尼辛 cinchonicine
辛可宁 cinchonine
辛辣气味 pungent odor
辛普森多样性指数 Simpson's diversity index
辛普森法 Simpson's rule
辛普森指数 Simpson's index
辛酸 caprylic acid
辛糖 octose
锌白 zinc white
锌钡白 lithopone
锌粉 zinc dust
锌黄 zinc yellow
锌焦油注入法 Allardyce process
锌盐 oxonium salt
新北界 Nearctic region
新北区 Nearctic province
新变态 neomorphosis
新冰川作用 Neoglaciation
新剥皮的爬杠(装车时以滑动代替滚动) slip skid
新布伦斯威克原木材积表 New Brunswick log rule
新陈代谢 metabolism
新陈代谢过盛 hypermetabolism
新成虫 callow adult
新成土 entisol
新达尔文主义 Neo-Darwinism
新伐材 fresh cut; fresh wood; freshly-felled timber
新伐倒材 green timber; green wood
新伐木 freshly felled tree
新分类学 new systematics
新截 fresh cut
新界 neogaea
新拉马克学说 neo-Lamarckism
新孟德尔学派 Neo-Mendelism
新模式(标本) neotype
新木材 neowood
新热带的 neotropical; newtropical
新热带区 neotropical region
新热带植物区 neotropical floral kingdom; neotropics
新梢 current shoot
新生 palingenesis
新生层 cambium (*plur.* -bia)
新生代 cenozoic era
新生代的 cenozoic
新生代演替系列 cenosere
新生代植物 cenophyte
新生代植物演替系列 angeosere
新生的 nascent; neonatal
新生界的 cenozoic
新生物发生 neobiogenesis
新生枝 innovation
新石器 neolith
新石器时代 New stone age
新式家具 new art style furniture; new quaint style furniture
新闻纸 newsprint (paper)
新闻纸用磨木浆 news grade groundwood pulp
新闻纸用亚硫酸盐木浆 news grade sulphite pulp
新西兰琥珀 kauri gum
新鲜空气滤网 fresh air screen
新鲜未发芽粒 fresh ungerminated seed
新鲜蒸汽 live steam
新鲜蒸汽管系 live steam piping
新效等位基因 neomorph
新型 neotype
新性发生 caenogenesis
新性染色体机制 neo-sex chromosomal mechanism
新叶 current leaf
新引入杂草植物 neophyte
新月形的 lunar; lunate; luniform
新月形沙丘 barchan; barchane; meniscus dune; crescentic dune
新造林地 newly afforested area; young afforested land
新着丝粒 neo-centromere
新种 neospecies
薪材 energy wood; faggot wood; fascine wood; fire wood; firewood;

fuel wood; kindling wood; short end; yule log
薪材采伐中的报酬递减 diminishing returns in cutting cordwood
薪材采用权 right to fire-wood
薪材场 firewood yard
薪材堆 rick
薪柴砍伐者 firewood cutter
薪金 emolument
薪金额 rate of pay
薪水 emolument
薪炭材 firewood
薪炭林 firewood forest; fuelwood forest
信贷 credit
信风 trade; trade wind
信风带 zone of trade wind
信号刺激 sign stimulus
信号灯 eye
信号员 flagman
信息 message
信息传递 communication
信息档案 information file
信息简缩变换 data reduction
信息交换 message switch
信息控制系统 information control system
信息论 information theory
信息识别符号 acknowledge (character)
信息素 ecto-hormone; pheromone
信息提取 information extraction
信息系统 information system
信息学 informatics
信息转储 dump; storage dump
信息转接 message switching
信息组 byte
信用 credit
信用卡 credit card
信用票据 clean bill
信用证 letter of credit
信誉 good will
星裂 spider shake; star check; star shake
星散薄壁组织 diffuse parenchyma; scattered parenchyma
星散-聚合薄壁组织 diffuse-in-aggregate parenchyma
星散排列 diffuse arrangement
星下点 sub-satellite point
星形胶合板 star plywood
星云 nebula
星状节 star-shaped knot
星状毛分枝 ray
星状树枝编层(坝前铺垫用) bough stars
星状中柱 actinostele
行车 overhead travelling crane
行程 run
行程时间 travel time
行道树 alee-tree; avenue tree; roadside tree; street tree
行道树栽植 roadside planting
行使价格 exercise price
行为(性)隔离 ethological isolation
行为多态现象 behavioural polymorphism
行为多型 polyethism
行为二态现象 behavioural dimorphism
行为干扰 behavioural interference
行为隔离 behavioural isolation
行为生态学 behavioral ecology
行为适应 behavioural adaptation
行为习性 behavior
行为学 ethology
行为主义 behaviourism
行星齿轮 planetary gear
行星齿轮式轮边减速 epicyclic hub reduction
行星齿轮式起重机 planetary hoist
行政管理 administrative management
行政管理办公室 administrative office
行政管理费 administrative expense
行政管理机构 administrative agency
行政管理局 administrative office
行政管理区 administrative district
行政施业区 administrative working circle
行政征用 eminent domain
T字形(的) tee
形成 establishment
形成层 cambium; cambium (*plur.* -bia); cambium layer
形成层带 cambial zone; cambium zone
形成层纺锤形原始细胞 cambial fusiform initial
形成层环 cambium ring
形成层阶段 cambial stage

形成层区 cambial region; cambium area
形成层射线 cambial ray
形成层射线原始细胞 cambial ray initial
形成层-树冠面积比 cambium-crown surface ratio
形成层细胞 cambial cell; cambium cell
形成层细胞状 cambiform
形成层原始细胞 cambial element; cambial initial
形成过程 establishment
形成粒 plastid
形点 form point
形点高 form-point height
形高 form height
形高系列 form height series
形级[木] form class
形率 form quotient; form ratio; quotient
形似小提琴的椅背板 fiddleback
形数 form factor; form-figure
形数表 form-factor table
形数测高法 Pressler's method of height determination
形态发生(形态建成) morphogenesis
形态渐变群 morphcline
形态可塑性 morphological plasticity
形态素 morphactin
形态突变体 morphological mutant
形态形成 morphosis
形态型 morphotype
形态休眠 morphological dormancy
形态学 morphology
形态演进 aromorphosis
形态种 morphological species; morphospecies
形体稳定性 dimensional stabilization
形质生长 increment due to quality; quality growth; quality increment; increase in quality
形质生长率 quality increment percent
形质疏伐 quality thinning
形质指数 quality index
形状(像片判读因子) shape
形状比 height-DBH ratio
形状参数 shape parameter
形状高 tree form height; volume-basal area ratio
形状高生长 form height increment
形状指数 form exponent; index of form; form index
型 caste
型材(面) pattern
型刨 four side moulder; too(u)lder
型式 type
型芯 core
兴趣中心 centre of interest
杏仁桉 giant gum; peppermint gum
杏仁油 apricot kernel oil; persic oil
杏仁状辉绿岩 dunstone
性孢子(锈菌) spermatium (*plur.* -ia); pycniospore; microspore
性孢子器(锈菌) clinosporangium; spermogonium; pycnium
性孢子器阶段(锈菌) pycnial stage; pycnial state
性别 sex
性别集群 sexual assemblage
性成熟 sexual maturity
性的 sexual
性反转 sex reversal; sex transformation
性分化 sex differentiation
性隔离 sexual isolation
性交 copulation
性连锁 sex linkage
性母 sexupara (*plur.* -rae)
性母细胞 auxocyte; meiocyte
性能分析 performance analysis
性能试验 performance test
性器官 sex organ
性嵌合体 sex mosaic
性染色体 accessory chromosome; allosome; heterosome; idiochromosome; sex chromosome
性外激素 sex pheromone
性细胞 sex cell; sexual cell
性腺 gonad
性信息素 sex pheromone
性选择 sexual selection
性蚜 sexuales
性因子 fertility factor
性引诱剂 sex attractant
性诱素 sex pheromone
性原细胞 gonocyte

性周期 sexual cycle
性转化 sex transformation
性状 trait; character
性状区异 character divergence
凶火(速行林火) running fire
胸 thorax
胸板 breast board
胸部 truncus
胸高 breast height; chest-height; height of chest
胸高处 chest

胸高定位 breast height locating
胸高断面积 basal area; basal area of breast-height; cross-sectional area at breast height
胸高断面积表 basal-area table; table of basal areas; table of sectional areas
胸高断面积平均木 mean basal area tree
胸高断面年龄 basal area-age
胸高干围 girth breast height; normal girth
胸高年龄 age at breast height
胸高形率(伪形率) breast-height form quotient; artificial form quotient; false form quotient
胸高形数(伪形数) breast-height form factor; artificial form factor; chest-height form factor; common form factor; form factor at breast height; false form factor
胸辊 breast roll
胸径 breast-height diameter; chest-height diameter; diameter at breast height; diameter breast-high (DBH); normal diameter
胸径分布 DBH distribution
胸径生长 chest-helght diameter increment
胸廓 thorax
胸围 breast-height girth
雄 male
雄单倍体 male haploidy
雄单性生殖 male parthenogenesis
雄核 male nuclei
雄核发育 androgenesis
雄花 male flower
雄花两性花同株的 andromonoecious
雄花两性花异株的 androdioecious
雄花无性花同株(菊科) agamandroecism
雄激素 androgen
雄配子 androgamete; male gamete; microgamete
雄配子体 male gametophyte
雄器 antherid; antheridium; spermary
雄器苞(苔鲜植物) perigone
雄球花 male cone
雄全同株 andromonoecy
雄全同株的 andromonoecious
雄全异体 androdioecy
雄全异株的 androdioecious
雄蕊 androecium; stamen
雄蕊不稔性 staminal sterility
雄蕊的 staminate
雄蕊花瓣状 andropetalous
雄蕊较花瓣少的 meiostemonous; miostemonous
雄蕊群 androecium; oecium
雄蕊外花盘 extrastaminal disc
雄蕊先熟 protandry
雄蕊先熟的 protandrous; proterandrous
雄蕊先熟花 protandrous flower
雄蕊异长现象 heteranthery
雄蕊与花瓣互生 alternipetalous
雄兽发情期(大型动物) rut season
雄穗 tassel
雄榫 carpentry tongue; cog; tenon; tongue
雄榫接台 cogged joint
雄榫上的斜肩 gain
雄榫斜嵌接合 cogged-scarf joint
雄系亲缘 agnation
雄细胞 male cell
雄性不育 male sterile
雄性不育法 sterile-male method
雄性不育性 male sterility
雄性不育症 male sterility
雄性的 male; mas; masculine
雄性交配素 androgamone
雄性两性同株 andromonoecy
雄性两性异体 androdioecy
雄性生殖系统 male reproductive system
雄性生殖细胞 male sex cell
雄性先熟 proandry; protandry

雄性先熟雌雄同体 protandrous hermaphrodite
雄性先熟的 protandrous; proterandrous
雄性植物 androecy
熊 bear
熊蜂类 bumble-bee; wild bee
熊阱式闸门 bear-trap gate
熊掌 bear claw
休伯区分求积法 sectional measurement by means of Huber's formula
休耕 fallow; fallowing
休耕地 fallow; fallow land
休眠 anabiosis; dormancy
休眠孢子 dormant spore; resting-spore
休眠插穗 dormant cutting
休眠火山 dormant volcano
休眠季节 dormant season
休眠解除 release of dormancy
休眠解除机制 dormancy release mechanism
休眠菌丝体 resting mycelium
休眠开始 dormancy onset
休眠卵 resting egg
休眠期 dormant period; rest period
休眠维持机制 dormancy-maintaining mechanism
休眠型 dormancy form
休眠芽 dormant bud; latent bud; resting-bud; sleeping bud; statoblast
休眠芽眼 dormant eye
休眠诱发机制 dormancy-inducing mechanism
休眠种子 dormant seed; latent seed
休眠状态 dormant state
休息室 lounge
休闲的 fallow
休闲地 fallow; fallow land
休闲地翻耕 lea ploughing
休闲地作物 fallow crop
休闲活动 leisure activity
休闲期 reforestation interval
休闲作物 fallow crop
休养地 health resort; recreation site; resort area
休止角 angle of repose
修边刨 edge trimming plane; jointing plane
修补基面涂层 wash coat
修齿 fit(ting)
修低枝 brush out
修干枝 dry pruning; dry prunning
修根 prune; root pruning
修根机 rootprunner
修活枝 green prunning
修剪 shearing; trimming; cutting back
修剪树冠 pollarding
修剪下部(枯死枝) low pruning
修剪下的枝 lop
修剪整形 prune
修剪作业 pruning practice
修建滑道工 axeman
修锯 filing; saw fitting; saw sharpening
修锯工 filer; saw doctor
修锯砧 saw bench
修理工具箱 repair kit
修理与更换判定(机械) repair-replace decision
修路 repair(ing)of road
修路工组 bull gang
修平刀 trowel
修齐[的] tumbled-in
修渠 ditching
修饰基因 modifying gene
修饰木工 finish carpenter
修饰涂层 finish coat(ing)
修饰因子 modifying factor
修树剪 secateur
修缘刨 edge trimming plane
修整(锯条) doctor
修整(刨平或装饰等) dress
修整(整形) shaping
修整机械 shaping equipment
修整装置(砂轮用) dresser
修正的决定系数 adjusted coefficient of determination
修正的收获蓄积比 modified yield-volume ratio
修正毛刺 briar dress
修正条款 amending clause
修正系数 correction factor
修正系数法 correction factor method

修正像片材积表 adjusting photo-volume table
修正因子 correction factor
修正值 allowance
修枝 branch pruning; lopping; lopping of branches; lopping off; prune; prunning; snedding; trim; trimming
修枝刀 billhook; lopping knife
修枝工具 pruning tool
修枝剪 branch-pruner; pruner; pruning shears; pruning-scissors
修枝锯 limbing saw; pruning-saw
修枝桠 lopping and topping
鸺鹠 owlet
髹漆 lacquering
朽节 punk knot; punky knot
朽木 frowy; mouldered wood; punk
朽心 punky heart
朽枝 branch-rot
袖珍程序计算器 programmable pocket calculator
袖珍罗盘 pocket compass
锈孢子 aecidiospore
锈孢子器 aeci(di)um (*plur.* -ia)
锈孢子器阶段(锈菌) aecidial stage; aecidial state
锈菌 rust
锈菌性孢子 pycnospore
锈皮虫 jewel beetle
锈色节 rusty knot
嗅觉标记 scent-marking
嗅觉器 olfactory organ
嗅味柱 scent post
嗅味装夹[法] scent set
溴化法(简易木材防腐防火处理法) bromination method
溴化银相纸 bromide paper
须 barba
须根 rootlet
须根病 hairy root
须根系 fibrous root system
虚股 watered stock
虚积 superficial volume
虚积材 stacked timber
虚积量 stacked content
虚假交易 wash sale
虚价 nominal price
虚列成本 inflated cost
虚卖 wash sale
虚拟 dummy
需光的 light-demanding
需光度 light-requirement
需光发芽 light germination
需光种子 light(-demanding) seed; light-requiring seed; positively photoblastic (seed)
需气土壤生物 aerophytobiont
需求变动因素 demand shifters
需求侧管理 demand-side management
需求疲弱 weakening in demand
需求预测 demand projection
需水量 water requirement
需锌植被 zinc vegetation
需盐种 salt-obligate species
需氧生活 aerobiosis
需氧生物 aerobiont
需氧脱氢 aerobic dehydrogenation
需氧微生物 aerobe
需氧细菌 aerobic bacteria
需阴的 shade-needing
徐变 creep
许可证 charter
许可证税 excise tax
序贯抽样 sequential sampling
序贯分析 sequential analysis
序号 ordinal
序列处理 list processing
序位 hierarchy
续燃时间 after flame time
絮浆 flocculate
絮凝 flocculation
絮凝沉淀物 flocky precipitate
絮凝剂 flocculant
絮凝胶体 flocculated colloid
絮凝物 floc
絮凝作用 flocculation
蓄电池 battery
蓄洪 flood storage
蓄洪坝 flood dam
蓄积 present yield; stock
蓄积保持 keep in stock
蓄积保育 management of the growing stock
蓄积比 volume ratio
蓄积构造 growing stock structure
蓄积价 value of growing stock

蓄积结构 growing stock structure
蓄积控制法 volume control method; volume regulation method
蓄积量表(估计定期蓄积生长量) stock table
蓄积平分法 volume frame work
蓄积调查 estimate survey; inventory of growing stock; valuation
蓄积推定 growing stock determination; timber determination
蓄积增长量 volume increment
蓄积贮备 volume reserve
蓄料槽 material chamber
蓄能器 accumulator
蓄水 impounding
蓄水坝 storage dam
蓄水槽 storage reservoir
蓄水层 aquifer
蓄水池 impounded basin; impounded body; lasher; reservoir
蓄水量(库塘) pondage
蓄水闸 flood dam
玄武岩 basalt
玄武岩刀辊 lava flybar
悬臂 bracket; hanger; lip
悬臂简支梁 free-free beam
悬臂梁 cantilever beam
悬臂梁式冲击机 Izod impact machine
悬臂梁式冲击强度 Izod impact strength
悬臂梁式冲击试验 cantilever beam impact test
悬臂梁式试验机 Izod impact machine
悬臂剖分锯(将工件顶在靠山上) line-bar resaw
悬臂起重机 boom hoist
悬臂起重台架 cantilever gantry
悬臂桥 cantilever bridge
悬臂式持久性试验 cantilever endurance test
悬臂式起重机 arm crane; bracket crane; cantilever crane
悬臂式真空伏辊 suction couch with cantilever arm
悬臂托梁 hammer beam
悬垂伐根 overhung stump
悬垂呼吸根 pendent respiratory root
悬垂花 penduliflory
悬垂花篮 hanging basket
悬滴 hanging drop
悬滴培养 hanging drop culture
悬吊式运材挂车 tug
悬浮固体 suspended solid
悬浮颗粒 suspended particle
悬浮泥沙量 suspended sediment load
悬浮式气流分选机 suspension sifter
悬浮式气流干燥机 suspension-type dryer
悬浮试验 flotation test
悬浮室 suspension chamber
悬浮体散射仪 nephelometer
悬浮物 seston; suspension
悬浮液 suspension
悬浮质泥沙 sediment in suspension
悬挂螺栓 hanger bolt
悬挂式防火犁 dependent plough
悬挂式干燥机 festoon dryer
悬挂式集材 suspension yarding
悬挂式输送机 underslung conveyer
悬挂索 suspension line; suspension rope
悬挂装置 bogie; hitch; hookup; suspender; suspension
悬果瓣 mericarp
悬空集材 suspension yarding
悬链线 catenary
悬梁式裁边机 beam log edger
悬铃木属 plane
悬木减速装置(滑道) wolfbrake
悬木制动器(滑道) deadener
悬溶胶体 suspensoid
悬索桥 cable-suspension bridge
悬线垂度 catenoid
悬崖 cliff
悬移质 suspended load
悬蛹 suspensi pupa
悬园 hanging garden
悬着地下水 perched ground water
悬着水位 perched water table
旋板机 rotary lathe; veneer lathe; veneer rotary lathe
旋臂锯 arm saw
旋臂起重机 swing crane
旋臂式龙门起重机 gantry slewing crane

X

旋臂圆锯 radial (arm) saw
旋边工具 bead-tool
旋床 turning-lathe
旋刀 lathe knife
旋刀式剥皮机 cutterhead barker
旋刀压板 knife cap
旋风 vortex
旋风分离器 cyclone separator
旋风抛土式梯田机 whirlwind terracer
旋风器 cyclone
旋耕地 tiller
旋耕机 rotavator; rotocultivator; rototiller
旋工工艺 turnery
旋管 coil
旋光的 optic
旋光度 optical activity; optical rotation
旋光改变作用 mutarotation
旋光计 polarimeter
旋光色散 optical rotatory dispersion
旋光性 optical activity; optical rotation
旋角 crab
旋卷 volution; vortex
旋切 peel; peeling
旋切材长度倍数 peeler multiple
旋切单板 rotary cut veneer; rotary cut(ting) veneer; rotary veneer
旋切单板松面 loose face
旋切机 lathe; peeler; peeling lathe; peeling machine; rotary cutter; rotary lathe; veneer peeling
旋切机上木装置 lathe charger
旋切角 knife pitch; knife setting angle
旋切紧密的单板 tight veneer
旋切裂隙(单板, 背面) lathe check
旋切缺陷 peeling defect
旋切软材单板 peeled softwood veneer
旋切式刨片机 lathe flaker
旋切试验 peeling test
旋切速度 peeling velocity
旋切用原木 peeler (lathe) log
旋切原木 lathe log; peeler log
旋塞 faucet
旋涡 vortex
旋涡纹理 swirl grain
旋圆 rounding(up)
旋凿 screw driver
旋制品 turnery-ware
旋转刀痕 revolution mark
旋转的(多叶或花卷迭式) convolute
旋转点 pivot
旋转阀 rotary valve
旋转风暴 revolving storm
旋转接头 shay swivel
旋转卡轴 live spindle
旋转槛门 stile
旋转力 turning force
旋转门 revolving door; pivoted door
旋转磨石环 rotating stone ring
旋转抛光 rotary finishing
旋转筛 gyratory screen; rotating screen
旋转式剥皮机 rotary peeling machine; rotor-peeler
旋转式干燥机 rotary dryer; rotation dryer
旋转式干燥筒 rotary dryer
旋转式割灌机 rotary slasher
旋转式耕耘机 rototiller
旋转式滚筒干燥机 rotary drum dryer
旋转式剪草机 rotary mower
旋转式剪草机具 rotary mowing equipmant
旋转式粒化机 rotary granulator
旋转式绿篱修剪机 rotary hedge cutter
旋转式喷气干燥机 rotary jet dryer
旋转式起重臂 swing boom
旋转式切片机 rotary microtome
旋转式湿度计 whirling psychrometer
旋转式食品架 dumb waiter
旋转式中耕机 rototiller cultivator
旋转式贮料仓 rotating storage bin
旋转式装卸机 swing-boom log loader
旋转书架 revolving bookcase
旋转塔式起重机 rotary tower crane
旋转型压碎机 rotary crusher
旋转椅 spring revolving chair; swivel chair
旋转主轴 live spindle
旋转锥形喷管 rotating conical nozzle tube
选板输送机 panel selecting conveyer
选别 reject

选材 assorting; sort; sorting
选材报号员 mark caller
选材场(陆上) sorting bay; sorting ground; sorting plant
选材场(水上) sorting pocket
选材导缏 sorting boom
选材缏 sorting jack
选材工 log sorter
选材机 grader; sorter
选材口(水上) sorting gap
选材平车 sorting carriage
选材输送机 sorting conveyor
选材台 assorting table; sorting deck; sorting table
选材通廊 pokelogan; sorting pocket
选材线 sorting line
选材员 culler; grader; sorter; timber grader
选材装置 log sorter
选材自动线 automated log-sorting line
选出 single out
选伐 economical selection method; selective cutting; selective felling; selective logging; selective method; thinning out
选伐林 culled standard
选分 classification
选良种 select fine seed strains; selection of good tree seeds
选留木 standard; teller
选留木林分 elite stand; selected crop; selected stand
选木打枝 selective high pruning
选线员 line locator
选型交配 assor(ta)tive mating
选型育种 assor(ta)tive breeding
选优间苗 selective thinning
选择 selection
K-选择 K-selection
选择差异 selection differential
选择成本 alternative cost
选择毒性 selective toxicity
选择反应 selection response
选择放牧 selective grazing
选择极限 selection limit
选择价值 selection value
选择理论 theory of preference
选择率(入选率) selection rate
选择能力 selectivity
选择品系 selection strain
选择强度 intensity of selection; selection intensity
选择式疏伐 extraction thinning
选择受精 selective fertilization
选择透性 selective permeability
选择吸附 preferential adsorption; selective absorption
选择系数 selection coefficient
选择效应 selection effect
选择性 selectivity
选择性采伐 selective exploitation
选择性抽样 selective sampling
选择性除草剂 selective herbicide
选择性培养基 selective media; selective medium
选择性杀虫剂 selective insecticide
选择性淘汰 selective elimination
选择压力 selection pressure
选择育种 selective breeding
选择增效 selective advance
选择指数 selection index
选择株 select plant
选种 seed sorting
选种机 seed cleaner
镟切的 rotary(-cut)
镟切裂隙(单板) cutting check
镟圆 round up
削度百分数 percentile taper
削度表 taper table
削度方程 taper equation
削度公式 taper formula
削度函数 taper function
削度留量 taper allowance
削度曲线 taper curve
削度允许范围 taper allowance
削端工(将木端削圆) sniper
削斧 scoring ax
削减产量 cutting output
穴播 dibble; hill dropping; hill-drop drilling; hill-drop planting; hole seeding; hole sowing; seeding in hole; sowing in hole
穴播器 dibble
穴侧栽植 side hole planting
穴居 pit dwelling
穴居动物 cave animal; cavernicolous animal

穴居动物的 cryptozoic
穴距 distance between hills
穴栖的 cavernicolous
穴施 spot application
穴心栽植 centre-hole planting
穴栽 dibble
穴植 hill-drop planting; hole planting; holing; pit planting; pitting; spot planting
穴植法 hole-method
穴植器 dibber
学理法 rational method; utilization percent method
学生化残差 studentized residuals
雪斑 snow bank; snowbank; snow-patch; snowpack
雪斑植被 snow patch vegetation
雪暴 blizzard
雪被 snowpack; snow bed
雪崩 avalanche; snow avalanche
雪崩防护林 protection forest for preventing snow avalanche
雪带 nival belt
雪的 nival
雪地装夹(法) snow set
雪堆 snow bank; snow pack; snowbank; snowdrift
雪害 snow damage
雪荷载 snow load
雪华 green snow
雪滑道 snow chute; snow slide; snowslip
雪犁 snow plough
V形雪犁 V-plough
雪路装夹(法) snow trail set
雪霉病 snow mould
雪面 nival surface
雪嵌部 snow-patch
雪嵌部植被 snow patch vegetation
雪橇 scoot
雪茄排 cigar raft
雪茄形海排 Benson raft
雪丘 snow bank
雪山 snow-capped mountain
雪蚀(作用) nivation
雪水 slush
雪松 Himalaya cedar; deodar cedar
雪松胶 cedar gum
雪松木油 deodar cedarwood oil; Himalayan cedarwood oil
雪松酮 cedrone; deodarone
雪松烯 himachalene
雪松叶油 cedar leaves oil
雪松油 cedar (wood) oil
雪线 snow line; nival line
雪压 crushing by snow
雪疫病 snow blight
雪栅栏 snow fence
雪折 snow breakage; snowbreak
雪折木 snow breakage
雪辙道 snow rut road
雪中生长的 nival
血柏木油 red cedar wood oil
血柏叶油 red cedar leaves oil
血粉 blood meal
血管的 vascular
血红素 heme
血红素蛋白 hemoprotein
血浆 blood plasma
血胶 blood adhesive; blood glue
血竭 draconis sanguis
血竭黄脂 dracoresene
血竭玫红素 dracorhodin
血竭玉红素 dracorubin
血蓝蛋白(昆虫) haemocyanin
血淋巴 blood lymph; h(a)emolymph
血腔 haemocoele
血清 blood serum; serum
血清测定 serological test
血清免疫 seroimmunity
血清球朊 seroglobulin
血清学 serology
血球 blood corpuscle
血色素 haemoglobin
血细胞 blood corpuscle; h(a)emocyte
血细胞计数器 cytometer
血纤维蛋白 fibrin
血型 blood group
血缘 kin
血缘的 genealogical
血缘关系 blood relationship
熏仓混剂(熏蒸剂) serafume
熏黑高度 scorch height
熏烟 smudging
熏烟处理 smoke treatment

熏烟干燥(法) smoking seasoning
熏烟干燥法 smoke drying
熏烟干燥窑 smoke (dry) kiln
熏烟器 smoker
熏衣草精油 lavender oil
熏蒸 fumigation; suffumigation
熏蒸剂 fumigant; fumigation agent
熏蒸器 smoker
熏蒸消毒剂 fumigant
薰衣草油 lavender oil
寻觅行为 searching behaviour
巡防员 lookout patrol(man)
巡回采伐 stage cut
巡逻区 beat
巡逻时间(防火) patrol time
巡木船 log patrol
循环 circle
循环采伐带 circular strip
循环带式干燥机 endless belt dryer
循环的 cyclic
循环钢带 endless steel belt
循环活化 cyclic activation
循环加压法 circulational pressure process
循环交配 circular mating
循环链 endless chain
循环链式输送机 endless chain conveyor
循环模式 cyclical pattern
循环期 circulation period
循环式采伐 circular cutting
循环式动力索道 endless cable system
循环式运材索道 continuously transporting cableway
循环索 continuous line; endless line
循环索道集材 moving cable yarding
循环索滑轮 endless pulley
循环索卷筒 endless drum
循环索系 moving cable system
循环油 turning oil
循环作用 circulation
循回病毒 circulative virus
循回期 celation period
训练区 training site
训练项圈(训犬) training collar
训鹰术 falconry
讯号 message
汛期 flood period; flood season
驯化 acclimation; domestication; habituation; naturalization; acclimatization
驯化种 naturalized species
驯化作用 acclimatization
驯鹿 Caribou; reindeer
驯养 domestication
蕈(xùn) mushroom
蕈毒 mushroom poison
蕈形地板 mushroom floor
压板 platen; press plate; bumping plate
压板机 plating machine
压板间距 daylight clearance
压板间距(压机) daylight opening
压边条 fillet
压草钩(耕地时犁土压倒杂草的) weeding hook
压尺 knife bar; nose bar; pressure bar
压尺垂高 pressure-bar lead
压尺垂直开度(压尺与刀刃的垂直距) vertical-bar opening
压尺导距(镟切) pressure-bar lead; vertical gap; vertical opening
压尺间距 horizontal gap; horizontal opening; nose-bar gap; pressure-bar gap
压尺开度(压尺与刀刃距离) nose-bar opening
压尺水平开度(压尺与刀刃的水平距) horizontal-bar opening
压电(现象) piezoelectricity
压电变质 piezo-metamorphism
压电常数 piezoelectric constant
压电结构 piezoelectric structure
压电晶体 piezoelectric crystal
压电模量 piezoelectric modulus
压电效应 piezoelectric effect
压电性质 piezoelectric property
压电学 piezoelectricity
压电张量 piezoelectric tensor
压钉机 stapling machine
压断 crushing
压缝栽植法 compression method
压盖 gland
压沟板 marking board
压沟滚筒 drill roller
压光 calendering

压光机 calender
压光机施胶 calender sizing
压辊(镟切机) pinch roll; press roll; pressure roll; back-up roll
压痕 imprint; indentation; pressmark
压痕试验 indentation test
压痕硬度 indentation hardness
压花 emboss
压花辊 design roller; embossing roll; engraved roller
压花辊涂布 design roller coating
压花机 embosser; embossing machine
压花胶合板 embossed plywood
压花纹板 marking board
压花性能 embossability
压花硬纸板 embossed hardboard
压花硬质纤维板 embossed hardboard; moulded hardboard
压簧杆 cocking lever
压簧杆槽 slot for cocking lever
压簧销孔 hold for cocking pin
压机 press
压机工作台 press bed
压机加压 loading of press
压机紧急停车 press emergency stop
压机开档 daylight clearance
压机排污泵 press pit sump pump
压机座 press bed
压脚子 log jack; raft hook; gaff
压紧 compaction; jam; pack
压紧辊 hold-down roll
压紧环 clamp ring
压紧块 shoe
压紧旋切单板 tight-cut veneer
压锯辊 stretcher roll
压块机(木屑) cuber
压溃压力 crushing pressure
压捆机 baler; trusser
压力表 pressure gauge
压力差 pressure difference
压力传感器 pressure-sensing device
压力单位 pressure unit
压力弹 pressure bomb
压力风速表 pressure anemoneter
压力罐 pressure tank
压力计压力表 pressure meter
压力解除 pressure relief
压力控制器 pressure controller
压力控制设备 pressure control equipment
压力流浆箱 pressure type head box
压力曲线 pressure curve
压力容器 pressure vessel
压力势 pressure potential
压力室 pressure chamber
压力送风 forced draft
压力梯度 pressure gradient
压力堰板 pressure slice
压梁木(装在车中的) bunk load
压料 swage set; swaging
压料齿 upset tooth
压料机 swage
压料矫正器 swage sharper
压料锯 swage saw
压料器(整锯) eccentric swage; jumper upset; jumper; swager
压路机 road roller; roll; roller
压轮 hanging roller; pinch roll
压滤机 filter(ing) press
压敏黏合 pressure-sensitive adhesion
压敏黏合剂 pressure-sensitive adhesive
压模 mould
压刨机 thicknesser; thicknessing machine
压膨胀 compression swelling
压片机 sheeter
压片技术(研究染色体) squash technique
压齐(短材横装后) bump up
压汽试验器 ompressometer
压强 intensity of pressure
压热器 autoclave
压刃 hook
压实 compaction; densify
压实器 packer
压碎 crush; crushing
压缩 compaction; compression; constriction
压缩板 compressed board
压缩比 compression ratio
压缩变定 compression set
压缩变形 compression deformation
压缩处理 compression treatment

压缩单板胶合板 densified veneer plywood
压缩断裂面 compressive fracture
压缩计(测量压缩变形) ompressometer
压缩假说 compression hypothesis
压缩剪切 compression shear
压缩空气发动机 air motor
压缩空气管 air pressure duct
压缩空气量 air supply
压缩空气输送管 blow pipe
压缩力 compressive force
压缩率 compressibility; compression rate
压缩木 compact wood
压缩喷雾器 backpack pump
压缩破坏 compression failure
压缩破裂 compressive fracture
压缩热固木 staypak
压缩系数 coefficient of compressibility; coefficient of compression; compressibility; compression coefficient
压缩应变 compression deformation; compression strain; compressive strain
压缩振动 compressional vibration
压条 layer; malleolus; propago; layerage
压条法 layering; planting of layer; layerage
压条萌蘖 layering
压条苗 layering plant
压条母校 parent layer
压条栽植 planting by layer
压铁 clamp iron; pressure bar
压头移动速率(试验机) rate of head movement
压涂法 extrusion method
压弯机 bulldozer
压弯木(被伐倒木压弯, 砍断时易弹回伤人) widow maker
压纹 emboss
压纹机 embosser; embossing machine; plating machine
压靴 shoe
压延机 calender
压印法 imprint method
压印辊 impression roll
压应力 compression stress; compressive stress; stress in compression
压榨部 press part; press section
压榨辊 press roll
压榨毛毡 press felt
压榨区(线)压力 nip pressure
压榨区挡板 nip guard
压榨纸板 millboard; pressboard
押汇证 cash against document
押金 cash bond; cash deposit as collateral; mortgage
押品保证的要求 margin requirement
鸦胆子苷 bruceolide
鸦雀 crowtit
鸭脚木碱 alstonine
鸭爪式锄铲 duck foot
牙签插 cutting held by toothpick
芽 bud; burgeon; shoot
芽苞叶 cataphyll (*plur.* -lla)
芽胞 brood cell
芽变 bud mutation; bud sport; bud variability; bud variation; sport
芽变品种 bud variety
芽插 bud cutting; eye cutting
芽腐 bud rot
芽接 bud grafting; budding
H形芽接 H-shaped budding; H-form budding; H-budding
T形芽接 T-shaped budding; T-form budding; T-budding
芽接刀 budding knife
芽接苗 budling; oculant
芽接者 inoculator
芽卷蛾 budworm
芽孔 germ pore
芽蕾 flower bud
芽鳞 bud scale; chlamydia; perula
芽鳞痕 scale scar
芽鳞脱落 deperulation
芽前 pre-emergence
芽前用量 pre-emergency rate
芽前中耕 pre-emergence cultivation
芽球 gemmule
芽生 blastogenesis; gemmation
芽生孢子 blastospore
芽条 bud stick; budling
芽条选择 bud selection

芽型 vernation
芽序 gemmation
芽眼(木材缺陷) pig eye
芽眼(植物器官) eye; oculus
芽眼嫁接法 eye grafting
芽殖 budding
芽肿病 bud hypertrophy
蚜虫 aphid; aphis; plant lice
蚜狮(草蛉科幼虫) aphis lion
崖柏油 morpankhi oil
崖屑坡 talus slope
哑变量 dummy variable
哑光装饰 flatting finish
雅各布森发芽装置 Jacobsen (germination) apparatus
亚胺硫磷(杀虫剂) imidan
亚北方期 subboreal period
亚北极期 subarctic period
亚北极针叶树 subarctic conifer
亚倍体 hypoploid
亚变群丛 lociation
亚变型(法瑞学派) subvariant
亚变种(分类单位) subvariety
亚表土 subsurface soil
亚表土径流 subsurface run-off
亚层 subhorizon; sub-layer
亚成虫 subimago
亚大西洋期 subatlantic period
亚单位 subunit
亚单位结构 subunit structure
亚等位基因 hypomorph
亚顶极(演替) subclimax
亚端着丝点染色体 S-chromosome
亚对生 subopposite
亚分子结构 sub-molecular structure
亚纲(群落或土壤分类) suborder; subclass
亚高山带 subalpine zone
亚高山林 subalpine forest
亚高山植物群落 orophytia
亚灌木 semishrub; semi-shrub
亚灌木的 suffruticose
亚光学显微镜 sub-light microscope
亚恒雪带 subnival belt; subnival region; subnival zone
亚茴香基丙酮 anisylidene acetone
亚基 subunit
亚基本纤丝 sub-elementary fibril
亚基因 subgene
亚级带 subpolar zone
亚级地区 subpolar region
亚极带 subarctic zone
亚极地森林 subpolar zone forest
亚己基二胺 hexamethylene diamine
亚糠基 furfurylidene
亚科 subfamily
亚颏(昆虫) submentum
亚类(分类单位) subgroup
亚历山大番泻叶 Alexandria senna
亚磷酸盐 phosphite
亚硫酸废液 spent sulfite liquor; sulfite waste liquor
亚硫酸镁法 magnesium sulphite process
亚硫酸氢铵法 ammonium bisulfite process
亚硫酸氢盐法 bisulfite process
亚硫酸盐(纸)浆 sulfite pulp
亚硫酸盐处理 sulfitization
亚硫酸盐法 sulfite process; sulphite process
亚硫酸盐废液 sulphite waste liquor
亚硫酸盐废液酒精 sulphite alcohol
亚硫酸盐黑液 sulfite liquor; sulphite liquor
亚硫酸盐化荆树皮栲胶 sulfited mimosa extract
亚硫酸盐浆 sulphite pulp
亚硫酸盐纸浆 sulfite pulp
亚麻 fibre flax
亚麻杆 flax shive
亚麻仁油 linseed oil
亚麻碎料板 flax shives board
亚麻下脚料 flax shive
亚麻下脚碎料板 flax shives board
亚门(分类单位) subdivision; subphylum
亚敏感性 hyposensitivity
亚目(分类单位) suborder
亚南极的 subantarctic
亚前缘脉(昆虫翅) subcosta
亚群丛(法瑞学派) subassociation
亚群丛(群落分类) faciation
亚群落 subcommunity
亚群属 suballiance
亚群系 subformation
亚染色单体 sub-chromatid

亚染色丝 subchromonema
亚染色体 sub-chromatid
亚染色线 subchromonema
亚热带 subtropical zone
亚热带常绿林 subtropical evergreen forest
亚热带风景 subtropical scenery
亚热带阔叶林 subtropical hardwood
亚热带林 subtropical forest
亚热带林区 subtropical forest region; subtropical forest zone
亚热带气候 subtropic climate
亚热带庭园 subtropical garden(ing)
亚热带夏绿林 subtropical summer green forest
亚热带雨林 subtropical rain forest
亚热带园林 subtropical garden(ing)
亚肉桂基 cinnamal; cinnamilidene
亚山地带 submontane zone
亚社会性的 subsocial
亚砷酸铜复盐 Schweinfurt(h) green
亚砷铜铵防腐剂 chemonite preservative
亚属(分类单位) subgenus
亚属名(分类单位) subgeneric name
亚铁螯合酶 ferrochelatase
亚铁黄素蛋白 ferroflavoprotein
亚铁氰化铁 Berlin blue
亚微观大小 submicroscopic size
亚微观多孔性 submicroscopic porosity
亚微观各向异性晶粒 submicroscopic anisotropic crystalline particulate
亚微观结构 submicroscopic structure
亚微观粒子 submicroscopic particle
亚微观特征 submicroscopic feature
亚微观微纤丝 submicroscopic microfibril
亚微观形态 submicroscopic morphology
亚稳离子 metastable ion
亚系统 subsystem
亚细胞结构 subcellular structure
亚纤丝 sub-fibril
亚香茅 Ceylon citronella grass
亚硝化作用 nitrosation; nitrosification
亚硝基氯化物 nitrosochloride
亚硝酸根离子 nitrite ions
亚硝酸细菌 nitrite bacteria
亚硝酸盐氧化细菌 nitrite-oxidizing bacteria
亚影响种 subinfluent; subinfluent species
亚优势的 subdominant
亚优势木 co-dominant; codominant tree; subdominant tree
亚优势树种 subdominant tree species
亚优势种 subdominant; subdominant species
亚优势主伐木 codominant crop tree
亚种(分类单位) subspecies
亚种群 subpopulation
亚总体(统计学) subpopulation
亚族(分类单位) subfamily; subtribe
砑光 glaze
砑光机 breaker stack; calender
咽 pharynx
咽侧体 corpus allatum
咽喉 gula
咽神经节 pharyngeal ganglion
咽下部 hypopharynx
胭脂红 carmine
胭脂树的果肉色质 orlean
烟囱 chimney
烟道 draft; draught; duct; flue; gas flue
烟道气 flue gas
烟道试验 tunnel test
烟害 damage by fume; fume damage; smoke damage; smoke injury; smoke nuisance
烟迹侦察 smoke detection
烟碱(杀虫剂) nicotine
烟缕 smoke plume
烟霉 sooty mould
烟密度速度 smoke density rating
烟幕 smoke curtain; smoke screen
烟示[测]气流法 smoke-test
烟酸 nicotinic acid
烟团 puff
烟雾 smog
烟雾发生器 smoke generator
烟雾剂 aerosol
烟雾喷射机 aerosol sprayer
烟雾器 smoke generator
烟洋椿木油 cedrela oil

烟柱探测仪(防火) haze meter; visibility meter
阉割 emasculation
淹没 submergence
淹水沼泽 fen
延迟 delay
延迟期 lag phase
延迟时间(加载和变形间的) incubation time (of strain load)
延迟效应 delayed effect
延后密度制约 delayed density dependence
延期付款 deferred-payment
延期付息股份 deferred stock
延期交货 back order
延伸率 elongation; unit elongation
延伸生长 excurrent growth
延伸仪 extensometer
延时 time-lag
延时信号 time retarding signal
延性 stretchability
延长器 extender
延滞性密度依存因子 delayed density-dependent factor
严格的自然保护区 strict nature reserve
严格需氧菌 strict aerobe
严格厌氧菌 strict anaerobe
严酷气候 vile weather
严霜 killing frost
严重变色 heavy discolouration
岩崩 rockfall; detritus avalanche
岩表植物 lithophyte
岩成土(松散母质) regosol
岩成土壤 endodynamomorphic soil
岩洞 grotto (*plur.* grotto(e)s)
岩浆 magma
岩浆流 lava flow
岩浆圈 pyrosphere
岩浆水 juvenile water; magmatic water
岩浆岩 magmatite
岩兰 vetivazulene
岩兰草酮 vetivone
岩兰草油 cuscus oil; oil of cuscus; vetiver oil
岩兰草薁 vetivazulene
岩栖的 saxicoline; saxicolous
岩蔷薇胶 labdanum(gum)
岩蔷薇油 labdanum oil
岩溶 karst
岩溶景观 karst landscape
岩生的 epipetreous; rupestral; saxicoline; saxicolous
岩生植物 rock plant
岩石花境 rock border
岩石火成论 plutonism
岩石侵蚀 rocky erosion
岩石圈 geosphere; lithosphere
岩石学 lithology; petrology
岩石园 rock garden
岩石植物 rock plant; rupicolous plant
岩下荒漠群落 petrideserta
岩屑 debris; detritus
岩屑堆 scree; talus deposite
岩屑滑动 debris slide
岩屑坡 talus slope
岩屑土 talus soil
岩屑锥 debris cone
岩芯 core
岩性 lithology
岩性单位 lithological unit
岩穴 abri
岩锥 talus cone
沿岸带 littoral zone
沿岸的 littoral
沿岸林 fringing forest; gallery forest
沿岸沙滩 barrier beach
沿岸生物群落 communities in littoral zone
沿海岸漂散木 long-shore drift
沿海的 coastal
沿海防护林 coast protection forest
沿海浮游生物 coastal plankton
沿海平原 coastal plain
沿海沙漠 coastal desert
沿海渔业 coastal fishery
沿海沼泽地 coastal swamp
沿河荒地 riverwash
沿路公园 passing-through park
沿脉变色症(病毒危害) vein-banding
沿石 curbstone
沿视线的 radial
沿四周锯切 sawing around
研光 satin finishing
研糊油 mull oil
研究浮标 research buoy

研究开发与示范 research, development and demonstration
研究用自然区 research natural area
研磨 abrasion; sharpening
研磨粉 abrasive powder
研磨机 grinder; knife grinding machine
研磨剂 abrasive compound; abrasive material; abrasive substance
研磨加工 abrasive machining
研磨角 grinding angle; wedge angle
研磨面 abrasive surface
研磨纸 emery paper
研制周期 lead time
盐[渍]土 saline soil
盐成土 halogenic soil; halomorphic soil
盐度障碍 salinity barrier
盐分累积 salt accumulation
盐分胁迫 salinity stress
盐杆菌 halobacteria
盐湖 salt lake
盐化草甸 saline meadow
盐化栗钙土 salinized chestnut soil
盐基饱和度 base cation saturation; cation saturation
盐剂混合罐 salt mixing tank
盐碱滩 salina
盐碱土 saline-alkali soil
盐结皮 salt crust; salt incrustatiou
盐浸作用 brining
盐壳 salt crust
盐类防腐剂 salt preservative
盐类化合物处理木材 salt-treated wood
盐敏性的 salt sensitive
盐浓度 salinity
盐盘 salt pan
盐生的 halophilic
盐生灌木 salt bush
盐生荒漠 haloeremion
盐生群落 salt community
盐生生物 halobiont
盐生演替系列 halosere
盐生植被 halophytic vegetation
盐生植物 halophilous plant; halophyte; halophytes
盐水 saline
盐水浮游生物 haloplankton
盐水侵入(侵蚀) saltwater intrusion (encroachment)
盐水贮木池 salt chuck
盐酸橡胶 rubber hydrochloride
盐土 salt soil; solonchak; white alkali soil
盐土草地 salt meadow
盐土草原 salt steppe
盐土荒漠 salt desert; solonchak desert
盐土沼泽 salt marsh; salt swamp
盐土植物 halic; halicole; salt plant
盐雾 salt spray
盐析洗提色谱 salting-out elution chromatography
盐液干燥法 salt seasoning
盐跃层 halocline
盐沼 salt marsh
盐沼区 salina
盐沼群落 limnodium
盐渍化 salinization
盐渍土 salinized soil
盐渍土壤 salt-affected soil
颜料 pigment
颜料黑 aniline black
颜色辨认 colour discrimination
颜色不正的 off-colour
颜色混合 colour mixture
颜色立体 colour solid
颜色匹配 colour match
颜色视觉 visual colour sensation
檐板 projection
檐槽板 gutter board
檐撑条 cave strut
檐稽 plate
檐口人字木 barge couple
檐状菌 bracket fungus
衍射 diffraction
衍射光栅 diffraction grating
衍射弧宽度 diffraction arc width
衍射图 diffraction diagram
衍射图像 diffraction pattern
衍生 derivation
衍生物 derivation; derivative
衍生稀树草原 dcrived savanna
掩蔽地 sheltered ground
掩蔽剂 masking agent
掩蔽效应 masking effect
眼点 ocellus (*plur.* ocelli); stigma

眼基线 eye base; inter-ocular distance
眼睑 lid
眼圈 orbital ring
演变树种 succession species
演化 evolution
演进变形 aromorphosis
演示场 simulator enclosure
演替 succession
演替顶极 climax
演替顶极理论 climax-pattern hypothesis
演替度 degree of succession
演替阶段 seral stage
演替进程 progression of succession
演替群丛 associes
演替生境 successional habitat
演替时序 chronosequence
演替树种 succession species
演替系列 sere
演替系列变群丛 facies
演替系列顶极(群落) serclimax
演替系列群落 seral community
演替系列微群落 associule
演替系列现象 seral phenomenon
演替系列小群落 locies
演替系列组合 socies
演替亚群丛 facies
厌气的 anaerobic
厌气性变质 anaerobic deterioration
厌气性浸泡 anaerobic waterlogging
厌压的 barophobic
厌氧处理 anaerobic treatment
厌氧的 anaerobic
厌氧细菌 anerobic bacteria
厌氧性的 oxyphohous
厌氧性生物 anaerobe
厌氧有机营养 anaerobic organotrophy
砚标 beacon
验收 acceptance check; acceptance inspection
验收程序 acceptance procedure
验收概率 probability of acceptance
验收合格证 acceptance certificate
验收试验 acceptance test; approval test; inspection test
验收台 live deck
验收条件 acceptance condition
堰 dam; weir
堰坝(流送木材用) logging dam
堰坝载荷 weir loading
堰板 gate
堰板缝隙 slice opening
堰塞湖 dammed lake
堰塞盆地 dammed basin
雁 goose
雁鸭类 waterfowl
燕麦 Avena sativa
燕鸥 tern
燕尾 forficated tail
燕尾榫 dovetail; dovetail tenon
燕尾状劈茬(伸出伐根) swallow tail stump
燕子 swallow
扬程 lift
扬声器 representative fraction
扬种机 seed blower
羊齿类温室 fern house
羊肚菌 morel; toadstool
羊角状花纹 ram's horn figure
羊毛染色剂 matching stain
羊毛脂 lanolin
羊皮纸 parchment paper
羊皮状涂饰 kid finish
羊圈(拦收未分类的赶羊木材) bull pen
羊腿漂子(简易式钢索浮木结构) sheep shank boom
阳光 helio- (*eg.* heliophilous, heliophobous)
阳光指标 light indicator
阳(茎)基(鞘翅目) tegmen (*plur.* tegmina)
阳极 anode
阳茎(昆虫) aedeagus; phallus
阳离子防腐剂 cationic preservative
阳离子交换 cation exchange
阳离子交换量(土壤) cation exchange capacity
阳离子交换能力 base exchange capacity
阳坡 adret; sunny slope
阳生叶 sun leaf
阳生植物 heliophyte; oread; intolerant species; sciophyte; sun plant; light-demander; light-demanding tree

阳性的 light-demanding; light-needing; masculine; shade-avoiding; heliophilous; heliophilic
阳性树 light-demander; light-demanding tree
阳性树冠 sun crown
阳性树种 intolerant tree; intolerant tree species; oread; heliophile; heliophyte; sciophyte; sun plant; light-demander; light-demanding tree
阳性植物 sciophyte; sun plant; oread; heliophile; heliophyte; intolerant species; light-demander; light-demanding species; heliophilous species
杨苷 populin
杨克式纸机 Yankee machine
杨梅 Chinese waxmyrtle; strawberry tree
杨梅栲胶 myrica extract
杨氏弹性模量 Young's modulus (of elasticity)
杨氏硬度 Janka hardness
杨属灵 populin
杨树叶锈病 poplar-rust
洋菜 agar; agar-agar
洋椿酮 cedrellone
洋红 fuchsin; magenta
洋槐 false acacia; black locust
洋槐黄素 robinetin
洋中脊 mid-ocean ridge
仰俯角 vertical angle
仰拱(管渠涵洞内壁下部) invert
仰卧的 procumbent
养虫室 insectary
养分缺乏症 food-deficiency disease
养分消耗 nutrient depletion
养分循环 nutrient cycle
养分转移 nutrient translocation
养花师 florist
养料 nutrient
养鱼 fish culture; fish farming
养鱼缸 aquarium
养育 nursing
氧的吸收速率 oxygen uptake rate
氧分压 oxygen tension
氧化巴西木素 brazilein
氧化变色 oxidation stain; oxidative stain
氧化处理 aerobic treatment
氧化还原电位 oxidation-reduction potential; redox potential
氧化还原电位-pH关系 Eh-pH relationship
氧化还原酶 oxido-reductase
氧化还原系统 oxidation-reduction system
氧化褐变 brown oxidation stain; chemical brown-stain
氧化乐果(杀虫剂) omethoate
氧化磷酸化 oxidative phosphorylation
氧化铝 alumina
氧化铝磨料 aluminium oxide mineral
氧化酶 oxidase
氧化偶合 oxidative coupling
氧化树脂 resene
氧化苏木素 haematein
氧化锑 antimony oxide
氧化铁黑 iron oxide black
氧化铁红 iron oxide red
氧化铁黄 iron oxide yellow
氧化铁棕(涂料) brown iron oxide
氧化物活动性 oxide activity
氧化纤维素 oxycellulose
氧化锌 zinc bloom; zinc white
氧化亚氮(N_2O) nitrous oxide (N_2O)
氧化作用 oxidation
氧卡系数 oxycalorific coefficient
氧量计 oxygen meter
氧桥聚合作用 oxolation
氧鎓盐 oxonium salt
氧吸收比速度 specific oxygen uptake rate
氧乙炔焦化材面防腐法 oxyacetylene process
氧杂萘邻酮 coumarin
氧债 oxygen debt
样板 mould; sample board; template
样板平尺 straightedge
样本 sample
样本大小(容量) sample size
样本单元 sampling unit
样本地法 fixed area sampling
样本点 sample point
样本分布 sampling unit distribution
样本检尺 sample scaling

样本均方差 mean square of the deviation
样本量 sample size
样本平均数 sample mean
样带 belt-transect; sample strip
样地 plot; sample area; sample plot
样地半径系数 plot radius factor
样地材积 plot volume
样地法 sample plot method
样地方差 plot variance
样地复位 plot recovery
样地估测法 plot estimate method
样地每木调查表 plot tally sheet
样地描述 site description
样地数据 sample plot data
样地调查 plot survey; sample-plot survey
样点 sampling point
样点观察法 point observation method
样点接触法 point contact method
样方 plot; quadrat; releve
样方法 quadrat method; square-foot method
样方目测法 ocular estimate-by-plot method; ocular plot-estimate
样方图 quadrat chart
样格法 grid method
样格取样 grid sampling
样规 mortise gauge
样木 model tree; sample tree
样品 sample
样品转换装置 sample changer
样区标准地 sample plot
样区调查法 inventory by sample plots
样条 transect; transect plot
样线(植被调查) line transect
样线法 line transect method; line-intercept method
样线调查 line survey
样芯 increment core
夭折 perish
腰部绷索(集材杆) buckle guy
腰果 cashew nut
腰果酚 anacardol; cardanol
腰果树香脂 acajou balsam
腰果树油 cardol
要求低温的种子 cold-requiring seed
姚金娘酮 benihione
窑顶 kiln roof
窑干 kiln seasoning; kiln-dry
窑干材 kiln-dried timber; kiln-dried wood
窑干材变色 kiln stain
窑干材堆 kiln charge
窑干材降等 kiln degrade
窑干操作 kiln operation
窑干堆材 kiln-dry lumber
窑干法 kiln drying; kiln seasoning
窑干褐变 kiln brown stain; kiln burn
窑干基准 kiln schedule
窑干基准表 kiln schedule; kiln-drying schedule
窑干控制板 controlling kiln sample
窑干杀菌 kiln sterilization
窑干试样 kiln sample
窑干装材车 kiln truck
窑干装材车架 kiln bunk
窑烧炭法 kiln process
摇摆 oscillation
摇摆式浮动履带 oscillating floating track
摇摆式输送机 swing conveyor
摇摆舞 locking
摇臂 rocker
摇尺 set
摇尺杆 set lever
摇尺工 block setter
摇尺机构 setworks
摇床 shaking table
摇动筛 shaker; shaking screen
摇杆 rocker
摇篮 cot
摇溶凝胶 thixotropic gel
摇台 shaking table
摇椅 digestive chair; rocker; rocking chair
遥测计 telemeter
遥测技术 telemetry
遥感 remote sensing
遥光度 transmittancy
遥控 remote control
遥控跑车(索道) remote controlled carriage
遥控索道跑车 tracer skyline-crane
遥控油锯 remote-control chain-saw
咬白剂 leucotrope
咬底(漆病) picking-up; lifting

咬起 lifting
药草园 herb garden; herbary
药粉拌种 seed dusting
药膏 salve
药害 spray injury
药剂驱除法 chemical control method
药剂注入法 impregnation
药室 theca (*plur.* thecae)
药筒 shell
药物抗性 chemoresistance
药用化学品 pharmaceutical chemicals
药用炭 medicinal carbon
药用植物 drug plant; medical plant; medicinal plant
要素 element; essence
要素费用 factor cost
钥孔锯 keyhole saw
钥匙卡 key card
鹞(yào) harrier
椰皮纤维 coir
椰子 copra
椰子壳 cocoanut shell
椰子林 coconut palm grove
椰子树 coco
椰子外壳 cocoanut husk
椰子纤维 coconut fibre
椰子油 coconut oil; copra oil; palm oil; palm-fat
椰子脂 cocoa butter
冶炼厂 refinery
野餐场地 picnic ground
野草控制 weed control
野花 wild flower
野化种群 feral population
野火 forest fire; wild fire; wildfire
野麻根素 datiscetin
野蛮 wildness
野栖的 campestral
野趣园 wild garden; wild park
野生的 feral
野生动物 wild animal; wildlife
野生动物保护 conservation of wildlife; protection of wild life
野生动物保护区 wildlife refuge
野生动物保养 preservation of games
野生动物管理 wildlife management
野生动物经济学 wildlife economics
野生动物类型自然保护区 nature reserve for wild animals
野生动物区系 wild fauna
野生动物生态应用(调查) wildlife ecology application
野生动物调查 wildlife censusing
野生苗 natural seedling; naturally sown young seedling; volunteer; wild seedling; wild stock; wildling; wilding
野生苗栽植 wild seedling (tree) planting
野生亲缘种 wild relative
野生群体 wild population
野生生物 wildlife
野生生物管理 wildlife management
野生生物类自然保护区 nature reserve for wildlife
野生物种 wild species
野生纤维植物 wild fibrous plant
野生橡胶植物 wild rubber plant
野生型 wild type
野生性 wildness
野生移植苗 wild transplant
野生原种 wild ancestor
野生植物 wild plant
野生植物类型自然保护区 nature reserve for wild plants
野生种 wild species
野生种群 feral population
野兽 wildlife
野兽害 damage caused by game
野兽林作业 wildlife system
野兽食害 browsing by game; damage by game
野兽园 game park; game cover; park
野兔 hare
野外 field; in situ
野外草图 field sketch
野外测量 field measurement
野外测量草图 survey field sketch
野外处理 field treatment
野外工作 field-work
野外记录 field sheet
野外记载簿 field-book
野外检查 field checking
野外鉴别 field identification
野外实验室 field laboratory
野外试验 field trial
野外数据 field data

Y

野外样地 field plot
野味 venison
野账(野外记录) measurement list
业务 transactions
业务费用 operating outlay
业务会计 activity accounting
业务容量 volume of business
业务准备金 operational reserve
业余花卉栽培 amateur floriculture
业余园艺 amateur gardening
叶 leaf; frondome; lobus; foliage
叶(蕨类、棕榈类、苏铁类) frond
叶白素 leueophyll
叶斑病 leaf spot
叶背 blade back
叶边界层 leaf boundary layer
叶表 phylloplane
叶柄 leaf petiole; leaf stalk; mesopodium; petiole; stipe; thecaphore
叶柄痕 scar
叶柄间芽 interpetiolar bud
叶柄上的 petiolar
叶柄下的 subpetiolar
叶柄状的 petiolar
叶部淋洗 foliage leaching
叶部喷雾 foliage spray
叶部诊断 foliar diagnosis
叶插 leaf cutting
叶蝉 leafhopper
叶齿 leaf teeth
叶赤素 phylloerythrin
叶刺 leaf spine; leaf thorn
叶耳 leaf auricle
叶蜂 sawfly
叶幅 leaf width
叶附生植物 epiphyllae
叶高 height of leaf
叶高度多样性 foliage height diversity
叶痕 leaf scar; phyllule
叶红素 erythrophyll
叶化石 phyllite
叶黄素 lutein
叶基 leaf base
叶级 leaf size class
叶际 phyllosphere
叶迹 folial trace; leaf trace
叶迹维管束 leaf trace bundle
叶甲 chrysomel; leaf-beetle
叶尖 leaf apex
叶焦 leaf scorch; scorch
叶界面层 leaf boundary layer
叶境 phyllosphere
叶卷叠式 foliation; frondescence
叶卷须 leaf tendril
叶孔 stoma
叶枯病 leat blight
叶宽度 leaf width
叶龄 leaf age
叶轮 impeller; runner
叶轮搅拌机 turbomixer
叶轮片 impeller blade
叶绿蛋白 chloroplastin
叶绿素 chlorophyll
叶绿素酶 chlorophyllase
叶绿素衍生物 phyllin
叶绿素荧光 chlorophyll fluorescence
叶绿素原 chlorophyllogen
叶绿体 chloroplast
叶脉 leaf vein; nerve; vein
叶脉型 venation
叶脉序 leaf venation
叶霉 leaf mould
叶霉病 leaf mould; leafmold
叶密度 leaf density
叶面附生植物群落 epiphyllous community
叶面积 leaf area
叶面积比 leaf area ratio
叶面积持续期 leaf area duration
叶面积密度 leaf area density
叶面积指数 leaf area index
叶面淋溶 foliage leaching
叶面喷施 foliage spray
叶面施肥 foliage dressing
叶疱病 leaf blister
叶片 blade; lamina (*plur.* laminae); leaf blade; limb; paddle
叶片分析 leaf analysis
叶片式搅拌器 blade agitator
叶片式空气制动器 fan brake
叶片体积 foliar volume
叶片诊断 foliar diagnosis; leaf diagnosis
叶片状 foliaceous
叶器官 leaf organ

叶鞘 blade sheath; leaf sheath; pericladium; vagina
叶肉 mesophyll
叶肉的 mesophyllous
叶上部着生的 superfoliaceous
叶舌 leaf ligule; ligule; strap; tongue
叶舌的 ligular; ligulate
叶舌痕 ligular scar
叶舌穴 ligular pit
叶身芽鳞 leafy scale
叶室 leaf chamber
叶突 lobe; lobule
叶围 phyllosphere
叶隙 folial gap; leaf gap
叶先端 leaf apex
叶形 leaf shape; phylliform
叶锈病 leaf rust
叶序 leaf arrangement; phyllotaxis; phyllotaxy
叶芽 gem; leaf bud
叶芽插 leaf-bud cutting
叶腋 leaf axil
叶疫 leat blight
叶原 phyllogen
叶原基 leaf primordium
叶缘 leaf margin; margin
叶缘类型 marginal type
叶缘萎黄病 marginal plant chlorosis
叶源地表腐殖质 foliogenous ecto-humus
叶诊断 foliar diagnosis
叶枕 leaf cushion; pad; pedestal; phyllula; pulvinus
叶重比 leaf weight ratio
叶轴 leaf axis; rachis (*plur.* raches); rhachis
叶状柄 frondescence; leaf-like petiole; phyllode
叶状的 foliaceous; phylloid; phylloidal
叶状茎 kladodium
叶状器官 leaflet
叶状体 pedalium; thallus
叶状体一年生植物 thallotherophytes
叶状体植物 thallophyte
叶状体状 thalliform
叶状突 foliole
叶状幼体 phyllosoma
叶状枝 cladode; cladophyll; phylloclade; phylloid
叶状枝束 phylloid truss
叶阻力 leaf resistance
叶座 leaf cushion; pulvinus; sterigma (*plur.* -mata); trichidium
页脚 footer
页岩 laminated rock; shale
页岩焦油 shale tar
页岩油 shale oil
曳材索 pulling line
曳槽 slip
曳木机 log haul(-up)
夜虫粪 nocturnal frass
夜出动物 nocturnal; nocturnal animal
夜蛾 cutworm
夜光云 luminous cloud
夜间出行的 nocturnal
夜间开放的 noctiflorous
夜间冷却 nocturnal cooling
夜间迁移 nocturnal migration
夜间巡逻员 night watch
夜莺 bulbul
液 juice
液胞 sap vesicle
液动装置 fluid power transmission
液封 wet seal
液固色谱法 liquid-solid chromatography
液化 liquidation
液化剂 liquefier; liquifier
液化酶 amylase; α-amylase
液化器 liquefier; liquifier
液胶 liquid adhesive; liquid glue
液胶状态 sol state
液晶 liquid crystal
液控卡木钩 hydraulic dog
液控冷压机 hydraulic cold press
液控调位装置 hydraulic offset
液控铡刀机 hydraulic clipper
液力喷射 hydro-jet
液力制动器 liquid brake
液料表面处理 surface dressing
液流(树干水分运输) sap flow
液泡 vacuole
液丝 brin
液态胶黏剂 liquid adhesive; liquid glue

液态矿脂 liquid petrolatum
液态石蜡 liquid paraffin
液态树脂 liquid resin
液态杂酚油 liquid creosote
液体催干剂 liquid drier
液体肥料 liquid fertilizer
液体回收量 liquid recovery
液体浆 pulp slurry
液体浸透 fluid penetration
液体冷却剂 liquid coolant
液体抛光剂 liquid polishing agent
液体培养基 liquid medium
液体倾角传感器 fluid-tilt sensor
液体闪烁计数器 liquid scintillation counter
液体生物燃料 liquid biofuel
液体松香 liquid rosin
液体填孔剂 liquid filler
液体温度计 liquid thermometer
液体压力 fluid pressure
液体研磨 liquid honing
液相色谱法 liquid chromatography
液相色谱分析法 liquid phase chromatography
液相色谱图 liquid chromatogram
液压 sap pressure
液压泵 fluid pump
液压传动装置 shuttle
液压动力链锯 hydraulically powered chain saw
液压堆垛机 hydraulic stacker
液压伐木楔 hydraulic wedge
液压机 hydraulic press
液压加压系统 hydraulic loading system
液压剪(伐木) hydraulic shear
液压进料链式榫槽机 hydraulic-feed chain mortise
液压锯 hydraulic action saw
液压控制装置 hydraulic(al) control device
液压链锯 hydrochain
液压马达 fluid motor
液压起重臂 hydraulic log loading boom
液压起重机 hydraulic crane
液压升降油缸 hydraulic lift cylinder
液压升降装置 hydraulie lift
液压试验 water test
液压梯 hydraulic ladder
液压推杆 timber tipper
液压挖掘机 hydraulic excavator
液压用油 hydraulic fluid
液压助力系统 power assist hydraulic system
液压抓木钩 hydraulic log grapple
液-液色谱法 liquid-liquid chromatography
腋 axil
腋的 alar
腋间生的 interaxillary
腋生技 axillary shoot
腋生聚伞花序 axillary cyme
腋生托叶 axillary stipule
腋下的 subaxile
腋下生的 infra-axillary
腋芽 axillary bud; lateral bud
靥窝纹理 dimpled grain
一般(地位级III) average
一般材积表 regional volume table
一般法正林 generalized normal forest
一般均衡理论 general-equilibrium theory
一般配合力 general combining ability
一般收获表 general yield table
一倍体 monoploid
一边刨光材 dressed one edge
一串缏漂木 string of stick
一雌多雄 polyandry
一雌一雄的 monogamous
一次繁殖生物 semelparity
一次集材 single haul
一次集材量 pull
一次集运量 turn
一次胶合 primary gluing
一次结果的 monocarpic plant
一次结实性 monocarpy
一次流送完毕 clcan drive
一次伞伐法 one-cut shelterwood method
一次生殖 semelparous
一次性能源 primary energy
一次选择 single selection
一次窑干材积 charge; kiln run
一次窑干过程 kiln run

一次渐伐法 one-cut shelterwood method
一次主伐的(轮伐内) monocyclic; monocyclic felling
一次装料量 charge
一代的(无机盐) primary
一等材 first grade timber; first prime quality
一端搬运 end haul(ing)
一段水解 one-stage hydrolysis
一堆木材(一次集材量) bunch(ing)
一对基因杂交 monohybrid cross
一对基因杂种 mono-hybrid
一对双爪链钩 double-coupler
一个基因一种酶假说 one-gene-one-enzyme hypothesis
一个基因一种作用假说 one-gene-one-function hypothesis
一行短原木(用于编排) sweep
一化[的] univoltine
一化昆虫 univoltine insect
一回羽状复叶 unipinnate compound leaf
一级的 prime
一级河流 first-class river
一级火 class A fires
一级胶合板 type-one plywood
一级结构 primary structure
一级水解 one-stage hydrolysis
一级压缩 single-stage compression
一级枝 primary branch
一级主伐木 first-class standard
一价酸 monobasic acid
一碱价酸 monobasic acid
一节平排 brail
一捆 bundle
一揽子交易 package deal
一类清查 national forest inventory; first-category forest inventory
一列废材 windrow
一零五九(杀虫剂) demeton
一硫化汞 cinnabar
一六零五(杀虫剂) parathion
一六零五粉剂(杀虫剂) parathion dust
一六零五可湿性粉剂(杀虫剂) parathion wettable powder
一卵双生 uniovular twins
一面光硬质纤维板 hard-board; one surface smooth hardboard
一面两边刨光材 dressed one side, two edges; surfaced one side, two edges
一面刨光材 dressed one side; surfaced one side
一面一边刨光材 dressed one side, one edge; surfaced one side, one edge
一年多代昆虫 multlvoltine insect
一年二代的 bivoltine
一年群体 annual colony
一年生草花园 annual garden
一年生的 annual
一年生地被物 annual ground covers
一年生溃疡 annual canker
一年生实生苗 yearling
一年生杂草 annual weed
一年生枝 annual branch
一年生植物 annual; annual plant; crop fiber; therophyte; therophyte (Th)
一年生子实体 annual fruit-body
一年一代的 univoltine
一品红 christmas flower
一齐林(高度密度均质) uniform stand
一日游公园 park day-use visitation
一束 bundle
一岁的 yearling
一维数组 linear array
一心皮的 monocarpellary
一雄多雌的 monarsenous
一元材积表 local volume table; one way volume table; single entry volume table; single variable volume table
一元材积模型 unitary volume model
一元材种出材率表 one-way merchantable volnme table
一元的 single-entry
一元酸 monoprotic acid
一致 coincidence
一致性 consistency
一致性削度方程 compatible taper equation
一昼夜(的) diel
伊福特剥皮机(铣削型多刀剥皮机) Efurd machine
伊利石 hydrous micas; illite
伊佐德冲击强度 Izod-value
衣橱 clothes press; wardrobe

衣橱式衣箱 wardrobe trunk
衣蛾 clothes moth
衣柜 clothes press
衣架 clothes-rack
衣壳(病毒的) capsid
衣帽架 clothes tree; hall stand
衣箱 dress case
依存细菌 dependent bacteria
依赖群落 dependent community
依兰浸膏 ylang-ylang concrete
依兰油 ylang-ylang oil
仪表 instrumentation
仪表板 meter board; panel
仪表化 instrumentation
仪表图 meter diagram
仪器 instrument
仪器柜 instrument locker
仪器盒 instrument case
仪器位置(放置点) instrument position
仪式化 ritualization
夷为平地的 stand-leveling
宜林地 suitable land for forest
移出 explantation
移动 movement
移动板 moving-slat conveyor
移动板链式输送机 moving-apron conveyor
移动常数 migration constant
移动带式输送机 moving-band conveyor
移动刀头单板齐边机 travel(l)ing head veneer jointer
移动的 erratic
移动滚柱式输送机 moving-roller conveyor
移动海滩 drifting beach
移动火 moving fire
移动链式榫槽机 portable chain-saw mortiser
移动苗圃 flying nursing; shifting nursery; wandering nursery
移动沙丘 temporary sandhill
移动式剥皮机 portable debarker
移动式成型铺装机 moving former
移动式打枝造材联合机组 limbing-bucking station
移动式垫板秤 portable scale
移动式伐木削片联合机 swathe-felling mobile chipper
移动式副集材杆 mobile back spar
移动式横截锯 flying cut-off saw; travel(l)ing cut-off saw
移动式后拉锚桩 walking anchor
移动式滑轮 travel(l)ing block
移动式集材杆 portable spar
移动式架杆起重机 portable jib crane
移动式架空采种塔 mobile aerial tower
移动式绞盘机 donkey; road donkey
移动式缆索起重机 mobile cable crane
移动式木材检尺机 vehicle scale
移动式喷涂柜 travel(l)ing spray booth
移动式拼合机构 moving butt-joint mechanism
移动式起重机 travel(l)ing hoist
移动式书架 moving bookstand
移动式数字转换装置 portable digitizer
移动式台面圆锯 circular saw with sliding table
移动式绞盘机 mobile skidder
移动式预压机 movable prepress
移动式圆锯 portable saw; running saw
移动式造材机 mobile slasher; slashmobile
移动式制材厂 peckerwood mill; woodpecker mill
移动索 setting line
移动台式锯机 rack saw
移动预备林 flowing reserve; mobile reserve
移动制材厂 portable saw mill
移动转向式装载机 skid-steer loader
移行区 transition region
移积土 allochthonous soil
移码 frameshift
移码突变 frameshift mutation
移苗开孔器 dibber
移位 shift
移位作用 translocation
移植 grafting; implant; implantation; inoculate; lifting; plant out; transfer; transplant; transplantation; transplanting
移植板 planting-board; transplant(ing) board

移植床 planting bed; transplant bed; transplanting bed
移植法 transfer method
移植环 transfer loop
移植机 transplanter; transplanting machine
移植犁 lining-out plough; transplanting plough
移植苗 nursery transplant; transplant; transplant stock; transplanting seedling
移植苗圃 planting-nursery
移植圃 lining out nursery; transplant nursery
移植箱 transplant box
移植休克 transplanting shock
移植针 transfer needle
遗本(遗产本金) corpus
遗传 heredity; inheritance
遗传保守性 hereditary conservation
遗传变异 hereditary variation
遗传变异性 hereditary variability
遗传成分 genetic element
遗传纯度 genetic purity
遗传单位 hereditary unit
遗传的 autochthonous; heritable
遗传度 degree of heritability
遗传多样性 genetic diversity
遗传反馈 genetic feedback mechanism
遗传方差 genetic variance
遗传防治 genetic control
遗传负荷 genetic load
遗传改良 evolution
遗传工程 genetic engineering
遗传公式 genetic formula
遗传机制 genetic mechanism
遗传渐变群 genetic cline
遗传交叉 genetic crossing over
遗传进度 genetic advance
遗传距离系数 coefficient of genetic distance
遗传可塑性 genetic flexibility; genetic plasticity
遗传控制 genetic control
遗传力(率) heritability
遗传密码 genetic code
遗传漂变 genetic drift
遗传漂移 genetic shift
遗传平衡 genetic equilibrium
遗传平衡假说 genetic equilibrium hypothesis
遗传生态学 genecology
遗传特性 emphytic character
遗传特征 hereditary feature
遗传体系 genetic system
遗传物质 genetic material; hereditary substance
遗传相关 genetic correlation
遗传相似系数 coefficient of genetic similarity
遗传信息 genetic information; hereditary information
遗传型 genotype
遗传型的 genotypic
遗传型方差 genotypic variance
遗传性 heredity
遗传性变异 genetic variation
遗传性不育 genetic sterility
遗传性的 heritable
遗传性优良的 elite
遗传性状 hereditary feature
遗传性阻隔遗漏 leakage of genetic block
遗传学 genetics
遗传因素 genetic factor
遗传因子 genetic factor; hereditary factor; inherited factor
遗传优势 genetic dominance
遗传有机体 genetic entity
遗传与环境(比喻) nature and nurture
遗传重组 genetic recombination
遗传资源 germplasm; germplasm stock; genetic resources
遗传资源保护 germplasm conservation
遗传组成 genetic make-up
遗存固有种 conservative endemism
遗痕器官 rudimentary organ
遗迹化石 ichnofossil; trace fossil
遗迹化石相 ichnofacies
遗留木 reserve; reserve cell
乙基纤维素 ethyl cellulose
乙酸 acetic acid
乙酸钙 acetate of lime
乙酸人造丝 acetate rayon
乙酸丝 acetate silk

乙酸铜 Schweinfurt(h) green
乙酸细菌 acetic acid bacteria
乙酸纤维 acetate rayon
乙酸纤维素 acetyl cellulose
乙酸纤维素泡沫(塑料) cellulose acetate foam
乙烯基树脂 vinyl resin
乙烯基塑料 vinyl plastic
乙酰丙酸钙 calcium levulinate
乙酰单宁 acetannin
乙酰丁子香酚 acetyl eugenol
乙酰辅酶A acetyl coenzime A
乙酰化 acetylation; acetylate
乙酰化材 acetylated wood
乙酰化产物 acetylate
乙酰基转移酶 acetylase
乙酰甲胺磷(杀虫剂) acephate
乙酰纤维素 acetyl cellulose
乙酰异丁子香酚 acetyl isoeugenol
乙酰转移酶 acetyl transferase
乙酰紫罗兰酮 acetyl ionone
已打浆纸料 beater furnish
已缴款股份 paid-up stock
已缴资本 paid-in capital; paid-up capital
已清理的采伐迹地 released slash
已筛浆料 screened stock
已生根的扦插苗 rooted cutting
已验明种源的 source-identified
以火攻火 counter fire; draft fire
以毛重作净重 gross for net
以木材为主要产品的林区 timber key area
以碳汇换排放 sink for emmission
以新代老 novation
蚁 emmet
蚁布植物 myrmecochore
蚁的 formic
蚁蛉 antlion; doodle bug
蚁狮 ant lion; antlion; doodle bug
蚁筑菌巢 comb
倚缘胚根 radicle accumbent
椅 chair
椅背 back rest; chair back
椅背顶横档 yoke
椅背横条 cross rail; slat
椅背桌面互换椅 settle table
椅垫 squab
椅横档 chair rail
椅后腿 back post
椅面 chair seat
椅前搁脚凳 ottoman footstool
椅状劈裂 barber chair
义务(救火)官员 obligation officer
义务护林员 fire co-operator
刈幅式伐木削片联合机 swathe-felling mobile chipper
刈割演替顶极 mowing climax
刈割样方 denuded quadrat
议价 bargaining
异胞质体 heterocyton; heteroeytosome
异倍体 heteroploid
异步计算机 asynchronous computer
异侧 antarafacial
异侧柏酮 isothujone
异侧压的 heterotasithynic
异常 abnormality
异常材 abnormal wood; anomalous wood
异常的 ataxinomic
异常地表径流 abnormal surface runoff
异常分裂 abnormal division
异常构造 anomalous structure
异常脉序 anomodromy
异常气候 abnormal climate
异常形成 malformation
异常循环 abnormal circulation
异常值 outlier
异常状态 abnormal state
异常组织 abnormal tissue
异狄氏剂(杀虫剂) endrin
异地的 allopatric
异淀粉酶(生化) isoamylase
异动态昆虫 heterodynamic insects
异发演替 allelochemics; allogen(et)ic succession; allogenic succession
异方差性 heteroscedasticity
异分子聚合 copolymerization; copolymerization
异粉性(胚乳) xenia
异粉性现象 xenity
异甘草苷 isoliquiritin
异构体 isomer
异固缩 heteropycnosis
异果性 allocarpy
异海松烷 isopimarane

异花传粉 cross-pollination
异花受精 cross fertilization
异花受精的 allogamous
异花授粉 allogamy
异花授粉的 allogamous
异花同株 polygamy
异化 dissimilation; catabolism
异化作用 catabolism; dissimilation
异基因的 allogenic; heterogenic; xenogeneic; xenogenic
异接合子 allozygote
异榄香素 elemecin
异泪柏烯 isomanoene
异类的 off-type
异类物质 extraneous material
异类种子 foreign seeds
异裂[反应] heterolytic fission
异龄的 uneven-aged
异龄混交法 highly structured forests
异龄混交林 mixed uneven-aged forest
异龄林 irregular forest; irregular plantation; many-aged forest; uneven-aged forest; uneven-aged wood
异龄林分 uneven-aged stand
异龄林经理 uneven-aged management
异龄林收获表 uneven-aged yield table
异龄林作业 uneven-aged system
异龄乔林 uneven-aged high forest
异密度材料 differential density material
异面叶 dorsi-ventral leaf
异木兰酚 isomagnolol
异泡桐素 isopaulownin
异配生殖 anisogametism; anisogamety; anisogamy; heterogamy; oogamy
异配生殖的 anisogamous
异配子型 digamety
异栖的 heteroxenous
异群落的种类 heterochore
异染色体 allosome
异染色质 heterochromatin
异染现象 heterochromaty
异生物群落 allobiocenose
异时的 allochronic
异时种 allochronic species
异数孢子 aneuspory
异四倍体 heterotetraploid
异速生长 allometry; heterogony
异速生长比例 allometric scaling
异速生长法 allometry method
异速生长方程 allometric equation
异速生长系数 allometry coefficient
异态不亲和 heteromorphic incompatibility
异态现象 heteromorphism
异态蛹类 amorpha; false in-digo
异体受精 allogamy
异体受精的 allogamous
异托的 heterocline
异位细胞壁层 heterotropic cell-wall layer
异位显性 epistasis; epistasy
异位显性基因 epistatic gene
异位抑制 allosteric inhibition
异温动物 heterotherm
异戊二烯 isoprene
异物 foreign matters
异系交配 exogamy
异细胞 idioblast
异香 by-note
异向性弹性系统 anisotropic-elastic system
异向性木梁 anisotropic wood beam
异形胞 heterocyst
异形变态 anisomorpha
异形股钢丝绳 flat strand rope
异形花被的 heterochlamydeous
异形木射线 heterogeneous ray
异形配子 heterogamete
异形射线 heterocellular ray; heterogeneous ray
异形射线组织 heterogeneous ray tissue
异形细胞 heterocyst
异形细胞射线 heterocellular ray
异型 off-type
异型不亲和性 heteromorphic incompatibility
异型分子 metagon
异型基因接合子 heteromerozygote
异型接合体 heterozygote
异型接合性 heterozygosity
异型接合子 heterozygote

异型酶 allozyme
异型配子的 anisogamous
异型生殖 heterogony
异型世代交替 heterogenesis; heterogeny
异型叶 heterophyll
异型杂交 outcrossing
异性接合 legitimate copulation
异亚硝基 hydroxyimino
异亚硝基樟脑 isonitrosocamphor
异养 heterotrophic nutrition
异养层 heterotrophic stratum
异养成分 heterotrophic component
异养的 heterotrophic
异养呼吸(非光合生物) heterotrophic respiration
异养湖 allotrophic lake
异养菌 heterotrophic bacteria
异养生物 heterotroph; heterotrophic organism
异养演替 heterotrophic succession
异养营养 heterotrophic nutrition
异养植物 heterotrophic plant
异叶绿素 allochlorophyll
异域分布 allopatry
异域分离 allopatric segregation
异域群种 allopatric population
异域现象(分布区不重叠) allopatric phenomena
异域型 allopatric form
异域种 allopatric species
异域种化 allopatric speciation
异源 heterogenesis; heterogeny
异源的 allogen(et)ic; heterogenic
异源多倍体 allopolyploid; dibasic
异源多倍性 allopolyploidy
异源二倍体 allodiploid
异源接合的 allozygous
异源结合 non-hemologous association
异源联会 allosyndesis; non-hemologous association
异源配对 allosynapsis
异源群体 heterogeneous population
异源染色体 chromosome nonhomologous; non-homologous chromosome
异源世代交替 antithetic alternation
异源四倍体 allotetraploid
异源四聚体 heterotetrameric
异源体类型 alloploid type
异源体起源 alloploid origin
异源外激素 allomone
异源植物区系 allogenous flora
异长叶烯 isolongifolene
异质 alloplasm
异质材料 heterogeneous material
异质的 heterogenic
异质多样生境 heterogeneous-diverse habitat
异质核仁 amphinucleolus
异质接合 allosyndesis
异质联会 allosynapsis
异质同晶 allomerism
异质土壤 heterogeneous soil
异质性 alloplasmy; heterogeneity
异质性系数 coefficient of heterogeneity
异种传播 heterochore
异种的 heterogenic; xenogeneic
异种集团 clan
异种嫁接 heteroplastic graft; xenoplastic graft
异种克生 allelopathy
异种信息素 allelochemic; allomone
异种移植 heterograft
异株克生物质 allelopathic substance
异株异花受精 xenogamy
异主寄生的 heteroxenous
异苎酮 isothujone
异状的 abnormis
异状发芽 abnormal germination
异状苗 abnormal seedling
异宗接合 heterothallism
异宗配合的 heterothallic
异宗配合现象 heterothallism
异祖发生 heterogenesis; heterogeny
抑菌剂 bacteriostatic agent; fungistat
抑菌素 ablastin
抑菌物质 fungistatic substance
抑菌作用 bacteriostasis; fungistasis
抑食剂 feeding suppressant
抑音器 damper
抑莠隆(除草剂) anisuron
抑藻剂 algae inhibitor
抑真菌剂 fungistat

抑真菌作用 fungistasis
抑制 check; inhibition; suppression
抑制病毒生长的 virostatic
抑制点 inhibition point
抑制基因 suppressor
抑制剂 inhibitor
抑制浓度 inhibition concentration
抑制效应 depressor effect
抑制因素 inhibitor
抑制栽培 retarding culture
抑制真菌能力 fungus-inhibiting capacity
抑制值(防腐剂) threshold value
抑制作用 inhibiting effect
译码器 encoder
易爆炸性 explosion hazard
易变的 labile
易变基因 labile gene
易地保护 ex-situ conservation
易飞散种子 light seed
易风化岩 eugeogenous rock
易感病的 amenable to disease; susceptible
易货贸易 barter
易接合性 autohesion
易罹霜害的 frost-tender
易裂性 fragility
易劈裂的 cleavable
易燃材料 combustible material; inflammable material
易燃木 lightwood
易燃碎物 inflammable debris
易燃物 fine fuel; flash fuel; fuel
易燃性 flammability; inflammability
易燃枝桠材 light slash
易溶胶黏剂 solvent-sensitive adhesive
易渗透材 receptive wood
易蚀性 erodibility
易受霜害地 frost hollow
易碎的 brash
易碎性 fragility
易损零件 wear parts
易脱屑种子 chaffy seed
易危种 vulnerable species
易位(染色体畸变) translocation
易位作用 metathesis
易杂交的 liable to cross
疫病 blight
益本率 benefit-cost ratio(s)
益本影价 benefit-cost shadow price
益他素 kairomone
逸出河梗的原木 prize log
逸散蒸汽 escaping steam
逸野[种] escape
翌年发芽种子 carryover seed
意外保险 accident insurance
溢出 overflow; run-off; spill
溢出量 outflow
溢出效应 spill-over effect
溢洪道 overflow; spillway; water escape
溢洪区 overflow area
溢流 effluence; overfall; overflow
溢流坝 spillway dam; weir dam
溢流挡板 over-pass
溢流段(断面) overflow section
溢流口 flow-off
溢流深度 overflow depth
溢流水舌 nappe
溢流箱 circulating stock chamber
溢流堰 flush weir; rolling dam
溢漫 spill
溢泌 exudation
溢泉 overflow spring
溢水坝 overfall; rafter-type splash dam
溢水道 spillway
溢水地 overflow area
溢水堰 overfalling weir; overflow weir
缢痕 constriction
缢扎 strangulation
翼坝 wing dam
翼瓣 ala (*plur.* alae); wing
翼的 alar
翼梁 spar
翼墙 wing wall
翼形插垛(由河中斜伸至一侧岸边) wing jam
翼羽 flight feather
翼状薄壁组织 aliform parenchyma
翼状打磨机 wing beater mill
翼状的 alary
翼状节 moustache-knot; wing-shaped knot
翼足类软泥 pteropod ooze
因变量 dependent variable

Y

因次 dimension
因次分析 dimensional analysis
因果分量 causal component
因果关系 cause-effect relationship
因果演化 consequential evolution
因果演替 consequent succession
因生境而异的 site-specific
因素 element
因子对性 gene allele
因子分解 factoring
因子实验 factorial experiment
因子型 genotype
阴地(阴影) shade; scio- (*eg.* sciophilous, sciophobous)
阴地群落 shade community
阴地植物群落 heliophobous community
阴干材 shed drying stock
阴极射线管 cathode-ray tube
阴极射线致发光扫描电镜显微照片 cathodoluminescence scanning electron micrograph
阴离子 anion
阴离子电泳 anaphoresis
阴离子防腐剂 anionic preservative
阴离子呼吸 anion respiration
阴离子交换剂 anionite
阴离子交换色谱 anion-exchange chromatography
阴历的 lunar; lunate; luniform
阴模 bed-die
阴坡 opaco; shady slope; ubac
阴生叶 sciophyll; shade leaf
阴天 overcast day; cloudy day
阴性的 shade-bearing; shade-demanding; shade-enduring; shade-requiring
阴性附生植物 shade epiphyte
阴性树种(阴性树种) shade species; heliophobe; sciophile; sciad; shade-loving plant; sciophilous species; heliophobous species
阴性植物(阴性植物) heliophobe; sciophile; shade species; sciad; shade-loving plant; sciophilous species; heliophobous species; sciophyte; shade plant
阴影 shadow; schatten
阴影法 shadow method
阴影长度 shadow length
音量 loudness
音频信号 acoustic signal
音响 sound
音响测定 sounding
音响检查 acoustic(al) inspection
音障 sound barrier
银白杨 abele; European white poplar; silver-leaf poplar
银枞油 silver pine oil
银光花纹 silver figure
银光纹理 silver-grain
银焊料 silver solder
银化薄层色谱 argentation thin-layer chromatograph
银桦树胶 silk oak gum
银色光泽 slivery sheen
引发刺激 releaser
引发二次演替 initiating secondary succession
引发剂 initiator
引洪漫地 irrigation with torrential flood
引火材 touchwood
引火柴 kindling wood
引火木材 spunk
引火盘 flash pan
引火人 firer
引火物(用复数) kindling
引火药 priming
引进 introduction
引进的 adventive
引进树种 exotic tree
引进外来植物 foreign-plant introduction
引理 lemma
引力势 gravitational potential
引流 water spreading
引路彩色带(防火标示) dope
引燃器 lighter
引入 introduction
引入物种 introduced species
引绳轮 operating line support
引水槽 flume
引水拉沙 diverting water for sluicing sand
引索 grass line; guinea line; pass line
引物 primer
引线 leader
引线孔 fairlead

引信线 fuse
引鸭犬 tollers-dog
引诱剂 attractant
引诱杀虫剂 attractant insecticide
引诱物 lure
引诱作用 attraction
引种 introduction; introduction of exotic species
引子 primer
吲哚丁酸 indolebutyric acid
吲哚乙酸 indoleacetic acid
饮用喷泉 drinking fountain
蚓状 vermiform
隐蔽抽屉 secret drawer
隐蔽处 abri; shelter
隐蔽的瑕疵 hidden defect
隐蔽价格 shadow price
隐蔽生物 cryptobion
隐蔽税 hidden tax
隐蔽所(野生动物保护设施) covert
隐蔽症状 masked symptom
隐藏色 concealing colour
隐藏盐土 disguised solonchak
隐翅虫 rove-beetle
隐伏 delitescence
隐含成本分析 implicit analysis of cost
隐含利息 implicit interest
隐含租金 implicit rent
隐花植物 cryptogam
隐花植物材 cryptogamic wood
隐花植物木质部 centripetal xylem
隐花植物群落 cryptogamic community
隐黄素 cryptoxanthin
隐灰壤 latent podzol
隐火 smouldering fire
隐节 bark-ringed knot; blind knot; enclosed knot; healed over knot; occluded knot; overgrown knot
隐晶岩 aphanite; cornean; cryptomere; kryptomere
隐居动物的 cryptozoic
隐没河 sinking creek
隐门(与墙齐平) jib door
隐秘种 cryptic species
隐棚(狩猎) blind
隐期(病毒) lag period
隐燃 smouldering
隐燃火 smouldering fire
隐伤 hidden defect; rind-gall
隐射点(狩猎) stand
隐头果 syconium
隐头果植物 sycon
隐头花序 hypanthium; hypanthodium; sycon; syconium
隐细腐殖质层 crypto-mull
隐心材树 light coloured heartwood tree
隐形腐殖质 crypto-humus
隐性病毒 latent virus
隐性纯合体 recessive homozygote
隐性纯合子 allozygote
隐性的 recessive
隐性基因 recessive gene
隐性基因纯合体 bottom recessive
隐性同型合子 bottom recessive
隐性同质结合 recessive homologous association
隐性性状 recessive character
隐芽 kryptoblast
隐芽植物 cryptophyte; cryptophytes
隐域群落 azonal community
隐域土 intrazonal soil
隐樟碱 cryptoçarine
荫蔽防火线 shaded firebreak
荫棚 shade-frame
荫亭 pavilion
印号槌 die-hammer
印痕 indentation
印花辊 embossing roll
印花机 embossing machine
印花税 stamp duty
印花税法 stamp act
印花乙烯薄膜 printed vinyl film
印画纸 developing paper
印记 imprinting
印茜草色素 munjistin
印刷 impression; imprint
印刷花纹 imprinted figuring
印刷机 press
印刷胶合板 printed plywood
印刷墨 printing ink
印刷输纸辊 feed roller
茚满型麝香 indane musk
英尺程 foot run
英尺秒(弹速) foot-second
英寸厚木板 inch stuff; inch wood

英国皇家检尺规程 empire grading rule
英国庭园 English garden
英国药典 British pharmacopoeia
英国药物处方书 British pharmacopoeia codex
英国园林 English garden
英吉利坐犬 English setter
英里英尺水量(1平方英里1ft深) mile foot
英亩 acre
英亩数 acreage
英亩-英尺(容积单位, 等于43560ft^{3}或1233.5m^3) acre-foot
英寻(水深单位, 6ft或1.829m) fathom
英制圆材材积计算单位(270ft^3) English standard
婴儿床 babycrib
婴孩摇椅 mammy's bench
婴孩椅 nursing chair
缨 tassel
缨状胶束 fringe(d) micelle
缨状胶束学说 fringed micell(ar) theory
缨状微束 fringe(d) micelle
缨状原纤维 fringed fibrils
樱核油 cherry stone oil
樱李类 plum (*Prunus*)
樱桃树胶 cherry gum
樱桃香精 cherry essence
鹰嘴(跑车上) dog
鹰嘴板(跑车上) backing board; dog board
应变 strain
应变度 strain intensity
应变功率 power of straining
应变率 strain-rate
应变能量 energy of strain
应变时效 strain aging
应变势能 potential energy of strain
应变速率 rate of straining
应变仪 strain gauge; strainometer
应变圆 circle of strain; strain circle
应变状态 state of strain
应答值(气相色谱等) response
应对范围 coping range
应付款 due
应付票据 bill payable
应付养恤金 benefit payable
应付账款 account payable
应归成本 imputed cost
应归利息 imputed interest
应激理论 stress theory
应急刹车 emergency brake
应计利息 accrued interest; implicit interest
应计收益 imputed income
应计折旧 accrued depreciation
应计资产 accrued asset
应缴额 assessment
应拉木纤维 tension wood fibre
应力 stress
应力表层结构 monocoque; stress(ed) skin construction; stressed shell construction
应力材 reaction wood
应力等值线 stress contour
应力范围 range of stress
应力分布 distribution of stress; stress distribution
应力分级 stress grading
应力分级材 stress-graded lumber
应力分级机 stress grader; stress grading machine
应力分级制 stressgrading system
应力分析 stress analysis
应力分析图解法 graphic method of stress analysis
应力集中 stress concentration
应力检验片 stress section
应力交变 reversal of stress
应力解除 release of stress
应力解除处理 relief treatment of stress
应力裂缝 tension crack
应力逆转 reversal of stress
应力强度 intensity of stress
应力试验 stress test
应力释放 release of stress
应力双折射 stress birefringence
应力松弛 relaxation of stress; stress relaxation; stress relief
应力松弛仪 relaxometer
应力速率 rate of stressing
应力塑性固定变形 drying stress set
应力系统 stress system
应力消除 stress relief
应力循环图 stress-cycle diagram

应力因子 stress factor
应力-应变关系 stress-strain relation
应力圆 circle of stress; stress circle
应力振幅 stress amplitude
应力状态 state of stress
应敏胶黏剂 reaction sensitive adhesive
应收账款 account receivable
应摊未摊折旧 accrued depreciation
应压材(木材缺陷) compression wood; compressed wood; glassy wood; hard streak(s); pressure wood; bull wood; tenar
应用程序设计 application programming
应用技术 application technique
应用技术卫星 application technology satellite
应用昆虫学 applied entomology
应用生态学 applied ecology
应用土壤学 edaphology
应征税货物 dutiable goods
迎春花属 Jasmine Jasminum
迎风 exposure to wind; wind exposure
迎风的 wind-exposed; windswept; windward
迎风面 windward side
迎风坡 windward slope
迎面火 back fire; back firing; counter fire
迎面灭火 back burn; back fire; back firing
迎面阻火 back stop
迎射 approaching shot
荧光 fluorescence; fluorescent light; luminescence
荧光灯 fluorescent lamp; fluorescent light
荧光灯管 fluorescent lamp
荧光分光光度法 fluorescence spectrophotometry
荧光分光光度计 fluorescence spectrophotometer; fluorospectrophotometer
荧光分析 fluorescence analysis
荧光光谱 fluorescence spectrum
荧光色谱 fluorescence chromatogram
荧光素 luciferin
荧光素酶 luciferase
荧光箱 fluorescent light box
盈亏分界点 break even point
盈亏分析 break even analysis
盈亏界限 margin for profit and risk
盈利企业 paying concern
盈余 margin; gain
盈余账 surplus account
萤石 liparite
营地 camp
营地单元调查法 camp unit system
营火 campfire
营救价值 wrecking value
营林段 beat
营林防治 management control; silvicultural control
营林费 silvicultural cost
营林机械 silviculture machinery
营林机械化 silviculture mechanization
营林机械系统 silviculture machine system
营林监督 control of silviculture
营林监督制度 supervisor system
营林局 district (forest-)office
营林农场 tree farm
营林强度 intensity of silviculture
营林区 range
营林区划 silvicultural division
营林区长 divisional officcr
营林署长 supervisor
营林所长 district officer
营林用除莠剂 silvicide
营林员 forest ranger; ranger
营林员管理区 ranger district
营期可稔性 bud-fertility
营养 nutrient; nutrition
营养比 nutritive ratio
营养钵 nutrition pot; nutritive pot
营养钵苗栽植 pot planting
营养钵压制机 soil pot press
营养不良 dystrophy; malnutrition; undernutrition
营养不良湖 dystrophic lake
营养不足 oligotrophy
营养不足的 nutrient poor
营养不足的植物 oligotrophic plant
营养充足的 nutrient rich
营养带(采脂割面间的树皮带) inter space
营养的 nutrient; trophic

营养动态 trophic dynamics
营养繁殖 vegetative propagation; vegetative reproduction
营养繁殖苗 planting stock (by vegetative propagation)
营养繁殖系 clone
营养分解带 tropholotic zone
营养分离 vegetative segregation
营养腐殖质 nutrient humus
营养根 feeder root; feeding root
营养过分 overnutrition
营养耗尽的 exhausted
营养核 trophonucleus; vegetative nucleus
营养基 food base
营养级 trophic level
营养价值高的饲料 heavy feeder
营养嫁接 vegetative grafting
营养阶段 vegetative phase
营养接 suckling grafting
营养结构 trophic structure
营养金字塔 trophic pyramid
营养菌丝 vegetative hyphae
营养菌丝体 vegetative mycelium
营养理论 nutritional theory
营养联系 trophic relationship
营养良好的 eutrophic
营养面积 growing space
营养喷雾 nutrient mist
营养期 vegetative period
营养琼脂 nutrient agar
营养缺乏 nutritional deficiency
营养缺乏病 deficiency disease
营养缺陷突变体 auxotrophic mutant
营养溶液 nutrient solution
营养肉汤 nutrient broth
营养生长 vegetative growth
营养生殖 vegetative propagation; vegetative reproduction
营养时期 vegetative phase
营养体[生长] vegetation
营养体时期 vegetative stage
营养体型 vegetative form
营养突变型 auxotrophic mutant
营养土块 peat cube
营养物质 nutritious substance; nutritive substance
营养性单性结实 vegetative parthenocarpy
营养性共生 nutritive symbiosis
营养学 dietetics; nutrition
营养学说 nutritional theory
营养芽 vegetative bud
营养叶(蕨类) sterile frond; sterile leaf
营养异常 anomalotrophy
营养因子 nutritional factor
营养杂种 vegetative hybrid
营养障碍 dystrophy
营养周期 vegetative cycle
营养砖 planting brick
营养状况 nutrient status
营业报表(损益表) operating statement
营业储备 operating reserves
营业额 business volume; turnover
营业公积金 operating reserves
营业利润 operating profits
营业税 sales tax; turnover tax
营业盈余 income from continuing operation
楹联 couplet on pillar
蝇类 circular-seamed flies
颖 glume
颖果 cariopside; caryopsis
颖花 glumous flower
颖壳 lepicena
颖毛 excoemum
颖片 glume; lepicena; peristachyum
颖状的 glumaceous
影相 shadow photograph
影响 impact
影响范围 incidence
影响评估 impact assessment
影响区(竞争个体的) influence zone
影响者(生态) influent
影响种 influent species
影响最小的采伐 reduced impact logging
影像 imagery
影像比值分析 image ratio analysis
影像处理 image processing
影像地图 photomap
影像模糊 blur
影像位移 blur; image displacement

影像增强 enhancement image; image enhancement
影像重叠 image overlap
影子 schatten; silhouette
影子定价 shadow pricing
影子价格 shadow price
瘿蜂 gall wasp
瘿螨 eriophyid mites; gall mite
瘿蚊 gall gnat; gall midge
映绘 tracing
映射(图像处理) mapping
硬币 hard cash
硬材 hardwood
硬材阔叶树 hardwood tree
硬垫群系 hard-cushion formation
硬吊河缏 rigid boom; stiff boom
硬吊排 fixed raft
硬度 hardness
硬度测定仪 hardness tester
硬度级 durometer level
硬度计 scale of hardness; sclerometer
硬度试验机 hardness tester
硬腐(初期腐朽) hard rot
硬化 sclerosis; hardening
硬化薄壁组织 sclerotic parenchyma
硬化表面 hardened surface
硬化层压木 densified laminated wood
硬化处理 hardening procedure
硬化剂 accelerator; hardener; hardening agent; stiffener; stiffening agent
硬化期 setting-up period
硬化侵填体 sclerotic tylosis
硬化时间 hardening time
硬化松香 hardened rosin
硬化温度 setting temperature
硬化细胞 sclereid; sclerotic cell
硬化纤维 sclereid fibre; sclerotic fibre
硬化组织 sclerotic tissue
硬件(计算机) hardware
硬浆 hard stock
硬浆蒸煮 hard cook
硬壳的 hardshell
硬壳结构 monocoque; stressed shell construction
硬壳种子 impervious seed
硬阔叶材 hard deciduous woods
硬阔叶树 hardwood tree
硬阔叶树材 hard deciduous woods
硬粒 hard grain
硬粒种子 hard seed
硬粒种子树种 hard seeded tree species
硬木家具 hardwood furniture
硬木结构 heavy-timber construction
硬木楔 glut
硬磐(土壤等) hardpan; orterde; pan
硬皮 hard bark; incrustate
硬韧皮部 hard bast
硬砂岩 graywacke; greywacke
硬伤 mechanical damage
硬实 hard seed
硬树皮 hard bark
硬树脂 animi; hard resin; solid resin
硬水 earth water; hard water
硬松类 hard pine
硬纤维 indurated fibre; stiff fibre
硬叶(植物) sclerophyll
硬叶草本群落 duriherbosa
硬叶常绿林 sclerophyllous forest
硬叶的 sclerophyllous
硬叶灌木 sclerophyllous shrub; true shrub
硬叶灌木群落 durifruticeta
硬叶剑麻纤维 sisal hemp
硬叶林 durisilvae; hard leaved forest
硬叶木本群落 durilignosa
硬叶乔木群落 durisilvae
硬叶树 sclerophyllous tree
硬叶小灌丛 sclerophyll scrub
硬叶植物 sclerophyllous plant; sclerophyte
硬枝插穗 hardwood cutting; straight cutting
硬脂萜(芳香油浸膏中之固形物) stearoptene
硬质合金 carbide alloy
硬质合金刀具 inserted tool
硬质合金锯链 carbide tipped chain
硬质碎料 hardcore
硬质纤维板 hard fibre board; hardboard
硬质橡胶 ebonite
硬质心土层松土器 pan-breaker
硬种皮 hard seed coat; sclerotesta
拥挤 crowding

拥挤效果 crowding effect
拥挤效应 effect of crowding
拥有水上选材设备的公司 boom company
壅高水位 back-up water level
壅水 backwater
壅水坝 barrage

Y

永冻 permafrost
永冻层 ever frozen layer; permafrost
永冻地区 permafrost
永冻面 permafrost table
永冻生态系统 permafrost ecosystems
永久保存林 permanent reserve forest
永久变形 permanent deformation; permanent set
永久标准地 growth plot; permanent growth plot; permanent sample area; permanent sample plot
永久磁铁 permanent magnet
永久凋萎含水量 permanent wilting dose
永久凋萎含水率 permanent wilting percent
永久冻土面 permafrost table
永久放牧地 permanent pasture
永久国有林 permanent state forest
永久积雪 firn
永久景观特征 permanent landscape attribute
永久苗圃 fixed nursery; permanent nursery
永久培养 permanent culture
永久栖息地 permanent habitat
永久切片 permanent slide
永久群落 permanent community
永久伸长 permanent elongation
永久生境 permanent habitat
永久塑性变形 permanent plastic deformation
永久萎蔫 irreversible wilting; permanent wilting
永久萎蔫百分率 permanent wilting percentage
永久萎蔫点 permanent wilting point
永久线 permanent way
永久样地 growth plot; permanent growth plot; permanent plot; permanent sample area; permanent sample plot
永久样方 permanent quadrat
永久样方法 permanent quadrat method
永久样区 continuous forest inventory with permanent samples
永久样区森林调查 continuous forest inventory
永久预备林 permanent reserve forest
永久杂种 permanent hybrid
永久组织 permanent tissue
永续 sustainablility
永续基准 sustainability criteria
永续经营 sustained-yield forest management; sustained-yield management
永续经营单位 sustained yield management unit
永续经营规整(调整) sustained yield regulation
永续经营区 sustained yield management unit
永续经营调整 sustained yield regulation
永续连年作业 sustained annual yield working
永续林 continuous cover forest; continuous forest
永续林经营 continuous cover forest management system
永续林思想 idea of continuous forest
永续林业 sustainable forest management
永续林作业 continuous forest method
永续森林经营 sustainable forest management
永续生产量 sustained yield
永续生产平均流量 sustained-yield-even-flow
永续生产作业 sustained-yield working
永续收获林 sustained-yield forest
永续收获评价 sustained-yield appraisal
永续性 sustainablility
永续准则 sustainability criteria
永续作业 intermittent sustained working; sustained-yield forest management; system of sustained working
蛹 pupa
蛹壳 puparium

蛹龄 pupal instar
蛹室 pupal cell
佣金 brokerage charge; commission
用材 sawlog
用材高度 timber height
用材林 saw-timber forest; saw-timber stand; timber forest; timber stand; timber wood
用材林地 timber key area; timber-land
用材林地估价 valuation of timberland
用材林面积 commercial acreage
用材树 timber tree
用材树种 commercial tree species
用材中央干围 mid-timber girth
用材中央直径 mid-timber diameter
用户 ultimate consumer; user
用户费 user charge
用户特征 user characteristic
用火许可证 burning permit
用火指标 burning index
用火指标尺 burning index meter
用碱量 alkali charge
用具 instrument
优等材 prime; select
优等的 prime
优等嵌装板 prime panel
优等原木 prime log
优点商品 merit goods
优葛缕酮 eucarvone
优化分配 optimum distribution
优化设计 optimum design
优惠价 concessional rate
优惠利率 prime rate of interest
优惠税率 preferential tariff
优惠条件 concessional terms
优惠条款 preferential clause
优级松节油 light oil; super turpentine
优良采种林分 elite seed stand
优良的 elite
优良林分 elite stand; superior plantation; superior stand
优良母树林 elite seed stand
优良种子 selected seed
优生的 eugenic
优生学 eugenics
优生学的 eugenic
优势(数量程度等) dominance; odds; prevailing
优势(杂交) potence
优势比率 potence ratio
优势的 dominant
优势顶极 prevail climax
优势度 cover degree; cover(age); degree of dominance; dominance
优势法则 law of dominance
优势个体 despot
优势木 dominant; dominant crop; dominant stem; dominant tree; dominating stem; prevailing stem; thriving tree
优势木高度 dominant height; top height
优势木平均高 average top height
优势木疏伐 thinning in dominant
优势年龄级 dominant year class
优势树木层 dominant tree layer
优势树种 dominant tree species
优势序位 dominance hierarchy
优势种 dominant; dominant species
优势主林木 dominant crop trees
优树 select tree; selected tree; superior tree
优树采集区 clonal archive
优树种子 selected tree seed
优先雇用会员企业 preferential shop
优先权 preemptive right; priority
优先燃烧 preferential combustion
优先顺序 priority
优先吸收 preferential absorption
优先制图系统 precedence charting system
优形学 euphenics
优型树 accent plant
优型学(遗传) euphenics
优选法 optimization
优选开工率 preferred operating rate
优质板 high-quality board
优质边材 bright sap(wood)
优质材 select
优质出口材 select export quality
优质大原木 big blue butt
优质单板 high-grade veneer
优质锯材 quality saw timber
优质林 high-grading forest

优质磨木浆 supergroundwood
优质木 primary timber
优质饲料 heavy feeder
优质原木 pay pole
优紫罗兰酮 euionone
尤德西林分经营法 Judeich's method of management; Judeich's working method
尤氏法 Urich's method
由体细胞发育的 somatogenic
由下向上锯截 undercutting
油斑 grease spot; oil spot; oil streak; scum mark
油泵膜片 fuel pump diaphragm
油槽 oil receptacle
油槽车 tank car
油茶 tea-oil tree; oil camellia
油茶皂苷 sasanquasaponin
油刀宽度 joint width
油道 oil canal; oil duct
油的聚合程度 oil length
油点 oil spot
油动绞盘机 motor winch unit
油分 cut fraction
油封的 oil-tight
油柑 oil orange
油柑栲胶 emblic extract
油膏剂 salve
油罐标尺 tank scale
油灰 clearcole; oil putty; putty; wood filling; wood putty
油回火 oil tempering
油基清漆 oil varnish
油基水乳化漆 oil bound water paint
油迹检测 oil test
油加热器 oil heater
油脚 oil foot
油浸 oil immersion
油浸系统(显微镜物镜) immersion system
油锯 gasoline powered chain saw
油锯插木齿 sticker
油料加入器 oil filler
油料林 oil-yielding trees
油毛毡纸容器 asphalt-coated felt paper container
油门定位栓(油锯) throttle latch
油门控制扳机 throttle control trigger
油门锁(油锯) throttle lock
油墨 printing ink
油囊 oil cyst
油漆 oil paint; paint
油漆标号 paint marking
油漆刀 spatula
油漆工 japanner; lacquerer
油漆起皮 skinning
油漆涂刷 brush painting
油漆未干 wet paint
油漆纤维板 painted fibreboard
油枪 gun
油热处理 oil tempering
油热处理硬质纤维板 oil tempered hardboard
油溶剂 oil solvent
油溶性防腐剂 oil-borne preservative
油溶性香精 oily perfume
油石 abrasive stick
油水处理 oil-water treatment
油水分离器 oil-water separator
油填充剂 oil extender
油萜(芳香油中液体部分) eleoptene
油涂料 oily paint
油污染 oil pollution
油雾润滑 mist lubrication
油细胞 oil cell
油箱 reservoir
油性腻子 oil putty
油性漆 oily paint
油性清漆 oil varnish
油性填料 oil filler
油压泵 oil-pressure pump
油压机 oil-hydraulic press
油烟 lamp black
油盐复合法 oil-salt combination
油眼 gum pocket; pitch pocket; resin cyst; resin pocket
油渣 oil foot
油着色(木材涂饰) oil-staining
油脂斑点 grease spot
油脂枪 grease gun
油脂状的 sebaceous
油质防腐剂 oil type preservative
油质坚果 oil nut
油中煮沸法(防腐) boiling-in-oil method
油注入 oil impregnation
疣 verruca (*plur.* -ae)
疣瘤状 clavate

疣状的 knobby
游标(尺) vernier
游标卡尺 caliper rule; sliding caliper; vernier caliper
游车 idler
游动孢子 swarm spore; zoosperm; zoospore
游动孢子囊 zoosporangium
游动精子 antherozoid; zoosperm
游动配子 zoogamete
游动位置 float(ing) position
游动压尺(镟切) floating pressure bar
游浆 slurry
游客 user
游客安全 user safety
游客教育 visitor education
游客来源 visitor origin
游客日 user day; visitor-day
游客容量 user capacity
游客特征 user characteristic
游客月分布 monthly distribution of visitation
游客中心 visitor centre
游客周分布 weekly distribution of visitation
游览列车 excursion train
游览区 sightseeing resort
游廊 porch; veranda(h)
游乐场 amusement ground; pleasure ground
游乐需求规划 planning recreational needs
游离度 freeness
游离度测定仪 freeness tester
游离度分析器 freeness analyzer
游离分子 free molecule
游离基聚合反应 radical polymerization
游离基因 episome
游离水 free water
游离水分 free moisture
游离松香 free rosin
游离松香胶 free rosin size
游离氧化物 free oxide
游离有机酸 free organic acid
游离状打浆 fast beating; free beating
游离状浆料 free beaten stock; free shellac
游离状木浆 free stock
游牧 nomadism
游牧民 nomad
游憩 recreation
游憩成本-利益分析 recreation cost-benefit
游憩程序 recreation program
游憩道路 recreation road
游憩地 recreation site; recreational forest
游憩地点 recreation area
游憩发展 recreation development
游憩方案 recreation program
游憩费用 recreation fee
游憩公园 recreation park
游憩供应 recreational accomodation
游憩规划 recreation planning
游憩行为 recreational behavior
游憩活动 recreation activity
游憩计划 recreation program
游憩价值 recreational value
游憩经费 recreation expenditure
游憩利润 recreational benefits
游憩领导权 recreation leadership
游憩旅行 recreational travel
游憩评价 valuation of recreation
游憩区 recreation area
游憩区的租借权 recreation area concession
游憩区分配定额 rationing of recreation areas
游憩区门票 recreation fee
游憩设施 recreation facility
游憩室 recreation room
游憩收益 recreational benefits
游憩态度 recreation-attitude
游憩团领导 recreation party leader
游憩性质 recreational property
游憩需求 recreation demand
游憩研究 recreation research
游憩招待设备 recreational accomodation
游憩政策 recreation policy
游憩支出 recreation expenditure
游憩装备 recreational equipment
游憩状态 recreation-attitude
游憩资源 recreational resource
游憩作用 recreational behavior
游人教育 visitor education
游人来源 visitor origin
游人每周分布 weekly distribution of visitation
游人日 visitor-day

游人容量 visitor-carrying capacity
游人中心 visitor centre
游丝 hairline
游隼 falcon
游艇 pleasure-boat; yacht
游戏理论决策 strategy in game theory
游泳的 natatorial
游泳生物 necton
游育土 gley
游资 floating capital; hot money
有杯状窝的 caliculate
有边的 marginate
有边无肩的(子弹) rimmed and shoulderless
有边有肩的(子弹) rimmed and shouldered
有柄的 pedicellate
有柄叶 stalked leaf
有侧护木的木轨道 lateral pole truck road
有撑档的框式结构门 framed and braced door
有齿的凿子 jagger
有翅工蚁 pterergate
有翅蚜 alatae
有翅种子 winged seed
有抽屉的写字桌 escritoire
有刺灌丛 thorn-bush
有刺毛的 setose
有错误的 faulty
有单节的 uninodal
有倒锯齿的 retroserrate
有抵抗力的 resistant
有垫矮凳 ottoman
有垫椅 matted chair
有毒的 toxic
有刚毛的 setose
有根水生植物 hydrophyta radicantia
有拱顶的走道 arcade
有沟的 plicate
有固定车立柱的承载梁 staked bunk
有故障的 faulty
有关森林事业 going forest concern
有害的 detrimental
有害废物 hazardous waste
有害基因 detrimental
有害排放物 emission
有害生物 pest
有害生物危害 damage from pests
有害野生动物 noxious wildlife
有害因子 detrimental
有害杂草 injurious weed
有核的 nucleated
有花瓣的 petalous
有花格平顶的 lacunar
有花植物 anthophyta; flowering plant
有喙的 rostrate
有机氨肥 ammoniate
有机防火剂 organic fire retardant
有机肥料 organic fertilizer; organic manure
有机腐朽 organic decay
有机合成杀虫剂 synthetic organic insecticide
有机矿物油杀虫剂 organic mineral insecticide
有机粒形物 organic particulate matter
有机磷杀虫剂 organophosphorus insecticide
有机氯杀虫剂 chlorinated insecticide; organochlorine insecticide
有机凝胶 organogel
有机气溶胶 organic aerosol
有机溶剂木素 organosol lignin
有机溶剂型防腐剂 organic-solvent type preservative
有机生物量 organic biomass
有机酸 organic acid
有机碎屑 organic debris; organic detritus
有机碎屑食物链 organic detritus food chain
有机体 biont; organism
有机体的 organismal
有机体概念 organismic concept
有机土 histosol
有机污染 organic pollution
有机质 organic matter
有机质层(土壤剖面A层含枯落物层) organic layer; organic horizon
有机质土 organic soil
有机阻燃剂 organic fire retardant
有甲板的驳船 flat-decked barge; flat-topped barge
有肩(子弹) shouldered
有节的 bossy

有节乳汁管 articulate lactiferous tube; laficiferous vessel
有菌盖的 pileate
有糠秕的 scurfy
有孔材 pored wood; porous wood
有孔胶纸带 perforated tape
有孔区 porous region
有理公式 rational formula
有力的 dynamic
有林地 timber stand; wooded area
有林地区 forested area
有鳞片的 paleace; paleaceous; scaly
有鳞脐的 umbonate
有霉斑的 pecky
有膜片的 chaffy
有胚乳种子 albuminose (albuminous) seed; endospermic seed
有胚植物 embryophyte
有皮鳞茎 tunicated bulb
有偏误差 biased error
有腔隙的 lacunose
有缺陷的 faulty
有色玻璃 potmetal glass
有生命力的种子 viable seed
有丝分裂 caryokinesis; karyokinesis; karyokinetic division; meiotic mitosis; mitosis; mitotic division
有丝分裂交换 mitotic crossing over
有丝分裂重组 mitotic recombination
有蹄类动物 ungulate
有细锯齿的 serrulate
有限单元法 finite element method
有限公司 company of limited liability
有限花序 definite inflorescence; determinate inflorescence
有限年金 terminating annuities
有限群体 finite population
有限染色体 limited chromosome
有限人口 finite population
有限元法 finite element method
有限自然增长率 finite rate of natural increase
有限总体 finite population; limited population
有效半径 effective radius
有效捕获区 effective trapping area
有效材积 effective volume
有效成分 active component; active constituent; active ingredient
有效持水力 available water-holding capacity
有效氮 available nitrogen
有效地压 active earth pressure
有效点估计式 efficient estimator
有效发芽能力 effective germinative capacity
有效辐射 effective radiation
有效功率 active power
有效估计量 efficient estimator
有效含水率 effective moisture content
有效积温 effective accumulative temperature
有效积温定律 law of total effective temperature
有效剂量 effective dose
有效碱 available alkali; effective alkali
有效可燃物能量 available fuel energy
有效孔径 effective aperture
有效孔隙度 active porosity
有效宽度 effective width
有效立木度 effective stocking
有效利率 effective (interest) rate
有效亮度温度 effective brightness temperature
有效量 effective dose
有效磷 available phosphorate
有效氯 available chlorine
有效面积 effective area
有效年龄(树木释压后) effective age
有效燃料 available fuel
有效深度 effective depth
有效时间 pot life
有效时间总额 total (amount) of effective time
有效实生苗 effective seedling
有效寿命 active life
有效数字 significant digit; significant figure
有效水分 available water
有效投入资本 available investment capital
有效土壤深度 effective soil depth
有效温度 active temperature; effective temperature

Y

有效温度带 zone of effective temperature
有效温区 effective temperature range
有效物质 active substance
有效吸收 active absorption
有效吸收剂 fertile absorber
有效系数 coefficient of efficiency
有效性 availability
有效需求 effective demand
有效养分 available nutrient
有效因子 effective factor
有效营养期 active nutritive period
有效应力 actual stress
有效载荷 payload
有效蒸散 actual evapotranspiration; effective evapotranspiration
有效中量 effective dose$_{50}$ (ED$_{50}$)
有效中心 active centre
有效装载量 operating capacity; operating load
有形资本 real capital
有形资产 tangible assets
有形资产价值 tangible value
有性雌蚜 gamic female
有性的 sexual
有性繁殖 generative propagation; reproduction; sexual propagation; sexual reproduction
有性过程 sexual process
有性后代 sexual progeny
有性阶段 perfect state
有性生殖 amphigenesis; generative propagation; reproduction; sexual reproduction
有性生殖细胞 sexual reproductive cell
有性生殖周期 sexual cycle
有性世代 gamobium; sexual generation
有性杂交 sexual hybridization
有性杂种 sexual hybrid
有须鸢尾(花类) bearded iris
有序无序转变 order-disorder transition
有焰燃烧阶段 flaming stage
有氧呼吸 aerobic respiration
有氧糖酵解 aerobic glycolysis
有叶的 lobed
有意识选择 conscious selection
有翼的 alate
有颖的 glumaceous
有硬节的 warted
有疣瘤的 warted
有辕滑橇 loose-tongued sloop
有证票据 documentary bill
有种子的 seed-bearing
有皱纹的 channeled
有贮木池的制材厂 wet mill
有蛀孔的 pecky
莠去净(阿特拉津除草剂) atrazine
右缠(茎) dextrorse
右捻 right hand lay; Z-lay
右偏(狩) cast off
右切齿(锯) right hand cutter
右绕 right hand lay; Z-lay
右式带锯机 right hand bandmill
右向反捻索 right lay rope
右旋的(攀缘茎等) dextral; dextro-rotatory; dextrorse
右旋扭曲 right-handed twist
右旋糖 dextrose
右旋糖酐 dextran(e); dextrase
幼材期 young timber stage
幼虫 larva
幼虫坑道 larval gallery
幼虫龄 larval instar
幼虫适应程度 larval adaptation
幼虫外膜 outer indusium
幼根 radicle
幼激素 juvenile hormone
幼林 sapling crop; thicket; young crop
幼林病害 immature forest disease
幼林除草松土机 young stand cultivator
幼林抚育 tending (after young plantation); tending of thickets; tending of young growth
幼林检查 checking (of young plantation)
幼林疏伐机 small wood harvester
幼龄材 juvenile wood
幼龄林 large sapling forest; stand of sapling; young forest; young growth; young stand
幼龄林抚育 tending of cultures
幼苗 youngling

幼苗出齐 established field; established stand
幼苗出土 emergence; field emergence; seedling emergence
幼苗活力 seedling vigour
幼苗靠接 seedling inarching
幼苗库 seedling bank
幼苗评价 seedling evaluation
幼年材 immature wood; pith wood
幼年的 juvenile
幼年阶段 juvenile phase
幼年期 brephic; juvenile period; juvenile stage
幼年生长 juvenile growth
幼年土 amorphic soil
幼年效应 juvenility effect
幼年性 juvenility
幼年因素 juvenility factor
幼年状态 juvenile form
幼树 sapling; juvenile tree; staddle; young growth; young tree; youngling
幼树阶段 sapling stage
幼树林 staddle
幼树生长 sapling growth
幼树试验(防腐) sapling test
幼态 juvenile form
幼态成熟 neoteny
幼态成熟的 neoteinic
幼态持续 neoteny
幼态持续的 neoteinic
幼体单性生殖 paedogenetic parthenogenesis
幼体发育 paedomorphosis
幼体两性生殖 paedogenesis bisexual
幼体配合 paedogamy
幼体生殖 paedogenesis
幼体死亡率 infant death rate; infant mortality (rate)
幼心皮 ovellum
幼芽 chit; germ; plumule
幼芽分生组织 plumular meristem
幼芽嫁接 sprout grafting
幼叶 young leaf
幼叶卷叠式 foliation; praefoliation; prefoliation
幼植物 plantling
柚木林 teak forest
诱变 induction
诱变基因 mutator gene
诱变剂 mutagen
诱变作用 mutagenic action
诱捕 trap
诱捕路线 trap line
诱捕网 trap baiting net
诱虫标志 vexillary
诱虫剂 insect attractant
诱导 derivation; induction
诱导漂子 glance boom; glancing boom; rudder boom; sheer boom; side boom; swing boom
诱导设施 sheer
诱饵 decoy
诱饵作物 bait crop
诱发变异 modification; modificational variation
诱发基因 modificator gene
诱发抗性 induced resistance
诱发突变 induced mutation
诱发物 evocator
诱发性休眠 induced dormancy
诱发性周期 induced periodicity
诱光灯 light trap
诱剂 lure
诱裂发生(染色体) Clastogenesis
诱裂剂 clastogen
诱杀 trapping
诱杀板 trap-billet
诱杀沟 insect-trench; trap-ditch; trap-trench
诱杀穴 trap-hole
诱树 trap log; trap tree
诱因 predisposition
诱致投资 induced investment
鼬鼠 weasel
迂回渠 round-about channel
淤地 warp
淤地坝 silt storage dam for farmland building; warp land dam
淤灌 basin irrigation
淤积 siltation
淤积谷 filled valley
淤积土 alluvial soil; warp soil
淤积物 alluvium
淤泥 mud; warp
淤塞 siltation
淤水 water logging

Y

淤渣 sludge
余额 balance
余甘栲胶 emblic extract
余甘子 oil orange
余火(余烬) ember
余量 allowance
余留木素 residual lignin
余汽 excess steam
余隙角 clearance angle
余渣 residue
余震 aftershock
鱼肝油烯 squalene
鱼骨法 herring-bone method
鱼骨式 fishbone system
鱼骨式采脂法 herring-bone method
鱼骨式伐倒法 fishbone system
鱼骨形伐倒 herringbone felling
鱼骨形伐倒法 herring bone system
鱼骨形切法 full-herringbone cut
鱼胶 fish gelatin; fish glue; isinglass
鱼椒粉(餐饮) isinglass
鱼类学 ichthyology
鱼类养殖 fish culture; fish farming
鱼鳞板 shiplap; shutter
鱼鳞坑 fish scale pit
鱼苗 fry
鱼群 school; shoal
鱼藤根 derris root
鱼藤属 jewel vine Derris
鱼藤酮(杀虫剂) rotenone
鱼网浮标 fish-net float
鱼尾板 fish plate; splice bar
鱼尾接 fished joint
鱼眼镜头 fish-eye lens
鱼秧 fry
娱乐公园 amusement park
娱乐区 resort area
娱乐室 recreation room
娱乐中心 amusement center
渔船 fishing boats
渔港 fishing harbor (port)
隅撑 knee brac(ing)
逾量压力 excess pressure
逾期费 demurrage
榆枯萎病 Dutch elm disease
榆木型管孔排列 ulmiform pore arrangement
虞美人 shirley poppy; corn poppy
与萼片互生的 alternisepalous
与花瓣互生的 alternipetalous
与集材线路平行伐木 falling to the lead
与距离无关的树木生长模型 distance-independent individual tree model
与距离有关的树木生长模型 distance-dependent individual tree model
与面积有关的 area-related
羽干 rhachis
羽根 calamus
羽管 pteryla
羽冠 crest
羽化(昆虫) emergence; eclosion
羽化孔 exit hole; mergence hole
羽化期 emergence period
羽毛 feather
羽片互生的 alternipinnate
羽区 pteryla
羽纹板 feather board
羽域 pterylae
羽状 pinniform
羽状的 pinnate; pinnulate
羽状多出复叶 pinnately decompound leaf
羽状复叶 pinnate compound leaf; pinnately compound leaf
羽状网脉 pinnately reticulately veined
羽状花纹 feather figure; swirl crotch figure
羽状接合 feather joint
羽状脉 pinnate vein
羽状脉的 penninerved
羽状浅裂的 pinnately lobed
羽状刃缘(锯或斧的) feather edge
羽状叶 pinnate leaf
雨冰 glaze
雨层云 nimbostratus
雨带 rain-band
雨滴 raindrop
雨滴光谱计 raindrop spectrograph
雨滴侵蚀 raindrop erosion
雨豆树 raintree; monkey pod
雨季 rainy season
雨季河 blind creek
雨夹雪 rain and snow mixed; sleet
雨量表 ombrometer

雨量计 hyetograph; hyetometer; hyetometrograph; rain gauge; udometer
雨量计防护罩 raingauge shield
雨量器 pluviometer; rain gauge
雨量区 hyetal region
雨量筒 pluviometer; rain measuring glass
雨量-温度图 rainfall-temperature diagram; rainfall-temperature graph
雨量系数 hyetal coefficient
雨量站 precipitation station
雨林 rain forest; rainforest
雨林灌木群落 hiemifruticeta
雨林木本群落 hiemilignosa
雨绿的 hiemal
雨绿灌木群落 hiemefruticeta
雨绿林 hiemisilvae; rain-green forest
雨绿木本群落 hiemelignosa
雨绿乔木群落 hiemisilvae
雨绿植物 rain green plant
雨期 pluvial; rain spell
雨区移地 storm transposition
雨日 rain day
雨水(暴雨水) storm water
雨水(广义) rain-water; pluviofluvial denudation
雨水截流 interception of rainfall
雨水流失 storm water run off
雨凇 glaze; glazed frost; silver thaw
雨养沼泽 ombrogenic bog
雨影 rain shadow
雨云 nimbus
语言 common business oriented language
玉桂 Chinese cinnamon
玉兰 white magnolia; white yulan
玉米棒子 corn cob
玉米黄质 cryptoxanthin
玉米素核苷 zeatin riboside
玉米穗 corn ear
玉米芯 corn cob
郁闭 close canopy; crown-contact
郁闭的 closed
郁闭度 canopy density; crown closure; crown density; degree of closeness; degree of closing; degree of closure; gloomy degree; percent crown cover; shade density
郁闭度标尺 crown density scale
郁闭冠层 closed canopy
郁闭林 complete stand
郁闭林分 close stand; fully stocked stand
郁闭群丛 closed association
郁闭群落 closed community
郁闭群系 closed formation
郁闭植被 closed vegetation
郁金香园 tulip garden
育成品种 breed cultivar
育成种 improved variety
育雏 parental care
育林 stand tending
育林采伐 silvical felling; silvicultural cutting
育林成熟 silvicultural maturity
育林费 expenses for silviculture
育林基金 silviculture fund
育林技术 silvicultural technique
育林目标 silvicultural goal; silvicultural objective
育林实务 silvicultural practice
育林收获 silvicultural yield
育林系统 forest system
育林学 silviculture
育林学家 silviculturist
育林原理 silvicultural principle
育林者 forest-grower
育林作业法 silvicultural method
育林作业最适投资 silvicultural operations optimum investment
育苗 raising seedling
育苗机械 seedling machinery
育苗浅箱 plant tray; seedling tray
育苗区 nursery garden
育苗台 bench
育苗箱 bench; transplant box
育培品种 breed cultivar
育幼室 brood chamber
育种 breeding
育种标本园 breeding arboretum
育种材料 breeding material
育种策略 breeding strategy
育种程序 breeding programme
育种方案 breeding strategy
育种幅度 breeding range

Y

育种规模 breeding scale
育种过程 breeding process
育种目标种 breeding objective
育种圃 breeding field
育种区 breeding zone
育种体系 breeding system
育种系统 system of breeding
育种小区 breeding plot
育种效率 breeding efficiency
育种者 breeder
育种值 breeding value
育种周期 breeding rotation
预报 prediction
预备伐 advance-cutting; preparation cutting; preparation felling; preparatory cutting; preparatory felling; preparatory stage; preparatory thinning; provisional cutting; provisional felling
预备工作 preliminary work
预备基金 reserve-fund
预备林 reserve; reserve cell; reserved forest; timber reserves
预备期 preparatory period
预备调查 pilot inventory; pilot survey; preliminary reconnaissance inventory
预测 forecast; prediction; prognosis
预测价格 forecast price
预测器 presumascope
预抽真空泵 forepump
预处理 preconditioning; pretreatment
预定采伐砍口标记 felling notch
预定日期(开始或结束) target date
预定售材伐区 budgeted chance for sale
预定位控制机构 preselector mechanism
预定压力 predetermined pressure
预防的 prophylactic
预防法 preventative measure; preventive measure
预防腐朽 preventable decay
预防剂 prophylactic
预防矫正法 prevention and remedy
预防密度 security density
预防性购买 precautionary purchase
预付 advance; advance payment; payment in advance
预付贷款 payment in advance
预付房租 key money
预付费用 advanced charge
预付款项 advance payment
预干室 predrier; predryer
预干燥 predrying
预干燥处理 preseasoning treatment
预干燥器 predryer
预购股票价格 exercise price
预估比例概率抽样 probability proportional to prediction sampling; sampling with probability proportional to prediction (3P sampling)
预固化 precure; precuring; prehardening
预计(一般的) projection (generic)
预计采伐量 future cut
预计立木价格增长 future increase in stumpage price
预计立木蓄积 future growing stock
预计使用期限 life expectancy
预加工 preframing
预加料 preload
预加应力 prestressing
预加载荷 preload
预架索道 pre-rigging
预紧力 preliminary tension
预浸 pre-soaking
预扣赋税 withholding tax
预捆 prechoking
预拉伸 prestressing
预冷处理 pre-chilling
预冷却 precooling
预期利润 anticipated profit
预期年限 expected life
预期培养木 potential crop tree
预期收益 expected income
预期寿命 expected life; life expectancy
预期调查 anticipation survey
预期需求 anticipated demand
预热 preheat(ing); preliminary heating; warming-up
预热法 preheating process
预热烘缸 predryer

预热磨木浆 thermomechanical pulp
预热器 preheater
预鞣 pretanning
预若虫 pronympha
预适应 preadaptation
预适应的 preadapted
预收货款 sales received in advance
预收收益 unearned revenue
预水解 prehydrolysis
预算草案 budget proposal
预算赤字 budget deficit; financial deficit
预算亏损 budget deficit
预算期 budget period
预算外的 extra-budgetary
预算限额 budget ceiling
预算盈余 budget surplus
预涂层 precoat
预涂装饰板 prefinished panel
预先安装 pre-setting
预先打捆 pre-setting
预先堆集 prepile
预先加料 preloading
预先调节 preconditioning
预向动作 intention movement
预选机构 preselector mechanism
预选装置 preselector
预压 form press; precompressure; prepressing
预压机 form press; prepress
预压木 precompressed wood
预压榨辊 pony press
预应力 initial stress
预硬化 prehardening
预蛹 prepupa
预制 prefabrication
预制干燥窑 prefabricated kiln
预制构件 prefabricated part
预制门 prefabricated door
预制装配式房屋 prefabricated house
预装 preloading
预装垫木 false bunk
预装工 preloader
预装设备 preloader
预装用垫木 false bunk
预钻孔 predrilling
域内流行病 endemic
欲求行为 appetitive behaviour
阈值 threshold; threshold value
御用工会 company union
愈疮木 guaiacum-wood
愈疮木二烯 guaiadiene
愈疮木粉 guaiacin
愈疮木基型木素 guaiacyl-type lignin
愈疮木烯 guaiene
愈疮木油 guaiac wood oil
愈疮树脂 guaiac; guajac; guaiac resin; guaiacum
愈合 coalesce; concrescent; healing; occlusion
愈合材 callus wood
愈合的 coalescent
愈合节 healed over knot; occluded knot; overgrown knot
愈合伤疤 scar cicatrice
愈伤 healed scar; healed wound
愈伤斑 callus fleck
愈伤组织 callus
愈伤组织形成 callus formation
愈伤组织形成期 callusing period
愈着 concrescent
薁(yù) azulene
薁类化合物 azulenoid
鸢尾树脂 orris resinoid
鸢尾酮 irone
鸢尾油 orris oil
鸳鸯 mandarin duck; mandarin teal
元分析 meta-analysis
元件 detail; unit
元数据 meta-data
元素 element
芫荽油 coriander oil
园 garden
园丁 gardener
园景公园 landscape park
园景树 specimen tree
园棱 eased arris; eased edge
园篱 garden fence
园林 park and garden
园林爱好 landscape preference
园林草坪 garden lawn
园林草图 garden sketch map
园林陈设 garden furniture
园林传统风格 traditional garden style
园林大道 boulevard
园林底视图 garden bottom view

园林底图 garden base map
园林地形 garden topography
园林地形改造 topographic treatment of garden
园林雕塑 garden sculpture
园林雕像 garden statuary
园林风格 garden style
园林更新 renovation in gardening
园林工程 landscape engineering
园林构图 composition of garden
园林管理 landscape management
园林灌木林 garden shrubbery
园林规划 landscape planning; park and garden planning
园林规划设计学 landscape architecture
园林化 forest gardenization
园林化郊区 garden suburb
园林画面 garden tableau
园林环境 garden circumstance
园林建设 park and garden construction
园林建筑 garden architecture; garden structure
园林郊区 garden suburb
园林空间 garden room
园林理水 water treatment of garden
园林露台 garden terrace
园林绿化 landscaping
园林美学 art of landscape garden(ing); esthetics of garden
园林铺地 garden pavement
园林入口 garden entrance
园林色彩构图 colour composition of garden
园林色彩配列 garden colour scheme
园林色彩调和 garden colour harmony
园林色彩效果 garden colour effect
园林色彩学 garden chromatics
园林设计 garden design; park and garden design
园林设计评选 landscape design judging
园林设计师 landscape architect
园林实施方案 garden lay-out
园林手用小喷雾器 garden automizer
园林树木 landscape tree and shrub; landscape woody plant
园林四季色彩 seasonal colours in gardening
园林坛庙 altar and monastery garden
园林拖拉机 grove tractor; orchard tractor
园林小品 garden ornament
园林小气候 garden microclimate
园林形式 garden style
园林学 landscape architecture
园林养护 park maintenance
园林艺术 art of landscape garden(ing); garden art
园林栅栏 garden fence
园林植物 garden plant; landscape plant
园林植物育种 breeding landscape plant
园林中心 garden centre
园林种植设计 garden planting design
园林周围情况 garden circumstance
园林装饰物 garden ornament
园林总平面图 garden general layout
园林总体规划 garden overall plan
园林座位 garden seat
园门 garden entrance
园墙 garden wall
园亭 arbour; garden house
园屋 garden house
园艺 gardening
园艺工作者 gardener
园艺学 horticulture; landscape gardening
园艺学家 horticulturist
园艺专家 gardener
园座 garden seat
垣篱式整枝 espalier training
原表皮层 protoderm
原卟啉 protoporphyrin
原材料 feed stock; raw material
原产地 provenance; provenience
原地处理 in-service treatment
原地截根 wrenching
原地土壤 autochthonous soil; sedentary soil
原地种 autochthonous species
原点 origin
原顶极群落 proclimax

原定模式标本 type by original designation
原儿茶鞣质 protocatechuic tannin
原发性休眠 primary dormancy
原分生组织 primordial meristem; promeristem
原果胶 protopectin
原果胶酶 protopectinase
原核生物 procaryote; prokaryote
原花粉 propollen
原花青素(定) procyanidin
原花色苷元 proauthocyanidin
原花色素 proanthocyanidin; proauthocyanidin
原基 primordium (*plur.* -dia)
原基因 protogene
原寄生物 primary parasite
原价与运费 cost and freight
原浆 protoplasm; virgin pulp; virgin stock
原菌丝 promycelium (*plur.* -lia)
原理 key card
原理图 schematic drawing
原粒体 elementary body
原料 feed stock; feedstock
原料补给 feedstock
原料指数 material index
原卵 ovule
原卵的 ovular
原木(端面不平) accordion
原木(广义) log
原木搬钩 log-handling peavy
原木板英尺 log-scale board foot
原木板英尺材积表 log rule
原木变色 log stain
原木标准材积表 standard rule
原木剥皮 log barking
原木材积 log volume
原木材积表 log rule; log scale; log volume table
原木材积记录 log tally
原木采伐 log-length logging; short-length logging
V形原木槽道 stringer road
原木测链(66ft或21.1168m) log chain
原木叉铲 log fork
原木插垛(赶羊流送) log jam
原木承枕 log bunk
原木池 log pond
原木尺 log scale
原木出材率 run-of-the-log
原木传送装置 log haul(-up)
原木串(原木大小头相连) string logs; trail
原木打捆 preloading
原木大头段 butt block
原木登记簿 log book
原木等级 log class; log grade
原木等级测定 log grade evaluation
原木垫 pad
原木吊车 log hoist
原木定位 log positioning
原木定心 log centring
原木定心机 block centering device
原木堆 log dump
原木垛 log pile
原木翻滚捆扎法 rolling bind
原木翻转 log-turning
原木翻转下锯 sawed around
原木反转下锯法 around-the-dog method
原木方向调转器 log direction changer
原木分等 grading of logs; grading of stems
原木分级 log grading
原木分级规格 log grading rule
原木分类(选材) log sorting; logs diversity
原木分选机 log-length sorting machine
原木供料装置 log delivery device
原木归堆机 log dozer
原木归楞 log handling
原木海田梁(运材车上) log bunk
原木号印 log mark; log stamp
原木横向吊装 gun a log
原木横向装车 cannon a log
原木滑道 log slide; log chute
原木滑橇 log boat
原木火灾 log fire
原木集材 log-length logging; short wood logging; short-length logging
原木几何形状测定 log evaluation for geometry
原木夹钳 log clamp

原木检尺 log scale; log scaling; measurement of felled timber; measurement of log
原木检尺手册 scaling handbook
原木截断规格 log break down rule
原木进料机构 log feeding mechanism
原木径级检测 calipering of log
原木锯架 log saw carriage
原木锯解 primary breakdown
原木卡钩 logging scissors
原木捆 log bundle; run of logs
原木蓝变 log blue
原木楞 log pile
原木楞场 log yard
原木楞堆 log deck
原木楞台 log deck; log dump
原木贸易 log trade
原木爬犁 log boat
原木盘状截头 lily pad
原木跑车 log carriage; log saw carriage; saw-carriage
原木劈开器 log splitter
原木平行下锯 through-and-through sawing
原木平行下锯法 sawing alive
原木评等 log grading
原木起吊装置 timber jack
原木起重器 log jack; timber jack
原木牵引机械 log hauler
原木清扫锯 barking saw; log-cleaner saw; rock saw
原木去皮 log inside bark
原木扫描器 log scanner
原木上木机 log charger
原木升降跑车 log raising and lowering carriage
原木市场 log market
原木输送 log-hauling
原木输送道 logway
原木输送机 log hoist
原木输送链 log chain
原木输送装置 hoist; jack ladder; jack works; log delivery device; log ladder; logway
原木四分下锯 quarter sawing
原木四分下锯的 quarter-sawed; quarter-sawn
原木四面锯切 slabbing
原木台账 log tally
原木提升机 log elevator
原木提升装置 log lift
原木调头(装车时) log-turning; change ends
原木调转索 roll
原木推河 log-rolling
原木未完全锯断 Irish coupling
原木芯 heart of log; log core
原木形状 shape of log
原木削度 log taper
原木削片制方机 chipping headrig
原木印号 log brand
原木预装 log transfer
原木运输槽道 log trough
原木着地面(拖曳时) ride; ride side
原木蒸煮 log cooking; log steaming
原木整理机 log dozer
原木整形 log rough rounding
原木制材量 log scale recovery; run-of-the-log
原木制方 slabbing
原木制家具 rustic furniture
原木制椅 rural chair
原木贮存台 storing table for log
原木抓钩 log grapple
原木转装 log transfer
原木转装台 log transfer deck
原木装车机 log grapple
原木装卸机 log handler
原木装载机 logger
原木自动储存送料装置 log magazine
原木纵横堆垛法 log cabin piling
原木纵滑(横滚装车时受阻) cannon a log
原木纵向装车 mountain style
原胚 proembryo
原群 parent stock
原染色体 original chromosome
原色 primary colour
原色的 bright
原生(始)林 primitive forest
原生的 indigenous
原生动物 protozoa
原生动物糖 paramylum
原生林 original forest; original growth; primary forest; primeval forest; primordial forest

原生木质部 protoxylem
原生木质素 protolignin
原生泥炭 autochthonous peat
原生群落 primary community
原生韧皮部 protophloem
原生生物 protist
原生水 connate water; juvenile water
原生体 cytoblast
原生土 autochthonous soil; primary soil; sedentary soil
原生纹孔 primordial pit
原生小体(病毒) elementary body
原生岩 parent rock
原生演替 primary succession
原生质 bioplasm; plasma; protoplasm
原生质融合 plasmogamy
原生质素 bioplasmin
原生质体 protoplast
原始薄壁组织 initial parenchyma
原始标准器 primary standard
原始材积毛生长 gross growth of initial volume
原始材料 parent material
原始成本 original cost
原始的 primary; seminal
原始腐殖质层 embryonic humus-layer
原始记录 initial entry
原始价值 initial value
原始景观 original landscape; wild landscape
原始阔叶树材 primitive hardwood
原始林 first growth; old growth; primary forest; primeval forest; savage forest; tall uncut; virgin forest; virgin growth; virgin stand; virgin timber; virgin wood-land; wild wood
原始密度 crude density
原始凭证 source document
原始期 initial stage
原始区 virgin area
原始群体 initial population
原始射线 initial ray
原始数据 raw data
原始土壤 embryonic soil
原始细胞 archaeocyte; initial cell
原始细胞壁 original cell wall
原始型 archetype; primitive type
原始叶 primordial leaf
原始雨林 primary rain forest
原始择伐 primitive selection felling
原始植被 original vegetation
原始植被区 primeval area; virgin area
原始种 initial species; original species
原始种群 initial population
原始资本 seed capital
原始资料 base-line
原始组型 archetype
原水 raw water
原酸 raw acid
原套 tunic
原体 elementary body
原条 log; bole; bole wood; long wood; pole with top; stem timber; stem length; tree length; whole log; whole stem
原条剥皮机 whole log barker
原条材积表 tree-length volume table
原条集材 long log system; stem-length skidding; tree-length logging; whole-stem logging
原条集运 long logging; long-length logging
原条捆 tree-length bundle
原条削片机 whole log chipper
原条作业 full-length logging
原脱植基叶绿素 protochlorophyllide
原位密度 density in situ
原位杂交 hybridization in situ
原细纤维 elementary fibril
原纤化作用 fibrillation
原纤晶体 fibrillar crystal
原纤取向 fibrillar orientation
原纤丝 elementary fibril; protofibril
原纤维 elementary fibril; fibril; fibrilla (*plur.* -llae); protofibre
原纤维定向角 fibril angle
原线圈 primary
原形成层 procambium (*plur.* -bia)
原形成层束 procambia strand
原型 primer type
原芽 gemmule
原野 wilderness (area); wold

原野保留区 wilderness; wilderness (area)
原野地游憩 wild land recreation
原野风景 wild landscape
原野公园 wild park
原野游客调查 wilderness visitor survey
原野游憩 wilderness recreation
原野园 wild garden
原油 crude; crude oil; crude petroleum
原予吸收光谱测定法 atomic absorption spectrometry
原育种子 breeder's seed
原原种(植物) breeder's stock
原植体 thallus
原纸 base paper; body paper; body stock; nurseryman; paper backing; raw paper
原质小体 elementary body
原种 elite; pedigree seed; stock; stock seed
原种的 elite
原种林(选种) elite stand
原种培养 stock culture
原种品系 foundation stock
原种圃 elite plot; nursery of elite; original seed farm; stock plot; stock seed farm
原种区 foundation block
原种植物 foundation seed
原株 etlite tree
原状土柱 intact soil cores
原子光谱 atomic spectrum
原子化 atomization
原子吸收光谱测定 atomic absorption spectrometry
原子吸收光谱学 atomic absorption spectroscopy
原子线 atom line
原紫胶 stick lac
原总体 parent population
原足 proleg
圆 circle
圆柏油 sabina oil; savin(e) oil
圆背椅 round back chair
圆材 round stuff; round wood; unhewn timber
圆材堆 camp run
圆齿 crena; crenel
圆齿状的 crenate
圆锉 fullround file
圆顶 dome
圆顶床 dome bed
圆度 circularity
圆二色性谱 circular dichroism spectrum
圆杆材(胸径5～10.5英寸) pole-timber; pole-wood
圆股(钢丝绳) normal strand
圆弧刨 capping
圆件砂光机 round stock sander
圆角柜 round corner cabinet
圆角接合 round corner joint
圆节 round knot
圆锯 buzz saw; circular saw; plate saw; rotary saw
圆锯裁边机 edger; round saw edger
圆锯锉锯机 circular saw sharpener
圆锯的 rotary(-cut)
圆锯伐木归堆机 circular saw feller buncher
圆锯机(具有小型跑车) knee bolter
圆锯片 circular saw blade
圆锯削片机 circular saw chipper
圆锯制材厂 circular mill
圆锯制材车间 circular mill
圆锯主锯机 circular saw headrig
圆棱加工 round process
圆犁刀 rolling coulter
圆砾岩 pudding stone
圆裂的 lobate; lobed
圆裂片 lobe
圆满度 circularity
圆面积 circular area
圆面积表 basal-area table; table of basal areas; table of sectional areas
圆木 log; round log; round timber
圆木端头 nose
圆木分级 log sorting
圆木轨道 pole tram-road
圆木料 spar
圆木匣(薄木制品) round chip box
圆排锯 circular gang-saw (*or* gang-mill); gang rip saw; rotary gang
圆盘 disc; disk
圆盘刀 cutting disc
圆盘方程 disc equation

圆盘粉碎机 disk mill
圆盘锯 disc saw
圆盘开沟器 rolling coulter
圆盘犁 disc machinery; disk plough
圆盘-立轴组合砂光机 disc and spindle sander
圆盘耙 disk harrow
圆盘耙耙地 disking
圆盘式割刀 circular disk spliter
圆盘式鼓风机 disc fans blower
圆盘松土耕种机 disk scarifier
圆盘削片机 disc type chipper
圆刨 circular plane
圆偏振光 circularly polarized light
圆锹 circular spade
圆丘 hummock; mamelon
圆筛 circular screen
圆石 cobble
圆桶 drum
圆筒 cylinder
圆筒剥皮机 barker drum; barking drum; rotary barker
圆筒法 cylinder method
圆筒摩擦式原木剥皮机 drum friction log barker
圆筒木箱试验机 tumbling drum
圆筒筛 drum screen; perforated cylinder
圆筒式枪筒 cylinder bore
圆筒形堆积法 heading
圆头 snap; snipe; snout
圆头的 obtuse
圆头工(将原木前端下侧削圆或削成斜面) log fixer
圆头木螺钉 roundhead woodscrew
圆头式(弹头) blunt-nosed
圆网单缸造纸机 cylinder Yankee machine
圆网式成型机 cylinder (sieve) machine
圆形背椅 loop back
圆形刀头 round cutterhead
圆形的 round
圆形浮雕 boss
圆形浮雕装饰板 medallion
圆形管胞 rounded tracheid
圆形广场 circus
圆形架(涡轮式干燥机主件) ring disk
圆形具缘纹孔 circular bordered pit
圆形靠背椅 hoop back chair
圆形料仓 circular bin
圆形气球 spherical balloon
圆形双折射 circular birefringence
圆形纹孔 circular pit
圆形旋转筛 circular and gyratory screen
圆形样地 circular plot
圆形样地面积 circular sample area
圆形样区法 sampling using a range of fixed radius plots
圆形凿齿链锯 chipper chain-saw
圆柚油 grapefruit oil
圆缘刨 astragal plane
圆凿 quick gouge
圆铡磨锯机 circular saw sharpener
圆直面接合 circular mitre
圆周速度 rim speed
圆周铣削 peripheral milling
圆柱材 roundwood
圆柱塞规 plug gauge
圆柱体 cylinder
圆柱形刀头 cylindrical cutterhead; round molding head
圆柱轴 cylindrical axis
圆柱钻 cylinder bit
圆锥花托 conopodium
圆锥花序 panicle
圆锥花序的 paniculiform
圆锥体 cone; ordinary cone
圆锥形旋切 cone-cutting
圆锥形整枝 cone-shaped training; pyramid training
圆锥形支腿 taper leg
圆锥状整形 cone-shaped training; pyramid training
圆桌 compass table
缘界效应 marginal effect
缘裂 marginal slit
缘毛(植) tricholoma; marginal seta
缘丝 periphysis
缘缨 fringe
源程序(计算机) source program
源汇关系 source-sink relation
源泉 fountain
源头(河流) headwater
源于生物的 biogenic

Y

远岸带 infralittoral zone; sublittoral zone
远程飘射 fight shooting
远程数据站 remote data station
远程信息处理 teleprocessing
远地点 apogee
远东春夏脑炎 Russian spring-summer encephalitis
远红外线 far infra(-)red (ray)
远红外线干燥法 far infra(-)red ray drying
远基细胞 distal cell
远交者 outbreeder
远景 distant view; long-range perspective
远期合同 froward contract
远期价 forward price
远亲交配 allogamy
远日点 aphelion
远系繁殖 outbreeding
远系繁殖材料 outbreeder
远洋带 pelagic zone
远洋的 pelagic
远洋浮游生物 oceanic plankton
远洋渔业 pelagic fishery
远缘嫁接 distant grafting
远缘染色体 alien chromosome
远缘杂交 distant crossing; distant hybridization; wide crosses
远缘杂种 distant hybrid
远轴傍管薄壁组织 abaxial paratracheal parenchyma
远轴薄壁组织 abaxial parenchyma
远轴种子 distal seed
远足 hike
约丹定律 Jordan's rule
约定价值 agreed value
约翰逊诱捕器 Johnson trap
约束条件 constraint
月洞门 moon gate
月钩纹 moonshine crotch
月桂 bay; Grecian laurel; true bay
月桂碱 laureline
月桂油 laurel leaf oil; laurel oil; bay oil
月积温 monthly cumulative temperature
月季花 everflowering rose; China rose
月季园 rosary
月亮的 lunar; lunate; luniform
月牙板(输送机用) traverser-link
月运节律 lunar rhythm
月运生殖周期 lunar reproductive cycle
月运现象 lunar phenomenon
月运周期 lunar cycle; lunar periodism
月运周期性 lunar periodicity
阅读椅 reading chair
悦目色 eye-rest colour
跃变 jump
跃变层 metalimnion
跃进 transition
跃迁 jump
跃障犁 stump jumper; stump-jump plough
越顶木 overtopping tree
越冬 overyear; hibernation; over wintering; overwintering; wintering
越冬孢子 resting-spore
越冬保护 winter protection
越冬场 wintering ground
越冬场所 hibernaculum
越冬洄游 wintering migration
越冬性 winter hardiness
越冬性好的 winter hardy
越冬一年生植物 hibernal annual plant; winter annual plant
越界火 breakaway; breakover
越橘蜡 ursolene
越年果 biennial fruit
越亲分离 transgression; transgressive segregation
越亲遗传 transgressive inheritance
越亲育种 transgression breeding
越夏 oversummering
越障能力(车辆) obstacle (climbing) capability
晕 halo
晕渲(地形图) plastic shading
云层分析 nephanalysis
云储存 cloud storage
云带 cloud belt
云高辐射计 cloud altitude radiometer
云计算 cloud computing
云量 amount of clouds

云母 mica; isinglass
云母黏土 micaceous clay
云母片岩 micaschist
云墙 wave-topped wall
云杉 dragon spruce; Chinese spruce
云杉栲胶 spruce extract
云杉型纹孔 plceoid pit; piciform pit
云杉型纹孔对 piciform pit pair
云杉叶锈病 spruce needle rust
云杉油 spruce oil
云实鞣花素 brevilagen
云实素 brevifolin
云雾林 cloud forest; mist forest
云雾林带 cloud forest belt; cloud forest zone
云消(日消云) burn-off
匀浆 homogenate
匀浆机 refiner
匀浆器 homogenizer
匀料机 refiner
匀染剂 level(l)ing agent
芸实荚 Divi-divi
芸香 herb of grace (grue)
芸香苷 barosmin; osyrit(r)in
芸香油 rue oil
耘锄 hoe
允许采伐材积 possibility by volume
允许采伐量 allowable cut; allowable cut amount; possibility
允许采伐面积 possibility by area
允许放牧证 leaseholder; permit holder
允许负荷量 carrying capacity
允许概率 probability of acceptance
允许利用度 allowable use
允许烧毁面积 allowable burn area
允许摄入量 acceptable intake
允许收获量 yield prescription
允许土壤流失量 soil loss tolerance
陨石 aerolite; aerolith; meteorite
陨星 meteor
孕花 flower initiation
运材 timber hauling; bringing timber; bringing wood; haul; log transportation; log-hauling; wood transport; log removal
运材便道 rack way
运材驳船 logging barge
运材岔线 spur road
运材车 hauler; log hauler; lorry; timber carrier
运材车防护棚 logging guard package
运材车驾驶员 log hauler
运材船 log ship
运材道 logging road; lumber road
运材道岔线 lateral log truck road
运材干线 truck line
运材轨道 logging-track
运材轨道车 bolster tram
运材机车 locie; logging locomotive
运材机械 log hauling machinery
运材距离 haul; hauling distance; transport distance
运材卡车 line truck; log truck; timber lorry
运材联橇 logging sled
运材列车 log truck with trailer; tandem truck
运材爬犁 log sledge; logging sleigh
运材平车 bolster tram
运材汽车 hauling truck; logging truck
运材手推车 logging dolly
运材双爬犁 logging sled
运材水道 log canal
运材四轮车 timber wagon
运材索道 transporting cableway
运材铁路 logging railroad (railway)
运材拖车 logging trailer; timber carriage
运材线 hauling route
运材小车 logging larry
运材小道 extraction lane
运材雪橇 log sledge; logging sleigh
运材支线 minor road
运材终点 terminal
运筹学 operation(al) research
运出量 removal
运动 locomotion
运动公园 play-fieid park
运动火 moving fire
运动黏度 kinematic viscosity
运短材汽车 short logger
运费 cost of transport
运费率 freight rate
运行机构 running gear
运行时间 working time
运行式索道 running skyline system

Z

运货滑盘 skid pan cargo carrier
运积土 transported soil
运价 freight rate
运生泥炭 allochthonous peat
运生土 allochthonous soil
运输便道 tote road; toting road
运输干燥材 shipping-dry lumber
运输干燥度 shipping-dry
运输机 conveyer; conveyor
运输机垫板 transport plate
运输设备 hauling equipment
运输索赔 transportation claim
运输作用 translocation
运算部件 arithmetic unit
运算速率 operating rate
运用资产 working assets
运载成本 carloading cost
运载工具 carrier
运载索具爬犁 rigging sled
运转 run; running; working
运转负荷 operating load
运转时间 operation time
韵律 rhythm
熨平作用(单板干燥) flatiron-effect
扎(zā) bundle
扎钩板 dog board
扎排 constructing raft; raft lashing
扎排木材上的砍口(放排梁) crease
杂斑(病毒病症状) mottle
杂倍体 dysploid
杂草 chomophyte; weed
杂草丛生 rank weed
杂草丛生的 weedy
杂草防除 weed control
杂草生态学 weed ecology
杂草植物 ruderal plant
杂费 incidental charge; incidentals; miscellaneous expense
杂酚煤焦油溶液 creosote coal tar solution
杂酚油(杀虫剂) coal tar; creosote; creosote oil
杂酚油防腐处理材 creosote wood
杂酚油加压法 creosote process
杂酚油乳剂 creosote emulsion
杂酚油石油液 creosote-petroleum solution
杂灌压碎机 tree crusher
杂合的 heterozygous
杂合体 heterozygote
杂合系统 heterological system
杂合性 heterozygosity
杂合性刺激作用(关于杂种优势) stimulation of heterozygosis
杂合子 heterozygote
杂交 allomixis; cross; crossing; hybridization; intercross; mistus; mixtus
杂交不结实的 cross-infertile
杂交不结实性 cross-infertility
杂交不亲和性 cross incompatibility
杂交不实的 cross-unfruitful
杂交不育 cross sterility
杂交不育的 cross sterile
杂交不孕的 cross-infertile
杂交不孕性 cross-infertility
杂交茶香月季 hybrid tea rose
杂交差型 taxocline
杂交第一代 first filial generation
杂交繁育 cross breeding; crossbreed; crossing
杂交果 allocarpy
杂交后代 filial generation
杂交后代退行 filial regression
杂交结实 cross-fertile
杂交结实性 cross-fertility
杂交界限群 commiscuum; comparium
杂交可实的 cross-fruitful
杂交可实性 cross-fruitfulness
杂交可孕 cross-fertile
杂交可孕性 cross-fertility
杂交类型 crossing form
杂交苗 cross seedling; hybrid seedling
杂交配合 xenogamy
杂交品系 cross-bred strain
杂交亲和性 cross compatibility
杂交受精 cross fertilization
杂交体 crossbreed
杂交育种 cross breeding
杂交育种法 breeding by crossing; breeding by hybridization
杂交育种工作者 hybridologist
杂交长春月季 hybrid perpetual rose
杂交种 cross-bred strain
杂交种子园 cross seed orchard
杂孔材 miscellaneous-porous wood

杂类草草甸 forb meadow
杂类草干草原 forb steppe
杂乱堆积的 jack-strawing
杂乱堆积木 jackknife
杂乱交配 promiscuity
杂乱物燃烧 debris burning
杂络物 mixed (ligand) complex
杂木 miscellaneous tree; weed tree
杂木林 coppice; copse; holt; scrub; spinny; thicket; wooded area
杂拼性 inhomogeneity
杂砂岩 graywacke; greywacke
杂食动物 omnivore; omnivore (*plur.* -ra); omnivorous animal
杂食性 omnivority; omnivory; pantophagy
杂食性的 heterophagous; omnivor(o)us
杂项开支 miscellaneous expense
杂项资产 miscellaneous asset
杂性的 polygamous
杂性花 polygamous flower
杂志 magazine
杂质 foreign matters; foreign particles; impurity
杂种 bastard; crossbreed; hybrid; metis; mistus; mixtus; mongrel; rogue
杂种第二代 F_2
杂种第一代 F_1
杂种繁茂 luxuriance
杂种核蛋白体 hybrid ribosome
杂种核糖体 hybrid ribosome
杂种后代 hybrid progeny
杂种圃 hybrid nursery; nursery of hybrid
杂种群集 hybrid swarm
杂种弱势 pauperization
杂种生境 hybrid habitat
杂种世代 hybrid generation
杂种衰退 breakthrough
杂种旺势 luxuriance
杂种无活性 hybrid inviability
杂种细胞 hybrid cell
杂种香水月季 hybrid tea rose
杂种性 hybridity
杂种优势 heterosis; heterotic vigor; hybrid vigour
杂种长春月季 hybrid perpetual rose
杂种指数 hybrid index
杂种制种 hybrid seed production
杂种状态 hybridism
灾变论 catastrophe theory
灾害保险 casualty insurance
灾后生长预测 growth prognosis
灾难性因素 catastrophic
灾难性因子 catastrophic factor
灾损估算 damage appraisal
栽培 culture; planting; raising
栽培草地 cultivated meadow
栽培措施 cultural practice
栽培品系 cultivar
栽培品种 cultivar; cultivated variety
栽培投资 cultural investment
栽培植物 cultivated plant; hemerophyte
栽培种 cultigen
栽培作业 cultural treatment
栽针保阔 recruiting conifers while retaining broad-leaved trees; reinforcing conifers while retaining broad-leaved trees
栽植 planting; planting out
栽植标志桩 planting-peg
栽植锄 planting shoe; planting spud; planting-hatchet
栽植带 planting strip
栽植地 planting site
栽植地区 planting area
栽植点 planting point
栽植斧 planting-hatchet
栽植镐 planting mattock
栽植更新 reproduction by planting
栽植工具 planting tool
栽植行 planting line; planting row
栽植机 plant setter
栽植计划 afforestation plan; establishment plan
栽植间隔 planting space; spacing of plants
栽植距离 planting distance; planting interval; planting space; spacing of plants
栽植面积 planted area; planting area
栽植枪 planting gun
栽植绳 planting-cord

栽植始 planting hammer
栽植楔 planting-wedge
栽植穴 plant(ing) hole
栽植造林 plantation
栽植杖 planting stick
栽植锥 planting borer; planting-dagger; planting-peg
栽种 plant out
载波频率 carrier frequency
载玻片 microscope slide; slide
载畜率 stocking rate
载荷 load; loading
载荷变形曲线 load-deformation curve
载荷测定器 loadometer
载荷传感阀 load sensing valve
载荷传感器 load cell
载荷点轨迹 locus of loaded point
载荷监测系统 load-monitoring system
载荷量 load capacity
载荷曲线 load curve
载荷伸长曲线 load-elongation-curve; load-extension curve
载荷试验 loading test
载荷因素 load factor
载荷-应变曲线 load-strain curve
载荷指示仪表(汽车) load-indicating apparatus
载泥船 mud-boat
载钯木炭 palladium charcoal
载频 carrier frequency
载气 carrier
载热剂 heat transfer mudium
载热体 heating medium
载入单元 loader module
载色剂 binding agent
载色体 chromatophore
载色体的 chromogenic
载体 carrier; support(er); vector
载体胶黏剂 supported adhesive
载体涂布 support coating
载液 carrier
载运费 carloading cost
载重标准 loading gauge
载重滑车 trolley
载重滑轮 depending block
载重卷筒 main drum
载重量 carrying capacity; load (carrying) capacity
载重爬犁 draft sled (*or* sledge)
载重跑车 hoisting carriage
载重索 load rope; running line
载重线 load line
再保险 reinsurance
再成岩 deuterosomatic
再处理 reconditioning
再次层积 restratifying
再次培养 subculture
再次侵染 secondary infection
再度蒸馏 rerun
再分带锯 band resaw
再分含晶细胞 subdivided crystalliferous cell
再分锯 slab saw; slabber
再分离 reisolation
再复制 reduplication
再干燥单板 redry veneer
再耕 over tillage
再活化 reactivation
再积化石 reworked fossil; secondary fossil
再加工材 shop lumber; shop timber
再建价 replacement value
再接种 reinoculation
再锯 resawing
再锯车间 resaw mill
再锯机 resaw
再剖 resawing
再剖锯 resaw
再迁入 remigration
再侵染 reinfection
再烧除 reburn
再烧区 reburn
再生(更新等) after-growth; reactivation; regeneration; regrow; regrowth; renewal; reproducer; revivification; aftermath
再生林 secondary woody growth
再生木材 reconstituted wood
再生能力 regenerative capacity
再生炭 regenerated carbon
再生纤维 fibre recovery; regenerated fibre
再生纤维素 regenerated cellulose
再生循环 reproductive cycle
再生盐渍土 regraded saline soil

再生源 renewable sources
再生长 regrowth
再生紫胶 rebuilt lac; rubulac
再碎机 disintegrater; disintegrator; mill; mill hog; reshredder
再贴现 rediscount
再投资 reinvestment
再涂重新涂饰 recoating
再现期 return period
再引入 reintroduction
再造林义务 legal obligation to replant; replant duty
再造木材(纤维分离后再黏合) reconstituted wood
再重复 reduplication
再煮锅 reboiler
在产品 goods in process
在伐木踏板上钉卡钉 shoe a springboard
在山场 at stump
在外面 ectad
在制品 goods in process; work in process
在制品周转 turnover of goods in process
暂居微生物 transient microbe
暂时变形 temporary set
暂时表面硬化 temporary casehardening
暂时收入 transitory income
暂时萎蔫 temporary wilting
暂时休眠 temporary dormancy
錾(zàn) paring chisel
藏红(碱性) safranine
藏花素 crocin
藏花油 saffron oil
藏茴香油 caraway oil
凿 chisel
凿石锤 bush hammer
凿榫机 mortiser; mortising machine
凿榫孔 mortising
凿形齿(木材防腐刻痕刀) chisel-point tooth
凿形犁 chisel plough
凿形松土器深松土 chisel ploughing
凿岩尖锤 grab skipper
早材 early wood; spring wood
早材带 zone of early wood
早材导管分子 early wood vessel element; spring wood vessel segment
早材管胞 early wood tracheid; spring tracheid
早材率 percent of early wood; percent of spring wood; proportion of spring wood
早材区 zone of early wood
早成雏 precocial
早成的 precocious
早春的 prevernal
早春景色 prevernal aspect
早第三纪 Paleogene
早固缩染色体 megameric chromosome
早花的 precocious
早花品种 early blooming cultivar
旱裂 drought crack
早落的 caducous
早皮 soft bark
早期采伐 prior cutting
早期成岩作用 syndiagenesis
早期干燥 ineipient drying
早期固化 precuring
早期疏伐 early thinning
早期树皮 early bark
早期修剪 early pruning
早期硬化 prehardening
早期预测 early prediction
早熟 early mast; thrifty mature
早熟品种 early variety
早熟现象 precocity
早熟性 earliness
早霜 early frost; first frost; first injurious frost
枣 jujube; common jujube
枣树 Chinese date tree; common jujube; jujube
枣子香精 jujube essence
藻被生态系 algal mat ecosystem
藻花 algal bloom
藻菌 phycomycete
藻类 alga (*plur.* algae); algae
藻类学 algology; phycology
藻酸 alginic acid
藻酸纤维 alginate fibre
皂草苷 saponin(e)

Z

皂荚碱 gleditschine
皂荚素 gleditsin
皂角苷 saponin(e)
皂脚 niger
皂石果榄 saponite
造孢细胞 sporogenous cell
造材 buck; bucking; cross cutting; dissecting; dissection; log making; logging off; saw buck; sawing jack; severing cut; slashing; transverse cut; wood buck
造材场 conversion site
造材垫枕 bucking loader
造材分离机 log rejector
造材工 bucker; cross cutter; log-maker
造材归堆工 chopper
造材归堆机 bucker-buncher
造材规程 bucking rule
造材滚道 bucking chute; bucking skid
造材后干材 dissected bole
造材机 bucker
造材架 spring pole
造材检尺 laying off
造材锯 buck saw; cross-cutting saw; swede fiddle; blocking saw
造材锯垫 undercutter
造材锯口偏斜 unsquare bucking
造材联合机 tree processor
造材链锯机 bucking chain saw
造材率 bucking percent; bucking recovery; conversion rate
造材木段 bucking cut
造材台 bucking deck; deck
造材梯形垫木 bucking ladder
造材楔 hanging wedge
造材斜道 bucking chute; bucking skid
造材因素 conversion factor
造材圆锯机 bucking circular saw
造材支杆 spring pole
造材自动线 automated bucking line
造船材 shipbuilding timber; ship-building timber; ship-timber
造景 landscaping
造景建筑业 landscape architecture
造酒厂 distillery
造林(育林) afforest; cultivation; forest culture; forestation; outplanting; crop establishment; establishment (of a stand); stand establishment
造林保存率 survival rate of plantation
造林成活率 stand establishment percentage
造林地 planted land; planting site
造林地区 afforested area; planting area
造林地整理 site preparation
造林方法 method of forestation
造林费 cost of reforestation
造林工程 afforestation project
造林规划 plantation layout
造林核定资本值 capitalized value of planting-cost
造林机械 afforesration and reforestation machinery
造林计划 afforestation plan; establishment plan; plan of silvicultral operations; planting plan; planting schedule
造林记录簿 planting record
造林密度 density of plantation; planting density
造林面积 planting area
造林难以成活的(立地) refractory
造林配置 planting composition
造林区划 silvicultural regionalization
造林日程 planting schedule
造林失败地 failed area
造林调查设计 plantation survey and project
造林学 silviculture
造林学家 silviculturist
造林学特性 silvicultural characteristics
造林学原理的 silvical
造林学原论 silvics
造林援助 reforestation assistance
造林运动 afforestation drive
造林者 forest-grower
造林政策 reforestation policy
造乳细胞 lactiferous cell; laticiferous cell
造山运动 mountain building; orogenesis; orogeny

造山作用 orogeny
造树锯 cut off saw
造丸 pelleting
造型 mould-making
造型机 moulding machine
造型树木园 topiary garden
造影剂 contrast agent
造园 garden making
造园技法 technique of garden-making
造园师 landscape architect
造园学 landscape architecture; landscape gardening
造纸 paper-making
造纸材 paper wood; pulpwood
造纸材堆 rick
造纸材伐木工 short log logger
造纸材框架 pulp rack
造纸厂 paper mill
造纸短圆材 pulpwood bolt
噪度 noisiness
噪声比 noise ratio
噪声标准 noise standard
噪声测量 noise measurement
噪声分析器 sonic noise analyzer
噪声级 noise level
噪声计 acoustimeter; noise meter
噪声控制 noise control
噪声量 noisiness
噪声评价数 noise rating number
噪声特性 noisiness
噪声污染 noise pollution
噪声系数 noise factor
噪声源 noise source
噪音防护套 hearing protector
噪音计 sound meter
责任 liability
责任会计 activity accounting
责任预算 responsibility budget
择伐 selection cutting; selection felling; selective logging
择伐更新 regeneration in the selection system; regeneration under selection system
择伐经营 uneven-aged management
择伐林 all-aged forest; plenter-wood; selection forest; selection stand
择伐率 selection percent
择伐式矮林 shelterwood coppice
择伐式疏伐 plenter-thinning; selection thinning
择伐体系(欧洲) pleater system
择伐周期 circulation period; selective cutting cycle
择伐作业 felling under selection system; forest selection system; selection method; selection system; single-tree selection system
择伐作业体系 forest selection system
择伐作业天然更新 natural regeneration by selection method
择优采伐 creaming cutting; high-grading cutting
择优间伐 selective thinning
择优取向 preferred orientation
增白剂 brightener
增充剂 extending agent
增稠过程 thickening
增稠剂 stiffener; stiffening agent; thickener; thickening agent; thickening material
增稠能力 thickening power
增稠器 thickener
增大 auxesis
增大母细胞 auxocyte
增伐 increased cutting
增感染料 sensitizing dye
增光剂 brightener
增红感 hyperpanchromatic
增厚(细胞壁) thickening
增汇 enhancement of carbon sequestration; enhancement of carbon sink
增加 recruitment; upward adjustment
增加单位产量 vertical expansion
增加值 value added
增进生长 advance growth
增力滑轮 block-and-a-half
增力式缆车道 Sessom's system
增力式索道 add-forced cable system; Tyler system
增力式索道集材 Tyler yarding
增量 increment
增量成本 incremental cost
增量剂 extender
增列 write up
增黏剂 tackifier

增黏系数 coefficient of wetness
增强薄片刨花板 waferboard-plus
增强剂 amplifier; fortifier; reinfurcer; strengthening agent
增强用胶料 strength increasing sizes
增韧剂 plasticizer; toughening agent
增溶剂 solubilizer; solubilizing agent
增色剂 toner
增生 proliferation
增湿处理 humidifying treatment
增湿机 dampening machine
增湿剂 wetter
增湿器 humidifier; wetter
增湿作用 humidification
增塑 plastify
增塑剂 plasticizer; plasticizing agent; softener
增塑效应 plasticising effect
增塑增容剂 plasticizer extender
增效剂 activator; synergist
增效作用 synergism
增雪 snow addition
增压扩散泵 booster diffusion pump
增益 gain
增益系数 gain factor
增援 follow-up
增长 accretion; build-up
增长截距 growth intercept
增长量 accretion
增长量估计 estimation of increment
增长量中央木 mean growth tree
增长率 growth rate
增长模型 growth model
增长曲线 increment curve
增长系数 increment coefficient
增长者 increaser
增支成本 incremental cost
增值 accrue; added value; appreciation; increment; write up
增值成本 incremental cost
增值曲线 reproduction curve
增值税 duty on value added; value-added tax
增值资产 accrued asset
增殖 hyperplasia; multiplication; reduplication
增殖病毒 propagative virus
增殖价值 added value
增殖率 rate of multiplication
增殖培养 enrichment culture
增殖区 increase block
增殖体 vegetation
增殖系数 growth coefficient
增殖者 multiplier
憎水剂 water repeller
憎液的 lyophobic
赠卷 coupon
轧边机 edger
轧花机 embossing machine
轧毛机 linter
轧籽机 seed crusher
闸 brake
闸板 dam board; splash boards; stop log
闸板槽 stop log recess
闸柄闩 blocking device
闸带 brake strap
闸阀 gate valve
闸门 gate; hatch
闸门底梁 stop log
闸门横梁 stop beam
闸门坎 tripsill
闸门木 drop log
闸式拦河绠 gate boom
闸水流送 sluice; splash
闸水流送原木 flood log
铡草机 straw cutter
铡刀机 guillotine
栅格数据 raster data
栅栏 barrier; fence
栅栏薄壁组织 palisade parenchyma
栅栏菌(担子菌纲) palisade fungus
栅栏细胞 palisade cell
栅栏柱 fence post
栅栏组织 palisade tissue
栅网河绠 fence boom
栅形椅背 bar back
栅状防护罩 grill(e) guard
栅状围绠 fence boom
炸药筒(劈大材) gun
炸药楔 powder wedge
蚱蜢 locust
摘 pick
摘除 ablation
摘钩 chase; raft hook
摘钩工 chaser; head loader; hook chaser; unhooker

摘挂 dis-lodging
摘挂工 tong unhooker
摘果机 picker
摘心 nipping; pinching; top removal
摘芽 bud pruning; debudding; disbudding; nipping
摘要 round up
摘叶 defoliation
宅园 home garden
窄板条 slat
窄板条间隔壁板 bar and batten paneling
窄带锯 narrow band-saw
窄带滤波器 narrow band filter
窄带砂光机 scroll sander
窄带芯板 banded core
窄带状极强度疏伐 very heavy thinning by narrow strip
窄带状皆伐 clear-cutting in narrow strip
窄带状择伐法 narrow strip-selection method
窄幅条播种 narrow-row drilling
窄幅条栽植 narrow-row planting
窄弓形锯 buhl saw
窄轨森林铁道 minor forest rail road
窄轨铁路 narrow gauge railroad
窄行播种 close drilling
窄行条播 sowing in close drill; sowing in narrow drill
窄林带 belt of trees
窄林冠 narrow crowned
窄轮的 narrow ringed
窄年轮的 close-grown; narrow ringed; slow-grown
窄片型刨花板 strandboard
窄生态幅度 narrow ecological amplitude
窄条手锯 coping saw
窄细胞法 empty-cell process
窄硬叶植物 narrow sclerophyll plant
窄长木排(装置编排机的) alley
债权 claim; financial claim
债券 bond
债券利率 bond rate
债券收益 bond yield
债券溢价 bond premium
债务 liability
债务待付数 debt service requirement
债务限额 debt limit
沾染区 contaminated zone
毡毛状的 pannose
蛅蟖食害 damage by caterpillars
展开(色谱法) developing
展开的 evolute
展开剂前沿 solvent front
展开模型 splay mould
展宽器 stretcher
展示林 demonstration forest; instruction forest; training forest; university forest
展叶 foliage expanding; leaf out; leafing
占地面积 floor space
占区(动物活动范围) territory
占用 appropriation
占有者 occupant
栈单 landing account
栈道 suspension chute
栈架结构 trestlework
栈架森铁岔线 piling track
栈架式水栅 trestle-boom
栈桥 framed trestle; jetty
栈桥森林铁路 piling railroad
栈台 scaffold
栈租 storage charges
战利品 trophy
战术(防火) tactics
战术计划 tactical plan
站杆(断头树) stub
站杆(枯立木) snag
站立位置(伐木时) footing
站台 landing board
站桩 snag
绽芽 bud bursting
张钩索 grapple-opening line
张紧绷索 spring guy
张紧度(锯) strain
张紧辊 tension roll
张紧滑轮 tightening block
张紧滑轮组 heel tackle
张紧卷筒 extension drum
张紧轮 jockey roller; tightener
张紧螺旋 tension screw
张紧索 extension rope; heel line
张紧索夹板 heel strap

张紧装置 stretcher; tension device; tightener
张紧状态 tension
张力 tension
张力计 tensiometer
张力裂缝 tension crack
张力调节辊 dancing roll
张力调整 stretch rolling; tensioning
张应力 stress in tension
章鱼 octopus
樟木油 camphor wood oil
樟脑酐 comphoric anhydride
樟脑烯 comphene
樟脑油 camphor oil
樟脑原油 camphor original oil
樟树 camphorwood; camphor tree; true camphor wood
蟑螂 cockroach
涨潮 flood
涨潮流 flood current
涨价 appreciation
涨湾 embayment
掌舵人 steerman
掌玫油 pamorusa oil
掌状半裂的 palmatifid
掌状的 palmate
掌状复出的 digitate
掌状复叶 paimately compound leaf; palmately compound leaf
掌状裂叶 palm-like lobe
掌状脉 palmate vein
掌状脉的 palmately nerved; palmately veined; palminerved
掌状脉序 palmate venation
掌状浅裂的 palmately lobed; palmatilobate
掌状全裂的 palmatisect
掌状深裂的 palmatipartite
掌状网状脉的 palmately reticulately veined
掌状叶 palmate leaf
掌状圆裂的 palmately lobed; palmatilobate
丈量 chaining
杖 stick
杖测疏伐(依最小空间疏伐) stick thinning
杖尺 scale-rule; scale-stick
杖式种子取样器 sticker
帐幕 veil
帐篷 lodge
账户 account
账面存货 book inventory
账面价值 carrying value
账面盘存 book inventory
胀量 bulk
胀式离合器 expanding type clutch
胀缩交变 movement
障视线 blind lead
辗压滚动 roll
辗压机 packer wheel; tamping machine
辗压机械 compacting equipment
辗压裂纹 roller check
辗压适张度(修锯) roll tensioning
朝开暮谢花 ephemeral flower
招虫害的 insect-prone
招风 exposure to wind
招风的 wind-prone; windswept
招火灾的 fire-prone
招贴纸 poster paper
着火 catching fire; ignition; inflammation
着火点 inflammation point; kindling point; point of ignition
着火时间 ignition time
着火温度 ignition temperature; kindling temperature
爪 claw; ungula; nail
爪垫(昆虫) pad; pulvillus
爪球形支腿 talon and ball foot
爪哇木棉天牛 kapok longhorned borer
爪形撬棍(起钉用) spike bar
爪形饰条 talon moulding
爪形支脚 paw foot
沼地 fen land
沼地泥炭 fen peat
沼地黏土 fen clay
沼腐殖质 bog mor
沼气 methane
沼生植物 helophyte
沼氏硬度试验 Rockwell-test
沼泽 bog; marsh; mose; quagmire; swamp
沼泽草本群落 emersiherbosa
沼泽草炭 drag turf
沼泽草原 grass moor

沼泽的 boggy
沼泽地 bog-land; everglade; glade; marshland; swamp; swamp-land; swampy area; wet soil
沼泽地的 marshy
沼泽地集材 fording
沼泽地推土机 swamp dozer
沼泽腐殖土 marsh muck
沼泽化的 water logged
沼泽灰壤 marsh podzol
沼泽犁 bog plow; marsh plough; moor plough
沼泽林 bog forest; moor forest; swamp forest
沼泽群落 fen
沼泽群落演替 bog succession
沼泽水泉群落 hel(e)okrene
沼泽土 bog soil; moor soil; swamp soil
沼泽性 bogginess
沼泽性植被 mire vegetation
沼泽园 bog garden; marsh garden; swamp garden
沼泽植物 bog plant; marsh plant; swamp-growing plant
兆吨 megaton (Mt)
兆克 megagram (Mg)
照查法 methods of control
照灯 illuminator
照度 illuminance; illumination; illumination (intensity)
照度计 lightmeter; luxmeter
照顾行为 epimeletic behaviour
照明 illumination
照明装置 illuminator
照片扭曲 distortion on photograph
照射百分比 percentage illumination; percentage irradiance
照射剂量 expose dose
照相底片 photographic plate
照叶灌木群落 laurifruticeta
照叶林 laurel forest; laurilignosa; laurisilvae
照准仪 alidade
罩光 glazing
罩光层 finish coat(ing)
罩光清漆 coating varnish
罩光涂层 overcoating
罩壳 lag
罩面清漆 finish varnish
遮板 shake
遮蔽 shadowing
遮盖力(油漆等) hiding power
遮盖力测定计 coverimeter; hidimeter
遮盖力计(涂料) cryptometer
遮盖物 veil
遮光板 orifice
遮光布 shade cloth
遮光涂料 opaque
遮光育苗室 lathhouse; shade house
遮框抽屉 overlap drawer; overlay drawer
遮视区 blind area
遮阳窗 awning window
遮阴 shading
遮阴度 shade density
遮阴灌木 shade-giving shrub
遮阴树 shade tree; shelter tree
遮阴网 shade mesh
折棒分布 broken-stick distribution
折臂式装载机 Knuckle-boom loader
折测线 meander line
折测线角 meander corner
折茬 beard; hinge wood; holding wood; sloven; tipping edge
折尺 folding (pocket) rule
折凳 campstool
折点 break point
折叠床 press bed
折叠刀 jackknife
折叠绠漂 jackknife boomsticks
折叠角(挂车与牵引车间因倒车形成的锐角) jackknife
折叠链状理论 folded-chain theory
折叠门 folding door
折叠屏风 folding screen
折叠式家具 folding furniture; nomadic furniture
折叠式门 folding door
折叠式起重臂 jill poke sizzer boom; Knuckle-boom
折叠式躺椅 deck chair
折叠椅 folding chair
折叠桌 folding table
折叠座椅 tilting seat
折断 breakage
折断数 folding number
折放市场 call money market

折光度 diopter
折减坡度 compensating grade
折旧 amortization; depreciation
折旧费 capital consumption allowance; depreciation expense
折旧公式 formula for depreciation
折旧估价法 depreciation-appraised method
折旧基金 depreciation funds
折旧准备 depreciation reserve
折旧准备率检验 depreciation-reserve-ratio test
折扣 agio; discount; rebate
折扣率 discount rate
折扣商店 discount houses
折款 call (able) loan; call money
折流板 baffle; baffle board; baffle plate
折门 accordion door; folding door
折面桌 flap and elbow table; flap table; pembroke table; snap table
折片 bridge
折扇状的 plicative
折射 refraction
折射角 angle of refraction
折射位移 refraction displacement
折算系数 discount factor
折损 breakage
折线索道 connecting curved cableway
折腰式车架 articulated frame
折腰式集材机 articulated frame steering skidder
折腰式原木装载机 articulated log loader
折叶板撑 fly rail
折椅 flap-seat
折余值 salvage value
折衷平分法 area and volume period method; combined frame work, area and volume period method
折转板 deflector
蛰伏 dormancy; torpor
摺面桌 tip-top table
摺扇状的 plicate
辙叉 frog
辙叉工 frogger
辙岔尖 point
辙岔枕木 point sleeper
辙沟 rut
褶层(胶合板缺陷) pleat
褶皱 fold
蔗糖 sucrose
蔗糖酶 invertase; saccharase; sucrase
蔗糖蜜 cane molasses
蔗渣 cane trash
蔗渣板 bagasse board
蔗渣干燥机 bagasse dryer
蔗渣刨花板 bagasse particleboard; bagasse shaving board
蔗渣软质纤维板 cellotex
蔗渣碎料板 bagasse particleboard
蔗渣纤维 bagasse fibre
针 pin; needle
针刺式堆垛机 stick stacker
针辊 picked roller; spiked roller
针节(直径小于$\frac{1}{4}$英寸) pin knot
针晶束 raphides
针晶体 raphides
针晶细胞 raphidian cell
针孔(径$\frac{1}{100}\sim\frac{1}{4}$英寸) pinhole
针阔叶混交林 mixed broad-leaf and coniferous forest
针形 aciculiform
针形闸门 needle gate
针叶 acerose leaf; needle
针叶变色 needle discolour
针叶材 softwood
针叶材板 softwood board
针叶材板料 deal
针叶材干馏 softwood distillation
针叶材厚板 softwood plank
针叶材焦油 softwood tar
针叶材焦油沥青 softwood tar pitch
针叶材木炭 softwood charcoal
针叶材纤维 softwood fibre
针叶层 needle layer
针叶纯林 pure coniferous stand
针叶的 needle-leaved
针叶灌木林 conifruticeta
针叶灌木群落 aciculifruticeta
针叶枯死病 needle blight
针叶林 aciculisilvae; coniferous forest; conisilvae; needle-leaved forest; softwood forest
针叶林土 taiga soil
针叶木本群落 aciculignosa; conilignosa

针叶乔木群落 aciculisilvae; conisilvae
针叶树 conifer; coniferous tree; softwood tree; needle-leaved tree
针叶树材 coniferous timber; coniferous wood
针叶树害虫 pests of coniferous trees
针叶树林 needle-leaved woodland
针叶树群系 coniferous formation
针叶树叶锈病 needle rust
针叶树种 coniferous species
针叶油 needle oil
针叶植被 aciculignosa
针叶植物 coniferophyte
针状的 acicular; needle-shaped
针状晶体 crystal acicular
针状纤维 raphioid fibre
针状隐色体 cryptostrobin
侦察点(防火) detection point
侦察机 reece plane
侦察哨 lookout point
侦察摄影 reconnaissance photography
侦察卫星 reconnaissance satellite
侦察长 general scout
珍贵材 high value assortments
珍贵物种 precious species; valuable species
珍奇自然风景区 unique natural area
珍珠灰 pearl ash
珍珠陶土 nacrite
珍珠岩 pearlite
帧频率 frame frequency
真北美草原 true prairie
真比重 real specific gravity; true specific gravity
真潮间带 eulittoral zone
真担子菌 eubasidii
真底栖生物 eubenthos
真方位 true azimuth
真浮游生物 euplankton
真海洋的 holopelagic
真核生物 eucaryote; eukaryote
真核细胞 eucaryotic cell; eukaryotic cell
真核细胞生物 eucary
真湖岸带 eulittoral zone
真菌 eumycetes; fungi; fungus; fungus (*plur.* -gi)
真菌病 mycosis
真菌病的 mycotic
真菌病毒 mycovirus
真菌病害 fungus disease
真菌毒素 fungaltoxin; mycotoxins
真菌纲 eumycetes
真菌共生体 fungal symbiont
真菌寄生物 mycoparasite
真菌菌落 fungus colony
真菌类 eumycetes
真菌瘤 fungus gall
真菌泌出酶 enzyme secreted by fungus
真菌上面生长的 epigeous
真菌性褐变(木材) fung(o)us brown-stain
真菌学 mycology
真菌学家 mycologist
真菌瘿 fungus gall
真菌状的 mycoid
真空泵 suction pump; vacuum pump
真空成型机 vacuum former
真空度 vacuum tightness
真空法(木材防腐) vacuum process
真空伏辊 suction couch
真空干燥 vacuum drying; vacuum seasoning
真空干燥法 vacuum drying
真空干燥炉 vacuum drying oven
真空干燥器 vacuum drier
真空格式压机(纤维板坯脱水成型) vacuum-gridpress
真空辊 suction roll
真空计 vacuum gauge; vacuum-meter
真空计数器 vacuum counter
真空加压法(木材防腐) vacuum pressure process
真空密闭的 vacuum-tight
真空铺装机 vacuum former
真空式捡拾器 vacuum pickup
真空输送托架 suction carriage
真空网笼 suction cylinder
真空吸板堆垛机 suction stacker
真空吸板进料机 suction feeder
真空吸盘 vacuum pickup
真空吸头 suction head
真空洗涤机 vacuum washer
真空熏蒸法 vacuum fumigation
真空压榨 suction press

Z

真空压榨辊 suction press roll
真空油煮法 boiling under vacuum
真空种子收集机 vacuum seed harvester
真空周期 vacuum cycle
真空贮藏 vacuum storage
真空贮藏罐 vacuum-storage tank
真密度 real density; true density
真皮的 hypodermal
真皮下的 subepidermal
真染色质 euchromatin
真社会性 eusociality
真实生长序列 real growth series
真实样品 authentic sample
真形率(正形率) normal form quotient; true form quotient
真形数(正形数) normal form factor; true form factor; Hohenadl's form factor
真休眠 true dormancy
真穴居动物 eucaval animal
真有丝分裂 eumitosis
真杂种优势 euheterosis
真正光合速率 true photosynthetic rate
真值 true value
真中柱 eustele
真种 euspecies
砧边接 side grafting
砧木 parent stock; rootstock; stock; tree stock
砧木蒙导 stock mentor
砧穗关系 rootstock-scion relationship; stionic relation; stock-scion relationship
砧穗接合体 stion
砧穗相互作用 stock-scion interaction
砧穗影响 stionic effect
砧状云 anvil cloud
榛角斑蚜 hazel aphid; filbert aphid
诊断 diagnosis
诊断层 diagnostic horizon
枕梁 saddler's wood
枕梁(与桥长成直角横设的枕木) corbel
枕木 railroad tie; crosstie; log for rail-road tie; sleeper; tie; wood sleeper; wood tie
枕木垫板 tie plate
枕木堆楞机 tie-piling machine
枕木归垛机 tie piler
枕木轻轨道 tie tram
枕木栓塞 tie plug
枕木下垫块 shim
枕资 short log show; sleeper block
阵风 gustiness; puff
阵列 array
阵列计算机 array computer
阵雨 shower
振荡 oscillation
振荡加压法 oscillating pressure process
振荡器 mechanical shaker; oscillator
振荡筛 shaking screen
振捣机 compactor
振动板 vibrating plate
振动采种机 tree shaker
振动槽式输送机 vibrating gutter conveyor
振动锄 vibro-hoe
振动光谱 vibronic spectrum
振动回转筛 vibrating rotary screen
振动片 vibrating plate
振动器 vibrator
振动强度 oscillation volume
振动切刀 vibration cutter
振动切削 vibration cutting
振动筛 oscillating screen; vibrating screen; vibrating sieve
振动式压实机 vibrating tamper
振动试验 shake test; vibration test
振动音量 oscillation volume
振幅 amplitude
震荡培养 shake culture
震动辗滚 tamping roller
震中 epicentre
镇 township
镇压 compacting
镇压器(田间) compactor; roller; land roll(er)
争夺型竞争 contest type of competition
征收 condemnation
征税 taxation
征税基数 tax basis
征用 condemnation
蒸发 evaporation; vapourization
蒸发表 evaporimeter
蒸发当量 equivalent evaporation

蒸发计 atmidometer; atmometer; evaporimeter
蒸发量测定术 atmidometry; atmometry
蒸发潜热 vapourization heat
蒸发强度 intensity of evaporation
蒸发热 heat of vaporization
蒸发仪 evaporimeter
蒸发蒸腾 evapo-transpiration
蒸发蒸腾计 evapotranspirometer
蒸发蒸腾作用 evapo-transpiration
蒸发作用 evaporation
蒸解 cooking
蒸解度 degree of cooking
蒸馏罐甑 retort
蒸馏木醋液 distilled pyroligneous acid
蒸馏室 distillery
蒸馏水 boiled-out water
蒸馏松节油 lightwood oil
蒸馏塔 hog still
蒸馏物 running
蒸馏甑 retort
蒸气压差 vapor pressure deficit
蒸汽挡板 vapour barrier
蒸汽干燥 steam seasoning
蒸汽干燥处理 vapour drying process
蒸汽干燥法 vapour drying; vapour seasoning
蒸汽管道 live steam piping
蒸汽管路 steam line
蒸汽光泽加工 steam finish
蒸汽集材机 steam yarder; steam-skidder
蒸汽加热干燥机 steam-heated dryer
蒸汽加热间歇式干燥机 steam-platen dryer
蒸汽加热压尺 steam-heated nosebar
蒸汽夹杂水 entrained water
蒸汽绞盘机 dolbeer; steam pot
蒸汽-浸渍作业 steaming and soaking process
蒸汽密度 vapour density
蒸汽灭菌器 steam sterilizer
蒸汽喷射法 steam blow; vapour injection process
蒸汽喷射器 steam jet
蒸汽劈材机 steam splitter
蒸汽牵引车 steam tractor
蒸汽热板式干燥机 steam-platen dryer
蒸汽软化 steam softening
蒸汽透过性 vapour permeability
蒸汽弯曲 steam bending
蒸汽相 vapour phase
蒸汽消耗量 steam consumption
蒸汽压亏缺 vapour pressure deficit
蒸汽压力 steam pressure; steaming pressure; vapo(n)r pressure
蒸汽压力差 vapour pressure difference
蒸汽预热法(防腐) Collmann method
蒸球 boiler; globe boiler
蒸去轻油 skimming
蒸散作用 evapotranspiration; transpiration
蒸腾比 transpiration ratio
蒸腾表 phytometer
蒸腾计 potometer
蒸腾流 transpiration current
蒸腾速率 transpiration rate
蒸腾系数 transpiration coefficient
蒸腾效率 transpiration efficiency
蒸腾作用 transpiration
蒸压速闭装置 quicklock autoclave closure
蒸煮 cooking; pulping
蒸煮不足 soft cook
蒸煮槽 cook vat
蒸煮程度 cooking degree
蒸煮池 cook vat; steaming pit
蒸煮池温度 pit temperature
蒸煮度 degree of cooking
蒸煮工 cooker
蒸煮锅 boiler; cooker; digester
蒸煮锅衬砖 digestor lining
蒸煮器 digester
蒸煮条件 cooking condition
蒸煮液 cooking liquor
蒸煮周期 cooking cycle; digester cycle
拯救伐(卫生伐或伐除受害木) salvage cutting; salvage logging; salvage logging
整倍的 euploid
整倍体 euploid
整车称重检尺机 vehicle scale

Z

整车载量 carload freight
整齿 tooth set
整齿工具 saw fitting tool
整齿规 jointing gauge
整齿机 saw-setting machine
整齿料 fit(ting)
整齿器 swage sharper
整齿钳 setting iron
整齿铁砧 set block
整齿砧 setting anvil
整地 clearing of the felling area; land preparation; preparation of soil; site clearing; site preparation; soil preparation
整地方法 method of soil preparation
整地费 cost of preparation of ground
整地机 grade shovel
整幅单板 full size veneer; full-sheet veneer
整合 conformity
整合分析 meta-analysis
整机使用重量(含驾驶员及油水等) operating weight
整锯 doctor
整锯齿(拨料或压料) grade stabilization structure; side set
整锯工 saw filer; saw fitter
整锯工长 head filer
整锯机 jointer
整理 rearrangement
整理伐 liquidation cutting
整理择伐 improvement-selection cutting
整料(修锯) square dress; equalizing; fit(ting)
整批购进 lump sum acquisition
整平道路用地 dirt work
整平机(单板) tenderizer
整平路基 dirt work
整平压光机 breaker calender
整齐的 regular
整齐林 regular forest
整齐同龄林 even-aged uniform forest
整群抽样 cluster sampling; block sampling
整群识别 cluster identification
整饰椽 trimming rafter
整饰横梁 trimming joist
整索(将脱出的绳索拨回到卷筒上) spool off
整体插垛(伸至两岸) solid jam
整体齿锯 solid tooth saw
整体齿圆锯 solid tooth circular saw
整体处理 bulk treatment
整体计划 intergrated planning
整体聚合 bulk polymerization
整体流 bulk flow
整体论 holism
整体木材 solid wood
整体式 integral arch
整体式导板头(油锯) solid nose
整体水平 level of integration
整体修饰 monolithic finishing
整形器 dresser
整形式园林 geometric style garden
整形修剪树 clipped tree
整修台 dressing table
整张化芯板 unitized crossband
整枝工人 limber
整枝剪 dresser
整株病 systemic disease
正常材 normal wood
正常防火期 normal fire season
正常分散 normal dispersion
正常根 normal root
正常含水量 normal moisture capacity
正常降水量 normal precipitation
正常接合 legitimate copulation
正常苗 normal seedling
正常年度 normal year
正常配偶 orthogamy
正常群丛分析 normal association analysis
正常群体 normal population
正常树脂道 normal resin canal; normal resin duct
正常速度 normal speed
正常相对湿度 normal relative humidity
正常型 on-type
正常芽 normal bud
正常演替系列 prisere
正常营养的 eutrophic
正常运行损失 normal operating loss
正常轴向胞间道 normal axial intercellular canal
正常状态 normal state

正反交 reciprocal crosses
正反交杂种 reciprocal hybrids
正反馈 positive feedback
正反配板法 book matching
正反射 normal reflection
正方形芯层 square cell core
正负突变型 sign mutant
正规采伐 regular cutting; regular felling; regular logging
正规传粉 legitimate pollination
正规的 regular
正规林 regular forest
正规森林经营 regulated forest management
正规种子林 normal seed stand
正号林分 plus stand
正号母树林 plus seed stand
正号树 plus tree; select tree
正极 anode
正交法 planing
正交晶系 rhombic system
正交面 normal surface
正交切削 orthogonal cutting
正交设计 orthogonal design
正交条件 orthogonality condition
正交异性弹性材料 orthotropic elastic material
正交直径 orthogonal diameter
正截原木 well-bucked end
正密度制约因子 positive density-dependent factor
正面 face side; felt side; top surface; front face; block front
正面图 body plan
正片 positive
正片麻岩 orthogneiss
正切 well cut
正切削 orthogonal cuttmg
正曲率 positive curvature
正入射 normal incidence
正三角形栽植 equilateral triangle planting
正射投影 orthogonal projection; orthographic projection
正射投影法 orthophoto graphy
正射投影图 orthophoto map
正射像片 orthophoto
正射像片比例尺 orthophoto scope
正射校正 ortho-rectification
正视投影图 orthophoto map
正视图 elevation
正态分布 normal distribution
正态分布规律 law of normal distribution
正态分布函数 normal distribution function
正态分布曲线 normal distribution curve
正态曲线 normal curve
正碳离子 carbonium ion
正统命名 legitimate name
正位异构体 anomers
正无级变速装置 positive infinitely variable
正弦曲线联锁拼接(砂带) sine-wave interlock
正相关 positive correlation
正向地性 positive geotropism
正向光性 positive phototropism
正向牵引(从绞盘机正前方集材) straight lead
正向突变 forward mutation
正向推板(推土机) straight dozer
正向演替系列 orthosere
正向杂种优势 positive heterosis
正形率(真形率) normal form quotient; true form quotient
正形数(真形数, 霍赫纳德形数) normal form factor; true form factor; Hohenadl's form factor
正盐 normal salt
正应力 normal stress
正长石 orthoclase
正长岩 syenite
证券 security
证券纸 bond paper
证书 certificate
政策和措施 policies and measures
政府间气候变化专门委员会 Intergovernmental Panel on Climate Change
政府经营的林业 forest-working through government officer
症状 symptom
之后 meta-
之形铁(Z形) Z-iron
之字线 zigzag
之字形的 zigzag
之字形管孔排列 zigzag pore arrangement

Z

之字形木杆道 skipper road
支臂式装车 heel(ing) boom loading
支臂式装车机 single-tong loader
支撑 strut; support
支撑板 poling board; support bracket
支撑木 tree-prop; tree-support
支撑腿 leg
支撑物 buttress
支撑轴承 axial bearing
支承 bear; middle support
L形支承板(爬犁) lug support
支承点 pivot
支承辊 supporting roll
支承接盘 fifth-wheel
支承接盘式半挖车 fifth-wheel trailer
支承面积 area of bearing
支承强度 bearing strength
支承索 supporting strand
支承托架 base unit main
支承小梁 apron piece
支承压辊 counterpressure roll
支承压力 force of support
支承应力 bearing stress
支持带 supporting belt
支持根 brace root
支持辊(砂光机) back-up roll; back-roll
支持架 supporting bar
支持细胞 supporting cell
支持组织 supporting tissue
支出 pay out
支道 feeder
支地槽 embayment
支点式起重臂 heel(ing) boom
支付 disbursement
支付书 documents against payment
支杆 spud
支化 branching
支化聚合物 branched polymer
支基因抗性 major gene resistance
支架 bent; carriage; gibbet; horse; lip; portal bracing; saddle; side mark
支架式打枝机 cradle type delimber
支架式起重杆 jill poke
支棱(天平) knife edge
支链淀粉 amylopectin
支链型高分子 branched polymer
支流 affluent; branch; confluent
支路 branch road; by-pass; leg
支配-从属 diominant-submissive
支配-从属关系 dominant-submissive
支配木 dominant stem; prevailing stem
支票 check
支票簿 check book; cheque book
支票利率 cheque rate
支渠 lateral canal
支枢点 pivot point
支索 side cable
支索滑轮 sail block
支腿间饰板 skirting piece
支腿折叠桌 gateleg table
支线 feeder; leg; spur
支柱 brace; leg; picket; pillar; prop; puncheon; stay; stick; strut; stull; support; upright
支柱根 brace root; prop root; stilt root
支柱起重机 pillar-crane
支柱气根 propaerial root
支柱托梁 saddle
支座 abutment; bearing
只读存贮器 read-only storage
汁液 juice; sap
汁液接种 sap inoculation
枝 branch; ramus
枝编篮形椅 croquet chair
枝编条的 wicker
枝杈 stub
枝插 shoot cutting
枝插穗 stem cutting
枝茬(断枝残基) branch stump; snag
枝的(昆虫) ramal; rameal
枝端嫁接 shoot-apex grafting
枝腐病 branch-rot
枝干高 clean length
枝高 branch height
枝痕 branch scar; limb marking
枝迹 branch trace
枝剪 lopper
枝接 peg graft; scion grafting; stem grafting; stub grafting
枝接法 whip-grafting
枝节 branch(ed) knot; knottiness
枝卷须 branch tendril

枝孔 branch hole
枝枯病 shoot blight
枝轮 branch whorl
枝梢材(较大) faggot wood; lap; lop and top; tops; lapwood
枝梢冻死 killing back
枝梢捆 faggot
枝条 branch
枝条材 branchwood; slender-branched
枝条材积 branch volume; shoot volume
枝条材积率 branch volume percent
枝条的 branchy
枝条扩展 branch spread
枝条形数(枝材形数) branch form factor
枝条性 branchiness
枝稀度 freedom from branches
枝隙 branch gap
枝下的 subaxile
枝下高 clear length; clear trunk; clear-bole height
枝下弯的 catacladous
枝序 branch order
枝丫切碎机 slash chopper
枝丫收集机 slash collector
枝桠剥皮机 branchwood rosser
枝桠材 branchwood; brushwood; fascine wood; lopwood
枝桠材削片机 brush chipper
枝桠处理 brush disposal
枝桠打捆机 branchwood baler
枝桠度(枝桠多少的程度) branchness
枝桠堆 brush pile
枝桠捆 fasciae
枝桠捆道 fascine work road
枝桠密度 limbiness
枝桠切碎机 brash chopper; brush chopper
枝桠清理费 cost of slash disposal
枝桠收集机 branchwood collecting machine
枝桠剔出器 rejector
枝桠推集机 forest rake
枝桠削片机 branchwood chipper
枝芽接 cion budding
枝叶饲料 browse
枝轴 main branch
枝桩(原木上的) snag; swell; horn
枝状的(昆虫) ramal
知更鸟 robin
肢 pes (*plur.* pedes); periopods
肢节 podite
织物 fabric; tissue
织物图案 drapery figure; drapery pattern
栀子花净油 gardenia absolute
栀子胶 kambi
栀子香精 gardenia compound
脂(动物) tallow
脂蛋白 lipoprotein
脂点病 grease spot
脂多糖 lipopolysaccharide
脂肪(动物) adipose
脂肪(固体) solid fat
脂肪分解 steatolysis
脂肪酶 lipase
脂肪树脂 fatty resin
脂肪体 fat-body
脂肪形成 lipogenesis
脂肪油 liquid fat
脂肪种子 fatty seed
脂肪组织 adipose tissue
脂化层压材 resin-treated laminated compressed wood
脂解作用 lipolysis
脂类 lipid(e); lipoid; lipin
脂类分解 lipolysis
脂酶 lipase
脂囊 gum check; resin sac
脂泡 pitch blister
脂朊 lipoprotein
脂色素 wear-and-tear pigment
脂松节油 gum turpentine
脂松香 gum rosin
脂檀素 amyrin
脂檀油 amyris oil
脂杨树脂 tacamahac
脂质 lipin
脂质体 lipoplast; liposome
脂状粗腐殖质 greasy mor
脂族树脂 aliphatic resin
蜘蛛 spider
执行 implementation
执行笔录 execution record
执行成本 implementation cost
执行稽核 control of prescriptions

Z

执照 charter
直板门 ledged door
直播机 direct seed machine
直播造林 direct seeding; direct sowing
直插(造林) straight cutting; vertical cutting; direct cutting
直插造林 direct slip planting
直齿链锯 scratcher chain saw
直齿耙 straight-tined harrow
直挡栅 straight boom
直刀修果机 vertical blade root pruner
直动态 orthokinesis
直堆法 vertical piling
直堆烧炭(法) stand heap charing
直方图 histogram
直方图规定化 histogram specification
直方图均衡化 histogram equalization
直方图线性化 histogram linearization
直感 xenia
直感现象 xenity
直干形 cylindrical
直根 peg root; taproot
直根系 taproot system
直观检查 ocular examination
直河绠 straight boom
直喙象鼻虫 straight-snouted weevil
直火式木材干燥法 drying of wood by direct fire
直减率 lapse rate
直交层压法 cross-laminating
直交单板 cross band (veneer); cross banding; crossing (veneer)
直角测量法 rectangular surveying system
直角尺 try square
直角接 heading joint
直角棱镜 right angle prism
直角刨削手押刨床 squaring planing-jointing machine
直角拼接 butt joint
直角器(测距离) cross-staff
直角器(测坡度) swede level
直角切削 orthogonal cuttmg
直角曲折处(面或饰条) return
直接成本 direct cost; prime cost; traceable cost
直接出售 direct sale
直接防治法 direct control
直接费用 direct charge
直接分层 direct stratification
直接辐射表 pyrheliometer
直接付款的租金 explicit rent
直接复型[法] direct replica (method)
直接辊涂 direct roll coating
直接集材 direct skidding; direct yarding
直接计数 direct count
直接加热干燥机 directly heated dryer
直接剪切试验 direct shear test
直接径流 direct runoff
直接拉力 direct tension
直接劳动 direct work (*or* labor)
直接灭火 direct fire suppression
直接灭火法 direct method of suppression
直接排水口 direct outlet
直接评定法 method of direct evaluation
直接扑灭法 method of direct attack
直接汽蒸 direct steaming
直接侵染 direct infection
直接侵入 direct penetration
直接燃气干燥机 direct fired dryer
直接燃气干燥窑 direct fired gas type kiln
直接生产工人 production worker
直接因子 direct factor
直接印刷 direct painting; direct printing
直接应力 direct stress
直接运材 forwarding
直接运出 direct haulage
直接蒸汽 live steam; open steam
直接蒸煮 direct cooking
直接正像 direct positive
直接装车 hot (deck) loading
直接自花受精 direct autogamy
直径测定 calipering; diamter measurement
直径尺 diameter gauge
直径分布 DBH distribution; diameter distribution

直径级 diameter class; diameter grade
直径铗 tree compass
直径阶 diameter grade; diameter class
直径界限 diameter limit
直径进级法 diameter step-up method
直径卷尺 diameter tape
直径括约 diameter rounding
直径量测 diameter measurement
直径率 diameter quotient
直径生长 diamemter growth; diameter growth; diameter increment; growth in thickness; secondary growth
直径生长率 diameter-increment percent; percentage diameter growth
直径-树高方程 DBH-height function
直径调查 calipering
直径下限 lower diameter limit
直径限制(小于该直径的立木限制采伐) minimum diameter
直径因子 diameter factor
直径约数 diameter rounding
直径整化 diameter rounding
直锯 log gang saw
直立的 upright
直立键 axial bond
直立卷轴集材机 spool donkey
直立蕨茎 caudex
直立木闸门 bracket gate
直立射线细胞 upright ray cell
直立生长的 upright-growing
直立炭窑 alpine kiln
直立圆柱体 right cylinder
直立植物 erect plant
直链淀粉 amylose
直流 co-current
直流堰 suppressed-weir
直馏油料 virgin stock
直木纹 edge grain
直拼撑门 ledged door
直拼斜撑框构门 braced door
直嵌板 straight panel
直闪石 anthophyllite
直上花药 vertical anther
直射光 direct light; undeviated light
直升飞机吊箱(灭火) helitank
直升飞机基地 base heliport
直升飞机集材 copter logging; helicopter logging; helicopter yarding; hover-place logging
直升飞机降落场 helispot
直升飞机空降灭火员 helijumper
直升飞机喷雾机 helicopter sprayer
直升飞机站 heliport
直升机集材 helicopter logging
直升机灭火装置 helitanker
直升机灭火作业班 helitack crew
直生说 orthogenesis
直饰条 straight banding
直条锯 web saw
直纹理 straight grain
直纹木材 straight grained wood
直纹凸榫 straight tongue
直线横截锯 straight line cross-cut saw
直线望远镜的经纬仪 theodolite with linear telescope
直线性强风 line blow
直线遗传 linear inheritance
直线英尺(一定断面的木材长度) running foot; linear foot
直线元素 vertical element
直线折旧法 depreciation-straight-line method; straight line depreciation
直线纵剖锯 straight line ripping saw
直向地性 orthogeotropism
直向选择 linear selection; orthoselection
直运加工厂的采运作业 logging to mill
直中柱 eustele
Q_{10}值 Q_{10} value
A值(同位素法测定土壤养分的指标) A-value
PF值 PF-value
pH值 pH-value
职能财政 functional finance
职能预算 performance budgeting
职位描述 job description
职位评定 job evaluation
职务说明 job description
职业捕兽者 professional trapper
植保素 phytoalexin

植被 plant cover; vegetal cover; vegetation; vegetation(al) cover
植被抽象单位 nodum
植被地带 vegetation zone; vegetated terrain
植被地理区 vegetation province
植被分析 vegetation analysis; vegetational analysis
植被覆盖百分比 percentage of vegetational cover
植被覆盖的 vegetated
植被恢复 revegetation; re-vegetation
植被均匀度系数 coefficient of vegetational homogeneity
植被类型 cover type
植被连续体概念 continuum concept of vegetation
植被剖面图 vegetation profile; vegetation profile chart
植被区划法 method of classifying vegetation
植被区域 vegetation region
植被圈 circle of vegetation
植被搔碎机 reforestation scarifier
植被生物量 vegetation biomass
植被省 vegetation province
植被碳 vegetative carbon
植被图 vegetation map
植被图形(像片判读) vegetational pattern
植被型 type of vegetation; vegetation form; vegetation type
植被型组 vegetation type group; group of vegetation type
植被亚型 vegetation subtype
植被指数 normalized difference vegetation index (NDVI)
植被重建 revegetation
植病细菌学 phytobacteriology
植草沟道 vegetated channel
植成土 phytogenetic soil
植距 espacement
植距指数 spacing index
植苗更新 reproduction by planting
植苗锥 dibble
T形柄植苗锥 T-grip dibble
植前处理 preplanting treatment
植前耕耘 preplanting cultivation
植前熏蒸 pre-planting fumigation
植鞣 vegetable tanning
植入 implant
植入物 implant
植绳 planting line
植食性 phytophagy
植树 afforest; plant trees; tree planting; treeplanting
植树镐 mattock; planting hoe
植树工具 planting tool
植树活动 tree-planting drive
植树机 plant setter; planter; planting machine; tree planter; tree planting machine
植树计划 planting design
植树季节 planting season
植树间隔 planting space; spacing of plants
植树节 Arbor-Day
植树锹 planting bar
植树日 tree-planting day
植树造林 forest plantation; forest planting; planting
植树种草绿化所有土地 green all available land with trees and grasses
植物 plant; ~phyte; phyto~
植物保护 plant protection
植物标本室 herbarium
植物病害流行 epiphytotics
植物病害流行预测 forecast of epiphytotic
植物病害体系 plant pathosystem
植物病理解剖学 pathologic plant anatomy
植物病理学 pathology of plant; phytopathology
植物病理学家 phytopathologist
植物病理掌 plant pathology
植物病原菌 pbytopathogen
植物残体 dead plant part; plant debris
植物虫害病毒 phytotoxemia
植物单宁 natural tannin
植物蛋白胶 vegetable protein glue
植物地理学 botanical geography; phytogeography; plant geography; vegetation geography
植物地区(区系) region
植物定居 plant colonization

植物冻胶 vegetable jelly
植物毒素 phytotoxin
植物毒性 phytotoxicity
植物发生的 phytogenetic
植物发生学 phytogenetic botany
植物芳香油 essential oil
植物防病注射 vaccination of plant
植物分布发展史学 epiontology
植物分类学 plant taxonomy; systematic botany
植物酚类 plant phenolics
植物附生物 epiphytic periophyton
植物覆盖的基质 vegetated substrate
植物化 vegetalization
植物化石 phytolith
植物化石学 phytopaleontology
植物化学 phytochemistry
植物化学生态学 phytochemical ecology
植物活性炭 vegetable active carbon
植物畸形学 plant teratology
植物激素 phytohormone; plant hormone
植物计 phytometer
植物计量法 phytometer method
植物寄生物 plant parasite
植物检疫 plant quarantine
植物碱 plant alkaloid
植物胶 pectic material; vegetable glue; vegetable gum
植物胶黏剂 vegetable adhesive
植物解剖学 phytotomy; plant anatomy; vegetable anatomy
植物界 vegetable kingdom
植物景观 vegetative landscape
植物抗毒素 phytoalexin
植物空间 vegetation space
植物类群 phytogroup; plant group
植物篱障 vegetative barrier
植物流行病 epiphytotic disease; epiphytotics
植物流行病学 epiphytology
植物面积指数 plant area index
植物苗 plantling
植物内生的 endophytic
植物内种皮 tegmen (*plur.* tegmina)
植物黏液 plant mucilage
植物配置 planting arrangement
植物屏障 vegetative barrier
植物气候 plant climate
植物气候学 phytoclimatology
植物器官学 plant organography
植物区 kingdom
植物区气候 biochore
植物区系 flora
植物区系成分 floristic element
植物区系的 floristic
植物区系地理学 floristic plant geography
植物区系名录 cybele
植物区系区 floristic region
植物区系组成 floristic composition
植物群丛 plant association
植物群落 flora; phytocoenose; phytocoenosis; phytocoenosium; phytocommunity; plant community
植物群落发生学 genetic synecology
植物群落学 phytocoenology; phytosociology; plant sociology
植物群系 plant formation
植物群系级 class of formation
植物群系团 group of formation
植物群形 phyad
植物染料 wood dye
植物认证 certification of planting; certification of plants
植物溶介素 phytolysine
植物鞣法 vegetable tanning
植物鞣革 vegetable tanned leather
植物鞣料 vegetable tanning material
植物散布 plant dispersal
植物色素 phytochrom; plant colouring matters
植物杀虫剂 vegetable insecticide
植物杀菌素 phytoncide
植物社会学 phytosociology; plant sociology
植物生理化学 physiological chemistry of plant
植物生理学 plant physiology
植物生态地理学 ecological plant geography; phytoecological geography
植物生态解剖学 ecological plant anatomy

Z

植物生态学 phytoecology; plant ecology
植物生物化学 plant biochemistry
植物生长素 auxin
植物生长调节剂 plant-growth regulator; plant-growth substance
植物体 plant body
植物体(藻类或苔藓) frond
植物调节剂 plant regulator
植物图解 phytograph
植物脱色炭 vegetable decolourizing carbon
植物外部形态学 plant external morphology
植物外貌(群落) plant physiognomy
植物物候学 phytophenology; plant phenology
植物烯 phytene
植物系统学 phylogenetic botany
植物细胞 vegetable cell
植物细胞分裂素 phytokinin
植物细胞学 plant cytology
植物纤毛 cilium (*plur.* -ia)
植物纤维 plant fiber; plant fibre; string; vegetable fibre
植物纤维素 plant cellulose
植物线虫病 nemas
植物小气候 phytoclimate
植物小群落 clan
植物形态学 plant morphology
植物性杀虫剂 botanical insecticide
植物学 botany; phytology
植物岩 phytolith; phytophoric rock
植物演替 plant succession
植物羊皮纸 vegetable parchment
植物养料 plant-food
植物药理学 phytopharmacology
植物遗传学 phytogenetics
植物蚁巢 myrmecodomatia
植物引种 plant introduction
植物油 vegetable oil
植物育种学 plant thremmatology
植物园 arboretum; botanical garden; botanical park; breeding arboretum
植物栅栏 vegetative barrier
植物脂 vegetable fat
植物志 flora
植物质单宁 vegetable tannin
植物致病污染物 plant pathogenic pollutant
植物组成 species composition
植物组合 plant society
植物组织学 plant histology
植穴 plant hole; plant pit
植穴挖掘机 hole digger
植株分布 distribution of plants
止点(树干出现实质性变化的点) stop
止动臂 clamp arm
止动夹钳(牵引索上的) button
止动器 claw; stop
止动楔 chock block
止回阀 back-stroke valve; load check; non-return valve
止回凝气阀 non-return trap
止菌作用 fungistasis
止轮垫(防止车子在斜坡往下滑) sprag
止轮器 sandal
止推滚珠轴承 ball thrust bearing
止推轴承 axial bearing; thrust bearing
纸板 board; paper board
纸板厂 cardboard mill
纸板车间 cardboard mill
纸板处理 board liner
纸板垫 pad
纸草 papyrus
纸层析 paper chromatography
纸带 paper tape
纸电泳 paper electrophoresis
纸幅 paper web
纸浆 Jiffy pot; paper-pulp; pulp; pulp stuff
纸浆材(短于10ft) pulp log; pulplog; short wood
纸浆厂 cellulose plant
纸浆回收机 pulp reclaimer; pulp saver; saveall
纸浆浓缩机 pulp thickener
纸浆筛 pulp screen
纸浆筛滤器 pulp strainer
纸浆脱水机 pulp thickener
纸巾 paper towel
纸料捕集机 stuff catcher
纸料柜 stuff chest
纸料收回机 saver; stuff catcher
纸料箱 stuff box

纸坯 raw paper
纸皮桦 paper birch
纸色谱法 paper chromatography
纸莎草 bulrush
纸素 paper factor
纸压切机 score cutter
纸质背衬(砂带) paper backing
纸质蜂窝芯 paper honeycomb
纸质塑料贴面 paper-plastic overlay
纸质贴面 paper overlay
指标 criterion (*plur.* -ria); index
指标林分 control stand; index stand; indicator stand
指标值 indicator value; indicators
指标种 index species
指定银行 authorized bank
指接 finger joint
指接机 finger jointing machine
指令 command
指令序列 sequence
指率 indicating percent
指射 pointer; pointing
指示(示意犬猎鸟) pointing
指示板 marking board
指示棒(防火) indicator stick
指示记录计 indicating and recording gauge
指示剂 indicator
指示菌株 indicator strain
指示林 index-forest
指示培养基 indicator medium
指示平均有效压力 indicated mean effective pressure
指示器 indicator
指示犬 pointer
指示群落 community indicator
指示生物 bioindicator; biological indicator; eucoen; indicator species
指示调节器 indicator-regulator
指示温度计 indicating thermometer
指示性价格 indicative price
指示压力 indicated pressure
指示样地 indicator plot
指示植被 indicator vegetation
指示植物 eucoen; indicator plant; plant indicator
指示植物谱 indicator spectra
指示种 indicator
指数 index; index number
指数分布 exponential distribution
指数分布函数 exponential distribution function
指数函数 exponential function
指数扩散 exponential dispersal
指数律 exponential law
指数曲线 exponential curve
指数生长 exponential growth
指数生长期 exponential growth phase
指数式增长 exponential growth
指数型 exponential type
指数型生长 exponential growth
指数选育 index breeding
指数因子 exponential factor
指数增长 exponetial growth
指数增长曲线 exponential growth curve
指相化石 facies fossil
指向植物 compass plant
指形搭接 finger-lap joint
指形接合 finger joint
指形刨(带有曲线底板的榫槽刨) thumb plane
指形拼接胶合板 finger jointed plywood
指形压紧器 doffer finger
指状的 digitate
指状分裂为五裂片的 pentadactylous
指状弦茬 finger of wood
趾甲 nail
酯胶 ester gum
酯酶 esterase
志留纪 Silurian period
制表机 tabulator
制材 lumber conversion; lumbering; wood conversion
制材产量 mill run
制材产品 lumber product
制材厂 log mill; lumber mill; sawmill; stud mill; timber mill; timber sawmill
制材厂工作者 lumber mill operator
制材厂用原木板材积表 mill-study log rule
制材厂原木检尺 mill scale
制材厂原木楞场 millyard
制材成本 milling cost
制材成品 sawn product

Z

制材废料 raw waste lumber; run-of-the-mill; sawmill refuse
制材费用 sawing cost
制材工段长 mill foreman
制材工人 lumberer
制材工业 lumber industry; lumber manufacturing industry
制材工艺 sawing technology
制材工作者 lumber manufacturer; lumberman
制材机械 lumbering machinery; sawing machine
制材记数 timber tally
制材量(原木) log run
制材量不足 underrun
制材领班 mill foreman
制材率 conversion rate
制材清单 stock-cutting list
制材商 lumber manufacturer
制材生产 saw leg production
制材用材 sawlog stage
制材用原木 mill-log
制材原木(圆木) sawlog
制材允许偏差 toleration in sawing
制材作业 lumber-job
制成品 finished goods; finished product
制弛卷筒 slack puller
制弛跑车 sharkey carriage
制弛索 slack puller; slack puller line; slack pulling line
制动 locking
制动测力计 brake dynamometer
制动带 brake band; brake strap
制动杆 friction lever; gooseneck
制动滑块 brake-block
制动距离 brake distance
制动力 braking effort
制动链轮 chain-bull block
制动轮 brake sheave; brake wheel
制动木块 brow log
制动器 arrestor; arrester; blocking apparatus; brake; drag; stop device
U形铁制动器(爬犁上) down-hill clevis
制动索 brake-rope; bridle
制动闸瓦 brake-block
制动装置 arresting device; brake device; friction device
制度(工作方法) regime (*eg.* thinning regime); schedule; system
制管厂 tube mill
制火器 fire stop
制剂 formulation
制浆 pulp; pulping
制浆材 pulpwood
制浆材抓钩 pulp hook
制浆厂 cellulose plant; pulp mill
制浆工艺 pulping process
制胶锅 glue cooker; glue pot
制模框 deckle
制木瓦碎片 spall; spalt
制盆材料 tubbing
制品规格 manufacturing characteristics
制品输送台 accumulation table
制品特性 manufacturing characteristics
制塞工人 plug cutter
制榫机 matcher
制炭场 Kiln-site
制桶业 cooperage
制图 platting
制图等级 mapping level
制图投影仪 mapograph
制图用纸 detail drawing paper
制药原液 raw acid
制造厂分销办事处 manufacture's sales branches office
制造厂分销机构 manufacturers sales branches
制造厂交货 ex factory
制造费用率 factory overhead rate
制造工艺 workmanship
制造精度 accuracy of manufacture
制枕木工 tie chopper; tie hack
制枕木机 tie chopper; tie hack
制止器 arrestor; arrester
制纸 paper-making
质壁分离 plasmolysis
质地 texture
质粒 plastid
质量 brand
质量标准 quality specification
质量测定 quality determination
质量级 quality class

质量检查 quality control
质量鉴定试验 qualification test
质量抗病性 qualitative resistance
质量控制 quality control
质量控制水准 quality control level
质量流 mass flow
质量系数 quality coefficient; quotient of quality
质量效应 mass effect
质量性状 qualitative character
质量指标 quality index
质量作用定律 law of mass action
质谱 mass spectrum
质谱分析 mass spectrometric analysis; mass spectroscopy
质谱分析法 mass spectrography; mass spectroscopy
质谱学 mass spectroscopy
质谱仪 mass spectrograph; mass spectrometer
质体 plastid
质体基粒 granum (*plur.* grana)
质体醌 plastoquinone
质外体 apoplast
质心抽样 centroid sampling
质型多角体病毒 cytoplasmic-polyhedrosis virus
治疗(病菌防治) cure; curing
治线磷(杀线虫剂) thionazin
栉(zhì) pecten; comb
致癌的 carcinogenic
致癌基因 oncogene
致癌物 cancer-causing agent
致癌物质 carcinogen
致病代谢物 pathogenic metabolite
致病的 pathogen(et)ic; pathogenous
致病机理 pathogenesis
致病菌系 pathogenic strain
致病力强的 virulent
致病生理型 pathogenic biotype
致病现象 pathogenism
致病小种 pathogenic race
致病型 pathologic form; pathotype
致病性 pathogenicity
致病源 etiology
致活酶 kinase
致密 densify
致密材 compact wood
致密木材 close-grained wood
致密片麻岩 dense gneiss
致密组织 close texture
致死的 lethal
致死低温带 zone of low fatal temperature
致死点 killing point
致死点死亡温度 death point
致死高温 fatal high temperature; lethal high temperature
致死基因 lethal gene
致死极限 lethal limit
致死剂量 lethal dosage; lethal dose
致死临界温度 zero point
致死率 lethality
致死浓度 killing concentration; lethal concentration
致死温度 killing temperature; lethal temperature; thermal death point
致死物 lethal
致死现象 lethality
致死限量 limes death
致死中量 lethal dose 50; median lethal dose
致死中浓度 50-lethal concentration (LC50)
致育因子 fertility factor
致增种 increaser
秩 rank
秩相关 rank correlation
窒息 asphyxia; suffocation
蛭石 vermiculite
滞洪坝 detention dam
滞洪区 detention basin
滞洪蓄水 detention storage
滞后 lag
滞后期 lag period
滞后曲线 hysteresis curve
滞后损耗 hysteresis loss
滞后现象 hysteresis
滞后现象原理 theory of hysteresis
滞后效应 delayed effect; hysteresis effect
滞留材 drift; driftwood
滞留成本 sunk cost
滞留费 demurrage
滞留锋 stationary front
滞留期 demurrage; residence time
滞留气旋 stationary cyclone

滞纳储备 provision for delinquency
滞纳税款 delinquent tax
滞燃时间 residence flame time
滞水 perched ground water
滞水面 perched water table
滞水岩层 aquiclude
滞育 diapause
滞育卵 resting egg
置产费用 capital expenditure
置换 displacement
置换漂白 displacement bleaching
置换洗涤 displacement washing
置换作用 metathesis
置石 installed stone; rock layout
置信界限 confidence limit
置信区间 confidence interval
置信水平 confidence level
雉 pheasant
稚羽 juvenal plumage
中凹侧突(橱) blockfront
中凹侧突橱 tub front
中班 swing shift
中板 core; cross band veneer; cross band veneer; cross banding; crossing veneer; crossing
中板木条 core strip
中部受精 mesogamy
中层(海洋) mesopelagic zone
中层(细胞) middle lamella; middle layer
中层原木(车上的) rib log
中肠 midgut; ventriculus
中程采伐计划 mid-term harvest planning
中尺度气候学 mesoclimatology
中尺度气象学 mesometeorology
中唇舌 glossa
中等 meso-
中等材质 fair average quality
中等级材 middlings
中等裂缝 medium check; medium shake
中等裂纹 medium torn grain
中等耐阴树种 intermediate shade-tolerant species; mid-tolerant species
中等劈裂 medium split
中等品 fair average quality
中等营养的(生境) mesotrophic
中等原木 average log
中点受精 mesogamy
中毒 poisoning
中毒的 toxic
中度的 moderate
中度干扰假说 intermediate disturbance hypothesis
中度火烧 moderate burn
中度耐阴树 mid-tolerant tree
中度疏伐 C-grade thinning; moderate thinning
中度休眠 intermediate dormancy
中段 midrib
中段原木 middle cut (log); middle log
中断 outbreak
中断洗脱 interrupted elution
中方材(约6~12英寸) medium square
中枋 die-square
中复羽 medium coverts
中副冠类(水仙等) medio-coronate
中腹林道 mid-slope read
中高草 midgrass
中高位芽植物 mesophanerophyte
中隔墙 mid-board
中耕 intermediate culture; intertillage
中耕铲 tine
中耕机 cultivator
中耕钻 culti-anger
中沟 median gutter
中国白苎麻 Chinese white ramie
中国大草原 China's vast grassland
中国红 Chinese red
中国蜡 Chinese insect wax
中国绿化基金会 Green China Fund
中国漆 Chinese lacquer
中国青苎麻 Chinese green ramie
中国肉桂 Chinese cinnamon
中国森林生态系统研究网络 Chinese Forest Ecosysterm Research Network (CFERN)
中国生漆 raw lacquer
中国生态系统研究网络 Chinese Ecosysterm Research Network (CERN)
中国压条 Chinese layering
中国紫胶 Chinese lac
中果皮 mesocarp; sarcocarp

中和 neutralization
中和面 neutral surface
中和平面 neutral plane
中和位置 neutral position
中和轴 neutral axis
中和作用 neutralization
中横档 middle rail
中护木(纵木板路) centre guard
中级的 medium (*plur.* -ia)
中级中密度纤维板 intermediate of medium density fibre board
中继集材杆(树)(分段集材) yarder tree
中继曲线索道 connecting curved cableway
中间 meta-
中间绷索 buckle guy
中间材 intermediate wood
中间层 central layer
中间产品 middlings
中间产物 intermediate product
中间代谢 intermediary metabolism
中间带 intermediate belt; intermediate zone
中间的 median; medium (*plur.* -ia)
中间反应 intermediate reaction
中间负片 inter-negative
中间工作台 centre table
中间辊 intermediate roll; middle roll
中间横档 centre rail
中间寄主 interhost
中间加热管 booster coil; reheat coil
中间阶段 intermediate stage
中间类型 intermediate form; passage form
中间楞场 intermediate landing; relay yard
中间楞场装车 reloading
中间利用 intermediate
中间料仓 in-process-storage-unit; intermediate bin
中间喷蒸 intermittent steaming
中间切面 middle section
中间缺失 interstitial deficiency
中间收获 intermediate yield
中间收获(间伐) intermediate felling yield
中间望远镜的经纬仪 theodolite with central telescope
中间维管束 intermediate bundle
中间纤维 intermediate fibre
中间斜支撑 middle shore
中间性植物 day neutral plant
中间性状 intermediate character
中间宿主 intermediate host
中间样品 mean sample
中间易位 interstitial translocation
中间砧木 intermediate stock
中间支架 intermediate support; middle support
中间支架式架空索道 supported line
中间支柱 intermediate post; intermediate-support spar; relay tree
中间支座(索道) tree hanger
中间纸板 mid-board
中间重复嫁接法 intermediate re-grafting
中间贮存和转运装置 surge and transfer station
中间状态 transition state
中景 middle ground
中距地区 middle ground
中孔 transitional pore
中跨 mid-span
中楞 cold deck; deck
中楞装车 cold deck loading
中砾石排水沟 cobblestone drain
中林 composite coppice; composite forest; compound coppice; coppice with standards; middle forest; sprout seeding forest
中林更新法 middle forest regeneration
中林作业 composite forest system; composite system; coppice with standard method; coppice with standards system; standard coppice system; storied coppice system
中林作业法 composite forest (method); middle-forest system
中林作业更新法 coppice-with-standards method
中龄林 half-mature forest; semimature forest
中流插垛(不及两岸) stream jam; centre jam
中脉(叶) midrib; primary vein
中脉出脉的 costal-nerved
中脉线 ramellus

中帽头 frieze rail
中密度板 medium density board
中密度刨花板 medium density particle board; medium density wood chipboard; medium-density particleboard; particle board of medium density
中密度碎料板 medium density particle board; medium density wood chipboard; particle board of medium density
中密度纤维板 intermediate density fibreboard; medium density fiberboard; medium density fibreboard; semi-hardboard
中欧山松针油 dwarf pine needle oil
中刨齿 centre raker
中胚轴 mesocotyl
中坡 midslope
中期 intermediate stage
中期采伐计划 mid-term harvest planning
中期分裂 metakinesis
中期腐朽 intermediate decay
中期计划 medium term planning
中期染色体 metaphase chromosome
中期预测 medium-term forecast
中期种 mesospecies
中乔木 medium tree
中日性植物 day intermediate plant
中深底栖生物 mesobenthos
中生(干湿度) meso-
中生的 mesic
中生高草群落 protoherbosa
中生化 mesophytization
中生林 mesophytic forest
中生生物区系 mesobiota
中生树种 mesophilous tree species
中生性 mesophil(e); mesophytism
中生性的 mesophil(e); mesophilus
中生性湿旱生植物 mesophilous tropophyte
中生性适季植物 mesophilous tropophyte
中生演替 measarch succession
中生演替的(生态) mesarch
中生演替系列 mesosere
中生植被 mesophyte vegetation
中生植物 mesophyte
中生植物的 mesophytic
中石器时代 Mesolithic age
中始式(植物解剖) mesarch
中始武木质部 mesarch xylem
中式盐 normal salt
中试车间 pilot plant
中室(昆虫翅) discal cell
中寿种子 medium lived seed
中枢神经系统 central nervous system
中枢销 kingpin
中庭庭园 patio
中头型的 mesocephalic
中纬线 equator
中位数 median
中温 mesotherm
中温带 midtemperate zone
中温固化胶 intermediate-temperature glue
中温固化胶黏剂 intermediate temperature setting adhesive
中温固化黏合剂 warm-setting adhesive
中温生物 mesophile
中温温室 warm intermediate house
中温硬化胶 intermediate-temperature glue
中温硬化胶黏剂 intermediate temperature setting adhesive
中温植物 mesotherm
中污水的 mesosaprobic
中污水生物 mesosaprobe
中物种 mesospecies
中线 median
中线配板法 centre matching
中小型长沙发 settee
中心板 plate nuclear
中心层硬化 case tension
中心插入伐木法 central cut
中心点 focus point
中心点火法 centre firing
中心法则 central dogma
中心环 centre ring
中心景 centre of interest
中心拼接 centre matching
中心投影 central projection
中心线 axis
中心线平均法(粗糙度测定) centre line average (method)

中心乡土区 region of climatic optimum
中心引爆(狩猎) central percussion
中心硬化 reverse casehardening
中新世 Miocene
中型防火犁 middlebuster plough
中型浮游生物 mesoplankton
中型生物 mexofauna
中型叶 mesomorphic leaf; mesophyll
中型叶的 mesophyllous
中型柱 intermediate column
中性 neuter; neutralism
中性的 eutrophic; neutral
中性肥料 neutral fertilizer
中性两性花同株 agamonoecism
中性滤色镜 neutral filter
中性面 neutral surface
中性平面 neutral plane
中性施胶剂 neutral size
中性树脂 neutral resin
中性说 total neutrality hypothesis
中性说辩论 neutrality controversy
中性土 neutral soil
中性蚁 neuter
中性油 neutral oil
中性轴 neutral axis
中胸侧板 mesopleuron
中修 medium maintenance
中央处理机 central processor
中央处理器(计算机) central processing unit
中央的 median
中央断面积求积式 Huber's formula
中央断面区分求积法 sectional measurement by means of Huber's formula
中央干围 mid-girth
中央公园 central park
中央计划经济 centrally planned economics
中央木 arithmetical mean sample tree; average stem; average tree; central stem; mean sample tree; mean tree
中央木平均生长 mean growth tree
中央木调查法 method of the arithmetical mean sample tree
中央数据库 central data bank
中央尾羽 central rectrices
中央圆周 mid-girth
中央载荷 centre load
中央直径 central diameter; centre point diameter; diameter at middle; half-height diameter; mid-diameter
中央中轴胎座 central axial placenta
中央周围 mid-girth
中央主枝 central leader
中央柱木 head spar
中营养湖 mesotrophic lake
中庸木 intermediate tree
中油(聚合程度中等) medium oil; intermediate oil
中长岩 andesinite
中值断面积直径 diameter of the central basal area tree; basal area central diameter
中值直径 median DBH
中种被 sarcoderm
中种皮 mesosperm; sarcoderm
中洲(河川中央沙洲) centre-field
中轴胎座式 axial placentation; axile placentation
中昼长植物 day intermediate plant
中柱(植物组织) stele; stela
中柱(柱子) filling post; intermediate column
中柱接头(螺旋梯的) newel joint
中柱鞘 pericycle
中柱鞘纤维 pericyclic fibre
中柱原 plerome
中柱轴 cylindrical axis
中转点 trans-shipment point
中转集材杆(索道跨越山脊) relay tree
中转税 transit duty
中转运输 portage
中转装车场 reloading landing
终变定 final set
终变期 diakinesis
终点站台 platform terminal
终端产品 end product
终端能源 final energy
终端支柱承索器 tail shoe
终伐(渐伐后) clearance; harvest cutting; main felling; major harvest; principal felling; final cutting; final felling
终伐材积 final volume
终伐期 final stage; late stage

Z

终伐蓄积表 final stock table
终含水率 final moisture content
终极群落 final community
终价 prolonged value
终结者 terminator
终期含水量 final moisture content
终期蒸汽处理 final steaming
终碛 terminal moraine
终渗能力 final infiltration capacity
终生浮游生物 holopelagic plankton; holoplankton; permanent plankton
终饰 finishing
终饰材 finish
终饰层 finish coat(ing)
终霜 last frost; latest frost
终宿主 final host
终应力 final stress
终值 end value; future value
终止反应 stopping reaction
终止年金 terminating annuities
终止区 terminator
钟乳石 stalactite
钟乳体 cystolith
钟乳体细胞 cystolith cell
钟状 mitriform
钟状的 campanulate
钟状蒴帽 mitriform calyptra
螽斯(产于美洲的昆虫) katydid
肿块 tumor
肿腿蜂 bethylids
肿胀 tumor; turgescence; turgescency
肿枝病 swollen shoot
种 race; species; stirps (*plur.* stirpes)
种柄 seed pedicel; seed stalk
种翅 seed wing
种畜群 breeding herd
种的多样性 species diversity
种的确限度 species fidelity
种的周转率(更替率) species turnover rate (replacement rate)
种肥 seed manure
种阜 calyptra; caruncle; strophiole
种复合体 species complex
种根 seminal root
种脊 raphe
种脊的 raphal
种加词 specific epithet
种间的 interspecific
种间关联 interspecific association
种间关系 interspecfic relationship
种间合作 interspecific cooperation
种间嫁接 heteroplastic graft
种间竞争 interspecific competition
种间外激素 allelochemic
种间杂交 interspecific crossing
种间种群 interspecies population
种胶 brood lac
种茎 seed stem
种类 category; species
种类鉴定 species identification
种类异质性 species heterogeneity
种类组成 floristic composition; species assemblage
种鳞 seminiferous scale
种-面积曲线 species-area curve
种苗 provenance; provenience
种苗分级机 seedling grading machine
种苗认证 certification of planting; certification of plants; certification of seed
种名 specific name
种名形容词 specific epithet
种内变异 intraspecific variation
种内的 intraspceific; intraspecific
种内关系 intraspecific relationship
种内嫁接 homoplastic graft
种内竞争 intraspecific competition
种内拟态 automimicry
种内群聚 intraspecies aggregation
种内协作 intraspecific cooperation
种内杂交 intraspecific crossing
种胚 embryo culture
种胚败育 embryo abortion
种胚嫁接 embryo grafting
种批 seed lot
种批均匀度 uniformity of seed lot
种皮 episperm; rind; seed coat; seed shell; seed vessel; spermoderm; testa (*plur.* tastae)
种皮引起的休眠 seed-coat dormancy
种脐 Hilum; hylum; hylus; naucum (*plur.* nauca); omphalodium

种圈 circle of species
种群 guild; population
种群暴发 population explosion; population irruption
种群变化 population change
种群波动 population fluctuation
种群大小 group size; population size
种群动态 dynamic of population; population dynamics
种群分布 population dispersion
种群分析 population analysis
种群过程 population process
种群过密 overpopulation
种群集度 species packing
种群零增长 zero population growth
种群密度 population density
种群密度过低 underpopulation
种群灭绝 population extinction
种群模型 population model
种群品质 population quality
种群平衡 population equilibrium
种群普查 census(ing)
种群趋势 population trend
种群趋势指数 index of population trend
种群曲线 population curve
种群生存力分析 population variability analysis
种群生态学 genecology; population ecology
种群生物学 population biology
种群调查 census
种群调节 population regulation
种群稳定性 population stability
种群系统 population system
种群相 groupy phase
种群循环 population cycle
种群压力 population pressure
种群增长 population growth
种群增长率 population growth rate
种群增长曲线 population growth curve
种群指数 population index
种群指数增长模式 exponential growth form
种生物量 species biomass
种属系数 generic coefficient
种数–多度数曲线 species-abundance curve
种特征 essential character
种外进化 macroevolution
种丸 pellet; pelletted seed
种系 germ line
种系发生 phylogeny
种小蜂 torymid
种源 provenance; provenience; seed origin; seed source; source
种源鉴定 determination of provenance
种源试验 provenance test; provenance trial
种源选择 provenance selection
种源种子源 provenance seed orchard
种质 germ plasm; germplasm; idioplasm
种质保存 germplasm preservation
种质保护 germplasm conservation
种质库 germplasm repository; germplasmbank
种质论 germplasm theory
种质遗传 blastogenesis
种质资源 germplasm resource; germplasm stock
种质资源库 germ plasma resource bank
种株 stock tree
种子(广义的) seed; semina; kernel
种子(苹果或梨) pip
种子安全含水量 safe seed moisture
种子包衣机 seed pelleting machine
种子饱满度 seed-filling
种子保证单 certification of planting; certification of plants
种子变异性 seed variability
种子采集处理机械 seed harvesting and processing machinery
种子采收 seed collecting; seed harvesting
种子层积催芽 seed stratification
种子掺杂 adulteration of seed
种子产量 seed yield
种子成熟 seed ripening
种子抽样器 seed trier
种子处理 seed processing; seed treatment
种子传播 seed dispersal; seed dissemination
种子传播的病害 seed-borne disease

Z

种子传播方式 modes of seed dispersal
种子纯度 purity of seed
种子催芽床 forcing bed
种子大小 seed mass
种子带(用两层富含营养的纸裹卷种子) seed tape
种子的 seminal; seminific
种子的碎屑 chaff
种子等级 seed class
种子电导性测定 seed conductivity test
种子顶端 seed end
种子锻炼 hardening of seed
种子发芽器 germinator
种子法规 seed law
种子繁育 seed growing
种子飞散极限 flight limit
种子肥料混播 combine drilling
种子分级 seed sorting
种子分级器 seed sorter
种子分离机 extractor
种子腐烂病 seed rot
种子覆土深度 seed depth
种子改良 seed amelioration
种子干燥 seed drying
种子干燥机 seed drying machine
种子干燥炉 seed drying oven
种子干燥室 seed kiln
种子根 seminal root
种子工作者 seedsman
种子害虫 seed bug
种子含水量 seed moisture content
种子活力 seed vigour; viability of seeds
种子机械 seed harvesting and processing machinery
种子机械破皮 mechanical scarification
种子计数器 seed counter
种子甲虫 seed beetle
种子嫁接 seed grafting
种子检索表 seed key
种子检验 certification of seed; seed certification; seed inspection; seed test(ing)
种子检验标准 seed certification standard
种子检验规程 rules for seed testing
种子检验机构 seed-certifying agency
种子检验人员 seed analyst
种子检验员 seed inspector
种子检验站 seed control (station); seed testing station
种子进料斗 seed hopper
种子净度 seed purity
种子库 seed bank; seed storage
种子老化 seed ageing; seed senescence
种子漏斗 seed hopper
种子目录 seed catalog(ue)
种子年 mast year; seed year
种子签证 seed certification
种子签证标准 seed certification standard
种子签证机构 seed-certifying agency
种子轻小的种类 light-seeded species
种子清选 clean-up of seed
种子群体 seed population
种子热水处理 hot-water seed treatment
种子认证 certification of planting; certification of plants
种子容重 seed volume-weight
种子散落 seed dispersal
种子筛 seed riddle
种子商 seedsman
种子生产区 seed production area
种子生活力 seed viability; seed vitality
种子生理学 seed physiology
种子生态学 seed ecology
种子生物学 seed biology
种子湿度电测计 electronic seed moisture meter
种子适用率(净度百分数×发芽百分率/100) pure live seed value
种子收集器 seed trap
种子寿命 seed longevity
种子条播机 tree seed drill
种子条例 seed act
种子调拨区 seed zone
种子丸(播种) seed ball; seed pellet
种子喂入箱 feed(er) box
种子无菌层积 aseptic seed stratification
种子无菌培育 aseptic seed culture
种子消毒 seed disinfection
种子消毒剂 seed disinfectant
种子休眠 seed dormancy

种子选择 seed selection
种子学 seed science; spermatology; spermology
种子易传播种类 light-seeded species
种子用价(等于净度百分数×发芽率百分数) utilization value of seed; stock-value
种子园 forest garden; seed orchard
种子站 seed station
种子真价(净度×发芽势) real value of seed
种子直播机 tree seed planter
种子直感 xenia
种子植物 carpophyte; seed bearing plant; seed plant; spermatophyta
种子质量 seed quality
种子质量检验 seed quality examination
种子贮藏 seed storage
种族 stirps (*plur.* stirpes)
种族发生 historical development; systematic development
种组 species group
踵状插穗 heel cutting
踵状扦插 heel cutting
种草绿化 green land with grasses
种花者 florist
种植 plant; grow; cultivate; growth
种植点配置 spacing of planting spots; spacing of seeding spots
种植规划 planting schema
种植机 planting machine
种植绿肥 green manuring
种植绿肥的休闲地 green(manured) fallow
种植密度 density of plantation
仲夏梢 midsummer-shoot
众数 mode
重材 heavy wood
重氮基(-N:N-或N:N=) diazo-
重氮基法 diazo method
重氮片 diazo film
重氮片印显机 diazo printer and developer
重刀打浆 dead beating; hard beating
重点植物(园林) accent plant
重点中和色谱(园林) accented neutral colour scheme
重革 heavy leather
重过磷酸钙 one-lime phosphate
重力 gravitation
重力坝 gravity dam
重力分级机 gravity separator
重力风 katabatic wind
重力风选机 gravity wind sifter
重力感受 graviperception
重力轨道(空车溜放) gravity spur
重力辊 gravity roll(er)
重力滑道 gravity chute; live chute; running slide
重力集材作业区 gravity chance
重力计量 weighing dosing
重力计量刨花进料机 gravimetric chip-feeder
重力溜槽 gravity chute
重力浓缩机 gravity thickener
重力侵蚀 gravitational erosion
重力侵蚀分级 classification of gravitational erosion
重力倾卸装置 gravity tipple
重力式进料削片机 gravity feed chipper
重力式缆车道 gravity incline-way; gravity system of cable railway
重力式凝汽阀 gravity trap
重力式抛木归堆机(选材) drop sorter and stacker
重力式抛木机 gravitational scrambler
重力式抛木选材机 drop sorter
重力式索道 gravity cableway
重力式索道集材 gravity skyline yarding
重力势 gravitational potential
重力水 gravitational water
重力松紧式索道 gravity slack line
重力索道集材 gravity cable logging
重力载荷 gravity loading
重力种子清选机 gravity seed cleaner
重力作业区 gravity chance
重量 weight
重量测定 weight determination
重量法 weight method
重量方程 weight equation
重量含水率 gravimetric moisture content
重量计 weightometer
重量克分子浓度 molal concentration

Z

重量控制台 weight control station
重量密度 weight density
重量平均分子量 weight-average molecular weight
重量平均聚合度 weight-average degree of polymerization (DP_w)
重量增加百分数 weight percent gain
重黏土 heavy clay
重燃物(慢燃物) heavy fuel
重壤土 heavy loam
重施胶 hard sized
重石油 pacura
重蚀地 badland
重型出河链 bull chain
重型单板刨切机 heavy duty veneer slicer
重型钩环(连接承载索) admiralty shackle
重型横铡机 cross-cut guillotine
重型集材滑轮(U形环可打开) shackle yarding block
重型锯木机 sawing rig; timber mill
重型梁(厚4英寸宽8英寸以上的成材) heavy joist
重型木建筑 heavy-timber construction
重型木结构 heavy-timber construction
重型四滚筒砂光机 heavy four drum roll feed sander
重型四面镶板机 heavy four-sided matchboarding machine
重型索道 heavy skyline cableway
重型原木铲叉 heavy-duty log fork
重型铡切锯 guillotine saw
重要天气 significant weather
重要值 importance value
重油 dead oil; heavy oil
重载钢索挠度 loaded deflection
重载行程(集材) inhaul trip
重载列车 drag
重载跑车 loaded carriage
重载牵引 inhaul
重质松节油 heavy turpentine
州立公园 state park
州林务员 state forester
州有林 provincial forest; state forest
州有林代办机构 state forest agencies
周壁的 parietal
周边形成层(根中柱最外部细胞层) pericambium
周边营养性菌根 peritrophic mycorrhizae
周丛生物(群体) periphyton
周分 cycles per minute (cpm)
周面磨削 cross grinding
周秒 cycles per second (cps)
周末使用区 weekend-use area
周年变化 annuation
周皮 periderm
周期 cycle time; return interval
周期变化 aspection; periodic change
周期变形 cyclomorphosis
周期单性生殖 cyclical parthenogenesis
周期低潮 trough
周期式干燥窑 box kiln; compartment kiln
周期式烘干窑 intermittent kiln
周期特性 cyclophysis
周期图 periodogram
周期性 periodicity; periodism
周期性爆发 cyclical outbreak
周期性波动 cyclic fluctuation
周期性大发生 periodic outbreak
周期性放牧 periodic grazing
周期性浮游生物 periodic plankton
周期性演替 periodic succession
周生鞭毛 peritrichous flagellum
周围的 circumferential
周围神经系统 peripheral nervous system
周围压力 ambient pressure
周位的 perigynous
周位花 perigynous flower
周限增长率 finite rate of increase
周缘的 parietal
周缘嵌合体 periclinal chim(a)era
周长 periphery
周转率 turnover rate
周转时间 turnover time
周转资金 working capital
轴 axis; pivot; shaft
轴柄 mandrel
轴衬 bush
轴承 bearing
轴键 axial bond

轴晶 axialite
轴可倾斜式多能圆锯 universal circular saw with tilting arbor
轴流风机 axial fan; axial flow fan; propeller fan
轴流涡轮 axial flow turbine
轴生式(性器官) axogamy
轴式滑轮 axle pulley
轴套 nave; sleeve
轴瓦 bush
轴下输出式削片机 spout undershaft chipper
轴线 axis
轴向 axial direction
轴向胞间道 axial intercellular canal; vertical intercellular canal
轴向薄壁细胞 axial parenchyma cell; cell of axial parenchyma
轴向薄壁组织 axial parenchyma
轴向壁 axial wall
轴向分子 axial element; vertical element
轴向高速搅拌器 axial high speed stirrer
轴向管胞 axial tracheid
轴向间隙 axial clearance
轴向拉力 axial tension
轴向力 axial force
轴向前角 axial rake angle
轴向束状薄壁组织 axial strand parenchyma
轴向树脂道 axial resin canal; vertical resin canal
轴向调整 axial adjustment
轴向推力 axial thrust
轴向位移 axial displacement
轴向压力 axial pressure
轴向压缩 axial compression
轴向应变 axial strain
轴向应力 axial stress
轴向载荷 axial load
轴向振动 axial vibration
轴心木 centre; center
轴心下锯法 centre sawing; cutting on centre
轴芯棒 axle
轴型鼓式刨片机 cutter spindle flaker; knife shaft flaker
轴用材 axial wood
肘部 elbow
肘钉 staple
肘管 elbow
肘节 binder
肘脉 cubitus
肘形连接杆 toggle
帚化 fibrillation
帚化纤维 fuzz fibre
绉纹垫 craped wadding
昼出动物 diurnal animal
昼行性 diurnalism
昼夜变动 diurnation
昼夜变化 diurnal change; diurnal variation
昼夜变异不规则 diurnal inequality
昼夜波动 diurnal fluctuation
昼夜活动节律 daily activity rhythm
昼夜节律 circadian rhythm; day-night rhythm
昼夜迁徙 diurnal migration
昼夜温差范围 diurnal temperature range
昼夜周期 daily periodicity
昼夜周期性 diurnal periodicity
昼夜转换定时器 day-night timer
昼长 day length
昼长处理 day-length treatment
皱的 crumpled
皱肤效应 skin effect
皱痕 crease
皱裂 alligatoring
皱缩 collapse; crimp; crinkle
皱缩种子 shriveled seed; wrinkled seed
皱缩作用 washboarding effect
皱纹 crinkle; ruga (*plur.* rugae); stria (*plur.* striae)
皱纹叶 rugose leaf
皱折变形 buckling deformation
皱折应力 wrinkling stress
皱褶 ruga (*plur.* rugae)
皱纸 crepe paper
骤减 die-off
骤冷 quench
骤冷器 expander
朱科测树仪 Jukohscope
朱林方程式 Jurin's equation
侏罗纪 Jurassic period
珠被 integument

Z

珠柄 funic(u)le; funiculus; ovule stalk; spermaphore; umbilical cord
珠柄钩 retinaculum
珠脊 raphe
珠脊的 raphal
珠孔 micropyle
珠鳞 cone scale
珠心 nucellus; nucella
珠芽 bulbil
珠缘 bead
珠缘嵌板 bead-flush panel
株距 distance between hills; individual spacing; spacing (between trees); tree spacing
株数 number of stems; number of trees
株系隔离 line isolation
硃砂 cinnabar
蛛网心裂 spider heart
竹编 bamboo weaving
竹编胶合板 bamboo mat plywood
竹鞭 bamboo rhizome
竹材 bamboo wood
竹材场 culm yard
竹材胶合板 bamboo plywood
竹材塑料结合物 bamboo-plastic combination
竹雕 bamboo sculpture
竹钉 bamboo nail
竹筏 bamboo raft
竹秆 bamboo culm
竹秆插穗 cutting of the culm
竹秆扦插 cutting of the culm
竹滑道 bamboo chute
竹蝗 bamboo locust
竹浆 bamboo pulp
竹节虫 phasmid; stick insect; walkingstick
竹类(泛指) cane
竹篱 bamboo fence
竹林 bamboo forest
竹林更新 bamboo-stand regeneration
竹林天然更新 natural reproduction by rhizome
竹林作业 bamboo stand system; working system for bamboo stand
竹笼(护岸) bamboo cylinder
竹篾 bamboo skin
竹排 bamboo raft; raft
竹刨花板 bamboo particle board
竹棚架(建筑) bamboo scaffold
竹皮 bamboo skin
竹片 bamboo strip
竹器 bamboo article; bambooware
竹签 tally
竹碎料板 bamboo particle board
竹荪 bamboo fungus
竹笋 bamboo shoots
竹索 bamboo cable
竹席 bamboo mat
竹制家具 bamboo furniture
竹制品 bamboo article; bambooware
竹子 bamboo
逐步回归分析 stepwise regression analysis
逐步聚合 transition polymerization
逐次分析 sequential analysis
逐渐变细 taper
逐日温度 daily temperature
逐选样方法 nested quadrat method
烛光[度] candlepower
烛果油 candle nut oil
主坝 barrage; key dam
主测 main survey
主产物 main product; major produce; major product; principal product
主产物收获 principal produce
主产物作业法 major product system
主成分 principal component
主成分分析 principal component analysis
主椽 principal rafter
主弹簧 main spring
主导基因 master gene
主导树种 leading species
主导因素 key factor
主道(板院) main alley
主道(广义的) main haul road; main road
主道(林区) main ride
主点 principal point
主电动机 main motor
主动轮 driving wheel
主动土压力 active earth pressure
主动微波 active microwave

主动吸收 active absorption
主动遥感 active remote sensing
主动轴 drive shaft
主伐 cut-over; felling; final clearing; final cutting; final felling; harvest cutting; main felling; major harvest; principal felling; regeneration cutting; regeneration felling; regeneratory cutting; reproduction cutting; reproduction felling
主伐成熟期 ripe for the harvest
主伐方式 system of cutting
主伐计划 harvest planning
主伐林 cutover forest; cut-over forest; final crop
主伐龄 final age
主伐木 crop tree; final crop tree; final stock; future tree; growing stock tree; principal stand; main crop
主伐年龄 cutting rotation age; felling age
主伐收获 final yield; principal yield; princrpal yield; yield of final crop; yield of final cutting
主伐收入 final income; financial returns from clear felling; financial returns from final cutting
主伐收益 return from final clearing
主分类(木材) main assortment; main category
主分量 principal component
主风 cardinal wind; prevailing wind
主干 main stem
主干顶生枝 leader
主干高 bole height
主干无枝度 cleanness of bole
主干型 central leader type
主干型整枝 central leader type training
主根 axial root; main root; principal root; taproot; top root
主根系 taproot system
主骨架 backbone
主管会计 comptroller
主管型加热器 header coil
主河 master river
主河缨 terminal boom
主桁架 main truss
主滑轮 head block
主基因 key-gene
主集材杆 head spar; head tree; main spar
主集材杆树 head spar tree
主夹板 bottom plate
主景 main feature
主景植物 accent plant
主锯 head saw; log saw
主锯边皮削片机 headrig slab chipper
主锯机 headrig
主卷筒 main drum
主控制程序 primary control program
主力锯 pony rig
主立柱 head spar; head tree; principal post
主联杆 master link
主链 backbone
主链片 master link
主梁 king post
主梁屋架 king post roof truss
主裂片 main lobe
主林班线 main compartment line; main frame
主林层 main storey; overstorey
主林产物 major forest produce
主林带 main forest belt
主林道 main forest road
主林分 main stand; principal stand
主林木 crop tree; dominant portion of forest; final crop tree; future tree; growing stock tree; leading tree; main crop; main crop after thinning; major forest tree; principal crop; principal forest crop; principal stand
主林木数量成熟 quantitative maturity of principal tree
主流 master stream
主脉(动物) main nerve
主脉(昆虫翅) primary vein
主脉(叶) rib
主苗 axial shoot
主模式标本 holotype
主排 crab
主平面 principal plane
主起重臂 main boom
主汽管系 live steam piping

Z

主切削方向 principal cutting direction
主切削角 primary cutting angle
主权 equity
主收获 principal yield
主收益 return from final clearing
主水解 main hydrolysis; principal hydrolysis
主索 logging line; main cable; main lead; main lead line; main line; traction cable; traction line; traction rope
主索滑轮 main-line block
主索具装配 mainline rigging
主塔(塔式起重机) king-tower
主题 key-note
主体 bulk
主体工程 main work
主脱木素作用 bulk delignification
主线 principal line
主销 kingpin
主效基因 major gene resistance
主型(染色体) primer type
主眼(射击时起主要作用的眼) master eye
主要变种 leading variety
主要部件 essential component
主要产品 key product; principal product
主要成分 chief constituent; essential component
主要反应 main reaction
主要害虫 primary pest
主要基因 major gene; oligogene
主要经营项目 major management subject
主要林产品 forest staple product
主要林木 main tree
主要零件 major parts
主要木材 primary timber
主要农作物 main crop
主要生产 principal produce
主要树种 main species; main tree species; principal species; chief species
主要特征 essential character; main feature
主要土类 main soil group
主要养分 macronutrient
主要用途 end use
主要优势种 leading dominant
主要指示植物 key indicator plants
主业(经营方案编制) major work; principal work
主应变 principal strain
主应力 principal stress
主优势木 head dominant
主运材 major transportation
主运料路 main haul road
主枝 axial shoot; leader branch; main branch
主枝延长枝 extension leader
主轴 arbor; caudex; main shaft; rhachis; spindle
主轴箱 headstock
主轴芽 leader bud
主子午线 principal meridian
主组分 principal component
煮沸试验(耐水胶着力试验) boiling water test
苎烯 cinene; citrine
助倒器(掌握树倒方向) gun; gun(ning) stick
助浮木 riser log
助根接 nurse-root grafting
助剂 adjuvant
助接 adjuvant grafting
助力扩散泵 booster diffusion pump
助力油泵 power assist pump
助留剂 retention aid
助滤剂(助滤器) filter aid
助溶剂 cosolvent
助色团 auxochrome group
助食素 feeding stimulant
助细胞 synergid
住房 lodging
住房建筑 housing
住宅林 residential forest
贮备 stockpile
贮备林 forest reserve
贮材 timber storage
贮藏 storage
贮藏管胞 storage tracheid
贮藏期 shelf life
贮藏期病害 storage disease
贮藏所 cache
贮藏细胞 storage cell
贮藏组织 storage tissue

贮槽 stock chest; sump
贮存 hoarding; stockpile
贮存腐朽 storage decay; storage rot
贮存绠 storage boom
贮存罐 accumulator tank; storage tank
贮存缺陷 piling defect; storage defect
贮存台 accumulation table
贮存有效期 storage life
贮放 seasoning
贮粉室 pollen chamber
贮浆池 stock chest; stuff chest
贮精囊 seminal vesicle
贮料仓 silo
贮料柜 stuff chest
贮料式磨木机 magazine grinder
贮料室 material chamber
贮料箱 stuff box
贮木 storage; timber storage
贮木场 log deposit; log depot; log dump; log yard; timber basin; timber depot; timber magazine; timber yard; wood yard; wood depot; logging depot; lower landing; service yard; storage area; storage yard; concentration yard; loading area
贮木场保管员 store manger
贮木场地设备 yard equipment
贮木场机械 timber yard machine
贮木场库存量 inventory of log yard
贮木场木材变色 yard stain
贮木场木材褐变(堆积变色) yard brown (stain)
贮木场木材蓝变(堆积变色) yard blue (stain)
贮木场内运输线 yard track
贮木场年吞吐量 annual input and output of log yard
贮木池 booming ground; log dump; log pond; mill pond; pond; storage pond; timber storage pond; timber-pond; water magazine; waterpool depot; yard pool
贮木池工人 short log cat
贮木池交货 at the pond
贮木池截锯 pond saw
贮木池造材 pond buck
贮木登记簿 log ledger
贮木能力 storage capacity
贮木区 storage area
贮木水渠 log canal
贮木作业 depoting
贮食传播 endochory
贮水池 storage reservoir; water-storage tank
贮水箱 water-storage tank
贮水堰 storage dam
贮索拖车 cable storage trailer
贮油槽 oil storage tank
贮油罐 accumulator tank; oil storage tank
贮脂池 gum reservior
注册支票 cancelled check
注册种子 registered seed
注满 fill
注入处理 injection treatment
注入剂量 injection dosage
注入口 influx
注入扩散法 gun-injection
注入系统 injection system
注射口 injection port
注射器 injection syringe; injector
注销 write-off
注销支票 cancelled check
注液截口 injection kerf
注油 lubrication
柱保留 column hold-up
柱材 pilaster; post; post pillar
柱层析 column chromatography
柱垫 cap block
柱顶过梁 architrave
柱基 plinth
柱基装饰 surbase
柱脚垫木 footer; footlid
柱廊 colonnade; portico
柱螺栓 fid
柱木 spar
柱塞 piston; trunk
柱塞泵 piston-type pump
柱色谱法 column chromatography
柱身 shaft
柱式锄铲中耕机 shank shovel cultivator
柱式压机 column press
柱头 stigma

Z

柱头垫木 cap block
柱轴形椅背 baluster-and-spindle back
柱状的 styloid
柱状花托 thalamus
柱状节理 columnar joint(ing)
柱状晶体 crystal styloid; elongated crystal; styloid crystal
柱状图 histogram
柱子(支柱或电柱) pole; column; post
蛀虫 borer
蛀道 tunnel
蛀道真菌 ambrosia fungus
蛀粪屑 frass
蛀干害虫 stem feeder
蛀孔 bore; borer hole; bore hole; tunnel; canker
蛀木虫 timber worm; wood borer
蛀木海蚤(海产甲壳类) gribble; chelura
蛀木害虫 xylophagous borer
蛀木昆虫 wood-destroying insect
蛀木筒蠹 timber beetle
蛀屑 bore dust; bore-powder; wormdust
铸锭 keel block
铸铁散热器 cast-iron radiator
筑坝 dam
筑巢 nest-building
筑巢处 rookery
筑成坡度 grading
筑床 bed-up
筑床机 bed former; bed shaper
筑堤 banking; mounding
筑路机 grader
筑丘 mounding
筑山 hill making
筑山园(日本式园林) hill garder
抓臂(伐木联合机的) grab arm
抓斗 clamshell; grapple
抓斗式起重机 grab(bing) crane
抓斗式挖土机 grab skipper
抓斗爪 rock rake
抓钩 bale hook; grapple; grapple hook; span dog
抓钩闭合索 dosing line
抓钩集材 grapple skidding
抓钩开闭索 grapple operating line
抓钩开度 grapple opening
抓钩起重容积 grapple lift volume
抓钩容量 grapple capacity
抓钩式集材(免捆木索) grapple yarding
抓钩式集材机 grapple skidder
抓钩式装载机 grapple loader
抓具 clamshell; grab; grab handle; grapple; grip
抓木器 log grapple; logging grapple
抓取量 grapple capacity
抓岩机 rock rake
专单性生殖 obligatory parthenogenesis
专化 specialization
专化寄生物 specialized parasite
专化型 specialized form
专化性病原 specialized pathogen
专类花园 specific flower garden
专利[权] patent
专利证 letters patent
专卖法 monopoly law
专门设计的 custom-designed
专题工作计划 thematic work program
专项贷款 tied loan
专性的 obligate; obligatory
专性腐生菌 obligate saprophyte
专性寄生菌 obligate parasite
专性需氧菌 strict aerobe
专性厌氧菌 strict anaerobe
专一性 specificity
专用林地 absolute forest land; absolute forest soil
专用绿地 exclusive green space
专用名 specific name
专有的 exclusive
专有性 despotism
专有种 exclusive
专账 separate account
砖坝 brick dam
砖衬 brick lining
砖隔状 muriform
砖红化作用 latosolization
砖红壤 laterite; latosol; latosols
砖红壤化 laterization
砖红壤性红壤 lateritic red earth; lateritic red loam
砖红壤性土 lateritic soil
砖红壤状土壤 lateritoid

转氨酶 aminotransferase; transaminase
转氨作用 transamination
转白剂 leucotrope
转变 conversion
转导 transduction
转导物 transducer
转导作用 transduction
转动惯量 moment of inertia
转动中心 pivot point
转化 conversion
转化酶 invertase; sucrase
转化糖 invert sugar; invertose
转化作用 transformation
转换 conversion; transformation; transition
转换成本 conversion cost
转换程序 conversion program
转换器 converter; transformer
转换系数 conversion factor
转绘镜 transferscope
转集楞 hot deck
转角 bight
转角滑轮 brush block
转节(虫足) trochanter
转口税 transit duty
转磷酸酶 transphosphorylase
转录 transcription
转录酶 transcriptase
转染 transfection
转让 assignment; transfer
转生 palingenesis
转羧基酶 transcarboxylase
转羧基作用 transcarboxylation
转糖苷作用 transglycosylation
转头运动(植物) nutation movement
转弯角柱 newel
转弯角柱接头 newel joint
转弯角柱柱顶装饰 newel cap
转酰基酶 transacylase
转线滑轮 transfer block
转向 orientation
转向车 swivel truck
转向处 turning place
转向河绠 swing boom
转向滑轮 knuckle sheave; out-of-lead block; side block; stump roller
转向滑轮系索 stump wire
转向架(铁路或采运机械用) orbital; bogie
转向架车轮的防滑履带套 bogie track
转向拉杆 tie rod
转向轮 fifth-wheel
转向设备 turn
转向销 kingpin
转向轴 steering axle
转写纸 transfer paper
转延侧面 return
转移 transfer; translocation
转移绞盘机滑轮 moving block
转移酶 transferase
转乙酰作用 transacetylation
转运材 removal
转运场 lower landing; relay yard
转运站 reloading landing; transit camp; trans-shipment point
转载 reloading
转折点 inflexion point
转折点桩 intersection point mark
转辙 passing-place
转辙杆 point lever
转辙器 derail; point
转主寄生 heteroecism
转主寄生的 heteroecious; metoecious
转主寄生现象 heteroecism
转主寄主 alternate host
转装 reloading; skyline swing
转板机 board turner
转臂 tumbler
转臂起重机 derrick; derrick crab; derrick crane
转臂起重架 guy derrick
转臂式卸车装置 donovan unloader
转臂式装载机 jib-type loader; swing-boom log loader
转材棍 roll-out bar
转车盘 turning-plate
转锄 rotary hoe
转钩 swivel hook
转鼓滚涂 barrelling
转环 swivel hook
转环滑车 monkey block
转环连接索 vanderpool
转环式剥皮机 rotary-ring barker
转矩 torque moment
转矩扳手 torque wrench

转矩压辊(木材旋切) torque backup roll
转门 swing door
转盘求积仪 rolling-disk planimeter
转球式求积仪 rolling-ball planimeter
转数 number of revolutions; number of speeds
转速表 tachometer
转速测定法 tacheometry
转速计 cycloscope
转筒 drum
转筒式剥皮机 rotary drum barker
转筒式干燥机 drum dryer
转椅 digestive chair; revolving chair; rotary chair
转子 plug; rotor
桩材 piling jack; post
桩箍 cap
桩基础 pile foundation
桩接 peg graft
桩帽 cap
桩木 pile; pile timber; piling; round timber
桩木林 piling show
桩子 ground prop; post
装板机 loader
装板机吊笼 loading cage
装板机液压泵 loader hydraulic pump
装捕兽夹 trapping
装材场 loading area
装材车 kiln bogie; kiln bunk
装材车台 bank
装材框架 load cradle
装材量 load capacity
装材索 drop line
装车 hoisting; loading; on-loading
装车场(河边) banking ground
装车场(陆上) brow; landing; log dump; ramp site; snatching point
装车场账目 landing account
装车短链(固定车立柱) bitch chain
装车杆绷索 jack guy; loading guy line
装车工(广义的) dock man; docker; jack; loader
装车工(绞盘机的) spool tender
装车工(用卡钩) tong hooker
装车工组长 head loader
装车机 loader
装车机牵引索 spotting line
装车架杆 bowleg loader; brow skid; gin pole
装车检尺材积 track scale
装车绞盘机 loading winch
装车卡钩 loading hook
装车楞场 skidway
装车木材压齐器 load squeezer
装车爬杠 loading skid
装车台 deck; elevated ramp; jack works; landing; loading deck; loading dog; loading jack; skid ramp; skidway; timber deck
装车台防护木 brow skid
装车台横垫木 loading skid
装车调节索 tongline
装车线挡木 brow log
装车用短链(固定架杆) bull chain
装肥机 manure loader
装锅器 packer
装夹路线 trap line
装绞盘的小木排 crab
装框 stow
装料 charge; fill
装料斗 feed hopper
装料工 dock man
装料机 filling machine
装料装置 charger
装门锁横档 lock rail
装排推木小船 boom boat
装配 assemblage; assembly; build-up; fit(ting); making up
装配干燥窑 prefabricated kiln
装配工序 assembly section
装配夹具 assembling clamp
装配胶 assembly glue
装配胶黏剂 assembly adhesive
装配孔 pilot-hole
装配面 surfacing
装配图 assembly drawing
装配压紧装置 assembling clamp
装拼 assemblage
装饰板 decorative board; decorative panel; decorative plaque
装饰薄板 decorative sheet(ing)
装饰薄膜 decorative film; decorative sheet(ing)

装饰层压板 decorative laminate
装饰层压纤维板 decorative laminated fibre board
装饰单板 decorative veneer; fancy veneer; figured veneer
装饰单板样片 swatch
装饰的 ornamental
装饰防腐剂 decorative preservative
装饰风格 motif
装饰工段 finishing section
装饰技艺 decorative art
装饰胶合板 decorative plywood; fancy plywood; finished plywood
装饰漆 decorative coating
装饰特点 motif
装饰图案 ornamental design
装饰椅背的薄板 spla; splad
装饰艺术 decorative art
装饰硬质纤维板 ornamental hardboard
装饰用观叶植物 greens
装饰植物 decorative plant
装饰纸 decorative paper
装死 death feigning; death mimicry
装锁横档 lock rail
装锁门梃 locking stile
装填 pack; stow
装填器 loader; packer
装桶 barrelling
装土机 soil filling machine
装屋面板 weather boarding
装卸 handle; handling
装卸点 landing
装卸机 loader and unloader
装卸机械 handling machine
装卸期间 lay-days
装卸桥 cantilever gantry; gantry crane; loading bridge
装卸踏板 loading ramp
装卸台 rollway
装卸曳索 landing line
装卸用三脚架 loading tripod
装卸作业 handling operation; off-and-on loading
装檐板 weather boarding
装有大型绞盘机的铁路平板车 donkey mover
装有绞盘的车 spool cart
装有绞盘机的木排 donkey float
装有螺旋桨的直升飞机 gyrodine
装运 shipment
装运港船边交货 free alongside ship
装运号印 shipping marks
装运清单 shipped bill
装运箱 shipping case
装载 charge; lade; load; on-loading; spot; spotting
装载程序 loading procedure
装载尺寸 load size
装载垫木 carrier block
装载过程 loading procedure
装载滑轮 pick-up block
装载量 shipment
装载区(索道集材) ducking area
装载索具(集材) loading jack
装载台 loading ramp
装载推土两用机 loader-dozer
装载斜坡 loading ramp
装载重量指示器 load-weight indicator
装置 aerial skidder; gear; plant; unit
壮材(直径大于3英寸) stout wood
壮花月季(大姊妹月季) grandiflora rose
壮年期 adolescent stage; maturity phase
壮枝 vigorous branch
状况(动态) regime (*eg.* moisture regime)
状态(动态) regime (*eg.* drying regime)
撞板 bumping plate
撞击震颤声 axial knock
撞针孔螺丝塞 screw plug
追出猎 hunt by chasing or beating
追肥 additional fertilization (fertilizer); after manuring; dressing; supplementary fertilizer; top dressing
追赶法 chasing method
追猎 battue; chase
追猎者 chaser
追踪 trace
追踪调查 periodic inventory
椎接 joggle
锥间谷 atrio
锥接 awl graft
锥面面板 tapered facing

锥形精磨机 conical refiner; jordan (mill)
锥形料仓 conical bin
锥形轮式打磨机 cone-type ring mill
锥形磨浆机 cone-mill; conical refiner
锥形取样器 thief-type trier
锥形物 taper
锥形镟切单板(圆桌面) cone-cut veneer
锥形栽植法 cone planting method
锥植 peg planting
锥足 conopodium
坠落(原木一侧悬空造材时) drop
坠落式分选装置 drop sorter
坠制铅丸 drop shot
准备金 appropriations reserve
准备时间 lead time
准顶极植物群落 quasiclimax
准各向同性 quasi-isotropic
准红壤 lateritoid
准灰壤 semipodzol
准静止锋 quasi-stationary front
准距播种 space drilling
准平原 erosion plain; peneplain; peneplane
准确度 accuracy
准星 fore sight
准星兜 sight hood
准蚁后补充蚁后 substitute queen
准蚁王 substitute king
准有机体 quasi-organism
准周期的 quasi-periodic
准棕壤 brown like soil; para braunerde
准租金 quasi rent
拙劣工作 jackpot
桌面 table top
桌上装饰 table decoration
桌椅两用家具 table chair; table chairewise
桌子 table
着床 embedding
着粉腺 retinaculum
着陆 landing
着陆基地(气球) bedding area
着色 colouring; stain; tincture; tintage; tinting
着色的 tinct
着色剂 colourant; colouring agent
着色力 tinting strength
着色木 stained wood
着色器 tinter
着色物 stainer
着色性 dyeability
着生根瘤的 noduli-ferous
着生植物 epiphyte
着丝点 centromere; kinetochore
着丝点并合 centric fusion
着丝点分裂 centric fission
着丝粒 kinetochore
灼伤(植物) heat asphyxiation; burn; scorch
浊度 turbidity
浊度计 haze meter; nephelometer
浊流层 turbidite
浊流沉积 turbidite
浊流岩 turbidite
浊溶性防腐剂 oil-solvent(-type) preservative
啄钩 floating hook; pike pole
啄痕 peck; peckiness
啄木鸟 creeper
啄食权 peck right
啄位 peck order
孜然油 cumin oil
咨询服务 advisory service
姿态 attitude
资本边际效率 marginal efficiency of capital
资本拨款 capital appropriation
资本财 capital goods
资本-产值比 capital-output ratio
资本负债 capital liability
资本划拨 capital appropriation
资本化 capitalization
资本化公式 formula of capitalization
资本化价值 capitalized value
资本货物 capital goods
资本积累 capital accumulation
资本集约的 capital-intensive
资本集约型工业 capital intensive industry
资本价 soil expectation value
资本价值 capital value
资本结构 capital structure
资本净值 net worth
资本密集度 capital intensity

资本设备 capital equipment
资本升值收入 capital gains
资本生产率 productivity of capital stock
资本收益 capital gains
资本消耗的补偿 capital consumption allowance
资本预算 capital budget
资本预算学 capital budgeting
资本增值税 capital gains tax
资本账户 capital account
资本征收 capital levy
资本支出 capital expenditure
资本重估价 revaluation of capital
资本周转 turnover of capital
资本总额 capitalization
资产 asset
资产负债表 balance sheet
资产和债务 assets and liabilities
资产重估价 revaluation of assets
资金 asset
资金筹集活动 fund raising activity
资金流 capital flow
资金配额 capital rationing
资金运用 application of funds
资料 data
资料卷代号 file number
资源 resource; resources
资源保护区 managed resource protected area
资源管理 resource management
资源管理制度 resource management system
资源量为基础 resource base
资源清查设计 inventory design
资源调度令 resource order
资源调令簿 resource order log
滋养共生 trophic symbiosis
滋养水 enriched water
滋养枝 nurse branch
滋养组织 nurse tissue
髭状节 moustache-knot; wing-shaped knot
子孢交配(真菌) merogamy
子虫道(小蠹虫) brood tunnel
子代 filial generation
子代测定种子园 progeny test seed orchard
子代的 filial
子代退行 filial regression
子弹模 bullet cast
子弹木 ausubo; bullet wood
子弹群射击 group shooting
子弹群中心 centre of group
子二代 F_2
子房 ovary
子房半下位 ovary semiinferior
子房上位 ovary superior
子房下位 ovary inferior
子根 daughter root
子宫 matrix
子核 daughter nucleus
子坑道 larvae gallery
子母五斗柜 chest-on-chest
子囊 ascus (*plur.* asci)
子囊孢子 ascospore
子囊果(真菌) ascocarp
子囊阶段 ascigerous stage
子囊菌 ascomycetous fungus; ascus-fungus
子囊菌类 ascomycetes
子囊壳 perithecium; verruca (*plur.* -ae)
子囊盘 apothecium (*plur.* -ia); discocarp
子囊腔 locule
子囊下层 hypothecium (*plur.* -ia)
子球 cormlet
子染色体 daughter chromosome
子实层 ascus layer; hymenial layer; hymenium (*plur.* -ia)
子实层体 hymenophore
子实体 basidiocarp; conk; fructification; fruit(ing)-body; sporocarp; sporophore
子实体柄 carpophore
子室 brood chamber
子系统 subsystem
子细胞 daughter cell
子线股(钢索) strand
子样品(样本) subsample
子叶 cotyledon; seed leaf; seed lobe; seminal leaf
子叶螺卷胚 embryo spirolobal
子叶下轴 hypocotyls
子叶折叠的 orthoplocal
子叶折叠胚 embryo orthoplocal

Z

子一代 F_1
子座 stroma (*plur.* -ata)
子座的 stromatic
姊妹染色单体 sister chromatid; sister chromosomes
姊妹染色体 sister chromosomes
姊妹体 sib; sibling
姊妹株授粉 adelphogamy
籽苗 plantling
籽苗嫁接 nurse-seed grafting
梓油 stillingia oil
梓樟油 tamala oil
紫变色 purple stain
紫丁香基型木质素 syringyl-type lignin
紫根腐 violet root rot
紫梗 lac encrusted twig
紫梗胶 stick lac
紫胶 lac
紫胶(经脱蜡漂白) wax-free white shellac
紫胶棒(修饰木材) stick shellac
紫胶虫 lacca; laccifer lacca; tachardia lacca
紫胶二氧化硅饼(唱片的原料) biscuit
紫胶封闭漆 seal lac
紫胶红色素 lac dye
紫胶蜡 lac wax
紫胶片 shellac
紫胶染料 lac dye
紫胶树脂 lac resin
紫胶原胶 lac
紫罗兰 gilliflower; common stock
紫罗兰色 florist violet; sweet violet
紫罗兰酮 ionone
紫铆黄酮 butin
紫铆因 butein
紫木兰 lily magnolia; purple yulan
紫苏花岗岩 charnockite
紫苏子油 perilla oil
紫檀素 pterocarpine
紫外光谱 ultraviolet spectrum
紫外-红外线漆膜固化设备 ultraviolet infrared film curing apparatus
紫外线 ultraviolet; ultraviolet ray
紫外线辐射 ultraviolet radiation
紫外线辐射(B区) ultraviolet-B radiation
紫外线光谱图 ultraviolet spectrogram
紫外线摄影 ultraviolet photography
紫外线吸收剂 ultraviolet absorbent
紫外线显微镜 ultraviolet microscope
紫外线像片 ultraviolet photography
紫外线荧光沉积物 ultraviolet fluorescent deposit
紫玉兰 lily magnolia; purple yulan
自变量 independent variable
自播植物 autochore; autochory
自差法 homodyne
自产绿色食品 foot diet
自成土 automorphous soil
自持系统 self-sustaining system
自导式自然小径 self-guiding nature trail
自动编程序 automatic programming
自动编码机 automatic coding machine
自动编码系统 automatic coding system
自动变速装置 automatic transmission
自动补偿 automatic compensation
自动测距镜 range-finder scope
自动测量 automatic gauging
自动程序控制 automatic program control; automatic sequence control
自动秤 automatic scale
自动打捆机 automatic tying machine
自动带式计量装置 automatic band weigher
自动带式砂光机 automatic belt sander
自动单板拼缝机 automatic veneer edge composer
自动登记站(游客) self-registration station
自动抵抗力 active resistance
自动电子计数器 automatic electronic counter
自动定位 automatic allocation
自动定心 automatic centering; self-centering
自动定心吊杆 self-centering boom
自动读数 automatic reading
自动堆垛机 unscrambler
自动堆料机 automatic stacker
自动对流 autoconvection
自动对准压带器 self-aligning slide
自动二进制电子计算机 automatic binary computer

自动跟踪分析数字计算机 automatic digital tracking analyzer computer
自动跟踪控制 automatic following control
自动挂钩 automatic hitch
自动灌溉系统 automatic watering system
自动辊式再锯机 auto-roller feed band resaw
自动恒温器 self-contained thermostat
自动划行器 automatic marker
自动化程度 degree of automation
自动换档 automatic transmission
自动回位起重臂 self-centering boom
自动计量装置 automatic scale
自动计数器 autoscaler
自动记录(游客) self-registration
自动记录测径规 recording caliper
自动记录监控器(温湿度) recorder controller
自动记录轮尺 automatic caliper
自动记录器 self-recorder
自动记录仪 automatic recording instrument; self-registering instrument
自动驾驶仪传动比 gain
自动检测 automatic gauging
自动检尺仪 autoscaler
自动检索 automatic request
自动剪板机 automatic clipper
自动截断锯 automatic cut-off saw
自动进料跑车大带锯 band saw with auto-feed carriage
自动进料台 automatic feeding deck
自动进样 automatic sampling
自动纠正仪 automatic rectifier
自动锯链润滑器 automatic chain oiler
自动抗病性 active resistance
自动控制 automatic control
自动控制干燥窑 automatically controlled kiln
自动控制量槽 auto immersion unit
自动控制器 self-operated controller
自动缆车道 self-acting tramway
自动免疫性 active immunity
自动灭火器 automatic fire sprinkler
自动磨刀机 automatic knife grinder
自动木材缺陷截除锯 automatic defecting saw
自动跑车 auto-feed carriage; automated carriage; automatic carrier
自动喷头 automatic nozzle
自动喷涂装置 automatic spray apparatus
自动喷雾器 automobile sprayer
自动平衡 automatic balance
自动平土机 road patrol
自动气象站 automatic weather station; robot weather station
自动侵染 active infection
自动倾卸装置 tipple
自动取样 automatic sampling
自动润滑装置 automatic lubricating device
自动三面刨 three side planing and moulding machine
自动砂光机 auto-sander
自动砂光生产线 automated sanding line
自动上木装置 automatic lathe charger
自动上树修枝机 tree monkey
自动湿度控制器 automatic humidity controller
自动式鸟枪 automatic shotgun
自动手 autohand
自动售货机 vendor
自动输入 automatic input
自动数据处理 automatic data processing
自动数据处理系统 automatic data processing system
自动数据传输装置 automatic data link
自动数据交换 automatic data exchange
自动数字编码系统 automatic digital encoding system
自动数字输入-输出系统 automatic digital input-output system
自动天平 automatic balance
自动天文导航 automatic celestial navigation
自动调节 automatic adjustment; automatic control; automatic regulation; autoregulation; self-regulation
自动调节恒温器 self-acting thermostat

自动调心供料输送机 self centering infeed conveyer
自动调增条款 escalator clause
自动调整 automatic adjustment
自动调整握紧器 self-aligning grip
自动停止装置 automatic stop device
自动推送机 auto-pusher
自动推送台 unscrambler
自动铣床 automatic shaper
自动校验 automatic check
自动校正 autocorrelation
自动校准 automatic gauging
自动卸板机 automatic press unloader
自动卸材车 dump car
自动信号 automatic signal
自动信息计算 automatic message counting
自动信息检索系统 automatic information retrieval system
自动悬挂 automatic hitch
自动选材机 auto-sorter
自动压力控制 automatic pressure control
自动遥控 automatic remote control
自动液压喷雾器 automatic hydraulic sprayer
自动译码机 automatic code translator
自动雨量计 pulviograph
自动原木归楞机 automatic log deck machinery
自动制动 automatic braking
自动转换输送装置 tipple
自动装板机 automatic press loader
自动装车臂 lift skid arm
自动装配 automatic assembly
自动装卸机 automatic loader and unloader
自动自花授粉 automatic self pollination
自动组装 automatic assembly
自毒的 autotoxic
自读标尺 self-reading rod; speaking-rod
自发的 spontaneous
自发林业 involuntary forestry
自发生长 autonomous growth
自发投资 autonomous investment
自发突变 spontaneous mutation
自发演替 autogenetic succession; autogenic succession
自沸石灰硫黄合剂 self-boiled lime sulphur
自复寄生现象 autoparasitism
自割 autotomy
自给林业 self-supporting forestry
自给农业 subsistence farming
自给生态系统 self-contained ecosystem
自给自足 autarchy
自根的 scion rooted
自根树 own-root tree
自根植物 self-root plant
自行固化胶 self-curing glue
自行固化胶黏剂 self-curing adhesive
自行润滑链锯 oilomatic chain
自行式绞盘机 mobile yarder; winch truck
自行式喷雾机 self-propelled sprayer
自行式起重机 self-propelled crane
自行式修枝机 self-propelled tree prunner
自行式旋耕机 self-propelled rototiller
自行式移植 self-propelled transplanter
自行式装载机 self-propelled loader
自行修枝 self pruning
自核交配 autogamy
自核交配的 autogamous
自花不稔 self-sterile
自花不育基因 self-sterility gene
自花不孕性 self-sterility
自花传粉 self-pollination
自花传粉植物 autogamic plant
自花可稔杂种 self-fertile hybrid
自花能孕的 self-fertile
自花受精 self-fertilization; autogamy
自花受精果 autocarp
自花授粉 selfing; self-pollination
自花授粉植物 self-fertilising plant; self-pollinated plant
自回归 autoregression
自回归变换 autoregressive transformation
自记风速计 recording anemometer
自记湖泊水位计 limnimeter
自记检潮仪 self-recording tide gauge

自记轮尺 recording cal(l)iper(s)
自记气压表 self-recording barometer
自记气压记录 barograph
自记生长计 autographic auxanometer
自记湿度计 recording hygrometer
自记水位计 self-recording tide gauge
自记图 reorder chart
自记温度计 recording thermometer; thermograph
自记温度图表 thermogram
自记雨量计 udomograph
自寄生 autoparasitism
自交 self; selfing
自交不亲和性 self-incompatibility
自交不育 self-sterile; self-sterility
自交不育基因 self-sterility gene
自交后代 selfed progeny
自交率 inbreeding rate
自交能孕的 self-fertile
自交能孕性 self-fertility
自交能孕植物 self-fertile plant
自交树 self
自交系 inbred line; self-bred line; selfed line
自接愈合 self-grafting
自紧绷索 self-tightening guyline
自紧式装车捆扎器 self-tightening load binder
自净作用 self purification
自绝防治 autocidal control
自来水 tap water
自立植物 aautonomous land plants
自流灌溉 gravity irrigation
自流含水层 artesian aquifer
自流井 artesian well
自黏的 self-bonded
自黏性 autohesion
自启通风器 self-opening ventilator
自切 autotomy
自然保护 conservation of nature; protection of nature
自然保护法 natural conservation law
自然保护区 landscape reservation; natural protected area; natural sanctuary; nature conservation; nature preserve; nature protection; nature reserve; nature sanctuary; wilderness area
自然保护区功能区划分 zoning of nature reserve
自然保护区管理计划 management plan of nature reserve
自然保护区管理有效性评估 assessment of nature reserve management effectiveness
自然保护区级别 level of nature reserve
自然保护区禁伐区 reservation
自然保护区类型 categories of nature reserve
自然保护区群 nature reserve complex
自然保护区适宜面积 suitable area of nature reserve
自然保护区体系 nature reserve system
自然保护区域 protected region
自然保护区总体规划 nature reserve master plan
自然保护小区 nature conservation spot
自然变异的 natural variable
自然产生的 spontaneous
自然成熟 physical maturity
自然传粉 natural pollination
自然单性结果 natural parthenocarpy
自然的龄级 stand development class
自然地理类型 physiographic type
自然地理图 physiographic map
自然地理学 physical geography; physiography
自然度 naturalness
自然发生 abiogenesis; autogenesis; autogeny
自然发生的 autogenetic
自然发生说 abiogeny
自然伐期龄 silvical final age
自然法则 natural law
自然防治 natural control
自然分布区 natural range
自然分化 natural differentiation
自然分类系统 natural system
自然丰度 natural abundance
自然风化 natural weathering
自然风景 natural landscape; scenery
自然干燥 air seasoning; natural drying; natural seasoning
自然干燥材 natural seasoned timber; natural seasoned wood

自然更新林 natural regenerated forest
自然公园 nature park
自然环境恢复局 Bureau of Reclamation
自然火 spontaneous fire
自然加热 spontaneous heating
自然嫁接 natural graft
自然交配 wide pairing
自然禁猎区 nature sanctuary
自然景色 landscape
自然控制 natural control
自然老化 natural ageing
自然利息 natural interest
自然林 indigenous forest
自然轮伐期 biological rotation; natural rotation; physical rotation
自然旅游资源 natural tourism resources
自然平衡 balance of nature; natural balance
自然奇观 natural wonders
自然侵蚀 natural erosion
自然区划 natural division
自然区划法 natural division method
自然群集 natural community
自然生态系统类自然保护区 nature reserve for natural ecosystem
自然式(园林) natural style
自然式园林 natural garden
自然淘汰 natural elimination
自然调节 natural regulation
自然通风干燥窑 natural ventilating dry kiln; natural-draft kiln; natural-draught kiln
自然突变 natural mutation
自然污染 natural pollution
自然稀疏 autothinning; law of natural thinning; natural thinning; self thinning
自然选择 natural selection
自然选择说 theory of natural selection
自然循环干燥窑 gravity circulation dry kiln; natural-circulation kiln; natural-draft kiln; natural-draught kiln
自然演替 natural succession
自然遗迹 natural monument
自然遗迹类自然保护区 nature reserve for natural monument
自然杂种 natural bastard
自然灾害 physical damage
自然再生林 natural regenerated forest
自然增长(利息) accrue
自然增长利润 accrued interest
自然增值 accrued value; unearned increment
自然整枝 autolopping; natural pruning; self pruning
自然植被 natural vegetation
自然种群 natural population
自然资本 natural capital
自然资源 natural resource; natural resources
自然资源管理 Natural Resource Management
自然资源信息系统 natural resource information system
自燃 self-combustion; spontaneous combustion
自燃温度 autoignition temperature
自燃着火点 self-ignition point
自燃作用 self-ignition
自溶(细胞) autolysis
自上而下模型 top-down model
自身寄生物 autoparasite
自身驯化 self-domestication
自生的 autogenic
自生林 indigenous forest
自生苗 naturally sown young seedling; volunteer
自生正片 autopositive
自生种 autochthonous species
负$\frac{2}{3}$自疏法则 thinning law of $-\frac{2}{3}$
自锁捆木索钩 buzzard hook
自锁跑车 automatic locking carriage
自锁索道跑车 self-locking skidding carriage
自锁套接 buzzard hook
自体传播 autochory
自体繁殖 autoreproduction
自体分化 self-differentiation
自体分裂 autotomy
自体嫁接 autograft; autoplastic graft
自体适应 autoadaptation
自体噬菌体 autobacteriophage

自体受精 autofecundation; autofertilization; self-fertilization; autogamy
自体受精的 autogamous
自体抑制物质 self-inhibitor
自体愈合 self-healing
自卫潜能 protective potential
自我登记(游客) self-registration
自我登记站(游客) self-registration station
自我破坏 self-destructive
自我吞噬作用(细胞) autophagy
自我修复 self-healing
自下而上模型 bottom-up model
自相关 autocorrelation
自相关年表 autocorrelation chronology
自卸车 dumper
自型土 automorphous soil; idiomorphic soil
自养 autotrophism; autotrophy
自养层 autotrophic stratum
自养成分 autotrophic component
自养代谢作用 autotrophic metabolism
自养的 autotrophic; holophytic
自养呼吸 autotrophic respiration
自养湖 autotrophic lake
自养类型 holophytic form
自养生物 autotroph; autotrophic organism
自养细菌 autotrophic becteria
自养演替 autotrophic succession
自养营养 autotrophic nutrition
自养植物 autophyte; autotrophic plant; holophyte
自营苗圃 home nursery
自用林 home-use forest
自由度 degree of freedom
自由伐 felling of wolf trees; free cutting; isolation cutting; liberation cut; liberation felling; severance felling
自由放牧地 open range
自由分子 free molecule
自由基引发聚合反应 radical polymerization
自由交配 pangamy
自由空间 free space
自由孔隙 free pore space
自由梁 free-free beam
自由蔓延火 free-burning fire
自由贸易 free trade
自由旁拉索道集材法 free end method
自由漂浮水生植物群落 free floating hydrophyte community
自由切割 free cutting
自由生活 free living
自由生活型 free-living form
自由湿气 free moisture
自由市场行情 free market scenario
自由授粉 free pollination; open pollination; random pollination
自由授粉实生种子园 open pollination seedling orchard
自由授粉子代 open pollination progeny
自由疏伐 free-thinning
自由树 open-growing tree
自由水 free water
自由水分 free moisture
自由支配的购买力 discretionary purchasing power
自由支配的开支 discretionary spending
自由支配的收入 discretionary income
自由柱容积 free column volume
自由转钮(连接索具) shay swivel
自由组合 independent assortment
自游生物 nekton
自有资本 equity capital
自愿捐款 voluntary contribution
自愿协议 voluntary agreement
自运节律 free-running rhythm
自载拖车回空 piggyback
自增长 self-propagation
自殖系统 selfed line
自重 dead load(ing); dead weight
自重挠度 unloaded deflection
自主变量 autonomous variable
自主投资 autonomous investment
自装集材机 self-loading skidder
自装家具 self-assemble furniture
自装运材车 crane car; crane truck; self-loading log truck
自组织 self-organization
字符 character

Z

字符发生器 character generator
字节 byte
字码处理 word processing
综合 integration
综合产量 combined production
综合防治 integrated control
综合分析 meta-analysis
综合工厂 integrated mill
综合公园 comprehensive park
综合管理方法 integrated management approach
综合立地指示 combination site indicator
综合利用 integrated utilization
综合品种 synthetic variety
综合评估 integrated assessment
综合区划法 combined division method; intergrated division method
综合设计 general design
综合生命维持系统 integrated life support system
综合疏伐 combined method of thinning
综合性采伐 integrated logging
综合休眠 combined dormancy
综合因素 integrative factor
综合应力 compound stress
综合优势比 summed dominance ratio
综合指数 aggregate index number; composite index number
综述 round up
棕腐酸 ulmic acid
棕钙土 brown soil
棕褐色 tan
棕榈 hemp palm; Chinese coin palm; windmillpalm
棕榈材 palm wood
棕榈核油 palm-nut oil
棕榈科树木 palm
棕榈蜡 caranday wax; palm wax
棕榈仁油 palm kernel oil
棕榈温室 palm house
棕榈叶蜡 raphia wax
棕榈油 palm butter; palm oil
棕漠草原土 brown desert-steppe soil
棕漠土 brown desert soil
棕壤 brown earth; brown soil; brunisolic soil; burozem
棕壤化 browning
棕壤型土 brown like soil; para braunerde
棕色草甸草原土 brown meadow steppe soil
棕色冲积土 alluvial brown soil
棕色腐殖质钙质土 brown humus calcareous soil
棕色钙质土 brown calcareous soil
棕色火山灰土 brown andean soil
棕色森林土 brown forest soil
棕色石灰土 brown rendzina
棕色土 umber
棕色团粒作用 brown granulation
棕色氧化硅铝土 brownish oxisiallitic soil
棕尾毒蛾 yellow-tail moth
鬃 bristle
总苞 anthodium; envelope; involucre; perianth; periclinium; periphoranthium
总苞的 involucral
总保持量 gross retention
总材积 gross scale; gross volume
总材积增益百分率 gross volume increase percentage
总采伐计划 general cutting-plan
总参数 population parameter
总产量 gross output; gross product; ultimate production
总产值 gross product
总成 unit
总成本 total cost
总尺寸 overall size
总抽出物 total extractive
总初级产物 gross primary production
总初级生产力 gross primary productivity (GPP)
总传动比 overall gear ratio
总的 entire
总店 home office
总断面积 total basal-area
总额 lump sum
总发芽率 total germination; total germination percent
总发芽势 total germinability
总费用 all-in cost
总负荷 total load
总干缩率 total shrinkage
总公司 home office

总功 total work
总光合速率 gross photosynthetic rate
总含量 total content
总和 resultant
总花梗 peduncle
总花托(用于菊科) clinanthium; clinium; coenanthium
总花序梗 common peduncle
总环流 general circulation
总环流模式 general circulation model
总计 lump sum
总价采购 lump sum acquisition
总价格生长率 total value increment percent
总检修 major overhaul
总碱 total alkali
总经理 managing director
总经营计划 general working-plan
总开支 overhead cost
总科(分类单位) superfamily
总括盈余概念 all-inclusive concept
总利润 total profit
总连年生长量 total annual growth
总量 population parameter
总林分表 total stand table
总平均年生长量 gross mean annual increment
总平均生长量 mean gross increment; total mean increment
总平均值 general average; population mean
总平面布置 general arrangement
总热 integral heat
总热量 total heat
总生产量 gross production
总生产率 gross production rate
总生长 gross growth
总生长量 gross increment; total growth; total increment
总生殖率 gross reproductive rate
总收额 gross revenue
总收获表 general yield table
总收获材积 total volume
总收获量 total yield
总收获最多的轮伐期 highest gross yield rotation
总收入 gross income
总收缩率 total shrinkage
总收益 total revenue
总受器 catch-all
总数 population parameter
总死亡率 crude death rate
总所得税 gross income tax
总体(统计学) population
总体布置 general arrangement
总体参数 parameters of population
总体单元 units of population
总体规划 general plan; master plan
总体流动 bulk flow
总体统计 population statistic
总图(机械与建筑) general plan; key map; general drawing
总吸收 total absorption
总吸收量 gross absorption; gross retention; gross uptake
总吸收热 integral heat absorption
总效率 total efficiency
总需求 aggregate demand
总需氧量 total oxygen demand
总悬浮颗粒物 particulate matter less than 100 μm in diameter (PM100)
总压力 gross pressure; total pressure
总叶柄 common petiole
总叶面积 summed leaf area
总叶重量 summed leaf weight
总应力 total stress
总有机碳含量 total organic carbon
总账 ledger
总照明度 integrated illumination; integrated illumination intensity
总值 lump sum
总重 gross weight
总状的 racemose
总状分枝式 monopodial branching
总状花序 botrys; raceme; racemose inflorescence
总状聚伞花序 cymo-botryoid; cymo-botryose
纵挡板 longitudinal wall
纵断面 edge grain; longitudinal profile; longitudinal section; longitudinal surface; vertical section
纵断面曲线 vertical curve
纵割 slit
纵割刀 slitting knife

Z

纵横交叉倾斜堆积法 crosswise sloped piling
纵横截两用锯 combination saw
纵横锯边机 sizing saw system
纵横两用锯 mitre saw
纵横两用圆锯 circular combination saw
纵火 incendiary; incendiary fire
纵火剂 incendiary agent
纵锯 rip
纵连钩链(串集) coupling grab
纵梁 stringer piece
纵裂(木材) simple check; lengthwise check
纵裂木材 splinter wood
纵木滑道 pole chute
纵木滑道杆滑道 pole slide
纵剖 ripping
纵剖锯 pit saw; rip saw; ripper; ripping saw
纵剖开 rip removed
纵剖面 longitudinal profile
纵剖图 elevation
纵切 rip; rip cutting
纵切面 longitudinal section; longitudinal surface
纵切排锯 gang rip saw
纵切圆锯机 bolter
纵视图 longitudinal view
纵条(虫) stripe
纵纹 side grain
纵向薄壁组织 longitudinal parenchyma
纵向薄壁组织束 longitudinal strand parenchyma
纵向壁 longitudinal wall
纵向成型铺装头 machine direction forming head
纵向创伤树脂道 longitudinal traumatic gum canal
纵向弹性压带器 longitudinal pad
纵向道 fore-and-aft road
纵向定向结构刨花板 oriented structural board-machine direction
纵向堆垛法(材长与干燥窑长平行) end-piled loading; lengthwise type piling
纵向堆积 end stacking
纵向分子 longitudinal element
纵向干缩 longitudinal shrinkage
纵向管胞 longitudinal tracheid
纵向间隙 axial clearance
纵向剪切 longitudinal shear
纵向胶合 end gluing
纵向接缝 longitudinal joint
纵向锯解 rip cutting; ripping; saw with the grain
纵向锯开 rip removed
纵向开裂 longitudinal check; longitudinal crack
纵向空气流动 longitudinal movement of air
纵向扩张 vertical expansion
纵向联合 vertical integration
纵向木薄壁组织 longitudinal wood parenchyma
纵向木材分子 longitudinal wood element
纵向木质部细胞 longitudinal xylary cell
纵向排水道 longitudinal culvert
纵向膨胀 longitudinal swelling
纵向拼缝机 longitudinal splicer
纵向强度 longitudinal strength
纵向翘皓 spring
纵向翘曲 longitudinal warping
纵向切削 longitudinal cutting
纵向侵蚀 longitudinal erosion
纵向生长 elongation; longitudinal growth
纵向收缩 kick; longitudinal shrinkage; vertical-contraction
纵向输送机 longitudinal conveyor
纵向树脂道 longitudinal resin canal; vertical resin canal
纵向弯曲 longitudinal bending
纵向位移 longitudinal displacement
纵向形成层原始细胞 longitudinal cambial initial
纵向压缩过程 longitudinal compression process
纵向应变 longitudinal strain
纵向应力 longitudinal stress
纵向载荷 longitudinal load(ing)
纵向造材 endwise bucking
纵向振动 longitudinal vibration
纵向中耕 length(wise) cultivation
纵向重叠 end lap

纵向装堆 end-piled loading
纵形炭窑 alpine kiln
纵压 end compression; endwise compression
纵枕木 stringer piece
纵轴型干燥窑 longitudinal shaft kiln
纵坐标 ordinate
走触性 stereokinesis
走刀 feed
走道 walkway
走火通道 fire escape
走廊 passage; veranda(h)
走廊林 galleria
走线材 miscut lumber
走向 orientation
租地制度 land tenure
租金 rent
租金和维修费 rental and maintenance
租金价 rental value
租赁 leasehold; rent
租赁者 leaseholder
租用者 permit holder
足 pes (*plur.* pedes)
足尺造材 cut full
足迹味 foot-scent
足丝蚁 web-spinner
足味 foot-scent
足浴 foot bath
族(分类) tribe; race
族间交配 racial crossing
族聚 aggregation; granulation
阻碍 blocking
阻碍素 deterrents
阻挡效应 blocking effect
阻风门(化油器) choke
阻火位置 anchor point
阻截火头(防火) front stop
阻抗计 impedometer
阻拦 detain
阻力矩 moment of resistance
阻尼 quench
阻尼能力 damping capacity
阻尼器 damper
阻燃 fire prevention; fire-proofing; flame retardation; glow retardant
阻燃处理 fire retardant treatment
阻燃处理材 fire-retardant-treated wood
阻燃防腐剂 fire-retardant preservative
阻燃剂 fire proofing agent; fire retardant; fire retardant agent; fire retardant chemical; flame retardant
阻燃时间 duration of fire resistance
阻燃涂层 fire retardant coating
阻燃盐剂 fire retardant salt
阻塞 clogging; jam; stem
阻湿处理 moisture-retarding treatment
阻湿剂 moisture repellent; moisture retardant
阻湿涂料(层) moisture retardant coating
阻食剂 feeding deterrent
阻水层 aquiclude
阻抑 depression
阻止 damp
阻止作用 interception
阻滞剂 retardant; retarding agent
阻滞木 dragger
组 battery; branch; class; section
组氨酸 histidine
组成 build-up; composition; making up
组成成分 constituent
组成代谢 anabolism
组成代谢现象 anabolic phenomenon
组成酶 constitutive enzyme
组蛋白 histone
组点 point of aim
组分 component
组分结构 components structure
组分可加性 additivity of components
组分区分(器官及大小等) comoponent fractionation
组合 resultant; society
组合薄木 reconstituted veneer
组合材堆单元 unit package
组合齿 combination teeth
组合单板 reconstituted veneer
组合刀头 dado head
组合堆积法 packaged stacking
组合辊 line up roll
组合家具 combination furniture; unit furniture
组合件 package
组合件的 nodal

组合胶 assembly glue
组合胶黏剂 assembly adhesive
组合宽度 aggregate width
组合梁 built-up beam
组合砂光机 abrasive planing-contact sanding and satining machine
组合时间 assembly time
组合式砂光机 built-up type sander
组合梯 sectional ladder
组合铣刀 cutter block
组合选择 combined selection
组合原木板英尺材积表 combination log rule
组件 assembly; layup
组距 class interval
组模 gang mould
组内相关系数 intraclass correlation coefficient
组坯 layup; lay-up
组坯机 lay-up machine
组筛(成套筛子) sizing screens
组台堆积法 package piling
组织 fabric; tissue
组织层次 level of organization
组织发生 histogenesis
组织分化 histological differentiation
组织分解 histolysis
组织化学 histochemistry
组织解剖学 histological anatomy
组织内寄生的 endobiotic
组织排列 orientation of tissue
组织培养 explantation; tissue culture
组织切片 microsection
组织亲和性 histocompatibility
组织溶解 histolysis
组织图 organization chart
组织走向 orientation of tissue
组装 solid content
组装胶合 assembly gluing; constructional gluing; secondary gluing
祖先 progenitor
钻床 drilling machine; power drill
钻杆(生长锥) borer stem
钻孔 auger boring; bore
钻孔毒杀(用液状或粉状毒药) bore-hole poisoning
钻孔防腐法 bore hole method
钻孔机 borer; boring machine; drilling machine
钻孔纤维板 peg board
钻孔枕木 bored tie
钻刃 cutting lip
钻头(生长锥等) auger borer; drill bit; power drill; borer bit
钻头螺纹(生长锥) bit thread
钻销孔 draw-bore
钻缘角 lip angle; lip clearance
嘴(喷口) spout
嘴峰 culmen
嘴甲 nail
最初的 primary
最初日产置 initial daily production
最大材积轮伐期 rotation of maximum volume production; rotation of the greatest volume
最大材积收获轮伐期 rotation of maximum volume yield
最大承载力 Maximum carrying capacity
最大持水量 maximum water holding capacity
最大持续收获量 maximum sustainable yield; maximum sustained yield
最大出生率 maximum natality
最大负载 top load
最大工作概率值 maximum working probit
最大荷重功 work to maximum load
最大降水带 zone of maximum precipitation
最大可见距离 visual range
最大立木度 maximum stocking
最大流量 peak flow
最大流最 peak discharge
最大毛管持水量 maximum capillary capacity
最大毛管力 maximum capillarity
最大年间保蓄量 maximum annual sustained yield
最大容许剂量 maximum permissible dose
最大容许误差 maximum allowable error
最大森林净收益轮伐期 rotation of maximum income; rotation of the highest net income

最大似然估计值 maximum likelihood estimator
最大似然判断法 maximum likelihood decision
最大似然性 maximum likelihood
最大收益轮伐期 maximum cash flow rotation; rotation of maximum cash flow
最大树高 maximum tree height
最大田间持水量 maximum field capacity
最大土地净收益轮伐期 rotation of maximum soil rent
最大稳定期 maximum stational phase
最大纤维应力 extreme fibre stress
最大限度利润 profit maximization
最大蓄积林分 fully stocked stand
最大运算概率值 maximum working probit
最大载荷 peak load
最大自然生产力轮伐期 rotation of maximum physical productivity
最低光强 minimum light-intensity
最低光照要求 minimum light requirement
最低价 bottom price
最低生长温度 minimum growth temperature
最低售价 upset price
最低限价 bottom price
最低养分法则 law of minimum nutrient
最低有效温度 minimum effective temperature
最低致死温度 minimum fatal temperature
最高纯收益原理 theory of the highest netrevenue
最高存活温度 maximum survival temperature
最高点 zenith
最高分辨度 peak resolution
最高价 ceiling price
最高利润 profit maximization
最高毛收获原理 theory of maximum gross yield
最高收入轮伐期 rotation of the highest income
最高收益轮伐期 largest income rotation
最高收益原理 theory of highest income
最高温度 maximum temperature
最高限价 ceiling price
最高有效温度 maximum effective temperature
最高致死温度 maximum fatal temperature
最高自记温度表 maximum recording thermometer
最高总收获原理 theory of maximum gross yield
最后产量衡值法则 law of constant final yield
最后产物 end product
最后加工 finish
最后价值量 ultimate quantity of value
最后一批木材推河 tail-down
最后致死带 zone of eventual death
最佳程序栽植 optimum programming plantation
最佳抽样设计 optimization of sampling deign
最佳分布状态 optimum distribution
最佳干燥条件 optimum drying condition
最佳门限值 optimum threshold
最佳设计 optimum design
最佳条件选择 optimization
最佳下锯法 best opening face
最佳阈值 optimum threshold
最近邻法 nearest neighbour method
最近亲杂交 most closed crossing
最少样本数 required sample size
最适捕获量 optimum catch
最适捕食理论 optimal foraging strategy
最适度 optimum
最适含水量 optimum moisture content
最适密度 optimum density
最适气候分布区 region of climatic optimum
最适曲线 optimum curve
最适生长率 optimum growth rate
最适温度 optimum temperature
最适因子 optimum factor
最小持水量 least water-holding capacity
最小存活面积 minimum viable area

Z

最小二乘法 least square method
最小干围 minimum girth
最小空气容量 minimum air capacity
最小量定律 law of the minimum
最小面积 minimal area
最小损失防火理论 minimum-damage fire-control theory
最小田间持水量 minimum field capacity
最小需光度 minimum light-intensity
最小样方面积 minimum quadrat area; minimum quadrate area
最小样方数目 minimum quadrat number
最小因子 minimum factor
最小因子法则 law of the minimum
最小直径 minimum DBH (by inventory); minimum diameter
最小值 valley value
最小致死量 minimum lethal dose
最小转弯半径 minimum turning radius
最优产量法则 theory of the optimal yield
最优方式 optimal fashion
最优化 optimization
最优配置 optimum allocation
最优生长 best growth
最优收获量 optimum yield
最优政策 optimal policy
最终产品 end product; end-use product
最终处理 final treatment
最终发芽率 final germination percent
最终间距 final spacing
最终结果 end value
最终平均年生长量 final mean annual increment
最终渗水量 final infiltration capacity
最终消费者 ultimate consumer
最终因子 ultimate factor
最终用途 end use
最终值 end value
最终贮木场 terminal yard
最终状态(热硬化性树脂) C-stage
左聚糖 levan
左螺旋纹理 left-handed spiral grain
左捻 left-hand lay; S lay
左捻索 left lay rope
左偏(狩) cast on
左切刀 left-hand cutter
左绕 S lay
左绕钢丝绳 left lay rope
左视图 left view
左手带锯木机 left-hand sawmill
左手带锯制材厂 left hand bandmill; left-hand sawmill
左旋扭曲 left-handed twist
左旋糖 laevoglucose; laevulose
左旋铣刀 left-hand cutter
左移 paramagnetic shift
左右对称 lateral symmetry
左右双筒 side-by-side double barrel
左右调节手轮(大带锯的) right and left adjustment hand wheel
佐剂 adjuvant
柞蚕 perny silk moth; perny silk worm; tusser worm; wild silk worm
作床 bedding
作床机 bed former; bed shaper
作功系数 working factor
作垄 bedding; earthing up; ridge plowing; ridging
作垄器 shaper
作畦 bedding
作畦打埂器 shaper
作坛 bedding
作土堆 mounding
作物 crop
作物生产潜力 crop productivity potential
作物生态型 agroecotype
作穴 dibbling
作业 chance; job; process; system; working
作业槽 operating tank
作业队 side
作业法 silvicultural system; working system; regime; system of management
作业方便的伐区 short haul
作业费 operating cost
作业费用 working cost; working-expenses
作业分析 job analysis
作业工段 working-block
作业规程 job specification

作业规划 works planning
作业稽核 control of operation; work tracking
作业级 working circle; management class; system class; working group; working section; working unit
作业级标准采伐量 standard cutting amount
作业级更新面积 standard planting area
作业计划 works program
作业计划期 forest management period; planning period
作业进度图 operation map
作业坑 operating pit
作业困难的伐区 one-choker show
作业目标林分 exploitable stand
作业评价 job evaluation
作业区 allotment; camp; chance; crops; felling unit; parcel
作业设计簿(按小班) working project sheet
作业深度(耕挖深度) operating depth; depth of work
作业生产计划 operative productive planning
作业说明书 job specification
作业索 operating line; operating rope
作业调查 forest operational inventory
作业图 flow diagram
作业现场 working site
作业线 operating line
作业线能力 line capacity
作业小组 logging crew
作业研究 operation(al) research
作用 working
作用半径 radius of action
作用光谱 action spectrum
作用面 active surface
作用区 active region
作用物 substrate
作用子 cistron
坐标 coordinate; coordinates
X-Y坐标定心 X-Y charger
坐标几何计算机程序 coordinate geometry computer program
坐标仪 coordinatograph
坐标纸 coordinate paper
坐标制图器 coordinatograph
坐茬(伐而不倒) skybound
坐果 fruit setting
坐坡 jump; kickback
坐犬 setter
坐式(射击) sitting
坐卧两用椅 chair bed
坐卧长椅 davenport bed
座板 bed plate; stay plate
座框 seat-frame

英文索引

A-flute, 495
A-frame, 231, 396
A-frame derrick, 231
A-frame loader, 231
A-frame spar, 223
A-frame yarder, 231
A-grade thinning, 222, 444
A-layer, 151
A-value, 615
A_0-horizon, 103
A_0-layer, 42, 278
A_2-layer, 210
Aapa moor, 162, 382
aapamoor, 162, 382
aautonomous land plants, 645
abacus, 105, 283
abandoned assets, 15
abandoned field, 139
abandoned land, 297, 371, 446
abandonment, 297, 371
abatement, 16, 139, 233, 277, 531
abaxial paratracheal parenchyma, 495, 594
abaxial parenchyma, 495, 594
Abbe refractometer, 1
abdomen, 155
abdominal ganglion, 155
abdominal leg, 155
abele, 570
aberration, 24, 221
abieninic balsam, 286
abies balsam, 286
abietic-type acid, 71
abietin, 465
abietine, 465
abietyl, 71, 465
abietylamine, 71, 465
abiogenesis, 510, 645
abiogenic, 138
abiogenic organic molecule, 138
abiogeny, 138, 645
abiological, 138
abioseston, 138
abiosis, 422, 462
abiotic, 138
abiotic component, 138
abiotic control, 138
abiotic environment, 138
abiotic factor, 138
abiotic substances, 138
ablactation, 271, 276
ablastin, 568
ablastous, 510, 511
ablation, 327, 377, 532, 602
abloom, 268
Abney level, 1
abnormal circulation, 566
abnormal climate, 566
abnormal cutting amount, 136
abnormal division, 566
abnormal final age, 136
abnormal forest, 136, 138
abnormal germination, 568
abnormal overstocked stand, 289
abnormal seedling, 568
abnormal spoilage, 138
abnormal state, 34, 566
abnormal surface runoff, 566
abnormal tissue, 566
abnormal wood, 566
abnormal working section, 136
abnormality, 24, 129, 221, 566
abnormis, 568
aboospore, 86
aboriginal, 19, 491, 527
abortion, 7, 32, 116, 125, 306
abortive egg, 7
abortive pollen, 7
aboveground biomass, 97
abradability, 326, 327
abradant, 326
abrade, 326, 327
abrader, 326
abrading test, 326
abrasion, 31, 326, 327, 555
abrasion hardness, 327, 337
abrasion platform, 283
abrasion resistance, 270, 337
abrasion test, 326, 337
abrasion testing machine, 326, 337
abrasive, 326
abrasive band, 412
abrasive belt, 412
abrasive belt size, 412
abrasive belt tension roll, 412
abrasive brush, 412, 451
abrasive cloth, 412
abrasive compound, 153, 555
abrasive disk, 326, 327, 412
abrasive dresser, 412
abrasive grain, 326, 351
abrasive hardness, 326, 337
abrasive machining, 327, 555
abrasive material, 326, 555
abrasive paper, 412
abrasive particle, 326
abrasive planer, 326, 412
abrasive planer-contact sander, 244
abrasive planing, 326, 327, 412
abrasive planing-contact sanding and satining machine, 652
abrasive powder, 555
abrasive resistance, 270, 337
abrasive stick, 326, 327, 578
abrasive substance, 326, 555
abrasive surface, 326, 555
abrasive wear, 327
abrasive wheel, 326, 412
abration, 379
abri, 21, 554, 571
abric acid, 197
abrin, 197, 528
abrupt ecospecies, 486
abruptly pinnate, 349
abscisic acid, 494
abscisin, 494
absciss layer, 287

A

absciss phelloid, 334
abscissa, 503
abscission layer, 287
abscission period, 314
abscissphelloid, 334
absconded swarm, 474
absconding swarm, 474
absinthe oil, 278
absolute alcohol, 510
absolute altitude, 265
absolute atmosphere, 265
absolute cost, 68, 264
absolute cover, 264
absolute density, 265
absolute dry weight, 265, 385
absolute error, 265
absolute even-aged stand, 265
absolute forest land, 265, 636
absolute forest soil, 265, 636
absolute form factor, 265
absolute form quotient, 265
absolute humidity, 265
absolute lethal gene, 498
absolute maximum bending moment, 265
absolute maximum fatal temperature, 265
absolute maximum gradient, 265, 526
absolute maximum shearing force, 265
absolute minimum fatal temperature, 265
absolute moisture content, 265
absolute oil, 251, 257
absolute orientation, 265
absolute parallax, 265
absolute pressure, 265
absolute price increment, 265
absolute programming, 264
absolute reproductive value, 265
absolute return, 264, 265
absolute screw-holding power, 265
absolute strain value, 265
absolute taper, 265
absolute temperature, 265
absolute temperature scale, 265, 269
absolute value, 265
absolute value computer, 265
absolute viscosity, 265
absolute weight, 265, 436
absolutely dried condition, 265, 385
absolutely dry wood, 265, 385
absorbance, 514
absorbance index, 514, 515
absorbancy, 514, 531
absorbed energy, 514
absorbed water, 515
absorbent, 513, 514
absorber, 233, 514
absorbing capacity, 514
absorbing inheritance, 492
absorbing quality, 515
absorptance, 514
absorptiometer, 514
absorptiometry, 514
absorption, 514
absorption ability, 514
absorption band, 514
absorption coefficient, 515
absorption costing, 183, 385
absorption curve, 514
absorption-desorption kinetics, 514
absorption hygrometer, 514
absorption of moisture, 513, 514
absorption period, 514
absorption rate, 515
absorption ratio, 514
absorption spectrum, 514
absorption water, 515
absorption zone, 514
absorptive crossing, 492, 515
absorptive power, 514
absorptivity, 514, 515
abstinence element, 246
abstract community, 61
abstract growth series, 61
abundance, 116, 144
abundance-frequency ratio, 116
abundance ratio, 144
abundant year, 144
abutment, 172, 377, 612
abutment pier, 3, 377
abutting edge, 114
abutting joint, 111, 114
abutting plate, 114
abyss, 420
abyssal, 420
abyssal benthos, 420
abyssal zone, 420
Abyssinian gum, 1
abyssopelagic zone, 420
A-C soil, 246
Ac-zol, 484
acacatechin, 285
acacetin, 71, 248
acacia, 248
acacia extract, 194
acacia gum, 1, 248
acacia oil, 248
acacia savanna, 248
acaciin, 71
acacin, 248
acacinin, 248
acajou balsam, 558
acalcerosis, 387
acalcicosis, 387
acantha (*plur.* -thae), 71
acanthaceous, 260, 266, 462
acanthon, 71
acarid mite, 144
Acarina, 357
acaroid gum, 191
acaroid resin, 191
acarology, 357
acaryotic, 509
acaulescent, 509
acauline, 260, 509
accelerant, 73
accelerated ageing, 228, 394
accelerated air-drying, 228, 376
accelerated cement, 279
accelerated creep, 228
accelerated depreciation, 228
accelerated erosion, 228
accelerated germination test, 279
accelerated growth, 228
accelerated gum, 466
accelerated increment, 228
accelerated test, 228

A

accelerated weathering, 228
accelerating agent, 73
accelerating belt, 228
accelerating culture, 466
accelerating death phase, 228
accelerating germination, 73, 74
acceleration belt conveyor, 80
acceleration conveyor, 228
accelerator, 73, 74, 228, 244, 575
accent plant, 577, 629, 633
accented neutral colour scheme, 629
acceptable daily intake, 397
acceptable intake, 595
acceptance certificate, 556
acceptance check, 556
acceptance condition, 556
acceptance inspection, 556
acceptance procedure, 556
acceptance test, 556
accepted chip, 192
accepted stock, 192
acceptor, 443
access door, 24, 233
access hole, 233
access mechanism, 74
access plate, 178, 232
access road, 24, 481
access time, 74, 111, 384
accessibility, 251, 273, 421
accessibility class, 96, 272
accessibility of forest, 408
accessibility of timber, 301
accessible forest, 272
accessible forested land, 273
accessory, 150, 152
accessory branch, 154
accessory bud, 154
accessory calyx, 154
accessory cambium, 154
accessory cell, 154
accessory chromosome, 120, 154, 541
accessory equipment, 150
accessory frame, 154, 383
accessory fruit, 152, 154, 230
accessory gland, 154
accessory pigment, 150
accessory pollination, 150
accessory product, 154
accessory risk, 152
accessory seed, 214
accessory species, 71, 150
accessory system, 150
accident insurance, 438, 569
accident yield, 304
accidental, 348
accidental cutting, 304, 513
accidental error, 348
accidental shut down, 348, 438, 486
accidental species, 348
accidental sport, 32, 348
accidental variation, 348
acclimation, 440, 549
acclimatization, 369, 549
accommodation, 440, 479
accompanying species, 11
accompanying table, 152
accordion, 274, 589
accordion door, 606
account, 225, 278, 604
account day, 246
account payable, 572
account receivable, 573
account sales, 532
accounting, 225, 278
accounting approach, 194, 225
accounting item, 278
accounting method, 194, 225
accounting procedure, 278
accounting record, 279
accounting system, 279
accounting transaction, 278
accounting voucher, 278
accretion, 53, 114, 497, 515, 602
accretion borer, 430
accretion cutting, 429
accretion thinning, 429
accrue, 602, 646
accrued asset, 306, 572, 602
accrued depreciation, 572, 573
accrued interest, 572, 646
accrued value, 646
accumulated amortization, 285
accumulated deficit, 285
accumulated depletion, 285
accumulated depreciation, 285
accumulated error, 285
accumulated loss, 285
accumulated temperature, 219
accumulated total punch, 285
accumulation horizon, 103
accumulation table, 620, 635
accumulation terrace, 114
accumulational platform, 114
accumulative raingauge, 285
accumulator, 74, 212, 285, 545
accumulator clamp, 183
accumulator forwarder, 112
accumulator tank, 635
accuracy, 254, 640
accuracy of adjustment, 480
accuracy of dimension, 57
accuracy of manufacture, 620
accuracy of measurement, 44
accuracy rating, 254
aceessary, 150, 152
acentric chromosome, 512
acentric fragment, 512
acentric inversion, 512
acentric ring, 512
acentrics, 512
acephate, 566
acer tannin, 371

A

aceric acid, 371
acerose leaf, 606
acervulus(*plur.* -li), 143
acescent, 501
acetannin, 566
acetate of lime, 73, 565
acetate rayon, 565, 566
acetate silk, 565
acetic acid, 73, 565
acetic acid bacteria, 566
acetone oil, 29
acetone yeast, 29
acetyl cellulose, 566
acetyl coenzime A, 566
acetyl eugenol, 566
acetyl ionone, 566
acetyl isoeugenol, 566
acetyl transferase, 566
acetylase, 566
acetylate, 566
acetylated wood, 566
acetylation, 566
acheb, 113
achenc, 444
achenodium, 454
achievement in afforestation, 315
achlamydeous, 508
achromatic, 136, 391
achromatic colour, 136
achromatic colours, 531, 532
achromatic fiber, 138
achromatic figure, 138
achromatic spindle, 138
achromatin, 138
achromatoplasm, 138
achromic period, 532
achromogenic, 35
achromosomal, 138
acicular, 607
aciculiform, 606
aciculifruticeta, 606
aciculignosa, 606, 607
aciculisilvae, 606, 607
acid alpine meadow soil, 467
acid-base balance, 467
acid-base equilibrium, 467
acid bog, 467
acid brown forest soil, 467
acid brown soil, 467
acid clay, 467
acid cooking, 467
acid deposition, 467
acid equivalent, 467
acid fast bacteria, 337
acid fast stain, 337
acid fastness, 337
acid feeder, 228
acid gland, 467
acid hardener, 467
acid humification, 467
acid humus, 467
acid-insoluble lignin, 467
acid medium, 467
acid number, 467
acid orange, 467
acid plant, 337
acid precipitation, 467
acid-proof coating, 337
acid proofing treatment, 337
acid rain, 467
acid resistance, 337
acid-resisting tile, 337
acid rosin size, 467
acid scarification, 467
acid size, 6, 467
acid smut, 467
acid soil, 467
acid-soluble lignin, 467
acid spring, 467
acid sulfite process, 467
acid sulfite semichemical pulp, 467
acid-tolerant, 337
acid value, 467
acid wood, 157, 467
acidation, 467
acidic lava, 467
acidic resin, 467
acidic rock, 467
acidification, 467
acidimetric standard, 25
acidite, 184, 467
acidity, 467
acidoid, 467
acidophilous, 441
acidophilous indicator plant, 467
acidophilous vegetation, 516
acidophobous, 524
acidosis, 467
acidotheca, 47
acidotrophic, 467
acidulant, 467
aciduric, 337
acknowledge (character), 66, 540
aclimatic soil formation, 138
aconite, 507
aconitine, 507
acoradiene, 48
acorn, 201, 531
acorn cup, 377, 531
acorn nut, 531
acorn planter, 531
acotyledon, 512
acoustic absorption coefficient (factor), 514
acoustic absorptivity, 514
acoustic fibre board, 514
acoustic filter, 316, 532
acoustic impedance, 430
acoustic insulating material, 166
acoustic reflection coefficient, 430
acoustic resistance, 430
acoustic signal, 481, 570
acoustic tile, 514
acoustic(al) board, 514
acoustic(al) conductivity, 66, 430
acoustic(al) energy, 430
acoustic(al) fibreboard, 514
acoustic(al) inspection, 376, 570
acoustic(al) material, 514
acoustic(al) permeability, 486
acoustical property, 430
acoustimeter, 430, 601
acquinite, 315
acquired character, 200, 217
acquired immunity, 217
acquired resistance, 217
acquisition cost, 384
acquisition of land, 419, 441
acre, 95, 488, 572
acre-foot, 572
acreage, 488, 572
acreage requirement formula, 323
acrisol, 376
acrocarpous, 105, 252
acrocentric, 250
acroconidium, 106
acropetal, 530
acrophyta, 163
acrophytia, 163
acrosporangium, 106
acrospore, 106
acrosporous, 106
across-the-board-cut, 386

A

acrox, 162, 376
acrylamide, 29
acrylamide gel, 29
Act of soil and water conservation, 459
actic, 52
actic association, 52
actidione, 135
actinic absorption, 181
actinism, 181
actinodaphnine, 210
actinodrome, 261
actinometer, 17, 159
actinometry, 148, 181
actinomorphic, 135, 148
actinomycete, 135
actinomycete-antagonist, 135
actinomycin, 135
actinophage, 135
actinostele, 540
action, 375
action body, 218
action bolt, 218, 278
action face, 218, 288
action potential, 110
action program, 436
action spectrum, 655
activated carbon, 215
activated carbon fibre, 215
activated char(coal), 215
activated silica gel, 214
activated sludge, 215
activated sludge method, 215
activated sludge process, 215
activating oven, 214
activation, 214, 221
activation analysis, 135, 214
activation energy, 214
activation pathway, 214
activator, 74, 214, 215, 221, 602
active absorption, 582, 633
active acidity, 215
active adaptation, 214
active assets, 214
active bud, 214
active carbon, 215
active centre, 215, 582
active char(coal), 215
active component, 215, 581
active constituent, 215, 581
active dune, 306
active earth pressure, 581, 632
active ferment, 215
active humus, 215
active immunity, 643
active infection, 643
active ingredient, 215, 581
active layer, 214, 398
active life, 437, 581
active microwave, 632
active nutritive period, 582
active porosity, 581
active power, 581
active region, 215, 655
active remote sensing, 633
active resistance, 642, 643
active sludge, 215
active substance, 215, 582
active surface, 214, 655
active temperature, 581
activity, 214, 215
activity accounting, 422, 560, 601
activity gradient, 214
activity rhythm, 214
actual area, 435
actual budget, 264
actual cost, 435
actual cut, 38, 435
actual density, 435
actual discharge, 435
actual efficiency, 435
actual error, 435
actual evapotranspiration, 435, 582
actual forest, 526
actual growing-stock, 435
actual incidence, 435
actual site, 526
actual stress, 435, 582
actual stumpage price, 435
actual supply, 435
actual use, 435
actual utilization, 435
actual vegetation, 526
actual vegetation map, 526
actual volume, 435
actual yield, 435
actual yield table, 435
actuary, 254, 484
actuating device, 383, 480
actuating gear, 66, 383
actuating unit, 66
aculeate, 71, 260
aculeus (*plur.* -lei), 71
acuminate, 237
acuminose, 250
acuminulate, 518
acutate, 500
acute, 222
acute dose, 222
acute intoxication, 222
acute toxicity, 222, 261
acutilingues, 231
acutungular, 401
Aczol, 484
ad valorem, 3, 71
ad valorem tariff, 3, 72
ad valorem tax, 72
adamic earth, 198
adaptability, 440
adaptation, 440
adaptation assessment, 440
adaptation cost, 440
adaptation norm, 440
adaptation reaction, 440
adaptation tolerance, 440
adapter sleeve, 249, 292
adaptiogenesis, 440
adaptive capacity, 440
adaptive character, 440
adaptive colouration, 440
adaptive compensation, 440
adaptive complex, 440
adaptive convergence, 440
adaptive differentiation, 440
adaptive enzyme, 440
adaptive evolution, 440
adaptive faculty, 440
adaptive fitness, 440
adaptive form, 440
adaptive migration, 440
adaptive peak, 440
adaptive phase, 440
adaptive radiation, 440
adaptive selection, 440
adaptive value, 440
adaptive variation, 440
adaptivity, 440

adaxial paratracheal parenchyma, 251, 339
adaxial parenchyma, 251, 339
add-forced cable system, 601
added value, 602
adder, 227
addition line, 152, 228, 478
additional cost, 120
additional equipment, 150
additional fertilization (fertilizer), 639
additional paid-in capital, 120
additional peak, 152
additionality, 120
additive, 227, 228, 478
additive colour formation, 228
additive colour mixture, 228
additive colour process, 228
additive factor, 228, 285
additive gene, 228
additive genetic effect, 220, 228
additive genetic variance, 228
additive variance, 228
additive viewer, 228
additivity, 273
additivity of components, 651
additivity of luminance, 296
adduct, 227
adelphia (*plur.* -phiae), 287
adelphic, 287
adelphogamy, 482, 642
adelphoparasitism, 484
adelphous, 287
adenosin(e), 527
adephagid, 399
adequate retention, 14
adequate stocking, 440
adequate variability, 440
adhere, 240, 344
adherend failure, 240, 344
adherent, 241, 345
adhering knot, 214, 292
adherometer, 152, 344
adhesiometer, 240, 344
adhesion, 152, 240, 344, 345
adhesion failure, 240, 241
adhesion test, 241
adhesive, 240, 241, 345
adhesive dispersion, 241
adhesive failure, 240
adhesive felt, 53, 240
adhesive force, 240, 345
adhesive glassine tape, 485
adhesive lacquer, 241
adhesive layer, 240
adhesive paper, 241
adhesive root, 152
adhesive solvent, 241
adhesive strength, 240, 241
adhesive tape, 240
adhesive varnish, 344
adhesiveness, 241
adhesivity, 240, 241
adiabatic(al) evaporation, 265
adierstein, 144
adipo-cellulose, 189, 452
adipose, 120, 613
adipose tissue, 613
adjacency analysis, 298
adjacent chromosome, 528
adjacent peak, 528
adjacent segregation, 298
adjoining cleared strip, 298
adjoining wood, 298
adjunct account, 152
adjusted coefficient of determination, 543
adjusted treatment effect, 106
adjusted volume, 243
adjusting device, 480
adjusting photo-volume table, 544
adjusting range, 480
adjustment time, 480
adjuvant, 150, 634, 654
adjuvant grafting, 634
administered price, 41, 179
administration cost, 179
administrative agency, 540
administrative district, 540
administrative expense, 540
administrative management, 540
administrative office, 540
administrative working circle, 540
administratively assigned shubland, 474
admiralty shackle, 630
admission passage, 482
admix, 87
admixture, 47, 213
adnascent, 152
adnate, 294, 480
adobe, 59, 491
Adohero gauge, 152
adolescent, 54
adolescent river, 379, 417
adolescent stage, 639
adret, 556
adsere, 152
adsorbability, 513, 514
adsorbed water, 514
adsorbed water layer, 514
adsorption, 513
adsorption bed, 513
adsorption chromatography, 513, 514
adsorption column, 514
adsorption complex, 514
adsorption curve, 514
adsorption isotherm, 513
adsorption stripping, 200, 248, 370
adsorptive water, 514, 515
adult, 54
adult emergence, 54
adult feeding, 199
adult instar, 54
adult leaf, 55
adult phase, 54, 55
adult tree, 54, 55
adult wood, 54
adulteration of seed, 627
advance, 103, 476, 586
advance-cutting, 373, 586
advance decay, 63
advance felling, 373
advance germination, 74
advance growth, 127, 373, 522, 601

A

advance of sand dunes, 411
advance payment, 586
advance planting, 127
advance reproduction, 127, 373
advance-reproduction system, 127
advance rot, 63
advanced charge, 586
advanced decay, 199, 419
advanced generation, 199
advanced generation cross, 199
advanced genetics, 162
advanced linear programming system, 162
advanced reconnaissance satellite, 162
advanced stage of decay, 151, 420
advanced thinning, 199
advanced unit in afforestation, 315
advanced wastewater treatment, 162, 419
advanced workers in afforestation, 315
advection, 362
advection fog, 362
advection radiation frost, 362
advective frost, 362
adventitious bud, 32
adventitious embryo, 31
adventitious embryony, 31
adventitious inflorescence, 31
adventitious plant, 496
adventitious root, 31
adventitious shoot, 32
adventitious species, 496
adventive, 32, 496, 570
adventive root, 31
adverse grade, 343, 416
advertising colour, 255
advertising cost, 182
advice of drawing, 213, 476
advisory service, 640
adze, 19, 23, 177, 196
adzing, 48, 177, 269
adzing and boring machine, 532
adzing machine, 19, 533
aecidial stage, 544
aecidial state, 544
aecidiospore, 544
aeci(di)um (*plur.* -ia), 544
aedeagus, 556
aelophilous, 144, 145
aeolian accumulation, 144, 145
aeolian basin, 144
aeolian deposit, 144
aeolian landform, 144
aeolian rock, 144
aeolian sediment, 144
aeolian soil, 144
aeolianite, 144
aerate, 17, 59, 514
aerating root, 200, 482
aeration, 59, 67, 481, 482
aeration-cooling, 481
aeration-drying, 481
aerator, 59
aerenchyma, 482
aerial biomass, 447
aerial cableway, 231
aerial camera, 191
aerial contamination, 275
aerial detection, 276
aerial dressing, 136
aerial exposure index, 191
aerial film, 190
aerial forest fire protection, 190
aerial forest survey, 407
aerial fuels, 276
aerial hypha(e), 370
aerial layering, 164, 276
aerial line, 231
aerial logging, 276
aerial mapping, 190
aerial monorail-tramway, 231
aerial mycelium, 370
aerial oblique photograph, 191
aerial patrol, 276
aerial photo coverage, 191
aerial photo interpretation, 191
aerial photo mensuration, 191
aerial (photo) survey, 191
aerial photo triangulation, 276
aerial photogrammetry, 190, 191
aerial photograph, 191
aerial photograph index, 191
aerial photograph scale, 191
aerial photograph type, 191
aerial photography, 191
aerial plant, 370
aerial planting, 136
aerial plot measurement, 276
aerial population, 276, 370
aerial reconnaissance, 276
aerial root, 370
aerial ropeway, 231
aerial seeding, 136
aerial shoot, 97
aerial skidder, 231, 276, 639
aerial skidding, 231
aerial smoke jumping, 275
aerial snubbing system, 470
aerial sowing, 136
aerial sprayer, 136
aerial spraying, 136
aerial stand volume table, 190
aerial surveying, 191
aerial termite, 275
aerial tramway, 231
aerial transmission, 275
aerial (tree) volume table, 190
aerial triangulation, 276
aerial vertical photograph, 276
aerial view, 507
aerial volume table, 190
aerial wireropeway, 231
aeriductus, 369, 402, 503
aero control, 136
aero sketchmaster, 191
aero spraying, 136
aero-yarder, 370
aerobe, 191, 544
aerobic bacteria, 191, 517, 544
aerobic culture, 191
aerobic decomposition, 191

A

aerobic dehydrogenation, 544
aerobic glycolysis, 582
aerobic nitrogen-fixing bacteria, 191
aerobic respiration, 582
aerobic treatment, 557
aerobiology, 275
aerobiont, 191, 544
aerobiosis, 191, 544
aerocartography, 191
aeroclimatology, 162
aerodynamic control, 368
aerodynamic method, 77
aerogams, 524
aerogenic bacteria, 48
aerolite, 308, 532, 595
aerolith, 308, 532, 595
aerological theodolite, 44
aeromorphosis, 370
aerophyte, 370
aerophytobiont, 544
aeroplane timber, 136
aeroplankton, 276
aeroplanktophyte, 77
aerosat, 191
aerosol, 370, 553
aerosol deodorant, 370
aerosol effect, 370
aerosol fragrance product, 370
aerosol sprayer, 355, 553
aerosols, 370
aerosor, 32
aerosphere, 369, 370
aerostat, 136, 370
aerosynusia, 370, 483
aerotaxis, 383
aerotolerant, 338
aerotropism, 530
aerovane, 146
aesculin, 365
aestatifruticeta, 522
aestatisilvae, 522
aesthetic forest, 145
aesthetic forestry, 145
aesthetic point, 145, 320, 420
aesthetic property, 320
aesthetic value, 178, 320
aestheticism, 502
aestiduriherbosa, 522
aestidurilignosa, 522
aestilignosa, 522
aestival, 521
aestival annual plant, 522
aestival aspect, 522
aestival bud, 522
aestivation, 202, 522
aetiolation, 210, 492
aetiology, 29
affected timber, 378, 443
affiliate, 143
affinis, 379
affinity, 379
affinity chromatography, 379
affinity labeling, 379
affluent, 213, 612
afflux, 213, 307
afforesration and reforestation machinery, 600
afforest, 600, 616
afforest desert, 315
afforestation, 209, 315
afforestation and reforestation, 209
afforestation by cuttings, 372
afforestation drive, 315, 600
afforestation for carbon sequestration, 473
afforestation plan, 597, 600
afforestation project, 600
afforestation project in the flat-land areas, 362
afforestation standard, 315
afforested area, 600
afforested areas, 315
African copaiba balsam oil, 139
African sandalwood oil, 139
after cost, 199, 444
after-culture, 32, 33
after-cure, 121
after-effect, 200
after effect of elasticity, 472
after flame time, 544
after flaming, 152
after floating, 307
after-growth, 61, 71, 199, 598
after manuring, 639
after-regeneration, 125
after-reproduction, 125
after-ripening, 200
after-ripening seed, 200
after-ripening treatment, 199
after-ripening type of dormancy, 200
after treatment, 199
aftermath, 598
aftershock, 584
Ag horizon, 27
agalite, 205, 433
agamandroecism, 542
agameon, 510, 511
agamic complex, 510, 511
agamic reproduction, 137, 510
agamo-pseudopolyploidy, 511
agamobium, 511
agamogenetic, 510, 511
agamogony, 510
agamogynaecism, 69
agamogynomonoecism, 403, 508
agamogynomonoecy, 510
agamohermaphroditism, 511
agamomonoecy, 510
agamonoecism, 625
agamont, 138, 298, 510
agamospecies, 510, 511
agamospermy, 510
agar, 381, 557
agar-agar, 381, 557
agar-block decay test, 381
agar-block test, 381
agar deep culture, 381
agar medium, 381
agar oil, 53
agar plate, 381
agar plate method, 381
agar slant, 381
agar-wood block method, 381
agar-wood oil, 53
agaric (fungus), 404
Agarofuran, 53
agarwood, 53
agave scale, 198
age and area hypothesis, 344
age-area system, 301
age at breast height, 542
age calss distribution, 305
age class, 305
age-class distribution, 305

age-class factor, 305
age-class interval, 305
age class method, 305
age-class period, 305
age-class space, 301
age-class table, 305
age-class-transition probability, 301
age-class vector, 301
age classes in regulated forest, 480
age composition, 344
age determination, 343
age distribution, 344
age gradation, 305
age grade, 305
age-grade, 305
age group, 305
age of seed-bearing, 245
age of stand, 300, 301
age pyramid, 344
age ratio, 343
age specific life table, 474
age specific mortality, 474
age specific survival rate, 474
age structure, 344
aged seed, 175
aged soil, 284
ageing, 53, 284, 451
ageing condition, 284
ageing process, 284
agent of disease, 29
agglomerate, 152, 224, 346, 491
agglutinant, 241, 346, 417
agglutination, 346
agglutinin, 346
aggradation, 388
aggregate, 264, 491
aggregate demand, 649
aggregate error, 285
aggregate flowers, 322
aggregate fruits, 263, 264
aggregate genotype, 263
aggregate impact, 285
aggregate index number, 648
aggregate loss, 285
aggregate parenchyma, 263
aggregate ray, 263
aggregate resin cell, 263
aggregate width, 358, 652
aggregated distribution, 263
aggregated soil, 491
aggregation, 263, 491, 651
aggregation index, 263, 264
aggregative response, 263
aggressive hardener, 428
aggressive individuals, 379
aggressive species, 379
aggressive strain, 379
aggressive type, 379
aggressivity, 379
agilawood, 53
agilawood oil, 53
aging, 55, 451
aging quality, 284
aging time, 284, 451
agio, 480, 606
agitator, 243
agitator shaft, 243
aglycon(e), 473
agmatoploidy, 230
agmogenesis, 62, 86, 510
agnation, 542
agonistic behaviour, 115
agraphitic carbon, 137, 138, 509
agreed period, 414
agreed value, 594
agreement year, 536
agri-silviculture, 214
agric horizon, 168
agricultural antibiotic, 348
agricultural chemical, 347
agricultural chemical pesticide, 347
agricultural eco-engineering, 347
agricultural ecology, 347
agricultural economical zoning, 347
agricultural indicator, 347
agricultural resources, 348
agricultural timber, 348
agriculturist, 347
agro-forestry, 214
agrocide, 309
agroecology, 347
agroecosystem, 347
agroecotype, 654
agroforestrial geology, 347
agroforestry, 214, 347
agroforestry system, 214
agromedical, 347
agromedicine, 347
agrometeorological forecast, 347
agronomic variety, 348
agronomy, 347
agrosilvopastoral system, 214
agrostological index, 191
agrostology, 191
agynous, 508
Ahig process, 1
aid cooking, 467
aid-to-planting, 33
aid-to-sowing, 32
aigret(te), 178
aiphyllium, 281
aiphytia, 48
air accumulator, 65, 276
air attack boss, 276
air base, 276
air binding, 276
air bladder, 370
air-blast, 175, 369
air blast sprayer, 145
air blast system, 175
air-bleed, 61, 135
air blower, 175
air-blowing, 175
air-borne, 276
air brake, 276, 370
air bubble level, 370
air bunk, 56
air capacity, 275
air-carrier sprayer, 145
air circular classifier, 369
air circulation, 275, 276
air-circulation system, 275
air clamp, 368
air classification, 369
air cleaner, 275
air cleaner traverse, 195
air clipper, 368
air compression machine, 276
air compressor, 276
air-conditioned channel, 276
air-conditioning, 276
air-conditioning unit, 276
air-conductivity, 486
air contaminant, 276

air cooling system, 275
air-crack, 157
air current, 369
air curtain, 370
air damper, 144, 245, 276
air debarker, 368
air diaphragm valve, 368
air dog, 368
air-drain, 369, 481, 482
air drainage, 135, 350
air-dried, 144, 368
air-dried condition, 368
air-dried lumber, 368
air-dried wood, 144, 368
air drilling, 136
air drive, 368
air drum sander, 370
air-dry, 144, 296, 368
air-dry condition, 368
air-dry lumber, 368
air-dry moisture content, 368
air dry weight, 144
air-dry wood, 144, 368
air-dry (wood) density, 368
air-drying, 368, 477
air-drying loss, 144, 368
air dusting, 67, 136
air embolism, 370
air exhaust, 61, 350
air feeder, 67, 171, 250
air-felting, 369
air-felting sheet, 369
air filter, 275
air flow, 369
air-flow sprayer, 145, 369
air-gauge, 276, 368, 370
air governor, 275
air grading spreader, 369
air heater, 275, 392
air hoar, 447, 455
air hoist, 144, 368
air hole, 146, 482
air hose, 482
air humidity, 275
air inlet flue, 250
air inlet port, 250
air jet, 275, 354
air jet dryer, 354
air knife coater, 368
air layering, 164, 276
air leakage test, 309
air level, 370
air lift logging, 276
air mass, 370
air-meter, 369
air motor, 368, 551
air oil separator, 276
air-operated automatic control, 368
air-operated clipper, 368
air-operated dog, 368
air-operated humidity controller, 368
air-operated puffer, 368
air-operated recorder regulator, 368
air-operated temperature controller, 368
air-operated thermostat, 368
air outlet flue, 350
air-oven, 393
air permeability, 482, 486
air photo mosaic, 191
air pit, 481
air plankton, 77
air pollutant, 276
air pollution, 78, 275
air pollution observation station, 275
air pollution potential, 276
air-powered pump, 368
air preheater, 276
air preheating, 276
air pressure, 370
air pressure duct, 551
air pressure gauge, 370
air pruning, 276
air-purification equipment, 275
air-purification system, 275
air purifier, 275
air reducing valve, 275
air register, 275, 479
air regulator, 275, 479
air resistance, 276, 485
air root, 368
air sac, 370
air scale system, 370
air-screen cleaner, 145, 369
air seal, 368
air-seasoned, 368
air-seasoned timber, 368
air seasoning, 144, 368, 645
air-seasoning condition, 368
air seeding, 136
air separation, 369
air separator, 369
air shrinkage, 158
air-sowing, 136
air-spraying, 136
air station, 276
air stream, 369
air suction pipe, 514
air suction port, 514
air supply, 171, 275, 370, 465, 551
air supply duct, 171
air supply line, 171
air support, 368
air survey, 190
air suspension, 276, 368
air tanker, 324
air tight joint, 370
air tight seal, 370
air-to-air heat exchanger, 275
air valve, 368, 369
air velocity, 275, 369
air volume, 275
airborne disease, 275
airborne infection, 275
airborne profile recorder, 190
airborne radiometer thermometer, 218
aircraft plywood, 190
aircraft sounding, 136
aircraft sprayer, 136
aircraft stock, 191
airflow dryer, 369
airfoil section, 503
airing, 275, 296, 481
airless spray, 509
airless spray gun, 509
airless sprayer, 164, 509
airless spray(ing), 164, 510
airplane lumber, 136
airplane observation, 136
airplane plywood, 190
airproof, 32
aithalium, 48
akaryote, 509
akazgine, 139
akee oil, 1
akineton, 34, 510
ala (*plur.* alae), 58, 569
Alabastrum, 203
Alameda, 303
alangine, 4
alanine, 29
alantin, 259, 489
alar, 58, 562, 569

alarm call, 255
alarm pheromone, 255
alarm reaction, 249, 255
alarm substance, 255
alary, 569
alaskite, 6
alatae, 580
alate, 582
Alban(an)e, 1
albedo, 129
albedo-effect, 129, 130
albert lay, 461
Albertol, 1
albic horizon, 6, 230, 358
albic layer, 6
albic soil, 6
albinism, 6
albino, 6
albite, 336
alboll, 358
albumen, 88, 352
albumen glue, 88
albuminose (albuminous) seed, 581
alburnum, 21
alburnum tree, 21
alcohol resistance, 336
alcove, 4, 21, 295
aldicarb, 477
aldrin, 2
aldrin dust, 2
aldrin emulsion, 2
alee-tree, 540
aletophyte, 310
aleuriospore, 143
aleurone, 201
Alexandria senna, 232, 552
aleyrodiform, 144
alfalfa, 336
alfisol, 304
alga (*plur.* algae), 599
algae, 599
algae inhibitor, 568
algal bloom, 599
algal mat ecosystem, 599
algarobilla, 465
algarobilla extract, 465
algarobilla tannin, 465
algarobin, 465
alginate fibre, 599
alginic acid, 189, 599
algivore, 437
algology, 599
algorithm, 467
algorithmic language, 467
algorithmic oriented language, 467
alidade, 605
alien, 183, 496, 528
alien chromosome, 496, 594
alien plant, 496
alien species, 496
aliform-confluent parenchyma, 264
aliform parenchyma, 569
aligned mat, 107
ali(g)ning board, 243
alignment (alinement) chart, 297
alignment device, 474
alignment of strip, 37
alinement chart method, 45, 297
aliphatic resin, 613
alkali charge, 577
alkali consumption, 234
alkali extractive, 234
alkali fusion, 234
alkali hydrolysis, 235
alkali lake, 234
alkali lignin, 234
alkali-lime series, 234
alkali make-up tank, 353
alkali pulp, 234
alkali recovery, 234
alkali resistance, 337
alkali resisting cellulose, 270
alkali soil, 234
alkali-sulfite process, 235
alkaline cooking, 234
alkaline glue, 234
alkaline hydrolysis, 228, 234
alkaline meadow soil, 234
alkaline patche, 234
alkaline plant, 191, 439, 516
alkaline pulp, 234, 235
alkaline rocks, 235
alkaline semichemical pulp, 234
alkaline s(i)erozem, 234
alkaline soil, 234, 235
alkaline solvolysis, 235
alkaline treatment, 234
alkalinity, 234
alkaliotropism, 530
alkaliplant, 235
alkaloid, 427
all-aged, 386
all-aged forest, 118, 386, 601
all-aged stand, 386
all-axle drive, 387
all bark particle board, 386
all-heart, 387
all-in cost, 648
all-inclusive concept, 649
all-male group, 387
all-metal prefabricated greenhouse, 386
all-purpose gun, 386
all-purpose language, 482
all risk insurance, 386, 484
all-veneer construction, 386
all-veneer plywood, 385
all weather, 387
all-weather foundation, 387
all-weather grapple loader, 387
all-weather road, 381, 387
all-year grazing, 293
Allardyce process, 1, 539
allautogamia, 217
Allee's law, 1
Allee's principle, 1
allele, 92
allelic gene, 92
allelic pair, 527
allelic recombination, 92
allelic series, 92
allelism, 92, 115, 483
allelochemic, 568, 626
allelochemical, 205
allelochemical(s), 497
allelochemics, 566
allelogenous, 47
allelomimetic behaviour, 118
allelomorph, 92, 115
allelomorphism, 92, 115, 483
allelopathic substance, 470, 568
allelopathy, 205, 422, 568
allelosomal, 92
Allen's curve, 2
Allen's rule, 1
allergen, 187

A

allergic crisis, 187
allergic reaction, 187
allergy, 187
alley, 286, 482, 533, 534, 603
alleyway, 223
alliance, 389
alliance association, 294, 389
allidochlor, 42, 121
alligator, 58, 121, 349, 533
alligator chain, 58
alligator imitation paper, 134
alligatoring, 121, 183, 631
allite, 155, 314
allitic weathering, 155
allivalite, 159
allobiocenose, 567
allocarpy, 566, 596
allocated cutting method, 142
allocation of felling, 38
allochlorophyll, 568
allochronic, 422, 567
allochronic isolation, 435
allochronic species, 567
allochthonic ground water, 350, 497
allochthonous, 496
allochthonous flora, 496
allochthonous peat, 71, 496, 596
allochthonous soil, 564, 596
allochthonous species, 496
allodiploid, 568
allogamous, 567
allogamy, 567, 594
allogen(et)ic, 568
allogen(et)ic plankton, 497
allogen(et)ic succession, 566
allogenic, 484, 497, 567
allogenic succession, 497, 566
allogenous flora, 568
alloiobiogenesis, 438
allomerism, 568
allometric equation, 567
allometric scaling, 567
allometry, 567
allometry coefficient, 567
allometry method, 567
allomixis, 596
allomone, 568
allomorphism, 484
alloocimene, 27
allopatric, 140, 566
allopatric form, 568
allopatric phenomena, 568
allopatric population, 568
allopatric segregation, 568
allopatric speciation, 568
allopatric species, 568
allopatry, 568
allophane, 314, 457
alloplasm, 568
alloplasmy, 568
alloploid origin, 568
alloploid type, 568
allopolyploid, 568
allopolyploidy, 568
alloprene, 1, 27, 315
allosome, 541, 567
allosteric inhibition, 567
allosynapsis, 568
allosyndesis, 568
allotetraploid, 152, 568
allotment, 37, 142, 655
allotment for felling area, 127
allotment logging, 142
allotment method, 142, 360
allotrope, 483
allotrophic lake, 568
allowable bearing capacity, 397
allowable burn area, 273, 595
allowable cut, 272, 595
allowable cut amount, 272, 595
allowable cut effect, 225
allowable cut formula-Grosenbaugh, 165
allowable cutting unit, 397
allowable defect, 398
allowable error, 398
allowable load, 398
allowable load capacity, 397, 398
allowable loaded deflection, 398
allowable periodic cut, 106
allowable pressure, 398
allowable resistance, 398
allowable strength, 397
allowable stress, 398
allowable use, 225, 595
allowance, 17, 33, 170, 249, 397, 398, 544, 584
allowance for bark, 449
allowance for defect, 388
allowance for fault, 388
allowance for trimming, 196
allowed value, 398
allozygote, 567, 571
allozygous, 568
allozyme, 484, 568
alluvial apron, 59
alluvial brown soil, 648
alluvial clay, 59
alluvial community, 59
alluvial cone, 59
alluvial deposit, 59, 358
alluvial fan, 59
alluvial flat, 59, 193
alluvial formation, 59
alluvial horizon, 59
alluvial layer, 59
alluvial plain, 59
alluvial slope, 59
alluvial soil, 59, 583
alluvial terrace, 59
alluvial valley, 59
alluvium, 59, 583
alluvium period, 100
allyl resin, 515
almaciga, 316
almond moth, 143, 156
alnusenone, 58, 365
aloewood oil, 53
alpha-cellulose, 523
alpha-helix conformation, 313
alpha humus, 151
alpha ray, 418
alpine, 163
alpine belt, 163
alpine fauna, 163
alpine flora, 163
alpine garden, 163
alpine hemlock, 76
alpine humus soil, 163
alpine kiln, 615, 651
alpine landscape, 163
alpine lichen, 163
alpine mat, 163
alpine meadow, 163
alpine meadow region, 163
alpine meadow soil, 163

alpine moss, 163
alpine plant, 163
alpine region, 163
alpine relict, 163
alpine soil, 163
alpine species, 163
alpine spruce, 76
alpine steppe soil, 163
alpine stony desert, 163
alpine sward, 163
alpine tundra, 163
alpine vegetation, 163
alpine zone, 163
alpinone, 295
alstonine, 1, 219, 551
altalf, 194
altar and monastery garden, 588
alternate clear strip system, 239
alternate clearcutting method, 239
alternate felling, 239
alternate host, 312, 637
alternate pitting, 202
alternate plough, 454
alternate strip system, 239
alternate (year) bearing, 166, 240
alternating bending test, 238
alternating deformation, 238
alternating displacement, 238
alternating forced cross circulation, 238, 273
alternating impact machine, 129, 238
alternating load(ing), 23, 238
alternating motion, 238, 500
alternating pressure, 238
alternating pressure method, 129
alternating pressure process, 228, 239, 359
alternating spiral grain, 239
alternating stress, 238
alternating temperature, 24
alternating tension stress, 238
alternating testing machine, 238
alternation of generations, 438
alternation of slope and terrace, 166
alternative cost, 272, 477, 547
alternative development path, 477
alternative energy, 477
alternative host, 312
alternative investment, 305
alternative prey, 477
alterne, 388
alternipetalous, 542, 584
alternipinnate, 584
alternisepalous, 202, 584
altherbosa (altoherbosa), 161
altimeter, 44, 188
altimetry, 44
altitude, 162, 188
altitude acclimatization, 163, 164
altitude-compensating turbocharger, 162
altitude surveying, 161
altitudinal acclima(tiza)tion, 188
altitudinal adaptation, 188
altitudinal belt, 68, 188
altitudinal vicariad, 68, 188
altitudinal zonality, 67
altitudinal zone, 68
alto∼, 162
alto*sim*, 161
altocumulus, 162
altoherbiprata, 161
altoherbosa (altherbosa), 161
altoll, 194
altostratus, 161
altricial, 499
altruism, 290
alula (*plur.* -lae), 58
alum, 325
alum earth, 325
alum tanning, 128
alumina, 128, 314, 557
aluminium can, 314
aluminium oxide mineral, 160, 557
aluminium paint, 314
aluminium-paper laminated bag, 45
aluminium tag, 314
aluminium tanning, 128
amalgamated balance sheet, 191
amargosa, 278
amargosa bark, 278
amaroid, 278
amateur floriculture, 560
amateur gardening, 385, 560
amber, 201
ambergris, 309
ambergris tincture, 309
ambient pressure, 208, 496, 630
ambient temperature, 208, 248, 441
ambisexual, 70, 296
ambrosia beetle, 50, 485, 533
ambrosia borer, 50, 485, 533
ambrosia fungus, 522, 636
ambrosia oil, 489
ambulatory chattel, 108
amelioration with protection forest, 406
amenability, 461
amenable to disease, 569
amending clause, 543
amenity forest, 145
amenity resources, 444
amenity value, 255
amensalism, 357
ament, 399
amentaceous plant, 399
American frankincense, 320
American lumber standard, 320
American Society for Testing Materials, 320
American standard of testing materials, 320
American storax, 17
American white moth, 320
American wormseed oil, 320
amerospore, 509
ametabola, 508
ametoecious, 87
amidase, 523
amidation, 523
amide, 523
amidin, 103
amidithion, 402
aminate, 3
amine, 3
amine preservative, 3

A

amino acid, 3
amino resin adhesive, 3
aminocarb, 324
aminocellulose, 3
aminoglucose, 3
aminolipoid, 3
aminotransferase, 637
aminotriazole, 3, 410
amiton, 3
amitosis, 510
amitrole, 410
ammonia liquor, 3
ammoniacal nitrogen, 3
ammoniate, 3, 580
ammoniated fertilizer, 3
ammonification, 3, 53, 227
ammonifying bacteria, 3
ammonite, 157, 259
ammonium, 3
ammonium bisulfite process, 552
ammophilous, 516
amnicolous, 411
amnios, 64
amobam, 78
amoebocyte, 24
amoeboid, 462
amorph, 511
amorpha, 567
amorphic soil, 62, 64, 583
amorphous, 508, 512
amorphous area, 137, 509
amorphous cellulose, 137
amorphous portion, 137, 508
amorphous region, 137, 509
amortization, 471, 606
amortization formula-Meyer, 317
amount of clouds, 594
amount of inbreeding, 250
amount of infection, 378
amount of precipitation, 238
amphibian, 296, 457
amphibiotica, 296
amphibious, 296
amphibious plant, 296
amphibole, 414
amphibrya, 87
amphidiploid, 152, 452
amphidiploid combination, 452
amphiecious, 182
amphigamy, 246, 443
amphigenesis, 296, 582
amphigenetic, 296
amphigonic female, 296
amphigony, 296
amphigous individual, 296
amphilepsis, 453
amphimictic population, 296
amphimixis, 296
amphinucleolus, 452, 568
amphion, 296
amphiont, 244
amphiphyte, 296
amphiplasty, 391, 468
amphiploid, 452
amphipolyploid, 452
amphiprotic, 296
amphistomatic, 295
amphitene stage, 296
amphitoky, 47
amphitropic, 196
amphitropous, 196
ampholytoid, 296
amphophilic, 454
amphoteric, 296
amphoteric colloid, 296
amphoteric tannin, 296
amplifier, 135, 602
amplifying recorder, 135
amplitude, 23, 608
amrad gum, 1
amusement center, 584
amusement ground, 579
amusement park, 584
amygdalase, 278
amylase, 103, 561
α-amylase, 103, 561
β-amylase, 103, 317, 473
amylin, 201
amylo process, 320
amyloid, 103, 240
amylolytic enzyme, 103
amylopectin, 612
amylose, 615
amyradiene, 6
amyrenone, 6
amyrin, 613
amyris oil, 613
amyrolin, 6
anabasin, 324
anabatic wind, 413, 416
anabiont, 48
anabiosis, 153, 212, 230, 431, 543
anabolic phenomenon, 651
anabolism, 651
anacardol, 230, 558
Anaconda process, 1
anadromous fish, 466
anaerobe, 556
anaerobic, 556
anaerobic bacteria, 524
anaerobic cellulose-decomposing bacteria, 524
anaerobic culture, 388
anaerobic deterioration, 556
anaerobic organotrophy, 556
anaerobic pectin decomposing bacteria, 524
anaerobic respiration, 511
anaerobic treatment, 556
anaerobic waterlogging, 556
anaerophytobiont, 524
anagenesis, 249
anaglyph, 33
anaglyphic principle, 33
anal angle, 492
anal comb, 492
anal leg, 503
analbite, 86, 495
analcime, 130, 352
analcimite, 130
analcite, 130, 352
analog computer, 325
analog-digital, 325
analog-digital converter, 325
analog-digital integrating translator, 325
analog-digital recorder, 325
analogue, 325
analogue computer, 325
analogue computing device, 325
analogy-digital-analog converter, 325
analyser, 143
analysis of covariance, 172, 201, 536
analysis of variance, 130, 287
analyte, 18
analytical aerial triangulation, 248
analytical engine, 143

A

analytical model, 248
analytical purity, 143, 256
analyzer, 143
anamorphic curve, 24
anamorphosis, 24, 236, 431
anaphalanx, 348
anaphase, 199
anaphoresis, 570
anastomosis, 294, 391, 499, 506
anatomical area, 248
anatomical feature, 248
anatomical structure, 247
anatomy, 247, 248
anatropous, 90
anauxite, 155, 399
anchor, 227, 319
anchor block, 96, 319
anchor bolt, 96, 319
anchor cable, 176, 319
anchor line, 319, 367
anchor log, 319
anchor point, 107, 133, 651
anchor rod, 319
anchor root, 176
anchor stump, 319
anchorage, 176, 319, 481
ancient erosion, 174
ancient garden, 174
ancient organisms remains, 174
ancient relicts, 40
ancillary benefits, 150
ancillary equipment, 150
ancillary investment, 150
ancillary resource, 150
andesinite, 625
andesite, 2
ando soil, 381
andosol, 156
andrena, 96
androcyte, 254
androdioecious, 542
androdioecy, 542
androecium, 542
androecy, 543
androgamete, 542
androgamone, 542
androgen, 542
androgenesis, 86, 542
androgenetic parthenogenesis, 48
androgynal, 70
androgynism, 70
androgynophore, 70
androgynous, 70
androgyny, 70
andromonoecious, 542
andromonoecy, 542
andropetalous, 542
anemochore, 144
anemochory, 144
anemoentomophily, 144
anemogamy, 145
anemogram, 145
anemograph, 145
anemometer, 145, 146
anemophic plant, 145
anemophilae, 145
anemophilous flower, 145
anemophilous plant, 145
anemophilous pollination, 145
anemophily, 145
anemosporae, 144
anemotaxis, 383
anemotropism, 530
anemovane, 146
anerobic bacteria, 524, 556
aneroid, 275, 511
aneroid barograph, 275, 511
aneroid barometer, 275, 511
anethole, 114, 212
aneuploid, 138
aneuspory, 567
angelic acid, 88
angelica oil, 88
angelicin, 88
angelicone, 88
angeosere, 539
angiosperm, 19
angiospermous forest, 282
angle bar, 242, 360
angle block, 242
angle board, 242
angle count cruising, 242
angle count method, 242
angle count plot, 242
angle count sample, 242
angle count sampling, 242
angle gauge, 242
angle gauge sampling, 242
angle grain, 537
angle measurement, 242
angle method, 141
angle mirror, 44
angle of attack, 59, 171
angle of depression, 149
angle of deviation, 357
angle of difference, 46, 357
angle of drift, 191
angle of field, 439
angle of incidence, 400
angle of inclination, 380, 537
angle of obliquity, 380
angle of orientation, 107
angle of reflection, 129
angle of refraction, 606
angle of repose, 258, 543
angle of shearing resistance, 270
angle of slope, 363
angle of sun, 471
angle of view, 439
angle of vision, 439
angle pipe, 497
angle plane, 242
angle planting, 377
angle post, 242
angle-summation method, 242
angledozer, 536
anglemeter, 44
angocopaloresene, 2
Angola copal, 2
angolensin, 2
angostura bark oil, 2
angostura extract, 2
angostura tincture, 2
angular aperture, 268, 276
angular deflection, 242
angular dispersion, 242
angular joint, 242
angular leaf spot, 242
angular plane surface, 537
angular surface, 537
angular test, 498
angulometer, 44, 296
angustifoliol, 2
anhydrobase, 494
anhydrodicatechin, 494
anhydroglucose, 364
anhyetism, 388
aniline, 19
aniline black, 19, 555
animal adhesive, 109
animal black, 175, 444
animal charcoal, 175, 444
animal community, 109
animal ecology, 109

animal forage, 109
animal furrow, 109
animal glue, 109
animal size, 109
animal skidding, 65
animal society, 109
animal sociology, 109
animal species with high values for wellfare, economy and science, 403
animal unit, 109
animi, 575
anion, 570
anion-exchange chromatography, 570
anion respiration, 570
anionic preservative, 570
anionite, 570
anisacetone, 76, 114
anise camphor, 114, 212
anise oil, 212
aniseed, 4, 76, 212
aniseed oil, 212
anisketone, 114, 212
anisogametism, 567
anisogametismanisogamety, 353
anisogamety, 567
anisogamonty, 353
anisogamous, 567, 568
anisogamy, 353, 567
anisogeny, 70
anisole, 19, 212, 230
anisomorpha, 567
anisoploid, 366
anisopolyploid, 366
anisotropic-elastic system, 567
anisotropic wood beam, 567
anisotropism, 166
anisotropy, 166
anisotropy of wood, 329
anisoxide, 212
anisum, 212
anisuron, 568
anisyl, 212, 230
anisyl acetone, 212
anisyl methyl ketone, 114, 212, 230
anisylidene acetone, 552
ankaflavin, 198
annelid, 207
annellospore, 207
annex I country, 152
annex II country, 152
annosum root rot, 167
annual, 563
annual amplitude, 343
annual bearing, 320, 343
annual branch, 563
annual budget, 344
annual budget formulation directive, 343
annual canker, 563
annual change, 343
annual charge, 344
annual colony, 563
annual coupe, 343
annual cut, 343
annual cutting, 292
annual cutting area, 343
annual cutting percent, 343
annual discharge, 344
annual earning, 320
annual equivalent, 99, 488
annual erosion, 344
annual felling, 292
annual felling area, 343
annual flood, 344
annual flow, 344
annual fruit-body, 563
annual garden, 563
annual gross earning, 344
annual ground covers, 563
annual growth, 344
annual (growth) layer, 344
annual (growth) ring, 344
annual increment, 344
annual input and output of log yard, 635
annual interest rate, 343
annual isotherm, 343
annual loss, 344
annual operation plan, 343
annual output, 343
annual overhaul, 343
annual plant, 563
annual precipitation, 343
annual production, 344
annual range, 343
annual range of temperature, 344
annual rent for bare land, 313
annual rent for forest, 408
annual ring, 344
annual ring analysis, 344
annual ring boundary, 344
annual ring breadth, 344
annual ring density, 344
annual ring width, 344
annual runoff, 343
annual shoot, 88
annual stumpage return, 292
annual supervisory return, 292
annual variation, 343
annual weed, 563
annual working, 292
annual working days, 343
annual yield, 344
annual yield by area, 85
annual zone, 344
annuation, 630
annuity formulae, 343
annuity in arrears, 493
annular bit, 208
annular budding, 209
annular cell, 208
annular nozzle, 208
annular vessel, 208
annulus, 267, 506
anobium beetle, 229
anode, 556, 611
anogamic propagation, 510
anomalotrophy, 574
anomalous structure, 566
anomalous succession, 34
anomalous wave, 129
anomalous wood, 566
anomaly, 261
anomers, 611
anomodromy, 566
anonaceine, 128
anorthite, 156
anorthoclase, 495
anorthoclasite, 495
anorthose, 495
anorthosite, 537
anorthositic gabbro, 537
anoxic, 388
anoxic condition, 388
anoxic treatment, 511
ant lion, 566
ant plant, 440, 517
ant-proof course, 134
antagonism, 536
antagonist, 115, 130

A

antagonistic action, 115, 536
antagonistic effect, 130, 536
antagonistic organism, 536
antarafacial, 566
antarctic, 338
antarctic circle, 338
antarctic desert, 338
antarctic life zone, 338
antarctic region, 338
antarctic zone, 338
antecedent precipitation, 373
antecedent stream, 522
anteclise, 310, 471
antenna, 65, 478
antennae, 65
antennal nerve, 65
anteposition, 373, 527
anterior wing, 372
anthemia, 202
anthemy, 202
anther, 204
anther cell, 204
anther culture, 204
antherid, 41, 254, 542
antheridium, 41, 254, 542
antheromania, 204
antherophylly, 204
antherozoid, 579
anthesis, 203, 268
anthesis isolation, 203
anthesis regulation, 203
anthobian, 365
anthocyan, 204
anthocyanase, 204
anthocyanidin, 204
anthocyanin, 204
anthocyanophore, 204
anthodium, 485, 648
antholysis, 202
anthophilous, 516
anthophorids, 203, 479
anthophyllite, 615
anthophyta, 524, 580
anthoxanthin, 203, 210
anthracene, 121
anthracene oil, 121
anthracnose, 472
anthraconite, 194, 291
anthraglucorhein, 76
anthropic factor, 396
anthropochore, 396
anthropogenic, 395
anthropogenic-alluvial soil, 168, 180, 395
anthropogenic emission, 395
anthropogenic erosion, 396
anthropogenic factor, 396
anthropogenic succession, 396
anthropogenic vegetation, 396
anthropophyte, 396
anti-blooming agent, 271
anti-blue stain dip, 133
anti-checking clamp, 133
anti-checking coating, 133
anti-checking iron, 133
anti-derailing rail, 202
anti-ecdysone, 271
anti-feedant, 260
anti-feeding effect, 260
anti foaming agent, 133
anti form, 18, 129
anti-movement treatment, 271
anti-noise paint, 532
anti-shrinkage treatment, 271
anti-skew protection, 134
anti-split iron, 133
anti-stain chemical, 131
anti-stain treatment, 131
anti-subsidy duty, 129
anti-tracking varnish, 133
anti-transpirant, 271
anti-weathering chemical, 270
antialexin, 270
antiantibody, 270
antiantitoxin, 270
antiar, 237
antiauxin, 271
antibiogram, 270
antibiont, 270, 271, 528
antibiosis, 271
antibiotic, 270, 271
antibiotic activity, 270
antibiotic microbe, 271
antibiotic spectrum, 270
antibiotic substance, 271
antibody, 271
antiborer plywood, 131
antichlor, 308, 494
anticipated demand, 586
anticipated profit, 586
anticipation survey, 373, 586
anticlinal cambial cell division, 68
anticlinal division, 68
anticline, 18
anticoagulant, 270
anticode, 129
anticodon, 129
anticonsequent stream, 343
anticorrosive agent, 133, 134
anticorrosive paint, 132, 134
anticrustator, 27
anticyclone, 129, 163
antidetonant, 270
antidetonator, 270
antidote, 247
antidromal torsion, 343
antienzyme, 270
antiferment, 133
antifertilizin, 271
antifouling coating, 134
antifouling paint, 66, 134
antifreeze, 131
antifreezing, 131
antifriction property, 270
antifungal, 271
antigen, 271
antigenic action, 271
antigenicity, 271
antigeny, 70
antigrowth hormone, 271
antimellin, 319, 365
antimetabolite, 270
antimony oxide, 476, 557
antimorph, 130
antimutagen, 271
antimycotic, 271
antioxidant, 271
antiparallel, 129, 343
antipodal cell, 130
antipodal nucleus, 130
antiport, 130
antipredator response, 129
antique finishing, 134, 174
antiradiation effect, 270
antirust paint, 134
antiscorch(ing), 270
antisepsis, 131
antiseptic, 131, 132

antiseptic chemical, 132
antiseptic effect, 132
antiseptic process, 131
antiserum, 271
antishrink(age) efficiency, 271
antiskid capability, 132
antiskid chain, 132
antithetic alternation, 568
antitoxic property, 270
antitoxin, 270
antitrade (wind), 130
antitriptic wind, 326
antler, 397
antlion, 566
antu, 2
anucleate, 509
anucleate cell, 509
anucleate mutation, 509
anus, 160
anvil, 481
anvil cloud, 608
aorta, 78, 109
apartment kiln, 143
aperiodicity, 31
aperture, 499, 506
aperwind, 247
apex, 58, 105, 446, 449, 522
aphanite, 138, 571
aphelion, 594
aphid, 552
aphidicide, 411
aphis, 552
aphis lion, 552
aphotic zone, 509
aphrodisiac, 74
aphyllous, 511
apical, 105, 106, 530
apical abolition, 105
apical bud, 106
apical cell, 105
apical dominance, 105
apical grafting, 105
apical growing point, 105
apical meristem, 105
apical reduction, 105
apinclum, 62
apinol, 62
apiology, 323
aplanat, 366, 532
aplanatic magnifier, 91, 532
aplanogamete, 32, 258
aplasia, 125, 522
Apmew (centrifugal) screen, 1
apneustic, 510
apobornylene, 1, 71, 72
apocarpous pistil, 287
apocarpy, 287
apochromatic lens, 154
apocrine gland, 105, 106
apocyte, 117
apode, 512
apoenzyme, 320, 493
apogamy, 510
apogee, 594
apogeoesthetic, 18
apogeotropism, 18
apoid, 146
apolar cell, 509
apomict, 510
apomixes, 510
apomixia, 510
apomixis, 86, 173
apophysis (*plur.* -hyses), 175, 305, 338, 461
apophyte, 177
apoplast, 138, 621
aporetin, 76
aposematic colouration, 255
aposematism, 255
apospory, 508
apothecium (*plur.* -ia), 641
apotracheal banded parenchya, 287
apotracheal diffuse-in-aggregate parenchyma, 287
apotracheal parenchyma, 287
apparent availability, 26, 438
apparent consumption, 27
apparent cord, 25
apparent creep, 26
apparent density, 26
apparent distance, 439
apparent elastic modulus, 26
apparent expansion, 26
apparent height, 439
apparent lignin, 26
apparent photosynthesis, 26, 524
apparent photosynthetic rate, 26
apparent specific gravity, 26
apparent stress, 26
apparent volume, 26, 439, 465
apparent weight, 26, 319, 439
appeasement substance, 209
appendage, 152
appetitive behaviour, 587
apple tree, 363
application for export permit, 445
application for share, 175, 396
application of funds, 641
application programming, 573
application technique, 573
application technology satellite, 573
applicator, 100, 147, 402, 431, 487
applicator roll, 431
applied ecology, 573
applied entomology, 573
applied force, 496
applied stress, 496
applique, 152, 375
apposition, 147
appraisal, 173
appraisal method, 173
appraisal of aesthetic value, 320
appraisal of bare land, 313
appraisal of stumpage, 289
appraisal procedure, 363
appraised stumpage price, 94, 363
appraised value, 173
appreciation, 363, 602, 604
appreciation of scenery, 145
appressorium, 152
approach, 250, 482
approach flow of stock, 237
approach grafting, 271
approaching direct shot, 362
approaching high shot, 164
approaching low shot, 94
approaching shot, 573

appropriation, 29, 252, 603
appropriations reserve, 252, 640
approval sale, 438
approval test, 556
approved budget, 193
approved estimate, 193
approximate calculation method, 250
approximate value, 250
approximated real growth series, 250
apricot kernel oil, 541
apron, 69, 90, 131, 202, 205, 388, 446
apron board, 69, 88, 388
apron conveyor, 294
apron lining, 53, 165, 388, 534
apron piece, 612
apron plain, 28
apsorption spectrum, 514
aptera(e), 508
apterous, 508, 511
apterous form, 508
aqua ammonia, 3
aquaculture, 456
aquapulper, 457
aquar garden, 457
aquarium, 460, 557
aquatic, 458–460
aquatic animal, 459
aquatic animal pest control, 459
aquatic community, 459
aquatic ecosystem, 459
aquatic insect, 459
aquatic macrophyte vegetation, 459
aquatic mosses, 459
aquatic plant, 459
aquatic vegetation, 459
aquent, 52
aqueous bulking agent, 458
aqueous preservative, 458
aqueous rock, 456
aqueous vapour, 458
aquept, 52
aquert, 52
aquiclude, 622, 651
aquiculture, 456
aquifer, 460, 545
aquifuge, 32
aquiherbosa, 57, 459
aquiprata, 456, 459
aqult, 52
aquod, 52
aquoll, 52
aquox, 52
arabic gum, 1
arable, 272, 440
arable land, 272
arable land minimum, 168
arable layer, 168, 272
araliene, 71
aralin, 71
aramite, 411
arasan, 149
arathane, 94
araucaroid pit, 338
arbitrary value, 396
arbor, 88, 377, 448, 538, 634
Arbor-Day, 616
arbor form, 377
arbor species, 377
arbor-vitae, 528
arboreal, 377, 446, 449
arboreal adaptation, 449
arboreal animal, 302
arboreal epiphyte, 449
arboreal growth, 448
arboreal habitat, 449
arboreal species, 377
arboreous, 377
arboreous desert, 515
arborescent, 377, 449
arborescent phanerogamic type, 328
arboret, 179, 535
arboretum, 448, 618
arboriculture, 448, 449
arboriculturist, 448
arbour, 295, 588
arbuscle, 2, 179
arbuscular, 72, 180
arc shake, 201
arcade, 294, 580
arch, 170, 172
arch breaker system, 132
arch bridge, 172
arch cat, 79
arch dam, 172
arch hook, 172
arch line, 172
arch logging, 172
archaeal, 174
archaeocyte, 591
archaeophyte, 437
Archaeornithes, 174
arched axle logging wheel, 172
arched door, 172
arched stretcher, 172
archegonium, 41, 254
archetype, 591
archicarp, 47
architectural acoustics, 236
architectural colour effect, 236
architectural design, 236
architectural finish, 236
architectural furniture, 236
architectural style garden, 236
architecture, 236
architrave, 320, 635
architrave block, 321
arctic air mass, 17
arctic alpine, 17
arctic barren, 17
arctic circle, 17
arctic desert, 221
arctic lichen barren, 17
arctic life zone, 17
arctic mat grassland, 17
arctic meadow, 17
arctic plant, 17
arctic region, 17
arctic timber line, 221
arctic tundra, 17
arctic wind, 17
arctic zone, 17, 189
arcto-tertiary flora, 17
arctogaea, 17
area, 97, 300, 305, 323
area allotment method, 323
area and volume method combined, 323
area and volume period method, 323, 606
area book, 323
area calculator, 323
area classification, 383
area control method, 323
area control system, 323
area determination, 323
area district, 97
area division, 323, 408
area effect, 323
area estimate, 323
area frame work, 323
area grid ignition, 366
area ignition, 116, 279, 386
area-increment, 113

area-increment percent, 113
area load, 323
area measurement, 323
area-method, 323
area normalization method, 323
area of application, 431
area of bearing, 612
area of special risk, 474
area period method, 323
area redevelopment program, 383
area register, 323
area regulation, 323
area regulation method, 323
area-related, 584
area sampling error, 323
area suppression method, 279
area table, 323
area-to-perimeter ratio, 323
area under canopy, 300
area under (crown) cover, 300
area under management, 253
area-volume check, 323
area-wise felling, 127, 323
areae, 59
areca palm, 27
arenaceous rock, 412
arenaceous sediment, 412
arenite, 412
arenosol, 412
arenyte, 412
areolae, 59
arete, 525
argentation thin-layer chromatograph, 570
argillaceous cement, 342
argillaceous facies, 342
argillaceous limestone, 342
argillaceous red bed, 198
argillaceous slate, 342
argillic, 345
argillic horizon, 344
argillite, 200, 342
arhythmic type, 32
arhythmicity, 245
ari, 504
arid, 157, 359
arid area, 157
arid region, 157
arid soil, 157
arid zone, 156
aridisol, 190
aridity, 158
aridity index, 157
aril, 230
aristiform, 318
arithmetic instruction, 467
arithmetic mean, 467
arithmetic mean diameter, 467
arithmetic mean height, 467
arithmetic unit, 596
arithmetic(al) mean, 91, 467
arithmetic(al) mean diameter, 467
arithmetical mean sample tree, 467, 625
ark, 169
arkose, 50
arkosic conglomerate, 50
arkosite, 50
arm chair, 147
arm crane, 545
arm rail, 147
arm rest, 147
arm saw, 442, 545
arm-tie, 195, 536
armanene, 205
Armillaria root rot, 323
armor-plywood, 249
arm's length pricing, 171
arm's length transaction, 171
Arnold arboretum, 1
aromadendrin, 528
aromadendrone, 528
aromatic hydrocarbon metabolism, 131
aromatic plant, 131
aromorphosis, 541, 556
around-the-dog method, 589
Arprocarb, 40
array, 35, 451, 608
array computer, 608
arrester, 21, 620
arresting device, 620
arrestor, 21, 620
arrhenotoky, 48
arris, 22, 285
arris knot, 285
arris rail, 402
arrow back, 237
arrowhead scale, 437
arroyo, 156, 157, 190
arsenic salt, 419
arsenical creosote, 189
Arsenious oxide, 6
art of landscape garden(ing), 588
art paper, 320
artemisia ketone, 379
artemisia oil, 379
arterial drainage, 350
arterial forest road, 160
arterial highway, 171
artesian aquifer, 55, 645
artesian well, 354, 645
arthropod-borne virus, 60
arthrospore, 245
articulate lactiferous tube, 581
articulated boom, 243
articulated frame, 243, 606
articulated frame steering skidder, 243, 606
articulated loading shovel, 243
articulated log loader, 243, 606
articulated ripper, 243
artificial afforestation, 394, 395
artificial auxiliary energy, 394
artificial board, 396
artificial camphor, 396
artificial coast sand dune, 394
artificial community, 395
artificial compulsory pollination, 395
artificial control, 394
artificial conversion, 395
artificial crop, 395
artificial culture, 395
artificial direct regulation, 395
artificial division, 395
artificial division method, 395
artificial drying, 394
artificial enrichment ex-periment, 395
artificial eutrophication, 394
artificial fibre, 396
artificial forest, 395
artificial forestation, 395

A

artificial form factor, 503, 542
artificial form quotient, 503, 542
artificial grassland, 394
artificial grinding stone, 396
artificial heartwood, 395
artificial hill, 230
artificial hole set, 395
artificial hybrid, 395
artificial induced mutation, 395
artificial infection, 394
artificial information flow, 395
artificial inoculation, 394
artificial lake, 394
artificial leather, 396
artificial manure, 114
artificial measures promoting regeneration, 394
artificial nesting method, 52
artificial parthenogenesis, 394
artificial plantation, 395
artificial planting, 395
artificial pollination, 395
artificial pond, 394
artificial population, 395
artificial precipitation, 394
artificial propagation, 394
artificial pruning, 395
artificial pulpstone, 396
artificial rain device, 394
artificial rainfall, 394
artificial reef, 395
artificial reforestation, 394
artificial regeneration, 394
artificial regeneration with the aid of field crop, 394
artificial reproduction, 394
artificial resin, 396
artificial ripening, 394
artificial sea water, 394
artificial seasoning, 394
artificial seeding, 394
artificial selection, 395
artificial spiral grain, 395
artificial stand, 395
artificial stocking, 394
artificial storage, 395
artificial ventilating dry kiln, 395
artificial ventilation, 395
artificial water tank, 395
arvideserta, 307
aryl-glycerol-β-aryl ether structure, 131
asafoetida oil, 1
asbestine, 434
asbestos, 434
asbestos blanket, 434
asbestos board, 434
asbestos cardboard, 434
asbestos cement plate, 434
asbestos cement sheeting, 434
asbestos fibre, 434
asbestos fibreboard, 434
asbestos panel, 434
asbestos plate, 434
asbestos sheet, 434
asbestos-veneer plywood, 434
asbestos wallboard, 434
asbestus, 434
ascending chromatography, 415
ascending face, 416
ascending grade, 416
ascending shot, 416
ascending water, 416
ascigerous stage, 641
ascocarp, 641
ascomycetes, 641
ascomycetous fungus, 641
ascorbic acid, 270, 502
ascorbic acid oxidase, 270
ascospore, 641
ascuerdalith, 167, 315
ascus (*plur.* asci), 641
ascus-fungus, 641
ascus layer, 641
asebotin, 316
aseptic, 131, 132, 509
aseptic culture, 509
aseptic filtration, 509
aseptic seed culture, 628
aseptic seed stratification, 628
asexual cycle, 511
asexual flower, 510, 511
asexual generation, 511
asexual germ cell, 511
asexual hybridization, 511
asexual multiplication, 511
asexual propagation, 511
asexual reproduction, 511
asexual spore, 511
ash, 210
ash-free dry weight, 509
ash-free weight, 509
ashen-grey soil, 211
asiphonogamae, 509
asomate, 149
asparagin, 477
asparagines, 477
aspect, 165, 226, 363, 504, 528, 529
aspect ratio, 49
aspection, 226, 255, 630
aspergillosis, 384
aspergillus, 384
asphalt, 7, 291
asphalt-coated felt paper container, 578
asphalt size, 7
asphaltic process, 291
asphyxia, 621
asphyxy, 213, 230
aspirated bordered pit pair, 20
aspirated pit, 20
aspirated pit-pair, 20
aspirated psychrometer, 481
aspirated thermometer, 482
aspiration, 506
aspiration of bordered pit, 261
Asplund process, 1
Asplund's freeness-tester, 1
asporogenous, 33
asporous, 508
assay determination, 143
assay procedure, 143
assemblage, 222, 224, 638
assembling clamp, 638
assembling wood, 222, 329
assembly, 224, 353, 389, 638, 652
assembly adhesive, 638, 652
assembly drawing, 638
assembly glue, 638, 652

A

assembly gluing, 121, 173, 652
assembly section, 638
assembly time, 53, 114, 652
assessed program cost, 471
assessed valuation, 173
assessed value, 173
assessed versus market value, 173
assessment, 47, 173, 363, 471, 572
assessment of nature reserve management effectiveness, 645
assessment of yield, 441
assessment of yield-capacity, 422
assessment report, 363
assessment unit, 363
assessor, 173, 363
assessory system, 150
asset, 641
assets and liabilities, 641
assignable cause, 138
assignable credit, 274
assigned amount unit, 142
assignment, 637
assignment allowance, 496
assimilate, 482
assimilating tissue, 482
assimilation, 482
assimilation box, 482
assimilation chamber, 482
assimilation efficiency, 482
assimilation number, 482
assimilation product, 482
assimilation quotient, 482
assimilation ratio, 482
assimilation-respiration ratio, 482
assimilation substance, 482
assimilation system, 482
assimilative, 482
assimilatory, 482
assimilatory tissue, 482
assistance cutting, 150
associate, 246, 294, 353
associated costs, 152
associated species, 11, 214
associated tree species, 11
associates, 11
association, 172, 294, 353, 388
association coefficient, 178
association complex, 388
association forest, 388, 418
association fragment, 388
association group, 388
association individual, 388
association segregate, 388
association table, 388
association temperature, 388
association type, 388
association unit theory, 388
associes, 556
associule, 556
assor(ta)tive breeding, 547
assor(ta)tive mating, 547
assorting, 141, 359, 547
assorting table, 141, 547
assortment, 37, 141, 142, 359
assortment table, 37, 141
assortment yield table, 37, 290
assumed growing stock, 230
assurance factor, 2
astatic, 35
ASTM designation, 320
ASTM specification, 320
ASTM standard, 320
astragal, 11
astragal plane, 593
astragal tool, 11
astringent, 442
Astrom barking machine, 1
astronautics, 191
astronomical method, 478
asulam, 210
asymmetric absorption band, 32
asymmetric growth, 32
asymmetric method, 32
asymmetric state, 32
asymmetrical distribution, 32
asymmetry, 32
asymptote, 237
asynapsis, 32, 34
asynchronous computer, 566
at-railhead, 480
at stake, 46
at stump, 413, 599
at the base, 3
at the dock, 316
at the pond, 635
at the stump, 3
atactic polymer, 509
atactic polypropylene, 509
atactostele, 404
atavism, 130
ataxinomic, 221, 566
atlantone, 78
atlapulgite, 215
atmidometer, 371, 609
atmidometry, 609
atmidoscope, 432
atmolith, 368
atmometer, 371, 609
atmometry, 609
atmosphere, 77, 78
atmosphere pollution, 78
atmospheric agency, 78
atmospheric circulation, 77
atmospheric density, 78
atmospheric diffusion, 77
atmospheric drought, 77
atmospheric evaporation, 78
atmospheric feedback process, 77
atmospheric humidity, 78
atmospheric inversion, 78
atmospheric moisture, 78
atmospheric pollution, 78
atmospheric pressure process, 49
atmospheric radiation, 77
atmospheric refiner, 49
atmospheric refining, 49
atmospheric rerun, 49

A

atmospheric rock, 368
atmospheric stability, 78
atmospheric turbidity, 77
atmospheric turbulence, 78
atmospheric vortex, 78
atmospheric window, 77
atmospheric(al) pressure, 78
atokous, 32, 509
atoll, 207
atoll reefs, 207
atom line, 592
atomic absorption spectrometry, 592
atomic absorption spectroscopy, 592
atomic spectrum, 592
atomization, 513, 592
atomizer, 321, 354, 513
atractylis concrete, 41
atrazine, 582
atrio, 215, 639
atrophy, 503
atropin(e), 1, 100
attack a fire, 324, 364
attacked timber, 443
attendance time, 62
attenuance, 451
attenuated nutrient solution, 515
attenuation, 233, 451
attenuation index, 451
attitude, 640
attractant, 571
attractant insecticide, 571
attractant of ant, 317
attracting force, 515
attraction, 571
attribute, 446, 475
attribute sampling, 446
attribution, 183
attrition, 326
attrition mill, 327, 522
attrition rate, 191
attrition test, 326, 327
auction, 349
auction market, 349
auction sale, 25, 349
audio method, 430
auditory organ, 481
auger, 313
auger bit, 313, 316, 333
auger borer, 313, 652
auger boring, 313, 652
aurantiin, 56
aurantiol, 56
auraptene, 56
aurea, 210
aurone, 57, 348
aurora, 222
austerity program, 249
Australian gum, 528
Australian region, 4
australol, 4, 115
Austrian formula, 4
Austro-Malayan region, 4
Austroriparian life zone, 338
ausubo, 641
autarchy, 111, 644
autecology, 166
authentic sample, 274, 608
authority to purchase, 503
authorized bank, 444, 619
authorized capital, 127, 396
authorized (capital) stock, 127, 193
auto-feed carriage, 643
auto immersion unit, 643
auto-pusher, 644
auto-roller feed band resaw, 643
auto-sander, 643
auto-snatch block, 214
auto-sorter, 644
autoadaptation, 646
autoallopolyploid, 484
autobacteriophage, 646
autobiology, 166
autocarp, 644
autochore, 642
autochory, 642, 646
autochthon, 19, 527
autochthonous, 491, 522, 527, 565
autochthonous flora, 19
autochthonous peat, 591
autochthonous soil, 588, 591
autochthonous species, 19, 588, 646
autocidal control, 645
autoclave, 164, 394, 550
autoclaving, 164
autocolony, 462
autoconvection, 642
autocorrelation, 644, 647
autocorrelation chronology, 647
autodecaploid, 484
autoecious, 87
autoecious fungus, 87
autoecism, 87
autoecology, 166
autofecundation, 647
autofertilization, 647
autogamic plant, 644
autogamous, 644, 647
autogamy, 644, 647
autogardener, 442
autogenesis, 645
autogenetic, 645
autogenetic plankton, 19
autogenetic succession, 644
autogenic, 646
autogenic succession, 341, 644
autogenus, 86
autogeny, 645
autograft, 483, 646
autographic auxanometer, 645
autohand, 218, 643
autohaploid, 484
autohesion, 569, 645
autoheteroploid, 484
autoignition temperature, 646
autolopping, 646
autolysis, 646
automated bucking line, 600
automated carriage, 643
automated log-sorting line, 547
automated sanding line, 643
automatic adjustment, 643, 644
automatic allocation, 642
automatic assembly, 644
automatic balance, 643
automatic band saw sharpener, 79
automatic band weigher, 642
automatic belt sander, 642
automatic binary computer, 642
automatic braking, 644
automatic caliper, 643
automatic carrier, 643

A

automatic celestial navigation, 643
automatic centering, 642
automatic chain oiler, 643
automatic check, 644
automatic clipper, 643
automatic code translator, 644
automatic coding machine, 642
automatic coding system, 642
automatic compensation, 642
automatic control, 643
automatic cut-off saw, 643
automatic data exchange, 643
automatic data link, 643
automatic data processing, 102, 643
automatic data processing system, 643
automatic defecting saw, 643
automatic digital encoding system, 643
automatic digital input-output system, 643
automatic digital tracking analyzer computer, 643
automatic electronic counter, 642
automatic feeding deck, 643
automatic fire sprinkler, 643
automatic following control, 643
automatic gauging, 642–644
automatic hitch, 643, 644
automatic humidity controller, 479, 643
automatic hydraulic sprayer, 644
automatic information retrieval system, 644
automatic injection planter, 495
automatic input, 643
automatic knapsack sprayer, 18
automatic knife grinder, 643
automatic lathe charger, 643
automatic loader and unloader, 644
automatic locking carriage, 646
automatic log deck machinery, 644
automatic lubricating device, 643
automatic marker, 643
automatic message counting, 644
automatic nozzle, 643
automatic picture transmission, 487
automatic pitch, 256
automatic press loader, 644
automatic press unloader, 644
automatic pressure control, 644
automatic program control, 642
automatic programming, 56, 642
automatic reading, 642
automatic recording instrument, 643
automatic rectifier, 643
automatic regulation, 643
automatic remote control, 644
automatic request, 643
automatic sampling, 643
automatic scale, 642, 643
automatic self pollination, 644
automatic sequence control, 642
automatic shaper, 644
automatic shotgun, 643
automatic signal, 644
automatic spray apparatus, 643
automatic stacker, 642
automatic stop device, 644
automatic transmission, 642, 643
automatic tying machine, 642
automatic veneer edge composer, 642
automatic watering system, 643
automatic weather station, 643
automatically controlled kiln, 643
automimicry, 626
automobile sprayer, 643
automorphous soil, 642, 647
autonomous growth, 644
autonomous investment, 644, 647
autonomous variable, 647
autoparasite, 646
autoparasitism, 484, 644, 645
autopelagic plankton, 415
autophagy, 647
autophyte, 647
autoplasm, 484
autoplastic graft, 483, 484, 646
autoploid origin, 484
autopolyploid, 484
autopositive, 646
autoradiography, 135
autoregression, 644
autoregressive transformation, 644
autoregulation, 643
autoreproduction, 646
autoscaler, 643
autosome, 48
autosynapsis, 484
autosyndesis, 484
autotetraploid, 484
autothinning, 646
autotomy, 644–646
autotoxic, 644
autotroph, 647
autotrophic, 647
autotrophic becteria, 647
autotrophic component, 647
autotrophic lake, 647
autotrophic metabolism, 647
autotrophic nutrition, 647
autotrophic organism, 647
autotrophic plant, 647
autotrophic respiration, 647

A

autotrophic stratum, 647
autotrophic succession, 647
autotrophism, 647
autotrophy, 647
autotropism, 530
autozygote, 482
autumn circulation, 382
autumn flower-bed, 382
autumn frost, 382
autumn growth, 382
autumn normal growing method, 382
autumn planting, 382
autumn ploughing, 382
autumn seeding, 382
autumn sowing, 382
autumn wood, 382, 499
autumnal aspect, 382
autumnal circulation, 382
auxanographic method, 430
auxanometer, 429
auxesis, 518, 601
auxetic, 517
auxiliary, 149, 150
auxiliary activities (works), 152
auxiliary body, 152
auxiliary carriage, 150
auxiliary energy, 150
auxiliary equipment, 150
auxiliary forest belt, 154
auxiliary land, 303
auxiliary rafter, 149, 154
auxiliary road, 154
auxiliary solvent, 150
auxiliary species, 11
auxiliary tank, 150
auxiliary torque backup roll, 150
auxiliary winch, 150
auxin, 429, 618
auxochrome group, 634
auxocyte, 430, 541, 601
auxograph, 429
auxotrophic mutant, 574
availability, 274, 582
available alkali, 581
available balance, 274
available chlorine, 581
available depth of soil, 491
available fuel, 581
available fuel energy, 581
available investment capital, 581
available nitrogen, 581
available nutrient, 582
available phosphorate, 581
available resource, 274
available water, 274, 581
available water-holding capacity, 581
available width, 274
avalanche, 19, 471, 548
avalanche of loose snow, 144
avalanche of massive snow, 279
avaram, 121
Avena sativa, 556
avenue, 303
avenue planting, 90, 245
avenue tree, 540
average, 362, 562
average accretion, 361
average age, 361
average annual growth rate, 344
average annual stand depletion, 300
average basal area, 361
average climate, 361
average cost, 361
average deviation, 361
average diameter, 362
average diameter at breast height, 361
average dominance, 361, 362
average fiber length grit, 361
average forest area per capita, 395
average form class, 361
average frequency, 361
average growth, 361
average height, 361
average height of stand, 300
average increment, 361
average information content, 361
average intake, 361
average life, 361
average log, 622
average moisture content, 361
average outgoing quality limit, 361
average radius, 361
average ratio, 361
average regional fire danger, 361
average stand age, 361
average stand height, 361
average stem, 361, 625
average stocking, 361
average top height, 577
average tree, 361, 625
average variable cost, 361
average volume table, 361
average yield class, 127
averruncator, 39, 164
aviary, 346, 379
avicelase, 501
Avicennia type, 188, 483
avicide, 411
avifauna, 345
avirulent, 509
avirulent strain, 233, 509
avoidance, 120, 524
avoiding, 524
avoiding movement, 474
awful weather, 369
awl graft, 639
awn, 318
awning window, 355, 605
axe gauge, 149
axe handle, 149
axe marking, 75, 269
axe mattock, 149
axe wedge, 356
axeman, 126, 269, 543
axenic culture, 509
axial adjustment, 631
axial bearing, 612, 618
axial bond, 615, 630
axial clearance, 631, 650
axial compression, 631
axial direction, 631
axial displacement, 631
axial element, 631
axial fan, 631
axial flow fan, 631
axial flow turbine, 631
axial force, 631
axial high speed stirrer, 631
axial intercellular canal, 631
axial knock, 639
axial load, 631
axial parenchyma, 631
axial parenchyma cell, 631

A

axial placentation, 625
axial pressure, 631
axial rake angle, 631
axial resin canal, 631
axial root, 633
axial shoot, 633, 634
axial strain, 631
axial strand parenchyma, 631
axial stress, 631
axial tension, 631
axial thrust, 631
axial tracheid, 631
axial vibration, 631
axial wall, 631
axial wood, 631
axialite, 631
axil, 562
axile placentation, 625
axillary bud, 562
axillary cyme, 562
axillary shoot, 562
axillary stipule, 562
axin, 60
axis, 624, 630, 631
axle, 52, 631
axle load, 52
axle pulley, 631
axogamy, 631
azadarach, 278
azadarichta, 278
azalea, 314
azedarine, 278, 295
azelon, 88
azeotropic drying, 172
azimuth, 97, 131
azimuth-elevation, 131
azimuthal angle, 131
azinphos methyl, 175
azobenzene, 348
azonal, 130, 136
azonal climax, 136
azonal community, 130, 571
azonal formation, 130
azonal soil, 130
azonal vegetation, 136
azophoska, 88
azotase, 175
azotification, 175
azotobacter, 175
azotobacteria, 175
azotobacterin, 175
azotogen, 175
azulene, 156, 587
azulenoid, 587
azygote, 86
azymia, 320
B-C soil, 246
B-grade thinning, 401
B-horizon, 103
B-layer, 103
B-stage, 469
bablah, 248
babool, 248
babul, 248
baby dryer, 535
baby square, 533
babycrib, 572
bacca, 237
baccate, 237
baccausus, 264
bacciferous, 260
bachelor's chest, 533
Bachert process, 17
baciilaris, 159
bacillar, 159
bacilliform, 159
bacillin, 159
bacillus, 159
bacillus (*plur.* -lli), 159
bacilysin, 398
back, 17, 18, 57, 88, 94, 95, 261
back action lock, 90, 130
back azimuth, 129
back bevel, 200
back bevel angle, 200
back-blading, 90
back board, 17
back bulb, 284
back burn, 343, 573
back chair, 271
back crossing, 17, 18
back cupping, 18
back cut, 11, 415, 524
back diffusion, 130
back draft, 343
back-extraction, 129
back face, 18, 57, 415
back fire, 573
back firing, 573
back-flap, 7
back-flap hinge, 325
back flow, 212
back focus, 199
back gauge, 18, 22
back ground, 18, 200
back guy, 199
back iron, 202
back knotter, 150, 503
back leg, 18, 199
back line, 57, 212
back lining, 18, 94
back log, 92, 451
back marsh, 318
back mutant, 211
back mutation, 211
back order, 554
back panel, 17
back post, 566
back rail, 199
back rest, 566
back rigging, 154, 503
back-roll, 612
back roll assembly, 134
back rule, 18
back saw, 200, 529
back-sawing, 524
back-sawn, 524
back shore, 27, 105
back slat lower, 18
back spar, 154, 503
back stop, 573
back stroke, 130
back-stroke valve, 618
back swamp, 200
back swimmer, 465
back-to-back letter of credit, 114, 115
back tracking, 129
back-up data, 17
back-up roll, 550, 612
back-up roller, 134
back-up water level, 212, 576
back veneer, 17
back wardation, 239, 525
back yard, 8, 200
backbone, 309, 456, 633
backcross, 211
backcross breeding, 211
backcross hybrid, 212
backcross method of breeding, 211
backcross parent, 211
backer loader, 129
backfill, 212
backfiller, 212, 479
background adaptation, 18
background equivalent activity, 19
background luminance, 18
background noise, 18
background planting, 18
background processing, 200
background radiation, 18, 19
background spectrum, 19
backhoe, 129
backhoe loader, 129

B

backing board, 17, 53, 165, 572
backing paper, 18, 27
backing pump, 63, 373
backing stage, 373
backing wire, 95, 288
backlog of unfilled orders, 504
backpack pump, 18, 551
backreflection, 200
backscattering, 200
backside, 18
backslope, 18, 199
backward linkage, 129, 343
backward roll, 106
backwash, 129, 212, 503
backwash effect, 212
backwater, 6, 212, 576
backwoods, 357, 504
bacteria-activator, 518
bacteria contamination, 518
bacterial chlorin, 518
bacterial cloning, 518
bacterial decomposition, 518
bacterial disease, 518
bacterial genetics, 519
bacterial insecticide, 518
bacterial ooze, 519
bacterial shot-hole, 519
bacterial toxin, 518
bacterial virus, 518
bactericide, 411
bactericidin, 411
bacteriform, 519
bacteriochlorin, 267
bacteriochlorophyll, 519
bacteriocin(e), 518
bacteriology, 519
bacteriolysin, 398
bacteriolysis, 398
bacteriophage, 441
bacteriophagia, 441
bacteriophagic, 441
bacteriophagic lysis, 441
bacteriopurpurin, 267
bacteriorhiza, 267
bacterioscopy, 518
bacteriosis, 518
bacteriostasis, 568
bacteriostatic agent, 568
bacteriotoxin, 518
bacteriotrophic, 519
bacterium (*plur.* -ia), 159, 518
bacteroid, 519
bacterorrhiza, 518
baculiform, 159
badan, 5
badger, 287
badian oil, 5
badland, 209, 297, 630
badly formed stem, 221
baffle, 88, 209, 606
baffle board, 88, 129, 166, 606
baffle plate, 88, 165, 606
baffled stripping column, 88
bag, 297
bag barker, 81
bag boom, 81, 338
bag moulding, 81
bag-worm moth, 469
bagasse board, 606
bagasse dryer, 606
bagasse fibre, 606
bagasse particleboard, 606
bagasse shaving board, 606
bagillumbang oil, 403
bagworm, 81
bahada, 413
bai-u, 319
bailing, 80
Bailing's hydrometer, 5
bailment, 227, 503
baisacki, 7
bait crop, 583
baiting method, 436
bajada, 413
baker raft, 360
baking finish, 197, 271
baking varnish, 197, 381
balance, 46, 246, 360, 584
balance cut-off saw, 360
balance due, 246
balance matching, 114, 360
balance of nature, 646
balance of trade, 319
balance piston saw, 360
balance pressure, 360
balance sheet, 246, 360, 641
balance sheet date, 246
balance weight, 360, 361
balanced construction, 114, 360
balanced forest, 127
balanced growth, 480
balanced hypothesis, 360
balanced lethal gene, 361
balanced nutrient solution, 361
balanced polymorphism, 360
balanced sash, 360
balanced selection stand, 128, 266
balanced structure, 114, 360
balanced type, 360
balanced unevenaged stands, 128, 266
balancing selection, 360
balancing sheet, 32, 360
balanophora, 417
balata, 5, 481
bald, 486
bald mountain, 486
bald tree, 511
bale breaker, 47, 75
bale hook, 636
bale opener, 47
bale pulper, 237
bale room, 75
baler, 75, 550
baling, 75, 114
baling paper, 12
balk, 73, 76, 173, 295, 309, 478
balk board, 132, 165
balk-ploughing, 340
balk-timber, 295
ball and ring method, 208
ball bearing, 184
ball catch, 321, 382
ball hooting, 184, 205
ball mill pulverizer, 382
ball-plant, 80
ball planting, 12, 80
ball thrust bearing, 492, 618
ballast, 5, 90, 310, 361, 364
ball(ed) planting, 80
balled seedling, 80
balled stock, 80
balled tree, 80
ballhooter, 184, 205
ballochore, 297
balloon drag, 370
balloon framing, 380
balloon logging, 370
balneotherapy, 280
balometer, 149
balsa, 380
balsa wood, 380

B

balsam, 529
baluster, 283
baluster-and-spindle back, 636
baluster back, 283
balustrade, 147, 283
bamboo, 632
bamboo article, 632
bamboo brake, 131
bamboo cable, 324, 632
bamboo chute, 632
bamboo culm, 632
bamboo cylinder, 632
bamboo fence, 632
bamboo forest, 632
bamboo fungus, 632
bamboo furniture, 632
bamboo grows, 531
bamboo locust, 632
bamboo mat, 632
bamboo mat plywood, 632
bamboo nail, 632
bamboo particle board, 632
bamboo plantation, 395
bamboo-plastic combination, 632
bamboo plywood, 632
bamboo pulp, 632
bamboo raft, 632
bamboo rhizome, 632
bamboo scaffold, 632
bamboo sculpture, 632
bamboo shoots, 632
bamboo skin, 632
bamboo splitting machine, 356
bamboo-stand regeneration, 632
bamboo stand system, 632
bamboo strip, 632
bamboo weaving, 632
bamboo wood, 632
bambooware, 632
ban forest, 251
band application, 81
band brake, 80
band carriage, 352
band clamp, 79, 80
band conveyer, 80
band dendrometer, 80, 81
band drier, 79
band girdling, 80, 207
band mill, 79
band mill edger, 79
band pass, 80
band-pass filter, 80
band planting, 79, 81
band resaw, 363, 598
band resaw with horizontal roll, 458
band saw, 79
band saw air-operated shaper, 368
band saw blade, 79
band saw hand-operated shaper, 442
band saw mill, 79
band saw sharpener, 79
band saw stretcher, 79
band saw with auto-feed carriage, 643
band sawing of veneer, 81
band-sawn, 79
band seeding, 79
band spread, 359
band tailing, 365
band veneer dryer, 79
bandage process, 12, 19
bandage treatment, 12, 19
banded apotracheal parenchyma, 80
banded core, 146, 603
banded paratracheal parenehyma, 80
banded parenchyma, 80
banded pores, 80
banding, 19, 146, 206, 209
banding method, 12, 19
banding plane, 526
banding treatment, 313, 502
banister back, 283
bank, 3, 183, 192, 285, 471, 534, 638
bank erosion, 94
bank planting, 94
bank protection, 202
banker, 353, 495
bankfull stage, 317
banking, 183, 479, 537, 636
banking ground, 285, 492, 638
baptist cone, 223
bar, 5, 11, 184, 196, 262, 412, 479
bar and batten paneling, 280, 603
bar back, 602
bar chair, 259
bar clamp, 492
bar gravel, 412
bar planting, 376
bar slit method, 48
bar stock, 11
barb, 90, 173
barba, 318, 390, 544
barbate, 261
barbed, 260
barbellate, 260
barber chair, 75, 566
barchan, 417, 539
barchane, 539
bardane oil, 346
bare area, 34, 313
bare-drum, 181, 275
bare faced tenon, 310, 325
bare fallow, 265, 508
bare ground, 313
bare hill, 209, 313
bare ion, 313
bare land, 34, 313
bare machine, 313
bare measure, 318
bare mountain, 486
bare rock, 313
bare-root(ed) planting, 313
bare-rooted seedling, 313
bare seashore, 182
bare size, 183, 318
bare wood, 355
bareroot seedling, 313
bargain money, 106
bargaining, 238, 566
bargaining counter, 62
bargaining power, 474
bargaining strength of producer, 422
barge boom, 360
barge couple, 555
baric topography, 370
bark, 30, 448
bark adjustment factor, 449
bark allowance, 449
bark bar, 165
bark beetle, 533
bark blazer, 177, 269, 532
bark blister, 356, 449
bark board, 79
bark borer, 449
bark breaking machine, 449
bark coppice, 31
bark crusher, 449
bark cutter, 449

B

bark deduction, 356, 449
bark dieback, 356
bark extract, 449
bark factor, 449
bark-free, 385
bark gauge, 448, 449
bark grafting, 356
bark grinding mill, 449
bark growth, 449
bark hack, 39, 165, 532
bark hog, 449, 468
bark increment, 449
bark inhabiting bryophyte, 447
bark injury, 356
bark inversion, 449
bark mark, 21
bark measuring instrument, 44
bark miner, 449
bark necrosis, 356
bark peeler, 31
bark peeling machine, 31
bark pellet, 449
bark percent(age), 448
bark pocket, 229, 449
bark removal, 30
bark-ringed knot, 356, 571
bark ringing, 209
bark scaler, 30
bark scorch, 6, 356
bark scorching, 449
bark scraper, 177
bark seam, 229, 356
bark shaver, 177
bark shaver face, 177
bark shaving face, 177
bark shredder, 449
bark side, 328
bark slip(ping), 449
bark speck, 448
bark spud, 30
bark stripping, 449
bark stripping machine, 449, 479
bark tanner, 449
bark thickness, 449
bark-thickness measurer, 449
barked log, 31, 385
barked wood, 30, 385
barker, 31
barker drum, 175, 593
barker of rosser-head type, 31
barker of rotary-ring type, 208
barkie, 79
barking, 30, 385
barking drum, 175, 593
barking in strip, 479
barking iron, 30
barking loss, 31
barking machine, 31
barking saw, 31, 590
barkometer, 5, 399
barlike thickening, 479
barnyard manure, 259
barogram, 370
barograph, 370, 645
barometer, 370
barometric pipe, 370, 456
barophilic, 440, 441
barophilic bacteria, 440
barophobic, 524, 556
Baroque garden, 5
barosmin, 35, 595
barosmoid, 35
barothermograph, 370
barothermohygrometer, 370
barotolerant, 338
Barr dendrometer, 5
barrage, 283, 576, 632
barras, 319, 401
barrel, 89, 175, 184, 375, 484, 485
barrel-butted, 76, 160, 167
barrel head, 484
barrel method, 485
barrel saw, 485
barrelling, 637, 639
barren, 34
barren land, 34
barren (land), 209
barrenness, 27
barrier, 166, 363, 602
barrier ant tape, 134
barrier beach, 554
barrier effect, 363
barrier film, 166
barrier reef, 94
barrier sheet, 166
bars of sanio, 319
barter, 569
Bartrev-press, 5
basal area, 113, 542
basal area-age, 542
basal area at stocking, 3
basal area central diameter, 625
basal area control, 113
basal area factor, 113, 242
basal area of breast-height, 542
basal area regulation, 113
basal-area table, 113, 542, 592
basal bark treatment, 160
basal body, 220
basal cover(age), 220
basal cut, 219
basal cutting, 219
basal diameter, 95, 113, 220
basal dressing, 219, 431
basal fertilizer, 219
basal log, 95
basal medium, 219
basal metabolic rate, 219
basal metabolism, 219
basal-nerved, 219
basal placenta, 219
basal placentation, 219
basal stem girdle, 160
basal wounding, 160
basal zone, 219
basalt, 545
base, 95, 219, 220, 234, 310, 413
base capital, 220
base cation saturation, 555
base coat, 95
base current, 220
base diameter, 167, 447
base exchange capacity, 556
base face, 221
base flow, 219
base height, 221
base-height ratio, 220
base heliport, 615
base length, 220
base level, 221
base level of erosion, 378
base lifting device, 492
base-line, 18, 115, 220, 591
base log, 205
base of erosion, 378
base pair, 234
base pairing, 234
base paper, 592
base peak, 219
base period, 220, 221
base plywood, 471
base population, 219
base price, 220

base run-off, 219, 278
base saturation, 234
base saturation degree, 234
base sequence, 234
base sheet, 95, 219
base status, 234
base substitution, 234
base tank, 94
base unit, 94, 95, 219
base unit main, 612
base year, 221
baseboard, 476
baseline, 221
baseline correction, 220
baseline data, 220
baseline drift, 220
baseline noise, 220
baseline shift, 220
basement, 95, 98, 218, 219
basement membrane, 219
basic density, 219
basic dye, 235
basic-level market, 63
basic-level production, 63
basic line, 220
basic manure, 219
basic metabolic rate, 219
basic metabolism, 219
basic number, 220
basic operating system, 219
basic rock, 220
basic seed, 219
basic shrinkage, 219
basic stress (value), 219
basic volumetric weight, 171, 219
basicity, 234
basicole, 234
basidial layer, 81
basidio lichen, 81
basidiocarp, 81, 641
Basidiomycetes, 81
basidiospore, 81
basidium (*plur.* -ia), 81
basiflory, 220
basifuge, 21, 524
basigamy, 219, 220
basikaryotype, 219
basil oil, 312
basilaire, 497
basilar style, 220
basin, 308, 355
basin dam, 318
basin irrigation, 317, 534, 583
basin recharge, 308
basin storage, 308
basing-point system, 219
basipetal, 530
basiphile, 516
basiphil(ous), 439, 516
basisternum, 219
basket chair, 308, 475
basket hook, 474
basket plant, 283
basket planting, 283
basket weave matching, 23, 310
basket willow, 191
basophilous vegetation, 439
bass, 113
basset, 5, 297
basset table, 350
bast, 396
bast fiber, 396
bast fibre, 396
bast parenchyma, 396
bast ray, 396
bast zone, 396
bastard, 70, 72, 73, 597
bastard acacia, 71
bastard face, 524
bastard fallow, 10, 112
bastard grain, 524
bastard-lumber, 92, 524
bastard-sawn, 524
bastard-sawn grain, 114, 524
batch culture, 54, 142
batch digester, 236
batch mixer, 142, 236
batch number, 355
batch process, 54, 142, 236, 355
batch production, 54, 355
batch scale, 142, 236
batch still, 142, 236
batcher, 225
Batesian mimicry, 17, 255
bathophilous, 516
bathoxenous, 349, 524
bathyal zone, 420
bathybic, 420
bathymetric contour, 188
bathymetry, 44, 188
bathypelagic zone, 419
bathyphytia, 93
bathyscaph, 420
bathysphere, 473
bathythermograph, 420
batten, 479, 533, 535
batten door, 8
battenboard, 279
batter, 376, 380, 537
battery, 141, 544, 651
battery edger, 34
battery of extractor, 251
battery of saws, 263
battue, 639
Bauer defibrator, 15
Bauer double disk refiner, 15
Bauer mill, 15
Bauer-refiner, 15
baulk, 73, 231
bau(l)k, 76
Baume hydrometer, 30
bauxite-laterites, 314
Bavarian group system, 5
bavistin, 19, 117
bay, 594
bay cutting, 189, 498
bay felling, 189, 498
bay ice, 189
bay oil, 594
baycurine, 218
bayonet topped, 71
BCa-layer, 155
be covered with, 155
beach deposit, 188
beach erosion, 188
beach ice, 188
beach plant, 188
beach strand, 188, 471
beacon, 248, 556
bead, 281, 486, 632
bead-and-batten, 67
bead-and-quirk, 4
bead and reel, 11
bead-butt, 362
bead-butt and square, 362
bead-flush, 67
bead-flush panel, 11, 632
bead-jointed, 486
bead-plane, 486
bead router, 274
bead saw, 529
bead sight, 324
bead-tool, 546
beaded ray, 345
beading, 11, 67
beagle, 19, 297
beak, 213, 231
beam, 182, 196, 288, 292, 295, 418

B

beam log edger, 545
beam simultaneous closing device, 161
beam with simply supported ends, 365, 485
bear, 225, 276, 317, 543, 612
bear claw, 281, 543
bear market, 317
bear-trap gate, 452, 508, 543
beard, 57, 71, 113, 523, 605
bearded iris, 582
beardless iris, 511
bearer, 55, 103, 492, 493
bearing, 107, 131, 245, 530, 612, 630
bearing age, 245
bearing branch, 245
bearing capacity, 56, 245
bearing habit, 245
bearing strain, 56
bearing strength, 612
bearing stress, 612
bearing tree, 25, 107, 245, 521
beat, 75, 383, 549, 573
beatability, 75
beater, 75, 217, 243
beater additive, 75
beater bar, 75
beater furnish, 75, 566
beater roll, 75, 136
beater sizing, 75
beating, 32, 75, 243
beating coefficient, 75
beating degree, 75, 277
beating machine, 75
beating pressure, 75
beating up, 32
beating (up), 33
Beaufort (wind) scale, 364
beaver tail, 193
becket, 208, 431
becket hook, 113
bed, 67, 126, 309, 310, 364, 366, 472, 521
bed density, 324
bed-die, 95, 570
bed former, 324, 636, 654
bed knife, 94, 175
bed load, 94
bed moulding, 4
bed of furnace, 310
bed of packing, 479
bed piece, 94, 103, 217
bed pillar, 67
bed plate, 94, 133, 655
bed rock, 95, 220
bed shaper, 324, 636, 654
bed step, 67
bed-up, 636
bed with storage, 79
bedding, 204, 375, 654
bedding area, 640
bedding culture, 67
bedding plant, 204
bedding wood material, 103
bedplate bar, 94, 184
bedplate box, 94
bedplate of beater, 237
bedpost, 67
bedroom suites, 507
bedstead, 67
bee plant, 323
beech nut, 413, 458
beef steak fungus, 347
beegle iron, 316, 480
beehive kiln, 52
bee's wing figure, 146
beeswax, 146
beetle, 230, 331
before reproduction system, 127
beginning of regeneration-cutting, 373
behavior, 540
behavioral ecology, 540
behavioural adaptation, 540
behavioural dimorphism, 540
behavioural interference, 540
behavioural isolation, 540
behavioural polymorphism, 540
behaviourism, 540
beheading, 247, 384
Beija-Jung defibrator, 17
Belfast truss, 17, 201
bell, 205
bell-shapled butt, 139
Bellmer bleacher, 17
bellows, 30, 145, 326
belly cut, 263
below grade, 34, 92
below-the-line expenditure, 485
belt, 66, 79, 95
belt balance, 80
belt bottom feeder, 94
belt conveyer, 80
belt drive, 356
belt dryer, 79
belt of soil water, 490
belt of trees, 315, 603
belt of weathering, 144
belt planting, 79, 132
belt precompressor, 80
belt pulley, 356
belt sander, 80
belt sowing, 79
belt speeder, 356
belt-transect, 558
belt-type caulles loader, 80
belt under regeneration, 169
belt veneer dryer, 499
belt weigher, 80
bench, 170, 262, 585
bench grafting, 3, 411, 440
bench grafting on root, 167
bench hook, 332
bench in greenhouse, 505
bench lathe, 471
bench plane, 471
bench run, 439, 471
bench saw, 471
bench screw, 170
bench shaper, 471
bench stop, 332
bench table, 50
bench terrace, 243, 458, 471
benchmark, 221, 460
bender, 497, 498
bending, 339, 384, 497
bending deflection, 497, 498
bending failure, 498
bending flexure, 339, 497
bending moment, 498
bending moment diagram, 497
bending planer, 384
bending radius, 497
bending rigidity, 271
bending stiffness, 271, 498
bending strain, 498
bending strength, 271
bending stress, 498

B

bending test, 498
bending vibration, 498
bending-wood, 498
bending yield point, 271
Bendtsen-method, 19
bendwood furniture, 498
beneficiary, 443
benefit-cost analysis, 54
benefit-cost procedure, 536
benefit-cost ratio(s), 54, 569
benefit-cost shadow price, 569
benefit payable, 572
benefit value, 290, 443
benihione, 17, 558
benlate, 19
benomyl, 19
Benson raft, 548
bent, 334, 350, 497, 612
bent gouge, 384
bent log, 498
bent plywood, 498
bent wood, 498
benthal, 188, 456
benthic, 188, 456
benthic animal, 95
benthic fauna, 95
benthic flora, 188
bentho-potamous, 515
benthoal, 95
bentho(n)ic, 95
benthophyte, 456
benthos, 95
benthos feeder, 436
bentwood furniture, 498
benzene, 19
benzene hexachloride, 309
benzidine stain test, 294
benzine, 211, 371
benzoin resin, 2
benzylaminopurine, 23
benzylcoumaranone, 23
benzylidenecoumaranone, 19, 57, 348
bergamot leaf oil, 528
bergamot oil, 147, 528
bergamot petitgain oil, 528
bergamotene, 528
bergamottin, 528
bergaptol, 147, 528
bergy bit, 533
Bering land bridge, 6
Berlin blue, 7, 365, 553
berm, 202, 310
Bernoulli equation, 31
berry, 237
berth throughput, 31
best growth, 654
best opening face, 653
beta cellulose, 523
beta distribution, 17
betelnut palm, 27
bethal zone, 10
Bethel (full-cell) process, 17
bethylids, 626
bettom hot platen, 521
betulin(ol), 207
between-row space, 190
bevel, 262, 537
bevel angle, 536, 537
bevel cutting, 537
bevel edge, 536
bevel indicator flap, 380
bevel saw, 85, 537
bevelled-dress(ed), 29
bevelled halving, 537
bevelled joint, 75, 537
bevelled tenon, 537
Bezner floating cutterhead peeler, 17
BFCA salt, 355
Bfe-layer, 13
bhatta, 398
bhilwaya, 282
bi-directional vane, 454
bi-linear equation, 454
biacuminate, 453
bialatus, 452
biangulate, 122
biapocrisis, 207
bias, 357
biased error, 357, 581
biaxial compression, 455
biaxial stress, 455
bibasic acid, 122
bicable, 454
bicable aerial tramway, 454
bicable system, 454
bicableway, 454
bicarpellary, 454
bicentric distribution, 455
bicollateral bundle, 454
bicoloured flower, 123
biconvex cavity, 454
bicrenate, 261
bicuspid, 452
bicycle, 352, 470
bid, 15, 62, 485
bid price, 25, 317
bidding decision, 485
bidding process, 485
biennial, 296
biennial bearing, 78, 166
biennial fruit, 122, 594
biennial plant, 296
biennial (plant), 122
Biffar-mill, 19
bifoliolate leaf, 29
bifollicular, 452
big blister figure, 77
big blue butt, 577
big cup shake, 76
big data, 78
big game, 77, 260
big slip, 46, 378
big square, 76
big-stick loader, 76
bigarade oil, 467
bigarade orange, 278, 467
bigener, 446
bigeneric, 296, 446
bigeneric cross, 446
bight, 384, 431, 637
bight angle, 12
bigness scale, 72, 318
bilabiate, 121
bilateral symmetry, 295, 452
bileucofisetinidin, 123
bilge saw, 484
bilinear hybrid, 296
bill for collection, 493
bill of exchange, 213
bill of health, 233, 236
bill of lading, 476
bill of sufferance, 323
bill payable, 572
bill to fall due, 358
bill to mature, 159, 358
billet, 113, 356, 385
billet timber (wood), 112, 113
billhook, 497, 544
bilobate, 260, 295
bilocular, 260, 296
Bilterlich-blade, 20, 242
Biltmore stick, 20
bimetal, 153, 453
bimodal distribution, 123, 452
bin, 296
Bin-layer, 481
bin wall, 296
binary, 122, 123, 452
binary cell, 122
binary chain, 122
binary code, 122
binary counter, 122

binary digit, 122
binary loader, 122
binary multiplier, 122
binary notation, 122
binary number, 122
binary symmetry, 123
binato-palmate, 122
binato-pinnate, 122
bind, 229, 281, 344, 437, 446
binder, 227, 241, 246, 249, 281, 344, 631
binder content, 241
binding, 229, 281, 344, 481
binding agent, 241, 598
binding-beam, 294
binding chain, 280, 281
binding genera, 172
binding log, 146, 281
binding material, 344
binding nature of soil, 490
binding power, 246
binding species, 246, 294
binding strap, 280, 281
binocular, 454
binocular microscope, 453
binocular shooting, 454
binocular vision, 454
binomial coefficient, 123
binomial distribution, 123
binomial nomenclature, 453
binomial system, 453
bio-massy, 428
bio-measurement, 426
bio-oxidation, 428
bio-reduction, 427
bio-science, 427
bioassay, 215, 426
bioavailability, 427
biocatalyst, 426
biochemical deterioration, 422
biochemical engineering, 422
biochemical evolution, 427
biochemical fuel cell, 422
biochemical genetics, 422
biochemical oxygen demand, 422
biochemical staining test, 422
biochemical treatment, 422
biochemistry, 427
biochore, 423, 426, 427, 617
biochrome, 428
biochronology, 427
biochronometer, 428
biocidal, 411
biocide, 411, 427
bioclimate, 427
bioclimatic, 427
bioclimatic law, 427
bioclimatic zonation, 427
bioclimatic zone, 427
bioclimatics, 427
bioclimatology, 427
biocoenology, 428
biocoenose, 428
biocoenosis, 428
biocoenotics, 428
biocompatibility, 428
bioconcentration, 427
biocontrol system, 427
biocybernetics, 427
biocytoculture, 215
biodegradable, 427
biodegradable pollutant, 273
biodegradation, 427
biodemography, 425, 428
biodeterioration, 427
biodiesel, 426
biodiversity, 426
biodiversity conservation region, 426
biodiversity hotspots, 426
biodiversity monitoring, 426
biodynamic agriculture, 426
biodynamics, 426
bioecology, 428
bioeconomics, 427
bioelectricity, 426
bioenergetics, 427
bioenergy, 428
bioengineering, 427
bioerodible, 272
bioethanol, 427
biofacies, 428
biofacies map, 428
biofeedback control, 426
biofog, 428
biofuel, 428
biogalvanic source, 426
biogecochemical cycle, 426
biogenesis, 427, 429
biogenetic law, 426
biogenic, 593
biogenous, 48, 423
biogeochemical circulation, 426
biogeochemical cycle, 426
biogeochemistry, 426
biogeoclimatic zone, 426
biogeocoenosis, 425, 426
biogeographic barrier, 426
biogeographic province, 426
biogeographic realm, 426
biogeographic region, 426
biogeography, 426
bioindicator, 619
bioisolation, 427
biolite, 427, 428
biolith, 428
biological accumulation, 427
biological activity, 427, 428
biological admixture, 428
biological agriculture, 427
biological assay, 426
biological attack, 427
biological breakdown, 426
biological cleaning, 427
biological clock, 428
biological concentration, 427
biological conditioning, 426
biological constant, 426
biological control, 426
biological corridor, 427
biological decay, 428
biological effect, 428
biological efficiency, 428
biological element, 428
biological energy efficiency, 427
biological energy subsidies, 427
biological enrichment, 427
biological equilibrium, 427

B

biological factor, 428
biological fitness, 428
biological geography, 426
biological half-life, 426
biological index of pollution, 428
biological indicator, 619
biological invasion, 428
biological isolation, 427
biological magnification, 426
biological micro-cycle, 428
biological minimum temperature, 428
biological monitoring, 427
biological nitrogen fixation, 427
biological oxidation, 428
biological oxygen demand, 422
biological periodism, 428
biological population, 428
biological process, 422, 428
biological production, 428
biological production system, 428
biological race, 423, 428
biological resources, 428
biological rhythm, 427
biological rotation, 423, 646
biological species, 426, 428
biological spectrum, 423
biological strain, 423, 427
biological system, 428
biological test, 428
biological time, 428
biological type, 428
biological variability, 426
biological weathering, 426
biological zero, 428
biological zero point, 428
biologicals, 426, 428
biologics, 426, 428
biologization, 428
biology, 428
bioluminescence, 426
bioluminescent organism, 124
biolysis, 426
biomagnification, 426
biomass, 427, 428
biomass boiler, 428
biomass density, 427
biomass expansion factor, 427
biomass feedstock, 427
biomass harvester, 179, 428
biomass increment, 427
biomass method, 427
biomass power plant, 428
biomass productivity, 427
biomass-respiration ratio, 427
biomass yield, 427
biomathematics, 428
biome, 428
biome type, 428
biomechanism, 427
biometeorology, 427
biometer, 427
biometer method, 427
biometric method, 427
biometric procedure, 428
biometrical genetics, 426
biometrics, 428
biometry, 426, 428
biomimesis, 427
biomolecule, 426
bion, 249, 423, 428
bionics, 134
bionomic strategy, 424
bionomics, 422, 425
biont, 426, 580
biophage, 215
biophore, 429
biophotometer, 427
biophysics, 428
bioplasm, 591
bioplasmin, 591
biopolymer, 123, 427
bioramous, 122
biorhythm, 427, 428
bios, 427, 429
biosafety, 426
bioses, 123
bioseston, 428
biosocial facilitation, 427
biosociology, 428
biosphere, 427
biosphere reserve, 427
biostatics, 427
biostatistics, 428
biostratigraphy, 426
biostratonomy, 426
biosynthesis, 427
biosystem, 428
biosystematics, 428
biosystematry, 426, 428
biota, 427, 428
biota-biotop system, 427, 428
biotelemetry, 428
biotemperature, 428
biotest, 426
biotic, 426
biotic agency, 428
biotic agent, 427
biotic balance, 427
biotic climax, 426, 428
biotic community, 428
biotic component, 428
biotic control, 426
biotic equilibrium, 427
biotic factor, 428
biotic formation, 428
biotic index, 428
biotic pesticide, 427
biotic potential, 427, 428
biotic pressure, 427, 428
biotic province, 426
biotic region, 427
biotic succession, 428
biotite, 195
biotope, 389, 423, 428, 534
biotron, 427
biotronics, 427
biotroph, 215
biotype, 69, 428, 483
biovolume, 485
biozone, 426
bipalmate, 122
biparasitic, 60, 227
biparental progeny, 453
biparous branching, 121
bipentaphyllus, 535
bipinnate, 122
bipolar distribution, 452
birainy, 454
birch bark, 207
birch bark tar, 207
birch black, 207
birch bud oil, 207
birch conk, 206
birch oil, 207
birch tar oil, 207
bird-banding, 209
bird-bath, 346

bird migration, 200
bird of passage, 111, 200
bird of prey, 321, 437
bird peck, 346
bird repellent, 21, 133, 383
bird sanctuary, 345
bird shot, 372, 534
bird watching (birding), 415
birdmouth, 55, 521
bird's eye, 346
bird's eye figure, 346
bird's eye grain, 346
bird's-eye view, 345
birds-eye view, 345
birds' eye wood, 346
bird's mouth, 521
birds' nest fungus, 345
birdwatcher, 345
birefringence, 455
Birmingham wire gauge, 31
birth control, 429
birth-death ratio, 424
birth rate, 430
birthwort, 316
bisabolene, 131, 197
bisacki, 7
biscuit, 72, 466, 642
bisect, 115, 360, 364
bisect method, 364
biserial ray, 453
biseriate, 122, 123
biserrate, 60, 260
bisexual, 70, 296
bisexual group, 296
bisexualism, 70
Bismarck brown method, 20
Bison-combi-dryer, 20
bisphenol A, 452
bisulfite process, 467, 552
bit, 20, 58, 149, 504, 529
bit profile, 149
bit thread, 652
bitch chain, 277, 638
bitch hook, 470
bitch link, 470
bite, 57, 65, 249, 378
bitter almond oil, 278
bitter principle, 278
bitter substance, 278
Bitterlich-relascope, 20, 300
Bitterlich sampling, 20, 242
Bitterlich variable radius plot, 20
Bitterlich's sector fork, 20
bitumen, 291
bitumen-bonded insulating board, 291
bitumen-impregnated insulating board, 291
bituminous board, 291
bituminous cement, 291
bituminous compound, 291
bivalent, 122
bivane, 454
bivariate normal distribution, 123, 452
bivoltine, 122, 563
bivoltinism, 121, 122
Bjorkman's lignin, 20, 327
black alkali soil, 194
black alkaline soil, 194
black and white infrared film, 194
black-body emission, 195
black-body radiation, 195
black-box model, 195
black carbon, 195
black carpenter ant, 194
black charcoal, 195
black check, 195
black cotton soil, 194
black dammar, 194
black earth, 195
black fallow, 297
black fungus, 194
black gang, 126, 215, 312, 461
black gun-powder, 194
black heart, 195
Black Hills formula, 195
black knot, 194, 462
black leg, 194
black light, 34
black line, 195
black liquor, 139, 195
black locust, 71, 557
black lye, 194
black pepper oil, 201
black prairie soil, 195
black root rot, 194
black rot, 194
black soil, 195
black spots, 194
black strap, 58
black streak, 195
black turf soil, 195
black wattle extract, 194, 254
black wood charcoal, 194
blackbody, 195
blackout position, 386
blade, 88, 149, 238, 263, 492, 560
blade accumulator cylinder, 48
blade agitator, 238, 560
blade back, 88, 261, 560
blade bracket, 492
blade coating, 177
blade grader, 48, 492
blade holder, 88, 263, 326
blade machine, 360
blade sheath, 561
blade straining device, 79
blank, 16, 73, 277, 304, 355, 388, 504, 519
blank spot, 309
blank test, 115, 277
blank value, 115
blanket, 155, 487
blanking, 8, 33
blansand, 156
blast blower, 175
blast drier, 175
blasticidin, 324
blastochore, 321
blastoderm, 352
blastogenesis, 551, 627
blastokinesis, 352
blastospore, 551
blaze, 165, 269
blcached earth, 358
bleach, 358
bleach pulp, 358
bleachability, 273, 358
bleached brown forest soil, 358
bleached forest soil, 358
bleached lac, 358
bleached layer, 358
bleached mineral soil, 358
bleached pulp, 358
bleached soils, 358
bleaching clay, 358
bleaching podzolisation, 358
bleaching process, 358
bleaching wood, 358
bled steam, 140
bled timber, 385

B

bleed-through, 420, 485
bleed valve, 135, 349
bleeding, 62, 287, 321, 405, 414, 420, 485
blemish, 387, 469, 520
blend, 47, 213, 353
blender, 11, 213, 243
blending inheritance, 398
blight, 103, 569
blighted area, 209, 451
blimp logging, 136
blind, 27, 571
blind area, 296, 318, 605
blind conk, 221
blind creek, 157, 235, 584
blind dovetail, 387
blind drain, 3, 318
blind holed fibre board, 20
blind knot, 146, 318, 571
blind lead, 604
blind mortise, 4
blind pit, 318
blind seed, 27, 462
blind tenon, 4, 32
blind tracery, 67, 436
blind tree, 318
blind-your-eye-tree, 188
blinding tree, 188
blister, 12, 175, 352
blister blight, 352
blister detector, 175
blister figure, 352
blister grain, 352
blister rust, 352
blizzard, 16, 548
block, 113, 142, 205, 333, 371, 383, 402, 432
block-and-a-half, 205, 601
block and tackle, 205
block bill, 279
block board, 519
block brake, 279, 495
block centering device, 332, 589
block charger, 416
block cutting, 61, 279
block density, 272, 279
block diagram, 131
block floor, 333, 358
block front, 611
block hot-water treating, 332
block house, 332
block mottle figure, 279
block-piled, 322, 435, 436
block plane, 112, 196
block planting, 279, 389
block record, 55, 383
block resistance, 270, 344
block sampling, 610
block setter, 558
block sheave, 205
block shell, 205
block stacking, 322, 435
block steaming, 332
block strap, 205
block system, 20, 383
block tackle, 153, 205
block tender, 248, 280
blockboard, 335
blocked account, 110
blocked group, 18
blocked shellac, 246
blockfront, 622
blocking, 14, 20, 131, 146, 335, 345, 651
blocking apparatus, 620
blocking device, 20, 602
blocking effect, 146, 651
blocking saw, 600
blockstrop, 205
blocky structure, 279
Blodgett log rule, 31
blonde, 375
blood adhesive, 548
blood corpuscle, 548
blood feeder, 515
blood glue, 548
blood group, 548
blood lymph, 548
blood meal, 548
blood plasma, 548
blood relationship, 548
blood serum, 548
bloom, 202, 210, 367
bloom-food, 203
bloom oil, 367
blooming, 6, 254, 268, 367
blooming date, 203, 268
blooming period, 203
blooming stage, 268
blossom bud, 204
blossom cluster, 202
blossom end, 186, 202
blossom fall, 203, 204
blow back, 16, 129
blow-by, 32, 309, 420
blow-down, 90
blow down heat recovery system, 135
blow of mat, 8
blow-off, 67, 177, 349, 354
blow-off pipe, 67, 349, 354
blow-off valve, 350
blow-out, 15
blow-out loss, 354
blow-out pressure, 354
blow pipe, 135, 275, 481, 551
blow pit, 354, 538
blow sand, 136
blow-up, 15
blow up fire, 15, 215
blow valve, 354, 465
blowdown, 135, 144
blowdown tree, 144
blower, 145, 175
blower (dry) kiln, 175, 465
blower sprayer, 145
blowing, 16, 67, 135, 309, 354
blowing-up cutter, 530
blowing-up stump, 16
blowmark, 367, 370
blown glass, 67
blowout, 145, 178, 309, 354, 465
blue camphor oil, 283
blue-collar workers, 283
blue-green algae, 283
blue lumber, 283, 379
blue ointment, 460
blue-rot, 379
blue sapstain, 283, 379
blue sapstain fungus (*plur.* -gi), 283, 379
blue stain, 283, 379
blue stain fungus, 283, 379
blue stained wood, 283, 379
blue vitriol, 87, 283, 512
blueing, 283, 379
blueprint, 283, 418
bluge, 308
Blume-leiss altimeter, 35
Blume-leiss hypsometer, 35
blunt counter knife, 94
blunt-nosed, 115, 593
blur, 574
blushing, 123, 513
board, 8, 153, 328, 618
board-and-brace work, 529

board-caul separator, 140
board-chute, 8
board cooler, 286
board dog, 371
board ejector, 491
board fence, 8
board foot-cord conversion, 8
board foot-cubic foot conversion, 8
board foot-cubic foot ratio, 8
board foot (feet), 8
board-foot log rule, 200
board-foot measure, 8
board-foot scaling, 8
board-foot volume table, 8
board hole, 471
board industry, 396
board iron, 471
board joint, 359
board liner, 202, 214, 618
board measure, 8
board rule, 8
board (saw) mill, 8
board scale, 8
board separating device, 140
board splitter, 140, 147
board turner, 128, 637
boarding, 165, 364, 500, 529
boat conformation, 66
boat-tailed, 66, 503
bob, 493, 534
bob-tail system, 87, 387
bob-tailing, 87, 387
bobber, 9, 147
bobbin, 264, 526
bobbin spool, 412, 452, 526
bobbin square, 179
bobbin wood, 179, 412
bodhi, 364
bodied mayonnaise-type emulsion, 157
body, 288, 289, 301, 529
body cavity, 476
body length, 477
body paper, 592
body plan, 611
body size, 477
body stock, 592
body wall, 476
bodying, 62, 227
bodywood, 159, 512
bog, 342, 467, 604
bog forest, 605
bog garden, 605
bog-land, 605
bog mor, 604
bog plant, 605
bog plow, 605
bog soil, 605
bog succession, 605
bogginess, 342, 605
boggy, 17, 605
bogie, 103, 410, 443, 453, 533, 545, 637
bogie track, 637
boiled hide, 356
boiled oil, 381, 446
boiled-out water, 609
boiler, 185, 609
boiler horsepower, 185
boiler plant, 185
boiling and cooling process, 393
boiling-in-oil method, 578
boiling-in-water method, 460
boiling point, 140
boiling temperature, 140
boiling under vacuum, 233, 608
boiling water test, 634
boiling-without-vacuum, 49
bole, 72, 125, 198, 591
bole area index, 159
bole form factor, 447
bole height, 633
bole stage, 159
bole wood, 72, 125, 591
bolection moulding, 258, 486
bolochory, 297
bolster, 103, 195, 196
bolster arm, 147
bolster tram, 595
bolster truck, 196
bolt, 46, 113, 312, 451
bolt action repeating shotgun, 452
bolt-bearing strength, 312
bolt-bearing stress, 312
bolt displacement, 312
bolt-handle, 282, 451
bolt-hole treater, 312
bolt loop, 372
bolter, 112, 650
bolting, 61, 312, 412
Boltzmann constant, 30
bombycinous, 461
bond, 237, 240, 344, 603
bond bending, 237
bond energy, 237, 246
bond line, 240, 241
bond paper, 241, 611
bond premium, 603
bond quality, 241
bond rate, 603
bond strength, 237, 344
bond stress, 152, 240
bond twisting, 237
bond water, 246, 514, 515
bond yield, 603
bonded goods, 15
bonded-phase chromatography, 237
bonding agent, 241
bonding strength, 241
bone char, 175
bone-dry, 158, 265
bone-dry shellac, 358
bone dry weight, 265
bone glue, 175
bonsai, 355
bonus, 197, 238
bonus dividend, 197
book account, 417, 500
book inventory, 604
book matching, 114, 444, 611
book rack, 444
bookcase, 444
booking list, 106
boom, 168, 193, 333, 358, 367
boom auger, 168
boom boat, 638
boom bridle, 104
boom buoy, 358
boom cable, 168
boom chain, 168
boom company, 576
boom cylinder, 368
boom haul back line, 367
boom hoist, 545
boom hold-up strap, 367
boom pin, 168, 349
boom poke, 135, 193
boom raft, 169
boom sprayer, 159
boom stay, 358
boom stick, 168, 358
boom strap, 367
boom-suspension guy, 368

boom swing lead block, 367
boom swing line, 367
boom tree block, 104
boomage, 193, 442, 458
booming, 23
booming dog, 168
booming engineering, 442
booming ground, 168, 635
booster, 150, 171, 414, 421
booster coil, 33, 623
booster diffusion pump, 602, 634
booster solution, 367
Bordeaux mixture, 30
border, 258, 303, 366, 481, 520
border-cutting, 22, 79
border effect, 22
border irrigation, 80, 366, 502
border of forest, 303
border plane, 22
border thickening, 22
border tree, 258, 303
bordered pit-pair, 61, 261
borderline tree, 22, 258
bore, 276, 375, 636, 652
bore dust, 60, 636
bore hole, 60, 636
bore hole method, 652
bore-hole poisoning, 652
bore-hole treatment, 65
bore hyphae, 65
bore-powder, 60, 636
bore sighting, 375
boreal, 17, 189
boreal coniferous forest, 471
boreal forest, 190, 471
boreal forest zone, 17
Boreal period, 17
bored tie, 652
borer, 636, 652
borer bit, 652
borer clogging, 402
borer handle, 430
borer hole, 60, 636
borer-proof plywood, 131
borer stem, 652
Borggreve thinning, 12
boring cut(ting), 46, 473
boring defect, 60
boring direction, 400
boring machine, 652
Borneo camphor oil, 309
borneol flake, 357
bornyl, 309
bornyl alcohol, 28, 309
borovina, 195, 462
borowina, 195, 462
borrowing, 248
borrowing capital, 248
borzoi, 305
boscage, 41, 180, 446
bosk, 1, 72, 446
boskage, 41, 180, 446
bosket, 1, 72, 446
bosquet, 1, 72, 446
boss, 593
bossy, 117, 580
bostrix, 313
bostrychid, 49
bostrychoid cyme, 313
Bosun's chair, 104
boswellic resin, 400
botanical garden, 618
botanical geography, 616
botanical insecticide, 618
botanical park, 618
botany, 97, 618
botfleneck effect, 363
both surface smooth hardboard, 295
botry-cymose, 264
botryoblastospore, 73
botrys, 649
bottle-butted, 76, 160
bottle grafting, 46, 363
bottleneck, 2, 169, 363, 520
bottleneck check, 20
bottom bed, 471
bottom belt, 94
bottom-board, 61, 94, 95
bottom coat, 95
bottom coater, 95
bottom coating, 95
bottom colour, 95
bottom community, 95, 456
bottom-diameter, 95, 167
bottom discharge chipper, 94
bottom dog, 94
bottom dumper, 94
bottom dyeing, 94
bottom end, 78, 167
bottom fauna, 456
bottom felt, 521
bottom flora, 456
bottom knife coater, 94
bottom lip (of slice), 520
bottom log, 160
bottom plate, 94, 633
bottom population, 95
bottom press roll, 521
bottom price, 653
bottom recessive, 571
bottom rib, 94
bottom sampler, 95
bottom season, 94
bottom slope, 95
bottom-up model, 647
bottom wear, 94
bottombind, 520
bough, 449
bough stars, 540
boulder clay, 342
boulevard, 171, 303, 587
Boulton drying, 31
Boulton process, 31, 233
bound moisture, 246, 446
bound water, 246, 446
boundary, 22
boundary correction, 237, 258
boundary description, 248, 258
boundary layer, 22
boundary limit, 22
boundary line, 258
boundary log, 22
boundary-map, 258
boundary-mark, 258
boundary of property, 469
boundary of servitude, 437
boundary-register, 258
boundary settlement, 248, 269
boundary stone, 248, 433
boundary surface, 22
boundary tree, 258
bounty, 238, 415
bouquet, 204
Bourdon tube, 35
bourse, 240
bouse block, 223
bow back, 170
bow handle, 170
bow-saw, 170
bow top, 170
bowed wood, 112, 170
Bowen's ratio, 15
bowguide, 170

B

bowing, 170, 461
bowleg loader, 638
bowstring truss, 170
box, 184, 263, 529
box beam, 194, 529
box board, 529
box-condenser, 529
box culvert, 529
box dolly, 111
box dovetail, 529
box(-end) stacking, 529
box heart sawing, 61
box hooks, 455
box kiln, 529, 630
Box-layer, 481
box lock, 519, 529
box lumber, 529
box pile, 131, 288, 529
box planting, 529
box-section loader arm, 529
box shook timber, 529
box stacking, 529
box stool, 529
box-system, 529
box tenon, 529
box the heart, 61
box timber, 529
box trap, 527
boxed frame, 529
boxed heart check, 262, 538
boxed-heart (timber), 80, 189
boxed pith, 468
boxing shutter, 529
brace, 271, 384, 612
brace back chair, 53
brace root, 612
braced door, 294, 615
brachelytra, 112
brachiation, 104
brachyblast, 113
brachycephalic, 113
brachypterous form, 112
brachysclereid, 113
bracing wire, 228, 282
bracken heath, 266
bracket, 492, 493, 545
bracket boom, 117
bracket crane, 545
bracket fungus, 201, 555
bracket gate, 615
bracket guy line, 223, 367
bracket light, 20, 493
brackish water, 10
brackish water meadow, 10, 501
brackish water swamp, 10, 501
braconidfly, 232
bract, 12
bract leaf, 12
bract scale, 12
bracteal leaf, 12
bracteole, 533
bractlet, 533
brad setter, 526
bradawl, 75, 536
Bragg angle, 35
Bragg diffraction, 35
Bragg direction, 35
Bragg reflection, 35
braided river, 24
braided sling, 23
braided stream, 24, 499
brail, 349, 536, 563
brake, 72, 179, 602, 620
brake band, 620
brake-block, 620
brake device, 620
brake distance, 620
brake dynamometer, 620
brake ratchet wheel, 222
brake-rope, 620
brake sheave, 620
brake sled, 80
brake strap, 602, 620
brake wheel, 620
braking effort, 620
branch, 36, 143, 612, 613, 651
branch angle, 143
branch bundler, 281
branch copic method, 485
branch cutting, 46, 75
branch forest road, 154
branch form factor, 613
branch gap, 613
branch height, 612
branch hole, 245, 613
branch litter, 278
branch order, 613
branch-pruner, 544
branch pruning, 544
branch road, 71, 154, 612
branch root, 43
branch-rot, 544, 612
branch scar, 612
branch spread, 613
branch stump, 612
branch tendril, 612
branch trace, 612
branch volume, 613
branch volume percent, 613
branch whorl, 613
branch(ed) knot, 140, 612
branched polymer, 612
branchiness, 613
branching, 75, 143, 612
branching habit, 143
branching system, 143
branchlessness, 512
branchlet, 535
branchline forest road, 298
branchness, 613
branchwood, 613
branchwood baler, 613
branchwood chipper, 613
branchwood collecting machine, 613
branchwood rosser, 613
branchy, 120, 613
brand, 25, 91, 191, 359, 620
brand name, 414
branding, 75, 284
branding hammer, 191
branding iron, 75, 191, 284
brash, 38, 74, 569
brash chopper, 613
brash fracture, 74
brashness, 74
brash(y) wood, 74
brass shell, 484
brattice, 7, 165
Braun's native lignin, 35, 478
Brayn preservative, 35
brazer, 244
Brazil wax, 5
brazilein, 557
Brazilian copal, 5
Brazilian rosewood oil, 5
Brazilian sassafras oil, 5
brazilin, 5
breach of contract, 502
break, 24, 124, 268, 363, 486
break a jam, 47
break a landing, 47
break a rollway, 492
break cut, 378
break-down hoist, 47
break-down landing, 329, 492
break-even, 35, 442

break even analysis, 361, 573
break even chart, 14, 469
break even point, 469, 573
break-in, 326, 438, 439
break out, 268, 349
break point, 113, 605
break-through curve, 65
breakage, 113, 605, 606
breakaway, 594
breakaway cable anchor, 161
breakdown, 177, 262, 364
breakdown saw, 364
breakdown timber, 247, 364
breaker, 209, 363
breaker calender, 610
breaker plough, 268
breaker stack, 47, 553
Breaking-down test, 113, 363
breaking joint, 45, 113
breaking load, 113, 363
breaking moment, 113
breaking of dormancy, 363
breaking strength, 113, 363
breaking stress, 113, 363
breaking-up plough, 268
breakover, 136, 594
breakthrough, 125, 301, 597
breakwater, 131
breakwind, 131
breast, 57, 58, 262, 288
breast bench, 262, 263
breast board, 542
breast box, 499
breast height, 542
breast-height diameter, 542
breast-height form factor, 503, 542
breast-height form quotient, 503, 542
breast-height girth, 542
breast height locating, 542
breast line, 223
breast log, 21
breast roll, 542
breast-summer, 77, 196, 493
breast-wall, 202
breast-work log, 205, 223
breasting, 57
breather drier, 200
breathing, 135
breathing pore, 200
breathing rate, 200
breathing root, 200
breathing veneer drier, 200
breccia, 242
breech, 375
breech bolt, 95, 375
breech loader, 200
breed, 128, 353, 359
breed cultivar, 585
breed true, 69
breeder, 586
breeder's seed, 592
breeder's stock, 592
breeding, 128, 585
breeding arboretum, 448, 585, 618
breeding area, 128
breeding behaviour, 430
breeding by crossing, 596
breeding by hybridization, 596
breeding by isolation, 141
breeding by separation, 166
breeding display area, 123
breeding efficiency, 129, 586
breeding field, 586
breeding herd, 129, 626
breeding landscape plant, 588
breeding material, 585
breeding migration, 128, 430
breeding objective, 586
breeding ornamental plant, 178
breeding place, 128
breeding plot, 586
breeding population, 129
breeding potential, 129
breeding process, 586
breeding programme, 585
breeding range, 128, 585
breeding rate, 129
breeding rotation, 586
breeding scale, 586
breeding season, 128
breeding strategy, 585
breeding success (rate), 128
breeding system, 128, 586
breeding value, 586
breeding zone, 586
breedy, 116, 155, 245
brephic, 125, 583
Brereton (log) scale, 35
Breuil mercury volume-meter, 35
brevifolin, 595
brevilagen, 595
brew, 345, 357
brewing kettle, 345
briar, 126, 196
briar dress, 29, 543
briar dressing, 29
brick dam, 636
brick lining, 636
bridge, 306, 377, 523, 606
bridge crane, 377
bridge grafting, 377
bridge method, 377
bridging, 233, 377
bridging blue, 114, 285
bridle, 113, 199, 227, 292, 620
bridle iron, 161, 349
bridle joint, 346
bridle path, 316
bridling (out), 43
Brigalow scrub, 35
bright, 590
bright bark pocket, 523
bright field, 325
bright sap wood, 31
bright sap(wood), 510, 577
brightener, 325, 601
brightening, 68, 351
brightness, 5, 182, 296
brightness range, 296
brightness reversion, 130, 212
brightness temperature, 296
brilliance, 182, 296
brin, 85, 561
brindle, 7, 201
Brinell hardness, 35
bringing in log, 222
bringing timber, 595
bringing wood, 595
brining, 555

briquet(te), 320, 390, 472
bristle, 160, 648
bristled grain, 343
bristly, 160, 260
British gum, 201
British pharmacopoeia, 572
British pharmacopoeia codex, 572
Brittany spaniel, 32
brittle heart, 74
brittle point, 74
brittle wood, 74
brittleness, 74
Brix, 6
Brix hydrometer, 6
broad axe, 269, 279
broad-base terrace, 401
broad hatchet, 279
broad-leaf forest, 281
broad-leaf tree, 282
broad leafed, 281
broad-leafed forest, 282
broad leaved, 281
broad-leaved evergreen, 48
broad-leaved forest, 281
broad-leaved herb, 281
broad-leaved tree, 282
broad-leaved woodland, 281
broad line seeding, 279
broad-ovate, 182
broad ray, 279
broad-ringed, 279
broad-sclerophyll plant, 280
broad spectrum, 182, 183
broad spectrum fungicide, 182
broad spectrum insecticide, 182
broad sword, 77, 281
broad-zoned, 279
broadcast burning, 386, 405
broadcast seeder fertilizer, 401
broadcast seeding, 401
broadcast sowing, 401
broken knot, 35
broken-stick distribution, 605
broken stripe figure, 113
broken timber, 469
broker, 78, 252
brokerage charge, 252, 577
bromacil, 64
bromide paper, 159, 544
bromination method, 544
Bronnert process, 35
Bronswerk-dryer, 35
brood, 15, 147, 483
brood bud, 129
brood cell, 551
brood chamber, 15, 585, 641
brood lac, 626
brood tree, 15, 60
brood tunnel, 641
broomed, 112
broomrape, 297
brother-sister mating, 483
brotochore, 395
brow, 222, 638
brow log, 620, 638
brow skid, 205, 286, 638
brow tine, 319
brown andean soil, 648
brown atrophy, 194
brown calcareous soil, 648
brown calcium acetate, 194
brown camphor oil, 198
brown cubical rot, 279
brown desert soil, 648
brown desert-steppe soil, 648
brown dry forest soil, 157
brown earth, 648
brown forest dark soil, 4
brown forest soil, 410, 648
brown granulation, 648
brown humus calcareous soil, 648
brown illimerized soil, 211
brown iron oxide, 557
brown like soil, 640, 648
brown meadow steppe soil, 648
brown mottled rot, 479
brown oxidation stain, 557
brown packing paper, 12, 347
brown pocket rot, 147
brown podzolic soil, 211
brown rendzina, 648
brown ring stringy rot, 209
brown rot, 194
brown-rot fungus, 194
brown size, 194
brown soil, 648
brown spongy rot, 188
brown stain, 194
brown stock chest, 210, 358
brown wood pulp, 194
browning, 648
brownish oxisiallitic soil, 648
browse, 274, 341, 613
browse line, 274
browsing, 57, 274
browsing by game, 559
browsing level, 274
bruceolide, 551
bruising, 532
brunisolic soil, 648
brunizem, 432
brush, 73, 180, 487
brush-and-bog plough, 180
brush and weedcutter, 165
brush block, 637
brush breaker, 64, 165
brush breaker plough, 180
brush chipper, 613
brush chopper, 613
brush cleaner, 64
brush crew, 180
brush cutter, 64, 165
brush cutting saw, 165
brush disposal, 179, 613
brush drag, 254
brush-eating device, 165
brush-field, 180
brush fire, 179
brush form factor, 73, 448
brush hook, 165
brush (land) plough, 180
brush out, 269, 381, 543
brush painting, 578
brush pasture, 179
brush pile, 139, 613
brush polishing, 351, 451
brush rake, 39
brush roll, 451
brush saw, 165
brush scythe, 76
brush shade, 46
brush spreader, 451

brush-stage, 180, 377
brush stripping, 64
brush treatment, 487
brush-type dewinger, 451
brushing lacquer, 451
brushkiller, 179
brushland, 179
brushmaster, 179
brushwood, 180, 613
brushy, 117
bryocoenology, 471
bryophyte, 471
bucca (*plur.* buccae), 42, 229, 277
buccal appendage, 277
buck, 11, 263, 600
buck for grade, 3, 192
buck saw, 196, 600
bucker, 600
bucker-buncher, 600
bucker-limber, 76
bucker's coupling, 348
bucket conveyer, 110
bucket crane, 110
bucket elevator, 110
bucket excavator, 48
bucket steam trap, 148
bucket valve, 214
bucket wheel, 202, 417
buckhorn chain, 310
buckhorn feed conveyor, 310
bucking, 600
bucking allowance, 199
bucking bar, 84
bucking chain saw, 600
bucking chute, 600
bucking circular saw, 600
bucking cut, 600
bucking deck, 600
bucking ladder, 600
bucking loader, 600
bucking percent, 600
bucking recovery, 600
bucking rule, 600
bucking skid, 600
buckle guy, 558, 623
buckling, 24, 338, 384, 497
buckling deformation, 497, 631
buckling limit, 498
buckling load, 498
buckling resistance, 271
buckling strain, 498
buckling strength, 271
buckling stress, 498
buckling stress coefficient, 498
buckrake, 224
buckskin, 104, 310, 385
buckwheat, 75
buckwheater, 126
bud, 551
bud bursting, 603
bud cutting, 551
bud-fertility, 285, 573
bud grafting, 551
bud hypertrophy, 552
bud mutation, 551
bud off, 62, 124
bud pollination, 285
bud pruning, 247, 603
bud rot, 551
bud scale, 551
bud selection, 551
bud sport, 551
bud stick, 551
bud variability, 551
bud variation, 551
bud variety, 551
budding, 62, 551, 552
budding knife, 551
budelium, 327
budget ceiling, 587
budget deficit, 587
budget estimate, 156
budget period, 587
budget proposal, 156, 587
budget regulation, 442
budget surplus, 587
budgeted chance for sale, 586
budling, 551
budwood, 245, 384
budworm, 264, 551
Buehler process, 19
buffer, 209, 532
buffer action, 209
buffer coat, 209
buffer effect, 209
buffer memory unit, 209
buffer solution, 209
buffer species, 209, 477
buffer strip, 14, 209
buffer zone, 14, 166, 209
buffered hardener, 209
buffering bin, 209
buffering gene, 209
buffet, 40, 184
buffing machine, 351
buffing wheel, 351, 412
bug, 62, 533
buggy, 443, 470
bugle weed, 98
buhl saw, 280, 420, 603
build-in end, 46
build-in (fitted) furniture, 375
build-up, 219, 292, 602, 638, 651
build-up index, 238, 285, 292
building board, 236
building construction, 134
building log, 236
building lumber, 236
building phase, 236
building poria, 236
building-rot fungus, 236
building timber, 236
built terrace, 30, 59
built-up beam, 652
built-up type sander, 219, 652
bulb, 33, 305
bulb pan, 382
bulb planter, 382
bulbil, 632
bulblet, 534
bulbous geophyte, 382
bulbous plant, 382
bulbul, 561
bulge, 486
bulk, 77, 464, 476, 604, 634
bulk conductivity, 476
bulk delignification, 77, 634
bulk density, 398, 464
bulk elasticity, 476
bulk flow, 610, 649
bulk lot of seeds, 404
bulk method of breeding, 213
bulk modulus, 476
bulk operation, 54, 355
bulk pile, 322, 404, 508
bulk polymerization, 19, 610
bulk progeny test, 224
bulk selection, 224
bulk soil, 491
bulk specific gravity, 398
bulk stacking, 322
bulk stress, 476
bulk system, 213, 404
bulk treatment, 476, 610
bulk viscoelasticity, 476
bulked down, 435
bulking, 355, 401
bulking agen, 355, 401
bulking effect, 355, 401

B

bulking property, 464
bulkmeter, 397
bull block, 372
bull buck(er), 126
bull chain, 66, 233, 630, 638
bull gang, 470, 543
bull hook, 223, 281
bull line, 10, 493
bull-nose, 497
bull pen, 556
bull screen, 73, 77
bull stick, 352
bull team logging, 65, 347
bull wood, 573
bulldozer, 492, 551
bulldozer blade, 492
bullet, 87
bullet cast, 641
bullet planting, 87, 152
bullet shape, 87
bullet weight, 87
bullet wood, 641
bull's eye fairleader, 499
bulrush, 91, 279, 619
bumble-bee, 543
bummer, 93, 453
bump, 175, 245, 308, 486
bump up, 550
bumper, 15, 209
bumper crop, 144
bumping knot, 377
bumping plate, 366, 549, 639
bunch grass, 72
bunch planting, 72, 73
buncher, 74, 75, 183, 192
bunchgrass steppe, 72
bunch(ing), 72, 75, 183, 281, 534, 563
bunching hand line, 130
bundle, 280, 502, 563, 596
bundle bucking, 54
bundle floating, 333
bundle load, 281
bundle rafting, 333
bundle scar, 502
bundle sheath, 502
bundle skidding, 54
bundle-survival ratio, 333
bundled stock, 54, 333
bungalow, 169, 360
bunion, 333
bunk, 56, 103, 188, 195, 224, 446, 536
bunk bed, 452
bunk block, 88
bunk chain, 88, 479
bunk hook, 52, 88
bunk jaw, 56
bunk jaw skidder, 55, 415
bunk load, 83, 94, 550
bunk log, 94
bunked long, 50
bunked short, 113
buoy, 147
buoyancy, 148
buoyancy theory, 148
buoyant force, 148
buprestid-beetle, 221
burden, 151
burdo, 231
bureau bookcase, 81
Bureau for the Revision of Working Plans, 407
Bureau of Forestry, 302
Bureau of Reclamation, 320, 646
bureau scale, 3
burgeon, 341, 551
Burgundy mixture, 465, 473
burl, 446
burl figure, 448
burl grain, 448
burlap planting, 80
burlapping, 12
burn, 216, 417, 640
burn a stump, 47
burn-off, 191, 416, 513, 595
burn-out test, 417
burn pattern, 215
burned area, 216
burned lime, 424
burner, 190, 390
Burnett process, 31, 315
Burnettizing, 31, 315
Burnett's fluid, 31
burning block, 390
burning edge, 390
burning-in knife, 177
burning index, 390, 577
burning index meter, 216, 577
burning index rating, 390
burning index system, 390
burning intensity, 390
burning of chip, 333
burning off, 416
burning out, 416
burning period, 390, 431
burning permit, 577
burning pile, 390
burning point, 390
burning-power, 390
burnish brush, 351
burnt-wood, 417
burnup analysis, 390
burozem, 648
burr, 103, 308, 318, 333, 403, 446, 474, 535
burr wood, 448
burrow, 110
burrowing animal, 110
bursa copulatrix, 239
burst, 16, 363
burst check, 201
bursting strength, 16, 337
bursting tester, 337
burying storage, 317
bush, 53, 179, 630, 631
bush breaker, 64, 165
bush cleaner, 64, 165
bush combine, 294
bush cutter, 165
bush fallow, 179, 297
bush fire, 179
bush hammer, 57, 599
bush hedge, 179, 180
bush kauri, 523
bush killer, 64
bush knife, 165, 269, 497
bush nursery, 301
bush ranger, 126
bushel, 225, 365
bushfire, 72
bushiness, 142
bushman, 126
bushveld, 42, 180
bushy tree, 2, 411
business entity, 367
business failure, 367
business forecasting, 415
business management, 367, 415
business planning, 179, 253
business volume, 574
butadiene dioxide, 123
butcher-bird, 7, 31
butein, 642
butin, 642
butt, 78, 159, 167, 237, 247, 377
butt and top grapple, 168

B

butt block, 159, 589
butt carriage, 359
butt chain, 281, 292
butt cut, 159, 167
butt diameter, 160
butt edge joint, 114
butt end, 72, 78, 167
butt end diameter, 78, 167
butt-end reducer, 167
butt figure, 167, 448
butt flare, 282
butt hook, 104, 178, 292
butt joint, 111, 114, 614
butt length, 159, 167
butt log, 159, 167
butt matching, 111
butt off, 247, 366
butt plate, 375
butt rigging, 113, 292
butt root, 76
butt rot, 160, 167
butt splice, 111, 114, 361
butt swelling, 76, 355
butt treatment, 167
butt veneer, 167
butte, 174
butten, 533
butter of zinc, 315
butterfly table, 201, 455
butting saw, 247, 366
button, 3, 618
button lac, 254, 347
button mushroom, 105
buttress, 8, 228, 486, 612
buttress flare, 8, 309, 319
buttress root, 8
buttress (root), 4
buttressed root, 4, 8, 9
butyl benzyl cellulose, 105
butyl cellulose, 105
buy at retail, 305
buy-back price, 211
buying capacity, 173
buying order, 106
buzz saw, 592
buzzard hook, 646
by contract, 55
by-law, 171, 183
by-note, 567
by-pass, 142, 351, 612
by-pass road, 154, 232, 534
by-path, 533
by-product, 154
by-road, 154, 534
by-track, 17
byte, 504, 540, 648
C-clamp, 170, 229
C-grade thinning, 444, 622
C-horizon, 328
C-iron, 133
C-layer, 328
C-stage, 654
C_3 plant, 473
C_4 plant, 473
caatinga, 267
cabin hook, 320
cabin trunk, 41, 78
cabinet, 64
cabinet finish, 7
cabinet hardware, 229
cabinet scraper, 519
cabinet wood, 519
cabinet-work, 64, 519
cabinetmaker, 229, 519
cabinetmaker's wood, 229, 519
cabinetmaking, 229, 519
cable, 101, 160, 161, 469
cable boom with floats, 148, 283
cable capacity, 397
cable car, 283
cable change, 56
cable chuting, 283
cable conveyor, 283
cable crane, 283
cable extraction, 161, 231
cable grapple, 470
cable guide pulley, 89
cable hauling, 470
cable holder, 283, 476, 493
cable lift, 283
cable log haul, 161, 470
cable logging, 161, 469
cable-railroad, 283
cable reel, 264
cable saddle, 161
cable skidder, 281
cable skidding, 469
cable sling, 104
cable step, 161
cable storage trailer, 635
cable strap, 104, 161
cable-suspension bridge, 161, 545
cable system, 470
cable tram, 103, 283, 469
cable tramway, 469
cable-tramway of single rope, 85
cable wiper, 161
cable work, 161
cable yarding, 161
cable yarding system, 161, 470
cableway, 276, 469
cableway bucket, 470
cableway carriage, 470
cableway hauling, 470
cableway tension device, 249, 470
cableway winch, 470
cabling, 493
cacao-butter, 273
cacao-fat, 273
cacao-tallow, 273
cache, 634
cacodylic acid, 122, 267
cacogenenesis, 297
cactin, 522
cactus house, 522
cadastral map, 488
cadastre, 488
cadinene, 20, 111
caducous, 494, 599
caecum, 318, 352
caeno-monoecious, 69, 84
caenogenesis, 539
caeoma, 313
caespitosa, 72, 73
caespitose, 72, 73
caespitose plant, 73
cajan, 332
cajaput oil, 6
cajeput oil, 6
cajuput oil, 6
caking of hygroscopic salt, 514
calamene, 5, 267
calameone, 5
calamondin orange, 248
calamus, 179, 194, 584
calamus camphor, 5
calamus oil, 5
calandria pan, 349
calantas wood oil, 528
calarene, 5, 456
calcareous, 156, 434
calcareous algae, 156
calcareous sandstone, 156
calcareous soil, 434
calcareous spar, 131
calcic, 156, 434
calcic horizon, 155
calcicole, 155, 156

C

calcicole plant, 156
calcicolous, 155
calcification, 155
calcifuge, 21, 524
calcifuge plant, 524
calcifugous plant, 524
calcilutite, 156, 211, 342
calcipete, 516
calcipetrile, 434
calciphile, 439, 516
calciphilous, 516
calciphilous plant, 439, 516
calciphobe, 21, 524
calciphobous plant, 21, 524
calciphyte, 156
calcite, 131
calcium acetate, 73
calcium carbonate, 473
calcium clay, 155, 434
calcium levulinate, 566
calcium soil, 156
calciumresinate, 450
calculation of cutting and filling, 489
calculation-period, 226
calculator, 226
calender, 312, 550, 551, 553
calender particleboard process, 184
calender sizing, 550
calendering, 549
calibration, 106, 243
calibration of prism, 285
caliche, 155, 433, 532
caliculate, 260, 580
California red scale, 198
caliper, 44, 311
caliper (calliper), 267
caliper finish, 200
caliper gauge, 44
caliper measure, 311
caliper rule, 44, 579
caliper scale, 36, 311
caliper square, 44
caliper-stick, 311
calipering, 614, 615
calipering of log, 590
calix, 121
call, 104, 243, 441
call (able) loan, 81, 214, 606
call money, 113, 214, 606
call money market, 214, 605
call premium, 441, 446
calliper, 44, 311, 371
cal(l)iper, 497
callitrisoid (*or* callitroid) bordered pit, 4
callitrisoid (*or* callitroid) thickening, 4
callose, 357
callow adult, 539
calls, 200
callus, 357, 587
callus fleck, 587
callus formation, 587
callus wood, 587
callusing period, 587
calm, 509
calomorphic soil, 155
caloric power, 268, 394
caloric value, 268, 394
calorimeter, 296, 393
calorimetric measurement, 393
calorimetry, 393
calotropin, 347
calvescent, 182, 251
Calvin cycle, 267
calycanthemy, 202
calycanthine, 522
calyciform, 17, 121
calycled, 260
calyculate, 260
calycule, 154
calyptra, 167, 319, 626
calyx, 121, 202
cam, 158, 486
CAM plant, 255
camber, 170, 172, 371
camber-beam, 170
cambial cell, 541
cambial element, 541
cambial fusiform initial, 540
cambial initial, 541
cambial ray, 541
cambial ray initial, 541
cambial region, 541
cambial stage, 540
cambial wall, 63
cambial zone, 540
cambiform, 134, 541
Cambio-barker, 267
cambium, 540
cambium area, 541
cambium (*plur.* -bia), 539, 540
cambium cell, 541
cambium-crown surface ratio, 541
cambium layer, 540
cambium ring, 540
cambium zone, 540
Cambrian period, 190
camellia flower, 413
camellia oil, 413
camelliagenin, 413
camera lucida, 485, 525
camera orientation, 419
camera station, 276
camera tube, 419
camouflage, 503
camouflage detection film, 472, 503
camp, 125, 169, 573, 655
camp ground, 311
Camp process, 269
camp run, 3, 592
camp site, 169, 311
camp unit system, 573
campanion cell, 11
campanulate, 626
campestral, 478, 522, 559
campestrian, 17
campfire, 173, 573
camphene hydrate, 456
camphor oil, 604
camphor original oil, 604
camphor tree, 604
camphor wood oil, 604
camphorwood, 604
campo, 5
campo cerrado, 5
campo sujo, 5
campstool, 605
camptothecin, 516
campylodromous, 201
Canada balsam, 228
Canadian garden, 228
Canadian log rule, 228
Canadian Standard Freeness, 228
canal, 173, 456
canary, 173
cancelled check, 151, 635
cancer, 1
cancer-causing agent, 621
cancer-tree, 1
cancerous, 1
candelabra-topped tree, 278
candelila wax, 536
candle nut oil, 632
candlepower, 632
cane, 159, 632
cane molasses, 606

cane trash, 606
canella, 6, 369
caney, 475
canine tooth, 387
canker, 277, 280, 636
canker stain, 280
canker-worm, 57
cannabene, 77
cannibalism, 484
cannon a log, 589, 590
canoe, 111, 204
canoe paddle, 111
canoe rib, 533
canonical correlation, 100
canonical (correlation) analysis, 100
canonical variate, 25
canons of taxation, 274
canopy, 300, 447
canopy analizer, 178
canopy class, 300
canopy closure, 300
canopy cover, 300
canopy cover class, 300
canopy density, 585
canopy drip, 302
canopy fire, 447
canopy interception, 300
canopy silhouette, 447
canopy surface, 300
canopy tree, 300
cant, 1, 130, 318, 537
cant board, 477, 537
cant flipper, 318, 538
cant setter, 349
canter, 532
cantilever beam, 545
cantilever beam impact test, 545
cantilever bridge, 545
cantilever crane, 545
cantilever endurance test, 545
cantilever gantry, 545, 639
canvas covered plywood, 128
canvas retainer plate, 128
canyon, 519
canyon carriage, 519
canyon wind, 519, 520
caoutchouc, 319, 428
cap, 156, 167, 223, 266, 638
cap-and-trade, 526
cap block, 309, 635, 636
cap cloud, 413
cap-fungus, 404
cap moulding, 319
cap plasmolysis, 178, 319
capacitor paper, 101
capacity building, 341
capacity of heat, 393
capacity utilization rate, 47, 417
capactiance moisture meter, 101
cape bulb, 188
capillarity, 318, 319
capillary, 319
capillary action, 319
capillary attraction, 319
capillary condensation, 318, 319
capillary condensed water, 319
capillary force, 319
capillary form, 319
capillary fringe, 318, 319
capillary gas chromatography, 319
capillary meniscus, 319
capillary moisture capacity, 319
capillary penetration, 319
capillary percolation, 319
capillary pipe, 319
capillary porosity, 318
capillary potential, 318
capillary-transient water, 319
capillary tube, 319
capillary viscosimeter, 319
capillary water, 319
capital account, 268, 641
capital accumulation, 640
capital appropriation, 640
capital budget, 641
capital budgeting, 23, 641
capital consumption allowance, 606, 641
capital equipment, 176, 641
capital expenditure, 220, 622, 641
capital felling, 125
capital financing, 398, 476
capital flow, 641
capital gains, 641
capital gains tax, 641
capital goods, 176, 422, 640
capital growing stock, 290
capital intensity, 640
capital-intensive, 77, 640
capital intensive industry, 640
capital levy, 37, 641
capital liability, 49, 176, 640
capital-output ratio, 640
capital rationing, 641
capital rental value of normal growing stock, 128
capital rental value of soil, 488
capital repair, 78
capital stock, 175, 220
capital structure, 640
capital value, 640
capitalization, 485, 640, 641
capitalized cost, 193
capitalized value, 193, 640
capitalized value of planting-cost, 600
capitate antenna, 68
capitulum, 485
capoeira, 5
capping, 592
capping method, 290, 314
capreolate, 260
capricorn-beetle, 477
caproic acid, 224
caprylic acid, 539
capsanthin, 282
capsella, 444
capsicin, 128, 282
capsid, 564
capsill, 195, 415
capsomer(e), 88, 272
capsorubin, 282
capstan, 242, 290
capstan spool, 243
capstan winch, 290
capsular, 461
capsular polysaccharide, 229
capsule, 12, 229, 461
captafol, 94
captan, 274
capture-recapture method, 33

C

capturer, 305
caput, 12
car mover, 471
car retort, 52
car spotter, 451
car stake, 52
carabidoid, 36
carajurin, 382
caramel test, 242
caranday wax, 648
caraway oil, 166, 210, 599
carbachol, 473
carbamate, 3
carbamide, 346, 473
carbamide resin adhesive, 346
carbamylcholine chloride, 87
carbaryl, 513
carbendazim, 19
carbendazim (*or* carbendazol), 117
carbendazol, 19
carbide alloy, 575
carbide tipped chain, 575
carbofuran, 147, 267
carbohydrase, 473
carbohydrate, 473
carbolineum, 121, 242
carbon-14 dating, 473
carbon allowance, 473
carbon assimilation, 473
carbon black, 465, 473
carbon brick, 472
carbon consumption, 473
carbon credit, 473
carbon cycle, 473
carbon debt, 473
carbon dioxide (CO_2) fertilization, 123
carbon-dioxide compensation point, 123
carbon disulfide, 122
carbon efflux, 473
carbon emission, 473
carbon emmission, 473
carbon fiber, 473
carbon-fibre composite, 473
carbon flux, 473
carbon footprint, 473
carbon gain, 473
carbon leakage, 473
carbon loss, 473
carbon neutral, 473
carbon-nitriding, 473
carbon-nitrogen ratio, 473
carbon offset, 473
carbon pool, 473
carbon reduction cycle, 473
carbon reservoir, 473
carbon sequestration, 177, 473
carbon sink, 473
carbon source, 473
carbon stock, 473
carbon storage, 473
carbon tax, 473
carbon tetrachloride, 463
carbon tissue, 154, 473
carbon tracker, 473
carbon trading market, 473
carbonaceous aerosol, 189
carbonaceous shale, 473
carbonaceous soil, 189
carbonate rock, 473
carbonated mist, 59
carbonation, 473
carbondioxide enrichment, 123
carbondioxide fertilizer, 123
carbonhydrate, 473
Carboniferous period, 434
carbonium ion, 473, 611
carbonization, 242, 473
carbonization of wood, 331
carbonizing rot, 194
carboxin, 503
carboxybiotin, 469
carboxykinase, 469
carboxylase, 469
carboxylation resistance, 469
carboxylic acid, 469
carboxymethyl cellulose, 469
carburett, 206
carburetter, 206, 371
carburettor, 371
carburization, 420
carburizer, 420
carcass, 173
carcass-flooring, 318
carcass-saw, 77
carcassing timber, 173
carcinogen, 621
carcinogenic, 621
carcinology, 1, 230
card collator, 267
card key, 267
card per minute, 320
Card process (treatment), 267
card punching unit, 267
card reader, 111
card sorting key, 141
cardamon oil, 533
cardanol, 558
cardboard, 200, 267
cardboard mill, 618
cardinal altitude, 219
cardinal wind, 431, 633
cardol, 376, 558
cards per minute, 320
care, 149
care of plantations, 395
care of stands, 299
care of woods, 406
Caribou, 242, 462, 549
carina (*plur.* carinae), 225, 309
carinate, 260, 309
cariopside, 574
carissin, 230
carload freight, 610
carloading cost, 596, 598
carmine, 553
carnauba-wax, 5, 267
carnivore, 436, 437
carnivor(o)us, 437
carnivor(o)us insect, 399
carnivorous plant, 436, 437
carnivorus plant, 399
carnose, 399
carob gum, 207, 242
carotenoid, 285
carotin, 201
carotinase, 201
carp, 538
carp (*or* carpel), 186
carpaine, 128
carpel, 538
carpellary, 538
carpellary bundle, 538
carpellate strobilus, 69
carpenter, 332
carpenter ant, 335
carpenter bee, 332
carpenter moth, 332
carpenter's jack (trying) plane, 332
carpenter's level, 332
carpenter's pincers, 201, 332

carpenter's rule, 332
carpenter's square, 242, 259, 332
carpenter's wood, 332
carpenter's wood saw, 332
carpentry, 332, 334
carpentry tongue, 332, 542
carpet bed, 319
carpet bedding, 319
carpet (flower) bed, 319
carpet herb, 319, 364
carpet laying, 98
carpet-like alpine meadow, 98
carpodermis, 186
carpology, 186
carpophore, 186, 538, 641
carpophyll, 76, 538
carpophyte, 76, 524, 629
carpopodium, 186, 257
carpospore, 186
carr, 267
carriage, 88, 104, 309, 352, 612
carriage feed, 352
carriage offset, 352
carriage rider, 178, 352
carriage stop, 352
carriage track cleaning device, 352
carriage way, 52, 352
carrier, 29, 52, 66, 79, 352, 596, 598
carrier block, 639
carrier cable, 56, 283
carrier for auxin, 429
carrier frequency, 598
carrier roll, 89
carring capacity, 151
Carr's disintegrator, 267
carry-over, 246, 250
carry-over of stock, 211
carrying capacity, 595, 598
carrying charge, 37, 253
carrying value, 184, 246, 604
carryover seed, 569
cart, 122, 443
cart track, 184, 222
carthamin, 197
cartier cable logging, 470
cartography, 487
carton paper, 267
cartridge rim, 87
cartwrights' wood, 52
caruncle, 626
carvene, 528
carvenone, 166, 528
carving, 103
carving machine, 103
carving wood, 103
carvomenthene, 528
carvomenthone, 528
carvotanacetone, 528
caryogenetics, 194
caryogram, 194
caryokinesis, 193, 235, 581
caryopsis, 574
caryotype, 194, 391
cascade, 104, 307, 534
cascade system, 66
cascara-sagrada, 17
cascarilla oil, 267, 528
case, 194, 474, 529
case-hardening, 26, 27, 382
case-hardening stress, 26
case side, 184
case tension, 343, 624
case wood, 529
casebearers, 7, 377, 457, 469
cased frame, 529
casein, 157, 284
casein adhesive, 284
casein cement, 284
casein fibre, 284
casein glue, 284
casein-paste, 284, 400
casein-soybean glue, 284
cash, 115, 525, 526
cash against document, 363, 551
cash bond, 526, 551
cash deposit as collateral, 15, 551
cash on delivery, 217, 239
cash purchase, 525, 526
cashew nut, 230, 558
casing, 12
cassava glue, 334
cassava starch, 334
cassia (bark) oil, 399
cassia leaf powder, 399
cassia oil, 399
cassia scrapped, 184
cassiamin, 413
cassideous, 280
cast-iron radiator, 636
cast off, 582
cast on, 654
castalin, 291
castaneous, 291
castanogenin, 4
castanozem, 291
caste, 221, 541
caste differentiation, 91
caste system, 91
castine, 328
castonin, 291
castration, 385
casual species, 348
casualty insurance, 597
casurin, 333
cat, 223
cat bummer, 493
cat eye knot, 318
cat face, 318, 423
cat hooker, 223, 493
catabolism, 141, 567
catacladous, 613
catadromous, 238
catadromous fish, 238
catalase, 187
catalysis, 74
catalyst, 74
catamaran, 53, 349
cataphalanx, 286
cataphoresis, 102
cataphyll (*plur.* -lla), 92, 305, 551
catastrophe theory, 597
catastrophic, 597
catastrophic factor, 260, 597
catathermometer, 157, 286
catch, 33
catch-all, 142, 247, 649
catch boom, 283
catch crop, 236
catch curve, 33
catch mark, 21
catch per unit effort, 85
catch shackle, 207
catch up box, 244
catchability, 272
catching fire, 390, 604
catching frame, 245
catching reaction, 33
catching trestle, 283
catchment, 224, 308, 350
catchment area, 224, 443
catchment basin, 224
catechol tan, 121
catecholase, 121, 298
catechotannin, 121
catechu, 121
cateehin, 121

categories of nature reserve, 645
categories of threatened species, 513
category, 285, 626
category of forest, 304
catena, 489
catenary, 545
catenoid, 545
caterpillar, 104, 314, 318, 536
caterpillar grinder, 293, 295
caterpillar tractor, 314
catharometer, 89, 392
cathead, 242, 319, 326
cathetometer, 44
cathode-ray tube, 570
cathodoluminescence scanning electron micrograph, 570
cation exchange, 556
cation exchange capacity, 556
cation saturation, 555
cationic preservative, 556
catkin, 399
catologue price, 415
catpiece, 479
catty logger, 446
cattyman, 342
cauda, 503
caudex, 252, 615, 634
caul board, 53, 103, 351
caul plate, 103
caul pusher, 103
caul speed-up conveyor, 103
caulescent, 260
caulicole, 252
cauliflorous, 160, 252
cauliflory, 284, 447
cauline bundle, 252
caulking, 479
caulless forming line, 508
caulless loader, 508
caulless prepress, 508
caulless pressing system, 508
caulless system, 508
causal component, 570
causative agent, 29
cause-effect relationship, 570
causticity, 234, 271
causticlye, 271
caustobiolith, 273
cave animal, 547
cave community, 110
cave strut, 555
cavernicolous, 110, 548
cavernicolous animal, 547
cavitas, 375
cavity, 375
cavity forming rot, 277
caxd code, 267
Cayenne linalor oil, 269
Cayenne rosewood oil, 269
ceanothine, 320
ceara rubber, 513
ceara rubber plant, 334, 402
cebur balsam, 402
cecidium, 60, 267
cecidogenous, 428
cecological civilization (eco-civilization), 425
cedar gum, 548
cedar leaves oil, 7, 548
cedar oil, 7
cedar (wood) oil, 7, 548
cedrat oil, 173, 529
cedrela oil, 553
cedrellone, 557
cedrenone, 7
cedrone, 7, 548
ceiling, 477
ceiling baffle (board), 477
ceiling board, 477
ceiling coil, 105
ceiling joist, 360, 477
ceiling price, 653
celastrine, 338
celation period, 549
cell, 59, 85, 86, 154, 208, 517
cell-bound enzyme, 13
cell cavity, 518
cell content, 518
cell corner, 518
cell differentiation, 517
cell division, 517
cell-dump suspension, 518
cell enlargemet, 518
cell fractionation, 517
cell-free extract, 511
cell-free system, 511
cell fusion, 518
cell genetics, 518
cell heredity, 518
cell inclusion, 518
cell lethality, 518
cell lumen, 518
cell lysis, 13, 518
cell membrane, 518
cell of axial parenchyma, 631
cell organ, 518
cell organell, 518
cell pattern, 518
cell plate, 517
cell sap, 518
cell separating enzyme, 517
cell-shaped strip, 147
cell space, 518
cell space ratio, 518
cell structure, 518
cell transformation, 518
cell-type reforestation, 534
cell wall, 517
cell wall check, 13, 313
cell wall component, 13
cell wall layer, 13
cell wall sculpturing, 13
cell wall substance, 13
cell-wall thickening, 13
cellar rot, 96
cellobiase, 522
Cellon process, 402
cellophane, 30, 402
cellotex, 606
cellotriose, 523
cellular, 517, 530
cellular automata, 518
cellular board, 146
cellular construction, 146
cellular element, 517
cellular honeycomb material, 146
cellular morphology, 518
cellular necrosis, 517
cellular physiology, 518
cellular structure, 146, 518
cellular texture, 146
cellular tissue, 518
cellular type, 518
cellulase preparation, 523
cellule, 277, 452, 535
cellulosan, 153, 523
cellulose, 523
α-cellulose, 523
β-cellulose, 523
γ-cellulose, 523
cellulose acetate foam, 566
cellulose adhesive, 523

cellulose base fibre, 523
cellulose biodegradation, 523
cellulose bundle, 523
cellulose chain, 523
cellulose degradation, 523
cellulose derivative, 523
cellulose-dissolving fungus, 523
cellulose esters, 523
cellulose ethers, 523
cellulose fiber, 523
cellulose fibre, 523
cellulose fibril, 523
cellulose hydrate I, 457
cellulose hydrate II, 457
cellulose lacquer, 523
cellulose matrix, 523
cellulose molecular chain, 523
cellulose plant, 618, 620
cellulosic exchanger, 523
cellulosic fibre, 523
cellulosic layer, 523
cellulosic strand, 523
celluloslysis, 522
Celsius degree, 419
Celsius' thermometric scale, 419
cembrene, 513
cement, 241, 457
cement-bond building panel, 458
cement-bonded particleboard, 458
cement particle board, 458
cementation, 27, 241
cementation index, 241
cementation zone, 241
cementing, 240, 344, 345, 398
cementing agent, 241
cemetery, 171, 305
cenanthy, 275
cenophyte, 539
cenosere, 539
cenospecies, 172, 263, 389
cenozoic, 539
cenozoic era, 539
census, 395, 627
census tract, 365
census(ing), 395, 627
center, 46, 335, 538, 631
center of abundance, 116
center of dispersal, 367, 404
center of distribution, 140
center sawn, 256
centering, 332
centigrade, 6, 419
centigrade degree, 6, 419
centipede, 512
central area, 194
central axial placenta, 625
central cut, 624
central data bank, 451, 625
central diameter, 625
central dogma, 624
central layer, 623
central layer of secondary wall, 70
central leader, 625
central leader type, 633
central leader type training, 633
central nervous system, 624
central park, 625
central percussion, 625
central processing unit, 625
central processor, 625
central projection, 624
central rectrices, 625
central stem, 625
centrally planned economics, 625
centre, 46, 335, 631
centre board, 385
centre-field, 237, 625
centre firing, 624
centre guard, 623
centre hole, 160, 276
centre-hole planting, 548
centre jam, 623
centre line average (method), 624
centre load, 625
centre matching, 624
centre of curvature, 384
centre of group, 641
centre of impact, 419
centre of infection, 66
centre of interest, 541, 624
centre of origin, 124, 367
centre of percussion, 59
centre plank, 385
centre point diameter, 625
centre rail, 623
centre raker, 624
centre ring, 624
centre rot, 538
centre-sawed, 256
centre sawing, 631
centre table, 623
centr(e)ing, 107, 115
centr(e)ing device, 107, 115
centr(e)ing machine, 107
centr(e)ing splitting (*or* ripping) saw, 115
centric fission, 640
centric fusion, 640
centrifugal atomizing head, 287
centrifugal blower, 287
centrifugal cleaner, 287
centrifugal dehydrator, 287
centrifugal effort, 287
centrifugal fan, 287
centrifugal (force) drying, 287
centrifugal glue sprayer, 287
centrifugal inflorescence, 287
centrifugal jet, 287
centrifugal protoxylem, 341
centrifugal seasoning, 287
centrifugal spraying method, 287
centrifugal tube, 287
centrifugal xylem, 287
centrifugation, 287
centrifuge, 287
centripetal inflorescence, 530
centripetal protoxylem, 497
centripetal xylem, 530, 571
centroid sampling, 621
centromere, 640
cenuglomerate, 342
cephalanthin, 146
cephalanthium, 485
cephalic index, 485
cephalophorum, 204
cephalotaxine, 72, 402
cephalothorax, 485
cera, 146
ceraceous, 282
cerambycid-beetle, 477
cerambycoid, 477

ceratium, 49, 242
cere, 75, 282, 487
cereal grass, 191
cereal straw, 175
ceresin, 513
ceresin wax, 96
certificate, 611
certificate of origin, 47
certification, 192, 372
certification of planting, 617, 626–628
certification of plants, 617, 626–628
certification of seed, 626, 628
certified, 192, 252
certified seed, 233, 372
cervix, 254
cespitose, 72, 73
Ceylon cinnamon bark oil, 462
Ceylon citronella grass, 462, 553
chaeta, 71, 160, 461
chaetoplankton, 242
chaetotaxy, 160, 319
chafer, 248
chaff, 175, 270, 628
chaffy, 72, 581
chaffy seed, 569
chafing, 36, 327
chain, 44, 207, 295
chain banking, 292
chain bar, 295
chain-belt conveyer veneer dryer, 499
chain blade, 295
chain board profile, 294
chain-bull block, 620
chain conveyor, 295
chain dancer, 249
chain discount, 292
chain drive, 294
chain file, 262
chain-flail barker, 295
chain-flail delimber, 295
chain flight conveyer, 295
chain grinder, 262, 295, 326
chain guide, 295
chain hand saw, 295
chain hoist, 46, 104
chain index, 292
chain line, 100, 295
chain loader, 295
chain molecule, 295
chain mortiser, 295
chain of sand dunes, 411
chain oiler, 262
chain peeler, 295
chain plate, 176, 202
chain press, 295
chain printer, 295
chain pulley, 295
chain saw, 295
chain saw felling head, 295
chain-saw mortising machine, 295
chain shifter, 295
chain skidding, 295
chain sprocket, 295
chain store, 292, 294
chain tender, 159, 281, 349, 469
chain-terminating codon, 295
chain transfer, 295
chain tray sorter, 295
chain type barker, 295
chaining, 295, 493, 604
chainlike cell, 295
chair, 355, 492, 566
chair back, 566
chair bed, 655
chair rail, 454, 566
chair seat, 566
chair table, 274
chalaza, 192
chalcid-fly, 226
chalcid wasp, 533
chalk, 5
chalk-loving, 516
chalking, 5, 433
chalky, 5
chalky deposit, 5
chalybeate spring, 481
chamaecin, 23
chamaephyta gramminidea, 191
chamaephyta lichenosa, 99
chamaephyta reptantia, 364
chamaephyta scandentia, 351
chamaephyta sphagnoides, 342
chamaephyta succulenta, 399
chamaephyta suffruticosa, 9
chamaephyta velantia, 317
chamaephyte, 97
chamaephyte climate, 97
chamber, 57, 58, 87, 184, 440, 529
chamber table, 444
chambered, 141
chambered crystalliferous cell, 143
chambered parenchyma, 143
chambered pith, 141
chamene, 23
chamenol, 23
chamfer, 90
champaca leaf oil, 210
champaca oil, 210
champagne, 528
champion tree, 254, 475
chance, 127, 224, 479, 654, 655
chance error, 218, 348
change detection technique, 23
change ends, 161, 243, 590
change of species, 450
change road, 209
change speed gear box, 24
changeability, 272
changing weight, 23
channel, 4, 172, 307, 369, 383, 481
channel fill deposit, 193, 384
channel net, 460
channel sedimentation, 384
channel storage, 193
channel straightening, 193
channeled, 260, 582
channeling, 173
chanootin, 202
chap, 108, 183
chaparral, 17, 411
chaparrinone, 47
char, 242, 334, 417, 472
char length, 472
character, 475, 542, 647
character displacement, 475
character divergence, 542
character generator, 648
character release, 475
character-seed, 47
characteristic absorption peak, 475

C

characteristic adhesion constant, 240
characteristic animal, 475
characteristic curve, 475
characteristic curve of stand density, 299
characteristic function, 475
characteristic scale, 475
characteristic species, 475
characteristic spectrum, 475
charcoal, 334, 472
charcoal-burner, 124, 334, 417
charcoal-burning, 417
charcoal dust, 472
charcoal kiln, 472
charcoal-making in pit, 275
charcoal oven, 472
charcoal standard, 334
charcoal-wood, 472
charcoalman, 417
charge, 101, 140, 228, 441, 562, 563, 638, 639
charge account, 173, 305, 417
charge-and-discharge statement, 442
charge-coupled devices, 101
charge coupled diode, 101
charge end, 249, 432
charge membrane, 79
charge off, 59, 475
charge sale, 417
charger, 415, 416, 638
charges for remittance, 213
charging hopper, 228
charging tank, 171, 225
charnockite, 642
charophyta, 312
Charpy (impact) machine, 7
Charpy impact strength, 7
Charpy pendulum machine, 7
charred wood, 472
charring, 417, 472
charring-and-spraying treatment, 417
charring in heaps, 114
charring in pit, 275
charring-making in pit, 275
charring rate, 472
chart quadrat, 487
charter, 544, 614
chase, 248, 297, 602, 639
chase mortise, 41
chase-mortising, 268
chaser, 248, 602, 639
chasing method, 639
chasmogamy, 268
chasmophyte, 434
chatki, 363
chattel mortgage, 108
chatter, 48, 480
chatwood, 1, 179
chaulmoogra oil, 52, 76
chavicine, 194
chavicol, 194
che shu lacquer, 366, 424
cheap money, 93
cheat stick, 232
cheater, 233, 328
check, 115, 193, 243, 277, 298, 569, 612
check book, 193, 612
check cruise, 193
check dam, 175, 283
check experiment, 114
check-free veneer, 510
check method, 193
check of zero, 305
check plot, 115
check scaler, 152, 232
check scaling, 152, 153
check system, 193, 232
check throat, 67
check-up flap, 232
check variety, 115
check weigher, 225
checkcrow, 366
checked knot, 297
checker, 226, 233, 420
checkerboard protection, 411
checkered seeding, 366
checking (of young plantation), 582
check(ing) point, 232
checking solution concentration, 44
check(ing) station, 232
checklist, 193, 232
cheek pouch, 229
cheese block, 88, 132
cheirolin, 184
chelate, 4
chelate substance, 4
chelation, 4
chelura, 636
chemi-finer, 163
chemi-ground wood pulp, 206
chemi-mechanical pulping, 206
chemi-thermo-mechanical pulping, 206
chemi-thermo-mechanical puping, 206
chemical attack, 206
chemical barking, 206
chemical bending, 206
chemical brown-stain, 206, 557
chemical cleaning, 206
chemical composition, 206
chemical control, 206
chemical control method, 206, 559
chemical debarking, 206
chemical defense, 206
chemical degradation, 206
chemical drying, 206
chemical ecology, 206
chemical erosion, 206
chemical evolution, 206
chemical fossil, 206
chemical-ground pulp, 206
chemical in fire suppression, 206
chemical induced breeding, 206
chemical-mechanical pulping, 206
chemical modification, 206
chemical oxygen demand, 206
chemical permanent set, 206
chemical potential, 206
chemical processing of forest products, 298
chemical pruning, 206
chemical pulp, 206
chemical seasoning, 206
chemical sediment, 206
chemical shift, 206
chemical stain(ing), 206
chemical sterilization, 206
chemical storage, 206

chemical thinning, 206
chemical treatment, 206
chemical weathering, 206
chemical weed control, 206
chemical weed killer, 64
chemical weeding, 206
chemical wood, 206
chemical (wood) pulp, 206
chemical wood-pulp, 206
chemiluminescence, 206
chemimechanical pulp, 206
chemisorption, 206
chemoautotroph, 206
chemolithotroph, 205, 206
chemomorphosis, 206
chemonite preservative, 553
chemophysiology, 423
chemoreceptor, 206
chemoresistance, 559
chemosensory organ, 206
chemosphere, 181
chemostat, 195
chemosterilant, 206
chemosynthesis, 205
chemosynthetic, 206
chemotaxis, 206, 383
chemotaxonomy, 206
chemotherapeutant, 206
chemotherapeutics, 206
chemotherapy, 206
chemotrophic, 206
chemotropism, 530
chenopodium oil, 489
cheque book, 612
cheque rate, 612
chequer, 130
chequered seeding, 402
chernozem, 194
cherokee rose extract, 197
cherry essence, 572
cherry gum, 572
cherry laurel oil, 184
cherry-picker, 52, 307, 380
cherry stone oil, 572
chersophyte, 157
chesnatin, 291
chest, 41, 184, 529, 542
chest-height, 542
chest-height diameter, 542
chest-height form factor, 503, 542
chest-helght diameter increment, 542
chest of drawers, 61, 512
chest-on-chest, 452, 641
chest saw, 534
chesterfield, 78
chestnut bark tannin, 291
chestnut blight, 291
chestnut earth, 291
chestnut extract, 291
chestnut pole borer, 291
chestnut soil, 291
cheval glass, 65
chev(e)ron burn, 216
chewing mouthpart, 259
chi-square distribution, 140, 267
Chi-square (*or* χ^2) test, 267
chi-square test, 233, 267
chianophile, 441
chianophobe, 524
chiasmata, 238
chiasmata deviation, 238
chibou, 297
chicle, 473
chicot, 10
chief constituent, 634
chief species, 634
children park, 121
chilled (glue) joint, 286
chilled seed, 286
chilled shot, 286
chilling, 93, 286
chilling injury, 110
chim(a)era, 231, 375
chimney, 482, 553
chimney in pile, 36
chimonanthus fragrans concrete, 282
chimonochlorous, 13
chimonophilous, 108, 516
china cabinet, 53, 69
china case, 53, 69
china clay, 5, 162, 358
China rose, 594
China wood-linseed varnish, 484
China wood oil, 484
China's vast grassland, 622
Chinese angelica tree, 71
Chinese arbor-vitae, 42
Chinese balsam, 182, 286
Chinese cinnamon, 585, 622
Chinese cinnamon oil, 399
Chinese coin palm, 648
Chinese date tree, 599
Chinese Ecosysterm Research Network (CERN), 622
Chinese Forest Ecosysterm Research Network (CFERN), 622
Chinese gall, 512
Chinese gall aphid, 512
Chinese gallnut, 512
Chinese gallotannin, 512
Chinese green ramie, 622
Chinese insect wax, 6, 622
Chinese lac, 622
Chinese lacquer, 424, 622
Chinese layering, 276, 622
Chinese neroli oil, 204
Chinese red, 622
Chinese red pine, 316
Chinese soapberry, 509
Chinese spruce, 595
Chinese tallow, 507
Chinese tannin, 259, 512
Chinese water-lily, 193, 294
Chinese waxmyrtle, 557
Chinese weeping cypress, 7
Chinese white ramie, 622
Chinese (wood) oil, 484
Chinese yam, 446
chinook, 158
chip, 13, 16, 333
chip-axe, 120
chip barking method, 532
chip bin, 334
chip blower, 333
chip breaker, 16, 333
chip bruising, 318, 469
chip budding, 375
chip cake, 16
chip cap, 333
chip contact dryer, 334
chip container, 333, 334
chip conveyor, 334
chip crusher, 334
chip disintegrater, 334

C

chip distribution pipe, 333
chip distributor, 333, 334
chip dryer, 16
chip feeding screw, 334
chip filler, 334
chip flaker, 334
chip grinder, 334, 532
chip loft, 333
chip mark, 318, 469
chip metering, 16
chip-N-saw, 263, 532
chip packing machine, 333
chip plug, 33, 402, 468
chip screen, 334
chip silo, 334
chip-spreading machine, 16
chip steaming, 334
chip thickness screening slicing system, 334
chip-throwing roll, 334
chip van, 334
chip washer, 334
chip washing, 334
chip wood, 532
chipboard, 16, 468
chipharvestor, 532
chipidium, 414
chipless cutting, 511
chipless wood-cutting, 331
chipped grain, 378, 468
chipper, 16, 262, 532
chipper-blower, 145, 532
chipper canter, 532
chipper chain-saw, 593
chipper disc, 532
chipper-edger, 532
chipper-mill, 532
chipper spout, 532
chipping, 16, 39, 377, 378, 532
chipping canter, 532
chipping edger, 532
chipping headrig, 532, 590
chipping machine, 16, 532
chiral molecule, 443
chirality, 357, 443
chirarity, 443
chisel, 131, 599
chisel plough, 420, 599
chisel ploughing, 34, 599
chisel-point tooth, 599
chisel saw chain, 131
chisel test of plywood, 240
chisel tooth, 131
chisel tooth saw, 131
chisley soil, 291, 434
chit, 124, 341, 583
chitin, 221, 272
chitosamine, 272, 364
chitting, 62, 74
chlamydia, 202, 551
chlamydo-, 200
chlamydospore, 200
chlorbufam, 315
chlordane, 315
chlordimeform, 411
chlorenchyma, 315
chloride paper, 315
chlorinated insecticide, 580
chlorinated lignin, 315
chlorinated rosin, 315
chlorinated rubber adhesive, 315
chlorine number, 316
chlorinity, 315
chlormequat chloride, 2
chlorofluorocarbon, 315
chlorofos, 94
chlorophenasic acid, 512
chlorophoria, 390
chlorophyll, 560
chlorophyll fluorescence, 560
chlorophyllase, 560
chlorophyllogen, 560
chloropicrin, 315, 403
chloroplast, 560
chloroplastin, 560
chloroprene rubber, 315
chlorosis, 210, 387, 492
chlorosity, 476
chlorothalonil, 7
chlorotic dwarf, 503
chloroxuron, 278
chlorpyrifos, 111
chlortetracycline, 248
chlorthion, 316
chlorzinc process, 315
chock, 103, 402
chock block, 88, 132, 618
chocker, 281
choice of tree species, 450
choke, 281, 375, 651
choke disease, 529
choke lever, 145
choker, 281
choker gun, 281
choker-hole, 65
choker hook, 281
choker man, 281
choker rester, 281
choker setter, 280, 281
choker skidder, 281
choker socket, 281
choking limit, 222
choline sterase, 87
cholla gum, 522
chomophyte, 434, 596
chop, 75, 125, 269
chopper, 75, 126, 269, 378, 600
chopper lump tubes, 149
chopper plough, 149
choppers' tool, 126
chopping axe, 269
chopping boss, 38
chopping platform, 126
chord-foot, 523
chorion, 311
chorisepalous, 287
chorology, 140, 426
Christen hypsometer, 274
christmas flower, 563
Christmas tree, 431
Chroma, 405
chromatic aberration, 405
chromatic adaptation, 405
chromatic colour, 39
chromatic pigment, 40
chromaticity, 391, 405
chromaticity diagram, 405
chromatid, 390
chromatin, 391
chromato-, 405
chromatogram, 45, 405
chromatograph, 405
chromatographia, 405
chromatographic, 405
chromatographic column, 405
chromatographic fractionation, 405
chromatographic grade, 405
chromatography, 45, 405
chromatophore, 405, 598
chromatoplate, 405
chrome azurol S reagent, 167
chrome tanned leather, 167

chromed hide powder, 167
chrominated hide powder, 167
chromogen, 124, 405
chromogene, 405
chromogenic, 598
chromomere, 390
chromometer, 20
chromone, 405
chromonema, 391
chromonucleoprotein, 391
chromophore, 124, 424
chromophoric, 124
chromoresin, 405
chromosin, 390
chromosomal aberration, 390
chromosomal breakage, 390
chromosomal damage, 391
chromosomal disjunction, 390
chromosomal fibrilla, 391
chromosomal loci, 391
chromosomal mechanism, 391
chromosomal nondisjunction, 390
chromosomal race, 391
chromosomal rearrangement, 391
chromosome, 390
chromosome aberration, 390
chromosome arm, 390
chromosome banding, 391
chromosome basic number, 390
chromosome bridge, 390
chromosome complex, 390
chromosome configuration, 390
chromosome fragment, 390
chromosome homologous, 484
chromosome mutation, 391
chromosome nonhomologous, 568
chromosome number, 390
chromosome pairing, 390
chromosome recombination, 391
chromosome repeat, 391
chromosome set, 391
chromosome sterility, 390
chromosome tetrad, 391
chromosome theory, 391
chromosome theory of heredity, 391
chromosome theory of inheritance, 391
chromosomoid, 285, 342
chromoxide green, 166
chronic dose, 318
chronic toxicity, 318
chronobiology, 428
chronological age, 3, 435
chronology, 343
chronology of cutting, 38, 39
chronosequence, 343, 556
chrysanthemine, 259
chrysomel, 560
Chuball, 382
chuck, 229, 267
chuckless veneer lathe, 509
chump, 113
chunk, 40, 381, 468, 491
chunk out, 269, 292
chunk truck, 40
chunk up, 442
churn-butted, 76, 125, 160
chute, 205, 222, 365, 536
chute cutoff, 536
chute stick, 205
chuting, 205
cicatrization, 246
cicatrize, 246
cigar raft, 189, 548
ciliate, 261
ciliato-dentate, 261
cilium (*plur.* -ia), 519, 618
cill, 67, 103, 320
cinch line, 249, 442
cinch rope, 249, 442
cinchona, 248
cinchonicine, 248, 539
cinchonidine, 248, 539
cinchonine, 248, 539
cincinnal cyme, 536
cincinnus, 536
cinene, 346, 634
cinnabar, 563, 632
cinnamal, 19, 553
cinnamic amide, 399
cinnamilidene, 19, 553
cinnamon, 399
cinnamon bark, 399
cinnamon bark oil, 184
cinnamon leaves oil, 399
cinnamon oil, 399
cinnamon soil, 194
cinnamon water, 399
cinnamyl, 399
cion budding, 613
circadian rhythm, 423, 631
circannian rhythm, 343
circannual rhythm, 343
circinal, 387
circinate, 387
circle, 432, 438, 549, 592
circle of microspecies, 536
circle of species, 627
circle of strain, 572
circle of stress, 573
circle of vegetation, 388, 616
circular and gyratory screen, 593
circular area, 592
circular bin, 593
circular birefringence, 593
circular bordered pit, 593
circular chromosome, 209
circular combination saw, 650
circular cutting, 208, 311, 549
circular dichroism spectrum, 592
circular disk spliter, 593
circular gang-saw (*or* gang-mill), 118, 592
circular integration, 208
circular mating, 549
circular mill, 592
circular mitre, 593
circular pit, 593
circular plane, 593
circular plot, 593
circular sample area, 593
circular saw, 592
circular saw blade, 592
circular saw chipper, 175, 592

circular saw feller buncher, 592
circular saw headrig, 592
circular saw sharpener, 592, 593
circular saw with sliding table, 564
circular screen, 593
circular-seamed flies, 574
circular spade, 593
circular strip, 549
circular veneer saw, 82
circularity, 592
circularly polarized light, 593
circulating asset, 306
circulating capital, 306
circulating stock chamber, 569
circulation, 208, 423, 549
circulation of closed system, 322
circulation of open system, 268
circulation of sap, 449
circulation period, 311, 549, 601
circulational pressure process, 549
circulative virus, 549
circumferential, 208, 630
circumgenital gland, 502
circumocular, 502
circumpolar, 221
circumposition, 164, 276
circus, 503, 593
cirrhus (*plur.* cirri), 264
cirriform, 264
cirrus, 65, 264, 518
cis-trans-isomer, 461
cissing, 442
cistron, 461, 655
citric acid cycle, 346
citriculture, 156
citrin, 346, 502
citrine, 346, 634
citriodora oil, 346
citron leaf oil, 173, 529
citron oil, 173, 529
citron petitgrain oil, 529
citronella oil, 528
citronellene, 528
citronellyl, 528
citroxanthin, 346
citrus white fly, 156
citrylidene acetone, 230, 346
city forestry, 56
city garden, 56
city garden distribution, 56
city park, 56, 438
city planning, 56
city planning area, 56
city Planning law, 56
city planting, 56
city water supply, 56
city zoning, 56
civic landscape, 56
civilian conservation corps, 405
cladode, 561
cladogenesis, 143
cladogram, 249, 250
cladophyll, 561
cladosiphonic, 261
claim, 469, 470, 603
clam-bunk skidder, 229
clamp, 229
clamp arm, 618
clamp connection, 470
clamp iron, 229, 551
clamp release, 229
clamp ring, 134, 550
clamp-type rope wheel, 229
clamping device, 229
clamp(ing) time, 228
clamshell, 202, 636
clan, 224, 568, 618
clap post, 64
clapboard, 202, 484
clapnet, 33
clasolite, 469
claspers, 15
clasping, 15
class, 160, 221, 651
class A fires, 563
class formation, 389
class insecta, 280
class interval, 221, 652
class mean sample tree method, 141
class of formation, 617
class width, 255
classic architecture, 174
classic style, 174, 342
classical garden, 174
classical genetics, 252
classical style, 174
classification, 140–142, 221, 285, 547
classification mark, 141
classification of gravitational erosion, 629
classification of site quality, 98
classifier, 141, 143
clastic, 469
clastogen, 583
Clastogenesis, 583
clavate, 12, 578
clave (*plur.* clavae), 11, 12
clavicle, 264
clavicornia, 68
claviform, 12
claw, 173, 604, 618
clay, 345
clay brown loam, 345
clay content, 345
clay impoverishing, 345
clay loam, 345
clay mineral, 345
clay pan, 345
clay pan chernozem, 345
clay pot, 474
clay sandstone, 345
clay shale, 345
clay slate, 344
clay soil, 345
clayed podzol, 345
claypan, 345
clcan drive, 562
clean, 246, 508, 509, 512
clean bark, 256, 385
clean bill, 181, 540
clean-bole, 509, 512
clean brush, 451
clean burn, 386, 416
clean cultivation, 381, 509
clean cut, 69, 181
clean-cut timber, 508
clean cutting, 244
clean-cutting method, 244
clean deal, 508
clean development mechanism, 381
clean felling, 244
clean length, 257, 612
clean logging, 244
clean room, 509
clean seed, 69
clean-stemmed, 512
clean summer fallow, 522
clean timber, 415, 509
clean-up of seed, 628
clean wood, 509
cleaned seed, 69
cleaner, 381
cleaning, 64

C

cleaning cartridge, 257
cleaning-cutting, 64
cleaning in small growth, 64
cleaning-machine, 381
cleaning of seed, 257
cleaning symbiosis, 381
cleaning-up of seed, 257
cleaning-up operation, 381
cleanness of bole, 278, 445, 633
clear, 256, 381, 415, 508, 509, 512
clear air turbulence, 381
clear-bole, 509
clear-bole height, 613
clear-burning strip, 386
clear-cut area, 244
clear-cut logging, 244
clear cutting, 244
clear cutting board, 257
clear-cutting coppice method, 1
clear cutting design, 244
clear cutting forest, 244
clear-cutting high-forest system, 377
clear-cutting in narrow strip, 603
clear cutting method, 244
clear cutting operation, 244
clear cutting system, 244
clear face board, 181, 257
clear fell ban, 252
clear-felled area, 244
clear felling, 244
clear felling method, 244
clear-felling system, 244
clear internode, 245
clear length, 613
clear lumber, 415, 508
clear-strip felling, 80, 239
clear-strip system, 80
clear teeth, 232, 349
clear trunk, 509, 613
clear wood, 508
clearance, 304, 381, 625
clearance angle, 57, 199, 584
clearcole, 75, 578
clearcut area, 244
clearcutting in alternate strip, 239
clearcutting in patches, 279
clearcutting in progressive strip, 293
clearcutting in strips, 80
clearcutting method with natural regeneration by seeds, 244
clearcutting sprout system, 321
cleared area, 244, 381
clearer (tooth), 16
clearer(tooth), 48, 381
clearfell, 199, 244
clearfell area, 244
clearing, 246, 381
clearing axe, 356, 381
clearing machine, 299
clearing of a cutting area, 127, 304
clearing of the felling area, 429, 610
clearing permit, 39
clearstrip method, 80
clearwing moth, 485
cleat, 133, 335, 536
cleavability, 356
cleavable, 569
cleavage, 311, 356
cleavage resistance, 270
cleavage strength, 356
cleave timber, 356
cleaver, 126, 269, 333, 335, 356
cleaving-axe, 356
cleaving-saw, 77
cleft, 297, 420
cleft cutting, 165, 371
cleft grafting, 356
cleft-grafting chisel, 165
cleft inarching, 165
cleft timber, 356
cleft wood, 355
cleistocarp, 20
cleistogamous, 20
cleistogamy, 20
cleistothecium (*plur.* -ia), 20
clench nailing, 497
cleptoparasitism, 90
clevis, 173
click-beetle, 277
cliff, 377, 545
climacteric, 200, 201
climagraph, 369
climate, 368
climate change, 368
climate climax, 369
climate feedback, 369
climate financing, 369
climate model (hierarchy), 369
climate negotiation, 369
climate prediction, 369
climate scenario, 369
climate school, 369
climate sensitivity, 369
climate system, 369
climate variability, 369
climatic chamber, 395
climatic change, 368
climatic climax, 369
climatic climax community, 369
climatic cycle, 369
climatic factor, 369
climatic fluctuation, 368
climatic indicator, 369
climatic life form, 369
climatic material, 369
climatic migration, 369
climatic optimum, 369
climatic plant formation, 369
climatic race, 369
climatic rhythm, 369
climatic snowline, 369
climatic soil formation, 369
climatic stability, 369
climatic succession, 369
climatic vacillation, 369
climatic variety, 369
climatic zone, 369
climatize, 440
climatizer, 369
climatogenic, 369
climatograph, 369
clim(at)ograph, 369
climatological forecast, 369
climatype, 369, 425
climax, 556
climax adaptation number, 105
climax community, 105
climax complex, 105
climax forest, 105
climax forest stage, 105
climax formation, 105
climax-pattern hypothesis, 105, 556
climax pattern theory, 105
climax species, 105
climax vegetation, 105
climb cutting, 461

climb milling, 461
climb sawing, 461
climber, 78, 349–351
climbing, 351
climbing-iron, 349, 351
climbing plant, 351
climbing rope, 349
climbing rose, 317
climbing shrub, 351
climbing stem, 351
climbing tool, 350
clime, 368
climosequence, 369
clinal variation, 236
clinanthium, 649
cline, 236, 380, 425
cling, 345
clinium, 649
clinometer, 45, 380
clinosol, 363
clinosporangium, 541
clip, 507
clip magazine, 87, 229
clipped tree, 610
clipper, 165, 233, 234
clipper cleaner, 212
clipping, 234
clisere, 369
clockwise rotation, 461
clockwise spiral growth, 461
clod, 489
clod crusher, 468
clogging, 111, 651
clohal test, 511
clonal, 511
clonal archive, 577
clonal growth, 274
clonal preservation, 511
clonal seed orchard, 511
clonal selection, 511
clonal test, 511
clonal trial, 511
clonal variant, 511
clone, 511, 574
clone propagation, 511
clone trial site, 511
close, 75, 178, 246, 320
close breeding, 250
close canopy, 249, 585
close-centered training, 20
close-contact adhesive, 322
close-crowned tree, 322
close cultivation, 250
close cycle, 146
close drilling, 603
close fertilization, 20, 250
close grain, 519
close-grained wood, 519, 621
close-grown, 318, 603
close herding, 224
close pile, 435
close piling, 183, 435
close planting, 322
close-plied, 322, 435, 436
close pollination, 250
close population, 20, 322
close relative plant, 250
close stacking, 322
close stand, 322, 585
close stracking, 261
close texture, 518, 519, 621
close time, 252
closed, 146, 585
closed assembly time, 20
closed association, 322, 585
closed bark-pocket, 340
closed canker, 146
closed canopy, 300, 585
closed circulation, 146
closed coat, 326
closed community, 585
closed conduit, 322
closed corporation, 146
closed crossing, 250
closed economy, 20
closed-end investment company, 175
close(d) fire-season, 132, 146
closed forest, 146, 322
closed formation, 585
close(d) game season, 251
closed-loop setworks, 20
closed pitting, 322
closed season, 133, 146, 252
closed shop, 34
closed socket, 146
closed split, 20
closed stand, 322
closed steaming, 146, 322
closed system, 146
closed traverse, 20
closed vegetation, 585
closed vista, 146
closely spaced planting, 322
closeness, 249
closing inventory, 365
closing layer, 146
closing membrane, 146
closing of standing-crop, 301
closing temperature, 20
closing the land for reforestation, 146
closing time, 20
cloth backing, 35
cloth house, 35
clothe bare mountains in verdure, 486
clothes line, 520
clothes moth, 564
clothes press, 563, 564
clothes-rack, 564
clothes-rail, 178
clothes tree, 564
cloud altitude radiometer, 594
cloud belt, 594
cloud computing, 594
cloud forest, 595
cloud forest belt, 595
cloud forest zone, 595
cloud storage, 594
cloudbuster, 363
cloudiness, 214
clouds of dust, 52
cloudy, 120
cloudy day, 570
clove, 194, 519, 534
clove bud oil, 105
clove bud oleoresin, 105
clove leaf oil, 105
clove oil, 105
clove-stems oil, 105
clovene, 105
clover, 42, 52, 403
clump, 72, 237, 238, 446
clumped, 224
clumped distribution, 224
cluster, 73, 146, 322
cluster analysis, 263
cluster bedding plant, 202
cluster crystal, 72, 73
cluster flower-bed, 202
cluster identification, 263, 610
cluster knot, 72, 73
cluster sampling, 285, 389, 610
clustered, 73
clustered distribution, 224
clutch drum, 287

clutch lever mechanism, 287
clutch size, 507
clutch spring, 287
clypeus, 69, 116
co-benefit, 172
co-benefits, 172, 536
co-crystallization, 153, 172
co-current, 29, 615
co-dominance, 92, 172
co-dominant, 92, 553
co-dominant species, 92
co-dominant tree, 92
co-dynamic, 528
co-existence, 172
co-ferment, 149
co-generation, 392
co-plasticizer, 149
CO_2-equivalent, 123
CO_2 fertilization, 123
coactions, 172, 528
coactor, 528
coadaptation, 528
coadapted system, 536
coak, 278, 335
coal gas tar, 320
coal off, 38, 126
coal tar, 319, 596
coal-tar creosote, 319
coal tar pitch, 319
coalesce, 192, 587
coalescent, 471, 587
coalescent pit aperture, 192
coarctate pupa, 502
coarse aggregate, 73
coarse bark, 73
coarse fibre, 73
coarse grain, 73
coarse-grained, 72
coarse-grained environment, 72
coarse grid, 73
coarse-grown, 279
coarse humus, 72
coarse para, 73
coarse ploughing, 72
coarse root, 72
coarse sand, 73
coarse sand soil, 73
coarse sandy loam, 73
coarse screen, 73
coarse splinter, 200
coarse structure, 72
coarse texture, 72
coarse-textured landscape, 72
coarse vacuum, 94
coarse woody debris, 72
Coase's theory, 163
coast, 3
coast dune, 188
coast line, 188
coast protection forest, 554
coastal, 188, 554
coastal community, 188
coastal deposit, 188
coastal desert, 554
coastal dune, 188
coastal effect, 188
coastal fishery, 554
coastal forest, 188
coastal plain, 554
coastal plankton, 554
coastal sand dune afforestation, 188
coastal sand dune fixation, 188
coastal swamp, 554
coastal windbreak, 188
coasting, 66, 205
coasting grade, 205
coatability, 487
coated abrasive belt, 412
coating, 27, 366, 487
coating resin, 487
coating roll, 487
coating varnish, 605
Cobb tester, 272
cobble, 120, 593
cobblestone drain, 623
cobra process, 71, 417
cocaine, 174
cocciferous, 260
coccus, 382, 533
cochlea, 264
cock-bead, 486
cockchafer, 248
cocking, 268
cocking lever, 550
cockroach, 604
coco, 559
cocoa butter, 273, 559
cocoanut husk, 559
cocoanut shell, 559
coconut fibre, 559
coconut oil, 559
coconut palm grove, 559
cocoon, 232, 311
cod, 229
code, 23, 78, 322
code converter, 322
code forestier, 406
code of hand signal, 57
codominant crop tree, 553
codominant tree, 553
codon, 322
coedificator, 172
coefficient, 517
coefficient of annual direct runoff, 344
coefficient of association, 178
coefficient of coincidence, 29, 148
coefficient of community, 389
coefficient of community similarity, 389
coefficient of competition, 257
coefficient of compressibility, 551
coefficient of compression, 551
coefficient of contraction, 442
coefficient of correlation, 528
coefficient of determination, 264
coefficient of difference, 528
coefficient of differentiation, 24, 141
coefficient of dispersion, 404
coefficient of efficiency, 536, 582
coefficient of elasticity, 472
coefficient of expansion, 355
coefficient of frequency, 359
coefficient of genetic distance, 565
coefficient of genetic similarity, 565
coefficient of germination, 124
coefficient of grade, 91
coefficient of growth, 430
coefficient of heat conduction, 89
coefficient of heat conductivity, 89
coefficient of heat transfer, 66
coefficient of heat transmission, 66
coefficient of heterogeneity, 568

C

coefficient of homogeneity, 484
coefficient of permeability, 421
coefficient of rolling friction, 184
coefficient of roughness, 72
coefficient of runoff, 256
coefficient of shear, 234
coefficient of shrinkage, 158
coefficient of similarity, 528
coefficient of sliding friction, 205
coefficient of sound absorption, 514
coefficient of thermal conductivity, 89
coefficient of thermal transfer, 66
coefficient of thermal transmission, 66
coefficient of variability, 24
coefficient of variation, 24
coefficient of vegetational homogeneity, 616
coefficient of velocity in germination, 124
coefficient of wetness, 401, 602
coelome, 476
coenanthium, 649
coenobiology, 389, 428
coenobiosis, 389
coenobium, 107, 246, 267, 388
coenocarpium, 263
coenocline, 388, 389
coenocytic, 117, 509
coenogametangium, 117
coenology, 389
coenomonoecia, 403
coenosis, 388
coenosium, 388
coenospecies, 250
coenzyme, 149
coenzyme A, 149
coevolution (co-evolution), 536
coextract, 172
cofactor, 149, 150
coffer dam, 134, 502
coffin, 178, 335
cog, 115, 542
cogged joint, 542
cogged rail, 375
cogged-scarf joint, 536, 542
cogging caulking, 268
cognation, 250, 483
cograft, 172
coherence, 292, 340, 527
coherent image radar, 527
coherent knot, 232, 292
coherent noise, 527
cohesion, 152, 294, 340, 344
cohesion failure, 244, 340
cohesion theory, 340
cohesive, 340, 344
cohesive action, 340, 345
cohesive end, 345
cohesive soil, 345
cohesive species, 340
cohort, 483
cohort life table, 483
coil, 526, 546
coiling, 264, 351, 415
coin larch, 248
coincidence, 29, 148, 563
coincidence index, 60, 172
coinsurance, 172
coir, 559
coke-oven coal tar, 242
coke-oven tar, 242
col, 3, 4, 413
cola, 273
cold adaptation, 286
cold-application creosote, 49, 287
cold bath treatment, 287
cold checking, 93, 110
cold-crack resistance, 336
cold cracking, 93, 110
cold curing, 93
cold current, 190
cold damage, 190
cold deck, 183, 286, 623
cold deck loading, 286, 623
cold desert, 190
cold dormancy, 93
cold endurance, 337
cold forest, 162
cold forest zone, 189
cold frame, 234, 286
cold-glued block, 286
cold gluing, 48, 286, 287
cold ground pulp, 286
cold groundwood pulp, 286
cold hardener, 287
cold hardening, 270, 287
cold hardiness, 337
cold injury, 190
cold junction, 286
cold moist storage, 93
cold pole, 190
cold-prepressed, 287
cold press, 287
cold pressing, 287
cold-requiring seed, 558
cold resistance, 270
cold setting adhesive, 286
cold setting glue, 286
cold skidding, 34, 295
cold snap, 190
cold soak(ing), 286
cold-soaking method, 286, 287
cold-soda pulp, 286
cold-soluble extract, 286
cold storage, 286
cold stratification, 93, 286
cold test, 93, 286
cold tolerance, 270, 337
cold treatment, 93
cold veneering press, 82
cold water cosmopolitan, 286
cold-water species, 286
coldness index, 190, 286
coleoptile, 352
coliform, 76
collagen, 241
collapse, 19, 280, 470, 631
collapse treatment, 134
collapsing hill, 19
collar, 167, 254, 353
collar borer, 167, 530
collar diameter, 167, 220
collar roof, 402
collarless circular saw, 511
collateral, 152
collateral accessory buds, 29
collateral bud, 29
collateral bundle, 29, 496
collateral security, 94, 152
collateral vascular bundle, 496
collecting channel, 224

collecting drain, 224
collecting place, 222
collecting yard, 217, 222
collection, 39
collection for others, 78
collective fruit, 263
collective tree farm, 224
collectively owned forest, 224
collector, 148, 442, 513
college forest, 243, 436
collenchyma, 200
collenchymatous cell, 200
colliery timber, 280
Collipress, 272
collipress process, 529
Collmann method, 272, 609
collodion, 215, 241
collodion cotton, 92, 215
collodion silk, 241
colloid, 241
colluvial, 19
colluvial deposit, 19
colluvial soil, 19
colluvium, 19
colonial breeding, 389
colonial insect, 388
colonization, 108, 224
colonization curve, 108
colonizing species, 269, 379
colonnade, 297, 635
colony, 224, 267, 389
colophany, 465
colophonium, 465
colophony, 465
Colorado white fir, 6
colour association, 405
colour composite, 40
colour composition of garden, 588
colour correction filter, 40
colour discrimination, 555
colour effect, 405
colour grades of rosin, 465
colour infrared film, 40
colour match, 555
colour matching, 359, 480
colour mixture, 555
colour negative, 40
colour perspective, 405
colour reaction, 524
colour reproduction, 40
colour retention, 15
colour reversal film, 40
colour reversion, 212
colour sapstain, 21
colour sensitivity, 40, 159
colour solid, 555
colour temperature, 405
colour transparency, 40
colour-washing tank, 516
colourant, 405, 640
colouration, 405
coloured paint, 405
colourimeter, 20, 405
colourimetric method, 20
colourimetry, 20, 405
colouring, 640
colouring agent, 640
colouring matter, 391, 405
Columbian pine, 204
column, 401, 636
column chromatography, 635
column hold-up, 635
column press, 635
columnar joint(ing), 636
comate, 72, 73
comb, 146, 399, 444, 566, 621
comb back chair, 444
comb-grained, 256, 444
comb knife, 444
comb structure, 444
combat forest fires, 133
combed joint, 444
comber, 358
combination backing, 153
combination edger, 294
combination furniture, 651
combination log rule, 652
combination of convection and infrared oven, 115
combination of forest and farm crop, 347
combination sale, 294
combination saw, 650
combination site indicator, 648
combination teeth, 153, 213, 651
combine drilling, 213, 628
combined cropping system, 377
combined division method, 648
combined dormancy, 648
combined frame work, area and volume period method, 606
combined method of thinning, 648
combined production, 648
combined selection, 652
combined system of field crop and coppice, 1
combined system of field crop and high forest, 377
combined transport bill of lading, 294
combined transport document, 294
combined trenching and transplant board, 268
combined water, 246, 446
combining ability, 353
combustibility, 273
combustible material, 569
combustion, 390
combustion phases, 390
combustion process, 390
combustion system, 390
combwebbing, 354
comeback, 212
comfort index, 444
command, 325, 619
command-directed economy, 277
commence, 269
commence growth, 269
commencing of growth, 269
commensal, 172, 484
commensalisms, 172
commercial acreage, 414, 577
commercial clear cutting, 415
commercial district, 415
commercial floriculture, 415
commercial forest, 367
commercial forest land, 415
commercial forestry, 415
commercial height, 273, 414

C

commercial plant, 170
commercial rotation, 252
commercial-sized timber, 415
commercial thinning, 415
commercial timber, 170, 252, 414
commercial timberland, 414
commercial tree species, 577
commercial veneer, 13, 415
commiscuum, 274, 596
commission, 78, 140, 577
commode, 61, 512
commode table, 61
commodity, 48, 217, 414
common apple, 363
common business oriented language, 482, 585
common dovetail, 365
common feeding ground, 172
common forest, 171
common form factor, 365, 503, 542
common foxglove, 318
common furniture beetle, 229
common grafting, 365, 377
common hyacinth, 146
common joist, 365
common jujube, 599
common mango, 318
common of estovers, 328
common of pasture, 135
common osier, 191
common peduncle, 649
common petiole, 649
common rafter, 365
common species, 48
common spruce, 348
common stock, 642
common walnut, 201
common(s), 484
communal, 388
communal courtship, 388
communal facility, 171
communal forest, 171
communal habitat, 389
communal land, 171
communication, 482, 540
communication console, 482
communication satellite, 482
communication technology satellite, 482
communities in littoral zone, 554
communities in the limnetic zone, 201
communities in the profundal zone, 420
community, 388, 418
community biomass, 389
community co-management, 418
community complex, 388
community composition, 389
community continuum, 389
community dynamics, 388
community ecology, 389
community ecosystem, 389
community equilibrium, 389
community forest, 171, 418
community function, 388
community gross prodution rate, 388
community indicator, 619
community integration, 389
community matrix, 389
community mosaic, 389
community organization, 389
community park, 259
community regulation, 388, 389
community respiration, 388
community ring, 388
community stability, 388, 389
community structure, 388
community succession, 389
community surface, 388
community table, 388
community type, 389
community value, 418
commutation of rights, 385
comoponent fractionation, 651
comose, 72
comospore, 260
compact crown, 322
compact fragipan horizon, 249
compact planting, 322
compact wood, 551, 621
compacted soil, 249
compacting, 608
compacting equipment, 604
compaction, 550
compactness, 232, 249
compactor, 75, 190, 608
companion cell, 11
companion crop, 236
companion cropping, 236
companion planting, 11
companion species, 11
company of limited liability, 581
company of unlimited liability, 511
company standard, 171
company union, 171, 587
comparative advantage, 19, 527
comparative anatomy, 19
comparative cost, 19
comparium, 342, 596
compartment, 165, 235, 298, 440
compartment boundary, 298
compartment cutting, 3
compartment felling, 3
compartment furnace, 235
compartment history, 298
compartment kiln, 143, 235, 630
compartment line, 298, 383
compartment management method, 299, 533
compartment map, 298
compartment register, 298

C

compartment stand and stock table, 290, 298
compartment system, 298, 305, 404
compartment under regeneration, 141, 169
compartment uniform method, 127, 386
compartmentalization, 383
compartmentation, 383
compass, 312
compass bearing, 312
compass man, 312
compass plane, 4, 384
compass plant, 619
compass rebate plane, 384
compass roof, 11
compass saw, 247, 384, 537
compass surveying, 312
compass table, 593
compass-timber, 384, 498
compatibility, 214, 232, 272, 353, 379
compatibility testing, 440
compatible, 379
compatible information system, 232
compatible reaction, 379
compatible taper equation, 563
compensating balance, 32
compensating grade, 606
compensating planimeter, 33
compensating volume, 32
compensation, 32, 360, 480, 532
compensation cylinder, 33
compensation depth, 33
compensation of employee, 177, 353
compensation point, 32
compensation temperature, 33
compensation trade, 32
compensator, 33, 360
compensator sheet, 360
compensatory financing, 32
compensatory fiscal policy, 33
compensatory predation, 33
competing tree, 257
competition, 257
competition coefficient, 257
competition curve, 257
competition density effect, 257
competition exclusion principle, 257
competition factor, 257
competition hypothesis, 257
competition index, 257
competition pressure, 257
competitive ability, 257
competitive advantage, 257
competitive association, 257
competitive bidding, 257
competitive capacity, 257
competitive coefficient, 257
competitive coexistence, 257
competitive displacement, 257
competitive exclusion, 257
competitive exclusion principle, 257
competitive inhibition, 257
competitive interaction, 257
competitor, 257
competive lottery, 61
comphene, 604
comphoric anhydride, 604
compilation, 23, 213
compile, 23
compiler, 23
compiling computer, 23
complemental reproductive, 33
complementary action, 201
complementary assay, 201
complementary base, 201
complementary base sequence, 201
complementary colour, 201
complementary factor, 201
complementary gene, 201
complementary pit, 201
complementary pollination, 150
complete barking, 498
complete canopy, 498
complete cultivation, 386
complete cutting, 244
complete diallel cross, 498
complete dominance, 498
complete enumeration, 320, 386
complete equational segregation, 498
complete equipment, 55
complete exploitation (felling), 244
complete failure, 385
complete fertilizer, 498
complete flower, 498
complete inhibition, 498
complete intermittent working, 236
complete leaf, 498
complete medium, 498
complete metamorphosis, 498
complete mint oil, 13
complete plant material, 498
complete set of equipment, 55
complete spare parts, 55
complete stand, 585
complete-tree utilization, 386
complete tree utilizer, 386
complete working co-operative society, 498
completed compartment, 385
completely automated heating and cooling greenhouse, 387, 421
complex, 153, 314, 391
complex tissue, 153
complex whip grafting, 153
complexity theory, 154

C

compliance, 274, 398, 399
complication, 29
component, 54, 142, 651
component of combustion, 390
components structure, 651
composite, 153
composite beam, 153
composite board, 152
composite coppice, 521, 623
composite crossing, 153, 154
composite factor, 153
composite fault scarp, 153
composite forest, 623
composite forest (method), 623
composite forest system, 1, 623
composite index number, 192, 648
composite lumber, 153
composite panel, 153
composite particle board, 153
composite photograph, 192
composite plywood, 153
composite rock, 224
composite system, 623
composite volcano, 153
composite wood, 153
composite wood-concrete beam, 329
composition, 651
composition board, 152
composition of furnish, 353
composition of garden, 588
composition rate by volume, 397
compost, 114, 213
compost bacteria, 114
compound annual rate of growth, 153
compound ball mill, 118
compound body, 153
compound coppice, 623
compound corymb, 153
compound cyme, 153
compound dichasium, 152
compound eye, 154
compound fertilizer, 153
compound frequency distribution, 153
compound geared donkey, 153
compound growth rate, 153
compound head, 153
compound inflorescence, 153
compound interest, 153
compound layering, 30
compound leaf, 154
compound (medullary) ray, 153
compound microscope, 154
compound middle lamella, 153
compound miter joint, 154
compound oil, 153
compound ovary, 153
compound pistil, 152
compound shake, 153
compound sieve-plate, 153
compound spike, 153
compound storied forest, 152
compound stratification, 213
compound stress, 153, 648
compound tariff, 153
compound tissue, 153
compounding, 353
comppensatory planting, 33
compreg, 251, 450
compregnated wood, 251, 450
comprehensive control, 386
comprehensive park, 648
compressed board, 550
compressed fibre board, 228
compressed wood, 573
compressibility, 274, 551
compression, 228, 322, 550
compression area, 443
compression coefficient, 551
compression debarker, 345
compression debarking, 345
compression deformation, 550, 551
compression failure, 551
compression hypothesis, 551
compression member, 443
compression method, 549
compression rate, 551
compression ratio, 550
compression set, 550
compression shear, 551
compression strain, 551
compression strength, 271
compression strength parallel to grain, 461
compression strength perpendicular to grain, 196
compression stress, 551
compression swelling, 550
compression test, 271
compression testing machine, 271
compression treatment, 550
compression wood, 443, 573
compressional vibration, 551
compressive force, 551
compressive fracture, 551
compressive strain, 551
compressive strength, 271
compressive stress, 551
comptroller, 420, 633
compulsory crossing, 376
computer, 101, 225, 226
computer aided design, 225
computer analog input, 226
computer assisted analysis, 225
computer compatible tape, 225
computer identification of hardwood species, 101
computer oriented language, 323
computer program, 225

C

computer simulation analysis, 225
computer truss estimating, filing and quoting system, 329
computerized linear programing projection model, 226
computerized wood identification, 330
concave, 4
concave plane, 4, 341
concave plasmolysis, 4
concave saw, 4
concavus, 4
concealed tenon, 4
concealing colour, 571
concede the land to forestry, 492
concentrated extract, 348
concentrated load(ing), 224
concentrated-solution-dip, 348
concentrated stress, 224
concentration, 224, 348
concentration factor, 155, 348
concentration gradient, 348
concentration yard, 635
concentric, 483
concentric exfoliation, 209
concentric included phloem, 483
concentric jointing, 483
concentric lamina, 483
concentric pore arrangement, 483
concentric ring, 483
concentric sample plot, 483
concentric stria, 483
concentric structure, 483
concentric vasculer bundle, 483
conceptual model, 156
concessional rate, 391, 577
concessional terms, 577
concrescent, 244, 587
concrete, 213, 251
concrete casing, 214
concrete community, 192, 219
concrete dam, 213
concrete form, 214
concrete of neflier, 356
concrete oil, 251, 257
concrete shuttering, 214
concrete slab, 214
concrete tie, 214
concretion, 246
concretionary laterite, 246
concretionary texture, 246
condemnation, 608
condensate, 286
condensate collecting vat, 286
condensate feed back, 286
condensation, 286, 346
condensed planting, 323
condensed tannin, 346
condensed tannin extract, 346
condensed water, 286, 346
condenser (dry) kiln, 286
condenser paper, 265
condensing (dry) kiln, 286
condensing factor, 346
condition class map, 301
condition of habitat, 423
conditional climatology, 440, 479
conditional constant, 479
conditional linear regression, 479
conditional multiple regression, 479
conditional reflex, 479
conditional response, 479
conditional sale, 152
conditioned reflex, 479
conditioning, 385, 479
conditioning chamber, 480
conditioning (treatment), 479, 480
conducting bundle, 445, 502
conduction, 66, 89
conductive tissue, 445
conduit, 178, 350, 456
conduplicate, 115
conduplicate cotyledon, 115
Condux mill, 269
condyle, 245, 272, 309
cone, 144, 146, 277, 382, 593
cone bearer, 382
cone bearing tree, 382
cone-bract, 382
cone breaker, 382
cone collector, 382
cone-cut veneer, 640
cone-cutting, 593
cone harvesting machine, 382
cone kiln, 382
cone-mill, 640
cone picking machine, 382
cone planting method, 640
cone rust, 382
cone scale, 186, 632
cone serotiny, 382
cone shaker, 382
cone-shaped training, 593
cone-sowing, 382
cone-type ring mill, 640
conelet, 534
conference of the parties, 100
conference table, 213
confidence interval, 622
confidence level, 622
confidence limit, 622
configuration, 173
confined aquifer, 18, 55
confining stratum, 32, 166
confirmed credit, 14
confiscatory taxation, 327
conflagration, 76, 387, 486
conflagration fire, 280
confluence basin, 213
confluent, 213, 612
confluent parenchyma, 263, 264
conformation, 173
conformity, 610
confronting, 115
confusant, 321
congealer, 286
congealing temperature, 110
congelation, 110
congeneric element, 483
congeneric species, 483, 484
congenial graft, 379

C

Conger method, 269
conglomerate, 54, 73, 117, 277, 291, 294, 469
Congo copal, 160
Congo pea, 332
conical bin, 640
conical refiner, 640
conidendrin, 481
conidial layer, 142
conidiophore, 143
conidiospore, 142
conidium (*plur.* -ia), 143
conifer, 607
Coniferales, 464
coniferophyte, 382, 464, 607
coniferous, 261
coniferous forest, 606
coniferous formation, 607
coniferous plant, 464
coniferous species, 607
coniferous timber, 607
coniferous tree, 464, 607
coniferous wood, 464, 607
conifruticeta, 606
conilignosa, 606
conisilvae, 606, 607
conivium, 166
conjoint bundle, 192
conjugant, 244
conjugate, 54, 172, 246, 294
conjugate cell, 292
conjugate parenchyma, 244
conjugate parenchyma cell, 292
conjugate-pinnate, 54
conjugate tracheid, 311
conjugated cellulose, 153
conjugation, 244
conjugon, 244
conjunctive symbiosis, 192
conjunctive tissue, 246, 292
conk, 332, 641
connate water, 478, 591
connecting curved cableway, 606, 623
connecting fittings, 292
connecting line, 292
connecting link, 292
connecting rod, 214, 292
connecting rod gear, 292
connecting rope, 292
connecting shaft, 292
connecting sleeve, 292
connectivity, 292
connector, 292
connnon dahlia, 77
conocarpium, 263
conometer, 106
conopodium, 593, 640
consanguineous mating, 250
conscious selection, 582
consequent drainage, 461
consequent succession, 570
consequential evolution, 570
consere, 482, 483
conservancy, 14
conservation, 14
conservation biology, 15
conservation culture, 425
conservation district, 14
conservation law, 14
conservation of nature, 645
conservation of wildlife, 559
conservative endemism, 565
conservative grazing, 14, 439
conservative management, 14, 224
conservative thinning, 14
conservator, 438
conservatory, 505
conserve water and soil, 14
consign, 227, 503
consigned forest, 503
consignee, 55
consistency, 61, 62, 344, 348, 528, 563
consistency coeffcient, 61
consistency controller, 61
consistency curve, 61, 528
consistency index, 62
consistency regulator, 61
consistency variable, 61
consistometric value, 62
consociation, 86
consocies, 86
consociule, 533
consortism, 172
conspecific, 484
conspicuousness, 525
constancy, 108, 195, 506
constant, 31, 48, 49, 195
constant cost, 31, 175
constant forest-labourer, 176
constant hybrid, 108
constant rate of drying, 92
constant species, 195
constant speed belt conveyor, 92
constant wind, 48, 506
Constantine log rule, 270
Constantine measure, 270
constituent, 54, 651
constitution, 300, 407
constitution isolation, 246
constitutive dormancy, 436
constitutive enzyme, 651
constraint, 594
constriction, 550, 569
constructing raft, 596
construction block, 223, 236
construction line, 223
construction timber, 236
construction timber for earth work, 489
construction timber for water-work, 456
constructional gluing, 121, 173, 652
constructive delivery, 491
constructive species, 236
consumer, 384, 531
consumer credit, 531
consumer demand, 531
consumer finance company, 531
consumer price index, 531
consumer sovereignty, 531
consummatory act, 498
contact adhesive, 65
contact agent, 65
contact angle, 244
contact area, 244
contact chemicals, 244
contact dryer, 244
contact heating, 244
contact herbicide, 244

C

contact insecticide, 244
contact poison, 244
contact print, 244
contact roll, 244
contact sander, 244
contact sticker, 244
contact thermometer, 244
contact transmission, 244
contact wheel sander, 3
contagion, 317
contagious distribution, 66
contagious population, 224
contain a fire, 502
contain fire, 215
container, 224
container-growing, 397
container-grown seedling, 397
container nursery, 397
container planting, 397
container seedling, 397
container veneer, 397
containerized nursery system, 397
containerset making machine, 397
containning method, 12, 146
contaminant, 508
contaminate, 508
contaminated zone, 508, 603
contamination, 103, 390
contemporary furniture, 88, 525
contest competition, 157
contest type of competition, 608
context, 267
contiental drift hypothesis, 77
continental, 77
continental air (mass), 77
continental bridge hypothesis, 310
continental climate, 77
continental drift, 77
continental forest, 77
continental glacier, 77
continental hemisphere, 310
continental island, 77
continental margin, 77
continental migration, 77
continental rift, 77
continental rise, 77
continental shelf, 77
continental slope, 77
continentality, 310
contingency correlation, 348
contingency table, 297
continuous areal, 293
continuous band hot press, 293
continuous barking drum, 293
continuous beater, 293
continuous bucket excavator, 295
continuous canopy, 387
continuous chain grinder, 295
continuous circulating cableway, 294
continuous community, 293
continuous cover forest, 195, 576
continuous cover forest management system, 195, 576
continuous cropping, 294
continuous cruise, 293, 294
continuous culture, 293
continuous curved cableway, 293
continuous deformed micelle, 293
continuous digester, 293
continuous (dry) kiln, 293, 373
continuous feed debarker, 293
continuous feeding, 293
continuous fibreboard dryer, 293
continuous-flow drier, 293
continuous forest, 195, 576
continuous forest inventory, 293, 410, 576
continuous forest inventory with permanent samples, 293, 576
continuous forest method, 576
continuous forming belt system, 293
continuous furrow planter, 293
continuous gluing, 293
continuous grazing, 293
continuous inventory method, 293
continuous inventory system, 294
continuous layering, 294
continuous line, 549
continuous loading, 294
continuous mist, 293
continuous multi-deck roller-dryer, 116
continuous one-deck roller-dryer, 82
continuous plankton recorder, 148
continuous prepress, 293
continuous press, 293
continuous quadrat, 293
continuous random variable, 294
continuous ribbon, 293
continuous sterilization, 293
continuous stream, 48
continuous strip camera, 293
continuous tree cover, 293
continuous (type) mixer, 293
continuous variable, 292, 293
continuous variation, 293
continuous wave, 293
continuously stable community, 294
continuously transporting cableway, 549
continuously working feed system, 293
continuum, 293, 389
continuum concept of vegetation, 616
continuum index, 293
contour cropping, 91
contour cultivation, 91
contour feather, 282
contour furrow, 458
contour-line, 91, 92
contour (line) map, 91

C

contour (line) road, 91
contour logging system, 91
contour map, 91
contour microclimate, 99
contour planting, 91
contour plowing, 91
contour ridge, 458
contour sander, 312
contour sod strip, 91
contour strip cropping, 91
contour tillage, 91
contour trench, 458
contouring, 91
contract authorization, 106
contract the project out to individuals, 5
contractile root, 372, 442
contraction, 442
contramarque, 21
contrast, 114, 129
contrast agent, 53, 601
contrast method, 114
contrast on photograph, 530
contrast ratio, 114
contrast test, 114
contrasting character, 527
contravention of forest-police, 502
contribution, 142, 143, 264
control, 115, 134, 277
control area, 115
control-book, 47
control bulb, 277
control console, 41, 277
control desk, 41, 277
control flap, 277
control line, 133
control method, 193, 232, 277
control of a fire, 216
control of mixture, 450
control of operation, 221, 253, 655
control of prescriptions, 435, 613
control of silviculture, 573
control piece, 115, 277
control plot, 115, 297, 381
control point, 277
control pollution, 277
control risk, 145
control skyline, 277
control stand, 619
control strip, 277
control threshold, 134
control time, 277
control unit, 277, 420
control-variate, 277
controlled-atmosphere storage, 370
controlled availability fertilizer, 50
controlled burn, 277
controlled burning, 277
control(led) burning, 277
controlled mosaic, 277
controlled pollination, 395
control(led)-pollination progeny test, 277
controlled prescribed burning, 225, 277
controlled-release fertilizer, 50, 209
controlled-release herbicide, 209
controller, 233, 277, 278, 420, 479
controlling gate, 246
controlling gene, 277
controlling kiln sample, 558
convection, 66, 115
convection column, 115
convection dryer, 115
convectional current, 115
convectional rain, 115
convenience, 171, 422
conventional competition, 481
conventional costing, 66
conventional furniture, 66
conventional kiln-drying, 48
conventional milling, 343
conventional open-loop setwork, 365
convergence, 148, 213, 383, 442
convergent adaptation, 383
convergent evolution, 383
convergent improvement, 263
convergent oscillation, 383
conversation algebraic language, 114
conversion, 23, 115, 227, 477, 637
conversion chart, 209
conversion cost, 637
conversion cut, 23
conversion factor, 600, 637
conversion per pass, 83
conversion period, 23
conversion plant, 227
conversion program, 637
conversion rate, 600, 620
conversion return, 227
conversion site, 600
conversion system, 23, 487
converted product, 227
converted timber, 227
converter, 637
convex plasmolysis, 486
conveyer, 66, 446, 596
conveyer (veneer) dryer, 293, 499
conveyor, 66, 446, 596
convivum, 96
convolute, 264, 546
conyolution, 264
cook vat, 609
cooker, 394, 609
cooking, 460, 609
cooking condition, 609
cooking cycle, 609
cooking degree, 609
cooking liquor, 609
cool house, 93, 286
cool intermediate house, 93
cool-white fluorescent lamp, 286
cool-white fluorescent tube, 286
cooling degree days, 286
cooling device, 286
cooling duct, 286
cooling tower, 286
cooling water, 286
cooling water pipe, 286
coom, 263, 320
coop and cask timber, 485
cooperage, 335, 620
cooperage bolt, 484
cooperage stock, 484

C

cooperating public agency, 536
cooperating sustained yield unit, 536
cooperation, 192, 536
cooperation farm forestry act, 347
cooperative, 192
cooperative forest management act, 192
cooperative forest management program, 408
cooperative hunting, 192
cooperative interaction, 536
cooper's-wood, 484
coordinate, 253, 353, 655
coordinate geometry computer program, 655
coordinate paper, 655
coordinates, 655
coordinates method, 253
coordinatograph, 655
copaene, 173
copaiba balsam, 278
copal, 271
copal oil, 175, 271
copal oil varnish, 174, 271
Copenhagen table, 165
coping range, 572
coping saw, 170, 526, 603
copolymer, 172
copolymerization, 172, 566
copper-coated shot, 484
copper pentachlorophenate, 512
copper pyridine, 20
copper-soap, 484
copper-tolerant fungi, 337
copperized boliden salt, 484
copperized chromated zinc arsenate, 484
coppice, 1, 321, 597
coppice conversion, 1
coppice coupe, 1
coppice forest, 1, 321
coppice-forest system, 1
coppice grown stock, 321
coppice-land, 1, 321
coppice method, 1
coppice of two rotations, 295
coppice selection forest, 1
coppice selection method, 1
coppice selection system, 1
coppice shoot, 321, 460
coppice sprout, 321, 460
coppice stand, 1
coppice system, 1
coppice system by clearcutting, 244
coppice with field crops system, 1
coppice with pollard, 247
coppice with reserve, 15, 122
coppice with standard method, 623
coppice with standards, 1, 321, 623
coppice-with-standards method, 623
coppice with standards system, 1, 623
coppice wood, 1, 321
coppicing, 1
copra, 559
copra oil, 559
coproduct, 154
coprolite, 144
copropalynology, 144
coprophagy, 436
coprophilous vegetation, 144
copse, 1, 321, 597
copse regeneration, 321
copsy, 180
copter logging, 615
coptisine, 210
copulation, 239, 240, 541
copulation chamber, 239
copulatory behaviour, 239
copy error, 154
copying lathe, 134
copying machine, 134
copying planer, 134
coral, 414
coral bleaching, 414
coral limestone soil, 414
coral reef, 414
corbel, 53, 295, 608
corcassian walnut, 201
corcule, 352
cord, 271
cord-measure, 45, 271
cord-tariff, 45
cord volume, 45, 120
cord volume unit, 45
cordage, 271, 431, 470
cording, 45
corduroy (road), 196, 333
cordwood, 45, 112
cordwood saw, 214
cordwood splitter, 356
core, 96, 160, 335, 538, 541, 554, 622
core area, 194
core blender, 539
core bracer assembly, 335
core composer, 539
core density, 539
core extractor, 344, 430
core layer, 539
core of dam, 94
core sample, 344
core spreader, 539
core stock, 539
core strip, 539, 622
core veneer, 539
core veneer splicer, 539
core wood, 468
core zone, 194
coreboard, 538
coreline, 539
coreline ratio, 240
coremium (*plur.* -ia), 12, 446
corer locking, 229
corer sticking, 229
coriaceous, 376, 530
coriander oil, 201, 587
coriarine, 316
cork, 334, 401, 452
cork block, 401
cork board, 334, 401
cork cambium, 334
cork-carpet, 334
cork filler, 401
cork oak, 334
cork product, 334, 401
cork sheet, 401
cork tissue, 334
corkscrew-growth, 313, 347
corky, 334
corky bark, 73
corm, 382
cormlet, 534, 641
cormus, 252

C

corn cob, 585
corn ear, 585
corn poppy, 584
cornean, 571
Cornell Peat-Lite Mix, 272
corneous, 242
corner, 46, 212, 503
corner bind, 94
corner block, 371, 402
corner brace, 242
corner chair, 21
corner cupboard, 21
corner cutting, 178, 351, 403
corner defect, 310, 469
corner knot, 285
corner locked joint, 444
corner man, 269
corner table, 21, 376
cornering, 178, 351
cornice, 136, 204, 415, 474
cornice plane, 202, 526
corol, 203
corolla, 203
corolla limb, 178
corolla throat, 203
corona, 154, 167, 178
corona discharge, 102
corpa, 254
corporation forest, 418, 491
corpus, 220, 565
corpus allatum, 553
corpuscle, 291, 501, 517, 535
corral a fire, 502
corral time, 502
correcting apparatus, 243
correction factor, 543, 544
correction factor method, 543
correlation, 114, 528
correlation character, 528
correlation-coefficient, 528
correlation index, 115, 528
correlation of growth, 430
correlogram analysis, 528
corridor, 223, 458, 481
corridor planting, 283
corrosion, 150, 378, 398
corrosion inhibitor, 150, 209, 270
corrosion prevention, 132
corrosion preventive, 133
corrosion testing, 150
corrosive effect, 150
corrosivene, 150
corrugated board, 30, 495
corrugated fastener, 30
corrugated nail, 41, 495
corrugated plywood, 495
corrugated sheet material, 495
corrugated surface, 495
corrugating medium, 495
cortex (*plur.* -ces), 339, 356, 448
cortical bundle, 356
cortical cell, 356
corticolous bryophyte, 449
corticose, 118, 448
cortification, 356
corvestrene, 528
corymb, 404
corymbose, 404
corymbose cyme, 404
corymbous, 404
cosere, 482, 483
cosmetic raw material, 206
cosmopolitan plant, 438
cosmopolitan (species), 140, 438
cosmopolite, 438
cosmopolitic species, 438
cosolvent, 374, 634
cost accounting, 53
cost analysis, 53
cost and freight, 53, 589
cost-benefit analysis, 54
cost-benefit ratio, 54
cost centre, 54
cost control, 53, 54
cost depletion, 53
cost-effective, 431
cost effectiveness, 54
cost-effectiveness ratio, 54
cost estimate, 53, 140
cost flow, 54
cost keeping, 53
cost of gene recombination, 220
cost of harvesting, 126
cost of hauling, 223
cost of mating, 239
cost of meiosis, 233
cost of planting stock, 324
cost of preparation of ground, 610
cost of reforestation, 600
cost of slash disposal, 613
cost of transport, 595
cost-plus-fixed-fee contract, 53
cost-plus method, 285
cost price, 53
cost record, 53, 54
cost-recovering, 441
cost residue, 504
cost value, 54
cost value of forest, 301
cost value of forest property, 406
cost value of forest soil, 298, 409
cost value of normal growing stock, 128
cost value of soil, 488
cost value of stand, 301
cost value of standing timber, 289
cost, insurance and freight, 90
cost, insurance, freight and commission, 90
costal-nerved, 623
costal vein, 373
costate, 260, 261
costene, 182
costing, 53, 173
costing procedure, 53
costs fixed per m.b.f., 320
costus root oil, 335
cosy corner, 21
cot, 103, 558
cotton asbestos, 434
cotton linters, 323
cotton seed hull, 323
cotton tree, 333
cotyledon, 641
cotyloid receptacle, 351
couch, 50, 67, 460
couch press, 147, 196
couch roll, 147
coujugate-palmate, 54
coulisse felling, 239
coulter, 288
coumaranone, 528
coumarin, 528, 557

C

coumarone resin, 19
count quadrat method, 225
counter, 225
counter balance, 32, 94, 360, 493
counter current air, 343
counter current type (dry) kiln, 343
counter-dam, 202, 271, 372
counter fire, 566, 573
counter-flap hinge, 18
counter-floor, 318
counter rope, 115, 353
counter-rotating double-stream ring refiner, 115
counter wedging, 454
counter weight jack, 56
counterbalance system, 110
counterbearing, 353
countercyclical action, 130
counterfort, 147, 172, 202
counterknife, 94
counterpart, 115, 353
counterpart expenditure, 115
counterpart funds, 114
counterpressure roll, 612
countervailing duty, 94, 129
counterweight, 353, 360
counting board, 225
counting device, 225
country kiln, 472
country of dispatch, 123
country park, 185
country-road, 185
country turned furniture, 527
county board of forestry, 97, 525
county forest, 383
county ranger, 525
coupe, 38, 127
coupe inventory, 127
coupe inventory record, 127
couple, 101, 288, 349, 353
couple-close, 376, 508
coupled action, 348
coupled log, 348
coupled respiration, 348
coupler, 52, 292, 294, 349
couplet on pillar, 574
coupling, 294
coupling agent, 348
coupling grab, 650
coupling series, 528
coupon, 152, 515, 602
course, 45, 186, 191, 310, 329
court, 171, 481, 499
courtship, 382
courtship behavior, 382
courtship behaviour, 382
courtship display, 382
courtship feeding, 382
courtyard, 481
covariance, 536
covariance matrix, 536
covariant, 536
cove, 339, 413, 535
cove moulding, 4
cove strip-cutting, 498
covenant, 371
cover, 300, 447
cover class, 156
cover crop, 139, 155
cover degree, 156, 577
cover fillet, 529
cover leaf, 12
cover-overheaded, 415
cover plant, 155
cover the country with trees, 315
cover type, 302, 616
cover type map, 302
coverage, 155
cover(age), 156, 577
covered land, 18
coverer, 155
coverimeter, 605
covering, 128, 155, 353
covering depth, 155
covering element, 155
covering harrow, 155
covering structure, 12
covering timber, 105
covering works, 155
covering works with fascine, 47
covert, 571
cow-day, 347
cow-month, 347
Cowan screen, 278
coxa, 220
coxa suture, 220
cozymase, 149
cpicranial, 485
crab, 103, 333, 350, 357, 506, 546, 633, 638
crack, 157, 269, 297, 298, 469
cracked wood, 269
cracker, 144
cracker pulp, 363
cracking, 269, 297
cradle, 23
cradle type delimber, 294, 612
craft raft clause, 333
crag, 377
cramped plasmolysis, 257
cranab, 108
cranberry wax, 198, 317
crane, 103, 198, 367, 421
crane car, 467, 647
crane carrier, 367
crane truck, 367, 647
cranemobile, 371
cranking saw, 384
craped wadding, 631
crassulacean acid metabolism, 255
crate, 8
crater, 216
crater lake, 215
cratering, 310
crawler, 314, 349
crawler crane, 314
crawler feller-buncher, 314
crawling, 349
craze, 298
crazing, 123, 501
crazy paving, 404
crbon storage, 473
creaming cutting, 601
creasability, 338
crease, 447, 596, 631
crease resistance, 271, 338
creasing strength, 338
creche, 287
credit, 81, 372, 540
credit card, 540
crediting period, 225
creek, 533, 535
creep, 75, 205, 317, 318, 399, 466, 544
creep compliance, 399
creep curve, 399
creep experiment, 399
creep flow, 399
creep limit, 399
creep of wood, 330
creep rate, 399

C

creep recovery, 399
creep relaxation, 399
creep resistance, 271, 399
creep rupture, 399
creep rupture test, 399
creep speed, 399
creep strength, 271
creep test, 399
creep-time curve, 399
creep unit, 399
creeper, 107, 349, 364, 640
creeping fire, 147, 317
creeping plant, 364
creeping stem, 364
creeping surface fire, 506
cremocarp, 454
crena, 592
crenate, 261, 592
crenatin, 397
crenel, 592
crenellate, 115, 261
crenium, 458
crenophilous, 440, 516
crenoxenous, 348, 496
creosote, 333, 596
creosote coal tar solution, 596
creosote emulsion, 596
creosote oil, 596
creosote-petroleum solution, 596
creosote process, 596
creosote wood, 596
crepe paper, 631
crepuscular, 210
crepuscular period, 53
crepuscular periodicity, 500
crescentic dune, 539
cresol resin adhesive, 230
crest, 5, 134, 146, 306, 446, 584
crest profile, 199
crest stage, 198
crested, 219
Cretaceous period, 5
cretanin, 397
crevasse of glacier, 28
crevice, 298, 356
crib, 23, 168, 283, 333, 349, 434, 533
crib bridge, 333
crib dam, 333
crib log, 81, 168
crib stacking, 403
crib test, 332
crib unstacker, 280
cribbing, 362
cribral parenchyma, 413
cribriform pit, 152, 413
cribwork, 316, 333
cricket, 328
cricket table, 403
crimp, 112, 161, 225, 264, 280, 631
crinkle, 631
crinkle iron, 30
cripple, 73, 243, 381
cripple tree limit, 2
criss-cross inheritance, 239
criss-crossed structure, 239
cristate, 219
criterion (*plur.* -ria), 619
criterion (*plur.* -rial), 25, 351
crithmene, 188, 465
critical altitude, 304
critical buckling stress, 304
critical compression ratio, 304
critical day-length, 304
critical depth, 304
critical experiment, 304
critical friction velocity, 304
critical height, 304
critical height sampling, 304
critical level, 304
critical load, 304
critical moisture content, 304
critical night length, 304
critical-path analysis, 304, 351
critical path method, 304, 351
critical photoperiod, 304
critical point, 304
critical pressure, 304
critical regions of biodiversity, 426
critical relaxation time, 304
critical temperature, 304
critical tractive force, 304
critical value, 304
critical velocity, 304
critical wind velocity, 304
crocin, 599
cron, 274
crook, 498
crooked cable way, 384
crooked log, 139, 498
crooked stem, 498
crooked tree, 498
crook(ing), 497
crop, 299, 301, 441, 466, 654
crop age, 300
crop density, 289, 299, 445
crop description, 299, 406, 409
crop establishment, 299, 600
crop expense, 39
crop fiber, 563
crop of formed tree, 236, 377
crop of small poles, 156
crop productivity potential, 654
crop rotation, 312
crop tree, 633
crop-tree thinning, 415
crop type, 302
crop year, 39, 441
cropland, 168
cropping system, 168
crops, 37, 298, 655
croquet chair, 612
cross, 238, 279, 596
Cross and Bevan cellulose, 274
cross arm, 195
cross band, 538
cross band veneer, 538, 622
cross band (veneer), 614
cross banding, 538, 614, 622
cross-beam, 196
cross bedding, 239
cross break, 195–197
cross-breaking strength, 271
cross-bred strain, 596
cross breeding, 596
cross check, 196, 204
cross compatibility, 596
cross conveyor, 197
cross current, 74
cross current type dry kiln, 196, 197
cross-cut guillotine, 630
cross-cut saw, 196

C

cross cutter, 196, 247, 600
cross cutting, 196, 600
cross-cutting saw, 196, 600
cross dimension, 195
cross direction forming head, 197
cross drain, 197
cross feed splicer, 196
cross-fertile, 596
cross-fertility, 596
cross fertilization, 567, 596
cross field, 238
cross field pit(ting), 238
cross fold, 104, 197
cross-fruitful, 596
cross-fruitfulness, 596
cross-grafting, 129
cross grain, 537
cross grained plywood, 196
cross-grained wood, 537
cross grinding, 196, 365, 630
cross-hair reticule, 433
cross incompatibility, 596
cross infection, 239
cross-infertile, 596
cross-infertility, 596
cross inoculation, 239
cross-laminating, 614
cross lap joint, 197
cross leg chair, 240
cross link, 239
cross link a chain, 442
cross linkage, 239
cross-linking agent, 239
cross-measurement, 238
cross member, 195
cross-over value, 239
cross piece, 195, 242
cross pile, 239
cross-piled loading, 196, 197
cross piling, 196, 197, 433
cross-pollination, 567
cross product, 46, 56
cross protection, 238
cross rail, 196, 566
cross-reaction, 239
cross-recovery method, 239
cross resistance, 239
cross-ride, 196
cross roll, 197, 537
cross section, 195, 196, 364
cross-section of a stand, 299, 300
cross-section rod, 195
cross-sectional area, 195, 446
cross-sectional area at breast height, 542
cross-sectional surface, 196
cross-sectioning, 196
cross seed orchard, 596
cross seedling, 596
cross-shaft (dry) kiln, 113, 197
cross shake, 255
cross-shaped budding, 433
cross-shooting, 239
cross skid, 364
cross sleeper, 47
cross slot, 196
cross sowing, 238
cross stacking, 196, 197, 238
cross-staff, 433, 614
cross sterile, 596
cross sterility, 596
cross stratification, 239
cross-stream impact mill, 197
cross-tongue, 196
cross tracheid, 256, 418
cross-unfruitful, 596
cross walk, 433
cross yarding, 197
crossability pattern, 273
crossbar, 195, 196, 367
crossbreed, 596, 597
crosscut end, 196
crossed classification, 239
crossed girdle, 239
crossed pit, 238
crosser, 53, 103, 166
crosser stain, 103, 166
Crossett Experimental Forest, 274
Crossett Lumber Company, 274
crosshaul(ing), 197
crossing, 538, 596, 622
cross(ing) ability, 273
crossing aperture, 239
crossing form, 596
crossing-over, 239
crossing sleeper, 47
crossing timber, 47
crossing veneer, 538, 622
crossing (veneer), 614
crossover, 65, 111, 141, 238, 247, 377
crossover suppressor, 239
crosstie, 184, 608
crosswise concaving, 377
cross(wise) cultivation, 197, 239
crosswise piling, 196, 197
crosswise sloped piling, 650
crotch, 380, 449
crotch chain, 140
crotch figure, 446
crotch grab, 140
crotch line, 79, 140
crotch line loading, 140
crotch swirl, 446
crotch tongue, 349
crotch veneer, 446
crotching, 269
crotovina, 479
crow, 507
crowded trees, 322
crowding, 575
crowding effect, 576
crown, 154, 172, 178, 447, 486
crown architecture, 448
crown-area index, 447
crown breadth, 178
crown bud, 178
crown budding, 178
crown canopy, 447
crown characteristics, 447
crown class, 429, 447
crown closure, 585
crown competition, 447
crown competition factor, 447
crown-contact, 585
crown count, 447
crown cover, 447
crown coverage, 447
crown-cut, 523, 524
crown density, 447, 585
crown density scale, 585
crown depth, 447
crown diameter, 178
crown diameter ratio, 178
crown diameter wedge, 448
crown dimension, 447
crown fire, 447

C

crown forest, 186, 499
crown form, 178
crown-formed wood, 447
crown gall, 167
crown grafting, 167, 178
crown height, 447
crown interception, 447
crown interception loss, 447
crown land, 171, 499
crown layer, 447
crown length, 448
crown length ratio, 178
crown mapping, 447
crown mean breadth, 361
crown mean diameter, 447
crown molding, 105, 309
crown openess, 300
crown percent (ratio), 447
crown post, 195
crown projection, 447
crown projection area, 447
crown projection diagram, 447
crown pruning, 415, 448
crown ratio, 447
crown root ratio, 178
crown rot, 178, 254
crown scorch, 447
crown size, 447
crown slope, 310
crown spread, 447
crown structrue, 447
crown structure, 447
crown surface, 447
crown texture, 447
crown thinning, 415
crown timber land, 499
crown wash, 178
crown width, 178
crown wire, 161
crown wood, 499
crowned, 260
crown(ing) fire, 447
crown(mean)width, 361
crownmeter, 447
crowntree, 172, 453
crowtit, 551
cruciferous, 261
cruciform, 433
cruck, 498
crude, 72, 477, 592
crude birth rate, 72
crude death rate, 649
crude density, 591
crude fiber, 73
crude Japan tallow, 423
crude mint oil, 13
crude oil, 592
crude peppermint oil, 13
crude persimmon tannin, 439
crude petroleum, 592
crude pine tar, 73
crude product, 9, 73
crude pyroligneous acid, 73
crude rubber, 428
crude tall oil, 72
crude tallow, 73
crude turpentine, 73
crude wax, 73
crude wood vinegar, 73
cruise, 44, 298, 471
cruiser, 408, 409
cruiser's stick, 44
cruising data, 104
cruising design, 104
cruising error, 104
cruising party, 269
cruising pattern, 104
cruising percentage, 471
crumb, 491
crumb structure, 491
crumpled, 351, 368, 631
crumpling resistance, 337
crush, 550
crusher, 363
crushing, 19, 549, 550
crushing by snow, 548
crushing limit, 271
crushing pressure, 550
crushing strength, 271
crust, 96
crust mor, 246
crust soil, 246
crustacean borer, 230
crustal movement, 96
cryofixation, 28
cryogen, 286
cryogenics, 93
cryophilous, 439, 516
cryophyta, 28
cryophyte, 28
cryoplankton, 28
cryoplanktophyte, 28
cryopump, 93
cryosphere, 28
cryostat, 93
cryotropism, 530
cryoturbation, 28, 110
cryptic species, 483, 571
crypto-humus, 571
crypto-mull, 317, 571
cryptobion, 571
cryptocarine, 200, 571
cryptoclimate, 440
cryptogam, 571
cryptogamic community, 571
cryptogamic wood, 571
cryptogram, 322
cryptojaponol, 308
cryptomere, 571
cryptomerione, 308
cryptomerone, 308
cryptometer, 605
cryptophyte, 571
cryptophytes, 571
cryptopore, 521
cryptostrobin, 607
cryptoxanthin, 571, 585
cryptozoa, 98
cryptozoic, 548, 571
crystal, 254
crystal acicular, 607
crystal cell, 189
crystal conglomerate, 72
crystal druse, 254
crystal dust, 254
crystal grain, 254
crystal grating, 254
crystal idoblast, 189
crystal locule, 189
crystal orientation, 254
crystal plane, 246
crystal repository, 254
crystal sand, 254
crystal sclerenchyma, 189
crystal styloid, 636
crystal unit cell, 254
crystalliferous cell, 189
crystalliferous element, 189
crystalliferous parenchyma, 189
crystalliferous xylary parenchyma, 189
crystalline area, 254
crystalline bloom, 310
crystalline cone, 254
crystalline core, 246, 254
crystalline material, 254
crystalline region, 246
crystallinity, 246
crystallite, 64, 254, 500
crystallizability, 273
crystallization, 246
crystallization tendency, 246
crystallize, 246

C

crystallized region, 246
crystallographic axis, 254
crystalloid, 254
crystalogeneous cell, 246
crystals in chain, 246
cubeb oil, 20
cubeba oil, 413
cubebhol, 20
cuber, 550
cubic-foot scaling, 289
cubic foot volume table, 289
cubic measure, 289
cubic meter of dense timber, 330
cubic meter of piled wood, 329
cubic meter of trunk wood, 330
cubic meter stacked, 45
cubic ton, 289
cubical content, 289
cubical rot, 279
cubing formula, 382
cubitus, 631
cuckoo-spit insect, 327
cuckoo wasp, 379
cucujid beetle, 23
cuesta, 84, 298
cuesta landform, 84, 86
cull, 92, 474, 475
cull factor, 92, 388
cull log, 92
cull lumber, 92
cull percentage, 139
cull tree, 34, 92
cull wood, 34
culled paper, 70, 475
cull(ed) seedling, 92, 297
culled standard, 547
culled wood, 21, 139
culler, 44, 547
culm, 276
culm base, 159
culm stalk, 159
culm yard, 632
culmen, 652
culti-anger, 622
cultigen, 597
cultivar, 597
cultivate, 629
cultivated cutting, 371
cultivated land, 168
cultivated landscape, 395
cultivated meadow, 396, 597
cultivated plant, 597
cultivated soil, 168
cultivated variety, 597
cultivation, 168, 353, 600
cultivator, 622
cultural control, 168
cultural cost, 353
cultural expense, 353
cultural investment, 597
cultural landscape, 506
cultural operations, 353
cultural park, 506
cultural practice, 149, 597
cultural tourism resource, 396
cultural treatment, 149, 597
cultural work, 149
culture, 353, 597
culture dish, 353
culture diversity, 506
culture medium, 353
culture of excised embryos, 287
culture solution, 353
cultures test, 353
Cumberland method, 269
cumin oil, 278, 640
cumulated temperature, 219
cumulative diffusion index, 285
cumulative error, 285
cumulative frequency curve, 285
cumulative growth curve, 285
cumulative number, 285
cumulative quantity discount, 285
cumulative selection, 285
cumulative volume tally sheet, 285
cumuliform cloud, 219
cumulonimbus calvus cloud, 486
cumulonimbus cloud, 219
cumulose soil, 151
cumulus cloud, 219
cuneus, 403, 536
cunit, 7
cuoxam lignin, 484
cup, 377, 444
cup frilling, 17
cup frills, 17
cup fungus, 351
cup shake, 201, 208, 312
cup system, 17
cuparene, 202
cuparone, 202
cuparophenol, 202
cupboard, 436, 499
cupped wood, 196, 495
cupping, 39, 197
cupping axe, 39
cuprammonium rayon, 484
cuprammonium silk, 484
cuprammonium solution, 484
cupressene, 7
cupressoid pit, 7
cupriammonium solution, 484
cupula, 377
curb the sand, 134
curbstone, 310, 554
curcuma, 237
curcuma oil, 237
curcumin, 237
curd, 346
cure, 176, 241, 308, 621
curf, 262
curing, 65, 176, 308, 446, 621
curing agent, 73, 176, 446
curing rate, 176
curing time, 176, 308, 446
curl, 30, 264
curly figure, 264
curly grain, 30, 264
curly top, 264
current accretion, 526
current and mean annual growth, 293
current annual, 292
current annual forest percent, 292
current annual growth, 292
current annual increment, 292
current annual rate of interest, 292
current annual uptake, 88
current annual value increment, 292
current annual value increment percent, 292

C

current annual volume growth, 292
current annual yield, 292
current asset, 306
current fire danger, 106
current fund, 306
current gross volume growth, 526
current growth, 88
current increment, 526
current investment, 112, 306
current leaf, 539
current liability, 112, 306
current meter, 307
current order, 526
current outlay cost, 525
current ratio, 306
current ruling price, 435, 438
current shoot, 539
current slash, 88
current value, 525, 526
current yield, 292
current yield system, 292
curtain coater, 304
curtain coating, 294, 336
curtain log scanner, 363
curve, 497
curve setting, 384
curve skidding, 49, 384
curve strip-cutting system, 384
curved cross-section, 384
curved culm, 498
curved dam, 384
curved plywood, 384
curved timber, 497
curvilinear, 384
curvimeter, 384
curving, 227
curving linear structure, 208
cuscus oil, 528, 554
cushion moss, 103
cushion plant, 103
cushion power saw, 233
cusp, 57, 231
custom-built truck, 108
custom-designed, 636
custom (kiln) drying, 48
customs duty, 178
cut, 25, 262, 274, 310, 378, 494
cut-and-mitred-string, 537
cut area, 38, 127
cut area method, 226, 323
cut-away view, 339, 364
cut flower, 377
cut fraction, 308, 578
cut full, 651
cut-grafting, 377
cut-in, 46, 228, 263, 378
cut log, 332
cut-off, 133, 193, 247, 377
cut-off and get-out, 283
cut off date, 247
cut-off drain, 224, 247
cut off saw, 196, 247, 601
cut-out and get-off, 283
cut-over, 38, 127, 244, 633
cut-over area, 244
cut-over forest, 633
cut over land, 244
cut-through, 298, 301
cutch, 121
cuticle, 27, 242
cuticula, 27
cuticular infection, 27
cuticular transpiration, 242
cuticularization, 242
cutin, 242
cutinization, 242
cutlery cabinet, 40
cutocellulose, 242
cutover, 37, 125
cutover area, 38
cutover forest, 38, 633
cutover land, 38
cutover nonrestocking, 38
cutover restocking, 38
cuttage, 371
cuttage grafting, 46
cutted land, 38
cutter, 88, 89, 126, 262, 380
cutter bar, 263, 295
cutter block, 88, 89, 516, 652
cutter cylinder chipper, 175
cutter disk, 88
cutter head, 88, 89, 516
cutter-lifter, 377
cutter link, 377
cutter mark, 378
cutter mill, 89
cutter sharpener, 74, 326
cutter spindle chipper, 88
cutter spindle flaker, 631
cutter tooth, 377
cutter-type particle, 16, 378
cutterhead barker, 546
cutting, 37, 46, 125, 262, 371, 378, 521
cutting age, 38
cutting and planting record, 38
cutting angle, 88, 378
cutting area, 38, 127
cutting assumption, 38
cutting auger, 127
cutting back, 247, 543
cutting bed, 371
cutting blank, 38, 263
cutting block, 127
cutting budget, 38, 39
cutting by group method, 279
cutting by group selection method, 279
cutting carriage, 88
cutting chance, 127, 223
cutting check, 547
cutting circle, 378
cutting class, 37
cutting compartment, 38
cutting consumption, 378
cutting control, 441
cutting cross, 378
cutting cutter, 378
cutting cycle, 211, 311
cutting date, 38
cutting direction, 38, 378
cutting disc, 592
cutting edge, 378
cutting edge displacement, 378
cutting force, 378
cutting form, 127
cutting gauge, 378
cutting graftage, 46
cutting grafting, 46
cutting height, 38, 125
cutting held by splinter, 331
cutting held by toothpick, 551
cutting in advance, 373
cutting in block, 279

C

cutting in strip, 80
cutting inarching, 46
cutting indicator, 39
cutting interval, 211, 311
cutting jig, 378
cutting leak, 149, 262
cutting limit, 39
cutting line, 127
cutting lip, 378, 652
cutting list, 39, 126
cutting-lopping-machine, 126
cutting map, 39
cutting material, 371
cutting medium, 371
cutting method, 39
cutting nick, 520
cutting of stand, 38
cutting of the culm, 632
cutting off, 269
cutting off centre, 357
cutting on centre, 631
cutting orchard, 39
cutting order, 39
cutting output, 378, 547
cutting percent, 37
cutting percent method, 38
cutting period, 39, 126, 225
cutting plan, 38
cutting planter, 46
cutting plate, 295
cutting plot, 371
cutting point, 521
cutting policy, 38
cutting practice, 39
cutting procedure, 39
cutting prohibited forest, 251
cutting propagation, 371
cutting quota, 39
cutting rate, 38
cutting ratio, 37, 38
cutting record, 38
cutting record of cultural operation, 353
cutting register, 37
cutting report, 38
cutting resistance, 378
cutting right, 38
cutting ring, 207
cutting rotation, 211, 311
cutting rotation age, 127, 633
cutting rule, 38
cutting saw, 247
cutting schedule, 38
cutting season, 38
cutting section, 37, 38
cutting sequence, 37
cutting series, 38, 39
cutting side-grafting, 46
cutting speed, 378
cutting strip, 37
cutting succession, 39
cutting surface, 57, 377
cutting system, 37, 39, 263, 378
cut(ting) test, 377
cutting throat, 90, 149
cutting tip, 378
cutting trajectory, 378
cutting treatment standard, 37
cutting unit, 37
cutting velocity, 378
cutting volume, 38
cutting wood, 46
cutworm, 377, 561
cyanide insensitive respiration, 381
cyanide resistant respiration, 270
cyanide sensitive respiration, 381
cyanidin, 204
cyanidine-3-glucoside, 204
cyanoethyl cellulose, 381
cyanoethylation of wood, 330
cyanogenetic glucoside, 424
cyanogenous plant, 424
cyathium, 17
cybele, 617
cybernetics, 277
cycad cycas, 466
cycle of erosion, 379
cycle of infection, 378
cycle of tending operations, 149
cycle time, 252, 630
cycles per minute (cpm), 630
cycles per second (cps), 630
cyclic, 207, 209, 312, 549
cyclic activation, 549
cyclic bond, 208
cyclic fluctuation, 630
cyclic timer, 312
cyclical outbreak, 630
cyclical parthenogenesis, 630
cyclical pattern, 549
cyclocaoutchouc, 207
cycloheximide, 135
cyclomorphosis, 630
cyclone, 370, 546
cyclone separator, 546
cyclonic precipitation, 370
cyclonic rain, 370
cyclophysis, 630
cycloscope, 638
cylinder, 264, 368, 593
cylinder bit, 593
cylinder bore, 593
cylinder door, 180
cylinder gluing machine, 184
cylinder grinder, 184
cylinder method, 593
cylinder saw, 485
cylinder (sieve) machine, 593
cylinder Yankee machine, 593
cylindrical, 614
cylindrical axis, 593, 625
cylindrical cutterhead, 593
cyma reverse, 30, 416
cyme, 264
cymelet, 534
cymo-botryoid, 649
cymo-botryose, 649
cymose, 264
cymose inflorescence, 485
cymose umbel, 264
cymule, 534
cyperon, 528
cyperone, 469, 528
cyperus oil, 469, 528
cypress oil, 7
cypress swamps, 7
cypsela, 292
cyrrhus, 264
cyst, 12, 13
cystidium (*plur.* -ia), 165
cystocarp, 338
cystolith, 626
cystolith cell, 626
cytidine, 13
cytidylic acid, 13
cyto-morphology, 518
cytobiology, 518
cytoblast, 517, 591

cytochemical staining, 517
cytochimera, 518
cytochrome, 518
cytochrome oxidase, 518
cytochrome peroxidase, 518
cytochrome reductase, 518
cytococcus, 443
cytode, 509
cytodifferentiation, 517
cytoecology, 518
cytogenetics, 518
cytohet, 13
cytokinesis, 13
cytokinin, 517
cytology, 518
cytometer, 548
cytomorphosis, 517, 518
cyton, 518
cytoplasm, 518
cytoplasmic gene, 518
cytoplasmic heredity, 518
cytoplasmic inheritance, 518
cytoplasmic male sterile, 518
cytoplasmic-polyhedrosis virus, 621
cytosine, 13
cytotaxonomy, 517
cytotype, 518
cytula, 192, 443
Czapek's medium, 47
D-grade thinning, 376
D-layer, 328
daconil, 7, 463
dacrene, 285
dacrydene, 285
dacrydioid pit, 285
dadap shield scale, 405
dado, 195, 202, 269
dado-and-lip joint, 195
dado-and-rabbet joint, 195
dado-and-tenon joint, 195
dado head, 268, 651
dado-head saw, 516
dado plane, 360
dado rail, 202
dahlia, 77
daily activity rhythm, 397, 631
daily compensation point, 397
daily cumulative temperature, 397
daily flow, 397
daily mean discharge, 397
daily output, 397
daily periodicity, 397, 631
daily precipitation, 336
daily production capacity, 397
daily range, 397
daily rhythm, 397
daily succession, 397
daily temperature, 48, 632
daily variation, 397
daily wage, 225
dalbergin, 210
dale, 515, 533
dalmon fly, 381
dam, 455, 460, 556, 636
dam board, 88, 602
damage, 414, 469
damage appraisal, 597
damage by caterpillars, 603
damage by drought, 190
damage by forest-insects, 60
damage by frost, 455
damage by fume, 553
damage by game, 559
damage by hail, 13
damage by hoar frost, 29
damage by insects, 60
damage by lighting strike, 414
damage by lightning, 414
damage caused by game, 559
damage from pests, 29, 580
damage from rats, 446
damaged knot, 469
damaged merchandise, 469
damaged tree, 18, 443
damaged wood, 469
Damascus barrel, 75
dammar, 75
dammar resene, 75
dammar varnish, 75
dammed basin, 556
dammed lake, 556
damp, 52, 433, 651
damp ground, 432
damp-proof fungicidal paint, 131
damp-proof paint, 131
damp rot, 432
damp-wood termites, 432
dampening machine, 602
damper, 144, 233, 370, 568, 651
damping, 211, 401
damping capacity, 651
damping-off, 73, 289
damping-off fungus, 73
damping stretch, 433
damsel flies, 110
dancing roll, 90, 604
dandy mark, 460
danger board, 216
danger class, 216
danger index, 216
danger line, 500
danger meter, 216
danger table, 216
Danish thinning, 81
daphne oil, 401
daphniphylline, 201, 240
dark adaption, 4
dark brown organic soil, 4
dark brown soil, 4
dark cell content, 420
dark cell inclusion, 420
dark conifer, 4
dark false heartwood, 420
dark-felling, 348, 521
dark field, 4
dark-field microscope, 4
dark forest soil, 4
dark germination, 4
dark germinator, 3
dark gray gleysolic soil, 3
dark-growth reaction, 4
dark heart, 4
dark reaction, 3
dark red latosol, 3
dark red soil, 3
dark respiration, 3
Darwin theory, 75
Darwinian fitness, 75
Darwinism, 75
Darwnism, 75
dashpot, 209, 233
dasyphyllous, 200, 260, 322
data, 450, 641
data acquisition, 450
data bank, 451

D

data collecting, 450
data collection system, 451
data compression, 451
data downloading, 451
data handling, 450
data library, 451
data logger, 450
data logging, 450
data processing, 450
data processing centre, 450
data processing equipment, 450
data processing machine, 450
data processing subsystem, 450
data reduction, 450, 540
data storage, 450
data transfer, 450
datatron, 450
date of maturity, 55
date of ripening, 55
date of seeding, 31
datiscetin, 531, 559
datum level, 161, 221
daughter cell, 641
daughter chromosome, 641
daughter nucleus, 641
daughter root, 641
daumen, 66
dauricine, 23, 413
davenport, 50, 535
davenport bed, 655
davenport table, 411, 520
day bed, 411
day degree, 397
day intermediate plant, 624, 625
day length, 631
day-length treatment, 631
day neutral plant, 181, 397, 623
day-night rhythm, 631
day-night timer, 631
day(-to-day) worker, 100, 225
daylight, 397
daylight clearance, 549, 550
daylight opening, 549
DBH distribution, 289, 301, 446, 542, 614
DBH-height function, 615
de-knotter ring, 384
deactivation, 115, 233
deactivator, 115, 233
dead-and-down, 90
dead-and-down tree, 278
dead and dying tree, 278
dead arm, 278
dead bark, 462
dead beating, 462, 629
dead branch, 278
dead centre, 462
dead-end of logging road, 224
dead face, 157
dead finish, 360, 509
dead fuel, 462
dead horizon, 462
dead knife, 94
dead knot, 462
dead lake, 462
dead level, 265, 362
dead-line road, 139, 462
dead load stress, 258
dead load(ing), 258, 647
dead lock, 362, 462
dead maturity, 498
dead oil, 132, 630
dead-piled, 322, 435, 436
dead plant part, 616
dead-ripe stage, 278
dead roller conveyor, 510
dead rollers, 258
dead shore, 53
dead size, 183, 462
dead soil covering, 278, 462
dead sowing, 511
dead stage, 278
dead standing tree, 278
dead timber, 278
dead-top, 278
dead tree, 462
dead trees, 278
dead weight, 258, 647
dead-wood, 52, 53, 278
deadener, 532, 545
deadening, 166
deadhead, 52, 53, 75, 278, 332
deadman, 319, 507
deadman anchor, 319, 507
deadning, 207
deadwood, 278
deafener, 532
deal, 200, 606
deal flame saw, 500
dealation, 493
deaminase, 493
deamination, 384
death feigning, 639
death mimicry, 639
death point, 621
death rate, 462
debacle, 19, 247, 413
debark, 30, 538
debarker, 31
debarking, 30
debarking abrader, 385
debarking machine, 31
debeardor, 64
debit, 248
deblade, 314
debouch, 193
debrancher, 76
debranch(ing), 75, 385
debris, 38, 40, 468, 554
debris barrier, 283
debris burning, 139, 597
debris cone, 59, 554
debris dam, 283
debris flow, 342
debris flow erosion, 342
debris slide, 554
debt limit, 603
debt service requirement, 603
debt-to-net-worth ratio, 151
debudding, 247, 327, 385, 603
decagynian, 433
decagynous, 433
decalcification, 493
decapetalous, 261
decaphyllous, 261
decapitation, 384
decaploid, 433
decarboxylase, 494
decarboxylation, 385
decasepalous, 261
decaspermial, 261
decay, 141, 150, 363, 451
decay control, 151
decay control by water spray, 354
decay durability, 337, 338
decay fungus, 151
decay hazard, 151
decay hazard classification, 150
decay of living tree, 289
decay period, 451
decay pocket, 151
decay product, 451
decay proof, 270, 337

D

decay requirement, 150
decay resistance, 270, 337
decay-resistant tissue, 270
decay scheme, 451
decay test, 151
decay type, 151
decayed knot, 151
decayed seed, 150
decayed timber, 151
deceiving colouration, 366
deceptive spectrun, 230
decibel, 140
deciduilignosa, 314
deciduous broad-leaved forest, 314
deciduous coniferous forest, 314
deciduous forest, 314
deciduous seedling, 314
deciduous shrub, 314
deciduous species, 314
deciduous tree, 314
deciduous wood, 281, 282
decimal digit, 433
decimal log rule, 433
decimal multiplier, 433
decimal rule, 433
deck, 156, 286, 323, 600, 623, 638
deck-bridge, 415, 471
deck chair, 605
deck layer, 26
deck loader, 183, 285
deck plank, 229, 377, 465
deck saw, 286
deck scaler, 286
deck stop, 286
decker, 62, 159, 183, 184, 348, 480, 494
decker loader, 480
decking, 229, 364, 471
decking aisle, 286
decking up, 114, 183
deckle, 106, 280, 620
deckle board, 106, 280
deckle box-method, 280
deckle frame, 106
deckle pully, 106
deckle strap, 106, 280
declared value, 419
declination, 58, 69, 357
decline, 363, 380, 451, 537
decline disease, 451
declining, 187, 231, 269
decoated seed, 385
decolourant, 494
decolouration, 494
decolourized shellac, 494
decolourizer, 494
decolourizing ability, 494
decolourizing carbon, 494
decolourizing char(coal), 494
decolourlzed (tannin) extract, 494
decomposed dung, 150
decomposed manure, 150
decomposer, 141
decomposing, 141, 150
decomposition, 141, 150
decompression timer, 238
decorative art, 639
decorative board, 638
decorative coating, 320, 639
decorative film, 638
decorative laminate, 639
decorative laminated fibre board, 639
decorative panel, 638
decorative paper, 639
decorative plant, 178, 639
decorative plaque, 638
decorative plywood, 639
decorative preservative, 639
decorative sheet(ing), 638
decorative veneer, 639
decortication, 30
decorticator, 30, 385
decoy, 319, 583
decreaser, 233
decreasing distribution, 99
decurrent, 521
decussate, 239
dedifferentiation, 129, 493
deduction for defect, 388
dee, 280, 292, 493
deep bead, 88
deep-columnar solonetz, 420
deep dormancy, 420
deep erosion, 419
deep percolation, 419
deep planting, 420
deep planting auger, 420
deep ploughing, 420
deep plowing, 420
deep-rooted, 420
deep rooted plant, 420
deep-rootedness, 420
deep scattering layer, 420
deep-sea bacteria, 420
deep-sea fish, 420
deep shake, 420
deep soil, 200
deep taproot system, 420
deep tillage, 420
deep water muck, 420
deep worm hole, 419
deepsea fishery, 420
deepwater formation, 420
deer, 310
deer damage, 310
deer foot, 173, 310
deer foot hook, 452
deer yard, 310
deerhound, 297
def(a)ecation, 481
defect, 387
defect deduction, 388
defect grading, 387
defect in growth, 429
defect system, 387
defected wood, 387
defective length, 388
defective log, 388
defective wood, 387
defectoscope, 161, 387, 388, 473
defended territory, 134
defense behaviour, 134
deferred income, 99
deferred-payment, 554
deferred stock, 554
defibering, 522
defibering machine, 522
defiberizer, 522
defibrater, 393, 522
defibration, 522, 523
defibration temperature, 523
defibrator, 393, 522
defibrator method, 393
defibrator second, 316, 393
defibrination, 327, 523
deficiency, 81, 280, 387
deficiency disease, 574

deficiency heterozygote, 387
deficiency homozygote, 387
deficiency symptom, 387
deficient chromosome, 40
deficient-duplicated chromosome, 387
deficient stocking, 289
deficit, 58, 280, 343
deficit financing, 58
defile, 2, 519
definite education, 107
definite inflorescence, 581
definite variability, 388
definite variation, 107
definition, 129, 140, 325, 381
deflaker, 287, 468, 469
deflation, 67, 145
deflect board, 357
deflecting block, 357
deflection, 338, 339, 357
deflection hood, 497
deflection nozzle, 357
deflectometer, 338
deflector, 88, 89, 357, 606
deflocculation, 130, 248
defloration, 314
defoamer agent, 532
defoliant, 314
defoliate, 314, 511
defoliation, 385, 494, 603
defoliator, 437
deforest, 213, 269
deforestation, 269, 283
deformation, 24, 221
deformed, 24, 221
deformed seedling, 221
deformity, 24, 221
defrother, 532
degassing, 135, 350, 383, 494
degeneracy, 234, 492
degenerate code, 234
degeneration, 24, 451, 492
degenerative character, 492
deglaciation, 28
degleyfication, 494
degradable pollutant, 273
degradation, 99, 238, 297
degradation of preservative, 132
degradation resistance, 270
degrade, 238
degraded allite, 24, 492
degraded chernozem, 492
degraded forest, 492
degraded soil, 492
degraded solonetz, 24, 492
degreasing, 64, 494
degree of anchor, 72
degree of atomization, 513
degree of automation, 643
degree of beating, 75
degree of bleaching, 358
degree of branching, 143
degree of closeness, 585
degree of closing, 585
degree of closure, 585
degree of contrast, 114
degree of cooking, 609
degree of crystallinity, 246
degree of density, 322
degree of dispersion, 142
degree of dominance, 525, 577
degree of drought, 156
degree of erosion, 59
degree of fineness, 518
degree of freedom, 647
degree of hardiness, 271, 337
degree of heritability, 565
degree of hydration, 456
degree of orientation, 107
degree of overlap, 60
degree of polymerization, 263
degree of purity, 256
degree of ripeness, 55
degree of risk index, 216
degree of similarity, 528
degree of stocking, 289, 445
degree of substitution, 384
degree of succession, 556
degree of swelling, 355, 398, 401
degree of tanning, 399
degree of thinning, 445
degree of viscosity, 344
degree of water erosion, 459
degree of wear, 327
degree of wetness, 401
degree of whiteness, 5
degree of wind, 145
degree of wind erosion, 145
degree Schopper-Riegler, 531
degressive mutation, 130, 183
degressive tax, 99
degummed ramie, 493
degumming, 493
dehiscence, 269
dehiscent fruit, 297
dehorn, 263, 269
dehorning, 247
dehumidification, 233, 385
dehumidifier, 158
dehusker, 30, 385
dehydrase, 494
dehydratase, 494
dehydration, 494
dehydrodicatechin, 494
dehydrogenase, 494
dehydrogenated rosin, 494
dehydrogenation, 494
dehydroxylation, 494
deinking, 494
deionization, 384
deionized water, 384, 493
delaminating, 287, 493
delamination, 30, 45, 140, 240, 367, 493
delamination strength, 30
delay, 481, 554
delayed density dependence, 554
delayed density-dependent factor, 554
delayed dominance, 57, 209
delayed effect, 554, 621
delayed elastic deformation, 209
delayed germination, 124
deletion, 387
delignification, 385, 494
delimber, 76
delimber and slasher, 76

D

delimber-bucker, 76
delimber-bucker-buncher, 76
delimber buncher, 75
delimber topper, 76
delimbing, 64, 75
delimbing gate, 76
delimitation, 205, 258
delineation on photograph, 530
delinquency, 213, 493
delinquent tax, 622
deliquescent branching, 72
deliquescent growth, 366
delitescence, 373, 571
delivered power, 445
delivered price, 239
delivery at landing, 222
delivery expense, 465
delivery hose, 445
delivery pressure, 349, 445
delivery season, 142
dell, 533, 535
delphinidin, 74, 136
delta, 403
delta deposit, 403
delta-wing balloon, 402
delta wood, 116
deltaic fan, 403
deluge, 16, 199
deluge fire control system, 217
deluging, 293
deluvium, 199, 363
demand, 516
demand for wood, 331
demand note, 222, 243
demand projection, 544
demand shifters, 544
demand-side management, 544
demanding, 516
demarcated forest, 14
demarcation, 205, 258
deme, 129, 468
dementholised peppermint oil, 13
demersal fish, 94
demeton, 341, 563
demi-circle, 44
demineralized water, 384, 493
demister, 64
demography, 389, 395
demonstration forest, 243, 436, 438, 439, 603
demonstration plot, 438
demulsify, 129, 494
demurrage, 584, 621
denature, 24
denatured alcohol, 24
denaturing agent, 24
dendricola, 449
dendriform, 450
dendritic drainage, 449
dendritic drainage pattern, 449
dendro-climatology, 344
dendro-entomology, 448
dendrochronograph, 344
dendrochronology, 344, 448
dendrocola, 449
dendroecology, 344
dendrograph, 448
dendroid(al), 448
dendrolite, 333, 449
dendrologist, 448
dendrology, 448
dendrometer, 44
dendrometric data, 44
dendrometry, 44
dendropathology, 446
dendrophilous, 227, 516
dendropoion, 392
dene, 411, 515
dengue Fever, 91
denitrification, 130
denitrifying bacteria, 130, 493
dens (*plur.* dentes), 57, 58
dense crop, 322
dense forest, 322
dense gneiss, 621
dense pasture, 224
dense planting, 323
dense powder, 348
dense sowing, 322
dense stand, 322
dense stocking, 162, 163
densification, 322
densified core, 322
densified laminated wood, 376, 575
densified veneer plywood, 551
densified wood, 322
densiflorene, 58
densify, 322, 550, 621
densimeter, 322
densitometer, 181, 322
densitometry, 181, 322
density, 32, 322
density class, 322
density-dependent, 322
density-dependent controls, 322
density-dependent factor, 322
density dependent mortality, 322
density-dependent theory, 322
density effect, 322
density-governing factor, 322
density-governing reaction, 322
density gradient, 322
density-gradient centrifugation, 322
density in situ, 591
density-independent, 137
density-independent controls, 137
density-independent factor, 137
density independent mortality, 137
density index, 322
density of crop, 299
density of crown cover, 447
density of forest stand, 299
density of infection, 378
density of initial stocking, 63
density of plantation, 600, 629
density of stocking, 289, 299
density of trees, 448
density of wood substance, 330
density ratio, 322
density slicing, 322
densometer, 322, 486
dentate, 58
dentate margin, 58
dentate ray tracheid, 58
dentate thickening, 58
dentation, 58
dentel, 58
denticle, 58
dentil, 58
denudation, 31, 313
denuded land, 313
denuded plantable land, 274
denuded quadrat, 566

D

denuded unplantable land, 34
deodar cedar, 548
deodar cedarwood oil, 548
deodarone, 548
deoxyerthrolaccin, 494
deoxysantalin, 494
Department of Forestry, 303
departure, 261, 357
dependable process technology, 55
dependent bacteria, 564
dependent community, 564
dependent crown fire, 506
dependent plough, 545
dependent variable, 569
depending block, 56, 598
depending sheave, 372
deperulation, 551
depigmentation, 494
depithing machine, 64
depithing mill, 64
depleted stand, 93
depletion, 191, 387, 451, 469
depletion allowance, 191, 469
depletion effect, 451
depletion mutant, 387
deposit, 53, 280
deposit feeder, 436
deposit-refund system, 15
depot, 126, 222
depoting, 635
depreciation, 23, 606
depreciation-annuity method, 343
depreciation-appraised method, 606
depreciation expense, 606
depreciation funds, 606
depreciation on replacement value, 61
depreciation reserve, 606
depreciation-reserve-ratio test, 606
depreciation-straight-line method, 362, 615
depression, 53, 92, 451, 495, 532, 651
depression angle, 149, 238
depressor effect, 233, 569
depth contour, 91
depth gauge, 420
depth gauge clearance, 57, 58
depth of field, 255
depth of penetration, 65, 179, 421
depth of sowing, 31
depth of work, 655
depthometer, 44, 420
depulper, 186
depulping, 64, 248
derail, 62, 493, 637
deratization, 324
derepression, 247
derivation, 555, 583
derivative, 121, 555
derivative rock, 89
derived air pollutant, 71
derived demand, 350
derived savanna, 555
dermatogen, 27
dermatoplasm, 13
dermatosome, 27
derno podsolic soil, 421
derrick, 470, 637
derrick barge, 147, 458
derrick crab, 637
derrick crane, 637
derris root, 584
desalinization, 494
descendance, 200
descending chromatography, 520
descending face, 520
descending grade, 520
descending inflorescence, 520
descending paper partition chromatography, 520
descending shot, 520
description of forest, 301
description of locality, 289
description record form, 104
descriptive statistics, 324
desert, 209, 411
desert advancing, 411
desert climate, 411
desert garden, 411
desert scrub, 209
desert soil, 209, 411
desert vegetation, 209, 411
desertification, 411
desertion of moisture, 494
desertization, 209
desiccant, 158
desiccated coconut, 158
desiccating wind, 156, 190
desiccation, 158, 385, 494
desiccation fissure, 158
desiccation-sensitive seed, 32, 157
desiccator, 158
design flood, 417
design in detail, 530
design roller, 184, 550
design roller coating, 184, 550
design test, 237
designed slope, 225, 417
desirable tree, 336
desired normal structure, 365
desired normal volume, 366
desk and bookcase, 81
desk-size electronic computer, 471
desolation, 209
desorber, 248, 493
desorption, 248
desorption isotherm, 248
desorption of moisture, 248
despot, 577
despotism, 423, 636
Dessemond process, 90
destcription of site, 96, 289, 429, 488
destructional landscape, 378
destructive cutting, 283
destructive distillation, 141, 157, 213
destructive distillation of wood, 329
destructive hydrogenation, 380
destructive lumbering, 283
destructive metabolism, 141
destructive test, 363
desulfurization, 494
deswelling, 492
desynapsis, 294

detachable maturity, 39
detachable time, 39, 272
detached truck, 410
detail, 305, 518, 587
detail drawing paper, 620
detail paper, 42, 324
detail part, 305
detail plan, 529
detail point, 468, 518
detail tracing paper, 324
detailed account, 461
detailed timber data, 529
detain, 277, 651
detanning, 494
detar, 493
detection air-ground system, 96
detection and attribution, 472
detection fixed point, 176
detection of forest fire, 407
detection point, 607
detector, 232, 472
detector element, 472
detention basin, 621
detention dam, 283, 621
detention storage, 621
detergent, 516
deterioration, 7, 24, 120, 238, 451
deterioration hazard, 24
deterioration indicator, 120
determinant, 108, 190, 264
determinate inflorescence, 581
determinate service, 176
determinate variation, 107
determination of cut, 38
determination of moisture content, 189
determination of provenance, 627
determination of volume, 36, 382
determination of yield, 441
deterministic model, 264
deterrents, 651
detrial food chain, 468
detrimental, 580
detrital deposit, 342, 468
detrital food chain, 150
detrital laterite, 40, 469
detrital rock, 469
detrition, 191, 327
detritivore, 437, 468
detritus, 151, 468, 554
detritus avalanche, 342, 554
detritus feeder, 437
detritus food chain, 151, 469
deuterogene, 199
deuteromycete, 11
deuteroparasite, 99
deuterosomatic, 598
deuterotoky, 47
deutoxylem, 199
devastated land, 209
devastated torrent, 209
devastation of forest, 408
developed land, 268
developer, 525
developer solution, 525
developing, 525, 603
developing paper, 525, 571
development, 125, 268, 525
development of a stand, 299, 300
development organ, 125
development period, 125
development phases of (natural) forests, 478
developmental branch, 125
developmental cycle, 125
developmental genetics, 125
developmental rate, 125
developmental stage, 125
developmental threshold, 125
developmental unit, 125
developmental variation, 125
developmental zero, 125
devernalization, 384
deviate at random, 468
deviation, 287, 357
deviation from average, 261
deviation from randomness, 468
deviation of grain, 506
devibrated handle bars, 134
devillicate, 143
Devonian period, 342
dew, 311
dew duster, 433
dew point, 311
dew point control, 311
dew-point temperature, 311
dewatering, 494
dewatering machine, 494
dewaxed decolourized shellac, 493
dewaxed shellac, 493
dewaxing, 493
dewing, 384
dewinger, 384
dexon, 94
dextral, 582
dextran gel, 364
dextran(e), 364, 582
dextrase, 364, 582
dextro-rotatory, 582
dextrorse, 582
dextrose, 364, 582
diabase, 211
diabatic, 66, 137
diacytic type, 196
diad, 83, 122
diadelphian, 123
diageic, 364
diageotropism, 197
diaglyph, 4
diagnosis, 237, 608
diagnostic horizon, 237, 608
diagnostic peak, 237
diagnostic species, 237
diagonal, 114, 537
diagonal grain, 114
diagonal member, 537
diagonal plywood, 537
diagonal wood, 537
diagram log rule, 487
diagram of tree, 289, 449
diagrammatic life table, 487
diagrammatic(al) view, 234, 487
diakinesis, 625
dial, 274
dial gauge, 274
dial indicator, 27
diallel cross, 453, 454
dialysis, 421, 486
diamemter growth, 139, 615
diameter at base, 220
diameter at breast height, 542

diameter at butt end, 78, 160
diameter at foot, 167
diameter at height of the eye, 336
diameter at middle, 625
diameter at small end, 417, 535
diameter at top end, 417
diameter band, 44
diameter breast-high (DBH), 542
diameter class, 255, 615
diameter class limit, 255
diameter class-specific sample tree, 255
diameter-classes method, 255
diameter classification, 255
diameter distribution, 614
diameter factor, 615
diameter gauge, 311, 614
diameter grade, 255, 615
diameter-grade interval, 255
diameter growth, 615
diameter-height curve, 255
diameter increment, 615
diameter-increment percent, 615
diameter inside bark, 385
diameter limit, 255, 521, 615
diameter limit cupping, 448
diameter-limit cutting, 255
diameter limit selection cutting, 255
diameter measurement, 615
diameter of mean basal area tree, 113
diameter of the central basal area tree, 625
diameter of the tree with the mean volume, 361
diameter outside bark, 79
diameter over bark, 79
diameter quotient, 615
diameter rounding, 615
diameter step-up method, 615
diameter tape, 502, 615
diameter under bark, 385
diameter with bark, 79
diameter without bark, 385
diametral pitch, 33, 255, 256
diamond impregnated circular saw, 375
diamond knife, 248
diamond lead, 43, 537
diamond line, 140
diamond matching, 305
diamonding, 305
diamter measurement, 614
dianthus garden, 434
diapause, 622
diaphan(o)us, 250, 485
diaphragm, 166, 326
diaphragm sprayer, 166
diapositive, 209, 485
diarch, 123
diaspore, 66, 404
diastase, 103
diastereoisomer, 136
diastrophism, 96
diatom, 183
diatom ooze, 183
diatomaceous earth, 183
diatomaceous silica, 183
diatomite, 183
diatropic, 197
diazinon, 122
diazo-, 629
diazo film, 629
diazo method, 629
diazo printer and developer, 629
dibasic, 122, 123, 568
dibber, 100, 495, 548, 564
dibble, 442, 547, 548, 616
dibble board, 495
dibble transplanting, 268
dibbler, 100, 495
dibbling, 100, 654
dicamba, 94, 317
dicarpellary, 260
dicatechin, 122
dice-square log, 131
dicentric chromosome, 455
dichasial cyme, 122
dichasium, 122
dichogamous, 70
dichogamous flower, 70
dichogamy, 70
dichotomous, 121, 122
dichotomous branching, 121
dichotomous cyme, 122
dichotomous key, 114, 121
dichotomous system, 121
dichotomy, 121, 122
dichotomy method, 122
dichroic grating, 123
diclesium, 467
diclinism, 70
diclinous, 70
dicliny, 70
dicollateral bundle, 29
dicotyledon, 455
dicotyledonous, 455
dicotyledonous tree, 455
dicotyledonous wood, 455
dicotylous, 455
dicotyls, 455
dicrotophos, 7
dicyclic, 452, 455
dicyme, 453
die-back, 278
die-hammer, 571
die-off, 265, 462, 631
die-square, 463, 622
dieback, 106, 417
diel, 563
dielectric, 101
dielectric coefficient, 248
dielectric constant, 248
dielectric constant moisture meter, 248
dielectric heating, 101
dielectric loss, 248
dielectric property, 248
dielectric strength, 248
Diels-Alder adducts, 94
diesel skidder, 47
dietetics, 574
diethyl-aminoethyl cellulose (DEAE-cellulose), 123
difference method, 243
difference of co-ordinate, 253
difference parallax, 439
differential adsorption, 46
differential colouration, 237, 390

D

differential density material, 567
differential gear, 46
differential host, 237
differential method, 500
differential quotient, 89, 501
differential reaction, 46
differential resistance, 475
differential species, 382, 435
differential swelling, 500
differential thermal analysis, 46
differential thermal analysis gas chromatography-mass spectrography, 46
differentially permeable membrane, 140
differentiation, 141, 237, 382, 500
differentiation center, 141
difffuse reflection, 317
diffhsion potential, 281
diffluence, 142
difflusivity, 281
diffraction, 391, 555
diffraction arc width, 555
diffraction diagram, 555
diffraction grating, 555
diffraction of X-ray, 418
diffraction pattern, 555
diffusate, 420
diffuse arrangement, 540
diffuse canker, 281
diffuse competition, 142
diffuse-in-aggregate parenchyma, 540
diffuse light, 318
diffuse parenchyma, 540
diffuse porous wood, 404
diffuse ray, 404
diffuse sky radiation, 477
diffuse-zonate parenchyma, 312
diffused light, 404
diffuser washing, 281
diffusibility, 281
diffusion, 281, 318, 321, 404, 420
diffusion gradient, 281
diffusion index, 281
diffusion phenomena, 281
diffusion pressure deficite, 281
diffusion process, 281
diffusion resistanee, 281
diffusion velocity, 281
diffusivity ratio, 281
diflubenzuron, 324
diflusion pressure, 281
digamety, 567
digamous, 296
digenesis, 438
digenetic, 296
digester, 251, 609
digester cycle, 609
digester silo, 184
digestibility, 532
digestive chair, 558, 638
digestive efficiency, 532
digestor lining, 609
digger, 494
digger blade, 494
digging auger, 495
digging machine, 495
digital computer, 451
digital display counter, 451
digital elevation model (DEM), 451
digital facsimile, 451
digital feed system, 451
digital image, 451
digital image processing, 451
digital indicator, 451
digital load cell, 102
digital positiometer, 451
digital response, 451
digital surface model (DSM), 451
digital tape recorder, 451
digital terrain mould, 451
digitalin, 318
digitate, 604, 619
digitinervate, 261
digitization, 451
digitizer, 451
digitoxose, 318
dihybrid, 454
dihydroeugenol, 122
dihydromyrcene, 122
dihydrophenanthrene, 122
dihydropyran, 122
dihydroquercetin, 122
dihydrorobinetin, 453
dihydrostilbene, 122
diisosafrole, 123
dika fat, 96
dike, 94
dilatation, 355
dilatation fissure, 355
dilatation joint, 355, 419
diluent, 515
dilutability, 274
dilute-solution-soaking, 515
dilution count, 515
dilution effect, 515
dilution end point, 515
dilution plate count, 515
diluvial soil, 199
diluvial upland, 199
dimension, 57, 111, 183, 296, 502, 570
dimension board, 183
dimension cutting, 255
dimension-lumber, 183
dimension saw, 519
dimension shingle, 183
dimension stock, 183
dimensional analysis, 570
dimensional stability, 476
dimensional stabilization, 57, 541
dimerous, 122
dimethoate, 284
dimethyl benzene, 122
dimethyl sulfoxide, 122
dimethylacetylamide, 122
dimethylformamide, 122
dimethylol urea, 266
dimictic lake, 452
dimilin, 324
diminished stile, 34
diminishing method, 99
diminishing returns, 442
diminishing returns in cutting cordwood, 540
dimorphantis, 71
dimorphism, 123
dimpled grain, 4, 562
dimyrcene, 122
dingy yellow horizon, 3
dining (room) suites, 40
dinitrophenol, 123
dinkey, 535
dioecious, 70
dioecism, 70
dioecy, 70
diominant-submissive, 612
diopter, 606
diorite, 414

D

dip, 251, 380, 521
dip angle, 380
dip cleat, 537
dip-coating, 251
dip iron, 177
dip-method, 251
dip section, 537
dip slope, 381
dipentene, 122, 454, 465
dipentene resin, 454
diphenyl oxide, 121
diphyllous, 260
dipinene, 453
diploid, 121
diploid gametophyte apomixes, 121
diploid organism, 452
diploidization, 121
diplonema, 454
diploparasitism, 123
diplophase, 121
diplophyte, 12, 121
diplosis, 18
diplospory, 121
diplotegia, 467
diplotene stage, 454
diploxylic, 454
dipped seat, 4
dipper, 48, 177, 193, 417, 442
dipping, 177, 251, 442
dipping axe, 265
dipping tank, 251
dipping treatment process, 251
dipterex, 94
diquat, 410
direct autogamy, 614
direct charge, 614
direct control, 614
direct cooking, 614
direct cost, 614
direct count, 614
direct-current property of wood, 331
direct cutting, 614
direct factor, 614
direct fire suppression, 614
direct fired dryer, 614
direct fired gas type kiln, 614
direct haulage, 614
direct infection, 614
direct light, 615
direct line, 107
direct method of suppression, 614
direct outlet, 614
direct painting, 614
direct penetration, 614
direct positive, 614
direct printing, 614
direct replica (method), 614
direct roll coating, 107, 614
direct runoff, 97, 614
direct sale, 614
direct seed machine, 614
direct seeding, 614
direct shear test, 614
direct skidding, 614
direct slip planting, 614
direct sowing, 614
direct steaming, 614
direct stratification, 614
direct stress, 614
direct tension, 614
direct-vernier, 461
direct work (*or* labor), 614
direct yarding, 614
directed application, 107
directed education, 107
directed hybridization, 107
directed mutation, 107
direct(ed) pressure, 107
directed variation, 107
direction finder, 107
direction finding, 107
direction indicator, 131
direction of check, 298
direction of fall, 90, 125
direction of felling, 125
direction of helix, 313
direction of regeneration, 169
direction of slope, 131, 363, 537
direction of spiral, 313
direction of tilt, 220, 380
direction of twisting, 347
direction of view, 178
directional felling, 90, 107
directional selection, 107
directional shrinkage, 107, 526
directive breeding, 107
directly heated dryer, 614
director, 107
dirt-catcher, 513
dirt chute, 489
dirt path, 489
dirt road, 489
dirt slide, 489
dirt work, 489, 610
dirtness, 52
dis-lodging, 603
disaccharides, 123
disafforestation, 125
disamenities, 32
disaster prevention forest, 134
disbark(ing), 30
disbarking machine, 31
disbranch, 64
disbudding, 247, 327, 385, 603
disbursement, 151, 612
disc, 88, 447, 592
disc and spindle sander, 593
disc equation, 297, 592
disc fans blower, 593
disc machinery, 593
disc saw, 593
disc type chipper, 593
discal cell, 624
discard, 139
discharge, 307, 349
discharge area, 187
discharge coefficient, 256, 307
discharge curve, 307
discharge end, 62, 156
discharge measurement, 307
discharge of ground water, 98
discharge onto the wire, 135
discharge outlet, 350
discharge site, 307
discharge spiked roller, 350
discharging device, 538
discharging expense, 538
discharging roll, 62
dischbank, 50, 170
disclimax, 357
discocarp, 641
discolouration, 23, 492
discolouration effect, 24, 492
discoloured heart, 24
discoloured wood, 24, 492
discomycetes, 351
disconcordance, 528
disconnected trolley, 122
disconnected truck, 410

D

discontinued operation, 481
discontinuity, 34
discontinuity layer, 235
discontinuous distribution, 235
discontinuous (growth) ring, 235
discontinuous phyllotaxy, 235
discontinuous sterilization, 236
discontinuous variation, 34
discopodium, 351
discount, 480, 606
discount factor, 480, 606
discount houses, 480, 606
discount rate, 480, 606
discounted cash flow, 525
discounted investment, 480
discounted marginal cost, 480
discounted marginal revenue, 480
discounted value, 480
discounted volume, 480
discrete, 287
discrete correlation, 287
discrete generation, 287
discrete method, 137, 141
discrete model, 287
discrete random variable, 287
discrete spectrum, 142, 287
discrete variable, 287
discretionary income, 647
discretionary purchasing power, 647
discretionary spending, 647
discriminant function, 351
discriminatory analysis, 351
disease, 29
disease complex, 29
disease dynamics, 29
disease-enduring, 336
disease-escaping, 21
disease forecasting, 29
disease forecasting model, 29
disease-free, 236, 508
disease-free certification, 508
disease index, 29
disease potential, 29
disease prediction, 29
disease reaction scale, 29
disease-resistance, 270
disease-resistant, 270
disease-treated seed, 531
diseased wood, 29
disengagement cutting, 64, 149, 247, 381
disepalous, 260
disforest, 125
disguised solonchak, 571
disharmonic biota, 35
disharmonic lake type, 35
dish(ing), 104
disiccation-tolerant seed, 337
disinfectant, 531
disinfection, 531
disintegrater, 144, 247, 522, 599
disintegration, 495
disintegrator, 144, 247, 522, 599
disintoxication, 247
disinvestment, 151
disjunctive area, 137, 235
disjunctive cell, 311
disjunctive distribution, 235
disjunctive parenchyma, 311
disjunctive parenchyma cell, 311
disjunctive tracheid, 311
disk, 447, 592
disk barker, 351
disk clearance indicator, 327
disk fan, 351
disk harrow, 593
disk mill, 351, 593
disk operating system, 69
disk planer, 351
disk plough, 593
disk refiner, 351
disk sander, 351
disk scarifier, 593
disk-type chipper, 351
disk type electrostatic sprayer, 351
disking, 593
dislocation, 24, 113, 504
dismantling raft, 47
dismutation, 366
disoperation, 378
disoperative competition, 528
dispatch sheet, 104
dispatcher, 104, 142
dispatcher's meter, 104
dispatching guide, 104
dispermy, 122
dispersal, 65, 404
dispersal unit, 66
dispersed soil, 142
dispersion, 142, 287, 288, 405
dispersion agent, 142
dispersion in marketing, 438
dispersive process, 142, 287
disphotic zone, 401
displacement, 504, 622
displacement activity, 384
displacement behaviour, 384
displacement bleaching, 622
displacement on photograph, 530
displacement washing, 622
disposal system, 139
disproportionated rosin, 366
disproportionation, 35, 366
disroot, 64
disrupt, 142
disruptive selection, 142
dissaving, 109
dissected bole, 600
dissecting, 600
dissection, 377, 600
dissemination, 31, 65
disseminule, 66
disseminule form, 66
dissimilation, 567
dissipative structure, 191
dissipative theory, 191
dissolved gum, 398
dissolved organic matter, 273
dissolved oxygen, 398
dissolved tar, 398
dissolving oil, 398
dissolving pulp, 206, 398

D

distal cell, 594
distal seed, 594
distance between hills, 548, 632
distance between rows, 190
distance-dependent individual tree model, 584
distance effect, 261
distance-independent individual tree model, 584
distance method, 235, 261
distance of dissemination, 65
distance-related, 261
distant crossing, 594
distant grafting, 594
distant hybrid, 594
distant hybridization, 594
distant view, 594
distichous, 122
distilled pyroligneous acid, 609
distillery, 600, 609
distinct variety, 475, 506
distorting disease, 221
distortion, 24, 221, 347, 431, 495
distortion on photograph, 605
distractile anther, 287
distrene, 94
distress selling, 280, 294
distributed knot, 404
distributed lag, 142
distributed load, 140
distributing disk, 140
distributing roller, 142
distributing scraper band, 140
distributing scraper chain, 140
distribution, 140
χ^2 distribution, 267
distribution area, 140
distribution coefficient, 142
distribution fringe, 140
distribution function, 140
distribution of age-classes, 305
distribution of cut, 38
distribution of plants, 618
distribution of stress, 572
distribution of three phases, 403
distribution pattern, 140
distribution range, 140
distribution screw, 313
distribution series, 140
distribution type, 140
distributional barrier, 140
distributor, 35, 89, 142, 349
district, 300, 382, 431
district administration, 300
district (forest-)office, 302, 573
district forester, 302
district of forest-ranger, 300
district of forest-warden, 300
district of guard, 14
district officer, 573
district park, 382
district-ranger system, 202
disturbance agent, 157
disturbance climax, 357
disturbance indicator, 120
disturbance regime, 157
disturbed profile, 391
disutility, 130
ditch, 172, 383
ditch cutter, 268, 494
ditch irrigation, 173
ditch planting, 173
ditch plough, 268
ditcher, 268, 494
ditching, 268, 543
ditching blade, 268
ditching machine, 268
diterpene, 454
dithane Z-78, 78
ditopogamy, 204
ditreme, 453
diurnal animal, 631
diurnal change, 631
diurnal fluctuation, 631
diurnal inequality, 397, 631
diurnal migration, 631
diurnal periodicity, 631
diurnal temperature range, 631
diurnal variation, 631
diurnalism, 631
diurnation, 631
diuron, 94
divagation, 130
divalent, 122
divarication, 140, 143, 269
divergence, 383
divergent adaptation, 383
divergent allele, 141
divergent angle of check, 297
divergent oscillation, 383
diverginervius, 148
diversification, 119
diversified disposal, 141
diversified farm, 120
diversified slash disposal, 127
diversity, 119
diversity gradient, 119
diversity index, 119
diversity of species, 513
diverter, 142, 491, 538
diverting system, 142, 491
diverting water for sluicing sand, 570
Divi-divi, 595
divide, 143
divided leaf cutting, 10
dividend, 175, 197
dividend in kind, 436
dividend yield, 175
diving duck, 374
division, 77, 143, 216, 298, 320, 432
division into annual coupes, 383
division into compartment, 298, 408
division into regeneration areas, 383
division of forest area, 408
division of forestry, 303
division of labour, 141
divisional forest-office, 406
divisional officcr, 432, 573
dlagenesis, 55
dlffusion coefficient, 281
doat, 150, 451
dock man, 638
dockage, 37, 330

docker, 196, 247, 638
docking, 247
docking saw, 196, 247
doctor, 177, 363, 543, 610
doctor blade, 177
doctor roll, 106
doctrine of forest rent, 304
doctrine of soil rent, 99
document against acceptance, 476
document against payment, 151
documentary basis, 506
documentary bill, 168, 582
documents against payment, 612
dodder, 491
dodecaploid, 433
doe, 69
doffer finger, 619
doffing roll, 314
dog, 5, 52, 88, 105, 173, 401, 572
dog board, 572, 596
dog eared, 264
dog hole, 267
dog iron, 5, 316
dog knocker, 367
dog leg crane, 384
dog spike, 173
dog warp, 79
dog wedge, 79, 536
dogger, 106, 178
dogger jaw, 371
dog's work, 65, 316
dolabradiene, 312
dolabrine, 312
dolabrinol, 312
dolbeer, 609
doldrums, 58
dolichocephalic, 50
dolina, 434
dolly, 1, 89, 103, 111, 113, 534
dolomite, 6
dolomitite, 6
dolomitization, 205
dolostone, 6
dolphin, 227, 389
dolphin foot, 188
dolphin hinge, 188
domain, 48, 305
dome, 172, 592
dome bed, 592
dome-like structure, 172
domestic fungus, 229
domestic timber, 185
domesticated alpine animal, 163
domestication, 549
dominance, 525, 577
dominance hierarchy, 577
dominance hypothesis, 525
dominant, 577
dominant character, 525
dominant crop, 577
dominant crop trees, 577
dominant feature, 525
dominant gene, 525
dominant height, 577
dominant lethal, 525
dominant mutant, 525
dominant portion of forest, 633
dominant species, 577
dominant stem, 577, 612
dominant-submissive, 612
dominant tree, 51, 220, 577
dominant tree layer, 577
dominant tree species, 577
dominant variance, 525
dominant year class, 577
dominated crop, 71, 154
dominated stem, 297
dominating stem, 51, 577
dominigene, 525
dominule, 501
donkey, 242, 564
donkey float, 639
donkey mover, 639
donkey skidding, 243
donkey snubber, 470, 520
donor, 171
donovan unloader, 637
doodle bug, 566
door carrier, 321
door case, 320
door cheeks, 320
door frame, 320
door lining, 320
door panel, 321
door post, 321
door sill, 320
door skin, 320
door strip, 320
dope, 40, 570
doping agent, 47
Doppler effect, 118
Doppler location, 118
Doppler radar, 118
Doppler ranging, 118
dormancy, 108, 543, 606
dormancy-breaking measure, 363
dormancy form, 543
dormancy-inducing mechanism, 543
dormancy-maintaining mechanism, 543
dormancy onset, 543
dormancy release mechanism, 543
dormant bud, 543
dormant cutting, 543
dormant eye, 543
dormant period, 543
dormant season, 543
dormant seed, 543
dormant spore, 543
dormant state, 543
dormant volcano, 543
dorsal, 17, 18
dorsal line, 18
dorsal suture, 18
dorsal vessel, 18
dorsi-ventral leaf, 567
dorsiventrality, 18
dorsopleural line, 18
dorsum (*plur.* dorsa), 17, 18
dosage bunker, 225, 353
dosage effect, 226
dose, 226
dose-response relationship, 226
dosing band, 225, 353
dosing bin, 225, 353
dosing line, 636
dosing silo, 225, 353
dot grid, 499
dot grid sample, 499
dote, 150
doted tree, 151
doty, 150
doty wood, 151, 320
doub bole, 452
double-acting door, 454
double-action shear, 452
double (annual) ring, 152, 453
double arbor gang, 455
double arbor resaw, 455
double back, 454
double-back tape, 453
double band (saw) mill, 452
double bar gauge, 452

double bark thickness, 453
double barrel shotgun, 454
double-base powder, 452
double-bead, 454
double block, 453
double block edger, 453
double bridging, 233
double budding, 61, 123
double carriage, 455
double circular (saw) mill, 454
double-column bandmill, 455
double-conical saw, 454
double-coupler, 463, 563
double cross, 455
double cross grain, 154
double crossing over, 453
double crossover, 238
double-crotch grab, 455
double-curvature arch bridge, 453
double cut-off saw, 153, 453
double cut twin band-mill, 453
double-cutting band saw, 453
double-cutting bands, 453
double cylinder machine, 454
double deck fairleader, 454
double-declining-balance, 227
double desk, 453
double-diffusion treatment, 455
double-diflusion process, 453
double dish, 353
double dividend, 455
double dormancy, 455
double dovetail key, 453
double dray, 452, 453
double-dryer, 452
double edge cutting band resaw, 454
double edge knife, 453
double edger, 454
double end tenoner, 454
double end tenoning machine, 454
double endless cableway, 454
double endless system, 454
double-entry bookkeeping, 153
double fertilization, 455
double float, 152
double flower, 60
double-frill girdling, 454
double gang edger, 455
double gate leg table, 454
double grafting, 123, 455
double-handed saw, 453
double-headed nail, 5, 316, 454
double headed spray gun, 454
double header, 117, 452
double heart, 454
double heart-wood, 153, 454
double heterozygote, 455
double-hook detacher, 453
double hung sash window, 453
double hybrid, 453
double-image photogrammetry, 454
double jawed trap, 60
double knot, 47
double line, 455
double-line incline-way, 454
double line system, 454
double margined door, 452
double match joint, 454
double membrane, 452
double mill, 452
double monosomic, 452
double mould board plow, 452
double of multiple annual ring, 153
double pass rotary dryer, 452
double pith, 454
double plane iron, 452
double point sampling, 455
double-pointed hoe, 296
double Poisson distribution, 455
double profile, 60, 123
double quirk-head, 452
double-rack sled, 453
double refraction, 455
double revolving disc refiner, 453
double ring, 453, 503
double roll dissolution head, 452
double roller bar, 452
double roller machine, 452
double-rope cableway, 454
double-row drilling, 452
double-row planting, 452
double Rueping process, 121
double samara, 452
double sampling, 455
double sampling technique, 455
double sampling with regression, 455
double sap wood, 339
double saw, 453
double set saw, 452, 454
double shearing strength, 452
double-shield budding, 452
double-sided sprayer, 452
double sides planer, 453
double sizer, 453
double sizing, 453
double slackline yarding, 454
double sled, 454
double sledge, 453
double spindle lathe, 453
double spindle shaper, 455
double spread(ing), 453
double standing line, 454
double stream mill, 453
double summit, 153, 154
double surface planer, 453
double swage saw, 452, 454
double taxation, 455
double teeth saw, 452
double tenons, 454
double the hill, 452
double-tongue graft, 61, 454
double-topped tree, 452
double-tree, 29, 453

D

double trigger, 452
double trisomic, 454
double trunk, 452
double vacuum cycle, 455
double vacuum process, 455
double wall, 60
double wavy grain, 452
double-winch tractor, 453
double-wire tramway, 454
double-work, 123, 455
double yard, 122, 141
doublet, 54, 102, 154, 348, 452, 455
doubling chromosome, 227
doubling time, 18
Douglas fir, 204
Douglas fir oil, 204
douglas-tree, 204
dove, 165
dove tree, 172
dovetail, 258, 469, 556
dovetail angle, 258
dovetail bit, 258
dovetail cutter, 258
dovetail groove, 258
dovetail joint, 258
dovetail lap joint, 258
dovetail saw, 258
dovetail slot, 258
dovetail tenon, 258, 556
dovetailed halving joint, 10
dovetailed housing joint, 258
dovetailer, 258
dovetailing machine, 258
Dow process, 90
dowel, 4, 46, 229, 452, 469
dowel bar, 532
dowel bit, 57, 417
dowel edge joint, 469
dowel hole boring machine, 332
dowel inserting machine, 469
dowel pin, 4, 296
dowel saw, 278, 452
dowelling joint, 375, 532
dowelling machine, 332
down, 112, 113
down budding, 521
down dead wood, 278
down draft, 520, 521
down-hill clevis, 620
down log, 53
down milling, 461
down payment, 106, 443
down-stroke hotpress, 521
down-timber, 125, 278
down time, 177, 481, 507, 521
down tree, 90
downgrade, 238, 521
downhill cable logging, 521
downhill north bend system, 17
downhill (side), 413, 521
downhill skidding, 67, 461, 521
downhill timber, 521
downpipe, 289, 520
downstream, 461, 521
downward grade, 461, 520, 521
downwind, 461, 520
downy mildew, 455
Doyle log rule, 119
Doyle-Scribner log rule, 119
Doyle tog rule, 119
dozer, 492
dozer boat, 492
dozer rotter, 492
dozy, 63
drab clay, 4
drab soil, 194
dracoalban, 309
draconis sanguis, 548
dracoresene, 548
dracorhodin, 548
dracorubin, 548
draft, 61, 282, 369, 372, 481, 513, 553
draft chair, 213
draft control hitch, 282
draft fire, 566
draft sled (*or* sledge), 598
drag, 16, 88, 284, 360, 372, 453, 493, 620, 630
drag cart, 453
drag chute, 316
drag conveyer, 177
drag hole, 493
drag line, 493
drag-lizard, 223, 493
drag road, 212, 222
drag saw, 201
drag sled, 372
drag spot, 237
drag turf, 604
dragger, 46, 165, 651
dragging, 97
dragging cap, 223
drag(ging) chain, 493
dragging dog, 223
drag(ging) in, 222
dragging pin, 223
dragging rack, 196
dragging shoe, 223
dragging slip, 223
dragging track, 196, 222
dragon fly, 381
dragon spruce, 595
dragon's blood, 309
dragsaw, 201, 493, 500
drain, 191, 350, 531
drain digger, 268
drain flux, 350
drain pipe, 350
drain port, 538
drain tank, 350
drainability, 316
drainage, 350
drainage area, 308, 350
drainage basin, 308, 350
drainage ditch on slope, 413
drainage divide, 143, 308
drainage error, 94
drainage hole, 350
drainage line, 308
drainage map, 308
drainage ratio, 256
drainage system, 350
drainage terrace, 247
drainage time, 350
drainage tray, 350
drake, 171
drapery figure, 613
drapery pattern, 613
drastic thinning, 222, 376
draught, 61, 282, 369, 372, 481, 513, 553
draw-bore, 652
draw hoe, 64, 348
draw knife, 177
draw shear, 282
draw table, 282
drawbar, 372
drawbar pull, 372
drawbar skidding, 372
drawer, 61
drawer back, 61
drawer bottom, 61
drawer front, 61
drawer guide, 61

D

drawer runner, 61
drawer side, 61
drawer slide, 61
drawing board, 213
drawing box, 61
drawing diagram, 487
drawing room suites, 274
drawing-table, 213
drawshave, 177
dray, 223
dray road, 65, 223
dredge, 284, 495
dregs, 315
drepanium, 294
dress, 543
dress case, 564
dressed and matched, 16
dressed board, 16
dressed four sides, 463
dressed lumber, 16
dressed one edge, 562
dressed one side, 563
dressed one side, one edge, 563
dressed one side, two edges, 563
dressed pole, 385
dressed timber, 16
dressed two edges, 295
dressed two sides, 295
dressed two sides, one edge, 296
dresser, 437, 543, 610
dressing, 11, 27, 141, 381, 431, 639
dressing table, 610
dri-vac tank, 158
dribbling, 94
dried blood, 158
dried fig moth, 143, 156
dried flower, 157
dried joint, 157
dried-out peatland, 157
dried top, 278
drier, 158
drift, 28, 191, 357, 358, 621
drift bed, 28
drift boulder, 358
drift clay, 28
drift dammed lake, 28
drift deposit, 28
drift epoch, 358
drift-road, 307
drift sand, 136, 307
drift terrace, 28
drift topography, 28
drifting beach, 564
drifting epoch, 28
drifting organism, 358
drifting sand waste, 307
driftwood, 358, 621
drill, 190, 309
drill bit, 652
drill-harrow, 190
drill-hoe, 190, 479
drill machine, 31, 479
drill planting, 190, 366, 479
drill roller, 549
drill seeder, 479
drill seeding, 366, 479
drill sowing, 479
driller, 479
drilling, 31, 350, 479
drilling machine, 652
Drilon process, 81
drimenene, 33
drimenone, 33
drinking fountain, 571
drip irrigation, 94
drip-line, 94, 184
drip-watering, 94
dripstone, 94
drive, 66, 84, 159, 307, 383
drive flank, 383
drive force, 383
drive link, 383
drive lug, 66
drive shaft, 383, 633
driven shaft, 71
driven wheel, 71
driver, 307, 383, 461
driveway, 52, 223, 430
driving effort, 383
driving road, 307
driving sprocket, 383
driving station, 470
driving wheel, 383, 632
drogue, 147
drone, 510
drop, 640
drop-leaf hinge, 215
drop-leaf table, 215
drop line, 110, 638
drop log, 103, 602
drop method, 94
drop pocket, 238
drop shot, 640
drop siding, 75, 201
drop softening-point method, 400
drop sorter, 238, 629, 640
drop sorter and stacker, 629
drop trimmer, 367
drop tube, 446
dropped seat, 4
dropping, 31, 350
drought, 156, 157, 190
drought crack, 157, 599
drought damage, 190
drought dormancy, 190
drought hardening, 157
drought hardiness, 337
drought index, 157
drought resistance, 270, 337
drought ring, 190
drought tolerance, 337
drouth, 156, 157, 190
drove, 65
drug plant, 559
drug store beetle, 378
drum, 175, 264, 484, 593, 638
drum barker, 175, 184
drum brake, 175
drum capacity, 264
drum chipper, 175
drum debarker, 175, 184
drum debarker system, 175
drum dryer, 175, 638
drum friction log barker, 593
drum narrower, 264
drum-peeler, 175, 184
drum polishing machine, 184
drum sander, 175, 184
drum sanding, 184
drum screen, 175, 593
drum skidding, 243
drum-system cable railway, 283
drum-system of inclineway, 283
drum winch, 184
drumlin, 175
drunken saw, 268, 279, 357
drupe, 193
drupel, 533
drupelet, 533
drupetum, 263
druse, 254
dry-and-wet bulb hygrometer, 157
dry basis moisture content, 160
dry bin, 157
dry breaking length, 159
dry bulb, 157

D

dry-bulb temperature, 157
dry-bulb thermometer, 157
dry chernozem, 157
dry chip, 157
dry chip bin, 157
dry chute, 157
dry climate, 157
dry clip, 159
dry code, 157
dry cold front, 157
dry deciduous forest, 158
dry desert, 157
dry distillation, 157
dry distillation of wood, 329
dry end, 62, 156
dry face, 156
dry fallen wood, 278
dry felt, 157, 197
dry fiberfelter, 156
dry-floated timber, 350
dry flume, 157, 489
dry fruit, 156
dry gluing, 157
dry gluing strength test, 48, 158
dry heat, 157
dry kiln, 158
dry kiln for plywood, 240
dry knot, 157
dry-laid masonry dam, 157
dry line, 460
dry mat, 156
dry matter content, 158
dry matter production, 158
dry method, 156
dry (*or* drier) part, 158, 197
dry-packing (method), 156, 157
dry podzolic soil, 157
dry practice, 276
dry-process, 156
dry pruning, 278, 543
dry prunning, 543
dry rear, 165
dry resistance, 270
dry ring crush, 157
dry roll, 237
dry rot, 156, 194
dry rot fungus, 160, 285
dry-rot of wood, 329
dry run, 158
dry (saw) mill, 156
dry season, 157
dry shrinkage, 158
dry slide, 157, 489
dry sloop, 110, 454
dry spray booth, 157
dry sterilization, 158
dry storage, 156
dry strength, 158
dry sunny slope, 158
dry tack, 158
dry tally, 156
dry tensile strength, 158
dry through, 385
dry to handle, 435
dry-topped, 278
dry topped tree, 278
dry type plate clutch, 157
dry-type rotary drum barker, 158
dry type wide belt sander, 158
dry valley, 156
dry wash, 156, 157
dry weight, 159
dry wood, 158, 278
dry wood beetle, 143
dry wood cutting, 278
dry wood storage, 157
dry-wood termite, 157
dryer, 158
dryer emission, 158
dryer (*or* drier) end, 158
dryer (*or* drier) unloader, 158
drying bed, 158
drying by boiling in oily liquid, 140
drying by dehumidification, 64, 233
drying condition, 158, 159
dry(ing) curve, 158
drying cycle, 159
drying defect, 158
drying element, 159
drying finger, 103
drying in hydrophobic liquid, 524
drying in paraffin(ic) oil, 434
drying mechanism, 158
drying medium, 158
drying of wood by direct fire, 329, 614
drying of wood in smoke and firing gas, 331
drying oil, 158
drying oven, 158, 197
drying period, 158, 159
drying power, 158
drying rate, 158
drying regime, 159
drying room, 158, 197
drying schedule, 158
drying shed, 158
drying shrinkage, 158
drying split, 157
drying stress, 159
drying stress set, 159, 572
drying temperature, 158
drying time, 158
drying time automatic recorder, 158
drying tree, 278
drying velocity, 158
drying with absorptive chemical, 494
drying with superheated steam, 187
drying yard, 368
dryness, 158
dryness index, 158, 159
dual-bath treatment, 393
dual control bulb, 453
dual fliter-agglomerator, 66
dual-function, 296, 455
dual infection, 455
dual spindle lathe, 453
dual thickness feed system, 451
dubbing, 7, 400, 470
Dublin standard, 110
duck foot, 454, 551
ducking area, 515, 639
duct, 89, 144, 446, 553
dudler, 108, 223
dudley, 108, 223
due, 140, 213, 460, 572
duff, 9, 72, 278
duff fire, 520
duff horizon, 72
duff hygrometer, 9
duff layer, 72
duff moisture code, 151
duff mull, 9
dug-hole methods, 269
duke, 282
dull edge, 115
dulling, 115, 169
dumb waiter, 546
dummy, 544
dummy tree, 150

D

dummy variable, 552
dump, 285, 492, 538, 540
dump car, 128, 644
dump chest, 538
dumper, 538, 647
dumping, 380, 381, 538
dumping bunk, 381
dumping fork, 46
dumping ground, 538
dumping ramp, 538
dune, 411
dune fixation, 411
dune fixation afforestation, 176
dune plant, 411
dune-sand, 382, 411
dunging, 431
dunite, 68
dunnage, 139
dunnage (wood), 103
dunstone, 320, 541
duopoly, 454
duplex, 153, 454
duplex and triplex mixture, 452
duplex flyer yarding, 454
duplex loader, 454
duplex loading system, 453
duplicase, 154
duplicate effect, 60
duplicate gene, 60
duplicating paper, 154, 475
duplication, 60, 154
duplicato-pinnate, 61, 122
duplicato-serrate, 60
duplicato-ternate, 60
duprene, 315
durability, 337
durability of germination capacity, 124
durability test, 337, 357
durable species, 337, 338
durable wood, 337, 338
duramen, 538
duration of fire resistance, 651
duration of life, 443
durifruticeta, 575
duriherbosa, 575
durilignosa, 575
Durio type, 308
duripan, 184
durisilvae, 575
Duro, 108
durometer-hardness, 225
durometer level, 575
dust, 144, 500
dust accumulator, 513
dust arrestor, 513
dust board, 88, 131
dust bowl, 52, 145
dust catcher, 513
dust collecting duct, 224
dust collecting hood, 513
dust collection, 224
dust collector, 224
dust counter, 225
dust devil, 52
dust dispipe, 52
dust-filter, 64, 316
dust-free, 35, 508
dust guard, 131
dust guard plate, 131
dust hoop, 131
dust mask, 131
dust panel, 88, 131
dust proofing, 131
dust screw conveyer, 52
dust seal, 131
dust separator, 140, 513
dusting, 354
Dutch bulb, 193
Dutch elm disease, 584
Dutch garden, 193
Dutch hyacinth, 146
dutchman, 103, 112, 177, 195, 227, 248, 351
dutchman line, 351
dutchman spipe, 316
dutiable goods, 573
dutiable value, 498
duty on value added, 602
duty on wood, 329
dwang, 103
dwarf bamboo, 2
dwarf branchlet, 113
dwarf culture, 1
dwarf effect, 1
dwarf forest, 1, 2
dwarf mistletoe, 1
dwarf pine needle oil, 2, 624
dwarf shoot, 113
dwarf shrub, 1
dwarf shrub heath, 2
dwarf tree limit, 2
dwarf(ing), 1
dwarf(ing) stock, 1, 2
dwarfism, 2
dyad, 122, 452
dyclesium, 467
dyeability, 273, 640
dyeing capacity, 390
dyestuff, 390
dyewood, 390
dying tree, 27, 67, 278
dying tree cutting, 278
dynamic, 108, 109, 581
dynamic balance, 109
dynamic characteristic, 109
dynamic classification, 109
dynamic community, 109
dynamic-composite life table, 109
dynamic creep, 109
dynamic ecology, 109
dynamic equilibrium, 109
dynamic life table, 109
dynamic load, 109
dynamic model, 109
dynamic of population, 627
dynamic price model, 109
dynamic programming, 109
dynamic sampling, 109
dynamic site class, 109
dynamic strain, 109
dynamic stress, 109
dynamic test, 109
dynamic viscosity, 108
dynamic viscosity coefficient, 108
dynamics of drying, 158
dynamometer, 44, 171
dyplotegia, 467
dyrene, 94
dysbolismus, 79
dysembryoplasia, 352
dysfunction, 218
dysgenesis, 32, 430
dysgenic, 297, 492
dysontogenesis, 125
dysphotic zone, 401
dysploid, 138, 596
dystrophic, 359
dystrophic lake, 573
dystrophy, 573, 574
E-grade thinning, 444
E-horizon, 304
E-layer, 304
eaglewood, 53
eaglewood oil, 53
ear, 103, 121, 175
ear covert, 121

ear-to-row selection, 469
earliness, 599
early bark, 599
early blooming cultivar, 599
early frost, 599
early mast, 599
early prediction, 599
early pruning, 599
early stage of decay, 63
early summer budding, 64
early thinning, 599
early variety, 599
early wood, 599
early wood tracheid, 599
early wood vessel element, 599
earth and underground working timber, 489
earth auger, 491, 495
earth boring machine, 495
earth chute, 489
earth crust, 96, 488
earth dam, 488
earth debris flow, 538
earth fall, 488
earth flow, 342, 489
earth kiln, 472, 491
earth pressure, 491
earth pressure cell, 491
earth resource observation satellite, 97
earth resource technology satellite (ERTS), 97
earth road, 489
earth slide, 489
earth water, 575
earth work, 489
earthflow, 342
earthing, 155, 353
earthing up, 353, 654
earthquake, 99
earthworm mull, 382
earwig, 368, 384
eased arris, 587
eased edge, 587
easement zone, 209
easterlies, 108
easy chair, 2, 385
eater-eaten interaction, 33
ebonite, 241, 575
ebonize, 134
ebpyriform, 90
ebullition, 140, 367
ecad, 440
eccentric adjustment, 357
eccentric drive, 357
eccentric force, 357
eccentric growth, 357
eccentric load(ing), 357
eccentric pith, 357
eccentric ring, 357
eccentric stress, 357
eccentric swage, 550
eccentric wear, 357
eccentricity, 357
ecdysis, 492
ecdysoid, 343
ecdysone, 492
ecesis, 106
echard, 511
echelon pore arrangement, 476
echinuline, 188, 211
echo sounding, 212
echolocation, 212
eclosion, 147, 584
eco-catastrophe, 426
eco-chemical, 424
eco-economical zoning, 424
ecobiogeography, 428
ecoclimate, 424
ecocline, 424
ecocycling, 425
ecodeme, 425
ecoenergetics, 424
ecogenesis, 424–426
ecogenetics, 426
ecogeographical divergence, 424
Ecolodge, 424
ecologic-phytocoenostic series, 426
ecological age, 424
ecological agriculture, 424
ecological agriculture model (pattern), 424
ecological amplitude, 424
ecological backlash, 424
ecological balance, 424
ecological balance sheet, 424
ecological benefit, 425
ecological biogeography, 425
ecological bonitation, 424
ecological climax, 424
ecological compensation, 424
ecological compensation mechanism, 424
ecological complex, 424
ecological concentration, 424
ecological condition, 425
ecological crisis, 425
ecological density, 424
ecological distribution, 424
ecological divergence, 424
ecological dominance, 426
ecological dominant, 426
ecological effect, 425
ecological efficiency, 425
ecological embryology, 424
ecological energetics, 424
ecological engineering, 424
ecological equilibrium, 424
ecological equivalence, 424
ecological equivalent, 424
ecological equivalent (species), 424, 425
ecological factor, 426
ecological force, 424
ecological genetics, 426
ecological geography, 424
ecological gradient, 425
ecological group, 425
ecological growth efficiency, 425
ecological history, 425
ecological homeostasis, 424
ecological homologue, 425
ecological impact assessment, 426
ecological indicator, 426
ecological invasion, 425
ecological isolation, 424
ecological light unit, 424
ecological location, 424
ecological longevity, 425
ecological morphology, 425
ecological mortality, 425
ecological natality, 424

ecological niche, 425, 534
ecological optimum, 426
ecological overlap, 426
ecological paradigm, 425
ecological performance, 425
ecological physiology, 425
ecological plant anatomy, 617
ecological plant geography, 617
ecological polymorphism, 424
ecological process, 425
ecological pyramid, 424, 426
ecological race, 424–426
ecological red line, 424
ecological release, 425
ecological safety, 424
ecological sequence number, 425
ecological serial method, 425
ecological series, 425
ecological setting, 424, 425
ecological significance, 426
ecological speciation, 425, 426
ecological stability, 425
ecological structure, 424
ecological subsystem, 425
ecological succession, 425
ecological survey method, 425
ecological sustainability, 424
ecological system, 425
ecological technique, 424
ecological threshold, 424, 426
ecological time, 425
ecological tolerance, 424
ecological tourism, 424
ecological turnover, 426
ecological type, 425
ecological valence, 424
ecological valency, 424
ecological value, 424
ecological variation, 424
ecological vicariad, 425
ecological zero, 424
ecological zoogeography, 424
ecologist, 425
ecology, 425
econometric modelling, 225
econometrics, 225, 252
economic age, 252
economic ailments, 252
economic balance, 252
economic density, 252
economic diameter limit, 252
economic downturn, 252
economic effect, 252
economic efficiency, 252
economic externality, 252
economic fire-control theory, 252
economic forest, 252
economic girth limit, 252
economic intensity, 252
economic life, 252
economic maturity, 252
economic plant, 252
economic potential, 252
economic rotation, 252
economic sanction, 252
economic selection cutting, 252
economic space, 252
economic take-off, 252
economic threshold, 252
economic trees of special use, 475
economic unit, 252
economic value of biodiversity, 426
economic volume, 252, 290
economical consideration, 252
economical maturity, 252, 288
economical rent, 252, 253
economical selection cutting, 252
economical selection method, 252, 547
economics of thinning, 445
economies in transition, 253
economies of forestry, 303
economy of scale, 183
economy-wide, 386
ecophene, 424, 425
ecophenotype, 424
ecophysiology, 425
ecoregion, 424
ecoregulation, 425
ecorticate, 510
ecospecies, 426
ecosphere, 424, 425
ecosystem, 425, 428
ecosystem development, 425
ecosystem ecology, 425
ecosystem energetic, 425
ecosystem function, 425
ecosystem functioning, 425
ecosystem health, 425
ecosystem integrity, 425
ecosystem interface, 425
ecosystem management, 425
ecosystem modeling, 425
ecosystem prediction, 425
ecosystem services, 425
ecosystem stability, 425
ecosystem type, 425
ecosystem value assessment, 425
ecosysterm research nNetwork, 425
ecotonal, 388
ecotonal community, 187, 239
ecotone, 239, 388, 424
ecotone hypothesis, 424
ecotope, 424
ecotopic adaptation, 424
ecotourism, 424
ecotype, 425
ecotypic differentiation, 425
ecotypic variation, 425
ecozoogeography, 424
ectad, 497, 530, 599
ectendotrophic mycorrhiza, 341
ecto-hormone, 496, 540
ectocarp, 496
ectocrine substance, 495
ectoderm, 495, 496
ectodynamic soil, 495
ectodynamic succession, 495
ectodynamomorphic, 495
ectodynamomorphic soil, 495

E

ectogeny, 88, 186, 497
ectognathous, 496
ectomycorrhiza, 496
ectomycorrhizal fungi, 496
ectoparasite, 496
ectoparasitic, 496
ectoparasitism, 477, 496
ectophagous, 497
ectorganic humus layer, 95
ectostroma, 497
ectotheca, 495, 496
ectotherm, 24
ectotroph, 497
ectotrophic, 477, 496
ectotrophic mycorrhiza, 496
ecumene, 273
edaphic, 443, 489–491
edaphic climax, 489
edaphic climax community, 489–491
edaphic ecotype, 490
edaphic factor, 98, 491
edaphic indicator, 488, 491
edaphology, 168, 348, 490, 491, 573
edaphon, 489, 490
edaphon(e), 490
eddy, 507
eddy correlation method, 507
eddy covariance, 507
edge, 36, 239, 262, 303, 396
edge angle, 22, 285, 357
edge banding machine, 146
edge belt sander, 42, 412
edge bend, 22, 43, 196
edge bonding machine, 146
edge butt joint, 21, 22
edge chipper, 532
edge community, 22
edge cross lap joint, 22
edge effect, 22
edge fibre chest, 22
edge fire, 22
edge firing, 22
edge gluer, 22
edge-gluing veneer, 22, 43
edge grain, 10, 256, 333, 615, 649
edge-grain lumber, 256, 463
edge-grain surface, 256
edge joint, 22, 358
edge knot, 22
edge load stress, 22
edge-mark, 21
edge mitre joint, 22, 42
edge of a stand, 299, 303
edge of forest, 303
edge piling, 42, 43
edge position control, 22, 43, 352
edge protection, 299, 303
edge rabbet joint, 22, 42
edge sander, 42, 412
edge-sawn, 256
edge seal(ing), 146
edge sorter, 22, 43
edge species, 22
edge split, 22
edge stacking, 42
edge tailer, 37
edge tree, 258, 303
edge trim saw, 22, 366
edge trimming plane, 543
edge-veneer, 146
edge, glue and rip process, 366
edger, 16, 37, 326, 366, 592, 602
edgewise distortion, 22
edging, 22, 37, 262, 529
edging grinder, 22
edging plant, 22, 529
edible fungus, 437
edible maturity, 437
edificator, 236
edudesmin, 2
Eelworm, 526, 535
effect of crowding, 576
effect of neighbours, 298
effective accumulative temperature, 581
effective age, 581
effective alkali, 581
effective aperture, 581
effective area, 581
effective brightness temperature, 581
effective demand, 582
effective depth, 581
effective dose, 581
effective dose 50 (ED50), 10
effective dose$_{50}$ (ED$_{50}$), 582
effective evapotranspiration, 582
effective factor, 582
effective germinative capacity, 581
effective (interest) rate, 435, 581
effective moisture content, 581
effective radiation, 581
effective radius, 581
effective seedling, 581
effective soil depth, 581
effective stocking, 581
effective temperature, 581
effective temperature range, 582
effective trapping area, 581
effective volume, 581
effective width, 581
effervescence, 319, 352, 367
efficiency, 536
efficiency of assimilation, 482
efficiency of biological system, 428
efficiency of labor, 284
efficiency variance, 284, 536
efficient estimator, 581
effloresce, 144, 268
efflorescence, 143, 144, 268
effluence, 124, 306, 569
effluent, 135, 139, 306, 349, 516
effluent disposal, 139
effluent seepage, 363, 420
effluent standard, 349, 350
efflux, 306, 307, 404, 418, 496, 497
Efurd machine, 563
egesta, 350
egestion, 350
egg, 311
egg batch, 311
egg-case, 311
egg cell, 311
egg deposition, 47
egg-larva, 63
egg-laying, 47
egg mass, 311
egg parasite, 311

E

egg-parasite wasp, 311
Eh-pH relationship, 557
eitolation phenomenon, 210
ejector gun, 354, 480
ejector tumbler, 480
ektodynamic soil, 495
ektodynamomorphic soil, 495
el Niño, 120
el Niño Southern Oscillation, 120
elaminate, 511
elastic after effect, 472
elastic caoutchouc, 472
elastic consequence, 472
elastic deformation, 472
elastic demand, 472
elastic energy, 472
elastic failure, 472
elastic fatigue, 472
elastic feed roll, 472
elastic force, 472
elastic gum, 428, 472
elastic hysteresis, 472
elastic limit, 472
elastic medium, 472
elastic modulus, 472
elastic-plastic property, 472
elastic-plastic strain, 472
elastic potential energy, 472
elastic recovery, 472
elastic resilience, 472
elastic resistance, 472
elastic rubber, 472
elastic strain, 472
elastic strain energy, 472
elastic strength, 472
elastic stress, 472
elastic vibration, 472
elastic-viscoplastic body, 472
elastic-viscous system, 472
elasticity, 305, 419, 472
elastomer, 87, 192, 472
elastoplasticity, 472
elbow, 147, 222, 498, 631
elbow chair, 2
elbow table, 215
elder flower concrete, 244
elecampane oil, 65, 489
electric bottom-beat cable, 94
electric-bulb oven, 100
electric-capacity moisture meter, 101
electric clipper, 101
electric conductivity, 89, 100, 392
electric dry oven, 101
electric drying, 101, 163
electric field lens, 100
electric hand plane, 443
electric mixer, 101
electric power, 101
electric powered chain saw, 101
electric precipitator, 101, 257
electric resistance, 102, 103
electric-resistance moisture meter, 102, 103
electric-resistance thermometer, 103
electric resistivity, 101–103
electric saw, 101
electric seasoning, 101
electric set-work, 101
electric soil-heating cable, 101, 490
electric tensioner, 101
electric thermo-couple, 392
electric truck, 100, 101
electric tying machine, 101
electric(al) accounting machine, 100–102
electric(al) conductivity, 89, 100
electric(al) heating system, 101
electric(al) membrane potential, 101, 326
electric(al) moisture meter, 100, 456
electric(al) potential, 101
electrically conducting coat, 89
electrically conductive adhesive, 89
electrically conductive glue, 89
electrically scanning microwave radio-meter, 102
electrochemical gradient, 101
electrochemical potential, 101
electroconductibility, 89, 100
electroconductivity, 89, 100
electrodialysis, 101
electrolyte, 101
electromagnetic brake, 100
electromagnetic energy, 100
electromagnetic radiation, 100
electromagnetic spectrum, 100
electromagnetic vibrator, 100
electromagnetics, 100
electromagnetism, 100
electromotor, 101, 123, 316
electron acceptor, 102
electron beam accelerator, 102, 376
electron beam curing, 102
electron beam probe, 102
electron carrier, 99, 102
electron density, 102
electron donor, 102, 167
electron flow, 102
electron micrograph, 102
electron microscope, 102
electron microscopy, 101, 102
electron nuclear double resonance, 102
electron paramagnetic resonance, 102
electron spin resonance, 102, 460
electron transport chain, 102
electronic computer, 101, 102, 225
electronic data processing, 102
electronic data processing system, 102
electronic gluing, 102
electronic ignition, 102
electronic leaf, 102
electronic moisture detector, 102
electronic moisture meter, 102

electronic position sensor, 102
electronic scale, 102
electronic seed counter, 102
electronic seed moisture meter, 102, 628
electronic spectroscopy, 102
electronic weighbridge, 102
electronic weighing apparatus, 102
electronically steerable array radar, 102
electrophile, 379
electrophilic reagent, 379
electrophoregram, 102
electrophoresis, 101, 102
electrophoretogram, 102
electrostatic coating, 257, 258
electrostatic hand gun, 258, 443
electrostatic nebulizer, 258
electrostatic spray gun, 257, 258
electrostriction, 101, 102, 130
elemecin, 567
elemene, 283
elemenone, 283
element, 36, 85, 86, 100, 218, 305, 559, 570, 587
element of forestry, 303
element of production, 422
elementary body, 589, 591, 592
elementary cell, 85, 352
elementary fibril, 219, 221, 591
elementary species, 219
elemi, 283
elemi oil, 283
elemi resin, 283
elemicin, 283
eleoptene, 578
elephant, 530
elepidote, 510
elevated microsite, 500
elevated panel, 175
elevated ramp, 162, 638
elevation, 161, 188, 309, 421, 476, 611, 650
elevation instrument, 44, 188
elevation screw, 421
elevator, 101, 367, 421
elfin forest, 1, 163
elfin tree, 163
elfin wood, 2, 163
elimination reaction, 531, 532
eliminator, 142, 257, 316, 531
elite, 254, 565, 577, 592
elite breeding, 295
elite plant, 475
elite plot, 592
elite seed, 254
elite seed stand, 577
elite stand, 328, 475, 547, 577, 592
elite tree, 254
ellagitannin, 399
ellipsoid particle, 494
elliptic growth, 494
elongated crystal, 636
elongated octahedron, 519
elongating stage, 5, 419
elongation, 282, 419, 554, 650
elongation of tree stem, 163
elongation percentage, 419
elongation period, 419
elongation rate, 419
elongation set, 419
elongation strain, 419
elongation tensor, 419
Eltonian pyramid, 2
eluant, 516
eluent, 516
elution, 474, 516
elution analysis, 474
elutropic, 516
eluvial horizon, 304
eluvial-hydromorpbic soil, 304
eluvial layer, 304
eluvial soil, 304
eluviation, 304
eluvium, 40, 144, 304
elytra, 18, 377
emasculation, 385, 554
embankment, 94, 310, 479
embargo, 252
embayment, 3, 55, 188, 193, 498, 604, 612
embedding, 12, 317, 640
ember, 211, 584
emblic extract, 578, 584
emboss, 147, 148, 486, 550, 551
embossability, 147, 550
embossed hardboard, 550
embossed plywood, 147, 550
embosser, 147, 550, 551
embossing, 147
embossing machine, 147, 550, 551, 571, 602
embossing roll, 550, 571
embouchure, 193, 519
embryo, 352, 353
embryo abortion, 352, 626
embryo cavity, 352
embryo culture, 352, 626
embryo dormancy, 352
embryo grafting, 352, 626
embryo nucleus, 352
embryo (*or* embroyonic) bud, 352, 374
embryo orthoplocal, 641
embryo phase, 352
embryo ratio, 352
embryo ripening, 352
embryo sac, 352
embryo spirolobal, 641
embryoid, 353
embryon, 352, 353
embryonic axis, 352
embryonic development, 352
embryonic humus-layer, 352, 591
embryonic implantation, 352
embryonic root, 352
embryonic soil, 591
embryophyte, 581
emerged plant, 481
emergence, 62, 124, 438, 524, 583, 584
emergence curve, 124
emergence period, 124, 584
emergency backfire, 249
emergency brake, 249, 572
emergency cutting, 249
emergency district, 251
emergency sport, 486
emergent, 62, 123, 418, 481, 486
emergent evolution, 486
emergent layer, 486
emergent tree, 486

E

emergy, 342
emersiherbosa, 604
emersion zone, 148, 251
Emerson beater, 1
emery, 160, 248
emery belt, 248, 412
emery cloth, 248, 412
emery fillet, 248, 412
emery paper, 248, 412, 555
emery tape, 248, 412
emery wheel, 248, 412
Emhorn tube, 1
emigrant, 372
emigration, 372
eminent domain, 185, 488, 540
emission, 123, 124, 135, 148, 354, 580
emission inventory, 349
emission line, 124
emission reduction unit, 233
emission spectrography, 124
emission spectrum, 124, 135
emissions, 349
emissions permit, 349
emissions quota, 349
emissions reduction unit, 349
emissions scenario, 349
emissions tax, 349
emissions trading, 349
emittance, 124, 135, 148
emmet, 317, 566
emmision reduction, 233
emolument, 15, 540
emphytic character, 565
empire grading rule, 572
empirical yield table, 253, 526
employed labour force, 177
employmem benefit cost analysis, 259
empty cable sag, 161
empty-cell impregnation, 276
empty-cell process, 35, 106, 276, 603
empty seed, 27, 275
emulator, 134, 325
emulsifier, 400
cement, 344
emulsion, 400
emulsion characteristic curve, 400
emulsion concentrate, 400
emulsion preservative, 400
enamel, 128, 473
enamelled hardboard, 69, 473
enantiomer, 115
enation, 121, 486
encased knot, 356, 462
enclave, 40, 136
enclosed belt conveyor, 322
enclosed knot, 339, 571
enclosed pocket, 229
enclosing of game, 502
enclosure, 492, 502, 534
encoder, 23, 569
encroachment, 378
encrusting material, 12, 246
encystment, 13, 54
end, 112, 247, 249, 327
end beam, 111
end-boring method, 111, 327
end-bud, 106
end-butt(ing), 247, 366
end check, 111
end check(ing), 111
end-coating, 112
end compression, 112, 461, 651
end-contraction, 43, 112
end cutting, 247
end diameter, 112
end dog, 111
end dogging carriage, 295
end dumper, 199
end-for-end treatment, 295
end gluing, 112, 650
end-grain cutting, 112, 196
end-grain section, 196
end hardness, 112
end haul(ing), 563
end hook, 111
end joint, 111, 112
end lap, 111, 191, 327, 373, 650
end lap joint, 111
end load(ing), 327
end mark, 112, 113, 246
end mast, 199, 503
end matched specimen, 112
end matcher, 120
end matching, 111, 112
end milling, 112, 290
end peak, 503
end penetration method, 112
end-piled loading, 650, 651
end piling, 450
end pressure, 112
end product, 54, 625, 653, 654
end-rack, 52, 112
end racking, 46, 239
end sealed, 112
end shake, 111, 112
end split, 111
end stacking, 450, 650
end stop, 112
end strip, 529
end surface, 112, 196
end-to-end joint, 114, 361
end trimming, 247, 366
end use, 634, 654
end-use point, 437
end-use product, 654
end-use test, 437
end value, 199, 626, 654
end wall, 111
end-way piling, 450
endangered species, 27, 304
endarch, 340
endarch protoxylem, 340
endarch xylem, 340
endecagynian, 433
endecagynous, 433
endecandrous, 433
endemic, 95, 96, 99, 145, 474, 475, 491, 587
endemic density, 96, 362
endemic plant, 475
endemic species, 88, 177, 475, 527
endemism, 96, 475, 491
endergonic reaction, 514
ending inventory, 365
endlap, 111
endless bed drum sander, 80
endless belt dryer, 549
endless cable system, 549
endless chain, 549
endless chain conveyor, 549
endless chain saw, 295, 509
endless drum, 549

E

endless line, 549
endless pulley, 549
endless saw, 79
endless steel belt, 549
endless tong, 104
endless Tyler system, 471
endo-humus, 489
endobasidium (*plur.* -ia), 155, 340
endobenthos, 95
endobiont, 340
endobiose, 340
endobiotic, 427, 476, 652
endocamphene, 340
endocarp, 339
endochory, 635
endoconidium (*plur.* -ia), 340
endodermis, 340
endodynamic soil, 339
endodynamic succession, 341
endodynamomorphic soil, 339, 554
endoecism, 201
endoenzyme, 13, 340
endogamy, 483
endogenetic rock, 339, 340
endogenous, 340, 341
endogenous migration, 341
endogenous rhythm, 341
endogenous spore, 339, 340
endolithic, 434
endolithophyte, 434
endomitosis, 193, 194
endomycorrhiza, 340
endoparasite, 339, 340
endoparasitism, 340
endopelic, 342
endophagous, 340
endophloem, 340
endophloic, 340
endophyte, 340
endophytic, 340, 617
endoplasm, 339–341
endoplasmic reticulum, 341
endopolyploidy, 193, 339
endopsammon, 411
endopterygote, 339
endoreduplication, 193, 194, 341
endorganic humus layer, 489
endorheic basin, 340
endosarc, 341
endosmosis, 340
endosomatic, 476
endosperm, 352
endospermic seed, 581
endospermous, 352
endospore, 339, 340
endotesta, 341, 377
endotherm, 195
endothermic point, 340, 514
endothermic reaction, 514
endothermy, 341
endotrop(h)ic, 340, 518
endotrophic mycorrhiza, 340
endotrop(h)ic mycorrhiza, 340
endowment cost, 264
endozoic, 109, 429, 482
endozoochore, 109
endrin, 566
endurance limit, 57, 356
endurance strength, 57, 337, 356
endurance test, 337, 357
endwise bucking, 650
endwise compression, 461, 651
endwise piling, 450
endwise tensile strength, 461
endwise tension test, 461
endy veneer, 537
energid, 215
energy analysis, 341
energy balance, 341
energy barrier, 341
energy budget, 341
energy change, 341
energy consumption, 191, 341, 342
energy conversion, 342
energy crop, 342
energy derived from non-fossil-fuel source, 137
energy-dispersive X-ray analysis, 342
energy dissipation, 341
energy equilibrium in forest, 408
energy flow, 341, 342
energy flow path, 342
energy flow structure, 342
energy flux, 341
energy forest, 342
energy input, 341
energy intensity, 342
energy metabolism, 341
energy of activation, 214
energy of desorption, 248, 493
energy of strain, 572
energy plantation, 342
energy release component, 341
energy resource, 109, 342
energy-rich bond, 163
energy-rich compound, 163
energy service, 342
energy sink, 341
energy source, 342
energy subsidy, 149
energy tax, 342
energy transfer, 341
energy transformation, 341
energy transformer, 341
energy wood, 342, 539
enfleurage, 204, 286
enforced dormancy, 376, 536
enforced self-pollination, 376
engine sizing, 75
engineered particle, 170
engineered wood product, 169
engineered wood products, 169
Engler degree, 121
Engler number, 121
Engler viscosimeter, 121
English garden, 572
English plane tree, 123
English setter, 572
English standard, 572
engraved roller, 103, 550
engraver beetle, 533
engraving, 103
enhancement image, 575
enhancement of carbon sequestration, 601
enhancement of carbon sink, 601
enhancement of multi-spectral satellite image, 117
enlarged cell, 355
enneagynian, 260
enneandrous, 260
enneaploid, 259

E

enneaploidy, 259
enomosphore, 131
enriched water, 348, 641
enrichment, 32, 33, 155, 348
enrichment culture, 155, 602
enrichment horizon, 155
enrichment seeding, 32
ensate, 236
ensign-fly, 366
entailed forest, 438
entailed forest property, 438
enterprise, 367, 438
enthalpy, 189, 392, 398
entine, 339
entire, 22, 68, 385, 387, 498, 648
entire leaf cutting, 387
entisol, 539
entomogenous, 60, 280
entomological factor, 60
entomologist, 280
entomology, 280
entomoparasitism, 280
entomopathogenic, 280, 437
entomophagous, 436, 437
entomophagous parasite, 280, 436
entomophagy, 436
entomophilous, 60, 436, 516
entomophilous flower, 60
entomophily, 60
entomophyte, 60, 108
entomophytous, 60
entomosis, 280
entophagous, 340
entophagous insect, 436
entrained water, 229, 609
entrepreneur, 55, 367, 438
entropy, 415
entrusting substance, 12, 246
enucleated, 384, 509
enumeration, 300
enumeration survey, 320
envelope, 12, 18, 474, 496, 648
enveloping, 261
environment, 207
environment complex, 208
environment deterioration, 207
environment hormone, 207
environment monitoring, 207
environment signal, 208
environment survey satellite, 207, 208
environmental agency, 208
environmental assessment, 207
environmental biology, 208
environmental capacity, 207, 208
environmental climate, 208
environmental complex, 207, 208
environmental condition, 208
environmental conservation, 207
environmental consideration, 208
environmental covariance, 207, 208
environmental crisis, 208
environmental degradation, 207, 208
environmental density, 207
environmental deviation, 207
environmental disaster, 208, 426
environmental disruption, 171, 208
environmental ecology, 208
environmental education, 207
environmental engineering, 207
environmental ethics, 207
environmental factor, 208
environmental goods, 207
environmental gradient, 208
environmental greening, 207
environmental heterogeneity, 208
environmental horticulture, 208
environmental impact assessment, 208
environmental indicator, 208
environmental noise, 208
environmental physiology, 208
environmental pollution, 207, 208
environmental preference, 208
environmental problem, 208
environmental protection, 207
environmental quality, 208
environmental resistance, 207, 208
environmental resource patch, 208
environmental science, 207
environmental sensitivity, 207
environmental sociology, 208
environmental variation, 207
environmental white paper, 207
environmentalism, 207
environmentalist, 207
environmentally sound technologies, 208
enzootic, 96
enzootic disease, 96
enzymatic oxidation, 320
enzymatically liberated lignin, 243, 320
enzyme, 320
enzyme activity, 320
enzyme catalysis, 320
enzyme-linked-immuno-sorbent-assay, 320
enzyme resin, 320
enzyme secreted by fungus, 607
enzymology, 320
enzymolysis, 320
Eocene, 438
eoclimax, 174
eolation, 144
eolian deposit, 144
eolian deposition, 144

eolian erosion, 145
eolian rock, 144
eonsistometer, 61
eosere, 522
eosin, 446, 463
eotic, 307
epharmone, 440
ephedroid perforation, 316
ephedroid perforation plate, 316
ephedroid vessel perforation, 316
ephemeral, 112
ephemeral desert, 112
ephemeral flower, 604
ephemerophyte, 112, 113
ephippium, 3
epi-, 415, 496
epiachene, 292
epiafzelechin, 26
epibenthic, 94
epibenthic plankton, 94
epibenthos, 94
epibiont, 27, 152
epibiotic, 264, 477, 496
epibiotic endemic species, 174
epibiotic endemism, 174
epibiotic plant, 174
epibiotic species, 40
epibiotics, 264
epicalyx, 121, 154
epicarp, 496
epicarpanthous, 416
epicarpius, 416
epicarpous, 416
epicatechin, 26
epicentre, 608
epichil, 415
epichile, 415
epicole, 496
epicormic, 154, 487
epicormic branch, 341, 487
epicormic knot, 341, 487
epicotyl, 416
epicotyl dormancy, 416
epicubehol, 26
epicyclic hub reduction, 540
epidemic, 306
epidemiology, 306
epidermis, 27, 416
epidosite, 315
epifauna, 27
epiflora, 27
epigaeal, 62, 429
epigaec, 97
epigallocatechin, 26
epigeal, 62, 97, 375, 429, 430, 480
epigeal germination, 62, 97
epigean, 62, 97, 429
epigeous, 62, 97, 429, 430, 607
epigynous, 416
epigynous disc, 416
epigynous flower, 416
epigyny, 416
epilimnion, 201
epilithic, 434
epilithon, 434
epimeletic behaviour, 605
epimer, 27, 46
epimeric form, 46
epimerization, 26, 46
epimeron (*plur.* -era), 199
epinasty, 357
epineuston, 456
epiontology, 617
epiorganism, 51, 152
epiparasitism, 60, 477, 496
epipedon, 26, 27, 45
epipelagic zone, 415
epipelic, 508
epipetalous, 202, 203, 416
epipetalous stamen, 178, 203
epipetreous, 554
epipharynx, 339, 416
epiphyllae, 560
epiphyllous community, 560
epiphyta arboricola, 449
epiphyte, 152, 476, 640
epiphyte-quotient, 152
epiphytic, 152
epiphytic orchid, 152
epiphytic periophyton, 617
epiphytology, 617
epiphyton, 152
epiphytotic disease, 617
epiphytotics, 616, 617
epiplankton, 415
epipleurite, 415
epipodium, 162, 323, 373, 416
epipsammic, 411
epipsammon, 411
episepalous, 114, 121
episome, 152, 579
episperm, 626
epistasis, 416, 567
epistasy, 416, 567
epistatic deviation, 416
epistatic effect, 416
epistatic gene, 416, 567
episternum, 372, 415, 416
epithelial cell, 322, 416
epithelial layer, 322, 416
epithelial parenchyma, 321, 416
epithelium, 322, 416
epithermal, 51
epizoan, 94, 152
epizoic, 109, 476, 496
epizoite, 477
epizoi(ti)c, 476
epizoochore, 109
epizoochory, 109
epizootic, 109, 477
epizootiology, 109, 444
epjgynous calyx, 416
epoekie, 476
epoxidation, 209
epoxy resin adhesive, 209, 499
equal loudness, 92
equalization treatment, 360
equalize, 247
equalized annual sustained working, 266
equalized area, 155
equalizer, 112, 366
equalizing, 247, 360, 610
equalizing block, 454, 469
equalizing layer, 360
equalizing roller, 266
equalizing vat, 225, 266
equation of value, 231
equator, 58, 624
equatorial low-pressure, 58
equatorial plate, 58
equatorial rain forest, 58
equi-productive area, 155, 422
equiconditional area, 482
equiformal progressive area, 92
equilateral triangle planting, 611
equilibrium, 266, 360, 527

equilibrium and transient climate experiment, 360
equilibrium density, 360
equilibrium modulus, 360
equilibrium moisture content, 360
equilibrium paradigm, 360
equilibrium population, 360
equilibrium slope, 360, 478, 506
equilibrium strength, 360
equilibrium theory, 360
equinoctial flower, 107
equipluve, 91, 92
equipotential temperature, 92
equitability, 266
equity, 385, 431, 634
equity capital, 175, 385, 647
equivalent acidity, 88
equivalent basicity, 88
equivalent bending moment, 91, 92
equivalent CO_2 (carbon dioxide), 123
equivalent evaporation, 91, 608
equivalent exchange, 91
equivalent load, 88, 92
equivalent porosity, 88, 92
equivalent species, 92
equivalent uniform load, 88, 92
eradicant, 48, 167
eradicant insecticide, 169
eradicating of climber, 64
eradication, 48, 167, 257, 381
eradication of pests, 29
eramose, 34, 512
erasibility, 336
erect plant, 615
eremacausis, 311, 318
eremium, 209, 411
eremophilous, 209, 365, 439
eremophyte, 209, 411
eremus, 209, 411
ergeron, 434
Erichsen tester, 1
ericifruticeta, 348
ericilignosa, 348
erinaceous, 118, 260, 399
eriodictyol, 318, 431
erion, 323
eriophyid mites, 575
ermine moth, 52, 90
eroded, 346, 387
eroded land, 378
eroded soil, 379
erodibility, 378, 490, 569
eromoparasitism, 111
erosion, 59, 150, 378, 379
erosion basis, 378
erosion by water, 459
erosion by wind, 145
erosion control, 59, 378
erosion control planting, 133
erosion cycle, 379
erosion modulus, 378
erosion of glacier, 28
erosion plain, 379, 640
erosion plateau, 378
erosion terrace, 378
erosion(al) basin, 378
erosion(al) lake, 378
erosion(al) landform, 378
erosion(al) scarp, 379
erosion(al) surface, 59, 144, 378
erosion(al) valley, 378
erosive power, 59, 378
erosiveness, 59, 378, 379
erratic, 358, 564
erratic block, 260, 358
error nomogram, 513
error of mean square, 266
error of mistake, 187
error propagation, 513
error term, 40, 74, 513
error variance, 218, 513
eruptive rock, 215, 354
erythro form, 58
erythrolaccin, 198
erythrophyll, 204, 560
erythrose, 58
escalator clause, 230, 419, 480, 644
escape, 474, 569
escape clause, 291, 323
escape mechanism, 474
escaping steam, 569
escritoire, 580
escutcheon, 115, 277, 470, 533
eskar, 417
esker, 417
espacement, 616
espalier, 288, 448, 449
espalier training, 588
espinal, 116, 190
esquamate, 510
essence, 254, 436, 528, 559
essential aggregation, 202
essential amino acid, 20
essential character, 627, 634
essential component, 634
essential element, 20
essential oil, 254, 528, 617
essential organ, 20, 219
essential parameter, 19, 219
established field, 54, 366, 583
established stand, 54, 366, 583
establishment, 54, 106, 366, 540, 541
establishment (of a stand), 299, 600
establishment period, 54
establishment phase, 106
establishment plan, 597, 600
ester gum, 465, 619
esterase, 619
esthetic function, 320, 420
esthetics of garden, 588
estidurilignosa, 522
estimate survey, 36, 44, 545
estimated average life, 173
estimating by tree, 320
estimating changes, 23
estimation by eye, 336
estimation of increment, 602
estimation of stand characteristics, 300
estimator, 173
estipulate, 510
estival, 521
estival aspect, 431
estivation, 202, 522
Estrade process, 1
estuarine, 193

estuarine plankton, 193
estuary, 188, 193
etaerio, 264
etchant, 150, 251, 437
Etesian climate, 99, 129
ether extractive, 321
Ethiopian region, 1
ethnobotany, 324
ethological isolation, 540
ethology, 109, 420, 425, 540
ethyl cellulose, 523, 565
etiolation, 210
etiolation effect, 210
etiolation method, 210
etiology, 29, 621
etlite tree, 254, 475, 592
eubasidii, 607
eubenthos, 607
eubiosis, 424
eucalyptene, 2
eucalyptus citriodora oil, 346
eucalyptus oil, 2, 3
eucarvone, 577
eucary, 607
eucaryote, 607
eucaryotic cell, 607
eucaval animal, 608
euchromatin, 48, 608
euchromosome, 48, 476
euchrozems, 420
eucoen, 619
eudesmene, 3
eugenic, 577
eugenics, 577
eugenin, 105, 128
eugenitin, 105, 128
eugenitol, 105, 128
eugenol benzyl ether, 105, 128
eugenol type basil oil, 105, 128
eugenone, 105, 128
eugenyl, 105, 128
eugeogenous rock, 569
euhalabous, 429
euhermaphrodite, 48
euheterosis, 49, 608
euionone, 578
eukaryote, 607
eukaryotic cell, 607
eulittoral zone, 607
eumitosis, 49, 326, 608
eumycetes, 607
euonymin, 503
euonymus, 503
euphenics, 577
euphorbium, 76
euphotic stratum, 485
euphotic zone, 485
euplankton, 607
euploid, 609
eupodzolic, 100
eurendzina, 100, 139
Euro-Asia plate, 348
European plane, 123
European spruce, 348
European walnut, 201
European white poplar, 570
European wood-block test, 348
eurybaric, 183
eurybathic, 182
eurychoric, 182, 183
euryecious, 182
euryhadric, 182
euryhaline, 183
euryhalinous, 183
euryhydric, 182
euryhygric, 182
euryoecic, 182
euryoecious, 182
euryokous, 182
euryoky, 182
euryoxybiotic, 183
euryphagic, 182
euryphagous, 182
euryphagy, 182
euryphotic, 182
euryspecies, 182
eurythermal, 183
eurythermic, 183
eurytomid, 182
eurytopic, 182
eurytopic species, 182
euryvalent, 182
euryzonal, 183
euryzonous, 182
eusociality, 608
euspecies, 608
eustatic sea-level change, 188
eustele, 608, 615
eutric arenosol, 14
eutric lithosol, 14
eutrophic, 139, 155, 182, 187, 191, 574, 610, 625
eutrophication, 155
eutrophy, 155
eutrophyte, 139, 155
eutropic, 530
evapo-transpiration, 609
evaporation, 608, 609
evaporimeter, 608, 609
evapotranspiration, 609
evapotranspirometer, 609
even-aged, 483
even-aged artificial mixed forest, 483
even-aged crop, 483
even-aged forest, 483
even-aged group stands, 483
even-aged high forest, 483
even-aged stand, 483
even-aged system, 483
even-aged type, 483
even-aged uniform forest, 610
even colour, 266
even-end, 366
even grain, 266, 344
even load, 266
even pinnate, 349
even shade, 266
even tail, 362
even texture, 266
even year, 362, 508, 509
evenness, 266
evenpolyploid, 349
ever frozen layer, 576
ever-wet forest, 48
everblooming, 48, 293, 463
everflowering rose, 594
everglade, 605
everglades national park, 185
evergreen, 48
evergreen broad-leaved forest, 48
evergreen community, 48
evergreen coniferous forest, 48
evergreen cutting, 48
evergreen fire-break, 48
evergreen forest, 48
evergreen hardwood forest, 48
evergreen needle-leaved forest, 48
evergreen sclerophyllous forest, 48
evergreen seasonal forest, 48
evergreen shrub, 48
evergreen silvae, 48
evergreen thicket, 48
evergreen undershrub, 48
evocator, 221, 583
evolute, 129, 603

E

evolution, 124, 249, 556, 565
evolution ecology, 249
evolution force, 249
evolution rate, 250, 349
evolutionary biology, 249
evolutionary ecology, 249
evolutionary retardation, 250
evolutionary reversion, 343
evolutionary time, 249
eversion, 201, 507
ex dock, 316
ex factory, 169, 620
ex pier, 316
ex quay, 316
ex-situ conservation, 569
exalbuminous, 510
exalbuminous seed, 510
examination method, 193, 221, 232
exanthium, 121
exarch, 497
exarch protoxylem, 497
excavated pith, 276
excavation, 266, 277, 494, 495
excavator, 494, 495
excelsior, 333, 334, 519
excelsior board, 334
excelsior cutting machine, 16, 263, 334
excelsior plate, 334
excelsior-shredding machine, 334
excentric embryo, 357
excentric growth, 287
excess height, 51
excess length, 51
excess pressure, 51, 584
excess profit, 51
excess reserve, 50, 51, 431
excess steam, 584
excess weight, 52, 187
excessive cutting, 51, 187
excessive felling, 187
excessive height, 51
excessive tree felling, 187
exchange acidity, 239
exchange base, 239
exchange capacity, 239
exchange constant, 239
exchange-value, 239, 317
exchangeable base, 239
exchangeable cation, 239
excise tax, 186, 217, 475, 544
excised embryo, 287
excised embryo test, 287
excised seed, 364
excitation of spectra, 181
exclosure, 252, 502, 526
exclusion, 35, 64, 349
exclusion chromatography, 350, 412
exclusive, 636
exclusive green space, 636
exclusive species, 388
exclusive territory, 111
exclusiveness, 388
excoemum, 574
excoriation, 30, 36, 327
excrescence burl, 308
excreta, 142, 350
excreting cell, 142, 350
excretion, 350
excurrent growth, 87, 470, 554
excursion train, 579
execution record, 436, 613
exendospermous, 510
exercise price, 192, 396, 540, 586
exergonic reaction, 135
exfoliate, 31, 357
exhaust, 134, 139, 350
exhaust fan, 61, 349, 350
exhaust gas desulfurization, 139, 350
exhaust noise, 350
exhaust outlet, 136, 350
exhaust pipe, 350
exhaust port, 350
exhaust purification, 139
exhaust steam, 139
exhausted, 139, 275, 504, 512, 574
exhausted soil, 191, 444
exhaustion of soil, 488, 490
exhaustion stage, 532
existing forest, 525
exit hole, 62, 584
exoadaptation, 477
exobasidium, 495, 496
exobiology, 496
exocarp, 496
exocuticule, 495
exodermis, 496
exogamy, 567
exogenous, 496, 497
exogenous dormancy, 497
exogenous migration, 497
exohormone, 496
exomycorrhiza, 496
exons, 497
exorbitant profit, 51
exoskeleton, 496
exosmosis, 420, 496
exospore, 12, 495, 496
exotesta, 497
exothermic reaction, 124, 135
exotic, 495, 496
exotic plant, 496
exotic species, 496
exotic stream, 497
exotic timber, 496
exotic tree, 496, 570
exotic wood, 496
exotoxin, 495
exotrophic, 496
exotrophic mycorrhiza, 496
expander, 53, 281, 631
expanding bit, 281
expanding type clutch, 281, 604
expanse, 59
expansion bar, 355
expansion-bath treatment, 355
expansion factor, 209, 281
expansion fissure, 355
expansion joint, 33, 355, 419
expansion moulding, 281
expansion of production, 422
expansive clay, 355
expectation, 365, 366
expectation value, 366
expectation value of forest, 408
expected income, 586
expected life, 586
expected life-period, 366
expected regression line, 366
expected value, 366
expected-value-model, 366
expediting setting, 279

expenses for silviculture, 585
experiment design, 439
experiment forest, 243, 436
experiment-station, 439
experiment zone, 436
experimental area, 438, 439
experimental basin, 436
experimental ecology, 436
experimental form factor, 436
experimental plot, 436, 439
experimental population, 436
experimental site, 438, 439
experimental taxonomy, 438
experimental unit, 436
experimental vegetation, 436
experlmental error, 439
explant, 142, 497
explantation, 287, 497, 564, 652
explicit analysis of cost, 497
explicit interest, 325, 371
explicit rent, 325, 614
exploitability, 268, 290
exploitable, 272, 273
exploitable age, 252, 272
exploitable diameter, 273
exploitable girth, 273
exploitable size, 272
exploitable stand, 272, 655
exploitable value, 268, 290
exploitation, 37, 253, 290
exploitation age, 127, 269, 311
exploitation ban, 251, 252
exploitation cutting (felling), 83, 311
exploitation management, 311
exploitation method, 37, 39, 268, 291
exploitation percent, 38, 62, 291
exploitation value, 268, 290
exploitive competition, 291
exploratory sampling, 269
explosion hazard, 16, 569
explosion process, 16
explosive disease, 16
explosive epidemic, 16
explosive evolution, 16
explosive wedge, 16
exponential curve, 323, 619
exponential dispersal, 619
exponential distribution, 619
exponential distribution function, 619
exponential factor, 619
exponential function, 619
exponential growth, 619
exponential growth curve, 619
exponential growth form, 627
exponential growth phase, 115, 619
exponential law, 619
exponential type, 619
exponetial growth, 619
expose dose, 605
exposed bark-pocket, 496
exposed rock, 313
exposure, 16, 17, 524
exposure factor, 17
exposure interval, 17
exposure meter, 17
exposure plot, 16
exposure test, 16, 17, 441
exposure to wind, 573, 604
expressivity, 27, 220
exsiccata, 156, 282
extant vegetation, 525
extended (pit) aperture, 497
extender, 33, 281, 554, 601
extending agent, 601
extending table, 282
extensible language, 273
extensin, 419
extension, 183, 281, 419
extension drum, 603
extension fitting, 419
extension growth, 419
extension ladder, 419
extension leader, 419, 634
extension rope, 603
extension slash, 38
extension stake, 49
extensional strain, 282
extensive agriculture, 72
extensive cultivation, 72
extensive forest survey, 408
extensive forestry, 72
extensive management, 72
extensive pasture, 72
extensive pressure, 355
extensive repair, 78
extensive selection cutting, 255
extensive silviculture, 72
extensive treatment, 72
extensometer, 24, 419, 554
exterior defect, 495
exterior door, 496
exterior finish, 495, 497
exterior logging road, 302
exterior plywood, 336, 441
exterior varnish, 497
exterior view, 496, 497
external benefit, 495
external circulating kiln, 497
external coating, 27, 497
external condensing-coil kiln, 495
external condition, 495, 496
external cost, 54
external decay, 21, 495
external defect, 27, 495
external diameter, 496
external dimension, 497
external environment, 496, 497
external factor, 496, 497
external-fan kiln, 496, 497
external fibrillation, 27
external force, 496
external forcing, 376
external integument, 497
external medulla, 468

E

external parasitism, 496
external phloem, 497
external pith, 468
external sizing, 218, 495
external stress, 495, 497
external torque, 496
external yarding distance, 497
externality, 495
externally fired retort, 496
extinction, 324
extinction coefficient, 531, 532
extinction curve, 324
extinction rate, 324
extinction threshold, 324
extinguish, 248, 364, 441
extirpation, 213
extra banking, 120, 152
extra-budgetary, 587
extra-cutting, 120
extra dividend, 120, 152
extra-fascicular cambium, 446
extra heavy particleboard, 163
extra-metabolites, 495
extra-nuclear inheritance, 518
extra-pliable hoisting rope, 474
extra-stress, 120, 152
extracellar polysaccharide, 13, 518
extracellular enzyme, 13, 518
extracellular space, 13
extract, 61, 74, 251, 476
extract wood, 74, 254
extractable material, 274
extracting, 126, 222, 251, 476
extracting seeds from cone, 382
extraction, 37, 39, 74, 224, 251, 384, 493
extraction battery, 251, 476
extraction damage, 39, 62
extraction distance, 7, 223
extraction drum, 493, 494
extraction fan, 61, 350
extraction-free wood, 510
extraction lane, 223, 595
extraction method, 61, 74, 251
extraction of stump, 265
extraction rack (road), 7, 223
extraction rate, 61, 211, 251, 476
extraction thinning, 415, 547
extractive, 74, 251, 476
extractor, 203, 251, 384, 494, 628
extractory, 493
extragenetic mutation, 220
extramatrical, 27, 429
extramatrical mycelium, 27
extraneous component, 61
extraneous material, 495, 567
extraneous substance, 74, 251, 476
extraordinary felling, 138, 304
extraordinary harvest, 138, 304
extraordinary yield, 120
extrapolating growth percent curve, 497
extrapolation, 495, 497
extrastaminal disc, 542
extratropical belt, 392, 505
extratropical zone, 392, 505
extraxylary, 335
extreme fibre stress, 26, 653
extreme position, 222
extreme pressure, 222
extreme value, 222
extreme weather event, 222
extremely high frequency, 222
extrinsic cycle, 497
extrinsic factor, 495, 497
extrinsic superiority, 497
extron, 220, 497
extruded particle board, 225
extruder, 224, 225
extrusion, 225, 394, 419
extrusion coater, 225
extrusion die, 225
extrusion method, 225, 551
extrusion moulding, 225
extrusion press, 225
extrusion-pressing, 225
extrusion pressure, 225
extrusion system, 225
extrusive rock, 216, 354
exudate, 420
exudation, 142, 420, 569
exuviae, 40, 275, 492
exuviation, 492, 494
eye, 207, 276, 288, 400, 405, 540, 552
eye base, 556
eye cutting, 86, 551
eye grafting, 552
eye hitch, 294, 372
eye of flower, 204
eye-rest colour, 594
eye-splice, 23, 207, 430, 470
eyelet, 178, 277, 534
F-horizon, 123
F-layer, 123
F_1, 597, 642
F_2, 597, 641
Fabian log rule, 140
fabric, 246, 613, 652
fabric unit, 500
face, 37, 39, 149, 165, 323, 520, 521
face angle, 58, 536
face cord, 323
face crossing, 26, 27
face cut, 221
face joint, 26, 37
face knot, 26, 37
face lathe, 314, 362
face layer, 26
face-mark, 37, 323
face milling, 112
face plate, 323, 362
face ply, 26
face side, 8, 26, 294, 611
face value, 358
face veneer, 26, 323
face veneer (composer), 26
faced plywood, 480
facet, 274, 534, 535
facial disk, 323
faciation, 24, 389, 552
facies, 389, 556
facies fossil, 619
facing, 26, 323, 438, 480
facing of pile, 286
facing point, 115

F

factor cost, 422, 559
factor of erosion, 379
factor of production, 422
factor of risk, 145
factor of safety, 2
factor of soil formation, 55
factorial experiment, 515, 570
factoring, 78, 570
factory burden, 50, 169
factory garden, 169
factory lumber, 227
factory management, 169
factory overhead rate, 169, 620
factory planting, 169
factory price, 62
factory seasoning, 50
factory shutdown loss, 169, 481
factory tax, 62
factory timber, 227
facultative aerobe, 232
facultative anaerobe, 232
facultative anaerobic bacteria, 232
facultative autotroph, 232
facultative halophyte, 232
facultative parasite, 232
facultative parthenogenesis, 232
facultative saprophyte, 232
fade, 451, 492
fading variability, 451
faeces, 144, 350
faee measure, 37
faee spreader, 26
fagetum, 458
faggot, 47, 613
faggot wood, 539, 613
fagine, 458
Fahrenheit (temperature) scale, 204
Fahrenheit thermometer, 204
fail-year, 375
failed area, 600
failure, 177, 363, 387, 431
fair average quality, 622
fairlead, 89, 570
fairlead arch, 89
fairleading, 89, 242, 534
fairy ring, 327
falcate, 294
falcon, 297, 469, 580
falconry, 136, 297, 549
fall a quarter, 530
fall block, 108, 367, 421
fall block system, 464
fall in rounds, 142
fall-off, 231
fall place, 304
fall planting, 382
fall-plough, 382
fall seeding, 382
fallen, 365
fallen dead wood, 278
fallen tree, 90, 125
fallen trunk, 90, 125
faller, 126
falling, 125
falling ball impact test, 314
falling block yarding, 421
falling knot, 494
falling mould, 384
falling plate, 126
falling rate period, 233
falling table, 215
falling to the lead, 360, 584
falling wedge, 126
fallout, 53
fallow, 543
fallow crop, 543
fallow land, 543
fallowing, 543
false acacia, 71, 557
false annual (growth) ring, 230
false annual ring, 230
false bunk, 587
false colour image, 230
false dichotomy, 230
false duramen, 230
false form factor, 503, 542
false form quotient, 503, 542
false-heart, 230, 503
false heartwood, 230, 503
false hybrid, 230, 503
false identification, 503
false in-digo, 567
false leg, 155, 503
false mildew, 455
false pleiotropy, 230
false point, 230
false powderpost beetle, 49
false raceme, 313
false ray, 230, 503
false ring, 230, 503
false sapwood, 230, 339
family, 229, 272
family dominant, 229
family heritability, 229
family selection, 229, 365
family sere, 484
family tree, 229, 517
famous historical site, 324, 325
famous site, 324
fan, 59, 145, 414, 465
fan blower, 145, 175
fan blower kiln, 175, 414
fan brake, 145, 560
fan deposit, 414
fan dryer, 175
fan drying, 376
fan-shaped yarding, 414
fan tail, 414
fan terrace, 59
fan training, 414
fan type kiln, 145, 376
fancy pelargonium, 76
fancy plywood, 639
fancy veneer, 639
fanglomerate, 414
fanner, 101, 144, 146, 481
fanning, 146
fanning treatment, 67, 481
fantail towing, 414
far infra(-)red (ray), 594
far infra(-)red ray drying, 594
farm forest, 347
farm forest cooperative, 347
farm forest owner, 347
farm forestry, 232, 347
farm shelterbelt, 347
farm woodland, 347
farm woodlot, 347
farm woods, 347
farming, 168
farmland, 347
farmland ecosystem, 347
farmland shelter-belt, 347
farmstead windbreak, 347
farmyard manure, 114, 259, 347

F

farnesyl-, 127, 248
farrerol, 111
fasciae, 47, 613
fasciae wicker work, 23
fasciation, 19, 23, 79
fascicle, 73, 322, 446
fascicled, 55, 73
fascicular, 73, 446, 502
fascicular cambium, 446
fascicular system, 502
fascicular xylem, 446
fasciculate, 72, 73, 446
fasciculated root, 72
fascine covering, 47, 417
fascine wood, 539, 613
fascine work, 114, 281
fascine work road, 47, 613
fast beating, 579
fast-drying lacquer, 279
fast-germinating, 124
fast-growing and high-yield plantation, 466
fast-growing and high-yielding forest, 466
fast growing species, 466
fast growing tree, 466
fast growing wood, 466
fast grown, 279, 466
fast technology, 279
fastening, 249, 280
fastness, 32, 232, 337
fastness to light, 32, 337
fat-body, 613
fat soil, 139, 507
fatal high temperature, 621
fathom, 289, 572
fathometer, 212, 459
fatigue behavior, 357
fatigue-bending test, 498
fatigue damage, 357
fatigue deformation, 356
fatigue durability, 337, 356
fatigue failure, 356
fatigue fracture, 356
fatigue limit, 356, 357
fatigue resistance, 270
fatigue strength, 356
fatigue stress, 357
fatigue test, 357
fatty resin, 613
fatty seed, 613
fatwood, 120, 325
faucet, 46, 309, 546
fault, 69, 113, 177, 213, 387
fault strike, 113
fault-tolerant computer, 397
faulty, 221, 580, 581
fauna, 109
faunal barrier, 109
faunal province, 383
faunal region, 109
faunal zoogeography, 383
faunation, 109
faunula, 109
Faustmann's hypsometer, 149
Fauteuil, 2, 147
faux swirl, 230, 325
favorable grade, 461, 521
fawn foot, 310
feasibility study, 272
feather, 584
feather board, 584
feather-checking, 123, 124
feather edge, 13, 532, 584
feather edge board, 13, 536
feather figure, 584
feather joint, 469, 584
feather mitre joint, 13
feather veined, 261
feathering out, 64
feature, 474, 475
feature extraction technique, 475
feces, 144, 350
fecundation, 66, 443, 444
fecundity, 429
fecundity schedule, 429
federal forest, 186, 294
federal forest service, 294
federal revenue law, 294
Federal Seed Act, 294
federal-state fire control system, 294
federation, 294
fee, 140, 443
feed, 228, 249, 250, 280, 651
feed attachment, 250
feed chute, 250
feed conveyor, 250, 464
feed ditch, 179, 250
feed head, 33, 250, 280, 319, 505
feed hopper, 167, 228, 250, 505, 638
feed rate, 249, 250
feed roll, 250
feed roller, 250, 446, 571
feed speed, 249, 250
feed stock, 588, 589
feed supplement, 464
feed table, 167, 250
feed yeast, 464
feedback, 129, 212
feedback control, 129, 212
feedback inhibition, 129
feedback loop, 129
feedbed, 250
feeder, 171, 228, 250, 464, 465, 612
feed(er) box, 143, 250, 280, 505, 628
feeder pipe, 250
feeder root, 514, 574
feeder strip, 353
feeder suspension cable, 250
feedforward control, 373
feeding, 167, 171, 250, 465, 505
feeding adjusting lever, 250
feeding and centering unit, 416
feeding apparatus, 250
feeding behaviour, 419
feeding belt, 250
feeding bin, 250
feeding chain, 250
feeding chain conveyor, 250
feeding current, 419
feeding deterrent, 651
feeding device, 250, 505
feeding funnel, 250
feeding habit, 437
feeding incitant, 73
feeding migration, 470
feeding niche, 384
feeding place, 419
feeding root, 514, 574
feeding single-log mechanism, 30, 83
feeding site, 419
feeding stimulant, 634
feeding suppressant, 568
feedstock, 589
feedway, 171, 446
feedworks, 250

feeling for fire, 217, 232
fell, 125, 126
fell field, 190
fell limit, 39
felled sale, 125
felled strip, 37
felled timber, 125
felled tree, 125
feller, 126
feller-boom, 125
feller-buncher, 126
feller-buncher delimber, 126
feller-buncher head, 126
feller-chipper, 126
feller-delimber, 126
feller-delimber-bucker-buncher, 126
feller-delimber-buncher, 126
feller-delimber-slasher-buncher, 126
feller-delimber-slasher-forwarder, 126
feller-director, 126
feller-forwarder, 126
feller-limber-bucker, 126
feller-processor, 126
feller-skidder, 126
fellfield, 209
felling, 37, 125, 633
felling age, 127, 269, 311, 633
felling and mortality record, 38
felling area, 38, 127
felling area line, 127
felling axe, 126
felling blank, 38
felling budget, 38
felling bunching machine, 126
felling by compartment, 3
felling by group, 279, 389
felling class, 37
felling contract, 38
felling contractor, 37
felling control, 38
felling cost, 37, 38
felling crew, 38
felling cut, 126
felling cycle, 39, 311
felling decision, 38
felling direction, 126
felling dog, 46, 371
felling face, 90, 149, 520, 521
felling for group system, 279, 389
felling for shelter-wood system, 404
felling front, 38
felling gap, 125
felling in alternate strips, 165, 239
felling in progressive strips, 293
felling in strip, 80
felling interval, 211, 311
felling key, 38
felling licence, 39
felling machine, 126
felling machinery, 406
felling notch, 126, 521, 586
felling of wolf trees, 16, 111, 416, 647
felling operation, 127
felling order, 39
felling period, 38
felling permit, 39
felling plan, 38, 39
felling point, 125, 446
felling quota, 39
felling rate, 38
felling record, 38
felling register, 126
felling return, 126
felling-road, 37
felling rotation, 311
felling rule, 126
felling saw, 126
felling season, 126
felling section, 37, 38, 125
felling sequence, 39
felling series, 38, 39
felling shake, 126
felling under selection system, 601
felling unit, 37, 39, 126, 655
felling volume, 38
felling volume assessment, 38
felling wedge, 126
felling with axe, 149, 269
felling yield, 38
felling yield forecast, 38
felloe wood, 312
felt board, 265, 401
felt grain, 256
felt side, 318, 611
felting, 55
felting machine, 364
female cell, 328
female cone, 69
female flower, 69
female parthenogenesis, 173
femur, 78, 175, 492
fen, 554, 605
fen clay, 604
fen land, 604
fen peat, 93, 604
fen soil, 92, 93
fence, 271, 288, 502, 602
fence boom, 602
fence for heaping sand, 133
fence for protection against avalanche, 134
fence holding boom, 283
fence month, 252
fence post, 288, 502, 602
fenchyl, 146
fender, 21, 88, 209
fender boom, 89
fender skid, 88, 132, 209
fenestriform pit, 67
fenitrothion, 411
fennel oil, 212, 534
fennel seed oil, 534
feral, 559
feral population, 559
ferallitic soil, 481
ferment, 123
fermentation, 123
fermentation horizon, 123
fermentation layer, 123
fermented soil, 446
fermention, 123
fern, 266
fern house, 266, 556
fern leaf, 266
ferrallitic soil, 480, 481
ferredoxin, 481
ferric-siallitic soil, 481
ferrochelatase, 553
ferroflavoprotein, 553
ferruginous lateritic soil, 481
ferrule, 281, 474
ferti-seeding, 213, 431
fertile, 139, 274, 341, 342
fertile absorber, 582
fertile branch, 245, 342
fertile ecosystem, 139
fertile flower, 69, 342

F

fertile meadow, 139
fertile pasture, 139
fertile pollen, 342
fertility, 96, 128, 139, 342, 396
fertility factor, 541, 621
fertility gradient, 139
fertility table, 430
fertilization, 431, 443
fertilizer, 139, 431
fertilizer apparatus, 431
fertilizer application, 431
fertilizer applicator, 431
fertilizer blower, 145, 431
fertilizer distributor, 205, 401, 431
fertilizer formula, 139, 171
fertilizer guarantee, 139
fertilizer injector, 139
fertilizer irrigation, 431
fertilizer practice, 431
fertilizer ratio, 139
fertilizer spreader, 139, 401
fertilizing, 431
fertilizing machinery, 431
festmeter(fm), 435
festoon dryer, 208, 545
festoon-like, 202
festoon porous wood, 202
Feulgen reaction, 149
fiber, 522, 523
fiber board, 522
fiber bundle, 523
fiber classification, 522
fiber particleboard, 523
fiber refining, 523
fiber saturation point, 522
fiber sticking, 344
fiberization, 523
fiberize, 523
fibre, 522, 523
fibre backing, 522
fibre block, 523
fibre bonding, 523
fibre building board, 236
fibre bundle, 182, 522, 523
fibre cap, 523
fibre classification, 522
fibre cleaning, 523
fibre core, 316, 523
fibre flax, 523, 552
fibre fractionation, 523
fibre furnish, 523
fibre membrane, 523
fibre orientation, 522, 523
fibre pot, 522
fibre recovery, 523, 598
fibre saturation point, 522
fibre separator, 522
fibre sheath, 523
fibre spreader head, 523
fibre strength, 523
fibre stress, 523
fibre stuff, 237, 523
fibre suspension, 523
fibre texture, 523
fibre tracheid, 523
fibre-water-suspension, 523
fibreboard, 522
fibreboard core plywood, 522
fibreglass, 30
fibrestock, 237, 523
fibrid, 53, 285
fibriform vessel element, 523
fibriform vessel member, 523
fibril, 501, 519, 522, 591
fibril angle, 522, 591
fibril layer, 522
fibrilla (*plur.* -llae), 522, 535, 591
fibrillar crystal, 501, 591
fibrillar orientation, 522, 591
fibrillar pattern, 522
fibrillation, 519, 522, 591, 631
fibrin, 522, 548
fibro-vascular bundle, 502
fibro-vascular tissue, 502
fibrocortex, 523
fibrograph, 523
fibrous cell, 523
fibrous peat, 523
fibrous root, 319, 523
fibrous root system, 544
fibrous tracheid, 523
fictitious ring, 503
fid, 312, 532, 635
fid hook, 23
fiddle butt, 476
fiddleback, 476, 518, 541
fiddleback chair, 379, 476
fiddleback figure, 476, 518
fiddleback grain, 476, 518
fiddleback mottle, 476, 518
fidelity, 15, 19, 388
fiducial interval, 274
fiducial limits, 274
fiducial mark, 98, 221
fiducial point, 219, 221
field, 50, 478, 559
field abandoned stocking, 297
field-book, 559
field capacity, 478
field checking, 559
field contouring, 91
field data, 497, 559
field design, 478, 497
field emergence, 62, 583
field experiment, 478
field germination, 50
field grafting, 311, 441
field identification, 525, 559
field ionization, 50
field laboratory, 302, 559
field layer, 41
field maximum moisture capacity, 478
field measurement, 435, 559
field moisture capacity, 478
field moisture deficiency, 478
field moisture equivalent, 478
field-note, 497
field nursery, 301, 304
field of microscope, 525
field of view, 439
field performance, 478
field planting, 62, 108
field planting value, 478
field plot, 438, 560
field reconnaissance, 525
field resistance, 478
field-run-seedling, 62
field scope, 85
field setting, 62, 108
field sheet, 559
field sketch, 559
field survey, 435
field survival, 478
field test, 478
field treatment, 525, 559
field trial, 559

field velocity of ground water, 98
field water capacity, 478
field-work, 478, 559
fielded panel, 529
fifth girth method, 512
fifth-wheel, 9, 17, 612, 637
fifth-wheel trailer, 612
fight shooting, 594
figure, 204, 487
figure in wood, 329
figured butt, 204
figured stumpwood, 204
figured veneer, 204, 639
figured wood, 204
figuring, 204
filament, 204, 462, 523
filamentary fibril, 462
filamentous fungus, 462
filbert aphid, 608
file, 74, 506
file area, 506
file holder, 74
file number, 506, 641
filer, 74, 543
filial, 199, 641
filial generation, 596, 641
filial regression, 596, 641
filiform, 462, 523, 526
filing, 23, 74, 543
filing angle, 58, 262, 327
filing clamp, 74
filing machinery, 74
filings, 74
fill, 479, 635, 638
filled seed, 14
filled valley, 583
filler, 228, 306, 309, 479
filler strand, 73, 479
fillet, 166, 520, 529, 538, 549
fillet strength, 538
filling, 14, 32, 478, 479
filling and sowing equipment for containerset, 397
filling in, 229
filling (in), 33
filling machine, 59, 74, 75, 638
filling of blanks, 33, 229
filling post, 625
filling-up, 32, 33
fillister, 4, 16
fillistered joint, 41
film, 13, 241, 401
film adhesive, 13, 241
film base, 401
film-builder, 54
film force, 13
film former, 54
film glue, 13, 241
film yeast, 47
filter, 187, 316, 357
filter aid, 634
filter feeder, 316
filter feeding, 316
filter food, 316
filter medium, 187, 316
filterable membrane, 316
filter(ing) press, 550
filtrable virus, 187
filtrate, 316
filtrated air, 187, 509
filtration, 187, 316, 420
fin boom, 55, 366
final age, 127, 633
final age of the maximum forest rent, 301
final air press, 199
final clearing, 381, 633
final community, 626
final crop, 633
final crop tree, 441, 633
final cutting, 625, 633
final cutting are, 127
final decay, 327
final diameter, 127
final energy, 625
final felling, 625, 633
final germination percent, 654
final host, 626
final income, 127, 633
final infiltration capacity, 626, 654
final inventory, 365
final mean annual increment, 127, 654
final moisture content, 626
final sander, 254
final set, 625
final spacing, 654
final stage, 625
final statement, 264
final steam bath, 199
final steaming, 626
final stock, 633
final stock table, 626
final stress, 626
final treatment, 654
final vacuum, 199
final volume, 625
final water sander, 433
final yield, 127, 633
finance, 37, 171, 248
financial claim, 603
financial deficit, 37, 587
financial maturity, 252
financial plan, 37
financial position, 37
financial record, 37, 279
financial returns from clear felling, 633
financial returns from final cutting, 633
financial returns from thinnings, 235, 252, 445
financial returns of forestry, 303
financial rotation, 252
financial value, 217, 248
financial yield, 252
financing, 62, 248, 476
finder, 472
fine bird's eye (figure), 519
fine cellulosic strand, 501
fine clay, 519
fine crumb, 519
fine earth, 519
fine fibre, 519
fine filter, 254, 519
fine fuel, 569
fine fuel moisture code, 519
fine fuels, 519
fine grain, 519
fine-grained, 519
fine-grained environment, 519
fine-grained soil, 519
fine-grained wood, 519
fine grid, 519
fine-grit belt, 519
fine mesh wire, 519
fine mull, 518
fine particle, 519
fine root, 518
fine sand, 519
fine sandy loam, 519
fine scale, 533
fine setting, 254
fine shake, 518
fine shellac, 415
fine structure, 51, 519
fine texture, 518, 519
fine-textured wood, 518
fineness, 68, 518
fineness ratio, 291
finger joint, 619

finger jointed plywood, 619
finger jointing machine, 619
finger-lap joint, 619
finger of wood, 619
finger plate, 492
fingerling, 535
finish, 27, 180, 181, 326, 487, 626, 653
finish carpenter, 543
finish coat(ing), 27, 543, 605, 626
finish varnish, 605
finished goods, 54, 620
finished plywood, 639
finished product, 54, 620
finished size, 16, 54, 256
finished stock, 227, 256
finishing, 16, 26, 27, 254, 487, 626
finishing beater, 54
finishing press, 121, 438
finishing section, 639
finite element method, 581
finite population, 581
finite rate of increase, 630
finite rate of natural increase, 581
finned pipe, 59
Finnish parabolic caliper, 143
fir, 414
fir balsam, 228, 286
fir cones oil, 286
fir (needle) oil, 286, 465
fir resin, 286
fir tussock moth, 77, 210
fire-adapted, 440
fire alarm, 217
fire-alarm indicator, 217
fire-alarmbox, 217
fire analysis, 215
fire annihilator, 324
fire appraisal, 215, 217
fire area, 133, 215
fire atlas, 133
fire barrier, 133
fire beater, 75, 324
fire behavior, 215, 216
fire behaviour, 215, 216
fire belt, 132, 531
fire blight, 217
fire boss, 133, 364
fire box, 216, 390
fire brand, 217
fire break, 132, 133
fire break belt, 132
fire break forest, 133
fire break tree belt, 133
fire brigade, 531
fire bucket, 531
fire calibration, 216
fire cause class, 217
fire chamber, 216, 390
fire climax, 216
fire climax community, 216
fire co-operator, 566
fire concentration, 217
fire containment, 166
fire control improvement, 133
fire control organization, 133, 217
fire control plan, 132, 133
fire control station, 133
fire control tactics, 132
fire-damage class, 216
fire damaged, 216
fire danger, 216
fire-danger board, 216
fire danger class, 217
fire danger division, 216
fire danger index, 216
fire danger map, 216
fire-danger measurement station, 216
fire danger meter, 216
fire-danger meter method, 217
fire danger rating, 216
fire danger rating system, 216
fire danger scale, 216
fire danger season, 216
fire-danger table, 216
fire danger weather, 216
fire department, 531
fire detection, 217
fire detection by satellite remote sensing, 503
fire detective device, 217
fire devil, 217
fire discovery, 217
fire district, 133
fire drill, 215, 259, 531
fire ecology, 216
fire edge, 217
fire engine, 259, 324, 531
fire escape, 2, 651
fire extinguisher, 324, 531
fire fighter, 364, 531
fire fighting, 75, 364, 531
fire fighting crew, 133, 531
fire-finder, 217
fire-finder map, 217
fire flank, 217
fire foam, 324, 352
fire frequency, 216
fire guard, 132, 133, 394
fire hazard, 216, 217
fire hazard index, 216, 217
fire hazard reduction, 238
fire head, 216
fire in stem, 447
fire injury tree, 216
fire insurance (seed) tree, 133
fire intensity, 216, 217
fire killed timber, 216
fire kiln, 215
fire-lane, 132
fire line, 133, 166
fire line construction equipment, 133
fire load index, 216, 390
fire management, 215, 531
fire-mantle, 133
fire margin, 217
fire master, 531
fire net, 259
fire occurrence map, 217
fire office, 217
fire origin, 217
fire patrol, 217, 301
fire place, 21, 216, 310, 390
fire plotting map, 300
fire plough, 133
fire plow, 133
fire presuppression, 133
fire prevention, 132, 217, 651
fire-prevention forest, 133
fire-prevention plan, 217
fire-prevention project, 217
fire prevention tree, 133
fire-prone, 604
fire-proofing, 132, 651
fire proofing agent, 133, 337, 651
fire proofing chemical, 133, 337
fire proofing material, 132, 337

fire proofing treatment, 132
fire propagation, 215
fire protecting construction, 133
fire protecting zone, 132
fire protection, 132, 531
fire protection break, 133
fire protection tree, 133
fire rapidly spreading, 136
fire rear, 216
fire resistance, 133, 337
fire-resistance test, 337
fire-resistant plywood, 133, 337
fire resistant tree, 337
fire resisting finish, 337
fire retardant, 133, 338, 651
fire retardant agent, 651
fire retardant chemical, 651
fire retardant coating, 133, 651
fire retardant paint, 133
fire-retardant preservative, 651
fire retardant salt, 651
fire-retardant-treated wood, 651
fire retardant treatment, 651
fire-retarding of wood, 331
fire return interval, 217
fire ring, 215, 216
fire risk, 216, 217
fire risk map, 216
fire scan, 198, 215
fire scar, 215, 216
fire season, 217
fire severity, 215
fire simulator, 217
fire size class, 217
fire-slash, 217
fire spread, 216, 301
fire stop, 88, 132, 620
fire storm, 215, 280
fire strategy, 324
fire suppression, 324, 364
fire suppression system, 324
fire swatter, 75, 324
fire-tool cache, 324
fire tower, 133
fire-trace, 132, 216
fire trail, 259
fire trench, 132
fire-tube test, 215
fire type, 217
fire warden, 97, 133
fire watch tower, 133
fire weather, 216
fire weather forecast, 216
fire weather index, 216
fire-weather station, 133
fire whirl, 216
fire whirlwind, 216
fire wood, 539
fire wound, 216
firearm, 215, 297, 375
firebreak, 132
fireclay, 337
fireman, 531
fireplace fungus, 441
fireproof plywood, 133, 337
fireproofed wood, 133
fireproofing wood, 133
firer, 100, 135, 570
firetriangle, 215, 217
firewood, 390, 539, 540
firewood cutter, 377, 540
firewood forest, 540
firewood yard, 540
firm knot, 232, 292
firm mull, 506
firm red heart, 63, 232
firm wood cutting, 46
firmer chisel, 130, 332
firmer gouge, 11, 332
firmo-viscosity, 176
firn, 28, 291, 576
first-category forest inventory, 185, 563
first-class river, 563
first-class standard, 563
first coat, 95
first coat of paint, 95, 100
first cutting, 443
first felling-line, 62
first filial generation, 596
first filter, 63, 72
first frost, 64, 455, 599
first grade timber, 563
first growth, 65, 591
first in first out, 522
first injurious frost, 64, 599
first-instar larva, 63
first level carnivore, 63, 485
first log, 167
first periderm, 100
first prime quality, 563
first snow, 64
first stage, 100
first stage suspension drying, 63
first thick branch, 100
first thinning, 63
first working period (5 years), 253
fiscal year, 37, 278
fish culture, 557, 584
fish-eye lens, 584
fish farming, 557, 584
fish gelatin, 584
fish glue, 584
fish-net float, 584
fish plate, 229, 584
fish pond-dike system, 278
fish scale pit, 584
fishbone system, 396, 584
fished joint, 229, 584
fishing boats, 584
fishing harbor (port), 584
fishing intensity, 33
fishtail, 80, 520
fissility, 356
fission, 142, 297
fissure, 297
fissure network soil, 500
fissured bark, 297
fit, 30
fit grafting, 192
fit hook, 207, 249
fitch, 2
fitness, 439, 440
fitness of environment, 208
fitted furniture, 375
fitting, 30
fit(ting), 269, 296, 543, 610, 638
fitting height curves, 447
five-leaved pine, 512
five-needle pine, 512
fixation of shifting sand, 307
fixative, 15, 107, 176
fixed-area plot sampling, 176
fixed area sampling, 176, 557
fixed assets, 176
fixed beam, 176
fixed bed, 175

F

fixed capital, 176
fixed cost, 31, 175, 176
fixed dune, 176
fixed forest, 175, 176
fixed guide, 176
fixed head jointer, 176
fixed horizontal dryer, 176
fixed input, 176
fixed knife planer, 176
fixed knot, 232, 292
fixed lateral boom, 176
fixed load, 176
fixed nosebar, 176
fixed nursery, 176, 576
fixed phase, 176
fixed price, 176
fixed program computer, 175
fixed raft, 575
fixed regeneration block, 176
fixed reserve, 175, 176
fixed satellite, 97, 258
fixed-size plot, 175, 176
fixed stone ring, 176
fixed tan, 246
fixed wire cableway, 56, 85
fixed-wire logging, 107
fixed yearly cut, 176
fixer solution, 108
fixing reserve, 176
fixing solution, 107, 108, 176
fixings, 152, 375
fixity factor, 506
fjeld, 28
fjord, 519
flabellate, 414
flabellate dichotomy, 414
flabellinerved, 414
flag-shaped tree, 366
flagellum (*plur.* -lla), 23
flagman, 461, 540
flagship species, 366
flagstone, 8, 13, 364
flail type delimbing device, 295
flake, 13, 16, 140, 367
flake bark, 358
flake board, 400
flake-faced particle board, 13, 519
flake orienter, 16
flake-producing machine, 16, 532
flake-shaving machine, 16, 532
flake shellac, 357, 358
flakeboard, 13, 16, 468
flaker, 16, 532
flaking, 16
flaky wood, 358
flame cleaning, 216
flame figure, 217
flame front, 215
flame gun, 100
flame ionization detector, 216, 380
flame-like arrangement, 217
flame-like pore arrangement, 217
flame-penetration test, 216
flame photometer, 216
flame-proof, 337
flame propagation, 216
flame reaction, 216
flame resistance, 337
flame retardant, 651
flame-retardant fibre board, 337
flame retardation, 651
flame spectrometry, 216
flame spread, 216
flaming, 216
flaming porous wood, 216
flaming stage, 582
flammability, 273, 569
flamy figure, 217
flange, 8, 127, 184, 309, 486
flange joint, 127
flank, 285
flank fire, 43, 217
flank fire suppression, 43
flank of fire, 43, 217
flanking backfire, 43
flanking fire, 43
flanking fire suppression, 43
flap, 75, 88, 128, 214
flap and elbow table, 606
flap-seat, 606
flap table, 606
flare root, 8
flash drying, 222, 461
flash evaporator, 414
flash fuel, 569
flash pan, 570
flash tube dryer, 279
flash(ing) point, 414
flat, 362, 375
flat bark beetle, 23
flat bed, 359, 360, 362
flat bed truck, 359
flat boom, 82, 362
flat-bottom gullet, 360
flat-bottom rail, 360
flat-bottomed valley, 360
flat cut, 524
flat (cut) grain, 524
flat-decked barge, 580
flat finish, 362
flat form training, 23
flat garden, 362
flat grain face, 524
flat grain lumber, 524
flat headed borer, 221
flat helix, 361
flat mitre joint, 362
flat molding process, 362
flat nose tool, 362
flat-nosed, 362
flat peach, 351
flat piling, 360
flat plate radiometer, 359
flat-platen-pressed particle board, 362
flat-pressed board, 362
flat raft, 362
flat rate, 484
flat sanding, 362
flat-sawed, 524
flat sawing, 524
flat-sawn, 524
flat-sawn grain, 524
flat-sawn surface, 524
flat-sawn timber, 524
flat screen, 362
flat screen classifying system, 362
flat spring, 8, 23, 357
flat stacking, 360
flat strainer, 359, 362
flat strand rope, 567
flat top peak, 360
flat-top saw, 184
flat-topped barge, 360, 580
flat type piling, 360
flat vibrating screen, 359
flathead, 126
flathead (wood) borer, 221
flathead woodscrew, 362
flatiron-effect, 596
flatpressing, 362
flattened pit orifice, 458

F

flatting, 13, 16, 23, 140, 361, 509, 531
flatting agent, 531
flatting finish, 531, 552
flatting varnish, 509
flavanol, 210
flavin(e), 210
flavodoxin, 210
flavon(e), 210
flavonoid, 210
flavonoid tannin, 210
flavoprotein, 210
flavorant, 437
flavour, 437, 480, 528
flaw, 298, 387
flaw-detector, 473
flaw-piece, 22
flax shive, 552
flax shives board, 552
flea beetle, 480
fleam, 262
fleck, 7, 501
fledgling, 287
fleet angle, 357
flesh liquor, 6
flesh weight, 399
fleshy, 399
fleshy fruit, 399
fleshy root, 399
fleshy seed, 399
flexibility, 339, 398, 399
flexibility coefficient, 339, 398
flexibility in cut, 305, 472
flexibility mechanism, 305
flexible bent surface, 398
flexible conduit, 400, 417
flexible contact platen, 472
flexible coreboard, 399
flexible coupling, 339
flexible door, 264
flexible hose, 339
flexible metallic conduit, 249
flexible metallic tubing, 249
flexible pipe, 339, 400, 417
flexible plastic caul, 339
flexible plywood, 339
flexible price, 472
flexible rubber impeller pump, 401
flexing, 273
flexural displacement, 497, 498
flexural failure, 498
flexural rigidity, 271, 339
flexural strain, 339, 498
flexural strength, 271, 339
flexural test, 498
flexural vibration, 498
flexure strength, 271
flexure stress, 498
flight altitude, 136, 190
flight behavior, 136
flight conveyor, 177, 294
flight distance, 136
flight feather, 136, 569
flight height, 136, 190
flight level, 136
flight limit, 136, 628
flight line, 136, 191
flight lock gun, 216
flight map, 191
flight path, 136, 190
flight strip, 190, 191
flip-flop, 65
flipper, 88, 128, 351
flitch, 17, 81, 318
flitch band saw, 318
float, 307, 358, 534
float gauge, 147, 148
floatability, 358
floatable waterway, 307
floatage, 330, 404
floatation test, 148
floatation tire, 147
floated bright, 257, 307, 350, 381
floater, 147, 148, 307
floating, 123, 307, 358
floating assets, 306
floating-boom, 147, 148
floating capital, 306, 580
floating-channel, 307
floating channel realignment, 307
floating conveyer, 62, 148
floating crane, 148, 458
floating dryer, 306
floating facility, 307
floating hook, 75, 173, 307, 640
floating island, 358
floating lateral boom, 42, 43, 362
floating-leaf water plant, 148
floating-leaved aquatic plant, 148
floating leaved plant, 148
floating-leaved vegetation, 148
floating log path, 307, 358
floating log set, 148
floating mark, 147
floating meadow, 358
floating plant, 358
floating plant community, 358
float(ing) position, 147, 579
floating pressure bar, 579
floating rake, 148
floating rear, 273, 307
floating root, 148
floating-scale calculation, 147
floating transporter, 148
floating wood, 307
floc, 241, 544
floccosoids, 5
flocculant, 544
flocculate, 544
flocculated colloid, 544
flocculation, 544
flocks, 113, 397
flocky precipitate, 544
flood, 199, 604
flood absorption capacity, 283
flood basin, 130
flood coater, 251
flood control, 132, 199
flood control forest, 132
flood control project, 132
flood control reservoir, 132
flood current, 199, 604
flood dam, 307, 544, 545
flood damage, 199
flood damage prevention forest, 132
flood discharge, 199, 538
flood erosion, 199
flood gate, 132
flood height, 199
flood land, 193, 198
flood level, 199
flood log, 602
flood mark, 199
flood peak, 198

F

flood peak discharge, 198
flood peak rate, 198, 199
flood period, 199, 549
flood plain, 59, 193, 198
flood-plain accumulation, 318
flood-plain bench, 318
flood-plain marsh, 318
flood plain meadow, 193
flood plain terrace, 318
flood retention, 283
flood season, 199, 549
flood stage, 199, 500
flood storage, 544
flood tolerance, 338
flood trash, 307, 358
flooded solonchak soil, 130
flooding, 317
flood(ing) irrigation, 317
floodplain forest, 193
floor, 95
floor baffle, 95
floor beam, 95, 196, 309
floor-ceiling construction, 95
floor cramp, 249
floor joist, 309
floor layer, 302
floor load, 95
floor plank, 95, 309
floor plant, 302
floor plate, 95, 103
floor space, 603
floor stratum, 302
floor strutting, 95
floor system (of bridge), 377
floor-type belt sander, 314
floor vegetation, 302
flooring, 95, 258
flooring block, 131, 364
floor(ing) board, 95, 309
flooring saw, 58, 95
flooring timber, 95
flora, 617, 618
floral arrangement, 46, 203
floral axis, 204
floral bud, 204
floral design, 203
floral envelope, 202
floral formula, 202
floral induction, 54
floral initiation, 203, 204
floral nectar, 203
floral preservative, 203, 377
floral stimulus, 54
flord, 519
florence flask, 254, 360
florescence, 203, 268
floret, 533
floretum, 204
floribunda rose, 144, 263
florigen, 54
florist, 557, 629
florist industry, 203
florist violet, 642
floristic, 617
floristic composition, 383, 617, 626
floristic element, 617
floristic plant geography, 617
floristic region, 617
florulent, 202
floscular, 202, 533
flosculous, 202, 533
flotation, 148, 358
flotation test, 148, 358, 545
flow, 306, 307, 466
flow approach, 237
flow area, 307
flow box, 306, 307, 499
flow chart, 56, 306, 307
flow coater, 240, 307
flow coating, 240, 306, 307
flow control, 307
flow cup, 344
flow curve, 306, 308
flow diagram, 306, 655
flow duration curve, 307
flow figure, 306
flow measurement, 307
flow meter, 307
flow of gum, 306
flow of veneer, 293
flow-off, 256, 569
flow resistance, 270, 306, 308
flow sheet, 306
flow stress, 306
flowage, 130, 306, 398
flower, 202, 254, 268
flower abscission, 203
flower arrangement, 46, 203
flower basket, 203
flower bed, 203, 204
flower border, 203, 258
flower branch, 204
flower bud, 204, 551
flower bug, 202
flower centre, 203
flower cluster, 202, 491
flower colour chart, 203
flower decoration, 203
flower form, 204
flower garden, 203, 204
flower hedge, 203
flower induction, 54, 268
flower initial, 204
flower initiation, 203, 268, 595
flower of ice, 28
flower plot, 174
flower pot, 202, 203
flower primordium (*plur.* -ia), 204
flower show, 203, 204
flower trellis, 203
flower tub, 204
flower type, 204
flower vase, 203
flowering hormone, 54, 268
flowering period, 203, 268
flowering plant, 580
flowering shrub, 178, 203
flowering tree, 203
flowering wood, 204
flowing method, 240, 307
flowing reserve, 564
flowstone, 307, 308
fluctuating drying schedule, 30
fluctuating load, 30, 35, 317
fluctuating-pressure process, 30
fluctuating temperature, 24, 30
fluctuation, 30
fluctuation of water table, 459
fluctuation pressure, 30, 317
fluctuation variation, 45
flue, 350, 481, 553
flue gas, 553
fluff, 367, 445
fluffer, 445, 464, 522
fluffiness, 367
fluid capital, 306
fluid motor, 562
fluid penetration, 562
fluid power transmission, 561

fluid pressure, 562
fluid pump, 307, 562
fluid resistance, 306, 307
fluid-tilt sensor, 562
fluidics, 306, 418
fluid(i)meter, 306, 344
fluidity, 306
fluidization, 306
fluidized bed, 306
fluidized bed dryer, 306
flume, 456, 570
flume box, 456
fluorescence, 573
fluorescence analysis, 573
fluorescence chromatogram, 573
fluorescence spectrophotometer, 573
fluorescence spectrophotometry, 573
fluorescence spectrum, 573
fluorescent lamp, 397, 573
fluorescent light, 573
fluorescent light box, 573
Fluorising process, 147
fluoro gum, 147
Fluoroacetamide, 147
fluorospectrophotometer, 573
flush, 59, 124, 237, 486
flush bead, 4
flush control, 282
flush door, 290, 362, 385
flush handle, 3, 59, 362
flush panel, 362
flush panelled door, 362
flush point, 383
flush weir, 538, 569
flute, 4, 172, 495
flute budding, 529
fluted bed roll, 41
fluting plane, 4
fluvent, 59
fluvial, 192, 193
fluvial erosion, 193
fluvial terrace, 192, 193
fluviatile deposit, 192, 193
fluviatile loam, 59
fluviogenic soil, 59
fluvioglacial, 28, 192
fluvioglacial erosion, 28
fluvisol, 59
flux adjustment, 482
fly-bar, 88
fly camp, 304, 531
fly rail, 606
fly rollway, 110
flycatcher, 5, 507
flyer carriage, 470
flyer slackline yarding, 464
flying crane, 136
flying cut-off saw, 8, 279, 564
flying dutchman, 44, 133
flying height, 136, 190
fly(ing) knife, 136
flying machine, 470
flying nursing, 304, 564
flying-shore, 195
flying spot scanner, 136
flying survey, 191
flyway, 372
flywheel horsepower, 136
fm (festmeter), 435
foam breaker, 363, 532
foam core, 352
foam glue, 123, 352
foam gluing, 123, 352
foam killer, 532
foam-like structure, 352
foam rubber, 188, 352
foamed adhesive, 123, 352
foamer, 123, 352
foaming agent, 123, 352
foamy adhesive, 367
focal distance, 242
focal length, 242
focal plane, 242
focal region, 194
focus of infection, 66, 159
focus point, 242, 263, 624
fodder, 464
fodder crop, 464
fodder plant, 464
fodder tree, 464
fodder yeast, 464
foehn (wind), 143
fog, 513
fog belt, 513
fog chamber, 513
fog desert, 119
fog drip, 513
fog droplet, 513
fog forest, 513
fog prevention forest, 134
fog prevention forest zone, 134
fogger, 355
fogging and spraying machine, 355
fog(ging) machine, 355
foil sampler, 207
fold, 606
folded-chain theory, 605
folding chair, 274, 605
folding door, 605, 606
folding endurance, 338
folding furniture, 605
folding number, 338, 605
folding (pocket) rule, 605
folding screen, 605
folding strength, 271, 338
folding table, 605
folding tester, 338
foliaceous, 560, 561
foliage, 73, 560
foliage dressing, 560
foliage expanding, 603
foliage height diversity, 560
foliage leaching, 560
foliage plant, 178
foliage spray, 168, 560
foliaged tree, 282
folial gap, 561
folial trace, 560
foliar diagnosis, 560, 561
foliar volume, 560
foliation, 560, 583
foliogenous ecto-humus, 561
foliole, 535, 561
follicle, 174
follicular, 174
follow-up, 467, 602
follow-up device, 467
food additive, 437
food base, 437, 574
food chain, 437
food chain structure, 437
food consumption, 419
food cycle, 437
food-deficiency disease, 557
food flavor, 437
food flavoring, 437
food gatherer, 39
food habit, 437
food insecurity, 295
food intake, 250, 419
food niche, 437

F

food patch, 464
food plant, 437
food relationship, 33
food-seeking behaviour, 321
food web, 437
food-yeast, 437
foot bath, 651
foot board, 243, 470
foot bolt, 320
foot-candle, 57
foot diet, 642
foot of the bed, 67
foot path, 36, 470, 533
foot pedal, 243
foot run, 57, 571
foot-scent, 651
foot screw, 96, 219
foot-second, 571
foot switch, 243
footage, 8, 57
footer, 243, 561, 635
foothill, 382, 413
footing, 220, 221, 603
footlid, 635
footstool, 243
forage, 464
forage acre, 273
forage crop, 336, 464
forage factor, 464
forage plant, 336, 464
foraging, 321
foraging behaviour, 321
foraging component, 514
foramen (*plur.* -mina), 276, 352, 404, 534
foraminate included phloem, 404
foraminate perforation, 404
foraminate perforation plate, 404
foraminate type, 404
foraminoid, 342
forb, 137, 281
forb meadow, 597
forb steppe, 281, 597
force of cut, 378
force of friction, 326
force of support, 612
forced-air drying, 376
forced burning, 376
forced circulation kiln, 376
forced cutting, 376
forced draft, 376, 550
forced draught kiln, 376
forced feed, 376
forced germination, 376
forced recirculating dry kiln, 376
forceps, 346, 373
forcible separation method, 376
forcing bed, 74, 505, 628
forcing culture, 73
forcing house, 73, 505
forcing of sprouting, 73, 74
forcing structure, 74
ford, 187
fording, 605
fore-and-aft road, 650
fore-gut, 372
fore plane, 73
fore sight, 324, 640
fore-stock, 157, 159
fore trigger, 372
fore wing, 372
forecast, 586
forecast of abundance, 60
forecast of epiphytotic, 29, 616
forecast price, 586
forecourt planting, 373
foredune community, 455
foreground, 373, 486
foreground processing, 373
foreign matters, 229, 567, 597
foreign particles, 597
foreign plant, 496
foreign-plant introduction, 570
foreign pollen, 496
foreign seeds, 567
forepump, 373, 586
foreshock, 373
foreshore, 188, 373
forest, 405
forest accessibility, 408
forest act, 406
forest administration, 407
forest aerial survey, 407
forest amelioration, 298
Forest and Rangeland Renewable Resources Planning Act, 407
forest animals, 406
forest appraisal, 408
forest area, 406
forest area under management, 409
forest assessment, 407
forest assets, 407, 410
forest association, 302
forest attraction, 314
forest-beat, 405
forest belt, 298
forest biology, 409
forest biomass, 409
forest biotope mapping, 409
forest bird, 302
forest bookkeeping, 406
forest border, 303
forest botany, 410
forest boundary, 407, 448
forest budget, 410
forest business, 410
forest by-product, 300
forest cadastral register, 406
forest cadastre, 406
forest calamity, 410
forest canopy, 300
forest capital, 410
forest catastrophe, 409
forest category, 304
forest chair, 73
forest charge, 289, 406
forest chemical product, 298, 300
forest chemistry, 298, 407
forest circle, 432, 438
forest-clad, 406
forest classification, 406
forest clearing, 268
forest climate, 408
forest climax, 406
forest co-operative society, 303
forest college, 303
forest colony, 302
forest combined with agriculture, 214
forest commodities, 298
forest community, 408
forest compartment, 298, 408
forest composition, 410
forest concession, 39, 409
forest condition, 301
forest conservation, 405
forest conservation law, 405
forest consisting of even-aged stand, 483
forest-cost-capital, 406, 408

F

forest council, 408
forest cover, 406
forest cover percent, 406
forest coverage, 406
forest credit, 410
forest criminal law, 410
forest critical temperature, 409
forest crop, 299, 409
forest cruise, 407, 409
forest cultivation machinery, 406
forest culture, 408, 600
forest cut, 406
forest degradation, 409
forest density, 408
forest denudation, 283
forest depot, 222, 413
forest description, 301, 409
forest destroyer, 213
forest destruction, 213, 408
forest deterioration, 409
forest devastation, 213, 407
forest development, 406
forest devil, 90, 266
forest disease, 405
forest distribution map, 406
forest district, 302, 408, 431
forest division, 408
forest easement, 409
forest ecology, 408
forest economics, 303
forest economy, 303, 407
forest economy goal, 407
forest ecosystem, 408
forest edge, 303
forest effect, 409
forest encephalitis, 408
forest encroacher, 302
forest engineering, 406
forest enterprise, 303, 408, 409
forest entomology, 408
forest environmental management, 407
forest estate, 406
forest esthetics, 408
forest evaluation, 408
forest examiner, 407
forest expectation value, 408
forest experiment officer, 407
forest experiment station, 409, 436
forest experimental station, 303
forest exploitation, 408
forest extension, 303
forest extension service, 303
forest farm, 298
forest fatal shade, 410
forest fauna and flora, 406
forest fertilization, 299, 409
forest finance, 406
forest fire, 300, 407, 559
forest fire communication system, 301
forest fire control, 301
forest fire danger meter, 407
forest fire danger rating, 407
forest fire destructive power, 301
forest fire detection, 301
forest fire-fighting caravan, 409
forest fire insurance, 407
forest fire law, 407
forest fire meteorology, 407
forest fire prediction, 301
forest fire presuppression, 408
forest fire prevention, 407
forest fire prognosis and prediction, 301
forest fire protection, 300
forest fire protection compact, 406
forest fire size class, 300
forest fire suppression, 301
forest floor, 298, 406
forest for carbon sequestration, 473
forest for community, 171
forest for protection against soil denudation, 133
forest for special use, 475
forest form, 302
forest form factor, 300
forest fragmentation, 409
forest fringe, 303
forest fuel, 408
forest functions map, 407
forest garden, 407, 629
forest gardenization, 588
forest genetics, 410
forest glade, 304
forest-grass measures for soil and water conservation, 459
forest grazing, 302
forest-grower, 585, 600
forest growth and yield science, 409
forest growth region, 409
forest guard, 202
forest-guard system, 202
forest hamlet, 304, 406
forest harvesting, 406
forest harvesting combine, 406
forest harvesting simulation model, 409
forest highway, 302
forest homestead, 303, 410
forest humus, 406
forest husbandry, 407
forest hydrology, 409
forest hygiene, 409
forest hygienic, 409
forest-income tax, 303
forest indicator, 410
forest industry, 406
forest inflammability, 408
forest influence, 410
forest information retrieval, 303
forest inheritance tax, 410
forest insect, 408
forest inspection, 410
forest inspection bureau, 407
forest inspectorate, 407, 409
forest insurance, 405
forest interior, 302
forest inventory, 409
forest inventory by using sampling method, 406
forest inventory design, 409

F

forest inventory for planning and design, 407, 410
forest inventory objective, 409
forest inventory sheet, 409
forest inventory specialist, 409
forest laboratory, 302
forest-labourer, 303
forest land, 298
forest land amelioration, 299
forest land capability, 299
forest-land tax, 299
forest land use, 407, 408
forest land-use class, 299, 407
forest land-use maping, 299
forest land-use planning, 299
forest landscape, 406, 408
forest landscape management, 406, 408
forest law, 406
forest limit, 302, 407
forest limit temperature, 407
forest line, 302
forest liquidation, 408
forest litter, 406, 409
forest management, 407, 408
forest management by individual trees, 82, 84
forest management inventory, 407, 410
forest management inventory; forest inventory for planning and design; second-category forest inventory, 122
forest management objective, 408
forest management office, 408
forest management period, 225, 408, 655
forest management plan, 407
forest management planning, 408
forest management purpose, 407
forest management regulation, 409
forest management rules, 407
forest management scale, 407, 408
forest management table, 407
forest management unit, 407
forest mantle, 302, 303
forest map, 302, 409
forest mapping, 410
forest margin, 303
forest master, 407
forest mathematics, 303, 407, 409
forest maturity, 406
forest measurement, 44
forest mensuration, 44
forest meteorology, 408
forest microclimate, 409
forest mnagement plan, 408
forest moss peat, 409
forest multiple use, 406
forest name, 170, 406
forest nature trail, 405
forest net yield, 406
forest nursery, 408
forest offence, 502
forest officer, 302, 407
forest on land liable to inundation, 130
forest operational inventory, 403, 655
forest ordinance, 406
forest organization, 407, 410
forest outlier, 174
forest owner, 410
forest owner's association, 304
forest ownership, 302
forest park, 407
forest pasture, 302
forest pathology, 405
forest peat, 408
forest percent, 298, 406, 410
forest photogrammetry, 408
forest planning, 407
forest plant, 410
forest plantation, 616
forest planting, 616
forest plot, 409
forest police, 408
forest policy, 303
forest practice, 303, 409
forest practice rule, 409
forest prductivity, 408
forest privilege, 409
forest produce, 298
forest product, 298
forest production, 298, 408, 421
forest productivity, 408
forest products, 298
forest profitability, 409
forest property, 406, 409
forest protection, 405
forest protection act, 405
forest protection machinery, 405
forest railroad, 409
forest railroad transportation, 410
forest railway, 409
forest rake, 613
forest range, 302, 405
forest ranger, 202, 301, 573
forest reclamation, 409
forest recreation, 410
forest recreation facility, 410
forest regeneration, 406
forest regeneration from seeds, 410
forest regeneration survey, 406
forest region, 302, 406, 410
forest regulation, 407, 409
forest regulatory technique, 409
forest renewal, 406
forest rent (al), 406, 410
forest rent theory, 406
forest rent(al), 301, 304
forest-rental value, 406
forest reserve, 410, 634
forest reserve area, 405, 410
forest reserve funds, 406
forest reserves, 17
forest residue, 299
forest resource, 410
forest resource archives, 410
forest resource code, 410
forest resource data base, 410

forest resource distribution map, 410
forest resource information system, 410
forest resource record, 410
forest resources, 410
forest resources appraisal, 410
forest resources evaluation system, 410
forest returns, 406, 410
forest right, 302, 409
forest right system reform, 302
forest road, 298
forest road contruction plan, 298
forest road density, 298
forest road net, 298
forest road network, 298
forest road system, 298
forest sanitation, 409
forest school, 303
forest science, 303
forest sector, 405
forest seedling, 315
forest selection system, 601
forest service, 302
forest service manual, 407
forest servitude, 409
forest settlement, 48, 406
forest settlement officer, 302
forest site, 408
forest site type, 408
forest soil, 409
forest soil productivity, 409
forest soil science, 409
forest spring encephalitis, 408
forest squatter, 302, 406
forest-staff, 303, 410
forest stand, 299
forest staple product, 410, 634
forest statics, 407, 409
forest statistics, 409
forest steppe, 406
forest stocking, 408
forest store, 406
forest structure, 407
forest subdivision, 298, 408
forest succession, 410
forest supervisor, 302
forest supervisor's headquarter, 302
forest survey, 303, 408, 409
forest surveying, 406
forest swamp, 410
forest system, 585
forest tariffs, 407
forest tax, 409
forest taxation, 408, 409
forest technology, 407
forest-tramway, 407
forest transition stability, 410
forest-tree, 301
forest tree breeding, 302
forest tree improvement, 301
forest-tree nursery, 301
forest tree nutrition, 302
forest tree seed, 302
forest-tundra, 406
forest type, 302, 410
forest typeology, 302
forest typology, 302
forest under common ownership, 171
forest under intermittent management, 236
forest under management, 253
forest use, 407, 408
forest utilization, 408
forest valuation, 301, 407
forest value, 301, 407
forest vegetation, 410
forest village, 304, 406
forest walk, 405
forest-warden, 410
forest-watershed experiment, 407
forest weather station, 408
forest-weeds, 410
forest working plan, 409
forest-working through government officer, 611
forest year, 409
forest yield, 409
forest zone, 406
forest zoology, 406
forestation, 600
forested area, 581
forested land, 298
forester, 302, 303
forestry, 302, 303
forestry act, 409
forestry association, 302
forestry capital, 303
forestry commission, 303
forestry cost, 303
forestry economics, 303
forestry extension agent, 303
forestry fund, 303
forestry gantry crane, 303
forestry herbicide, 303
forestry history, 303
forestry land, 298, 303
forestry machinery, 303
forestry on short rotation, 112
forestry on small scale, 533
forestry overhead travelling crane, 303
forestry planning system, 303
forestry planting, 303
forestry plough, 303
forestry program committee, 303
forestry railroad car, 410
forestry regulation, 303
forestry sector, 303
forestry taxation system, 303
forestry torch, 303
forficated tail, 556
fork, 41, 45, 46, 121, 140, 446
fork arm, 45, 46
fork caliper, 46
fork lift loader, 46
fork lift operator, 46
fork lift truck, 46
fork lifter, 46
fork loader, 46
forked growth, 140
forked mycorhiza, 46
forked tenon, 46
forked tree, 140
forked wood, 46
form bridge, 75
form class, 160, 541
form class volume table, 160, 403
form exponent, 541
form factor, 541
form factor at breast height, 503, 542
form-factor table, 541

F

form-figure, 541
form height, 300, 541
form height increment, 449, 541
form height series, 541
form increment, 449
form index, 160, 541
form line, 99, 312
form of coupe, 38
form of crop, 302
form point, 541
form-point height, 541
form press, 587
form quotient, 541
form ratio, 541
form work, 325, 327
form(a), 24, 285
formal garden, 183
formal style, 183
format, 78, 165, 530
formation, 310, 389
formation group, 389
formation period, 54
formative period of enterprise, 367
formative tissue, 57
formazan, 230
formed contact wheel, 55
formed plywood, 55, 326
former, 55, 327
former management, 373
formic, 230, 506, 566
forming belt, 8
forming conveyor, 8
forming die, 55, 254
forming head, 55
forming line, 55
forming machine, 55
forming roll, 55
forming table, 55, 499
forming unit, 55
Formosau hinoki leaf oil, 471
formula for depreciation, 606
formula for determining the volume, 382
formula method, 171, 451
formula method of periods by area and volume combine, 451
formula of capitalization, 640
formula of forest condition, 301
formula transformation, 171
formula translation, 171
formula translator, 56, 171
formulation, 353, 620
forset function, 407, 408
fortified rosin, 376
fortified rosin size, 376
fortifier, 33, 602
forward business, 365
forward integration, 200
forward mutation, 373, 611
forward price, 365, 594
forwarder, 112, 387
forwarding, 387, 614
forwarding machine, 112, 387
fossil, 206
fossil biocoenosis, 206
fossil CO_2 (carbon dioxide) emissions, 206
fossil erosion surface, 174, 206
fossil fuel, 206
fossil pollen, 203
fossil resin, 201, 206
fossil soil, 174
fossil species, 206
fossil zone, 206
fossilization, 206
foul, 161, 258, 470, 480
fouling organism, 508
foundation, 95, 219
foundation beam, 96, 219
foundation bearer, 36, 114
foundation block, 592
foundation bolt, 96
foundation pile, 96, 221
foundation plant, 219, 376
foundation planting, 219, 376
foundation seed, 69, 592
foundation stock, 592
foundational niche, 219
founder effect, 236
founder principle, 236
fountain, 354, 593
fountainhead, 387
four boundaries, 463
four-chamber short-retention-time blender, 463
four cutter, 463
four-faced sawing, 463
four-lane traffic way, 462
four-o'clock, 4
four-piston metering pump, 463
four side moulder, 463, 541
four-side tree planting, 463
four-sided matchboarding machine, 463
four sides planing and moulding machine, 463
four-strand crossing over, 463
four-way cross, 463
fourdrinier, 50
fourdrinier board forming machine, 50
fourdrinier machine, 50
fourdrinier-method, 50
fourdrinier part, 50
fourdrinier section, 50
fourdrinier table, 50
Fourier analysis, 154
fourier transform, 154
fowling net, 33
fowling piece, 297, 346
fox-tail saw, 201, 231
fox wedged tenon, 201
foxglove, 318
foxhound, 297
foxiness, 194
foxtail, 201
foxty-tree, 267
fractal dimension, 143
fractal geometry, 143
fraction, 142, 308, 468
fraction collector, 308
fractional cover, 156
fractionation, 141, 142, 382
fracture, 269
fracture length, 297
fragility, 74, 569
fragipan, 74
fragipan soil, 74
fragmental rock, 469
fragmental soil, 72, 291
fragmentation, 363
fragmentation of forest holdings, 409
fragmospore, 117
fragrance (material), 131, 528
fragrant substance, 131
frame, 173, 218, 280, 505
frame culture flower, 505

F

frame frequency, 607
frame girder, 173
frame of roof, 508
frame press, 280
frame saw, 280, 349
frame (saw) mill, 280
frame saw sharpener, 280
frame-sawn, 280
frame-type furniture, 280
frame unloader, 231
frame-work, 173, 218, 280
frame work method, 360
framed and braced door, 280, 580
framed door, 280
framed floor, 280
framed grounds, 320
framed trestle, 173, 349, 603
frameshift, 564
frameshift mutation, 564
framework Convention on Climate Change, 369
framework substance, 175
framing, 280
framing scaffold, 243
framing square, 332
framing timber, 173
frangula-emodin, 76
frankincense oil, 283, 400
franking, 537
frass, 59, 280, 636
fraternity, 264
Fraunhofer line, 147
Fraunhofer line discriminator, 147
Fraxetin, 6
Fraxin, 6
Fraxinol, 6
fraying, 310
free alongside ship, 66, 639
free beam, 234
free beaten stock, 579
free beating, 579
free-burning fire, 647
free central placentation, 142, 474
free column volume, 647
free cutting, 647
free end method, 647
free floating hydrophyte community, 647
free-free beam, 545, 647
free growing, 504
free growth management, 466
free hand curve, 487
free-hand section, 487
free living, 647
free-living form, 647
free log, 509
free market scenario, 647
free moisture, 579, 647
free molecule, 579, 647
free on board, 66, 287
free on car, 371
free on lorry, 217
free on plane, 136
free on quay, 316
free on rail, 215
free on truck, 217
free on vessel, 66
free organic acid, 579
free oxide, 579
free planting stock, 323
free pollination, 647
free pore space, 510, 647
free pupa, 287, 314
free rosin, 579
free rosin size, 6, 579
free-running rhythm, 647
free-setting (of) the crown, 515
free shellac, 404, 579
free space, 647
free stock, 579
free-thinning, 647
free-to-grow, 504
free trade, 647
free water, 579, 647
freeboard, 50, 51, 257
freedom from branches, 613
freehand fitting, 487
freely cutting machine, 93
freely supported beam, 234
freeness, 316, 579
freeness analyzer, 579
freeness tester, 579
freeze, 110
freeze drying, 286
freeze etching, 28
freeze-thaw action, 110
freeze-thaw erosion, 110
freezing, 110
freezing damage, 110
freezing injury, 110
freezing microtomy, 28
freezing resistance, 270, 336
freezing temperature, 110
freight and cartage, 457
freight rate, 595, 596
French leg, 453
French method of thinning, 127, 415
French polish, 127
French poodle, 127
French thinning, 127, 415
frequency, 359
frequency class, 359
frequency curve, 359
frequency diagram, 359
frequency distribution, 359
frequency law, 359
frequency modulation, 480
frequency of infection, 378
frequency percentage, 359
frequency-response curve, 359
frequency spectrum, 359
frequency table, 359
fresh air screen, 539
fresh cut, 539
fresh gale, 376
fresh mass, 523
fresh ungerminated seed, 539
fresh water, 87
fresh water community, 87
fresh water plant, 87
fresh water succession, 87
fresh weight, 433, 523
fresh wood, 421, 432, 539
freshet, 16, 68, 474
freshly-felled timber, 539
freshly felled tree, 539
freshwater ecology, 87
freshwater ecosystem, 87
freshwater lens, 87
fret, 165
fret saw, 160, 526
friction, 326
friction debarker, 326
friction device, 326, 620
friction drive, 326

F

friction feeding, 293, 326
friction-gearing, 326
friction lever, 326, 620
friction nigger, 326
friction of moment, 109
friction of rest, 258
friction receder, 326
friction resistance, 326, 337
friction separator, 326
friction velocity, 326
frictional force, 326
frictional strength, 326
frictional unemployment, 112, 326
frieze panel, 321, 446
frieze rail, 103, 624
frigid belt, 189
frigid forest-region, 189
frigid zone, 189
frigorideserta, 190
frigorigraph, 93
frill cut, 450
frill girdling, 450
fringe, 22, 307, 404, 593
fringe area, 140
fringe woodland, 22
fringed, 260, 261
fringed fibrils, 572
fringed micell(ar) theory, 572
fringe(d) micelle, 572
fringing forest, 49, 554
fringing reef, 3
froe, 355, 356, 536
frog, 46, 205, 288, 606
frogger, 606
frond, 560, 618
frondescence, 560, 561
frondome, 560
frons, 120
front, 58, 88, 120, 127, 298, 373
front bevel angle, 373
front-end dumper, 373
front-end lift, 372, 373
front-end loader, 373
front face, 372, 373, 611
front fire, 215
front frame, 372
front guide shoe, 372
front handle, 373
front loader, 373
front-mounted cutter bars power cuitlvator, 373
front-mounted scrubclearing equipment, 373
front rail, 373
front stop, 373, 651
frontal analysis, 373
frontier period of forest economy, 407
frost, 455
frost action, 28, 110, 455
frost-canker, 110, 455
frost cleft, 110
frost crack, 110, 455
frost cracking, 110
frost damage, 455
frost erosion, 110, 455
frost-free period, 510
frost-free season, 510
frost hardiness, 336, 337
frost-hardy, 337
frost heart (wood), 110
frost heave, 110, 455
frost heaving, 110, 455
frost hole, 455
frost hollow, 455, 569
frost-injury, 455
frost killing, 110
frost level, 455
frost lift(ing), 110, 455
frost line, 28, 110, 455
frost locality, 118, 455
frost pocket, 455
frost pool, 455
frost prevention, 133
frost protection, 133
frost resistance, 270
frost-resisting power, 270, 337
frost-rib, 110, 455
frost-ring, 110
frost-shed, 455
frost split, 110, 455
frost spot, 455
frost-tender, 569
frost weathering, 28
frostbite, 110, 455
frosting, 367
frostless season, 510
frosty-zone, 455
frother, 123, 367
frotrude, 143
froward contract, 365, 594
frowy, 544
frozen-dehydration method, 110
frozen road, 110
frozen soil, 110
frozen wood, 110
fructification, 245, 641
fructosan, 186
frugivorous, 437
frugivorous animal, 437
fruit, 186
fruit bearing, 245
fruit-bearing in alternate years, 166
fruit bearing percentage, 245
fruit-bearing shoot, 245
fruit branch, 186
fruit bud grafting, 204
fruit character, 245
fruit-eating, 437
fruit-formation period, 186
fruit grafting, 186
fruit hedge, 186
fruit-processing, 186
fruit setting, 245, 655
fruit thinning agent, 445
fruit tree, 186
fruitful, 118, 139, 273
fruitfulness, 245
fruit(ing)-body, 641
fruiting year, 245
fruitlet, 533
frustum, 247, 361
frustum form factor, 247
frustum of neiloid, 247
frustum of paraboloid, 247
frutescence, 462
frutescens, 180
frutescent, 250
fruticeta, 180
fruticose lichen, 180
fruticous desert, 180
fruticous fundra, 180
fruticulose, 533
fry, 584
fubo, 359
fuchsin, 235, 359, 557
fucugetine, 149
fuel, 273, 390, 569
fuel and induction system, 390
fuel bed, 390
fuel gas, 273
fuel hazard, 273
fuel loading, 273
fuel metering diaphragm, 390
fuel moisture content, 273
fuel pump diaphragm, 578
fuel reduction, 238

fuel-reduction burning, 381, 416
fuel stick, 273
fuel switching, 390
fuel type, 273
fuel-type classification, 273
fuel value, 390, 394
fuel wood, 416, 540
fuel(l)er, 229
fuelwood forest, 540
fuge, 524
fugous, 524
fuil dovetail dado, 386
full-bodied, 160
full-bole, 498
full-boled stem, 498
full-cell impregnation, 317
full-cell process, 59, 317, 387
full circle crane, 387
full cone, 498
full-cost pricing, 385
full crown cover, 385
full cultivation, 386
full cut, 50, 306, 318
full-forest utilization, 386
full grown larva, 284
full-herringbone cut, 386, 584
full immersion method, 317, 386
full-length logging, 385, 591
full-length tree, 386
full light, 385
full maturity, 498
full measure, 50, 306, 318
full nutrient solution, 498
full of staybolts, 117
full overhead shade, 386
full-round cut(ting), 48, 387
full scale, 318, 385
full-scale plant, 169
full seed, 14
full seed year, 245
full seeding, 386
full-sheet veneer, 610
full sib, 387
full size, 50, 306, 318
full size veneer, 610
full spiral cut, 386
full-stocked, 289
full-stocked stand, 322, 498
full stocking, 498
full sun, 385, 386
full sun light, 385
full time drying system, 293
full track, 385, 386
full trailer, 385, 387
full-tree chipper, 125, 386
full-tree field chipping, 125
full-tree logging, 125, 386
full-tree skidding, 125, 386
full-tree utilization, 386
full view, 386
fuller's earth, 358
fullness, 144
fullround file, 592
fully regulated forest, 498
fully ripened seed, 498
fully stocked even-aged stand, 498
fully-stocked plantation, 510
fully stocked stand, 585, 653
fully stocked woods, 322, 498
fully stocking percent, 498
fumarole, 354
fume damage, 553
fume-protection forest, 134
fume resistance, 271, 338
fume-tolerant plant, 338
fumigant, 549
fumigation, 549
fumigation agent, 549
function response, 171
functional design, 171
functional finance, 615
functional group, 171
functional guild, 171
functional niche, 171
functional resistance, 171
functional response, 171
fund raising activity, 641
fundamental gneiss, 219
fundamental meristem, 219
fundamental niche, 219
fundamental rock, 219
fundamental species, 219
fundamental tissue, 219
fundatrix, 157
funded debt, 50
funded reserve, 220
funereal cypress, 7
fungal decay, 266, 267
fungal symbiont, 172, 607
fungaltoxin, 607
fungi, 607
fungi imperfecti, 11, 35
fungicidal effectiveness, 411
fungicidal efficiency, 411
fungicidal paint, 133, 411
fungicidal preservative, 411
fungicide, 411
fungistasis, 568, 569, 618
fungistat, 568
fungistatic substance, 568
fungivore, 437
fungivorous, 437
fung(o)us brown-stain, 607
fungous gall, 267
fungus, 607
fungus colony, 607
fungus damage, 267
fungus disease, 607
fungus filament, 267
fungus gall, 267, 607
fungus garden, 267
fungus (*plur.* -gi), 267, 607
fungus-inhibiting capacity, 569
fungus spread, 267
funic(u)le, 470, 632
funiculus, 470, 632
funnel-form, 309
furan(e) resin, 147
furan(e) resin adhesive, 147
furanose, 147, 512
furca, 45, 46, 472
furfurylidene, 147, 552
furnace-heated kiln, 310
furnish weighing station, 468
furnishes, 171, 282
furnishing department, 353

F

furniture, 229
furniture back, 229
furniture beetle, 229
furniture board, 229
furniture components, 229
furniture dimension stock, 229
furniture industry, 229
furniture plywood, 229
furniture varnish, 229
furnos process, 472
furnos treatment, 472
furoyl, 147, 270
furrow and ridge tillage, 173
furrow drilling, 172
furrow irrigation, 173
furrow planting, 172, 173
furrow slice, 127, 128
furrow transplanting, 288
furrower, 268
furrowing, 268
fuse, 15, 89, 398, 571
fused aperture, 192
fused calcium-magnesium phosphate, 155
fused tricalcium phosphate, 398
fuseous soil, 4
fusible tape, 273
fusiform body, 134
fusiform (cambial) initial, 134
fusiform cell, 134
fusiform initial (cell), 134
fusiform radical, 134
fusiform ray, 134
fusiform (wood) parenchyma cell, 134
fusion nucleus, 29, 263, 398
fusion of hyphae, 267
fuste, 210
fustic wood, 210
future cost, 504
future cut, 586
future growing stock, 586
future increase in stumpage price, 289, 586
future market, 365
future tree, 441, 633
future value, 199, 626
fuzz, 367, 397
fuzz fibre, 367, 631
fuzz grain, 319
fuzzy clustering, 325
fuzzy grain, 318
G-horizon, 374
G-layer, 241, 374
gabbro, 211
gabion, 333, 434
gabion boom, 434
gable roof, 396, 402
gad, 53
gaff, 75, 79, 482, 550
Gaia hypothesis, 76
gain, 135, 469, 542, 573, 602, 643
gain factor, 602
gaining, 217, 238, 268, 442
galactane, 241
galactase, 10
galactoglucomannan, 10
galactolipid, 10
galactomannan, 10
galactosidase, 10
galactoside, 10
galban, 30, 174
galbulus, 237
gale, 76
gale damage, 144
galea, 495
galilee, 321, 373
galipot (resin), 188, 465
gall, 60, 87, 308, 512
gall aphid, 334, 382
gall apple, 512
gall fly, 512
gall gnat, 575
gall midge, 575
gall mite, 575
gall nut, 12, 327, 512
gall rust, 308
gall wasp, 512, 575
galleria, 651
galleried table, 486
gallery, 49, 95, 144, 206, 275
gallery forest, 49, 554
gallery pattern system, 275
gallnuts extract, 12
gallocatechin, 12, 327
gallotannin, 12, 327
galvanometer, 101
game, 444
game animal, 297, 444
game area, 444
game bird, 272
game cover, 252, 297, 559
game cropping, 297
game habitat, 297
game land, 444
game law, 444
game-licen, 444
game-license, 444
game management, 444
game manager, 444
game park, 252, 297, 444, 559
game pasture, 444
game preserve, 252, 297
game-preserver, 297
game protector, 444
game refuge, 252
game species, 272
game theory, 31, 114
game warden, 444
gametangium, 353
gamete, 353
gamete abortion, 353
gametes, 353
gametic nucleus, 353
gametic number, 353
gametophore, 353
gametophyte, 353
gamic, 443
gamic female, 582
gamma distribution, 155
gamma radiation, 148, 155
gamma-ray, 155, 418
gamma-ray-control-system, 418
gamobium, 438, 582
gamodeme, 166, 240
gamogastrous, 192
gamomery, 192
gamopetalous, 191
gamosepalous, 192
gang borer, 350
gang mould, 652
gang rip saw, 592, 650
gang saw, 280, 349
gang sawyer, 349
ganglion, 237, 420
gang(saw)mill, 349
gangway, 62
gantry, 309, 321, 471
gantry crane, 321, 639
gantry slewing crane, 545
gap, 235, 287, 304, 309, 330, 387
gap clear-cutting system, 277

gap-cover, 33, 88, 304
gap cutting, 277, 491
gap cutting system, 277, 491
gap felling, 277, 491
gap-filling adhesive, 358, 479
gap-filling glue, 358, 479
gap-filling hardener, 358, 479
gap gauge, 200, 236, 496
gap joint, 79, 358
gap-phase replacement, 302
gap-phase species, 302
gap size, 235, 387
gap stick, 88, 146
gape grafting, 213
gapping, 235
gapping machine, 235, 445
garage, 371, 493
garden, 481, 587
garden architecture, 481, 588
garden art, 588
garden automizer, 588
garden balsam, 147, 222
garden base map, 588
garden bottom view, 587
garden centre, 204, 588
garden chromatics, 588
garden circumstance, 588
garden city, 204, 478
garden colour effect, 588
garden colour harmony, 588
garden colour scheme, 588
garden dahlia, 77
garden design, 481, 588
garden entrance, 588
garden fence, 587, 588
garden for tea, 47
garden furniture, 481, 587
garden general layout, 588
garden house, 204, 588
garden lawn, 587
garden lay-out, 481, 588
garden making, 601
garden microclimate, 588
garden ornament, 481, 588
garden overall plan, 588
garden path, 204
garden pavement, 588
garden plant, 588
garden planting design, 588
garden room, 481, 588
garden sculpture, 481, 588
garden seat, 588
garden shrubbery, 588
garden sketch map, 587
garden statuary, 481, 588
garden structure, 588
garden style, 588
garden suburb, 56, 588
garden tableau, 588
garden terrace, 481, 588
garden topography, 588
garden wall, 588
gardener, 587, 588
gardenia absolute, 613
gardenia compound, 613
gardening, 72, 588
Gardner colour, 227
Gardner dryer, 227
garnet, 248, 434
garnet lac, 434
garret, 105, 165
Garrigue, 228
gas chromatograph-mass spectrometer, 370
gas chromatography, 370
gas chromatography-mass spectrography, 370
gas chromatography-mass spectrography computer, 370
gas conductivity, 370, 486
gas evolution, 370
gas exchange quotient, 370
gas-filled tube, 59
gas fired kiln, 370
gas flue, 553
gas generator, 319, 370
gas handling system, 250, 370
gas hold-up, 370
gas-house tar, 320
gas-liquid chromatography, 370
gas-liquid partition chromatography, 370
gas permeability, 486
gas pin, 371
gas-solid chromatography, 368
gas sweetening, 370
gas treatment, 370
gas vacuole, 368, 370
gaseous cycle, 370
gaseous pollutants, 370
gasification, 369
gasohol, 371
gasoline powered chain saw, 578
gasoline skidder, 371
gasometry, 370
gasteromycetes, 155
gastric caeca, 504
gate, 262, 556, 602
gate boom, 602
gate saw, 280, 349
gate strut, 321
gate valve, 205, 602
gateleg table, 612
gateway road, 2
gatherer, 442
gathering agent, 465
gathering ground, 263, 264
gauge, 25, 183, 184, 262, 296
gauge constant, 242
gauge insulation, 184
gauge lath, 178
gauge line, 184
gauge of track, 184
gauging station, 44, 459
gauging weir, 44, 296
gauging works, 44, 459
gaultheria oil, 6, 108
gaultherilene, 6
Gause's axiom, 163
Gause's principle, 163
Gauss-Newton, 163
gear, 58, 66, 417, 639
gear bogie, 58
gear drive, 58
gear rack, 58
gear shift, 24, 209
gear-(type) pump, 58
gear-type rope wheel, 58
gear-(type)oil pump, 58
gearbox, 24, 58
geijerene, 221
geijerin, 221
geitonogamy, 484
gel, 346
gel chromatography, 346
gel consistency, 346
gel exclusion chromatography, 346

G

gel filtration chromatography, 346
gel peptization, 346
gel permeation chromatography, 346
gel state, 241, 346
gel strength, 346
gelatin, 109, 325, 346
gelatination, 241, 346
gelatinization, 346
gelatinize, 241, 487
gelatinizer, 240, 241
gelatinous fibre, 241, 346
gelatinous fibre tracheid, 241
gelatinous layer, 241
gelatinous matrix, 241
gelatinum, 6
gelation, 346
gelation time, 241
geling power, 346
gellike, 285
gel(ling) point, 241
gem, 561
gem-faced civet, 186, 203, 471
gemma, 13
gemmation, 551, 552
gemmiferous, 124, 261, 428
gemmule, 352, 535, 551, 591
gena, 229
gene, 220
gene action, 221
gene allele, 92, 570
gene block, 220, 221
gene complex, 221
gene conversion, 220
gene dosage, 220
gene flow, 220
gene frequency, 220
gene locus, 220
gene mutation, 220
gene order, 220
gene pool, 220
gene recombination, 220
gene redundancy, 220
gene segregation, 220
gene system, 220
genealogical, 517, 548
genealogical tree, 517
genecology, 565, 627
general adaptation syndrome, 440
general agreement on tariff and trade, 178
general arrangement, 649
general average, 172, 649
general board, 7
general circulation, 649
general circulation model, 649
general combining ability, 365, 562
general cultivation, 386
general cutting-plan, 648
general design, 648
general drawing, 649
general-equilibrium theory, 562
general histology, 365
general infection, 24, 60
general overhaul, 78, 386
general plan, 225, 649
general plan of working, 432
general problem solver computer (GPS), 482
general purpose computer, 482
general purpose language, 482
general purpose machine, 482
general scout, 607
general supervision of the forest, 410
general Tarifs, 482
general volume table, 123, 482
general working-plan, 432, 649
general yield table, 562, 649
generalized normal forest, 183, 562
generating tissue, 143
generation, 438
generation time, 438
generative nucleus, 430
generative propagation, 582
generative tissue, 143, 430
generator cell, 430
generator gas, 124, 310
generic coefficient, 627
generic equation, 365, 482
genesclinic, 357
genesis, 124, 367, 430
genetic advance, 565
genetic cline, 565
genetic code, 220, 565
genetic control, 565
genetic correlation, 565
genetic crossing over, 565
genetic diversity, 565
genetic dominance, 565
genetic drift, 220, 565
genetic element, 565
genetic engineering, 565
genetic entity, 565
genetic equilibrium, 565
genetic equilibrium hypothesis, 565
genetic factor, 565
genetic feedback mechanism, 565
genetic flexibility, 565
genetic formula, 565
genetic hermaphrodite, 124
genetic information, 565
genetic load, 565
genetic make-up, 565
genetic material, 565
genetic mechanism, 565
genetic plasticity, 565
genetic purity, 565
genetic recombination, 220, 565
genetic resources, 221, 565
genetic shift, 565
genetic sterility, 565
genetic synecology, 388, 617
genetic system, 565
genetic variance, 565
genetic variation, 565
genetical classification of soils, 489
genetics, 565
genic balance, 220
geniculate, 516
geniculatus, 516
genital chamber, 430
genitalia, 430, 497
genoholotype, 446
genoid, 285, 518
genolectotype, 446
genom(e), 221, 391
genomere, 220
genonema, 220
genophore, 220
genosome, 220
genosyntype, 446
genotype, 220, 446, 565, 570

G

genotype-environment correlation, 220
genotype-environment interaction, 220
genotypic, 220, 565
genotypic frequency, 220
genotypic selection, 220
genotypic value, 220
genotypic variance, 220, 565
genovariation, 220
Gentan probability method, 233
gentiin, 309
gentle, 360, 361
gentle ground, 361
gentle slope, 209
genus cross, 446
genus (*plur.* genera), 446
geo-engineering, 97
geobenthos, 201, 310
geobiocoenosis, 426
geobiont, 490
geobotanics, 99
geobotany, 99
geochemical cycle, 97
geochronology, 97
geocline, 96, 99
geocode, 96
geocoenosium, 96
geocryptophyte, 98
geodesy, 76
geodetic control point, 76
geodetic coordinate, 44, 76
geoecotype, 96
geographic area, 96
geographic isolation, 96
geographic race, 96
geographic range, 96
geographic space, 96
geographic variant, 96
geographic variation, 96
geographical distribution, 96
geographical indicator, 96
geographical variation, 96
geographic(al) variety, 96, 99
geographical vicariad, 96
geographtype, 96
geography, 96
geoisotherms, 91
geologic age, 99
geologic interpretation, 99
geologic structure, 99
geologic time scale, 99
geological erosion, 99
geological formations, 99
geological function, 99
geological map, 99, 169
geological science, 99
geological succession, 99
geology, 99
geometric distribution, 224
geometric growth, 224
geometric mean, 91, 224
geometric solid, 224
geometric style garden, 224, 610
geometrical centr(e)ing, 224
geometrical principle, 224
geometrical style, 224
geomicrobiology, 97
geomorphic cycle, 96
geomorphology, 96
geomorphology (landform), 99
Geoorkiantz form point method, 246
geophagous, 437
geophile, 516
geophysical airborne survey system, 97
geophyta bulbosa, 305
geophyta rhizomatosa, 167
geophyte, 98
geosere, 99
geosphere, 97, 310, 377, 554
geostatics, 98
geostationary meteorological satellite, 97
geostrophic wind, 99
geosyncline, 95, 98, 310
geotaxis, 383
geothermal, 97
geothermal gradient, 98
geothermal power, 97
geothermy, 97
geotropism, 530
geovore, 437
geoxene, 349
geozoology, 95
geranene, 318, 528
geraniin, 318
geraniolene, 318, 528
geranium oil, 284, 528
geranyl chloride, 315
germ, 352, 501, 518, 583
germ cell, 430
germ-free box, 509
germ line, 627
germ plasm, 627
germ plasma resource bank, 627
germ pore, 124, 551
germ tube, 321
germacrone, 76
German cocoa powder, 90
German thinning, 90
germicidal lamp, 411
germicide, 411
germifuge, 411
germinability, 124
germinable, 341
germinal bud, 352
germinant, 124, 321
germinate, 124, 321
germinating ability, 124
germinating-apparatus, 124
germinating bed, 124
germinating-box, 124
germinating energy, 124
germinating (germination) capacity, 124
germinating power, 124
germinating quality, 124
germinating-ripe, 124
germinating test, 124
germination, 124, 321
germination blotter, 124
germination by repetition, 121
germination cabinet, 124
germination consant, 124
germination ecology, 124
germination energy index, 124
germination energy period, 124
germination (germinating) behaviour, 124
germination index, 124
germination inhibiting substance, 124
germination inhibitor, 124
germination number, 124

G

germination percent, 124
germination promoting substance, 124
germination rate, 124
germination resistance, 124
germination speed, 124
germination stimulator, 124
germination tray, 124
germination value, 124
germination viability, 124
germinative, 124
germinative force, 124
germinator, 124, 628
germplasm, 430, 565, 627
germplasm conservation, 565, 627
germplasm preservation, 627
germplasm repository, 627
germplasm resource, 627
germplasm stock, 565, 627
germplasm theory, 627
germplasmbank, 627
gerontomorphosis, 474
gettama gum, 111
Gewecke pressure and suction method, 165
geyser, 235, 236
gharcoal trap, 472
ghatti gum, 229
ghedda wax, 269
ghost peak, 178, 230
ghosting effect, 178, 230
giant cell, 260
giant chromosome, 260
giant gum, 541
giant mistletoe, 201
giant redwood, 260
giant sequoia, 260
giant silkworm moth, 477
giant sized board, 76
giant tree, 260
gibberellin, 58
gibbet, 103, 367, 612
Gibson raft, 221
gig-back, 90
gig run, 90
giga-, 371, 433
gigagram, 371, 433
gigantism, 260
gigas, 260
gigaton (Gt), 433
gill, 7, 228, 267, 404
gill guy, 149, 351
gill poke, 458, 464
gilliflower, 642
gimlet, 443
gin, 76, 242, 403
gin block, 84, 367
gin pole, 231, 368, 638
ginger oil, 237
Girard form class, 221
Girard form quotient, 221
Girard's rule of thumb for tree volume, 221
girder, 41, 53, 77, 195
girdle, 289
girdling, 207–209
girth, 160, 502
girth above buttress, 281
girth breast height, 542
girth class, 160
girth increment, 160
girth limit, 160
girth-limit cutting, 502
girth over bark, 79
girth quotient, 160
girth tape, 160
girth under bark, 385
gizzard, 218, 412, 466
glacial, 28
glacial acetic acid, 28, 510
glacial age, 28
glacial anticyclone, 29
glacial conglomerate, 28
glacial debris, 28
glacial delta, 28
glacial deposit, 28
glacial drift, 28
glacial epoch, 28
glacial erosion, 28
glacial flora, 28
glacial lake, 28
glacial landform, 28
glacial period, 28
glacial recession, 28
glacial refuge, 28
glacial relics, 28
glacial relicts, 28
glacial retreat, 28
glacial rock, 28
glacial scratch, 28
glacial striae, 28
glacial striation, 28
glacial terrace, 28
glacial till, 28
glaciation, 28
glacier, 28
glacier-avalanche, 28
glacier cascade, 28
glaciofluvial activity, 28
glacis, 209, 537
glade, 301, 304, 605
glance boom, 583
glancer, 205
glancing boom, 583
gland, 479, 527, 549
gland lure, 527
glandular, 261
glandular hair, 527
glandular pubescent, 261
glandular punctate, 261
glandular scale, 527
glans, 201
glareous, 430
glass house, 30, 505
glass knife, 30
glass roof, 30
glassine paper, 30, 485
glassy wood, 573
glaucescent, 79
glaucous, 210, 260
glaze, 123, 326, 351, 553, 584, 585
glazed door, 30
glazed frost, 28, 585
glazed joint failure, 30
glazed paper, 162
glazed rain, 28
glazier's wood, 28
glazing, 182, 432, 605
glazing bar, 30, 529
glazing bead, 529
glazing machine, 326, 351
gleditschine, 600
gleditsin, 600
glei alluvial brown soil, 374
glei forest grey soil, 374
glei horizon, 374
glei process, 211, 374
gleization, 211, 374
gley, 374, 580
gley horizon, 211, 374
gley-like podzol, 374
gley-like rice soil, 374
gley podxol, 210, 211
gley-podzol, 374
gley-prairie soil, 374
gley process, 211, 374
gley soil, 374
gleyed forest soil, 374
gleyed soil, 374
gleying humus layer, 374

gleysol, 374
glide direction, 205
glide lamella, 205
glide mirror, 205
glide plane, 205
gliding growth, 205
Global Atmospheric Research Program, 386
global change, 386
global ecology, 386
Global environmental monitoring system, 386
global position system, 386
global surface temperature, 386
global warming, 386
global Warming Potential, 386
globe boiler, 609
globulin, 382
glochid, 173
glochidiate, 260
glochidium, 173
Gloger's rule, 165
glomerule, 491
gloomy degree, 585
gloss, 181, 351
gloss agent, 182
gloss finish, 296, 351
gloss oil, 182, 296, 465
gloss reduction, 182, 431
gloss retention, 14
glossa, 622
glow retardant, 651
glow-type combustion, 58
glowing of wood, 329
glowing time, 58
glucan, 364
glucanase, 364
glucaric acid, 364
glucase, 364
glucogallin, 217, 364
glucolipid(e), 474
glucomannan, 364
glucoprotein, 473
glucopyranose, 20
glucoresin, 473
glucosan, 364
glucosidase, 364
glucoside, 364, 473
glucosidic bond, 473
glucosum, 364
glucuronide, 364
glue, 240, 241
glue and filler bond, 241
glue applicator, 431, 487
glue ball, 241, 242
glue blending, 431
glue bond, 241
glue coater, 487
glue container, 240
glue cooker, 620
glue dissolver, 353, 479
glue drum, 479
glue film, 157, 240, 241
glue gun, 354
glue joint, 241
glue joint failure, 241
glue-joint-ripsaw, 241
glue-joint strength, 240
glue jointer, 241
glue lam, 240
glue laminated wood member, 240
glue line, 240
glue line cleavage test, 240
glue-line heating, 240
glue line strength, 240
glue line viscosity, 240
glue machine, 11
glue mixer, 11, 479
glue penetration, 420, 485
glue pot, 240, 620
glue preparation, 353, 479
glue preparing station, 353
glue shear test, 240
glue sprayer, 354
glue spreader, 487
glue spread(ing), 487
glue stain, 240, 241
glue strength, 240
glueability, 240
glued joint, 241
glued lamination board, 240
glued lamination member, 240
glueless dry process, 509
gluing, 240, 241
gluing pressure, 240, 241
glulam, 240
glumaceous, 574, 582
glume, 574
glumelle, 238, 339
glumous flower, 574
glut, 345, 575
glutamine, 175
glutelin, 175
gluten, 175, 323
glutenin, 317
glycan, 264
glycerin gelatine, 156
glycerin gum, 156
glycerin jelly, 156
glycogen, 474
glycogenase, 474
glycolipid, 474
glycolipin, 474
glycolysis, 243, 474
glycolytic enzyme, 473
glycolytic ferment, 473
glycolytic pathway, 473
glycoprotein, 473
glycose, 85, 364
glycosidase, 473
glycosidation, 156
glycoside, 353, 473
glycosylation, 473
glycyrrhiza, 156
glyptal resin, 156
glyptostroboid pit, 459
gnarl, 245, 333
gnawed, 346, 387
gneiss, 357
go back road, 212
go-devil, 122, 179, 223, 309
gobi, 164
goggle, 202
going forest concern, 580
going market price, 525
gold lacquer, 87
gold painted lacquer ware, 324
golden larch, 248
Goldschmidt barometer, 165
gonad, 430, 541
gondang wax, 398
gondola, 50, 296, 360, 480
goneoclin, 525
goniometer, 44, 45, 242
gonocyte, 430, 541
gonophore, 70, 430
good two sides, 295
good will, 415, 540
goodness of fit, 342
goods credit, 217
goods in bond, 15
goods in process, 599
goods-out on consignment, 227
goods shelf, 217, 415
goose, 120, 556
goose saw, 7, 104
gooseneck, 120, 205, 294, 620

gooseneck boom, 120, 384
gooseneck trailer, 120
goosepen, 216
gopher, 65, 210, 338, 535
gopher hole, 65, 352, 446
Gordon setter, 165
gorge, 466, 519
gossypetin, 323
Gothenburg standard, 165
gouging, 16, 355
government square, 125
governor, 479, 480
grab, 173, 205, 372, 636
grab arm, 636
grab driver, 106
grab handle, 507, 636
grab hook, 295, 367
grab link, 295
grab maul, 106
grab ring, 295
grab skipper, 599, 636
grab(bing) crane, 636
Grabinski yarding, 56
gradation, 236
gradation of age classes, 305
gradation of moisture, 456
grade, 91, 359, 362, 363, 380
grade destroyer car, 112
grade hydrolysis, 141
grade limit, 363
grade-line, 179, 363
grade of locality, 98
grade of rosin, 465
grade resistance, 363
grade separation, 91, 290
grade shovel, 362, 610
grade stabilization structure, 176, 363, 610
graded bench terrace, 243, 458
graded density, 236
graded-density particleboard, 236, 322
graded terrace, 475
graded timber, 91
graded topocline, 475
grader, 141, 233, 360, 362, 547, 636
gradient, 363, 380, 475
gradient analysis, 475
gradient capability, 349, 475
gradient hypothesis, 475
gradient preservative, 132
gradient thin-layer chromatography, 475
grading, 141, 362, 363, 636
grading analysis, 291
grading certificate, 91
grading log, 91, 365
grading of logs, 37, 589
grading of stems, 37, 589
grading regulation, 140
grading rule, 141, 363
grading station, 363
grading table, 141
gradoni, 129, 340
gradual dial, 274
graduated-particle board, 236
graduated reticle, 274
graduated tail, 486
graduating curve, 274
graft, 231
graft copolymer, 245
graft copolymerization, 245
graft hybrid, 231
graft incompatibility, 231
graft on root, 167
graft reaction, 245
graft union, 231, 244
graft with cutting, 213
graftage, 231
grafting, 231, 244, 564
grafting affinity, 231
grafting by approach, 271
grafting case, 231
grafting chimaera, 231
grafting plant, 231
grafting saddle, 3
grafting wax, 244
grain, 175, 272, 291, 506
grain angle, 506
grain character, 506
grain direction, 506
grain orientation, 254, 272, 506
grain printer, 335
grain printing, 335
grain raising, 367
grain rupture, 506
grain size, 254, 272, 291
grain weight, 291
grained paper, 335, 506
graining, 134, 291, 367, 515
graining brush, 335
graining paint, 335
grainlifting, 318, 367
Gram-negative, 165
Gram-positive, 165
Gram-stain, 165
gramineous, 191
graminivorous, 42
graminoid, 191
Gram's stain, 165
grand period of growth, 78, 429
grandiflora rose, 639
granite, 203
granite sand, 203
granodiorite, 203
granosan, 175
granular activated carbon, 272
granular aggregate, 491
granular amphibolite, 291
granular gneiss, 291
granular limestone, 291
granular mor, 491
granular structure, 291
granular (tannin-)extract, 291
granularity, 272, 291
granulated fertilizer, 272
granulated peat, 491
granulation, 263, 291, 491, 651
granulator, 54
granule, 272, 412, 491
granule size, 291
granulose, 260, 272
granulosis, 272
granulosis virus, 272
granum (*plur.* grana), 621
grape, 364
grapefruit oil, 593
graphic interpolation, 487
graphic method, 487
graphic method of stress analysis, 572
graphic method of valuation, 37, 384
graphical estimation, 487
graphical presentation, 487
grapple, 636
grapple-bunk skidder, 56

G

grapple capacity, 636
grapple hook, 367, 636
grapple lift volume, 636
grapple loader, 636
grapple opening, 636
grapple-opening line, 603
grapple operating line, 636
grapple saw, 229
grapple skidder, 636
grapple skidding, 636
grapple yarding, 636
grappling iron, 120, 480
grass, 41
grass-covered land, 41
grass eliminator, 64
grass fen, 432
grass fire, 95
grass glade, 303
grass grub, 248
grass heath, 42
grass hook, 42
grass line, 150, 570
grass mold, 150
grass moor, 42, 421, 604
grass rubber, 41, 42
grass savanna, 191, 402
grass tundra, 42
grass walk, 42, 364
grass waterway, 42
grasshopper, 210
grassland, 41, 42
grassland diminishes, 42
grassland ecology, 41
grassland ecosystem, 41
grassland improvement, 42
grassland indicator, 41
grassland management, 42
grassland vegetation, 42
grassplot, 302
grating, 165
grating for stopping floating wood, 283, 499
grating spectrograph, 182
grauser, 314
gravel, 291, 412
gravel culture, 412
gravel ground, 291
gravel path, 412
gravel soil, 291
gravelly loam, 291
gravelly sandy loam, 291
gravelly soil, 291
graveyard test, 489
gravid, 207, 396
gravimetric chip-feeder, 629
gravimetric moisture content, 629
graviperception, 629
gravitation, 499, 629
gravitational erosion, 629
gravitational potential, 570, 629
gravitational scrambler, 629
gravitational water, 629
gravitator, 53, 64
gravity cable logging, 277, 629
gravity cableway, 629
gravity chance, 629
gravity chute, 629
gravity circulation dry kiln, 646
gravity dam, 629
gravity feed chipper, 629
gravity incline-way, 629
gravity irrigation, 645
gravity loading, 629
gravity roll(er), 629
gravity seed cleaner, 629
gravity separator, 629
gravity skyline system, 509
gravity skyline yarding, 629
gravity slack line, 629
gravity spur, 629
gravity system of cable railway, 629
gravity thickener, 629
gravity tipple, 629
gravity trap, 629
gravity wind sifter, 629
gray acetate of lime, 210
gray-brown desert soil, 211
gray-brown forest soll, 211
gray desert soil, 211
gray forest soil, 211
gray heart, 211
gray humic acid, 210, 211
gray lime acetate, 210
gray scale, 210
gray shade, 211
gray soil, 211
gray solodic soil, 211
gray tropical clay, 392
gray warp soil, 211
grayish-black stain, 210
grayness after rubbing, 36
graywacke, 575, 597
grazed woodlot, 135
grazer, 336
grazing, 135
grazing allotment, 135
grazing animal, 135, 436
grazing capacity, 135
grazing food chain, 135
grazing forest, 135, 213
grazing ground, 135
grazing habit, 135
grazing height, 419
grazing herbivore, 135
grazing in forest, 302
grazing indicator, 135
grazing land value, 135
grazing line, 135
grazing meadow, 135
grazing period, 135
grazing preference, 135
grazing rate, 39, 419
grazing season, 135
grazing system, 135
grease dauber, 90, 229, 401
grease gun, 205, 578
grease proofness, 338
grease resistance test, 271, 338
grease spot, 343, 578, 613
greaser, 90, 229, 401
greasy mor, 613
great circle, 76, 78
greater covert, 76
greaved, 261
Grecian laurel, 594
greek fret, 165, 515
green, 315, 421, 432, 504
green all available land with trees and grasses, 616
green area, 315
green-belt, 80, 315, 510
green biomass, 427
green branch, 215
green building, 315
green chain, 432
green-chain tally, 421
Green China Fund, 622
green consumer, 315
green cover percentage, 315
green coverage, 315
green density, 421
green end, 249, 432

G

green fodder, 379
green food, 315
green fuel, 214
green glue-line, 423, 504
green gross domestic product, 315
green house, 203, 348, 505
green jobs, 315
green lacewing, 42
green land with grasses, 629
green liquor, 234, 315, 379
green lumber, 421, 432, 504
green lumber storage, 432
green lumber yard, 432
green manure, 315
green manure crop, 315
green manuring, 432, 629
green moisture content, 421, 432
green mold, 315, 379
green pruning, 215, 315, 430
green prunning, 543
green ramie, 379
green revolution, 315
green root, 214
green-rot, 315, 379
green salt, 315, 463
green snow, 315, 548
green sorting deck, 432
green space, 315
green-stain, 315
green table, 432
green the country, 315
green timber, 214, 421, 432, 539
green veneer, 432
green weight, 421, 432, 523
green wood, 421, 432, 539
green-wood cutting, 315
greenery, 42, 315, 348, 379, 505
greenheart, 315
greenhouse, 505
greenhousc budding, 505
greenhouse cluster, 505
greenhouse culture, 505
greenhouse effect, 505
greenhouse flower, 505
greenhouse gas, 505
greenhouse headhouse, 505
greenhouse heating, 505
greenhouse management, 505
greenhouse plant, 505
greening, 315
greening campaign, 315
greening commission, 315
greenish cast, 315
greenland, 315
green(manured) fallow, 629
greens, 315, 639
gregarious, 264, 388
gregarious animal, 388
gregarious flowering, 224
gregarious habit, 388
gregarious insect, 388
gregarious parasitism, 388
gregarious parasitoid, 388
gregarious phase, 388
gregariousness, 388, 389
gregarization pheromone, 263
grey-brown podzolic soil, 211
grey desert soil, 211
grey ferruginous soil, 211
grey forest soil, 211
grey mould, 211
grey soil, 211
grey stain, 210
grey tropical clay, 211
grey wooded soil, 211
greywacke, 575, 597
gribble, 188, 636
grid, 165
grid co(-)ordinate, 165
grid line, 499
grid method, 558
grid sampling, 558
gridironing method of strip survey, 81
grill(e), 165, 203
grill(e) guard, 132, 602
grind stone, 326, 412
grinder, 326, 327, 412, 555
grinder pocket, 327
grinding angle, 396, 536, 555
grinding chamber, 327
grinding disk, 327, 412
grinding face, 327
grinding material, 326
grinding pan, 327
grinding pressure, 327
grinding surface, 327
grip, 15, 29, 267, 636
gripping device, 229
grisein, 211
griseofulvin, 211
grit, 73, 291, 326, 412
grit number, 326
grit size, 326
grog table, 259, 353
groin, 104, 131, 239, 381
groined roof, 381
grooming, 444
groove, 4, 172
grooved and tongued joint, 366, 469
grooved bit, 75
grooved-faced, 41
grooved plywood, 8
groover, 268
grooving cutter, 268, 516
grooving machine, 268
grooving plane, 268
grooving saw, 268, 516
gross absorption, 319, 649
gross anatomical feature, 198
gross current annual increment, 526
gross domestic product, 186
gross feature, 198
gross for net, 566
gross growth, 318, 649
gross growth of initial volume, 63, 591
gross income, 73, 649
gross income tax, 649
gross increment, 73, 318, 649
gross knot, 77
gross loss, 319
gross mean annual increment, 649
gross national product, 185
gross output, 648
gross photosynthetic rate, 649
gross pressure, 649
gross primary production, 648
gross primary productivity (GPP), 648

G

gross product, 648
gross production, 649
gross production rate, 649
gross profit, 290, 318
gross reproductive rate, 649
gross retention, 648, 649
gross revenue, 649
gross scale, 72, 318, 648
gross structure, 198
gross uptake, 319, 649
gross volume, 72, 318, 648
gross volume increase percentage, 648
gross weight, 319, 649
grotto (*plur.* grotto(e)s), 172, 394, 554
ground, 94, 95, 97, 132, 219, 327, 336, 488
ground avalanche, 470, 488
ground-beetle, 36
ground board test, 8
ground-cable logging, 387
ground check, 97
ground chute, 489
ground clearance, 52, 95, 287, 299
ground cloth, 97, 99
ground coat, 95
ground control, 97
ground control map, 97
ground cork, 452
ground cover, 95
ground cover plant, 95
ground disinfection, 97
ground distance, 97
ground drain, 50, 98
ground drainage, 50, 98
ground fertilizer, 219
ground fire, 97, 98
ground flora, 95
ground frost, 97
ground fuel, 97
ground game, 534
ground hauling, 97
ground-hog, 83, 495
ground information, 97
ground lead block, 97
ground-lead (cable) logging, 97
ground(-lead) hauling, 97
ground-lead(cable) logging, 387
ground-line, 96, 220
ground-line area, 96
ground-line (cable) logging, 97
ground(-line) logging, 97, 387
ground(-line) skidding, 97, 387
ground-line treatment, 96
ground-litter, 278
ground meristem, 219
ground nadir, 97
ground-off saw, 13
ground-penetrating implement, 489
ground-penetrating shank implement, 257
ground plan, 94, 95
ground plate, 244, 317
ground pressure, 97, 244
ground prop, 231, 638
ground pulp, 218, 327
ground rent, 99
ground resolution, 97
ground resolved distance, 97
ground sill, 96, 507
ground skidder, 10, 242
ground sledge, 488, 489
ground-slide, 66, 489
ground sliding, 66
ground snigging, 97, 387
ground state, 98, 220
ground station, 97
ground stratum, 97
ground tissue, 219
ground truth, 97
ground vegetation, 95, 214
ground visibility, 97
ground water, 98
ground water flow, 98
ground water laterite, 374
ground water lateritic soil, 374
ground water level, 98, 374
ground water podsol soil, 374
ground water podzol, 374
ground water podzolization, 374
ground water resource, 98
ground water rice soil, 98, 374
ground water runoff, 98
ground water soil, 374
ground water table, 98, 374
ground-weir, 53, 251
ground wood pulp, 327
ground (wood) pulp, 218
ground work, 219, 310, 489
ground yarding, 97, 387
groundling, 364
groundwood core, 327
ground(wood) pulp, 327
group, 388, 449, 534
group clearcutting method, 389
group cutting, 389
group cutting method, 389
group effect, 224
group felling, 389
group formation, 389
group hunting, 389
group knot, 388
group management, 534
group mixture, 279
group of formation, 389, 617
group of sampling unit, 61
group of trees, 388, 449
group of vegetation type, 616
group planting, 279
group regulation, 389
group seed tree method, 306
group selection, 389
group-selection cutting, 389
group selection felling, 389
group selection management, 389
group selection method, 389
group selection system, 389
group-selection (system), 389
group-shelter wood cutting, 279
group shelterwood, 389
group shelterwood cutting, 389
group shelterwood felling, 389
group shelterwood system, 389
group shooting, 641

group size, 418, 627
group spacing, 389
group system, 389
group system of natural regeneration, 390
grouped pores, 179
grouping bed, 224
groupy phase, 627
grouse, 464
grouse ladder, 120
grove, 446, 535
grove tractor, 588
grow, 629
grow saplings, 353
growing facility, 353
growing period, 429
growing point, 429
growing season, 429
growing season grafting, 429
growing season length, 429
growing space, 429, 574
growing-space ratio, 429
growing stock, 290
growing stock determination, 37, 545
growing stock regulation, 290
growing stock structure, 544, 545
growing stock tree, 633
growing stock value, 410
growing stock volume classification, 300
growth, 299, 301, 429, 629
growth acceleration, 429
growth analysis, 429
growth and forest yield science, 409
growth (annual) increment, 344
growth band, 429
growth cabinet, 395
growth chamber, 395
growth coefficient, 602
growth correlation, 430
growth curve, 429
growth-cut ratio, 429
growth cycle, 430
growth data, 429
growth effect, 430
growth efficiency, 430
growth equation, 429
growth factor, 430
growth form, 429, 430
growth function, 429
growth habit, 423, 430
growth hormone, 429
growth in elongation, 419, 447
growth in height, 447
growth in length, 447
growth in storeys, 140
growth in thickness, 615
growth in volume, 37
growth intercept, 602
growth interruption, 430
growth layer, 429
growth-limiting substrate, 430
growth method, 429
growth model, 429, 602
growth of cross-sectional area, 113
growth of stand, 300
growth of tree value, 448
growth onset, 269
growth parameters, 44, 429
growth percent, 55, 429
growth percentage, 429
growth period, 429
growth plot, 176, 576
growth prediction, 430
growth prognosis, 430, 597
growth projection, 430
growth-promoting auxin, 73
growth rate, 429, 602
growth-rate function, 429
growth region, 429
growth regulator, 430
growth retardant, 2, 430
growth ring, 429
growth-ring boundary, 429
growth series, 430
growth space, 429
growth stress, 429, 430
growth study, 430
growth table, 429
growth test, 429
growth type, 430
growth zone, 429
groyne, 104, 131, 239
grub, 265, 366
grub-breaker, 265
grub felling, 265, 292
grub-hoe, 64, 164, 265
grub-hoe-slit-method, 64
grub hole, 60, 366
grub puller, 265
grubber, 5, 265, 420, 495
grubbing axe, 265
grubbing-instrument, 266
grumosol, 391
grumosolic soil, 391
guaiac, 587
guaiac resin, 587
guaiac wood oil, 587
guaiacin, 587
guaiacum, 587
guaiacum-wood, 587
guaiacyl-type lignin, 587
guaiadiene, 587
guaiene, 587
guajac, 587
guano, 345
guanosine, 345, 346
guarana tannin, 352
guard cell, 15
guard plot, 132
guard row, 14
guard stone, 202
Gugel process, 174
guibortacacidin, 184
guide, 89
guide bar, 89
guide bar nose, 89
guide block, 89
guide curve, 89
guide plate, 89
guide roll, 89, 243
guide track, 89
guide tree, 222, 231, 478
guiding curve, 89, 221
guild, 285, 389, 483, 627
guillotine, 233, 602
guillotine saw, 630
guillotine shears, 233, 492
guinea line, 209, 570
gula, 497, 553
gullet, 57, 58, 88, 520, 521
gullet depth, 57
gullet file, 57
gullet saw, 57, 173
gullet sawdust capacity, 58
gullet-to-chip area ratio, 58
gulleting machine, 268
gully, 172, 173
gully density, 173
gully erosion, 173
gully erosion control forest, 172
gully head protection, 173
gully type, 59
gum, 448

gum arabic, 1
gum benzoin, 2
gum canal, 448
gum cavity, 448
gum check, 448, 613
gum crusher, 465
gum duct, 448
gum formation, 242
gum kino, 220
gum-lac, 450, 531
gum mastic, 316, 400
gum navel stores, 465
gum of pine, 465
gum pocket, 448, 578
gum reservior, 635
gum resin, 448
gum rosin, 613
gum spot, 240, 448
gum streak, 448
gum syrup, 1, 448
gum tragacanth, 210
gum turpentine, 613
gum up, 345, 450
gum vein, 240, 448
gumbo, 232
gummase, 366
gum(med) tape, 240
gummeline, 201
gummer, 57, 74, 268, 431
gumming, 241, 487
gummosis, 306
gummous reaction, 306
gun, 375, 401, 578, 602, 634
gun a log, 589
gun boat, 53
gun cotton, 217, 532
gun-hour, 375
gun-injection, 375, 635
gun-injection process, 375
gun powder, 217
gunite method, 354
gunning, 195, 197, 324, 418
gun(ning) stick, 634
gunstock, 375
gurgan, 174, 494
gurjun, 174, 494
gurjun balsam, 174, 309
gurjun balsam oil, 174, 309
gurjunene, 174, 309
gurjunene ketone, 174, 309
gurjuresene, 174, 309
gustiness, 280, 608
gutta, 111, 174
gutta percha, 111, 174
guttation, 491
gutter, 41, 68, 90, 172
gutter board, 479, 555
gutter trench, 132
guy, 19
guy cable, 19, 282
guy derrick, 7, 637
guy line, 19, 176
guy-line loading, 19
guy-line post, 19
guy-line spacing, 19
guy-line system, 19
guy rope, 19
guy-spike, 19, 175
gymnosperm, 314
gynandrian, 192
gynandromorph, 70
gynandrophore, 296
gynandrous, 70
gynandrous spike, 70
gynobase, 69
gynodioecious, 69
gynoecium, 69
gynogenesis, 69, 173
gynomonoecious, 69
gynophore, 69
gynospore, 69, 76
gynostemium, 192
gypsum, 433
gypsum rock, 433
gypsy, 242, 264
gypsy spool, 242, 264
gypsy yarder, 264
gyratory screen, 357, 546
gyrodine, 639
gyrodozer, 48
gyttja, 150, 201
gyttja soil, 150, 201
H-budding, 551
H-form budding, 551
H-horizon, 342
H-layer, 342
H-shaped budding, 551
habal zone, 420
habit, 516
habit of growth, 430
habitat, 365, 423
habitat diversity, 423
habitat factor, 423
habitat form, 423
habitat fragmentation, 423
habitat group, 423
habitat island, 423
habitat isolation, 423
habitat managemen, 365
habitat management, 423
habitat management area, 513
habitat niche, 423
habitat preference, 423
habitat restoration, 365
habitat rstoration, 423
habitat segregation, 423
habitat selection, 365, 423
habitat type classification, 423
habituation, 549
hack, 177, 269, 433, 443
hack-saw, 160, 170
hack tie, 269, 356
hadal zone, 51
hadopelagic zone, 51
hadromase, 334
hadrome, 511
hadromestome, 335
haematein, 557
haematin, 376
haematoxylin, 466
haemocoele, 548
haemocyanin, 548
h(a)emocyte, 548
haemoglobin, 548
h(a)emolymph, 548
haemolytic, 398
haemophagous, 437, 515
hag, 37, 125, 127, 187, 269
hail, 13, 27
hair check, 123, 519
hair crack, 519
hair hygrograph, 318
hair hygrometer, 318
hair split, 519
hairline, 324, 518, 580
hairline crack, 123, 519
hairy root, 123, 544
half balk, 11, 115
half-blind dovetail, 10, 11
half bog soil, 11
half bordered pit, 11
half bordered pit-pair, 11
half-cellulose, 10
half choke, 9, 11
half-desert, 10
half-diallel crossing, 10
half dovetail, 10
half-dovetail dado joint, 10
half-fallow, 10
half-finished goods, 9

half-finished product, 9
half-gantry crane, 83
half-hard (fibre) board, 11
half hardy, 10
half hardy annuals, 10
half hatchet, 9
half-height diameter, 625
half-herringbone cut, 11
half-lap joint, 9
half-life, 9, 10
half-log, 11
half-mature forest, 623
half mutant, 10
half-mutation, 10
half negative slope, 10
half-parasite, 9
half positive slope, 10
half-principal, 9, 504
half ringshake, 10
half-rip saw, 518
half-rotary-cut veneer, 11
half-rotary-cutting, 11
half-round, 11
half-round cutting, 11
half-round method, 11
half round post, 11
half round profile board, 11
half-round sleeper, 11
half-round veneer, 11
half-round veneer lathe, 11
half-round wood, 11
half-saprophyte, 9
half seed year, 245
half shaded slope, 10
half-shrub, 9
half-sib, 10
half-sib covariance, 10
half size veneer, 9
half-sized, 380
half spiralcurve cut, 10
half-stock, 9, 10
half strength nutrient solution, 10
half stuff, 9, 10
half sunny slope, 10
half-tide-level, 9
half timber, 9
half-timbering, 10
half-track, 10
half truss, 9
half-wrought timber, 9
halfmoon tie, 11
halfround cutting veneer, 11
halic, 555
halicole, 555
halide lamp, 233, 249, 310
hall seat, 78
hall stand, 564
halloysite, 1, 118
halo, 276, 594
halobacteria, 555
halobion, 440, 516
halobiont, 555
halocline, 555
haloeremion, 555
halogenated fire retardant, 310
halogenation, 310
halogenic soil, 555
halomorphic soil, 555
halophile, 440, 441, 517
halophilic, 555
halophilous plant, 555
halophyte, 440, 517, 555
halophytes, 555
halophytic vegetation, 555
haloplankton, 555
halosere, 555
halt desertification, 134
halt soil erosion slow down, 134
halter, 360
halticid beetle, 480
halving joint, 10, 41
hamabiosis, 511
hamamelin, 17
hamamelitannin, 248
hamlet forest, 74
hammada, 434
hammer, 68, 376
hammer basket screen mill, 68
hammer beam, 66, 492, 545
hammer crusher, 68
hammer-head crane, 68, 470
hammer hog, 68, 468
hammer mark, 68, 269
hammer mill, 68
hammer post, 66
hammer type barker, 68
hammer type shredder, 68
hammered gun, 68, 325
hammerless gun, 3, 340
hamulus, 58, 173, 533
hand axe, 442
hand-bag, 395
hand-bag (hand-bank), 395
hand-bank, 395
hand banker, 395
hand bar, 320
hand barker, 394
hand barking, 394
hand block sander, 395, 443
hand brake, 443
hand cart, 443
hand charging, 395
hand clamp, 442
hand drill, 442, 443
hand-feed, 394
hand feed band-saw, 394
hand-feed grinder, 443
hand (feed) planer, 395
hand incisor, 442
hand jig, 443, 526
hand-lens identification, 135
hand lever, 442
hand loading, 395
hand logger, 65, 395
hand log(ging), 65, 395
hand mill, 442
hand-operated winch, 442, 443
hand peeler, 394
hand pike, 112
hand plane, 442
hand-power, 442
hand prehauling, 395
hand rail, 147, 283
hand reflected level, 442
hand sander line, 443
hand-saw, 442
hand screw clamp, 332, 442
hand seeder, 442
hand set, 442
hand skid(ding), 65, 395
hand sprayer, 442
hand tool, 442
hand trencher, 442
hand weeding, 395, 442
hand winch, 443
hand yarding, 65, 395
hand yoller, 395
handkerchief tree, 172
handle, 5, 7, 170, 639
handling, 41, 65, 502, 639
handling expense, 7, 65, 179
handling machine, 7, 639
handling operation, 639

H

handscrew, 442
handsheet, 442
hang stick, 147
hang up, 75, 223, 268, 307, 470
hang-up tree, 75
hanger, 104, 545
hanger bolt, 104, 545
hanging basket, 104, 545
hanging drop, 545
hanging drop culture, 545
hanging garden, 276, 545
hanging rail, 231
hanging roller, 550
hanging sash, 103
hanging stile, 243
hanging tree, 75
hanging-up, 178, 367, 473
hanging wedge, 600
hangover fire, 373, 374
hank shellac, 243
Hanzlik formula, 190
Hanzlik's formula, 190
haplite, 518
haplochromosome, 82
haplodiplont, 82
haploid, 82
haploid apogamy, 82
haploid gametophyte apomixis, 82
haploid number, 82
haploid parthenogenesis, 82
haplometrosis, 83
haplomict, 82
haplomitosis, 82
haplont, 82
haplontic sterility, 82
haplopetalous, 82
haplostemony, 84
hapto, 314
haptotropism, 530
hard bark, 499, 575
hard bast, 575
hard beating, 629
hard-board, 295, 563
hard cash, 525, 575
hard cook, 575
hard-cushion formation, 575
hard deciduous woods, 575
hard fibre board, 575
hard grain, 575
hard leaved forest, 575
hard pine, 575
hard-pruned, 376
hard resin, 575
hard rime, 455
hard rot, 575
hard seed, 575
hard seed coat, 575
hard seeded tree species, 575
hard sized, 630
hard stock, 575
hard streak(s), 573
hard water, 575
hard wood, 281
hardboard, 575
hardcore, 310, 575
hardened off, 324, 335
hardened rosin, 575
hardened surface, 575
hardener, 575
hardening, 113, 575
hardening agent, 575
hardening of seed, 628
hardening of the soil, 491
hardening off, 113
hardening procedure, 575
hardening time, 575
hardiness, 94, 271, 337
hardness, 575
hardness tester, 575
hardpan, 575
hardshell, 89, 575
hardware, 249, 512, 535, 575
hardwood, 282, 575
hardwood cutting, 46, 282, 575
hardwood forest, 282
hardwood formation, 282
hardwood furniture, 575
hardwood tree, 575
hardy, 337
hardy annuals, 337
hardy crop, 270
hardy perennials, 337
hare, 559
harem, 264
harmonic biota, 479
harmonic lake type, 479
harmonic mean, 479
harmonic peak, 537
harmonious growth, 536
harmonization of curve, 384, 536
harmonized curve, 479
harpes, 15
harrier, 559
harrow, 5, 349
harrower, 5
harrowing, 5
harsh climate, 369
Hart moisture gauge, 187
Hartig net, 187
Hartignet, 187
Hartig's method, 187
Hartman process, 187
harvest cut with seed trees, 306
harvest cutting, 55, 625, 633
harvest felling, 37
harvest loss, 441
harvest management, 38
harvest maturity, 441
harvest method, 441
harvest percent, 291, 441
harvest plan, 126
harvest plan map, 38
harvest planning, 633
harvest volume, 38, 441
harvest waste, 441
harvestable timber, 272
harvested wood product, 335
harvester, 38, 39
harvester termite, 41
harvesting, 37, 39
harvesting cost, 38
harvesting costs, 125, 441
harvesting cycle, 211, 442
harvesting insurance, 441
harvesting platform, 39
harvesting system, 37
Hasa altimeter, 187
Hasserman process, 187
hastening germination, 73
hatch, 41, 125, 165, 223, 268, 534, 602
hatch crane, 41
hatchability, 147
hatched grafting, 529
hatchery, 147
hatchet, 19, 247, 269, 442
hatchet planimeter, 234
hatching, 147
hatching factor, 147
hatchling, 534
hati, 122, 352, 471
haul, 70, 222, 474, 595

H

haul ahead line, 372
haul-in line, 372
haul line, 372
haul-out line, 212
haul road, 298
haul(age) cable, 372
haulback, 212
haulback block, 212
haulback drum, 212
haulback lead block, 212
haulback line, 212
haulback side block, 212
haulback tail block, 212
hauler, 595
hauling capacity, 372
hauling cost, 224
hauling distance, 595
hauling equipment, 372, 596
hauling route, 595
hauling truck, 595
hauling up logs from water, 329
haulmiser, 252
haunch, 469
haunch mortise and tenon joint, 229
haunch rabbet joint, 229
haunched tenon, 229
haunching, 269
haustorium (*plur.* -ia), 514
hawk moth, 477
Haworth formula, 217
hawser, 283, 319, 372
hay production, 464
hay wire, 2, 150
hayrack (boom) loading, 475
haywire boom, 150, 304
hazard rating, 500
hazardous waste, 580
haze, 317
haze catting filter, 513
haze meter, 554, 640
haze-penetrating filter, 486
haze, smog, 513
hazel aphid, 608
head, 88, 89, 105, 149, 216, 231, 314, 459
head block, 53, 243, 286, 352, 633
head box, 250, 306, 459, 499
head capsule, 485
head dominant, 634
head erosion, 173, 466
head filer, 610
head fire, 461, 485
head gate, 250, 384
head lean, 290
head loader, 52, 269, 602, 638
head log, 373, 417
head of bed, 67
head of fire, 216
head rigger, 280, 470
head saw, 364, 485, 633
head spar, 223, 625, 633
head spar tree, 223, 633
head stick, 350
head tree, 633
head trip block, 212
head water conservation forest, 460
head way, 52, 257
head woodcutter, 126
headboard, 67, 491
header coil, 633
heading, 131, 191, 485, 593
heading back, 212, 247, 327
heading joint, 111, 614
heading off, 247
head(log) grab, 443
headrig, 364, 485, 633
headrig slab chipper, 633
headroom, 257
headstock, 67, 218, 294, 485, 634
headward erosion, 173, 466
headwater, 593
headwood system, 485
headwork, 242, 250, 384, 413, 444
healed over knot, 571, 587
healed scar, 587
healed wound, 587
healing, 587
healing duct, 228
health resort, 296, 543
healthy tree, 236
heap charring, 114
hearing protector, 601
heart board, 468
heart centre, 468
heart chamber, 538
heart check, 538
heart of log, 590
heart plank, 468
heart ring-shake, 538
heart rot, 538
heart shake, 538
heart split, 449
heart stain, 538
heart substance, 538
heartwood, 538
heartwood formation, 538
heartwood rot, 538
heartwood tree, 538
heat-absorbing reaction, 514
heat absorption, 394
heat asphyxiation, 640
heat balance, 393
heat budget, 393
heat capacity, 393
heat conduction, 66, 391
heat conductivity, 89
heat consumption, 191
heat content, 189, 392
heat convection, 392
heat convertible resin, 394
heat curtain door, 166
heat decomposition of wood, 330
heat deformation, 391
heat degradation, 392
heat distortion, 391
heat drop, 392
heat exchanger, 392
heat flux, 393
heat input, 171, 393
heat insulator, 15, 393
heat island, 392
heat isolated door, 166, 265
heat kill, 393
heat lesion, 392
heat loss, 393
heat of adsorption, 514
heat of condensation, 286, 346
heat of swelling, 355
heat of vaporization, 371, 609
heat of wetting, 401
heat output, 124, 392–394
heat penetration, 391, 393, 486
heat pollution, 394
heat press, 394
heat probe, 473
heat radiation, 392
heat reclaim, 139, 392
heat recovery, 392
heat resistance, 270, 337
heat-sensitive equipment, 393

heat set, 391, 392
heat-setting resin, 392
heat stabilization, 394
heat sum, 219
heat supply pipeline, 171
heat therapy, 393, 394
heat tolerance, 337
heat transfer, 66, 393
heat transfer by contact, 244
heat transfer by convection, 115
heat transfer by radiation, 148
heat transfer coefficient, 66
heat transfer mudium, 66, 598
heat transfer surface, 66
heat transmissibility, 391
heat-treated board, 391
heat-treated wood, 391
heat treatment, 391
heat value, 394
heated-air seasoning, 393
heated nebulization chamber, 228, 394
heated roll, 228
heater, 228
heath, 348
heath peat, 434
heath plough, 180
heath podzol, 434
heathland, 348
heating by infrared rays, 198
heating by radio frequency, 418
heating coil, 228, 392, 404
heating cycle, 228, 394
heating degree days, 228
heating drum, 228
heating effect, 394
heating element, 228
heating medium, 393, 598
heating period, 228
heating platen, 394
heating power, 124, 268, 390, 394
heating rate, 228
heating temperature, 228
heating time, 228
heating unit, 228
heating-up, 228, 421
heavily cut, 375
heavily cutover area, 375
heavily thinned stand, 376
heaviness of felling, 38
heavy bodied, 344
heavy clay, 630
heavy discolouration, 554
heavy dressing, 77
heavy duty grinding, 376
heavy-duty log fork, 630
heavy duty plough, 420
heavy duty veneer slicer, 630
heavy feeder, 574, 578
heavy four drum roll feed sander, 630
heavy four-sided matchboarding machine, 630
heavy fuel, 47, 163, 318, 630
heavy grading, 110
heavy irrigation, 77, 186, 318
heavy joist, 78, 630
heavy leather, 200, 629
heavy loam, 630
heavy market, 438
heavy oil, 630
heavy patternmaker's gap lathe, 187
heavy pruning, 376
heavy seeding, 322
heavy skyline cableway, 630
heavy soil, 345
heavy texture, 345
heavy thinning, 376
heavy-timber construction, 575, 630
heavy torn grain, 420
heavy turpentine, 630
heavy wood, 629
hedge, 315, 423, 448
hedge cutter, 315
hedge shears, 315
hedgemaker, 315
hedger trimmer, 315
hedgerow, 288, 315
hedging, 315
heel, 57, 88, 168, 444
heel block, 103, 153
heel cutting, 356, 629
heel line, 603
heel strap, 66, 104, 603
heel tackle, 603
heel wear, 295
heelin, 230
heel(ing) boom, 105, 612
heel(ing) boom loading, 612
heel(ing)-in, 230, 317
heerabolene, 327
height according to the 40% rule, 6
height accretion, 447
height accumulation, 285
height age curve, 447
height board, 475
height class, 447
height class volume table, 447
height classification, 447
height curve, 447
height-DBH ratio, 541
height determination, 447
height development curve, 447
height-diameter curve, 447
height finder, 44, 162
height gauge, 161
height grade, 447
height growth, 447
height growth curve, 447
height increment, 447
height index, 447
height-measure, 162
height measurement, 161, 447
height measures of stand, 299
height meter, 44
height of chest, 542
height of dominant trees, 415, 416
height of leaf, 560
height of mean basal area tree, 113
height of tooth, 57
height of tree, 447
height sampling methods, 447
hekistotherm, 439
hel(e)okrene, 605
heleoplankton, 57
helical cavity, 313
helical conveyer, 313
helical hinge, 453
helical orientation of microfibril, 501

helical stump cutter, 313
helical surface, 313
heliciform pore arrangement, 313
helicoid cyme, 313
helicopter logging, 615
helicopter-mounted broadcaster, 136
helicopter sprayer, 615
helicopter yarding, 615
helics, 313
helijumper, 615
helio- (*eg.* heliophilous, heliophobous), 556
heliophile, 517, 557
heliophilia, 516
heliophilic, 557
heliophilous, 516, 557
heliophilous species, 557
heliophobe, 21, 517, 524, 570
heliophobous, 517, 524
heliophobous community, 570
heliophobous species, 517, 570
heliophyte, 517, 556, 557
heliotherm, 397
heliotropic, 530
heliotropism, 530
heliport, 615
helispot, 615
helitack crew, 275, 615
helitank, 456, 615
helitanker, 615
helix, 313
helix angle, 313
hell hook, 104
helm wind, 120, 414
helophyte, 604
helve, 29, 149, 169
hemartine extract, 466
heme, 548
hemera, 222
hemerophyte, 597
hemiascomycete, 11
hemibasidiomycete, 9
hemibiotroph, 9
hemicellulose, 10
hemicellulose biodegradation, 10
hemicolloid, 10
hemicryptophyte, 10
hemicryptophyte climate, 97
hemicryptophytes (He), 97
hemicyclic, 10
hemielytra, 10
hemiendobenthos, 10
hemiepiphyte, 9, 10
hemikaryon, 82
hemilignin, 10
hemimetabola, 9, 35
hemin, 316
hemiparasite, 9, 232
hemiplankton, 9
hemisaprophyte, 9
hemispherical photography, 10
hemitropal, 196
hemitropous, 196
hemizygote, 9
hemlock extract, 481
hemlock spruce oil, 481
hemoprotein, 548
hemp palm, 648
hemp seed oil, 77
hendecaploid, 433
hepialid moth, 23
heptachlor, 365
heptaploid, 365
heptet, 365
heptose, 168
herb, 41
herb garden, 559
herb layer, 41
herb of grace (grue), 595
herbaceous, 41
herbaceous border, 42, 203
herbaceous layer, 41
herbaceous stem, 41, 42
herbaceous stratum, 41
herbaceous vegetation, 41
herbaceous walk, 42, 203
herbage, 41, 336
herbarium, 282, 616
herbary, 41, 559
herbcide, 64
herbicidal oil, 64
herbicide, 64, 410
herbivore, 436
herbivorous, 436
herbivorous animal, 436, 437
herbland, 41
herbosa, 41
hercules, 56, 161, 442, 538
Hercules drop method, 194
hereditability, 226, 274
hereditary conservation, 565
hereditary factor, 565
hereditary feature, 565
hereditary information, 565
hereditary substance, 565
hereditary unit, 565
hereditary variability, 565
hereditary variation, 565
heredity, 565
heritability, 565
heritable, 272, 273, 565
hermaphrodite, 70
hermaphrodite flower, 296
hermaphroditic, 70, 296
hermaphroditic generation, 70, 296
hermaphroditism, 70
hermetical storage, 322
herpetology, 349
herring-bone figure, 139
herring-bone method, 139, 584
herring bone system, 396, 584
herringbone face, 139
herringbone felling, 396, 584
hesperidium, 156
heteranthery, 542
heterobeltiosis, 51
heterocellular ray, 567
heterocephalous, 260
heterochlamydeous, 121, 567
heterochore, 567, 568
heterochromatin, 567
heterochromaty, 237, 567
heterochronogenous soil, 71
heterocline, 70, 567
heterocyst, 567
heterocyton, 566
heterodynamic hydrid, 357
heterodynamic insects, 566
heterodyne, 495
heteroecious, 637
heteroecism, 637
heteroeytosome, 566
heterogamete, 567
heterogamety, 353
heterogamy, 70, 353, 567
heterogeneity, 34, 119, 568
heterogeneity value, 35

H

heterogeneous-diverse habitat, 568
heterogeneous material, 34, 137, 568
heterogeneous population, 137, 568
heterogeneous ray, 567
heterogeneous ray tissue, 567
heterogeneous soil, 137, 568
heterogenesis, 486, 568
heterogenic, 567, 568
heterogeny, 486, 568
heterogonous, 204
heterogony, 204, 438, 567, 568
heterograft, 568
heterological system, 596
heterolytic fission, 567
heteromerozygote, 567
heterometabola, 35
heteromorphic incompatibility, 567
heteromorphism, 118, 567
H
heterophagous, 597
heterophyll, 568
heteroplastic graft, 568, 626
heteroploid, 566
heteropycnosis, 518, 566
heteroscedasticity, 566
heterosis, 597
heterosomal aberration, 390
heterosome, 541
heterostyly, 204
heterotasithynic, 33, 566
heterotetrameric, 568
heterotetraploid, 567
heterothallic, 70, 568
heterothallism, 70, 568
heterotherm, 567
heterothermic animal, 24
heterotic vigor, 597
heterotroph, 568
heterotrophic, 568
heterotrophic bacteria, 568
heterotrophic component, 568
heterotrophic nutrition, 568
heterotrophic organism, 568
heterotrophic plant, 568
heterotrophic respiration, 568
heterotrophic stratum, 568
heterotrophic succession, 568
heterotropic cell-wall layer, 567
heteroxenous, 567, 568
heterozone organism, 118
heterozygosity, 567, 596
heterozygote, 567, 596
heterozygous, 596
heulandite, 357
hew, 269, 356
hewed mountain square, 413
hewer, 269
hewing axe, 19
hewing crosstie, 19
hewn square, 130
hewn surface, 356
hewn tie, 19
hewn timber, 356
hewn tree, 125
hewn-tree skidding, 125
hexagon cell core, 309
hexahydroretene, 309
hexahydroxy diphenyl, 309
hexamerous, 309
hexamethylene diamine, 552
hexandrian, 309
hexaploid(6n), 308
hexokiuase, 224
hexosan, 224
hexose, 224
heyderiol, 531
Hi-lo pulper, 162
hiba leaf oil, 312
hiba wood oil, 312
hibernaculum, 594
hibernal, 108
hibernal annual plant, 594
hibernation, 108, 594
hidden defect, 571
hidden fire scar, 216
hidden solonchak, 374
hidden tax, 235, 571
hidden tenon, 4
hidden tree, 18
hide, 356, 424, 466
hide glue, 356
hide humus layer, 317
hide powder, 356
hide powder method, 356
hidimeter, 605
hiding edge, 146
hiding power, 605
hiemal, 108, 190, 585
hiemal aspect, 108
hiemefruticeta, 585
hiemelignosa, 585
hiemifruticeta, 585
hiemilignosa, 585
hiemisilvae, 585
hierarchy, 544
hierarchy theory, 91
higb-lead skidder, 10
high analysis fertilizer, 161, 348
high-ball logging, 10, 469
high bog, 164
high clearance digging tractor, 162
high clearance tractor, 162
high climber, 162, 349
high cover, 162, 416
high density digital tape, 163
high-density particleboard, 163
high density plywood, 163, 164
high density stock pump, 163
high-energy jet, 163
high face, 163
high-flelding seed, 161
high-flying logging, 370
high forest, 377, 436
high forest compartment system, 377
high forest system, 377
high forest with reserves (system), 14, 122
high-forest with soil protecting wood, 459
high forest with standards (system), 14, 122
high free rosin size, 164
high frequency, 163
high frequency (dielectric) drying, 163
high-frequency drying, 163
high frequency equipment, 163

high frequency gluing, 163
high frequency hardening, 163
high frequency heating, 163
high frequency power, 163
high frequency titration, 163
high-frequency ultrasonic wave, 163
high frequency vacuum drying, 163
high frequency vibration, 163
high-grade stock, 415
high-grade veneer, 162, 577
high grading, 311
high-grading cutting, 601
high-grading forest, 577
high grass, 161
high-humidity treatment, 163
high-impact material, 336
high-lead, 10
high-lead block, 10
high-lead cable logging, 10
high-lead hauling, 10
high-lead logging, 10
high-lead spar tree, 10
high lead yarding, 243
high-lead yarding system, 10
high level laterite, 162
high moor, 164, 359
high moor peat, 164
high mountain forest, 163
high oblique photo, 163
high-pass filter, 164
high peat-bog, 164, 359
high performance liquid chromatography, 164
high-planting, 114
high pole (wood), 78
high polymer, 162
high-population fitness, 164
high porous board, 162
high power objective, 161
high-pressure air receiver, 164
high pressure gas chromatography, 164
high-pressure laminating system, 164
high pressure liquid chromatography, 164
high pressure mass spectrometry, 164
high-pressure process, 164
high pressure steam sterilizer, 164
high-priced furniture, 161
high productive chromatography, 161
high pruning, 164, 415
high-quality board, 577
high-quality seed, 295
high relief, 162
high resolution, 162
high resolution infrared radiation sounder, 162
high resolution infrared radiometer, 162
high-road, 171
high side, 161, 164, 415
high spar, 161
high speed cooking, 279
high-speed mixing machine, 163, 279
high stand, 164
high straight-away shot, 163
high taper, 231, 417
high-temperature adhesive, 164
high-temperature dormancy, 164
high-temperature drying, 164
high-temperature kiln drying, 164
high-temperature pasteurization, 164
high-temperature processing, 164
high-temperature setting glue, 164
high thinning, 415
high topping, 162
high tree, 377
high value assortments, 162, 184, 607
high-velocity superheated steam dry kiln, 187
high water bed, 161, 163, 199
high-way, 171
high-wheeler, 162
high wheels, 162
high wind, 76, 222
high yield, 161
high-yield grass, 161
high yield pulp, 161
highboy, 162
higher-fungus, 161
higher harmonic wave, 161
higher-monosaccharide, 164
highest forest rent rotation, 406
highest gross yield rotation, 319, 649
highest soil expectation value rotation, 299
highest soil rent rotation, 298
highly mycotrophic, 162
highly structured forests, 213, 567
highwall, 22
highway ditch, 171, 202
highway tree, 171
hike, 240, 487, 594
hill, 31, 381, 489, 534
hill culture, 309, 363
hill-drop drilling, 547
hill-drop planting, 547, 548
hill dropping, 547
hill garder, 636
hill making, 636
hill planting, 382
hill side, 413
hilling, 114, 167, 353
hillock, 534
hillside channel works, 413
hillside covering works, 413
hillside erosion control, 413
hillside stone masonry, 413
hillside terracing, 413
hillside wicker works, 413
hillside works, 413
hillslope, 413
hilly land, 363, 381
hilly zone, 382
Hilum, 366, 626
himachalene, 548
Himalaya cedar, 548

H

Himalayan cedarwood oil, 548
hinge, 243
hinge hardware, 243
hinge mounting plate, 243
hinge pin, 243
hinge-type jib boom, 243
hinge-type tower, 243
hinge wood, 306, 605
hingebound door, 74
hinged bearing, 243
hinged end, 243
hinged hammer pulverizer, 177
hinged joint, 243
hinged stake, 243
hingeless arch, 509
hinoki cedar, 396
hinoki ketone, 396
hinoki oil, 396
hinokiic acid, 396
hinokiol, 202, 396
hip rafter, 236
hipping, 353
histic epipedon, 342
histidine, 651
histochemistry, 652
histocompatibility, 652
histogenesis, 652
histogram, 614, 636
histogram equalization, 614
histogram linearization, 614
histogram specification, 614
histological anatomy, 652
histological differentiation, 652
histology of wood, 331
histolysis, 652
histone, 651
Historia animalium, 109
historic relic, 174
historic site, 174
historical biogeography, 288
historical development, 517, 629
historical garden, 288, 437
historical plant geography, 288
history book, 253
history of forestry, 303
history of stand, 299
histosol, 580
hit-and-miss, 309
hitch, 104, 545
hitching force, 178
hitching post, 227
ho oil, 131
hoading, 8
hoar frost, 6, 28
hoarding, 635
hodometer, 310, 312
hoe, 64, 595
hoeing, 64, 190
hog-back(ed) sleeper, 11
hog gum, 363
hog still, 609
hogarth chair, 497
hogged slab, 218
hogger, 468
hog(ger), 144, 468
hogging, 363
hogging machine, 144, 468
Hohenadl's form factor, 608, 611
hoist, 264, 367, 590
hoist cable, 104, 368
hoist hook, 104
hoist line, 104, 368
hoister, 242, 367, 421
hoisting, 367, 638
hoisting block, 367
hoisting carriage, 598
hoisting drum, 367
hoisting rope, 368
hoisting tackle, 367
holarctic floral kingdom, 130, 385
Holarctic Region, 385
holarctis, 385
holard, 491
hold-down roll, 550
hold for cocking pin, 550
hold over, 15
hold water in the soil, 15
holdback carrier, 130
holder, 65, 221, 229, 231
holder shank, 262
holdfast, 177, 513
holding, 442
holding area, 168, 442
holding boom, 283, 442
holding company, 57, 175, 277
holding ground, 168, 442
holding paddock, 502, 534
holding power, 507
holding project, 442
holding wood, 306, 605
holding work, 442
hole, 110, 276
hole borer, 495
hole digger, 495, 618
hole-method, 548
hole planting, 548
hole seeding, 547
hole sowing, 547
hole transplanting, 269
holeuryhaline, 183
holing, 548
holism, 610
Holland log rule, 193
Hollander (beater), 193
Hollander roll, 193
hollow, 4
hollow-backed saw, 4
hollow butt, 125
hollow chisel, 276
hollow cone, 5
hollow-core board, 276
hollow-core construction, 276
hollow-core particleboard, 276
hollow cutting bit, 276
hollow dam, 276
hollow-ground saw, 452
hollow heart, 339
hollow homing, 146
hollow-horned, 146
hollow knot, 275, 494
hollow pith, 468
hollow plywood, 276
hollow seat, 4
hollow seed, 275
hollowed out carving, 309
hollowed out tenon, 486
hollowness, 117
holobasidium, 385, 509
holocellulose, 387
Holocene, 387
holocoen, 425
holocoenotic, 425
holoenzyme, 386
holometabola, 385
holometabolic insect, 385
holomietic lake, 387
holoparasite, 498
holopelagic, 386, 607
holopelagic plankton, 626
holophyte, 647
holophytic, 647
holophytic form, 647
holoplankton, 626
holopulping process, 387

holotype, 387, 498, 633
holozic, 385
holozygote, 385
holt, 302, 534, 597
home garden, 229, 481, 603
home gardening, 229
home market, 186
home nursery, 229, 647
home office, 648
home range, 52, 214
home-use forest, 229, 647
homeohydric plant, 195
homeoplastic graft, 484
homeorhesis, 109
homeostasis, 341
homeostatic organism, 341
homeostatic process, 341
homeotherm, 195
homeothermal animal, 195
homeotype, 91, 483
homoannular diene, 482
homocamphor, 164
homocellular, 483
homocellular ray, 483
homochlamydeous, 483
homodiploid, 68
homodyne, 642
homoeosis, 484
homoeotype, 91, 483
homoepicamphor, 161
homogamete, 483
homogametic, 483
homogamous, 70
homogamy, 70, 483
homogenate, 595
homogeneity, 266, 484
homogeneous, 266, 483, 484
homogeneous cover, 266
homogeneous fibre board, 266
homogeneous hydrolysis, 266
homogeneous particle board, 266
homogeneous pith, 266
homogeneous plywood, 86
homogeneous population, 366, 484
homogeneous ray, 483
homogeneous shear, 266
homogeneous soil, 266
homogeneous stress, 266
homogeneous wood beam, 266
homogeneousness, 266
homogenizer, 266, 595
homogenizing mill, 266
homogony, 204
homoiohydric plant, 195
homoiosmotic, 92
homoiosmotic animal, 195
homoiotherm, 195
homoiothermal animal, 195
homoiothermic, 195
homo(io)thermism, 505
homoiothermy, 195
homokaryotic, 482
homologous, 483, 484
homologous association, 484
homologous chromatids, 484
homologous chromosome, 484
homologous series, 484
homologue, 115, 484
homology, 482, 484
homolysis, 266
homomenthone, 161
homopimanthrene, 162
homoplastic graft, 484, 626
homoplasy, 383
homoploid, 482
homopolar, 483
homopolar crystal, 172
homopolymer, 266
homopolymerization, 483
homoptercarpin, 164
homoretene, 163
homoscedasticity, 130
homosomal aberration, 390
homospore, 483
homosporic, 261
homostatic period, 100
homoterpenylmethylketone, 163
homothallic, 484
homothallism, 484
homothermal animal, 195
homotype, 483
homotypic division, 483
homotypic spindle, 483
homoxylous wood, 483
homozygote, 68, 483
homozygous, 68, 483
hon-sho oil, 19
honey fungus, 323, 537
honey mushroom, 323, 537
honey plant, 323
honeycomb check, 146
honeycomb core, 147
honeycomb rot, 146
honeycomb stacking, 147, 277
honeycomb structure, 147
honeycombing, 146
honeydew, 323
hood, 88, 132, 218, 496
hook, 173, 178, 281, 292, 550
hook-aroon, 16, 484
hook catch tongue, 173
hook chaser, 602
hook-lever, 7, 173
hook rebate, 4
hook scale, 173
hook teeth, 173
hook tender, 223, 280
hooked joint, 322
Hooke's law, 201
hooking-on point, 160
hooklet, 533
hookman, 7, 178
hookup, 178, 545
hoop back chair, 593
hoop wood, 484
hoop(ing), 174
hopper, 167, 228
hopper feeds, 309
hopper joint, 537
Hoppus foot, 217
Hoppus measure, 217
Hoppus Rule, 217
Hoppus string measure, 217
Hoppus ton, 217
horizon, 45, 97, 458
horizon of soil, 488
horizontal air flow, 458
horizontal band resaw, 507
horizontal band-saw, 507
horizontal-bar opening, 549
horizontal boom, 135, 458
horizontal branching, 458
horizontal closure, 458

H

horizontal conveyor, 197, 458
horizontal disk type chipper, 507
horizontal disk type flaker, 507
horizontal distribution, 458
horizontal ditch, 458
horizontal element, 256, 458
horizontal extruder, 507
horizontal feed chipper, 458
horizontal feeder, 507
horizontal flue, 196
horizontal furnace test, 507
horizontal gang borer, 507
horizontal gang saw, 507
horizontal gap, 458, 549
horizontal grilled dam, 165
horizontal integration, 483
horizontal intercellular canal, 256
horizontal interception, 458
horizontal kiln, 196
horizontal line sampling, 458
horizontal migration, 458
horizontal mixture, 458
horizontal mortiser, 507
horizontal multi-spindle wood borer, 507
horizontal multiple-deck chip screen, 507
horizontal opening, 458, 549
horizontal point sampling, 458
horizontal pressure, 458
horizontal push-type centrifuge, 507
horizontal resin canal, 197, 256
horizontal resistance, 458
horizontal retort, 507
horizontal root, 458
horizontal shear(ing), 458
horizontal slicer, 507
horizontal soil zonality, 458
horizontal spindle tenoner, 507
horizontal splitter, 507
horizontal stand structure, 300
horizontal storage bin, 458, 507
horizontal tube-bundle dryer, 458
horizontal veneer gangsaw, 507
horizontal veneer slicer, 507
horizontal visibility, 458
horizontal wood borer, 507
horizontal zonality, 458
hormone, 221
hormone silvicides, 193
hormonogenesis, 221
horn, 613
horn knot, 231, 242
horn tail, 446
horn worm, 477
hornblende schist, 242
hornet, 210
hornets, 210
hornfels, 242
horntail, 446
horny, 242
horny knot, 242
horphenathrene, 238
horse, 612
horse-hoofs leg, 316
horse-road, 65, 347
horse skidding, 65
horse tree, 103, 263
horsetail pine, 316
horticulture, 588
horticulturist, 588
hose-in-hose flower, 60, 474
hose-lay, 456
host, 221, 227
host density, 227
host-parasite combination, 227
host-parasite complex, 227
host-parasite interaction, 227
host-parasite relationship, 227
host plant, 227
host range, 227
host resistance, 227
host-selection, 227
host-specific pathogen, 227
host specificity, 227
host suitability, 227
hostility, 94
hot air drying, 393
hot air drying kiln, 393
hot air predryer, 393
hot air seasoning, 392
hot air sterilization, 157
hot and cold bath method, 393
hot and cold bath process, 393
hot-and-cold open-bath treatment, 393
hot and cold open tank process, 393
hot bath treatment, 394
hot choker, 304
hot choking, 304
hot coal-tar coating, 393
hot-cold circle, 393
hot deck, 304, 307, 637
hot (deck) loading, 467, 468, 614
hot desert, 391
hot dip, 394
hot filtration process, 393
hot flue gas, 394
hot gas duct, 393
hot gluing, 228
hot grind(ing), 393
hot-ground wood pulp, 392
hot hardener, 392, 394
hot-hot circle, 393
hot-house, 505
hot junction, 392
hot logging, 307
hot melt adhesive, 393
hot money, 580
hot pile, 304
hot plate, 394
hot-plate dryer, 393
hot plate press, 394
hot press, 394
hot pressing, 394
hot-rolling, 392, 394
hot-set, 391, 392
hot setter, 164, 392
hot-setting, 164, 392
hot-setting adhesive, 392
hot shortness, 391
hot skidding, 293, 307
hot skidway, 307
hot soaking, 393, 505
hot-spray lacquer, 393
hot spring, 505
hot spring ground, 505

hot-spring organism, 505
hot stack, 393
hot-water coil, 393
hot-water platen dryer, 393
hot-water seed treatment, 628
hot-water treatment, 393, 460
hot-wire anemometer, 394
hotbed, 505
hotel garden, 314
hotspot, 58
hound, 297
hour-control standard, 435
hours on stream, 41
house-block cap, 6
house garden, 440
house plant, 441
housed dovetail joint, 387
housed joint, 529
housing, 156, 496, 529, 634
hover-place logging, 615
hub, 53, 312
hub block, 312
Huber's formula, 201, 625
hue, 405
hula saw, 200
hull, 186
huller, 493, 494
hulling, 384, 493
human capital, 395
human demography, 395
human disturbance, 395
human ecology, 395
human excrement, 394
human factor, 396
human resource, 395
human settlement, 395
human system, 395
humate, 151
Humboldt notch, 129, 521
humeral angle, 232
humic acid, 151, 194
humic alkali soil, 151
humic allophane soil, 151
humic carbonated soil, 151
humic fertilizer, 201
humic gley soil, 151
humic iron pan, 151
humic latosol, 151
humic-like substance, 285
humic siallite, 151
humic substance, 151
humid climate, 433
humid limestone brown loam, 433
humid volume, 432, 433
humidification, 433, 602
humidifier, 433, 602
humidifying treatment, 602
humidistat, 195
humidity, 432
humidity chart, 432
humidity control system, 277
humidity controller, 432
humidity regulated dry kiln, 480
humidity resistance, 337
humification, 151
humified organic matter, 151
humin, 201
humivore, 437
hummock, 381, 413, 593
humo-fulvic acid, 201
humocarbonate soil, 151
humod, 151
humox, 151
hump, 175, 309, 486, 494
humpback, 19, 52
humulene, 315
humulith, 151
humult, 151
humus, 151
humus-calcic pseudogleyed soil, 151
humus horizon, 151
humus-layer, 151
humus podzol, 151
humus soil, 151
humus theory, 151
Hundeshagen's formula, 195
hunk, 40, 468
hunt by chasing or beating, 159, 639
hunt by stalking, 374
hunt units, 502
hunter, 444
hunting, 444
hunting-box, 297
hunting cabin, 297
hunting camp, 444
hunting-field, 297, 444
hunting-ground, 297, 444
hunting-instrument, 297
hunting licence, 444
hunting-license, 444
hunting-right, 444
hunting season, 47
hunting tackles, 297
huntsman, 444
huon pine wood oil, 285
hurricane, 261
husk, 218, 262, 496
hyacinth concrete, 146
hyaline, 485, 510
hybrid, 597
hybrid cell, 597
hybrid cost system, 213
hybrid generation, 597
hybrid habitat, 597
hybrid index, 597
hybrid inviability, 597
hybrid nursery, 597
hybrid perpetual rose, 596, 597
hybrid progeny, 597
hybrid ribosome, 597
hybrid seed production, 597
hybrid seedling, 596
hybrid swarm, 597
hybrid tea rose, 596, 597
hybrid vigour, 597
hybridism, 597
hybridity, 597
hybridization, 596
hybridization in situ, 591
hybridologist, 596
hydathodal cell, 350, 491
hydathodal hair, 491
hydathode, 350
hydatogen rock, 456
hydatogenous rock, 456
hydnaceous fungus, 58
hydnocarpus oil, 76
hydrabrusher, 457
hydraclone, 64
hydrafiner, 164, 457
hydram, 3
hydrant, 531
hydrapulper, 457
hydrarch succession, 459
hydrase, 457
hydratase, 456
hydrate, 456
hydrated cellulose, 457
hydration, 456, 457
hydration number, 456

hydrature, 456
hydraulic action saw, 562
hydraulic barker, 457
hydraulic clipper, 561
hydraulic cold press, 561
hydraulic crane, 562
hydraulic debarker, 457
hydraulic debarking, 457
hydraulic debarking machine, 457
hydraulic dog, 561
hydraulic dynamometer, 457
hydraulic engineering, 457
hydraulic excavator, 562
hydraulic-feed chain mortise, 562
hydraulic fluid, 562
hydraulic gradient, 457, 459
hydraulic ladder, 562
hydraulic lift cylinder, 562
hydraulic loading system, 562
hydraulic log grapple, 562
hydraulic log loading boom, 562
hydraulic offset, 561
hydraulic peeler, 457
hydraulic potential, 459
hydraulic press, 562
hydraulic radius, 457
hydraulic regulator, 457
hydraulic ring barker, 457
hydraulic shear, 562
hydraulic slab debarker, 457
hydraulic splitter, 457
hydraulic stacker, 562
hydraulic wedge, 562
hydraulic(al) control device, 562
hydraulically powered chain saw, 562
hydraulics, 457
hydraulie lift, 562
hydric, 459
hydro-jet, 561
hydro-junction, 457
hydro-melioration, 457
hydroabietylamine, 380
hydrobiology, 459
hydrocellulose, 457
hydrochain, 562
hydrochemical metamorphism, 456
hydrochore, 456
hydrochory, 457
hydroclimate, 460
hydroclimograph, 460
hydroconductibility coefficient, 89
hydrocracking, 380
hydrodynamometer, 307, 457
hydroexpansivity, 515
hydrofluorocarbons, 380
hydrofuge, 131, 133
hydrogel, 458
hydrogen acceptor, 380
hydrogen bond, 380
hydrogen clay, 380
hydrogen donor, 380
hydrogen flame ionization detector, 380
hydrogen ion activity, 380
hydrogen ion concentration, 380
hydrogenated rosin, 380
hydrogenation, 380
hydrogenator, 380
hydrogenic rock, 456
hydrogenolysis, 380
hydrogenolytic cleavage, 380
hydrogeology, 459
hydrograph, 459
hydrographic net, 193, 459, 460
hydrographic station, 459
hydrographical condition, 459
hydrography, 459
hydroid, 460
hydrol humic latosol, 456
hydrolase, 457
hydrolignin, 380
hydrolith, 456
hydrologic balance, 456
hydrologic budget, 456, 457, 459
hydrologic cycle, 456
hydrologic equation, 456
hydrologic regime, 456
hydrological balance, 456
hydrological basin, 460
hydrological condition, 459
hydrological cycle, 456
hydrological divide, 459
hydrological station, 459
hydrological statistics, 459
hydrology, 459
hydrolysable tannin, 457
hydrolysable tannin extract, 457
hydrolysate, 457
hydrolysis, 457
hydrolysis constant, 457
hydrolysis reactor, 457
hydrolyst, 457
hydrolyte, 457
hydrolytic acidity, 457
hydrolytic action, 457
hydrolytic degradation, 457
hydrolytic lignin, 457
hydrolyzable tannins, 457
hydrolyzate, 457
hydrolyze, 457
hydrolyzed shellac, 457
hydrolyzer, 457
hydrome, 89
hydromechanics, 307
hydrometeorology, 459
hydrometric station, 459
hydromicas, 457
hydromorphic soil, 456
hydronasty, 159
hydronium, 456
hydrophile, 379
hydrophilic, 379
hydrophilic colloid, 379
hydrophilic fibre, 379
hydrophilic gum, 379
hydrophilous, 379, 457, 459
hydrophobe, 445
hydrophobic, 445
hydrophobic colloid, 445
hydrophobic effect, 445
hydrophobic liquid, 524
hydrophyta adnata, 177
hydrophyta natantia, 358
hydrophyta radicantia, 580
hydrophyte, 459
hydroponics, 398, 460
hydrosafroeugenol, 380
hydrosera, 459
hydrosere, 459
hydrosol, 458
hydrosphere, 457, 458, 460

H

hydrostatic clipper, 258
hydrostatic level, 258
hydrostatic pressure, 258, 307
hydrostatic skeleton, 460
hydrostatics, 307
hydrostratigraphic unit, 459
hydrotechnical amelioration, 457
hydrotechnical construction, 457
hydrotherm figure, 505
hydrotropic pulping, 458
hydrotropic solvent, 458
hydrotropism, 530
hydrous micas, 457, 563
hydrous oxide, 457
hydroxy, 376
hydroxy complex, 376
hydroxyapatite, 380
hydroxyimino, 507, 568
hydroxylamine, 187, 376
hydroxylase, 376
hydroxylation, 376
hydroxyquinoline, 376
hydroxyresveratrol, 376
hydroxystibene, 376
hyetal coefficient, 585
hyetal region, 585
hyetograph, 585
hyetometer, 585
hyetometrograph, 585
hygric, 433
hygro-, 433
hygro-stability, 433
hygrocole, 433
hygrogram, 432
hygrograph, 432
hygrometer, 432
hygrometric moisture meter, 514
hygrophil(e), 516
hygrophilous, 516
hygrophilous tree species, 433
hygrophorbium, 93, 432
hygrophyte, 433
hygroscope, 432
hygroscopic agent, 514
hygroscopic behavior, 514
hygroscopic coefficient, 514
hygroscopic limit, 514
hygroscopic material, 514
hygroscopic moisture, 514
hygroscopic moisture meter, 514
hygroscopic nature, 514
hygroscopic property, 514
hygroscopic substance, 514
hygroscopic treating agent, 514
hygroscopic water, 514, 515
hygroscopicity, 514
hygrostat, 44, 195
hygrothermograph, 505
hylum, 366, 626
hylus, 366, 626
hymen, 326
hymenial layer, 641
hymenium (*plur.* -ia), 641
hymenophore, 641
hypanthium, 571
hypanthodium, 571
hyper-focal distance, 51
hyper-geometric distribution, 51
hyperbola, 453
hyperboloid, 453
hyperconjugation, 51
hyperendemic, 162
hyperinfection, 162
hypermetabolism, 79, 539
hypermetamorphosis, 152
hypermorph, 51
hypernic, 198
hyperpanchromatic, 51, 601
hyperparasite, 60
hyperparasitism, 60
hyperpermeability, 51
hyperplasia, 518, 602
hyperploid, 50
hyperploidy, 50
hyperpolyploid, 50
hypersaline water, 164
hypersensibility, 187
hypersensitive, 187
hypersensitive resistance, 187
hypersensitivity, 187
hyperspace, 119
hyperspectral, 162
hyperspectrum, 162
hypertonic, 163
hypertonic solution, 163
hypertrophic, 186
hypertrophy, 139, 186
hypervolume, 119
hypha (*plur.* hyphae), 267
hyphal hole, 267
hyphal strand, 267
hypo, 78, 188, 512
hypo-, 521
hypo- (*eg.* hyposensitivity), 35
hypoachene, 521
hypobasidium, 520
hypocarpium, 186
hypocotyls, 521, 641
hypodermal, 356, 521, 608
hypodermis, 521
hypogaeic, 98
hypogaeous cotyledon, 306
hypogeal, 98, 306
hypogeal germination, 98, 306
hypogean, 98, 306
hypogeous, 98, 306
hypoglycemia, 93
hypognathous, 521
hypogynous flower, 521
hypogyny, 521
hypolimnion, 201
hypomorph, 552
hyponeuston, 520
hypopharynx, 417, 521, 553
hypophloedes, 340
hypoplankton, 520
hypoplasia, 125, 517
hypoplastic, 125, 517
hypoplasty, 125, 517
hypoploid, 552
hyposensitivity, 552
hypostatic, 521
hypostatic gene, 521
hypothecation, 81, 94
hypothecium (*plur.* -ia), 338, 641
hypothesis of pangenesis, 130
hypotonic, 93
hypotonic solution, 93
hypsithermal interval (period), 164, 505
hypsometer, 44
hypsometric curve, 162
hypsometric formula, 44
hypsometrical measurement, 447
hysteresis, 621
hysteresis curve, 621
hysteresis effect, 621

hysteresis loss, 621
hystrella, 186, 538
hyther, 505
hythergraph, 505, 506
hytherograph, 505
I-joist, 170
I-shaped budding, 170
iarovization, 68
ice age, 28
ice-and-snow belt (zone), 28
ice avalanche, 28
ice ball method, 110
ice-break, 29
ice cap, 28
ice chute, 28
ice climate, 28
ice columns in ground, 489
ice damage, 28
ice field, 28
ice guard, 193
ice period, 28
ice-pit storage, 28
ice ribbon, 80
ice road, 28
ice rut road, 29
ice sheet, 28
ice shelf, 28
ice-slide, 28
ice storm, 28
ice wag(g)on, 240
iceberg, 28
iceborne sediment, 28
ichneumon fly, 219
ichnofacies, 206, 565
ichnofossil, 565
ichthyology, 584
ichthyovorous animal, 437
icing, 28, 240
icy shower, 28, 110
idea of continuous forest, 195, 576
ideal cylinner, 230
ideal fluid, 288, 510
ideal forest, 127, 288
ideal forest model, 128, 288
ideal growing stock, 288
ideal moisture condition, 288
ideal profile, 288
ideal selection system, 128, 288
ideal standing volume, 288
idealized profile, 288
idempotent matrix, 323
identification, 237, 435
identification card, 237, 435
identification key, 233, 435
identification number, 435, 448, 486
identity, 85, 195, 266, 483
identity matrix, 85
identity test, 237
idiobiology, 166
idioblast, 567
idiochromosome, 541
idioecology, 166
idiogram, 391
idiomorphic soil, 647
idiophase, 141
idioplasm, 627
idiothermal, 195
idiotrophic, 475
idiotype, 166
idiovariation, 486
idle discharge, 81
idle fund, 523
idle pulley, 120, 276
idle speed, 81
idle wheel, 120, 276
idler, 120, 276, 579
idling consumption, 276, 509
igneous rock, 215
ignitability, 273
igniting-chamber, 100
ignition, 100, 604
ignition loss, 417
ignition point, 123, 390
ignition probability, 273
ignition source, 217
ignition temperature, 390, 604
ignition time, 604
ignition wire, 100
illegal cutting, 136
illite, 563
illuminance, 605
illumination, 182, 605
illumination (intensity), 605
illuminator, 123, 605
illuvial clay, 103
illuvial horizon, 103
illuvial layer, 103
illuviation, 103
illuvium, 103
ilyotrophe, 437
image compression, 487
image data digitizer system, 487
image description, 487
image displacement, 574
image encoding, 487
image enhancement, 487, 575
image-motion compensator, 487
image overlap, 575
image point, 530
image processing, 487, 574
image ratio analysis, 574
image representation, 487
image restoration, 487
image segmentation, 487
image transformation, 487
imagery, 55, 487, 574
imago, 54
imbibed water, 246, 515
imbibition, 251, 433, 514, 515
imbricate, 155
imidan, 552
imitation Japanese vellum, 134
imitation leather, 230, 396
imitation parchment, 134
imitative wood, 134
immaterial assets, 511
immature, 503
immature forest disease, 582
immature longitudinal tracheid, 504
immature longitudinal xylary cell, 504
immature soil, 426, 504
immature stage, 504
immature stand, 504
immature wood, 503, 504, 583
immaturity, 503, 504
immediate assets, 306
immersed aquatic plant, 53
immersion depth, 251
immersion system, 49, 50, 578
immersion test, 251
immersion treatment, 251
immigrant, 372
immigration, 372
immigration curve, 372
immigration rate, 372

I

immiscibility, 35
immobile liquid, 176
immobile phase, 176
immobilization, 175
immune, 323
immune response, 323
immune serum, 323
immunity, 323
immunization, 323
immunochemistry, 323
immunology, 323
immutable landscape attribute, 31
imp, 231, 244, 321, 341
impact, 574
impact assessment, 574
impact bending, 59
impact bending strength, 59
impact bending test, 59
impact coefficient, 59
impact disk mill, 59
impact energy, 59
impact extrusion, 59, 286
impact force, 59
impact hammer, 59
impact hardness, 59
impact head, 59
impact height, 59
impact load(ing), 59
impact pressure, 59
impact shearing strength, 59
impact strength, 59
impact stress, 59
impact stroke, 59
impact test, 59
impact tester, 59
impact value, 59
imparipinnate, 366
impedometer, 651
impeller, 243, 560
impeller blade, 109, 560
imperatorin, 373, 499
imperfect flower, 35
imperfect fungus, 11, 35
imperfect heart wood, 35, 446, 504
imperfect heartwood tree, 35, 446, 504
imperfect stage, 35, 511
imperfection, 387
imperforate tracheary cell, 508
imperial forest, 210
impermeability, 32
impermeable bed, 32
impermeable layer, 32
impermeable wood, 338
impervious layer, 32
impervious seed, 575
impervious surface, 32, 166
imping, 244
implant, 564, 616
implantation, 564
implement stock, 169
implementation, 613
implementation cost, 613
implicit analysis of cost, 571
implicit interest, 4, 571, 572
implicit rent, 571
imporosity, 32, 137, 509
importance value, 630
imported timber, 250
imposed dormancy, 376
impounded basin, 545
impounded body, 457, 545
impounding, 283, 545
impoverished soil, 359
impreg, 251
impregnant, 251
impregnatability, 251
impregnated paper, 251
impregnated shell, 251
impregnated timber, 251
impregnating process, 251
impregnation, 251, 421, 443, 444, 559
impregnation from end, 111, 474
impregnation liquid, 251
impregnator, 251
impression, 94, 327, 366, 571
impression roll, 551
imprest fund, 106
imprint, 550, 571
imprint method, 551
imprinted figuring, 571
imprinting, 571
improved variety, 155, 585
improved wood, 155
improvement, 149
improvement cutting, 149, 155
improvement cutting schedule, 149
improvement felling, 149, 155
improvement of forest, 301
improvement of site condition, 274
improvement of soil condition, 274
improvement of stand condition, 274
improvement planting, 155
improvement seed orchard, 155
improvement-selection cutting, 149, 610
improvement-thinning, 155
impulse, 59, 317, 492
impulse oscillating sprinkler, 59
impulse switch, 317
impulse wheel, 59
impulsion, 59
impurity, 33, 229, 508, 597
imputed capital flow, 173
imputed capital value, 173
imputed cost, 230, 339, 572
imputed income, 173, 572
imputed interest, 173, 572
in-bark DBH, 385
in-line resaw moulder system, 66
in-line seed cleaner, 66
in-line system, 293
in-place inventory, 533
in-process-storage-unit, 360, 623
in-row, 190
in-row weeder, 190
in-service treatment, 588
in situ, 525, 559
in-situ conservation, 259
in the swing, 473
in the water, 458
in the white, 504
in vitro, 215, 287, 395, 438
in vivo, 215
inaccessible forest, 34, 504
inaccessible forested land, 34, 504
inactivation, 115, 238
inactive porosity, 511

inadapted river, 35, 401
inadmissible assets, 32
inarching, 271
inarching by bark incision, 356
inarching by cleaving, 165
inarching by inlaying, 375
inarching by tongueing, 417
inarching by veneering, 377, 529
inarching grafting, 271
inarching with a branch, 86
inarching with an eye, 86
inarticulate, 34, 509
inbark, 229, 356
inborn error of metabolism, 522
inbred line, 250, 645
inbred-variety cross, 105
inbreeding, 250
inbreeding coefficient, 250
inbreeding depression, 250
inbreeding rate, 645
incendiary, 135, 390, 650
incendiary agent, 650
incendiary fire, 650
incentive payment, 238
inception phase, 367
inceptisol, 437
incest breeding, 250
inch stuff, 571
inch wood, 571
inchi grass oil, 6, 379
incidence, 123, 124, 400, 574
incidence of molding, 320
incident angle, 400
incident beam, 400
incident energy, 400
incident intensity, 400
incident light, 400
incident-particle distribution, 400
incident ray, 400
incidental charge, 596
incidental cost, 152
incidental felling, 304
incidental species, 348
incidentals, 596
incineration, 113
incinerator, 113
incipient decay, 63
incipient gullying, 63
incipient plasmolysis, 64
incipient podzolization, 63
incipient rot, 63
incipient species, 112
incipient stage, 63
incised, 261, 401
incised groundline, 96
incised lacquer, 103
incised meander, 420
incising, 274
incising tooth, 274
incision, 103, 377, 387
inclination angle, 380
incline, 97, 283, 413, 537
incline grain, 537
inclined cableway, 97, 283
inclined cutting, 537
inclined haulage, 283
inclined piling, 537
inclined plane, 380
inclined rope railway, 97, 283
inclined stacking, 537
inclinel cutting, 381
included phloem, 339
included (pit) aperture, 339
included sapwood, 339
inclusion, 12, 40, 339
inclusion body, 12, 339
incoherent knot, 34
incoherent noise, 138
income-capital ratio, 442
income cycle, 442
income distribution, 442
income from continuing operation, 574
income rotation, 442
income tax allocation, 469
income value, 442, 469
income velocity, 442
incoming charge, 250
incoming stock, 250
incompatibility, 32, 34, 35
incompatible reaction, 35
incompletae, 35
incomplete combustion, 35
incomplete cyme, 84
incomplete diallel cross, 35
incomplete dominance, 35
incomplete flower, 35
incomplete inhibition, 35
incomplete metamorphosis, 35
incomplete plant, 40
incomplete ripeness, 504
incomplete working co-operative society, 36
incompressibility, 34
incorporated business, 171
incorporation, 191
incorporation procedure, 171
increase block, 602
increase in growth, 429
increase in growth rate, 429
increase in quality, 541
increased cutting, 601
increaser, 602, 621
increasing cost, 99
increment, 429, 601, 602
increment borer, 430
increment borer extractor, 430
increment coefficient, 429, 430, 602
increment core, 344, 558
increment curve, 429, 602
increment cutting, 429, 443
increment due to increased light, 443
increment due to quality, 541
increment due to quantity, 37
increment felling, 429
increment gauge, 430
increment parameters, 44, 429
increment percent, 429
increment study, 429
increment table, 429
increment tax on land value, 488
increment thinning, 429
incremental cost, 601, 602
incross, 68, 359
incrust, 32, 50, 246
incrustate, 5, 30, 261, 575

incrustation, 246
incubating carrier, 373
incubation, 147, 353, 373, 506
incubation period, 147, 353, 373
incubation time (of strain load), 554
incubator, 147, 195, 353
incumbent anther, 340
incumbent cotyledon, 18
incumbent radicle, 340
incurvation, 340, 497
incyathescence, 263
indane musk, 571
indefinite bud, 32
indefinite corymb, 511
indefinite inflorescence, 511
indehiscent fruit, 20, 34
indent cylinder seperator, 507
indentation, 4, 274, 384, 387, 550, 571
indentation hardness, 550
indentation test, 550
indented ring, 262
indented scarfed joint, 58, 536
indenture, 4, 371
independent assortment, 111, 647
independent culling level, 111
independent variable, 642
independent wire rope core, 83
independent wire strand core, 160
index, 315, 470, 619
index breeding, 619
index-forest, 619
index fossil, 25
index mosaic, 470, 529
index number, 619
index of aridity, 158, 159
index of concentration, 155
index of dispersion, 140
index of diversity, 119
index of form, 541
index of fungicidal effectiveness, 411
index of growth, 430
index of moisture, 433
index of population trend, 627
index of similarity, 528
index of species diversity, 513
index species, 25, 619
index stand, 619
indexation of price, 3, 231
indicated mean effective pressure, 619
indicated pressure, 619
indicating and recording gauge, 619
indicating percent, 292, 619
indicating thermometer, 619
indicative price, 619
indicator, 25, 524, 619
indicator medium, 619
indicator plant, 619
indicator plot, 619
indicator-regulator, 619
indicator species, 619
indicator spectra, 619
indicator stand, 619
indicator stick, 619
indicator strain, 619
indicator value, 619
indicator vegetation, 619
indicators, 24, 470, 619
indifferent, 468
indifferent plant, 182
indifferent species, 182
indigenous, 19, 491, 527, 590
indigenous forest, 478, 646
indigenous peoples, 491
indigenous plant, 19, 527
indigenous species, 19, 527
indigenous tree species, 527
indigo yellow, 103
indirect (cell) division, 235
indirect cooking process, 235
indirect cost, 235
indirect effect, 235
indirect estimate, 235
indirect fire suppression, 235
indirect heating, 235
indirect labour, 235
indirect method of suppression, 235
indirect reduction division, 235
indirect replica (method), 235
indirect selection, 235
indirect steaming, 235
indirect utility of forest, 407
indirect vegetative division, 235, 511
indirubin, 103
indiscriminate cutting, 283, 363
individual, 87, 166, 474
individual cell, 83
individual converse rate, 166
individual density, 166
individual ecology, 166
individual fibre, 83
individual growth rate, 166
individual habitat, 166
individual micellar theory, 83
individual ontogenesis, 166
individual proprietorship, 111, 166
individual selection, 87, 166
individual site, 166, 474
individual spacing, 632
individual station method, 83, 224
individual tree, 84
individual tree competition index, 84
individual-tree mixtue, 87
individual tree model, 84
individual tree-selection, 301
individual tree valuation, 84
individual variation, 166
individualizing method, 299, 533
indoleacetic acid, 571
indolebutyric acid, 571
indoor flower decoration, 440
indoor furniture, 440
indoor garden, 440, 441
indoor grafting, 440
indoor plant, 441
indoor planting, 440

indoor seasoning, 440
indoor seeding, 440
indoor weather, 440
induced dormancy, 583
induced investment, 583
induced mutation, 583
induced periodicity, 583
induced resistance, 583
induction, 159, 183, 583
inductive method, 183
indurated fibre, 575
indurated red earth, 232
industrial cut stock, 170
industrial finish, 170
industrial forest, 170, 367
industrial forest management, 170, 367
industrial forestry, 170, 367
industrial forestry association, 406
industrial forestry enterprise, 170
industrial leather, 170
industrial lumber, 227
industrial melanism, 170
industrial nitrogen fixation, 170
industrial revolution, 170
industrial safety, 170
Industrial Safety and Health Law, 170
industrial sewage, 170
industrial waste, 170
industrial wood, 170
industrialization, 170
industrial(ized) society, 170
ineipient drying, 599
inert filler, 120
inert humus, 120
inert matter, 229, 510
inertia, 179
inertia balance, 109, 179
infant death rate, 583
infant mortality (rate), 583
infauna, 95
infected site, 378
infected timber, 159, 443
infection, 378
infection center, 378
infection court, 378
infection gate, 378
infection hypha, 378
infection peg, 400
infection process, 378
infection thread, 378
infection type, 159, 378
infectious disease, 378
infectious organism, 378
infective, 378
infective center, 378
infective dose, 159
infective stage, 378
infectivity, 378
infeed roll, 250
infer-coat, 122
inferior calyx, 521
inferior ovary, 521
inferior species, 93
inferior stand, 93
inferior tree, 297
infertile forest land, 359
infertile land, 359
infertile soil, 444
infertility, 32, 34, 509
infestation, 66, 317, 378, 379, 500
infilling, 32, 33
infiltration, 251, 420, 421
infiltration capacity, 420
infiltration rate, 420, 421
infiltration velocity, 421
infiltrometer, 421
infinite, 511
infinite population, 511
infinite reflux, 386
infinitely variable speed, 509
inflammability, 273, 569
inflammable debris, 569
inflammable material, 569
inflammation, 390, 604
inflammation point, 604
inflate, 355
inflated cost, 283, 544
inflating agent, 123, 424
inflexibility, 137, 160
inflexion point, 178, 637
inflorescence, 204
inflow-storage-discharge curve, 250
influence zone, 574
influent, 307, 574
influent species, 574
influx, 193, 307, 635
infoliate, 18
informal style, 34
informatics, 540
information control system, 540
information extraction, 540
information file, 540
information-oriented language, 323
information system, 540
information theory, 540
infra-axillary, 504, 562
infrabathyal zone, 520
infracapillary space, 318
infracutaneous, 27
infralittoral zone, 52, 594
infrared absorption, 198
infrared absorption spectroscopy, 198
infrared absorption spectum, 198
infrared detection system, 198
infrared drying, 198
infrared film, 198
infrared fire detector, 198
infrared frequency, 198
infrared (IR), 198
infrared line scanner, 198
infrared photography, 198
infrared radiation, 198
infrared radiometer, 198
infrared radiometer spectrometer, 198
infrared ray, 198
infrared scanner, 198
infrared sensing photographic, 198
infrared spectrometry, 198
infrared spectrum, 198
infrared transmission, 198
infrastructure, 219
infructescence, 186
infusion kettle, 251
infusion method, 251
infusorial earth, 183
ingestion, 419
ingraft, 231, 245
ingraftment, 231, 244
ingrown bark, 229, 356
ingrowth, 250, 530
ingrowth volume, 250
inhalant current, 514
inhalant stream, 514
inhaul, 630
inhaul line, 372
inhaul trip, 630

I

inherent fertility, 374, 477
inherent stress, 177
inherent vice, 177
inheritance, 565
inheritance of acquired character, 217
inherited factor, 565
inhibiting effect, 569
inhibition, 324, 569
inhibition concentration, 569
inhibition point, 569
inhibitor, 569
inhomogeneity, 34, 119, 597
initial absorption, 63
initial air pressure, 63
initial angle, 63
initial boiling point, 63
initial budget estimate, 62
initial capital, 63, 67
initial capital investment, 67
initial cell, 591
initial cost, 67, 268
initial creep, 63
initial daily production, 652
initial density, 64
initial dry weight, 63
initial entry, 591
initial grinding factor, 63
initial harvest, 63
initial inventory, 63
initial load, 64
initial meiosis, 100
initial moisture content, 63
initial outlay, 67
initial parenchyma, 591
initial pit border, 63
initial population, 591
initial position, 64
initial ray, 591
initial retention, 63
initial set, 62, 63
initial spacing, 63
initial species, 591
initial spread index, 63
initial stage, 63, 64, 591
initial stational phase, 443
initial stock table, 63
initial stress, 64, 587
initial tension, 64
initial toxicity, 63
initial uptake, 64
initial vacuum, 372, 373
initial value, 64, 591
initial volume, 63
initiating secondary succession, 570
initiation, 123, 367
initiative community, 522
initiator, 367, 570
injection dosage, 635
injection kerf, 635
injection port, 250, 635
injection syringe, 635
injection system, 635
injection treatment, 635
injector, 635
injurious beast, 189
injurious bird, 189
injurious weed, 580
injury by bail, 13
injury by frost, 455
injury symptom, 414
ink absorption, 514
ink disease, 327
ink receptivity, 514
ink resistance, 270
inlaid work, 529
inland drainage, 340
inland inundation, 340
inland quarantine, 185
inland-sand, 340
inland water, 339
inlay, 46, 358, 529
inlay approach grafting, 46
inlaying grafting, 46
inlet, 46, 188, 250, 400, 459
inlet needle, 250
inlet pressure, 250
innate, 95, 317, 522
innate behaviour, 522
innate dormancy, 522
innate releasing mechanism, 177
innately dormant, 177
inner bark, 340, 396
inner capacity increase, 339
inner capillary water, 340
inner checking, 340
inner circle of bundles, 340
inner coat, 341
inner core, 341
inner diameter, 340
inner diffusion, 339
inner endodermis, 340
inner-environment, 340
inner evaporation of soil, 490
inner integument, 341
inner lamella, 339
inner layer, 339
inner layer of secondary wall, 70
inner overlap, 339
inner periodicity, 341
inner pit aperture, 506
inner ring, 339
inner spindle, 341
inner surface, 339, 340
inner surface of wood, 339
inner web, 340
innovation, 539
inoculate, 245, 564
inoculating hood, 245
inoculating loop, 245
inoculation, 245
inoculator, 245, 551
inoculum, 245
inoperable area, 34
inorganic adhesive, 509
inorganic colloid, 509
inorganic fertilizer, 509
inorganic fire retardant, 509
input-output analysis, 485
input-output control system, 445
input-output framework, 485
input speed, 250
input translator, 445
inquiline, 227
inquilinism, 227
inscribed tablet, 23
insect attractant, 583
insect born(e) disease, 60
insect born(e) virus, 60
insect control, 60
insect damage, 60
insect fauna, 280
insect gall, 60
insect mull, 60
insect parasite, 226
insect-paste, 33
insect pest, 189
insect pollination, 60
insect powder, 410
insect-prone, 604
insect repellent paper, 131, 383

insect resistance, 270, 336
insect society, 280
insect transmission, 280
insect-trench, 33, 583
insect vector, 280
insect wax, 60
insectary, 557
insecticidal check method, 411
insecticidal oil, 411
insecticidal smoke, 411
insecticide, 411
insecticide paint, 411
insecticide resistance, 411
insectivorous, 436
insectivorous plant, 436
insert, 375, 529
insert piece, 46
inserted tool, 529, 575
inserted tooth, 375, 529
inserted tooth circular-saw, 375
inserted tooth saw, 375
insertion, 46, 375
insessorial, 440
inshore, 250
inside defect, 339
inside diameter, 340, 385
inside-light of canopy, 300
inside-light of tree crown, 447
inside-light of woods, 302
inside red spot, 339
inside rot, 339, 538
inside sapwood, 339
inside surface, 339, 340
insolation, 397, 413
insoluble redwood, 35
insoluble tar, 53
insolvency, 363, 510
inspection, 232, 473
inspection certificate, 233
inspection conveyor, 233
inspection district, 232
inspection door, 233
inspection hole, 178, 232
inspection plate, 232
inspection rule, 232
inspection standard, 233
inspection test, 556
inspector, 233, 367
instability, 35
installation diagram, 2
installed stone, 622
installment, 142
installment and interest charge, 142
installment credit, 142
instantaneous birth rate, 461
instantaneous death rate, 461
instantaneous field of view, 461
instantaneous growth rate, 461
instantaneous mortality rate, 461
instantaneous rate of increase, 461
instar, 305
instinct, 19
instinctive behaviour, 522
institutional forest, 436
institutional land, 218
instruction forest, 243, 436, 439, 603
instrument, 130, 169, 192, 442, 564, 577
instrument case, 169, 564
instrument locker, 564
instrument position, 564
instrumentation, 564
insular species, 166
insulated electrode, 265
insulating board, 265
insulating coating, 166, 265
insulating fibre board, 265, 401
insulating layer, 265
insulating paper, 265
insulating property, 265
insulating-type particleboard, 265
insulation board, 265, 401
insulation potential, 265
insulator, 265
insurable risk, 272
insurance policy, 15
insurance premium, 15
insusceptibility, 32
intact forest, 498
intact soil cores, 592
intangible assets, 511
intarsia, 359
integral agroforestry ecosystem, 347
integral arch, 223, 610
integral heat, 219, 649
integral heat absorption, 649
integrated assessment, 648
integrated control, 648
integrated control of insect pests, 189
integrated forest utilization, 410
integrated illumination, 649
integrated illumination intensity, 649
integrated leaf amount, 285
integrated life support system, 648
integrated logging, 648
integrated management approach, 648
integrated mill, 648
integrated pest management, 29
integrated utilization, 648
integrated veneer, 224
integration, 224, 294, 648
integrative factor, 648
integrifolious, 86, 387
integument, 27, 476, 631
intensely podzolized soil, 375
intensity, 375
intensity axis, 376
intensity curve, 375
intensity of evaporation, 609
intensity of growth, 429
intensity of investment, 485
intensity of labour, 284
intensity of management, 253
intensity of pressure, 550
intensity of radiation, 148
intensity of sampling, 61
intensity of selection, 547
intensity of silviculture, 573
intensity of sound, 430
intensity of stress, 572
intensity of thinning, 445

I

intensity of utilization, 291
intensity of water erosion, 459
intensity of wind erosion, 145
intensive afforestation, 76
intensive culture, 224
intensive drying, 222, 376
intensive forest management, 224
intensive forestry, 224
intensive managed forest, 224
intensive management, 224
intensive selection cutting, 254
intensive selection felling, 224
intensive silviculture, 224
intention movement, 587
inter annual, 343
inter annual change, 343
inter fibre bonding, 523
inter-negative, 623
inter-ocular distance, 296, 556
inter-planting, 301
inter-row, 190
inter-row cultivation, 190
inter space, 235, 573
inter-tree competition, 448
inter-tree cultivation, 448
interact, 528
interaction, 528
interaction deviation, 239, 416
interaction effect, 239
interaction term, 239
interaction variance, 202, 239
interaxillary, 562
interbiotic, 427
interblend, 528
interbreed, 24, 359
intercalary growth, 245, 259
intercalary growth point, 245
intercalary meristem, 259
intercalation, 375
interceilular duct, 13
intercellular canal, 13
intercellular cavity, 13
intercellular fluid, 13
intercellular layer, 13
intercellular mycelium, 13
intercellular passage, 13
intercellular space, 13
intercellular substance, 13
intercept, 247
intercepting ditch, 247
intercepting safety, 247
intercepting safety gear, 247
interception, 247, 651
interception of precipitation, 238
interception of rainfall, 585
interception storage, 247
interchange, 201, 239
interchangeability, 201
intercommunications system, 339
intercompany profits, 171
intercropping, 235, 236
intercross, 201, 596
intercrystalline swelling, 246
interdeme selection, 388
interest-bearing note, 152
interest rate, 290
interface, 244, 248
interfacial surface tension, 248
interfamily grafting, 272
interfascicular cambium, 446
interfascicular medullary ray, 446
interfascicular phloem, 446
interfascicular ray, 446
interfascicular region, 446
interfascicular xylem, 446
interfelting, 240
interference, 157
interference competition, 157
interference filter, 157
interferon, 157
interfertile, 201
interflow, 45, 213, 239, 489
interfluve, 193, 237
interfoyles, 12, 305, 493
intergeneric grafting, 446
interglacial period, 232
Intergovernmental Panel on Climate Change, 611
intergradation, 232, 388
intergrated division method, 648
intergrated planning, 386, 610
intergroup relation, 388
intergrown knot, 214
interhost, 623
interior, 339, 340
interior blue stain, 339
interior check, 339
interior decoration, 441
interior detail, 383
interior drainage, 340
interior finish board, 441
interior glue, 440
interior landscape, 441
interior plywood, 441
interior sap stain, 339
interior species, 339
interior stain(ing), 339
interior tension, 340
interior trim, 441
interior varnish, 441
interlaced chair back, 80
interlaminar strength, 45
interlayer, 122
interlock, 202, 294
interlock arrangement, 294
interlocked grain, 239
interlocking directorate, 232, 294
interlocking drum, 294
interlocking gear, 294
interlocking skidder, 294
interlucation, 443, 485
intermediary metabolism, 623
intermediate, 235, 373, 623
intermediate belt, 186, 623
intermediate bin, 623
intermediate bundle, 623
intermediate character, 623

intermediate coat, 122
intermediate column, 625
intermediate culture, 236, 622
intermediate cutting, 149, 235
intermediate decay, 624
intermediate density fibreboard, 624
intermediate disturbance hypothesis, 622
intermediate dormancy, 622
intermediate felling yield, 235, 623
intermediate fibre, 623
intermediate form, 187, 623
intermediate host, 623
intermediate landing, 623
intermediate layer, 186
intermediate of medium density fibre board, 623
intermediate oil, 625
intermediate post, 623
intermediate product, 623
intermediate re-grafting, 623
intermediate reaction, 623
intermediate revision, 533
intermediate roll, 623
intermediate shade-tolerant species, 622
intermediate stage, 235, 623, 624
intermediate stock, 623
intermediate support, 3, 623
intermediate-support spar, 623
intermediate-temperature glue, 624
intermediate temperature setting adhesive, 624
intermediate tree, 236, 625
intermediate wood, 186, 623
intermediate yield, 166, 235, 623
intermediate zone, 623
intermicellar reaction, 241
intermicellar space, 241
intermicellar swelling, 241
intermicrofibrillar substance, 501
intermingle, 213, 214
intermittent barking drum, 236
intermittent furrow planter, 235
intermittent kiln, 236, 630
intermittent lake, 236
intermittent managemmt, 166, 236
intermittent mist, 236
intermittent production, 236
intermittent river, 236
intermittent selection, 236
intermittent selective cutting, 166, 235
intermittent steaming, 236, 623
intermittent stream, 236
intermittent sustained working, 166, 576
intermittent type grinder, 236
intermittent working, 166
intermittent yield, 166, 235
intermittent yield system, 166, 235
intermittently working feed system, 236
intermix application, 213
intermontane basin, 413
intermontane plain, 413
internal adjustment, 340
internal audit, 339
internal bond strength, 340
internal check, 339
internal circulating kiln, 341
internal clock, 476
internal combustion engine, 340
internal condensing-coil kiln, 340
internal decay, 339
internal defect, 339
internal distribution pattern, 339
internal dormancy, 341
internal drainage, 339
internal environment, 341
internal exposure, 476
internal-fan kiln, 341
internal fibrillation, 339
internal friction, 340
internal gas-heated retort, 340
internal gauge, 340
internal integument, 341
internal mixing gun, 340
internal normalization, 339
internal parasitism, 340
internal phloem, 340
internal plasticization, 341
internal quarantine, 185, 339
internal rate of return, 288, 339
internal rhythm, 339
internal sapwood, 339
internal shake, 339
internal sizing, 339
internal soil drainage, 490
internal standard, 339
internal standard normalization, 339
internal strain, 341
internal stress, 341
internal variability, 339
International Association for Ecology, 185
International Association of Landscape Ecology, 185
International Biological Programme, 185
International Code of Botanical Nomenclature, 185
International Code of Nomenclature for Cultivated Plants, 185
International Council for Bird Preservation, 185
international Energy Agency, 185

International Geographical Year, 185
International Hydrologic Decade, 185
international log rule, 185
international (log) rule, 185
international rules for seed testing, 185
international seed analysis certificate, 185
international textural grade, 185
International Union for the Conservation of Nature and Nature Resources, 185
International Union of Biological Sciences, 185
internodal elongation, 245
internode, 245
interpenetrating samples, 528
interpetiolar bud, 560
interphase, 142
interplant, 301
interplanting, 236
interpolation for local volume table, 339
interposition growth, 378
interpretation aids, 351
interpretation key, 351
interpretation of agricultural natural conditions, 348
interpretation of agricultural regionalization, 348
interpretation of bedding surface types of river basin, 308
interpretation of forest distribution, 406
interpretation of grassland types, 41
interpretation of land features, 488
interpretation of land resources appraisal, 488
interpretation of land types, 488
interpretation of landform, 96
interpretation of linear features, 527
interpretation of man made, 395
interpretation of river system, 460
interpretation of soil erosion, 490
interpretation of soil groups, 490
interpretation of the present state of land use, 488
interpretation statistical analysis, 351
interpretation stereograms, 351
interpretation template, 351
interpreter, 248, 351
interpreter for photograph, 530
interprocess movement, 170
interquartile, 463
interrupted elution, 622
interrupted fiddle-back, 235
interrupted inflorescence, 235
interrupted wavy grain, 235
interruptedly pinnate, 45, 349
intersecting line, 196, 239
intersection photogrammetry, 239
intersection point, 238
intersection point mark, 238, 637
interseeding, 301
intersegment pit, 245
intersex, 70, 232
intersowing, 301
interspecfic relationship, 626
interspecies population, 626
interspecific, 626
interspecific association, 626
interspecific competition, 626
interspecific cooperation, 626
interspecific crossing, 626
interstitial, 235
interstitial animal, 235
interstitial canal, 13
interstitial cavity, 13
interstitial deficiency, 623
interstitial fauna, 235
interstitial fraction, 519
interstitial space, 13
interstitial translocation, 623
interstock hybrid, 359
intertillage, 622
intertrachear(y) pitting, 179
interval, 235
intervalometer, 107, 435
intervarietal crossing, 359
intervarietal free cross-pollination, 359
intervarietal free crossing, 359
intervascnlar, 179, 502
intervascular pitting, 179
intervessel pit, 89
interwall space, 13
interwoven grain, 239
interwoven texture, 239
interxylary, 333
interxylary cork, 333
interxylary phloem, 333, 339
intestinal symbiont, 339
intimate mixing, 266
intine, 339
intolerance, 32
intolerant, 32
intolerant species, 517, 556, 557
intolerant tree, 32, 557
intolerant tree species, 516, 557
intracellular, 518
intracellular canaliculus, 13
intrachromosomal, 390
intraclass correlation coefficient, 652
intracrystalline swelling, 246
intragenic change, 220
intraprocess movement, 170
intraspceific, 626

intraspecies aggregation, 626
intraspecific, 626
intraspecific competition, 626
intraspecific cooperation, 626
intraspecific crossing, 626
intraspecific relationship, 626
intraspecific variation, 626
intraxylary, 335
intrazonal soil, 571
intrinsic factor, 341
intrinsic rate of increase, 339
intrinsic rate of natural increase, 339
intrinsic superiority, 341
introduced disease, 496
introduced species, 570
introduction, 570, 571
introduction of exotic species, 571
introgression, 237
introgressive hybridization, 237, 389
introne, 339
intrusion, 378
intrusive growth, 378
intumescent coating, 123, 355
intumescent paint, 355, 394
intussusception, 340, 469
intussusception growth, 340
intussusception theory, 340
inulase, 259
inulin, 259, 489
inundated plain, 130
inundation, 199
invader, 378
invader species, 378
invading water, 379
invariance, 31
invasion, 378
invasion stage, 378
invasive species, 400
inventory, 23, 37, 74, 278, 351, 381
inventory by sample plots, 558
inventory design, 641
inventory method, 408, 409
inventory of growing stock, 545
inventory of log yard, 635
inventory of stand, 300
inventory tag, 74, 351
inventory unit, 61, 104
inventory valuation adjustment, 74
inverse density-dependent factor, 343
inverse J-shaped curve, 90
inverse regression coefficient, 129
inverse stratification, 343
inversed grafting, 129
inversely density dependent, 129
inversion, 90, 129, 130, 343
inversion crossover, 90
inversion of altitudinal zone, 67
invert, 557
invert sugar, 637
invertase, 606, 637
invertebrata, 509
inverted cone-shaped opening, 90
inverted image, 90
inverted siphon, 90
inverted skyline, 129
inverted stereo, 129
inverted T-form budding, 90
inverted V-match, 90
inverted V-planting, 90
invertose, 637
invested capital, 485
investigation of locality, 96
investigation of stand-conditions, 301
investment guarantee, 485
invidual tree valuation, 84
invierno, 392
invigorating rootstock, 376
invisible hinge, 3
involucral, 648
involucrate, 261
involucre, 648
involucred, 261
involucret, 536
involuntary forestry, 644
involution, 340, 341, 492
involution form, 492
iodine number, 100
iodine staining, 100
iodine staining angle, 100
iodine staining method, 100
iodine value, 100
ion antagonism, 288
ion exchange, 288
ion exchange cellulose chromatography, 288
ion-exchange chromatography, 288
ion-exchange reaction, 288
ion hydration, 288
ion pump, 288
ion substitution, 288
ionic activity, 288
ionic balance, 288
ionic flow, 288
ionic fragmentation reaction, 288
ionic gradient, 288
ionic mobility, 288
ionic radius, 288
ionic regulation, 288
ionic strength, 288
ionization, 101, 288
ionizing radiation, 101
ionogen, 272
ionone, 642
ionophore, 288
ionophoresis, 102
iorn tag, 481
iridescent cloud, 198
Irish bridge, 196, 318
Irish coupling, 590
Irish setter, 2
iron age, 481
iron and aluminum oxide deposition layer, 481
iron bacteria, 481
iron clamp, 480
iron clay, 481
iron concretion, 480
iron crust soil, 481
iron degradation, 23
iron drier, 391
iron hardpan, 481
iron-humus ortstein, 480
iron jaw, 229, 480
iron lateritic soil, 481

iron oxide black, 480, 557
iron oxide red, 480, 557
iron oxide yellow, 480, 557
iron-pipe monument, 480
iron podzol, 481
iron sheet bandage, 133
iron stake, 481
iron strap, 481
iron-tannate stain, 291
iron trap, 33
ironbark dropper, 186
irone, 587
irradiance, 149
irradiancy, 149
irradiation, 149, 181
irradiation-cured, 182
irregular all-aged forest, 34
irregular cutting, 137
irregular forest, 152, 567
irregular grain, 34
irregular plantation, 152, 567
irregular shelterwood system, 34
irregular stand, 34, 35
irregular stocking, 137, 404
irregular style, 533
irregular twist, 34
irregular unevenaged stand, 34
irregular wavy grain, 34
irregularly stocked open woodtract, 404
irreversibility, 34
irreversible coagulation, 34
irreversible colloid, 34
irreversible diffusion, 34
irreversible inhibition, 34
irreversible wilting, 576
irrigating net, 179
irrigation, 179
irrigation channel, 179
irrigation injector system, 179
irrigation pump, 179
irrigation with torrential flood, 570
irrigator, 354
irruptive rock, 378
isadelphous, 295
isallobar, 91
isallohypsic wind, 91
isallotherm, 91
isinglass, 6, 584, 595
island, 90, 174
island biogeography theory, 90
island ecology, 90
islands, 90
iso-allele, 482
iso-B-chromosome, 91
iso-humic belt, 91
iso-x-chromosome, 91
isoamylase, 566
isoanth, 91
isobar, 92
isobaric chromatography, 92
isobaric map, 92
isobarometric line, 92
isobath, 91
isobathyic line, 91
isobilateral, 91, 121
isobilateral leaf, 91
isobiochore, 92
isocenter, 91, 191
isochion, 92
isochromosome, 91
isocost curve, 91, 92
isocotylous, 92
isodiametric cell, 91
isoelectric fractionation, 91
isoelectric pH, 91
isoelectric point, 91
isoelectrofocusing, 91
isoenzyme, 482
isogamete, 483
isogamy, 353, 483
isogenes, 91, 484
isogenetic, 482, 484
isogenic, 91
isogram, 92
isogynous, 91, 482
isohaline, 92
isohel, 91
isohume, 92
isohydric suspension, 91
isohyet, 92
isoiocus break, 92
isolate, 166
isolated planting, 174
isolated stant, 174
isolated tree, 174
isolateral, 91
isolaterality, 91
isolation, 142, 166, 247, 287, 515
isolation cutting, 287, 647
isolation fibre board, 166
isolation medium, 142, 265
isolation strip, 166
isolation theory, 166
isoliquiritin, 566
isolongifolene, 568
isomagnolol, 567
isomanoene, 567
isomer, 482, 484, 566
isometry, 91
isomorphous, 483
isonephe, 92
isonitrosocamphor, 568
isonome, 91
isopaulownin, 567
isoperimetric deficit, 92
isophene, 91, 92
isophyte, 91
isopiestic method, 92
isopimarane, 566
isopleth, 91, 92
isopleth map, 92
isopoll, 91
isopollen line, 91
isoprene, 531, 567
isoquants, 91
isosporic, 483
isospory, 12
isostasy, 96
isostatic land movement, 266
isostere, 91, 482
isosteric heat of adsorption, 91
isostomous, 121
isostress, 92
isostyled, 91
isostylous, 91
isoterra, 92
isotheral, 91
isothere, 91
isotherm, 92
isotherm linearity, 92
isothermal compression, 92
isothermal gas chromatography, 92
isothermal layer, 92
isothermal line, 92
isothermal map, 92
isothermal process, 92, 107
isothermal sorption, 92
isothermohyps, 92
isothujone, 566, 568
isothyme, 92
isotonic, 92

isotonic solution, 92
isotope, 483
isotope dilution method, 483
isotope gauge, 483
isotope tracer, 483
isotopic tracer, 438, 483
isotropic, 166
isotropic fibre, 166
isotropic layer, 166
isotropic material, 92, 166
isotropic wood beam, 92
isotropism, 166
isotropy, 166, 266, 511
isotype, 483
isozyme, 482
isthmian link, 98
isthmus, 98
iteration, 104
iteroparity, 116
iteroparous, 116
iva oil, 419, 433
ivy leaf oil, 48, 215
Izod impact machine, 545
Izod impact strength, 545
Izod-value, 563
J-shaped distribution, 140
jack, 46, 66, 109, 263, 368, 371, 638
jack chain, 295, 368
jack guy, 638
jack ladder, 334, 470, 590
jack plane, 73, 77
jack slip, 67
jack-strawing, 597
jack works, 590, 638
jacket cooling, 474
jacketed bullet, 12
jacketed kettle, 229, 474
jacketed still, 229, 474
jacketed syringe, 474
jackknife, 597, 605
jackknife boomsticks, 605
jackpot, 34, 150, 285, 485, 640
Jacobsen (germination) apparatus, 552
jagger, 161, 274, 580
jalap, 382, 538
jam, 46, 177, 186, 550, 651
jam cracker, 47
jam crew, 47
jam pier, 46
jamb, 44, 280
jammer, 157, 161, 223
Janka hardness, 234, 557
Janka indentation test, 234
Janka-test, 234
jap-A-lac, 396
Japan tallow, 397
Japan wax, 397
Japanese camellia, 413
Japanese cypress, 396
Japanese lacquer, 397
Japanese Peace and Friendship Forest, 396
japanner, 578
japanning, 487
jardiniere, 203
jarovization, 68
Jasmine Jasminum, 327, 573
java wax, 398
Jeffrey's maceration method, 246
jelly, 110, 186, 241
jelly strength, 241
jelly time, 346
jellyfish, 457
jequiritin, 197, 528
jet, 354, 355, 418
jet air cleaner, 354
jet angle, 354
jet barker, 457
jet barking, 457
jet blower, 354
jet cleaning device, 354
jet dry kiln, 354
jet dryer, 354
jet impregnation, 354
jet method, 354
jet nozzle, 355
jet-stream, 354
jet tube, 354
jet-tube dryer, 354
jet turbine bundle dryer, 354
jet veneer dryer, 354
Jethuri, 365
jetty, 131, 603
jewel beetle, 221, 544
jewel vine Derris, 584
Jew's ear, 194, 332
jib door, 362, 571
jib-strut crane, 108, 243
jib-type loader, 637
jib(arm), 195, 367
Jiffy pot, 324, 466, 618
jig-back cableway, 500
jig-back system, 500
jig saw, 309, 450, 519
jigger, 65, 177, 309, 533
jill poke, 491, 612
jill poke sizzer boom, 605
jim binder, 249, 281
jim crow load, 83
job, 170, 654
job analysis, 170, 654
job card, 170
job description, 615
job evaluation, 363, 615, 655
job specification, 654, 655
jobber, 252, 304, 355
jockey roller, 89, 120, 603
joe-log, 205, 349
joggle, 346, 639
joggle beam, 358
joggle joint, 346
Johns wurtfamilyr-hypericum famiy), 249
Johnson boom, 433
Johnson trap, 594
joiner, 244, 519
joinery, 332, 519
joining rafts, 192
joint, 244, 245, 359, 469
joint aging time, 244, 344
joint conditioning time, 244
joint demand, 292
joint factor, 240
joint fertilization, 172, 244
joint flooring, 95, 366
joint implementation, 294
joint occurrence, 172
joint product, 292
joint sleeve, 292
joint stock company, 175
joint strength, 241
joint venture, 112, 192, 294
joint width, 501, 578
jointed veneer, 359
jointer, 101, 244, 359, 610
jointing, 37, 82, 89, 190, 262, 359, 501
jointing cutting teeth, 377
jointing gauge, 610

jointing plane, 245, 543
jointless floor, 509
jointworm, 182
joist, 95, 165, 493
Jojner's chisel, 519
jojnt flange, 292
jojoba wax, 513
Jonson's form point, 319, 381
jordan (mill), 93, 640
Jordanon, 401
Jordan's rule, 594
Judas's ear, 194
Judeich's method of management, 578
Judeich's working method, 578
judicious mating, 107, 192
jugate, 29, 260
juglandin, 201
juglansin, 194, 201
juglone, 201
jugum, 58, 115, 186, 499
juice, 561, 612
jujube, 599
jujube essence, 599
jujubogenin, 467
jujuboside, 467
Jukohscope, 631
julaceous, 261
juliform, 261
jumbo, 351, 453
jumbo hopper, 78
jump, 129, 480, 486, 594, 655
jump a tree, 112
jump-butt, 167, 247
jump cut, 247, 480
jump saw, 421
jumper, 160, 278, 349, 480, 550
jumper upset, 550
jump(ing) fire, 136
junction point spectrum, 246
june-struck cutting, 9, 522
junene, 71
jungle, 391, 392
jungle form, 179
juniger forest, 71
junipene, 50, 71
juniper berry oil, 71
juniper wood oil, 71
juniper(us) oil, 71
junk value, 41, 139
Jurassic period, 631
Jurin's equation, 631
just price, 171
jute stalk, 210
juvabione, 15, 342
juvenal plumage, 64, 622
juvenescent phase, 379
juvenile, 379, 583
juvenile form, 583
juvenile growth, 583
juvenile hormone, 15, 379, 582
juvenile hormone analong, 15
juvenile period, 417, 583
juvenile phase, 583
juvenile stage, 583
juvenile tree, 583
juvenile water, 554, 591
juvenile wood, 582
juvenility, 33, 583
juvenility effect, 583
juvenility factor, 583
juvenoid, 15
juxtaposition, 29, 239, 537
K-band radar, 284
K-selection, 547
K-strategist, 114
K-strategy, 114
K value, 162
kadsura oil, 338
kairomone, 569
kakogenic, 32, 297
kali saltpeter, 230, 532
kalloplankton, 241
kalotermite, 328
kambi, 613
kame, 28
Kamyr digester, 267
Kamyr grinder, 267
kaolin, 162, 474
kaolinite, 162
kaolinization, 162
kaolinton, 162
kapok, 333
kapok longhorned borer, 604
Kappa number, 267
karaya gum, 71
karoo, 157, 510
karpose, 357
karroo, 267, 338
karst, 267, 554
karst landscape, 267, 554
karyapsis, 193
karyoclasis, 193
karyogamy, 194
karyogram, 194, 391
karyokinesis, 193, 581
karyokinetic division, 581
karyolemma, 193
karyology, 517
karyoplasm, 138, 193, 194
karyostenosis, 517
karyotype, 194, 391
kasugamycin, 68
katabatic wind, 520, 629
katharobic, 381
Katki, 267
katydid, 162, 626
kauri copal, 17
kauri gum, 17, 271, 539
kauri oil, 17
kauri resin, 17, 271
kaya oil, 139
keel, 66, 309
keel block, 309, 636
keel towing boom, 502
keeled, 260
keep in stock, 544
keeping quality, 14, 338
Kelvin temperature, 265, 269
Kelvin temperature scale, 265, 269
keratin, 242
keratinase, 242
kerf, 262, 356
kernel, 175, 193, 396, 627
kernel fruit, 193, 396
kerosene emulsion, 434
ketolysis, 248, 484
key, 233, 237, 306, 521, 523, 536
key age, 221
key area, 178
key card, 233, 559, 589
key dam, 632
key factor, 178, 632
key factor analysis, 178
key fruit, 58
key-gene, 633
key indicator plants, 634
key map, 237, 470, 649
key money, 120, 533, 586
key-note, 220, 634
key product, 634
key sort cards, 233
key species, 178
key stimulus, 178
keyed joint, 237
keyer, 225, 277
keyhole saw, 237, 452, 559

keypunch, 237
keystone species, 178
kick, 129, 167, 650
kickback, 61, 129, 130, 167, 199, 211, 212, 655
kickback guard, 131
kickback of preservative, 132
kicker, 103, 351, 492, 538
kicker system, 351, 492
kickxia rubber, 461
kid finish, 556
Kidney-form, 420
kidney shaped, 420
kieselguhr, 183
kiganene, 336
killig, 492
killing back, 613
killing concentration, 621
killing frost, 554
killing point, 621
killing temperature, 621
kiln, 158, 417
kiln bogie, 638
kiln brown stain, 558
kiln brown stam, 158
kiln bunk, 558, 638
kiln burn, 558
kiln-burning, 114, 472
kiln charge, 158, 558
kiln charge run, 158
kiln degrade, 558
kiln-dried timber, 558
kiln-dried wood, 558
kiln-dry, 558
kiln-dry lumber, 558
kiln drying, 558
kiln-drying schedule, 558
kiln operation, 558
kiln process, 558
kiln roof, 558
kiln run, 562
kiln sample, 558
kiln schedule, 558
kiln seasoning, 558
Kiln-site, 620
kiln stain, 558
kiln sterilization, 558
kiln track, 158
kiln truck, 558
kilnboy, 158
kimma sled road, 333, 349
kin, 229, 548
kin selection, 379
kinase, 221, 621
kind of timber, 37
kindling, 100, 570
kindling point, 390, 604
kindling temperature, 390, 604
kindling wood, 540, 570
kindred plant, 251, 484
kinds offorest fire, 301
kinematic viscosity, 595
kinesis, 109
kinetic adherence, 109
kinetic elasticity, 108
kinetic energy, 109
kinetic friction, 109
kinetics, 109
kinetin, 147, 221
kinetochore, 110, 640
king orange, 156, 411
king post, 104, 195, 633
king post roof truss, 87, 633
king-tower, 634
kingdom, 617
kingpin, 624, 634, 637
kinin, 221, 517
kink, 207, 215, 347, 497
kino, 220, 221
kinship, 379
kinsho oil, 397
kiosk, 295, 538
kiree, 81
kitchen cabinet, 40, 436
kitchen furniture, 64
kitten, 534, 535
Kjeldahl method, 220, 269
kladodium, 561
Klason lignin, 274
klausner hypsometer, 274
kleistogamous, 20
klinokinesis, 104, 537
klinotaxis, 104
knag, 245, 310, 332, 446
knaggy, 4, 117
knaggy wood, 117, 448
knapsack atomizer, 18
knapsack piston sprayer, 18
knapsack sprayer, 18
knar(l), 333
knead, 3, 213, 243, 346
kneader, 346, 469
kneading, 346
knee, 52, 168, 352, 371, 498, 516
knee bolter, 592
knee brac(ing), 242, 536, 584
knee-hole desk, 295, 296
knee table, 403
kneed timber, 498
kneel, 309
kneeling, 184
kneewood, 498
knephoplankton, 524
knife angle, 2, 249
knife bar, 88, 377, 549
knife barker, 88
knife beam, 88
knife block, 88, 89
knife cap, 88, 546
knife carriage, 88
knife chamber angle, 231, 536
knife check, 88
knife coater, 177
knife coating, 177
knife(-cutting) machine, 89
knife debarker, 88
knife edge, 88, 612
knife face, 88
knife grinder, 184, 326
knife grinding machine, 326, 412, 555
knife head, 88, 89
knife height, 88
knife hog, 89
knife mark, 88
knife pitch, 249, 546
knife ring flaker, 452
knife setting angle, 249, 546
knife shaft flaker, 631
knife test, 48
knifebelts limbing device, 295
knob, 3, 12, 29, 282, 446, 448, 474
knobby, 117, 579
knobby-bole, 117
knock-down block, 268
knock-down fittings, 272
knock-down furniture, 47
knocked down, 47, 125, 248
knocker screen, 64, 384
knockmeter, 16
knockout, 143, 354, 494
knockout shackle, 273
knockout tank, 143, 209
Knoop-test, 348
knopper gall, 512
knot, 245, 448, 504
knot borer, 384
knot bumper, 76

knot catcher, 384
knot cluster, 153, 245
knot cut lengthwise, 479
knot-free, 509
knot free lumber, 509
knot group, 245
knot hole, 245
knot hole boring machine, 245
knot horn, 231, 479
knot-root disease, 167
knot rot, 245
knot screen, 64
knot sealing, 333
knot(ted) garden, 385, 430
knotter, 64, 75, 76, 165, 384
knottiness, 117, 612
knotting, 146, 246, 478, 487
knotting varnish, 333
knotty, 117
knotty core, 245
knotty diameter, 246
knotty wood, 117
Knoweles tar, 348
Knuckle-boom, 178, 605
Knuckle-boom loader, 605
knuckle sheave, 243, 637
Koch's postulates, 271
Koch's steam-sterilizer, 272
Koetitze pile armour, 269
Koetitze pile armour process, 160, 269
koku, 397
kolinsky, 210
kolinsky yellow weasel, 210
kollaplankton, 241
Kolle flask, 274
konaki, 1
kondang wax, 398
Konig's hypsometer, 274
konimeter, 225, 275
koosmie, 278
Koppen's climatic province, 271
Koyamaki oil, 397
kraft, 347
kraft process, 308, 347
kraft pulp, 308, 347
kraft semi chemical pulp, 308
Kraft's tree classification, 274
krasnozem, 198
Krebs cycle, 274, 346, 403
krummhol(t)z, 163
kryptoblast, 571
kryptoclimate, 440
kryptomere, 571
Kulba salt, 278
kunhi, 316
Kuntz process, 276
kuromoji oil, 78
kuroshio, 194, 471
kurtosis, 146, 377
kusmi lac, 278
kyan process, 315
kyanising, 315
kyanizing process, 315
Kyoto Mechanism, 252
Kyoto Protocol, 252
L-band radar, 284
L-form bacteria, 518
L-joint, 244
L-root, 167
L-type colony, 267
la Niña, 282
labdanum oil, 282, 554
labdanum(gum), 282, 284, 554
label, 25, 69
label(l)ed compound, 25
labellum (*plur.* labella), 69
labiate, 69
labiate corolla, 69
labile, 35, 569
labile gene, 35, 569
lability, 35
labium (*plur.* labia), 520
laboratory beater, 436
laboratory decay test, 436
laboratory germination percentage, 436
labour agreement, 284
labour capacity, 284
labour cost, 284, 394
labour day, 284
labour force, 284
labour in forestry, 303
labour physiology, 284
labour productivity, 284
labour saving, 245
labour slow down, 81
labour trouble, 284
labour turnover, 284
labourer-insurance, 284
labrum, 69, 277, 415
labyrinth, 165, 321
lac, 60, 642
lac dye, 642
lac encrusted twig, 642
lac resin, 60, 642
lac varnish, 60, 181
lac wax, 642
lacca, 642
laccase, 60, 366
laccifer lacca, 642
lace bug, 499
lace veneer, 26
lace wing, 42, 317
lacerate, 444, 462
laciniate, 55, 479
lacinilene, 379
lacinulate, 261
lacinule, 535
lack of rainfall, 388
lacker, 77, 211, 354, 478, 532
lacquer, 77, 211, 354, 366, 478, 532
lacquer enamel, 211, 473, 532
lacquer tree, 366
lacquer ware, 366
lacquer wax, 366
lacquerer, 578
lacquering, 416, 487, 544
lactacidase, 400
lactase, 400
lactation, 33, 321, 424, 444
lacteal, 399, 400
lactic fermentation, 400
lacticiferous system, 400
lacticiferous vessel, 400
lactiferons duct, 400
lactiferous canal, 400
lactiferous cell, 600
lactiferous tube, 400
lactonase, 341
lactone, 341
lactose, 400
lacunar, 375, 580
lacunarity analysis, 277
lacunose, 117, 119, 581
lacustrine, 201
lacustrine deposit, 201
lacustrine soil, 201
ladder ditcher, 476
ladder-hording, 224
ladder-like perforation, 476
ladder stock, 162
lade, 193, 456, 639
lady-beetle, 358
lady bird, 358
lady Washington geranium, 76

laevigate, 361
laevoglucose, 186, 654
laevulose, 186, 654
laficiferous vessel, 581
lag, 8, 57, 132, 484, 605, 621
lag period, 571, 621
lag phase, 481, 554
lageniform, 201, 417
lagging, 275, 280, 333, 500
lagoon, 242, 524, 538
lagos silkrubber, 461
lake chalk raw soil, 201
lake effect, 201
lake loess, 201
lake terrace, 201
lake type, 201
Lamarckian effect, 282
Lamarckian inheritance, 282
Lambert conformal projection, 283
Lambiotte retort, 283
lamella (*plur.* -llae/-llas), 13, 267
lamellate, 45, 260, 358
lamellate antenna, 402
lamellated pith, 358
lamellation, 45
lamiid, 173
lamina (*plur.* laminae), 13, 45, 560
laminar-turbulent transition, 45
laminate, 13, 45, 480
laminate molding, 116
laminated arch, 45
laminated beam, 45, 104
laminated board, 8, 45, 480
laminated clay, 45, 358
laminated fibre board, 45
laminated member, 45
laminated moor, 45
laminated particleboard, 45
laminated plywood, 45
laminated rock, 45, 561
laminated stock, 45
laminated structure, 45
laminated surfacing equipment, 480
laminated thermosetting plastics, 45
laminated veneer, 45
laminated veneer lumber, 81
laminated wood, 45, 240
laminating plastics, 45
lamination, 13, 45
laminboard, 8, 45
lammas shoot, 226, 382
lamp black, 90, 578
lampstand, 90, 91
lanate, 260, 323
lance tooth, 231, 402
lanceate, 462
lanceolate, 319, 355
land, 488
land amelioration, 488
land appraisal, 488, 490
land breeze, 310
land bridge, 310
land bridge hypothesis, 310
land-capability class, 96
land capital, 488
land carrying capacity, 488
land class zoning, 488
land clearing machine, 299
land cover map, 488
land-depot, 310
land description, 96
land-drainer, 350
land expectation value, 488
land feature, 96–98
land for forestry, 298, 303
land form, 96
land improvement, 488
land law, 488
land leveling, 488
land leveling machine, 360
land-locked species, 310
land looker, 301
land management, 488
land map, 488
land mass, 77
land ownership, 488
land parcel, 96, 299, 488
land parcelling, 488
land parcelling legislation, 488
land preparation, 610
land reclamation, 274, 488
land register, 488
land registry office, 488
land rent, 99
land rent value, 99
land resource, 488
land retirement, 488, 490
land roll(er), 608
land scraper, 360
land settlement project, 488
land slide, 205
land storage, 310
land survey, 96, 488
land survey office, 488
land tax, 98, 488
land tenure, 488, 651
land use, 488
land-use capability, 488
land-use capability classification, 488
land-use capability survey, 488
land-use change, 488
land-use classification, 488
land-use cost, 488
land-use map, 488
land-use model, 488
land-use pattern, 488
land-use planning, 488
land-use policy, 488
land-use type, 488
land utilization map, 488
land value, 96, 488
land variety, 19, 488
land-use planning committee, 488
landcreep, 471
landform, 96, 98
landing, 224, 413, 638–640
landing account, 367, 603, 638
landing barge, 359
landing board, 603
landing line, 310, 639
landmark forest, 95, 96
Landsat, 310
landsat imagery, 310
landscape, 145, 255, 646
landscape analysis, 145, 255
landscape architect, 145, 588, 601
landscape architecture, 588, 600, 601
landscape control point, 145, 255
landscape depression, 255
landscape design, 145

landscape design judging, 255, 588
landscape development, 145, 255
landscape diversity index, 255
landscape ecology, 255
landscape element, 255
landscape engineering, 145, 255, 315, 588
landscape evenness index, 255
landscape fragmentation, 255
landscape gardening, 588, 601
landscape grain, 255, 303, 414
landscape judging, 145, 255
landscape management, 145, 255, 588
landscape mosaic, 255
landscape park, 478, 587
landscape pattern, 255
landscape planning, 145, 255, 588
landscape plant, 178, 588
landscape preference, 145, 587
landscape protection, 96, 255
landscape reservation, 145, 255, 645
landscape richness index, 255
landscape structure, 255
landscape style, 145, 255
landscape tree and shrub, 145, 588
landscape type, 145, 255
landscape unit, 255
landscape woody plant, 178, 588
landscaped city, 145
landscaping, 588, 600
landslide, 413, 470, 489
landsliding, 488
landslip, 205, 470, 488
lane, 52, 191, 298, 481
lane of traffic, 52
Lanewood particleboard process, 282
lang-lay, 461
lang lay rope, 461
lanolin, 556
lanose, 260, 323
lantern-fly, 64, 180
lanuginose, 261, 445
lap, 21, 39, 60, 75, 156, 168, 366, 487, 613
lap dovetail, 201
lap-joint strength, 75
lapillus (*plur.* -lli), 216
lappa, 71, 346
lappaceous, 71, 173, 318
lap(ped) joint, 75
lapse rate, 99, 233, 614
lapse rate to temperature, 505
lapwood, 39, 613
larch (bark) extract, 314
larch canker, 314
larch leaf oil, 314
larch-needle-rust, 314
larch turpentine, 314
large area clearcutting, 77
large carpenter bee, 77
large cleaning, 77
large clear-cutting, 77
large coupe, 76
large dimension timber, 76, 77
large end, 72, 78
large fire control, 76
large flake board, 76
large forest fire, 406
large hewn square, 356
large knot, 77
large plot, 78
large sample theory, 78
large sapling, 78
large sapling forest, 582
large sawtimber, 77
large scale, 76
large-scale afforestation drive, 76
large-scale production, 76
large square, 76
large stretches of forests, 77
large strip, 78
large timber, 77
large tree, 78
large worm hole, 76
largest income rotation, 653
larva, 227, 582
larva-fly, 227
larvae gallery, 641
larval adaptation, 582
larval gallery, 582
larval instar, 582
larval plankton, 148
laser, 221
laser beam detector, 221
laser cutting, 221
laser dendrometer, 221
laser detection and ranging, 221
laser guide, 221
laser image processing scanner, 221
laser radar, 221
laser range finder, 221
laser scanner, 221
lash pole, 27, 65, 350
lasher, 227, 283, 356, 545
lashing, 23, 281, 470
last frost, 626
last in first out, 199
last injurious frost, 499
last runnings, 503
lasting manure, 57
lata, 118, 486
lata type, 118
latch hook, 472
late bark, 499
late burning, 157
late-flushing groups, 499
late-flushing strain, 499
late frost, 499
late glacial deposit, 199
late-maturity, 199, 499
late stage, 625
late weed control, 199
late wood, 499
late wood percentage, 499
late wood vessel element, 499
late wood zone, 499
latency, 373
latent bud, 373, 543
latent heat, 374
latent heredity, 373
latent image, 374
latent infection, 373
latent learning, 373
latent period, 373
latent podzol, 571
latent root primordium, 373
latent seed, 543
latent virus, 373, 571
later thinning, 199
lateral, 42–44
lateral accessory bud, 43
lateral bending, 43, 197
lateral boom, 196
lateral branch, 44
lateral branch-root, 44
lateral bud, 43, 562

L

L

lateral canal, 43, 612
lateral corrosion, 43, 197
lateral cutting, 43
lateral deformation, 196
lateral division, 43
lateral dragging, 43
lateral drift of shot, 87, 418
lateral erosion, 43, 197, 351
lateral expansion, 43, 197
lateral fibre, 42
lateral floating boom, 148
lateral force, 43, 197
lateral irrigation, 43
lateral lap, 43, 351
lateral log truck road, 595
lateral meristem, 43
lateral natural seeding, 43
lateral nerve, 43
lateral ocelli, 42
lateral order, 43
lateral overlap, 43, 351
lateral petal, 42
lateral pole truck road, 580
lateral pressure, 43, 197
lateral ramification, 43
lateral road, 154
lateral root, 43
lateral sepal, 43
lateral shelter, 43
lateral shoot, 43, 44, 154
lateral spine, 42
lateral strain, 197
lateral stress, 197
lateral subsidiary cell, 43
lateral surface area, 43
lateral symmetry, 295, 654
lateral trace, 43
lateral vein, 43
lateral vibration, 197
lateral yarding, 43
laterite, 198, 636
lateritic red earth, 636
lateritic red loam, 636
lateritic soil, 636
lateri(ti)zation, 198
lateritoid, 636, 640
laterization, 198, 636
laterosternite, 43
laterotergite, 42
latest frost, 626
latex canal, 400
latex cell, 400
latex duct, 400
latex (*plur.* latices), 241, 400, 531
latex trace, 400
latex tube, 241, 400
lath, 8, 211
lath back, 335
lath-screen, 8, 335
lathe, 546
lathe charger, 546
lathe check, 88, 546
lathe flaker, 546
lathe for eccentric peeling, 11
lathe knife, 546
lathe log, 240, 546
lathhouse, 605
laticifer, 241, 400
laticiferous cell, 600
laticiferous system, 400
laticiferous tube, 400
latifoliate, 281
Latin-square, 282
latitude, 503
latitude and departure, 253
latitudinal zonation, 502
latosol, 636
latosolization, 636
latosols, 636
lattice back, 165
lattice boom, 165, 195
lattice constant, 100, 254
lattice fence, 165
lattice structure, 254
lattice-work, 165
latticed dune, 165
latticed perforation plate, 476
launching, 492, 521, 538
launching site, 124, 492
laundry board, 516
laurel forest, 605
laurel forest zone, 48
laurel leaf oil, 594
laurel oil, 594
laureline, 284, 594
laurifruticeta, 281, 605
laurilignosa, 48, 281, 605
laurisilvae, 281, 605
lava, 398
lava flow, 398, 554
lava flybar, 398, 545
lavender oil, 549
law of chance, 218
law of compensation, 32
law of conservation of energy, 341
law of constant final yield, 653
law of diminishing marginal utility, 22
law of diminishing returns, 15, 442
law of dominance, 525, 577
law of equi-marginal productivity, 22
law of equi-marginal utility, 22
law of filial regression, 199
law of geminate species, 311
law of increasing returns, 15
law of linkage and crossover, 292
law of mass action, 621
law of minimum nutrient, 653
law of natural thinning, 646
law of non-specialized, 138
law of normal distribution, 611
law of probability, 217, 221
law of segregation, 142
law of the minimum, 654
law of tolerance, 337
law of total effective temperature, 581
law of variability, 24
lawn, 41, 267, 512, 519
lawn-mower, 165, 233
lawn plant, 42
lawn renovation, 42
lawn sprinkler, 41, 42
lay, 45, 126, 391, 446
lay-days, 639
lay out young trees, 149
lay panel, 362
lay-up, 652
lay-up machine, 652
laydown, 41, 53, 165
layer, 13, 45, 54, 75, 140, 551
layer ploughing, 140
layer-shear, 45
layer society, 45
layerage, 551
layerage by girdling, 207

layerage by notching, 378
layerage by twisting, 347
layered pile, 45
layering, 45, 140, 551
layering plant, 551
laying, 2, 74, 147, 345, 363, 364
laying in, 378
laying off, 481, 521, 600
laying-out, 35, 147, 204
layup, 244, 353, 652
lea ploughing, 543
leach, 251, 291, 304
leach resistance, 337
leachability, 420
leached chernozem, 304
leached layer, 304
leached wood ashes, 251
leaching, 304
leaching erosion, 304
leaching field, 187, 291
leaching liquor, 251, 291
leaching of wood, 330
leaching test, 251, 291
leaching zone, 304
lead, 89, 125, 224, 372
lead block, 89
lead-lag relationship, 51
lead line, 372
lead log, 205, 286, 443
lead square, 281
lead time, 91, 239, 476, 523, 555, 640
leader, 89, 350, 570, 633
leader branch, 634
leader bud, 634
leader fungus, 165
leading dominant, 634
leading end, 161, 373
leading fossil, 25
leading indicator, 51, 305
leading peak, 373
leading roller, 89
leading shoot, 105, 106
leading species, 632
leading tree, 633
leading variety, 634
leaf, 560
leaf abscission, 314
leaf age, 560
leaf analysis, 560
leaf apex, 560, 561
leaf area, 560
leaf area density, 560
leaf area duration, 560
leaf area index, 560
leaf area ratio, 560
leaf arrangement, 561
leaf auricle, 560
leaf axil, 561
leaf axis, 561
leaf base, 560
leaf-beetle, 560
leaf blade, 560
leaf blister, 560
leaf boundary layer, 560
leaf bud, 561
leaf-bud cutting, 561
leaf cast, 314
leaf chamber, 561
leaf conopy, 300
leaf-cover, 278
leaf curl, 469
leaf cushion, 561
leaf cutting, 560
leaf density, 560
leaf developing period, 269
leaf diagnosis, 560
leaf drop, 314
leaf emergence, 62
leaf-fall, 314
leaf-fodder, 449
leaf gap, 561
leaf gold, 248
leaf ligule, 561
leaf-like petiole, 561
leaf-litter, 278
leaf-losing, 314
leaf margin, 561
leaf miner, 374
leaf mould, 151, 560
leaf organ, 560
leaf out, 62, 603
leaf petiole, 560
leaf primordium, 561
leaf resistance, 561
leaf roll, 264
leaf-roller, 264
leaf-rolling insect, 264
leaf rust, 561
leaf scar, 560
leaf scorch, 560
leaf shape, 561
leaf sheath, 561
leaf size class, 560
leaf skeletonizer moth, 7
leaf spine, 560
leaf spot, 560
leaf stalk, 560
leaf succulent, 399
leaf teeth, 560
leaf tendril, 560
leaf thorn, 560
leaf-tier (tyer), 246
leaf trace, 560
leaf trace bundle, 560
leaf tree, 282
leaf vein, 560
leaf venation, 560
leaf weight ratio, 561
leaf width, 560
leaf wood, 282
leafhopper, 560
leafiness, 119
leafing, 124, 603
leafit, 535
leafless, 511
leaflet, 154, 341, 401, 535, 561
leaflet trace, 535
leafminer fly, 374
leafmold, 560
leafy, 119, 281
leafy cutting, 80
leafy scale, 203, 561
leak detector, 233
leak-free, 322
leakage, 309, 538
leakage of genetic block, 565
leaking window, 309
lean clay, 444
leaner, 351, 381
leaning, 380
leaning stem, 381
leaning tree, 381
leap year, 401
leapfrogging, 480
learned behaviour, 516
leasehold, 651
leaseholder, 56, 595, 651
least-cost fire-control theory, 252
least square method, 654
least water-holding capacity, 653
leat blight, 560, 561
leather furniture, 356
leather-like, 165
leather waste, 139
leave, 306
leave strip, 14
leave tree, 15
leaving air, 287
Lebacq process, 288
lecithotrophic, 88
ledged-and-braced door, 53
ledged door, 614, 615
ledger, 94, 141, 196, 507, 649
lee, 17, 18, 520
lee eddies, 18

lee side, 18
lee slope, 18
lee wave, 18
leeward, 18
leeward slope, 18
left hand bandmill, 654
left-hand cutter, 654
left-hand lay, 654
left-hand sawmill, 654
left-handed spiral grain, 654
left-handed twist, 654
left lay rope, 654
left-over area, 209
left view, 654
leg, 243, 288, 612
legal entity, 127
legal highway load, 127
legal minimum reserve, 127
legal obligation to replant, 599
legal person, 127
legal standard, 127
legal tender, 127
legend system, 98
leggy, 119, 444, 487
legitimate combination, 483
legitimate copulation, 568, 610
legitimate genus, 192
legitimate name, 192, 611
legitimate pollination, 483, 611
legume, 110, 229
legume bacteria, 167
legumelin, 110
legumin, 110
leguminous green manure, 110
leguminous plant, 110
leguminous seed, 110
leiocarpus, 181
leiogynus, 181
leisure activity, 523, 532, 543
lek, 520, 521
lemma, 496, 570
lemnian earth, 457
lemon lac, 346
lemon oil, 346
lender, 62, 81
lenetic, 258
length, 49
length class, 49, 233
length-measuring control dial, 45
length measuring mechanism, 45
length of day, 397
length scanner, 49
lengthwise check, 650
length(wise) cultivation, 190, 650
lengthwise type piling, 650
lenific, 258
Leningrad standard, 297
lens examination, 135, 399
lens key, 399
lentic ecosystem, 258
lenticel, 356
lenticel infection, 356
lenticellate, 261
lenticelle, 356
lenticular pit aperture, 453
lenticular ray, 134, 485
lenticular transpiration, 356
lepal, 24
lepicena, 574
lepidosome, 305
leptomorphic rhizome, 13, 87
leptophyll, 501
leptotene, 519
lesion, 29
less electron dense portion, 515
less than carload lot, 94
lessee, 56
lessivation, 304
lessive, 304
lessive pseudogley, 304
lessor, 62
let, 310
let-down drop sorter, 521
lethal, 621
lethal concentration, 621
50-lethal concentration (LC50), 621
lethal dosage, 621
lethal dose, 621
lethal dose 50, 621
lethal gene, 621
lethal high temperature, 621
lethal limit, 621
lethal temperature, 621
lethality, 621
letter of credit, 540
letters patent, 475, 636
leuco-anthocyanidin, 6, 510
leuco-anthocyanin, 6, 510
leuco base stain, 510
leucocrate, 87
leucocyanidin, 6, 510
leucodelphinidin, 5, 510
leucofiseltinidin, 6, 510
leucoguibourtinidin, 510
leucopelargonidin, 510
leucoplast, 6
leucorobinetinidin, 510
leucotrope, 558, 637
leueophyll, 560
levan, 186, 654
Levant storax, 465
levee, 94, 132, 477
level, 221, 362, 458, 460
level bench, 458
level book, 226
level bottom, 359
level culture, 363
level cutting, 317, 458
level fork position, 48
level of integration, 610
level of nature reserve, 645
level of organization, 652
level of scientific understanding, 272
level of significance, 525
level trench, 458
level(l)er, 360, 460
level(l)ing, 266, 460, 480
level(l)ing agent, 266, 595
level(l)ing bench, 458
level(l)ing block, 204, 243, 458
level(l)ing-instrument, 460
level(l)ing machine, 243
level(l)ing (rod-)staff, 460, 470
level(l)ing tube, 460
lever, 41, 161, 282
lever-action, 161
lever arm, 161
leverage, 161, 347
leverblock, 156, 161, 442
levomenthone, 62
Lewellyn setter, 310
ley, 41, 336
liability, 151, 601, 603
liable to cross, 569
liana, 475
liane, 475

lianoid, 475
lianoid form, 475
liber, 142, 287, 396, 444
liberal pupa, 109, 287, 314
liberate carpel, 287
liberate ovary, 287
liberate pistil, 287
liberation cut, 16, 111, 416, 647
liberation cutting, 247, 443
liberation felling, 16, 111, 247, 416, 443, 647
liberation of virus, 29
libocedrene, 74, 531
libocedrol, 74, 531
library case, 78
libriform fibre, 396
libriform wood fibre, 396
licence, 444
license sales, 192
licensee, 57, 217
lichen, 99, 379
lichen bog, 99
lichen-quotient, 99
lichen tundra, 99
Lichenes, 99
lichenology, 99
lid, 203, 556
lie, 90, 446
Liebig theory of decay, 288
Liebig's law of minimum, 288
Liebig's law of the minimum, 290
lien, 278, 306
life arena, 422
life belt, 2, 426
life cycle, 422, 423
life expectancy, 173, 586
life form, 423
life-form class, 423
life-form dominance, 423
life form spectrum, 423
life habit, 422
life history, 422, 423
life history strategy, 422
life saving equipment, 259
life saving ring, 259
life span, 422, 437, 443
life stage, 422
life-support system, 423
life system, 423
life table, 423
life zone, 423, 426
lifespan, 443
lifetime, 443
lift, 367, 421, 556
lift boom, 103, 108, 367
lift bridge, 421
lift form, 423
lift gate, 421
lift skid arm, 644
lift truck, 45, 421
lifter, 367, 421
lifting, 366, 367, 476, 558, 559, 564
lifting by frost, 110
lifting capacity, 368, 476
lifting line, 368, 421
lifting of seedlings, 367
lifting operation, 367, 368
lifting platform, 260, 421
lifting table, 421
lifting the canopy, 300
lifting the canopy base, 300
lifting trailer, 222, 223
lifting transporter, 476
liftor, 367, 476
ligand, 353
ligase, 292
ligation, 246, 314
light, 180, 214, 445
light adaptation, 182
light and dark bottle method, 194
light break, 181
light brown steppe soil, 87
light burn, 380
light burning, 380
light cableway, 380
light chestnut earth, 87
light clay, 380
light coloured heartwood tree, 571
light compensation point, 180
light conifer, 296
light culture, 90
light cutting, 443
light-dark cycle, 180
light degradation, 181
light-demander, 516, 517, 556, 557
light-demanding, 516, 544, 557
light demanding guild, 516
light(-demanding) seed, 544
light-demanding species, 517, 557
light-demanding tree, 516, 556, 557
light discolouration, 380
light duration, 182
light extinction coefficient, 531
light factor, 182
light fastness, 337
light-favored seed, 516
light felling, 443
light flux, 182
light flux density, 182
light forest fire, 380
light forwarding vehicle, 380
light fuel, 380
light germination, 544
light green, 375
Light-growth inhibition, 181
light hardening, 182
light hardwoods, 400
light increment, 443
light independent, 35
light indicator, 556
light-induced bleaching, 182
light inducing, 182
light-inhibited seed, 182
light inhibition, 376
light injury, 181
light intensity, 181
light interception, 181
light knot, 232, 292
light line, 325
light loam, 380
light microscope, 182
light middle, 73, 380
light-needing, 557
light oil, 380, 577
light open, 445
light particleboard, 380
light permanency, 337
light quality, 182
light railway, 380
light rain, 535
light reaction, 180
light reflection, 180
light-requirement, 544
light-requiring seed, 544
light resistance, 180, 337
light-resistant, 337
light respiration, 181
light saturation, 180
light saturation point, 180
light scattering, 181

L

light seed, 380, 569
light-seeded species, 628, 629
light-sensitive seed, 181
light skyline, 380
light slash, 569
light soil, 380
light solvent treatment, 380
light source, 182
light stain, 380
light supply, 167
light surface fire, 500
light tar, 380
light thinning, 401
light transmission percent, 485
light trap, 583
light wood, 59, 325
lighter, 31, 75, 570
lighthouse, 90
lighthouse cat, 79
lighting of tree, 445, 485
lightmeter, 180, 605
lightning, 414
lightning-caused fire, 284
lightning-damaged tree, 284
lightning detection system, 284
lightning fire, 284
lightning risk, 284
lightning shake, 284
lightwood, 189, 325, 380, 569
lightwood oil, 380, 609
ligneous, 335
ligneous plant, 328
ligneous shrub, 335
ligneous tissue, 336
ligniferous, 335
lignification, 332, 335
lignificationlignifying, 335
lignified, 335
lignified cell wall, 335
lignified fiber, 335
lignified tissue, 335
ligniform, 331, 336
lignify, 332, 335
lignifying, 332
lignin, 335
lignin biodegradation, 336
lignin building stone, 336
lignin building unit, 336
lignin-carbohydrate complex, 334
lignin-decomposing fungus, 335
lignin-dissolving fungus, 336
lignin plastic, 336
lignin precursor, 334
lignin precursors, 334, 336
lignin skeleton, 335
ligninase, 336
lignivorous, 437
ligno-cellulose, 334, 336
ligno-humus, 334
ligno-protein complex, 334
ligno-proteinate(s), 334
lignocellulosic filler, 336
lignolysis, 334
lignometer, 44
lignosa, 328
lignous, 335
lignum, 328
ligular, 417, 561
ligular pit, 561
ligular scar, 561
ligulate, 417, 561
ligulate corolla, 417
ligulate flower, 417
ligulate vessel-element extension, 417
ligule, 69, 417, 561
liliaceous corolla, 6
lily garden, 6
lily magnolia, 642
lily pad, 327, 460, 590
lily-pool, 6, 460
limacid, 281
limb, 11, 75, 121, 178, 449, 560
limb marking, 612
limber, 75, 610
limber boom, 399, 401
limber bucker, 76
limbiness, 143, 613
limbing, 64, 75
limbing-bucking station, 564
limbing machine, 76
limbing saw, 544
limby tree, 120
lime bast, 113
lime carbonate, 473
lime concretion, 433
lime-constant association, 156
lime-fusion method, 434
lime hardpan, 434
lime ignition method, 434
lime oil, 6, 288
lime plant, 156
lime sulphur (solution), 434
limed rosin, 155, 434, 450
limed wood, 251
limes death, 621
limestone, 434
limestone red loam, 198, 434
limette, 6, 288
limette oil, 6, 288
limit of allowable error, 398
limit of elasticity, 472
limit of fatigue, 356
limit of fire resistance, 337
limit of plastic flow, 466
limit of rupture, 113, 363
limit of tolerance, 337
limited chromosome, 581
limited population, 581
limited range provenance trial, 259
limited use forest, 526
limiting factor, 526
limiting similarity, 526
limiting water level, 526
limiting water velocity, 526
limnic peat, 201, 434
limnimeter, 644
limnodium, 555
limnology, 201
limnoplankton, 87
limonin, 346
limy soil, 434
lindane, 161, 298
Lindeman's efficiency, 298
Lindeman's ratio, 298
linden concrete, 113, 364
linden oil, 113
line, 160, 161, 248, 310, 359, 517
line-bar resaw, 79, 545
line blow, 615
line boss, 133, 364
line breeding, 359, 517
line camp, 216, 304
line capacity, 161, 397, 655
line cross, 359

line drilling, 350, 479
line hook, 227, 281, 372
line-intercept method, 558
line intersect sampling, 526
line isolation, 632
line leaf, 527
line locator, 547
line log, 183, 242
line mixture, 190
line of attachment, 150
line of vision, 439
line pattern, 526
line per millimeter, 320
line-planting, 190, 297
line-plot, 458
line-plot method, 527
line-plot survey, 458, 527
line retainer, 161
line sampling, 526
line scout, 215
line seeding, 190, 479
line sowing, 190, 479
line speed, 372, 526
line spotter, 25
line survey, 558
line swivel, 161
line tensioner, 249
line thinning, 81, 166
line transect, 558
line transect method, 558
line truck, 365, 595
line up roll, 197, 651
lineage, 365
linear array, 527, 563
linear elasticity, 526
linear erosion, 527
linear expansion, 526
linear expansivity, 526
linear factor, 527
linear flame propagation, 216
linear foot, 57, 615
linear gravity, 527
linear inheritance, 615
linear leaf, 479
linear measure, 49
linear measurement, 526
linear model, 526
linear molecule, 526
linear order of genes, 220
linear parameter, 526
linear pit, 526
linear plot, 190, 527
linear prestressing, 527
linear programming, 526
linear ray, 84
linear regression, 526
linear regression estimate, 526
linear (sample) plot, 526
linear sampling, 526
linear selection, 517, 615
linear shrinkage, 107, 526
linear strain, 527
linear survey, 527
linear transformer strain gauge, 526
linear variable displacement transducer, 526
linear velocity, 526
linearity, 526
linebar, 261, 271
linen shelf, 67
linepull, 264
lingula, 69
lingy, 180
lining, 53, 184, 340
lining out, 54
lining out nursery, 565
lining-out plough, 565
link, 237, 262, 295
link lock, 214, 470
link peened, 295
linkage, 208, 282, 292
linkage coefficient, 207, 292
linkage computer, 293, 294
linkage editor, 292, 295
linkage group, 292, 293
linkage map, 292
linkage relationship, 292
linker, 292
Linnean species, 302
linseed oil, 552
lint, 323, 356
linter, 31, 323, 602
linter pulp, 323
liorhizal, 360
lip, 69, 88, 90, 378, 545, 612
lip angle, 69, 536, 652
lip clearance, 18, 199, 652
lip drawer, 74
liparite, 183, 308, 573
lipase, 613
lipid(e), 285, 343, 613
lipin, 285, 613
lipogenesis, 613
lipoic acid, 308
lipoid, 285, 343, 613
lipolysin, 398
lipolysis, 613
lipophilic, 379
lipophobic, 445
lipoplast, 613
lipopolysaccharide, 613
lipoprotein, 613
liposome, 613
lippia oil, 187
lipping, 69, 146
liptocoenosis, 426
liquefier, 561
liquid adhesive, 561
liquid asset, 306
liquid biofuel, 562
liquid brake, 561
liquid capital, 306
liquid chromatogram, 562
liquid chromatography, 562
liquid coolant, 286, 562
liquid creosote, 562
liquid crystal, 561
liquid cutting, 457
liquid drier, 562
liquid fat, 613
liquid fertilizer, 562
liquid filler, 562
liquid glue, 561
liquid honing, 457, 562
liquid-liquid chromatography, 562
liquid manure, 144, 259
liquid medium, 562
liquid paraffin, 562
liquid petrolatum, 128, 562
liquid phase chromatography, 562
liquid polishing agent, 562
liquid recovery, 562
liquid resin, 562
liquid rosin, 333, 464, 494, 562
liquid scintillation counter, 562
liquid-solid chromatography, 561
liquid thermometer, 562
liquidambar resinoid, 146, 241
liquidation, 49, 381, 398, 561
liquidation cutting, 381, 610
liquidation value, 381

liquidene, 146, 465
liquidity, 306
liquifier, 515, 561
liriodendrin, 120
list, 381
list price, 25, 106
list processing, 544
list sampling, 297
lister cultivator, 268
lister-planter, 172, 268
lister planting, 288
lister plougher, 268
listing, 268
lithic contact, 328
litho-morphic brownish calsiallitic soil, 434
litho-soil, 291, 434
lithological unit, 554
lithology, 554
lithophyte, 434, 554
lithophytic, 434
lithopone, 539
lithosere, 434
lithosol, 291, 434
lithosphere, 310, 554
lithotrophy, 509
litmus, 434
litse citrata oil, 413
litse cubeba oil, 20, 333, 413
litsea citrata oil, 413
litsea cubeba oil, 413
litter, 278, 409
litter decomposition, 278
litter feeder, 437
litter interception loss, 314
litter layer, 103, 278, 462
litter of weed, 278
litter size, 507
litter supply, 462
litter temperature, 278
litter trap, 103
little ice age, 533
little leaf disease, 535
littoral, 27, 188, 554
littoral dune, 188
littoral forest, 188
littoral zone, 554
live borer, 215
live branch, 215
live burning, 430, 433
live chute, 629
live crown ratio, 215
live crown surface, 215
live deck, 62, 113, 556
live decoy, 214
live discharge roll, 383
live end of loop, 208
live fence, 315
live hose, 55
live knot, 214
live labour, 214
live load(ing), 108, 170, 215
live pruning, 215, 315
live roll, 108, 383
live roller conveyer, 184, 383
live skyline, 464
live spindle, 546
live steam, 307, 539, 614
live steam piping, 539, 609, 633
live tandem, 455
live wood, 214
liver fungus, 347
liverwort, 471
livery board, 436
livery cupboard, 534
livestock, 229
living bamboo, 290
living cell, 215
living clock, 428
living environment, 422
living fire break, 133
living firebreak, 133
living fuelbreak, 133
living ground cover, 214
living model, 427
living pattern, 422, 423
living prototype, 428
living resources, 428
living tree, 214
lizard, 84, 93, 285, 516
ljght growth reaction, 181
llano, 76, 282
Lmnarckism, 282
load, 151, 226, 312, 445, 598, 639
load aligning, 197
load at failure, 113, 363
load baffle, 36
load bearing, 11, 55, 56
load binder, 249, 280, 281, 470
load brake, 51
load capacity, 74, 226, 598, 638
load (carrying) capacity, 56, 598
load cell, 598
load check, 618
load cradle, 224, 638
load curve, 598
load-deformation curve, 598
load-elongation-curve, 598
load-extension curve, 598
load factor, 151, 598
load-indicating apparatus, 598
load line, 66, 598
load-monitoring system, 598
load of long duration, 57
load pendulum, 469
load rope, 598
load sensing valve, 598
load size, 639
load squeezer, 366, 638
load-strain curve, 598
load-weight indicator, 639
loaded carriage, 630
loaded deflection, 151, 630
loaded skyline cable, 56
loader, 445, 638, 639
loader and unloader, 639
loader-digger, 495
loader-dozer, 639
loader hydraulic pump, 638
loader module, 226, 445, 598
loader-tractor, 223
loading, 327, 479, 514, 598, 638
loading area, 488, 635, 638
loading block, 367
loading bridge, 180, 377, 639
loading cage, 638
loading deck, 62, 638
loading dog, 62, 638
loading flipper, 128
loading gauge, 296, 598
loading guy line, 638
loading hook, 104, 638
loading jack, 62, 638, 639
loading of press, 550
loading procedure, 639
loading ramp, 639
loading shovel, 110, 492
loading skid, 638
loading test, 598
loading tripod, 639
loading winch, 638

loading without car stake, 508
loadometer, 44, 598
loam, 391
loam clay, 391
loamy coarse sand, 391
loamy fine sand, 391
loamy sand, 391
loamy soil, 391
loamy texture, 391
loan capital, 248
lobate, 142, 297, 375, 592
lobe, 30, 121, 297, 375, 561, 592
lobed, 142, 375, 582, 592
lobed foot, 11
lobelet, 534, 535
lobose, 260
lobulate, 261
lobule, 155, 534, 535, 561
lobulus (*plur.* lobuli), 58, 535
lobus, 297, 560
local, 95, 259
local abundance, 97, 259
Local Agenda 21, 95
local apparent time, 96
local association, 96
local character species, 97
local climate, 259
local cultivar, 96
local endemic species, 96
local hour angle, 96
local immunity, 259
local infection, 259
local irradiation, 259
local lesion, 259
local magnetic attraction, 96, 259
local nature reserve, 95
local necrosis, 259
local necrotic lesion, 259
local parasite, 88
local population, 259
local race, 96, 99
local seed source, 259
local solar time, 96
local species, 96, 177
local symptom, 259
local tarifs, 96
local time, 88, 96
local variation, 97
local variety, 96
local volume table, 95, 563
local yield class, 96
local yield table, 96
locality, 47, 95, 207, 288, 504
locality class, 98
locality factor, 289
localized fertilization, 107
localized incompatibility, 259
localized infection, 259
locant, 504
location, 310, 504
location map, 107, 504
location parameter, 504
location survey, 106, 107
lociation, 552
locie, 595
locies, 556
lock, 210, 451, 460, 470
lock bar, 294, 470
lock corner, 242
lock dowel, 470
lock down, 187, 350
lock-in technologies and practice, 251
lock joint, 277, 470, 532
lock mitre, 537
lock mitre joint, 229, 470
lock rail, 2, 638, 639
lock ring, 322, 470
lock time, 464
lock wedge, 470
locker, 80, 286, 470
locking, 178, 470, 558, 620
locking stile, 639
locomotion, 595
locomotive crane, 480
locomotive yard, 217
locule, 440, 641
loculicidal, 13, 440
loculicidal dehiscence, 440
loculus, 13, 134, 534
locus, 184, 504
locus of loaded point, 598
locust, 210, 602
locust tree, 71
lodge, 42, 75, 409, 444, 604
lodged tree, 75
lodging, 75, 90, 227, 268, 634
lodgment of stock, 103, 507
lodicule, 238, 305
loess, 210
loess deposit, 210
loess-like soll, 342
loessal, 210
loft, 165
loft dryer, 158, 529
log, 112, 159, 170, 328, 589, 591, 592
log allotment, 331
log assemblage, 222
log assortment, 37
log band saw, 79, 208
log barking, 589
log bin, 143
log blue, 590
log boat, 589, 590
log book, 589
log brand, 590
log break down rule, 590
log bundle, 590
log bunk, 589
log cabin, 111, 335
log cabin piling, 590
log canal, 595, 635
log carriage, 465, 590
log centring, 589
log chain, 589, 590
log chain conveyor, 295
log charger, 590
log charging, 332
log chute, 589
log clamp, 589
log class, 589
log-cleaner saw, 590
log cooking, 590
log core, 590
log cut-off saw, 493, 500
log dam, 328
log deck, 590
log delivery device, 589, 590
log deposit, 635
log depot, 635
log direction changer, 589
log dog, 207, 371
log dogging jaw, 371
log dozer, 492, 589, 590
log driver, 159, 307
log dump, 171, 492, 589, 590, 635, 638
log ejecter, 351
log elevator, 197, 590
log evaluation for geometry, 589
log feeding mechanism, 590
log fire, 589
log fixer, 593
log for rail-road tie, 608

log fork, 589
log fork-lift truck, 329
log frame, 68, 117
log frame-saw, 280
log gang saw, 349, 615
log getter, 223
log grab, 207, 371
log grade, 589
log grade evaluation, 589
log grading, 589, 590
log grading rule, 589
log grapple, 590, 636
log handler, 283, 590
log handling, 331, 589
log handling and conveying machinery, 330
log-handling peavy, 589
log handling portal crane, 321, 331
log haul(-up), 535, 538, 561, 589
log hauler, 329, 590, 595
log-hauling, 590, 595
log hauling machinery, 595
log hoist, 242, 283, 589, 590
log inside bark, 590
log jack, 7, 62, 550, 590
log jam, 34, 589
log kicker, 351, 476, 492
log ladder, 590
log land transportation, 330
log ledger, 330, 635
log-length logging, 589
log-length sorting machine, 589
log lift, 590
log loader, 331
log loading, 331
log lock, 307
log-lumber fork, 329
log magazine, 590
log-maker, 600
log making, 600
log mark, 589
log market, 330, 590
log mill, 76, 619
log pass, 127
log patrol, 202, 549
log pile, 589, 590
log pile area, 286
log pond, 589, 635
log positioning, 589
log preservation and management, 328
log-raft, 332, 333
log raising and lowering carriage, 590
log rejector, 600
log removal, 222, 224, 595
log-rolling, 590
log rotater, 128
log rough rounding, 590
log rule, 589
log run, 620
log saw, 77, 364, 633
log saw carriage, 590
log sawyer, 126, 262
log scale, 115, 589, 590
log-scale board foot, 589
log scale recovery, 62, 590
log scaling, 590
log scanner, 590
log series distribution, 115
log ship, 595
log sledge, 595
log slide, 589
log slip, 332
log sluice, 456
log sorter, 547
log sorting, 589, 592
log splitter, 355, 590
log stack, 286
log stacker, 113, 183
log stain, 589
log stamp, 191, 589
log standard, 328
log steaming, 590
log storage, 331
log storage deck, 171, 285, 538
log tally, 589, 590
log taper, 590
log trade, 590
log tramway system, 184, 231
log transfer, 331, 590
log transfer deck, 590
log-transformed model, 115
log transportation, 595
log transportation by truck, 371
log transportation by water, 331
log trough, 590
log truck, 595
log truck with trailer, 595
log turner, 128
log-turning, 589, 590
log type, 37
log volume, 589
log volume table, 589
log-wagon, 463
log watch, 446
log yard, 590, 635
logan, 304, 492
loganin, 128, 316
logarithmic death phase, 115
logarithmic distribution, 115
logarithmic growth phase, 115
logarithmic sprayer, 115
logarithmic volume curve, 115
loggable timber, 272
logged-crib abutment, 333
logged(-off) area, 38
logged (over) area, 38
logger, 39, 126, 590
logging, 37, 39, 222
logging accessory, 39
logging arch, 223
logging area, 38, 127
logging band, 126
logging barge, 360, 595
logging block, 268, 469
logging boss, 126
logging budget, 39
logging camp, 126
logging case, 39, 224
logging chance, 39
logging clearing, 127, 381
logging contractor, 126
logging crane, 224
logging crew, 7, 114, 655
logging cycle, 38
logging dam, 556
logging debris, 38
logging depot, 39, 635
logging dolly, 122, 595
logging engineering, 406
logging enterprise, 39
logging grapple, 223, 636
logging guard package, 595
logging headquarter, 125
logging hoist, 224
logging jobber, 126
logging larry, 595
logging line, 223, 634
logging locomotive, 595
logging loss, 39

L

logging machine system, 406
logging machinery, 39, 330
logging mechanization, 406
logging off, 328, 600
logging operation, 127
logging period, 38
logging plan, 37
logging plant, 38
logging practice, 127
logging procedure, 126
logging railroad (railway), 409, 595
logging regulation, 39
logging residue, 38
logging road, 222, 595
logging rule, 38
logging scissors, 126, 590
logging shack, 38, 126
logging shed, 126
logging show, 39, 224
logging site, 38, 125, 248, 371, 385
logging slash, 38, 125
logging sled, 595
logging sledge, 223
logging sleigh, 595
logging superintendent, 126, 127
logging to mill, 615
logging tongs, 223
logging-track, 409, 595
logging tractor, 223
logging trailer, 595
logging truck, 44, 410, 595
logging unit, 39, 127, 224
logging-wheels, 162
logical instruction, 312
logical language, 312
logical unit number, 312
logistic curve (Sigmoid curve), 312
logistic equation, 312
logistic growth, 312
lognormal distribution, 115
logs diversity, 589
logway, 62, 127, 358, 590
loment, 245
lomentaceous, 245, 260
London lane tree, 123
long axis, 50
long-boled, 49
long butt, 49
long column, 50
long cord, 49
long corner, 22
long-cut wood, 50
long day plant, 50
long-day treatment, 50
long flowering period, 49
long grained plywood, 461
long-handled pruning shear, 49
long-horned beetle, 477
long-length, 49, 50
long-length logging, 49, 591
long-lived seed, 50
long log system, 49, 591
long log with slender top, 79
long logging, 49, 591
long oil, 50, 376
long-period group system, 49
long plane, 49
long-range forestation program, 50
long-range perspective, 594
long-retention-time blender, 318
long rotation, 49
long run, 49, 50
long shoot, 50
long-shore drift, 307, 554
long slope, 49
long splice, 49
long stem cutting, 50
long sunshine period, 50
long-term, 49
long-term bending test, 49
Long-Term Ecological Research, 49
long-term effectiveness, 57
long term fire danger, 49
long-term retardant, 50
long term static stress, 49
long term tensile test, 49
long tie, 50
long timber, 49
long-time load, 50
long wood, 49, 159, 591
long wood column, 49
longevity, 443
longhorn beetle, 147
longipinene, 50
longitude, 49, 252
longitudinal bending, 650
longitudinal boom, 461
longitudinal cambial initial, 650
longitudinal check, 650
longitudinal compression process, 461, 650
longitudinal conveyor, 650
longitudinal crack, 650
longitudinal culvert, 650
longitudinal cutting, 650
longitudinal displacement, 650
longitudinal distortion, 170, 461
longitudinal element, 650
longitudinal erosion, 650
longitudinal growth, 163, 650
longitudinal increment, 447
longitudinal joint, 650
longitudinal load(ing), 650
longitudinal movement of air, 650
longitudinal pad, 650
longitudinal parenchyma, 650
longitudinal profile, 649, 650
longitudinal resin canal, 650
longitudinal section, 649, 650
longitudinal shaft kiln, 50, 651
longitudinal shear, 650
longitudinal shrinkage, 650
longitudinal slope of canal, 383
longitudinal splicer, 650
longitudinal strain, 650
longitudinal strand parenchyma, 650
longitudinal strength, 650
longitudinal stress, 650
longitudinal surface, 649, 650
longitudinal swelling, 650
longitudinal tracheid, 650

L

L

longitudinal traumatic gum canal, 650
longitudinal vibration, 650
longitudinal view, 650
longitudinal wall, 649, 650
longitudinal warping, 170, 650
longitudinal wood element, 650
longitudinal wood parenchyma, 650
longitudinal xylary cell, 650
look-up table, 47
lookout, 296
lookout cabin, 178, 296
lookout cupola, 178, 296
lookout dispatcher, 296
lookout fireman, 296
lookout house, 296
lookout observatory, 296
lookout patrol(man), 549
lookout point, 296, 607
lookout room, 296
lookout station, 178, 296
lookout system, 232, 296
lookout tower, 296
looming, 416
loop, 21, 160, 208, 212
loop back, 212, 593
loop bar, 474
loop chromatid, 208
looper, 57, 280
loose, 464
loose bark pole, 68, 464
loose-cut veneer, 464
loose driving, 159, 404
loose face, 18, 464, 546
loose floating, 159, 404
loose granular structure, 404
loose knot, 464, 494
loose raft, 81, 400
loose rock check dam, 468
loose rock dam, 468
loose seat, 272
loose side, 18, 464
loose stone dam, 157
loose-tongued sloop, 80, 582
loose veneer, 464
loose volume, 404
loosen, 374, 465
loosened grain, 464
loosener, 374, 465
loosening of the soil, 445, 490
loosening wheel, 464
lop, 125, 234, 247, 269, 520, 543
lop and top, 613
lophos, 225
lopper, 75, 164, 247, 612
lopping, 75, 234, 247, 544
lopping and scattering, 247
lopping and topping, 544
lopping-axe, 247
lopping forest, 247
lopping knife, 544
lopping of branches, 234, 247, 544
lopping off, 234, 247, 544
lopping regeneration, 247
lopping reproduction, 247
lopping shears, 164
lopping system, 247
lopping tool, 247
lopping tree, 247
lopsided tree, 357, 381
lopwood, 613
Lorey's mean height, 361
lorry, 595
loss, 280, 469
loss by solution, 398
loss leader, 294
loss of gloss, 431
loss of head, 314, 459
loss of soil nutrient, 491
loss-on-ignition, 417
loss tangent, 469
lost line, 431
lot, 95, 286, 332, 533
lot production, 77, 355
lotic ecosystem, 307
lotic water, 307
lotus, 193, 294
lotus root, 294, 349
loudness, 530, 570
lounge, 50, 474, 543
lounge chair, 50
louver board, 7, 404, 479
love seat, 453
Lovibond tintometer, 314
loving, 516
low-angle plane, 93
low back chair, 92
low-bed, 92, 93
low-bed trailer, 93, 359
low-branching tree, 94
low-carbon economy, 93
low-carbon lifestyle, 93
low-density particleboard, 93
low-density plywood, 93, 380
low-density wood chipboard, 93
low finish, 93
low forest, 1
low frequency, 93
low-frequency shock wave, 93
low humic gley soil, 93
low-land, 92, 495
low-lead skidding, 97
low level laterite, 93
low moor, 93
low moor peat, 93
low moor soil, 92
low moor wood peat, 93
low noise sawblade, 94
low oblique photo, 93
low-pass filter, 93
low pole, 156, 536
low pole wood, 1
low polymer, 93
low post bed, 92, 93
low-pressure laminating system, 94
low pressure short cycle processing, 94
low pruning, 543
low ring roll, 314
low-risk forest management, 93, 253, 408
low-statured, 92
low-stemmed, 1
low straight-away shot, 93
low-sun angle aerial photography, 93
low-temperature drying, 93
low-temperature setting, 93
low-temperature setting adhesive, 93
low-temperature storage, 93
low thinning, 520
low-thinning method, 520
low-tower system, 93
low-voltage heating, 93

low volume spray, 93
low volume spray treatment, 93
low volume sprayer, 93
low yield pulp, 92
lowboy, 80, 112
lowboy trailer, 93
lower bed, 92, 93
lower branch, 520
lower chord, 521
lower cord, 195
lower couch roll, 520
lower course, 93, 521
lower critical temperature, 93
lower crown height, 520
lower crown length, 520
lower density fibreboard, 93
lower diameter limit, 615
lower glume, 497
lower landing, 635, 637
lower plant, 92
lower plastic limit, 466
lower press roll, 521
lower-priced furniture, 92
lower reach, 521
lower storey, 520, 521
lower storey of forest, 302, 521
lower subsoil, 96, 538
lower tail covert, 503
lower wire, 521
lowerable load hook, 273
lowland moor, 93
Lowry process, 9, 10, 348
lubaloy-coated bullet, 305
lubaloy-coated shot, 305
lube-axial fan, 179
lubricant, 401
lubricate, 401
lubrication, 401, 635
luciferase, 573
luciferin, 573
luciferous, 123
lucumin, 310
luffer board, 7
luffing, 52, 367
lug, 262, 314, 486
lug hook, 161, 471
lug support, 121, 492, 612
lukabro oil, 471
lumbang oil, 434, 484
lumber, 54, 262, 328
lumber buggy, 453
lumber carrier, 15, 465
lumber checker, 331
lumber class, 262
lumber code, 328, 329
lumber consumption, 262
lumber container car, 15
lumber conversion, 619
lumber-core board, 479
lumber core composer, 8, 335
lumber-core construction, 8, 200, 262
lumber core edge gluer, 8
lumber core plywood, 262, 519
lumber dry kiln, 329
lumber drying, 262
lumber drying kiln, 262
lumber elevator, 8, 113, 262
lumber fork, 262
lumber gauge, 262
lumber grade, 262
lumber grade recovery, 91
lumber industry, 620
lumber inspection bureau, 330
lumber jack, 113, 126
lumber-job, 127, 620
lumber kiln, 262, 329
lumber manufacturer, 620
lumber manufacturing industry, 620
lumber market, 330
lumber merchant, 330
lumber mill, 125, 262, 619
lumber mill operator, 126, 619
lumber product, 619
lumber recovery, 262
lumber road, 222, 298, 595
lumber scale, 262
lumber stacker, 262
lumber storage room, 262
lumber tally, 262
lumber trade, 330
lumber transportation, 262
lumber value differential, 262
lumber volume determination, 262
lumber yard, 8
lumberer, 39, 620
lumbering, 8, 126, 262, 619
lumbering camp, 38
lumbering concession, 475
lumbering machinery, 620
lumberman, 126, 330, 620
lumen (*plur.* lumina), 307, 375, 518
luminescence, 123, 286, 573
luminescence bacteria, 123
luminescent organ, 123
luminosity, 296, 325
luminous cloud, 17, 358, 561
luminous emittance, 123, 180
luminous line, 325
luminous organism, 123
lump, 165, 246, 446, 486
lump sum, 648, 649
lump sum acquisition, 610, 649
lunar, 539, 570, 594
lunar cycle, 594
lunar periodicity, 594
lunar periodism, 594
lunar phenomenon, 594
lunar reproductive cycle, 594
lunar rhythm, 594
lunate, 539, 570, 594
lunated, 261
luniform, 539, 570, 594
lure, 571, 583
Lusitanian oak, 364
luster, 182
lusterless paint, 509
lustre, 182
lute in, 146, 375, 466
lutein, 210, 560
luteolin, 318, 335
luvisol, 304
luxmeter, 284, 605
luxuriance, 597
luxury absorption, 417
lying charcoal kiln, 507
lying dead tree, 278
lying meiler, 507
lying panel, 362
lying retort, 507

L

M

lyme disease, 282, 408
lyophilic, 379
lyophilic colloid, 379
lyophilization, 93, 286, 421
lyophilize, 286
lyophilizing, 110
lyophobe sol, 445
lyophobic, 445, 602
lyophobic colloid, 445
lyrate, 78, 450
lyre back chair, 365
lysate, 398
lysigenetic, 398
lysigenic, 398
lysigenous, 398
lysigenous cell, 398
lysigenous duct, 363, 398
lysigenous traumatic canal, 398
lysimeter, 304, 398, 420
lysin, 518
lysine, 282
lysis, 142, 398
lysosome, 398
lysozyme, 398
lytic response, 398
M-tooth, 262
M-toothed saw, 58
macadam pavement, 468
macadam road, 468
macadamizing, 468
macaroni, 49
MacArthur's brocken stick model, 316
macchia, 316
macerate, 75, 251, 287
macerated wood, 247, 287
macerating fluid, 142, 251
maceration, 251, 287
macerator, 75, 251, 385
machete, 269
machilus leaf oil, 338
machilus oil, 338
machinability, 273, 378
machine bum, 218
machine code, 218
machine direction forming head, 650
machine down-time, 481
machine drill, 218
machine grading, 218
machine hour, 218, 471
machine idle-time, 218, 481
machine instruction, 218
machine language, 218
machine-made paper, 218
machine operation code, 218
machine oriented language, 323, 440
machine running-time, 218
machine saw, 218
machine sensible information, 218
machine separation, 218
machine waiting time, 218
machinery layout, 218
machining, 218
machining defect, 218
machining precision, 218
maclurin, 405
Maco template, 316
macro-analysis, 48, 198
macro-consumer, 78
macrobenthos, 78
macrobiota, 78
macrobiotic seed, 50
macrochore, 76
macrochromosome, 78
macroclimate, 77
macroclimatology, 77
macroconidium, 76
macroconstituent, 48
macroelement, 48, 77, 198
macroevolution, 77, 198, 627
macrofauna, 78
macrofibril, 73, 78
macroflora, 78
macrogamete, 77
macrography, 198, 399
macrometeorology, 78
macromolecular structure, 76
macromolecule, 76, 162
macromutation, 78, 260
macronutrient, 77, 634
macrophages, 260
macrophanerophyta, 76
macrophyll, 78
macrophyllous, 78
macrophyte, 78
macrophytoplankton, 78
macroplankton, 78
macropore, 77
macroprecipitation, 48
macropterous form, 49, 76
macroscopic, 198, 399
macroscopic examination, 92, 198, 399
macroscopic feature, 198
macroscopic structure, 198
macroscopic symptom, 198
macroscopy, 399
macrospecies, 78
macrosporangium, 76
macrospore, 76
macrosporophyll, 76
macrostructure, 198
macrostylous, 49
macrotherm, 164
macrotin, 421
macular, 260
maculose, 260
madarin orange, 411
madder, 375, 396
Maeule reaction, 319
magaphanerophytes, 377
magazine, 87, 296, 401, 597
magazine follower, 491
magazine grinder, 41, 457, 529, 635
magazine spring, 87, 492
magenta, 359, 557
maggot, 276, 383
maggot chamber, 383
magma, 201, 554
magmatic water, 554
magmatite, 554
magnesium sulphite process, 552
magnet separator, 69
magnetic aberration, 69
magnetic catch, 69
magnetic cleaner, 69
magnetic defectoscope, 69
magnetic ink character reader, 69
magnetic ink character recognition, 69
magnetic pole, 69
magnetic tape, 69
magnetic valve, 100
magnetosphere, 69
magni-coronate, 49
magnifier, 135
magnifying power, 135
magnolia, 333
mai yang, 494
maiden, 1
mailer method, 114
main alley, 632

main assortment, 331, 633
main boom, 633
main branch, 613, 634
main cable, 56, 368, 372, 634
main category, 331, 633
main compartment line, 383, 633
main crop, 15, 633, 634
main crop after thinning, 306, 633
main crown class, 415
main drum, 598, 633
main feature, 633, 634
main felling, 625, 633
main forest belt, 633
main forest road, 633
main frame, 383, 633
main haul road, 632, 634
main hydrolysis, 634
main key, 220
main lead, 102, 372, 634
main lead block, 372
main lead line, 368, 372, 634
main line, 77, 160, 368, 372, 634
main-line block, 368, 372, 634
main lobe, 633
main motor, 632
main nerve, 633
main product, 632
main reaction, 634
main ride, 632
main road, 160, 632
main root, 633
main rope, 56, 368, 372
main shaft, 634
main soil group, 78, 634
main soil type, 78
main spar, 633
main species, 634
main spring, 632
main stand, 633
main stem, 633
main storey, 633
main survey, 435, 632
main tree, 634
main tree species, 634
main truss, 633
main work, 634
mainline rigging, 372, 634
maintenance, 15, 397, 422
maintenance load, 502
maintenance of canopy, 14
maintenance of stand, 15
major element, 77
major forest produce, 633
major forest tree, 633
major gene, 634
major gene resistance, 612, 634
major habitat, 78
major harvest, 625, 633
major management subject, 634
major overhaul, 78, 649
major parts, 634
major produce, 632
major product, 632
major product system, 632
major ride, 383
major transportation, 634
major tree planting drive, 76
major work, 634
major works, 78
make bed, 360
make coat, 412
make good, 32, 54, 353
make time, 20, 244
makeshift bridge, 24, 304
making up, 33, 147, 173, 638, 651
malacology, 401
malacophilous flower, 507
maladaptation, 440
malamine resin, 403
malaria, 348
malathion, 316, 463
male, 542
male cell, 542
male cone, 542
male flower, 542
male gamete, 542
male gametophyte, 542
male haploidy, 542
male nuclei, 254, 542
male parthenogenesis, 174, 542
male reproductive system, 542
male sex cell, 542
male sterile, 542
male sterility, 542
maleated rosin, 316
maleation, 316
malengket, 401
maletto extract, 2
maletto tannin, 2
malformation, 221, 566
malfunction, 177, 431
malice soil, 2, 464
mall, 303
mallee, 2
malleolus, 207, 551
mallet, 331
mallet cutting, 68
mallet tapping, 68
mallisol, 234
malnutrition, 573
Malpighian tube, 316
malpractice, 498, 502
malt, 317
malt agar, 317
malt-agar culture test, 317
malt-agar method, 317
malt extract, 317
malt extract medium, 317
malt process, 317
maltase, 317
Malthusian parameter, 316
maltobiose, 317
maltose, 317
maltotriose, 317
malvidin, 122, 249
malvin, 249
mamelon, 204, 593
mammal, 33
mammy's bench, 572
Man and the Biosphere Programme, 396
man-caused fire occurrence index, 395
man-caused risk, 396
man-hour, 169
man-made, 394, 396
man-made climate, 396
man-made community, 396
man-made crossing, 395
man-made ecosystem, 395
man-made environment, 396
man-made fibre, 206, 396
man-made forest, 395
man-made forest edge, 395
man-made pastrre, 395

M

M

man-made reforestation, 394
managed forest, 253, 432
managed resource protected area, 641
management, 253
management ability, 179, 253
management archives of forest farm, 298
management-book, 431
management by compartment, 299
management by objectives, 336
management by result, 54
management categories of protected area, 14
management class, 253, 655
management control, 573
management function, 179
management fund, 179
management goal, 253
management information system, 179
management instructions, 225, 253
management intensity, 253
management map, 254, 407
management objective, 253, 432
management of park, 171
management of the growing stock, 544
management on a stand basis, 299
management-period, 253
management plan, 253, 431
management plan of nature reserve, 645
management plan revision, 253, 431
management planning unit, 226, 253
management policy, 253
management potential class, 253
management prescription, 253, 254
management prescriptions, 225, 253
management record, 221, 253, 431
management requlations, 225, 253
management selection cutting, 254
management subdivision, 253
management survey, 253
management type, 253
management unit, 253
management-volume inventory, 254
management without plan, 509
managerial organization, 254
managing director, 649
managing staff, 179
mandarin duck, 587
mandarin oil, 259
mandarin teal, 587
mandible, 415, 520
mandibulate type, 259
mandrel, 263, 538, 630
maneb, 78
maneuverability, 218, 272, 274, 305
manganese deficiency, 387
manganese fertilizer, 321
mango, 318
mango gum, 318
mangrove, 198
mangrove extract, 198
mangrove forest, 198
mangrove formation, 198
mangrove soil, 198, 467
mangrove swamp, 198
mangrove tannin, 198
manifold, 119, 120, 153, 179, 366
manifold effect, 118, 119
manifold paper, 76, 154
manihot caoutchouc, 334, 402
Manila rope, 72, 316
manipulator, 41, 218
manna, 156, 254
manoxylic, 417, 445
mantel-board, 21
mantid, 474
mantle community, 22
mantle rock, 27, 144
manual cable winch, 395, 443
manual clippers, 442
manual collecting, 442
manual debarking, 442
manual gain control, 395
manual knapsack sprayer, 18
manual skidding, 395
manually operated kiln, 394
manufacture damp, 170
manufactured abrasive, 395
manufacturers sales branches, 620
manufacture's sales branches office, 169, 620
manufacturing characteristics, 620
manufacturing lumber, 227
manure, 139, 144
manure juice, 259
manure loader, 638
manure receiver, 144
manure spreader, 401
manuring, 431
many-aged forest, 118, 567
many headed, 153
many storied forest, 116, 152
many storied high forest, 116
many storied high forest system, 116
map coordinate system, 98
map distance, 98, 487
map of harvest operations, 38
map of potential natural vegetation, 374
map reference, 98
map unit, 486, 487
maple cream, 371
maple honey, 371
maple sugar, 371
maple syrup, 371
maple-wax, 371
mapograph, 620
mapping, 55, 575
mapping level, 620
mapping of chromosome, 391
maquis, 316
Marathon rigging, 50
marble, 77
marble polecat, 201

marble rot, 77
marble table, 77
marble top, 77
marcescent, 103, 278
marcottage, 164, 276
marcotting layerage, 164, 276
Margalef's model of succession, 316
Margary's process, 308
margin, 46, 290, 561, 573
margin for profit and risk, 290, 573
margin of safety, 2
margin requirement, 22, 103, 551
margin strip, 22
margin tolerance, 170, 397
marginal analysis, 21
marginal benefit, 22
marginal coefficient of production, 22
marginal community, 22, 303
marginal cost, 21
marginal cost of acquisition, 21
marginal cost pricing, 21
marginal cutting, 21
marginal distribution, 22
marginal effect, 22, 593
marginal efficiency of capital, 640
marginal-factor cost, 22
marginal habitat, 22
margin(al) knot, 22
marginal land, 22, 248
marginal log, 21
marginal meristem, 22
marginal method, 21
marginal outlay curve, 22
marginal ovule, 22
marginal parenchyma, 22, 312
marginal pit, 22
marginal placentation, 22
marginal plant chlorosis, 561
marginal product, 21, 22
marginal productivity, 21
marginal productivity theory, 22
marginal profit, 21
marginal propensity to consume, 22
marginal propensity to import, 21
marginal propensity to save, 21
marginal purchase, 21
marginal rate of return, 22
marginal rate of substitution, 21
marginal ray, 22
marginal ray cell, 22
marginal ray tracheid, 22
marginal revenue product, 21, 22
marginal rot, 21
marginal seeding, 43
marginal seta, 593
marginal slit, 593
marginal tracheid, 22
marginal tree, 91, 127, 303
marginal type, 561
marginal utility, 22
marginal-value product, 21
marginal vein, 22
marginate, 260, 580
margined flush door, 529
margo, 402
mariculture, 188
marine, 189
marine algae, 189
marine (algae) vegetation, 189
marine bacteria, 189
marine borer, 188
marine buoy, 188
marine clay, 188
marine climate, 189
marine deposit, 189
marine ecology, 189
marine ecosystem, 189
marine erosion, 188
marine fungus, 188
marine humus, 188
marine ooze, 189
marine palynology, 189
marine pile, 188
marine plain, 188
marine preservative, 66
marine solonchak, 27, 188
marine succession, 189
marine system, 189
marine zonation, 189
maritime forest, 188, 189
maritime vegetation, 188
mark, 25, 195, 226, 248, 475, 526
mark-and-recapture method, 25
mark-and-release method, 25
mark caller, 547
mark on, 53, 227
mark-recapture method, 25
mark-sense character reader, 148
mark sensing, 24, 25
mark up, 53, 476
marked individual, 25
marker, 75, 204
marker gene, 25
market barrier, 438
market-based incentive, 438
market economy, 438
market house, 240, 438
market impact, 438
market information institution, 190
market penetration, 438
market place, 240, 438
market potential, 438
market price, 438
market profile, 438
market rate, 438
market share, 438
market specification, 415
market value, 438
market wage, 438
marketable, 274, 438
marketing efficiency, 438, 532
marketing weight, 20, 416, 438, 532
marking, 24, 75, 178
marking-arrow, 45
marking axe, 269
marking behaviour, 25
marking board, 25, 549, 550, 619
marking cog, 30
marking gauge, 204
marking gun, 354
marking hammer, 191
marking iron, 191, 474
marking keel, 24
marking knife, 120, 204, 274

M

M

marking log, 25
marking method, 25
mark(ing)-pin, 44
marking roll, 25
marking rule, 25
marking trees for felling, 38
Markov chain, 316
Markovian model, 316
marl, 342
marl soil, 342
marlin(e) spike, 65
marly soil, 342
marquetry, 375, 529
marresistance, 270, 336
marri kino, 320
marrum, 480
marsh, 432, 604
marsh garden, 605
marsh muck, 605
marsh plant, 605
marsh plough, 605
marsh podzol, 605
marshland, 605
marshy, 459, 605
marten, 103
marvel of Peru, 4
Marvin sunshine recorder, 316
mas, 542
masculine, 542, 557
mash, 23, 90, 237, 317, 400, 480
mask, 294, 323
masked civet, 186, 203, 471
masked polecat, 2
masked symptom, 571
masking agent, 555
masking effect, 555
Masonite, 16, 316
Masonite-fibre gun, 316
Masonite hardboard, 16, 316
Masonite-process, 16, 316
masonry dam, 433, 507
mass-accretion, 37
mass breeding, 213
mass collection, 77
mass effect, 76, 621
mass fire, 76
mass flow, 491, 621
mass infiltration, 77
mass method of breeding, 213, 224
mass morphology, 388
mass movement, 279, 389, 413
mass pedigree method, 213, 224
mass production, 54, 77
mass provisioning, 171
mass rearing, 77
mass selection, 213
mass spectrograph, 621
mass spectrography, 621
mass spectrometer, 621
mass spectrometric analysis, 621
mass spectroscopy, 621
mass spectrum, 621
massif, 413
massive colony, 279
massive planting, 72, 224, 389
massive seeding, 72
massive structure, 279
Masson's pine, 316
massulae, 203
mast, 223, 232, 377, 502
mast crane, 502
mast timber, 502
mast-tree, 223
mast year, 144, 628
master eye, 191, 634
master fire operations map, 133
master gene, 632
master link, 114, 633
master pin, 314
master plan, 283, 649
master river, 160, 633
master stream, 160, 633
masterlist, 220
mastic, 241, 343, 400
mastic gum, 241, 400
mastication, 259
mastix, 400
masty, 4, 464, 500
mat, 8, 72, 73, 509
mat assembly belt, 8
mat compressibility, 8
mat finish, 325, 531
mat formation, 8, 334
mat former, 8, 55
mat forming machine, 8
mat-forming plant, 285
mat-laying, 8
mat preloader, 8
mat reject station, 8
match block, 215
match board(ing), 366, 417
match lock gun, 215, 216
match plane, 41, 366
match plank, 215
match sawing, 453
match splint, 215
match stick, 215
match stock, 215
match wood bolt, 215
match-wood log, 215
match(ed) joint, 366
matched piece, 115
matched sampling, 353
matched veneer, 115
matcher, 463, 620
matchet, 77, 165, 497
matching, 115, 353, 366
matching species with the site, 439
matching stain, 467, 556
matching surface, 353
mate, 5, 108, 239, 353, 359
material balance, 513
material budget, 513
material chamber, 545, 635
material cycling in forest, 409
material damage, 37
material index, 37, 589
material yield table, 513
maternal character, 327
maternal form, 327
maternal half-sib, 328
maternal influence, 328
maternal inheritance, 328
maternal line selection, 328
maternal plant, 327, 328
maternal progeny, 328
maternal sex determination, 328
mathematical climate, 451, 478
mathematical ecology, 451
mathematical economics, 451
mathematical log rule, 451
mathematical model, 451
mathematical statistics, 451
matico oil, 520
mating, 239, 240
mating behaviour, 239
mating continuum, 239
mating season, 239
mating system, 240
mating type, 240, 244

matricyte, 328
matrilinear inheritance, 328
matrix, 53, 221, 260, 326, 328, 641
matrix chromosome, 390
matrix (*plur.* matrixes/matrices), 220
matrix substance, 221, 479
matroclinous hybrid, 357
matroclinous inheritance, 357
Matsumura pine scale, 464
mattae, 364
matted chair, 580
matter cycle, 513
mattock, 19, 194, 433, 616
mattock-slit method, 164
mattress, 8, 42, 47
mattress dam, 47, 308
maturation, 54, 206
maturation period, 55
maturation process, 55
mature, 54
mature association, 55
mature bamboo, 55
mature community, 55
mature crop, 55
mature egg, 55
mature forest, 55
mature habitat, 55
mature soil, 55
mature stand, 55
mature timber, 54, 55
mature tree, 55
mature wood, 54
mature xylary cell, 55
maturing temperature, 440
maturity, 49, 54, 55, 90, 127
maturity index, 55
maturity of largest labor productivity, 284
maturity of maximum forest net income, 69
maturity of the highest income, 69
maturity phase, 55, 639
maturity stage, 55
maul, 49
Maule's reaction, 326
Maule's test for lignin, 326
maxilla, 415, 520
maximum allowable error, 652
maximum annual sustained yield, 652
maximum capillarity, 652
maximum capillary capacity, 652
Maximum carrying capacity, 652
maximum cash flow rotation, 248, 653
maximum crushing strength, 271
maximum effective temperature, 653
maximum fatal temperature, 653
maximum field capacity, 653
maximum likelihood, 653
maximum likelihood decision, 653
maximum likelihood estimator, 653
maximum natality, 652
maximum permissible dose, 652
maximum recording thermometer, 653
maximum stational phase, 653
maximum stocking, 652
maximum survival temperature, 653
maximum sustainable yield, 652
maximum sustained yield, 652
maximum temperature, 653
maximum tree height, 653
maximum water holding capacity, 652
maximum working probit, 652, 653
Mayerl process, 317
maze, 321
Mcfarlane slackline yarding, 317
McLean boom loading, 317
McLeod vacuum gauge, 317
meadow, 41, 93, 336, 456
meadow bog, 42
meadow burozem, 42
meadow chernozem, 41
meadow chestnut soil, 42
meadow mouse, 98, 478
meadow podzolic soil, 42
meadow saline soil, 42
meadow soil, 42
meadow solonchak, 42
meadow steppe, 41
mealy-wing, 144
mealybug, 144, 457
mean accretion, 361
mean annual forest percent, 344
mean annual increment, 361
mean annual uptake, 361
mean annual yield, 361
mean area, 361
mean basal area tree, 113, 542
mean crowding, 362
mean crowding-mean density method (m*-m method, 362
mean daily germination, 361
mean DBH according to the 40% rule, 6
mean deviation, 266, 361
mean diameter, 362
mean dominant height, 362
mean effective pressure, 362
mean error, 361
mean generation time, 361
mean girth, 361
mean gross increment, 362, 649
mean growth tree, 361, 602, 625
mean height, 361
mean increment, 361
mean length of incubation time, 361
mean of annual net productivity, 361
mean of ratio, 361
mean predominant height, 362
mean product, 266, 361

M

mean sample, 362, 623
mean sample tree, 25, 361, 625
mean sea level, 361
mean square deviation, 266
mean square of the deviation, 266, 558
mean square value, 266
mean tree, 361, 625
mean tree method, 361
mean value, 362
mean velocity formula, 361
meander, 384
meander corner, 89, 605
meander line, 89, 605
measarch succession, 624
measurable class, 367
measurable size, 272
measure and a half, 9
measurement, 44, 296
measurement error, 44
measurement list, 104, 232, 320, 325, 560
measurement location, 44
measurement mark, 44, 48
measurement of felled timber, 126, 590
measurement of log, 590
measurement weir, 44, 296
measurer, 233
measuring, 44, 232, 296
measuring cylinder, 296
measuring for bucking, 296
measuring mark, 44
measuring off, 108, 205, 296
measuring pin, 45
measuring staff, 232
measuring stick, 57, 296
measuring-tape, 264
measuring unit, 44
measuring worm moth, 57
meat block, 112, 167
mecca balsam, 317
mechanical adhesion, 218
mechanical analysis, 218
mechanical barking, 218
mechanical conditioning, 218
mechanical damage, 218, 575
mechanical defect, 218
mechanical degradation, 218
mechanical digger, 217, 367
mechanical efficiency, 218, 288
mechanical girdling, 218
mechanical glue spreader, 487
mechanical handling equipment, 218
mechanical injury, 218
mechanical lever strain system, 218
mechanical logging, 218
mechanical nonequilibrium, 288
mechanical peeling, 218
mechanical piling, 218
mechanical property, 218, 288
mechanical protection of wood, 329
mechanical pulp, 218, 327, 333
mechanical pulping, 218
mechanical resistance, 218, 288
mechanical sampler, 218
mechanical sampling, 218
mechanical scarification, 218, 628
mechanical sediment, 218, 468
mechanical seeder, 31
mechanical shaker, 218, 608
mechanical spreader, 487
mechanical stacker, 113, 114
mechanical strain, 218
mechanical strength, 218
mechanical stress, 218
mechanical thinning, 218
mechanical tissue, 218
mechanical treatment, 218
mechanical vibration, 218
mechanical wear, 218
mechanism, 218
mechanism of absorption, 514
mechanism of penetration, 421
mechanism of timber drying, 329
medallion, 593
medial lethal dose, 11
median, 623–625
median DBH, 113, 625
median gutter, 622
median height, 113
median lethal dose, 621
median tolerance limit, 10
medical plant, 559
medicinal carbon, 559
medicinal plant, 559
medio-coronate, 622
mediolittoral zone, 52
Mediterranean climate, 99
Mediterranean oak cork, 99
mediterranean region, 99
medium check, 622
medium coverts, 622
medium density board, 624
medium density fiberboard, 624
medium density fibreboard, 624
medium density particle board, 624
medium-density particleboard, 624
medium density wood chipboard, 624
medium hewn square, 19
medium (*plur.* -ia), 221, 248, 319, 353, 623
medium lived seed, 624
medium maintenance, 625
medium oil, 625
medium seed year, 245
medium shake, 622
medium split, 622
medium square, 622
medium-term forecast, 624
medium term planning, 624
medium torn grain, 622
medium tree, 624
medulla (*plur.* -llae), 267, 468

medullary, 468
medullary blemish, 468
medullary bundle, 468
medullary cavity, 175, 468
medullary ray, 468
medullary spot, 468
meeting of the Parties (to the Kyoto Protocol), 100
mega-, 7
megabreccia, 260
megafauna, 260
megagaea, 17
megagram (Mg), 7, 115, 605
megaloplankton, 260
megameric chromosome, 599
megaphanerophyte, 76, 78
megaphyll, 260
megatemperature, 164
megatherm, 164
megatherm climate, 164
megatherm plant, 164
megathermal climate, 164, 392
megaton (Mt), 7, 605
MeGiffert skidder, 317
meiler method, 114
meiler process, 114
meiobenthos, 535
meiocyte, 541
meiophyllous, 260
meiosis, 55, 233
meiosis gametic, 353
meiostemonous, 542
meiotic disturbance, 233
meiotic division, 55, 233
meiotic drive, 233
meiotic mitosis, 55, 581
melaleuca oil, 6
melanism, 194
melanite, 194, 391
melanization, 194
melioration, 489
melittophilae, 146
melliphagous, 437
melliphagy, 57
mellow earth, 188, 446
mellow-soil plough, 465
melon tree, 128
melt, 398
melting point, 398
member, 173
membrane, 326
membrane potential, 326
membrane resistance, 326
membranous, 326
memorandum entry, 17
memorandum sale, 438
memorial park, 226
memorizer, 74
memory, 74, 339
memory microprocessor, 74
Mendelian character, 321
Mendelian genetics, 321
Mendelian hybrid, 321
Mendelian inheritance, 321
Mendelian population, 321
Mendelian ratio, 321
Mendel's law, 321
Mendel's population, 321
Mendel's principle, 321
meniscus, 498
meniscus dune, 539
menopetalous, 191
menotaxis, 195
mensuration, 44
Mercator projection, 327
mercerization, 234, 251, 461
mercerized cellulose, 461
merch top, 273, 415
merchandise balance, 414
merchantability specification, 414, 532
merchantable, 273
merchantable age, 414
merchantable bole, 415
merchantable diameter, 414
merchantable diameter limit, 414
merchantable forest disease, 415
merchantable form factor, 414
merchantable height, 414
merchantable length, 414
merchantable log, 414
merchantable timber, 414
merchantable timber increment, 273, 290
merchantable timber volume, 54, 273
merchantable volume, 274, 414
mercurial level, 460
mercurialine, 413
mercury fulminate, 284
mergence hole, 584
mericarp, 141, 545
mericlinal chimaera, 22, 36
meridional cell, 256
meristem, 142, 143
meristem arthrospore, 143
meristem culture, 143
meristem insects, 502
meristematic cell, 143
meristematic tissue, 143
meristematic zone, 143
meristyle, 142
merit goods, 52, 149, 577
merocyte, 120, 431
merogamy, 35, 353, 535, 641
merogony, 311
merokinesis, 259
meromictic lake, 10
meromixis, 10, 259
meroplankton, 226
merozygote, 9, 36
Merriam's life zone, 319
Merritt hypsometer, 319
merry-go-round system, 212
mersosin, 278
mesa, 360, 471
mesarch, 624
mesarch xylem, 624
mesenchyma, 232
mesenchyme, 232
mesh, 41, 346, 412, 499
mesh analysis, 412
mesh-belt conveyor, 499
mesh-belt drier, 499
mesh dryer, 499
mesic, 624
meso-, 622, 624
mesobenthos, 624
mesobiota, 624
mesocarp, 622
mesocephalic, 624
mesoclimatology, 622
mesocotyl, 624
mesogamy, 622
Mesogean Sea, 174
Mesolithic age, 624
mesometeorology, 622
mesomitosis, 194
mesomorphic leaf, 625

M

mesopelagic zone, 622
mesophanerophyte, 622
mesophile, 624
mesophil(e), 624
mesophilous tree species, 624
mesophilous tropophyte, 624
mesophilus, 429, 624
mesophorbium, 163
mesophyll, 561, 625
mesophyllous, 561, 625
mesophyte, 624
mesophyte vegetation, 624
mesophytic, 624
mesophytic forest, 624
mesophytism, 624
mesophytization, 624
mesoplankton, 625
mesopleuron, 625
mesopodium, 560
mesopsammic, 411
mesopsammon, 411
mesosaprobe, 624
mesosaprobic, 624
mesosere, 624
mesospecies, 624
mesosperm, 625
mesotherm, 624
mesotherm climate, 505
mesotherm plant, 505
mesothermophilous, 440, 516
mesotrophic, 11, 622
mesotrophic lake, 625
mesquite gum, 336
message, 381, 540, 549
message switch, 540
message switching, 540
messenger line, 212
meta∼, 23
meta-, 611, 623
meta-analysis, 213, 587, 610, 648
meta-community, 224
meta-data, 587
metabiosis, 9, 199
metabola, 24
metabolic activity, 79
metabolic block, 79
metabolic by-pass, 79
metabolic chain, 79
metabolic feed back, 78
metabolic pathway, 79
metabolic pattern, 79
metabolic pool, 79
metabolic process, 78
metabolic product, 78
metabolic pump, 78
metabolic rate, 79
metabolic route, 79
metabolic sequence, 79
metabolic sink, 78
metabolic source, 79
metabolic stage, 79
metabolic symptom, 79
metabolic tissue, 79
metabolic transport, 79
metabolic type, 79
metabolism, 79, 539
metabolite, 78
metacentric, 261
metacentric chromosome, 91
metacollenchyma, 199
metacommunity, 224
metagenesis, 438
metagenetic cycle, 438
metagon, 24, 200, 567
metakinesis, 624
metal brace, 249
metal caul, 249
metal channel, 249
metal clip, 249
metal detector, 249
metal-faced plywood, 249
metal fitting, 249
metal foil, 249
metal foil overlay, 249
metal-fouling, 484
metal furniture, 249
metal hinge, 249
metal-impregnated wood, 249
metal shadowing, 249
metal thermometer, 249
metal-treated wood, 249
metalimnion, 594
metallic wood borer, 221
metallized plywood, 249, 376
metallized wood, 249, 376
metalloenzyme, 249
metalloflavoprotein, 249
metamorphic landscape, 24
metamorphic rock, 24
metamorphism, 24
metamorphosed leaf, 24
metamorphosed rock, 24
metamorphosis, 24
metanotum, 200
metapedes, 200
metaphase chromosome, 624
metaphloem, 199
metapleuron, 200
metapopulation, 224
metasomatite, 239
metastable ion, 553
metastructure, 71
metathesis, 152, 569, 622
metathorax, 200
metatracheal diffused parenchyma, 235
metatracheal narrow parenchyma, 235
metatracheal parenchyma, 235
metatracheal wide parenchyma, 235
metaxenia, 88, 186, 199
metaxylem, 199
meteor, 78, 308, 595
meteorite, 595
meteorological element, 370
meteorological phenomena, 370
meteorological symbol, 370
meteorology, 370
meteorology satellite, 370
meter board, 564
meter diagram, 564
metering device, 226
metering pump, 106, 225
metering roll, 106
metes and bounds survey, 258
methane, 230, 604
method by area, 323
method by volume, 36
method of advance reproduction, 127
method of bandage, 12
method of classifying vegetation, 616
method of control, 193
method of deflection angle, 357
method of direct attack, 614
method of direct evaluation, 614
method of double crossing, 153, 455
method of experimental design, 439
method of forestation, 600

method of hastening germination, 124
method of hot and cold bath, 393
method of indirect evaluation, 235
method of indirect suppression, 235
method of intersection, 220, 239
method of management, 253
method of mean sample tree, 361
method of mean tree, 361
method of measurement, 232, 296
method of periods, 106
method of periods by area, 106
method of periods by area and volume combined, 106, 107
method of periods by volume, 106
method of regeneration, 169
method of regulating yield (by comparing actua with normal crops), 128, 451
method of relative diameter, 527
method of repeated pollination, 116
method of soil preparation, 610
method of successive thinning, 237
method of superposition, 104
method of the arithmetical mean sample tree, 467, 625
method of vegetative approachment, 511
method of volume table, 36
method of yield regulation, 442
methods of control, 221, 605
methods of yield regulation, 442
methoxyl group, 230
methylolation, 376
metis, 597
metoecious, 637
metric, 111
metric system, 321
metrical character, 111
metromorphic hybrid, 328
metrophorum, 69
mexofauna, 625
Meyer ratio, 319
Meyer's coefficient of precipitation, 319
mica, 595
micaceous clay, 595
micanoparasitism, 498
micaschist, 595
micella(*plur.* -llae), 143, 241, 500, 501
micellar cavity, 241
micellar hypothesis, 241
micellar network theory, 241
micellar strand, 241, 500
micellar string, 241
micellar structure, 241
micell(e), 143, 241, 500, 501
Michaelis constant, 321
Michurin genetics, 321
micro budding, 502
micro cell, 500–502
micro-creep, 501
micro crystal, 500
micro-dissection, 525
micro-dissection apparatus, 525
micro-dissection technique, 525
micro-environment, 500, 533
micro-extensometer, 525
micro-fungus, 501
micro lamella, 500
micro-manipulative technique, 524
micro-nutrient element, 501
micro-polarimeter, 44
micro-propagation, 501
micro-sealing process, 500
micro-succession, 535
micro topography, 500
micro-veneer, 500
microaerophile, 502
microbe, 501
microbenthos, 501
microbial control, 501
microbial decomposition, 501
microbial ecology, 501
microbial genetics, 501
microbial pesticide, 501
microbicidal, 411
microbicide, 411
microbiological analysis, 501
microbiological assay, 501
microbiological degradation, 501
microbiology, 501
microbiota, 501
microbivore, 437
microcapillary, 501
microchromosome, 534
microclimate, 501, 534
microclimatology, 534
microcolony, 501
microcommunities, 534
microcomputer, 501
microconidium, 533
microconsumer, 535
microcosm, 439
microcosmos, 439
microcrystalline cellulose, 501
microcrystalline structure, 501
microdendrometer, 524
microdiodange, 203
microdiode, 203
microecosystem, 501
microelement, 501
microevolution, 500, 534
microfauna, 535
microfibril, 501
microfibril axis, 501
microfibril distribution, 501
microfibril parameter, 501
microfibril strand, 501
microfibril(lar) angle, 501
microfibril(lar) network, 501
microfibril(lar) organization, 501
microfibril(lar) orientation, 501
microfibril(lar) orientation pattern, 501
microflora, 502
microfossil, 501
microgamete, 534, 542
microgenetics, 501, 525
micrografting, 500
micrography, 525

M

M

microhabitat, 534
microheterochromatic, 535
microincineration, 211, 525
microl, 533
microlitic structure, 501
micrology, 525
micromanipulation, 524
micromelittophilae, 533
micrometeorology, 501, 534
micrometer, 44, 371
micrometer caliper, 44, 371
micrometer eyepiece, 44
micromethod, 44
micromyiophilae, 535
micronekton, 501
micronutrient, 501
micronutrient deficiency, 501
microorganism, 501
microphagous, 437
microphanerophyte, 533
microphotographic apparatus, 525
microphotography, 469
microphotometer, 525
microphyll, 535
microphyllous, 535
microphyte, 501
microplankton, 501
microplex, 190
micropolariscope, 357
micropore, 501
microporous, 501
micropowder, 51
microprecipitation test, 500, 501
microprocessor, 500
microprogramming, 500
microprojection apparatus, 525
micropterous, 533, 534
micropyle, 311, 632
microradiography, 418
microrelief, 501, 533
microscope, 525
microscope camera, 525
microscope slide, 598
microscopic cell-wall structure, 517
microscopic check, 501
microscopic examination, 500, 525
microscopic feature, 500, 525
microscopic morphology, 525
microscopic observation, 525
microscopic structural fracture, 500
microscopic structure, 500, 525
microscopic symptom, 500
microscopic(al) analysis, 525
microscopy, 525
microsection, 652
microsere, 535
microsite, 501
microsome, 501
microspecies, 535
microspectrophotometry, 524
microspectroscope, 524
microspore, 533, 541
microstrain, 502
microstratification, 500
microstructure, 500, 525
microstylous, 112
microsubspecies, 502
microsymbiont, 500, 533
microsyringe, 501
microtechnique, 525
microtherm, 93
microthermal climate, 93
microthermal plant, 93
microtome, 377
microtomed section, 525
microtomy, 378
microtonometer, 501
microtubule, 501
microvisco(si)meter, 501
microwave, 500
microwave drying, 500
microwave sensor, 500
microwave sounder, 500
microwave spectrometer, 500
microwave spectroscopy, 500
micrurgical technique, 524
mid-board, 235, 622, 623
mid-diameter, 625
mid-girth, 625
mid-ocean ridge, 557
mid-slope read, 413, 622
mid-span, 278, 623
mid-term harvest planning, 622, 624
mid-timber diameter, 577
mid-timber girth, 577
mid-tolerant species, 622
mid-tolerant tree, 622
midcycle data updating, 366
middle breaker plough, 268
middle buster, 268
middle cut (log), 622
middle forest, 623
middle forest regeneration, 623
middle-forest system, 623
middle gouge, 201
middle ground, 623
middle lamella, 13, 622
middle layer, 13, 622
middle layer chip, 539
middle log, 622
middle rail, 623
middle roll, 623
middle section, 623
middle shore, 623
middle support, 612, 623
middlebuster plough, 625
middlings, 622, 623
midget tractor, 535
midgrass, 622
midgut, 622
midparent, 453
midrib, 622, 623
midslope, 624
midsummer-shoot, 629
midtemperate zone, 624
migrant, 200, 372
migrant bird, 200
migration, 372
migration constant, 516, 564
migration mechanism, 372
migration rate, 372
migratory bird, 200
migratory bird conservation act, 14
migratory birds, 200
migratory dune, 306
migratory songbird, 372
migrule, 66
migrule form, 129
migrule type, 129
milacre, 371
mild humus, 400, 505
mild steel, 93, 400
mildew, 5, 144

mile foot, 572
milk-ripe stage, 400
milkiness, 124, 400
milking, 224, 398, 476
milkvetch, 210
milky maturity, 400
mill, 52, 75, 169, 254, 261, 326, 516, 599
mill cull, 92, 520
mill deck, 286
mill-end, 327
mill foreman, 620
mill grade, 262
mill grapple, 8
mill hog, 75, 599
mill-log, 620
mill pond, 635
mill run, 62, 619
mill scale, 54, 619
mill-size chip, 170, 534
mill-study log rule, 619
mill tally, 262, 421
mill waste, 139
millboard, 444, 551
milled edge joint, 516
milled wood, 516
milled wood lignin, 327
millenium assessment, 371
millenium plan, 371
milling, 254, 277, 327, 516
milling cost, 619
milling head debarker, 516
milling machine, 332
milling zone, 75
millipede, 316, 371
millwork, 169, 326
millwork plant, 227
millwright, 218
millyard, 619
mimesis, 343
mimic, 343
mimic gene, 343, 483
mimicry, 343
mimosa (bark) extract, 254
mimosa extract, 254
mimosine, 189
mine, 275
mine-prop, 275, 280
mine tie, 280
mine timber, 280
mineral, 280
mineral ash, 280
mineral black, 434
mineral-constituent, 280
mineral cycling, 280
mineral cycling in forest, 408
mineral deficiency, 280
mineral element, 280
mineral manure, 280, 509
mineral matter, 280
mineral nutrition, 280
mineral pitch, 7, 96
mineral soil, 280
mineral stain, 280
mineral streak, 280
mineralization, 280
mineralized-cell process, 280, 518
mingle, 213, 214
mingled forest, 213
mini-computer, 535
mini electro printer, 535
mini-size chain saw, 535
miniature gardening, 501
miniature landscape, 355
miniature potted-tree, 450
miniaturization, 535
minifundium, 533, 534
minig timber, 275
minimal area, 654
minimal medium, 219
minimum air capacity, 654
minimum-damage fire-control theory, 654
minimum DBH (by inventory), 367, 654
minimum diameter, 615, 654
minimum effective temperature, 653
minimum factor, 654
minimum fatal temperature, 653
minimum field capacity, 654
minimum girth, 502, 654
minimum growth temperature, 653
minimum lethal dose, 654
minimum light-intensity, 653, 654
minimum light requirement, 653
minimum quadrat area, 654
minimum quadrat number, 654
minimum quadrate area, 654
minimum threshold of farmland area, 168
minimum turning radius, 654
minimum viable area, 653
mining bee, 96
mining timber, 280
minirhizotron tube, 500
minisome, 535
Ministry of Forestry, 303
minium, 198, 372, 463
minivet, 413
minor base, 515
minor branch road, 71
minor community, 534
minor component, 417
minor control point, 92
minor element, 501
minor feature, 533
minor forest produce, 300
minor forest rail road, 603
minor gene, 71, 501
minor gene resistance, 501
minor nutrient element, 501
minor overhaul, 535
minor produce, 154
minor product, 154
minor ride, 154, 383
minor (river) bed, 533
minor road, 595
minor species, 71, 535
minor transportation, 222, 485
mint oil, 13
minus stand, 151, 297
minus tree, 151, 297
minute, 17, 140, 391, 461
minute egg parasite, 58
minute feature, 519
minute of angle, 140
minute structure, 519
Miocene, 625
miombo, 158
miostemonous, 542
mirage method, 90
mire vegetation, 605
mirene, 321
mirex, 324
mirror finish, 258
mirror polish, 258
mirror relascope, 129

M

mirror stereoscope, 129
misappropriation, 90, 283
miscellaneous asset, 597
miscellaneous expense, 596, 597
miscellaneous full-cell treatment, 119
miscellaneous-porous wood, 596
miscellaneous tree, 597
Mischerlich's law of plant growth, 321
Mischerlich's pot experiment, 321
miscut lumber, 352, 651
miscut(ting), 513
misfit river, 32, 401
mismanufacture, 227
mismatching, 335, 353
misogamy, 430
miss cut lumber, 352
miss matching, 335, 353
miss-spread mat, 34, 364
misshapen log, 221
misshapen tree, 221
missing plot, 387
mist, 354, 513
mist bed, 354
mist blower, 321
mist forest, 595
mist lubrication, 578
mist nozzle, 355
mist propagating installation, 354
mist propagation, 354, 460
mist sprayer, 354
mist system, 355
mistletoe, 201
mistus, 214, 596, 597
mite, 317
miticide, 411
mitigation, 233
mitigation of carbon emission, 233
mitigative capacity, 233
mitochondrion, 526
mitosis, 581
mitosis aberrant, 221
mitotic crossing over, 581
mitotic division, 581
mitotic recombination, 581
mitra, 280
mitragynine, 319
mitre, 537
mitre block, 537
mitre box, 149, 537
mitre clamp, 537
mitre cutting, 537
mitre gauge, 537
mitre joint, 537
mitre saw, 537, 650
mitre square, 537
mitre-with-end-lap joint, 111
mitre-with ribbet joint, 375
mitriform, 626
mitriform calyptra, 626
mix by single individuals, 87
mixed adhesive, 213
mixed border, 213
mixed boreal forest, 17, 190
mixed broad-leaf and coniferous forest, 606
mixed bud, 213
mixed by clumps, 72
mixed by individual tree, 87
mixed by stands, 299
mixed by strips, 80
mixed crop, 213
mixed cropping, 214
mixed economy, 213
mixed fertilizer, 153
mixed flock, 214
mixed-flow fan, 213
mixed flow turbine, 213
mixed forest, 213
mixed forest in groups, 279
mix(ed) glue, 213
mixed grafting, 213
mixed grain, 10, 213
mixed infection, 213
mixed inflorescence, 213
mixed inheritance, 214
mixed layer, 213
mixed (ligand) complex, 213, 597
mixed paint, 479
mixed pixel, 213
mixed plant community, 213
mixed planting, 214
mixed plywood, 213
mixed pollination, 213
mixed prairie, 213
mixed rot, 213
mixed seeding, 213
mixed seeding and sprout forest, 521
mixed-species population, 214
mixed stand, 213
mixed uneven-aged forest, 567
mixed wood, 213
mixed woodland, 213, 214
mixen, 259
mixer, 11, 243
mixing preservative, 480
mixing ratio, 213
mixing table, 353
mixing tank, 213, 243
mixing truck, 213
mixochromosome, 348
mixture, 214, 450
mixture by group, 279
mixture by isolated plants, 404
mixture by line, 80
mixture by single trees, 87
mixture distribution, 213
mixture in group, 279
mixture plantation, 394
mixture screw, 213
mixtus, 214, 596, 597
mlercepting safety stop, 247
mobile aerial tower, 564
mobile back spar, 564
mobile cable crane, 564
mobile dune, 306
mobile loader, 371
mobile phase, 306
mobile reserve, 564
mobile (saw) mill, 306
mobile skidder, 564
mobile slasher, 564
mobile yarder, 644
mobilideserta, 307
moccasin, 349
mock-up furniture, 229
mode, 131, 629
model, 325, 326
model area, 326
model compound, 326
model experiment, 325
model forest, 325
model garden, 326
model hierarchy, 326
model substance, 325
model system, 325
model tree, 25, 558
moder, 317, 467
moderate, 622
moderate burn, 622

moderate drying, 209, 439
moderate opening of canopy, 439
moderate thinning, 439, 622
modern architecture, 525
modern furniture, 525
modern garden, 525
modes of seed dispersal, 628
modification, 155, 438, 583
modification for migration, 372
modificational variation, 438, 583
modificator gene, 583
modified cellulose, 24, 155
modified choke, 155
modified mass method, 155
modified pedigree system, 155
modified rind graft, 155
modified rosin, 155
modified wood, 155
modified yield-volume ratio, 543
modifying factor, 23, 543
modifying gene, 23, 543
modular organism, 173
modulator, 480
module, 173
modulus, 325, 326
modulus of elasticity, 472
modulus of elasticity in bending, 497
modulus of rupture, 113, 339
modulus of section, 113
modulus of shearing elasticity, 234
Mohr's circle, 327
moiety, 11
moire calender, 30, 335
moist air, 432
moist-chilling, 433
moist forest, 52
moist gluc linc, 432, 504
moist heat, 432
moist heat sterilization, 433
moist land, 52
moist meadow, 432
moist prechilling, 433
moist soil, 433
moist wood, 432
moist-wood termites, 432
moisture, 432
moisture catena, 432
moisture content, 189
moisture (content) determination, 189
moisture (content) meter, 432, 456
moisture (content) section, 189
moisture detector, 456
moisture distribution section, 189
moisture driving force, 456
moisture equivalent, 57
moisture excluding efficiency, 134, 432
moisture gradient, 432, 456
moisture holding capacity, 57, 456
moisture-holding material, 15
moisture-holding medium, 15
moisture-laden air, 432
moisture meter, 189
moisture movement, 456
moisture of the capillary bond disruption, 318
moisture potential, 432
moisture-proof, 131, 133
moisture proof adhesive, 131
moisture proofing, 133
moisture-proof(ing) coating, 131
moisture range, 432
moisture regain, 211, 212, 514
moisture regime, 456
moisture register, 432
moisture repellent, 651
moisture-resistant, 131
moisture-resisting paint, 337
moisture-retaining, 15
moisture retardant, 651
moisture retardant coating, 651
moisture-retarding treatment, 651
moisture retention curve, 57
moisture stress, 387, 456
moisture suction, 456
moisture tension, 456
moisture transmitting, 486
moisture travel, 456
molal concentration, 629
molamma, 291
molar concentration, 476
molasses, 473
mold-cover, 151
moldability, 55, 274
molder, 353
molding-up, 353
mole burrow, 60
mole crickets, 309
mole fraction, 326
molecular architecture, 143
molecular biology, 143
molecular diffusion, 143
molecular disease, 143
molecular ecology, 143
molecular evolution, 143
molecular genetics, 143
molecular hybridization, 143
molecular sieve, 143
molecular sorption, 143
molle, 160
mollisol, 401
molluscan borer, 401
molluscicide, 411
molly hogan, 67, 83, 208
molting fluid, 494
molting hormone, 494
moment of flexure, 497
moment of inertia, 179, 637
moment of resistance, 651
momentum, 109
monadelphous, 85
monader, 85
monadical plot sampling, 86
monadnock, 40
monandreous, 260
monarsenous, 563
monascorubrin, 198
monascus, 198
monastery garden, 464
monazite, 111
monesin, 249
monetary value, 217, 248
money yield, 37, 288

M

M

money yield of financial rotation, 311
money yield table, 217
mongrel, 597
mongrelism, 359
mongrelize, 359
moniliform, 345
monitor, 53, 232
monitoring pollution, 508
monitoring program, 232
monkey, 76, 199, 214, 367
monkey block, 88, 292, 637
monkey pod, 584
monkey tail, 264
monkey wrench, 214
mono cable, 85
mono-hybrid, 86, 563
mono-phase sampling, 85
monobasic acid, 563
monocable aerial tramway, 85
monocambial, 86
monocarpellary, 86, 563
monocarpellary ovary, 86
monocarpellary pistil, 86
monocarpic, 245
monocarpic plant, 562
monocarpous, 245
monocarpous ovary, 86
monocarpous pistil, 86
monocarpy, 562
monochasium, 84
monochasy, 84
monochlamydeous, 82
monochlamydeous flower, 82
monochogamous, 70
monochogamous flower, 70
monochogamy, 70
monochrom, 85
monochromator, 85
monochrome image, 85
monoclimax, 86
monoclimax theory, 83
monoclinic, 86, 296, 498
monoclinous, 296, 498
monoclinous flower, 296
monocoque, 84, 572, 575
monocot, 87
monocotyledon, 87
monocotyledonous, 87
monocotyledonous tree, 87
monocotyledonous wood, 87
monocotylous, 87
monocrotophos, 259
monocular microscope, 84
monoculture, 69, 86
monocultute, 86
monocyclic, 83, 84, 86, 563
monocyclic felling, 563
monocyclic monoterpene, 83
monodelphous stamen, 85
monodispersity, 83
monodominant community, 86
monoecious, 70
monoecism, 70
monoecy, 70
monoembryonic, 84
monoembryony, 84
monofacial, 84
monofactorial, 86
monofil(ament), 84, 85
monogamous, 84, 562
monogamy, 84
monogenic, 83, 86
monogenic character, 83
monogenic resistance, 83
monogenomic species, 84
monogeny, 84
monogynaecial, 83
monogynian, 83
monogynous, 83, 260
monogyny, 83
monohaploid, 86
monohybrid cross, 86, 563
monokaryon, 83
monolayer, 82, 83, 85
monolithic finishing, 610
monolithic storage technology, 84, 219
monolocular, 85
monomer, 83, 85
monomerical inheritance, 83
monomerous flower, 83
monometrosis, 83
mononuclear, 83
monoparasitism, 83
monophage, 85, 87
monophagous, 85
monophagous animal, 85
monophagy, 85
monophenol oxidase, 83
monoploid, 82, 562
monoploidy, 82
monopodial, 86, 87
monopodial branching, 649
monopodial growth, 87, 470
monopodial inflorescence, 87
monopodium, 86
monopoly law, 636
monoprotic acid, 563
monopsony, 317
monopterous, 83
monorail, 83, 84
monorail cableway, 85
monorail transfer, 83
monosaccharide, 85
monosaccharose, 85
monose, 85
monoseeding, 84
monosepalous, 192
monosome, 83, 84
monosomic, 84, 85
monosomic diploidy, 85, 121
monospermic egg, 84
monospermous, 82, 86
monospermy, 84
monostrosity, 221
monoterpene, 85
monotocous animal, 85
monotony, 85
monotrophic, 85
monotropic, 82, 83
monotropitin, 108
monotropy, 82
monotypic genus, 84, 86
monovalent chromosome, 83
monoxenous, 87
monpohagous, 85
monsoon, 226
monsoon forest, 226
monsoon rain forest, 226
monster, 221
montane, 413
montane belt, 413
montane bog, 413
montane coniferous forest, 413
montane flora, 413
montane forest, 413
montane indicater species, 414
montane rain forest, 413
montane zone, 413

Monte Carlo stock, 261, 321
monthly cumulative temperature, 594
monthly distribution of visitation, 579
montmorillonite, 321
Montreal Protocol, 321
monument, 248
monumental square, 226
moon gate, 594
moonshine crotch, 594
moor, 342, 433
moor forest, 605
moor plough, 605
moor soil, 605
mooring buoy, 227
mooring post, 227
moorland, 164
mop-stick hand-rail, 360
mop up stray, 159
mopping up, 381
mor, 72
moraine, 28
moraine garden, 433
morbidity, 123
morel, 556
morgan, 326
morindone, 405
morpankhi oil, 552
morphactin, 541
morphcline, 541
morphogenesis, 541
morphological dormancy, 541
morphological mutant, 541
morphological plasticity, 541
morphological species, 541
morphology, 541
morphology of fibre, 523
morphoplankton, 55
morphosis, 541
morphospecies, 541
morphotype, 541
mortality, 278, 462
mortality curve, 462
mortality model, 278
mortality rate, 462
mortar stone check dam, 238
mortarless stone check dam, 157
mortgage, 94, 551
mortgage debt, 94
mortice, 469
mortice and tenon joint, 529
mortise, 469
mortise and tenon joint, 529
mortise chisel, 469
mortise gauge, 558
mortise joint, 469
mortise lock, 46
mortiser, 599
mortising, 46, 599
mortising machine, 599
mortmass, 72
mosaic, 204, 529
mosaic bedding plant, 529
mosaic discase, 204
mosaic dominance, 529
mosaic hybrid, 529
mosaic plywood, 375
mosaic soil, 529
mosaic surface, 358
mosaic vegetation, 529
mosaic virus, 204
mosaiculture, 529
mose, 604
moss, 525
moss layer, 471
moss-moor, 525
moss-quotient, 471
moss stage, 471
moss-tundra, 525
mossy forest, 471, 525
moss(y) forest, 471
mossy forest zone, 471
most closed crossing, 653
moth, 120
mother bamboo, 328
mother block, 327
mother cell, 328
mother culm, 328
mother gallery, 328
mother liquor, 328
mother nucleus, 328
mother nuclide, 328
mother tree, 328, 379
mothproof chest, 134
motif, 486, 639
motoplanter, 109
motor auger, 108
motor cultivator, 109
motor cutter, 108
motor-driven drill, 101
motor grader, 108
motor-manual cleaning, 442
motor mule, 379
motor scraper, 108
motor truck hauling, 217, 371
motor winch unit, 101, 578
motorized dust gun, 109
motorized knapsack sprayer, 18
motorized recreation, 218
motorized rossing tool, 218
motorway timber, 371
mottle, 7, 596
mottled figure, 7
mottled grain, 7
mottled rot, 7
mould, 285, 320, 326, 375, 384, 526, 550, 557
mould fungus, 320
mould-making, 236, 601
mould release agent, 494
mould resistant, 270
mould-stain, 320
mouldboard plough, 204, 205
moulded hardboard, 550
mould(ed) timber, 326, 327
mouldered wood, 544
mouldering, 150
moulding, 55, 202, 326, 526
moulding and shaping machine, 55
moulding cutter, 526
moulding machine, 526, 601
moulding plane, 526
moulding sander, 55
moullded plywood, 55
moult, 209, 492
moulting, 492
mound, 94, 202, 489
mound layering, 114
mound planting, 382
mound storage, 114
mounding, 636, 654
mountain, 413
mountain and water garden, 414
mountain building, 600
mountain chernozem, 413
mountain chestnut soil, 413
mountain climate, 413
mountain forest, 413
mountain hemlock, 76

M

mountain meadow soil, 413
mountain pasture, 413
mountain peat, 413
mountain podzolic soil, 413
mountain style, 590
mountain timber yard, 413, 414
mountain wind, 413
mountainous sawmilling, 302, 413
mountains wildlife refuge, 413
mounting test, 514
moustache-knot, 569, 641
mouth cavity, 277
movable bridge, 214
movable dam, 214
movable prepress, 564
movable quadrat, 214
movement, 504, 564, 604
moving-apron conveyor, 564
moving average, 205
moving-band conveyor, 564
moving block, 637
moving bookstand, 564
moving butt-joint mechanism, 564
moving cable system, 109, 549
moving cable yarding, 109, 549
moving endless line, 109
moving fire, 564, 595
moving former, 564
moving loading, 110, 184
moving main line, 109
moving-roller conveyor, 564
moving-slat conveyor, 295, 564
mower, 165
mowing, 233
mowing climax, 566
mu oil, 484
mucilage, 240, 241, 345
mucilage canal, 345
mucilage cell, 345
mucilage duct, 345
mucilaginous fibre, 345
muck, 151, 508
muco-cellulose, 345
mucopolysaccharide, 344
mucosa, 345
mud, 508, 583
mud-boat, 93, 512, 598
mud flow, 342
mud hole, 342, 350, 527
mud-road, 491
mud-rock flow, 342
mud-stone flow, 342
muddy shores, 342
mudflat, 188
mudflow, 342
mudsill, 95, 219
muffler, 532
mulberry, 405
mulberry scale, 405
mulberry silk worm, 229, 405
mulch, 155
mulcher, 27
mulching, 155
mule cart, 65, 380, 463
mule ear, 231, 242
mule-ear knot, 231, 242
mull, 150, 518
mull oil, 554
Müllerian mimicry, 321
mullion, 280, 320
multi-blade fan, 119
multi channel magnetic tape, 116
multi channel system, 119
multi-daylight press, 116
multi-head tenoner, 119, 120
multi-head wood borer, 120
multi-industry company, 117
multi-knife disk chipper, 116
multi-layer, 116
multi-layer adsorption, 116
multi-layer particle board, 116
multi-layered board, 116
multi-lens camera, 117
multi loaded cableway, 117
multi-male troop, 119
multi-open(ing) press, 116
multi-opening press-line, 116
multi-phase sampling, 119
multi-platen press, 116
multi-ply (plywood), 116
multi-salt preservative, 119
multi-saw edger, 117
multi-saw trimmer, 117
multi-seasonal, 117
multi-species mature ecosystem, 119
multi-spindle wood borer, 120
multi-stem delimber, 120
multi-storey reeling rack, 116
multi-storeyed high forest, 152
multi-storied forest, 152
multi-storied stand, 152
multi-strand chain drive, 118
multi-stranded cable, 117
multi-temporal, 118
multi-use desk, 119
multi-use management, 119
multiband camera, 116, 117
multiband colour enhancement, 116
multiband photography, 116
multiband system, 116
multicellular, 119
multicolour, 118
multicoloured flower, 118
multicostate, 120
multidimensional hypervolume (niche), 119
multidimensional niche, 119
multifid, 118
multiflorous, 117
multiforcasting, 119
multifunction system, 117
multigenic control, 117
multigenic resistance, 117
multihybrid, 117
multijugate, 116
multilacunar, 119
multilayer photography, 116
multilineal variety, 119, 214
multilinear heredity, 119
multilocular ovary, 118

multimer, 118
multimeter, 118, 499
multimolecular, 117
multinet growth hypothesis, 153
multinet theory, 153
multinodal, 117, 118
multinomial distrihution, 119
multiparasitism, 117
multiparasitization, 117
multiparous cyme, 116
multiphase sampling, 107, 119
multipie-cycle disease, 154
multipinnate, 117
multiple allele, 152
multiple allelic series, 152
multiple allelism, 154
multiple allelomorph, 152
multiple (annual) ring, 153
multiple axle vehicle, 120
multiple backfire, 119
multiple band dryer, 116
multiple bandsaw, 118, 153
multiple borer, 120
multiple bud, 154
multiple chain-saw mortiser, 120
multiple chromosome, 153
multiple coating, 116
multiple continuous circulating cableway, 118
multiple corolla, 60
multiple correlation, 154
multiple crossing, 117, 153
multiple cutterhead peeler, 516
multiple daylight press, 116
multiple-drop check dam, 117, 175
multiple-entry card key, 117
multiple entry system, 153
multiple entry volume table, 119
multiple factor inheritance, 119
multiple fruit, 152, 263
multiple gene, 117
multiple (growth) ring, 153
multiple-head sander, 119
multiple hollow-chisel mortising machine, 120
multiple hybrid, 153
multiple inheritance, 119
multiple leaders, 152
multiple parasitism, 120
multiple-pass conveyor dryer, 116
multiple perforation, 152
multiple planting, 72
multiple-pocket peeler, 116
multiple pore, 152
multiple purpose river development, 193
multiple-purpose (use) forestry, 119
multiple radial pore, 255
multiple regression, 120
multiple resonance, 152
multiple-row sled tree lifter, 117
multiple sampling inspection plan, 117
multiple saw, 117
multiple scale templet, 152
multiple seedling sprout, 72
multiple sex chromosome, 154
multiple single plant selection, 116
multiple spindle carving machine, 120
multiple-track kiln, 117
Multiple Use-Sustained Yield Act, 119
multiple-yield table, 153
multiplex, 116
multiplex projector, 116
multiplex system, 116
multiplication, 128, 602
multiplicative effect, 18
multiplicity, 120
multiplied dominance ratio, 153
multiplier, 18, 56, 281, 602
multiplier phototube, 180
multiploid, 116
multiploidy, 116
multiprocessing machine, 119
multipurpose chair, 119
multipurpose crawler tractor, 119
multipurpose timber-harvesting machine, 119
multipurpose tree chipper, 499
multiramose, 117
multiscope, 116
multiseptat, 117
multiseriate heterogeneous ray, 118
multiseriate homogeneous ray, 118
multiseriate pitting, 118
multiseriate ray, 118
multisheave heel tackle, 153
multishift operation, 116
multisomic analysis, 118
multispan, 117
multispan skyline, 117
multispectra, 117
multispectral photography, 117
multispectral scanner, 117
multistage compressor, 117
multistage inventory, 117
multistage sampling, 117
multistaminate, 119
multistratal, 116
multivalent chromosome, 117
multivariable histogram, 116
multivariable population, 116
multivariate analysis, 116, 120
multivoltinism, 116, 117
multiway-layout method, 120
multlvoltine insect, 116, 117, 563
mumal fund, 172
municipal forest, 438
munjistin, 571

N

Munsell colour system, 318
munting, 321
Murashige and Skoog high salt medium MS, 164
Murashige and Skoog medium MS, 353
muriform, 636
murolineum, 329
muscardine, 6
muscular stomach, 218
mush log, 271, 495
mushroom, 327, 404, 549
mushroom bullet, 268
mushroom floor, 207, 549
mushroom hot house, 327
mushroom poison, 549
mushroom spawn, 267, 327
musk, 419
musk yarrow oil, 419
muskatel, 399
muskeg, 342, 467, 515
Muskey, 377
mustard oil, 248
mutagen, 583
mutagenic action, 583
mutant, 24, 486
mutant gene, 486
mutarotation, 24, 546
mutase, 24
mutation, 486
mutation load, 486
mutation pressure, 486
mutation rate, 486
mutator gene, 486, 583
mutibo, 275, 471
mutiparous branching, 116
muton, 486
mutual company, 202
mutual interference constant, 528
mutual pollination, 528
mutualism, 172, 201
mutualistic symbiosis, 201
muzzle, 355, 375
muzzle bandage, 375
muzzle blast, 375
muzzle energy, 63
muzzle loader, 373
muzzle rest, 375
muzzle velocity, 64, 375
mycelial, 267
mycelial fan, 267
mycelial felt, 267
mycelial mat, 267
mycelial pad, 267
mycelial phage, 267
mycelial strand, 267
mycelium (*plur.* -lia), 267
mycelium mat, 267
mycetocole, 365
mycetophagous, 437
mycetophagy, 437
mycocecidium, 267
mycocide, 411
mycoid, 607
mycologist, 607
mycology, 607
mycoparasite, 607
mycophage, 441
mycoplasma, 267
mycoplasma like organism, 285
mycorhiza, 267
mycorrhiza, 267
mycorrhizae, 267
mycorrhizal association, 267
mycorrhizal fungi, 267
mycorrhizal fungus, 267
mycorrhizal plant, 267
mycorrhizal tree, 267
mycorrhyzal association, 267
mycosis, 607
mycotic, 607
mycotoxins, 607
mycotrophy, 267
mycovirus, 607
Myriapoda, 120
myrica extract, 557
myrmecochore, 566
myrmecocole, 365
myrmecodomatia, 618
myrmecophilous plant, 440
myrmecophobic, 524
myrmecophobous, 524
myrtle oil, 528
myxomycete, 345
n-dimensional hypervolume, 51
naa oil, 465
nacrite, 607
nadir point, 477
nail, 604, 619, 652
nail-holding ability, 507
nail-holding capacity, 507
nail-holding power, 507
nail test, 507
nailed joint, 105
nail(ed) joint, 105
nailing machine, 106
naked, 313, 508
naked bud, 313
naked bulb, 510
naked flower, 508
naked-rooted plant, 313
naked root(ed) planting, 313
namatium, 515
nanism, 2
nanmu oil, 338
nannofossil, 51
nannoplankton, 501
nanophanerophyte, 1
nanophyll, 501
nanous, 1, 2
nap, 112, 397, 519
naphthalene acetic acid, 338
nappe, 496, 569
nappy, 261
narrow band filter, 603
narrow band-saw, 526, 603
narrow board, 519
narrow crowned, 603
narrow ecological amplitude, 603
narrow face, 36
narrow gauge railroad, 603
narrow ringed, 603
narrow-row drilling, 603
narrow-row planting, 603
narrow sclerophyll plant, 603
narrow strip-selection method, 603
narrow-zoned, 520
nascent, 63, 539
nascent tissue, 143
nastic movement, 159
nasute, 49
natality, 62, 430
natatorial, 580
natatorial bird, 458
national afforestation, 185
national bank, 185
national battlefield park, 185
national bison range, 185
national capital park, 185

national compulsory tree-planting activities, 385
national fire danger rating system, 185
national flower, 185
national forest, 186
national forest advisory council, 185
national forest area, 186
national forest board of review, 185
national forest inventory, 185, 563
national forest land, 186
national forest recreation resource review, 185
national forest recreation survey, 185
national forest reservation commission, 185
National Forestation Commission, 185
national forestation drive, 385
national historic park, 185
national labour relations act, 385
national lakeshores, 185
national lumber manufactures association, 185
national memorial park, 185
national military park, 185
national nature reserve, 185
national park, 185
national recreation area, 185
national resources committee, 185
national resources planning board, 185
national scenic river ways, 185
national scenic trails, 185
national scientitle reserves, 185
national tree, 186
national tree-planting festival, 385
national waterway, 186
national wildlife refuge system, 185
native, 19
native cellulose, 478
native cellulose fibre, 478
native fiber, 478
native fibre, 478
native lignin, 478
native rock, 328
native rubber, 478
native species, 19, 527
native tree, 527
native tree species, 527
native variety, 96
native wood, 19, 185
natural abundance, 477, 645
natural afforestation, 477
natural ageing, 646
natural background, 477
natural balance, 646
natural bastard, 646
natural capital, 646
natural-circulation kiln, 646
natural colour, 19, 478
natural community, 646
natural conservation law, 645
natural control, 645, 646
natural defect, 478
natural differentiation, 645
natural division, 646
natural division method, 646
natural-draft kiln, 646
natural-draught kiln, 646
natural drying, 477, 645
natural durability, 478
natural durability test, 478
natural dye, 478
natural elimination, 646
natural enemy, 477
natural erosion, 646
natural flavour, 478
natural food, 315
natural forest, 478
natural formation of wood, 477
natural garden, 646
natural glue, 110, 478
natural graft, 646
natural grassland, 477
natural growth hormone, 429
natural high polymer, 477
natural history, 31
natural hybrid, 478
natural immunity, 478
natural ingredient, 477, 478
natural interest, 646
natural landscape, 478, 645
natural law, 645
natural layering, 478
natural monument, 478, 646
natural mutation, 646
natural park, 478
natural parthenocarpy, 645
natural pasture, 478
natural pigment, 478
natural pollination, 645
natural pollution, 478, 646
natural population, 646
natural preservation, 477
natural protected area, 645
natural pruning, 646
natural pulpstone, 478
natural range, 645
natural regenerated forest, 646
natural regeneration, 477
natural regeneration by clearcutting method, 244
natural regeneration by group seed tree method, 389
natural regeneration by marginal seeding, 43
natural regeneration by scattered seed tree method, 404
natural regeneration by seed-shedding, 478
natural regeneration by seeds, 478
natural regeneration by selection method, 601
natural regeneration by self-sown seeds, 478
natural regeneration by shelterwood method, 404

N

N

natural regeneration by shoots, 321
natural regeneration by shoots and suckers, 1
natural regeneration by sprouts, 321
natural regeneration under shelterwood, 415
natural regeneratiov by seed tree method, 306
natural regulation, 646
natural reproduction, 477
natural reproduction by rhizome, 98, 632
natural resin, 478
natural resin varnish, 478
natural resistance, 478
natural resource, 646
natural resource information system, 646
Natural Resource Management, 646
natural resources, 646
natural rotation, 646
natural sanctuary, 645
natural sandstone, 478
natural scenery resource, 477
natural seasoned timber, 368, 645
natural seasoned wood, 368, 645
natural seasoning, 368, 645
natural seeding, 478
natural seedling, 559
natural selection, 646
natural stand, 478
natural strand forest, 478
natural style, 646
natural succession, 646
natural system, 645
natural tannin, 477, 478, 616
natural thinning, 646
natural tourism resources, 646
natural variable, 645
natural vegetation, 646
natural ventilating dry kiln, 646
natural weathering, 645
natural wonders, 646
natural wood filler, 328
naturalist, 31
naturalization, 183, 549
naturalized plant, 183
naturalized species, 183, 549
naturally managed forest, 250, 251
naturally sown young seedling, 559, 646
naturalness, 645
nature and nurture, 522, 565
nature conservation, 645
nature conservation spot, 645
nature park, 646
nature preserve, 645
nature protection, 645
nature reserve, 645
nature reserve complex, 645
nature reserve for ancient organisms remains, 174
nature reserve for desert ecosystem, 209
nature reserve for forest ecosystem, 408
nature reserve for geological formations, 99
nature reserve for inland wetland and water area ecosystem, 340
nature reserve for natural ecosystem, 646
nature reserve for natural monument, 646
nature reserve for ocean and seacoast ecosystem, 189
nature reserve for steppe and meadow ecosystem, 42
nature reserve for wild animals, 559
nature reserve for wild plants, 559
nature reserve for wildlife, 559
nature reserve master plan, 645
nature reserve system, 645
nature sanctuary, 645, 646
naucum (*plur.* nauca), 193, 626
Naudet's barometer, 461
nautical mile, 188, 288
naval stores, 465
naval tank, 66
nave, 312, 631
navel, 366
navigation, 89
Navy abrader, 341
Navy-type wearometer, 341
neap, 165, 281
neap tide, 533
neap(ed) timber, 165, 281
near-infrared radiation, 250
near infrared (ray), 250
near-mature forest, 250
near to nature forest, 250, 251
near-zero slump, 250
Nearctic province, 539
Nearctic region, 539
nearest neighbour method, 653
nebula, 540
nebulization, 513
neck, 167, 254
neck-down, 254
neck rot, 254
necktie, 281
necrobiosis, 237
necrocoenosis, 431
necrogenous reaction, 207
necrology, 431
necrophag(o)us, 437
necrophagy, 437
necroplankton, 462
necrosis, 207, 278
necrotic, 207
necrotic depression, 207
necrotic disease, 207
necrotic lesion, 207
necrotic local lesion, 259
necrotic spot, 278
necrotic tissue, 207
necrotrophic, 462
necrotrophic pathogen, 462
Nectar plant, 323
nectary, 323
necton, 580
needle, 606
needle blight, 606
needle cast, 314
needle discolour, 606

needle gate, 606
needle hole, 533
needle layer, 606
needle-leaved, 606
needle-leaved forest, 606
needle-leaved tree, 607
needle-leaved woodland, 607
needle oil, 607
needle roller, 264
needle rust, 607
needle-shaped, 607
needle shedding, 464
negate, 136, 382
negative adsorption, 151
negative binominal distribution, 151
negative correlation, 151
negative cumulated temperature, 151
negative curvature, 151
negative density-dependent factor, 151
negative diageotropism, 151
negative feedback, 151
negative geotropism, 151
negative heterosis, 151
negative interest, 90, 343
negative investment, 151, 233
negative orthogeotropism, 152
negative phototropism, 151
negative plagiogeotropism, 151
negative staining, 151
negative-tax plan, 151
negative tropism, 151
negatively photoblastic (seed), 226
neighborhood, 298
neighborhood park, 245, 298
neighbourhood garden, 245
neighbourhood unit, 259
neiloid, 4
nekton, 458, 647
nema, 204, 526
nemas, 618
nematocide, 411
nematode, 526
nematode gall, 526
nematodiasis, 526
nematology, 526
neo-centromere, 539
Neo-Darwinism, 539
neo-Lamarckism, 539
Neo-Mendelism, 539
neo-sex chromosomal mechanism, 539
neobiogenesis, 539
neogaea, 539
Neoglaciation, 539
neolith, 539
neomorph, 539
neomorphosis, 539
neonatal, 63, 539
neophyte, 539
neoprene adhesive, 315
Neornithes, 248
neospecies, 539
neoteinic, 33, 583
neoteny, 583
neotropical, 539
neotropical floral kingdom, 539
neotropical region, 539
neotropics, 539
neotype, 539
neowood, 539
nephanalysis, 594
nephelometer, 545, 640
nepheloscope, 45
nephometer, 296
nephoscope, 45
nephroid, 420
neptunian theory, 456
neptunic rock, 188, 456
neptunism, 456
nervation, 317
nerve, 58, 560
nervose, 118, 261
nervulose, 261
neryl, 56
nest, 52
nest-building, 636
nest-hording, 110
nest of drawers, 61
nest planting, 73
nested quadrat method, 632
nested regression, 375
nested table, 474
nesting-box, 52
nesting looks, 52
nestling, 64
Nestos snubbing yarding, 340
net absorption, 257
net assimilation, 257
net assimilation rate, 257
net cash flow, 525
net community productivity, 389
net discounted revenue, 3
net domestic product, 186
net dry salt retention, 156
net ecosystem exchange, 425
net ecosystem production, 257
net ecosystem productivity, 425
net growth, 257
net-income, 69, 257
net increase, 257
net interception, 257
net investment, 257
net-like stone soil, 500
net national product, 185
net photosynthesis, 257
net photosynthetic rate, 257
net-plankton, 499
net precipitation, 257
net primary production, 256
net primary productivity, 256
net proceeds, 257
net production, 257
net production rate, 257
net profits, 257
net radiation, 257
net reproduction rate, 257
net reproductive rate, 257
net retention, 256
net revenue, 257
net scale, 256
net volume, 256
net volume increase, 36
net volumetric absorption, 257
net weight, 257
net worth, 257, 640
net yield, 256
netted venation, 500
nettle caterpillar, 71
network analysis, 499
network of accessory frame, 299
network of cutting-series, 38, 39
network of forest, 302
network pattern, 499

N

network structure, 499
neural, 420
neuration, 317
neuron(e), 420
neurotoxin, 420
neuston, 358
neuter, 625
neuter flower, 510, 511
neutral, 511, 625
neutral axis, 172, 623, 625
neutral fertilizer, 625
neutral filter, 625
neutral flower, 510, 511
neutral oil, 625
neutral plane, 623, 625
neutral position, 275, 623
neutral resin, 625
neutral size, 625
neutral soil, 625
neutral surface, 623, 625
neutralism, 625
neutrality controversy, 625
neutralization, 94, 437, 623
new art style furniture, 539
New Brunswick log rule, 539
new quaint style furniture, 539
New stone age, 539
new systematics, 539
newel, 309, 637
newel cap, 309, 637
newel joint, 309, 625, 637
newground plough, 268
newly afforested area, 539
news grade groundwood pulp, 539
news grade sulphite pulp, 539
newsprint (paper), 539
Newton's formula, 346
newtropical, 539
nexine, 495
Neyman's distribution, 336
niche, 425, 501, 534
niche breadth, 425
niche compression, 425
niche dimension, 425
niche overlap, 425
niche particulate, 137
niche separation, 425
niche shift, 425
niche theory, 425
niche width, 425
nick, 521
nicked, 261, 501
nicker, 281
nicotine, 342, 553
nicotine sulphate, 308
nicotinic acid, 553
nidifugity, 287
nidulariaceous fungus, 345
nieshout, 354
niger, 600
nigger, 128, 492
niggerhead, 35, 317, 507
night soil, 144
night stand, 67
night watch, 561
nimbostratus, 584
nimbus, 585
nimbus-cumuliformis, 219
ninhydrin reaction, 456
nip guard, 551
nip pressure, 551
nip roll, 225
nipping, 603
nipple, 313, 355
nirit, 123
nitosol, 376, 488
nitrate, 532
nitrate bacteria, 532
nitrate ions, 532
nitrate plant, 532
nitrate respiration, 532
nitration, 532
nitric bacteria, 532
nitrification, 532
nitrifier, 532
nitrifying bacteria, 532
nitrite bacteria, 553
nitrite ions, 553
nitrite-oxidizing bacteria, 553
nitro-cellulose, 532
nitro-lacquer, 532
nitrocellulose emulsion, 532
nitrocellulose lacquer, 532
nitrochloroform, 315, 532
nitrofen, 64
nitrogen content, 189
nitrogen cycle, 88
nitrogen cylinder, 88
nitrogen fertilization, 88
nitrogen fixation, 175
nitrogen-fixing bacteria, 175
nitrogen-fixing root nodule, 175
nitrogen free extract, 508
nitrogen metabolism, 88
nitrogen oxide, 88
nitrogen oxides, 88
nitrogen-phosphate, 88
nitrogen starvation, 88
nitrogenase, 175
nitrogin, 167
nitroglycerine, 532
nitrolignin, 532
nitrophile, 439, 516
nitrophilous plant, 439
nitrophilous vegetation, 439
nitrophyte, 439
nitrosation, 553
nitrosification, 553
nitrosochloride, 553
nitrous oxide (N_2O), 557
nival, 119, 548
nival belt, 195, 219, 548
nival flora, 28
nival line, 548
nival surface, 548
nivation, 548
nivometer, 296
NMR spectrometry, 193
no load, 276
no-regret option, 509
no-regrets opportunity, 509
no-regrets policy, 509
no-regrets potential, 509
no-seed year, 375
no-shatter feller, 510
no-spin front axle, 132
no tillage, 323
Noble germinating dish, 348
noctiflorous, 561
nocturnal, 561
nocturnal animal, 561
nocturnal cooling, 561
nocturnal frass, 561
nocturnal migration, 561
nodal, 245, 246, 651
nodal diaphragm, 245
nodal wood, 246
node, 239, 245, 246
nodular end wall, 67
nodular thickening, 67
nodule, 167, 246

nodule bacteria, 167, 168
noduli-ferous, 640
nodum, 616
noise control, 601
noise factor, 601
noise level, 601
noise measurement, 601
noise meter, 430, 601
noise pollution, 601
noise rating number, 601
noise ratio, 601
noise-shielded, 134
noise source, 601
noise standard, 601
noise temperature ratio, 505
noisiness, 601
nomad, 579
nomadic furniture, 605
nomadism, 579
nomenclature, 324, 325
nomenclature type, 325
nominal, 325
nominal density, 325
nominal diameter, 24, 171
nominal load, 120
nominal measure, 325
nominal pressure, 24, 171
nominal price, 325, 544
nominal size, 171, 325
nominal volume, 24, 171
nominal wage, 217, 325
nominal weight, 24, 171
nomograph, 226, 297, 348
nomography, 297
non-accessible forest, 34
non-adaptive character, 138
non-adhering knot, 462
non-aggressive strain, 138
non-allelic gene, 136
non-assimilation system, 138
non-available water, 511
non-calcareous soil, 138
non-cellulose, 138
non-cellulosic constituent, 138
non-cellulosic polysaccharide, 138
non-commercial forest land, 138
non-commercial thinning, 138, 149
non-compressed (fibre) board, 138, 401
non-conducting material, 265
non-conductor, 136
non-cyclic benzyl aryl ether structure, 137
non-cyclic photophosphorylation, 138
non-degradable pollutant, 137
non-destructive stress, 137
non-destructive test(ing), 137
non-digital computer, 138
non-dimension timber, 34, 92
non-disjunction, 34
non-dormant, 138, 511
non-drying oil, 137
non-ejector gun, 32
non-endospermic seed, 510
non-equilibrium theory, 137
non-essential element, 136
non-fibrous component, 138
non-fibrous matter, 138
non-fibrous type, 138
non-forest area, 137
non-forest land, 137
non-forest soil, 137
non-forested area, 510
non-forestry land, 137
non-frame construction, 510
non-frost period, 510
non-game birds, 252
non-genic variation, 137
Non-Governmental Organization, 138
non-grain-raising stain, 137
non-heel cutting, 510, 512
non-hemologous association, 568
non-hereditary variability, 35
non-hereditary variation, 138
non-heritable agency, 138
non-heritable variation, 35
non-homeostatic organism, 137
non-homologous chromosome, 138, 568
non-ignitable, 34, 35
non-infectious disease, 138
non-infective disease, 138
non-leaching treatment, 137
non-light-sensitive seed, 137
non-linearity, 138
non-living environmental condition, 138
non-log factor, 138
non-luminated, 137
non-market impact, 138
non-market value, 138
non-mechanical physical property, 137
non-operable forested land, 34
non-operable stand, 34
non-orthogonal design, 138
non-overlapping niche, 139
non-overlapping triplet, 138
non-parasitic, 137
non-parasitic disease, 137
non-passing sight distance, 481
non-pathogenic, 32, 136
non-periodic cruise, 508
non-physical loss, 511
non-point-source pollution, 136
non-polar adherent, 137
non-polar adhesive, 137
non-polar solvent, 137
non-pored wood (timber), 509
non-porous wood (timber), 509
non-pressure process, 49
non-pressure treatment, 49
non-productive forest land, 64, 137, 138
non-random distribution, 138

N

N

non-random sampling, 138
non-recurrent parent, 137
non-recurring expenses, 304
non-return trap, 618
non-return valve, 85, 618
non-sampling error, 136
non-saturated air, 503
non-secretory intercellular space, 137
non-selective filter, 138
non-silvicultural cost, 138
non-soluble matter, 35
non-spinning rope, 509
non-stocked forest land, 510
non-stocked land, 510
non-stratified meristem, 136
non-subterranean termite, 136
non-sulfur pulping, 510
non-symbiotic, 137
non-symbiotic nitrogen fixation, 137
non-tan, 138
non-tannin, 138
non-tanning substance, 138
non-tapering, 137, 144
non-taperness, 498
non-timber product forest, 252
non-transplanting, 306
non-uniform flow, 138, 506
non-volatile matter, 137
non-wood based panel, 137
non-woody plant, 137
Non-Annex I Country (Party), 137
noncalcic brown soil, 138
noncompatible taper equation, 138
noncompetitive inhibition, 137
noncontrol cableway, 137
noncrystalline matrix, 137
noncrystalline region, 137
nondurable goods, 137
nonflexible price, 510
nonforested land, 504, 510
nonhost crop, 137
nonhypersensitive resistance, 137
noninheritedness, 34
noninterbreeding, 138
nonlignin compound, 137
nonpersistent virus, 136
nonpolysomatic species, 136
nonresistant, 509
nonrigid road, 137
nonsense mutation, 511
nonspecific reaction, 139
nonsporing, 35
nonsporous, 33
nonstandard equipment, 136
nonsteady deformation, 35
nonstratified crown, 508
nonstratified form, 508
nonsymbiotic culture, 509
nontronite, 315, 338
nonuniform sampling, 137
nonuniformity, 34
nonuploid, 259
noose, 470, 474
Nordheim process, 348
normal age-class, 128
normal age-class arrangement, 128
normal age-class distribution, 128, 325
normal annual mortality, 292
normal annual yield, 25, 128
normal arrangement of stand, 299
normal association analysis, 610
normal axial intercellular canal, 610
normal bud, 108, 610
normal curve, 48, 611
normal curve of error, 513
normal cut, 25, 127
normal cutting ratio, 127
normal diameter, 26, 542
normal dispersion, 610
normal distribution, 611
normal distribution curve, 611
normal distribution function, 611
normal distribution of age classes, 305
normal distribution of stand, 299
normal empty-cell treatment, 325
normal erosion, 48
normal even-aged forest, 128
normal final age, 127, 128
normal fire season, 610
normal forest, 127, 325
normal forest model, 128, 288
normal form factor, 26, 608, 611
normal form quotient, 26, 608, 611
normal-fully stocked stand, 128
normal girth, 25, 542
normal growing stock, 127, 128, 325
normal growth rate, 128
normal (ideal) age classdistribution, 128
normal incidence, 611
normal increment, 128, 325
normal moisture capacity, 610
normal operating loss, 610
normal peeling lathe, 26, 365
normal plywood, 365
normal population, 610
normal precipitation, 610
normal reflection, 611
normal relative humidity, 610
normal resin canal, 610
normal resin duct, 610
normal root, 106, 610
normal rotation, 128
normal salt, 611, 624
normal seed stand, 328, 611
normal seedling, 610
normal series of age gradations, 128, 325

normal size, 25
normal solution, 88
normal spectrum, 48
normal speed, 610
normal stand, 25, 127
normal state, 610
normal stock, 128
normal stock method, 128
normal stocking, 128, 288
normal strain, 127
normal strand, 592
normal stress, 127, 611
normal succession of cutting, 25
normal succession of felling, 25
normal surface, 68, 611
normal temperature-pressure, 26
normal timber life, 331
normal uneven-aged forest, 128
normal volume, 127
normal volume method, 128
normal wood, 610
normal year, 362, 610
normal yield, 128
normal yield table, 25, 128
normality, 127
normalized difference vegetation index (NDVI), 315, 616
north Atlantic oscillation, 17
north bend skyline system, 17
north bend yarding, 17
north-facing slope, 17
north pole, 17
northern coniferous forest biome, 17
Norway spruce, 348
nose, 89, 349, 377, 592
nose angle, 88
nose bar, 549
nose-bar gap, 549
nose-bar opening, 549
nose guy, 367
noseboard, 19
nosing, 11, 349, 486
notate, 260
notch, 41, 149, 274, 377, 519
notch board, 4
notch-girdling, 207
notch graft, 231
notch planting, 377
notch weir, 387
notched and cogged joint, 41
notched joint, 41
notching, 207, 268
notching spade, 147, 376
note regarding future treatment, 253
notogaea, 338
notum (*plur.* nota), 17
noubirefringent layer, 138
novation, 78, 566
novelty saw, 16
novriable, 510
noxious weed, 121
noxious wildlife, 580
noxtane, 512
nozzle, 350, 355
nozzle coefficient, 355
nozzle control, 355
nozzle headbox, 355
nozzle type slice, 355
nozzle velocity coefficient, 355
nubbin, 12, 474
nucella, 632
nucellus, 632
nuciferous, 260
nuclear-cytoplasmic incompatibility, 194
nuclear family, 194
nuclear gene, 193
nuclear magnetic resonance, 193
nuclear membrane, 193
nuclear-polyhedrosis virus, 194
nuclear spindle, 193
nuclear type, 194
nuclease, 194
nucleated, 580
nucleic acid, 194
nucleic acid cycle, 194
nuclein, 194
nucleo-cytoplasmic interaction hypothesis, 194
nucleocapsid, 29, 193
nucleoid, 29, 285
nucleolar organizer, 194
nucleolus (*plur.* -li), 194
nucleophile, 379
nucleophilic addition, 379
nucleophilic reagent, 379
nucleophilic substitution, 379
nucleophilicity, 379
nucleoprotein, 193
nucleosidase, 193
nucleoside, 193
nucleotidase, 193
nucleotide, 193
nucleus (*plur.* nuclei), 193, 517
nude, 313
nudicaulous, 313
nugget variance, 279
nuisance control, 171
nulliplex, 305, 511
nullisomic, 387
nullisomic diploidy, 387
nullozygous gene, 305
number-average degree of polymerization, 451
number-average molecular weight, 451
number-marking hammer, 75
number of individuals, 166
number of revolutions, 638
number of speeds, 638
number of stages, 221
number of stems, 289, 290, 632
number of steps of speeds, 24
number of strockes, 59
number of trees, 632
number of trees per hectare, 171
number tag, 191
numbering hammer, 75
numerical analysis, 451
numerical chromosome change, 391
numerical mutation, 451
numerical response, 451
numerical taxonomy, 451
nunatak, 28
nuptial chamber, 239
nuptial colouration, 213
nuptial flight, 213
nuptial gift, 213
nuptial plumage, 213
nurse branch, 641
nurse-crop, 14
nurse-root grafting, 202, 634
nurse-seed grafting, 642

N

O

nurse species, 33, 71, 154, 252
nurse tissue, 641
nurse tree, 15, 20
nurse-wood, 14
nursery, 324
nursery bed, 323
nursery bed former, 324
nursery disease, 324
nursery garden, 324, 585
nursery gardening, 324
nursery in field, 311
nursery in forest, 301
nursery inventory, 324
nursery land, 324
nursery of collection, 39
nursery of elite, 592
nursery of hybrid, 597
nursery of scion, 39
nursery operation, 324
nursery planning, 324
nursery planter, 324
nursery practice, 324
nursery row, 324
nursery selection, 324
nursery soil, 324
nursery stock, 324
nursery-stock growing, 324
nursery tarp roller, 324
nursery transplant, 565
nurseryman, 324, 592
nursing, 149, 557
nursing chair, 572
nut (*plur.* nuces), 232
nut oil, 201
nutation movement, 637
nutgall, 327, 512
nuthook, 173
nutlet, 534
nutmeg butter, 399
nutmeg oil, 399
nutmeg tallow, 399
nutrient, 557, 573
nutrient agar, 574
nutrient broth, 574
nutrient cycle, 557
nutrient depletion, 557
nutrient humus, 574
nutrient medium, 353
nutrient mist, 574
nutrient poor, 359, 573
nutrient rich, 139, 573
nutrient solution, 574
nutrient status, 574
nutrient translocation, 557
nutrition, 573, 574
nutrition pot, 573
nutritional deficiency, 574
nutritional factor, 574
nutritional theory, 574
nutritious substance, 574
nutritive pot, 573
nutritive ratio, 573
nutritive substance, 574
nutritive symbiosis, 574
nutty structure, 194
nuxvomica poisonnut, 316
nyctinastic movement, 159, 259
nyctinasty, 159
nylon brush roll, 342
nymph, 401
nymphal instar, 401
Nymphalidae, 229
O-layer, 42, 278, 462
oak apple, 291
oak-bark coppice, 291
oak bark extract, 201, 291
oak bark tannage, 201, 291
oak coppice-wood, 291
oak grove, 291
oak line pattern, 531
oak parquet, 291
oak-type ray, 291
oak wood, 291, 379
oak (wood) extract, 291
oakery, 291
oakwood, 531
oar, 204, 238
oasis, 315
oasis (*plur.* oases), 208, 315, 507
obakunone, 210
obakurone, 210
obconic, 90
obcordiform, 90
obdeltoid, 90
obdiplostemonous, 496
obiotic component, 138
object micrometer, 513
object program, 336
objective, 336, 513
objective forest, 336
objective function, 336
objective lens, 513
oblanceolate, 90
oblate, 23
obligate, 636
obligate parasite, 636
obligate saprophyte, 636
obligation officer, 566
obligatorily coloured heartwood tree, 525
obligatory, 636
obligatory parthenogenesis, 636
oblique, 357, 536
oblique arch, 537
oblique bedding, 536
oblique blade root pruner, 536
oblique boom, 536, 537
oblique cleft grafting, 537
oblique cutting, 536, 537
oblique joint, 537
oblique notching, 269
oblique photograph, 381
oblique planting, 537
oblique pore arrangement, 537
oblique profile, 537
oblique root, 43, 537
oblique shear, 537
oblique tenon, 537
oblique tooth, 536
oblique wedge grafting, 537
obovoid, 90
observation station, 178
observation window, 178
observatoinal component, 178
observatory, 178, 519
observed profile, 178
observed shrinkage, 178, 435
obsolete, 34, 64, 125
obsolete character, 236
obstacle (climbing) capability, 594
obstruction by slime, 345
obtect pupa, 18
obtuse, 115, 593
obtuse ground cypress, 396
occasional crossing, 348
occasional species, 348
occasional table, 380
occiput, 200
occluded knot, 571, 587
occlusion, 587
occlusion of wounds, 414
occupant, 259, 603
occupied fallow, 10
occurrence, 62
ocean, 189
ocean basin, 189
ocean conveyor belt, 189

ocean-going raft, 189
ocean raft, 189
oceanic, 189
oceanic climate, 189
oceanic island, 189
oceanic plankton, 594
oceanic province, 78
oceanic region, 78
oceanity, 189
oceanodromous migration, 189
oceanology, 189
ocellus (*plur.* ocelli), 86, 555
ochrea, 493
ochric epipedon, 87
ocimene, 312
ocimenone, 312
octagon hewn lumber, 5
octahydro retene, 5
octal, 5
octal digit, 5
octamerous, 4
octoploid, 4
octopus, 604
octose, 5, 539
oculant, 551
ocular, 336
ocular estimate, 336
ocular estimate-by-plot method, 558
ocular examination, 614
ocular measurement, 336
ocular plot-estimate, 558
ocular volume estimation, 336
oculation, 244
oculus, 552
odd corner, 127
odd lengths, 366, 396
odd lot, 305
odd-number lengths, 366
odd-pinnate, 366
odd pinnately compound leaf, 366
odd sheets, 34
odds, 577
odoriferous material, 528
odoriphor group, 124
odorousness, 370
odour of wood, 330
oecium, 542
oecology, 425
oesophafus, 436
oesophageal ganglion, 436
oestrous cycle, 123
off-and-on grazing, 32
off-and-on loading, 639
off-bearer, 262, 521
off-center, 357
off-colour, 23, 492, 555
off-cut, 40, 227, 521
off-gas, 62, 139, 503
off-grade, 34, 92
off-grade seed, 92
off-grade timber, 92
off hand, 468
off-highway hauling, 51, 137
off-highway truck, 51, 137
off-load capability, 538
off-load table, 538
off-loading, 538
off-machine sizing, 218
off-peak period, 137
off products, 154
off-road hauling, 127, 302
off-saw size, 262
off-season, 87, 138
off-season nursery, 137
off-size, 34, 137
off-stream storage pond, 192
off-type, 136, 138, 567
off year, 375, 534
offal, 109, 150
offal timber, 40, 70, 139
offcut, 22, 521
offcut timber, 22, 333, 521
offer, 15, 123, 236
offloader, 538
offout, 521
offset, 40, 143, 492
offset mitre joint, 375
offset sawing, 357
offshoot, 44, 143
offshore, 287
offshore current, 287
offspring, 199, 200
offspring-parent regression, 379
offspring-parent relationship, 379
offtake, 350, 538
often cross-pollinated crop, 49
ogee, 384
ogee sticking, 308
ohort life table, 483
oidlum (*plur.* -ia), 142, 143
oikos, 423
oil-borne preservative, 578
oil bound water paint, 458, 578
oil camellia, 578
oil canal, 578
oil cell, 578
oil cyst, 578
oil drying, 394
oil duct, 578
oil extender, 434, 578
oil filler, 578
oil foot, 578
oil gland, 503
oil heater, 578
oil-hydraulic press, 578
oil immersion, 578
oil impregnation, 578
oil length, 189, 578
oil miscible concentrate, 400
oil number, 401, 515
oil nut, 578
oil of cuscus, 528, 554
oil orange, 578, 584
oil paint, 578
oil pollution, 578
oil-pressure pump, 578
oil proof, 134, 338
oil putty, 578
oil receptacle, 578
oil-recovery tank, 212
oil-salt combination, 578
oil seasoning, 394
oil solvent, 578
oil-solvent(-type) preservative, 640
oil spot, 578
oil-staining, 578
oil storage tank, 635
oil streak, 578
oil tar, 242
oil tempered hardboard, 578
oil tempering, 578
oil test, 578
oil-tight, 32, 578
oil type preservative, 578
oil varnish, 578
oil-water separator, 578
oil-water treatment, 578
oil-yielding trees, 578
oilomatic chain, 644
oily paint, 578
oily perfume, 578
Okal chipboard, 4, 290

O

Okal-extrusion process, 4
Okazaki fragments, 160
Okazaki piece, 160
olation, 353, 376
old bamboo, 284
old burns, 259
old crop, 284
old field, 371
old field stand, 284
old growth, 55, 591
old height crop, 284
old individual, 284
old stand, 284
old standard, 187
old thinning, 199
old timber, 284
old timber stage, 284
oleander oil, 348
oleoresin, 189, 254, 465
olfactory organ, 544
olibanoresene, 400
olibanum oil, 400
olibanum resin, 400
olibanum resinoid, 400
oligandrous, 417
Oligocene, 237
oligoforate, 417
oligogene, 177, 634
oligogenic character, 177
oligogenic resistance, 177
oligohaline, 177
oligomerous, 417
oligomictic lake, 417
oligophagous, 177
oligophagy, 177
oligophyllous, 417
oligosaprobic, 177
oligospermous, 417
oligostemonous, 417
oligotaxy, 203
oligothermal, 519
oligotrophic, 359
oligotrophic lake, 359
oligotrophic plant, 177, 573
oligotrophic water, 359
oligotrophy, 359, 573
olive oil, 159
olivine-fused phosphate, 159
olivline, 159
Olsen-type wearometer, 4
Olympic National Monument and National Park, 4
ombre table, 534
ombrogenic bog, 585
ombrometer, 502, 584
ombrophile, 440, 517
ombrophilous, 517
ombrophobe, 21, 524
ombrophyte, 440, 517
ombrotrophic bog, 342
ombrotropism, 530
ombrotype, 238
omethoate, 557
omnivore, 597
omnivore (*plur.* -ra), 597
omnivority, 597
omnivor(o)us, 597
omnivorous animal, 597
omnivory, 597
omphalodium, 626
ompressometer, 469, 550, 551
on-board load-measuring system, 52
on-grade, 91, 183
on hand, 278, 525
on-highway hauling, 171
on-line message processing, 294, 526
on line operation, 294
on-line real time processor, 294
on-loading, 638, 639
on-off switch, 209
on-road hauling, 298
on-site penetration test, 107
on-stream storage pond, 192
on-type, 610
on year, 77, 88
oncogene, 621
one-chamber type vertical shaft mill, 84
one-choker show, 655
one-component lacquer, 87
one cut per rotation, 320
one-cut shelterwood method, 562, 563
one-deck dryer, 82
one-dimensional chromatography, 85
one-donkey show, 535
one-eye cutting, 86
one-factor difference, 86
one-gene heterosis, 86
one-gene-one-enzyme hypothesis, 563
one-gene-one-function hypothesis, 563
one-handed saw, 82, 84
one-hinged arch, 83
one-legged ladder, 87
one-lime phosphate, 629
one-male troop, 86
one-man saw, 84
one match burn, 158
one-node cutting, 84
one-parent progeny test, 84
one-sided inheritance, 526
one-sided spike, 82
one-stage hydrolysis, 563
one-storey forest, 82
one surface smooth hardboard, 563
one-way circulation, 86
one-way disk plough, 86, 454
one way-drum dryer, 83
one-way incline, 85
one-way merchantable volnme table, 563
one way skidding, 85
one-way tramway, 85
one way volume table, 95, 563
onocerin, 318
onocol, 318
onset (*eg.* dormancy onset), 269
onshore current, 530
ontogenetic theory, 166
ontogeny, 166
ooapogamy, 121, 477
oocyst, 311
oocyte, 311
ooecium, 311
oogamete, 69
oogamy, 567
oogonium, 311
ookinesis, 311
oomycetes, 311
oopod(eus), 47
oosperm, 443
oosphere, 311
oosphere nucleus, 311
oospore, 18, 311, 443
oosporen, 311
ootid, 311
ootypus, 311
ooze, 401
opacimeter, 3, 32
opacity, 3, 32, 213
opaco, 570

opalescence, 400
opalwax, 399
opaque, 3, 32, 33, 605
open air seasoning, 311, 441
open assembly time, 269
open association, 269, 515
open bucket type steam trap, 148
open canal, 325
open canker, 268
open canopy, 515
open cell process, 276
open channel, 325
open circulation, 268
open coat, 444, 445
open community, 269, 515
open crop, 445
open-crowned tree, 445
open culture, 268, 293
open cycle, 136, 269
open defect, 27, 325
open ditch, 325
open (ditch) drainage, 325
open drum sander, 443
open fire, 214, 325
open forest, 445
open forest land, 445
open glue joint, 287
open grain, 73
open-growing tree, 647
open-grown, 279, 344
open grown tree, 174
open infusion method, 50
open intercrossing, 478
open jack, 43
open joint, 179, 240, 268
open knot, 297
open mortise, 268
open mortise-and-tenon joint, 268
open out, 445
open panicle, 269
open pile, 45
open piling, 45, 316, 360, 362, 515
open pollination, 268, 647
open pollination progeny, 647
open pollination seedling orchard, 647
open pore coating, 6, 19
open-pore finishing, 6
open range, 647
open root, 313
open-root(ed) planting, 313
open-root(ing) planting, 313
open season, 75, 444
open shrubland, 515
open-sided jack, 43
open sight, 387
open single lane, 269
open socket, 268
open-space, 277
open space in pile, 36
open space percentage, 315
open stacking, 306, 444, 485, 515
open stand, 445
open-stand increment, 443
open-stand system, 443
open steam, 614
open storage, 268
open sunny slopes, 269
open system, 268
open-tank method, 50
open-tank process, 50, 286
open time, 269
open-top culvert, 310, 325, 485
open up, 268
open vegetation, 269, 404, 515
open venation, 268
open waiting time, 74, 269
open wood system, 443
opener, 268
openess, 445
opening, 304
opening by valves, 11
opening line, 104
opening of canopy, 300
opening of the shelter wood, 15, 443
opening-out of leaf canopy, 300
opening time, 113, 269
opening up, 75, 268, 445
opening-up of a stand, 300
openside, 81, 82
openside block, 268
openside trolley, 268
operable area, 274
operable guide, 432
operable land, 274
operable plan, 432
operable stand, 274
operable timber, 24, 272
operating cable, 368, 372
operating capacity, 435, 582
operating cost, 654
operating depth, 655
operating instruction, 41
operating line, 41, 372, 655
operating line support, 372, 570
operating load, 170, 582, 596
operating outlay, 560
operating pit, 655
operating plan, 432
operating pressure, 170, 277
operating profits, 574
operating rate, 268, 596
operating reserves, 574
operating rope, 372, 655
operating statement, 574
operating tank, 654
operating weight, 610
operation map, 655
operation time, 41, 596
operational performance, 437
operation(al) research, 595, 655
operational reserve, 560
operative productive planning, 655
operator gene, 41
opercule, 156, 242, 276, 461
operon, 41
ophiolite, 417
ophioxylin, 417
ophite, 211, 414
opportunity, 218
opportunity cost, 218, 477
opposing wind, 343
opposite, 115
opposite bud, 115
opposite leaves, 115
opposite-pinnate, 115
opposite pitting, 115
oppositifolious, 115
oppositional allele, 115
oppositional factor, 528
oppositional gene, 115
oppositipetalous, 114

O

oppressed stem (tree), 18
optic, 182, 439, 546
optical activity, 546
optical calliper, 182
optical dendrometer, 182
optical density, 182
optical-mechanical scanner, 181
optical processing, 182
optical reader, 180
optical rotation, 357, 546
optical rotatory dispersion, 546
optical wedge, 182, 285
optimal fashion, 654
optimal foraging strategy, 653
optimal policy, 654
optimization, 577, 653, 654
optimization of sampling deign, 653
optimum, 653
optimum allocation, 654
optimum catch, 653
optimum curve, 653
optimum cut, 439, 440
optimum cut period, 440
optimum density, 653
optimum design, 577, 653
optimum distribution, 577, 653
optimum drying condition, 653
optimum factor, 653
optimum growth rate, 653
optimum moisture content, 653
optimum programming plantation, 653
optimum stocking, 440
optimum temperature, 653
optimum threshold, 653
optimum yield, 654
oral cavity, 277
orange lac, 56
orange oil, 57
orange peel, 56, 259
orbit, 184, 214
orbital, 184, 637
orbital parameter, 184
orbital period, 184
orbital ring, 556
orchard, 186
orchard spray, 186
orchard tractor, 588
orchid house, 283
orchids, 282
order, 71, 106, 336, 389
order correlation, 461
order-disorder transition, 582
order of architecture, 236
order of felling, 39
order of life, 423
order of living beings, 426
order of magnitude, 451
ordinal, 544
ordinary back up device, 49
ordinary cone, 593
ordinary lay, 160
ordinary spade, 365
ordinary splice grafting, 365
ordinary surveying, 365
ordinary thinning, 520
ordinary water level, 48
ordinate, 651
ordination, 350
Ordovician period, 4
oread, 556, 557
Oregon pine, 204
oreharding, 186
oreithalion tropicum, 391
organ culture, 371
organgenesis, 371
organic acid, 580
organic aerosol, 580
organic biomass, 580
organic debris, 580
organic decay, 580
organic detritus, 580
organic detritus food chain, 580
organic fertilizer, 580
organic fire retardant, 580
organic horizon, 580
organic layer, 580
organic manure, 580
organic matter, 580
organic mineral insecticide, 580
organic particulate matter, 580
organic pollution, 580
organic soil, 580
organic-solvent type preservative, 580
organism, 428, 580
organismal, 580
organismic concept, 580
organization by area, 323
organization by volume, 36
organization chart, 652
organization costs, 268
organization of administration, 179
organization of forest cash-office, 406
organization of forest management, 407
organization of forest-protection, 405
organization of labour, 284
organization of production, 422
organization of worker's position, 170
organochlorine insecticide, 580
organogel, 580
organogenesis, 371
organoid, 518
organophosphorus insecticide, 580
organosol lignin, 580
organotropy, 530
Oriental arbor-vitae, 42
oriental region, 108
orientating function, 107
orientation, 107, 183, 350, 384, 637, 651
orientation effect, 107, 384
orientation of cellulose molecule, 523
orientation of fibril, 522
orientation of tissue, 652
oriented adsorption, 107
oriented region, 107, 384
oriented strand board, 107
oriented structural board, 107
oriented structural board-cross direction, 197
oriented structural board-machine direction, 650
oriented waferboard, 107

orifice, 165, 245, 276, 355, 412, 605
origin, 47, 282, 367, 588
origin of life, 423
origin of species, 513
origin of stand, 300
original cell wall, 591
original chromosome, 590
original cost, 591
original forest, 590
original growth, 590
original landscape, 591
original seed farm, 592
original species, 591
original vegetation, 591
original wages, 219
orizabin, 382
orlean, 553
ormosine, 197
ornamental, 178, 639
ornamental bamboos, 178
ornamental design, 178, 639
ornamental fruit trees and shrubs, 178
ornamental hardboard, 639
ornamental horticulture, 178
ornamental species, 178
ornamental trees and shrubs, 178
ornamental vessel, 438
ornate, 73, 343
ornithologist, 345
Ornithology, 345
ornithophilous flower, 345
ornithophilous plant, 346
ornithophily, 345
oroboside, 528
orogenesis, 600
orogeny, 600, 601
orographic effect, 99
orographic factor, 99
orographic precipitation, 99
orographic rain, 99
orographic rainfall, 99
orography, 98, 414
orophytia, 552
orris oil, 587
orris resinoid, 587
orterde, 575
ortet, 511
ortho-dichlorobenzene, 298
ortho-effect, 298
ortho-phenylphenol, 298
ortho-rectification, 611
orthoclase, 611
orthod, 211
orthogamy, 610
orthogenesis, 107, 615
orthogeotropism, 615
orthogneiss, 215, 611
orthogonal cutting, 68, 611
orthogonal cuttmg, 611, 614
orthogonal design, 611
orthogonal diameter, 611
orthogonal projection, 611
orthogonality condition, 611
orthographic projection, 611
orthokinesis, 614
orthophoto, 611
orthophoto graphy, 611
orthophoto map, 611
orthophoto scope, 611
orthoplocal, 641
orthoquartzite, 215
orthoselection, 107, 615
orthosere, 611
orthotropic elastic material, 611
ortsrein, 481
Osborne fire finder, 4
oscillating-bit mortiser, 500
oscillating floating track, 558
oscillating mortising chisel, 500
oscillating pressure cycle, 7
oscillating pressure process, 7, 608
oscillating saw, 493, 500
oscillating screen, 608
oscillating table separator, 471
oscillation, 558, 608
oscillation volume, 608
oscillator, 7, 124, 228, 608
oscillograph, 310, 438
oscillography, 438
oscillometer, 438
osier, 308
osier willow, 191
osmanthus concrete, 184, 335
osmometer, 421, 528
osmometery, 421
osmophore, 124
osmoplastic method, 421
osmoregulation, 421
osmosar, 421
osmose preservative, 421
osmose process, 421
osmose treatment, 421
osmosis, 421
osmotaxis, 383
osmotic coefficient, 421
osmotic concentration, 421
osmotic potential, 421
osmotic pressure, 421
osmotic pressure gradient, 421
osmotic water, 421
osmotic work, 421
osmotica, 421
osmotroph(s), 421
osmotropism, 530
osseous, 175
ossiculus, 533
ostiole, 276
Ostwald viscometer, 4
osyris oil, 411
osyrit(r)in, 411, 595
otoba tallow (butter), 4
ottoman, 509, 580
ottoman footstool, 566
oued, 156, 278
ouffeed device, 62
ounterveneer, 94
ouricury wax, 533
out-of-lead block, 637
out-of-round log, 34
out-of-standard, 34
out-of-standard board, 34
out-of-wind, 32, 362
out-put, 62
out-turn, 47, 62
outbark, 496
outboard drum, 150
outbreak, 76, 113, 306, 363, 622
outbreak of insect, 189
outbreeder, 594
outbreeding, 594
outcrop, 311
outcrossing, 138, 568
outdoor flower, 311
outdoor furniture, 441
outdoor grafting, 202

O

O

outdoor recreation, 202
outdoor recreation model, 202
outdoor recreation resource availability, 202
outdoor recreational value, 202
outdoor seasoning, 441
outer aperture, 506
outer bark, 497
outer bark pocket, 496
outer boom, 154
outer coat, 497
outer glume, 497
outer indusium, 496, 582
outer integument, 497
outer layer, 495
outer layer of secondary wall, 70
outer lobe, 496, 497
outer margin, 497
outer part, 495
outer perianth, 496
outer pericycle, 497
outer perigone, 496
outer pit aperture, 506
outer rectrix (*plur.* -trices), 495
outer spathe, 495
outer spindle, 496
outer surface, 495
outer web, 496
outer wood, 54
outfall, 193, 350
outfeed guide, 140
outfeed roll, 62
outflow, 306, 569
outfold, 90
outhaul, 212
outhaul block, 212
outhaul line, 212
outhaul trip, 211
outlet, 350, 445
outlet conduit, 538
outlet drain, 350
outlier, 566
outplanting, 62, 600
outpouring, 306, 538
output, 47, 62, 422, 445, 536
outrigger, 55, 419
outrigger leg, 496
outside bark, 496
outside buffer zone, 497
outside diameter, 79
outside frost cleft, 495
outside gouge, 496
outside lining, 280
outside surface, 495
outskirt, 240, 496, 500
outthrows, 139, 183, 468
outturn control, 36, 37
outturn of stand, 299
outturn of the felling, 38, 62
outturn table, 62
outwash plain, 28
oval knot, 311, 494
oval wheel type meter, 494
ovariole, 311
ovary, 311, 641
ovary inferior, 641
ovary semiinferior, 641
ovary superior, 641
ovate, 311
ovellum, 583
oven, 158, 197
oven-dried condition, 197, 265
oven-dried weight, 197, 265
oven-dried wood, 197, 265
oven-dry, 197, 265, 310
oven-dry density, 265
oven-dry weight, 197
oven-dry weight of moisture section, 189
oven-dry wood, 197, 265
oven-drying method, 197
oven for dry distillation, 157
over-all application, 386
over-all control, 386
over bark, 496
over-bark diameter, 79
over-beating, 75
over cover system, 404
over cut, 187
over development, 186
over-dispersion, 51
over-dry veneer, 187
over-exploitation, 186, 187
over-fishing, 186
over grazing, 186
over-grown, 72, 186
over irrigation, 317
over laying sander, 480
over light, 415
over-mature timber stage, 186
over-pass, 569
over-potting, 203
over production, 421
over-specialization, 186
over tillage, 152, 598
over tire track, 474
over wintering, 594
overall gear ratio, 648
overall limit, 385
overall project of forest enterprises, 303
overall size, 312, 497, 648
overall soil preparation, 386
overarching wall, 172
overarm closure, 121
overarm opening, 121
overcast day, 570
overcoating, 14, 323, 327, 496, 605
overconditioning, 186
overcook, 19, 187
overcrowded, 187
overcrowding, 186
overcut size, 306, 318
overcutting, 187, 352
overdominance, 51
overdominance hypothesis, 51
overdraft, 186, 187, 486
overdrive transmission, 51
overexploitation, 186
overfall, 88, 211, 569
overfalling weir, 569
overfell, 187
overfelling, 51, 187
overflow, 187, 569
overflow area, 569
overflow dam, 184, 485
overflow depth, 569
overflow land, 193, 456
overflow section, 569
overflow spring, 569
overflow weir, 569
overfold, 90
overgrown knot, 571, 587
overgrowth, 146, 186, 317, 416, 429, 487
overhaul, 78, 232, 282
overhead beam edger, 162
overhead cable, 231
overhead cableway skidding, 231
overhead cost, 179, 235, 649
overhead cover, 415
overhead crane, 162, 377

overhead discharge chipper, 67
overhead door, 162
overhead internal-fan kiln, 36, 162
overhead irrigation, 354, 415
overhead-light, 415
overhead loading system, 162
overhead oscillating sprinkler, 231
overhead protection, 415
overhead release felling, 443
overhead saw, 104, 455
overhead shading, 415
overhead skidding, 231
overhead spray, 105
overhead spray system, 105
overhead sprinkling, 415
overhead travelling crane, 162, 377, 477, 540
overhead trimmer, 104
overhead yarding, 231
overhung stump, 496, 545
overlaid fibreboard, 480
overlaid veneer, 480
overland flow, 95, 317
overlap, 60, 104
overlap area, 60
overlap drawer, 605
overlap of photograph, 531
overlap splice, 75
overlapped peaks, 60
overlapping, 104
overlap(ping), 104
overlapping gene, 60
overlapping generations, 438
overlapping niche, 60
overlapping of sheet, 104
overlapping pair, 60
overlay, 26, 43, 155, 480, 487
overlay drawer, 605
overlength, 51, 187, 199
overload, 51
overload relief valve, 187
overload thermostat, 187
overluxnriant growth, 187
overmantel, 21
overmature forest, 187
overmature stand, 187
overmature wood, 187
overnutrition, 574
overplanting, 33
overpopulation, 627
overrun, 50
overrunning, 51
overscale, 36, 232
overshot wheel, 416
oversize, 50, 51, 136, 412
overspray, 186, 354
overstand wood, 187
overstock, 187
overstocked, 187
overstocked forest, 322
overstocked planting, 323
overstocked stand, 289
overstocking, 186
overstorey, 416, 447, 633
overstorey tree, 416
overstrain, 51
overstress, 51
overstuffing, 474
oversummering, 594
overthrust, 343, 415
overtime, 227
overtone, 130, 348, 405
overtopped stem, 18
overtopping tree, 594
overtreating, 65
overtrim, 199
overturn circulation, 459
overturned tree, 90
overturning moment, 380
overuse, 186
overweight, 187
overwind, 160, 416
overwintering, 594
overworking, 187, 283
overyear, 594
overyeared wood, 187
ovicide, 411
oviduct, 445
oviparity, 311
oviparous reproduction, 311
oviposition, 47
oviposition period, 47
ovipositor, 47
ovist, 311
ovogenesis, 54, 311
ovoid, 311
ovolo, 311
ovotestis, 311
ovoviviparity, 311
ovular, 311, 353, 589
ovulation, 350
ovule, 311, 353, 589
ovule abortion, 353
ovule stalk, 632
ovum (*plur.* ova), 311
owl, 318, 531
owlet, 544
own-root tree, 644
ox bow, 90, 193
ox-skidding, 347
oxalic acid, 42
oxaluria, 42
oxarch, 467
oxbow, 347
oxbow lake, 347
oxharrow, 65
oxidase, 557
oxidation, 557
oxidation-reduction potential, 557
oxidation-reduction system, 557
oxidation stain, 557
oxidative coupling, 557
oxidative phosphorylation, 557
oxidative stain, 557
oxide activity, 557
oxido-reductase, 557
oxo reaction, 473
oxolation, 557
oxonation, 473
oxonium salt, 539, 557
oxyacetylene process, 557
oxycalorific coefficient, 557
oxycellulose, 557
oxygen consumption, 191
oxygen debt, 557
oxygen deficient lake, 388
oxygen meter, 557
oxygen tension, 557
oxygen uptake rate, 557
oxygenation, 229
oxylophyte, 440, 516
oxyphile, 440, 516
oxyphilic, 440, 516
oxyphilous, 440, 516
oxyphobe, 21, 524
oxyphobous, 21, 524
oxyphohous, 556
oxyphyte, 440, 516
oxysere, 467
oyster cup fungus, 73, 187
oyster-knife tooth, 17

oystering, 81, 359
ozonation, 62
ozone hole, 62
ozone layer, 62
ozone-resistant vegetation, 270
ozone seasoning, 62
P-line, 151
P-sampling, 61
P/E ratio, 238
pace method, 35, 36
pachycarpous, 200
pachynema, 73
pachyphyllous, 200
pachytene, 73
pachytene stage, 73
Pacific Lumber Inspection Bureau, 471
Pacific plate, 471
pacing, 35
pack, 12, 114, 280, 446, 478, 550, 639
package, 12, 36, 280, 446, 651
package deal, 55, 563
package immersion treatment, 54
package loading, 54
package logging, 54
package pile, 165
package piling, 54, 652
package transfer, 55
packaged furniture, 47
packaged stacking, 54, 651
packager, 75
packaging veneer, 12
packed bed, 479
packed capillary column, 479
packed column, 479
packer, 75, 366, 550, 638, 639
packer wheel, 604
packing, 12, 53, 322, 478, 479
packing board, 12
packing box, 12
packing case, 12
packing case wood, 224
packing density, 74, 114
packing timber, 12, 529
pacura, 630
pad, 103, 561, 589, 604, 618
pad foot, 143
pad saw, 375, 519
padding, 479, 487
paddle, 238, 243, 470, 560
paddle fan, 287
paddle type agitator, 238
paddle-type blender, 238
paddock, 502, 534
paddy hull, 90
paddy soil, 456
paedogamy, 583
paedogenesis, 583
paedogenesis bisexual, 583
paedogenetic parthenogenesis, 583
paedomorphosis, 583
paeonidin, 417
pagoda, 470
paid-in capital, 566
paid-in surplus, 137
paid-up capital, 566
paid-up stock, 566
paid vacation, 80
paimately compound leaf, 604
paint, 487, 578
paint-and-batten method, 487
paint coating, 354, 487
paint film, 366, 487
paint gun, 354
paint marking, 578
paint naphtha, 480
paint nozzle, 354
paint remover, 494
paint rock, 481
paint scrubber, 516
paintability, 487
painted fibreboard, 578
painting, 487
painting characteristic, 487
pair of photographs, 530
paired with unwarmed plots, 137
pairing, 239, 353
pairing chromosome, 54, 353
palace park, 171
Palaearctic province, 174
Palaearctic region, 174
palaeo-endemic species, 174
palaeobiogeography, 174
palaeobiology, 174
palaeobotany, 174
palaeocormophyta, 174
palaeoecology, 174
palaeoentomology, 174
palaeogenetics, 174
palaeolith, 259
palaeolithic age, 259
palaeontology, 174
palaeopathology, 174
palaeophytology, 174
pal(a)eosol, 174
palaeotropical, 174
Palaeozoic era, 174
palar, 9
palate, 199, 417, 520
pale, 305, 339, 492
pale cutch, 1, 121
pale rosin, 375
palea, 305, 339, 492
paleace, 147, 261, 581
paleaceous, 147, 261, 581
Palearctic realm, 174
paleobiocoenosis, 174
paleoclimatology, 174
paleoecology, 174
Paleogene, 599
paleogeography, 174
paleoichnology, 174
paleola, 238
paleolate, 260
paleoliferous, 260, 261
paleopedology, 174
paleophyte, 174
paleosoil, 174
paleosynchorology, 174
paleosynecology, 174
paleotemperature, 174
paleotropical disjunction, 174
paleotropical floral kingdom, 174
paleotropical region, 174
palet, 305, 339, 492
palinactinodromous, 260
palingenesis, 312, 539, 637
palingenetic character, 60, 61
palingenetic metamorphosis, 61
palingenetie process, 61
palisade cell, 602
palisade fungus, 602
palisade parenchyma, 602
palisade tissue, 602
palladium charcoal, 355, 598
pallet, 120, 223, 229, 493, 533
palletize, 103, 136, 316

palm, 648
palm butter, 648
palm-fat, 559
palm house, 648
palm kernel oil, 648
palm-like lobe, 604
palm-nut oil, 648
palm oil, 559, 648
palm wax, 648
palm wood, 648
palmarosa oil, 108, 319
palmate, 604
palmate leaf, 604
palmate vein, 604
palmate venation, 604
palmately, 399
palmately compound leaf, 604
palmately lobed, 604
palmately nerved, 604
palmately reticulately veined, 604
palmately trifoliolate, 261
palmately veined, 604
palmatifid, 604
palmatilobate, 604
palmatipartite, 604
palmatisect, 604
palminerved, 604
palpus, 65
palyareal plot sampling, 118
palynology, 12, 203
palynotaxonomy, 12, 203
pamorusa oil, 528, 604
pampa, 365
pampa grassland, 42
pampas grass, 365
pampas grassland, 365
pampiniform, 264, 317
pan, 58, 160, 351, 489, 493, 575
pan back chair, 375
pan-breaker, 420, 575
pan logging, 493
pan skidding, 103, 160
Panama angle gauge, 5
panchromatic, 386
panchromatic emulsion, 130, 386
panclimax, 130
pandemic, 365
Pandia continuous digester, 350
panel, 8, 20, 277, 375, 564
panel-and-frame construction, 375
panel back chair, 375
panel board, 353, 358, 375, 529
panel dry kiln, 359
panel furniture, 8
panel gauge, 20
panel length, 245
panel point, 195, 245
panel products, 8
panel selecting conveyer, 546
panel shear test, 8
panel-shears, 8
panel unit, 13, 396
panel(ed) door, 529
panelled door, 165
panelling, 20, 321, 529
pangamy, 130, 647
pangen, 130, 352
panicle, 154, 404, 593
panicled spike, 404
panicled thyrsoid cyme, 322
paniculate, 261
paniculiform, 593
panmictic population, 468
panmictic unit, 468
panmixia (panmixis), 468
pannose, 603
panorama, 214, 386, 387
panoramic profile map, 386
pantanal, 87
panteplankton, 441
pantograph, 469
pantometer, 242
pantomictic plankton, 214
pantomorphism, 385
pantoon, 53
pantophagy, 597
pantropic, 130
pantropical, 130
pantropical plant, 130
papaw, 5
papaya, 128
paper backing, 592, 619
paper birch, 619
paper board, 618
paper chromatography, 618, 619
paper electrophoresis, 618
paper-face overlay, 251
paper factor, 342, 619
paper honeycomb, 619
paper laydown station, 364
paper-making, 601, 620
paper mill, 601
paper overlay, 619
paper-plastic overlay, 619
paper-pulp, 618
paper tape, 618
paper towel, 618
paper web, 618
paper wood, 601
papilionaceous corolla, 104
papilionaceous plant, 104
papilla (*plur.* papillae), 400
papillar(y), 400
papillate, 261, 400
papillose, 118, 261
pappiform, 178
pappus, 178, 397, 398
papreg, 466
paprika, 157
papyrus, 618
par-tox, 67
par value, 358
para braunerde, 640, 648
para caoutchouc tree, 531
para rubber, 5
para rubber tree, 531
parabiont, 292
parabiosphere, 154
parabola, 351
parabolic curve, 351
paraboloid, 351
paracarpium, 32
paracentric inversion, 21
paracorolla, 154
paracrystal, 70, 285
paracrystalline cellulose, 35
paracrystalline cortex, 35
paracrystalline region, 35
paradichlorobenzene, 115
paraffin emulsion, 434
paraffin section, 434
paraffin size, 434
paraffined paper cup, 434
paraffin(ic) oil, 434
paragenesis, 211, 232

P

paraglossa, 42
paraguay tea, 5
paragynous, 43
parallax, 439
parallax bar, 439
parallax difference, 439
parallax drainage pattern, 360, 439
parallax height equation, 439
parallax method, 439
parallax micrometer, 439
parallax wedge, 439
parallel boom of net type, 500
parallel community, 360
parallel construction, 89
parallel current type kiln, 483
parallel evolution, 360
parallel holding ground, 461
parallel-to-grain stress, 461
parallel veins, 360
parallelinervate, 360
parallelism, 360
parallelodromous, 360
paramagnetic shift, 461, 654
parameter, 40
parameter prediction model, 40
parameter recovery model, 40
parameter retrieval, 40
parameterization, 40
parameters of population, 649
paramn-impregnated wood, 434
paramyelin, 154
paramylum, 154, 590
paranotal theory (hypothesis), 42
parapatry, 305
parapetalum, 154
paraphysis, 42, 43, 166, 219
paraplectenchyma, 154
paraschist, 154, 456
paraselene, 209
parasite, 227
parasite food chain, 227
parasite-host interaction, 227
parasite resistance, 227
parasite scurrula, 405
parasite vascular plants, 227
parasites, 227
parasitic, 226
parasitic diptera, 227
parasitic fungi, 215
parasitic hymenoptera, 226
parasitic insect, 227
parasitic plant, 227
parasitic potential, 227
parasitic process, 226
parasitic root, 226
parasitic seed plant, 227
parasitic wasp, 226
parasitism, 226, 227
parasitoid, 227, 342
parasitoidism, 342
parasitology, 226, 227
parasol mushroom, 404
parasol pine oil, 396
parastamen, 32
parastemon, 32
parathion, 115, 563
parathion dust, 563
parathion wettable powder, 563
paratracheal, 12
paratracheal banded parenchyma, 12
paratracheal-confluent parenchyma, 12
paratracheal parenchyma, 12
paratracheal-vasicentric parenchyma, 12
paratracheal zonate parenchyma, 12
paratype, 154
parbuckle, 196
parbuckling, 196, 197
parcel, 37, 114, 355, 439, 655
parcell, 299
parchment paper, 556
parenchyma, 13
parenchyma cell, 13
parenchyma strand, 13
parenchymatous cells, 399
parenchymatous disease, 13
parent, 379
parent bed rock, 220, 328
parent cell wall, 379
parent company, 327
parent layer, 551
parent material, 328, 379, 591
parent material horizon, 328
parent-offspring correlation, 379
parent-offspring mating, 379
parent peak, 327
parent population, 592
parent rock, 328, 591
parent root, 327
parent stand, 39, 328
parent stock, 168, 328, 590, 608
parent tree, 328
parental behaviour, 453
parental care, 379, 585
parental investment, 379
parental pair, 379
parer, 324, 374
parhelion, 209
parietal, 43, 630
paring chisel, 13, 532, 599
paring gouge, 201
paring plough, 374
paripinnate, 349
paripinnate leaf, 349
Paris green, 5
Paris green preservative, 5
parish forest, 243
parity price, 91, 361
park, 171, 303, 559
park administration, 171
park and garden, 587
park and garden construction, 588
park and garden design, 588
park and garden planning, 588
park attractiveness, 171
park day-use visitation, 563
park engineering, 171
park forest, 407
park maintenance, 588
park management, 171
park nursery, 171
park planning, 171
park system, 171
park way, 171
parkine, 350
parking leg, 481
parkland, 445, 515
parkway, 171, 204, 303
parkway system, 303

parlour chair, 274
parqnet strip, 359, 529
parquet, 358, 529
parquet block, 358, 529
parquet border, 358, 529
parquet floor, 358, 529
parquetry, 529
parquetry stave, 358
part, 36, 141, 305
part per hundred parts of rubber, 6
part per million (ppm), 7
part time drying system, 236
parthenocarpic fruit, 86
parthenocarpic fruit-set, 86
parthenocarpic(al), 86
parthenocarpy, 86
parthenogeneic reproduction, 86, 174
parthenogenese, 86
parthenogenesis, 86, 174
parthenogenesis facultative, 348
parthenogenetic, 86, 174
parthenogenetic cleavage, 86
parthenogenetic diploid, 86
parthenogenetic egg, 86
parthenogenetic female, 86, 174
parthenogenetic generation, 86, 174
parthenogenetic propagation, 86, 174
parthenogonidium, 86
parthenomixis, 86, 174
partial adhering knot, 36
partial compression test, 36
partial correlation, 257, 357
partial cutting, 142, 259
partial deep decay, 36
partial diallel crossing, 36
partial disposal, 259
partial dominance, 35, 36
partial-equilibrium theory, 36
partial fertilization, 36, 259
partial forest survey, 259
partial habitat, 259
partial intergrown knot, 9
partial mast, 534
partial pressure gradient, 143
partial regression, 357
partial seed year, 245
partial seeding, 259
partial sex-linkage, 36
partial shallow decay, 36
partial shipment, 142
partial slash disposal, 127, 381
partial soil preparation, 259
partial sowing, 259
partial sterility, 36
partial sterilization, 36
particle, 272, 468
particle board, 16, 468
particle board core plywood, 16, 153
particle board core stock, 16, 468
particle board of medium density, 624
particle density, 272
particle moulding, 16
particle separator, 468
particle size, 111, 272, 291, 517
particulate material, 272
particulate matter, 501
particulate matter less than 100 μm in diameter (PM100), 649
particulate matter less than 2.5 μm in diameter (PM2.5), 273
particulate matter of 2.5–10 μm in diameter (PM10), 274
parting agent, 142, 494
parting bead, 166
parting slip, 166
partition, 141, 165, 235
partition chromatography, 142
partition coefficient, 142
partition column, 142
partition liquid, 142
partition wall, 166
partly bleach pulp, 10
partly finished goods, 9
partners desk, 115
partnership, 192
Paschke (drum-peeler), 349
pass, 90
pass chain, 2
pass line, 2, 570
pass-out head, 350
pass point, 66, 252
pass-shooting, 136
passage, 481, 651
passage cell, 481
passage form, 187, 623
passblock, 231
passing knot, 65, 179
passing migrant, 314
passing-place, 637
passing-through park, 554
passivating agent, 69
passive dispersal, 18
passive earth pressure, 18
passive immunity, 18
passive infection, 18
passive remote sensing, 18, 512
passive soil former, 120
passive warming, 18
passometer, 36
past costs, 187, 226
paste, 201, 237
paste paint, 200
paste resin, 201
paster, 240, 241, 487
pasteurization, 5, 228
pasteurizer, 5
pasturage, 336
pasture, 135, 336
pasture field, 135
pasture land, 135
pasture management, 336
pasture protection forest, 202
pasture range, 135
patch, 33, 277, 318, 402
patch bark, 279
patch budding, 33, 480
patch burning, 259, 279
patch clear-cutting system, 277, 279
patch logging, 239, 279
patch method, 277
patch planting, 279
patch-scarifier, 279
patch-sowing, 279
patcher, 81
patchiness, 7
patching, 375
patching machine, 81, 82
patching up, 32, 33

P

patchouli oil, 182
patent, 475, 636
paternal form, 151
paternal inheritance, 151
path, 395, 534
path analysis, 482
path coefficient, 482
path diagram, 482
pathochrome, 405
pathogen-free, 508
pathogen-free stock, 508
pathogen(e), 29
pathogenesis, 123, 621
pathogen(et)ic, 29, 621
pathogenic bacterium, 29
pathogenic biotype, 621
pathogenic fungus, 29
pathogenic metabolite, 621
pathogenic organism, 29
pathogenic parasite, 29
pathogenic race, 621
pathogenic strain, 621
pathogenicity, 29, 621
pathogenism, 621
pathogenous, 29, 621
pathologic form, 621
pathologic plant anatomy, 616
pathologic reaction, 29
pathologic(al), 29
pathological factor, 29
pathological resin, 29
pathologic(al) rotation, 29
pathology of plant, 616
pathotype, 621
pathradiance, 181
patina, 484
patio, 624
patobiont, 298
patowle, 298
patoxene, 299
patriliny, 151
patroclinous, 357
patroclinous inheritance, 357
patrogenesis, 174
patrol time, 549
pattern, 165, 204, 325, 327, 506, 541
pattern class, 487
pattern classification, 325, 487
pattern discrimination, 325, 487
pattern line sawing, 204
pattern maker, 333
pattern maker's lathe, 326, 333
pattern of mixture, 213
pattern recognition, 325, 487
pauperization, 597
paurometabola, 236
pave, 94, 155, 364
pavement, 364
pavi-coronate, 533
pavilion, 21, 295, 571
paving block, 364
paving design, 364
paw foot, 604
pawpaw, 5
pay-basis by cubic feet, 3
pay-basis by running feet, 3
pay-basis by the piece, 225
pay for stacking, 219
pay off, 32, 135, 151
pay off matrix, 442
pay off period, 212
pay out, 135, 207, 612
pay pole, 578
payee, 443
payer, 151
paying concern, 573
payload, 151, 170, 257, 582
payment in advance, 586
payroll, 169, 170
payroll tax, 169
pbytopathogen, 616
pcercent of forest cover, 406
pe-la insect, 6
peach gum, 474
peak day, 146
peak discharge, 198, 652
peak flow, 198, 652
peak germination percent, 146
peak load, 151, 653
peak resolution, 653
peak to valley method, 146
peak value, 146
peaker, 146, 367
pearl ash, 73, 607
pearlite, 607
peary hook, 7
peat, 342
peat block, 342
peat bog, 342
peat bog fire, 342
peat cube, 342, 574
peat horizon, 342
peat moss, 342
peat plant block, 342
peat soil, 342
peat tar, 342
peat wax, 342
peatland, 342
peaty clay, 342
peaty gley soil, 342
peaty soil, 342
peavie, 7
peavy, 7
peavy dog, 7
peavy (peavie), 173
pebble, 291, 434
pebble tool, 311
pecan oil, 413
pecies, 450
peck, 81, 640
peck order, 640
peck right, 640
peckerwood mill, 413, 564
peckiness, 81, 640
pecky, 7, 581, 582
pecky dry rot, 81
pectase, 186
pecten, 444, 621
pectic acid, 186
pectic body, 241
pectic material, 186, 617
pectic matrix, 186
pectin, 241
pectin substance, 186
pectinase, 186
pectin(e), 186
pecto-cellulose, 186
pectose, 186
ped, 246, 491
pedalfer, 304, 480
pedalium, 561
pedate, 346
pedestal, 103, 219, 221, 471, 561
pedicel, 169, 186, 203
pedicellate, 260, 261, 580
pedigree, 365
pedigree culture, 365, 517
pedigree method, 365
pedigree seed, 592
pedigree selection, 365
pedigree system, 365
pedigree trial method, 365
pedocal, 155
pedoclimax, 489

pedogenesis, 55, 489
pedogenic environment, 55
pedogenic process, 55, 489
pedology, 219, 491
pedometer, 288
pedon, 83, 491
pedonic, 489
peduncle, 29, 203, 649
pedunculagin, 203
peel, 30, 496, 546
peel adhesive test, 240
peeled billet-wood, 30
peeled log, 385
peeled softwood veneer, 546
peeled wood, 30, 385
peeler, 31, 240, 546
peeler bar, 31
peeler block, 113, 240, 534
peeler grade, 240
peeler (lathe) log, 546
peeler log, 240, 546
peeler multiple, 546
peeler picker, 240
peeling, 30, 385, 546
peeling axe, 31
peeling damage, 31
peeling defect, 31, 546
peeling iron, 30
peeling lathe, 31, 546
peeling lathe with telescopic spindle, 453
peeling machine, 31, 546
peeling(-off) reaction, 30
peeling strength, 30
peeling test, 30, 546
peeling velocity, 546
peep hole, 486
peepul, 364
peepul tree, 364
peet, 342
peewee, 534
peg, 105, 332, 334, 349
peg board, 106, 515, 652
peg graft, 612, 638
peg planting, 69, 640
peg-raker saw, 153, 453
peg root, 614
peg saw, 402
peg tooth, 402
peg-tooth harrow, 105
peg-tooth saw, 402
pegging, 107, 288
pelagic, 594
pelagic division, 456
pelagic egg, 148
pelagic fish, 415
pelagic fishery, 594
pelagic zone, 594
pelagium, 188
pelagos, 456
pelite, 342
pellet, 87, 139, 212, 291, 464, 627
pelleting, 601
pelletted seed, 498, 627
pellicle, 13, 148, 267, 326, 401, 492
pellucidity, 485
peltate, 116
pembroke table, 606
pen, 254, 353
penal rate, 221
pencil tracing, 372
pendant block, 67
pendent respiratory root, 545
penduliflory, 545
pendulum bar, 7
pendulum cross cut saw, 7, 104
pendulum dynamometer, 108
pendulum hammer, 7
pendulum impact test, 7
pendulum impact testing machine, 7
pendulum saw, 7, 104
penehlorol, 512
peneplain, 640
peneplane, 640
penetrability, 421
penetrance, 497
penetrant, 421
penetrating ability, 421
penetrating power, 421
penetration, 378, 421
penetration hypha, 378
penjing, 355
penner oil, 201
penninerved, 584
pennyroyal oil, 188
penta, 226, 512
penta prism caliper, 512
pentachaenium, 512
pentachiorophenol, 512
pentachloronitroben-zene, 512
pentacotyl, 512
pentacyclic, 512
pentacyclic flower, 512
pentadactylous, 512, 619
pentadelphous stamen, 512
pentafoliate, 512
pentagynous, 512
pentakenium, 512
pentaploid, 512
pentarch, 512
pentosan, 118, 119, 264, 512
pentose, 512
pentose phosphate pathway, 304
pentoside, 512
peonidin, 417
peonin, 417
pepo, 202
pepper oil, 201
peppermint gum, 541
peppermint oil, 13
peptize, 241
peptone, 88
per capita growth rate, 395
peracid, 187
perambulator, 44, 443
peraphyllum, 493
percent cover, 156
percent crown cover, 585
percent forest cover, 300, 408, 410
percent of autumn wood, 382
percent of bark, 449
percent of early wood, 599
percent of late wood, 499
percent of sapwood, 21
percent of spring wood, 599
percent of summer wood, 499
percentage absence, 6
percentage basal area growth, 113
percentage depletion, 191
percentage diameter growth, 615
percentage error, 513
percentage gross volume increase, 72, 318
percentage humidity, 14, 527
percentage illumination, 527, 605
percentage irradiance, 527, 605
percentage net growth, 257

P

P

percentage of absence, 504
percentage of cut-turn, 62
percentage of forest cover, 406
percentage of out-turn, 62
percentage of purity, 256
percentage of vegetational cover, 616
percentage of volume, 36
percentage point, 6
percentage relative humidity, 527
percentage shrinkage, 158, 442
percentage similarity, 6
percentage swelling, 355
percentage volume growth, 37
percentile, 6
percentile taper, 547
perception of landscape, 145
perched ground water, 258, 545, 622
perched water table, 258, 545, 622
perclimax, 372
percolate, 187, 251, 420
percolation, 420
percolation method, 420
percolation rate, 420, 421
percolation ratio, 420
percolator, 420
percussion-cap, 123, 284, 484
perdominant, 49
perennating bud, 386
perennial, 118
perennial canker, 118
perennial colony, 118
perennial forb, 118
perennial form, 118
perennial fruit body, 118
perennial fungus, 118
perennial grass, 118
perennial herb, 118
perennial plant, 118
perennial plants, 118, 467
perennial river, 48
perfect combustion, 498
perfect floret, 498
perfect flower, 260, 498
perfect leaf, 498
perfect state, 582
perfect stuff, 498
perfectly density-dependent factor, 498
perfluorocarbons, 385
perfoliate leaf, 15, 179
perforated card, 65
perforated cylinder, 117, 412, 593
perforated hardboard, 65
perforated index card, 65
perforated panel type absorber, 65
perforated plate, 65
perforated steam pipe, 117
perforated tape, 75, 581
perforation, 65
perforation plate, 65
perforation rim, 65
performance, 41, 536
performance analysis, 169, 541
performance budgeting, 615
performance test, 541
perfume, 528
perfume garden, 528
pergamyn paper, 338
pergola, 295, 475
perhydrogenated rosin, 386
perianth, 202, 461, 648
perianth lobe, 202
perianth segment, 202
perianthial, 202
peribion, 22
periblem, 356
pericambium, 630
pericarp, 186, 338
pericarpial, 186
pericarpic, 186
pericentric inversion, 21
pericladium, 561
periclinal cambial cell division, 363
periclinal chim(a)era, 630
periclinal division, 363
periclinium, 648
pericycle, 625
pericyclic fibre, 625
periderm, 27, 496, 630
peridermium, 352
peridium, 12, 186, 495
perigee, 250
periglacial, 29
perigone, 202, 203, 542
perigonium, 202
perigynous, 630
perigynous flower, 630
perigynous with inferior ovary, 521
perigynous with superior ovary, 416
perihelion, 250
perilla oil, 642
period, 142, 254
period of return, 211, 311
periodic annual growth, 106
periodic annual inerement, 106
periodic annual yield, 106
periodic area allotment method, 106
periodic attendant counts, 107
periodic average increment, 107
periodic block, 107
periodic change, 630
periodic coupe, 142, 312
periodic cruise, 107
periodic cut, 106
periodic cutting, 142
periodic cutting-area, 142
periodic cutting plan, 106
periodic forest inventory method, 107
periodic grazing, 630
periodic growth, 107
periodic increment, 107
periodic intermittent working, 106
periodic inventory, 107, 639
periodic maintenance, 107
periodic mean annual increment, 106
periodic mean annual intermediate yield, 106
periodic method by area and volume combined, 107
periodic method by volume, 106
periodic mortality, 106
periodic outbreak, 630

periodic plankton, 630
periodic sale, 106
periodic sample, 107
periodic selection system, 107
periodic steaming, 106, 236
periodic succession, 630
periodic working, 107
periodic yield, 107
periodical tapping, 39, 106
periodicity, 235, 630
periodicity of germination, 124
periodicity of seed bearing, 245
periodism, 630
periodogram, 630
periopods, 36, 613
peripheral milling, 593
peripheral nervous system, 497, 630
periphery, 502, 630
periphoranthium, 648
periphysis, 593
periphyte, 177, 459
periphyton, 630
perish, 150, 278, 558
perisperm, 496
peristachyum, 574
perithecium, 641
peritrichous flagellum, 630
peritrophic, 502
peritrophic membrane, 502
peritrophic mycorrhizae, 630
perivisceral sinus, 502
periwinkle, 48
permafrost, 48, 576
permafrost ecosystems, 576
permafrost table, 576
permanence, 57
permanency, 57
permanent bird, 306
permanent cage, 175
permanent community, 506, 576
permanent consumption, 49
permanent culture, 576
permanent deformation, 40, 576
permanent elongation, 40, 576
permanent expansion, 57
permanent flow, 506
permanent growth plot, 176, 576
permanent habitat, 576
permanent hybrid, 576
permanent income, 49
permanent landscape attribute, 576
permanent layer transect, 107
permanent magnet, 576
permanent nursery, 176, 576
permanent pasture, 576
permanent periodic block, 176
permanent plankton, 626
permanent plastic deformation, 576
permanent plot, 107, 576
permanent quadrat, 107, 576
permanent quadrat method, 576
permanent reserve forest, 576
permanent sample area, 176, 576
permanent sample plot, 576
permanent set, 40, 576
permanent slide, 576
permanent state forest, 576
permanent succession, 49
permanent tissue, 576
permanent way, 576
permanent wilting, 576
permanent wilting dose, 576
permanent wilting percent, 576
permanent wilting percentage, 576
permanent wilting point, 576
permanganate number, 163
Permasan, 349
permeability, 65, 421, 486
permeability barrier, 486
permeability coefficient, 486
permeable layer, 421
permeable membrane, 486
permeable seed, 273
permeant water, 421
Permian period, 122
permissible defect, 398
permissible load, 2, 397
permissible revenue, 398
permissible stress, 398
permit holder, 595, 651
permitted hunter-kill ratio, 397
permo-treatment, 421
perny silk moth, 654
perny silk worm, 654
peroxidase, 187
peroxisome, 187
perpendicular cutting pattern, 68
perpendicular heating, 68
perpendicular shot, 530
perpendicular to grain load at failure, 196
perpetual bloom, 293, 463
perpetual flowering, 293, 450, 463
perron, 311, 441
perseverance, 15
Persian walnut, 201
persic oil, 541
persimmon oil, 439
persistence, 57, 337
persistent calyx, 467
persistent pathogen, 467
persistent virus, 57
personal lag, 130
personate corolla, 230
perspective, 373, 486
perspective drawing, 486
perspective view, 486
perthite, 479
perthotroph, 462
perturbation lifetime, 34
perula, 551
pervious, 65, 421, 486
pes (*plur.* pedes), 613, 651
pessimum, 297
pest, 29, 580
pest control, 29
pest management, 189
pesticide, 347
pesticide contamination, 347

P

Pestox-14, 230
Pestox-15, 29
Pestox-III, 4
pests of broad-leaf trees, 282
pests of coniferous trees, 607
petal, 202
petal bundle, 202
petalode, 230
petaloid, 202
petaloid calyx, 202
petalous, 580
Petersburg standard, 20
petiolar, 560
petiole, 29, 155, 560
petioled, 261
petiolule, 535
petitgrain oil, 57
petra-, 371
petragram (Pg), 371, 433
petrefaction, 150
Petri dish, 353, 362
Petri-dish-agar method, 353
Petri-dish method of determining toxicity, 44
Petri dish test, 353
petrideserta, 554
petrification, 433
petrified forest, 433
petrium, 291
Petrograd standard, 20
petroleum emulsion, 434
petroleum engine, 434
petroleum moter, 434
petroleum-paraflin rosin mixture, 320
petrology, 554
petrophyte, 434
petty forestry, 533
petunidin, 372
pew, 8, 243
PF-salt, 123
PF-value, 615
pfanerophytes, 450
Pfister process, 139
pH-value, 467, 615
phaenerophyta herbaceae, 41
phaenerophyta scandentia, 475
phaenerophyta succulenta, 399
phaenerophyte, 164
phaeozem, 211, 406
phage, 441
phagocyte, 441
phagocytosis, 441
phallus, 556
phaltan, 324
phaner~, 164
phanero~, 164
phanerogam, 524
phanerophyta herbacea, 42
phanerophyta scandentia, 328
phanerophytes (Ph), 164
phaoplankton, 516
pharmaceutical chemicals, 559
pharyngeal ganglion, 553
pharynx, 553
phase, 243, 365, 530
phase-contrast microscope, 530
phase deviation, 530
phase diagram, 360, 530
phase gregaria, 388
phase of cell-wall thickening, 517
phase of lignification, 335
phase polymorphism, 243, 527
phase solitaria, 111
phase theory, 24
phase variation, 471, 530
phasic development, 243
phasmid, 632
pheasant, 622
phellem, 334, 452
phellem layer, 334
phellodendron, 210
phelloderm, 452
phellogen, 334
phelloid cell, 342
phellum, 452
phene, 19, 27
phenetics, 27
phenocopy, 27, 342
phenogam, 524
phenogenetic correlation, 27
phenogenetics, 27, 125
phenol, 19, 143, 434
phenol thermosetting resin, 19
phenolic-impregnated wood, 19
phenological calendar, 512
phenological isolation, 512
phenological observation, 512
phenological phase, 512
phenological spectrum, 512
phenological succession, 513
phenology, 512
phenology avoidance, 512
phenology mosaic, 512
phenophase, 512
phenotype, 27
phenotype covariance, 27
phenotype test, 27
phenotype value, 27
phenotype variance, 27
phenotypic selection, 27
pheophytin, 494
pheromone, 496, 540
phialospore, 363
~phile, 516
philia, 516
philic, 516
philous, 516
phinatas oil, 189
phlegma, 286
phloem, 396
phloem bundle, 396
phloem fiber, 396
phloem fibre, 396
phloem mother cell, 396
phloem parenchyma, 396
phloem parenehyma cell, 396
phloem ray, 396
phloem sclereid, 396
phobe, 524
phobi, 524
phobia, 524
phobic, 524
phobous, 524
pholadobiose, 519
phonological, 245
phonotaxis, 383
phonotropism, 530
phorate, 230
phoresy, 537
phosphamidon, 304
phosphatic sandstone, 305
phosphatic shale, 305
phosphatide, 305
phosphite, 304, 552
phosphorescence of wood, 330

phosphorus absorption coefficient, 514
phosphorus cycle, 304
phosphorylation, 304
photic zone, 485
photo base, 419
photo center, 531
photo controlled inventory, 531
photo-cured coating, 180
photo-electric direct reading spectrometer, 180
photo-electric effect, 180
photo interpretation, 530
photo-interpretation key, 530
photo map, 191
photo mapping radiometer, 530
photo measurement, 530
photo plot, 530
photo scale reciprocal, 530
photo timber cruise, 530
photo tone, 530
photo volume, 530
photo volume estimate, 530
photo volume table, 530
photoactivation, 181
photoassimilation, 182
photoautotrophic, 182
photoautoxidation, 182
photobacterium, 123
photobiology, 181
photoblastic seed, 180
photocatalyst, 180
photochemical, 181
photochemical air pollutant, 181
photochemical atmosphere, 181
photochemical complex, 181
photochemical induction, 181
photochemical pollution, 181
photochemical process, 181
photochemical reaction, 181
photochemical smog, 181
photochemical smog complex, 181
photochemistry, 181
photocuring agent, 180
photoelastic, 180
photoelastic test, 180
photoelectric analytical balance, 180
photoelectric cell, 180
photoelectric colourimetry, 180
photoelectric glossmeter, 180
photoelectric turbidometer, 180
photogrammetric cruising, 419
photogrammetry, 419
photograph coordinate, 531
photographic base, 419
photographic exposure, 419
photographic map, 191
photographic plate, 419, 605
photographic surveying, 419
photoinactivation, 180, 181
photoinhibition, 182
photointerpreter, 530
photokinesis, 180
photolysis, 181
photomap, 574
photometer, 180
photometric titration, 180
photometry, 180
photomorphogenesis, 182
photomultiplier, 180
photomultiplier tube, 180
photon flux density, 181
photonasty, 159
photooxidation, 182
photoperception, 180
photoperiod, 182
photoperiodic effect, 182
photoperiodic induction, 182
photoperiodic response, 182
photoperiodicity, 182
photoperiodism, 182
photophase, 182
photophilous, 441, 516
photophobic, 21
photoreaction, 180
photoreception, 180
photoreceptor, 443
photoreceptor cell, 159
photoreduction, 180
photorespiration, 181
photosensitivity, 181
photosensitization, 181
photosensitizer, 181
photostage, 182
photostimulation, 180
photosynthate, 180
photosynthesing cell, 181
photosynthesis, 181
photosynthetic ability, 180
photosynthetic active wave length, 181
photosynthetic area, 180
photosynthetic bacteria, 181
photosynthetic capacity, 180
photosynthetic effect, 182
photosynthetic efficiency, 181
photosynthetic organ, 181
photosynthetic parenchyma, 180
photosynthetic part, 180
photosynthetic photon flux density, 180
photosynthetic productivity, 181
photosynthetic quotient, 181
photosynthetic rate, 181
photosynthetic ratio, 180
photosynthetic surface, 181
photosynthetic system, 181
photosynthetic yield, 180
photosynthetically active radiation (PAR), 181, 423
photosystem II, 182
phototactic, 383
phototactic rhythm, 383
phototaxis, 383
phototaxy, 383
phototheodolite, 419
phototroph, 181
phototropism, 383, 530
phreatic surface, 98, 374
phreatic water, 98, 374
phreatophyte, 374, 524

phrygana, 48
phunki, 55
phyad, 617
phycochrome, 87
phycology, 599
phycomycete, 599
phyletic evolution, 526
phyllade, 305
phylliform, 561
phyllin, 560
phyllite, 371, 560
phyllocaline, 55
phylloclade, 230, 561
phyllocladene, 23, 230
phyllode, 230, 561
phyllody, 24
phylloerythrin, 560
phyllogen, 106, 561
phylloid, 561
phylloid truss, 561
phylloidal, 561
phylloplane, 560
phyllosoma, 561
phyllosphere, 560, 561
phyllotaxis, 561
phyllotaxy, 561
phyllula, 561
phyllule, 560
phylogenetic botany, 618
phylogenetic series, 517
phylogeny, 517, 627
phylum, 320
phyplankton, 148
physical age, 423
physical assets, 436
physical climate, 513
physical damage, 646
physical dormancy, 513
physical dryness, 513
physical ecology, 513
physical factor, 513
physical geography, 645
physical market, 436, 525
physical maturity, 423, 645
physical phenomena, 513
physical property, 513
physical resistance, 513
physical rotation, 423, 646
physical units, 436, 513
physical weathering, 513
physiognomic classification of community, 389
physiognomic dominance, 255
physiognomy, 496
physiognomy of forest, 409
physiographic climax, 99
physiographic map, 156, 645
physiographic plant formation, 96
physiographic type, 645
physiography, 98, 645
physiologic species, 423
physiological, 423
physiological acid fertilizer, 423
physiological basic fertilizer, 423
physiological chemistry, 423
physiological chemistry of plant, 617
physiological combustion, 423
physiological deterioration, 423
physiological disease, 423
physiological disorder, 423
physiological dormancy, 423
physiological drought, 423
physiological drying, 423
physiological dryness, 423
physiological ecology, 423
physiological genetics, 423
physiological insect-pest, 423
physiological isolation, 423
physiological life history, 423
physiological longevity, 423
physiological maturity, 423
physiological moisture, 423
physiological natality, 423
physiological plasticity, 423
physiological polymorphism, 424
physiological race, 423
physiological resin, 423
physiological resistance, 423
physiological rhythm, 423
physiological species, 423
physiological taxonomy, 423
physiological timing, 423
physiological water, 423
physiological zero, 423
physiology, 423
~phyte, 616
phytene, 618
phyto~, 616
phytoalexin, 615, 617
phytobacteriology, 616
phytobenthos, 95, 456
phytochemical ecology, 617
phytochemistry, 617
phytochrom, 617
phytochrome, 181
phytocide, 64
phytoclimate, 618
phytoclimatology, 617
phytocoenology, 617
phytocoenose, 617
phytocoenosis, 617
phytocoenosium, 617
phytocommunity, 617
phytoecological geography, 617
phytoecology, 618
phytoedaphon, 490
phytogenetic, 617
phytogenetic botany, 617
phytogenetic soil, 616
phytogenetics, 618
phytogeny, 517
phytogeography, 616
phytograph, 618
phytogroup, 617
phytohormone, 617
phytokinin, 618
phytolith, 617, 618
phytology, 618
phytolysine, 617
phytometer, 609, 617
phytometer method, 617
phytoncide, 617
phytopaleontology, 174, 617
phytopathologist, 616
phytopathology, 616
phytophaga, 437
phytophagous, 437

phytophagous animal, 437
phytophagy, 42, 616
phytopharmacology, 618
phytophenology, 618
phytophoric rock, 618
phytoplankton, 148
phytoplankton bloom, 148
phytosociology, 617
phytotomy, 617
phytotoxemia, 616
phytotoxicity, 617
phytotoxin, 617
phytotron(e), 395
picaroon, 75, 173
picaroon hole, 173
piciform pit, 595
piciform pit pair, 595
pick, 245, 254, 479, 602
pick-haue, 194, 433
pick mattok, 194, 231
pick strength, 27
pick the rear, 307
pick-up block, 367, 639
pick-up grain, 462
pick-up head, 44
pick-up man, 442
pickaroon, 112
pickaroon hole, 75, 173
pickaxe, 433
picked roller, 445, 606
picker, 186, 603
picket, 275, 336, 612
picking-up, 558
pickup, 380
picnic ground, 559
pictorial drawing, 255
picture-backing board, 486
picture-field, 206, 487
piece of work, 169
piece product, 225
piece rate, 225
piece-rate system, 225
piece wage, 225
piece wage rate, 225
piece-work, 225
piece-work agreement, 225
piece worker, 225
piedmont alluvial plain, 413
piedmont zone, 413
pier, 8, 131, 377
pier bedroom, 79
pier dam, 89, 333, 434, 446
pier table, 271
piercing saw, 537
piercing type, 71
piezo-metamorphism, 549
piezoelectric constant, 549
piezoelectric crystal, 549
piezoelectric effect, 549
piezoelectric modulus, 549
piezoelectric property, 549
piezoelectric structure, 549
piezoelectric tensor, 549
piezoelectricity, 549
pig eye, 552
pigeon, 165
pigeon pea, 332
piggy back tank, 456
piggyback, 647
pigment, 405, 555
pigment figure, 405
pike pole, 49, 484, 640
pilaster, 10, 21, 635
pile, 36, 76, 183, 286, 638
pile bent, 76, 350
pile burn, 114
pile discolouration, 114
pile drawer, 5
pile driver, 76
pile extractor, 5
pile foundation, 638
pile of lumber, 8, 262
pile of wood, 219
pile timber, 638
pile trestle, 76
pileate, 581
piled fathom, 114
piled lot, 114, 285, 286
piled measure, 114
piled volume, 45, 114
pileorhiza, 167
piler, 183
piles-rhiza, 167
pileus, 266, 267
piliferous, 18, 260
piliferous layer, 168
piling, 76, 183, 638
piling-and-burning, 183
piling blue, 114, 285
piling defect, 114, 635
piling hook, 333
piling jack, 183, 638
piling lumber for kiln drying, 394
piling machine, 76
piling railroad, 316, 603
piling show, 39, 638
piling site, 76, 285, 286
piling stick, 103, 286
piling track, 316, 603
piling up, 114
pillar, 221, 612
pillar-crane, 612
pillarwood, 280
pilot, 231, 480, 522
pilot forest, 439
pilot-hole, 107, 522, 638
pilot inventory, 522, 586
pilot motor, 150
pilot operated back-up roll, 522
pilot plant, 438, 624
pilot survey, 522, 586
pilotless, 510
pimaradiene, 188
pimaric type acid, 188
pin, 46, 444, 532, 606
pin-hole, 533
pin knot, 518, 606
pin-knot group, 518
pinch, 229
pinch roll, 229, 550
pinching, 75, 603
pine, 464
pine beauty (*Noctu piniperda*), 465
pine beetle (*Myelophilus piniperda* L.), 465
pine-black, 465
pine caterpillars, 464
pine-cluster cups, 465
pine greedy scale, 464
pine gum, 465
pine-leaf-cast, 464
pine leaf oil, 465
pine logger (Lidgerwood), 282
pine-moth (*Dendrolimus pini* L.), 464
pine needle oil, 465
pine-needle rust, 465
pine nut, 464, 465
pine oil, 465
pine oleoresin, 465
pine root oil, 464
pine-seed oil, 465
pine-soot, 465
pine tar, 464
pine tar pitch, 464
pine-tree oil, 464
pine-weevil, 464
pine wood, 464
pine-wood creosote, 464

P

pine wood nematode, 464
pine wood oil, 464
pine wood tar, 464
pinery, 464
pinetum, 464
pinger, 430
pingo, 534
pinheiro, 338
pinhole, 60, 606
pinhole borer, 50, 485, 533
pinic, 464
pink garden, 139
pinnate, 584
pinnate compound leaf, 584
pinnate leaf, 584
pinnate vein, 584
pinnately compound leaf, 584
pinnately decompound leaf, 584
pinnately lobed, 584
pinnately reticulately veined, 584
pinnately trifoliolate, 261
pinniform, 584
pinnulate, 260, 261, 584
pinnule, 122, 535
pinocembrin, 464
pinoid pit, 465
pinoid pit pair, 465
pinomyricetin, 464, 465
pioneer, 522
pioneer community, 522
pioneer enterprise, 522
pioneer guild, 522
pioneer plant, 522
pioneer species, 522
pioneer tree species, 522
pioneering, 268
pip, 627
pipe model theory, 178
pipe stock, 332
piped log, 178, 276
pipe(d) (rot), 179
pipeline transport, 178
pipework, 178
pipit, 309
piracy, 120
pirate stream, 120
pirce range, 230
pirn, 522
pisang wax, 5
piscivore, 437
pistil, 69
pistillate flower, 69
pistillody, 69, 538
pistol butt, 498
pistol grip, 443
pistol-grip dibble, 375
piston, 214, 635
piston displacement, 214
piston-type pump, 214, 635
pistorius condenser, 205
pit, 21, 41, 240, 275, 280, 506
pit-and-mound topography, 275
pit annulus, 506
pit aperture, 506
pit area, 506
pit aspiration, 506
pit border, 506
pit canal, 506
pit cavity, 506
pit chamber, 506
pit dwelling, 547
pit-fall trap, 527
pit field, 506
pit kiln, 275
pit kiln process, 275
pit loading, 92, 164
pit man, 521
pit membrane, 506
pit membrane pore, 501
pit orifice, 506
pit-pair, 506
pit planting, 275, 548
pit prop, 280
pit saw, 275, 453, 650
pit storage, 243
pit temperature, 609
pit torus, 21, 506
pit wood, 275, 280
pit zone, 506
pitch, 16, 58, 245, 291, 449, 468
pitch blister, 613
pitch canker, 450
pitch circle diameter, 245
pitch coke, 291
pitch defect, 165
pitch down, 521
pitch hole, 450
pitch pocket, 450, 578
pitch seam, 450
pitch shake, 450
pitch streak, 449, 450
pitch stump, 189, 325
pitch trouble, 450
pitch tube, 449
pitched pile, 380
pitched roof, 285, 537
pitching, 309, 364
pitfall, 527
pith, 468
pith cavity, 468
pith check, 538
pith fleck, 468
pith knot, 275, 468
pith ray, 468
pith ray fleck, 468
pith sheath, 468
pith wood, 468, 583
pithshaken, 468
pitted cell, 506
pitted element, 506
pitted perforation, 506
pitted sap rot, 275
pitted secondary wall, 506
pitted tracheid, 506
pitted vessel, 506
pitting, 266, 276, 506, 548
pivot, 243, 292, 383, 546, 612, 630
pivot point, 243, 612, 637
pivoted door, 546
pixel, 531
pjarapatric speciation, 298
placenta, 471
placic horizon, 13
plagioclase rock, 537
plagioclimax, 357
plagiogeotropism, 537
plagiosere, 357
plagiotropism, 537
plague, 506
plain, 362
plain cut figure, 524
plain edge joint, 114
plain grain face, 524
plain joint, 114, 361
plain-sawed, 524
plain-sawed lumber, 523, 524
plain sawing, 524
plain-wood, 366
plain wood of lacquerware, 366
plan of management, 432
plan of operations, 253
plan of silvicultral operations, 600
planchette, 519
Planck constant(b), 365
Planck's law, 365
plane, 16, 362, 545

plane-castes, 16
plane grating spectrograph, 362
plane-iron, 16
plane panel, 363
plane polarizer, 362
plane scarf joint, 362
plane stock, 16, 17
plane-surveying, 362
plane table, 359
plane table surveying, 359
plane type raft, 362
plane wood, 16
planed lumber, 16
planer, 16, 362
planer and matcher, 463
planer saw, 16
planer splitting, 16
planetary gear, 540
planetary hoist, 540
planetary powershift transmission, 108
planimeter, 323, 382
planimetry, 44
planing, 16, 611
planing machine, 16
planing mill, 16, 17
plank, 200
plank board, 200
plank buttress, 9
plank buttresses root, 9
plank chute, 8
plank-floored bridge, 328
plank road, 328
plank root, 9
plank-slide, 8
planker, 8, 284, 493
planking, 8, 66, 95, 364
plankon, 148
plankter, 148
plankton, 148
plankton net, 148
planktonic, 148
planktophile, 436
planktotrophic, 436
planned obsolescence, 225
planning horizon, 183
planning machine, 16
planning of production, 422
planning period, 225, 408, 655
planning recreational needs, 579
planning theory, 183
planning, programming and budgetary system, 225
planophyre, 45
planosol, 345
plansol, 432
plant, 52, 169, 324, 417, 616, 629, 639
plant alkaloid, 617
plant anatomy, 617
plant area index, 617
plant association, 617
plant biochemistry, 618
plant body, 618
plant box, 324
plant capacity, 417
plant cellulose, 618
plant climate, 617
plant colonization, 616
plant colouring matters, 617
plant community, 617
plant cover, 616
plant cytology, 618
plant debris, 616
plant dispersal, 617
plant division, 143
plant eater, 437
plant ecology, 618
plant effluent, 169
plant external morphology, 618
plant fiber, 618
plant fibre, 618
plant-food, 618
plant formation, 617
plant gall, 18
plant geography, 616
plant group, 617
plant-growth regulator, 618
plant-growth substance, 618
plant histology, 618
plant hole, 618
plant hormone, 617
plant indicator, 619
plant introduction, 618
plant lice, 552
plant lift plough, 367
plant lifter, 367
plant mite, 317
plant morphology, 618
plant mucilage, 617
plant organography, 617
plant out, 564, 598
plant parasite, 617
plant pathogenic pollutant, 618
plant pathology, 616
plant pathosystem, 616
plant percent, 54, 90
plant phenolics, 617
plant phenology, 618
plant physiognomy, 618
plant physiology, 617
plant pit, 618
plant protection, 616
plant quarantine, 617
plant regulator, 618
plant-school, 77, 324
plant setter, 597, 616
plant society, 618
plant sociology, 617
plant succession, 618
plant taxonomy, 617
plant teratology, 617
plant thinning, 235
plant thinning machine, 235
plant thremmatology, 618
plant-top removing machine, 378
plant tray, 585
plant trees, 616
plant trees on river banks, 192
plant woody cell, 328
plantable seedling, 274
plantation, 395, 598
plantation fill in gap, 33
plantation layout, 600
plantation management, 395
plantation survey and project, 600
planted area, 597
planted forest, 395
planted land, 600
planter, 31, 616
planting, 31, 597, 616
planting area, 597, 600
planting arrangement, 617
planting bar, 616
planting bed, 565
planting-board, 564
planting borer, 598
planting brick, 574
planting by cuttings, 46
planting by layer, 551
planting by layerings, 317
planting by rootcuttings, 141
planting by single tree, 87

P

planting by sucker, 142
planting by tillers, 142
planting by vegetative propagation, 142
planting composition, 600
planting-cord, 597
planting-dagger, 598
planting density, 600
planting design, 616
planting distance, 597
planting gun, 597
planting hammer, 534, 598
planting-hatchet, 597
planting hoe, 616
plant(ing) hole, 598
planting hole machine, 495
planting in advance, 127
planting in notch, 147
planting in notches, 147
planting in row, 190, 297
planting in slit, 147
planting in triangle, 403
planting interval, 597
planting line, 597, 616
planting machine, 616, 629
planting mattock, 597
planting-nursery, 565
planting of layer, 551
planting of slip, 46
planting out, 597
planting-peg, 597, 598
planting plan, 600
planting point, 597
planting record, 600
planting row, 597
planting schedule, 600
planting schema, 629
planting season, 616
planting shoe, 597
planting site, 597, 600
planting space, 597, 616
planting spud, 597
planting stick, 598
planting stock, 108, 324
plant(ing) stock, 108
planting stock age, 108
planting stock (by vegetative propagation), 574
planting strip, 597
planting tool, 597, 616
planting tube, 397
planting value, 31
planting-wedge, 598
planting width, 190
plantlet, 352, 534, 535
plantling, 583, 617, 642
plants supply, 324
plantule, 352
plaque assay, 398
plasficizing wood, 466
plasma, 237, 591
plasmagene, 518
plasmase, 522, 523
plasmodesma (*plur.* -mata), 13
plasmogamy, 13, 591
plasmolysis, 620
plasmon, 13, 518
plaster board, 376
plaster ceiling, 211
plaster figure, 433
plastering, 487
plastic, 466
plastic antichecking clamp, 466
plastic bending, 466
plastic clay, 466
plastic covered greenhouse, 466
plastic crack, 466
plastic deformation, 466
plastic extensibility, 273
plastic faced plywood, 466
plastic film, 466
plastic flow, 345, 466
plastic flow limit, 466
plastic furniture, 466
plastic house, 466
plastic limit, 466
plastic material, 466
plastic overlay, 466
plastic pot, 466
plastic property, 466
plastic range, 344, 466
plastic shading, 594
plastic shrinkage, 466
plastic state, 345, 466
plastic strain, 466
plastic track, 466
plastic wood, 466
plasticising effect, 602
plasticity, 274, 339, 345, 466
plasticity index, 274
plasticizer, 466, 602
plasticizer extender, 149, 602
plasticizing agent, 466, 602
plasticizing wood with ammonia, 328
plastid, 541, 620, 621
plastify, 466, 602
plastisol, 466
plastispray, 466
plastochrone, 235
plastochrone index, 235
plastoelastic deformation, 472
plastomer, 466
plastometer, 466
plastometer constant, 466
plastometry, 466
plastoquinone, 621
plat, 98, 362
plat link chain, 8, 23
plat sowing, 279
platband, 23
plate, 8, 13, 55, 359, 555
plate basal, 8
plate budding, 357, 358
plate count, 359
plate-culture, 362
plate dryer, 103, 236
plate girder bridge, 8
plate glass, 30
plate glazing, 8
plate nuclear, 624
plate-press dryer, 359
plate saw, 592
plate shear test, 233
plate shears, 233
plate techtonics, 8
plate tectonics, 8
plateau, 164
platen, 549
platen drier, 200
platen press, 362
platen-pressed particle board, 362
platform, 471
platform balance, 12, 471
platform dryer, 362
platform scale, 12, 471
platform starting, 437
platform terminal, 625
platform truck, 359
platina, 31, 72
plating efliciency, 62, 362
plating machine, 549, 551
platting, 620
platy structure, 358
platyphyllous, 279
platypodid beetles, 50
play a vital role, 367
play-fieid park, 595
playa, 157

plceoid pit, 595
pleasure-boat, 580
pleasure ground, 579
pleat, 606
pleater system, 491, 601
plectenchyma, 322
pleiade, 528
pleiochasium, 118
pleiocyele, 118
pleiophyllous, 119
pleiotropism, 220
pleiotropy, 220
Pleistocene, 169, 199
plenter-thinning, 601
plenter-wood, 601
plenum chamber, 59, 465
pleomorphic, 118
plerome, 625
pleura, 42
pleuston, 457
pliability, 273, 399
pliability coefficient, 398
pliability of wood, 330
pliable, 398
plicate, 261, 580, 606
plicative, 606
plinth, 21, 635
Pliocene, 416
plot, 96, 533, 534, 558
plot estimate method, 558
plot radius factor, 558
plot recovery, 558
plot sowing, 279
plot survey, 558
plot tally sheet, 558
plot variance, 558
plot volume, 558
plotless cruising, 511
plotless sampling, 242, 511
plough, 22, 41, 288
plough in, 339
plough land, 168
plough layer, 168
plough line, 288
plough out, 495
plough pan, 288
ploughing, 128, 168
ploughing profitability, 439
ploughless farming, 34, 510
ploughshare, 288
ploughshare mixer, 288
ploughshoe, 288
ploughstaff, 288
ploughtail, 288
plover, 195
plow, 22, 41, 288
plow steel wire, 372
plug, 33, 46, 179, 334, 335, 638
plug cutter, 334, 620
plug gauge, 340, 593
plug saw, 278, 452
plug tenon, 46
plum mottle, 288
plum (*Prunus*), 572
plum-pudding design, 288
plum rains, 319
plumb point, 67
plumbing bar, 114
plumular, 352
plumular meristem, 583
plumule, 352, 583
plumule axis, 352
plunge cut, 46
plurilobed, 118
plurivorous parasite, 120
plus modifier, 99
plus seed stand, 611
plus stake, 229
plus stand, 611
plus tree, 611
plutonism, 554
pluvial, 585
pluvial age, 119
pluvial erosion, 199
pluvial lake, 119
pluvial period, 119
pluviifruticeta, 49
pluviilignosa, 49
pluviisilvae, 49
pluviofluvial denudation, 193, 585
pluviometer, 585
ply, 45, 192, 328
plyboard, 82, 153
plymax, 249, 529
plymetal, 249
plymold, 327
plypanel, 529
plywall, 20
plywood, 191, 240
plywood caul, 240
plywood core veneer, 240
plywood flooring, 240
plywood for external use, 441
plywood for general use, 365
plywood for internal use, 441
plywood industry, 240
plywood of balanced construction, 360
pneumatic clamps, 368
pneumatic control, 368
pneumatic control pressure, 369
pneumatic conveying, 368, 369
pneumatic conveying dry, 369
pneumatic conveying dryer, 369
pneumatic conveyor, 369
pneumatic conveyor fan, 368
pneumatic drive, 369
pneumatic drum sander, 368, 369
pneumatic dryer, 369
pneumatic extinguisher, 145
pneumatic injection, 370
pneumatic sectional plate base, 369
pneumatic separation, 369
pneumatic separation arrangement, 368
pneumatic system, 368
pneumatic tired roller, 59
pneumatic tube, 369
pneumatogenic, 368
pneumatophore, 200, 370
pneumatophoric root, 200, 370
poacher, 485
poaching, 374, 485
pochard, 374
pocket, 23, 52, 81, 458, 504
pocket boom, 81, 442
pocket compass, 544
pocket debarker, 81, 535
pocket dry rot, 81
pocket edger, 296
pocket grinder, 535
pocket leaf, 81
pocket of shaving machine, 16
pocket rot, 81
pocosin, 375
pod, 41, 110, 229
podite, 613
podocarpoid pit, 312
podocarptorene, 312
podsol, 210, 211
podsolic soil, 210

P

podsolization, 211
podzol, 210, 211
podzolic, 210
podzolic horizon, 210
podzolic soil, 210
podzolization, 210
pogonip, 110
poikilosmotic, 24
poikilosmotic animal, 24
poikilotherm, 24
poikilothermal animal, 24
poikilothermy, 24
point, 57, 90, 606, 637
point angle, 57
point-centered quarter method, 100
point contact method, 558
point density, 100
point estimator, 100
point lever, 637
point load, 224
point mutation, 220
point observation method, 558
point of aim, 324, 651
point of flammability, 414
point of fusion, 398
point of ignition, 390, 604
point of impact, 87
point of inflexion, 178, 421
point of measurement, 44
point quadrat analysis, 100
point sampling, 100, 242
point sleeper, 47, 606
point-source pollution, 100
pointer, 619
pointing, 619
pointing-out method, 179
poison bait, 110
poison cell, 111
poison exponent, 111
poison exponent method, 111
poison girdling, 111
poison gland, 111
poison hair, 111
poison sac, 111
poison seta, 111
poison thinning, 111, 206
poison-vetch, 210
poisoning, 111, 622
poisonous dust, 110
Poisson distribution, 31
Poisson's ratio, 31, 196
poiyacrylonitrile fiber, 263
pokelogan, 304, 492, 547
poker-picture, 284, 417, 438
polachena, 512
polackena, 512
polar adherent, 222
polar adhesive, 222
polar circle, 222
polar climate, 221
polar compound, 222
polar flagellum, 222
polar front, 222
polar lake, 221
polar method, 222
polar molecule, 222
polar moment of inertia, 222
polar movement, 222
polar nuclei, 222
polar planimeter, 106
polar solvent, 222
polar wandering, 222
polar zone, 221
polarimeter, 357, 546
polarity, 222
polarization, 222, 357
polarization analyzer, 233
polarizing filter, 222, 357
polarographic scanner, 222
polder, 269
pole, 24, 30, 46, 156, 223, 636
pole bondaging, 101
pole calliper, 79
pole carriage, 159
pole chain saw, 159
pole chute, 156, 332, 650
pole circular saw, 159
pole crop, 156, 534
pole derrick, 231
pole drain, 519
pole drying, 46
pole lathe, 331
pole line, 101, 231
pole line inspection, 101
pole piling, 537
pole press, 76, 290
pole retard curtain, 332
pole road, 332
pole saw, 164
pole size timber, 156, 284
pole slide, 332, 650
pole-slide with plank-bottom, 8
pole stage, 156, 534
pole stage forest, 156
pole stock, 156
pole switch, 349, 538
pole tie, 156, 175
pole-timber, 592
pole trailer, 49, 159, 179
pole tram-road, 332, 592
pole wag(g)on, 159
pole with top, 591
pole with tops, 79
pole-wood, 156, 592
policies and measures, 611
poling, 288
poling board, 612
polished caul plate, 351
polished press plate, 351
polishing, 326, 351
polishing varnish, 351
polishing wax, 415
polishing wheel, 351
pollard, 247, 384, 508
pollard method, 485
pollard shoot, 247
pollarded tree, 247, 508
pollarding, 247, 280, 485, 543
pollarding-forest, 485
pollard(ing) system, 485
pollen, 203
pollen analysis, 203
pollen-bearing, 79
pollen carrier, 203
pollen chamber, 203, 635
pollen diagram, 203
pollen extract, 203
pollen flower, 47
pollen grain, 203
pollen (grain) culture, 203
pollen gun, 203
pollen killer, 411
pollen killer gene, 411
pollen mentor, 203
pollen mitosis, 203
pollen mother cell, 203
pollen parent, 151, 203
pollen plant, 144, 203
pollen prevention forest, 132
pollen profile, 203

pollen sac, 203
pollen spectrum, 203
pollen sterility, 203
pollen substitute, 203, 394
pollen supplier, 151, 203
pollen tube, 203
pollen zone, 203
pollenin, 203
pollenizer, 444
pollination, 66, 444
pollinator, 66
pollina(tor), 444
pollinia, 77, 203
pollinium, 203
pollinizer, 444
pollinosis, 203
pollshing of outer skin, 27
pollutant, 508
polluted atmosphere, 508
polluted water, 508
polluter, 508
pollution, 508
pollution abatement, 134
pollution-free food, 315
pollution index, 508
pollution load, 508
pollution source, 508
poly methylolurea, 263
poly parasitism, 117
polyacid base, 118, 120
polyacrylamide, 263
polyacrylonitrile, 263
polyaddition, 227
polyadelphian, 118
polyadelphous, 118
polyadelphous stamens, 118
polyallel-cross, 117, 118, 120
polyandrous, 119
polyandry, 119, 562
polybenzimidazole, 263
polycambial, 119
polycarpellary, 119
polycarpellary ovary, 119
polycarpellary pistil, 119
polycarpic, 116
polycarpic plant, 116
polycarpicus, 116
polychromatic, 118
polychrome lacquerware, 39
polychronic origin, 118
polyclimax, 119
polyclimax theory, 119
polycondensation, 469
polyconic projection, 120
polycony, 118
polycotyl, 120
polycotyledoae, 120
polycotyledon, 120
polycotyledonous, 120
polycotyledony, 120
polycotylous, 120
polycross, 119
polycross method, 119
polycross test, 117
polycrystalline materials, 117
polycyclic, 118
polydemic, 116, 182
polydispersity, 34, 116, 263
polyembryonate, 118
polyembryonic, 118
polyembryony, 118
polyethism, 540
polyethylene, 264
polyethylene container, 264, 466
polyethylene film, 264
polyethylene greenhouse, 466
polyflavonoid, 263
polyfoam, 352
polyfunctional group, 117
polygamous, 118, 597
polygamous flower, 597
polygamy, 567
polygene, 117, 501
polygenic inheritance, 117
polygenic resistance, 117
polygon mitre joint, 117
polygonal tundra, 116, 183
polygyny, 116, 119
polyhaline, 119
polyhaploid, 116
polyhedron, 117
polyhedrosis, 117
polyhybrid, 116, 119
polylayer forest, 116
polymer, 263
polymerase, 117, 263
polymeric genes, 92
polymeric homologue, 483
polymeride, 263
polymerism, 263
polymerization, 263
polymerized rosin, 263
polymerous flower, 116
polymictic lake, 119
polymictic plankton, 119
polymorphic curves, 118
polymorphic site curve, 119
polymorphism, 118, 119
polyneme, 119
polyoecious, 70
polyoecism, 70
polyose, 118, 264
polyovulation, 118
polyoxybiotic, 119
polyparasitism, 117
polypeptide, 118
polypetalae, 287
polypetaly, 287
polyphagous, 118
polyphagy, 118, 120
polyphenol, 117
polyphenol oxidase, 117
polyphenols, 117
polyploid, 18, 116, 120
polyploidy, 116
polypod type, 120
polypore fungus, 117
polypot, 466
polypropylene, 263
polyribesome, 117
polysaccharase, 118, 264
polysaccharide, 118, 162, 264
polysaccharide skeleton, 118
polysaccharoid, 264
polysepalous, 287
polyseriate ray, 118
polysome, 117
polysomic, 118
polysomy, 118
polyspermy, 117
polystyrene, 263
polytene chromosome, 119
polyteny, 119
polythermal, 164
polytopic, 117
polytropic, 117
polytypic evolution, 119
polyvalency, 117
polyvinyl adhesive, 264
polyvinyl chloride, 263
polyvinylidene chloride, 263
polyvoltinism, 117
polyxylose, 263
pomade, 528, 529

pome, 288, 396
pomegranate peel, 434
pomegranate tannin, 434
Pomolio-Celdecor process, 234, 315
pompon rose, 87
pond, 41, 57, 241, 487, 499, 635
pond buck, 635
pond-dozer, 284
pond man, 23, 62
pond saw, 635
pond succession, 57
pondage, 545
ponding, 460, 520
pontoon, 148, 283, 458
pontoon bridge, 148
pony band mill, 535
pony band-saw, 533
pony mill, 535
pony press, 587
pony rig, 535, 633
pony rig bandmill, 533
pony rig bandsaw and carriage, 352
pony truss, 1
pony truss bridge, 534
pool, 53, 191, 192, 294, 456
pool garden, 57
poophyte, 42
poor, 120
poor stocked, 289
poor visibility, 401
pop strength, 337
poplar-rust, 557
population, 60, 387, 627, 649
population analysis, 627
population biology, 627
population change, 627
population curve, 627
population cycle, 627
population density, 627
population dispersion, 627
population dynamics, 627
population ecology, 627
population equilibrium, 627
population explosion, 627
population extinction, 627
population fluctuation, 627
population genetics, 389
population growth, 627
population growth curve, 627
population growth rate, 627
population index, 627
population irruption, 627
population mean, 649
population model, 627
population parameter, 648, 649
population pressure, 389, 627
population process, 627
population quality, 627
population regulation, 627
population size, 627
population stability, 627
population statistic, 649
population system, 627
population trend, 627
population variability analysis, 627
populin, 557
popycarpic plant, 116
porch, 321, 579
porch column, 283
porcupine, 112, 191
pore, 124, 179, 318, 369, 501
pore arrangement, 179
pore capsule, 277
pore chain, 179
pore cluster, 179
pore diameter, 276
pore flame-like arrangement, 179
pore figured arrangement, 202
pore fungus, 117
pore multiple, 152
pore of soil, 490
pore pattern, 179
pore pattern arrangement, 179
pore ring, 276
pore size, 276
pore size distribution, 276, 519
pore solitary, 83
pore space, 277
pore volume, 277
pore volume distribution, 277
pore water pressure, 277
pore zone, 179
pored wood, 282, 581
porometer, 369
porosimeter, 276, 277
porosity, 117, 277
porosity of soil, 490
porospore, 276
porous circle, 276
porous region, 581
porous (spongy) pith, 188
porous vessel, 83
porous wet-bulb sleeve, 117
porous wood, 282, 581
porphyrin, 32
porphyrinogen, 32, 207
porphyrite, 27
porphyry, 7
portable chain-saw mortiser, 564
portable debarker, 564
portable digitizer, 564
portable electric router, 443
portable jib crane, 564
portable knife sharpener, 443
portable log splitter, 380
portable mortising machine, 443
portable nailing machine, 443
portable peeler, 379
portable planing machine, 380
portable power hole digger, 380
portable radio-controlled winch, 380
portable saw, 380, 564
portable saw mill, 564
portable scale, 564
portable spar, 564
portable sprayer, 443
portable spraying unit, 443
portable stapling machine, 442
portable tank, 380
portable tramway, 47, 380
portage, 457, 625
portal bracing, 321, 612
portal-to-portal pay, 3
portico, 321, 635
Portuguese oak, 364
position control, 107
position effect, 504

position pseudoalleles, 504
position pseudoallelism, 504
positive, 611
positive correlation, 611
positive curvature, 611
positive density-dependent factor, 611
positive displacement, 435
positive economics, 436
positive feedback, 611
positive geotropism, 611
positive heterosis, 611
positive infinitely variable, 611
positive phototropism, 611
positively photoblastic (seed), 544
possibility, 25, 595
possibility by area, 25, 595
possibility by volume, 25, 595
possible rates of return, 192
post, 635, 636, 638
post-adaptation, 199
post adjustment, 147
post-and-paling, 335
post classification, 147
post climax, 199
post drive, 199
post-egg stage, 47
post-emergence application, 62
post-frontal fog, 146
post-glacial forest, 28
post-glacial period, 28
post-glacial rebound, 28
post-irrigation cultivation, 179
post-logging, 199
post-mortum, 462
post notum, 199
post pillar, 275, 635
post-type dam, 332
postcambial, 199
postcambial stage, 199
postclimax, 50, 51
postcure, 199, 200
posted price, 25, 350
postemergence tipburn, 199
poster paper, 604
posterior intestine, 199
posterity, 199
postern, 24
postforming, 121, 392
posthydrolysis, 200
posting, 107
posting of account, 187
posting plumb, 107
posting square, 107
pot culture, 355
pot experiment, 355
pot layerage, 164, 276
pot life, 440, 581
pot planting, 30, 573
pot seedling setting machine, 397
pot shot, 75
pot transplanting, 355
potamic, 193, 237
potamium, 193
potamology, 192
potamoplankton, 193
potassium pump, 230
potato-dextrose agar, 489
potence, 577
potence ratio, 577
potency, 536
potential, 101, 374, 438
potential acidity, 374
potential area, 374
potential biota, 374
potential break, 374
potential climax, 374
potential crop tree, 273, 586
potential deformation, 374
potential distribution area, 374
potential energy, 438, 504
potential energy of strain, 572
potential evapotranspiration, 374, 504
potential fertility, 374
potential food, 374
potential growing stock, 374
potential growth, 374
potential impact, 374
potential increase, 374
potential longevity, 374
potential natural forest community, 374
potential natural landscape, 374
potential natural terminal community, 374
potential natural vegetation, 374
potential pest, 374
potential photosynthetic capacity, 374
potential production, 374
potential production line, 374
potential rupture surface, 273
potential variability, 373
potentiometer, 101
pothook, 129, 480
Potlatch tram, 454
potmetal glass, 581
potometer, 609
potted miniature landscape, 414
potted miniature tree, 450
potted plant, 355
potter wasps, 210
potting, 415, 416
potting mixture, 355
potting plant, 355
Poulain process, 30
powder-pest beetle, 143
powder-post beetle, 143
powder-post borer, 143
powder-post damage, 143
powder-post defect, 143
powder resin, 144
powder sweeper, 405
powder wedge, 602
powdered activated carbon, 144
powdered adhesive, 144
powdered fire retardant, 144
powdery glue, 144
powdery mildew, 5, 6
Powell process, 15, 419
powell process Powell, 328
Powellizing process, 15, 419
power, 108
power articulated-frame, 109
power assist hydraulic system, 562
power assist pump, 634
power brake, 109, 218

P

power cable transporting, 109
power closing, 108
power consumption, 109, 171
power cultivator, 218
power cutter, 108
power dozer, 109
power drill, 108, 109, 652
power-driven increment borer, 101
power duster, 218
power equation of common density effect, 257
power factor, 171
power ladder for tree, 217
power logging, 108
power loss factor, 171
power monkey, 108
power of sprouting, 321
power of sprouting form stools, 321
power of straining, 572
power plant, 123
power rating, 120
power saw, 108
power skidding, 108
power sprayer, 108
power steering, 109
power tool, 108
power transmission, 108
power wagon, 171
powered backup roll, 109, 383
power(ed) cable logging, 109
Power's method, 15
powersupply sensing device, 108
practical capacity, 435
practical fire precaution, 436
practical germination percent, 435
praefloration, 202
praefoliation, 119, 583
prairie, 17, 365
prairie breaker, 268
prairie-forest region, 42
prairiebuster plough, 268
prammophyte, 411
prata, 41
praying mantis, 474
pre-chilling, 586
pre-emergence, 62, 551
pre-emergence application, 62
pre-emergence cultivation, 62, 551
pre-emergency rate, 62, 551
pre-existing pole, 373
pre-existing tree, 373
pre-logging, 127, 373
pre-planting fumigation, 616
pre-rigging, 586
pre-setting, 587
pre-soaking, 586
pre-stratification, 141
pre-thicket, 395, 396
preadaptation, 587
preadapted, 587
Precambrian (period), 373
precautionary purchase, 586
precedence charting system, 577
preceding crop, 373
preceding table, 442
prechoking, 586
precious material, 184
precious species, 607
precipitate, 346, 458
precipitated wood tar, 53
precipitation, 53, 238
precipitation aloft, 162
precipitation cell, 194
precipitation effectiveness index, 238
precipitation-effectiveness ratio, 238
precipitation evaporation index, 238
precipitation-evaporation ratio, 238
precipitation gauge, 296
precipitation mechanism, 238
precipitation per annum, 343
precipitation station, 585
precipitus, 238
precision, 254
precision drilling, 254
precision grinder, 254
precision kiln drying, 254
precision roll coater, 254
precision seeder, 254
preclimacteric rise, 200
preclimax, 372, 373
preclimax community, 372
precoat, 75, 95, 415, 587
precocial, 174, 599
precocious, 599
precocity, 599
precogenic horizon, 421
precommercial thinning, 503
precompressed wood, 587
precompressure, 587
preconditioning, 586, 587
precooling, 586
precure, 586
precuring, 586, 599
precursor, 328, 373, 522
predacious insect, 399
predating, 33
predation, 33
predation hypothesis, 33
predation pressure, 33
predator, 33, 266, 399
predator food chain, 33
predator-prey interaction, 33
predator-prey system, 33
predetermined pressure, 586
predetermined variable, 372
prediction, 586
predictor, 44
predisposition, 466, 583
predistillation, 62
predominant, 51
predominant tree, 51
predrier, 586
predrilling, 587
predrive, 62, 522
predryer, 586
predrying, 586
preemptive right, 577
preening, 444
prefabricated door, 587
prefabricated house, 214, 587
prefabricated kiln, 587, 638
prefabricated part, 587
prefabrication, 65, 587
prefectural forest, 525

prefelling regeneration, 127
preferent, 440
preferential absorption, 577
preferential adsorption, 547
preferential clause, 577
preferential combustion, 577
preferential shop, 577
preferential tariff, 474, 577
preferential wetting, 440
preferred operating rate, 577
preferred orientation, 601
prefinished panel, 587
prefloration, 202
prefoliation, 583
preformed resistance, 177
preformed root initial, 522
preframing, 65, 586
pregeneration, 127
pregermination, 74
pregermination treatment, 124
preglacial period, 28
prehardening, 586, 587, 599
preharvest spray, 441
prehaul, 387
prehauler, 222, 387
prehauling, 387
preheater, 587
preheat(ing), 586
preheating process, 586
prehistoric age, 437
prehistoric naturalized plant, 437
prehistoric plant, 437
prehydrolysis, 587
preliminary air, 63
preliminary heating, 586
preliminary plan, 234
preliminary reconnaissance inventory, 586
preliminary steaming (treatment), 63
preliminary survey, 41
preliminary tension, 586
preliminary thinning, 155
preliminary vacuum, 63, 373
preliminary work, 586
preload, 586
preloader, 587
preloading, 587, 589
premating behaviour, 239
prematuration period, 55
premature drop, 39, 504
prementum, 373
premium, 15, 152, 238, 421, 480
preoviposition period, 47
preparation cutting, 586
preparation felling, 586
preparation of soil, 610
preparation of working plan, 431
preparatory cutting, 586
preparatory felling, 586
preparatory period, 586
preparatory stage, 586
preparatory thinning, 586
prepared paint, 479
prepared tar, 254
prepile, 587
preplanting cultivation, 31, 616
preplanting treatment, 616
prepress, 587
prepressing, 587
prepupa, 587
preregeneration, 127
preregeneration cleaning, 373
preregeneration system, 127, 237
prescap process, 155
prescribed burning, 183
prescribed fire, 183, 225
prescribed yield, 15, 25, 225
prescutum, 373
preseasoning treatment, 586
preselector, 587
preselector mechanism, 586, 587
presence degree, 74
present net worth, 480, 526
present reproductive value, 88
present value, 217, 526
present value cost, 526
present yield, 526, 544
presenting, 382
preservation, 14, 131
preservation chemicals, 132
preservation economics, 132
preservation mechanism, 132
preservation of fertility, 96
preservation of games, 559
preservation of wood, 328
preservation order, 251, 252
preservation process of wood, 328
preservation timber, 131
preservative, 132
preservative bandage, 131
preservative cartridge, 132
preservative classification, 132
preservative combination, 132
preservative detoxification, 132
preservative gradient, 132
preservative in bored hole, 132
preservative oil, 132
preservative paper, 132
preservative penetration, 132
preservative permanence, 132
preservative reliability, 132
preservative retention, 132
preservative selection, 132
preservative test method, 132
preservative volatility, 132
preserve, 14, 251
preserved plywood, 132
preserved tree, 443
preserved wood, 443
preserving agent, 132
preserving timber, 131
presizing, 478
presowing cultivation, 31

press, 64, 184, 550, 571
press bed, 550, 605
press cycle timer, 229
press emergency stop, 550
press felt, 551
press part, 228, 551
press pit sump pump, 550
press plate, 549
press platen temperature, 394
press roll, 550, 551
press section, 228, 551
pressboard, 551
pressed active carbon, 55, 108
pressed-fibre board, 522
pressing, 228
pressing cycle, 228
pressing period, 228
pressing temperature, 228
pressing time, 228
Pressler method, 500
pressler reference height, 500
Pressler reference point, 500
Pressler's growth percent, 365
Pressler's (increment) borer, 365
Pressler's method of height determination, 44, 541
Pressler's method of volume, 365
pressmark, 550
pressure altimeter, 370
pressure anemoneter, 550
pressure bar, 549, 551
pressure-bar gap, 88, 549
pressure-bar lead, 549
pressure bomb, 550
pressure chamber, 550
pressure control equipment, 550
pressure controller, 550
pressure cooker, 164, 394
pressure curve, 550
pressure cycle stage, 229
pressure difference, 550
pressure gauge, 550
pressure gradient, 550
pressure meter, 550
pressure period, 228
pressure potential, 550
pressure process, 228
pressure reducing period, 238
pressure reducing valve, 233
pressure reduction treatment, 233
pressure relief, 550
pressure roll, 550
pressure-sensing device, 550
pressure-sensitive adhesion, 550
pressure-sensitive adhesive, 550
pressure slice, 550
pressure sterilizer, 164, 394
pressure-stroke cycle, 228
pressure-stroke process, 228
pressure tank, 228, 550
pressure treatment, 228
pressure treatment in bored hole, 65
pressure type head box, 550
pressure unit, 550
pressure vessel, 550
pressure wood, 573
pressurized refiner, 228
pressurized refining, 228
presternum, 373
prestressing, 586
presumascope, 173, 586
pretanning, 587
pretreatment, 372, 586
prevail climax, 577
prevailing, 577
prevailing stem, 577, 612
prevailing westerlies, 431
prevailing wind, 431, 633
prevent log, 202
prevent water loss and soil erosion, 134
preventable decay, 586
preventative measure, 586
prevention and remedy, 586
prevention guard, 132, 232
prevention patrolman, 132, 202
preventive measure, 586
prevernal, 599
prevernal aspect, 599
previous generation, 372
prewood, 251
prey, 33, 297
preyarder, 183, 534
preyarding, 183, 534
prezygotic mechanism, 192
price-earnings ratio, 230
price formation, 231
price growth percent, 231
price increment, 231
price increment percent, 231
price index, 231, 513
price limit, 230
price mechanism, 230
price pushing, 197
price relative, 230
price-support program, 231
price uncertainty, 230
price variation, 230
price wage, 225
price wages, 225
price wise, 231
pricing, 106, 225
pricking out, 534
prickle, 71, 356
prima, 181
primaries, 63, 372
primary, 31, 63, 100, 219, 292, 563, 591, 652
primary bark, 63
primary bast, 63
primary branch, 563
primary breakdown, 364, 590
primary carnivore, 63
primary (cell) wall, 63
primary climax, 95
primary colour, 220, 590
primary community, 591
primary consumer, 63
primary control program, 633
primary conversion, 63
primary cortex, 63
primary creep, 63
primary cultivation, 63, 219
primary cutting angle, 634
primary deposit, 525
primary dormancy, 589
primary effects of gene, 220

primary energy, 562
primary exporting country, 63
primary extraction, 62
primary forest, 590, 591
primary gluing, 62, 562
primary growth, 63
primary host tree, 227
primary infection, 62
primary insect, 63
primary lamella, 63
primary meristem, 63
primary mortgage market, 63
primary mycelium, 63
primary parasite, 589
primary pathogen, 63
primary pest, 634
primary phloem, 63
primary pit-field, 63
primary producer, 63
primary production, 63
primary productivity, 63
primary rain forest, 591
primary ray, 63
primary road, 160
primary root, 63
primary sampling unit, 63
primary sex organ, 63
primary soil, 591
primary standard, 219, 591
primary structure, 63, 563
primary succession, 591
primary timber, 578, 634
primary tissue, 63
primary transportation, 62, 112, 222
primary tumor, 63
primary vascular bundle, 63
primary vascular tissue, 63
primary vein, 623, 633
primary wall, 63
primary wall texture, 63
primary wood, 63
primary xylem, 63
prime, 563, 577
prime-coated board, 95, 487
prime cost, 614
prime log, 415, 577
prime panel, 577
prime rate of interest, 577
primer, 63, 95, 123, 368, 437, 570, 571
primer coating, 487
primer type, 591, 634
primeval area, 591
primeval forest, 590, 591
primine, 497
priming, 570
priming foil, 75, 95
priming paint, 95
priming system, 367
primitive forest, 590
primitive hardwood, 591
primitive selection felling, 591
primitive type, 591
primordial forest, 590
primordial leaf, 63, 591
primordial meristem, 589
primordial pit, 591
primordium (*plur.* -dia), 589
principal, 19, 220
principal component, 632–634
principal component analysis, 632
principal crop, 633
principal cutting direction, 634
principal felling, 625, 633
principal forest crop, 633
principal forest product, 410
principal hydrolysis, 634
principal line, 634
principal meridian, 634
principal plane, 633
principal point, 632
principal post, 633
principal produce, 632, 634
principal product, 632, 634
principal rafter, 632
principal root, 633
principal species, 634
principal stand, 633
principal strain, 634
principal stress, 634
principal sum, 19
principal work, 634
principal yield, 633, 634
principle of allocation, 142
principle of competitive exclusion, 257
principle of the mathematics of large numbers, 78
princrpal yield, 633
print of high resolution, 162
printability, 440
printed plywood, 571
printed vinyl film, 571
printing ink, 571, 578
prior cutting, 599
priority, 577
prisere, 610
prism, 285
prism calibration, 285
prism calliper, 285
prism relascope, 285
prism spectrograph, 285
prismatic compass, 403
prismatic structure, 285
prismatic wood chip, 285
prismoidal formula, 361
private cost, 462
private forest, 462
private forest management, 462
private forest owner, 462
private forestry, 462
private forestry grants, 462
private landscape, 462
private recreation, 462
private timber manager, 462
prize log, 509, 569
proandry, 542
proanthesis, 522
proanthocyanidin, 589
proauthocyanidin, 589
probability, 156
probability-density, 156
probability density function, 156
probability of acceptance, 556, 595
probability of discrepancy, 287
probability proportional to prediction, 61
probability proportional to prediction sampling, 586
probability proportional to size (pps), 221
probability proportional to size sampling, 61

probability sampling, 156
probable maximum precipitation, 273
probe, 384, 473
probit, 156, 201
proboscis, 213
proboscypedia, 277
procambia strand, 591
procambium (*plur.* -bia), 591
P
procaryote, 589
proceeds, 436, 442
process, 56, 130, 170, 186, 443, 654
process chart, 170
process costing, 140
process flow diagram, 170
process production, 140
processor, 65, 76
prochromosome, 373
proclimax, 588
procreation, 128, 429
procumbent, 360, 362, 364, 557
procumbent cell, 196
procumbent ray cell, 196
procyanidin, 589
produce, 48, 441
produce check, 48
producer, 422
product class, 48
product design, 48
product difierentiation, 48
product life cycle, 48
product life history, 48
product management, 48
product method, 56
productibility of soil, 96, 488
production, 422
production-biomass ratio, 47
production context, 421
production cost, 421
production ecology, 422
production efficiency, 422
production evapotranspiration, 422
production forest, 422
production parameter, 421
production path, 422
production planning, 422
production pyramid, 47
production quota, 421
production rate, 422
production-respiration ratio (P/R ratio), 47
production target, 422
production time, 422
production worker, 421, 614
productive capacity on area, 299
productive factor of forestry, 303
productive structure, 422
productive structure diagram, 422
productive time, 422
productiveness, 422
productivity, 422
productivity class, 98
productivity clause, 422
productivity index, 98, 376, 422
productivity of capital stock, 641
productivity of ecosystem, 425
productivity of forestry, 303
productivity of labour, 284
productivity of soil, 96
productivity power of soil, 96, 488
proembryo, 590
proenzyme, 320
professional trapper, 615
profile, 42, 364
profile density, 247
profile diagram, 364
profile of equilibrium, 361
profiler, 55
profit and loss statement, 469
profit area, 290
profit-bearing investment, 79
profit margin, 290, 318
profit maximization, 653
profit of forestry, 303
profit of forestry for annual working, 48
profit of forestry for intermittent working, 166
profit ratio in stumpage appraisal, 289
profit sharing, 141, 290
profit squeeze, 290
profitability, 217
profitability of forestry, 409
progenitor, 379, 522, 652
progeny, 199
progeny test, 199
progeny test seed orchard, 641
progeny trial, 199
prognosis, 586
program evaluation and review technique, 56, 225
program evaluation research task, 56
program growth, 225
program of production, 422
program temperature controller, 56, 505
programmable calculator, 272
programmable pocket calculator, 544
program(me) control, 56
programmer, 56
programming language, 56
progress map, 249
progression of succession, 556
progressive burning, 467
progressive-burning powder, 99
progressive clear-strip system, 293
progressive disturbance, 249
progressive (drying) kiln, 373
progressive drying-kiln, 373
progressive (drying)kiln, 294
progressive evolution, 237
progressive gluing, 293
progressive improvement, 237
progressive method, 237
progressive mutation, 250, 373
progressive plate dryer, 293

progressive shelterwood felling, 169, 237
progressive succession, 250
progressive tax, 285
prohibition of clear cutting, 252
prohibition of devastation, 252
prohibition of exploitation, 252
project area, 485
project benefits, 169, 225
project boundary, 530
project fire, 474
projection, 438, 485, 555
projection area, 485
projection drawing, 485
projection (generic), 586
projector plan, 41
prokaryote, 589
proleg, 155, 592
proles, 96
proliferation, 602
proliferation of funds, 142
proline, 365
prolonged frontal rain, 57
prolonged value, 199, 626
proluvium, 199
promeristem, 589
promiscuity, 597
promising tree, 366
promissory note, 365
promote forestry with science and technology, 272
promoter, 367
prompt cash payment, 222
promutagen, 373
promycelium (*plur.* -lia), 522, 589
pronate, 149
prone (position), 507
prong, 46, 444
prong budding, 46, 113
pronghorn, 140
pronotum, 373
pronympha, 587
proof grader, 232
prop, 280, 612
prop root, 612
propaerial root, 612
propagating case, 128
propagation, 65, 128, 430
propagation coefficient, 129
propagation house, 129
propagation of error, 513
propagative organ, 129
propagative virus, 602
propago, 382, 551
propagule, 129
propagulum (*plur.* -la), 129
propanil, 94
propeller fan, 313, 631
propensity to consume, 531
propensity to invest, 485
proper crown grafting, 377
proper stocking, 439
proper use factor, 439
proper valves, 147
property co-operative society, 37
property map, 488
property-tax, 37
prophase, 373
prophylactic, 586
prophyll, 372, 522
propleuron, 373
propollen, 589
proportion of late wood, 499
proportion of spring wood, 68, 599
proportion of summer wood, 499
proportional allocation, 20
proportional limit, 20
proportional sampling, 221
proportional tax, 20
proportional to size sampling, 156
proportionate area by age class, 305
propulsive effort, 372
prosenchyma, 50, 401, 445
prosenchymatous element, 401
prosenehymatous cell, 401
prosize, 76
prospect, 255, 258, 373
prospector, 44, 269
prostrate, 362, 364, 451
prostrate plant, 364
prostrate stem, 362, 364
protandrous, 542, 543
protandrous flower, 542
protandrous hermaphrodite, 543
protandry, 542
protease, 88
protect forests, 202
protect-range, 14
protectant, 14
protected area, 14
protected forest, 14, 251
protected landscape, 255
protected region, 645
protected species, 14
protected tree, 14
protecting colouration, 14
protecting cover, 202
protecting device, 15, 132
protection, 14
protection belt, 14, 209
protection forest, 14, 132, 251
protection forest against snow, 134
protection forest against tide water damage, 188
protection forest for drought control, 157
protection forest for erosion control, 134
protection forest for flood hazard, 132
protection forest for land slide prevention, 491
protection forest for preventing snow avalanche, 548
protection forest for shifting sand prevention, 133
protection forest for water conservation, 460
protection forest land, 132
protection forest system, 132
protection forest under prohibition of cutting, 251
protection forestry, 132
protection maturity, 132
protection of birds, 345

P

protection of game, 251
protection of nature, 207, 645
protection of trees, 448
protection of wild life, 559
protection strip, 14, 209
protective beam, 132
protective belt, 132
protective clothing, 132
protective coast-forest, 188
P
protective colour, 14
protective colouration, 14
protective dune, 133
protective forest, 132
protective fungicide, 14
protective layer, 14
protective measure, 14
protective mimicry, 14
protective potential, 647
protective reaction, 14
protective reflex, 134
protective shield, 202
protective substance, 14
protective trade, 14
protective zone, 14
protein adhesive, 88
protein glue, 88
proteinase, 88
proterandrous, 542, 543
proteropetalous, 340
proterosepalous, 496
prothorax, 373
protist, 591
proto-cooperation, 63
protocatechuic tannin, 589
protochlorophyllide, 591
protoderm, 588
protofibre, 591
protofibril, 591
protogene, 589
protogynous, 69
protogynous hermaphriodism, 70
protogyny, 69
protoherbosa, 624
protolignin, 591
protoparasite, 62
protopectin, 589
protopectinase, 589
protophloem, 591
protoplasm, 589, 591
protoplast, 591
protoporphyrin, 588
protoporphyrinogen, 373
protoxylem, 591
protozoa, 590
protuberance, 246, 308, 486
provascuar tissue, 502
provascular strand, 502
provenance, 588, 626, 627
provenance seed orchard, 47, 627
provenance selection, 627
provenance test, 627
provenance trial, 47, 627
provenience, 588, 626, 627
proventriculus, 373, 527
provincial forest, 431, 630
provincial park, 431
provision for delinquency, 622
provisional cutting, 586
provisional felling, 586
provitamin, 373, 502
proximal, 250, 251
proximal seed, 251
proxy, 477
proxy statement, 503
pruinate, 260
pruinous, 260
prune, 75, 543, 544
pruned log, 512
pruner, 544
pruning, 75
pruning practice, 543
pruning-saw, 544
pruning-scissors, 544
pruning shears, 544
pruning tool, 544
prunning, 544
Prussian blue, 365
psamment, 412
psammitic regosol, 412
psammolittoral, 411
psammolittoral organism, 411
psammon, 411
psammophile, 440, 516
psammophilic, 440, 516
psammophilous, 411, 516
psammophyte, 411
psammophytic vegetation, 411
psammosere, 411
psephite, 291, 468
psephyte, 291, 468
pseudo-alleles, 342
pseudo-carp, 230
pseudo colour, 230, 502
pseudo-epiphyte, 230
pseudo granular structure, 230
pseudo-invariant feature, 502
pseudo-linkage, 342
pseudo mycorrhiza, 503
pseudo-noise, 230
pseudo-orthogenesis, 230
pseudo scopic stereo model, 230
pseudoanticline, 230
pseudoaposematic, 342
pseudocambium (*plur.* -bia), 230
pseudocorolla, 230
pseudodominance, 230, 343
pseudoepiphytes, 503
pseudogamy, 230
pseudogleyzation, 230
pseudoparaphysis, 342
pseudoparenchyma, 230
pseudopurpurin, 230
pseudoring, 230
pseudosclerotium, 230
pseudosematic colour, 14
pseudostroma, 230
pseudosycline, 230
pseudoterminal, 230
pseudoterminal bud, 230
pseudotransverse cambial cell division, 342
pseudotransverse division, 342
psilophyte, 313
psilopsida, 313
psychrometer, 157
psychrometric calculator, 432
psychrometric room, 395
psychrophilic organism, 441
psychrophyte, 163, 190
psylla, 334
pterergate, 580
pteridophyta, 266
pterocarpine, 642
pterodium, 58
pteropod ooze, 569
pterosere, 174
pterospermum type, 59
pterostigma, 59
pterygium, 58, 213
pterygopous, 260
pteryla, 584

pterylae, 584
puberulent, 18
puberulous, 18
pubescence, 18, 397, 400
pubescent, 261
public camp ground, 171
public corporation-forest, 171
public forest, 171
public forest owner, 171
public hazard, 171
public nuisance, 171
public park, 171
public utility company, 171
public welfare forest, 171
puckered, 352, 399
pudding stone, 592
puddling treatment, 11
puff, 553, 608
puffball, 316
Pulaski, 19, 453
Pulaski tool, 164, 296
pull, 177, 237, 346, 372, 442, 562
pull around, 346
pull back (line), 212
pull down front, 530
pull felling, 248, 292, 372
pull-grader, 372
pull out, 346
pull-type rotary cutter, 372
puller, 5, 47
puller-buncher, 5
puller drive, 373
pulley, 205
pulley groove, 205
pulley mortise, 205
pulley (wheel), 205
pulling, 237, 282, 372, 493
pulling block, 89
pulling face, 162
pulling-in line, 372
pulling line, 372, 561
pulp, 186, 238, 618, 620
pulp bin, 237
pulp board, 237
pulp chest, 237
pulp cleaner, 412
pulp consistency, 237
pulp hook, 112, 620
pulp lap, 45
pulp log, 618
pulp mill, 620
pulp rack, 601
pulp reclaimer, 618
pulp saver, 618
pulp screen, 412, 618
pulp slurry, 562
pulp stock, 237
pulp stone, 327
pulp strainer, 618
pulp stuff, 237, 618
pulp thickener, 618
pulp washer, 516
pulp washing, 516
pulp web, 237
pulper, 468
pulping, 75, 468, 609, 620
pulping process, 620
pulplog, 618
pulpwater, 6
pulpwood, 601, 620
pulpwood bolt, 601
pulse, 317
pulse code, 317
pulse-echo method, 317
pulse generator, 317
pulse jet fog machine, 317
pulse length, 317
pulse width, 317
pulverization, 345, 465, 513
pulverizer, 144, 354, 355, 468
pulverizer harrow, 468
pulvillus, 604
pulvinus, 561
pulviograph, 644
pumice, 148
pump crew, 455
pump gun, 205, 219
pumped tree, 150
puna, 73, 365
punch bar, 492
punch tape, 65
punch(ed) card, 65
punch(ed) card key, 65
punch(ed) card system, 65
punched tape, 65
puncheon, 11, 88, 112, 113, 364, 378, 612
puncher, 59, 65, 327
punching, 59
punctate, 260, 519
puncticulate, 261
puncture, 71, 274, 519
puncture resistance, 65, 270
pungent odor, 71, 539
punk, 150, 544
punk heart, 150
punk knot, 544
punk wood, 151
punky heart, 544
punky knot, 544
punky wood, 150, 538
pup trailer, 535
pupa, 576
pupal cell, 577
pupal instar, 577
puparium, 502, 576
pupation, 206
purchase block, 153, 367
purchase of timber, 329
purchase of wood, 329
purchaser, 173, 250
purchaser after felling, 125
purchaser of standing tree, 289
purchasing power parity, 173
Purdey double bolt, 365
pure bred, 69
pure-breeding variety, 69
pure cellulose, 69
pure-clonic form, 69, 511
pure community, 69
pure coniferous stand, 606
pure crop, 69
pure culture, 69
pure forest, 69
pure intolerant stand, 516
pure line, 69
pure live seed value, 628
pure profit, 69
pure rate of interest, 69
pure seed, 32, 69
pure shear, 69
pure shear stress, 69
pure stand, 69
pure strain, 69
pure stress, 69, 86
purification, 476
purine, 358
purity, 68, 256, 359
purity board, 257
purity of seed, 628
purity of variety, 359
purlin and rafter, 305
purlin roof, 305
purline, 195, 305
purple stain, 642
purple yulan, 642
purposeful hybridization, 107, 277

purposive sampling, 336
purr lingo, 508
push button, 3
push-button control, 3
push car, 443
push hoe, 443
push stick, 492
pusher bar, 492
pusher cat, 492
pusher drive, 200
pushing-boat, 492
push(ing) pole, 492
put in, 239
putamen, 193
putrefaction, 150
putty, 578
putty retouch, 494
PVC film overlay, 263
pycnial stage, 541
pycnial state, 541
pycnidiospore, 142, 371
pycnidium (*plur.* -dia), 143
pycniospore, 541
pycnium, 541
pycnometer, 20
pycnospore, 544
pygidium, 492, 503
pygmy tree, 2, 163
pyramid of biomass, 427
pyramid of energy, 341
pyramid of number, 451
pyramid of productivity, 422
pyramid-shaped opening, 242
pyramid training, 593
pyramidal crown, 249, 470
pyrene, 141, 534
pyrenocarp, 193
Pyrenomycete, 193
pyrethrum, 64
pyrethrum dust, 64
pyrethrum insecticide, 64
pyrgeometer, 77, 97
pyrheliometer, 614
pyric factor, 217
pyroclimax, 216
pyrogenic, 124, 392
pyrogenic decomposition, 164, 392
pyrogenic distillation, 157, 164
pyrogenic reaction, 242, 424
pyrogenic succession, 215
pyrograph, 284, 417
pyrography, 284
pyrolignucs acid, 332
pyrolysis, 164, 392
pyrolysis gas chromatography, 392
pyrolytic, 392
pyrolytic reaction, 392
pyrolyzate, 157, 392
pyrometer, 164
pyrophilous fungus, 516
pyrophyte, 337
pyrosphere, 215, 554
pyroxylin, 92, 215
pyroxyline lacquer, 333, 532
pyroxyline silk, 532
pyrrhic, 216
pyrrhic succession, 215
pythmic, 201
pythogenesis, 150
pyxidium, 156
pyxis, 156
Q_{10} value, 505, 615
quack lingo, 173
quad band saw, 463
quadrangle, 462, 534
quadrant, 463
quadrant angle, 530
quadrant dividers, 530
quadrat, 558
quadrat chart, 558
quadrat method, 558
quadrate plate, 131
quadratic mean, 360
quadratic mean deviation, 121
quadratic mean diameter, 113, 122
quadratic relationship, 121, 360
quadric stress, 27
quadrifoliate, 463
quadrifoliolate, 261
quadrilobate, 463
quadrilocular, 463
quadrimerous flower, 462
quadrinate, 463
quadripinnate, 463
quadrivalent, 463
quadrivalvate, 261, 463
quadrivalvular, 261, 463
quadruple, 462, 463
quadruplets, 463
quadruplex, 463
quagmire, 432, 604
qualification test, 192, 621
qualitative character, 621
qualitative resistance, 621
qualitative spectrochemical analysis, 181
quality class, 98, 620
quality coefficient, 621
quality control, 621
quality control level, 621
quality cruise, 330
quality determination, 620
quality factor, 37
quality growth, 541
quality increment, 541
quality increment percent, 541
quality index, 541, 621
quality of locality, 98
quality of site, 98
quality of soil, 491
quality of stand, 300
quality of standing-crop, 410
quality of tolerance, 170
quality saw timber, 577
quality specification, 226, 620
quality thinning, 541
quamitative epidemiology, 106, 306
quantification of colour, 40
quantitative character, 451
quantitative ecology, 451
quantitative environmental ranking system, 106
quantitative genetics, 451
quantitative geography, 106
quantitative inheritance, 451
quantitative maturity, 451
quantitative maturity of principal tree, 633
quantitative resistance, 451
quantitative rheology, 106

Q

quantitative spectrochemical analysis, 181
quantitative thinning, 106
quantity of heat, 393
quantizing, 451
quantum, 296
quantum biology, 296
quantum efficiency, 296
quantum evolution, 296
quantum requirement, 296
quantum theory of gene, 220
quantum yield, 296
quarantine, 166, 233
quarantine inspection, 233
quarantine inspector, 233
quarantine law, 233
quarantine regulation, 233
Quarles process, 278
quarry, 39, 297, 434
quarry stone, 318, 434
quarter-circle cut, 463
quarter-cut, 256, 463
quarter-cut figure, 256
quarter girth, 463
quarter-girth measure, 463
quarter-girth method, 463
quarter-girth tape, 463
quarter-mast, 375
quarter point loading, 463
quarter-rotary-cutting, 256
quarter round, 463
quarter saw, 463, 530
quarter-sawed, 256, 590
quarter-sawed lumber, 256
quarter sawing, 256, 590
quarter-sawn, 256, 590
quarter-sawn figure, 256
quarter sawn grain, 255, 256
quarter-sawn stock, 256
quarter section, 255, 463
quarter stuff, 463
quarter surface, 255
quarter tie, 463
quarter timber, 463
quartered, 256, 462, 463
quartering, 462
quartering fire, 43
quartering shot, 537
quartering to the tree, 537
quartet, 462, 463
quartz porphyry, 434
quartz trachyte, 434
quartzite, 434
quasi-conjugation, 51, 462
quasi-isotropic, 640
quasi-organism, 640
quasi-periodic, 640
quasi rent, 640
quasi-stationary front, 640
quasiclimax, 640
quassin, 278
quassinoids, 278
Quaternary period, 100
quaternary structure, 463
quay, 3, 94, 316, 434
quay crane, 62, 316
quaywall, 3, 31
Quebec log rule, 280
quebracho, 232
queen, 199, 327
queen bolt, 282
queen-post, 455
queen post girder, 455
queen post roof truss, 455
queen post truss, 455
queen substance, 146
quench, 74, 233, 516, 631, 651
quenched region, 4, 73
quencher, 73, 74, 233, 286, 324, 532
quercetum (*plur.* -ta), 291
quescent point, 257
quest constant, 465
queuing theory, 349
quick assets, 466, 526
quick-burning powder, 466
quick cook(ing) process, 466
quick gouge, 535, 593
quick lime, 424
quick liquidation, 289
quick-locking door, 279
quick make, 279
quick return, 279
quick sand, 307
quick travel, 279
quicklock autoclave closure, 609
quiescence, 258
quiescent load, 176, 258
quilted figure, 190, 239
quinate, 512
quincuncial, 319, 512
quincunx planting, 319, 512
quincunxial, 319, 512
quinone, 19, 280
quinquecostate, 512
quinquefid, 297, 512
quinquefoliolate, 512
quinquejugate, 512
quinquelocular, 512
quinqueloculate, 512
quinquepartite, 140, 512
quinquevalent, 512
quinquevalvate, 261
quint band saw, 512
quirk, 215, 419, 519
quirk router, 519
quota system, 106
quotient, 414, 541
quotient of quality, 621
qurik moulding, 345
r-k continuum of strategy, 114
R-layer, 328
R-strategist, 45
R-strategy, 45
rabbet, 4, 16, 41, 375, 377, 469
rabbet and dado joint, 366
rabbet edge joint, 375
rabbet joint, 41, 366
rabbet-mitre joint, 537
rabbet plane, 22, 41
race, 184, 207, 310, 359, 385, 456, 535, 626, 651
race-nonspecific resistance, 536
race-specific resistance, 536
race track eftect, 184
racemation, 497
raceme, 649
racemization, 497
racemose, 364, 649
racemose cyme, 264
racemose inflorescence, 649
racemule, 536
rachilla, 533, 535
rachis (*plur.* raches), 204, 561
racial crossing, 651

rack, 57, 58, 89, 205, 218, 231, 534
rack-and-pinion drive, 58
rack car, 79
rack saw, 58, 564
rack way, 534, 595
rackheap, 81, 329
rad, 282
radar automatic navigation, 284
radar shadow, 284
radial, 135, 149, 182, 256, 554
radial alignment, 255
radial (arm) saw, 499, 546
radial-arm saw, 499
radial arrangement, 149, 255
radial bark thickness, 256
radial blade, 256
radial bundle, 149
radial check, 148, 255
radial clearance, 256
radial cut, 255
radial displacement, 256
radial distance, 148
radial division, 256
radial element, 255
radial fan, 256
radial growth, 256
radial impact hardness, 256
radial increment, 256
radial intercellular canal, 256
radial line, 149
radial line plot, 149, 277
radial line plotter, 149
radial line plotting, 149
radial-longitudinal division, 256
radial longitudinal section, 256
radial multiple, 255
radial parenchyma, 256
radial penetration, 256
radial pitting, 255
radial plane, 256
radial pore arrangement, 256
radial (pore) multiple, 255
radial porous wood, 148, 255
radial pressure, 256
radial rack angle, 256
radial resin canal, 256
radial row, 255
radial sawing, 256
radial-sawn, 256
radial section, 256
radial shake, 148, 255
radial shrinkage, 256
radial swelling, 256
radial symmetry, 148, 256
radial triangulation, 149
radial twist, 256
radial vascular bundle, 149
radial wall, 256
radial wing, 256
radial wood, 256
radial wood borer, 332
radiance, 148
radiancy, 148
radiant energy, 148
radiant intensity, 148
radiant power, 148
radiate, 148, 149
radiate veined, 148
radiate veins, 148
radiating surface, 135
radiation, 148, 149, 392
radiation adaptation, 383
radiation breeding, 149
radiation curing, 149
radiation dose, 148
radiation dust, 135
radiation ecology, 135
radiation field, 148
radiation frost, 149
radiation genetics, 135
radiation-induced aberration, 149
radiation intensity, 148
radiation pollutant, 135
radiation source, 149
radiation sterilization, 148
radiative balance, 148
radiative forcing, 148
radiative forcing scenario, 149
radiator, 149, 404
radical, 168, 220
radical emergence, 310, 352, 491
radical leaf, 220
radical polymerization, 579, 647
radicated, 260
radication, 168, 268, 422
radicel, 533
radicel (radicle), 352
radicicolous, 167, 429
radiciferous, 260
radiciform, 168
radicite, 206
radicle, 533, 582
radicle accumbent, 566
radicoid form, 98
radicular, 167, 352
radio directing and ranging, 511
radio frequency, 163, 418
radio-frequency curing, 163
radio-frequency drying, 163
radio-frequency gluing, 163
radio-frequency heating, 163
radio-frequency moisture meter, 418
radio-frequency preheat, 163
radio telemetry, 511
radio tracking, 511
radioactive dating, 135
radioactive fallout, 135
radioactive indicator, 135
radioactive isotope, 135
radioactive phosphorus, 135
radioactive pollution, 135
radioactive ray, 135
radioactive tracer, 135
radioactivity, 135
radioautograph, 135
radioautography, 135
radiodensitometry, 135
radioecology, 135
radiogram, 419, 511
radiography analysis, 418
radioisotope, 135
radioisotope detector, 135
radioisotope method, 135
radiolabelled, 135
radiometer, 148
radiometer-scattermeter, 149
radiomutant, 149
radionecrosis, 135
radiopaque, 148
radioparent, 149

R

radioresistance, 148, 337
radioselection, 149
radiosensitivity, 148
radiosonde, 511
radiosonde observation, 473
radiosymmetric, 148
radius, 10
radius of action, 214, 655
radius of curvature, 384
radius of gyration, 212
radius-thickness ratio, 255
raffia, 259
raffia wax, 259
raffinate, 40, 476
raffinator, 83
raffinose, 322, 323
raft, 23, 127, 333, 350, 632
raft bundle, 350
raft ground, 23
raft hook, 75, 550, 602
raft lashing, 596
raft nail, 349
raft pocket, 458
raft remaking, 155
raft section, 349
raft-section driving, 84, 349
raft stick, 168
raft-wood, 350
rafter, 66, 135, 307
rafter harbor, 309, 481
rafter-type splash dam, 66, 332, 569
rafting, 23, 350
rafting block, 350
rafting boom, 349, 350, 502
rafting channel, 127, 307
rafting dog, 350
rafting machine, 23
rafting operation, 332, 350
rafting reservoir, 350
rafting rod, 135
raftsman, 23, 53, 350
ragged leaf, 298
rail, 147, 160, 184, 195, 196, 283, 295, 480, 533
rail track, 184
railing, 146, 283, 502
railroad spike, 90
railroad tie, 608
railway logging, 480
railway sleeper, 480
rain and snow mixed, 584
rain-band, 584
rain day, 585
rain forest, 585
rain gauge, 585
rain-green forest, 585
rain green plant, 585
rain gush, 16
rain gust, 16
rain height, 238
rain measuring glass, 585
rain shadow, 585
rain spell, 113, 585
rain wash, 238
rain-water, 585
raindrop, 584
raindrop erosion, 584
raindrop spectrograph, 584
rainer, 354, 394
rainfall, 238
rainfall characteristic, 238
rainfall duration, 238
rainfall erosion, 238
rainfall excess, 238
rainfall frequency, 238
rainfall hours, 238
rainfall intensity, 238
rainfall regime, 238
rainfall reliability, 238
rainfall-temperature diagram, 585
rainfall-temperature graph, 585
rainfall type, 238
rainfall under tree crown, 302
rainfall variability, 238
rainforest, 585
raingauge shield, 585
raintree, 584
rainy season, 584
raised bed, 161
raised bog, 164
raised grain, 4
raised line, 309
raised panel, 486
raiser, 494
raising, 353, 597
raising seedling, 585
rake, 88, 349, 373, 380
rake angle, 57, 88, 373, 380
rake conveyer, 177
rake dozer, 5
rake finger, 349
rake roller, 444
raker, 58
raker clearance, 58
raker gauge, 58
raker (tooth), 232, 349
ram pike, 146, 278
ramal, 612, 613
Raman spectroscopy, 282
rambler, 317
rambling, 317, 404
rameal, 612
ramellus, 535, 623
ramentum, 534
ramet, 511
ramification, 143
ramiform pit, 143
ramiform pitting, 143
ramify, 62, 143
rammer, 75, 190
ramose, 120, 143
ramp, 205, 224, 285, 363, 536
ramp site, 285, 638
ram's horn figure, 556
Ramson, 285
ramulose, 119
ramulous, 119
ramulus (*plur.* ramuli), 154, 535
ramus, 612
rancidity, 62, 467
rand, 22, 486
random, 467, 509
random access, 467
random access memory, 467
random access sort, 467
random assembly, 468
random breeding unit, 468
random chromatid segregation, 390
random chromosome segregation, 391
random crack, 34
random cruising, 468
random dispersion, 468
random distribution, 468
random error, 348, 468
random fluctuation, 467
random forming, 396
random grain, 311
random length, 34, 136, 396
random length pile, 33
random matching, 468
random mating, 468

R

random mismatching, 468
random pairs method, 467
random pollination, 647
random sample, 468
random sampling, 467
random start, 468
random subsample, 467
random variable, 467
random variation, 467
randomized block, 468
randomness, 468
range, 23, 41, 130, 140, 148, 261, 296, 413, 418, 573
range appraisal, 41
range condition, 336
range finder, 44
range finder dendrometer, 439
range-finder scope, 642
range fire, 336
range-hammer, 75
range improvement, 336
range land, 135
range management, 41
range manager, 336
range of stress, 572
range officer, 336
range survey, 41
range table, 359
range utilization, 41, 464
rangeeni lac, 311
rangeland, 336
ranger, 300, 573
ranger district, 14, 573
rank, 91, 190, 336, 621
rank correlation, 91, 621
rank order, 91
rank weed, 336, 596
raphal, 225, 626, 632
raphe, 225, 626, 632
raphia wax, 259, 648
raphides, 537, 606
raphidian cell, 606
raphioid fibre, 607
rapid fire, 279, 466
rapid forest management, 73
rapid generation scheme, 438
rapid-peeler, 279
rapid zone, 222
rapidly growing tree, 466
rapine, 283
raptor, 321
rare, 515
rare species, 515
rarefaction, 515
raster data, 602
ratchet wheel, 222
rate, 315
rate earned, 442
rate gene, 466
rate of combustion, 390
rate of drying, 158
rate of fire spread, 215
rate of fire travel, 216
rate of flame spread, 216
rate of forestry interest, 303
rate of germination, 124
rate of growth, 429
rate of head movement, 551
rate of increment, 429
rate of interest, 290
rate of loading, 229
rate of lostmass after fire, 390
rate of multiplication, 602
rate of pay, 540
rate of production, 422
rate of return on invested capital, 485
rate of seeding, 31
rate of spread, 281, 317
rate-of-spread meter, 216, 317
rate of straining, 572
rate of stressing, 572
rate of survival, 54
rate of thinning, 445
rated cruise speed, 120
rated load, 120
ratio estimation, 19
ratio estimator, 20
ratio of root to shoot, 167
ratio of tangential to radial shrinkage, 523
rational formula, 438, 581
rational method, 291, 548
rationalization, 192
rationing of recreation areas, 579
ratoon, 167, 168, 247
rattan, 475
rattan furniture, 475
rattanware, 475
rattler, 321, 380, 410
raum-meter (rm), 45
Raunkiaer's frequency spectrum, 284
Raunkiaer's law of frequency, 283
Raunkiaer's life-form classification, 284
Raunkiaer's system of life-form, 283
ravine, 173, 413, 420, 519
ravine forest, 175
ravine wind, 519
raw acid, 591, 620
raw bark, 73
raw data, 591
raw estimate, 72, 156
raw humus, 72
raw lacquer, 424, 622
raw material, 588
raw paper, 592, 619
raw rubber, 428
raw waste lumber, 620
raw water, 424, 591
rawinsonde, 511
ray, 404, 418, 540
ray cell, 418
ray contact area, 418
ray crossing, 418
ray epithelial cell, 418, 419
ray fleck, 418
ray flower, 21, 417
ray initial, 419
ray mother cell, 418, 419
ray parenchyma, 418
ray parenchyma cell, 418
ray sieve tube, 418
ray spacing, 418
ray storied, 418
ray tracheid, 418
ray tracheid cross field, 418
ray-vessel, 418
ray-vessel pitting, 418
ray volume, 418
Rayleigh scattering, 401
Raymond mill, 284, 363
rayon cord, 396
razor-strop fungus, 206
re-allocation, 61
re-allotment, 61
re-logging, 212
re-vegetation, 616
reach, 292, 372, 419, 470
reaction, 130
reaction rate, 130
reaction sensitive adhesive, 130, 573
reaction time lag, 130

reaction wheel, 129
reaction wood, 572
reactivation, 598
reactive fire retardant, 130, 215
reactive species, 130
read-only storage, 175, 176, 612
reading chair, 594
readout, 111
ready mixed paint, 479
reafforest, 61
reafforestation, 61
real-balance-effect theory, 435
real capital, 436, 582
real density, 608
real estate, 32
real forest, 526
real growing-stock, 526
real growth series, 608
real income, 435
real property, 32
real specific gravity, 607
real time, 436
real-time high resolution mass spectro metry, 279
real value of seed, 629
real vegetation map, 525
real yield-table, 435
realization value, 62, 272
realize, 24
realized heritability, 526
realized mortality, 435
realized natality, 435
realized niche, 435
realm, 248, 305
ream, 305
rear, 216, 307, 327
rear guide shoe, 199
rear handle, 199
rear lump, 199
rear mount, 200
rear sight, 199
rearrangement, 60, 61, 610
reassessment, 61
reassociation, 61
rebate, 4, 161, 212, 469, 606
rebate plane, 242
rebated joint, 41, 375
reboiler, 599
rebound effect, 129
rebuilt lac, 599
rebulac, 60
reburn, 153, 598
recapitulation, 61
recapture, 60, 120
recapture method, 60
receding block, 212
receding line, 212
receiving boom, 74, 244
recent erosion, 525
receptacle, 204, 338, 430
receptive wood, 569
receptivity, 159, 274, 324, 443
receptor, 159, 443
recess, 4, 387
recess cabinet, 20
recessed pilaster, 21
recession, 451
recessive, 571
recessive character, 571
recessive gene, 571
recessive homologous association, 571
recessive homozygote, 571
rechipper, 334
reciprocal crosses, 201, 611
reciprocal equation of common-density effect, 257
reciprocal factor, 239
reciprocal grafting, 239
reciprocal hybrids, 611
reciprocal saw, 500
reciprocal translocation, 528
reciprocating (blade) saw, 500
recircula ing dry kiln, 116
reckless cutting, 283
reclaim line, 139, 212
reclaimed stock, 212
reclamation, 268
reclamation plough, 268
recoating, 599
recognize, 435
recoil, 200
recoil pad, 200
recombination, 61
recon, 61, 239
reconditioning, 61, 355, 598
reconnaissance, 269, 471
reconnaissance photography, 607
reconnaissance report, 156, 269
reconnaissance satellite, 607
reconnaissanee soil survey, 489
reconstituted veneer, 651
reconstituted wood, 598, 599
reconstitution, 60, 61
record of cutting, 37, 38, 126
record on forest operation, 410
recorder controller, 643
recorder magnification setting, 226
recorder response, 226
recording anemometer, 644
recording caliper, 643
recording cal(l)iper(s), 102, 645
recording hygrometer, 645
recording thermometer, 645
recovered stock, 212
recovery, 40, 62, 211, 212
recovery creep, 211
recovery logging, 70, 121
recovery of invested capital, 485
recovery vacuum, 212
recovery value, 211, 273
recreation, 314, 579
recreation activity, 579
recreation area, 579
recreation area concession, 579
recreation-attitude, 579
recreation cost-benefit, 579
recreation demand, 579
recreation development, 579
recreation expenditure, 579
recreation facility, 579
recreation fee, 579
recreation leadership, 579
recreation park, 579
recreation participation, 40
recreation party leader, 579
recreation planning, 579
recreation policy, 579
recreation program, 579
recreation research, 579

R

recreation road, 579
recreation room, 270, 579, 584
recreation site, 543, 579
recreational accomodation, 579
recreational behavior, 579
recreational benefits, 579
recreational equipment, 579
recreational forest, 410, 579
recreational property, 579
recreational resource, 579
recreational travel, 579
recreational value, 579
recrossing, 60
recruiting, 32, 33
recruiting conifers while retaining broad-leaved trees, 597
recruitment, 33, 250, 601
rectangular patch budding, 260
rectangular planting, 49
rectangular plot, 260
rectangular profile, 259
rectangular raft, 49, 260
rectangular setting, 259
rectangular surveying system, 614
rectification of boundaries, 237, 258
rectified aniseed oil, 254
rectified point, 258
rectifier, 258
rectifying, 232, 258
rectifying still, 254
recurrent (forest) inventory method, 410
recurrent formula, 312
recurrent inventory system in cut determination, 388
recurrent mutation, 359
recurrent parent, 312
recurrent selection, 312
recursive system, 99
recutting, 167
red cedar leaves oil, 548
red cedar wood oil, 548
red cinchona bark, 197
red data book, 198, 513
red desert soil, 198
red earth, 198
red forest ant, 197
red garden, 58, 198
red heart, 198
red ink log, 92
red knot, 58, 214
red lead, 198, 372, 463
red line of ecological protection, 424
red line of farmland area, 168
red list, 198
red loam, 198
red mangrove, 198
red pine, 198
red prairie soil, 198
red-ray drying, 198
red ring rot, 197
red root, 197
red root extract, 197
red rot, 197
red silver fir, 471
red snow, 198
red soil, 198
red-spider, 198
red spruce oil, 58
red stain, 197
red staining fungus, 197
red tide, 58
red-top, 197
red water, 58
red wood ant, 197
red yellow podzolic soil, 210
reddish brown forest soil, 198
reddish brown lateritic soil, 198
reddish brown (steppe) soil, 198
reddish chestnut soil, 197
rediscount, 599
redox potential, 557
redry veneer, 598
reduce soil moisture, 238
reduced cost, 46, 238
reduced impact logging, 574
reduced root, 492
reduced scale, 469
reduced temperature, 106, 114
reducer, 207
reducing emissions from deforestation and forest degradation (REDD), 233
reducing end group, 207
reducing factor, 435
reducing sugar, 207
reductase, 207
reduction division, 233
reduction gear, 233
reduction in stock, 289
reduction of pressure to a standard level, 25
reduction of temperature to mean sealevel, 361
redunca, 2
redundancy, 398
reduplication, 598, 599, 602
Redux-process, 284
reece plane, 607
reed swamp, 310
reeding, 486
reef coral, 414
reef flat, 242
reel, 184, 242, 264, 527
reeling, 264
reeling machine, 264
reeling system, 264
reeving, 161
refectory table, 520
reference age, 221
reference coler, 25
reference image, 40
reference scenario, 40
refilling, 32, 33
refined, 254
refined bleached lac, 493
refined rosin, 254
refined shellac, 254
refined turpentine, 254
refined wood rosin, 254
refiner, 254, 327, 595
refiner ground pulp, 254
refinery, 54, 254, 559
refining, 254, 381
reflectance, 129
reflecting projector, 129
reflection densitometer, 129
reflectivity, 129
reflectoscope, 51
refluence, 90, 343
reforestation, 61, 207, 226, 407
reforestation assistance, 600
reforestation by sowing, 31
reforestation interval, 543
reforestation policy, 600

reforestation problem, 406
reforestation program, 406
reforestation scarifier, 616
reforestation with grass culture, 213
reforestation with pasture culture, 213
reforestation work, 406
refraction, 606
refraction displacement, 606
refractory, 338, 600
refractory seed, 338
refractory timber, 337, 338
refractory wood, 338
refrigerated storage, 286
refuge, 2, 20, 21, 252
refuge area, 14, 252
refugium, 21
refugium (*plur.* -ia), 40
refusal point, 260
refuse, 139
refuse burner, 139
refuse conveyer, 139
refuse grinder, 139
refuse lac, 41
refuse sorting, 282
refuse wood, 38
regarder, 302
regenerated carbon, 598
regenerated cellulose, 598
regenerated fibre, 598
regenerated forest land, 169
regenerated forest under group system, 390
regenerated forest under pollarding method, 485
regenerated forest under shelterwood system, 404
regenerated forest under strip system, 81
regenerated forest under the system of successive regeneration fellings, 237
regeneration, 169, 598
regeneration after removal of old growth, 125
regeneration area, 169
regeneration barrier, 169
regeneration block, 169
regeneration by compartments, 237, 404
regeneration by groups, 390
regeneration by sprout, 321
regeneration by stem-shoots, 247
regeneration by stool shoots, 1
regeneration by strip-felling, 79
regeneration class, 169
regeneration coupe, 169
regeneration cutting, 169, 633
regeneration felling, 169, 633
regeneration from root suckers, 168
regeneration from seed disseminated from outside, 43
regeneration from sprouts, 321
regeneration in group, 259, 279
regeneration in the selection system, 601
regeneration interval, 169
regeneration maturity, 169
regeneration period, 169
regeneration target, 169
regeneration under clear cutting system, 244
regeneration under coppice system, 1
regeneration under group system, 390
regeneration under pollarding method, 485
regeneration under selection system, 601
regeneration under shelterwood, 404
regeneration under shelterwood system, 404
regeneration under sprout method, 321
regeneration under strip system, 80
regenerative capacity, 598
regenerative nutrition, 169
regeneratory cutting, 169, 633
regime, 654
regime (*eg.* disturbance regime), 325
regime (*eg.* drying regime), 639
regime (*eg.* management regime), 477
regime (*eg.* moisture regime), 639
regime (*eg.* precipitation regime), 183
regime (*eg.* temperature regime), 285
regime (*eg.* thinning regime), 130, 131, 183, 620
region, 79, 616
region of climatic optimum, 625, 653
region of growth, 429
region office, 77
regional assignment, 383
regional association, 97, 383
regional biogeography, 383
regional disparities, 97
regional forester, 77
regional geography, 383
regional park, 383
regional planning, 383
regional test, 383
regional volume table, 383, 562
register, 226
registered seed, 635
regma, 472
regolith, 148, 445
regosol, 554
regraded saline soil, 598
regrafting, 129
regress, 492
regression, 211
regression analysis, 211
regression coefficient, 211
regression double sampling, 455
regression equation, 211
regression estimation, 211

regression estimator, 211
regression-line, 211
regression parameter, 211
regression species, 492
regressive differerentiation, 343
regressive evolution, 343
regressive succession, 200, 343
regressive tax, 99, 285
regressive type, 532
regrow, 598
regrowth, 199, 598, 599
regueing, 5
regular, 82, 610, 611
regular bleached lac, 365
regular cutting, 611
regular cutting budget, 25
regular distribution, 183, 266
regular drilling, 91
regular felling, 611
regular forest, 610, 611
regular lay, 129
regular lay rope, 343
regular logging, 611
regular plantation, 183
regular planting, 91
regular quadrat method, 183
regular size, 183
regular style, 183
regular tie, 365
regular wind, 48
regulate climate, 479
regulated forest management, 192, 611
regulation, 179, 183, 479
regulation of consistence, 348
regulation of felling budget, 480
regulation period, 33, 155, 360
regulator gene, 479
regulatory measure, 183
regulatory period, 480
regulatory rotation age, 311
regur, 194
regurgitation, 129, 491
rehaul line, 212
reheat coil, 33, 623
reindeer, 242, 462, 549
reinfection, 598
reinforced plywood, 376
reinforcement, 32, 227, 228
reinforcement (planting), 33
reinforcing conifers while retaining broad-leaved trees, 597
reinforeed plastic, 376
reinfurcer, 602
reinoculation, 598
reinsurance, 598
reintroduction, 599
reinvestment, 599
reisolation, 598
reject, 34, 92, 474, 475, 546
reject mat hopper, 139
rejected log, 34, 92, 475
rejector, 47, 139, 613
rejuvenation, 154, 169
rejuvenator, 363, 465
rejuvenescence, 154
relascop, 299, 300
relascope, 466
related crossing, 379
relative aperture, 527
relative astringency, 527
relative azimuth, 527
relative cover, 527
relative density, 527
relative dominance, 527
relative durability, 19
relative even-aged stand, 527
relative flammability index, 527
relative frequency, 527
relative growth, 527
relative growth coefficient, 527
relative growth method, 527
relative growth rate, 527
relative heat balance, 527
relative humidity, 527
relative income hypothesis, 527
relative light intensity, 527
relative light requirement, 527
relative moisture content, 527
relative pluviometric coefficient, 527
relative reproductive value, 527
relative retention time, 527
relative screw-holding power, 527
relative sea level, 527
relative taper, 527
relativc thinning intensity, 527
relative transpiration, 527
relative variance, 527
relaxation balance, 464
relaxation curve, 464
relaxation modulus, 464
relaxation of stress, 572
relaxation test, 464
relaxation velocity, 464
relaxometer, 572
relay, 226, 244, 477
relay set, 226
relay tree, 623, 625
relay yard, 623, 637
release, 64, 247, 493
release agent, 494
release cutting, 64, 247
release felling, 64, 247
release-kill ratio, 136
release of dormancy, 543
release of pressure, 136, 238
release of stress, 572
release of tension, 282
release-recapture method, 25
released liquor, 212, 441
released slash, 566
released thinning, 64, 149
releaser, 441, 570
releasing agent, 133, 441
releve, 104, 389, 558
reliability, 273
reliable minimum estimate, 273
relic, 40, 245
relic area, 40
relic coenosinm, 40
relic community, 40
relic ecotype, 245
relic endemic species, 40
relic flora, 40
relic plant, 245
relic soil, 40
relic species, 40, 245
relict, 40, 245
relict ecotype, 40
relict method, 40
relief, 96, 97, 147
relief cutting, 247

relief displacement, 99
relief-map, 99, 290
relief of casehardening, 531
relief ornament, 147
relief treatment of stress, 572
relier, 147
relish, 492
reloading, 61, 623, 637
reloading landing, 625, 637
reloading tool, 61
relogging, 33
remaining stand, 40
remaining tree, 15, 40
reman edger, 155
reman saw, 155
remedy of casehardening, 531
remigration, 130, 598
remnants, 40
remontant flowering, 116
remote control, 558
remote-control chain-saw, 558
remote controlled carriage, 558
remote data station, 594
remote sensing, 558
removal, 38, 125, 381, 406, 595, 637
removal age, 125
removal census, 349, 384
removal cutting, 199, 443
removal felling, 443
removal method, 349, 384
removal of felling area, 127
removal of forest litter, 381
removal of stump, 265
removal of wing, 384
removal sod right, 298
remove, 224, 381
removing time, 223
renascent herbs, 118
rendzina, 195
renewable, 274
renewable resource, 272, 274
renewable resource recycling, 274
renewable resources, 274
renewable sources, 599
renewal, 169, 598
renewal of working plan, 253
renewal pruning, 169
renovation, 169
renovation in gardening, 588
renovation of variety, 359
rent, 651
rent of growing stock, 302
rent of soil, 489
rental and maintenance, 651
rental of stumpage, 302
rental value, 651
rentavalue of forest of normal working section, 128
reorder chart, 645
rep finish, 479
repair kit, 543
repair planting, 33
repair-replace decision, 543
repairing, 32, 33
repairing of plywood, 240
repair(ing)of road, 543
repeatability, 60
repeated crossing, 60
repeated infection, 60
repeated sampling, 60
repeated selection, 60
repeated stress, 60
repeated tillage, 294
repeating method, 152
repeating theodolite, 152
repeating unit, 60
repellent, 132, 383
repetition rate, 60
repetitive gene, 60
repetitive stress, 60
replacement cost, 61, 169
replacement of air, 276
replacement of ecological factors, 426
replacement planting, 33
replacement resource, 79
replacement value, 477, 598
replant duty, 599
replanting, 33
repletum, 467
replica, 154
replica method, 154
replicase, 154
replication technique, 154
replicative form, 154
report time, 15
repotting, 209
repp finish, 479
representation sample, 100
representative area, 26, 100
representative fraction, 78, 100, 556
reproducer, 152, 154, 598
reproducibility, 61
reproducing class, 169
reproducing period, 169
reproduction, 154, 169, 582
reproduction by planting, 597, 616
reproduction by root cutting, 141
reproduction by sprout, 321
reproduction by sprout with stumpbark, 356
reproduction by stool shoot, 321
reproduction by stump division, 125
reproduction by sucker, 168
reproduction by tending treatment, 149
reproduction capacity, 129, 169
reproduction curve, 129, 602
reproduction cutting, 169, 633
reproduction effect, 430
reproduction felling, 169, 633
reproduction method, 128, 169
reproduction physiology, 129
reproduction plot, 169
reproduction potential, 129, 169
reproduction power, 128
reproduction rate, 169, 430
reproduction speed, 169
reproduction survey, 169
reproductive behaviour, 430

reproductive bud, 129, 204
reproductive capacity, 128
reproductive cost, 128
reproductive cycle, 430, 598
reproductive effort, 129
reproductive isolation, 430
reproductive pattern, 430
reproductive phase, 128, 129
reproductive potential, 430
reproductive power from stool, 168
reproductive rate, 129
reproductive system, 430
reproductive time lag, 430
reproductive value, 430
reptant, 364
repulsion, 58, 527
repulsion series, 527
require, 516
required sample size, 20, 653
requiring, 516
rerafting, 155
rerun, 598
resaw, 598
resaw mill, 36, 598
resaw waste conveyor, 36
resawing, 155, 598
research buoy, 554
research natural area, 478, 555
research, development and demonstration, 555
resection, 36, 199, 238, 247, 377
reseeding, 60
resene, 234, 557
reservation, 645
reservation cost, 14
reservation price, 14
reserve, 14, 565, 586
reserve area, 14, 252
reserve cell, 14, 565, 586
reserve coppice method, 14
reserve cutting, 14, 521
reserve-forest, 14
reserve-form stands, 14
reserve-fund, 586
reserve high forest system, 15
reserve-seed forest, 14, 328
reserve seed method, 15
reserve-seed tree method, 15, 306
reserve sprout forest, 15
reserve sprout system, 15
reserve substance, 65
reserve tissue, 65
reserve tree method, 15
reserve-tree system, 15
reserved forest, 14, 586
reserved tree, 15
reserves, 272
reserving of stand, 15
reservoir, 457, 460, 545, 578
reservoir protection forest, 457
reshredder, 599
residence flame time, 622
residence time, 40, 621
resident bird, 306
residential district green area, 259
residential district park, 259
residential forest, 634
residential region planting, 259
residual, 139
residual chronology, 46
residual crop, 125
residual deformation, 40
residual elasticity, 431
residual heredity, 40
residual impact, 41
residual lignin, 40, 584
residual luminescence in wood, 330
residual material, 40
residual products, 40, 154
residual reproductive value, 431
residual resistance, 431
residual retention, 431
residual shrinkage, 40
residual soil, 40
residual stand, 40, 125
residual strain, 41
residual stress, 41
residual stumpage price, 431
residual toxicity, 40
residual tree, 15
residual (value), 431
residual variance, 431
residue, 40, 316, 584
residue harvesting machine, 431
resilience, 211, 472
resiliency, 211, 472
resilient modulus, 472
resilient stability, 211
resin, 449
resin adhesive, 449
resin-blaze, 39
resin bleeding, 322
resin-bonded plywood, 449
resin boxing, 39
resin bulb, 449
resin canal, 449
resin cavity, 450
resin cell, 450
resin channel, 68
resin-coated particle, 11
resin content, 449
resin coverage, 449
resin crop, 39, 465
resin cyst, 450, 578
resin defect, 450
resin duct, 449
resin film, 241, 449
resin flow, 308, 322
resin flux, 308, 322
resin gall, 449, 450
resin gland, 450
resin-impregnated paper, 450
resin impregnation, 450
resin modified wood, 449
resin over glue bond, 487
resin over resin bond, 487
resin passage, 449
resin plate, 449
resin pocket, 450, 578
resin sac, 613
resin seam, 450
resin shake, 450
resin sizing, 416
resin-soaked wood, 59
resin soak(ing), 450
resin soap, 450
resin spread rate, 450
resin streak, 449, 450
resin tannage, 450
resin tanning, 450
resin-tapper, 449

resin tapping, 39
resin transfer, 450
resin-treated laminated compressed wood, 613
resin tube, 449
resin varnish, 450
resinaceous, 449
resinated mat, 450
resiniferous, 189
resiniferous system, 450
resinification, 449
resinified knot, 189
resinify, 23, 251, 449, 487
resinoid, 528
resinosis, 308, 322, 450
resinous, 449
resinous cancer, 450
resinous odor, 465
resinous tracheid, 449
resinous wood, 120, 189, 325
resinous wood distillation, 325
resinous wound, 450
resistance, 270, 271
resistance adaptation, 271
resistance breeding, 271
resistance due to curvature, 384
resistance heating, 102
resistance in debarking, 32
resistance of start, 367
resistance of surface to boiling water, 26
resistance of surface to cigarette burn, 27
resistance of surface to high temperature, 26
resistance of surface to stain, 27
resistance of surface to wear, 26
resistance prediction, 271
resistance to control, 338
resistance to decay, 270
resistance to trampling, 271, 337
resistance-type moisture meter, 103
resistant, 270, 580
resistant against termite, 270
resistant fungi, 338
resistant stability, 94
resistant stock, 271
resistant variety, 271
resistant wood, 338
resistibility, 94, 271
resistivity, 271
resolution, 140
resolution capacity, 140
resolution cell, 140
resolution of microscope, 525
resolving power, 140
resonance, 172, 537
resonance method, 172
resonance peak, 537
resonance wood, 172
resonant frequency, 172
resonant shell, 172
resort area, 314, 543, 584
resource, 641
resource base, 641
resource management, 641
resource management system, 641
resource order, 641
resource order log, 641
resources, 641
respiration, 201
respiration-biomass ratio (R/B ratio), 200
respiration loss, 201
respiratory chain, 200
respiratory current, 200
respiratory quotient, 200
respiratory rate, 200
respiratory root, 200, 370
respirometer, 200
response, 130, 305, 530, 572
response time, 530
responsibility budget, 601
responsivity, 130, 530
rest period, 258, 543
rest point, 258, 481
resting-bud, 373, 543
resting egg, 543, 622
resting metabolism, 258
resting mycelium, 543
resting-spore, 543, 594
resting stage, 258
restocking, 61, 169
restoration, 211
restoration ecology, 211
restoring gene, 211
restratifying, 598
restricted fund, 526
restriction enzyme, 526
restrictive rootstock, 526
resultant, 130, 192, 422, 649, 651
resultant mat, 55
resultant of forces, 192
resultant wind, 192
retail nursery, 305
retail price, 305
retail yard, 305
retained earnings, 15, 504
retained profit, 14, 257
retaining board, 53, 103
retaining bolt, 176
retaining ditch and embankment, 283
retaining nut, 176
retaining wall, 88, 202
retardant, 651
retarded tree, 18
retarding agent, 57, 651
retarding culture, 569
retention aid, 634
retention of state forest, 186
retention time, 14, 15
retentivity, 14
reticulate parenchyma, 499
reticulate perforation, 499
reticulate perforation plate, 499
reticulate scalariform perforation plate, 500
reticulate thickening, 499
reticulate tracheid, 499
reticulate vein, 500
reticulate vessel, 499
retinaculum, 632, 640
retirement area, 445, 492
retirement method of depreciation, 15
retort, 157, 384, 609
retractable chuck, 419
retraction, 212
retrieval, 129, 130
retriever, 233, 436, 465
retrogression, 90, 343
retrogression of succession, 343
retrogressive evolution, 492
retrogressive succession, 343, 492

R

retroserrate, 343, 521, 580
return, 15, 130, 211, 290, 442, 460, 492, 614, 637
return air duct, 212
return bead, 294
return beam vidicon, 129
return cargo, 211, 212
return chain, 212
return convection, 211, 286
return felt roll, 319
return from final clearing, 127, 633, 634
return from thinning, 445
return interval, 211, 235, 630
return line, 212
return line for caul plate, 165
return migration, 500
return period, 211, 599
return rate, 212, 290
return to scale, 183
return track cable, 56, 500
return trap, 212
return water, 212
returning cableway, 500
revaluation, 61, 421
revaluation of assets, 641
revaluation of capital, 641
revaluation surplus, 60
revegetation, 61, 315, 616
revenue, 442
revenue expenditure, 442
revenue recycling, 442
reverberant room, 214
reverberation of sound, 430
reverberation time of sonud, 430
reversal of stress, 572
reversal peak, 90
reversal transparency, 130
reverse acting spiral, 129
reverse casehardening, 343, 625
reverse circulation, 343
reverse circulation jet blower type kiln, 343
reverse creep, 130
reverse cross circulation, 273
reverse duplication, 129
reverse mutation, 211
reverse ogee, 30, 90
reverse osmosis, 129, 343
reverse roll coater, 273
reverse scraper, 211
reverse scraper chain, 211
reverse strain, 130
reverse transcriptase, 130
reversed budding, 343
reverse(d) casehardening, 343
reversed matching, 100
reversed phase chromatography, 129
reversed stress, 130, 238
reversible circulation, 273
reversible circulation kiln, 273
reversible forming line, 500
reversible inhibition, 273
reversible plough, 454
reversible system, 273
reversing air circulation, 343
reversion, 130, 211
reversion of agricultural lands to forest, 492
reversion phase chromatography, 129
revertant, 211
revision of forest management plan, 407, 408
revision of management plan, 253
revision of the management, 407
revision of working plan, 407, 408
revitalizer, 154, 169
revivification, 153, 598
revocable regeneration block, 273
revolute leaf, 496
revolution mark, 546
revolving bookcase, 546
revolving chair, 638
revolving door, 546
revolving storm, 546
rewetting, 211
rewinder, 153
reworked fossil, 598
Reynolds number, 285
Rf value, 20
rhabdovirus, 159
rhachis, 204, 561, 584, 634
rhegma, 472
rheological behaviour, 306
rheological proterty, 306
rheology, 193, 306
rheophilic vegetation, 439, 441
rheophyte, 307
rheostat, 24
rheotaxis, 383
rhimba wax, 76
rhipidium, 414
rhizine, 167, 230
rhizobium (*plur.* -ia), 168
rhizocaline, 54, 422
rhizocaul, 159, 167
rhizoid, 230
rhizomatic, 168
rhizome, 98, 168
rhizome bud, 167
rhizome cutting, 98
rhizome geophyte, 167
rhizome neck, 167
rhizome plant, 168
rhizome sheath, 98
rhizome-wood regeneration, 98
rhizomoid, 167, 230
rhizomorph, 168
rhizomorphoid, 168
rhizophore, 168
rhizoplane, 168
rhizosphere, 167, 168
rhizosphere effect, 168
rhizotaxis, 168
rhododendron garden, 111
rhombic system, 611
rhomboidal crystal, 305, 537
rhooki, 55
rhus lacquer, 366
rhusinic acid, 366
rhyolite, 308
rhythm, 245, 246, 596
rhythmic growth, 245
rhythmic timing, 245
rhythmicity, 245, 246
rhytidome, 73, 314, 496
rib, 285, 633
rib log, 89, 622
rib-type coil, 285

R

riband back chair, 80
ribbed cocoon maker, 373
ribbing, 30, 495
ribbon bed, 80
ribbon board, 479
ribbon figure, 80
ribbon flower-bed, 80
ribbon ice, 80
ribbon-like microfibril, 81
ribbon saw, 79
ribbon sowing, 81
ribbon spreading machine, 313
ribbon stripe, 80, 285
ribbon(-stripe) figure, 80
ribonuclease (Rnase/RNAse), 194
ribo(nucleo)side, 193
ribo(nucleo)tide, 193
ribosomal RNA, 194
ribosome, 194
ribulose, 194
rice paper, 481
rice soil, 456
rice straw, 90
Rich pruning tool, 288
richness, 144
rick, 45, 114, 183, 236, 540, 601
ricked measure, 45
Rickettsia(e), 285
Rickettsia(e) like bacteria, 285
Rickettsia(e) like organism, 285
riddle, 73, 165
ride, 52, 298, 301, 383, 590
ride side, 590
rider, 454
rider block, 84, 89
ridge, 168, 225, 305, 309
ridge beam, 110, 225
ridge culture, 309
ridge drilling, 309
ridge-piece, 225
ridge-planting, 168, 309
ridge plowing, 309, 353, 654
ridge roof, 508
ridger, 353, 367
ridgetop, 413
ridging, 353, 654
ridging loading method, 367
Riecker's lbrmula, 288
riffler, 53, 64, 479
riffles zone, 222
rifle, 36, 282
rifle cartridge, 282
rifle shooting position, 282
rifle sight, 282
rifle sling, 282
rifle stock, 282
rifling, 282
rift crack, 538
rift grain, 255, 256
rift saw, 364, 529
rift sawed, 255, 256
rift sawed lumber, 255, 256
rift sliced, 256
rift valley, 97, 297
rifting rod, 173
rig, 2, 231, 470
rig donkey, 2
rigger, 470
rigger's bar, 105, 377
rigger's block, 231
rigger's drum, 231
rigging, 470
rigging assembly, 470
rigging block, 231, 367
rigging cap, 223
rigging chain, 227
rigging choker, 281
rigging hardware, 470
rigging plate, 227
rigging rope, 2, 205
rigging sled, 596
rigging socket, 161
right and left adjustment hand wheel, 654
right angle prism, 614
right ascension, 58
right cylinder, 615
right hand bandmill, 582
right hand cutter, 582
right hand lay, 582
right-handed twist, 582
right lay rope, 582
right of bark, 448
right of forest use, 409
right of grass-cutting, 165
right-of-way grant, 482
right on the pin, 361
right to fire-wood, 540
right to litter, 95
right to pannage, 186
right tree on right site, 439
rigid boom, 160, 176, 575
rigid insulation board, 160
rigid structure, 160
rigidity, 160
rigidity modulus, 160
rill, 533, 535
rill erosion, 533
rim-fire, 22
rim-fire cartridge, 21
rim grove, 87
rim of sanio, 319
rim saw, 529
rim speed, 593
rime, 32, 318, 447, 513
rime fog, 465
rimless shouldered, 508
rimmed and shouldered, 580
rimmed and shoulderless, 580
rimming, 246
rimose bark, 183
rimu resin, 401
rimuene, 401
rind, 186, 448, 626
rind approach graft, 356
rind-gall, 414, 571
rind inarching, 356
ring, 267, 344
ring arrangement, 209
ring-bark(ing), 209
ring bivalent, 208
ring boundary, 344
ring budding, 209
ring chromatid, 208
ring chromosome, 208
ring count, 344
ring crack, 208, 312
ring crush compression resistance, 208
ring crush test, 209
ring debarker, 209
ring disk, 593
ring dog, 5, 207
ring failure, 209, 312
ring-formed clear-cutting, 209
ring-girdling, 207, 208
ring grinder, 208
ring-joint, 207
ring line, 208
ring-pored wood, 208
ring porous wood, 208
ring roll mill, 208
ring rot, 207, 312
ring (rotary) barker, 208
ring shake, 208, 312
ring shaken heart, 312

ring shaped selection system, 209
ring spot, 207, 312
ring-stripping, 209
ring-type delimber, 208
ring type grinder, 208
ring width, 344
ring width index, 344
ringed thickening, 208
ringed tracheid, 208
ringed vessel, 208
ringing, 207, 208, 228, 278
ringing (method), 207
rip, 516, 650
rip cutting, 650
rip removed, 650
rip saw, 650
riparian forest, 455
ripe for cutting, 37
ripe for the harvest, 633
ripe stand, 55
ripe wood, 54
ripe wood tree, 55
ripeness for felling, 37
ripening, 54
ripper, 5, 465, 650
ripping, 650
ripping saw, 650
ripple, 30
ripple mark, 30, 294
riprap, 351
rise, 231, 309, 416, 421
rise-and-fall saw, 421
risen moulding, 486
riser log, 634
rising and falling saw, 421
rising wind sifter, 416
risk capital, 145
risk factor, 216, 500
risk-frec interest, 509
risk-free rate, 509
risk of forest fire, 407
risk premium, 145
risk-taking, 145
ritualization, 564
river, 192, 356
river-basin, 308
river bed type, 193
river boss, 307
river crossing, 111, 195
river depot, 458
river driving, 159
river gauge, 459
river guardian, 141
river mouth or estuary, 193, 237
river scale, 3, 460
river stage, 459
river (surface) temperature, 193
river system, 193, 460
river terrace, 193
riverain forest, 192
riverine forest, 192
rivers die out, 193
riverside garden, 27, 192
riverside vegetation, 192
riverwash, 193, 554
riving, 356
riving hammer, 356
riving knife, 140, 356
rizoid, 230
road, 90, 121, 526
road block, 311, 364
road changing, 155
road donkey, 564
road intensity, 90
road junction, 90
road layout, 298
road-line, 222, 223
road network, 298
road patrol, 643
road pattern, 310, 526
road protection forest, 90
road roller, 222, 550
road scale, 222
road-slide, 223
road spacing, 298
road tread, 52
roading, 90, 121
roading plan, 298
roadless area, 510
roadside garden, 245, 310
roadside greenspace, 310
roadside planting, 540
roadside stockpile, 310
roadside strip, 310
roadside tree, 540
roaster, 19
robin, 383, 613
robinetin, 71, 557
robinetinidin, 71
robot weather station, 643
robtein, 71
rock blade, 492
rock bolt, 319
rock border, 554
rock desert, 434
rock-fill dam, 114
rock garden, 554
rock layout, 622
rock mantle, 27, 144
rock outcrop, 313
rock pavement vegetation, 433
rock picker, 39, 233
rock plant, 434, 554
rock rake, 434, 636
rock saw, 434, 590
rock stratum, 328
rocker, 558
rockery, 230
rockery and forest landscape, 434
rockery making, 104
rockfall, 554
rocking chair, 558
Rockwell-test, 604
rocky, 434
rocky erosion, 554
rocky-soil plough, 118
rocky tundra, 434
rod, 24, 533, 535
rod bivalent, 11, 159
rod chromosome, 159
rod-like crystal, 12
rod-shaped bacteria, 159
rodent repellent, 346
rodenticide, 346, 411
roe, 194
roe figure, 7, 310
roe stripe, 310
roestelium, 179
rogor, 284
rogue, 297, 384, 474, 597
rogued seed orchard, 385
roiling load, 184, 214
roll, 184, 550, 590, 604
roll coater, 184
roll front, 264
roll laminating, 184
roll-out bar, 637
roll press, 184
roll splice, 430, 470
roll spreader, 184
roll tensioning, 604
roll test, 264
roll-type bush breaker, 184
rolled towel test, 264
roller, 89, 184, 550, 608
roller bandsaw, 184
roller check, 604
roller coater, 184
roller conveyer veneer drier, 184
roller discharge head, 184
roller dryer, 184
roller fairlead, 89
roller frame for reel, 264

roller glue laying method, 184
roller gluing machine, 184
roller nose, 184
roller nosebar, 184
roller press, 184
roller spreader, 184
roller veneer drier, 184
rolling, 184, 492
rolling-ball planimeter, 638
rolling bind, 589
rolling boom, 401
rolling coulter, 592, 593
rolling dam, 184, 569
rolling device, 264
rolling-disk planimeter, 638
rolling off, 128, 480
rolling plain, 367
rolling readjustment, 367
rolling shear, 184
rolling shear test, 184
rolling stock, 217, 312, 353, 361
rolling-table saw, 184
rollover arm, 507
rollway, 171, 184, 492, 639
Röntgen ray, 311
Röntgenogram, 311, 419
Röntgenography, 311, 419
roof boarding, 508
roof-canopy, 447
roof corner test, 508
roof garden, 508
roof insulation board, 508
roof sheathing, 508
roof slab, 508
roof tree, 110, 225, 508
roof truss, 508
roofing, 155, 156, 508
roofing board, 508
rooftop garden, 508
rookery, 636
room garden, 440
room germinator, 124
room temperature, 441
room temperature setting adhesive, 441
roomy, 445
roost, 365
root, 168
root-and-shoot cutting, 167, 359
root apex, 167
root bit, 265
root breaker, 377
root buttress, 8, 281
root cancer, 167
root cap, 167
root climber, 168
root collar, 167
root crown, 167
root cutter, 247, 377
root cutting, 167
root density, 168
root division, 141
root dozer, 5, 265
root extractor, 494
root exudate, 168
root flare, 167
root-forming ability, 54, 422
root-forming substance, 54, 422
root frozen, 167
root grafting, 167
root grafting by approach, 167
root grubber, 265, 494
root hair, 168
root inhabiting fungus, 167
root initial, 168
root initiation, 123
root injury, 167
root knot, 167
root-knot eel-worm, 167
root-knot index, 167
root-knot nematode, 167
root leaf, 167
root maggot fly, 204
root mass, 168
root-mean-square, 266
root mean square deviation, 25, 266
root nodule, 167
root nodule bacteria, 167
root parasitism, 167
root planting, 141
root plate, 168
root plough, 494
root plow, 494
root pressure, 168
root primodium, 168
root-promoting, 73
root pruning, 168, 543
root-puddled, 324
root puller, 265
root rot, 167
root shear feller, 233
root-shoot, 167, 168
root-shoot ratio, 167
root spiralling, 168
root sprout, 167, 168
root-stem transition zone, 167
root stock, 63
root sucker, 167, 168
root suffocation, 167
root-swelling, 76, 159, 167
root system, 168
root thorn, 167
root timber, 167
root tuber, 279
root tubercle, 167
root tubercle bacteria, 167
root up, 167, 292
root volume, 167
root wad, 167
root wood, 167
root wrapping, 167
root zone, 168
root:shoot ratio, 167
rooted cutting, 566
rooter, 494
rooter bit, 494
rooting, 422
rooting ability, 123, 422
rooting bed, 371
rooting capacity, 123, 422
rooting frame, 46, 372
rooting habit, 168
rooting medium, 371, 422
rooting out of stump, 265
rooting percentage, 422
rooting zone, 422
rootlet, 518, 533, 544
rootprunner, 247, 377, 543
Rootrainer, 168
rootrake, 309, 494
Roo’ts blower, 310
rootstock, 167, 168, 608
rootstock form factor, 168
rootstock-scion relationship, 608
rope, 469
rope figure, 113, 431
rope ladder, 431
rope pitch, 431
rope print, 160
rope railway, 283
rope (storage) capacity, 264

R

rope-tightener, 249
rope transfer, 484
rope wheel, 470
ropeway, 469
ropeway driving wheel, 470
roping, 161, 431
rosary, 376, 594
rose and ribbon moulding, 376
rose-bengal, 319, 462
rose oil, 319
rosemary oil, 321
rosette, 72, 73, 294
rosette plant, 294
rosin, 465
rosin acid, 465
rosin amine, 465
rosin crystallization, 465
rosin derivative, 465
rosin nitrile, 465
rosin oil, 465
rosin pitch, 465
rosin resin, 465
rosin salt, 465
rosin size, 465
rosin soap, 465
rosin speck, 449, 465
rosin spot, 465
rosin varnish, 465
rosinate soap, 465
rosinol, 465
rosiny, 465
rosinyl, 465
ross, 30
rosser, 31
rosser head debarker, 516
rosserhead, 31
rossing, 39, 177
rossing-mill, 30
rossing tool, 177
rostrate, 213, 580
rostrum, 213
rot, 151
rot-resistant, 270
rotary ball digester, 212
rotary barker, 184, 593
rotary chain-saw, 170
rotary chair, 638
rotary clipper, 212
rotary crusher, 546
rotary(-cut), 547, 592
rotary cut veneer, 546
rotary cutter, 546
rotary cut(ting) veneer, 546
rotary drom kiln, 184
rotary drum barker, 638
rotary drum dryer, 546
rotary dryer, 546
rotary feeder, 212
rotary finishing, 546
rotary gang, 592
rotary granulator, 546
rotary hedge cutter, 546
rotary hoe, 637
rotary hot press, 212
rotary jet dryer, 546
rotary lathe, 545, 546
rotary microtome, 443, 546
rotary mower, 546
rotary mowing equipmant, 546
rotary nozzle dryer, 212
rotary peeling machine, 82, 546
rotary-ring barker, 637
rotary saw, 592
rotary slasher, 38, 546
rotary tower crane, 546
rotary valve, 212, 546
rotary veneer, 546
rotating conical nozzle tube, 546
rotating ring, 212
rotating screen, 212, 546
rotating stone ring, 546
rotating storage bin, 546
rotation, 209, 311, 312
rotation age, 311
rotation choice, 312
rotation determination, 312
rotation dryer, 546
rotation length, 311
rotation of crops and grass, 42
rotation of cuttings, 311
rotation of maximum cash flow, 248, 653
rotation of maximum income, 406, 652
rotation of maximum physical productivity, 653
rotation of maximum soil rent, 488, 653
rotation of maximum volume production, 652
rotation of maximum volume yield, 652
rotation of the greatest volume, 652
rotation of the highest income, 653
rotation of the highest net income, 406, 652
rotation pasture, 312, 347
rotation system, 312
rotational grazing, 312
rotational strain, 212
rotator, 212
rotavator, 546
rotenone, 584
rotocultivator, 546
rotor, 184, 313, 638
rotor-peeler, 546
rototiller, 546
rototiller cultivator, 546
rotten streak, 151
rotten timber, 150
rotten wood, 150
rottenness, 150
rough association table, 73
rough bark, 73
rough construction, 72
rough cut, 73
rough estimate, 72, 156
rough estimating, 72, 156
rough grain, 72
rough grazing, 72
rough hew, 73
rough log, 72, 73
rough lumber, 72, 504
rough mill, 73
rough patch, 33
rough plank, 504
rough sample plot, 72
rough sawn timber, 73
rough selection felling, 72
rough serrate, 72
rough staining fungus, 197
rough timber, 318, 466, 503, 504
rough wood, 72
roughage, 72, 73
roughness, 72
round, 45, 81, 124, 391, 593
round-about channel, 583
round back chair, 592
round-back(ed) sleeper, 11, 115
round-billet (wood), 535
round boom, 81
round-boom type raft, 81
round chip box, 592

round corner cabinet, 592
round corner joint, 592
round cutterhead, 593
round-headed borer, 477
round knot, 592
round log, 592
round measure, 435
round molding head, 593
round process, 592
round reducer, 16
round saw edger, 592
round shake, 208, 312
round stock sander, 592
round stuff, 592
round timber, 81, 156, 592, 638
round tree, 81
round up, 213, 269, 351, 547, 603, 648
round wood, 156, 592
rounded tracheid, 593
roundhead woodscrew, 593
rounding(up), 546
roundwood, 593
roundwood equivalent, 435
route survey, 310
router (machine), 309
router plane, 384
routine maintenance, 291, 397
routing, 309
rove-beetle, 571
roving agriculture, 312
row planting, 190
row seeding, 190, 479
row spacing, 190
row thinning, 166, 190
royalty, 385, 437
rub, 88, 89, 350
rub tree, 89
rubber bag moulding, 531
rubber-based adhesive, 66
rubber bearimg plant, 531
rubber contact roll, 531
rubber cushion glide, 531
rubber estate, 531
rubber hydrochloride, 555
rubber latex, 531
rubber plant, 531
rubber plantation, 531
rubber producing plant, 531
rubber roll, 531
rubber sectional infeed roll, 141
rubber softener, 531
rubber yielding plant, 531
rubbery wood, 472
rubbing varnish, 326, 337
rubby soil, 291
rubulac, 599
ruby lac, 14
ruby shellac, 14
ruby-tailed wasp, 379
rudder boom, 583
ruderal plant, 596
rudimentary, 40, 125, 492, 504
rudimentary embryo, 125
rudimentary organ, 125, 565
rudimentary seed, 125
rudite, 291, 468
rudyte, 291, 468
rue oil, 595
Rueping cylinder, 314
Rueping empty-cell cycle, 314
Rueping process, 106, 276, 314, 526
ruga (*plur.* rugae), 631
rugged terrain, 367
rugose leaf, 631
ruinous-exploitation, 283
rule joint, 215
rule of similarity, 528
rules for seed testing, 628
rules of forest police, 301
ruling gradient, 526
ruling price, 435, 438
ruling price timber, 331
rumen, 308
ruminant, 129
ruminate, 129, 242
run, 66, 161, 223, 301, 306, 438, 474, 540, 596
run-around, 208
run camp, 179
run cutter, 223, 269
run line, 44
run of logs, 590
run-of-the-log, 589, 590
run-of-the-mill, 620
run-off, 256, 306, 538, 569
runcinate, 90
Runkel coefficient, 283
runner, 55, 89, 205, 292, 349, 351, 364, 560
runner chain, 205, 349
runner dog, 349
runner plant, 148
running, 205, 364, 596, 609
running bamboos, 404
running cat, 223
running cost, 397
running crown fire, 281
running fire, 297, 466, 542
running foot, 57, 615
running gear, 66, 595
running guy, 21
running hook, 281
running line, 372, 598
running log, 461
running repair, 397, 535
running saw, 564
running skyline, 56, 231
running skyline system, 595
running slide, 629
running splice, 49
running tightline, 249
running time, 41
running water community, 307
runoff, 256
runoff analysis, 256
runoff coefficient, 256
runoff ratio, 256
runout, 161
runway, 533
rupestral, 554
rupicolous plant, 434, 554
rupideserta, 291, 434
rupture, 113, 297, 363
rupture strength, 113
rupture stress, 113
rural chair, 73, 590
rural forestry, 527
rural scenery, 240, 478
Russell effect, 312
Russian fathom, 120
Russian spring-summer encephalitis, 408, 594
rust, 544
rust-inhibiting paint, 134
rust preventative, 134

R

rust preventing agent, 134
rust resisting paint, 134
rustic furniture, 73, 590
rustic scenery, 478, 527
rusty knot, 544
rut, 52, 606
rut season, 542
Rütgers process, 314
rutter, 362
ruuudheaded wood borer, 477
rypophagous, 436
S-chromosome, 390, 552
S-colony, 181
S-curve, 130, 384
S-discharging device, 350
S-external, 70
S-helix, 313
S-hook detacher, 39
S-internal, 70
S-iron, 133
S lay, 654
S-type chip, 16
sabina oil, 592
sable, 194
sabre butt, 384
sabre leg, 266
sabulicole, 412
sabulose, 411
sac, 338
saccate, 261, 338
saccharase, 606
saccharide, 473
saccharification, 473
saccharimeter, 473
saccharimetry, 473
saccharin(e), 473
saccharol, 473
saccharometer, 473
saccus, 338, 370
sack, 12, 81, 307, 493
sack boom, 81
sack the banks, 465
sack the rear, 307
sack the slide, 232
sacker, 307
saddle, 3, 4, 88, 497, 612
saddle a skid, 269
saddle back, 3
saddle bag, 391
saddle block, 56, 402, 453
saddle fungus, 316
saddle grafting, 3, 366
saddle-planting, 3
saddle set, 3
saddler's wood, 377, 493, 608
safe guy angle, 19
safe handling, 2
safe-landing approach, 2
safe limit, 2
safe load, 2
safe moisture content, 2
safe operation, 2
safe seed moisture, 627
safe side, 2
safe sight distance, 2
safe stress, 2
safe working pressure, 2
safe working strength, 2
safe working stress, 2
safety code, 2
safety device, 2
safety factor, 2
safety gradient, 2
safety guy, 2
safety island, 2
safety level, 2
safety line, 2
safety measure, 2
safety saw, 2
safety slide, 2
safety strip, 2
saffron, 128
saffron oil, 599
safranine, 599
sag, 2, 4, 67, 306, 338, 520
sagging, 53, 306, 380
sagittal, 237, 437
sago, 513
sail block, 612
sail guy, 195, 196
sail-guy spreader line, 195
sailer, 136, 212
sailing line, 196
sale and leaseback, 317, 444
sale at fixed rate, 106
sale at tariff price, 106
sale by area, 289
sale by auction, 349
sale by private contract, 290
sale by royalty, 106
sale by sealed tender, 25
sale by successively increased bids, 19
sale depot, 253
sale lot, 532
sale-lot of numbers of pieces, 225
sale-lot of separate pieces, 166
sale of business, 62
sale of standing timber, 289
sale of standing trees, 289
sale of the whole lot, 385
sale on commission, 227, 503
sale on the stump, 289
sale value, 62
sal(e)able, 274
sal(e)able height, 273
sal(e)able plant, 273
sal(e)able volume, 273
sales discount, 532
sales of productive property, 422
sales on credit, 417
sales proceeds, 444
sales promotion expense, 492
sales received in advance, 587
sales tax, 574
salicortin, 460
salina, 555
saline, 189, 555
saline-alkali soil, 555
saline meadow, 555
saline resistance, 271
saline soil, 555
salineness, 189
salinity, 189, 524, 555
salinity barrier, 555
salinity stress, 555
salinity tolerance, 338
salinization, 555
salinized chestnut soil, 555
salinized soil, 555
salipurposide, 367
salisbury hook, 111
salivary duct, 494
salivary gland, 494
salivary gland chromosome, 494
salivation, 307, 494
sallon tree, 308
sallow flat body Depressaria conterminella Zeller, 308
salt accumulation, 555
salt-affected soil, 555
salt-and-pepper effect, 138
salt-avoiding species, 21

S

salt bush, 555
salt chuck, 555
salt community, 555
salt crust, 555
salt desert, 555
salt endurance, 338
salt enduring species, 338
salt incrustatiou, 555
salt lake, 555
salt marsh, 555
salt meadow, 555
salt mixing tank, 555
salt-obligate species, 544
salt pan, 555
salt plant, 555
salt preservative, 555
salt resistance, 271
salt seasoning, 555
salt sensitive, 555
salt soil, 555
salt spray, 555
salt steppe, 555
salt swamp, 555
salt tolerance, 338
salt-tolerant, 338
salt-tolerant plant, 338
salt-treated wood, 555
saltation, 34, 267, 389, 480, 486
salting-out elution chromatography, 555
saltpeter, 230, 532
saltwater intrusion (encroachment), 555
salty soil, 189
salvage, 141, 259, 376
salvage clipper, 468
salvage cutting, 609
salvage edger, 139
salvage logging, 64, 609
salvage timber, 216
salvage value, 40, 139, 606
salve, 147, 559, 578
salvene, 446
salviol, 446
samara (samera), 58
sample, 61, 384, 439, 557, 558
sample area, 25, 558
sample board, 438, 557
sample changer, 558
sample collector, 384
sample fraction, 61
sample mean, 558
sample number, 384
sample plot, 558
sample plot data, 558
sample plot method, 25, 558
sample-plot survey, 25, 558
sample-plot survey method, 25
sample point, 557
sample scaling, 557
sample size, 557, 558
sample strip, 80, 558
sample thinning, 25
sample tree, 25, 558
sample-tree age, 25
sampler, 233, 384
sampling, 61, 384
sampling at successive occasion, 293
sampling density, 61
sampling design, 61
sampling distribution, 61
sampling error, 61
sampling for attribute, 3
sampling for growing stock volume, 3
sampling fraction, 61
sampling inspection, 61
sampling intensity, 61
sampling methods, 384
sampling plot, 535
sampling point, 558
sampling proportional to size, 156
sampling rate, 384
sampling ratio, 61
sampling survey method, 61
sampling unit, 61, 557
sampling unit distribution, 557
sampling using a range of fixed radius plots, 593
sampling using variable radius plots, 23, 272
sampling with partial replacement, 36
sampling with probability proportional to prediction (3P sampling), 586
sampling with probability proportional to size, 156
sampling with varying probability, 34
sampson, 89, 492
san-mou oil, 414
sanctuary, 252, 346, 507
sand, 411, 412
sand bar, 412
sand-barren, 411
sand-beach community, 411
sand belt, 412
sand binder, 177, 411
sand-binding grasses, 176
sand binding plant, 177
sand break, 133
sand cloth, 412
sand collar method, 228
sand control hedge, 133
sand-creosote-collar method, 412
sand culture, 412
sand defence forest, 133
sand disk, 326, 412
sand-drift, 307
sand dune, 411
sand dune plant, 411
sand dune regosol, 411
sand dune vegetation, 411
sand fixation, 176
sand flea, 411
sand garden, 411
sand grass, 411
sand hill, 411
sand loam, 380
sand paper, 412
sand protecting plantation, 133
sand protection facility, 133
sand protection green, 133
sand run, 412
sand-shifting control forest, 131
sand soil, 412
sand solidification, 176
sand spit, 412
sand spreader, 155, 402
sand storm, 411
sand vegetation, 411
sand-washing machine, 516
sandal, 618
sandalwood oil, 472
sandbar, 412
sandbreak forest, 133
sanded one side, 84
sanded two sides, 453
sander, 326, 412
sander-dust, 332, 412

sanders oil, 472
sanding, 412
sanding cylinder, 327, 412
sanding dust, 332, 412
sanding machine, 326, 412
sanding pad, 412
sanding pressure, 412
sanding station, 326, 412
sandstone, 412
sandtrap, 53, 64
sandwich beam, 229
sandwich board, 229
sandwich composites, 153, 229
sandwich construction, 229
sandwich guide, 229
sandwich panel, 229
sandy loam, 412
sandy soil, 412
sandy soil vegetation, 412
sanguis draconis, 309
Sanio's law, 402
sanitary equipment, 503
sanitary insect, 503
sanitary waste, 207, 422
sanitation cutting, 503
sanitation felling, 503
sanitation in seasoning yard, 368
santal camphor, 472
santal oil, 472
sap, 449, 518, 612
sap acid-etching, 449
sap beetle, 311
sap displacement (treatment), 449
sap fir, 433
sap flow, 561
sap inoculation, 612
sap-peeling, 68, 521
sap pressure, 562
sap rot, 21
sap-season, 449
sap shake, 21
sap shoot, 321
sap stain, 21
sap stain control, 21
sap stain fungus, 21
sap streak, 21
sap-sucking insect, 71
sap vesicle, 561
sap (wood) rot, 21
sap-wood tree, 21
sapling, 583
sapling chopper, 222
sapling crop, 582
sapling growth, 583
sapling stage, 583
sapling test, 583
saponin(e), 599, 600
saponitc, 600
sapote gum, 396, 413
sappanin, 466
saprobia, 508
saprobic, 150, 508
saprobiont, 150, 508
saprobiotic, 508
saprogen, 150, 462
sapropel, 150
sapropelite, 150
sapropelith, 150
saprophage, 150
saprophagous, 436
saprophagous animal, 436
saprophagous organism, 436
saprophagy, 436
saprophile, 439, 516
saprophilous, 150, 439, 516
saprophyte, 150, 462
saprophyte-community, 150, 153
saprophytic, 150, 439, 516
saprophytic chain, 150
saprophytic form, 150
saprophytic fungus, 150
saprophytic nutrition, 150
saprophytism, 150
saproplankton, 508
saprotrophs, 150
saprotroph(s), 150
saprovore, 436
saproxenous, 348
saproxylobios, 462
sapstain prevention, 21
sapwood, 21
sapwood area, 21
Saran fabric, 412
sarcocarp, 186, 399, 622
sarcoderm, 625
sarcophaga, 437
sarcophagous, 437
sarcophagy, 437
sarcotesta, 237
sarment, 49, 522
sasanquasaponin, 578
sash, 67, 262, 280
sash bar, 280
sash centre, 67
sash cramp, 67
sash door, 375
sash fillister, 67
sash gang-saw, 280
sash mortise chisel, 67
sash pocket chisel, 67
sash saw, 280, 349
sash weights, 103
sassafras (wood) oil, 210
SAT-zone, 468
satellite, 468, 503
satellite data processing system, 503
satellite imagery, 503
satellite infrared spectrometer, 503
satellite inventory, 503
sathrophyta, 151
sathrophytia, 151
satin finishing, 351, 554
satining machine, 398
saturated air, 14
saturated humidity, 14
saturated pressure, 14
saturated steam, 14
saturated temperature, 14
saturated vapour, 14
saturated vapour pressure, 14
saturation, 14, 405
saturation adiabat, 432
saturation coefficient, 14
saturation deficit, 14
saturation density, 14
saturation point, 14
saucer peach, 351
Saug-Kappe process, 470
Saukey diagram, 405
saurochore, 417
sausage exercise, 290
savage forest, 591
savanna forest, 515
savanna woodland, 515
savanna(h), 515
saveall, 6, 88, 133, 618
saver, 618
savin(e) oil, 411, 592
saw, 261
saw adjustment, 263
saw alive, 318, 360
saw arbor, 263
saw bar, 262
saw bench, 263, 543
saw bill, 54
saw blade, 263
saw-blade projection, 263

S

saw block, 263
saw brazing clamp, 263
saw buck, 262, 263, 600
saw-carriage, 79, 590
saw carriage track, 326
saw chain, 262
saw chain filing machine, 262
saw cut, 262
saw doctor, 74, 262, 543
saw down, 262
saw draft, 262
saw filer, 74, 610
saw fitter, 74, 610
saw fitting, 543
saw fitting tool, 74, 610
saw frame, 262
saw guard, 263
saw guide, 262
saw gumming, 326
saw handle, 261
saw kerf, 262
saw kerf graft, 262
saw leg production, 620
saw line, 262, 263
saw log, 262
saw log grade, 262
saw log portion, 262
saw log (sawlog), 262
saw log short, 262
saw lumber, 54, 262
saw mandral, 263
saw pinching, 229
saw-plate projection, 263
saw protector, 295
saw punch, 268
saw set, 29, 262
saw-setting machine, 29, 610
saw shaft, 263
saw shaper, 268
saw sharpener, 326
saw sharpening, 326, 543
saw-sharper, 262
saw spindle, 263
saw tailer, 262
saw-teeth, 262
saw through and through, 318, 360
saw timber, 262
saw-timber forest, 577
saw-timber stand, 577
saw vice, 262, 263
saw wedge, 263
saw with the grain, 461, 650
saw-wrest, 30, 58
saw, dry and rip process, 262
sawblade sharpener, 326
sawblade strain system, 263
sawblade stretcher, 263
sawdust, 263, 335
sawdust and log waste conveyor, 263
sawdust molding product, 335
sawdust polymer, 263
sawed around, 589
sawed lumber, 54, 262
sawed timber, 54, 262
sawed veneer, 263
sawfly, 262, 560
sawhorse, 263
sawing, 262, 521
sawing alive, 318, 590
sawing around, 110, 554
sawing cost, 620
sawing jack, 263, 600
sawing machine, 620
sawing-off, 262
sawing pattern, 263, 521
sawing procedure, 262, 521
sawing rig, 630
sawing technology, 620
sawlog, 577, 620
sawlog class, 262
sawlog stage, 77, 620
sawmill, 262, 619
sawmill refuse, 620
sawn full, 263
sawn product, 262, 619
sawn square, 263
sawn stave, 263
sawn timber, 55, 262
sawn veneer, 263
sawnweod grading rule, 262
sawnwood, 54, 262
sawnwood price, 262
sawyer, 126, 262, 477
saxicoline, 554
saxicolous, 554
saxideserta, 291
saxitoxin, 17, 187
scab disease, 67
scab pitch pocket, 449
scab resin pocket, 449
scabrous, 72
scaffold, 243, 477, 603
scaffold board, 126
scalariform, 475, 476
scalariform bordered pit, 476
scalariform cell, 475
scalariform conjugation, 475
scalariform duct, 475
scalariform parenchyma, 476
scalariform perforation, 476
scalariform perforation plate, 476
scalariform pitting, 476
scalariform thickening, 475, 476
scalariform tracheid, 475
scalariform vessel, 475, 476
scale, 20, 36, 57, 184, 232, 248, 274, 305, 358
scale bark, 305
scale beam, 57
scale bill, 232, 233, 320
scale board, 17
scale book, 38, 127, 320
scale effect, 57
scale-insect, 248
scale measure, 232
scale of aerial photograph, 191
scale of hardness, 575
scale of operation, 399
scale-off, 305, 494
scale parameter, 57
scale-rule, 45, 233, 604
scale scar, 551
scale-stick, 45, 233, 604
scale ticket, 330
scaler, 30, 106, 233
scaling, 111, 232
scaling diameter, 232
scaling felled tree, 125
scaling handbook, 590
scaling law, 24
scaling of felled timber, 232
scaling ramp, 232
scaling rule, 232
scaling station, 233
scaling stick, 233
scalping, 279, 385
scalping roll, 177
scaly, 305, 581
scaly bark, 305
scaly bud, 305
scaly bulb, 510
scaly leaf, 305
scaly wood, 358
scalybark hickory, 73
scandent, 152, 351
scandent epiphyte, 351

scandent plant, 351
scandent shrub, 364
scanner, 405
scanner imagery, 405
scanning block centring, 332
scanning electron micrograph, 405
scanning electron microscope, 405
scanning radiometer, 405
scansorial, 350
scant, 112, 387
scant size, 32
scantentes, 351
scantling, 533, 535
scanty paratracheal, 515
scanty paratracheal parenchyma, 515
scape, 203, 204
scar, 5, 195, 560
scar (branch) knot, 414
scar cicatrice, 587
scar tissue, 7
scarab, 248
Scarab beetle, 248
scarabaeoid, 366
scarcely forested land, 404
scarf, 537
scarf joint, 537
scarfed plywood, 75
scarfjoint, 537
scarification, 414, 465
scarifier, 128, 414, 465
scaring, 284
scariosa, 116
scatter diagram, 404
scatter gun, 346, 404
scatter hording, 142
scattered felling, 404
scattered knot, 404
scattered light, 404
scattered mixed forest, 404
scattered parenchyma, 540
scattered periodic block, 404
scattered planting, 404
scattered seed-tree method, 404
scattered tree, 404
scattergram, 404
scattering, 404
scatterogram, 404
scavenger, 33, 257, 436
scenarios, 381
scene borrowing, 248
scenery, 144, 145, 645
scenery variation, 255
scenic analyses, 145
scenic area, 145
scenic beauty, 145, 146
scenic beauty forest, 145, 146
scenic beauty protection forest, 145
scenic forest, 145
scenic quality, 145, 146
scenic spot, 145
scent gland, 62
scent-marking, 544
scent post, 544
scent scales, 528
scent set, 544
schatten, 570, 575
schedule, 183, 336, 435, 620
schematic drawing, 438, 589
scheme of division, 408
Schiffel's formula, 515
schinus oil, 229
schizo-lysigenous, 297
schizo-lysigenous cell, 297
schizo-lysigenous duct, 297
schizo-lysigenous traumatic canal, 297
schizocarp, 141
schizocotyledon, 298
schizogenesis, 297
schizogenetic, 287, 297
schizogenous, 287, 297
schizogenous canal, 297
schizogenous cell, 297
schizogenous duct, 297
schizogenous intercellular cavity, 297
schizogenous intercellular space, 297
schizogenous resin canal, 297
schizogenous traumatic canal, 297
schizomycete, 298
schizophyte, 298
Schlyter test, 431
Schneider's formula, 431
school, 584
schooling, 388
Schopper-Riegler beating degree, 531
Schopper-Riegler beating degree tester, 531
Schopper-Riegler grade, 531
Schopper-Riegler-Mahlgrad, 531
schradan, 4
Schultze's maceration method, 444
Schweinfurt(h) green, 553, 566
Schweitzeros reagent, 484
sciad, 517, 570
sciadin, 397
sciadinone, 397
scintillator, 414
scio- (*eg.* sciophilous, sciophobous), 570
scion, 244
scion budding, 113
scion effect, 244
scion grafting, 612
scion grafting on root, 168
scion plucking nursery, 39
scion rooted, 644
scion-rooting, 244
scion-stock interaction, 244
scion wood, 244
sciophile, 440, 517, 570
sciophilic, 517
sciophilous, 517
sciophilous species, 517, 570
sciophobia, 21, 516
sciophyll, 570
sciophyte, 556, 557, 570
scission, 141, 142, 233, 377
scissors truss, 234
scleranthium, 467
sclereid, 232, 434, 575
sclereid fibre, 575
sclerenchymatous cell, 200
sclerometer, 575
sclerophyll, 575
sclerophyll scrub, 575
sclerophyllous, 575
sclerophyllous forest, 575
sclerophyllous plant, 575
sclerophyllous shrub, 575

S

sclerophyllous tree, 575
sclerophyte, 575
scleroscope, 212
sclerosis, 575
sclerotesta, 575
sclerotic cell, 434, 575
sclerotic fibre, 575
sclerotic parenchyma, 575
sclerotic tissue, 575
sclerotic tylosis, 575
sclerotiniose, 267
sclerotium, 267
scobs, 16, 263
scoop-net, 259
scooped basin, 378
scooped side graft, 57
scoot, 349, 548
scope, 439, 524
scope sight, 258
scopoline, 283
scorb, 204
scorch, 278, 279, 560, 640
scorch height, 548
scorching, 278
score, 274, 297, 414
score cutter, 619
scoring ax, 269, 547
scorpioid, 536
scorpioid eyme, 536
scorpioidal, 536
scotia, 4
scouring, 257, 378, 385, 476, 494
scramble competition, 291
scramble-type competition, 376
scramble type of competition, 143
scrambler, 351
scrap, 41, 468
scrap value, 40
scrap wood, 22, 40
scrape, 165, 177, 319
scrape iron, 177
scraper, 177, 362
scraper conveyor, 177
scraper plane, 177
scraping, 177
scraping machine, 177
scraping tool, 39, 177
scrapper, 177
scratch, 177, 205, 501, 534
scratch grading, 489
scratch resistance, 270
scratch-test, 204
scratcher, 177, 274
scratcher chain saw, 614
scratch(ing) hardness, 177
scree, 19, 307, 468, 534, 554
scree slope, 468
screeding, 177
screefing, 64
screen, 165, 363, 412
screen analysis, 412
screen-back, 499
screen classification, 412
screen door, 1, 145, 363
screen felling, 239
screen fractionation, 412
screen mesh, 412
screen planting, 363
screened pulp, 518
screened stock, 518, 566
screening, 363, 412
screening loss, 412
screw conveyer, 313
screw driver, 312, 546
screw extrusion press, 313
screw feeder, 313
screw feeding, 313
screw grinder, 313
screw-holding power, 346, 507
screw jack, 313
screw joint, 312
screw plug, 639
screw press, 313
screw-press process, 313
screw roller, 313
scribe, 75, 177, 204
scriber teeth, 403
scribing saw, 205
Scribner log rule, 462
Scribner log scale, 462
Scribner rule, 462
scroll, 507
scroll foot, 507
scroll leg, 507
scroll moulding, 507
scroll ornament, 507
scroll-over arm, 454
scroll sander, 339, 603
scroll saw, 384
scrolled marquetry, 507
scrolled pediment, 507
scrotiform, 338
scrub, 179, 597
scrub clearance, 180
scrub-clearing machine, 64, 165
scrub community, 179
scrub cutter, 165
scrub forest, 179
scrub plough, 180
scrub pulverizer, 180
scrub rake, 180
scrub slasher, 180
scrub (stand), 1, 179, 180, 449
scrub wood, 180
scrubber, 165
scuba, 460
scud, 136, 535
scuffing, 318
scuffle drill, 479
sculptor, 103
sculpture, 103
sculptured stone, 434
sculpturing, 103
sculpturing machine, 103
scum, 27, 148, 491
scum mark, 578
scurfy, 270, 581
scutate, 50, 260
scutel, 115
scutellum, 115, 352
scutiform, 116
sea drift, 188
sea ice, 188
sea level secular change, 188
sea-mount, 188
sea-side park, 188
seabird, 188
seal lac, 642
seal pattern wire rope, 146
sealant, 146, 343
sealed container, 322
sealed rope, 146
sealed storage, 322
sealer, 321
sealing, 146, 322
sealing bare wood, 328
sealing off, 30
seam, 244, 297, 319
seamless pipe, 509
sear, 7, 217
sear spring, 7
searching behaviour, 549
searching image, 465
SEASAT(sea satellite), 189
seasonal, 226
seasonal aspect, 226
seasonal change, 226
seasonal colours in gardening, 588

seasonal distribution, 226
seasonal fluctuation, 226
seasonal forest, 226
seasonal frequency, 226
seasonal garden, 463
seasonal grazing, 226
seasonal increment, 226
seasonal isolation, 226
seasonal migration, 226
seasonal periodicity, 226
seasonal prevalence, 226
seasonal reinfall, 226
seasonal rhythm, 226
seasonal succession, 226
seasonal swamp forest, 226
seasonal swamp thicket, 226
seasonal synusia, 226
seasonal variation, 226
seasonal vicariad, 226
seasonal water contents of living tree, 289
seasoned timber, 158, 368
seasoned wood, 158, 368
seasoning, 144, 158, 480, 487, 635
seasoning by leaching of wood, 449
seasoning check, 157
seasoning crack, 157
seasoning deflect, 158
seasoning degrade, 158
seasoning in saturated steam, 14
seasoning kiln, 158
seasoning schedule, 158
seasoning shake, 158
seasoning split, 158
seasoning yard, 368
seat-back, 271
seat-frame, 655
Seattle bunk, 470
Seattle car, 292
seawall, 188
seaweed marquetry, 507
sebaceous, 578
sebiferous, 260, 261
secateur, 543
seclusion type, 166
second-category forest inventory, 407, 410
second cutting, 121
second-generation, 99
second generation seed, 122
second grade timber, 122
second growth, 70, 71
second length, 99
second level carnivore, 70, 122
second log, 99
secondaries, 70
secondary adsorption, 99, 513
secondary bark, 71
secondary bast, 71
secondary benefits, 70, 350
secondary branchlet, 154
secondary breakdown, 363
secondary bud, 154
secondary carnivore, 70
secondary cell wall, 70
secondary center, 70
secondary community, 71
secondary compartment line, 383
secondary consumer, 70, 122
secondary conversion, 121
secondary cortex, 71
secondary crop, 152
secondary cyclone, 71, 154
secondary dormancy, 71, 122, 226
secondary embryo-cord, 352
secondary environment, 70
secondary felling, 199
secondary forest, 71
secondary fossil, 598
secondary gluing, 121, 652
secondary growth, 71, 615
secondary hardwood, 70
secondary infection, 226, 598
secondary insect, 70
secondary laterite, 71
secondary liber, 71
secondary low, 70, 154
secondary meristem, 70
secondary metabolites, 70
secondary parasite, 70
secondary particle reduction unit, 468
secondary periderm, 71
secondary phloem, 71
secondary ploughing, 152
secondary producer, 70
secondary product, 71, 99, 121
secondary production, 70
secondary productivity, 70
secondary rain forest, 71
secondary ray, 71
secondary road, 154
secondary root, 70
secondary saw, 122
secondary sere, 71
secondary soil, 71
secondary species, 71, 154
secondary spiral thickening, 71
secondary stand, 71
secondary structure, 70, 122
secondary succession, 71
secondary thickening, 70
secondary thickening growth, 71
secondary tillage, 152
secondary tiller, 121
secondary timber, 70
secondary tissus, 71
secondary transportation, 49, 122
secondary tumor, 71
secondary tundra, 70
secondary valence, 154
secondary wall, 70
secondary wall inner layer, 70
secondary-wall layer, 70
secondary wall middle layer, 70
secondary wall outer layer, 70
secondary woody growth, 71, 598
secondary xylem, 70
secret dovetail, 4
secret drawer, 571
secretaire, 444, 538
secretary, 444, 538
secretin, 142
secretion, 142
secretory cell, 142
secretory intercellular space, 142

secretory tissue, 142
section, 113, 247, 332, 364, 377, 382, 651
section analysis, 113
section line, 298
section modulus, 247
section plate, 292
section raft, 127, 349
section sampling, 378
section staining, 378
sectional air plate base, 141
sectional chip breaker, 141
sectional infeed roll, 141
sectional ladder, 359, 652
sectional measurement, 382
sectional measurement by means of Huber's formula, 543, 625
sectional measurement by means of Smalian's formula, 462
sectional measurement by means of volume, 382
sectional method, 382
sectional outfeed roll, 141
sectional plate, 141
sectional view, 113, 364
sectioned longitudinal pad, 141
sectioning, 377
sectionwise measurement, 382
sector, 534
sectorial chim(a)era, 382, 414
sectorial-periclinal chim(a)era, 382
secular trend, 49
security, 611
security density, 2, 586
sedentary, 40
sedentary soil, 106, 588, 591
sedge (*Carex*), 471
sediment, 53, 59
sediment charge, 189
sediment delivery ratio, 342
sediment discharge, 177, 446
sediment flux, 446
sediment in suspension, 545
sediment load, 446
sediment production, 53
sediment runoff, 177, 342
sediment slide, 53
sediment storage dam, 283
sediment storage dam with hole, 277
sediment trap, 64, 283
sediment yield, 177, 446
sedimentary cycle, 53
sedimentary landscape, 53
sedimentary phosplate, 456
sedimentary process, 53
sedimentary relict, 53
sedimentary rock, 53, 456
sedimentation, 53
sedimentation coefficient, 53
sedimentology, 53
sedoheptose, 255
sedoheptulose, 255
seed, 627
seed act, 628
seed ageing, 628
seed amelioration, 628
seed analyst, 628
seed aspirator, 370
seed ball, 628
seed bank, 628
seed barrow, 443
seed bearer, 328
seed-bearing, 246, 582
seed-bearing age, 245
seed bearing plant, 629
seed-bed, 31, 323, 521
seed beetle, 628
seed biology, 628
seed blower, 175, 556
seed-board, 31
seed-borne disease, 627
seed broadcaster, 401
seed bug, 628
seed capital, 591
seed carrying pest, 79
seed carrying pest and insect, 29
seed catalog(ue), 628
seed certification, 628
seed certification standard, 628
seed-certifying agency, 628
seed chalcid, 182
seed class, 628
seed cleaner, 257, 547
seed-cleaning, 257
seed coat, 626
seed-coat dormancy, 626
seed collecting, 39, 627
seed collection, 39
seed-collection area, 39
seed collection stand, 39
seed collection zone, 39
seed conductivity test, 628
seed control (station), 628
seed counter, 628
seed crusher, 602
seed depth, 31, 628
seed disinfectant, 628
seed disinfection, 628
seed dispersal, 627, 628
seed dissemination, 627
seed dormancy, 628
seed dressing, 11
seed drill, 31
seed drill for forestry, 303
seed driller, 479
seed drying, 628
seed drying machine, 628
seed drying oven, 628
seed dusting, 559
seed ecology, 628
seed end, 628
seed extraction, 493
seed extractor, 494
seed extractory, 494
seed-filling, 627
seed forest, 377, 436
seed grafting, 628
seed growing, 628
seed harrow, 31
seed harvesting, 627
seed harvesting and processing machinery, 627, 628
seed hopper, 628
seed-horn, 31, 521
seed huller, 384, 493
seed husking, 493
seed inspection, 628
seed inspector, 628
seed key, 628
seed kiln, 493, 628
seed lac, 272
seed law, 628
seed leaf, 641
seed lobe, 641
seed longevity, 628
seed lot, 355, 626

S

seed manure, 626
seed mass, 628
seed metering device, 350
seed moisture content, 628
seed orchard, 629
seed origin, 627
seed-pan, 31
seed pedicel, 626
seed pellet, 628
seed pelleting machine, 627
seed physiology, 628
seed plant, 629
seed plot, 323, 324
seed population, 628
seed processing, 627
seed production area, 39, 628
seed production stand, 39
seed purity, 628
seed quality, 629
seed quality examination, 629
seed-quantity, 31
seed rate, 31
seed release, 521
seed riddle, 257, 628
seed ripening, 627
seed rot, 628
seed science, 629
seed selection, 629
seed senescence, 628
seed separator, 257
seed setting, 245
seed setting percentage, 245
seed shedding, 521
seed shell, 626
seed sorter, 628
seed sorting, 547, 628
seed source, 627
seed sowing, 31
seed-sowing machine, 31
seed spot, 31
seed stalk, 626
seed stand, 39, 328
seed station, 629
seed stem, 626
seed storage, 628, 629
seed stratification, 627
seed tape, 628
seed test(ing), 628
seed testing station, 628
seed-time, 31
seed trap, 56, 628
seed treatment, 627
seed tree, 39, 328
seed tree cutting by the group method, 80, 279
seed-tree cutting method, 15
seed-tree cutting system, 15
seed-tree method, 306, 404
seed-tree method of strip cutting, 306
seed tree selection, 328
seed-tree system, 15
seed trier, 473, 627
seed variability, 627
seed vessel, 186, 626
seed viability, 628
seed vigour, 628
seed vitality, 628
seed volume-weight, 628
seed washer, 516
1000-seed weight, 371
seed wing, 626
seed year, 245, 628
seed yield, 627
seed zone, 39, 628
seedbed, 323
seedbed formation, 323
seedbed frame, 324
seedbed fumigation, 324
seedbed stake, 324
seedbed thinner, 235, 445
seedbed weeder, 324
seeder, 31
seeding, 31, 48, 74
seed(ing) cutting, 521
seed(ing) felling, 521
seeding-growth, 478
seeding in hole, 100, 547
seeding in line, 190, 479
seeding in spot, 100
seeding in strip, 79
seed(ing)-lath, 31
seeding nursery, 31
seeding on ridge, 309
seeding rate, 31
seeding stage, 31, 245
seeding time, 31
seeding trough, 521
seedless, 509, 512
seedlessness, 509, 512
seedling, 436
seedling age, 436
seedling bank, 583
seedling coppice, 436
seedling crop, 436
seedling emergence, 62, 583
seedling evaluation, 583
seedling-forest, 377
seedling forest system, 377
seedling grading machine, 626
seedling harvester, 367
seedling inarching, 583
seedling lifter, 367
seedling machine, 31
seedling machinery, 585
seedling packing, 324
seedling packing machine, 324
seedling percent, 54
seedling-plant, 436
seedling seed orchard, 436
seedling sorting, 324
seedling sprout, 321
seedling stage, 324
seedling stand, 436
seedling stand percent, 54
seedling storage, 324
seedling transplanter, 324
seedling tray, 585
seedling vigour, 583
seedling year, 62
seedsman, 31, 628
seemingly unrelated regression, 462
seen-area map, 296, 341
seep-off, 98, 420
seepage, 420
seepage area, 420
seepage flow, 420
seepage rate, 420
seepage water, 421
segment, 36, 245, 295, 386, 476, 519
segment belt, 359
segmental allopolyploid, 36
segmental arch, 170, 201
segmental interchange, 245, 390
segmental polyploid, 36
segmental waste gate, 170, 201
segmentation, 517
segmented distribution, 141
segmented polynomials, 140
segment(ed) saw, 245, 414
segregated coupe, 141

segregation, 141, 142
seiche, 230
seismonasty, 159
seize, 11
select, 577
select export quality, 577
select fine seed strains, 547
select plant, 547
select tree, 577, 611
selected crop, 328, 475, 547
selected seed, 577
selected stand, 328, 475, 547
selected tree, 577
selected tree seed, 577
selection, 141, 547
selection border cutting method, 81
selection coefficient, 547
selection cutting, 601
selection differential, 547
selection effect, 547
selection felling, 601
selection forest, 601
selection index, 547
selection intensity, 547
selection limit, 547
selection method, 601
selection of good tree seeds, 547
selection of route, 298
selection of tree, 302
selection percent, 601
selection pressure, 547
selection rate, 547
selection response, 547
selection stand, 601
selection strain, 547
selection system, 601
selection system of natural regeneration, 478
selection thinning, 601
selection value, 547
selective absorption, 547
selective absorption of wood, 329
selective advance, 547
selective breeding, 547
selective cupping, 127
selective cutting, 547
selective cutting cycle, 601
selective edger, 274
selective elimination, 547
selective exploitation, 547
selective felling, 547
selective fertilization, 547
selective grazing, 547
selective herbicide, 547
selective high pruning, 547
selective insecticide, 547
selective logging, 547, 601
selective media, 547
selective medium, 547
selective method, 252, 547
selective permeability, 547
selective sampling, 100, 547
selective thinning, 547, 601
selective toxicity, 547
selectivity, 547
selectivity of adsorption, 514
selerenchyma, 200
self, 645
self-acting thermostat, 643
self-acting tramway, 643
self-aligning grip, 644
self-aligning slide, 642
self-assemble furniture, 647
self-boiled lime sulphur, 644
self-bonded, 645
self-bred line, 645
self-centering, 642
self-centering boom, 642, 643
self centering infeed conveyer, 644
self-coloured flower, 85
self-combustion, 646
self-contained ecosystem, 644
self-contained thermostat, 643
self-curing adhesive, 644
self-curing glue, 644
self-destructive, 647
self-differentiation, 646
self-domestication, 646
self-fertile, 644, 645
self-fertile hybrid, 644
self-fertile plant, 645
self-fertilising plant, 644
self-fertility, 645
self-fertilization, 644, 647
self-grafting, 645
self-guiding nature trail, 642
self-healing, 647
self-ignition, 646
self-ignition point, 646
self-incompatibility, 645
self-inhibitor, 647
self-loading log truck, 647
self-loading skidder, 647
self-locking skidding carriage, 646
self-opening ventilator, 645
self-operated controller, 643
self-organization, 647
self-pollinated plant, 644
self-pollination, 644
self-propagation, 647
self-propelled crane, 644
self-propelled loader, 644
self-propelled rototiller, 644
self-propelled sprayer, 644
self-propelled transplanter, 644
self-propelled tree prunner, 644
self pruning, 478, 644, 646
self purification, 645
self-reading rod, 644
self-recorder, 643
self-recording barometer, 645
self-recording tide gauge, 644, 645
self-registering instrument, 643
self-registration, 643, 647
self-registration station, 642, 647
self-regulation, 643
self-reproducing, 477
self-root plant, 644
self-sowing, 477
self-sown seed, 478
self-sterile, 644, 645
self-sterility, 644, 645
self-sterility gene, 644, 645

self-supporting forestry, 644
self-sustaining system, 642
self thinning, 646
self-tightening guyline, 645
self-tightening load binder, 645
selfed line, 645, 647
selfed progeny, 645
selfing, 644, 645
selinene, 379, 417
selva, 58, 392
semelparity, 562
semelparous, 562
semet, 204
semi-amplexicaul, 9
semi-anatropal, 9
semi-anatropous, 9
semi-arid regions, 9
semi-automatic, 11
semi-automatic bundling device, 11
semi-automatic cableway carriage, 11
semi-bar lock, 9
semi-beaver tail, 9
semi-capsula, 377
semi-chemical pulp, 9
semi-chisel chain, 9
semi-circle cut, 115
semi-circular spade, 11
semi-conical spade, 11
semi-controlled mosaic, 10
semi-deciduous forest, 10
semi-desert, 9
semi-diffuse porous wood, 10
semi-drum, 11
semi dry process, 9
semi-dry pulp, 9
semi-drying oil, 9
semi-durable adhesive, 10
semi-evergreen forest, 9
semi-fast beating, 11
semi-finished products, 11
semi-fossil resin, 9
semi-frutex, 9
semi-gloss finish, 9, 398
semi-hammerless gun, 9
semi-hard board, 11
semi-hardboard, 11, 624
semi-hardwood cutting, 11
semi-hyaline, 10
semi-inferior ovary, 10
semi-isolation, 9
semi-lethal gene, 11
semi-mangrove, 9
semi-manufacture, 9
semi-manufactured commodity, 11
semi-mechanical pulp, 9
semi-mounted, 9, 36
semi-opaque, 10
semi-persistent virus, 9
semi-products, 11
semi-rigid insulation board, 9
semi-ring porous wood, 9
semi-shrub, 533, 552
semi-variable costs, 9
semiarid, 9
semiarid region, 9
semiboggy soil, 11
semibordered pit-pair, 10
semiconnate, 9
semidominance, 10
semidouble, 11
semifixed dune, 9
semilunar reproductive cycle, 11
semimature forest, 623
semimicro Kjeldahl method, 10
semina, 627
seminal, 63, 591, 628
seminal leaf, 641
seminal receptacle, 443
seminal root, 626, 628
seminal vesicle, 635
seminatural community, 11
seminatural vegetation, 10
seminature sowing, 10
seminiferous scale, 626
seminific, 628
semiparasite, 9
semiparasitic plant, 9
semipermeable membrane, 10
semipodzol, 9, 640
semisaprophytic, 9
semishrub, 533, 552
semispecies, 9
semisterility, 9
semivariance analysis, 9
semiverticillate, 10
sempervirentiherbosa, 48
sempervirentiprata, 48
send-up, 349
senesce, 451
senescence, 451
senile soil, 284, 451
senility, 162, 284
senna leaf wax, 413
sense organ, 159
sensibility, 305, 324
sensible heat, 524
sensible temperature, 159
sensillum chaeticum, 71, 159
sensillum trichodeum, 159, 319
sensing scanner, 111
sensing system, 44, 66
sensitive cell, 180
sensitive hair, 159
sensitive index, 324
sensitive species, 324
sensitiveness, 324
sensitivity, 159, 324
sensitivity analysis, 305
sensitizer, 181
sensitizing dye, 601
sensitometer, 17, 159
sensitometric step wedge, 159
sensor, 66, 324, 472
sepal, 121
sepalody, 121
sepaloid, 121
separate account, 111, 636
separating chamber, 142
separation, 141–143
separation layer, 287
separator of board and caul, 8, 140
separatory tank, 141
septate, 260
septate fibre, 141
septate fibre-tracheid, 141
septate parenchyma cell, 141
septate tracheid, 470
septate (wood) fibre, 141
septenate, 54, 365
septet, 365
septicidal, 440
septiferous, 260
septifragal, 441
septulate, 260
septum (*plur.* -ta), 166
septuploid, 365

S

sequence, 71, 461, 619
sequence rule, 461
sequential analysis, 544, 632
sequential decision making, 293
sequential forest inventory, 410
sequential inventory, 293
sequential sampling, 544
sequestration, 177
serafume, 548
seral community, 556
seral phenomenon, 556
seral stage, 556
serch, 402
serclimax, 556
sere, 556
serein, 381
serial, 54, 293
serial correlation coefficient, 517
serial map, 517
serial production, 54
sericeous, 18, 264, 462
sericite, 264
serine, 461
seringa, 531
serir, 291, 311
seroglobulin, 548
seroimmunity, 548
serological test, 548
serology, 548
serotinal, 57, 499
serotinal aspect, 57, 499
serotinous, 499
serous fluid, 238
serozem, 210
serpentine, 417
serpentine end matching process, 201
serpentine front, 486
serpentine-fused phosphate, 155
serpentine layering, 30
serpentine top, 486
serpentinite, 417
serrate, 262
serrature, 262
serrulate, 261, 581
serule, 502, 535
serum, 238, 381, 548
service charge, 443
service life, 437
service record, 437
service test, 437
service time, 437
service trial, 437
service yard, 8, 635
serviceability, 15, 274
serviceable life, 437
services, 284
servitude to by-product, 154
sesquicitronellene, 18
sesquidiploid, 18
sesquioxide, 18
sesquioxidic clay, 480
sessile, 177, 508, 509
sessile animal, 177
sessile leaf, 508
sessile organisms, 177
Sessom's system, 601
seston, 545
set, 23, 29, 49, 106, 176, 262, 280, 346, 558
set bead, 281, 474
set block, 610
set choker, 281, 474
set in fiberlayer, 523
set knob, 281
set lever, 558
set of chair, 55
set of chromosome, 391
set of stick, 168, 350
set time, 23, 346
set tooth, 29, 373
set-up, 231, 269, 304, 521
seta (*plur.* setae), 71, 160, 461
setaceous, 160, 260
set(ing) point, 346
setm inclination, 447
setose, 580
sett, 49, 278
settee, 49, 624
setter, 350, 405, 655
setting, 2, 29, 127, 304, 470
setting anvil, 610
setting back, 129, 200
setting bench, 29
setting free the crown, 447
setting gauge, 30, 107, 243
setting height of knife tip, 88
setting-in of bai-u, 400
setting iron, 30, 610
setting key, 30, 296
setting line, 2, 564
setting machine, 29
setting range, 480
setting temperature, 176, 575
setting time, 176, 346, 480
setting-up period, 176, 575
settle, 161
settle table, 566
settled tar, 53
settlement, 106, 224, 259, 263
settlement pattern, 224, 263
settling velocity, 53
setworks, 107, 558
severance cutting, 80, 287
severance felling, 80, 287, 445, 647
severance tax, 38, 142, 268
severing cut, 263, 600
severity thinning, 376
seville orange, 467
Sevin, 513
sewage, 139, 508
sewage farm, 508
sewage pollution, 508
sewage pump, 508
sewerage, 508, 521
sex, 541
sex attractant, 541
sex cell, 541
sex chromosome, 541
sex-controlled inheritance, 72
sex differentiation, 541
sex dimorphism, 296
sex-discrimination, 70
sex linkage, 11, 541
sex-linked gene, 11
sex-linked inheritance, 11
sex mosaic, 541
sex organ, 541
sex pheromone, 541
sex ratio, 70
sex reversal, 541
sex transformation, 541, 542
sexfarious, 309
sexing, 70
sexivalent, 309
sexless, 511
sextuploid, 308
sexual, 541, 582
sexual assemblage, 541
sexual cell, 541
sexual cycle, 542, 582
sexual dimorphism, 70
sexual generation, 582

S

sexual hybrid, 582
sexual hybridization, 582
sexual isolation, 541
sexual maturity, 541
sexual maturity of trees, 301
sexual organ, 430
sexual process, 582
sexual progeny, 582
sexual propagation, 582
sexual reproduction, 582
sexual reproductive cell, 582
sexual selection, 541
sexuales, 541
sexupara (*plur.* -rae), 541
sferics receiver, 77
shackle, 173, 207, 281, 431
shackle yarding block, 630
shade, 20, 570
shade area, 20
shade avoidance, 21
shade avoidance syndrome, 21
shade-avoiding, 557
shade-beare, 338
shade-bearer, 338
shade-bearing, 338, 570
shade bearing guild, 338
shade-bearing tree, 338
shade cloth, 605
shade community, 570
shade-demanding, 517, 570
shade density, 585, 605
shade-endurance, 338
shade-enduring, 338, 570
shade-enduring plant, 338
shade-enduring tree, 338
shade epiphyte, 570
shade-frame, 571
shade-giving shrub, 605
shade house, 605
shade intolarance, 516
shade-kind, 338
shade leaf, 570
shade loving, 517
shade-loving plant, 517, 570
shade mesh, 605
shade-needing, 544
shade plant, 570
shade-requiring, 517, 570
shade-requiring plant, 441
shade species, 517, 570
shade tolerance, 338
shade tolerant, 338
shade-tolerant species, 338
shade-tolerant tree species, 338
shade-tolerating tree, 338
shade tolorance, 338
shade tree, 605
shaded firebreak, 571
shadephobe, 21
shades of gray, 45, 210
shading, 605
shadow, 570
shadow length, 570
shadow line, 182
shadow method, 570
shadow method in photo, 530
shadow photograph, 574
shadow picture, 419
shadow price, 571, 575
shadow pricing, 575
shadowgraph, 419
shadowing, 354, 605
shadowing technique, 354
shadowless photograph, 511
shady slope, 570
shaft, 252, 446, 450, 630, 635
shaft-cleaning, 64, 381
shagbark hickory, 73
shaitan, 52
shake, 208, 312, 335, 605
shake bolt, 335
shake culture, 608
shake test, 608
shake tongue, 65
shaker, 186, 558
shaker bed, 110, 142
shaking board, 110
shaking screen, 558, 608
shaking table, 558
shaky butt, 208
shaky wood, 269
shale, 229, 272, 561
shale oil, 561
shale tar, 561
shallow, 374, 375
shallow convection, 374
shallow dormancy, 375
shallow ploughing, 374
shallow-rooted, 374
shallow rooted plant, 374
shallow rootedness, 374
shallow soil, 13
shallow tillage, 374
shank shovel cultivator, 635
Shannon-Weaver function, 528
Shannon-Weiner index, 528
Shannon's index of diversity, 528
shanty, 126
shape, 541
shape distortion, 24
shape of log, 590
shape of tree, 449
shape parameter, 541
shaped surface sander, 55
shaper, 55, 347, 516, 654
shaper head, 516
shaping, 16, 55, 543
shaping equipment, 55, 543
shaping lathe, 55
shaping machine, 16
share, 64, 268, 288
share plough, 205
shared forest, 172
shared natural resources, 172
sharing behaviour, 142, 172
sharing (of yield), 48
sharkey carriage, 314, 620
sharp corner, 222, 401, 533
sharp pitch, 363
sharpener, 74, 326
sharpening, 326, 555
sharpening number, 396
sharper, 74, 326
sharping machine, 326
sharpness angle, 58, 396, 536
shave, 16, 177, 477
shaver, 16, 177
shaving, 16, 177
shaving board, 16, 468
shaving chair, 206
shaving chute, 16
shaving knife, 16, 177
shaving machine, 16, 477, 532
shaving stand, 206
shaving table, 206

shaw, 292, 535
Shaw hardness, 531
shaw socket, 205
shay swivel, 546, 647
she-oak extract, 333
sheaf discharge, 351, 538
shear, 234, 396
shear along the grain, 461
shear area, 233, 443
shear elasticity, 234
shear failure, 234
shear force, 234
shear fracture, 234
shear gradient, 234
shear-legs, 108, 396
shear line, 377
shear modulus, 234
shear parallel to the grain, 461
shear perpendicular to the grain, 196
shear resistance, 270
shear stiffness, 270
shear strain, 234
shear strength parallel to grain, 461
shear strength perpendicular to grain, 196
shear tensile strength, 234
shear test, 234
shear-test block, 234
shear zone, 234
shearing, 543
shearing effect figure, 270
shearing force diagram, 233
shearing strain, 234
shear(ing) strength, 234, 270
shear(ing) stress, 234
sheath, 202, 377
sheath auricle, 494
sheath blade, 494
sheath cell, 377
sheathing, 20, 53, 66, 156, 500
sheathing paper, 53, 166, 265
sheathing screw, 20
sheaves, 535
shed drying stock, 355, 570
shed leaf, 314
shed piling, 355
shed seasoning, 355
shedding, 314, 494
sheep shank boom, 556
sheer, 396, 583
sheer arm, 537
sheer boom, 583
sheer legs, 396
sheer log, 88, 89
sheer rail, 328
sheer skid, 21
sheet, 8, 357
sheet erosion, 26, 357
sheet flood, 318, 357
sheet machine, 535
sheet material, 8
sheet-metal, 248
sheet-pile, 9
sheet pulp, 237, 357
sheeter, 550
sheeting, 5, 202, 480
shelf, 165, 231
shelf-board, 355
shelf fauna, 77
shelf hole hardware, 165
shelf life, 440, 634
shelf nog, 165
shelf supports, 231
shell, 26, 87, 186, 272, 346, 559
shell construction, 41, 384
shell gimlet, 474
shell moisture content, 495
shell rot, 495
shell shake, 208, 312
shellac, 54, 358, 642
shellac putty, 60
shellac varnish, 60
shellbark hickory (big shellbar hickory), 320
sheller, 31, 494
shellfish, 17, 230
shelling, 30, 72, 241, 384, 385
shelter, 14, 20, 21, 571
shelter-belt, 131, 132
shelter belts of protect farmland, 202
shelter belts to protect sea coasts, 188
shelter belts to protect waterhead, 460
shelter-forest, 132
shelter growth, 20, 373
shelter thermometer, 7
shelter tree, 605
shelter tube, 98
shelter-wood, 14, 132
shelter-wood compartment system, 127
shelter-wood cutting, 237, 404
shelter-wood felling, 237, 404
shelter-wood forest, 237, 404
shelter-wood group system, 389
shelter-wood method, 237, 404
shelter-wood selection system, 237, 404
shelter-wood strip system, 80, 81
shelter-wood system, 237, 404
shelter-wood uniform system, 386
shelter-wood wedge system, 536
shelterbelt, 14, 131
sheltered ground, 363, 555
sheltered tree, 19
sheltering effect, 20
shelterwood coppice, 404, 601
shelterwood cutting, 237
shelterwood felling by groups, 389
shelterwood group system, 389
shelterwood method, 404
shelterwood regeneration, 237, 404
shelterwood-strip felling, 80
shelterwood system, 404
shelterwood uniform method, 386
shelterwood wedge system, 536
sheugh(ing), 230
shibuol, 439
shide, 335
shield-budding, 116
shield bug, 115
shield grafting, 116
shield shaped, 115
shield volcano, 115
shift, 23, 564
shift sand control forest, 307
shifting agriculture, 312
shifting cultivation, 312

shifting divide, 372
shifting mosaic, 109
shifting nursery, 304, 564
shifting sand, 307
shikimol, 210
shim, 103, 402, 479, 536, 608
shim hoe, 19, 196
shingle, 335, 508
shingle beach, 311
shingle bolt, 335
shingle crib, 335
shingle weaver, 78, 335
shiny beating, 345
ship, 66, 456
ship-building timber, 600
ship knee, 285
ship-lap joint, 161
ship lumber, 62
ship-plank, 230
ship-timber, 62, 600
ship-timber beetle, 485
shipbuilding timber, 600
shiplap, 74, 584
shiplap joint, 75
shipmast locust, 71
shipment, 639
shipped bill, 41, 639
shipper, 123
shipping case, 639
shipping-dry, 596
shipping-dry lumber, 596
shipping marks, 123, 639
shipping tally, 331
shirley poppy, 584
shiu oil, 131
shiv, 24
shives, 535
sho-gyu oil, 347
shoal, 375, 412, 584
shock load, 59, 486
shock resistance, 59, 270
shock strength, 59
shock stress, 59
shoe, 133, 314, 550, 551
shoe a springboard, 599
shoestring fungus, 537
shoofly, 304
shook, 55
shoot, 205, 252, 323, 341, 468, 551
shoot a jam, 16
shoot-apex grafting, 612
shoot blight, 613
shoot cutting, 341, 612
shoot mutation, 417
shoot-root ratio, 252, 479
shoot volume, 613
shooter, 418
shooting, 444
shooting license, 444
shooting preserve, 253
shooting range, 418
shop cost, 52, 62
shop-cutting panel, 37, 92
shop lumber, 9, 598
shop seasoning, 50
shop timber, 9, 598
shore, 3, 106, 450
Shore hardness, 212
shore hold, 3
shore plant, 188
shorebird (wader), 419
short, 112
short branchlet, 113
short column, 113
short cord, 112
short coupled, 113
short curly chip, 264
short cut, 24, 246
short-day, 112
short day plant, 112
short-day treatment, 112
short end, 112, 529, 540
short grain, 113, 381
short grass, 1
short ground, 366
short-handled prunning shear, 112
short haul, 112, 654
short-haul extraction, 112
short-length, 112
short-length logging, 589
short-lived seed, 113
short log, 113
short log cat, 23, 635
short log country, 464
short log logger, 39, 464, 601
short log show, 280, 464, 608
short logger, 595
short oil, 113, 263
short out, 112
short-period group system, 112
short ream, 534
short regeneration period, 112
short-retention(-time) blender, 279
short road, 212
short rotation, 112
short run, 112, 534
short sale, 317
short shoot, 113
short splice, 112
short split, 113
short stock, 113
short sunshine period, 112
short-term forecast, 112
short-term retardant, 113
short ton, 112, 320
short wood, 618
short wood logging, 589
shortcovering, 33, 317
shortday plant, 112
shot-gun, 297, 346
shot-gun shell, 346
shot-gun-slug, 346
shot hole, 533
shot-string, 87
shoulder, 310
shoulder-clasp joint, 15
shoulder-inserted join, 46
shoulder plane, 377
shoulder sprayer, 232
shouldered, 580
shoulders, 165
shovel, 48, 481, 494
shovel reach, 494
show, 37, 39, 127, 223
show pelargonium, 76
show through, 8
show wood, 496
showcase, 53
shower, 608
shower bath, 304
shragging, 75
shredded bark, 468
shredder, 378, 523
shrike, 7, 31
shrine forest, 464
shrink, 368, 442
shrinkage, 158, 442, 469
shrinkage crack, 158, 442
shrinkage in percent, 158, 442
shrinkage mass, 158, 442
shrinkage on drying, 158
shrinkage ratio, 158, 442
shrinkage retarding treatment, 134
shrinkage strain, 158, 442

shrinkage stress, 158, 442
shriveled seed, 631
shrub, 179, 180
shrub border, 180
shrub desert, 180
shrub land, 180
shrub layer, 179
shrub savanna, 180
shrub species, 180
shrub stage, 180
shrub steppe savanna, 515
shrub zone, 180
shrubbery, 180
shrubbery mass, 180
shrubby, 180
shrubby vegetation, 180
shrubland, 179
shrubland assigned as forest, 474
shrubland specified as forest, 474
shrublet, 533
shucker, 30, 31
shudder, 378
shut-down period, 481
shutter, 7, 88, 264, 279, 584
shuttle, 370, 469, 562
shuttle belt conveyor, 500
shuttle hauling, 65, 312
shuttle skidding, 65
siamese cat, 454
sib, 482, 642
sib-mating, 250, 482
sib-pollination, 251
sibling, 482, 642
sibling species, 379, 482
siccative, 74, 157, 158
siccative base, 158
siccideserta, 157
sickle cyme, 294
sickle-like cyme, 294
side, 22, 127, 149, 286, 351, 654
side-attached, 43
side axe, 84
side bark-grafting, 356
side bind, 229
side block, 104, 212, 637
side blocking, 43
side board, 21, 524
side boom, 202, 583
side branches, 44
side-by-side double barrel, 654
side cable, 19, 612
side chair, 509
side cleft graft, 377
side compression, 43
side cut, 21, 43, 510
side cut piece, 21, 510
side cutting, 43, 177
side cutting edge, 377
side cutting edge angle, 377
side ditch, 43
side drain, 43
side dressing, 43, 57, 190, 453
side-edge angle, 42
side effect, 154
side fork lift truck, 43
side grafting, 155, 608
side grafting by inlaying, 375
side grain, 362, 650
side grain surface, 42, 524
side guards, 43
side gutter, 43
side gutter angle, 43
side hardness, 43
side heavy, 357
side-hill, 10, 413
side hole planting, 547
side inlaying, 375
side jam, 3
side lap, 197, 351
side lean, 289
side lining, 43
side link, 43
side load stress, 22
side log, 21, 205
side looking airborne radar, 43
side mark, 21, 356, 612
side notching, 177
side panel, 42, 351
side penetration, 43
side pier, 446
side plane, 22
side planting, 309
side plate, 23, 42, 52
side-plate angle, 42
side rake, 43
side regulator, 143, 245
side relief, 43
side rind graft, 155
side set, 610
side shading, 43
side shoot, 44, 154
side slope, 22
side spool, 495
side stick, 333
side stretchers, 43
side surface, 43
side table, 271
side tongue grafting, 417
side track, 23, 43
side veneer-grafting, 529
side view, 43
side walk, 395
side wear, 43
sideboard table, 40
sidehill logging, 67
sidehill yarding, 67
sidelap, 43
sideline, 12, 43, 391
sideswipe, 331, 448
sidewise bucking, 197
sidewise compression, 43
sidewise compressive strength, 43
sidewise tension test, 43
siding, 20, 42, 88
siding board, 20, 21
sierozem, 210
sierra club, 414
sieve, 316, 412
sieve analysis, 412
sieve area, 412, 413
sieve cell, 412
sieve disk, 412
sieve element, 412
sieve field, 412
sieve machine, 412
sieve mesh, 412
sieve mill, 412
sieve pitting, 413
sieve plate, 412
sieve-plate column, 412
sieve-plate tower, 412
sieve pore, 412
sieve selection, 412
sieve selection hypothesis, 412
sieve test, 412
sieve tube, 412
sieve-tube element, 412
sieving method, 412
sieving zone, 412
sift, 187, 412
sifting, 187, 412
siftings, 316, 412
sight gun, 90
sight hood, 640
sightseeing, 178, 314
sightseeing resort, 314, 579
sigmoid curve, 384
sigmoid growth, 429
sigmoid growth curve, 429
sign, 29

sign mutant, 611
sign stimulus, 540
significance, 525
significance test, 525
significant deviation, 525
significant digit, 581
significant figure, 581
significant weather, 630
sika deer, 319
silent feeding, 510
silent gear, 510
silhouette, 312, 575
silica gel, 183
silicate, 183
silice cell, 183
siliceous clay, 183
siliceous sandstone, 183
silicicolous plant, 439
silicle, 112
silicon carbide (mineral), 248, 473
silicone resin, 183
siliculose, 260
silique, 49
siliquiform, 49
silk moth, 41, 405
silk oak gum, 570
silk worm, 229, 405
silky-sheen, 398
sill, 67, 95, 103, 320
sill course, 320
silo, 485, 635
silo outfeed device, 296
silt, 144
silt concentration, 189
silt discharge, 350, 446
silt storage dam for farmland building, 583
silt transportation, 342
siltanalysis, 474
siltation, 583
silty clay loam, 144
silty loam, 144
Silurian period, 619
silva, 408, 410
silver figure, 201, 570
silver-grain, 201, 570
silver halide, 310
silver-leaf poplar, 570
silver pine oil, 570
silver solder, 570
silver thaw, 585
silvi-pastoral system, 302
silvical, 600
silvical felling, 585
silvical final age, 645
silvical habit, 303, 425
silvichemicals, 298
silvicide, 573
silvicolous, 259
silvics, 302, 303, 600
silvicultural characteristics, 600
silvicultural control, 573
silvicultural cost, 573
silvicultural cutting, 585
silvicultural cutting method, 253
silvicultural division, 573
silvicultural goal, 585
silvicultural maturity, 585
silvicultural method, 585
silvicultural objective, 585
silvicultural operations optimum investment, 585
silvicultural practice, 585
silvicultural principle, 303, 585
silvicultural regionalization, 600
silvicultural rotation, 169
silvicultural system, 410, 654
silvicultural technique, 585
silvicultural yield, 585
silviculture, 408, 585, 600
silviculture fund, 585
silviculture machine system, 573
silviculture machinery, 573
silviculture mechanization, 573
silviculture taungya plantation, 214
silviculturist, 585, 600
Simalian's formula, 361, 513
simazine, 513
similarity, 528
similarity matrix, 528
simple aerial cableway, 234
simple allometric growth, 234
simple average growth rate, 84
simple batch distillation, 83
simple beam, 234
simple beam impact machine, 7, 84
simple bending, 69, 83
simple but, 86
simple cauliflory, 234
simple check, 255, 650
simple check dam, 234
simple coppice system, 1, 234
simple coppice (system), 83
simple cutting, 365
simple-cycle disease, 86
simple factorial experiment, 86
simple growth rate, 84
simple heart shake, 86
simple inheritance, 234
simple interest, 84
simple layering, 86, 365
simple leaf, 86
simple logistic curve, 234
simple method, 82
simple perforation, 83
simple perforation plate, 83
simple pit, 85
simple pit-pair, 85
simple poer, 85
simple pore, 83, 84
simple random sampling, 234
simple ray, 84
simple sieve plate, 85
simple strain, 234
simple stress, 234
simple sugar, 85
simple tariff, 86, 234
simple terracing with fascine, 234
simple terracing with miscanthus, 234
simple terracing works, 234
simple theodolite, 234
simple torsion, 69
simple translocation, 234
simple whip grafting, 85, 365
simple wire log-slide, 161
simplex, 85
simplex algorithm, 83
simplex donkey, 533

simplex group of chromosomes, 87
simplex method, 83
Simpson's diversity index, 539
Simpson's index, 539
Simpson's rule, 539
simulated inventory data updating, 325
simulation, 325
simulation games for business, 436
simulation language, 325
simulation model, 325
simulator, 325
simulator enclosure, 325, 556
simultaneous closing units, 483
simultaneous diffusion, 483
simultaneously closing device, 483
sine-wave interlock, 611
single-action shear, 83
single back fire, 85
single band-mill, 83
single band saw, 84
single barrel repeating shot-gun, 85
single barrel single shot gun, 85
single-base powder, 83
single bit, 84
single-bladed axe, 84
single-boring assay, 84
single-cell culture, 85
single cell protein, 85
single cell variant, 85
single-celled, 85
single-conical saw, 486
single coppice system, 83
single cord, 84
single cross, 83
single crossing over, 83
single cut quad band-mill, 463
single-cutting band saw, 84
single-entry, 563
single entry volume table, 563
single floor, 84
single-gang edger, 87
single gene heterosis, 83
single glue joint, 84
single haul, 83, 562
single head sander, 85
single-hinged arch, 83
single-hook detacher, 84
single-hull balloon, 85
single knot, 84
single layer particle board, 82
single layered, 82
single-line inclineway, 85
single linkage method, 84
single-load cableway, 86
single machine, 235
single oblique cut, 86
single open press, 82
single opening hydraulic cold press, 83
single out, 83, 87, 547
single periodic block, 86
single periodic block method, 83
single revolving disc refiner, 83
single rope cableway, 85
single rotating disc refiner, 87
single-seed planting, 84, 100
single seedling sprout, 83
single selection, 83, 562
single shearing strength, 69
single side planer, 84
single-sided edge bander, 84
single slackline yarding, 85
single sliding door, 85
single span, 84
single-span skyline, 84
single-spindle carving machine, 87
single-spindle mortiser, 87
single-spindle shaper, 87
single spread(ing), 84
single-stage compression, 83, 563
single stand, 174
single stem, 84
single-stem delimber, 84
single-step suspension dryer, 83
single stick boom, 83
single-storey tapping, 82
single-storied forest, 82
single-storied stand, 82
single surface glue spreading, 84
single surface planer, 84
single swinging door, 85
single tax, 86
single-tong loader, 612
single-tong system, 87
single track-cable, 85
single track kiln, 83
single tree, 84, 493, 499
single-tree mixture, 87
single-tree model, 84
single tree sample, 87
single tree selection, 87
single-tree selection felling, 87
single tree (selection) method, 87
single-tree selection system, 87, 601
single tree selective cutting, 87
single trigger, 81
single variable volume table, 563
single wall, 85
single-wire logging, 85, 176
single wire-tramway, 85
single yard, 83
singler, 235, 445
singlet, 83
singlet state, 85
singling, 235, 445
sink, 194, 213
sink for emmission, 566
sinker, 52, 53, 71, 521
sinker boat, 53
sinker-stock, 53
sinking creek, 147, 571
sinking fund, 49
sinking fund table, 49
sinking hole erosion, 527
sinuous, 30, 497
sinus, 498
siphon, 198, 497
siphon raingauge, 198
siphonaceous, 179
sir sweeper, 275
sisal hemp, 575
sistens type, 481
sister chromatid, 642
sister chromosomes, 642
site, 50, 288, 423
site apparaisal, 96, 289, 429, 488
site apparaisement, 96, 289, 429, 488
site class, 98
site class estimation, 289
site class table, 98

S

site classification, 98
site clearing, 429, 610
site description, 96, 289, 429, 488, 558
site destruction, 50
site determination, 289
site evaluation, 289
site factor, 98
site index, 289
site index class, 289
site index curve, 289
site index table, 289
site indicator, 98
site map, 289
site mapping, 96, 289, 429, 488
site preparation, 600, 610
site preparation machinery, 299
site productivity, 98, 422, 429, 489
site quality, 289
site quality class, 98
site-quality table, 98
site-species combination, 289
site-specific, 474, 570
site-specific allometric equation, 96
site-specific volume table, 95
site survey, 96, 289, 429, 488
site type, 289
site-type class, 289
site type classification, 289
site type map, 289, 429
sitting, 655
sitting shot, 75
six-log timber, 309
six-row transplant machine, 309
six-tree sample, 309
sixtant, 308
sixty percent choke, 309
sizability, 431
size, 3, 57, 183, 241, 247, 431
size class, 78
size-class distribution model, 255
size degradation, 144, 327
size fastness, 240, 431
size-graded particleboard, 236
size of a sample, 384
size range, 57
size speck, 240
size test, 412
sizer, 141, 142, 332, 412, 415, 463, 479
sizing, 106, 241, 412, 431, 479
sizing agent, 133, 431
sizing box, 431
sizing roll, 431
sizing saw, 37
sizing saw system, 650
sizing screens, 652
sizing specification, 57
sizing tub, 431
sizing vat, 431
skeet, 342, 461
skeet choke, 342, 462
skein, 391
skeletal, 359
skeletal cellulose, 175
skeletal hyphae, 175
skeletal limestone soil, 72
skeletal soil, 72
skeletal substance, 175
skeleton, 175
skeleton forest road, 160
skeleton soil, 72
sketch map, 438
sketchmaster, 42
skew rebate plane, 536
skewness, 357
skiagram, 419
skiagraph, 419
skid, 103, 222, 243, 286, 349, 364
skid cat, 223
skid frame, 205
skid grease, 332
skid loader, 223
skid log, 205, 349
skid pan cargo carrier, 596
skid pole, 184
skid ramp, 537, 638
skid road, 7, 223
skid-steer loader, 564
skid timber, 286
skid tine, 48
skid up, 183, 332
skidder, 93, 223
skidder line, 223, 372
skidder-loader, 223
skidder-piler, 223
skidder yarding system, 470
skidding, 222, 364
skidding coefficient, 224
skidding cone, 223, 224
skidding dish, 223
skidding extension rope, 292
skidding lane, 223
skidding line, 223, 372
skidding line drum, 372
skidding machinery, 223
skid(ding) pan, 223
skidding reach, 223
skidding resistance, 224
skidding road, 196, 222
skidding rope, 223, 372
skidding scissors, 223
skidding shoe, 223
skidding sled, 223
skid(ding) trail, 222
skidding trip, 223
skidding winch, 223
skidway, 222, 638
skim ploughing, 374
skimming, 148, 358, 609
skin, 26, 30, 66, 241, 356, 496, 532
skin dry, 26, 37
skin effect, 224, 631
skin glue, 356
skin stress, 27
skinning, 30, 532, 578
skip, 30, 128, 235, 276, 480, 487
skip-in-dressing, 309
skip-in-planing, 309
skip-in-surfacing, 309
skipper road, 196, 612
skirting (board), 21, 476
skirting piece, 612
skotograph, 419
skotoplankton, 420
sky piece, 146
sky-ward facing, 530
skybound, 75, 371, 655
skycar, 470
skycrane, 470
skyflyer, 470
skyhook, 469
skylight, 477
skyline, 56, 231, 477
skyline crane, 272
skyline-crane carriage, 470
skyline-crane yarding, 469
skyline-highlead yarding, 10
skyline jack, 56
skyline loaded, 444
skyline logging, 469
skyline road, 469

S

skyline sag, 56
skyline sheave, 470
skyline shoe, 56
skyline skidder, 470
skyline skidding, 469
skyline strip, 469
skyline stub, 56
skyline support, 470
skyline support line, 56
skyline swing, 121, 637
skyline system, 469
skyline tail stump, 470
skyline tension, 56
skyline tensioning device, 56
skyline winch, 470
skyline yarder, 470
skyline yarding, 469
skyroad, 231
skyscooter, 305
slab, 8, 200, 263, 359, 377
slab check dam, 328, 433
slab-chipper, 8
slab cut, 221
slab dam, 328, 433
slab edging, 8, 318
slab joint, 362
slab roof, 351
slab saw, 364, 598
slab tie, 23
slabbed, 356
slabbed cant, 318
slabbed log, 263
slabber, 8, 364, 598
slabbing, 263, 590
slabbing edger, 8
slabwood, 22
slack, 184, 209, 464
slack face, 81, 82
slack line system, 464
slack puller, 314, 360, 620
slack puller line, 314, 620
slack pulling block, 314
slack pulling carriage, 79
slack pulling drum, 314
slack pulling line, 314, 493, 620
slack side, 81, 82
slack water, 93, 209, 258
slacker, 464, 465
slacking, 464
slackline system, 464
slackline yarding, 464
slacklining, 464
slant agar, 537
slant band saw, 381
slant culture, 537
slap happy loading rig, 196
slash, 74, 125, 261, 269, 297, 304, 377, 431
slash and burn, 88
slash and burn agriculture, 417
slash area, 38
slash burn, 143
slash chopper, 613
slash collector, 39, 613
slash disposal, 39, 127, 139
slash disposal of chipping, 532
slash figure, 524
slash figure check, 524
slash fire, 40
slash grain, 524
slash-grain lumber, 524
slash knot, 231, 242
slash mill, 139
slash pile, 39, 139
slash utilization machinery, 38
slasher, 49, 87, 165
slasher saw, 87
slashing, 269, 521, 600
slashing saw, 87, 117
slashmobile, 564
slat, 8, 13, 566, 603
slat back, 476
slat conveyor, 8
slat crate, 8
slat door, 7, 8
slat saw, 8
slate, 8, 433
sled, 223, 243
sled hauling, 349, 377
sled iron, 349
sled plate, 292
sled runner, 349
sled shoe, 349
sled tender, 349
sledge, 78, 349, 376
sledge road, 349, 376
sledge runner, 205, 349
sledgewinch, 205
sledging, 349, 377
sleeper, 96, 533, 608
sleeper block, 608
sleeping bud, 543
sleepy hollow chair, 2
sleet, 28, 110, 584
sleeve, 179, 276, 474, 476, 631
sleeve-type trier, 474
sleigh, 205, 349, 376
slender-branched, 534, 613
slenderness ratio, 49, 50
slew skid, 537
slewing, 212
slewing angle, 212
slewing torque, 212
slice, 13, 16, 377
slice opening, 556
sliced, 16
sliced sheet, 377
sliced veneer, 16
sliced veneer overlay, 16
slicer, 16
slicing action, 205
slicing knife, 16, 377
slicing machine, 16
slicing veneer, 16, 357
slick spot, 182, 205
slide, 205, 598
slide-action shotgun, 205, 219
slide caliper, 205, 311
slide matching, 461
slide plane, 205
slide rule, 225
slided land, 19
sliding, 205
sliding angle, 326
sliding caliper, 579
sliding door, 282
sliding growth, 205
sliding hook, 205
sliding microtome, 205
sliding out, 205
sliding parity, 274
sliding track, 205
sliding weight, 205
slight imperfection, 380
slight shake, 533, 534
slight skip, 380
slight stain, 380
slime, 345
slime-flux, 345
slime-fungus, 345
slime-mold, 345
sliminess, 344
slimy exudate, 345
sling, 280, 474
sling cart, 104, 281
sling chain, 281
sling dog, 104
sling-fruit, 472
sling hook, 104
sling load, 281, 333
sling psychrometer, 443
sling rope, 104, 281, 474
slip, 46, 48, 62, 75, 147, 205, 561

slip erosion, 205
slip feather, 417
slip grab hook, 281
slip hook, 205, 470
slip joint, 205, 419
slip line, 205
slip pipe, 419, 474
slip plane, 205
slip skid, 539
slip-tongue cart, 162
slippage, 205, 240
slipper rocker, 376
slit, 147, 519, 649
slit dam, 147
slit-like bordered pit, 298
slit-like simple pit, 298
slit nozzle, 519
slit planting, 377
slitting knife, 649
sliver, 479, 519
sliver tooth, 177
slivery sheen, 570
sloop, 223
slop, 139
slop cut, 34, 139
slope, 363, 380
slope base, 363
slope bottom, 363
slope correction, 363
slope failure, 363
slope index, 363
slope measurement, 363
slope of compensation, 32
slope of grain, 506
slope of soil, 488
slope of stack, 114
slope pile, 380
slope piling, 537
slope protection forest, 413
slope stake, 22
slope wash, 363, 537
slope wash soil, 363
sloping screen filter, 537
sloping terrace, 279, 363
slops, 139, 508
slot for action bolt, 278
slot for cocking lever, 550
slot mortise, 268, 519
slot mortiser, 516
slotted hole, 412
slotted plate, 49, 79
slotted side graft, 46
slotted templet, 41
sloven, 75, 306, 523, 605
slow beating, 345
slow cooking process, 209
slow fire, 318
slow flame, 318
slow-germinating, 124
slow-grown, 318, 603
slow-release fertilizer, 50
slow-rooting, 123
slow stock, 345
sludge, 342, 508, 584
sludge treatment, 508
slue line, 7
slug, 281
slug-caterpillar, 71
sluggish year, 78
sluice, 307, 456–458, 538, 602
sluice dam, 306
sluice gate, 460
slump, 451
slurry, 515, 579
slurry seal process, 237
slush, 548
slush pulp, 348
slusher, 445
slushing, 271, 487, 494
smack winch, 380
Smalian's formula, 462
small blister figure, 535
small blue horntail, 78
small board, 533
small clear specimen, 26, 508
small clear timber test, 26, 508
small-crowned tree, 448
small-dimension stock, 533
small door, 233
small end, 417, 535
small-end diameter, 327, 417, 535
small ermine moth, 52
small-format aerial photograph photo, 533
small gap felling, 534
small hewn square, 356
small pit-prop, 535
small pitch streak, 535
small pole, 156, 535
small pole forest, 156
small pomele figure, 535
small quadrat, 535
small sample theory, 535
small sapling, 535
small scantling, 534
small segregated coupes, 141
small square, 533
small stature of stock, 1, 2
small timber, 519, 534
small watershed management, 534
small wood, 533
small wood harvester, 582
small(er) end, 535
smearing culture, 487
smog, 553
smog-sensitive, 115
smoke curtain, 553
smoke damage, 553
smoke density rating, 553
smoke detection, 553
smoke (dry) kiln, 549
smoke drying, 549
smoke generator, 553
smoke injury, 553
smoke jumper, 275, 480
smoke nuisance, 553
smoke plume, 553
smoke protection forest, 134
smoke resistance, 271
smoke screen, 553
smoke-test, 553
smoke treatment, 548
smoker, 124, 549
smoking seasoning, 549
smooth bark, 181
smooth face, 181
smooth grain, 361, 519
smooth side convex, 361
smooth texture, 266, 361
smoothed two sides, 295
smooth(ing), 363
smoothing harrow, 363
smoothness, 361
smothering, 113
smouldering, 571
smouldering fire, 320, 571
smudging, 548
smut, 194, 195
smut-fungus, 194, 195
snag, 251, 278, 603, 612, 613
snag-felling law, 40
snag pusher, 278
snagged, 75, 117, 430
snagging, 40, 278
snaggy, 117, 430
snake, 30, 54, 417, 431, 497

S

snake block, 19
snake wood, 316
snakefly, 417
snaking, 97, 387
snaking line, 223, 372
snaking trail, 223
snap, 351, 593
snap guy, 372
snap line, 231
snap table, 606
snapshot, 222, 466
snatch block, 89, 268
snatch team, 150
snatch yoke, 268
snatching point, 638
snedding, 75, 544
sneezewood, 354
snib, 134
snig chain, 223
snig track, 223
snigging, 97, 387
snigging pan, 223
snig(ging) trail, 223
snipe, 269, 306, 351, 593
sniper, 547
snipper saw, 37
snippers, 234
snorkel boom, 419
snout, 351, 593
snout beetle, 530
snout moth, 325
snow accumulation gradient, 219
snow addition, 602
snow avalanche, 548
snow bank, 548
snow bed, 548
snow blight, 548
snow breakage, 548
snow-capped mountain, 548
snow catch, 134
snow chute, 548
snow cover duration, 219
snow cover line, 219
snow damage, 548
snow fence, 548
snow line, 548
snow load, 548
snow melt erosion, 398
snow mould, 548
snow pack, 219, 548
snow-patch, 548
snow patch vegetation, 548
snow plough, 405, 548
snow precipitation line, 238
snow rut road, 548
snow set, 548
snow shield, 134
snow slide, 548
snow trail set, 548
snowbank, 548
snowbreak, 134, 548
snowbreak forest, 134
snowdrift, 548
snowfall, 238
snowmelt duration, 398
snowpack, 548
snowslip, 548
snowy, 119
snub, 431, 538
snub check, 431
snubbing device, 277
snubbing guy, 107
snubbing line, 372, 451
snub(bing) yarding, 110
soakage, 515
soaking, 251
soaking method, 251
soaking pit, 251
soap nut tree, 509
sociability, 388
social aggregation, 418
social animal, 388, 418
social behaviour, 418
social benefit, 418
social benefits of forests, 407, 408
social bond, 418
social cost, 418
social distance, 388, 418
social dominance, 418
social ecology, 418
social-economic-natural complex ecosystem, 418
social facilitation, 418
social group, 418
social habit, 388
social hierarchy, 418
social insect, 418
social parasitism, 388
social structure, 418
social travel, 418
socialization, 418
sociation, 220
socies, 556
society, 418, 651
socio-biology, 418
socio-economic potential, 418
socio-economics, 418
socioeconomic status, 418
sociogram, 418
socket, 46, 55, 474
socket wedge, 46, 474
sod, 42, 364, 421
sod layer, 42
sod lyher, 42
sod seeding, 301
sod shoveling for fuel, 48
sod webworm, 42
soda bordeaux, 308, 465
soda-lime, 234
soda process, 271, 465
soda pulp, 235
soda recovery, 234
soda slag, 189
sodded check dam, 42
sodded spillway, 42
soddy forest soil, 421
sodic soil, 336, 465
sodium methyl dithiocarbamate, 500
sodium penta chloro phenate, 512
sodium silicate adhesive, 183
soffit, 172
soft bark, 401, 599
soft bast, 399
soft cook, 400, 609
soft deciduous wood, 400
soft fibre board, 401
soft fir, 17, 204
soft gel-like matrix, 401
soft hammer, 377
soft heart, 401
soft pine, 401
soft pulp, 401
soft roll, 401
soft rot, 400
soft-rot fungus, 400
soft rot tolerance, 400
soft scale, 400
soft stem cutting, 401
soft tissue, 13
soft wire-rope, 400
soft x-ray, 400
soft x-ray machine, 400
softboard, 401
softener, 400, 602
softening point, 400
softness, 399
software engineering, 400
softwood, 400, 606
softwood board, 400, 606
softwood charcoal, 400, 606

softwood distillation, 400, 464, 606
softwood fibre, 400, 606
softwood forest, 400, 606
softwood grafting, 401
softwood plank, 606
softwood (stem) cutting, 341
softwood tar, 400, 464, 606
softwood tar pitch, 400, 464, 606
softwood tree, 607
sog mor, 460
soi, 488
soil acidity, 490
soil aeration, 490
soil aggregate, 490
soil air, 490
soil alkalinity, 490
soil amelioration, 489
soil amendment, 489
soil analysis, 489
soil and water conservation, 459
soil and water conservation benefit, 459
soil and water conservation engineering, 459
soil and water conservation forest, 459
soil and water conservation measures, 459
soil and water conservation planning, 459
soil and water conservation regionalization, 459
soil and water conservation tillage measures, 459
soil and water losses, 459
soil animal, 489
soil anti-erodibility, 490
soil anti-scouribility, 490
soil association, 491
soil auger, 491
soil biology, 490
soil biota, 490
soil block test, 490
soil-borne disease, 489
soil borne fungus, 489
soil-building property, 55
soil catena, 489
soil class, 489
soil classification, 489
soil climate, 490
soil colloid, 490
soil colour, 491
soil compactness, 491
soil conservation, 489
soil conserving crop, 459
soil consistence, 489, 490
soil consistency, 490
soil constitution, 490, 491
soil core, 491
soil-covering, 95
soil crust, 488
soil cultivation, 490
soil disinfectant, 490
soil disinfection, 490
soil drainage, 350, 490
soil dressing, 274
soil drifting, 489
soil drought, 489
soil erosion, 459, 489, 490
soil erosion type, 490
soil expectation value, 298, 488, 640
soil family, 489, 491
soil fauna, 489
soil fertility, 489
soil filling machine, 639
soil formation, 490
soil formation process, 490
soil former, 490
soil forming factor, 490
soil forming process, 490
soil fraction, 489
soil fumigation, 491
soil genesis, 489
soil geography, 489
soil great group, 489
soil horizon, 488
soil humidity, 490
soil improvement, 489
soil improving tree, 139, 489
soil improving tree species, 155
soil indicator, 491
soil inhabitant, 490
soil inhabiting fungus, 490
soil-inhabiting termite, 489
soil injector system, 491
soil inoculant, 490
soil insect, 489, 490
soil insecticide, 490
soil invader, 490
soil invertebrate, 490
soil layer, 488
soil-lifting frost, 110, 455
soil loss, 490
soil loss amount, 437
soil loss tolerance, 595
soil macrofauna, 489
soil management, 490
soil mantle, 488, 489
soil map, 490
soil material, 490
soil mesofauna, 491
soil microbe, 490
soil microbiology, 490
soil microclimate, 490
soil microfauna, 490
soil microorganism, 490
soil microsite preparation, 279
soil mineral, 490
soil mixture, 491
soil moisture, 490
soil moisture characteristic curve, 490
soil moisture constant, 490
soil monolith, 491
soil morphology, 490
soil mulch, 489
soil mycostasis, 491
soil order, 489
soil organic matter, 491
soil particle, 490
soil permeability, 490
soil phase, 491
soil pick, 488
soil pit, 489
soil pore, 490
soil pore space, 490
soil porosity, 490
soil pot press, 573
soil preparation, 610
soil productivity, 490
soil profile, 490
soil protection, 489
soil protection and water regulation, 459
soil pulverizer, 128, 465
soil reaction, 489
soil regionalization, 490
soil renewal, 489, 491
soil rent, 488
soil rent value, 96, 489

S

soil rental, 488
soil-rent(al), 99
soil respiration, 490
soil salinity, 491
soil saving dam, 15, 247
soil science, 491
soil separate, 490
soil series, 491
soil solution, 490
soil sourness, 490
soil species, 491
soil split preparation, 490
soil sterilant, 490
soil sterilization, 490
soil sterization, 490
soil stratigraphy, 489
soil structure, 490
soil subgroup, 491
soil suborder, 491
soil survey, 490
soil suspension, 490
soil system, 490
soil temperature, 490
soil temperature overturn, 98
soil test, 489, 490
soil texture, 491
soil tilth, 490
soil treatment, 489
soil type, 489, 491
soil value, 488
soil-vegetation relationship, 491
soil water, 490
soil wounding, 375
soil zonality, 489
soil zone, 489
soiling crop, 139
soiling plant, 139
soilless bed, 458, 510
soilless culture, 510
sol, 398
sol state, 561
solar activity, 471
solar climate, 451, 478
solar constant, 471
solar drying, 397
solar emjoules, 471
solar encrgy, 471
solar-energy drying, 471
solar lumber pre-dryer, 331
solar radiation, 471
solar radiation curve, 471
solar transformity, 471
solarimeter, 397
solarization, 397
solder, 190
soldier ant, 29
sole, 288
sole leather, 94
sole plate, 94
sole timber, 103
solfatara, 308
solid, 177, 232, 245, 249, 436
solid billet, 436
solid bulb, 382
solid bullet, 436
solid content, 177, 278, 386, 435, 652
solid cubic content, 435
solid culture, 177
solid fat, 613
solid fertilizer, 177
solid floor, 436
solid fuel, 177
solid glued-up stock, 436
solid jam, 610
solid koji, 177
solid lift gate, 436
solid logic technology, 177
solid measure, 435
solid medium, 177
solid meter, 435
solid modified wood, 436
solid nose, 610
solid panel, 200
solid penetration, 498
solid pile, 435
solid-piled, 435, 436
solid piling, 322, 435
solid pith, 436
solid-plate circular saw, 177
solid resin, 177, 575
solid tooth, 435
solid tooth circular saw, 435, 610
solid tooth saw, 435, 610
solid train, 381
solid type piling, 322
solid volume, 435
solid waste, 177
solid wood, 436, 610
solid wood furniture, 435
solid wood products, 435
solidified oleoresin, 319
solidified resin, 176
solidnerve, 179
solifluction, 342
solitary, 85, 174
solitary animal, 111
solitary parasitism, 83
solitary pore, 83
solitary terminal, 327
solod, 493
solod soil, 493
solodic soil, 493
solodization, 493
solonchak, 555
solonchak desert, 555
solonetz, 234
solonetz soil, 234
solonization, 234
solonized brown soil, 234
soloth soil, 493
soloti soil, 492, 493
solubility, 398
solubilizer, 602
solubilizing agent, 602
soluble tar, 398
soluble wood tar, 398
solum, 488, 491
solution culture, 398
solvation, 398
solvency, 49
solvent-activated adhesive, 398
solvent adhesive, 398
solvent dried wood, 398
solvent drying, 398
solvent front, 398, 603
solvent recovery, 398
solvent-sensitive adhesive, 569
solventless epoxy varnish, 510
solventless varnish, 510
solvolysis, 398
soma, 476, 477
somatic, 477
somatic cell, 477
somatic cell hybrid, 477
somatic (cell) hybridization, 477
somatic crossing over, 477
somatic doubling, 477
somatic mitosis, 477
somatic mutation, 477
somatic pairing, 477
somatic prophase, 477
somatic reduction, 477
somatic synapsis, 477
somatogenic, 578
somatogensis, 477
somatoplasm, 477
somatostatin, 429
somatotype, 477
sombric horizon, 151
someiyoshine, 397

somite, 476
sonar, 430, 459
song-bird, 325
sonic noise analyzer, 601
sooty mould, 320, 553
sophorine, 207, 248
sorb, 204
sorbing agent, 513, 515
sorema, 538
sorose, 421
sorption, 515
sorption hysteresis, 514, 515
sorption isotherm, 515
sorption water, 515
sorptive nature, 515
sort, 130, 141, 262, 285, 547
sorted outturn of production, 140
sorter, 54, 141, 547
sorting, 141, 547
sorting bay, 547
sorting boom, 141, 547
sorting carriage, 547
sorting conveyor, 547
sorting deck, 547
sorting gap, 547
sorting ground, 547
sorting jack, 547
sorting lath, 31
sorting line, 547
sorting plant, 141, 547
sorting pocket, 547
sorting table, 141, 547
sound, 91, 236, 430, 570
sound absorbing coefficient, 514
sound absorption, 514, 515
sound arrester, 166
sound barrier, 430, 570
sound bridge, 430
sound cutting, 236
sound damper, 532
sound damping device, 532
sound-deadening plywood, 532
sound defect, 236
sound eliminator, 532
sound heart-wood, 236
sound-insulating board, 166
sound knot, 236, 423
sound-level meter, 430
sound location, 430
sound meter, 430, 601
sound on-grade, 236
sound pressure level, 430
sound proof board, 166
sound scale, 3
sound transmission, 430
sound wood, 236
sound wormy hole, 498
soundbird, 325
sounding, 570
sound(ing) board, 172
sour bleach, 228
sour humus, 467
sour log stain, 457
sour orange, 467
sour soil, 467
sour-telling, 126, 423
source, 168, 282, 627
source document, 591
source-identified, 566
source of inoculum, 245
source of pollution, 508
source program, 593
source-sink relation, 593
south bend yarding, 338
south-facing slope, 338
south pole, 338
southern Oscillation, 338
sow grass seeds, 31
sow tree seeds by plane, 275
sow tree seeds from aircraft, 275
sowbug, 52, 195
sowing, 31
sowing brick, 31
sowing depth, 31
sowing division, 31
sowing-drill, 31
sowing in broad drill, 279
sowing in close drill, 603
sowing in furrow, 288
sowing in hole, 547
sowing in line, 190
sowing in narrow drill, 603
sowing in patch, 279
sowing in row, 190
sowing in spot, 279
sowing in strip, 79
sowing-machine, 31
sowing on ridge, 309
sowing rate, 31
sowing-season, 31
sowing time, 31
sowing width, 31
soybean glue, 110
soybean meal glue, 110
space, 58, 232, 278
space block, 367
space coordinate, 275
space drilling, 640
space ecology, 275
space environment monitor, 275
space ordering, 275
space simulation, 275
spaced column, 287
spacer, 103, 166, 235, 275, 375
spacing, 232, 261
spacing (between trees), 160, 632
spacing figure, 235
spacing index, 616
spacing of planting spots, 629
spacing of plants, 597, 616
spacing of seeding spots, 629
spacing sticker, 166, 235
spacing thinning, 61, 106
spade, 48, 376
spade budding, 48, 116
spade foot, 232
spadiceous, 147, 291, 399
spadicose, 399
spadix, 147, 399
spado, 169, 170
spaghetti tube watering, 518
spall, 620
spalt, 30, 356, 620
span, 278
span dog, 636
span of useful life, 437
span piece, 196
span-rail, 278
span saw, 280
spaniel, 462
spanipelagic plankton, 515
spanworm, 57
spar, 223, 502, 569, 592, 635
spar cap, 223
spar foot, 223
spar pole, 223
spar tree, 223
spar varnish, 502
spare bunk, 103
spare parts, 17, 353
spare varnish, 17, 502
sparge, 353, 354

S

spark arrester-screen, 133
sparse planting, 515
sparse stand, 445
sparse vegetation, 515
sparsely paratracheal parenchyma, 515
spathaceous, 260
spathe, 147
spathulate, 57
spatial and temporal scale, 275
spatial art, 275
spatial autocorrelation analysis, 275
spatial autorelation analysis, 275
spatial distribution, 275
spatial explicit model, 275
spatial gradient, 275
spatial heterogenity, 275
spatial order, 275, 441
spatial pattern, 275
spatial resolution, 275
spatial scale, 275
spatial stress, 275
spatio-temporal, 435
spatiotemporal variability, 435
spattering, 237
spatula, 177, 578
Spaulding log rule, 462
spawn, 267
spawning, 47
spawning ground, 47
spawning migration, 47
spawning-time, 47
speaking-rod, 644
spear, 277, 506
spear edge bunk, 231
spearmint oil, 306
special assessment, 474, 475
special cutting, 474, 475
special densified hardboard, 475
special drying method, 475
special fiberboard, 475
special heterogeneity, 275
special map, 219
special plywood, 475
special tree, 14, 474
special vegetation geography, 475
specialization, 636
specialized form, 636
specialized parasite, 636
specialized pathogen, 636
speciation, 513
species, 513, 626
species-abundance curve, 627
species-area curve, 626
species assemblage, 626
species biomass, 627
species code, 450
species cohesion, 513
species complex, 626
species composition, 300, 450, 618
species constellation, 513
species conversion, 450
species density, 513
species diversity, 626
species diversity index, 513
species equilibrium, 513
species evenness, 513
species fidelity, 626
species group, 450, 629
species heterogeneity, 626
species identification, 450, 626
species mixture, 214, 450
species of trees, 450
species packing, 627
species richness, 513
species turnover rate (replacement rate), 626
specific activity, 19
specific adhesion, 19
specific beating pressure, 75
specific combining ability, 475
specific cutting resistance, 85
specific discharge, 20, 85, 307
specific dynamic action, 474
specific epithet, 626
specific flower garden, 636
specific gravity, 20
specific gravity in air dry, 368
specific gravity in green, 421, 504
specific gravity in over dry, 265, 310
specific gravity of wood substance, 330, 331
specific growth rate, 20
specific heat, 20
specific humidity, 20
specific impact strength, 85
specific inductivity, 101, 248
specific leaf area, 20
specific leaf mass, 20
specific leaf weight, 20
specific minimum retention, 183
specific mortality, 474
specific name, 626, 636
specific natality, 474
specific oxygen uptake rate, 557
specific pressure, 20, 85
specific pressure of grinding, 327
specific radioactivity, 19
specific reaction, 474
specific strength, 20
specific surface, 19, 27
specific surface area, 19
specific tariff, 72
specific volume, 20
specific volume resistance, 476
specific weight of wood, 328
specific work, 19
specification limit, 183
specification standard, 183
specificity, 475, 636
specified dimension, 171, 325
specified grain, 475
specified strength, 183
specimen, 24
specimen handling, 250
specimen tree, 24, 174, 587
speck, 7, 507
speckled wood, 7
speckness, 52
spectral analysis, 181
spectral background, 181
spectral band, 30, 181
spectral chromatography, 30, 181
spectral curve, 181

spectral interval, 30, 181
spectral reflectance, 181
spectral signature, 181
spectrogram, 30, 181
spectrograph, 181, 419
spectrometer, 141
spectrometric identification, 30
spectrometry, 181
spectrophotometric method, 141
spectrophotometriy, 141
spectrophotometry, 141
spectroscope, 141
spectrum projector, 181
spectrum (*plur.* -tra), 30, 181, 365
specular gloss, 258
specular reflectance, 258
speed change gear, 24
speed meter, 466
speed of attack, 152
speed of photographic plate, 159
speed of quarry, 297
speed the growth of the forests, 228
speed transmission, 24
speed variator, 24
spelean environment, 110
speleobiology, 110
spending unit, 531
spent sulfite liquor, 552
spent turpentine chip, 494
sperm cell, 254
spermaphore, 471, 632
spermary, 203, 542
spermatheca, 443
spermatium (*plur.* -ia), 541
spermatology, 254, 629
spermatophyta, 629
spermoderm, 626
spermogonium, 541
spermology, 254, 629
sphaeroblast, 308, 382
sphaeroid, 382
sphaeroma, 188
Sphagnum bog, 342
Sphagnum moor, 342
sphagnum moss, 342
sphagnum peat, 460
sphelrical swelling, 382
sphenoidal, 536
spherical, 382
spherical balloon, 593
spherical boiler, 382
sphinx moth, 477
spicate, 469
spicate cyme, 469
spice, 528
spicula, 535
spiculate, 261
spiculate spike, 153
spicy acrid taste, 528
spider, 613
spider heart, 632
spider shake, 540
spider web, 283
Spiegel relascop, 431
spike, 5, 76, 90, 469
spike bar, 604
spike (d-tooth) harrow, 105
spike harrow, 105
spike killing, 90
spike knot, 23, 231, 242
spike-top, 242, 278
spiked bell hook, 79
spiked roller, 71, 606
spiked skid, 79
spikelet, 535
spikenard oil, 156
spiling, 229, 336
spill, 313, 333, 569
spill-over effect, 569
spillage, 307, 538
spillway, 569
spillway dam, 184, 569
spin, 4
spindle, 44, 134, 268, 538, 634
spindle-and-bead, 11
spindle back chair, 134
spindle-like shape, 134
spindle moulder, 290
spindle sander, 290
spindle turning, 538
spine, 71, 222
spinescent, 71, 260
spinny, 1, 180, 597
spinule, 500, 533
spiracle, 369
spiracula, 369
spiral arrangement, 313
spiral bedroll, 313
spiral cavity, 13, 313
spiral checking, 313
spiral conveyor, 313
spiral crack, 313
spiral-curve cut, 313
spiral dowel, 313
spiral duct, 312
spiral grain, 313
spiral phyllotax, 313
spiral striation, 313
spiral system cut, 313
spiral texture, 313
spiral thickening, 313
spiral tracheid, 313
spiral vessel, 312
spiralling root, 351
spireme, 391
spirit(-based) stain, 69
spirit enamelling, 211
spirit-level, 259
spirit of turpentine, 464
spirit varnish, 466
spiroplasma, 313
spitzer, 231
spla, 50, 639
splad, 50, 639
splash, 602
splash boards, 602
splash dam, 219
splash erosion, 237
splash-lubrication, 136
splash zone, 283
splay knot, 23, 231, 242
splay mould, 377, 603
splayed foot, 5
splice, 104, 240–242, 345, 359
splice angle, 244
splice bar, 229, 584
splice grafting, 75, 192, 359
splice joint, 358
splice width, 359
splicer, 358, 508
splicing, 241
spline, 375, 402
spline edge joint, 58, 131
splint, 13, 215
splint cut, 177
splint-wood, 13, 215
splinter, 139, 468
splinter draw, 61, 75
splinter pull, 61
splinter pulling, 61, 356
splinter wood, 650
split, 115, 157, 356
split billet, 356
split cutting, 165, 356
split-design, 297
split face, 356
split fire-wood, 355, 356
split gene, 113
split guy, 295
split-level landing, 243
split-plot experiment, 297
split ring, 297
split-scion graft, 9
split stuff, 156

S

split tie, 356
split timber, 355
splitter, 355
splitting gun, 16
splitting machine, 356
splitting saw, 364, 454
splitting tool, 355
splitting wedge, 356
splitwood, 356
spodic, 210
spodic horizon, 210
spodic humus layer, 210
spodography, 211
spodosol, 211
spoil, 139, 371
spoil(age), 139
spoilage inhibitor, 132
spoke, 149, 312, 442, 475
spoke stock, 149
spokeshave, 148
sponge sander, 188
sponge sanding, 188
spongioplasm, 188
spongophyll, 188
spongy cortex, 188
spongy heart, 464, 538
spongy pith, 188
spongy rot, 188, 400
spongy tissue, 13, 188
spontaneous, 644, 645
spontaneous combustion, 646
spontaneous fire, 646
spontaneous heating, 646
spontaneous mutation, 644
spool, 242, 264, 412, 527
spool bar, 412
spool blank, 412
spool cart, 93, 639
spool diameter, 264
spool donkey, 615
spool length, 264
spool off, 135, 610
spool on, 89
spool tender, 30, 76, 89, 638
spool wood, 264, 412
spoon back, 57
sporangial vesicle, 70
sporangiole, 533
sporangiophore, 12
sporangiospore, 12
sporangium, 12
spore, 12
spore abortion, 12
spore ball, 12
spore-forming bacteria, 47
spore germination, 12
spore horn, 12
spore inoculum, 12
spore load, 12
spore mass, 12
spore print, 12
spore release, 12
spore reproduction, 12
spore shower, 12
spore trap, 12
sporidial stage, 81
sporidial state, 81
sporidium, 81, 533
sporocarp, 12, 641
sporodochium (*plur.* -ia), 143
sporogenous cell, 600
sporophore, 641
sporophyll, 12
sporophyte, 12
sporopollenin, 12
sport, 260, 486, 551
sporter, 297
sport(ing)-dog, 297
sporting rifle, 297
sports park, 477
sportsman, 444
sporulating mat, 12
sporulation, 12
spot, 7, 29, 405, 538, 639
spot application, 548
spot bark, 279
spot burning, 100
spot checking, 61, 100
spot coating, 100
spot disease, 7
spot felling, 100
spot fire, 136
spot gluing, 100, 259
spot heating, 259
spot planting, 548
spot price, 525
spot-seeding, 100
spot soil preparation, 279
spot sowing, 100
spotted deer, 319
spotting, 25, 106, 538, 639
spotting fire, 136
spotting line, 275, 451, 638
spotty bond, 100
spout, 41, 305, 309, 354, 446, 505, 652
spout undershaft chipper, 631
sprag, 618
spray, 73, 354
spray coating, 354
spray-cone angle, 513
spray cultivation, 355
spray drift, 354
spray dry, 354
spray dryer, 354
spray dust, 354
spray fan, 354
spray glue applicator, 354
spray gun, 354
spray head, 354
spray hose, 241, 354, 457
spray injury, 559
spray irrigation, 354
spray loss, 355
spray mist, 354
spray nozzle, 355
spray painting, 354
spray producer, 354
spray retention, 355
spray tunnel, 354
spray zone, 136
sprayer, 354
spraying, 354
spraying coater, 354
spraying machine, 354
spray(ing) treatment, 354
spread, 364, 415, 487
spread component, 317
spread-length, 364
spread of decay, 150, 151
spread of flame, 216
spreadable life, 274
spreader, 223, 281, 364, 368, 487
spreader bar, 103
spreader-bar loading, 104
spreader chain, 195
spreader for toxicant, 111
spreader head, 364
spreader truck, 401
spreader wheel, 140, 147
spreading agent, 281
spreading-angle, 364
spreading machine, 364, 487
spreading roll, 364, 487
sprig, 23, 535
sprig budding, 46
sprig grafting, 116
spring, 22, 29, 129, 351, 461, 650
spring-back, 211, 472

spring balance, 472
spring board, 74, 126
spring board notch, 126
spring circulation period, 68
spring clamp, 472
spring felling, 68
spring guy, 603
spring link yarder, 268
spring mattress, 472
spring normal growing stock, 68
spring outing, 68
spring overturning, 68
spring plant, 68
spring-planting, 68
spring pole, 600
spring pole falling, 472
spring revolving chair, 546
spring roll, 472
spring seeding, 68
spring set, 29
spring tracheid, 599
spring wood, 68, 599
spring wood vessel segment, 599
springer, 27, 242
springer-presser, 132
springer spaniel, 462
springhead, 387, 460
springing, 129, 224
sprinkle, 502
sprinkler, 354, 394
sprinkler apparatus, 354
sprinkler (sled), 401
sprinkling, 240, 354, 401
sprinkling can, 354
sprinkling norm, 240
sprocket-and-chain drive, 294
sprocket nose, 295
sprocket wheel, 295
sprout, 321
sprout clump, 321
sprout forest, 1, 321
sprout grafting, 583
sprout land, 321
sprout method, 321
sprout regeneration, 321
sprout reproduction, 321
sprout seeding forest, 623
sprout stand, 321
sprout-system, 1, 321
sprout system by clear cutting, 244
sprouting, 321
sprouting tree, 321
spruce extract, 595
spruce needle rust, 595
spruce oil, 595
spud, 30, 103, 612
spudder, 532
spunk, 570
spur, 4, 47, 113, 186, 261, 284, 349, 612
spur knife, 165, 284
spur road, 595
spurt, 354
squab, 566
squalene, 242, 403, 584
squall line, 26
squama, 305
squamaceous, 305
squamate, 260, 305
squamiform, 305
squamose, 118, 305
squamose erosion, 305
square, 130, 242
square around, 461
square beam, 130, 295
square budding, 131
square cell, 131
square cell core, 611
square-corner cabinet, 130
square cut, 262
square cutterhead, 131
square dress, 610
square-edge and sound, 79
square-edged, 131
square edged board, 463
square-edged timber, 130
square-end, 366
square-foot method, 558
square framed panel, 130
square grid, 130
square head, 131, 320
square joint, 362
square-jointed, 131
square log, 130, 131
square molding head, 131
square panelled door, 130
square planting, 462
square plot, 131
square profile board, 259
square ray cell, 131
square rebate plane, 130
square-sawn, 130, 463
square the break, 366
square type piling, 131
squared (squaring) log, 130
squared stuff, 130
squared tie, 131
squaring planing-jointing machine, 614
squash technique, 550
squaw log, 167, 325
squeegee coat, 399, 531
squeeze-out, 225
squeeze roll, 225
squids, 507
squirrel, 367, 464
squirrel block, 104, 212, 353
squirrel carriage slack puller, 352
squirrel line swing block, 367
squirrel suspension block, 475
squirrel tree, 120
stab culture, 65
stability, 506
stability of stand, 289
stability-resilience, 506
stabilization, 506
stabilization scenario, 506
stabilized dune, 176
stabilizer, 506
stabilizing reaction, 506
stabilizing selection, 506
stabilizing wood, 506
stable age distribution, 506
stable air (mass), 506
stable channel, 506
stable dung, 259
stable equilibrium, 506
stable isotope, 506
stable limit cycle, 506
stable manure, 259
stable vegetation, 506
stable water, 259
stachyose, 459
stack, 36, 183, 214, 286
stack bottom, 36, 120
stack device, 113, 114
stack elevator, 114
stack of lumber, 8
stack of timber, 329
stack stand, 36
stack wood, 114
stacked content, 114, 544
stacked cubic foot, 45
stacked cubic meter, 45

S

stacked measure, 45
stacked timber, 45, 544
stacked volume, 45
stacked wood, 114
stacker, 114
stacking, 114, 183
stacking area, 286
stack(ing) density, 114
stacking guide, 114
stacking in drift, 404
stacking machine, 113, 114
stacking stall, 329
stacking strip, 53, 103, 166
stacking wood, 45
stacking yard, 114, 286
staddle, 583
stadia hairs, 439
stadia rod, 439
stadia surveying, 439
stadia wire, 439
stadium, 305
staff, 24, 470
Stafford process, 462
stag, 76, 171
stag beetle, 376
stage, 60, 127, 435, 458, 459
stage cooking, 141
stage cut, 3, 237, 549
stage-duration curve, 459
stage of vernalization, 68
stagger joint, 74
staggered incision, 239
staggered lumber pile, 8
staggered piling, 239, 433
staggered setting, 165, 240
staggered stacking, 239
staghead, 278
stagheaded, 278
staging, 243
staging board, 471
stagnant air, 481
stagnant water, 462
stagnation, 481
stagnation period, 481
stagnoplankton, 258
stain, 23, 390, 507, 640
stain control, 134
stain fungus, 24
stainable, 273
stained heart (wood), 24
stained sap, 24
stained sap wood, 24
stained wood, 24, 640
stainer, 24, 640
staining, 23, 390
staining reaction, 390
staining wood, 24, 390
stair rail, 309
staircase, 309
stake, 25, 52, 107, 283
stake trial, 336
staked bunk, 580
stakeholder, 290
staking out, 288
stalactite, 626
stalactite cave landscape, 398
stalk, 29, 169
stalked leaf, 580
stalking, 365, 373
stamen, 542
staminal sterility, 542
staminate, 542
staminiferous, 261
staminode, 492
stamp, 75
stamp act, 571
stamp ax(e), 191
stamp duty, 571
stamping hammer, 191
stanchion, 52
stand, 299, 301, 533, 571
stand aerial volume table, 299
stand after thinning, 444
stand age, 300
stand age class organization, 299
stand area-age, 300
stand assessment, 299, 406, 409
stand basal area, 300
stand biomass, 300
stand boundary, 299, 533
stand-by column, 17
stand-by crew, 364, 497
stand characteristics, 300
stand-class, 299
stand classification, 299
stand climate, 408
stand composition, 300
stand condition, 301
stand condition map, 301
stand conversion, 299
stand data, 300
stand density, 299
stand density class, 299
stand density control diagram, 299
stand density index, 300
stand description, 299, 300
stand description factor, 300
stand description factors, 300
stand description key sort card, 299
stand development, 300
stand development class, 299, 645
stand diameter distribution, 300
stand diameter structure, 300
stand edge, 299, 303
stand establishment, 299, 600
stand establishment percentage, 62, 600
stand estimation, 300
stand expectation value, 300, 301
stand fire, 407
stand form factor, 300
stand-grow index, 300
stand growth, 300
stand heap charing, 450, 614
stand height, 299
stand height distribution, 300
stand height structure, 300
stand improvement, 299
stand-in, 79, 325
stand information, 300
stand inventory, 300
stand-leveling, 213, 564
stand management, 299
stand management planning, 299, 408
stand map, 300, 302
stand merchantable volume measurement, 299
stand-method, 298, 300, 386, 404
stand of sapling, 582
stand origin, 300
stand profile, 299, 300
stand projection in growth estimation, 429
stand quality, 300, 302
stand record, 406

S

stand register, 298
stand renewal equation, 299
stand-replacing, 298
stand sale, 290
stand size class, 299, 300
stand stability, 300
stand structure, 299, 300
stand survey, 300
stand table, 299
stand table method, 299
stand-table projection, 299
stand tending, 15, 149, 299, 585
stand top, 471
stand type, 299, 302
stand valuation, 300, 301
stand value, 289, 299, 301
stand volume, 289, 300
stand volume equation, 300
stand volume increment, 300
stand volume table, 289, 300
stand yield model, 300
standard, 25, 262, 547
standard-abrader, 25
standard atmosphere, 25
standard box, 26
standard chronology, 25
standard condition, 26
standard coppice system, 623
standard cut, 25
standard cutting amount, 655
standard deal, 25
standard density altitude, 25
standard deviation, 25
standard error, 26
standard error of estimate, 173
standard error the mean, 26, 362
standard felling volume, 25
standard hardboard, 26
standard height curve, 26
standard hundred, 6
standard knot, 25
standard log length, 25
standard of the first rotation, 100
standard of the fourth rotation, 99
standard of the second rotation, 99
standard of the third rotation, 99
standard performance, 47
standard planting area, 655
standard pressure, 26
standard ream, 25
standard rule, 589
standard specification, 25
standard target, 25
standard test, 25
standard treatment, 26
standard tree, 25
standard veneer lathe, 25
standard volume table, 25, 123
standardization, 25, 183
standardized field-test method, 25
standardized height curve, 25
stander, 38
standing, 290, 418
standing boom, 350
standing charcoal-kiln, 176
standing crop, 302
standing crop biomass, 525
standing current, 220
standing dead tree, 278
standing death, 289
standing line, 176, 249
standing panel, 50
standing place, 418
standing pole, 450
standing reserve, 176
standing skyline, 176
standing skyline yarding, 176
standing stock, 290
standing timber, 289, 301
standing timber fire insurance, 289
standing timber price, 289
standing timber value, 289
standing tree, 289
standing tree method, 215
standing volume, 289
staple, 105, 631
stapling machine, 106, 549
star anise(ed) oil, 5, 76
star check, 255, 540
star plywood, 540
star shake, 540
star-shaped knot, 540
starch adhesive, 103
starch glue, 103
starch grain, 103
starling, 295
starter solution, 367
starting bar, 367
starting value, 367
starvation show, 481
starve joint, 387
starved failure, 387
starved joint, 387
stat bending failure, 258
state farm, 186
state forest, 186, 630
state forest agencies, 186, 630
state forest economy, 186
state forest in the custody, 185
state forester, 630
state of strain, 572
state of stress, 573
state owned forest farm, 186
state park, 630
state-run logging, 186
state safety code, 185
state-wide land classification, 385
static, 258
static bending, 258
static bending machine, 258
static bending strength, 258
static bending test, 258
static compression, 258
static economy, 258
static elastic constant, 258
static endurance load in tension, 258
static equilibrium, 258
static hardness, 258
static life table, 258
static load(ing), 384

static modulus of elasticity, 258
static moment, 258
static pressure, 258
static rigidity, 258
static site class, 258
static stress, 258
static test, 258
static torsion, 258
statical load, 258
statics, 258
station pole, 24
stationary age distribution, 176
stationary cable crane, 176
stationary crane, 176
stationary creep, 506
stationary cyclone, 258, 621
stationary demand curve, 176, 258
stationary digester, 176
stationary drying kiln, 176
stationary fire, 258
stationary front, 258, 621
stationary (growth) phase, 506
stationary jammer, 176
stationary loading bridge, 176
stationary metal detector, 176
stationary motion, 506
stationary parasitism, 176
stationary (saw) mill, 176
stationary spreader, 176
stationary traction line, 176
stationary winch, 176
statistical analysis of inventory data, 451
statistical calculation, 484
statistical decision, 484
statistician, 484
statoblast, 282, 543
statolith, 360
statoscope, 305, 500
stature, 78, 162, 447
staubosphere, 52
stave, 8, 476, 484
stave bolt, 484
stave pipe, 8
stave-wood, 484
stay, 19, 52, 95, 175, 282, 612
stay boom, 150, 228
stay log, 371
stay-log cutting, 11
stay plate, 103, 655
stay rope, 19
staybwood, 391, 392
staypak, 391, 551
steady feed, 506
steady load, 506
steady state, 195, 506
steady-state phase, 506
steady wind, 48, 195
steam-and-quench treatment, 371
steam bending, 609
steam blow, 371, 609
steam consumption, 191, 609
steam distilled rosin, 251, 334
steam-distilled (wood) resin, 334
steam distilled wood turpentine, 251, 334
steam-distilled (wood) turpentine, 334
steam finish, 609
steam gauge, 371
steam-heated dryer, 609
steam-heated nosebar, 609
steam hose, 482
steam-injection knife, 354
steam jet, 609
steam jet blower kiln, 354
steam line, 609
steam-platen dryer, 609
steam pot, 609
steam pressure, 609
steam seasoning, 609
steam-skidder, 609
steam softening, 609
steam splitter, 609
steam sterilizer, 609
steam tractor, 609
steam trap, 346, 445
steam yarder, 609
steamed groundwood, 194
steaming, 355, 371
steaming and soaking process, 609
steaming and vacuum treatment, 355
steaming period, 371
steaming pipe, 355
steaming pit, 371, 609
steaming pressure, 609
steaming time, 371
steaming treatment, 355, 371
stearoptene, 575
steatolysis, 613
steel belt, 160
steel-blue wood-wasp, 78
steel link, 160
steel pontoon boom, 249
steel shim, 160
steel spar, 223
steel tie, 161
steel tower, 160
steel trap, 33, 160
steel wool, 160
steel yoke, 161
steep, 251, 380
steep grade engine, 110
steep helix, 110
steep land, 110
steep slope, 110, 222
steep slope tree planter, 363
steep(ing), 251
steeping tank, 251
steeping treatment, 286
steering axle, 373, 637
steering flap, 277
steerman, 135, 604
stela, 625
stele, 625
stem, 169, 252, 328, 446, 651
stem analysis, 447
stem axis, 160
stem class, 447
stem curve, 447
stem curve equation, 447
stem cutting, 612
stem density, 289
stem diameter, 447
stem dryness, 278
stem feeder, 636
stem fibre, 252, 396
stem-fire, 447
stem flow, 252
stem form, 447
stem form factor, 447
stem-formed wood, 447
stem grafting, 612
stem lean, 447
stem length, 447, 591
stem-length skidding, 591
stem-number, 302

S

stem profile, 447
stem profile models, 447
stem quality, 447
stem quality class, 447, 449
stem-root ratio, 178, 252
stem rot, 159, 252, 538
stem-shoot forest, 247
stem sprout, 447
stem strangulation, 252
stem-succulent, 399
stem timber, 446, 591
stem topwood, 417
stem-tuber, 98, 167
stem volume, 446
stem volume increment, 37
stemwood, 446
stenoecic, 520
stenoecious, 520
stenohaline, 520
stenohydric, 520
stenoky, 520
stenooxybiotic, 520
stenophagic, 520
stenophagous, 520
stenophagy, 520
stenophotic, 519
stenophyllism, 520
stenothermal, 520
stenothermic plant, 520
stenotopic, 519
stenozonal, 519
step allelomorph(ism), 243
step counter, 36
step felling, 243
step planting, 168, 309
step wedge, 210
stepladder, 463, 475
steppe, 42, 156
steppe and semideserty, 156
steppe chernozem, 42
steppe climate, 42
steppe forest, 156
steppe savanna, 445
steppe scrub, 42, 156
stepped scarf joint, 243
stepped stacking, 475
stepping stone, 36, 471
stepping treatment, 36
stepwise heated drying, 116
stepwise regression analysis, 632
stercorary, 259
stere, 289
stereo image alternator, 290
stereo plotting instrument, 290
stereo (scopic) pair, 290
stereo triplet, 403
stereo triplicate, 403
stereo zoom transfer scope, 290
stereogram method, 290
stereokinesis, 383, 651
stereometer, 290
stereomicrometer, 290
stereomicroscope, 290, 477
stereopair, 290
stereoscope, 290
stereoscope imagery, 290
stereoscope picture, 290
stereoscopic, 290
stereoscopic base, 290, 436
stereoscopic coverage, 290
stereoscopic effect, 290
stereoscopic examination, 290
stereoscopic measurement, 290
stereoscopic microscope, 290, 477
stereoscopic plotter, 290
stereoscopic view, 290
stereoscopic vision, 290
stereoscopy, 290
stereotaxis, 383
stereotriangulation, 290
stereotropism, 379, 530
stereotypic behaviour, 108
sterigma (*plur.* -mata), 81, 533, 561
sterilant, 206, 531
sterile, 32, 509
sterile board, 509
sterile caste, 511
sterile floral leaf, 32, 313
sterile flower, 32
sterile frond, 32, 574
sterile glume, 276
sterile land, 34, 359
sterile leaf, 32, 574
sterile lemma, 275
sterile-male method, 542
sterile nucleus, 32
sterile palea, 32
sterile soil, 359
steriling test, 324
sterility, 32, 35
sterilization, 265, 324, 531
sterilized box, 509
sterilizer, 324, 531
sterilizing oven, 157
sterilizing treatment, 411
sternum (*plur.* sterna), 155
stevioside, 417, 478
stick, 11, 112, 114, 168, 184, 345, 604, 612
stick back, 12
stick-culture, 65
stick furniture, 280
stick insect, 632
stick lac, 592, 642
stick-method, 71
stick placer, 114, 166
stick shellac, 642
stick stacker, 606
stick stacking, 515
stick thinning, 604
stick-type angle gauge, 159
stickability, 241, 345
sticker, 103, 114, 166, 345, 578, 604
sticker alignment, 166
sticker markings, 166
sticker spacing, 166
sticker stain, 166
stickered stacking, 79
sticking, 344, 345
sticking agent, 241
stickness, 241, 345
sticky point, 344
sticky point tester, 344
sticky trap, 345
sticky-trap method, 345
stiff adhesive, 74
stiff boom, 223, 575
stiff fibre, 232, 575
stiff-leg, 160, 202, 536
stiffener, 575, 601
stiffening agent, 575, 601
stiffness, 160, 232, 251
stiffness factor, 160
stiffness to weight, 376
stigma, 100, 369, 555, 635
stigma (*plur.* stigmata/stigmas), 59
stile, 67, 321, 450, 546
still target, 257
stillage, 112, 114
stillingia oil, 507, 642
stillingia tallow, 507

S

stilt root, 612
stimulant, 71
stimulation of heterozygosis, 596
stimuli (climate-related), 71
stimulus, 71, 534
stimulus-response chain, 71
sting, 441
stinger, 295, 350, 492, 493
stion, 608
stionic effect, 608
stionic relation, 608
stipe, 266, 560
stipel(la), 535
stipular scale, 493
stipulate, 261
stipule, 493
stipule scar, 493
stirps (*plur.* stirpes), 252, 474, 626, 629
stitching machine, 147
stochastic model, 468
stochastic process, 468
stochastic variable, 467
stochastics, 156
stock, 25, 168, 199, 278, 324, 328, 544, 592, 608
stock board, 183
stock book, 289
stock breeding, 295
stock capital, 302
stock chest, 65, 523, 635
stock culture, 65, 592
stock-cutting list, 620
stock density, 289
stock driveway, 52, 223, 430
stock-gang, 280
stock inlet, 250
stock line, 445
stock lumber, 183
stock map, 302
stock mentor, 608
stock outlet, 62
stock plant, 39
stock plot, 592
stock production, 324
stock pump, 237
stock rate, 85
stock route, 52, 223, 430
stock-scion interaction, 608
stock-scion relationship, 608
stock seed, 592
stock seed farm, 592
stock table, 299, 545
stock tree, 39, 627
stock-value, 301, 629
stock velocity, 237
stock volume, 171
stock yard, 114, 285, 286
stocked area, 289
stocked land, 289
stocked quadrat, 290
stockholders' equity, 175
stockholm tar, 464
stocking, 74, 136, 289, 290, 324, 388
stocking density, 289
stocking percent, 54, 289
stocking rate, 136, 598
stockpile, 183, 286, 634, 635
stocksaver, 33
stocktaking, 351
stolbor, 238
stole, 364
stolon, 243, 364, 430
stoloniferous plant, 364
stoma, 369, 560
stoma (*plur.* stomata/stomas), 200
stomach agent, 504
stomach insecticide, 504
stomata, 369
stomatal, 369
stomatal closure, 369
stomatal infection, 369
stomatal movement, 369
stomatal resistance, 369
stomatal transpiration, 369
stone, 186, 246, 433
stone boat, 205, 349
stone cell, 434
stone construction, 433
stone fruit, 193
stone furniture, 434
stone kiln, 434
stone lantern, 433
stone roll beater, 433
stone splash-dam, 433
stone weir, 114, 433
stoning, 57
stony, 118, 434
stony land, 434
stony soil, 434
stool, 92, 94, 165, 321
stool layering, 164, 276
stool shoot, 168
stool sprout, 168
stop, 494, 618
stop beam, 602
stop device, 620
stop log, 602
stop log recess, 602
stop moulding, 526
stoppe-lap dovetail joint, 3
stopping reaction, 506, 626
stopples, 203
storability, 336
storage, 74, 634, 635
storage area, 635
storage boom, 193, 635
storage capacity, 74, 397, 635
storage cell, 634
storage charges, 603
storage dam, 545, 635
storage decay, 635
storage defect, 114, 635
storage disease, 634
storage dump, 540
storage life, 65, 635
storage location, 74
storage parenchyma, 65
storage pocket, 74
storage pond, 635
storage potential, 336
storage premises, 278
storage pressure vessel, 164
storage reservoir, 545, 635
storage rot, 635
storage tank, 635
storage tissue, 634
storage tracheid, 634
storage yard, 635
storax oil, 466
store, 15, 302
store manger, 635
storey, 298, 300
storeyed high forest, 140
storeyed structure, 104
storied, 45, 104
storied arrangement, 45, 104
storied axial element, 104
storied building, 309
storied cambium, 104
storied coppice method, 140
storied coppice system, 1, 623
storied element, 104
storied high forest, 140
storied ray, 104
storied structure, 104
storing table for log, 590

storing water tillage, 270
storing wood in land, 310
storing wood in water, 460
storm damage, 144
storm surge, 144
storm transposition, 16, 585
storm water, 585
storm water run off, 585
storyed, 45, 104
storyline, 381
stout wood, 639
stove bolt, 41, 484
stovehouse, 505
stoving enamel, 197
stow, 114, 638, 639
straddle-carrier, 162, 278
straddle truck, 162
straggler bird, 321
straight-away shot, 362
straight banding, 615
straight boom, 103, 231, 614
straight coal-tar creosote, 69
straight cutting, 365, 575, 614
straight dozer, 611
straight forward shot, 465
straight front, 363
straight grain, 615
straight grained wood, 615
straight haul, 509
straight laminated wood, 461
straight lead, 611
straight line cross-cut saw, 615
straight line depreciation, 615
straight line ripping saw, 615
straight panel, 615
straight-snouted weevil, 614
straight-tined harrow, 105, 614
straight tongue, 615
straightedge, 88, 177, 557
strain, 72, 267, 359, 572, 603
strain aging, 572
strain circle, 572
strain energy of bending, 258
strain gauge, 572
strain intensity, 572
strain-linear hybrid, 105
strain-rate, 572
strain test, 359
strainer, 187
strainometer, 572
strake, 66, 132
strand, 165, 267, 404, 430, 446, 469, 641
strand lake, 188
strand-like structure, 470
strand parenchyma, 446
strand tracheid, 67, 446
strand vegetation, 188
strandboard, 479, 603
stranger, 349
stranger plant, 349
stranger species, 349
strangler, 243
strangulation, 569
strap, 227, 281, 417, 561
strap saw, 79
strap socket, 470
strapper, 12, 280
strategy in game theory, 580
stratification, 45, 54, 140, 411
stratification in aerial photograph, 191
stratified cambium, 104
stratified clip method, 140
stratified clip technique, 140
stratified community, 140
stratified crown, 140
stratified mixed stand, 152
stratified population, 140
stratified random sample, 140
stratified random sampling, 140
stratified ray, 104, 140
stratified sample, 140
stratified sampling, 140
stratified secondary wall, 45
stratosphere, 362
stratum (*plur.* strata), 13, 45
straw cutter, 602
straw drum, 2
straw line, 2
straw line block, 2
straw line fairleader, 2
straw line hook, 2
straw mushroom, 42
straw pulp, 42
strawberry tree, 557
stray, 358
streak, 43, 228, 405, 479
streak culture, 204
streaking, 165, 268
streaky structure, 479
stream, 515
stream bank protection forest, 202
stream barker, 418, 457
stream bed, 193
stream channel erosion, 193
stream community, 192
stream discharge, 307
stream driving, 159, 192
stream dryer, 369
stream flow, 192, 307
stream frequency, 192
stream jam, 623
stream-lined, 308
stream quadrant, 44
stream rating, 192, 199
streamflow, 307
streamline, 308
street garden, 245
street green area, 245
street planting, 90, 245
street tree, 540
strength factor, 376
strength increasing sizes, 602
strength of attack, 324
strength of wood, 330
strength perpendicular to the grain, 196
strength properties, 288, 376
strength test, 375
strength timber, 236, 246
strength-weight ratio, 376
strengthening agent, 602
streptolydigin, 290, 295
streptomycin, 295
stress, 343, 572
stress amplitude, 573
stress analysis, 572
stress at elasticity limit, 472

stress at proportional limit, 20
stress birefringence, 572
stress circle, 573
stress concentration, 572
stress contour, 572
stress-cycle diagram, 572
stress distribution, 572
stress factor, 573
stress germination, 343
stress-graded lumber, 572
stress grader, 572
stress grading, 572
stress grading machine, 572
stress in bending, 498
stress in biaxial, 455
stress in compression, 551
stress in tension, 604
stress in three dimension, 403
stress in triaxial, 403
stress relaxation, 572
stress relief, 572
stress section, 572
stress-strain relation, 573
stress system, 572
stress test, 343, 572
stress theory, 572
stressed shell construction, 572, 575
stress(ed) skin construction, 572
stressed skin panel, 443
stressfree wood, 512
stressgrading system, 572
stretch of pines, 77
stretch rolling, 604
stretchability, 282, 554
stretcher, 196, 263, 453, 603, 604
stretcher mounted sprayer, 81
stretcher roll, 263, 550
stria (*plur.* striae), 479, 631
striation, 479
strict aerobe, 554, 636
strict anaerobe, 554, 636
strict nature reserve, 554
strike, 46, 73
strikethrough, 420, 485
string, 249, 519, 618
string board, 309
string logs, 589
string (logs), 66
string measurement, 430
string of stick, 562
stringer, 295, 350
stringer piece, 195, 309, 350, 650, 651
stringer road, 328, 332, 589
stringy rot, 462
strip, 8, 30, 37, 53, 80, 103, 166, 470
strip and group shelter wood system, 80
strip and group system, 80
strip boom, 86
strip burning, 81, 461
strip catcher, 366
strip census, 81
strip clearcutting, 80
strip coating, 529
strip cropping, 80
strip cruise, 80, 81
strip cutting, 80
strip-cutting forest, 80
strip cutting system, 80
strip felled area, 80
strip felling, 80
strip-felling area, 80
strip felling system, 80
strip firing, 80, 81
strip heating, 335, 479
strip-method, 79, 81
strip planting, 81
strip plot, 81
strip road, 184, 489
strip-road processor, 298
strip sampling, 80
strip seeding, 79
strip-selection cutting, 81
strip selection method, 81
strip selection system, 81
strip-shelterwood cutting, 80
strip shelterwood method, 80, 81
strip soil preparation, 81
strip sowing, 479
strip spacing, 190
strip-stand method, 81
strip survey, 81
strip system, 79
strip-thinning, 81, 166
stripboard, 8
stripe, 479, 650
stripe figure, 80
striped veneer, 479
stripling, 324
stripped stacking, 79
stripper, 8, 30, 64, 177, 371
stripping, 80, 385, 444
stripping wheel, 451
strobile, 12, 382
strobilus, 12, 382
stroke, 59
stroke method, 204
stroke sander, 80
stroll, 114
stroma (*plur.* -ata), 232, 642
stromatic, 642
strong-light demander, 376
strong thinning, 376
strop, 205, 281
strophiole, 626
Stroud dendrometer, 462
struck moulding, 37
structural adhesive, 246
structural change, 246
structural colour, 246
structural component, 246
structural feature, 246
structural framework, 175, 246
structural gene, 246
structural hybrid, 246, 390
structural insulation board, 236
structural jointer, 246
structural material, 236, 246
structural panel, 236, 246
structural timber, 236, 246
structural veneer, 246
structural wood, 236, 246
structural work, 246
structure of wood, 329
struggle for existence, 422
strut, 271, 612
strychnos type, 117, 316
stub, 41, 74, 125, 146, 450, 603, 612
stub grafting, 612
stub line, 292

S

stub mortise, 375
stub out, 265, 292
stub pole, 113
stub puller, 450
stub snag, 450
stub tenon, 112
stub track, 410
stubbing, 64, 292
stubble, 41, 46
stubble mulch, 40
stuck moulding, 103
stud mill, 619
stud(ding), 20, 232
studentized residuals, 548
study plot, 438
stuff, 237, 333, 479
stuff box, 479, 618, 635
stuff catcher, 618
stuff chest, 618, 635
stuff knot, 237
stuff-over, 385
stuff pump, 237
stuffing box, 322, 479
stull, 275, 280, 612
stump, 40, 125, 168, 265, 359, 446
stump-age, 37, 125
stump analysis, 125
stump auger, 494
stump chipper, 125
stump crusher, 125
stump cutter, 125
stump deduct, 162
stump diameter, 125
stump extraction, 125, 265
stump extractor, 5
stump height, 125
stump-jump plough, 594
stump jumper, 162, 280, 594
stump-land, 168
stump mark, 450
stump out, 265, 292
stump plant, 321, 359
stump pull, 61, 381, 538
stump-puller, 5, 450
stump roller, 637
stump root system, 220
stump rot, 127, 160, 167
stump sale, 289
stump scale, 120, 167
stump scaler, 125
stump section, 38, 125
stump shoot, 168
stump splitter, 356
stump sprout, 168
stump tenon, 279
stump transplant, 359
stump turpentine, 325
stump veneer, 167
stump wire, 637
stump wood, 125, 167, 168, 325
stump wood chip, 125
stump wood figure, 168
stumpage, 289
stumpage appraisal, 301
stumpage charge, 290
stumpage conversion value, 301
stumpage cost, 301
stumpage formula, 301
stumpage margin, 301
stumpage market, 290
stumpage price, 289
stumpage ratio, 301
stumpage revenue per unit area, 85
stumpage sale, 290, 302, 317
stumpage value, 289, 301
stumper, 126, 265
stumping, 64
stumping powder, 125
stump(y) land, 38, 125
stunt, 1
stunted tree, 1, 112
stylar, 204
stylate, 260
style, 204
styler-borne virus, 277
stylet, 277, 535
styloid, 636
styloid crystal, 636
stylus (*plur.* styli), 503
styracin, 466
styrax, 465, 529
styrene-butadiene rubber, 104
sub-adult, 222, 503
sub-back, 154
sub-chromatid, 552, 553
sub-compartment, 533
sub-compartment boundary, 299, 533
sub-elementary fibril, 552
sub-fibril, 553
sub-irrigation, 98
sub-layer, 552
sub-light microscope, 552
sub-molecular structure, 552
sub-plot, 25, 154, 517, 535
sub-population, 70
sub-sampling, 70, 154
sub-satellite point, 540
sub-subcompartment, 517
subacuminate, 250
suballiance, 552
subalpine forest, 552
subalpine zone, 552
subantarctic, 552
subapical, 250
subapiculate, 311
subarborescent, 250
subarctic climate, 154
subarctic conifer, 552
subarctic period, 552
subarctic zone, 552
subassociation, 552
subatlantic period, 552
subaxile, 562, 613
subboreal period, 552
subcellular structure, 553
subchromonema, 553
subclass, 552
subclimax, 552
subcommunity, 552
subcompartment, 533
subcompartment investigation, 533
subcompartment management method, 533
subcosta, 552
subcrenate, 261, 502
subculture, 66, 598
subcutis, 356, 521
subdivided crystalliferous cell, 598
subdivision, 552
subdivision forest, 408
subdominant, 553
subdominant species, 553
subdominant tree, 553
subdominant tree species, 553
subdorsal line, 42
subdrain, 3
subdrainage, 98
subentire, 221
subepidermal, 27, 608
suber, 334, 452
suberification, 334, 401, 452
suberin, 334, 401

suberinlamella (*plur.* -llae), 452
suberization, 452
subfamily, 552, 553
subfloor, 94, 318
subflooring, 94
subformation, 552
subfreezing temperature, 305
subgene, 552
subgeneric name, 553
subgenus, 553
subgroup, 552
subhorizon, 538, 552
subiculum (*plur.* -la), 267
subimago, 552
subinfection, 380
subinfluent, 553
subinfluent species, 553
subirrigation, 98
sublabrum, 339
sublingual pouch, 417
sublittoral zone, 374, 594
submarginal land, 248, 253
submarginal log, 92, 534
submarine climate, 188
submarine photometer, 460
submaritime, 250
submentum, 552
submerged plant, 456
submerged reef, 3
submerged (water) plant, 53
submergence, 554
submersiherbosa, 53
submersion, 457
submicroscopic anisotropic crystalline particulate, 553
submicroscopic feature, 553
submicroscopic microfibril, 553
submicroscopic morphology, 553
submicroscopic particle, 553
submicroscopic porosity, 553
submicroscopic size, 553
submicroscopic structure, 553
submissive posture, 461
submontane zone, 553
subnival belt, 162, 552
subnival region, 552
subnival soil, 162
subnival zone, 552
subnormal-under stocked stand, 289
suboesophageal ganglion, 437
subopposite, 552
suborder, 552
subordinate, 18, 443
subordinate species, 72
subordinate tree, 297, 520
subordinates pecies, 71, 72, 154
subordination, 297
subpedunculate, 311, 342
subperiod, 142
subpetiolar, 560
subphylum, 552
subplot, 517
subpolar, 154
subpolar region, 552
subpolar zone, 552
subpolar zone forest, 552
subpopulation, 553
subsample, 122, 641
subsere, 71
subshrub, 9, 533
subsidence, 53, 520
subsidence (temperature) inversion, 520
subsidiary felling, 33, 150, 154, 212
subsidiary pollination, 150
subsidiary record, 150
subsidiary road, 154
subsidy, 33
subsistence activity, 423
subsistence economy, 422
subsistence farming, 644
subsistence herding, 423
subsistence level, 422
subsocial, 553
subsoil, 95, 538
subsoil plough, 420
subsoiler, 53, 538
subsoiling, 420, 538
subspecies, 553
substandard, 34
substitute fibre, 134
substitute king, 33, 640
substitute queen, 640
substitute tie, 79
substitute wood, 79
substitution forest, 477
substitution vegetation, 477
substorey, 70
substrate, 18, 95, 219, 353, 655
substratum (*plur.* -ta), 221, 353
substructure, 219, 520
subsurface irrigation, 98
subsurface run-off, 552
subsurface soil, 552
subsystem, 553, 641
subtending leaf, 12, 495
subterranean, 98
subterranean animal, 98
subterranean irrigation, 98
subterranean layer, 98
subterranean part, 98
subterranean river, 147, 374
subterranean shoot, 98
subterranean stem, 98
subterranean termite, 95, 489
subterranean water, 98
subtidal community, 52
subtidal zone, 52
subtractive colour formation, 233
subtractive colour process, 233
subtribe, 553
subtropic climate, 553
subtropical evergreen forest, 553
subtropical forest, 553
subtropical forest region, 553
subtropical forest zone, 553
subtropical garden(ing), 553
subtropical hardwood, 553
subtropical rain forest, 553
subtropical scenery, 553
subtropical summer green forest, 553
subtropical zone, 553
subunit, 552
subunit structure, 552
subvariant, 552
subvariety, 552
subzero temperature, 305

succession, 556
succession of cutting, 39
succession of felling, 39
succession species, 556
successional habitat, 556
successive cutting, 293, 461
successive felling, 293, 461
successive generation, 293
successive inventory, 293
successive regeneration, 293
successive regeneration cutting, 293
successive regeneration felling, 293
successive regeneration system, 293
successive strip, 293
succulence, 120, 399
succulent, 399
succulent chamaephyte, 399
succulent fruit, 120
succulent plant, 118, 399
suck stacker, 514
sucker, 167, 321, 514
sucker-down block, 521
sucker-stump, 521
suckering, 64, 247
suckling grafting, 574
sucrase, 606, 637
sucrose, 606
suction box, 515
suction box cover, 515
suction carriage, 514, 607
suction catcher, 513
suction couch, 607
suction couch with cantilever arm, 545
suction couch with cellular grating, 147
suction couch with cellular perforations, 147
suction couch with open chamber, 50
suction cylinder, 514, 607
suction drum drier, 513
suction duct, 514
suction feed type gun, 514
suction feeder, 607
suction force, 515
suction head, 607
suction pipe line, 446
suction press, 607
suction press roll, 608
suction pressure, 515
suction pump, 607
suction roll, 515, 607
suction stacker, 607
suction trap, 513
sudd, 77
suddenly applied load, 486
suffocation, 621
suffrutescent, 9
suffrutex, 2, 9
suffruticosa plant, 180
suffruticose, 9, 552
suffruticose chamaephytes, 9
suffruticulose, 9
suffumigation, 549
sugiol, 308
sugiresinol, 308
sugoyl, 308
suitable area of nature reserve, 645
suitable land for forest, 564
suitable moisture content, 440
suite, 55
sulcus, 41, 172
sulfate (cooking) liquor, 308
sulfate-reducing bacteria, 308
sulfation, 308
sulfite liquor, 552
sulfite process, 552
sulfite pulp, 552
sulfite waste liquor, 552
sulfited mimosa extract, 552
sulfitization, 552
sulfonated extract, 210
sulfonation, 210
sulfonator, 210
sulfur bacteria, 308
sulfur cycle, 308
sulfur hexafluoride, 308
sulky, 162, 223
sulky plough, 312
sulphate cook, 308
sulphate (cooking) liquor, 308
sulphate process, 308
sulphate pulp, 308
sulphate turpentine, 308
sulphide community, 308
sulphidity, 308
sulphite alcohol, 552
sulphite liquor, 552
sulphite process, 552
sulphite pulp, 552
sulphite waste liquor, 552
sulphur bacteria, 308
sulphur rain, 210, 308
sulphuric acid lignin, 274, 308
sum-of-the-year's-digits depreciation, 343
sumach extract, 366
sumac(h) tannin, 366
summed dominance ratio, 648
summed leaf area, 649
summed leaf weight, 649
summer annual (plant), 522
summer beam, 77, 319
summer bird, 521
summer diapause, 522
summer dormancy, 522
summer garden, 521
summer-green deciduous forest, 522
summer green forest, 522
summer-house, 21, 295
summer plumage, 522
summer resident, 521
summer resort, 21
summer show, 522
summer-spore, 521
summer stagnation period, 521
summer visitor, 521
summer wood, 76, 328, 499, 521
summer wood vessel segment, 499
summer wood zone, 499
summerwood cutting, 341
summit, 105, 413
sump, 57, 264, 508, 635
sun burn, 397, 413
sun-crack, 413
sun crown, 557
sun-eheck, 413
sun leaf, 556
sun plant, 517, 556, 557
sun power, 471
sun-scald, 397, 413
sun synchronous, 471
sundeala, 336
sunfleck, 180
sunk cost, 53, 621

S

sunk moulding, 520
sunk panel, 520
sunk pier, 53, 460
sunk top, 520
sunk wood, 53
sunken bed, 92
sunken flower bed, 52
sunken garden, 53
sunken joint, 4
sunken log, 52, 53
sunken stomata, 341
sunny, 182, 530
sunny slope, 556
sunscald, 397
sunscorch, 397, 413
sunscreening agent, 471
sunspot, 471
super bacteria, 51
super dominance hypothesis, 51
super elevation, 51
super ficial foot, 357
super fine paper, 90, 240
super-foot Hoppus, 217
super-foot true, 436
super guard saw chain, 51
super-hardboard, 474
super high frequency, 51
super low frequency, 50
super marginal, 50
super surface, 254
super turpentine, 577
superallele, 51
supercilium, 319
supercooling, 186
superdominant, 51
superfaced lumber, 16
superfamily, 649
superfemale, 50
superficial measure, 26
superficial treatment, 26
superficial volume, 26, 45, 496, 544
superfoliaceous, 561
supergroundwood, 334, 578
superheat, 187
superheated steam, 187
superheated steam drying, 187
superheated steam kiln, 187
superheated steam seasoning, 187
superheated vapor, 187
superheated vapor drying, 187
superinfection, 60
superior national forest, 162
superior ovary, 416
superior plantation, 577
superior stand, 577
superior tree, 577
supermale, 51
supermolecule, 241, 500
superorganism, 51
superparasitism, 60, 153
superparasitization, 60, 153
superposed bud, 104
superposed square planting, 512
superposition theory, 45
superpressed plywood, 164
supersex, 51
supersonic detector, 51
supersonic frequency, 51
supersonic wave, 51
superspecies, 52
superstoichiometric, 51, 186
superstructure, 415
supervisor, 232, 573
supervisor system, 573
supplemental cutting, 33
supplemental final cut, 33
supplemental nutrition, 33
supplementary fertilizer, 33, 639
supplementary ichneumon fly, 232, 534
supplementary illumination, 150
supplementary item, 33
supplementary natural regeneration, 32, 33
supplementary pollination, 150
supplementary reproduction, 150
supplementary windbreak, 152, 154
supply chute, 171
supply hot air duct, 393
supplying, 32, 33
support, 612
support bracket, 612
support carriage, 465
support coating, 598
support land, 303
support skyline, 56
supported adhesive, 598
supported line, 623
support(er), 598
supporting bar, 612
supporting belt, 612
supporting cell, 612
supporting roll, 612
supporting strand, 612
supporting tissue, 612
suppressed tree, 18
suppressed-weir, 615
suppression, 569
suppression action, 324, 364
suppression method, 324
suppressor, 243, 569
supralittoral zone, 3, 52
supraoesophageal ganglion, 437
supraorganism, 51
supratidal zone, 52
surbase, 635
surf-board, 133
surface active agent, 26
surface area, 26, 384
surface blemish, 27
surface bonding strength, 26
surface cableway, 97
surface carburization, 27
surface channel, 325
surface check, 26, 375
surface coating, 27
surface combustion, 27
surface-cover, 95
surface crack, 26, 375
surface cultivation, 27
surface decay, 26
surface defect, 27
surface densification, 27
surface drain, 95, 325, 350
surface dressing, 27, 147, 561
surface drill, 401
surface dry, 26
surface erosion, 323
surface evaporation, 27
surface expansion, 27
surface feature, 96, 98
surface film, 26, 455
surface finish(ing), 27
surface fire, 95
surface-float, 457
surface flow, 26, 95, 256
surface foot, 360
surface fuzz(ing), 27
surface ground cover, 95
surface growth, 27
surface hardening, 27

surface humus layer, 26
surface inspection, 26
surface inversion, 97
surface irregularity, 26
surface irrigation, 95
surface layer, 323
surface measure, 8
surface of rupture, 113, 363
surface planer, 16, 362
surface porosity, 26
surface profile, 26
surface quality, 27
surface reservoir, 97
surface roughness, 26
surface run-off, 95
surface runner, 95, 97
surface runoff, 95
surface seeding, 401
surface shake, 26
surface size(ing), 27
surface smoothness, 26
surface soil, 27
surface soundness, 26
surface sowing, 401
surface split, 26
surface spreader, 26
surface stain, 26
surface sterilization, 27
surface tension, 27
surface texture, 26
surface-treated particleboard, 26
surface treatment, 26
surface (veneer), 26
surface vibrator, 27
surface water, 95
surface worm hole, 26
surface yarding, 97, 387
surfaced, 16
surfaced four sides, 463
surfaced lumber, 16
surfaced one side, 563
surfaced one side, one edge, 563
surfaced one side, two edges, 563
surfaced timber, 16, 227
surfaced two sides, 295
surfaced two sides, one edge, 296
surfacer, 332
surfacing, 16, 412, 638
surfactant, 26
surfoyl, 496
surge and transfer station, 623
surge deck, 74, 81, 209
surging, 30, 352
surplus account, 573
surplus capacity, 187
surplus felling, 187
surplus stress, 41, 431
surplus value, 431
surtax, 50, 152
survey, 44, 104
survey field sketch, 559
survey for disease, 29
survey in detail, 468
survey line, 45
survey report, 104
survey unit, 61, 104
surveying, 44
surveying altimeter, 44
surveyor, 44, 104
surveyor's chain, 44
surveyor's compass, 44
surveyor's table, 44
survival, 40, 74, 422
survival curve, 74
survival of the fittest, 440
survival percent, 54, 74
survival percent of seeding, 54
survival percentage, 54, 74
survival potential, 422
survival rate, 54, 74, 422
survival rate of plantation, 600
survival ratio, 422
survival value, 74
survivor, 40, 74
survivorship, 74
survivorship curve, 74
suscept, 159
susceptibility, 159
susceptible, 569
susceptible reaction, 159
suspended cable yarding, 469
suspended load, 545
suspended particle, 545
suspended planimeter, 104
suspended sediment load, 545
suspended skyline, 231
suspended solid, 545
suspender, 104, 545
suspensi pupa, 67, 545
suspension, 104, 545
suspension block, 104
suspension bridge, 104
suspension chamber, 545
suspension chute, 231, 603
suspension door, 104
suspension feeder, 316
suspension guy line, 104
suspension line, 104, 545
suspension rod, 103
suspension rope, 104, 545
suspension sifter, 545
suspension-type dryer, 545
suspension yarding, 545
suspensoid, 545
suspensor, 338, 352
sustainability criteria, 576
sustainable agriculture, 57
sustainable development, 272
sustainable forest management, 272, 576
sustainable forestry, 272
sustainable growth, 57
sustainable use, 272
sustainablility, 15, 576
sustained annual yield working, 576
sustained contraction, 57
sustained lead, 57
sustained periodic yield, 107
sustained yield, 57, 576
sustained-yield appraisal, 576
sustained-yield-even-flow, 576
sustained-yield forest, 576
sustained-yield forest management, 576
sustained-yield management, 576
sustained yield management unit, 576
sustained yield of forest, 410
sustained yield regulation, 576
sustained-yield working, 576
Sutherlin side block, 402
suture, 147, 526
swage, 481, 550
swage saw, 85, 550
swage set, 550
swage sharper, 550, 610
swager, 550

S

swaging, 550
swale, 93, 304
swallow, 556
swallow tail stump, 556
swamp, 381, 604, 605
swamp dozer, 605
swamp forest, 605
swamp garden, 605
swamp-growing plant, 605
swamp hook, 222
swamp-land, 459, 605
swamp plough, 180
swamp soil, 605
swamped line, 125
swamper burning, 467
swamping, 125, 381
swampy area, 605
swan-neck, 120
swan-neck lock mortise chisel, 120
sward, 41
swarm, 142, 388
swarm spore, 579
swarming, 388
swatch, 30, 37, 639
swathe-felling mobile chipper, 564, 566
swather, 165, 364
sway grain, 537
sway sawback, 384
sway wood, 357
sweat gland, 190
sweating, 62, 420
swede fiddle, 196, 600
swede hook, 161
swede level, 143, 614
Swedish top height, 401
sweep, 7, 52, 177, 377, 405, 497, 538, 563
sweep front, 170
sweep sawing, 498
sweep the rear, 307
sweep time, 317
sweep wood, 53
sweeper, 177, 381, 498
sweeper harrow, 493
sweeping, 497
sweeping tree, 498
sweet basil oil, 197, 478
sweet birch oil, 207, 478
sweet extract, 399, 478
sweet orange oil, 478
sweet orange peel oil, 478
sweet violet, 642
swell, 160, 355, 479, 613
swell-butted, 76, 160
swell front, 170
swelling, 352, 355, 398, 401
swelling agent, 355, 401
swelling compressive stress, 355
swelling heat, 355
swelling pressure, 355, 401
swelling ratio, 355
swelling stress, 355, 401
swelling type cosolvent, 355
swept front, 170
swift-water community, 222
swifter, 168, 333
swifter gang, 23
swifter knot, 430, 470
swifter line, 350
swiftering machine, 23
swifts, 23
swing boom, 546, 583, 637
swing-boom log loader, 546, 637
swing charger, 7
swing conveyor, 558
swing crane, 212, 545
swing cut-off saw, 7, 104
swing dingle, 84, 205
swing dog, 362
swing donkey, 121, 150
swing door, 638
swing drum, 150
swing hot, 307
swing-jack, 196
swing line, 367
swing matching, 115
swing road, 121
swing saw, 7, 104
swing shift, 535, 622
swing tree, 223
swing up saw, 90
swing yard, 121
swing(ing), 7, 104, 121, 324
swinging blade, 292
swinging cross-cut saw, 7, 104
swinging in a bind, 125, 473
swingle bar, 499
swingle-tree, 499
swirl, 507
swirl crotch, 507
swirl crotch figure, 507, 584
swirl grain, 546
switch tie, 47
Switzerland selection forest, 401
swivel chair, 546
swivel hook, 637
swivel joint, 212, 243, 244
swivel truck, 637
swivel yoke, 205
swollen shoot, 626
sword, 295
sycon, 571
syconium, 571
syenite, 611
sylva, 408, 410
symbiont, 172
symbiosis, 172
symbiotic algae, 172
symbiotic bacteria, 172
symbiotic fungus, 172
symbiotic nitrogen fixation, 172
symbiotic relationship, 172
symbiotic saprophyte, 172
symbiotrophic, 172
symbolic algebra, 148
symmetrical axis, 114
symmetrical style, 114
symmetry, 114
symparasitism, 172
sympatric, 140, 483
sympatric phenomena, 172
sympatric speciation, 483
sympatric species, 483
sympatric subspecies, 483
sympatry, 483
sympetalous, 191
sympetalous corolla, 191
symphyllous, 294
symplasm, 172, 191
symplasmic movement, 172
symplast, 172, 191
symplastic growth, 536
sympodial branching, 192
sympodulospores, 192
symptom, 611
synapsing chromosomes, 390
synapsis, 390
syncarp, 192
syncarpous fruit, 192
syncaryocyte, 192
synchorology, 388

S

synchronic, 482
synchronization, 482, 483
synchronization factor, 482
synchronization regulation, 482
synchronology, 174
synchronous, 482
synchronous culture, 482
synchronous meteorological satellite, 97
synchronous satellite, 482
syncline, 530
synclinorium, 154
syncyanose, 172
syndesis, 390
syndiagenesis, 482, 599
syndiploidy, 172
syndrome, 153
syndynamic, 388
syndynamics, 388
synechthry, 376
syneclise, 310, 471
synecology, 389
synema, 204
syneresis, 241, 346
synergid, 634
synergism, 536, 602
synergist, 192, 536, 602
synergistic effect, 536
syngamy, 296, 353
syngenesis, 388, 389
syngenetic succession, 388
syngenetics, 389
synnema (*plur.* -ata), 143, 446
synoecium, 191
synoecy, 70, 274
synopsis of the forest resource, 410
synoptic chart, 477
synphyllous, 191
synphylogeny, 389
synphysiology, 389
synsystematics, 389
syntan, 192
syntaxonomy, 388
synthetic antenna, 192
synthetic aperture antenna (radar), 192
synthetic aperture radar, 192
synthetic camphor, 192
synthetic fibre, 192
synthetic glue, 192
synthetic medium, 192
synthetic organic insecticide, 580
synthetic resin, 192
synthetic resin adhesive, 192
synthetic rubber, 192
synthetic rubber adhesive, 192
synthetic size, 192
synthetic stone, 396
synthetic tannin, 191, 192
synthetic tanning agent, 191, 192
synthetic variety, 648
synusia, 45, 483
synzoochory, 109
syringyl-type lignin, 642
syrinx, 325
systellophytum, 152
system, 517, 620, 654
system administration, 517
system analysis, 517
system approach, 517
system class, 655
system design, 517
system ecology, 517
system loader, 517
system model, 517
system of after regeneration, 169
system of breeding, 586
system of classification, 141
system of cutting, 633
system of forest regulation, 407, 409
system of high forest with standards, 14
system of management, 253, 654
system of regeneration, 169
system of successive regeneration fellings, 237
system of sustained working, 576
system optimization, 517
system theory, 517
systematic, 141, 517
systematic bacteriology, 518
systematic botany, 517, 617
systematic development, 517, 629
systematic distribution, 517
systematic embryology, 517
systematic error, 517
systematic floriculture, 203
systematic plant sociology, 517
systematic plot sampling, 517
systematic sample, 517
systematic sampling, 517
systematic strip sampling, 517
systematic thinning, 517
systematic tree, 517
systematic zoology, 517
systematics, 141, 517
systematist, 141
systemic, 341, 386, 477, 517
systemic disease, 517, 610
systemic fungicide, 341
systemic infection, 517
systemic insecticide, 339, 341
systemic mutation, 477, 517
systox, 341
T-branch, 143
T-budding, 551
T-chromosome, 390
t-distribution, 140
T-form budding, 551
T-grip dibble, 616
T-lap joint, 75
T-notching, 147
T-shaped budding, 551
T-shaped hair, 318
T-shaped retaining wall, 88
t-test, 233
tabanid, 321
Taber-apparatus, 471
table, 26, 90, 126, 170, 362, 375, 469, 640
table band resaw, 471
table band-saw, 471
table chair, 640
table chairewise, 640
table decoration, 640
table lifting device, 170
table of basal areas, 113, 542, 592

T

table of quality of locality, 98
table of sectional areas, 113, 542, 592
table of site-class, 98
table roll, 3
table top, 640
tableland, 471
tablet, 534
tablet(-arm) chair, 444
tablet system, 359
tabouret, 533
tabular cell, 358
tabular ferric concretion, 9
tabular root, 9
tabulator, 619
tacamahac, 197, 613
tachardia lacca, 642
tacheometry, 439, 638
tachigaren, 489
tachograph, 44, 466
tachometer, 44, 638
tachymeter-survey, 439
tachymeter-transit, 439
tack, 63, 241, 332
tack dry, 158
tack range, 345
tack strength, 345
tackifier, 241, 601
tackiness, 241
tackiness agent, 241
tackle, 153, 205, 470
tackle block, 153, 205
tacky state, 344
tact system, 307
tactical plan, 324, 603
tactics, 603
tactile organ, 65
tag, 25, 178, 281, 292
tag chain, 112, 292
tag line, 104, 150, 282, 292
tagged atom, 25, 438
tagging, 178, 227, 483
tagulaway, 402
taiga, 190, 471
taiga soil, 284, 471, 606
tail block, 212, 503
tail chain, 360, 503
tail-down, 184, 653
tail dumper, 199
tail feather, 503
tail grip, 503
tail hold, 199
tail hook, 503
tail jack, 503
tail line, 212, 259
tail piece, 9, 503
tail rope, 212, 503
tail shoe, 625
tail spar, 154, 503
tail track, 275
tail tree, 154, 503
tailer, 244, 443
tailer-in, 184, 492
tailing, 41, 365, 493, 503
tailing pulp, 41, 503
tailwater, 503
taint, 150, 159
takakusa, 161
take-home pay, 435
takir, 183
takyr, 183
talc(k)ing, 205
talcum, 205
talisai oil, 283
tall field layer, 161
tall-grass prairie, 161
tall herbaceon layer, 161
tall herbaceous layer, 161
tall oil, 333, 494
tall oil derivative, 148
tall oil pitch, 148
tall oil rosin, 333
tall oil soap, 465
tall tree, 377
tall tree species, 377
tall uncut, 591
tallboy, 162
tallol, 333, 494
tallow, 347, 613
tally, 320, 334, 632
tally board, 25, 232
tally counter, 232
tally man, 233
tally sheet, 104, 232
Tallysurf measurement, 471
talon and ball foot, 604
talon moulding, 604
talus, 307, 363, 468
talus cone, 554
talus deposite, 554
talus slope, 552, 554
talus soil, 554
talus vegetation, 307
tamala oil, 642
tamping machine, 75, 604
tamping roller, 190, 608
tamponing, 36, 269
tan, 399, 648
tan-bark oak, 322
tan mill, 399
tan oak, 322
tan stain, 84
tanacetum oil, 2
tanbark, 449
tanbark coppice, 31
tandem drum, 453, 454
tandem duplication, 66
tandem mass spectrometer, 66
tandem truck, 493, 595
tandem twin bandmill, 453
tandem twin bandmill with remote controlled carriage, 80
tandem yarding, 141
tangent stick, 378
tangential arrangement, 524
tangential cut, 524
tangential direction, 524
tangential division, 524
tangential face, 524
tangential loading, 524
tangential longitudinal section, 524
tangential parenchyma, 524
tangential pitting, 524
tangential plane, 377, 524
tangential plank, 524
tangential porous wood, 524
tangential section, 524
tangential shrinkage, 524
tangential stress, 524
tangential surface, 524
tangential swelling, 524
tangential wall, 378, 524
tangerine oil, 197
tangeritin, 197
tangible assets, 435, 582
tangible value, 582
tangle wire rope, 160, 161
tank car, 578
tank scale, 578
tannage, 399
tannigen, 84, 123
tannin, 84, 399
tannin cell, 84, 399
tannin extract, 271, 399
tannin resin, 84
tanning agent, 399
tanning bark, 449
tanning extract, 271, 399
tanning material, 399

T

tanning materials extractor, 399
tanning matter, 84
tanning of leather, 399
tannometer, 5, 399
tanstuff, 399
tap, 135, 309
tap for resin, 465, 531
tap measure, 264
tap water, 645
tape, 240, 264, 310, 358
tape adhesive, 240
tape operating system, 69
tapeless splicer, 508
taper, 23, 231, 632, 640
taper allowance, 547
taper coefficient, 231, 232
taper curve, 547
taper equation, 547
taper formula, 547
taper function, 547
taper-ground saw, 454
taper leg, 593
taper plate, 536
taper rate, 232
taper sawing, 232
taper set, 480
taper table, 547
taper value, 231
tapered facing, 639
tapered log, 232
tapering, 231
tapering grade, 231
tapering lumber, 232
taperingness, 231
taperness, 231
taphocoenosis, 317
taphonomy, 317
taping machine, 240
tapioca flour, 334, 446
tapped-out timber, 450
tapping, 136, 165
tapping face, 165
tapping intensity, 39
tapping system, 165
taproot, 614, 633
taproot system, 614, 633
tar, 242
tar acid, 242
tar-base, 242
tar oil, 242
tar oil pitch, 242
tar-oil type preservative, 242
tar-works, 242
tara extract, 71, 470
tara tannin, 71
target, 5, 47, 336
target area, 336
target canker, 5
target date, 586
target diameter, 336
target forest, 336
target growing stock, 336
target-rod, 24
target shooting, 75, 336
targets and time table, 336
tariff, 36, 178, 441
Tarif(f) table, 470
tarp-layer, 364
tarp-laying device, 364
tarry residue, 242
tarsia, 359, 529
tarsus (*plur.* tarsi), 147
tassel, 204, 307, 542, 572
tattle tale, 371
taungya, 300
taungya method, 214, 300
taungya plantation, 214, 347
taungya system, 214, 216, 300
tax, 460
tax abatement, 233
tax administration, 460
tax appraisal, 460
tax assessment, 363
tax assessor, 173, 363
tax avoidance, 192
tax base, 225
tax basis, 608
tax burden, 460
tax credit, 94, 274
tax deduction, 274
tax equity, 274
tax evasion, 309, 474, 485
tax exclusion, 32
tax exemption, 323
tax holidays, 154, 323
tax in kind, 436
tax-interaction effect, 460
tax-loss carryback, 154
tax-loss carryforward, 154
tax on income, 469
tax penalties, 460
tax rate, 460
taxa, 141
taxable base, 225, 274
taxable year, 274
taxation, 608
taxation of cutover land, 37
taxis, 350, 383
taxite, 7
taxocline, 596
taxodine, 314
taxodioid pit, 414
taxodioid pit pair, 414
taxon (*plur.* taxa), 141
taxonomy, 141
tea-booth, 47
tea garden, 47
tea grove, 46
tea-kiosk, 46
tea oil, 47
tea-oil tree, 578
tea plantation, 46, 47
tea room, 46
tea-stall, 46
tea table, 47
TEAE-cellulose(triethyl aminoethyl cellulose), 403
teagle, 205, 242
teak forest, 583
team, 7
team snatch block, 79
tear fungus, 160, 285
tear speed, 462
tear strength, 462
tearing, 462
tearing resistance, 270
tearing strength, 462
tearing tester, 462
tearing tool, 30, 462
teat pipette, 531
technical cutting age, 170
technical economic index, 226
technical final age, 170
technical maturity, 170
technical measures of soil and water conservation on slope, 413
technical productive financial plan, 422
technical rotation, 170
technical schedule, 170, 226
technical specification, 226
technical supervision of management, 253
technique for photo-interpretation, 530

T

technique of garden-making, 601
technique of time-lapse remote sensing, 435
technological aspect, 170
technological potential, 226
technological process, 170
technological property, 170
technological tree, 170, 475
technology, 226
technology or performance standard, 226
technology transfer, 226
tectonic basin, 173
tectonic lake, 173
tedder, 471
tee, 105, 403, 540
teeth disk mill, 58, 351
teeth fungus, 58
teeth high, 57
teeth point, 57
tegmen (*plur.* tegmina), 155, 556, 617
tego (glue) film, 241
tegument, 27, 356, 495
tele-relascope, 500
telegoniometer, 102
telegony, 159, 522
telegraph pole, 101
telegraph pole cross arm, 101
telegraph-pole wood, 101
telegraph post, 101
telemeter, 44, 558
telemetry, 558
teleology, 336
telephone cabinet, 101
telephone pole, 101
teleprocessing, 594
telescope, 500
telescope level, 460, 500
telescopic boom, 419
telescopic pole, 419
telescopic spindle, 453, 474
telescoping jib, 419
telescoping spar, 419
teletype, 100
teleutosorus, 108
teleutospore, 108
television camera, 101
television detection, 101
telial stage, 108
telial state, 108
teliospore, 108
telium, 108
teller, 62, 233, 547
telocentric, 260
telome, 106
telotaxis, 383
temperate deciduous forest, 505
temperate forest, 505
temperate forest region, 505
temperate forest zone, 505
temperate hardwood, 505
temperate rain forest, 505
temperate warm region, 348
temperate warm zone, 348
temperate zone, 505
temperature anomaly, 505
temperature coefficient, 505
temperature conductivity, 89, 392
temperature controller, 480, 505
temperature cycle, 505
temperature dispersion, 505
temperature effect, 505
temperature-efficiency index, 505
temperature gradient, 505
temperature inversion, 343, 505
temperature limit, 505
temperature persistence, 15
temperature profile, 505
temperature programmer, 56
temperature regime, 505
temperature regulator, 505
temperature sensitive adhesive, 505
temperature sum, 219
tempered hardboard, 391
tempering, 211, 391, 395
tempering chamber, 391
tempering tower, 209
template, 55, 295, 326, 327, 333, 557
template paper, 327
temple forest, 464
temple garden, 420, 464
temporal isolation, 435
temporal order, 441
temporal scale, 435
temporary bridge, 24, 304
temporary casehardening, 599
temporary dormancy, 599
temporary mixture, 304
temporary nursery, 304
temporary parasitism, 304
temporary planting, 230
temporary quadrat, 304
temporary sample plot, 304
temporary sandhill, 564
temporary set, 472, 599
temporary wilting, 599
tenacious consistency, 345
tenacity, 270, 345, 396
tenar, 573
tender annuals, 32
tender price, 485
tenderizer, 341, 610
tending, 149
tending (after young plantation), 582
tending cost, 149
tending cutting (felling), 149
tending objective, 15, 149
tending of crop, 406
tending of cultures, 582
tending of thickets, 322, 582
tending of woods, 301, 406
tending of young growth, 582
tending operation, 149
tending period, 149
tending plan, 149
tending thinning, 149
tendril, 264
tendril climber, 264
tendrillous, 260
tenon, 486, 542
tenon and mortise joint, 70

tenon and slot mortise, 469
tenon dowel joint, 469
tenon saw, 269
tenoner, 269
tenoning, 269
tensile elasticity, 270
tensile force, 282
tensile strain, 282
tensile strength, 270
tensile strength parallel to grain, 461
tensile strength perpendicular to grain, 196
tensile stress, 282
tensile stretch, 419
tensile test, 282
tensile testing machine, 282
tensiometer, 282, 604
tension, 282, 419, 604
tension break, 282, 462
tension crack, 572, 604
tension device, 604
tension failure, 282
tension index, 282
tension member, 282, 443
tension parallel to grain, 461
tension perpendicular to the grain, 196
tension roll, 603
tension screw, 603
tension set, 282
tension shear test, 270
tension station, 470
tension strain, 282
tension test perpendicular to grain, 196
tension wood, 443
tension wood fibre, 572
tensioning, 604
tentative standard, 438
tentorium, 175, 336
tenure, 57
tepal, 18, 202
tephrochronology, 216
teppee (tipi) burner, 139
tera-, 499
teragram (Tg), 499
teratology, 221
teredo resistance, 336
term mortgage, 106
termi-repellent, 134, 338
terminal, 90, 106, 595
terminal anther, 106
terminal boom, 90, 633
terminal bud, 106
terminal cutting, 106
terminal deficiency, 111
terminal group, 111
terminal layer, 312
terminal moraine, 626
terminal oxidase, 327
terminal oxidation, 327
terminal parenchyma, 105, 312
terminal ray cell, 22
terminal yard, 654
terminating annuities, 581, 626
terminator, 626
termitarium, 6
termitary, 6
termite, 6
termite control, 6
termite damage, 6
termite field trail, 441
termite hill, 6
termite-proof, 134
termite-proof course, 131, 134
termite-proof particleboard, 131
termite resistance, 270
termite runway, 6, 98
termite savanna, 6
termite shield, 131, 134
Termiteol, 465
termitophilous, 516
tern, 556
ternary, 403
ternate compound leaf, 402
ternate pinnate, 402
ternately compound, 402
ternately trifoliolate, 402
terra rossa, 198
terrace, 243, 311, 362, 471, 475
terrace garden, 471
terrace ridge afforestation, 475
terraced building, 471
terraced field, 475
terracing, 243, 458
terrain, 95, 97, 98
terrain block, 89, 97, 99
terrain relief, 97
terras soil, 345
terrestrial, 97, 310
terrestrial animal, 310
terrestrial ecology, 310
terrestrial ecosystem, 310
terrestrial orchid, 97
terrestrial plant, 310
terrestrial whirl wind, 310
terricolous, 310
terrigenous sediment, 311
terriherbosa, 310
territorial behavior, 305
territorial circle, 432, 438
territorial division, 97, 431
territoriality, 305
territory, 130, 305, 603
tertiary (cell) wall, 403
tertiary consumer, 402
tertiary creep, 403
tertiary flora, 99
tertiary lamella, 403
Tertiary period, 99
tertiary plant, 99
tertiary split, 402
tertiary structure, 402
tertiary vein, 402
χ^2 test, 267
test bay, 439
test cross, 44, 439
test fire, 438
test of independence, 111
test of significance, 525
test plantation, 439
test plot, 25, 233
test rotation, 439
test tree, 25
testa (*plur.* tastae), 626
tested line, 18
tester, 44
tester strain, 44
tester strain line, 44
testicular envelop, 164
testicular feminization, 164
testicular follicle, 164, 423, 534
testing machine, 438
testing machine head, 439
test(ing) piece, 438
test(ing) procedure, 438
testis, 164, 254
testis spermatogonium, 164, 254
tether line, 104, 227
Tethys Sea, 475
tetrabasic acid, 463

T

tetrachlorvinphos, 411
tetrad, 462, 463
tetradymous, 12, 261
tetradynamous, 463
tetradynamous stamen, 463
tetrandrous, 261
tetraploid, 462
tetraplont, 462
tetraprotic acid, 463
tetrarch, 463
tetrasaccharide, 463
Tetraset process, 100
tetrasomic, 463
tetraspore, 462
tetraterpenes, 463
tetravalent, 463
tetrazolium staining, 463
texture, 218, 246, 506, 620
thalamifloral, 260, 492
thalamiflorous, 260, 492
thalamus, 636
thalassoplankton, 189
thalliform, 561
thallophyta, 267
thallophyte, 267, 561
thallotherophytes, 561
thallus, 267, 561, 592
thanatocoenosis, 431
thanatosis, 230
thaw, 247, 398
theca (*plur.* thecae), 204, 272, 377, 461, 559
thecaphore, 538, 560
theft of forest-produce, 298
thelephoraceous fungus, 165
thelykaryon, 69
thelytoky, 47
thematic work program, 636
theobromine, 273
theodolite, 253
theodolite surveying, 253
theodolite with central telescope, 623
theodolite with compass, 312
theodolite with excentric telescope, 43
theodolite with linear telescope, 615
theodolite with prism, 285
theoretical clearfell area, 127, 288
theory of alternation of species, 450
theory of cell-wall thickening, 517
theory of evolution, 249
theory of financial rotation, 252
theory of forest rent(al), 406
theory of highest income, 653
theory of hysteresis, 621
theory of island biogeography, 90
theory of maximum gross yield, 653
theory of natural selection, 646
theory of origin of species, 513
theory of preference, 547
theory of small sampling, 535
theory of soil rent, 488
theory of the highest netrevenue, 653
theory of the optimal yield, 654
theory of tolerance, 338
thermal activating point, 221
thermal adaptation, 505
thermal band, 391
thermal capacity, 393
thermal coefficient of expansion, 393
thermal conductance, 66, 391
thermal conductivity, 89, 392
thermal conductivity (cell) detector, 392
thermal content, 392
thermal cracking gas chromatography, 392
thermal death point, 621
thermal death time, 394
thermal decomposition, 392
thermal degradation, 392
thermal diffusivity, 393
thermal-efficiency index, 394
thermal energy, 393
thermal equilibrium, 393
thermal erosion, 393
thermal expansion, 393
thermal expansion coefficient, 393
thermal gravity-gas chromatograph-mass spectroscopy, 394
thermal imaging system for fire detecting, 473
thermal inertia, 392
thermal infrared, 392
thermal instability, 391
thermal insulating material, 265
thermal insulating property, 265
thermal insulating value, 265
thermal insulation, 15, 393
thermal insulation board, 265
thermal inversion, 343
thermal losses, 191, 393
thermal-mechanical pulping, 393
thermal pollution, 394
thermal radiation, 392
thermal resistivity, 394
thermal scanner, 393
thermal softening of wood, 330
thermal strain, 394, 443
thermal stress, 394
thermal treatment, 391
thermal unit, 393
thermal value, 124, 394
thermistor, 393
thermo-negative reaction, 514
thermo-positive reaction, 135
thermo-variable resistor, 393
thermobulb, 159, 505
thermocline, 506
thermocouple, 392, 505
thermocouple thermograph, 392
thermocouple thermometer, 392
thermodiffusion, 393
thermodormancy, 394
Thermodyn process, 402
thermodynamic efficiency, 393
thermodynamics, 393
thermoelectric couple, 392, 505

thermoelectric pyrometer, 392
thermogenic soil, 391
thermogram, 393, 505, 645
thermograph, 505, 645
thermogravimetric analysis, 393
thermohaline circulation, 505
thermoinsulation effect, 166
thermokarst, 28
thermolysis, 392, 404
thermomechanical pulp, 393, 587
thermometabolism, 391
thermometer, 505
thermometer-screen, 7
thermometer shelter, 7
thermometric conductivity, 505
thermometrograph, 505
thermonasty, 159, 505
thermoperiod, 506
thermoperiodicity, 394, 506
thermoperiodism, 506
thermophase, 68, 159
thermophile, 441
thermophilic, 441, 516
thermophilic organism, 441
thermophilous, 440, 516
thermophilous organism, 440, 516
thermophilous tree species, 516
thermophyte, 337, 516
thermoplast, 393
thermoplastic adhesive, 393
thermoplastic fibre, 393
thermoplastic foil, 393
thermoplastic plastic, 393
thermoplastic resin, 393
thermoplastic resin adhesive, 393
thermoplasticity, 393
thermoregulator, 393, 505
thermoresistance adhesive, 337
thermosetting, 392, 394
thermosetting adhesive, 392
thermosetting material, 394
thermosetting plastic, 392
thermosetting plasticizer, 392
thermosetting polymer, 392
thermosetting resin, 392
thermosetting resin adhesive, 392
thermosetting synthetic resin, 392
thermosetting varnish, 392
thermosol, 393
thermostability, 394
thermostage, 68, 159
thermostat, 195
thermostatic trap, 480
thermostatic type steam trap, 393
thermotaxis, 383
thermotropism, 530
thermoviscosimeter, 393
thermoviscosity, 393
therophyllous, 341, 522
therophyte, 563
therophyte (Th), 563
thertnoperiodism, 506
thick board, 200
thick forest, 322
thick shellbark hickory, 73
thick wall, 200
thick-walled crystal cell, 200
thick-walled fibre, 200
thickener, 53, 348, 601
thickener vat, 348
thickening, 348, 601
thickening agent, 601
thickening chest, 348
thickening material, 601
thickening power, 601
thickening proceed, 228
thicket, 179, 322, 582, 597
thicket stage, 180, 533
thickner, 348, 494
thickness gauge, 200
thickness of brush, 179
thickness of end wall, 111
thickness of stand, 301
thickness selector, 200
thickness swelling, 200
thicknesser, 44, 204, 550
thicknessing machine, 204, 550
thief stick, 232, 233
thief-type trier, 640
thigmotactic, 383
thigmotaxis, 383
thigmotropic reaction, 383
thigmotropism, 530
thill, 52
thimble, 219, 375, 474
thimble block, 375, 474
thimet, 403
thin baord, 13
thin-kerf blade, 13
thin-layer chromatography, 13
thin planting, 445, 515
thin sawdust board, 13
thin sowing, 444, 515
thin stocked land, 445
thin wall cell, 13
thin-walled fibre, 13
thin wood particleboard, 13
thinic, 411
thining yield, 235, 445
thinium, 411
thinner, 398, 515
thinner blade, 445
thinning, 235, 444, 515
thinning cycle, 445
thinning for tending, 149
thinning forecast, 445
thinning frequency, 445
thinning from above, 415
thinning from below, 520
thinning fruit, 445
thinning grade, 444
thinning in dominant, 415, 577
thinning in lower storey, 520
thinning in upper storey, 415
thinning intensity, 445
thinning interval, 149, 445
thinning law of $-\frac{2}{3}$, 646
thinning method, 235, 444, 445
thinning out, 235, 547
thinning period, 445
thinning periodicity, 445
thinning phase, 515
thinning plan, 444
thinning processor, 444
thinning rate, 444, 445

T

thinning regime, 149, 444, 445
thinning schedule, 444, 445
thinning series, 445
thinning system, 444, 445
thinning table, 444
thinning-to-waste, 138
thinning treatment, 444
thinning type, 445
thinning weight, 445
thinning winch, 445
thiolignin, 308
thionazin, 621
thioredoxin, 308
thiourea, 308
third-generation, 99
thiuram, 149
thixotaxis, 383
thixotropic agent, 65
thixotropic gel, 558
thixotropic preservative, 65
thixotropic system, 65
thixotropism, 65, 530
thixotropy, 65
thorax, 542
thorn-bush, 580
thorn forest, 391, 392
Thorne debarker, 81, 470
thousand-foot board measure, 371
thousand grain weight, 371
thrasher, 75, 376, 464, 494
thrashing, 75, 493
thread fungus, 462
thread rot, 527
threat behaviour, 500
threatened species, 27
threat(en)ing colouration, 500
three-bend leg, 403
three-cut shelterwood method, 402
three-dimensional appearance, 403
three dimensional-look, 403
three-drum skyline yarding system, 403
three-entry, 403
three faced sawing, 403
three heads wide belt sander, 403
three kneed carriage, 403
three layer, 402
three-layer particleboard, 402
three lick snipe, 403
three-log boom, 403
Three-North protection forest, 402
three-P sampling, 61
three phase distribution, 403
three planes of section, 403
three roll spreader head, 402
three sections of wood, 330
three side planing and moulding machine, 403, 643
three stage pressing cycle, 402
three-stage sampling, 403
three-way cross, 402, 403
three-way drum dryer, 402
three-way volume table, 403
thremmatology, 110
threonine, 465
threose, 466
threshold, 103, 320, 587
threshold moisture level, 189
threshold of hearing, 481
threshold return, 15, 367
threshold value, 304, 419, 569, 587
thrift, 128
thriftiness, 128
thrifty mature, 64, 279, 599
thrips, 227
thriving tree, 500, 577
throat, 57, 58, 62, 173, 205, 521
throat depth, 254
throat line, 58
throttle, 145, 245
throttle control trigger, 578
throttle latch, 578
throttle lock, 578
through-and-through sawing, 318, 590
through-bridge, 65, 520
through check, 179, 485
through dovetail, 179
through lump, 484
through plate girde bridge, 520
through shake, 179
through tenon, 179, 486
through truss, 65, 520
throughfall, 65, 257
throughput capacity, 422, 513
throughput speed, 482
throw, 353
throw line, 493
throw sifting, 351
throwing, 125, 184, 211, 237
throwing belt, 351
throwing roll, 351
throwing roll spreader head, 351
thrown chair, 52
thrust bearing, 618
thuja oil, 528
thujoid, 42
thumb plane, 619
thunbergene, 195
thunder shower, 285
thundercloud, 285
thunderstorm, 284
thylakoid, 285
thylosoid, 342, 462
thyme oil, 7
thyrse, 264
thyrsiferous, 264
thysanuriform, 29
tibia, 257
tibial spur, 257
tick, 42, 357
tick-borne encephalitis, 408
tidal flat, 52
tidal forest, 52
tidal marsh, 52
tidal prevention forest, 131
tidal swamp, 52
tidal vegetation, 52
tidal wave, 52, 189
tidal zone, 52
tide, 52
tide gauge, 232
tide water control forest, 131
tide-water-prevention forest, 131
tidebreak, 131
tie, 20, 227, 281, 282, 292, 608

tie beam, 227, 458
tie bolt, 90, 227
tie chopper, 620
tie-down line, 104, 176
tie hack, 620
tie in sale, 75
tie line, 104, 227
tie piler, 608
tie-piling machine, 608
tie plate, 608
tie plug, 608
tie point, 291
tie rod, 292, 517, 637
tie strap, 292
tie tram, 608
tie-up line, 176
tie water, 246
tieback angle, 199
tieback line, 199
tieback stump, 199
tied loan, 526, 636
tiedown rope, 104, 176, 227
tier, 45, 114, 140, 165, 190, 281, 349, 351
tier like arrangement, 45, 104
tiger grain, 201
tiger moth, 90, 201
tiger table, 201
tiger weasel, 201
tight cooperage stock, 484
tight-cut veneer, 550
tight knot, 214, 249
tight line, 249, 282
tight-line loading, 249
tight side (of veneer), 82
tight (sky) line yarding, 249
tight stacking, 322
tight veneer, 546
tightener, 603, 604
tightening block, 153, 603
tight(ening) skyline, 176, 249
tightlining, 249
tile cell, 495
tile of trend, 335
tiliacine, 113
till, 28
tillable acres, 272
tiller, 142, 168, 546
tiller cutting, 64
tillering, 140
tilling depth, 168
tilling width, 168
tilt, 380
tilt displacement, 381
tilt down conveyor, 381
tilted bandmill, 381
tilted circular saw, 381
tilter, 128, 381, 537
tilth, 168
tilting arbor variety saw, 381
tilting cylinder, 101
tilting seat, 605
timber, 66, 289, 301, 328, 333, 414
timber acquisition, 301
timber and lumber standard, 328
timber and lumber standardization, 328
timber and lumber standards, 328
timber appraisal, 301
timber assortment, 37, 329
timber assortment table, 62
timber basin, 635
timber beetle, 636
timber bind, 229
timber bob, 172
timber bonds, 331
timber borer, 331
timber bow, 455
timber carriage, 595
timber carrier, 278, 595
timber chute, 332
timber claim, 330
timber collecting, 224
timber compass, 126
timber concession, 39, 127
timber connection, 333
timber connector, 330, 333
timber construction, 333
timber consumption, 331
timber crew, 127
timber crop, 331
timber cruise, 289, 301
timber cruiser, 302
timber cruising cost, 302
timber cutter, 126
timber dam, 328, 335
timber decay, 329
timber deck, 286, 638
timber depot, 635
timber determination, 37, 545
timber direct photomeasurement, 302
timber dock, 329, 331
timber drying, 329
timber elevator, 330, 331
timber exchange, 330
timber export, 329
timber export trade, 329
timber famine, 330
timber fastener, 329
timber felling, 125
timber flooring, 95
timber for agricultural implement, 347
timber for bridge, 377
timber for vehicle, 52
timber forest, 377, 577
timber form factor, 37, 302
timber-framed building, 333
timber-framing machinery, 333
timber grade, 329
timber grader, 330, 547
timber growing stock, 302
timber harvest, 126, 127, 328, 331
timber harvest landing, 127
timber harvesting, 37, 331
timber haulage, 331
timber hauling, 595
timber height, 577
timber histology, 331
timber hitch, 281, 331
timber import quota, 330
timber import trade, 330
timber industry, 329, 406
timber inspection, 330
timber inspection bureau, 330
timber inspector, 77, 126, 330
timber inventory, 302
timber investment, 302
timber jack, 126, 590
timber key area, 566, 577
timber-land, 298, 577
timber-land lease, 299
timber lease, 301, 302
timber length, 447

timber licence, 301
timber limit, 301
timber line, 302
timber loans, 331
timber lorry, 595
timber lot, 329, 332
timber magazine, 330, 635
timber making, 25, 38
timber market, 330
timber marking, 301, 329
timber mechanics, 330
timber mill, 619, 630
timber pest, 329
timber physics, 331
timber pile, 336
timber plugger, 33
timber-pond, 635
timber port, 329, 330
timber price, 329
timber production, 330
timber purchaser, 330
timber rake, 135, 492
timber reserve, 146, 251, 252
timber reserves, 586
timber resources, 331
timber right, 38, 330
timber sale, 301
timber sale bidding record, 301
timber sale program, 331
timber sawmill, 619
timber scaling, 330
timber science, 330, 331
timber seasoning, 329
timber selling, 331
timber shave, 452
timber shaver, 31
timber shipment, 331
timber slide, 332
timber specification, 329, 330
timber stain, 328
timber stand, 577, 581
timber standard, 328
timber storage, 634, 635
timber storage pond, 635
timber supply, 329
timber sword, 99
timber tally, 620
timber tipper, 492, 562
timber-tract, 127, 302
timber tractor, 223
timber trade, 330
timber transaction, 330
timber transport, 331
timber treating plant, 329
timber tree, 54, 577
timber trend study, 330
timber trespass, 329, 485
timber truss, 332
timber type mapping, 302, 408
timber unloading, 331
timber used in children's toy, 499
timber volume, 36
timber wagon, 595
timber wood, 577
timber worm, 329, 636
timber yard, 8, 285, 635
timber yard machine, 635
timber yield, 301, 329
timbershed, 8, 54, 329, 330
timberwork, 332, 333
time barrier, 435
time cycle controller, 435
time device, 225
time discriminate analysis, 435
time-lag, 435, 554
time of migration (bird), 200
time officer, 226
time ordering, 435
time retarding signal, 554
time scale, 435
time schedule, 435
time series, 435
time series analysis, 435
time-specific life table, 474
time study, 169
time wage, 225
time wage rate, 225
timer, 107
timescale, 435
timing conveyor, 107
tin, 12, 111, 444, 515
tin nailer, 444
tinct, 390, 405, 640
tincture, 105, 251, 390, 405, 640
tinder, 89
tine, 45, 58, 64, 88, 465, 622
tine plough, 58
tine rake, 58
tine(d) harrow, 105
tint, 87, 375, 380, 405, 501
tintage, 390, 640
tinter, 391, 640
tinting, 390, 640
tinting strength, 405, 640
tintometer, 405
tip, 357, 417, 449
tip blight, 105
tip burn, 105, 231
tip cutting, 105, 417
tip grain, 461
tip growth, 105
tip layering, 106
tip-top table, 606
tip-up seat, 128
tipped, 529
tipped-tooth, 529
tipper, 381
tipping cart, 443
tipping edge, 306, 605
tipple, 643, 644
tire, 277, 312
tire-tube method, 111, 474
tire tube process, 111, 474
tirnao, 499
TIROS-N, 471
tissue, 14, 246, 613, 652
tissue culture, 652
tit, 413
titanium white, 471
title deed, 97, 371, 469
titration, 94
to-and-fro cableway, 500
to-and-fro system, 500
toad back moulding, 47
toadstool, 111, 556
toat, 471
toe ring, 417
toggle, 474, 631
toggle chain, 168, 281
toggle hook, 281, 474
toggler, 281
toilet chair, 444
toilet table, 444
Tokyo cherry, 397
tolerable-windows approach, 272
tolerance, 170, 337, 397
tolerance degree, 337
tolerance of shade, 338
tolerant fungi, 338
tolerant mesophytic tree, 338
tolerant species, 338

tolerant-tree, 338
tolerant understorey, 338
toleration ecology, 338
toleration in sawing, 620
tollers-dog, 571
tolu balsam, 491
tombstone, 75, 306
tombstone stump, 306
tomcat, 172
tomentellate, 18
tomentose, 18
tomentum, 323, 397
tompound umbel, 153
tone, 405
toner, 480, 602
tong choker, 178
tong hooker, 638
tong puller, 97
tong unhooker, 603
tonger, 178, 281
tongline, 178, 223, 368, 638
tongs, 229, 267
tongue, 89, 417, 469, 542, 561
tongue and groove joint, 469
tongue grafting, 417
tongue mitre, 537
tongue(d) and groove(d) joint, 366, 469
tongued approach grafting, 417
tongued-lap joint, 417
toning, 480
tonn, 115
toona Toona, 528
toonalol, 528
toonin, 528
tooth, 262
tooth angle, 58, 536
tooth harrow, 105
tooth pattern, 58
tooth pitch, 58
tooth scarifier, 58
tooth set, 29, 610
tooth setting, 262
tooth spacing, 58
tooth stop, 58
toothed disk-mill, 58
toothed gearing, 58
tooth(ed) harrow, 105
toothed plane, 58
toothed rail, 57
toothed roll, 57
toothed thickening, 58
toothlike projection, 58
too(u)lder, 55, 526, 541
top, 57, 105, 323, 417, 449
top-and-bottom saw, 416
top bevel angle, 57
top bind, 415
top chain, 105, 146
top chord, 416
top clamp, 52, 121
top coat, 27, 323
top-cross, 105
top-cross parent, 105
top-cross test, 105
top cut, 105
top dam, 105
top diameter, 105, 417
top discharge chipper, 416
top-down model, 646
top dressing, 27, 415, 639
top dry, 105, 278, 417
top drying, 278
top end, 417, 535
top (end) diameter, 327
top feed chipper, 415
top felt, 415
top fruit sprayer, 186
top grafting, 105, 162
top grass, 161
top guy, 156
top height, 449, 577
top layer chip, 26
top lip (of slice), 415
top load, 116, 652
top log, 42, 373, 417
top log chute, 205
top-lopping, 247, 385
top measurement, 105
top overhaul, 78
top-plate cutting angle, 105
top-plate (filing) angle, 105
top press, 416
top press roll, 416
top rail, 195
top removal, 75, 603
top rib, 105
top root, 633
top-root ratio, 178
top rot, 105, 417
top shoot, 105
top side wide belt sander, 416
top sizing, 27, 416
top soil, 27, 168
top stem, 417
top strut, 105, 106
top surface, 611
top wood, 417
topiary, 315, 448
topiary art, 448
topiary garden, 315, 601
topkill, 278, 447
topochemical reaction, 248, 259
topocline, 96, 99
topographic barrier, 99
topographic climax, 99
topographic feature, 99
topographic treatment of garden, 588
topographic unit, 98
topographic variation, 98
topographical condition, 99
topographical design, 99
topographical displacement, 99
topographic(al) divide, 143
topographical map, 99
topographical test, 107
topography, 97–99
topography refotnl, 99
topological dimension (vertical heterogeneity), 289
topology, 494
topophototaxia, 383
topophysis, 504
toposequence, 99
topotaxis, 383
topper, 378
topping, 57, 75, 247
tops, 613
topsin, 492
topsin-M, 230
topsoil, 27
topwood, 417
topwork(ing), 105
torchere, 91, 162
torn grain, 462
tornado, 309
torpor, 606
torque backup roll, 638
torque moment, 347, 637
torque wrench, 347, 637
torrent, 199, 222
torrent control, 209
torrent control works, 199
torrent erosion, 209
torrential flood drainage works, 413

T

torrential flood erosion, 413
torrent(ial) river, 222
torrential valley, 515
torreyene, 139
torrid zone, 391
torsion balance, 347
torsion constant, 347
torsion-fibre, 347
torsion micrometer, 501
torsion modulus, 347
torsion property, 270
torsion-shear strength, 270
torsion shear test, 270
torsion viscosimeter, 347
torsional bending, 347
torsional force, 347
torsional limit, 347
torsional load, 347
torsional moment, 347
torsional resonant vibration, 347
torsional rigidity, 270
torsional stiffness, 270
torsional strain, 347
torsional strength, 270
torsional strength tester, 347, 438
torsional stress, 347
torsional vibration, 347
torsionmeter, 347
tortoise beetle, 183
tortricid, 264
tortuosity, 384, 497
torus, 506
torus (*plur.* tori), 21, 204, 506
torymid, 627
total absorption, 649
total alkali, 649
total (amount) of effective time, 581
total annual growth, 649
total basal-area, 648
total content, 649
total cost, 648
total efficiency, 387, 649
total extractive, 648
total germinability, 648
total germination, 648
total germination percent, 648
total growth, 649
total heat, 189, 392, 649
total height, 385
total increment, 649
total inhibition point, 498
total load, 648
total loss, 386
total mean increment, 649
total neutrality hypothesis, 625
total organic carbon, 649
total oxygen demand, 649
total pressure, 649
total profit, 649
total revenue, 649
total shrinkage, 648, 649
total social labour, 418
total stand projection, 386
total stand table, 649
total station, 387
total stress, 649
total tree height, 386
total tree volume, 448
total value increment percent, 649
total volume, 649
total work, 649
total yield, 649
totara tree, 474
totarene, 401, 474
totarolone, 474
tote road, 596
toting road, 596
totipalmate foot, 386
touch dry, 244
touch hole, 244
touch-me-not, 147, 222
touch sanding, 380
touch-up coating, 33
touchwood, 570
tough glue, 396
tough wood, 232
toughening agent, 602
toughening fragile veneer, 396
toughness, 396
tour paws, 463
tour-piece butt matching, 463
tourism, 178, 314
tourism carrying capacity, 314
tourist environmental capacity, 314
tourist interest, 324
tourist map, 90
tourist trade, 314
tow, 372, 493
tow boat, 493
tow-way, 493
towage, 458, 493
towel rack, 318
tower cable way, 162
tower cupola, 178, 296
tower spar, 470
tower-yarder, 470
towering old tree, 40
towerman, 296, 500
towing bag (boom), 81
towing cable, 493
towing hook, 493
towing winch, 372, 493
town-forest, 74
town scape, 56
township, 608
township (corner), 382
toxaphene, 111
toxic, 110, 580, 622
toxic chemical, 111
toxic compound, 111
toxic dosage, 110
toxic fungus, 110
toxic hazard, 111
toxic limit, 111
toxic sill, 111, 132
toxic threshold, 111
toxic value, 111
toxicant, 110, 111
toxicant retention, 110
toxicity, 110, 111
toxicity determination, 111
toxicity limit, 111
toxicity of preservative, 132
toxicity screening test, 111
toxicity test, 111
toxin, 111
toxolysin, 247, 270
trabant, 468
trabecula (*plur.* -lae), 196, 255, 499
trabecular vessel, 196
trace, 195, 501, 639
trace element, 195, 501
trace fossil, 565
trace gap, 226
trace level, 195
trace routine, 168
traceable cost, 274, 614
tracer, 438
tracer analysis, 438
tracer shot, 438
tracer skyline-crane, 558
trachea, 89, 368
tracheal portion, 179
tracheary element, 179
tracheary tissue, 179

tracheid, 178
tracheid lumina, 178
tracheid wall, 178
tracheidal, 178
tracheidometry, 178
tracheobacteriosis (*plur.* -ses), 89
tracheomycosis, 89
tracheophyta, 89, 502
trachyte, 72
tracing, 324, 325, 575
tracing paper, 324
track, 184, 191, 312, 314
track barrow, 184
track cable, 231
track-laying tractor, 314
track line, 231
track loader, 314
track pad, 314
track plate, 314
track scale, 638
track scraper, 314
track shoe, 314
tracking, 438
traction cable, 372, 634
traction line, 372, 634
traction rope, 372, 634
tractive current, 372
tractor-cable, 493
tractor-donkey, 79
tractor hauling, 493
tractor logging, 493
tractor production data, 493
tractor shovel, 493
tractor skidder, 223
tractor skidding, 493
tractor sledge, 493
trade, 240, 319, 540
trade association, 190, 483
trade effects, 319
trade mark, 414
trade ring, 240
trade wind, 540
tradition furniture, 66
traditional flower, 66
traditional garden style, 587
traditional style, 66
tragacanth (gum), 210
trail, 66, 589
trail car, 177, 493
trail log, 503
trail setting, 66, 223
trail skidding, 54
trailed bush cutter, 177
trailer, 177, 349, 364, 493, 503
trailing, 54, 383
trailing chuting, 54, 205
trailing plant, 364
trailing slide, 54
training collar, 549
training forest, 243, 436, 439, 603
training site, 549
trait, 542
trajectory, 184
tram, 380
trama (*plur.* -ae), 267
tramway, 332, 359, 380
trans-shipment point, 625, 637
transacetylation, 637
transactions, 240, 560
transacylase, 637
transaminase, 637
transamination, 637
transboundary protected areas, 278
transcarboxylase, 637
transcarboxylation, 637
transcriptase, 637
transcription, 637
transducer, 66, 209, 637
transduction, 637
transect, 196, 520, 558
transect for area determination, 382
transect plot, 558
transection, 196
transfection, 637
transfer, 66, 391, 564, 637
transfer block, 637
transfer bridge, 111, 239
transfer coefficient, 66
transfer gantry, 309
transfer logs over dam, 329
transfer loop, 245, 565
transfer method, 565
transfer needle, 245, 565
transfer of detail, 518
transfer paper, 637
transfer unit, 66, 372
transferase, 637
transferscope, 637
transformation, 23, 637
transformation of data, 451
transformer, 24, 637
transfusion cell, 66, 445
transfusion strand, 66, 445
transfusion tissue, 66, 445
transgenosis, 221
transglycosylation, 364, 637
transgression, 188, 594
transgression breeding, 51, 594
transgressive inheritance, 594
transgressive segregation, 594
transhumance, 226
transient, 200, 461
transient capillary, 461
transient climate response, 461
transient creep, 461
transient distortion, 461
transient flow, 461
transient microbe, 187, 599
transistor pulse ignition, 100
transit, 253
transit camp, 637
transit device, 187
transit duty, 187, 625, 637
transition, 186, 594, 637
transition-creep, 187
transition curve, 187, 209
transition distance, 209
transition lamella, 186
transition phase, 186
transition polymerization, 632
transition region, 187, 564
transition state, 187, 623
transition tissue, 187
transition zone, 186, 187
transitional moor, 187
transitional moor soil, 187
transitional pore, 187, 623
transition(al) zone, 186
transitory income, 599
translocation, 155, 564, 569, 596, 637
translucent, 10
translucid, 10
transmission, 24, 65, 66, 486
transmission densitometer, 486
transmission electron micrograph, 486

T

transmission electron microscope, 486
transmission gear, 66
transmission of heat, 66, 391
transmission of light, 182
transmission of sound, 66, 430
transmissivity, 486
transmittance, 485, 486
transmittancy, 558
transmitter, 66
transmuted wood, 24
transocular stripe, 65
transoid, 129
transparence positive, 485
transparency, 485
transparency test, 485
transphosphorylase, 637
transpiration, 609
transpiration coefficient, 609
transpiration current, 609
transpiration efficiency, 609
transpiration rate, 609
transpiration ratio, 609
transplant, 564, 565
transplant bed, 565
transplant box, 565, 585
transplant nursery, 565
transplant stock, 565
transplantation, 564
transplanter, 565
transplanting, 564
transplanting bed, 565
transplant(ing) board, 564
transplanting machine, 565
transplanting plough, 565
transplanting seedling, 565
transplanting shock, 209, 565
transport caul, 446
transport chain, 446
transport distance, 595
transport plate, 596
transport roll, 66
transportable (saw) mill, 306
transportation, 7
transportation claim, 596
transported soil, 596
transporting cableway, 595
transversal pad, 197
transverse, 195, 196
transverse boom, 196
transverse cell, 256
transverse construction, 195
transverse conveyor, 197
transverse cut, 196, 600
transverse dehiscence, 196
transverse division, 197
transverse dune, 197
transverse elasticity, 197
transverse element, 256
transverse-flow fan, 196
transverse grade, 196
transverse heating, 68
transverse holding ground, 283
transverse intercellular canal, 256
transverse loading, 197
transverse-longitudinal cutting, 197
transverse parenchymatous element, 256
transverse piling, 238, 239
transverse resin canal, 256
transverse resonant vibration, 197
transverse section, 195, 196
transverse septum, 196
transverse shake, 196
transverse strength, 271
transverse stress, 197
transverse surface, 196
transverse thermal expansion, 197
transverse trench, 196
transverse vibration, 197
transverse wall, 111
transverse wall nodular, 246
transverse warping, 197
trap, 33, 346, 445, 583
trap baiting net, 583
trap-billet, 583
trap-ditch, 33, 583
trap drain, 247
trap-hole, 527, 583
trap line, 583, 638
trap log, 121, 583
trap tree, 121, 583
trap-trench, 33, 583
trapezoidal weir, 476
trapped bark, 229, 449
trapper, 229
trapping, 583, 638
trash gate, 139
trass, 148
trauma, 67, 496
traumatic axial intercellular canal, 67
traumatic duct, 67
traumatic heartwood, 67
traumatic intercellular canal, 67
traumatic parenchyma, 67
traumatic ray tracheid, 67
traumatic resin canal, 67
traumatic resin duct, 16
traumatic ring, 67
traumatic stimulation of flowering, 67
traumatic tylosis, 67
travel agency, 314
travel for pleasure, 314
travel time, 66, 487, 540
travel-time map, 487
traveler bird, 314
traveller, 314
travel(l)er's camp, 314
travel(l)ing belt screen, 79
travel(l)ing block, 564
travel(l)ing cut-off saw, 564
travel(l)ing head veneer jointer, 564
travel(l)ing hoist, 564
travel(l)ing laser beam, 66
travel(l)ing spray booth, 564
traverse, 65, 89, 196
traverse surface, 195
traverser-link, 197, 594
travois, 223
travois road, 223, 349
tray, 351, 374, 492
tray belt moulding system, 492
tray belt press loader, 493
tray top table, 47
trayless-system loader, 510
treatability, 65

T

treated tie, 132
treated timber, 131
treated wood, 131
treating cylinder, 65
treating living tree, 289
treating plant, 131
treating process, 131
treating tank, 65
treatment cycle, 65
treatment to refusal, 14
tree, 333, 377, 448, 450
tree aerial volume table, 448
tree age, 448
tree analysis, 447
tree annual ring, 448
tree architecture, 44, 448
tree bailing machine, 448
tree balling machine, 80
tree-base sprayer, 303
tree bicycle, 416
tree block, 89
tree boom cut radios, 294
tree boom loading, 104
tree botany, 410
tree breeding, 448
tree buttress, 8
tree-by-tree mixture, 87
tree canopy, 172, 447
tree class, 448, 449
tree classification, 448
tree climbing iron, 416
tree climbing ladder, 416
tree climbing spur, 349
tree clipper, 126
tree compass, 615
tree controller, 448
tree cover, 301
tree crasher, 448
tree crown, 447
tree crown class, 447
tree crusher, 448, 596
tree damage caused by forest fauna, 109, 310, 444
tree density, 301
tree digger, 367, 495
tree dog, 371
tree dozer, 492
tree eater, 492
tree extractive, 448
tree extractor, 5
tree farm, 573
tree feller, 126
tree fern, 448
tree forest, 377
tree form, 449
tree form factor, 448
tree form height, 449, 541
tree (gene) bank, 448
tree grades, 289
tree growth projection system, 301
tree-guard sprayer, 448
tree hanger, 623
tree harvester, 38, 126
tree height, 447
tree height measuring pole, 447
tree increment plot, 448
tree information, 302
tree injection, 448
tree iron, 12, 103
tree jack, 56, 126
tree knocker, 186
tree layer, 377, 446
tree length, 591
tree-length bundle, 591
tree-length logging, 591
tree-length volume table, 591
tree lifter, 367, 495
tree limit, 448
tree line, 409, 448
tree list file, 301
tree maintenance, 448
tree marking, 448
tree measuration, 44
tree monkey, 643
tree mover, 495
tree number, 448
tree nursery, 324
tree of economic value, 252
tree of forest border, 303
tree of overstorey, 416
tree of special use, 475
tree of understorey, 521
tree of upper storey, 416
tree pathology, 446
tree percent, 54
tree photo volume table, 448
tree physiology, 448
tree planter, 616
tree planting, 616
tree-planting day, 616
tree-planting drive, 616
tree planting machine, 616
tree plate, 12, 103
tree processor, 126, 600
tree-prop, 612
tree-protecting bird, 202
tree pruner, 164
tree puller, 266
tree pulling, 282
tree-pusher, 492
tree rack, 165
tree repair, 448
tree rig, 223
tree ring analysis, 448
tree ring analyzer, 344
tree ring calendar, 448
tree-ring climatology, 344
tree-ring core, 344
tree ring measurement system, 344
tree ring measuring device, 344
tree ring positiometer, 344
tree ring positioning table, 344
tree sap, 449
tree savanna, 515
tree scale, 289
tree-scriber, 31, 385
tree seed drill, 628
tree seed orchard, 302
tree seed planter, 629
tree seedling, 436
tree shaker, 608
tree shearing device, 234
tree shears, 126
tree shoe, 55
tree spacing, 160, 632
tree spade, 448
tree species, 450
tree species identification, 450
tree stage, 54
tree stem, 446
tree-stem theory, 447
tree stem volume, 447
tree stock, 608
tree stratum, 377
tree-support, 612
tree surgery, 448
tree tip, 446
tree transplanting machine, 448
tree veld, 392
tree vigour, 449
tree volume, 289
tree volume equation, 159, 301
tree volume function, 159
tree volume measure, 448
tree volume table, 289

tree ware, 334, 335
tree weight table, 448
tree well, 208
tree with coloured heartwood, 538
tree with faculatatively coloured heartwood, 348
tree with imperfect heartwood, 446
tree with retarded formation of heart-wood, 21
tree with torse fibre, 523
tree work, 448
tree zone, 377
treefall mound, 90
treefall pit, 90
treeless land, 510
tre(e)nail, 332, 334
treeplanting, 315, 616
treetop, 446, 449
treflan, 474
trellis, 203
trellis diagram, 165
trellis drainage pattern, 165
trelliswork, 165
trema (*plur.* tremata), 276, 321
trench, 132, 172, 366
trench cutting machine, 494
trench layering, 268
trench planting, 173
trench-sowing, 172
trench transplanting, 268
trencher, 268, 494
trenching, 268, 269
trenching plough, 268
trend surface analysis, 383
trestle-board, 76
trestle-boom, 603
trestle cable, 162
trestle for sawing, 263
trestle table, 46
trestlework, 603
tri-band resaw, 403
tri-merous flower, 402
triadelphous, 403
trial and error learning, 438
triander, 403
triangle-planting, 403
triangular division method, 402
triangular planting, 403
triangular profile board, 402
triangulation, 402
triangulation point, 402
triangulation station, 402
triarch, 403
Triassic period, 402
tribe, 651
tricarboxylic acid cycle, 403
trichasium, 403
trichidium, 533, 561
trichiferous, 260
trichlorfon, 94
trichogyne, 443
tricholoma, 277, 593
trickle irrigation, 94
tricostate, 261
tricussate, 239, 403
tridigitate, 403
tridynamous, 403
triethyl aminoethyl cellulose, 469
triethylaminoethyl cellulose, 403
trifluralin, 147
trigallic acid, 402
trigger, 7, 65
trigger guard, 7
trigger pull, 7
trigonometrical hypsometer, 402
trigonometrical leveling, 402
trihybrid, 402
trilacunar, 403
trilateral paper trimming machine, 402
trim, 247, 366, 377, 544
trimethyl glucose, 402
trimethylsilyl ethers, 402
trimethylsilylation, 402
trimmed, 196, 366
trimmed size, 54, 247
trimmed width, 54, 247
trimmer, 117, 196, 247
trimming, 75, 234, 247, 262, 543, 544
trimming charge, 359
trimming joist, 187, 610
trimming rafter, 610
trimming saw, 247
trimming section, 112
trinervate, 261
trinerved, 261
trinervious, 402
trinomial system, 403
trio table, 104
trioecious, 69, 84
triordinal, 403
trip line, 52, 212
trip timber, 268, 277
tripalmate, 402
tripalmately compound, 402
triparted, 402, 403
tripartite, 402, 403
tripennate, 402
tripinnate, 402
tripinnately compound, 402
triple band saw, 403
triple catch method, 402
triple deck chip screen, 402
triple disc refiner, 83
triple drum, 403
triple haul, 403
triple hybrid, 402
triplet, 403
triplet code, 403
triplet state, 403
triploid, 402
triploparasitism, 403
tripod, 403
tripod ladder, 403
tripod table, 403
tripsill, 602
tripton, 138
trisaccharide, 403
trisome, 403
trisomic, 403
trisomic diploidy, 121, 403
triternate, 402
triterpenes, 403
triterpenoid, 403
tritiated compound, 65
tritoparasite, 99
trivalent, 402
trivial name, 179, 466
trochanter, 637
trochometer, 52, 288, 312
trolley, 103, 443, 598
tropept, 392
trophallaxis, 238
trophi, 277
trophic, 277, 384, 437, 573
trophic dynamics, 574
trophic level, 437, 574
trophic pyramid, 574
trophic relationship, 574
trophic structure, 574

trophic symbiosis, 641
trophobiont, 437
trophobiosis, 384
trophotlotic zone, 574
trophonucleus, 574
trophy, 444, 603
tropic movement, 530
Tropic of Cancer, 17
Tropic of Capricorn, 338
tropic rain forest climate, 392
tropical, 391
tropical anticyclone, 391
tropical calm zone, 392
tropical cyclone, 392
tropical deciduous forest, 392
tropical forest, 392
tropical forest belt, 392
tropical forest region, 392
tropical forest zone, 392
tropical hardwoods, 392
tropical pedocal, 391
tropical plant, 392
tropical rain forest, 392
tropical red earth, 391
tropical red loam, 391
tropical red soil, 391
tropical shelterwood system, 392
tropical strand forest, 391
tropical timber, 391
tropical wood, 391
tropical zone, 392
tropism, 383, 530
tropolone, 207
tropopause, 115
tropophyte, 432, 439
tropophytic, 391
troposphere, 115
tropotaxis, 138
trough, 41, 172, 630
trough carousel, 212
trough method, 41
trough roof, 41, 508
trough(ed) belt conveyer, 41
trough(ed) chain conveyer, 41
trowel, 534, 535, 543
trowel adhesive, 163
truck hauling, 267
truck line, 595
truck logging, 267
truck-trailer log hauling, 371
trucking, 267
truckle bed, 243
true azimuth, 607
true bay, 594
true-breeding hybrid, 34
true camphor wood, 604
true density, 608
true dormancy, 608
true form factor, 608, 611
true form quotient, 608, 611
true humus calcareous soil, 100
true laterite, 100
true middle lamella, 13
true parasite, 69
true photosynthetic rate, 608
true prairie, 607
true saprophyte, 68
true shrub, 575
true specific gravity, 264, 607
true value, 608
true volume measure, 435
true-woodfibre, 100
truewood, 538
truffle, 98, 279, 464
truncated cone, 247
truncated normal distribution, 247
truncated paraboloid, 247
truncated plant, 247
truncheon, 46, 184, 284
trunciflory, 159
truncus, 383, 476, 542
trundle bed, 243
trunk, 72, 78, 160, 309, 446, 635
trunk rot, 159, 448
trunnel, 77, 332
truss (frame), 173, 195
truss member, 195
trussed beam, 195
trusser, 75, 550
try-gun, 439
try square, 242, 384, 614
trying plane, 49, 362, 519
tryon chute, 248
tshernosem, 194
tsugaresinol, 481
tub, 333, 484
tub chair, 11
tub front, 622
tub sizing, 41
tub sofa, 10, 485
tubbing, 484, 620
tube-bundle, 179
tube bundle dryer, 179
tube culture, 438
tube mill, 184, 485, 620
tube nucleus, 178
tube-plate, 178
tube sheet, 178
tube sprayer, 179
tuber, 279
tubercle, 167, 246, 486, 534
tubercular, 246, 308
tuberous plant, 382
tubicole, 179
tubicolous animal, 179
tubular board, 276
tubular bridge, 179
tubular budding, 179
tubular furniture, 160
tubular magazine, 65, 485
tubular particle board, 276
tubuli, 179, 267
tucking roll, 128
tuff, 346
tuft, 72, 264
tug, 493, 545
tug boat, 493
tug raft, 493
tugger, 242, 493
tulip garden, 585
tulipiferine, 120
tumbled-in, 366, 543
tumbler, 7, 184, 637
tumbling, 184
tumbling drum, 593
tumor, 626
tundra, 110
tundra climate, 110
tundra meadow, 110
tundra soil, 29, 110
tundra zone, 471
tung oil, 484
tung-tree, 484
tungsten-carbide, 473
tungsten carbide tip saw, 473
tunic, 305, 326, 591
tunicated bulb, 581
tunnel, 235, 469, 636
tunnel clearance, 469
tunnel kiln, 469, 492
tunnel test, 144, 553
turbidite, 640
turbidity, 640
turbidostat, 195
turbo-chipper, 59, 507

T

turbo-dryer, 507
turbo fan, 485, 507
turbocharger, 507
turbomixer, 560
turbosphere, 491
turbulence, 491
turbulent diffusion, 506
turbulent flow, 491, 506
turbulent jet, 491
turf, 41, 42, 73
turf culture, 42
turf planting, 42
turfing, 364
turgescence, 355, 626
turgescency, 355, 626
turgidity, 249
turgor, 249, 355
turgor pressure, 249, 355
turion, 260, 487
Turkey rhubarb, 489
Turkish gallotannin, 489
turmeric, 237
turmerone, 237
turn, 66, 211, 282, 500, 562, 637
turn-around, 211
turnbuckle, 249, 312, 464
turner, 128
turnery, 52, 546
turnery-ware, 546
turning bay, 211
turning force, 347, 546
turning-lathe, 290, 546
turning oil, 549
turning place, 211, 637
turning-plate, 637
turning radius, 212
turning saw, 170
turnings, 378
turnknife, 312
turnout, 21, 47, 213, 350, 422
turnover, 74, 305, 485, 574
turnover of capital, 641
turnover of circulating capital, 306
turnover of goods in process, 599
turnover of material resource, 513
turnover rate, 169, 630
turnover tax, 307, 574
turnover time, 630
turpentine, 464, 465
turpentine oil, 464
turpentine-orchard, 465
turpentine substitute, 464
turpentine-timber, 189
turpentining, 39, 476
turps, 464
turret veneer lathe, 212, 452
tusk and tenon joint, 529
tusk tenon, 119
tusser worm, 654
tussock, 41, 421
tussock moth, 110
tussock tundra, 421
tuzet, 492
twig, 214, 535
twig-budding, 113
twig grafting, 113
twilight migration, 53
twin band-saw, 153, 453
twin band (saw) mill, 153, 453
twin circular saw, 454
twin drilling, 452
twin edger, 454
twin-frame (saw) mill, 453
twin-hull balloon, 454
twin-sled, 292, 453
twine-tying device, 281
twiner, 47
twining, 47
twining plant, 47
twining stem, 47
twist, 4, 347
twist bit, 316
twist gimlet, 443
twisted barrel, 243
twisted fibre, 347
twisted grain, 347
twisted growth, 313
twisted state, 347
twister, 242, 249, 534
twisting, 347
twist(ing) angle, 347
twisting couple, 347
twisting moment, 347
twisting stem, 347
twitching, 222
two-bottom mouldboard plough, 454
two-chamber type vertical shaft mill, 453
two-component spray gun, 453, 455
two-cord grapple, 454
two-cut shelterwood method, 121, 122
two-deck conveyor, 452
two-deck dryer, 452
two-dimensional correlation, 123
two-dimensional paper chromatography, 454
two disk refiner, 453
two-entry, 123
two-faced planing machine, 453
two-faced sawing, 295
two-faced tie, 11, 115
two-factor method, 454
two-furrow plough, 452
two-handled saw, 452
two heads wide belt sander, 454
two-hinged arch, 453
two-lane traffic way, 452
two-legged ladder, 455
two level board pick-up, 452
two-line grapple, 454
two lined borer, 452
two-log boom, 453
two-man saw, 453
two-movement treatments with zinc chloride, 147
two-parent progeny test, 453
two-ply (plywood), 121
two-rail shear test, 452
two-rotation coppice (system), 295
two-row planter, 452
two-row planting, 453
two-saw edger, 454
two-saw trimmer, 453
two-side coater, 453
two sleigh, 292, 453
two-stage compression, 153, 295
two-stage cutting, 121
two-stage cycle, 122
two-stage drying plant, 122
two-stage sampling, 295
two-stage suspension dry, 122
two-stage suspension type dryer, 122
two step replica method, 121
two-storey face, 452
two-storey tapping, 452
two-storied, 121, 122
two-storied forest, 121
two-storied high forest system, 122

T

two-way cableway, 454
two-way cross, 123, 453
two-way merchantable volume table, 123
two-way method, 454
two-way plough, 454
two-way skidding, 452
two way system, 454
tychoplankton, 348
tying contract, 152
Tyler system, 471, 601
Tyler yarding, 471, 601
tylosis, 379
tylosis forming layer, 379
tylosis (*plur.* -se), 357, 379
tylosoid, 342, 462
tympanum, 175, 325
type, 285, 325, 541
type by original designation, 589
type culture, 25, 100
type genus, 325
type locality, 325
type map, 285, 299, 302
type of management objective, 253
type of mixture, 213
type of pit, 506
type of ray-component, 419
type of ray-width, 418
type of thinning, 235, 444, 445
type of tree height, 447
type of vegetation, 616
type-one plywood, 498, 563
type species, 326
type specimen, 325
type test, 100
type-three plywood, 402
type tree, 25
type-two plywood, 122
types of ownership, 37, 469
typhoon, 471
typical retention, 25
tyrosinase, 284
tyrosine, 284
U-blade cutter, 377
U-gully, 172
U-shaped gullying, 173
U. S. public land survey system, 320
U. S. rectangular surveying system, 320
ubac, 570
ubiquitist, 468
udometer, 585
udomograph, 645
ugly weather, 121
ulmic acid, 205, 648
ulmiform pore arrangement, 584
ultimate bearing capacity, 222
ultimate bending strength, 222
ultimate brightness, 222
ultimate compression strength, 222
ultimate consumer, 577, 654
ultimate factor, 654
ultimate load, 222
ultimate production, 648
ultimate quantity of value, 653
ultimate strength, 222
ultimate stress, 222
ultimate structure, 51
ultimate tensile strength, 222
ultisol, 284
ultra-hardboard, 51
ultra-high frequency, 51
ultra-high pressure method, 51
ultra-high purity, 51
ultra-low frequency, 50, 221
ultra-low-volume, 50
ultra-low-volume sprayer, 50
ultra-molecular structure, 51
ultra radio frequency, 51
ultra wide-angle camera, 51
ultracentrifugation, 51
ultracentrifuge, 51
ultramarine, 147, 389, 420
ultramicrochemistry, 51
ultramicrometer, 50
ultramicron, 51
ultramicropore, 51
ultramicroscope, 51
ultramicroscopic particle, 51
ultramicroscopic structure, 51
ultramicroscopy, 51
ultramicrostructure, 51
ultramicrotechnique, 51
ultramicrotome, 50
ultraplankton, 51
ultrared, 198
ultrared drying, 198
ultrared ray, 198
ultrasonic, 51
ultrasonic degradation, 51
ultrasonic flaw detector, 51
ultrasonic inspection, 51
ultrasonic pulse, 51
ultrasonic wave, 51
ultrasonic wave degradation, 51
ultrastructural lamella, 51
ultrastructure, 51
ultrathin section, 50
ultraviolet, 642
ultraviolet absorbent, 642
ultraviolet-B radiation, 642
ultraviolet fluorescent deposit, 642
ultraviolet infrared film curing apparatus, 642
ultraviolet microscope, 642
ultraviolet photography, 642
ultraviolet radiation, 642
ultraviolet ray, 642
ultraviolet-ray-resisting polyethylene, 271
ultraviolet spectrogram, 642
ultraviolet spectrum, 642
umbel, 404
umbellate cyme, 404
umbelliferone, 404
umbellule, 534
umber, 46, 242, 648
umbilical cord, 366, 632
umbo (*plur.* umbones), 305, 486
umbonate, 261, 581
umbrella roof, 404
umbrella species, 404
umbric epipedon, 4
unacceptable product, 34, 70
unaided eye, 399
unavailable water, 34, 511
unbalanced, 35

unbalanced distribution, 34
unbalanced growth, 35
unbarked wood, 503
unbias(s)ed estimate, 510
unbleached pulp, 504
unbonded plywood assembly, 240
unbudgeted cutting, 225
uncalled capital, 504
uncertainty, 32
unciform, 173
unclassed felling, 137, 509
uncollectible account, 78, 207
uncommitted amount, 504
uncontrolled cutting, 509
uncontrolled fire, 510
uncontrolled mosaic, 510
uncouple, 247
uncoupled respiration, 137
uncoupler, 247
uncrossed cheque, 137
uncultivated, 504
under bark, 340
under-bark volume, 64, 385
under bleached, 504
under coat(ing), 95, 323
under cook, 229
under cooked chip, 229
under cut trimmer saw, 243
under-developed area, 504
under-dispersion, 93
under grazing, 380
under-insurance, 15, 35
under seal, 146, 322
under side wide belt sander, 521
under size, 57, 412, 519
under stocking, 289, 445
under tanned, 504
underbark, 340
underbind, 104, 520
underbrace, 156, 520
underbrush, 520, 521
underbucking, 343
underconsumption, 531, 532
undercooked pulp, 229
undercrowded, 445
undercrowding, 186
undercure, 176
undercut, 37, 129, 262, 268, 269, 343, 520, 521, 526
undercutter, 269, 343, 494, 600
undercutting, 98, 343, 361, 578
underdevelopment, 125, 525
underdrain, 3, 4
underdrainage, 3
underexposure, 17
underfit river, 35, 401
underfit stream, 35, 401
underflow, 94, 374
under(ground) irrigation, 98, 508
underground reservoir, 98
underground rhizome, 98
underground river, 3, 98
underground runoff, 43
underground stratification, 98
underground water, 98
underground works, 98
undergrowth, 125, 302, 521
underlayer, 520
underlying rock, 220, 520
underlying surface, 520
undernutrition, 573
underplant, 521
underplant-operation, 521
underplanting, 302, 449
underpopulation, 627
underpowered, 93, 171
underrun, 94, 374, 375, 387, 620
undershrub, 9, 533
undersized chip, 187, 412
undersized log, 534
undersized tree, 534
underslung conveyer, 545
undersoil, 95, 520
understerilization, 94
understock, 74, 168
understocked, 289
understorey, 520, 521
understorey management, 302, 449, 521
understorey (tree) species, 520
undertone, 87, 95
underwind, 521
underwood, 520, 521
underwriter, 15, 55
underwriting limit, 55
undesirable species, 34, 137
undesirable tree, 137
undeviated light, 615
undistributed profit tax, 504
undressed timber, 72, 504
undulating ground, 367, 381
undulating topography, 99
undung, 413
unearned increment, 34, 137, 646
unearned revenue, 99, 587
uneconomic thinning, 137, 510
unedged, 318
unedged lumber, 318
unencrusted membrane, 313
unencumbered balance, 504
unequal-toothed, 35
unequally pinnate, 366
uneven-aged, 567
uneven-aged forest, 567
uneven-aged high forest, 567
uneven-aged management, 567, 601
uneven-aged stand, 567
uneven-aged system, 567
uneven-aged wood, 567
uneven-aged yield table, 567
uneven brightness, 182
uneven drying, 158
uneven grain, 34
uneven-span (green) house, 34
uneven texture, 34
unevenness of stand, 301
unexpected outcrossing, 348
unexpended balance, 504
unexploited forest, 504
unfloated timber, 34, 53

ungalvanized wire rope, 32
unglazed finish, 93
ungrazed woodlot, 137, 252
ungula, 476, 604
ungulate, 581
unhewn timber, 504, 592
unhooker, 248, 602
uniaxial compression, 87
uniaxial extension, 87
uniaxial orientation, 86
uniaxial tensile deformation, 87
unicarpellate, 86
unifacial leaf, 84
unified budget, 484
unified volume table, 484
unifoliate, 261
unifoliolate, 261
uniform, 266
uniform crop, 266
uniform distribution, 266
uniform grain, 266
uniform high forest, 266
uniform lay wire rope, 461, 483
uniform load, 266
uniform method, 266, 386
uniform pressure, 92, 266
uniform resistance, 266
uniform sampling, 266
uniform shelterwood cutting, 266
uniform shelterwood system, 77, 183
uniform stand, 266, 483, 484, 563
uniform strength, 266
uniform system, 266, 386
uniform texture, 266
uniform-textured wood, 266
uniformitarianism, 266
uniformity of seed lot, 626
uniformly distributed load, 266
uniformly varying load, 266
unilacunar, 86
unilateral pit, 82
unilaterally aliform parenchyma, 82
unilaterally compound pitting, 82
unilaterally compound pitting vessel, 82
unilaterally confluent parenchyma, 82
unilaterally paratracheal parenchyma, 82
unilaterally scanty paratracheal parenchyma, 82
unilayer particleboard, 82
uninodal, 320, 580
union drive, 294
uniovular twins, 563
unipalmate, 86
uniparental, 84
uniparous helicoid cyme, 84
uniparous scorpioid cyme, 86
unipinnate compound leaf, 563
uniprocessor, 83
unique and threatened system, 111
unique landscape, 111, 190
unique natural area, 111, 515, 607
uniseriate parenchyma band, 84
uniseriate ray, 84
unisexual, 70, 86
unit, 36, 85, 86, 121, 166, 218, 417, 587, 639, 648
unit cell, 86, 156, 165, 254
unit cost, 85
unit elongation, 85, 554
unit furniture, 55, 651
unit heater, 86
unit hydrograph, 85
unit leaf rate, 85
unit load, 85
unit membrane, 85, 166
unit of cost, 83
unit of measurement, 44, 111
unit of sampling, 61
unit package, 651
unit pressure, 20, 85
unit-price, 83
unit process, 86
unit safe stress, 85
unit strain, 85
unit stress, 85
unit tensile stress, 85
unit train, 381
unit value, 83
unit volume expansion, 85
unitary elasticity, 86
unitary organism, 85
unitary volume model, 563
United Nations Framework Convention on Climate Change, 294
unitized crossband, 610
unitized shipment, 224
units of population, 649
univalent, 83
universal abrasive wheel sander assembly, 118
universal automatic computer, 482
universal automatic map compilation equipment, 386
universal blender, 118
universal circular saw with tilting arbor, 631
universal circular saw with tilting table, 471
universal clamp, 118
universal computer oriented language, 482
universal dining table, 119
universal drill, 499
universal flaker, 118
universal lead, 499
universal log rule, 482
universal soil loss equation, 482
universal test machine, 499
universal testing machine, 482, 499
universal thinning, 365
universal time, 438
universal tool grinder, 499
universal tractor, 482
universal transverse Mercator (UTM), 484
universal umbel, 153
universal woodworking machine, 499
university forest, 243, 436, 439, 603
univoltine, 563
univoltine insect, 563

unlignified parenchyma, 137
unlimited swelling, 511
unloaded carriage, 276
unloaded deflection, 276, 647
unloaded skyline cable, 276
unloader, 538
unloader hydraulic pump, 538
unloader live-roller, 538
unloader valve, 538
unloading, 538
unloading cage, 538
unloading deck, 538
unloading deflector, 538
unloading gin pole, 110, 538
unloading platform, 538
unloading tank, 538
unmanaged forest, 504
unmanaged stand, 504
unmanned aerial vehicle, 510
unmanned aircraft, 510
unmanned plane, 510
unmerchantable log, 138
unmetalled road, 489
unmixed stand, 137
unmodified rosin, 478, 504
unoccupied land, 277, 313
unordinary yield, 474
unpalatable, 34
unpaved road, 489, 491
unpredictability, 34
unproductive forest land, 138
unproductive property, 138
unproductive stand, 138
unrecovered strain, 41
unreeling, 135
unreeling machine, 135
unreeling system, 464
unregulated felling, 137
unregulated fire, 396
unreserved forest land, 136
unresponsive supply, 138
unrigging, 47
unsaponifiables, 32
unsaturation, 33
unscrambler, 642, 644
unscreened stock, 72
unseasoned wood, 504
unseptate fibre, 509
unsized fabric, 509
unsound knot, 32, 464
unsound seed, 32, 33
unsquare bucking, 600
unstable gene, 35
unstable humus, 35
unstable vegetation, 35
unstack, 47
unstacker, 47
unsteady state, 35
unstiffened suspension bridge, 504
unstocked area, 277, 313, 504, 510
unstocked blank, 510
unstocked forest land, 510
unstocked stand, 510
unstoried cambium, 136
unstratified drift, 136
unsupervised classification, 509
unsupported adhesive, 512
unsurfaced road, 489
unsymmetrical bending, 32
unsymmetrical flower, 32
untreated board, 504
untreated timber, 466, 504
untrimmed size, 504
unutilized wood, 504
unweathered mineral soil, 504
unwinding, 135, 248
unwrought timber, 72
unwro(ugh)t timber, 504
up-and-over door, 162
up-budding, 416
up milling, 343
up-stream face, 416
upender, 104, 128, 450
upending, 100, 104
upfield, 161
uphill side, 416
uphill skidding, 343, 416
uphill sloping profile, 110
uphill timber, 416
uphill yarding, 343, 416
upholstered chair, 35, 400
upholsterer, 229, 441
upholstery, 229, 411, 441
upland forest, 162
upland game, 310
upland hickory, 73
upland meadow, 413
upland peat soil, 162
upland site, 164, 413
upland tundra, 163
upper, 415
upper air, 161, 162
upper atmosphere, 161
upper bound, 415
upper couch roll, 415
upper crown height, 415
upper crown length, 415
upper explosive limit, 16
upper felt, 415, 416
upper growth, 416
upper horizon, 26
upper landing, 222, 413
upper limit, 416
upper log, 157, 417
upper perturbation, 162
upper press, 416
upper press roll, 416
upper stem, 157, 417
upper-stem diameter, 447
upper stock, 415
upper storey, 415, 416
upper storey thinning, 415
upper stratosphere, 415
upper tail covert, 503
upper tree layer, 415
upper warm front, 162
uppermost layer, 415
uppermost soil, 27
Uppsala school, 17
upright, 320, 612, 615
upright-growing, 615
upright ray cell, 615
uproot, 167
uprooted stump, 5
uprooter, 494
uprooting, 167, 494
uprooting by frost, 110
uprooting machine, 494
upset, 128, 228, 469, 498
upset price, 349, 653
upset tooth, 550
upshear, 343
upslide surface, 415
uptake, 372, 416, 514
upturned sod, 127, 128
upvalley wind, 416
upward adjustment, 530, 601
upward growth, 163
upwelling, 416
urban air pollution, 56
urban area, 438
urban atmosphere, 56

urban construction, 56
urban decoration, 56
urban design, 56
urban ecology, 56
urban environment, 56
urban forest, 56
urban forestry, 56
urban green system planning, 56
urban heat island, 56
urban open space, 56
urban planning, 56
urban vegetation, 56
urbanization, 56
urceolate corolla, 472
urea, 346
urea resin, 346
urea-resin adhesive, 346
urea-resin treated wood, 346
uredial stage, 521
uredial state, 521
urediniospore, 521
uredinium, 521
ured(i)ospore, 521
uredium, 521
uredosorus, 521
Urich's method, 578
uridine, 346
uridylic acid, 346
urn stand, 135
uropod, 155, 503
uropygial gland, 503
Urotropine, 309, 507
ursolene, 594
ursolic acid, 507
urunday extract, 507
Urus logging, 507
urushi, 366
urushi tallow, 366
urushi wax, 366
urushin, 366
urushoil, 366
usable life, 274, 437, 440
use percent, 291
use tax, 437
user, 577, 579
user capacity, 579
user characteristic, 577, 579
user charge, 577
user day, 579
user safety, 437, 579
using percentage, 291, 441
utilisation, 290, 298
utility drum, 150
utility furniture, 119, 436
utility loader, 119
utilization, 290, 298
utilization coefficient, 291
utilization percent, 291, 441
utilization percent method, 291, 548
utilization period, 291, 432
utilization permit, 291
utilization value, 290
utilization value of seed, 629
utilized top, 273
V-grooved plywood, 240
V-gully, 378
V-lead, 372
V-matched figure, 396
V-plough, 452, 548
V-roof, 396
V-shaped gulling, 173
V-shaped incision, 377
V-shaped notch, 274
vac-vac process, 455
vacancy, 277
vaccination of plant, 617
vacuole, 275, 561
vacuum counter, 607
vacuum cycle, 233, 608
vacuum drier, 233, 607
vacuum drying, 607
vacuum drying oven, 607
vacuum former, 607
vacuum fumigation, 607
vacuum gauge, 607
vacuum-gridpress, 607
vacuum-meter, 94, 607
vacuum pickup, 607
vacuum pressure process, 607
vacuum process, 607
vacuum pump, 607
vacuum seasoning, 233, 607
vacuum seed harvester, 515, 608
vacuum storage, 608
vacuum-storage tank, 608
vacuum-tight, 322, 369, 607
vacuum tightness, 607
vacuum washer, 607
vadose, 420
vadose spring, 420
vadose water, 420
vagility, 404
vagina, 494, 561
vaginate, 261
vale, 113, 175, 193
valerian oil, 537
valine, 537
valley, 175, 413
valley bottom, 175
Valley size tester, 495
valley value, 175, 654
valley wind, 175
valonea extract, 531
valoneaic acid, 531
valoneaxanthone, 531
valonia, 499, 531
valonia extract, 531
valonia tannin, 531
valuable species, 607
valuation, 237, 359, 363, 545
valuation clause, 173
valuation of forest land, 299
valuation of land for forestry, 303
valuation of recreation, 579
valuation of second growth, 71
valuation of stumpage, 289
valuation of timberland, 577
value, 231
value added, 601
value-added tax, 602
value frame work, 231
value in land, 488
value increment, 231
value increment percent, 231
value of a stand, 289, 299, 301
value of forest, 407
value of growing stock, 301, 544
value of land, 96
value of normal growing stock, 128
value of soil, 488
value of stand, 289, 301
value standing, 289
valvate, 11, 346
valve, 11, 12, 127, 297
Van der Waals force, 130
vanderpool, 637
vanilione, 528
vanillic acid, 528
vanishing point, 532
Vapam, 500

vapo(n)r pressure, 371, 609
vapo(n)r pressure equilibrium method, 371
vapor pressure deficit, 609
vapour barrier, 371, 609
vapour density, 609
vapour drying, 609
vapour drying process, 609
vapour injection process, 371, 609
vapour permeability, 609
vapour phase, 371, 609
vapour pressure deficit, 14, 609
vapour pressure difference, 609
vapour seasoning, 609
vapourization, 371, 608
vapourization heat, 371, 609
vapourization without melting, 421
vapourizer, 354, 371
variability, 24, 272
variable, 23
variable cost, 23, 272
variable-density growth and yield model, 272
variable density yield table, 272
variable (fire) danger, 272
variable flow, 24
variable plot, 272
variable plot cruising, 272
variable plot method, 272
variable-probability-sampling, 272
variable radius plot sampling, 242, 272
variable wind, 31
variance, 130
variance analysis, 130
variance component, 130
variance-mean ratio, 266
variance ratio, 130
variance test, 130
variant, 24
variate, 23, 272
variation, 23, 24
variegated plant, 7
varietal character, 359
varietal characteristic, 359
varietal deterioration, 359
varietal resource, 359
varietalness, 359
variety, 24, 359
variety sander, 117, 118
variety saw, 118
varnish, 381, 487
varnish base, 381
varnish coating, 381, 487
varnish formation, 366
varnish paint, 381
varnish resin, 381
varnish tree, 159, 366, 397
varnished paper, 251, 487
varnishing, 416, 487
varved clay, 506
varying load, 23, 32
varying probability, 272
varying stress, 24
vascular, 89, 317, 502, 548
vascular bundle, 502
vascular bundle sheath, 502
vascular cambium, 502
vascular cylinder, 502
vascular disease, 89
vascular meristem, 502
vascular plant, 502
vascular ray, 502
vascular skeleton, 502
vascular system, 502
vascular tissue, 502
vascular tracheid, 89, 502
vase, 203, 363
vase baluster, 363
vase-form training, 17
vase splat, 363
vasicentric, 207
vasicentric abundant parenchyma, 207
vasicentric parenchyma, 207
vasicentric scanty parenchyma, 207
vasicentric tracheid, 207
vasiform, 179, 317
vectopluviometer, 107
vector, 65, 248, 530, 598
vector control, 66
vector data, 437
vector gauge, 107
vector of infection, 65, 378
Vee-balloon, 370
vegetable active carbon, 617
vegetable adhesive, 617
vegetable anatomy, 617
vegetable carbon, 334
vegetable cell, 477, 618
vegetable charcoal, 334
vegetable decolourizing carbon, 618
vegetable fat, 618
vegetable fibre, 618
vegetable glue, 617
vegetable gum, 617
vegetable insecticide, 617
vegetable jelly, 617
vegetable kingdom, 617
vegetable oil, 618
vegetable parchment, 230, 618
vegetable protein glue, 616
vegetable refuse, 150, 409
vegetable remains, 150, 409
vegetable tanned leather, 617
vegetable tannin, 477, 478, 618
vegetable tannin extract, 271
vegetable tanning, 616, 617
vegetable tanning material, 617
vegetal cover, 616
vegetalization, 617
vegetated, 616
vegetated channel, 616
vegetated substrate, 617
vegetated terrain, 616
vegetation, 574, 602, 616
vegetation analysis, 616
vegetation biomass, 616
vegetation form, 616
vegetation geography, 616
vegetation map, 616
vegetation period, 429
vegetation profile, 616
vegetation profile chart, 616
vegetation province, 616
vegetation region, 616
vegetation season, 429

vegetation space, 617
vegetation subtype, 616
vegetation type, 616
vegetation type group, 616
vegetation zone, 616
vegetational analysis, 616
vegetation(al) cover, 616
vegetational pattern, 616
vegetative barrier, 617, 618
vegetative bud, 574
vegetative carbon, 427, 616
vegetative cycle, 430, 574
vegetative embryo, 511
vegetative flower, 511
vegetative forest regeneration, 409
vegetative form, 423, 574
vegetative grafting, 574
vegetative growth, 574
vegetative hybrid, 511, 574
vegetative hybridization, 511
vegetative hyphae, 574
vegetative landscape, 617
vegetative mycelium, 574
vegetative nucleus, 574
vegetative parthenocarpy, 574
vegetative period, 574
vegetative phase, 574
vegetative progeny, 511
vegetative propagation, 511, 574
vegetative propagule, 511
vegetative reproduction, 511, 574
vegetative segregation, 511, 574
vegetative spread, 511
vegetative stage, 574
vehicle scale, 564, 609
vehicle stock, 52
veil, 267, 604, 605
vein, 58, 335, 506, 560
vein-banding, 554
vein-clearing, 325
veined paper, 335
veined wood, 204
veinlet, 519, 534
veld(t), 76, 140
vellum, 111, 134
velocity-height ratio, 466
velocity of air, 369
velocity of shot, 87
velometer, 466
velpar, 500
velvety, 18
venation, 59, 317, 560
vendee, 317
vendor, 317, 643
veneer, 13, 81, 240, 529
veneer band, 81
veneer bandsaw, 81
veneer belt dryer, 79
veneer bolt, 82
veneer budding, 84, 529
veneer chest, 82, 240
veneer-circular saw, 82
veneer clipper, 81, 82, 233
veneer composer, 82
veneer core, 335
veneer cutting saw, 82
veneer drier, 81
veneer dryer, 81
veneer drying, 81
veneer edge gluer, 81
veneer flitch, 81, 82
veneer gang saw, 82
veneer grafting, 357, 377, 529
veneer jointer, 82
veneer knife, 82
veneer lathe, 82, 545
veneer log, 82, 240
veneer log production, 82
veneer overlay, 36
veneer pack jointer, 81
veneer particleboard, 82
veneer patcher, 82
veneer peeling, 546
veneer plywood, 385
veneer preparation area, 82
veneer reel, 82
veneer reeling and unreeling machine, 82, 264
veneer ribbon, 81
veneer roller dryer, 184
veneer rotary lathe, 82, 545
veneer saw, 82
veneer sawing machine, 82
veneer score knife, 81
veneer slicer, 82
veneer sorter, 81
veneer splicer, 82
veneer splicing, 83
veneer stack, 81
veneer stacker, 81
veneer stitcher, 82
veneer stitching machine, 82
veneer storage magazine, 74, 82
veneer storage system, 82
veneer suction stacker, 515
veneer tapping machine, 81
veneer unitization, 82
veneer welding, 82
venetian blind, 214, 400
venetian blind slat, 214
Venice turpentine, 314, 500
venison, 310, 560
vent, 350, 481, 482
ventifact, 145
ventilated psychrometer, 481
ventilated thermometer, 482
ventilating damper, 370, 482
ventilating device, 482
ventilating dry kiln, 481
ventilating pit, 482
ventilation, 350, 481
ventilation hole, 146, 481, 482
ventilation tunnel, 481
ventilator, 350, 482
ventral, 155
ventral suture, 155
ventri-dorsal, 18
ventriculus, 504, 622
ventricumbent, 149, 323
venture capital, 145, 485
venulose, 116, 261
veranda(h), 579, 651
veranda(h) post test, 283
verbena oil, 131, 316
verbenene, 316
verdant zone, 510
verge board, 88, 146
verification, 194
verifier, 193, 233
vermiculite, 621
vermiform, 399, 571

V

Vermont log rule, 147
vernal aspect, 68
vernal equinox, 68
vernalization, 68
vernation, 119, 552
vernier, 501, 579
vernier caliper, 579
verruca (*plur.* -ae), 318, 578, 641
versatile, 105, 482
versatile reproduction, 414
vert-air-jet impingement dryer, 68
vertebral, 225
vertex, 105, 485
vertical, 67
vertical aerial photograph, 68
vertical anemometer, 372
vertical anemoscope, 372
vertical angle, 67, 557
vertical anther, 615
vertical band saw, 290
vertical-bar opening, 549
vertical blade root pruner, 614
vertical boom-nozzle system, 159
vertical boring machine, 290
vertical closure, 68
vertical-contraction, 650
vertical convection, 67, 372
vertical crown, 68
vertical curve, 450, 649
vertical cutting, 68, 614
vertical datum, 68, 161
vertical disk type chipper, 290
vertical disk type flaker, 290
vertical distribution, 68
vertical element, 615, 631
vertical exaggeration, 68
vertical expansion, 601, 650
vertical extruder, 290
vertical gap, 68, 549
vertical grain, 256
vertical-grain lumber, 256
vertical head sander, 288
vertical integration, 68, 650
vertical intercellular canal, 631
vertical interval, 67, 91
vertical labour mobility, 284
vertical line sampling, 68
vertical logging, 68, 276
vertical member, 68
vertical migration, 68
vertical opening, 68, 549
vertical photograph, 68
vertical photography, 68
vertical piling, 42, 614
vertical point sampling, 67
vertical resin canal, 631, 650
vertical resistance, 68
vertical retort, 290, 450
vertical ring-magazine, 68
vertical root, 68
vertical section, 649
vertical shear, 68, 196
vertical sketchmaster, 68
vertical slicer, 290
vertical spindle tenoner, 290
vertical stand structure, 299
vertical storage bin, 290
vertical stratification, 67
vertical structure, 68
vertical temperature profile radiometer, 505
vertical trade, 68, 338
vertical veneer slicer, 290
vertical vibration, 68
vertical zonality, 67
vertical zone, 68
verticil, 208, 312
verticillaster, 312
verticillate, 208, 312
verticillate leaf, 312
vertisol, 24
very dense forest, 222
very heavy thinning by narrow strip, 603
very high frequency, 222
very high resolution radiometer, 222
very low frequency, 221
vesicle, 352
vesicular-arbuscular mycorrhiza, 338
vesicular mycorrhiza, 352
vessel, 89
vessel element, 89
vessel element extension, 89
vessel line, 179
vessel member, 89
vessel perforation, 89
vessel segment, 89
vesselform tracheid, 89
vested interest, 226
vestigial tissue, 492
vestured pit, 152
veteran, 284
veteran stage, 284
vetivazulene, 554
vetiver oil, 528, 554
vetivone, 554
vexillary, 104, 583
vexillum, 266, 366
viability, 422
viability of seeds, 628
viability stain, 422
viable seed, 214, 274, 581
vibes-less, 233
vibrating gutter conveyor, 608
vibrating plate, 608
vibrating rotary screen, 608
vibrating screen, 608
vibrating sieve, 608
vibrating tamper, 608
vibration cutter, 608
vibration cutting, 608
vibration damper, 233
vibration test, 608
vibrator, 113, 608
vibro-hoe, 362, 608
vibronic spectrum, 608
vibroshock, 209, 233
vicariad, 477
vicarious community, 477
vicarious species, 477
vicarism, 477
vicinism, 348, 478
vicious circle, 121
Vickers (hardness) test, 502
video digitizer, 439
video frequency, 439
video tape recorder, 439
vidicon, 180
viewfinder, 384
viewing angle, 439
viewing equipment, 178

vigorous branch, 639
vigour, 429
vigour class, 214
vigour classification, 214
vigour evaluation, 214
vigour modification, 214
vigour rating, 214
vile weather, 554
villa, 27, 240
village forest, 527
villose, 18, 50
vinaceous figure, 364
vine, 351, 475
vine fibre, 475
vinyl plastic, 566
vinyl resin, 566
violent storm, 15
violet root rot, 642
viral, 29
viral infection, 29
viral pesticide, 29
virgin area, 591
virgin dip, 63
virgin forest, 591
virgin growth, 591
virgin land, 65, 422
virgin pulp, 589
virgin stand, 591
virgin stock, 589, 615
virgin strand forest, 188
virgin timber, 591
virgin wood-land, 591
viricidal, 410
virion, 29
viroid, 285
virology, 29
viroplasma, 29
virosin, 29
virosis, 29
virostatic, 569
virtual refusal, 260
virucidal, 410
virucide, 410
virulence, 110, 111
virulent, 110, 621
virulent phage, 297
virulent strain, 111, 375
viruliferous, 79, 341
virus, 29
virus complex, 29
virus crystal, 29
virus disease, 29
virus-free, 508
virus-free stock, 508
virus particle, 29
virus strain, 29
viscera, 341
visco-elastic body, 344
visco-elastic flow, 344
viscoelastic behavior, 344
viscoelasticity, 344
viscoelastometer, 344
viscogel, 345
viscoid, 345
viscoloid, 345
viscose, 345
viscose glue, 241, 345
viscose process, 345
viscose silk, 345
viscosimeter, 344
viscosity, 345
viscosity-average molecular weight, 345
viscosity unit, 344
viscous-elastic behaviour, 344
viscous flow, 345
visibility, 341
visibility angle, 273
visibility distance, 341, 439
visibility map, 296, 341
visibility meter, 341, 554
visible-area map, 296, 341
visible balance, 319
visible crack, 273
visible crown diameter, 273
visible (light) radiation, 273
visible line, 312, 497
visible mutation, 273
visible spectrum, 273
visible symptom, 273
visible tree height, 273
visitor-carrying capacity, 580
visitor centre, 579, 580
visitor-day, 579
visitor education, 579
visitor origin, 579
vista, 485, 486
vista cutting, 485
visual acuity, 439
visual and infrared radiometer, 273
visual angle, 439
visual colour sensation, 555
visual estimation, 173, 336
visual field, 439
visual inspection, 336, 399
visual interpretation, 336
visual organ, 439
visual range, 439, 652
visual vulnerability, 439
vital force, 422
vital index, 214
vital statistics, 395
vitalism, 423
vitality, 422
vitamin, 502
viviparity, 328, 471
vivipary, 471
vivipary plant, 471
vivosphere, 427
void, 277, 309
void content, 277
void volume, 276, 277
volatile content, 211
volatile matter, 211
volatile oil, 211, 254
volatile solvent, 211
volcanic ash, 216
volcanic ash soil, 216
volcanic chain, 216
volcanic desert, 216
volcanic lake, 215
volcanic mud flow, 216
volcanic product, 216
volcanic rock, 216
volcanic sand, 216
volcanic soil, 216
voltinism, 206
voluble plant, 47
voluble shrub, 475
voluble (volubile), 47
volume, 36
volume-alloting method, 36
volume allotment method, 36
volume and area allotment method, 36
volume and increment control method, 128
volume assessment, 36
volume-basal area ratio, 36, 449, 541
volume-basal area ratio table, 36
volume calculation, 36
volume check, 36
volume class, 36
volume control, 36, 37
volume control method, 545
volume curve, 36
volume-curve method, 384
volume deformation, 476

V

volume-derived biomass, 37
volume determination, 36
volume difference method, 36
volume elasticity, 476
volume equation, 36
volume estimation by sight, 336
volume expansion, 476
volume expansivity, 476
volume exponent method, 37
volume frame work, 36, 545
volume function, 36
volume growth, 37
volume increment, 545
volume increment percent, 37
volume index, 37
volume inside bark, 385
volume measure, 36
volume measurement, 36
volume mixing ratio, 476
volume modulus of elasticity, 476
volume of business, 560
volume of over bark, 79
volume of standing timber, 290
volume of wood, 36
volume outside bark, 79
volume per cent, 36, 397
volume-period method, 106
volume production, 37, 355
volume ratio, 397, 476, 544
volume ratio equation, 36
volume regulation, 37
volume regulation method, 545
volume reserve, 64, 545
volume rotation, 36
volume shrinkage, 476
volume shrinkage mass, 476
volume strain, 476
volume table, 36
volume viscoelasticity, 476
volume weight, 398
volume yield, 37
volumeter, 397, 476
volumetric callipers, 36
volumetric capacity, 397
volumetric heat capacity, 397
volumetric shrinkage, 476
volumetric strain, 476
volumetric swelling, 476
volumetric water content, 476
volumetric weight, 274, 397
volumetric yield, 37
voluminal resilience, 476
voluminosity, 397
volune measurement by sections, 382
voluntary agreement, 647
voluntary contribution, 647
volunteer, 477, 559, 646
volunteer growth, 373, 477
volution, 312, 313, 546
volva, 267
Von Mantel's formula, 147
vortex, 507, 546
vortex action pulp cleaner, 507
vortex cleaner, 507
voucher, 66, 363
voucher specimen, 363
vulcanized fibre, 161, 308, 379
vulnerability, 74
vulnerable species, 569
wad, 53, 103, 321
wade, 371
wader, 371
wafer, 13, 77, 157
wafer board, 204
waferboard, 13, 204
waferboard-plus, 602
waferizer, 77
wage, 170
wag(g)on, 463
wag(g)on hauling, 463
wag(g)on plank, 52
waggon retort, 52
wag(g)on sled, 453
Wagner's Blendersaumschlag system, 495
Wagner's border cutting, 495
Wagner's method, 495
waif (wood), 358, 512
wainscot, 202
wainscot chair, 375
wainscoting moulding, 20
waiting line theory, 349
wale, 195
walking anchor, 282, 564
walking guy, 160
walkingstick, 632
walkway, 395, 651
wall, 20
wall board, 20, 376
wall desk, 21
wall furniture, 375
wall garden, 376
wall panel, 20
wall plate, 55
wall post, 21
walling plant, 152
walnut oil, 194
wand, 341
wanded chair, 308
wandering bird, 358
wandering dune, 306
wandering heart, 357, 468
wandering nursery, 304, 564
wandoo extract, 2
wane, 115, 387
wane edge, 115, 387
waney edge, 115, 387
waney lumber, 115
want, 37
wany lumber, 115, 387
warden, 202, 269
wardrobe, 563
wardrobe handle, 64
wardrobe trunk, 564
warfarin, 411
warm colour, 348
warm current, 348
warm front, 348
warm front surface, 348
warm intermediate house, 624
warm-setting adhesive, 624
warm storage, 164, 348
warm stratification, 164, 348
warm temperate deciduous forest, 348
warm-temperate district, 348
warm temperate plant, 348
warm temperate rain forest, 348

warm-temperate zone, 348
warm water bath, 393
warm-water species, 348
warming-up, 197, 228, 348, 421, 586
warmth index, 505
warning colour, 255
warning colouration, 255
warning limit of arable land, 168
warning signal, 255
warp, 24, 136, 377, 497, 583
warp land dam, 583
warp soil, 136, 583
warped timber, 377
warping, 377
warrant, 173, 396
warranty, 15
Warren chain grinder, 507
wart, 308, 449, 502
wart-like structure, 308
warted, 80, 582
warty layer, 308
wash coat, 146, 516, 543
wash drum, 516
wash out pipette, 516
wash sale, 544
washboard, 74, 476, 516
washboarding, 30, 74, 495
washboarding effect, 495, 631
washer, 53, 103, 516
washing, 59, 416, 487, 516
washing loss, 460
washland, 130, 193
washstand, 294
wasp, 210
wastage, 139, 431, 469
waste, 139, 191, 431, 512
waste burner, 139
waste conveyor, 139
waste-gas stream dryer, 139
waste in felling and log-making, 126
waste in floating, 307
waste in wood, 38, 139
waste land, 209
waste liquor, 139
waste liquor recovery, 139
waste range, 139
waste rate, 469
waste steam, 139
waste steam coil, 139
waste stuff, 139
waste water chest, 139
waste wood, 139
wasted top, 139
wasteland, 209
wasteline, 112, 212
wastewater treatment, 139
wastewood, 139
wasting assets, 99, 233, 531
water-absorbing capacity, 515
water absorption, 515
water balance, 456
water balance diagram, 456
water bank forest, 455
water-bar, 310
water based paint, 460
water based stain, 460
water basin, 456
water bearing layer, 189
water bird, 458
water bloom, 456
water blush(ing), 460
water bombing, 136, 456
water-bound macadam, 457
water bronco, 492
water budget, 456
water capacity, 57, 397
water catchment, 224, 308
water cellar, 457
water cellulose, 59
water conductibility, 89
water-conducting tissue, 89
water conservation, 456, 460
water conservation forest, 460
water consumption, 191
water content, 189
water-core, 433
water course, 193, 456, 457
water culture, 398, 458
water cutting, 456
water cycle, 456
water deficit, 456
water divide, 143
water drain, 350
water drainage, 350
water dropping, 401
water equilibrium, 456
water erosion, 459
water escape, 569
water exchange, 456
water fall, 365
water filler, 460
water-film test, 457
water furrow, 325, 349
water garden(ing), 457
water-gas tar, 457
water-gas-tar creosote, 457
water gland, 460
water glass adhesive, 456
water glass paint, 456
water heart, 433
water-holding capacity, 57
water humus, 458
water-influence zone, 460
water-jacket oven, 459
water ladder, 459, 530
water lead, 307, 456
water-level, 458–460
water-level recorder, 459
water-lily pool, 460
water logged, 10, 251, 605
water logged land, 457
water logging, 59, 251, 457, 583
water-logging resistance, 270, 271
water logs, 307
water loss, 459
water magazine, 635
water mass(es), 459
water paint, 460
water-parting, 143
water path, 5
water permeability, 486
water plant, 459
water pollution, 460
water-pollution control, 460
water potential, 459
water proofing material, 133
water quality, 460
water quality control act, 460
water repellency agent, 133
water-repellent, 133
water-repellent preservative, 134
water-repellent size, 134
water repeller, 133, 602
water requirement, 544

W

water reservoir forest, 460
water resistance, 271, 337
water-resistance adhesive, 337
water-resistance glue, 337
water resource, 460
water resources application, 460
water retention, 15, 57
water root, 459
water rot, 432
water sac, 457
water saturation deficit, 455
water scale, 456, 458
water shield gum, 69
water slide, 456
water soak, 251, 433
water source, 460
water source protection forest, 460
water spout, 188, 350, 354, 534
water spray (dry) kiln, 354
water spray seeding, 354
water spreading, 141, 354, 570
water sprout, 167, 460, 487
water stain, 455, 458, 460
water storage, 456
water-storage capacity, 57
water-storage tank, 635
water stress, 460
water suspension, 460
water system, 193, 460
water table, 98
water table map, 459
water temperature mapping, 459
water test, 460, 562
water-transport elevator, 62
water treatment of garden, 588
water turbidity, 457
water use efficiency, 456
water-varnish, 460
water washer, 460
water-white, 455
water withdrawal, 457
water year, 307, 459
waterborne chemical, 458
waterborne salt, 458
waterborne(-type) preservative, 458
watered stock, 47, 544
waterfall erosion, 104
waterfall landscape, 365
waterfowl, 458, 556
waterfowl refuge, 458
watering, 240, 401, 492
watering machine, 61, 354
watering pot, 354, 401
waterlike solvent, 285
waterlogged wood, 53, 251
waterlogging, 284, 457
watermark, 356, 459, 460
watermark disease, 460
waterpool depot, 635
waterpour debarker, 433
waterproof adhesive, 134
waterproof glue, 134
waterproof glueline, 134
waterproof plywood, 337
watershed, 143, 224, 308
watershed area, 224, 308
watershed divide, 143
watershed leakage, 308
watershed management, 224, 308
watershed protection, 224
watershed variable, 308
waterside forest, 455
watersplash, 187
waterthinnable paint, 460
wattle, 288, 399
wattle extract, 254
wattle-fence, 23, 288
wattle-work, 23
wave board, 30
wave cyclone, 30
wave depression, 30, 146
wave erosion, 30, 283
wave moulding, 30
wave-topped wall, 595
wavelength, 30
wavelet analysis, 533
wavy figure, 30
wavy grain, 30
wavy veneer, 30
wax-free white shellac, 642
wax gland, 282
wax moth, 282
wax of bridelia leaf, 489
wax of kamala, 73, 139
waxed plastic sheet, 282
wax(en) maturity, 282
way leave, 482, 488
waypoint, 190, 310
ways, 89, 205
wayside pavilion, 90
weak dominant, 401
weakening in demand, 544
weakly alkaline soil, 401
weakly leached, 380
weakly podzolic soil, 401
wear, 327, 337, 469
wear-and-tear, 327, 469
wear-and-tear pigment, 451, 613
wear parts, 327, 569
wear plate, 133
wear testing machine, 326
wearability, 337
wearing resistance, 270
wearing surface, 326, 327
weasel, 210, 583
weather board, 146
weather boarding, 133, 639
weather evaluation, 477
weather map, 477
weather satellite, 370
weather shake, 145, 157
weather stain, 144, 145
weatherability, 337
weathered layer, 144
weathered material, 144
weathered wood, 144, 145
weathering, 144, 145
weathering residue, 144
weathering test, 144
weatherometer, 145
weatherproof plywood, 441
web, 13, 155, 404
web member, 155
web of life, 423
web saw, 160, 280, 357, 615
web-spinner, 651
webbed toe, 365
webbing, 282, 499
webspinning sawfly, 23
wedge, 333, 536
wedge angle, 536, 555
wedge cut, 90, 520
wedge felling, 228

wedge grafting, 536
wedge joint, 536
wedge planting, 536
wedge shaped sectioning, 536
wedge-shaped shelter-wood system, 536
wedge-shaped slit, 536
wedge socket, 536
wedge split, 536
wedge system, 536
wedging, 228
weed, 596
weed and loose soil machine, 64
weed clearing, 64
weed control, 559, 596
weed ecology, 596
weed killer, 64
weed oil, 64
weed puller, 64
weed sprayer, 64
weed tree, 93, 597
weeder, 64
weedicide, 64
weeding, 64
weed(ing) harrow, 64
weeding hook, 549
weed(ing) knives, 64
weeding machine, 64, 165, 233
weedy, 116, 596
weekend-use area, 630
weekly distribution of visitation, 579
weeping mulberry, 405
weevil, 530
Weibull distribution, 500
weigh hopper, 225
weigher, 53, 57
weighing dosing, 3, 629
weight, 629
weight apportionment, 385
weight-average molecular weight, 630
weight by volume, 397
weight control station, 630
weight density, 630
weight determination, 629
weight equation, 629
weight method, 629
weight percent gain, 630
weight scale, 53
weight-strength ratio, 376
weighted average, 228
weighted-average inventory method, 228
weighted mean, 228
weighting, 385
weighting of observation, 178
weightometer, 629
weir, 5, 460, 556
weir crest, 5
weir dam, 569
weir loading, 556
Weise's hypsometer, 502
welfare economics, 149
well-bucked end, 611
well cut, 611
well-drained, 350
well stocked, 278, 289, 445
Wellhouse process, 502
Werzalit process, 507
westerly wind, 513
wet abrasive paper, 458
wet and dry bulb hygrometer, 157
wet and dry-bulb recording hygrometer, 158
wet and dry-bulb thermometer, 157
wet basis moisture content, 432, 433
wet beating, 345, 432
wet bin, 432
wet bulb, 432
wet bulb depression, 157
wet-bulb potential temperature, 432
wet-bulb temperature, 432
wet-bulb thermometer, 432
wet-cemented, 432
wet chute, 456
wet damage, 432
wet-dry method, 9
wet-felting, 432
wet finishing, 432
wet flume, 456
wet-glued, 432
wet-glued plywood, 432
wet gluing, 432
wet heart, 433
wet lap, 432
wet machine, 50
wet mat, 432
wet-method hardboard, 432
wet mill, 582
wet oil, 189
wet paint, 578
wet podzolic soil, 432
wet pressing, 433
wet process, 432, 433
wet replica method, 432
wet rot, 400, 432
wet sanding, 432
wet seal, 561
wet sheet, 432
wet slide, 456
wet soil, 222, 605
wet stock, 345, 432, 433
wet storage, 432
wet strength, 432
wet strength paper, 337, 432
wet timber, 162, 432
wet type wide belt sander, 433
wet water, 401
wet web strength, 433
wet weight, 433, 523
wetland, 432
wetland mapping, 432
wettability, 273
wettable powder, 273
wetted perimeter, 433
wetter, 433, 602
wetting agent, 433
wetting angle, 433
wetting curve, 433
wetting heat, 433
wetting of surface, 27
wetting surface, 52
wetting tension, 433
wetwood, 162, 433
wetwood borer, 432
wharf borer, 316
wharfage, 316
wheatsheaf back, 317
wheel girdle, 132
wheel lock gun, 58
wheel lug, 52, 132
wheel-mounted small skidder, 535
wheel scraper, 312
wheel skidder, 312
wheeled cultivator, 312
wheeler raft, 189
wheelwright's tenon, 536
wheelwright's wood, 52
whiffle-tree, 499
whip, 23
whip-grafting, 417, 612
whip-poor-will, 89

whip saw, 173, 275, 453
whip sawing, 275
whipple-tree, 499
whirl, 507
whirling psychrometer, 546
whirlwind terracer, 546
whistle budding, 94, 179
white alkali soil, 6, 555
white ant, 6
white balsam fir, 6
white camphor oil, 6
white charcoal, 6
white fir, 6
white fly, 144
white ground-pulp, 6
white grub, 366
white liquor, 6, 417
white magnolia, 6, 585
white mottled rot, 5, 479
white mulberry, 405
white paste, 6, 372
white pocket rot, 147, 277
white resin, 6
white ring rot, 6
white rot, 6
white-rot fungus, 6
white rust, 6
white sandalwood, 472
white shellac, 358
white spongy rot, 188
white spot, 5
white stringy rot, 461
white tag seed, 5
white water, 6
white water chest, 6
white water driving, 222, 307
white water man, 446
white wax, 6, 59
white wax insect, 6
white-wood-free paper, 34, 90
white yulan, 6, 585
whiteness, 5
White's medium, 207
whitewash, 487
whitewood, 6, 504
whittle grafting, 75, 178
whole log, 591
whole log barker, 591
whole log chipper, 591
whole scene, 386
whole stand model, 386
whole stem, 591
whole-stem logging, 591
whole tree, 387
whole-tree chipper, 125, 386
whole-tree chipping, 125, 386
whole-tree harvesting, 386
whole-tree logging, 125, 386
whole-tree skidding, 125, 386
whole-tree utilization, 386
wholesale market, 355
wholesale nursery, 355
wholesale price, 355
wholesale price index, 355
wholesaler, 355
whorl, 312
whorl branch, 312
whorl foot, 313
whorl knot, 312
whorl-less, 174
whorle, 312
wickel, 83
wicker, 308, 612
wicker chair, 308
wicker-work, 23
wicker-work barrier, 23
wide angle, 182
wide band-saw, 279
wide band-saw blade, 279
wide belt sander, 279
wide-belt sanding, 279
wide belt sanding machine, 279
wide crosses, 594
wide ecological amplitude, 182
wide-field-of-view, 182
wide pairing, 646
wide planting, 515
wide range, 279
wide range provenance trial, 385
wide-ringed, 279, 344
wide scale, 76
widow maker, 500, 551
width, 279
width in the clear, 64
width of diameter class, 255
Wien's displacement law, 502
wiggle wire, 347
wigwam, 75, 355
wigwam burner, 139
wild ancestor, 559
wild animal, 559
wild apricot, 414
wild bee, 543
wild catting, 81
wild eye, 161
wild fauna, 559
wild fibrous plant, 559
wild fire, 559
wild flower, 559
wild garden, 559, 592
wild grain, 311
wild land, 209
wild land recreation, 592
wild landscape, 591, 592
wild park, 559, 592
wild peach, 414
wild pepper, 203
wild pile, 404
wild plant, 559
wild population, 559
wild relative, 559
wild rubber plant, 559
wild seedling, 436, 559
wild seedling (tree) planting, 559
wild silk worm, 654
wild species, 559
wild stand, 478
wild stock, 559
wild timber, 221
wild transplant, 559
wild type, 559
wild wood, 478, 591
wilderness, 592
wilderness area, 210, 645
wilderness (area), 209, 591, 592
wilderness hiker, 487
wilderness recreation, 209, 592
wilderness visitor survey, 592
wildfire, 209, 559
wilding, 559
wildland, 209, 303
wildlife, 559
wildlife censusing, 559
wildlife corridor, 109
wildlife ecology application, 559
wildlife economics, 559
wildlife management, 559
wildlife refuge, 559
wildlife reserve, 252
wildlife system, 559
wildlife under special state protection, 185
wildling, 559

wildness, 559
Wiley mill, 351, 500
Williams 2-deck vibrating screen, 500
Williamson's spruce, 76
willow, 308
willow extract, 308
willow pile check dam, 308
willow-rust, 308
willow spittle bug, 308
wilt disease, 278, 503
wilting, 503
wilting coefficient, 503
wilting percent, 503
wilting point, 503
winch, 242
winch drum, 243
winch line, 243
winch pull, 243
winch spool, 243
winch truck, 242, 644
Winchester, 505
winching, 242
wind belt, 131
wind bend, 145
wind blast, 144
wind-break, 131, 146
wind-breakage, 146
wind breaking forest, 131
wind breaks, 131
wind-crack, 145
wind current, 145, 369
wind damage, 144
wind direction, 146
wind eddy, 145
wind erosion, 145
wind-exposed, 573
wind exposure, 530, 573
wind fallen tree, 144
wind firmness, 15, 146, 270, 337
wind flower, 145
wind frequency, 145
wind furnace, 481
wind gap, 144, 145
wind gorge, 144
wind gully, 145
wind load, 146
wind loading, 144
wind-mantle, 131
wind-pollinated plant, 145
wind-pollination, 145
wind pressure, 146
wind-prone, 604
wind resistance, 270, 337
wind rose, 145, 146
wind shake, 145, 208
wind sifter, 145
wind sifting, 145
wind slash, 144
wind spreader, 369
wind-spreader chamber, 369
wind velocity, 145
windblow, 144
windbreak, 14, 131
windbreak forest, 131
windbreak trees, 131
windbreaking, 131
winder, 242, 264, 391
windfall, 144, 146
windfirm, 270
windfirm species, 270
windfirm tree, 270
winding, 264, 442, 526
winding mountain path, 351
windlean, 145, 146
windmillpalm, 648
window board, 67
window frame, 67
window garden, 67
window-like pit, 67
window-like pit pair, 67
window sash, 67, 416
window seat, 67
window stool, 67
window ventilator, 52, 368
windowsill, 67, 414
windrock(ing), 146
windrow, 114, 183, 190, 563
windrow planting, 190
windscreen, 146
windswept, 144, 573, 604
windthrow, 144
windthrow resistance, 15, 146, 270, 337
windward, 146, 530, 573
windward side, 530, 573
windward slope, 573
windy, 117, 144
wine grape, 364
wine table, 316, 534
wing, 58, 370, 569
wing beater mill, 569
wing bookcase, 78
wing chair, 161
wing dam, 446, 569
wing jam, 569
wing log, 22, 43
wing pad, 59
wing saw, 529
wing-shaped knot, 569, 641
wing venation, 59
wing wall, 569
wing without seed, 275
wingboom, 43
winged fruit, 58
winged seed, 580
wingless, 508
wink-dry, 461
Winkler's method, 505
winter annual plant, 108, 594
winter aspect, 108
winter bark-scorch, 108
winter bird, 108
winter bud, 108
winter callusing, 108
winter coat, 108
winter condition, 108
winter dormancy, 108
winter drying, 108
winter frost, 108
winter garden, 108
winter-green, 48, 108
winter hardiness, 270, 594
winter hardy, 270, 594
winter hardy tree species, 337
winter injury, 110
winter-kill, 110
winter-killing, 110, 190
winter mulch(ing), 108
winter plumage, 108
winter protection, 132, 594
winter resident, 108
winter resistance, 270
winter show, 108, 219, 387
winter stagnation, 108
winter sunscald, 108
winter visitor, 108
wintergreen oil, 108
winterin, 33, 108
wintering, 594
wintering ground, 594
wintering migration, 594
winternene, 108
wire, 469
wire belt dryer, 499
wire-covered table roll, 155
wire cylinder, 481
wire edge, 264, 527
wire fence, 481
wire frame, 499, 526

wire gauge, 481, 526
wire (leading) roll, 89
wire mark, 499
wire mesh belt conveyor, 249
wire part, 499
wire return roll, 484
wire rope, 160, 161
wire rope clip, 161
wire rope rigging, 161
wire rope sag, 160
wire rope sleeve, 160
wire rope slipper, 160
wire rope socket, 160
wire rope tensiometer, 160
wire rope thimble, 160
wire ropeway, 469
wire screen, 249
wire side, 499
wire side convex, 499
wire speed, 499
wire strand, 160, 161
wire stretch roll, 484
wire tramway, 380
wire worm, 277
wiring, 155, 160, 174, 245
Wirkkala hook, 502
Wirkkala (slackline) yarding, 502
Wirkkala system, 502
witches' broom, 72
witch's broom, 72
with, 281
with replacement, 135
with the grain, 461
withdrawal, 37, 61, 350, 476
withdrawal by notice, 482
withdrawal force, 5
withdrawal resistance, 5, 270
withe, 391
withholding, 277
withholding tax, 586
within family selection, 229
within-grade increment, 91
without replacement, 32
witness tree, 219
wobble saw, 268, 357
wold, 413, 591
wolf tree, 284
wolfbrake, 209, 545
wood, 301, 322, 328, 333, 335, 448, 534
wood anatomy, 330
wood arch, 332
wood ash, 333
wood assortment, 37, 329
wood barging, 329
wood-bark separation, 336
wood-base panel material, 335
wood-based panel, 335, 396
wood-based panel manufacturing machinery, 396
wood beam, 333
wood beetle, 333
wood bending, 331
wood biodegradation, 330
wood block, 103, 332, 333, 336, 364
wood block flooring, 333
wood block pavement, 333
wood block test, 333
wood boat, 223
wood boiler, 417
wood borer, 332, 636
wood boring insect, 331
wood boring lathe, 332
wood brick, 336
wood buck, 600
wood burning furnace, 330
wood capital, 331
wood carbonization, 331
wood carver, 103, 332
wood carving, 329, 332
wood carving by sand blast, 354
wood cellulose, 335
wood-cement board, 334
wood char (coal), 334
wood chemical stain, 456
wood chemistry, 329
wood chip, 16, 333, 335
wood chipboard, 335, 336
wood chopper, 126
wood coal, 334
wood colour, 37
wood column, 336
wood comprehensive utilization, 331
wood construction, 333
wood consumption, 331
wood contractor, 329
wood conversion, 329, 619
wood copying lathe, 332
wood core plywood, 200, 335
wood cotton-tree, 333
wood creosote, 333, 335
wood cull, 475
wood-cutter's implement, 126
wood cutting, 330
wood-dealer, 334
wood decay, 329
wood decay fungi, 332
wood-decaying fungus, 332
wood degradation, 330
wood density, 330
wood depot, 222, 635
wood destroy, 330
wood-destroying fungus, 332
wood-destroying insect, 636
wood destructive distillation, 329
wood deterioration, 328
wood distillation, 329
wood drier, 329
wood drying, 329
wood-dwelling termite, 365
wood dye, 330, 617
wood element, 329
wood engraver, 333
wood engraving, 328, 333
wood extractive, 448
wood failure rate, 330
wood fibre, 335
wood fibre board, 336
wood figure, 329
wood filler, 336
wood filling, 331, 578
wood fire retardant, 331
wood floatage, 330
wood floor, 332
wood flour, 332
wood for agricultural purpose, 348
wood for carving, 103
wood for children's toy, 498
wood for furniture, 229
wood for hydraulic works, 456
wood for match sticks and cases, 215

wood for musical instruments, 284
wood for pianoforte, 160
wood for telegraph-post, 101
wood for tool and implements, 169
wood for turnery, 52
wood form, 333, 335
wood frame, 333
wood frame construction, 333
wood-free writing paper, 34, 90
wood fuel, 330
wood gas, 333
wood gasification, 330
wood grain, 331
wood grain printing, 335
wood gum, 448
wood-gypsum board, 329, 330
wood histology, 331
wood hydrolysis, 331
wood identification, 330
wood impregnation, 330
wood inclusion, 330
wood industry, 329
wood-inhabiting fungus, 331
wood lathe, 332
wood limit, 409
wood loss, 40, 139
wood lot, 534
wood louse, 52, 195
wood machining, 330
wood market, 330
wood mast, 333
wood meal, 332
wood merchant, 334
wood molasses, 331
wood-mosaic, 529
wood moth, 332, 336
wood multi-cut lathe, 332
wood naphtha, 73
wood naval stores, 334
wood oil, 335, 484
wood organization, 331
wood painting, 331
wood paper, 335
wood parenchyma, 328
wood parenchyma cell, 328
wood parenchyma strand, 328
wood particle, 336
wood particleboard, 335, 336
wood pasture, 302
wood pavement, 364
wood paving, 364
wood paving block, 364
wood physics, 331
wood pipe, 332
wood pitch, 330
wood plant, 328
wood plastic, 331
wood plastic composite, 334, 466
wood plasticization, 331
wood plug, 334
wood-polymer, 466
wood polyose, 329, 332
wood powder, 332
wood preservation, 328, 329
wood preservation against fire, 331
wood preservative, 329
wood-preserving process, 329
wood price, 329
wood processing, 329
wood processing industry, 298, 329
wood processing machine, 332
wood procurement, 329
wood product, 329
wood production, 330
wood property, 331
wood protection, 328
wood protection in storage, 328
wood pulp, 333
wood putty, 343, 578
wood pyrolysis, 330
wood quality, 37, 331
wood raft, 332, 333
wood ray, 334
wood refuse, 139
wood residue, 139, 335
wood resource, 331
wood right, 330
wood road, 298
wood rosin, 251, 334
wood-rotting fungus, 332
wood-run, 298
wood saccharification, 331
wood sample, 328, 331
wood science, 331
wood screw, 333
wood sealer, 331, 343
wood seasoning, 329
wood shaping lathe, 332
wood shaving board, 335
wood shavings, 333, 335
wood shrinkage, 329, 331
wood (skid) trail, 302
wood sleeper, 608
wood slip, 53
wood sort, 37
wood species, 331
wood stabilization, 331
wood stack, 36
wood stain, 328, 330
wood staining, 328, 330
wood-staining fungus, 328
wood starch, 329
wood-steel construction, 160
wood strip, 8, 533
wood structure, 329, 333
wood substance, 331
wood sugar, 334
wood swelling, 330
wood tannin, 329, 330
wood tapping, 39
wood tar, 333
wood tar creosote, 333
wood tar oil, 333
wood tar pitch, 333
wood technology, 329
wood thermal property, 330
wood tie, 608
wood tissue, 331
wood transport, 595
wood transport by water, 331
wood transport(ation), 331
wood-turning lathe, 332
wood turpentine, 251, 334
wood ultra-structure, 329
wood vinegar, 331
wood volume, 328
wood ware, 334, 335
wood wasp, 446
wood waste, 139
wood-waste metering screw, 139
wood-wind, 332
wood with radial band, 148
wood wool, 333, 334, 519
wood-wool board, 334

wood wool cement board, 458
wood-wool machine, 334
wood wool machining, 334
wood-wool slab, 334
wood worker, 332
wood working, 329, 332
wood working machine, 329
wood working machinery, 332
wood yard, 222, 635
woodchecker, 331
woodcraft, 332
woodcut, 9, 333
woodcutter's axe, 126
woodcutter's crew, 126, 127
wooded area, 299, 581, 597
wooded land, 298, 405
wooden article, 334, 335
wooden beam, 333
wooden block, 332, 333
wooden building, 333
wooden clogs, 333, 335
wooden column, 336
wooden dam, 328, 335
wooden-famine, 333
wooden furniture, 335
wooden member, 332, 335
wooden mosaic, 529
wooden pin, 332, 334
wooden pipe, 332
wooden plate, 328
wooden plug, 332, 334
wooden product, 335
wooden-ship, 331
wooden sled, 223
wooden slide, 332
wooden spar, 333
wooden splash dam, 328
wooden stave, 484
wooden stringer road, 332
wooden tank, 335
wooden tower, 334
wooden tree, 333
wooden truss bridge, 332
wooden wedge, 335
woodenware, 334, 335
woodiness, 301
woodland, 2, 534
woodland belt, 406, 448
woodland park, 407
woodless, 510
woodllin, 333
woodman, 269, 302
woodpecker mill, 413, 564
woods, 328
woods cabin, 302, 304
woods contractor, 329
woods crew, 126
woods cull, 304
woods foreman, 126, 303
woods highway, 268, 302
woods laborer, 126, 303
woods operation, 39
woods run, 34
woodshop, 332
woodsman, 126, 303, 377
woodward, 302, 413
woodwork, 335, 519
woody, 328, 335
woody lianas, 328
woody part, 335
woody plant, 328
woody plant cell, 328
woody texture, 329, 330
woody tissue, 336
woody-plant cover, 301
woolly aphid, 323
woolly bear, 90, 201
woolly grain, 318, 319, 388
word processing, 648
woren wire dam, 161
work design, 431
work in process, 599
work life, 437
work measurement, 169
work output, 445
work status, 259
work to maximum load, 652
work to ultimate load, 363
work tracking, 221, 253, 655
workbench, 170
worked-out timber, 450
worker, 169, 170
working, 253, 378, 431, 596, 654, 655
work(ing) ability, 227, 431
working area, 431, 432
working assets, 306, 596
working bench, 170
working-block, 654
working capital, 214, 630
working circle, 432, 655
working cost, 654
working crew, 7
working cycle, 432
working division, 432
working-expenses, 654
working factor, 170, 654
working figure, 432
working group, 170, 253, 655
working life, 274, 437, 440
working load, 170, 437
working-map, 254
working of state-forest, 186
working-percentage, 253
working period, 253, 291
working-plan, 253, 431
working-plan area, 253, 431
working-plan control, 253
working-plan document, 253
working-plan period, 253, 432
working-plan renewal, 253
working-plan report, 253
working plan revision, 253, 432
working plans circle, 431
working plans conservator, 431
working plans officer, 431
working pressure, 170
working project sheet, 655
working quality, 227
working-rule, 41, 253
working scheme, 253, 432
working section, 39, 169, 431, 655
working site, 655
working strength, 170
working stress, 170
working system, 170, 654
working system for bamboo stand, 632
working table, 170, 471
working tank, 225
working thickness, 227
working time, 284, 595
working unit, 169, 432, 655
working width, 227

workmanship, 170, 620
workpiece, 169
works planning, 655
works program, 655
workshop, 52
World Wildlife Fund, 438
worm-eaten, 60
worm-eaten seed, 60
worm excretion, 60
worm grub, 60
worm hole, 60
worm roller, 313
wormdust, 60, 636
wormhole, 60
wormwood oil, 278
wormy grade, 60
worn-out road, 327
worse face, 243
worsening of environment, 207
wound, 67, 496
wound cambium, 67
wound cork, 67
wound dressing, 414
wound gum, 67
wound heartwood, 67
wound infection, 414
wound intercellular canal, 67
wound-invading organism, 414
wound parasite, 414
wound parenchyma, 67
wound periderm, 67
wound resin canal, 67
wound-rot, 414
wound stain, 67
wound tissue, 67
wound tumour, 414
wounding, 67
wounding cutting, 414
wounding improved response, 414
wove paper, 35
woven-wire fence, 249
wow, 384, 497, 498
wrack, 93, 297, 529
wrap, 12, 47
wrapper chain, 44, 280
wrapping paper, 12
WRE profile, 184
wreath, 203, 313
wreathed column, 313
wreathed hand-rail, 347
wrecking value, 573
wrench, 7, 347
wrenching, 324, 588
wrest of saw, 262
wrest tooth, 29, 373
wresting, 5, 373
wriggle nail, 30
Wright effect, 282
wrinkled seed, 631
wrinkling stress, 631
write-down, 238
write-off, 59, 635
write up, 476, 601, 602
writing arm, 538
writing-chair, 538
writing desk, 444, 538
writing table, 538
Wyssen system, 502
X-band radar, 284
X figure pit-pair, 239, 506
X-lifter, 421
X-radiation, 148
X-radiograph, 419
X-radiography, 418
X-ray, 418
X-ray analysis, 418
X-ray contrast method, 418
X-ray defectoscopy, 418
X-ray detection of defect, 418
X-ray diagram, 311, 418
X-ray diffraction, 418
X-ray diffraction pattern, 419
X-ray film, 418
X-ray fluorescence analysis, 419
X ray microscopy, 418
X ray pattern, 418
X ray photograph, 419
X-ray spectrograph, 418
X-ray spectrometer, 418
X-ray spectrum, 418
X-ray structure, 418
X-rayed, 418
X-raying, 418
X-rayogram, 418
X-shaped chair, 46
X-stretcher, 282
X-tree, 15
X-Y charger, 416, 655
xanthochroism, 210
xanthorhamnin, 210
xenia, 566, 614, 629
xenity, 566, 614
xenobiosis, 27
xenogamy, 568, 596
xenogeneic, 567, 568
xenogenic, 567
xenoplastic graft, 568
xeralf, 157
xerarch succession, 190
xeric, 190, 439
xeric community, 190
xeric succession, 190
xero-, 190
xerocline, 190
xerocole, 190
xeromorphic, 190
xeromorphic vegetation, 190
xeromorphism, 190
xerophil(e), 439, 516
xerophilous, 439, 516
xerophilous tree species, 190
xerophyte, 190
xerophytia, 190
xerophytic, 190
xeroplastic, 439
xerosere, 190
xerosol, 157
xerothermic interval (stage), 505
xylan, 333
xylanase, 333, 334
xylary cell, 335
xylary ray, 334
xylary ray cell, 334
xylem, 335
xylem bundle, 335
xylem cell, 336
xylem daughter cell, 335
xylem fibre, 336
xylem mother cell, 335
xylem parenchyma, 328
xylem ray, 334
xylem translocation, 335
xylematic element, 336
xylene, 122
xylenol, 122
xylocarpous, 335
xylochrome, 330, 335
xylodium, 335
xylogen, 335
xylogenesis, 335
xylography, 328, 333
xyloid, 336, 462
xyloketose, 335
xylol, 213
xylometer, 328, 329
xylometric measurement, 44
xylophagous, 437
xylophagous borer, 636
xylophilic, 334
xylophone, 334
xyloside, 334
xylotomy, 330
xyltile, 329

xylulose, 335
Y-cutting, 371
Y-level, 460
yacht, 279, 580
Yankee machine, 83, 557
yanon scale, 437
yard, 8, 222, 243, 285, 316, 480
yard blue (stain), 635
yard brown (stain), 635
yard carriage, 223
yard equipment, 9, 635
yard lumber, 9, 50
yard manure, 259, 385
yard pool, 635
yard preservative, 9
yard seasoning, 9
yard stain, 368, 635
yard stick, 172, 316
yard track, 50, 635
yarder, 223
yarder cable way, 223, 231
yarder-loader, 223
yarder mast, 243
yarder tree, 623
yarding, 222
yarding bell, 281
yarding block, 223
yard(ing) crane, 50, 223
yarding donkey, 223
yarding line, 223
yarding sled, 223
yarding span, 222, 223
yarding system, 223
yarding tower, 223
yarding trail, 223
yarding wheels, 223
yarovization, 68
yaw, 357
year class, 344
year-long grazing, 293
year-round logging, 48
year-to-year correlation, 343
yearling, 563
yearling face, 99
yearlong range, 386
yearly receipts, 344, 468
yearly ring, 344
yearly yield, 344
yeast extract, 243
yeast fermentation, 243
yellow-brown earth, 210
yellow-brown forest soil, 210
yellow camphor oil, 210
yellow earth, 210
yellow garden, 210
yellow jacket, 210
yellow locust, 71
yellow maturity, 210
Yellow Park forest reservation, 210
yellow podzolic soil, 210
yellow sandalwood, 472
yellow soil, 210
yellow stain, 210
yellow straw pulp, 90
yellow tag seed, 210
yellow-tail moth, 405, 648
yellow virus, 210
yellow weasel, 210
yellowing, 210
yellowish brown lateritic soil, 210
yellows, 210
Yellowstone National Park, 210
yerbine, 5, 108
yield, 15, 47, 170, 217, 291, 344, 421, 435, 441
yield capacity, 422
yield class, 441
yield coefficient, 47
yield curve, 441
yield-density effect, 47
yield determination, 442
yield determination methods, 25, 38
yield determination using harvest percent, 291
yield diagram, 47
yield estimate, 441
yield estimation, 441
yield forecast, 442
yield formula, 441
yield from felling, 38
yield function, 441
yield function estimate, 441
yield in material, 513
yield level, 422, 441
yield limit, 306, 383
yield model, 441
yield of charcoal, 442
yield of final crop, 127, 633
yield of final cutting, 633
yield of plantation, 395
yield of thinning, 445
yield percent, 291, 441
yield point, 217, 306, 383
yield potential, 98, 422, 429, 489
yield-power, 422
yield prediction, 442
yield prescription, 183, 595
yield regulation, 441, 442
yield regulation by area, 25
yield regulation by diameter classes, 255
yield sharing forest, 171
yield strain, 383
yield strength, 270, 383
yield stress, 383
yield table, 441
yield-table in material, 513
yield(-table sample) plot, 441
yield tax, 422, 441
yield uncertainty, 47
yield-value, 441
ylang-ylang concrete, 564
ylang-ylang oil, 528, 564
yoke, 67, 202, 207, 566
Yoshino cherry, 397
Yoshino-zakura, 397
young afforested land, 503, 539
young crop, 379, 582
young forest, 582
young growth, 582, 583
young leaf, 341, 583
young plantation, 395
young stand, 582
young stand cultivator, 582
young timber stage, 582
young tree, 583
youngling, 582, 583
Young's modulus (of elasticity), 557
yule log, 113, 540
Z-helix, 313
Z-iron, 133, 170, 611
Z-lay, 582
Zanzibar copal, 405
zeatin riboside, 585
zenith, 105, 264, 477, 653
zenith distance, 477
zeolite, 140
zerlate, 149
zero isotherm, 305
zero margin cutting cycle, 305

zero margin forestry, 305
zero margin logging, 305
zero margin tree, 305
zero net growth isoline, 305
zero point, 305, 621
zero population growth, 627
zero-sum game, 305
zeroed, 104, 232
zerotruncated distribution, 385
zeta-potential (ζ-potential), 101, 108
zheltozem, 210
zigzag, 262, 351, 611
zigzag felling, 384
zigzag pore arrangement, 611
zigzag tooth, 239
zigzag wind sifter, 262
zinc bloom, 460, 557
zinc chloride process, 315
zinc creosote process, 315
zinc dust, 539
zinc phosphide, 304
zinc tannin process, 315
zinc vegetation, 544
zinc white, 539, 557
zinc yellow, 539
zineb, 78
zingiberene, 237
zonal climax, 95
zonal distribution, 80, 525
zonal soil, 95, 525
zonal vegetation, 95, 525
zonality, 95
zonate parenchyma, 80
zonate pore, 179
zonation, 54, 140, 142
zonation of climate, 369
zone, 45, 79, 95, 130, 382, 385, 405
zone electrophoresis, 45, 382
zone line, 80
zone location, 383
zone of aeration, 482
zone of early wood, 599
zone of effective temperature, 582
zone of emergent vegetation, 481
zone of eventual death, 653
zone of heating, 228
zone of increasingly favorable temperature, 397
zone of late wood, 499
zone of low fatal temperature, 621
zone of maximum precipitation, 652
zone of overlap, 60
zone of saturation, 14
zone of trade wind, 540
zone time, 383
zoning, 95, 142, 382
zoning law, 142, 383
zoning of nature reserve, 645
zono biome, 427
zono ecotone, 388
zoo, 109
zoobenthos, 95
zoochlorella, 60
zoochoric, 109
zoochory, 109
zoocoenosis, 109
zoogamete, 579
zoogeographical realm, 109
zoogeographical region, 109
zoogeography, 109
zoom stereoscope, 272
zoom transfer scope, 272
zoometer, 109
zoophagous, 436, 437
zoophily, 109, 444
zoophobia, 109
zooplankter, 148
zooplankton, 148
zoosperm, 579
zoosporangium, 579
zoospore, 579
zootope, 109
zooxanthella, 60
zygomorphic, 295
zygomorphous, 295
zygomycete, 244
zygosis, 244
zygosome, 244
zygospore, 244
zygote, 192, 443
zygotene, 192, 349
zygotene stage, 192, 349
zygotic meiosis, 192
zygotic nucleus, 192
zygotic sterility, 192
zymase, 243, 259, 345
zymolysis, 320
zymoprotein, 320
zymurgy, 123, 345

主要参考书目

- Allaby M. 2001. 牛津生态学词典[M]. 上海：上海译文出版社.
- 测绘学名词审定委员会. 2002. 测绘学名词[M]. 2版. 北京：科学出版社.
- 地理信息系统名词审定委员会. 2002. 地理信息名词[M]. 北京：科学出版社.
- 林业部调查规划设计院. 1986. 英汉-汉英林业资源科技词汇[M]. 北京：中国林业出版社.
- 南京林业大学. 1988. 新英汉林业词汇[M]. 北京：中国林业出版社.
- 全国自然科学名词审定委员会. 1989. 林学名词[M]. 北京：科学出版社.
- 日本林学会林学検索用語編集委員会. 1990. 林学検索用語集（日文）. 東京：財団法人日本林学会.
- 日本学術振興会. 1988. 学術用語集-地理学編（日文）[M]. 東京：日本学術振興会.
- 日本学術振興会. 1994. 学術用語集-地学編（日文）[M]. 東京：日本学術振興会.
- 日本学術振興会. 1992. 学術用語集-農学編（日文）[M]. 東京：日本学術振興会.
- 日本学術振興会. 1987. 学術用語集-気象学編（日文）[M]. 東京：日本学術振興会.
- 日本学術振興会. 1990. 学術用語集-植物学編（日文）[M]. 東京：丸善株式会社.
- 沼田　真. 1995. 生態学辞典（日文）[M]. 東京：筑地書館.
- 芝　祐順，渡辺　洋，石塚智一. 1984. 統計学用語辞典（日文）[M]. 東京：新曜社.
- 中国科学院南京土壤研究所. 1984. 英汉土壤学词汇[M]. 北京：科学出版社.

策划编辑：肖基浒
责任编辑：张　佳　肖基浒
封面图片：刘琪璟　提供
封面设计：睿思视界视觉设计

ISBN 978-7-5038-8988-2
定价：168.00元